# Vector Optimization

The series in Vector Optimization contains publications in various fields of optimization with vector-valued objective functions, such as multiobjective optimization, multi criteria decision making, set optimization, vector-valued game theory and border areas to financial mathematics, biosystems, semidefinite programming and multiobjective control theory. Studies of continuous, discrete, combinatorial and stochastic multiobjective models in interesting fields of operations research are also included. The series covers mathematical theory, methods and applications in economics and engineering. These publications being written in English are primarily monographs and multiple author works containing current advances in these fields.

More information about this series at http://www.springer.com/series/8175

Christiane Tammer · Petra Weidner

# Scalarization and Separation by Translation Invariant Functions

## with Applications in Optimization, Nonlinear Functional Analysis, and Mathematical Economics

Springer

Christiane Tammer
Institute of Mathematics
Martin-Luther-University Halle-Wittenberg
Halle (Saale), Germany

Petra Weidner
Faculty of Engineering and Health
HAWK University of Applied Sciences
and Arts Hildesheim/Holzminden/Göttingen
Göttingen, Germany

ISSN 1867-8971 ISSN 1867-898X (electronic)
Vector Optimization
ISBN 978-3-030-44721-2 ISBN 978-3-030-44723-6 (eBook)
https://doi.org/10.1007/978-3-030-44723-6



This Springer imprint is published by the registered company Springer Nature Switzerland AG
The registered company address is: Gewerbestrasse 11, 6330 Cham, Switzerland

*To our parents*

# Preface

Our scientific collaboration began about 35 years ago, when both of us had just finished our Ph.D. Theses. The first article which we submitted together dealt with a certain type of nonlinear separating functionals and their application in vector optimization. Since that time, this function class has turned out to be of practical and theoretical importance in different fields of research. This inspired us to write a book about these functions, representing the related theory and showing the scope of applications. The studied functions serve as a tool for the separation of not necessarily convex sets and for the scalarization of problems in linear spaces. Their scope of application comprises, for example, the extension of classical results in nonlinear functional analysis, optimality conditions, the scalarization of solution sets in vector optimization and in decision making, risk measures in finance, and production models in mathematical economics. Writing this book became a long process since it was closely related to the development of new research results. The book is written for students as well as for researchers in mathematics and in mathematical economics. It covers topics in pure and applied mathematics, in mathematical finance, consumer and production theory. Mathematically interested engineers and researchers from other disciplines can benefit from the presented applications.

Halle (Saale), Germany — Christiane Tammer
Göttingen, Germany — Petra Weidner
December 2019

**Acknowledgments** First of all, we are grateful to Christian Rauscher of Springer Verlag for promoting our book project with limitless patience and to Johannes Jahn for his support throughout this process. We wish to express our gratitude to Rosalind Elster, Christian Günther, Elisabeth Köbis and Brian Winkler for carefully reading parts of the manuscript. Moreover, we would like to thank Christian Günther, Elisabeth and Markus Köbis for their technical support related to some of the figures in this book. Finally, the authors thank all colleagues who have contributed to the field of translation invariant functions and their applications.

The original version of the book was revised: In the original version author names were mentioned inadvertently and missing crosslinks in references of Chaps. 5 and 7 were fixed. The correction to the book is available at https://doi.org/10.1007/978-3-030-44723-6_16.

# Contents

# List of Symbols and Abbreviations

## Abbreviations

Assumptions which are used in more than one chapter:

($\mathrm{Sp0}_{A,k}$): $Y$ is a linear space, $A \subseteq Y$, and $k \in Y \setminus \{0\}$.

($\mathrm{Sp1}_{A,k}$): $Y$ is a topological vector space, $A$ is a proper subset of $Y$, $k \in Y \setminus \{0\}$,

$$\operatorname{cl} A - \mathbb{R}_{>} k \subseteq A.$$

($\mathrm{Sp2}_{A,k}$): $Y$ is a topological vector space, $A$ is a proper subset of $Y$, $k \in Y \setminus \{0\}$,

$$\operatorname{cl} A - \mathbb{R}_{>} k \subseteq \operatorname{int} A.$$

($\mathrm{H1}_{A,k}$): $Y$ is a topological vector space, $A$ is a proper closed subset of $Y$,

$$k \in -0^{+} A \setminus \{0\}.$$

($\mathrm{H2}_{A,k}$): $Y$ is a topological vector space, $A$ is a proper closed subset of $Y$,

$$k \in Y \setminus \{0\} \text{ with } A - \mathbb{R}_{>} k \subseteq \operatorname{int} A.$$

Set operations:

Assume that $A, B$ are nonempty subsets of a linear space $X$, that $\Lambda$ is a nonempty subset of $\mathbb{R}$, $x \in X$, and $\gamma \in \mathbb{R}$. Then

$$\begin{aligned}
&A + B := \{x + y \mid x \in A, y \in B\},\ A + \emptyset := \emptyset + A := \emptyset,\\
&x + A := A + x := \{x\} + A,\\
&\Lambda A := \Lambda \cdot A := \{\lambda x \mid \lambda \in \Lambda, x \in A\},\ \Lambda \cdot \emptyset := \emptyset,\\
&\gamma A := \gamma \cdot A := \{\gamma\} \cdot A,\ -A := (-1) \cdot A,\\
&A - B := A + (-B) = \{x - y \mid x \in A, y \in B\}.
\end{aligned}$$

Further abbreviations:

| | |
|---|---|
| B-l.s.c. | Berge-lower semicontinuous |
| B-u.s.c. | Berge-upper semicontinuous |
| Chap. | Chapter |
| Chaps. | Chapters |
| Cor. | Corollary |
| Fig. | Figure |
| n.v.s. | normed space |
| l.c.s. | locally convex space |
| l.s.c. | lower semicontinuous |
| Prop. | Proposition |
| resp. | respectively |
| Sect. | Section |
| t.v.s. | topological vector space |
| u.s.c. | upper semicontinuous |
| VVDS | vector optimization problem with a variable domination structure |
| w.l.o.g. | without loss of generality |
| w.r.t. | with regard to |

## Special Spaces

| | |
|---|---|
| $C[a,b]$ | space of functions which are continuous on the interval $[a,b]$ with $a,b \in \mathbb{R}$ |
| $C(T)$ | space of functions which are continuous on the set $T$ |
| $c_0$ | space of all real sequences in $\ell^\infty$ converging to zero |
| $e^j$ | unit vector in a Euclidean space with 1 in the j-th component |
| $\ell^p$ | space of all real sequences with $p$-norm $\|\cdot\|_p$, $1 \le p \le +\infty$ |
| $L^p(\Omega, M, \mu)$ | Lebesgue space on the measure space $(\Omega, M, \mu)$ $(1 \le p \le +\infty)$ |
| $L^p(\Omega, M, P)$ | Lebesgue space with some probability measure $P$ $(1 \le p \le +\infty)$ |
| $L^p(\Omega)$ | Lebesgue space on the nonempty set $\Omega \subseteq \mathbb{R}^\ell$ with the Borel $\sigma$-algebra and the Lebesgue measure $(1 \le p \le +\infty)$ |
| $L^p[a,b]$ | $L^p(\Omega)$ with $\Omega = [a,b]$ $(1 \le p \le +\infty)$ |
| $\mathbb{N}$, $\mathbb{N}_>$ | the set of nonnegative and of positive, resp., integers |
| $\mathbb{Q}$ | the set of rational numbers |
| $\mathbb{R}$ | the set of real numbers |
| $\mathbb{R}_+, \mathbb{R}_>$ | the set of nonnegative and of positive, resp., real numbers |
| $\overline{\mathbb{R}}$ | $\overline{\mathbb{R}} := \mathbb{R} \cup \{-\infty, +\infty\}$ |

| | |
|---|---|
| $\mathbb{R}^n$ | space of $n$-dimensional vectors of real numbers (Euclidean space if not stated otherwise) |
| $\mathbb{R}^n_+$ | nonnegative orthant of $\mathbb{R}^n$ |
| $\mathbb{R}^{m,n}, \mathbb{R}^{m,n}_+$ | (m × n)-matrix with elements from $\mathbb{R}$ and $\mathbb{R}_+$, resp. |
| $\operatorname{rank} W$ | rank of the matrix $W$ |
| $\mathbb{Z}$ | the set of integers |
| $(a,b), [a,b]$ | open and closed, resp., interval if $a, b \in \overline{\mathbb{R}}$ |
| $(a,b], [a,b)$ | half-open intervals if $a, b \in \overline{\mathbb{R}}$ |
| **1** | in a Euclidean space: vector with 1 in each component; in a function space: constant function with function value 1 |

## Linear and Topological Spaces

| | |
|---|---|
| lim inf | limit inferior |
| lim sup | limit superior |
| $B_{\lVert\cdot\rVert}(x^0, \varepsilon)$ | closed ball w.r.t. the norm $\lVert\cdot\rVert$ with the center $x^0$ and the radius $\varepsilon$ |
| $\overset{\circ}{B}_{\lVert\cdot\rVert}(x^0, \varepsilon)$ | open ball w.r.t. the norm $\lVert\cdot\rVert$ with the center $x^0$ and the radius $\varepsilon$ |
| $B_{\lVert\cdot\rVert}, \overset{\circ}{B}_{\lVert\cdot\rVert}$ | closed and open, resp., unit ball w.r.t. the norm $\lVert\cdot\rVert$ |
| $B_{X^*}$ | $B_{X^*} := \{x^* \in X^* \mid \lVert x^* \rVert_* \le 1\}$ |
| $\mathcal{N}_\tau(x), \mathcal{N}(x)$ | the set of neighborhoods of $x \in X$ in the topological space $(X, \tau)$ |
| $\mathcal{N}^{\mathrm{bal}}_\tau(x)$ | the set of all balanced neighborhoods of $x \in X$ in the topological vector space $(X, \tau)$; also denoted by $\mathcal{N}^{\mathrm{bal}}(x)$ |
| $Y^\#$ | algebraic dual space of the vector space $Y$ |
| $Y^*$ | topological dual space of the t.v.s. $Y$ |
| $\lVert\cdot\rVert_*$ | norm in the topological dual space of a n.v.s. $X$, defined by $\lVert x^* \rVert_* := \sup\{x^*x \mid x \in X, \lVert x \rVert = 1\}$ |
| $(x_i)_{i\in I}, \{x_i\}_{i\in I}$ | sequences if $I \subseteq \mathbb{N}$; nets otherwise |

## Sets

| | |
|---|---|
| $0^+A$ | recession cone of the set $A$ |
| $A \subseteq B$ | $A$ is a subset of $B$ |
| $A \subset B$ | $A$ is a subset of $B$ with $A \neq B$ |
| $A^\perp$ | orthogonal complement to the set $A$ |
| $A^+$ | (positive) dual cone of the set $A$ |
| $A^{++}$ | bipolar cone of the set $A$ |
| $\operatorname{aff} A$ | affine hull of the set $A$ |
| $\operatorname{bar} A$ | barrier cone of the set $A$ |
| $\operatorname{bd} A$, $\operatorname{r-bd} A$ | boundary and relative boundary, resp., of the set $A$ |

| | |
|---|---|
| $\mathrm{bd}_k(A)$ | $k$-directional boundary of the set $A$ |
| $\mathrm{bd}_{\mathrm{w}}A$, r-$\mathrm{bd}_{\mathrm{w}}A$ | boundary and relative boundary, resp., of the set $A$ in the weak topology |
| $\mathrm{bd}_{\mathrm{w}^*}A$, r-$\mathrm{bd}_{\mathrm{w}^*}A$ | boundary and relative boundary, resp., of the set $A$ in the weak* topology |
| $\mathrm{cl}\,A$, r-$\mathrm{cl}\,A$ | closure and relative closure, resp., of the set $A$ |
| $\mathrm{cl}_k(A)$ | $k$-directional closure of the set $A$ |
| $\mathrm{cl}_{\mathrm{w}}A$, r-$\mathrm{cl}_{\mathrm{w}}\,A$ | closure and relative closure, resp., of the set $A$ in the weak topology |
| $\mathrm{cl}_{\mathrm{w}^*}A$, r-$\mathrm{cl}_{\mathrm{w}^*}A$ | closure and relative closure, resp., of the set $A$ in the weak* topology |
| $\mathrm{conv}\,A$ | convex hull of the set $A$ |
| $\mathrm{cone}\,A$ | cone generated by the set $A$ |
| $\mathrm{core}\,A$ | algebraic interior (or core) of the set $A$ |
| $\mathrm{icr}\,A$ | relative algebraic interior (or intrinsic core) of the set $A$ |
| $\mathrm{int}\,A$, r-$\mathrm{int}\,A$ | interior and relative interior, resp., of the set $A$ |
| $\mathrm{int}_k(A)$ | $k$-directional interior of the set $A$ |
| $\mathrm{int}_{\mathrm{w}}A$, r-$\mathrm{int}_{\mathrm{w}}A$ | interior and relative interior, resp., of the set $A$ in the weak topology |
| $\mathrm{int}_{\mathrm{w}^*}A$, r-$\mathrm{int}_{\mathrm{w}^*}A$ | interior and relative interior, resp., of the set $A$ in the weak* topology |
| $\mathrm{lin}\,A$ | algebraic closure of the set $A$ |
| $\mathrm{N}(x;\Omega)$ | normal cone to $\Omega$ at $x$ |
| $\hat{N}(x;\Omega)$ | Fréchet normal cone to $\Omega$ at $x \in \Omega$ |
| $\hat{N}_{\varepsilon}(x;\Omega)$ | set of $\varepsilon$-normals to $\Omega$ at $x \in \Omega$ |
| $N_A(x;\Omega)$ | approximate (Ioffe) normal cone to $\Omega$ at $x$ |
| $\mathrm{N_C}(x;\Omega)$ | Clarke normal cone to $\Omega$ at $x$ |
| $N_M(x;\Omega)$ | basic, limiting, or Mordukhovich normal cone to $\Omega$ at $x$ |
| $\mathscr{P}(Y)$ | power set of $Y$ |
| $\mathscr{P}_C(Y)$ | $\mathscr{P}_C(Y) := \{D \subset Y \mid D$ is a $C-$proper set, and $D + C$ is closed$\}$ |
| $\mathrm{qint}\,A^+$ | quasi-interior of the dual cone $A^+$ |
| $\mathrm{span}\,A$ | linear hull of the set $A$ |
| $\mathrm{T_B}(x;S)$ | Bouligand tangent cone to $S$ at $x$ |
| $\mathrm{T_C}(x;S)$ | Clarke tangent cone to $S$ at $x$ |
| $\mathrm{vcl}(A)$ | vectorial closure of the set $A$ |

## Sets of Optimal Elements

| | |
|---|---|
| Be-PEff$(F,D)$ | Benson properly efficient point set of $F$ w.r.t. the set $D$ |
| $\mathrm{Eff}(F)$ | Pareto minimal point set of $F$ |
| $\mathrm{Eff}(F,D)$ | efficient point set of $F$ w.r.t. the set $D$ |
| $\mathrm{Eff}_f(S,D)$ | $\mathrm{Eff}_f(S,D) := \{x^0 \in S \mid f(x^0) \in \mathrm{Eff}(f(S),D)\}$ |

| | |
|---|---|
| $\mathrm{GEff}(F)$ | Geoffrion properly efficient point set of $F$ |
| $\text{He-PEff}(F,D)$ | Henig properly efficient point set of $F$ w.r.t. the set $D$ |
| $\mathrm{ND}(F,\mathscr{R})$ | nondominated element set of $F$ w.r.t. the binary relation $\mathscr{R}$ |
| $\text{NI-PEff}(F,D)$ | Nehse-Iwanow properly efficient point set of $F$ w.r.t. $D$ |
| $\mathrm{WEff}(F)$ | weakly Pareto minimal point set of $F$ |
| $\mathrm{WEff}(F,D)$ | weakly efficient point set of $F$ w.r.t. the set $D$ |

## Functions

| | |
|---|---|
| $f(A)$ | $f(A) := \{f(a) \mid a \in A\}$ for $f : X \to Y$ and $A \subseteq X$ |
| $f^{-1}(B)$ | $f^{-1}(B) := \{x \in X \mid f(x) \in B\}$ for $f : X \to Y$ and $B \subseteq Y$ |
| $f^*, f^{**}$ | (Fenchel) conjugate and biconjugate, resp., of $f$ |
| $\partial f(\bar{x})$ | (Fenchel) subdifferential of $f$ at $\bar{x}$ |
| $\partial f(\bar{x})(y^*)$ | limiting subdifferential of $f$ at $\bar{x}$ in direction $y^* \in Y^*$ |
| $\hat{\partial} f(\bar{x})(y^*)$ | Fréchet subdifferential of $f$ at $\bar{x}$ in direction $y^* \in Y^*$ |
| $\partial_A f(\bar{x})$ | approximate subdifferential of $f$ at $\bar{x}$ |
| $\partial_C f(\bar{x})$ | Clarke generalized subdifferential of $f$ at $\bar{x}$ |
| $\partial_M f(\bar{x})$ | basic, limiting, or Mordukhovich subdifferential of $f$ at $\bar{x}$ |
| $D^*f(\bar{x})$ | (basic, normal, Mordukhovich) coderivative of $f$ at $\bar{x}$ |
| $\hat{D}^*f(\bar{x})$ | Fréchet coderivative of $f$ at $\bar{x}$ |
| $\mathrm{epi} f$ | epigraph of $f$ |
| $\mathrm{graph} f$ | graph of $f$ |
| $\mathrm{hypo} f$ | hypograph or subgraph of $f$ |
| $\mathrm{Im} f$ | $\mathrm{Im} f := \{f(x) \mid x \in X\}$ for $f : X \to Y$ |
| $L(X,Y)$ | set of all continuous linear mappings from $X$ to $Y$ |
| $\mathscr{M}(\mathscr{P}_C(Y),\mathbb{R})$ | set of all functions from $\mathscr{P}_C(Y)$ to $\mathbb{R}$ |
| $P_C$ | metric projection mapping |
| $\mathrm{Pr}_X$ | $\mathrm{Pr}_X : X \times Y \to X$ with $\mathrm{Pr}_X(x,y) := x$ |
| $\iota_A$ | indicator function of the set $A$ |
| $\sigma_A$ | support function of the set $A$ |

## Extended Real-Valued Functions

| | |
|---|---|
| $\varphi_{A,k}$ | $\varphi_{A,k}(y) := \inf\{t \in \mathbb{R} \mid y \in A + tk\}$ |
| $\mathrm{argmin}_M \varphi$ | set of minimizers of the functional $\varphi$ on $M$ |
| $\mathrm{cl}\,\varphi$ | function defined by $\mathrm{epi}(\mathrm{cl}\,\varphi) = \mathrm{cl}(\mathrm{epi}\,\varphi)$ |
| $\mathrm{conv}\,\varphi$ | function defined by $\mathrm{epi}(\mathrm{conv}\,\varphi) = \mathrm{conv}(\mathrm{epi}\,\varphi)$ |
| $d(x,S)$ | distance from $x$ to $S$ |
| $\mathrm{dom}\,\varphi$ | (effective) domain of the functional $\varphi$ |
| $\mathrm{lev}_{\varphi,\mathscr{R}}(t)$ | $\mathrm{lev}_{\varphi,\mathscr{R}}(t) := \{x \in X \mid \varphi(x)\mathscr{R}t\}$ |

| | |
|---|---|
| $p_A$ | Minkowski functional of the set $A$ |
| $f \# g$ | inf-convolution of $f$ and $g$ |
| $\langle .,. \rangle$ | bilinear functional; scalar product if not stated otherwise |
| $\|\cdot\|_{C,k}$ | norm generated by an ordering cone $C$ and $k \in \operatorname{core} C$ (see Theorem 2.2.10) |

# Set-Valued Mappings

| | |
|---|---|
| $\Gamma : X \rightrightarrows Y$ | multifunction $\Gamma : X \to \mathscr{P}(Y) \setminus \{\emptyset\}$ |
| $\Gamma^{-1}(B)$ | $\Gamma^{-1}(B) := \{x \in X \mid \Gamma(x) \cap B \neq \emptyset\}$ for a multifunction $\Gamma$ and $B \subset \Gamma(X)$ |
| $\Gamma^{+1}(B)$ | $\Gamma^{+1}(B) := \{x \in X \mid \Gamma(x) \subset B\}$ for a multifunction $\Gamma$ and $B \subset \Gamma(X)$ |
| $D^*\Gamma(\bar{x}, \bar{y})$ | (basic, normal, Mordukhovich) coderivative of $\Gamma$ at $(\bar{x}, \bar{y}) \in \operatorname{graph} \Gamma$ |
| $D^*_A\Gamma(\bar{x}, \bar{y})$ | approximate (Ioffe) coderivative of $\Gamma$ at $(\bar{x}, \bar{y})$ |
| $D^*_C\Gamma(\bar{x}, \bar{y})$ | Clarke coderivative of $\Gamma$ at $(\bar{x}, \bar{y})$ |
| $\hat{D}^*\Gamma(\bar{x}, \bar{y})$ | Fréchet coderivative of $\Gamma$ at $(\bar{x}, \bar{y})$ |
| $\partial\Gamma(\bar{x}, \bar{y})(y^*)$ | (limiting, basic, normal) subdifferential of $\Gamma$ at $(\bar{x}, \bar{y})$ in direction $y^* \in Y^*$ |
| $\hat{\partial}\Gamma(\bar{x}, \bar{y})(y^*)$ | Fréchet subdifferential of $\Gamma$ at $(\bar{x}, \bar{y})$ in direction $y^* \in Y^*$ |
| $\mathscr{E}_\Gamma : X \rightrightarrows Y$ | epigraphical multifunction of $\Gamma$ |
| Lim sup | Kuratowski Painlevé upper limit |
| $\preceq^l_C$, $\preceq^u_C$ | lower and upper, resp., set less relation w.r.t. $C$ |
| $\preceq^s_C$ | set less relation w.r.t. $C$ |
| $\preceq^c_C$ | certainly less relation w.r.t. $C$ |
| $\preceq^p_C$ | possibly less relation w.r.t. $C$ |
| $\preceq^m_C$ | minmax less relation w.r.t. $C$ |
| $\preceq^{mc}_C$ | minmax certainly less relation w.r.t. $C$ |
| $\preceq^{mn}_C$ | minmax certainly nondominated relation w.r.t. $C$ |

# Chapter 1
# Introduction

Many theoretical results and applications in mathematics and mathematical economics are based on the separation of sets and scalarization. In this context, our book deals with functions of the following type. A function $\varphi_{A,k}$ is defined on a real vector space $Y$ by

$$\varphi_{A,k}(y) := \inf\{t \in \mathbb{R} \mid y \in A + tk\},$$

where $A$ is a subset of $Y$ and $k \in Y \setminus \{0\}$.

These functions have been used—generally under more restrictive assumptions—in separation theorems and scalarization statements for nonconvex vector optimization problems since the middle of the 1980s.

About fifteen years later, it became evident that the studied function class coincides with the class of translation invariant functions, which is frequently encountered in mathematical finance. The notion of translation invariance came into existence since the value of an extended real-valued function having this property alters by $t$ if the argument is translated by $tk$—where $k$ is a pre-specified direction in a linear space—and thus the shape of the sublevel sets is invariant w.r.t. the level. Via the representation as $\varphi_{A,k}$, a constructive approach to translation invariant functions became available.

The history of translation invariant functions, especially of their representation $\varphi_{A,k}$, will be described in Sect. 4.15.

The purpose of this book is to present the theory of translation invariant functions and the scope of their applications in order to serve the growing community of researchers who apply such functions and to connect research from different fields. These fields include

- nonlinear functional analysis, where separation theorems involve translation invariant functions,
- decision making and vector optimization, where translation invariant functions serve as scalarizing functions,



C. Tammer and P. Weidner, *Scalarization and Separation by Translation Invariant Functions*, Vector Optimization, https://doi.org/10.1007/978-3-030-44723-6_1

- nonlinear optimization including set optimization, where optimality conditions make use of translation invariant functions,
- mathematical finance, where risk measures turn out to be translation invariant,
- consumer and production theory, where benefit and shortage functions are translation invariant.

Many results can also be applied in discrete mathematics.

Note that each extended real-valued function can be extended to a translation invariant function (see Sect. 4.14), which gives us a tool for investigating arbitrary extended real-valued functions. One result is the characterization of continuity by properties of the epigraph.

In the past three decades, several authors have contributed to the theory of translation invariant functions. Chapter 4 contains the fundamental properties of translation invariant functions and foundations for their application. It is written in such a way that it can act as a toolbox for researchers, but also—together with Chaps. 2 and 3—as a course for undergraduates or graduate students. Numerous unpublished results are proved, and results from the literature are presented in a unified way.

Translation invariant functions have been used in optimization, functional analysis, and mathematical economics inside proofs or immediately in applications. This will be underscored by Chaps. 6–7 and 9–15. Chapter 4 indicates further applications in these fields.

The outline of this book is as follows, where Petra Weidner is the author of Chaps. 2–7, and Christiane Tammer is the author of Chaps. 8–15.

The functions on which our book focuses are defined on general linear spaces, which are usually equipped with some binary relation. This relation can be a partial order represented by some cone. The considered linear space is often endowed with a topology. In the first sections of Chap. 2, we provide well-known fundamentals of linear spaces and topological spaces, binary relations and cones. We will add some new results and details that are necessary for the understanding of the subsequent chapters and for the proofs therein. Furthermore, our notation will be introduced. In Sect. 2.3.2, we define the directional closedness of sets in linear spaces—together with related notions—and prove direction-depending properties of sets. In the cases where a set is not closed in a given topology, where no topology is assumed at all, or where the algebraic interior of a set is empty, the results of Sect. 2.3.2 build the basis for statements which usually require some topological assumptions.

Since translation invariant functions may have an effective domain which differs from the entire space, properties of extended real-valued functions will be presented in Chap. 3. Here we will also discuss problems that arise by admitting function values $\pm\infty$. Some properties which are stated as properties of translation invariant functions in the literature turn out to be properties of each extended real-valued function.

As mentioned before, Chap. 4 contains the fundamental properties of translation invariant functions and foundations for their application. Ways to construct translation invariant functions in general as well as with desired properties will be pointed out, and separation theorems for not necessarily convex sets in linear spaces are proved. The approximation of translation invariant functions by convex or concave functions is studied. Relationships between translation invariant functions on the one hand, and Minkowski functionals and norms on the other hand will be shown. Moreover, we will derive formulas for translation invariant extensions as well as for translation invariant envelopes of arbitrary extended real-valued functions.

Many assumptions, definitions and statements related to translation invariant functions have immediate practical interpretations in mathematical economics, especially in mathematical finance and production theory. Such interpretations will be given throughout Chap. 4. Section 15.1 contains applications in mathematical economics. Coherent risk measures in mathematical finance can be formulated using functionals of type $\varphi_{A,k}$ such that our results can be used to derive corresponding properties for coherent risk measures. Furthermore, we study the relationship between coherent risk measures and a strictly robust counterpart problem of an optimization problem under uncertainty. Sect. 15.1.2 deals with consumer and production theory. It is shown that the benefit function and the shortage function in mathematical economics are functions of type $\varphi_{A,k}$.

Since many applications are based on the minimization of translation invariant functions, this optimization problem is investigated in Chap. 5.

In Chaps. 6 and 11, vector optimization problems in linear spaces are studied. We have to point out that the notion of vector optimization will be used by us in a very general way. A vector optimization problem with a variable domination structure is equivalent to the problem of finding nondominated solutions in a set w.r.t. some arbitrary binary relation. Each decision problem can be shown to be of this type. We would like to mention applications in portfolio management, location theory and medical radiotherapy treatment. Section 6.1 contains the basic concept for variable domination structures. The case where the domination structure can be described by a single set—the domination set of the vector optimization problem—is studied in Chap. 6. There is an immediate correspondence between the solutions to such problems and the separation of sets. Statements related to efficient, weakly efficient and properly efficient points are proved, with an emphasis on the scalarization by translation invariant functions. Moreover, surrogates for the weakly efficient point set and problems with uncertainties or perturbations are considered. In Sect. 15.2, solutions to a vector-valued optimal control problem with PDE-constraints are described by means of functions $\varphi_{A,k}$, which can be used in a corresponding adaptive algorithm. Solution concepts for problems with variable domination structures are discussed in Chap. 11. In this context, we present several modifications of the functional $\varphi_{A,k}$ and characterize solutions to vector optimization problems with a variable domination structure by these scalarizing functions. The same functions are used in variational methods for vector optimization problems in Sects. 12.3.2 and 12.4 as well as in numerical algorithms for such problems in Sect. 13.6. Chapters 12 and 13 also contain results for other vector optimization problems, especially for multiobjective optimization problems.

The results of Chaps. 5 and 6 are applied to scalarization methods for multiobjective optimization problems in Chap. 7. In this way, a general framework for analyzing such methods is illustrated. Furthermore, Chap. 7 deals with properties of Geoffrion's properly efficient point set.

In Chap. 8, we recall well-known concepts of generalized differentiation which are relevant when deriving optimality conditions for vector-valued functions as well as for set-valued mappings.

In Sect. 4.8, functionals $\varphi_{A,k}$ are studied for the case where the involved set $A$ is given by a system of linear inequalities. Chapter 7 contains applications of this case for multiobjective optimization problems. In Sect. 9.1, results for functionals $\varphi_{A,k}$ of this type and related statements in multiobjective optimization are proved under the aspect that $A$ is described via block norms. These results are applied to multiobjective d.c. optimization problems in Sect. 9.1.4. Moreover, Sect. 4.8 contains statements for functions $\varphi_{A,k}$ where each element of $A$ has to fulfill at least one inequality of a given system.

Directional minimal time functions are modifications of functions $\varphi_{A,k}$. In Sect. 9.2, properties of such functions are shown. They are used as distance functions for—in general nonconvex—location problems in Sect. 15.3.

Relationships between the functional $\varphi_{A,k}$ and the oriented distance by Hiriart-Urruty are presented in Sect. 9.3.

Set optimization has become an important field in optimization theory as well as in various applications, especially in welfare economics, mathematical finance, optimization under uncertainty and medical image processing. The aim of Chap. 10 is to show that it is possible to give characterizations of certain set relations and corresponding solutions to set-valued optimization problems by means of the functional $\varphi_{A,k}$.

In Chap. 12, functionals $\varphi_{A,k}$ are applied to derive necessary optimality conditions in vector optimization, duality assertions, minimal point theorems, variational principles, necessary conditions for approximate solutions to vector optimization problems with a variable domination structure, and existence results for solutions of vector variational inequalities. Relationships to the Extremal Principle by Kruger and Mordukhovich are proved in Sect. 12.6.

Chapter 13 contains algorithms for the solution of optimization problems which are based on functions $\varphi_{A,k}$:

- Benson's outer approximation algorithm, a proximal-point algorithm as well as an adaptive algorithm for vector optimization problems,
- proximal-point algorithms for optimization problems and for control approximation problems,
- a descent method for set-valued optimization problems,
- algorithms for vector optimization problems with variable domination structures.

In Chap. 14, the functional $\varphi_{A,k}$ is used to describe various concepts of robustness and stochastic optimization which are well known from scalar optimization under uncertainty. Based on this unified approach, it is possible to derive new concepts of robustness in scalar optimization, and relationships to multiobjective optimization problems can be established.

# Chapter 2
# Sets and Binary Relations

In the subsequent chapters, we will investigate functions with uniform sublevel sets on general linear spaces and on topological vector spaces. These functions are usually applied on sets on which some binary relation is given. This relation is often a partial order represented by some cone. In the first sections of this chapter, we provide well-known fundamentals of linear spaces and topological spaces, binary relations and cones. We add some new results and details that are necessary for the understanding of the following chapters and for the proofs therein. Furthermore, our notation is introduced. Properties of recession cones are contained in Sect. 2.2. Norms which are defined by means of cones will be studied in Theorem 2.2.10 and its corollary. Statements about the algebraic interior and the algebraic closedness of sets are summarized in Sect. 2.3.1. In Sect. 2.3.2, we will introduce the directional closedness of sets in linear spaces together with related notions and will prove direction-depending properties of sets. In the cases where a set is not closed in a given topology, where no topology is assumed at all, or where the algebraic interior of a set is empty, the results of Sect. 2.3.2 build the basis for statements which usually require some topological assumptions. For terms used without definition, the reader is referred to [3] and to the references mentioned in the specific context.

## 2.1 Basic Definitions and Statements in General Spaces

### *2.1.1 Linear and Topological Spaces*

This section contains basic definitions in linear and in topological spaces. A space is always considered to be a nonempty set.

For our investigation, we need sets on which certain calculations are possible. The appropriate operators are given in linear spaces.



C. Tammer and P. Weidner, *Scalarization and Separation by Translation Invariant Functions*, Vector Optimization, https://doi.org/10.1007/978-3-030-44723-6_2

**Definition 2.1.1** A **(real) linear space** $X$ is a nonempty set which is equipped with a mapping $+\colon X \times X \to X$, called **addition**, and with a mapping $\cdot : \mathbb{R} \times X \to X$, called **multiplication by scalars**, satisfying the following conditions:

(a) $\forall x, y, z \in X : (x + y) + z = x + (y + z)$,
(b) $\forall x, y \in X : x + y = y + x$,
(c) $\exists 0 \in X\ \forall x \in X : x + 0 = x$,
(d) $\forall x \in X\ \exists x' \in X : x + x' = 0$,
(e) $\forall x, y \in X\ \forall \lambda \in \mathbb{R} : \lambda \cdot (x + y) = \lambda \cdot x + \lambda \cdot y$,
(f) $\forall x \in X\ \forall \lambda, \mu \in \mathbb{R} : (\lambda + \mu) \cdot x = \lambda \cdot x + \mu \cdot x$,
(g) $\forall x \in X\ \forall \lambda, \mu \in \mathbb{R} : \lambda \cdot (\mu \cdot x) = (\lambda \cdot \mu) \cdot x$,
(h) $\forall x \in X : 1 \cdot x = x$.

This space is also denoted by $(X, +, \cdot)$ and called a **(real) vector space** or a **set equipped with a linear structure**. The elements of a vector space are called **vectors** or **points**. The vector 0 in condition (c) is called the **zero element** of $X$ and is also denoted by $0_X$. The vector $x'$ in condition (d) is called the **additive inverse** of $x$ and denoted by $-x$. We will also write $\lambda x$ instead of $\lambda \cdot x$ for $\lambda \in \mathbb{R}$, $x \in X$.

A linear space $X$ is said to be **nontrivial** if $X \neq \{0\}$.

We will consider *real* linear spaces only and omit the word real later on.

Clearly, $\mathbb{R}^\ell$ with $\ell \in \mathbb{N}_>$ is a linear space when defining

$$x + y := (x_1 + y_1, \ldots, x_\ell + y_\ell)^{\mathrm{T}},$$
$$\lambda x := (\lambda x_1, \ldots, \lambda x_\ell)^{\mathrm{T}}$$

for $x = (x_1, \ldots, x_\ell)^{\mathrm{T}} \in \mathbb{R}^\ell$, $y = (y_1, \ldots, y_\ell)^{\mathrm{T}} \in \mathbb{R}^\ell$ and $\lambda \in \mathbb{R}$.

In linear spaces $F$ consisting of extended real-valued functions on some space $X$ or of functions mapping $X$ to some other linear space $Y$, addition and multiplication by scalars are defined by

$$\begin{aligned} (f + g)(x) &:= f(x) + g(x) \quad && \text{for each } x \in X, \\ (\lambda f)(x) &:= \lambda f(x) && \text{for each } x \in X \end{aligned}$$

for all $f, g \in F$ and $\lambda \in \mathbb{R}$.

Throughout the book, we will use the following notation. Assume that $A, B$ are nonempty subsets of a linear space $X$, that $\Lambda$ is a nonempty subset of $\mathbb{R}$, $x \in X$, and $\gamma \in \mathbb{R}$. Then

$$A + B := \{x + y \mid x \in A,\ y \in B\},\ A + \emptyset := \emptyset + A := \emptyset,$$
$$x + A := A + x := \{x\} + A,$$
$$\Lambda A := \Lambda \cdot A := \{\lambda x \mid \lambda \in \Lambda, x \in A\},\ \Lambda \cdot \emptyset := \emptyset,$$
$$\gamma A := \gamma \cdot A := \{\gamma\} \cdot A,\ -A := (-1) \cdot A,$$
$$A - B := A + (-B) = \{x - y \mid x \in A,\ y \in B\}.$$

We will now define some basic properties of sets.

**Definition 2.1.2** $A$ is said to be a **proper subset** of the space $X$ if $A \subset X$ and $A \neq \emptyset$. If $X$ is a linear space, then $A \subseteq X$ is **pointed** if $A \cap (-A) \subseteq \{0\}$.

**Definition 2.1.3** A subset $B$ of a nontrivial linear space $X$ is called a **(Hamel) basis** of $X$ if each finite subset of $B$ is linearly independent and each $x \in X$ is a linear combination of elements of $B$.

**Proposition 2.1.4** *Suppose that $X$ is a nontrivial linear space.*

(a) *$X$ has a Hamel basis.*
(b) *Each Hamel basis of $X$ has the same cardinality, which is called the **dimension** of $X$.*
(c) *For each $x \in X \setminus \{0\}$, the presentation of $x$ as a linear combination of elements of a given Hamel basis is uniquely determined.*
(d) *If $X$ is a Banach space, then each Hamel basis is finite or not countable.*

See e.g. [74] for (a)–(c), [5, p. 310] for (d).

In a linear space, the following substructures are considered.

**Definition 2.1.5** Let $X$ be a linear space and $X_0$ be a nonempty subset of $X$.

$X_0$ is a **linear subspace** of $X$ if $\lambda x + \mu y \in X_0$ for all $x, y \in X_0$ and $\lambda, \mu \in \mathbb{R}$.

$X_0$ is **affine** (or a **linear manifold**) if $\lambda x + (1-\lambda)y \in X_0$ for all $x, y \in X_0$ and $\lambda \in \mathbb{R}$.

If $X_0$ is a linear subspace of a vector space $(X, +, \cdot)$, then $(X_0, +, \cdot)$ is a linear space itself.

A nonempty set $M \subseteq X$ is affine if and only if there exist $x^0 \in X$ and a linear subspace $X_0$ of $X$ such that $M = x^0 + X_0$.

**Definition 2.1.6** Let $X$ be a linear space, $A \subseteq X$.
$A$ is **convex** if

$$\forall x, y \in A\colon\ [x, y] := \{\lambda x + (1-\lambda)y \mid \lambda \in [0, 1]\} \subseteq A.$$

$A$ is called **absorbing** if

$$\forall x \in X\ \exists \varepsilon \in \mathbb{R}_>\ \forall t \in \mathbb{R} \text{ with } |t| < \varepsilon\colon\ tx \in A.$$

$A$ is **balanced** (or **circled**) if

$$\forall a \in A\ \forall t \in [-1, 1]\colon\ ta \in A.$$

$A$ is said to be **starshaped** about $a^0 \in A$ if

$$\forall a \in A\ \forall \lambda \in [0, 1]\colon\ a^0 + \lambda(a - a^0) \in A.$$

$a \in A$ is an **algebraic interior point** of $A$ if

$$\forall x \in X\ \exists \varepsilon \in \mathbb{R}_>\ \forall t \in \mathbb{R} \text{ with } |t| < \varepsilon:\ a + tx \in A.$$

core $A$ is defined as the set of all algebraic interior points of $A$ and denoted as the **algebraic interior** or **core** of $A$. If $A = \operatorname{core} A$ holds, then $A$ is said to be **algebraically open**.

$a \in A$ is a **relative algebraic interior point** of $A$ if

$$\forall x \in \operatorname{aff}(A - A)\ \exists \varepsilon \in \mathbb{R}_>\ \forall t \in \mathbb{R} \text{ with } |t| < \varepsilon:\ a + tx \in A.$$

The set of all relative algebraic interior points of $A$ is denoted by icr $A$ and called the **relative algebraic interior** or **intrinsic core** of $A$.

$x \in X$ is said to be **linearly accessible** from $A$ if

$$\exists a \in A\ \forall \lambda \in (0, 1):\ a + \lambda(x - a) \in A.$$

The **algebraic closure** lin $A$ of $A$ consists of all elements of $X$ that are linearly accessible from $A$. The set $A$ is said to be **algebraically closed** if it coincides with its algebraic closure.

It is obvious that a linear subspace is affine and an affine set is convex. Moreover, any intersection of linear subspaces, affine sets or convex sets is a linear subspace, an affine set or a convex set, respectively. These properties allow us to introduce the **linear hull**, the **affine hull** and the **convex hull** of a set $A \subseteq X$ by

$$\begin{aligned}
\operatorname{span} A &:= \bigcap \{X_0 \subseteq X \mid A \subseteq X_0,\ X_0 \text{ linear space}\},\\
\operatorname{aff} A &:= \bigcap \{X_0 \subseteq X \mid A \subseteq X_0,\ X_0 \text{ linear manifold}\} \text{ and}\\
\operatorname{conv} A &:= \bigcap \{X_0 \subseteq X \mid A \subseteq X_0,\ X_0 \text{ convex set}\},
\end{aligned}$$

respectively.

A set $A \subseteq X$ is absorbing if and only if $0 \in \operatorname{core} A$. For a convex set $A \subseteq X$, this is equivalent to $X = \mathbb{R}_+ A$. Note that a nonempty set which is balanced must contain zero. A convex set is a set that is starshaped about each of its elements.

Properties of the algebraic interior and the algebraic closedness will be discussed in Sect. 2.3.1.

We will now present basic properties of sets and functions in topological spaces. Terms like continuity and limit are based on a topological structure of the set under consideration.

**Definition 2.1.7** Let $X$ be a nonempty set. A **topology** $\tau$ on $X$ is a family of subsets of $X$ satisfying the following conditions:

(a) every union of (finitely or infinitely many) sets of $\tau$ belongs to $\tau$,

(b) every intersection of finitely many sets of $\tau$ belongs to $\tau$,
(c) the empty set $\emptyset$ and the set $X$ belong to $\tau$.

The elements of $\tau$ are called **open** sets, and $X$ equipped with $\tau$ is called a **topological space** and is denoted by $(X, \tau)$. The elements of a topological space are also referred to as points.

For two topologies $\tau_1, \tau_2$ on $X$, $\tau_1$ is said to be **weaker** than $\tau_2$ if $\tau_1 \subset \tau_2$.

The weakest topology on each nonempty set $X$ is the so-called **trivial topology** $\{\emptyset, X\}$.

**Definition 2.1.8** Let $(X, \tau)$ be a topological space and $x \in X$.

A set $U \subseteq X$ is said to be a **neighborhood** of $x$ if there exists a set $O \in \tau$ such that $x \in O$ and $O \subseteq U$. We will denote the set of all neighborhoods of $x \in X$ by $\mathcal{N}_\tau(x)$ or $\mathcal{N}(x)$.

A subset $\mathcal{B}(x)$ of $\mathcal{N}_\tau(x)$ is called a **neighborhood base** at $x$ w.r.t. $\tau$ if, for every $U \in \mathcal{N}_\tau(x)$, there exists $V \in \mathcal{B}(x)$ such that $V \subseteq U$. A **neighborhood base** of $(X, \tau)$ is a union of neighborhood bases of all its points.

$X$ is said to be **first countable** if each point of $X$ has a countable neighborhood base.

$X$ is a **Hausdorff space** or **separated space** if, for all $x, y \in X$ with $x \neq y$, there exist disjunct neighborhoods of $x$ and $y$.

Each metric space is first countable.

Topologies are often defined via some neighborhood base. The following proposition (see [305]) shows which families of sets form a neighborhood base for some topology and in which way a topology can be constructed from the neighborhood base.

**Proposition 2.1.9** *Let $X$ be an arbitrary space. A system $\mathcal{B}$ of subsets of $X$ is a neighborhood base of some topology in $X$ if and only if $\mathcal{B}$ has the following properties:*

(i) *Each $x \in X$ is contained in at least one set of $\mathcal{B}$.*
(ii) *If $x \in X$ is an element of the intersection of two sets $N_1, N_2$ of $\mathcal{B}$, then there exists a set $N \in \mathcal{B}$ such that $x \in N \subseteq N_1 \cap N_2$.*

*If these properties are fulfilled, then the family of all unions of sets from $\mathcal{B}$ together with the empty set is a topology on $X$. We call this topology the **topology** on $X$ **generated** by $\mathcal{B}$.*

**Definition 2.1.10** Let $(X, \tau)$ be a topological space, $A \subseteq X$ and $x \in X$.

$A$ is called a **closed set** if $X \setminus A$ is an open set.

$A$ is said to be a **compact set** if every family of open sets whose union includes $A$ contains a finite number of sets whose union includes $A$.

$x$ is an **interior point** of $A$ if $A$ contains some neighborhood of $x$.

The set of all interior points of $A$ is denoted by $\operatorname{int} A$, the **interior** of $A$.

$x$ is a **boundary point** of $A$ if each neighborhood of $x$ contains points of $A$ as well as points of $X \setminus A$.

We denote the set of all boundary points of $A$ by $\mathrm{bd}\, A$, the **boundary** of $A$.

The **closure** of $A$ is defined as $\mathrm{cl}\, A := A \cup \mathrm{bd}\, A$.

$x$ is called a **cluster point** or an **accumulation point** of $A$ if $(V \setminus \{x\}) \cap A \neq \emptyset$ for each neighborhood $V$ of $x$.

The definition implies:

**Lemma 2.1.11** *Assume that $A$ and $B$ are subsets of a topological space $X$. Then the following statements hold.*

(a) $\mathrm{cl}\, A = \mathrm{int}\, A \cup \mathrm{bd}\, A$.
(b) *$A$ is closed* $\iff A = \mathrm{cl}\, A$.
(c) *$A$ is open* $\iff A = \mathrm{int}\, A$.
(d) $\mathrm{int}\,\mathrm{int}\, A = \mathrm{int}\, A$.
(e) $A \subseteq B \implies \mathrm{int}\, A \subseteq \mathrm{int}\, B$.
(f) $\mathrm{bd}(X \setminus A) = \mathrm{bd}\, A$.
(g) $\mathrm{bd}\,\mathrm{cl}\, A \subseteq \mathrm{bd}\, A$.
(h) $\mathrm{bd}\,\mathrm{cl}\, A = \mathrm{bd}\, A \iff \mathrm{int}\,\mathrm{cl}\, A = \mathrm{int}\, A$.
(i) $\mathrm{bd}\,\mathrm{int}\, A \subseteq \mathrm{bd}\, A$.
(j) $\mathrm{bd}\,\mathrm{int}\, A = \mathrm{bd}\, A \iff \mathrm{cl}\,\mathrm{int}\, A = \mathrm{cl}\, A$.
(k) $\mathrm{int}(A \setminus B) = \mathrm{int}\, A \setminus \mathrm{cl}\, B$.
(l) $\mathrm{cl}(A \setminus B) \subseteq \mathrm{cl}\, A \setminus \mathrm{int}\, B$.
(m) $\mathrm{int}(X \setminus A) = X \setminus \mathrm{cl}\, A$.
(n) $\mathrm{cl}(X \setminus A) = X \setminus \mathrm{int}\, A$.

For a proof of (f)–(l), see [472]. (k) yields (m). (n) follows from (m).

**Definition 2.1.12** Let $(X, \tau)$ be a topological space, $M$ a nonempty subset of $X$. $\tau_M := \{O \cap M \mid O \in \tau\}$ is called the **topology induced** by $\tau$ on $M$.

Obviously, $(M, \tau_M)$ is really a topological space, and we get:

**Lemma 2.1.13** *Let $(X, \tau)$ be a topological space, $M$ a nonempty subset of $X$.*

(a) *A subset of $M$ is compact in $(M, \tau_M)$ if and only if it is compact in $(X, \tau)$.*
(b) *A subset of $M$ is closed in $(M, \tau_M)$ if and only if it is the intersection of $M$ with a set that is closed in $(X, \tau)$.*

**Proposition 2.1.14** *Let $(X, \tau)$ be a topological space.*

(a) *Each closed subset of a compact set in $X$ is compact.*
(b) *The intersection of a compact set and a closed set in $X$ is compact.*
(c) *Each compact set in $X$ is closed if $X$ is a Hausdorff space.*

***Proof***

(a) follows immediately from the definition of compactness.

(b) The intersection of a nonempty compact set $A$ and a closed set $B$ is closed in the topology induced by $\tau$ on $A$. Hence, it is compact in this topology. This implies its compactness in $(X, \tau)$.
(c) See [307, §3.2].

□

An example showing that a compact set in a non-Hausdorff space need not be closed is given in [3, Example 2.33]. In the Euclidean space, the following characterization of compactness is well known.

**Proposition 2.1.15** *In the Euclidean space $\mathbb{R}^\ell$, the following properties of a subset $A$ are equivalent to each other:*

(a) *$A$ is compact.*
(b) *$A$ is closed and bounded.*
(c) *$A$ is closed, and there exist vectors $u, w \in \mathbb{R}^\ell$ such that $A \subseteq (u + \mathbb{R}^\ell_+) \cap (w - \mathbb{R}^\ell_+)$.*

In topological spaces, we can introduce continuity of functions.

**Definition 2.1.16** Consider two topological spaces $(X, \tau)$ and $(Y, \sigma)$ and a function $f: X \to Y$. $f$ is called **continuous** at $x^0 \in X$ if, for each $V \in \mathcal{N}_\sigma(f(x^0))$, there exists $U \in \mathcal{N}_\tau(x^0)$ such that $f(x) \in V$ for all $x \in U$. $f$ is called a **continuous function** if $f$ is continuous at each $x^0 \in X$. $f$ is said to be **continuous on** a nonempty set $D \subset X$ if it is continuous at each $x^0 \in D$ w.r.t. the topology induced by $\tau$ on $D$.

Since continuity on a subset refers to the topology induced on this set, a function can be continuous on $X_0 \subset X$ and on $X \setminus X_0$ without being continuous on $X$.

**Example 2.1.17** Consider the function $f: \mathbb{R} \to \mathbb{R}$ given by

$$f(x) := \begin{cases} 0 & \text{if } x < 0, \\ 1 & \text{if } x \geq 0. \end{cases}$$

$f$ is continuous on $\mathbb{R}_+$ and on $\mathbb{R} \setminus \mathbb{R}_+$, but not on $\mathbb{R}$.

From now on, given some function $f : X \to Y$ and a subset $A \subseteq X$, we will denote the image set of $A$ w.r.t $f$ by $f(A) := \{f(a) \mid a \in A\}$.

Essential operations in a linear space are defined on product spaces. Given two topological spaces $(X_1, \tau_1)$ and $(X_2, \tau_2)$, the **product topology** on $X_1 \times X_2$ is the weakest topology in which each of the projections $p_i : X_1 \times X_2 \to X_i$, $i \in \{1, 2\}$, given by

$$p_i((x_1, x_2)) := x_i \qquad \text{for all } (x_1, x_2) \in X_1 \times X_2, \tag{2.1.1}$$

is continuous.

We will always assume that the product of two topological spaces is equipped with the product topology. Some well-known properties of product topologies are summarized in the next proposition.

**Proposition 2.1.18** *Assume that* $(X_1, \tau_1)$ *and* $(X_2, \tau_2)$ *are two topological spaces and that the functions* $p_i \colon X_1 \times X_2 \to X_i$, $i \in \{1, 2\}$, *are given by* (2.1.1).

(a) *For each topological vector space* $Y$ *and each mapping* $g \colon Y \to X_1 \times X_2$, *we have:*
*$g$ is continuous if and only if $p_i \circ g$ is continuous for each* $i \in \{1, 2\}$.
(b) *For arbitrary subsets* $A \subseteq X_1$ *and* $B \subseteq X_2$, *we have:*

$$\operatorname{int}(A \times B) = \operatorname{int} A \times \operatorname{int} B \qquad \textit{and}$$
$$\operatorname{cl}(A \times B) = \operatorname{cl} A \times \operatorname{cl} B.$$

Continuity of the operations on a linear space $X$ which is equipped with some topology requires compatibility between the linear and the topological structure of $X$. The basic operations in $X$ are the addition, which is defined on $X \times X$, and the scalar multiplication on $\mathbb{R} \times X$. Here it is assumed that $\mathbb{R}$ is equipped with the Euclidean topology, and that the topologies on $X \times X$ and $\mathbb{R} \times X$ are the product topologies.

**Definition 2.1.19** Let $X$ be a linear space endowed with a topology $\tau$. We say that $(X, \tau)$ is a **topological vector space** (t.v.s.) if the addition and the multiplication by scalars in the linear space $X$ are continuous.

$X$ is a **locally convex space** if $X$ is a topological vector space in which every neighborhood of zero includes a convex neighborhood of zero.

Each linear space which is equipped with the trivial topology is a locally convex space. It is not a Hausdorff space if it is nontrivial.

The following lemma [286, 11.1.1] contains basic facts about open sets in topological vector spaces.

**Lemma 2.1.20** *Assume that* $(X, \tau)$ *is a topological vector space. Then*

(a) $\forall O \in \tau \setminus \{\emptyset\}\ \forall \lambda \in \mathbb{R} \setminus \{0\} : \ \lambda O \in \tau$.
(b) $\forall O \in \tau \setminus \{\emptyset\}\ \forall x \in X : \ x + O \in \tau$.
(c) $\forall O \in \tau$ *with* $0 \in O : \ \bigcup\{\lambda O \mid \lambda \in \mathbb{R}_>\} = X$.

**Corollary 2.1.21** *Assume* $X$ *is a topological vector space,* $A, B \subseteq X$ *with* $\operatorname{int} A \neq \emptyset$ *and* $x \in X$. *Then*

(a) $x + \operatorname{int} A = \operatorname{int}(x + A)$.
(b) $x + \operatorname{int} A \subseteq B \implies x + \operatorname{int} A \subseteq \operatorname{int} B$.

***Proof***

(a) results from Lemma 2.1.20(b).
(b) $x + \operatorname{int} A \subseteq B$ implies $x + \operatorname{int} A = \operatorname{int}(x + A) \subseteq B$ and thus $x + \operatorname{int} A = \operatorname{int}(\operatorname{int}(x + A)) \subseteq \operatorname{int} B$.

□

Lemma 2.1.20 yields a characterization of the neighborhoods in a topological vector space by the neighborhoods of zero.

**Lemma 2.1.22** *Assume that* $(X, \tau)$ *is a topological vector space.*

(a) *Each neighborhood base* $\mathscr{B}_\tau(0)$ *corresponds with a neighborhood base*

$$\bigcup \{x + \mathscr{B}_\tau(0) \mid x \in X\}$$

*of* $X$.

(b) *Every neighborhood of* $0$ *is absorbing.*

(c) *The zero element of* $X$ *has a neighborhood base formed by sets that are closed, balanced, and absorbing.*

A proof of Lemma 2.1.22(c) can be found in [3].

If $(X, \tau)$ is a topological vector space, then the set of all balanced neighborhoods of $x \in X$ will be denoted by $\mathscr{N}_\tau^{\text{bal}}(x)$ or $\mathscr{N}^{\text{bal}}(x)$.

Note that a topological vector space $(X, \tau)$ is a Hausdorff space if and only if, for every $x \in X \setminus \{0\}$, there exists $V \in \mathscr{N}_\tau^{\text{bal}}(0)$ such that $x \notin V$, which is equivalent to the set $\{0\}$ being closed.

**Definition 2.1.23** Assume that $(X, \tau)$ is a topological vector space and $A \subseteq X$.

$a \in A$ is a **relative interior point** of $A$ or a **relative boundary point** of $A$ if $a$ is an interior point or a boundary point, respectively, of $A$ in the topology induced by $\tau$ on the affine hull of $A$.

The **relative interior** r-int $A$ of $A$ consists of all relative interior points of $A$. The **relative boundary** r-bd $A$ of $A$ consists of all relative boundary points of $A$.

$A$ is called **bounded** if, for each $U \in \mathscr{N}_\tau(0)$, there exists $\lambda \in \mathbb{R}$ such that $A \subseteq \lambda U$.

A function $f : D \to X$, $D$ being some nonempty set, is **bounded** if $\{f(d) \mid d \in D\}$ is bounded.

We get the following from [289, 13.1(i)].

**Lemma 2.1.24** *Assume that* $A$ *is a convex subset of some topological vector space.*

(a) cl $A$ *and* int $A$ *are convex sets.*

(b) *If* $\operatorname{int} A \neq \emptyset$, *then* $\operatorname{cl} A = \operatorname{cl} \operatorname{int} A$ *and* $\operatorname{int} A = \operatorname{int} \operatorname{cl} A$.

Conditions under which $\operatorname{cl} A = \operatorname{cl} \operatorname{int} A$ or $\operatorname{int} A = \operatorname{int} \operatorname{cl} A$ holds for a nonconvex set $A$ are given in Lemmata 2.2.23 and 2.3.16.

**Definition 2.1.25** Let $X$ be a topological vector space and $A \subseteq X$.

$A$ is **strictly convex** if $(x, y) := \{\lambda x + (1 - \lambda)y \mid \lambda \in (0, 1)\} \subseteq \operatorname{int} A$ for all $x, y \in A$ with $x \neq y$.

The following statement will be used in different proofs.

**Lemma 2.1.26** *Assume that* $(X, \tau)$ *is a topological vector space,* $A \subset X$, $a \in A$, $x \in X \setminus \operatorname{cl} A$.

(a) *There exists* $t \in (0, 1]$ *such that* $ta + (1 - t)x \in \operatorname{bd} A$.

(b) *There exists* $t \in (0, 1]$ *such that* $ta + (1-t)x \in \operatorname{bd}_M A$, *where* $\operatorname{bd}_M A$ *denotes the boundary of* $A$ *w.r.t. the topology* $\tau_M$ *induced by* $\tau$ *on* $M := \{ta + (1-t)x \mid t \in [0, 1]\}$.

***Proof***

(a) $t_0 := \inf\{t \in \mathbb{R}_+ \mid ta + (1-t)x \in A\} = \inf\{t \in \mathbb{R}_+ \mid x + t(a-x) \in A\} \leq 1$ since $a \in A$.
To obtain a contradiction, suppose $t_0 = 0$. Then, for all $U \in \mathcal{N}(x)$, there exists $t \in \mathbb{R}_+$ with $x + t(a-x) \in A \cap U$. Hence, $x \in \operatorname{cl} A$, which contradicts the assumption. Thus, $t_0 > 0$. The definition of $t_0$ implies, for $b := t_0 a + (1-t_0)x$ and all $V \in \mathcal{N}(b)$, the existence of $t_1 \geq t_0$ and $t_2 < t_0$ such that $t_1, t_2 \in [0, 1]$, $x + t_1(a-x) \in A \cap V$ and $x + t_2(a-x) \in V \setminus A$. Therefore, $b \in \operatorname{bd} A$.
(b) can be proved in the same way as (a) when replacing the neighborhoods by their intersection with M.

□

Norms and seminorms are important for handling distances in linear spaces.

**Definition 2.1.27** A **seminorm** on a linear space $X$ is a mapping $p : X \to \mathbb{R}$ that has the following properties:

(a) $p(x^1 + x^2) \leq p(x^1) + p(x^2)$ for all $x^1, x^2 \in X$,
(b) $p(x) \geq 0$ for all $x \in X$,
(c) $p(\lambda x) = |\lambda| p(x)$ for all $\lambda \in \mathbb{R}, x \in X$.

If additionally $p(x) = 0$ implies $x = 0$, then $p$ is called a **norm** on $X$ and is denoted by $\|\cdot\|$. In this case, $X$ equipped with $\|\cdot\|$ is said to be a **normed space** and is also denoted by $(X, \|\cdot\|)$. Then we write $\|x\|$ instead of $p(x)$.

A norm defines a distance in a natural way.

**Definition 2.1.28** For a nonempty set $A$ in a normed space $X$, the **distance function** $d(\cdot, A)$ on $X$ is defined by

$$d(x, A) := \inf_{a \in A} \|a - x\|.$$

Each sublinear functional induces seminorms (see [3, Lemma 5.47] for (a)):

**Lemma 2.1.29** *Assume that* $X$ *is a linear space and* $p \colon X \to \mathbb{R}$ *is sublinear. Then each of the following functions is a seminorm on* $X$*:*

(a) $q_1 \colon X \to \mathbb{R}$ *defined by* $q_1(x) := \max\{p(x), p(-x)\}$,
(b) $q_2 \colon X \to \mathbb{R}$ *defined by* $q_2(x) := p(x) + p(-x)$.

We will now introduce the Minkowski functional of an absorbing set $A$. This is a special case of Definition 3.2.4, where the functional is not necessarily real-valued.

**Definition 2.1.30** For each absorbing subset $A$ of a linear space $X$, the **Minkowski functional** $p_A : X \to \mathbb{R}$ of $A$ is defined by

$$p_A(x) := \inf\{\lambda \in \mathbb{R}_> \mid x \in \lambda A\}.$$

**Lemma 2.1.31** *Assume that $p : X \to \mathbb{R}$ is a nonnegative function on a linear space $X$.*

(a) *$p$ is sublinear if and only if it is the Minkowski functional of a convex absorbing set $A \subseteq X$, in which case we may take $A = \{x \in X \mid p(x) \leq 1\}$.*
(b) *$p$ is a seminorm if and only if it is the Minkowski functional of a convex absorbing balanced set $A \subseteq X$, in which case we may take $A = \{x \in X \mid p(x) \leq 1\}$.*

**Lemma 2.1.32** *Suppose that $X$ is a topological vector space.*

(a) *If $A$ is a closed absorbing set that is starshaped about zero, then $A = \{x \in X \mid p_A(x) \leq 1\}$.*
(b) *A nonnegative real-valued function on $X$ is sublinear and lower semicontinuous if and only if it is the Minkowski functional of a closed convex absorbing set.*
(c) *A nonnegative real-valued function on $X$ is sublinear and continuous if and only if it is the Minkowski functional of a convex neighborhood of zero.*

For the statements in the previous two lemmata, see e.g. [3].

Seminorms induce topologies.

**Lemma 2.1.33** *Assume a seminorm $p\colon X \to \mathbb{R}$ on the linear space $X$.*

*Then there exists a unique topology $\tau$ with the following two properties:*

(a) *$\{\{x \in X \mid p(x) \leq \varepsilon\} \mid \varepsilon \in \mathbb{R}_>\}$ is a neighborhood base of $\tau$ at zero,*
(b) *$(X, \tau)$ is a topological vector space.*

*$(X, \tau)$ is a locally convex space.*

We say that a **topology** on a linear space $X$ is **generated by a family of seminorms**

$$\{p_i\colon X \to \mathbb{R} \mid i \in I\}$$

if it is generated by the system $\{x + \mathscr{B}(0) \mid x \in X\}$ with

$$\mathscr{B}(0) := \left\{\bigcap_{i \in I}\{x \in X \mid p_i(x) \leq \varepsilon\} \mid \varepsilon \in \mathbb{R}_>\right\}.$$

The topology in each locally convex space can be generated by a family of seminorms [3, Theorem 5.73].

**Definition 2.1.34** Suppose some normed space $(X, \|\cdot\|)$.

The topology on $X$ generated by the norm $\|\cdot\|$ is called the **norm topology** of $(X, \|\cdot\|)$.

For $\varepsilon \in \mathbb{R}_>$, the **closed ball** with the center $x^0 \in X$ and the radius $\varepsilon$ is given by

$$B_{\|\cdot\|}(x^0, \varepsilon) := \{x \in X \mid \|x - x^0\| \leq \varepsilon\}$$

and the related **open ball** with the center $x^0 \in X$ and the radius $\varepsilon$ by

$$\mathring{B}_{\|\cdot\|}(x^0, \varepsilon) := \{x \in X \mid \|x - x^0\| < \varepsilon\}.$$

We call the closed and the open ball with center 0 and radius 1 the closed and the open **unit ball** of the norm and denote it by $B_{\|\cdot\|}$ and $\mathring{B}_{\|\cdot\|}$, respectively.

A set $A \subseteq X$ is said to be **complete** if each Cauchy sequence of elements of $A$ converges to an element of $A$. $X$ is called a **Banach space** if $X$ is complete.

Normed spaces are always supposed to be equipped with the norm topology. In this way, they are locally convex spaces.

The topology in the product space $X \times Y$ of two normed spaces $(X, \|\cdot\|_X)$ and $(Y, \|\cdot\|_Y)$ can be generated by the norm $\|\cdot\|_{X\times Y}$ given by

$$\|(x, y)\|_{X\times Y} := \sqrt{\|x\|_X^2 + \|y\|_Y^2} \qquad \text{for all } x \in X,\ y \in Y.$$

Each Banach space is a separated locally convex space and first countable. The next proposition characterizes essential properties of Banach spaces (see [63] for these and further properties).

**Proposition 2.1.35** *Let $X$ be a Banach space. The following statements are equivalent:*

(a) *$X$ is finite-dimensional.*
(b) *Every linear subspace of $X$ is closed.*
(c) *Every linear real-valued functional on $X$ is continuous.*
(d) *Every convex real-valued functional on $X$ is continuous.*

We give some examples of Banach spaces (see [3, 272] for more details).

**Example 2.1.36**

(1) $\mathbb{R}^n$ is a Banach space when equipped with one of the norms $\|\cdot\|_p$, $1 \leq p < +\infty$, and also when equipped with the norm $\|\cdot\|_\infty$, where

$$\|x\|_p := \sqrt[p]{\sum_{i=1}^{n} |x_i|^p} \text{ and } \|x\|_\infty := \max_{i=1,\ldots,n} |x_i| \text{ for all } x = (x_1, \ldots, x_n)^T \in \mathbb{R}^n.$$

$\|\cdot\|_\infty$ is called the **Chebyshev norm** or **maximum norm** on $\mathbb{R}^n$.
$\|x\|_2$ is said to be the **Euclidean norm**. The related norm topology on $\mathbb{R}^n$ is the so-called **Euclidean topology** $\tau_0$. Equipped with this topology, $\mathbb{R}^n$ is called the **Euclidean space**. In $\mathbb{R}$, the open sets in this topology are the empty set and all unions of open intervals.

(2) Consider the space of real sequences $x = (x_i)_{i\in\mathbb{N}}$ with the norms $\|\cdot\|_p$, $1 \le p \le +\infty$, defined by

$$\|x\|_p := \sqrt[p]{\sum_{i\in\mathbb{N}} |x_i|^p} \text{ if } 1 \le p < \infty, \text{ and } \|x\|_\infty := \sup_{i\in\mathbb{N}} |x_i|.$$

Each space $\ell^p$, $1 \le p \le +\infty$, of real sequences $x$ with $\|x\|_p < +\infty$ is a Banach space when equipped with $\|\cdot\|_p$. The subspace $c_0$ of all sequences in $\ell^\infty$ converging to zero is a Banach space as well. This space contains each space $\ell^p$, $1 \le p < +\infty$.

(3) On the space $C[a, b]$ of all real-valued functions that are continuous on the interval $[a, b]$ ($a, b \in \mathbb{R}$, $a < b$), the **Chebyshev** or **supremum norm** $\|\cdot\|_\infty$ is defined by

$$\|x\|_\infty := \sup_{t\in[a,b]} |x(t)|.$$

$(C[a, b], \|\cdot\|_\infty)$ is a Banach space.

(4) Consider any measure space $(\Omega, M, \mu)$. Take any $p \in \mathbb{R}$ with $1 \le p < \infty$. The set of all measurable functions $f\colon \Omega \to \mathbb{R}$ with $\int_\Omega |f(x)|^p d\mu(x) < \infty$ is a linear space, where addition and scalar multiplication are defined in the usual way for function spaces. We denote this space by $\mathscr{L}^p(\Omega, M, \mu)$. A seminorm on $\mathscr{L}^p(\Omega, M, \mu)$ is given by

$$f \mapsto \left(\int_\Omega |f(x)|^p d\mu(x)\right)^{\frac{1}{p}}.$$

Equivalence classes of functions from $\mathscr{L}^p(\Omega, M, \mu)$ are built in the following way: Two functions belong to the same equivalence class if and only if they coincide almost everywhere on $\Omega$, i.e., if they can only be different on some set $N$ with $\mu(N) = 0$. Denote the set of these equivalence classes by $L^p(\Omega, M, \mu)$. On this space, the above seminorm becomes a norm, which is denoted by $\|\cdot\|_{L^p(\Omega,M,\mu)}$. Equipped with this norm, $L^p(\Omega, M, \mu)$ is a Banach space. Furthermore, the set of all measurable functions $f\colon \Omega \to \mathbb{R}$ with

$$\operatorname*{ess\,sup}_{x\in\Omega} |f(x)| := \inf_{\substack{N\in M\\ \mu(N)=0}} \sup_{x\in\Omega\setminus N} |f(x)| < \infty$$

is a linear space. We denote this space by $\mathscr{L}^\infty(\Omega, M, \mu)$.
A seminorm on $\mathscr{L}^\infty(\Omega, M, \mu)$ is given by

$$f \mapsto \operatorname*{ess\,sup}_{x\in\Omega} |f(x)|.$$

Equivalence classes of functions from $\mathscr{L}^\infty(\Omega, M, \mu)$ are built in the same way as above. Denote the set of these equivalence classes by $L^\infty(\Omega, M, \mu)$. On

this space, the seminorm becomes a norm, which is denoted by $\|\cdot\|_{L^\infty(\Omega,M,\mu)}$. Equipped with this norm, $L^\infty(\Omega, M, \mu)$ is a Banach space.
Take now any $p \in \mathbb{R}$ with $1 \le p \le \infty$. The space $L^p(\Omega, M, \mu)$ is called a (real) **Lebesgue space**.

If $\Omega \subseteq \mathbb{R}^\ell$, $M$ is the Borel $\sigma$-algebra on $\Omega$ and $\mu$ is the Lebesgue measure, then $L^p(\Omega, M, \mu)$ is denoted by $L^p(\Omega)$. In this case, we use the notation $L^p[a, b]$ if $\Omega$ is the interval $[a, b]$ with $a, b \in \mathbb{R}$, $a < b$.

Investigations in a vector space often make use of its dual space since linear functionals are appropriate for the separation of convex sets.

**Definition 2.1.37** The **algebraic dual space** $X^\#$ of a vector space $X$ is defined as the set of all linear mappings from $X$ into $\mathbb{R}$. The **topological dual space** $X^*$ of a topological vector space $X$ is defined as the set of all continuous linear mappings from $X$ into $\mathbb{R}$.

All linear real-valued functions defined on a finite-dimensional topological vector space are continuous [402]. By Proposition 2.1.35, $X^* = X^\#$ holds for a Banach space $X$ if and only if $X$ is finite-dimensional. In [190, 1.6], one can find an example of a linear real-valued function that is defined on some convex subset C of a Hilbert space, but is discontinuous at each point of some set that is dense in C.

When working with elements of $X^*$, $X$ is usually assumed to be a nontrivial separated locally convex space since this guarantees the existence of some $x^* \in X^* \setminus \{0\}$.

Many duality statements depend on the fact that a space is isomorph to the dual space of its dual space. This relationship can be described using the notion of a dual pair [67, 68].

**Definition 2.1.38** Given two linear spaces $X$ and $X'$, a function $\Phi$ is said to be a **bilinear functional** on $X \times X'$ if $\Phi\colon X \times X' \to \mathbb{R}$ fulfills the conditions (1)–(2).

(1) $\Phi(x, .)$ is linear for each $x \in X$.
(2) $\Phi(., x')$ is linear for each $x' \in X'$.

Such a functional is often denoted by $\langle ., . \rangle$, where $\langle x, x' \rangle$ is written instead of $\Phi(x, x')$.

A **dual pair** $(X, X', \Phi)$ is a pair of linear spaces $X$ and $X'$ together with a bilinear functional $\Phi$ on $X \times X'$ for which the conditions (3)–(4) are satisfied. A dual pair is also called a **dual system**.

(3) If $\Phi(x, x') = 0$ for all $x' \in X'$, then $x = 0$.
(4) If $\Phi(x, x') = 0$ for all $x \in X$, then $x' = 0$.

Each space of a dual pair can be interpreted as a set of linear functions on the other space which separates the elements of the other space.

Examples of dual pairs are:

- $(X, X^\#, \langle .,. \rangle)$, where $X$ is an arbitrary linear space, $X^\#$ is its algebraic dual space and $\langle x, x' \rangle = x'(x)$,
- $(X, X^*, \langle .,. \rangle)$, where $X$ is an arbitrary separated locally convex space, $X^*$ is its topological dual space and $\langle x, x^* \rangle = x^*(x)$,
- $(X, X, \langle .,. \rangle)$, where $X$ is an arbitrary linear space with a scalar product $\langle .,. \rangle$.

If notions and statements which we introduced for dual pairs will be used without reference to some dual pair, then they refer to the above dual pair $(X, X^*, \langle .,. \rangle)$.

Furthermore, the following dual pairs should be mentioned:

- $(\mathbb{R}^\ell, \mathbb{R}^\ell, \langle .,. \rangle)$ with $\langle x, x' \rangle = (x')^T x$,
- $(\ell^\infty, \ell^1, \langle .,. \rangle)$ with $\langle x, x' \rangle = \sum_{i=1}^\infty x'_i x_i$,
- $(C[T], M[T], \langle .,. \rangle)$, where $T$ is a compact subset of a separable Banach space, $C[T]$ is the space of continuous real-valued functions on $T$, $M[T]$ is the space of Radon measures on $T$, and $\langle f, \mu \rangle = \int_T f(x)\, d\mu(x)$,
- $(L^p(\Omega, M, \mu), L^q(\Omega, M, \mu), \langle .,. \rangle)$ with $1 < p < \infty$, $1 < q < \infty$, $\frac{1}{p} + \frac{1}{q} = 1$ and $\langle f, g \rangle = \int_\Omega f(x) \cdot g(x)\, d\mu(x)$,
- $(L^1(\Omega, M, \mu), L^\infty(\Omega, M, \mu), \langle .,. \rangle)$ with $\langle f, g \rangle = \int_\Omega f(x) \cdot g(x)\, d\mu(x)$.

Spaces $L^p(\Omega, M, \mu)$, $1 \le p \le \infty$, with some probability measure $\mu$ are used in risk analysis. In this case, $\langle f, g \rangle$ is the expected value of $f \cdot g$.

Given two dual pairs $(X^1, Y^1, \Phi_1)$ and $(X^2, Y^2, \Phi_2)$, a dual pair related to the product space is given by $(X^1 \times X^2, Y^1 \times Y^2, \Phi)$ with

$$\Phi((x^1, x^2), (y^1, y^2)) := \Phi_1(x^1, y^1) + \Phi_2(x^2, y^2).$$

Consider now an arbitrary dual pair $(X, X', \Phi)$. Then the functions $p_{x'} : X \to \mathbb{R}$ and $p_x : X' \to \mathbb{R}$ defined by

$$p_{x'}(x) := p_x(x') := |\Phi(x, x')| \quad \text{for all } x \in X,\ x' \in X'$$

are seminorms on $X$ and on $X'$, respectively.

The family $\{p_{x'} \mid x' \in X'\}$ of seminorms generates a topology on $X$, with which $X$ becomes a separated locally convex space. This topology is denoted by $\sigma(X, X')$ and called the **weak topology** w on $X$ generated by $X'$. It is the weakest topology on $X$ in which each function $\Phi(., x')$ with $x' \in X'$ is continuous.

The family $\{p_x \mid x \in X\}$ of seminorms generates a topology on $X'$, with which $X'$ becomes a separated locally convex space. This topology is denoted by $\sigma(X', X)$ and called the **weak* topology** w* on $X'$ generated by $X$. It is the weakest topology on $X'$ in which each function $\Phi(x, .)$ with $x \in X$ is continuous.

If $(X, X', \langle .,. \rangle)$ is a dual pair, then

$$(X, \sigma(X, X'))^* = X', \tag{2.1.2}$$

i.e., if $f: X \to \mathbb{R}$ is a $\sigma(X, X')$-continuous linear functional, then there exists a unique $x' \in X'$ such that $f(x) = \langle x, x' \rangle$ for each $x \in X$. Similarly,

$$(X', \sigma(X', X))^* = X. \tag{2.1.3}$$

Because of (2.1.2) and (2.1.3), statements which have been proved for the dual space $(X, X^*, \langle .,. \rangle)$ can be transferred to arbitrary dual pairs $(X, X', \langle .,. \rangle)$ if $X$ is equipped with the weak topology $\sigma(X, X')$ and $X'$ is equipped with the weak* topology $\sigma(X', X)$.

For a separated locally convex space $X$ and $X' = X^*$, (2.1.2) underscores that the $\sigma(X, X^*)$-continuous linear functionals on $X$ are the linear functionals on $X$ which are continuous in the original topology of the space.

If $A$ is a convex subset of a separated locally convex space $X$, then

$$\text{cl} A = \text{cl}_w A.$$

For details of dual spaces and dual pairs, we refer to [3, 198].

In Chap. 12, generalized differentiability will be studied in Asplund spaces. Calculus rules in such spaces can be used to derive optimality conditions based on limiting subdifferentials (see Sect. 12.1) and to prove extremal principles (see Sect. 12.6).

**Definition 2.1.39** A Banach space $X$ is called an **Asplund space** if every continuous convex real-valued function on any nonempty open convex subset $U$ of $X$ is Fréchet differentiable at the points of a dense subset of $U$.

The class of Asplund spaces contains each reflexive Banach space and each Banach space with a separable topological dual space. In particular, the spaces $\ell^p$ and $L^p[0, 1]$ are Asplund spaces for $1 < p < +\infty$. However, $\ell^1$ and $\ell^\infty$ are not Asplund spaces, although $c_0$ is an Asplund space. Asplund spaces play an important role in variational analysis and in the geometric theory of Banach spaces (see [60, 359, 361]).

### 2.1.2 Binary Relations

Binary relations are the basis for the definition of optimal elements and of convergence.

**Definition 2.1.40** Let $M$ be a set and $\mathcal{R}$ a subset of $M \times M$. Then $\mathcal{R}$ is called a **(binary) relation** on $M$ and the pair $(M, \mathcal{R})$ is called a set $M$ with binary relation $\mathcal{R}$. The containment $(x^1, x^2) \in \mathcal{R}$ will be denoted by $x^1 \mathcal{R} x^2$.

Two elements $x^1, x^2 \in M$ are said to be **comparable** if $x^1 \mathcal{R} x^2$ or $x^2 \mathcal{R} x^1$ holds.

The binary relation $\mathcal{R}$ is called:

(a) **reflexive** if $x \mathcal{R} x$ for every $x \in M$;
(b) **transitive** if, for all $x^1, x^2, x^3 \in M$, $x^1 \mathcal{R} x^2$ and $x^2 \mathcal{R} x^3$ imply $x^1 \mathcal{R} x^3$;

(c) **asymmetric** if, for all $x^1, x^2 \in M,\ x^1\mathscr{R}x^2$ implies $\neg(x^2\mathscr{R}x^1)$;
(d) **antisymmetric** if, for all $x^1, x^2 \in M,\ x^1\mathscr{R}x^2$ and $x^2\mathscr{R}x^1$ imply $x^1 = x^2$;
(e) **complete** if any two elements of $M$ are comparable;
(f) a **preorder** if $\mathscr{R}$ is transitive;
(g) a **quasiorder** if $\mathscr{R}$ is reflexive and transitive;
(h) a **partial order** if $\mathscr{R}$ is reflexive, transitive, and antisymmetric;
(i) a **linear** or **total order** if $\mathscr{R}$ is a complete partial order.

In the cases (f)–(i), the containment $(x^1, x^2) \in \mathscr{R}$ is also denoted by $x^1 \leq_{\mathscr{R}} x^2$ or simply by $x^1 \leq x^2$ if there is no risk of confusion.

The **inverse relation** of $\mathscr{R}$ is the binary relation

$$\mathscr{R}^{-1} := \{(x^1, x^2) \in M \times M \mid (x^2, x^1) \in \mathscr{R}\}.$$

The notions introduced in (e)–(h) may be defined different in the literature. For details, see e.g. [3].

We will now give two examples of partial orders.

**Example 2.1.41**

(1) The usual partial order defined on $\mathbb{R}^n$ is

$$\mathscr{R}_n := \{(x, y) \in \mathbb{R}^n \times \mathbb{R}^n \mid \forall i \in \{1, \dots, n\} \ :\ x_i \leq y_i\},$$

where $x = (x_1, \dots, x_n)^T$ and $y = (y_1, \dots, y_n)^T$.
It is a linear order if and only if $n = 1$.
On $\mathbb{R}^n$, we will denote $\mathscr{R}_n$ by $\leq$ and $\mathscr{R}_n^{-1}$ by $\geq$.
Furthermore, we consider on $\mathbb{R}^n$ the asymmetric transitive relation

$$\mathscr{R}_< := \{(x, y) \in \mathbb{R}^n \times \mathbb{R}^n \mid \forall i \in \{1, \dots, n\} \ :\ x_i < y_i\}.$$

We will denote $\mathscr{R}_<$ by $<$ and $\mathscr{R}_<^{-1}$ by $>$.

(2) A partial order on the set $M$ of subsets of some nonempty set $X$ is given by

$$\{(A, B) \in M \times M \mid A \subseteq B\}.$$

The following definition is based on [250].

**Definition 2.1.42** Let $\mathscr{R}$ be a binary relation on the nonempty set $M$ and $M_0 \subseteq M$. An element $x^0 \in M_0$ is called a **nondominated element** of $M_0$ w.r.t. $\mathscr{R}$ if

$$\forall x \in M_0\colon\ (x\mathscr{R}x^0 \implies x^0\mathscr{R}x).$$

The set of all nondominated elements of $M_0$ w.r.t. $\mathscr{R}$ is denoted by $\mathrm{ND}(M_0, \mathscr{R})$.

If $\mathscr{R}$ is an antisymmetric binary relation on $M_0$, then

$$\mathrm{ND}(M_0, \mathcal{R}) = \{x^0 \in M_0 \mid \forall x \in M_0 \colon (x\mathcal{R}x^0 \Rightarrow x = x^0)\}.$$

If $\mathcal{R}$ is an asymmetric binary relation on $M_0$, then

$$\mathrm{ND}(M_0, \mathcal{R}) = \{x^0 \in M_0 \mid \nexists x \in M_0 \colon x\mathcal{R}x^0\}.$$

For an arbitrary partial order $\mathcal{R}$ on $M$, the set $\mathrm{ND}(M, \mathcal{R})$ may contain none, exactly one or more than one element.

In Sect. 6.1, we will investigate sets of nondominated elements based on set-valued mappings which will be assigned to the relations.

**Definition 2.1.43** Let $\mathcal{R}$ be a binary relation on a set $M$ and $M_0$ be a nonempty subset of $M$.

1. $M_0$ is **bounded below** or **bounded above** (w.r.t. $\mathcal{R}$) if there exists $a \in M$ such that $a\mathcal{R}x$ or $x\mathcal{R}a$, respectively, holds for every $x \in M_0$, where $a$ is called a **lower bound** or an **upper bound**, respectively, of $M_0$ (w.r.t. $\mathcal{R}$).
2. Assume that $\mathcal{R}$ is a partial order. We say that $a \in M$ is the **infimum** of $M_0$ if $a$ is a lower bound of $M_0$ and we have $a'\mathcal{R}a$ for any lower bound $a'$ of $M_0$. We call $a \in M$ the **supremum** of $M_0$ if $a$ is an upper bound of $M_0$ and we have $a\mathcal{R}a'$ for any upper bound $a'$ of $M_0$.

The existence of nondominated elements w.r.t. binary relations is of basic importance for optimization problems. In general spaces, this requires the Axiom of Choice, from which we get **Zorn's lemma**.

**Lemma 2.1.44** (Zorn's lemma) *Let $M$ be a nonempty set and $\mathcal{R}$ a reflexive transitive binary relation on $M$. If every nonempty subset of $M$ on which $\mathcal{R}$ is complete is bounded below, then $M$ has at least one nondominated element.*

Zorn's lemma is used to prove the existence of a Hamel basis for a nontrivial linear space.

### 2.1.3 Convergence

We will now introduce convergence in arbitrary topological spaces. Here a point belonging to the closure of a set is not necessarily a limit of some sequence in this set. Examples in [3] illustrate this fact. However, the closure of a set can be described by limits of more general structures than sequences. This is done by extending the index set from $\mathbb{N}$ to a directed set and calling the new structure a net.

**Definition 2.1.45**

(1) A set $I$ is called **directed** w.r.t. a binary relation $\preceq$ given on $I$ if $\preceq$ is reflexive and transitive and if, for all elements $i_1, i_2 \in I$, there exists $i \in I$ such that $i_1 \preceq i$ and $i_2 \preceq i$.

(2) A **net** in a set $A$ is a function $f: I \to A$, where $I$ is a directed set. The directed set $I$ is called the **index set** of the net and each element of $I$ is an **index**. The net is also written as $\{a_i\}_{i\in I}$ or $(a_i)_{i\in I}$ with $a_i = f(i)$ or simply as $\{a_i\}$ or $(a_i)$. If $A$ is a subset of a topological vector space, then the net is said to be **bounded** if the function $f$ is bounded.

(3) A net $\{b_j\}_{j\in J}$ is a **subnet** of a net $\{a_i\}_{i\in I}$ if there exists a function $\varphi: J \to I$ with

   (i) $b_j = a_{\varphi(j)}$ for each $j \in J$,
   (ii) $\forall i_0 \in I \; \exists j_0 \in J \; \forall j \in J$ with $j \preceq j_0 : \varphi(j) \preceq i_0$.

Each sequence $(a_n)_{n\in\mathbb{N}}$ is a net, and the subnets of a sequence are its subsequences. For sequences, we will prefer the notation with parentheses.

Nets are used in a similar way as sequences to define convergence. In the following, we denote by $\succeq$ the inverse relation of $\preceq$.

**Definition 2.1.46** Let $X$ be a topological space, $\{x_i\}_{i\in I}$ a net in $X$, $x \in X$.

(a) $x$ is called a **cluster point** or an **accumulation point** of $\{x_i\}_{i\in I}$ in $X$ if

$$\forall U \in \mathcal{N}(x) \; \forall i_0 \in I \; \exists i \succeq i_0 : \; x_i \in U.$$

(b) $\{x_i\}_{i\in I}$ is said to **converge** to $x$ if

$$\forall U \in \mathcal{N}(x) \; \exists i_0 \in I \; \forall i \succeq i_0 : \; x_i \in U.$$

   In this case, we denote $x$ as a **limit** of $\{x_i\}_{i\in I}$ and write $x_i \to x$ or $x_i \overset{X}{\to} x$ to indicate that the elements of the net belong to $X$.

For sequences, the above notions coincide with the usual ones.

**Proposition 2.1.47** *Let $X$ be a topological space, $A \subseteq X$, $x \in X$.*

(a) *$x$ is a cluster point of a net in $X$ if and only if some subnet of this net converges to $x$.*
(b) *$x$ is a limit of a net in $X$ if and only if each subnet of this net converges to $x$.*
(c) *$x$ belongs to the closure of $A$ if and only if it is the limit of some net in $A$.*
(d) *If $X$ is first countable, then $x$ belongs to the closure of $A$ if and only if it is the limit of some sequence in $A$.*
(e) *$X$ is a Hausdorff space if and only if every net converges to at most one point.*

For the proof, see [3].

Consequently, the consideration of nets which are not sequences is not necessary if the topological space is first countable. This is the case for metric spaces.

Consider a bounded net $\{x_i\}_{i\in I}$ in $\mathbb{R}$. This net has a largest and a smallest cluster point, called the **limit superior** and the **limit inferior** of the net, respectively.

These cluster points are denoted by $\limsup_{i\in I} x_i$ and $\liminf_{i\in I} x_i$, respectively.

It can be shown that

$$\liminf_{i\in I} x_i = \sup_{i\in I}\inf_{j\succeq i} x_j \leq \inf_{i\in I}\sup_{j\succeq i} x_j = \limsup_{i\in I} x_i.$$

In $\mathbb{R}$, we have $x_i \to x$ if and only if $x = \limsup_{i\in I} x_i = \liminf_{i\in I} x_i$.

For the limit, the limit inferior and the limit superior of a sequence $(x_n)_{n\in\mathbb{N}}$, we will also write $\lim_{n\to\infty} x_n$, $\liminf_{n\to\infty} x_n$ and $\limsup_{n\to\infty} x_n$, respectively.

For a sequence $(t_n)$ of real numbers that is strictly monotonically increasing and converges to $t$, the notation $t_n \uparrow t$ or $t_n \nearrow t$ is used. For a sequence $(t_n)$ of real numbers that is strictly monotonically decreasing and converges to $t$, the notation $t_n \downarrow t$ or $t_n \searrow t$ is used. Moreover, $t_n \to +0$ as well as $t_n \to 0+$ indicates convergence to 0 from above.

Later, we will make use of the following well-known facts.

**Lemma 2.1.48** *Assume some dual pair $(X, X', \langle .,. \rangle)$.*

*A net $\{x_i\}_{i\in I}$ of elements from $X$ converges to $x \in X$ in the weak topology $\sigma(X, X')$ if and only if $\{\langle x_i, x'\rangle\}_{i\in I}$ converges to $\langle x, x'\rangle$ for each $x' \in X'$.*

*A net $\{x_i'\}_{i\in I}$ of elements from $X'$ converges to $x' \in X'$ in the weak* topology $\sigma(X', X)$ if and only if $\{\langle x, x_i'\rangle\}_{i\in I}$ converges to $\langle x, x'\rangle$ for each $x \in X$.*

**Lemma 2.1.49** *Assume $(a^n)_{n\in\mathbb{N}}$ to be a sequence of elements $a^n = (a_1^n, \ldots, a_\ell^n)^T$ from $\mathbb{R}^\ell$, $\ell \in \mathbb{N}_>$. If it is bounded, then it has at least one cluster point. The sequence has a limit $a$ if and only if it is bounded and has exactly one cluster point $a$. Moreover, we have:*

$$\lim_{n\to\infty} a^n = a \iff \forall i \in \{1, \ldots, \ell\} : \lim_{n\to\infty} a_i^n = a_i.$$

## 2.2 Cones

In linear spaces, partial orders that are compatible with the linear structure of the space can be represented by cones. We will investigate this fact and properties of cones. This section contains various examples of topological vector spaces and ordering cones.

When considering a binary relation on some vector space, usually a certain compatibility with the linear structure of the space is expected.

**Definition 2.2.1** Let $\mathcal{R}$ be a binary relation on the linear space $X$. We say that $\mathcal{R}$ is **compatible with the linear structure** of $X$ if the following conditions are satisfied:

$$\forall x^1, x^2 \in X \ \forall \lambda \in \mathbb{R}_> : x^1\mathcal{R}x^2 \implies (\lambda x^1)\mathcal{R}(\lambda x^2), \tag{2.2.1}$$

$$\forall x^1, x^2, x \in X : x^1\mathcal{R}x^2 \implies (x^1 + x)\mathcal{R}(x^2 + x). \tag{2.2.2}$$

$X$ is called an **ordered linear space** (w.r.t. $\mathcal{R}$) if $\mathcal{R}$ is a partial order that is compatible with the linear structure of $X$ and fulfills the following conditions:

(i) $\mathscr{R} \neq X \times X$,
(ii) $x^1 \mathscr{R} x^2$ holds for some $x^1, x^2 \in X$ with $x^1 \neq x^2$.

For $(\mathbb{R}^n, \leq)$, (2.2.1) and (2.2.2) are satisfied.

Relations on linear spaces can often be described using some subset of the space. A large number of relations that are compatible with the linear structure of the space can be defined by cones.

**Definition 2.2.2** A subset $C$ of a linear space $X$ is a **cone** if $\mathbb{R}_+ C \subseteq C$. The cone $C$ is said to be **reproducing** if $C - C = X$. The cone $C$ is called **nontrivial** if $C \neq \emptyset$, $C \neq \{0\}$ and $C \neq X$ hold.

Clearly, if $C$ is a nonempty cone, then $0 \in C$. Statements for reproducing cones related to their algebraic interior will be given in Sect. 2.3.1.

We now turn to binary relations which are defined by a fixed set.

**Theorem 2.2.3** *Suppose $X$ to be a linear space. For each subset $D$ of $X$,*

$$\mathscr{R}_D := \{(x^1, x^2) \in X \times X \mid x^2 - x^1 \in D\} \tag{2.2.3}$$

*defines a binary relation on $X$. This relation fulfills condition* (2.2.2). *We have:*

(a) *$\mathscr{R}_D$ is reflexive $\iff$ $0 \in D$.*
(b) *$\mathscr{R}_D$ is transitive $\iff$ $D + D \subseteq D$.*
(c) *$\mathscr{R}_D$ is asymmetric $\iff$ $D \cap (-D) = \emptyset$.*
(d) *$\mathscr{R}_D$ is antisymmetric $\iff$ $D \cap (-D) = \{0\}$.*
(e) *$\mathscr{R}_D$ satisfies condition* (2.2.1) *$\iff$ $D \cup \{0\}$ is a cone.*
(f) *$\mathscr{R}_D$ is a quasiorder that is compatible with the linear structure of the space if and only if $D$ is a nonempty convex cone.*
(g) *$\mathscr{R}_D$ is a partial order that is compatible with the linear structure of the space if and only if $D$ is a nonempty pointed convex cone.*

*The binary relations on $X$ with property* (2.2.2) *are the relations $\mathscr{R}$ that can be represented as $\mathscr{R}_D$ for some subset $D$ of $X$. If $\mathscr{R}$ is a binary relation on $X$ that fulfills* (2.2.2), *then $\mathscr{R} = \mathscr{R}_D$ for $D := \{x \in X \mid 0 \mathscr{R} x\}$.*

We will determine nondominated elements w.r.t. relations $\mathscr{R}_D$ in Chap. 6.

Theorem 2.2.3 yields:

**Corollary 2.2.4** *A linear space $X$ is an ordered linear space w.r.t some binary relation $\mathscr{R}$ if and only if $\mathscr{R} = \{(x^1, x^2) \in X \times X \mid x^2 - x^1 \in C\}$ for some nontrivial pointed convex cone $C \subset X$. The cone $C$ representing the partial order $\mathscr{R}$ is called the **ordering cone** or the **positive cone** of $X$, and $X$ is said to be **ordered** by $C$.*

If $X$ is a vector space ordered by a nontrivial pointed convex cone $C \subset X$, then we denote the corresponding partial order by "$\leq_C$", i.e.,

$$\forall x^1, x^2 \in X\colon\ (x^1 \leq_C x^2 \iff x^2 - x^1 \in C)$$

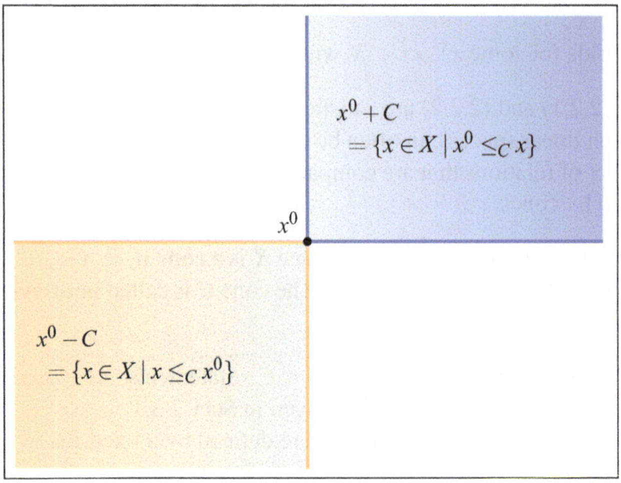


**Fig. 2.1** Partial order on a linear space $X$ with ordering cone $C$

(see Fig. 2.1). The subscript $C$ will often be omitted if no confusion can occur.

For a reproducing ordering cone $C$, we have:

$$\forall x \in X\ \exists c^1, c^2 \in C:\ -c^1 \leq_C x \leq_C c^2.$$

We will now give some statements and definitions related to cones, which are essential for the following chapters.

**Lemma 2.2.5** *Suppose that $C$ is a cone in a topological vector space $X$.*
*Then* $\operatorname{cl} C$ *and* $\operatorname{int} C \cup \{0\}$ *are cones.*

**Lemma 2.2.6** *Let $C$ be a nonempty subset of a linear space $X$.*

$$\begin{aligned} C \text{ is a convex cone} &\iff \mathbb{R}_+C + \mathbb{R}_+C \subseteq C,\\ &\iff C \text{ is a cone and } C + C \subseteq C,\\ &\iff C \text{ is convex, } C + C \subseteq C \text{ and } 0 \in C. \end{aligned}$$

**Lemma 2.2.7** *Suppose that $C$ is a convex cone in a topological vector space $X$.*

(a) $\operatorname{cl} C$ *and* $\operatorname{int} C \cup \{0\}$ *are convex cones.*
(b) *If* $\operatorname{int} C \neq \emptyset$, *then*

$$\operatorname{cl} C + \operatorname{int} C \subseteq \operatorname{int} C \qquad \text{and}$$
$$\operatorname{cl}(X \setminus C) - \operatorname{int} C \subseteq \operatorname{int}(X \setminus C). \tag{2.2.4}$$

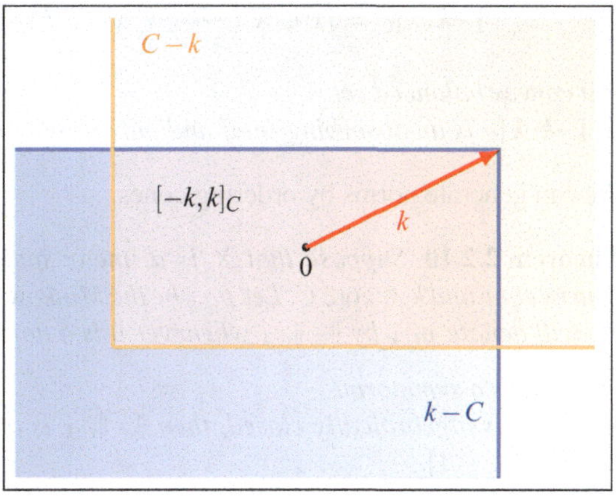


**Fig. 2.2** Order interval $[-k, k]_C$

***Proof***

(a) follows from Lemmata 2.2.5 and 2.1.24.
(b) Assume $\operatorname{int} C \neq \emptyset$. Lemma 2.1.24 implies $\operatorname{int} C = \operatorname{int} \operatorname{cl} C$. Since $\operatorname{cl} C$ is a convex cone, we get $\operatorname{cl} C + \operatorname{int} C = \operatorname{cl} C + \operatorname{int} \operatorname{cl} C \subseteq \operatorname{int} \operatorname{cl} C = \operatorname{int} C$.
Suppose, contrary to our claim, that (2.2.4) does not hold. Then there exist $x \in \operatorname{cl}(X \setminus C)$ and $c \in \operatorname{int} C$ such that $x - c \notin \operatorname{int}(X \setminus C)$. By Lemma 2.1.11(m), $X \setminus \operatorname{int}(X \setminus C) = \operatorname{cl} C$. Hence, $x - c \in \operatorname{cl} C$, which implies $x \in c + \operatorname{cl} C \subseteq \operatorname{int} C + \operatorname{cl} C \subseteq \operatorname{int} C$. This contradicts $x \in \operatorname{cl}(X \setminus C)$ since $X \setminus \operatorname{int} C = \operatorname{cl}(X \setminus C)$ by Lemma 2.1.11. □

**Definition 2.2.8** Let $X$ be a linear space and $C \subset X$ a nontrivial convex cone. A nonempty convex set $B \subset C \setminus \{0\}$ is called a **base** of $C$ if each $x \in C \setminus \{0\}$ has a unique representation $x = \lambda b$ with $\lambda \in \mathbb{R}_>$ and $b \in B$.

A nonempty convex subset $B$ of a convex cone $C$ is a base of $C$ if and only if $C = \mathbb{R}_+ B$ and $0 \notin \operatorname{aff} B$ [192, Theorem 2.1.15].

For all $x^1, x^2 \in X$ in a linear space $X$ with an ordering cone $C$,

$$[x^1, x^2]_C := \{x \in X \mid x^1 \leq_C x \leq_C x^2\}$$

denotes the related **order interval**. This interval may be empty. It is easy to verify that

$$[x^1, x^2]_C = (x^1 + C) \cap (x^2 - C).$$

The following properties of an order interval $[-k, k]_C$ (see Fig. 2.2) result immediately from its definition.

**Lemma 2.2.9** *Suppose that $X$ is a vector space ordered by a nontrivial pointed convex cone $C \subset X$ and $k \in C$. Then*

$$[-k, k]_C = \{x \in X \mid -k \leq_C x \leq_C k\} = (C - k) \cap (k - C)$$

*is a convex balanced set.*

$[-k, k]_C$ *is an absorbing set if and only if* $k \in \operatorname{core} C$.

We can generate norms by ordering cones.

**Theorem 2.2.10** *Suppose that* $X$ *is a linear space,* $C \subset X$ *a nontrivial pointed convex cone and* $k \in \operatorname{core} C$. *Let* $p_{C,k}$ *be the Minkowski functional of the set* $[-k, k]_C$. *We will denote* $p_{C,k}$ *by* $\|\cdot\|_{C,k}$ *whenever it is a norm.*

(a) $p_{C,k}$ *is a seminorm.*
(b) *If* $C$ *is algebraically closed, then* $\|\cdot\|_{C,k}$ *is a norm and* $[-k, k]_C = \{x \in X \mid \|x\|_{C,k} \leq 1\}$.
(c) *If* $X$ *is a separated locally convex space and* $C$ *has a weakly compact base, then* $\|\cdot\|_{C,k}$ *is a norm and* $(X, \|\cdot\|_{C,k})$ *is a reflexive normed space. In this case,* $(X, \|\cdot\|_{C,k})$ *is also a Banach space and an Asplund space.*
(d) *If* $X$ *is a finite-dimensional topological vector space, then* $p_{C,k}$ *is continuous.*
(e) *If* $X$ *is a topological vector space and* $C$ *is closed, then* $\|\cdot\|_{C,k}$ *is a lower semicontinuous norm. In this case,* $[-k, k]_C$ *is the closed unit ball w.r.t. the norm* $\|\cdot\|_{C,k}$, *and* $C$ *has a nonempty interior in the normed space* $(X, \|\cdot\|_{C,k})$.
(f) *If the assumptions of* (e) *and* $k \in \operatorname{int} C$ *hold, then* $\|\cdot\|_{C,k}$ *is continuous.*
(g) *If* $X$ *is a normed space,* $C$ *has a closed bounded base and* $k \in \operatorname{int} C$ *holds, then* $\|\cdot\|_{C,k}$ *is a norm which is equivalent to the norm given on* $X$.
(h) $p_{-C,-k} = p_{C,k}$.

***Proof***

(a) $A := [-k, k]_C$ is a convex absorbing balanced set. The Minkowski functional $p_{C,k} = p_A : X \to \mathbb{R}$ of $A$ is a seminorm by Lemma 2.1.31.
(b) was proved by Jahn [272, Lemma 1.45].
(c) The first statement was proved by Jahn [272, Lemma 1.45], the other statements immediately result from it.
(d) follows from the sublinearity of $p_A$.
(e) The norm property and the unit ball result from (b). Since $A$ is the closed unit ball w.r.t. the norm $p_A$ and $k + A \subseteq C$, $k$ is an interior point of $C$ w.r.t. the topology generated by the norm $p_A$. Furthermore, $p_A$ is lower semicontinuous by Lemma 2.1.32.
(f) Since $A$ is a convex neighborhood of zero, $p_A$ is continuous by Lemma 2.1.32.
(g) Assume that $(X, \|\cdot\|_X)$ is a normed space. Since $\|\cdot\|_{C,k}$ is continuous at 0, there exists $\alpha \in \mathbb{R}_>$ such that $x \in X$ and $\|x\|_X < \alpha$ imply $\|x\|_{C,k} < 1$. For each $x \in X \setminus \{0\}$, we have $\|\frac{\alpha}{\|2x\|_X} x\|_X = \frac{\alpha}{\|2x\|_X}\|x\|_X = \frac{\alpha}{2} < \alpha$, and thus $\|\frac{\alpha}{\|2x\|_X} x\|_{C,k} = \frac{\alpha}{2\|x\|_X}\|x\|_{C,k} < 1$. We obtain $\|x\|_{C,k} \leq \frac{2}{\alpha}\|x\|_X$ for all $x \in X$.
Since $C$ has a bounded base, $V := (C - k) \cap (k - C)$ is bounded [379, p. 62]. Thus, there exists $\beta \in \mathbb{R}_>$ such that $\|x\|_X \leq \beta$ holds for all $x \in V$. For each $x \in$

$X \setminus \{0\}$, $\|\frac{1}{\|x\|_{C,k}} x\|_{C,k} = \frac{1}{\|x\|_{C,k}} \|x\|_{C,k} = 1$ implies $\|\frac{x}{\|x\|_{C,k}}\|_X \leq \beta$ by (b). Therefore, $\|x\|_X \leq \beta \|x\|_{C,k}$ for all $x \in X$. Hence, the norm $\|\cdot\|_{C,k}$ is equivalent to $\|\cdot\|_X$.

(h) holds since the order interval generated by $C$ and $k$ coincides with the order interval generated by $-C$ and $-k$. □

Seminorms of the constructed type are contained in Example 2.2.13. We will prove further properties of $\|\cdot\|_{C,k}$ in Sect. 4.13.

In a linear space $X$ with an ordering cone $C$, $k \in X$ is a so-called order unit of $X$ if and only if $k \in \operatorname{core} C$. The **order unit norm**, which is often denoted by $\|\cdot\|_\infty$, coincides with $\|\cdot\|_{C,k}$. For details related to order units, see [3, 4]. The relationship between order units in complete lattices and the construction of norms was used by Nachbin in [363].

**Corollary 2.2.11** *If $X$ is a topological vector space and $C \subset Y$ a nontrivial closed pointed convex cone with nonempty core, then this core is the interior of $C$ in some norm topology of $X$.*

**Remark 2.2.12** Theorem 2.2.10 and Corollary 2.2.11 were proved in [481]. The related seminorm and norm was constructed in the same way by Jahn [272, Lemma 1.45]. The result that Theorem 2.2.10 contains in part (e) and (f) for locally convex spaces with $k \in \operatorname{int} C$ was given in [291, Example 2.2.13].

We will now collect a few examples of cones.

**Example 2.2.13**

(1) In $\mathbb{R}^n$, the **nonnegative orthant**

$$\mathbb{R}^n_+ := \{(x_1, \ldots, x_n)^T \in \mathbb{R}^n \mid \forall i \in \{1, \ldots, n\} \colon x_i \geq 0\}$$

is a nontrivial pointed convex reproducing cone, which is the usual ordering cone. We have

$$\operatorname{int} \mathbb{R}^n_+ = \{(x_1, \ldots, x_n)^T \in \mathbb{R}^n \mid \forall i \in \{1, \ldots, n\} \colon x_i > 0\}.$$

For $K := \mathbb{R}^n_+$, the relation $\mathscr{R}_K$ is the relation "$\leq$". A base of $K$ is given by $\{(x_1, \ldots, x_n)^T \in \mathbb{R}^n_+ \mid \sum_{i=1}^n x_i = 1\}$. For $k := \mathbf{1} \in \mathbb{R}^n$, the norm $\|\cdot\|_{K,k}$ introduced in Theorem 2.2.10 becomes the maximum norm $\|\cdot\|_\infty$.

(2) Another nontrivial pointed convex reproducing cone in $\mathbb{R}^n$, $n > 1$, is given by

$$C := \left(\bigcup_{i=1}^n C^i\right) \cup \{0\},$$

where $C^i := \{(x_1, \ldots, x_n)^T \in \mathbb{R}^n \mid x_i > 0, x_j = 0 \text{ for all } j < i\}$.
The relation $\mathscr{R}_C$ is a linear order called the **lexicographic order** on $\mathbb{R}^n$.

In $X = \mathbb{R}^2$, we have $C = \{(x_1, x_2)^T \in \mathbb{R}^2 \mid x_1 > 0\} \cup \{(x_1, x_2)^T \in \mathbb{R}^2 \mid x_1 = 0,\ x_2 \geq 0\}$ and $k := (1, 0)^T \in \operatorname{core} C$. For the seminorm $p_{C,k}$ defined in Theorem 2.2.10, we get $p_{C,k}((x_1, x_2)^T) = |x_1|$ for each $(x_1, x_2)^T \in \mathbb{R}^2$. The convex cone $C$ is not closed, and $p_{C,k}$ is not a norm.

(3) In $C[a, b]$,

$$C[a, b]_+ := \{x \in C[a, b] \mid \forall t \in [a, b]\colon x(t) \geq 0\}$$

is a nontrivial pointed convex reproducing cone, which has a nonempty interior in the topology generated by the supremum norm $\| \cdot \|_\infty$.

(4) Let $C^1[a, b]$ be the real vector space formed by all continuously differentiable real-valued functions defined on the interval $[a, b] \subset \mathbb{R}$ and equipped with the norm

$$\|f\|_1 := \left( \int_a^b (f(t))^2 \,\mathrm{d}t + \int_a^b (f'(t))^2 \,\mathrm{d}t \right)^{1/2}$$

for any $f \in C^1[a, b]$. The natural ordering cone

$$C^1[a, b]_+ := \{f \in C^1[a, b] \mid f \geq 0\}$$

has a nonempty interior [412].

(5) In $\ell^p$, $1 \leq p \leq +\infty$,

$$\ell^p_+ := \{(x_i)_{i \in \mathbb{N}} \in \ell^p \mid \forall i \in \mathbb{N}\colon x_i \geq 0\}$$

is the usual ordering cone. The cone $\ell^\infty_+$ is closed, pointed, convex, and reproducing. It has a nonempty interior [379, p. 186]. However, the cones $\ell^p_+$ with $1 \leq p < +\infty$ have an empty interior [272, Example 1.48]. It is well known that even the relative algebraic interior of these standard cones is empty.

(6) For a nonempty subset $\Omega$ of $\mathbb{R}^\ell$ and any $p$ with $1 \leq p \leq +\infty$, the ordering cone of the Lebesgue space $L^p(\Omega)$ is

$$L^p(\Omega)_+ := \{f \in L^p(\Omega) \mid f(x) \geq 0 \text{ almost everywhere on } \Omega\}.$$

It has a nonempty interior for $p = +\infty$, but an empty interior for any $p$ with $1 \leq p < +\infty$ [272, Example 1.51]. Moreover, $L^p(\Omega)_+$ has an empty relative algebraic interior if $1 \leq p < +\infty$.

In order to derive optimality conditions or duality statements in general spaces (see Chap. 12), the ordering cone is often required to have a nonempty interior. However, the ordering cone may have an empty interior even in some Asplund spaces. In Example 2.3.5, we will prove the existence of a nontrivial convex cone with empty relative algebraic interior in each infinite-dimensional linear space.

Sets are often assumed to fulfill certain conditions that depend on the ordering cone.

**Definition 2.2.14** Suppose $X$ to be a linear space, $A$ a nonempty subset of $X$ and $C \subseteq X$ a nonempty convex cone. $A$ is called **C-proper** if $A + C \neq Y$. $A$ is said to be **full** with respect to $C$ if $A = [A]_C$, where

$$[A]_C := (A + C) \cap (A - C).$$

Note that $[A]_C$ is full w.r.t. $C$ for every set $A \subseteq X$.

If $C$ is an ordering cone in $X$, then $[A]_C = \{x \in X \mid \exists a^1, a^2 \in A : a^1 \leq_C x \leq_C a^2\}$. If additionally $A = \{-k, k\}$ for some $k \in C$, then $[A]_C = [-k, k]_C$ is an order interval.

**Definition 2.2.15** Let $(X, \tau)$ be a topological vector space and let $C \subset X$ be a nontrivial convex cone. Then $C$ is said to be **normal** (relative to $\tau$) if $0_X$ has a neighborhood base formed by sets which are full w.r.t. $C$.

Properties of cones which are normal were discussed in [291].

**Definition 2.2.16** Assume that $X$ is a topological vector space, $A$ a nonempty subset of $X$ and $C \subseteq X$ a nonempty convex cone. $A$ is called **C-closed** if $A + C$ is a closed set. $A$ is said to be **C-bounded** if, for each neighborhood $U$ of $0 \in X$, there exists $t \in \mathbb{R}_>$ such that $A \subseteq tU + C$. We say that $A$ is **C-compact** if any cover of $A$ of the form $\{U_\alpha + C \mid U_\alpha \text{ open}\}$ admits a finite subcover.

In a separated topological vector space, every $C$-compact set is $C$-closed and $C$-bounded [338, Proposition 3.3].

When dealing with duality or generalized differentiability, cones in dual spaces are relevant.

**Definition 2.2.17** Assume some dual pair $(X, X', \langle .,. \rangle)$ and a nonempty set $A \subseteq X$.

$$A^+ := \{x' \in X' \mid \forall a \in A \colon \langle a, x' \rangle \geq 0\}$$

is said to be the **(positive) dual cone** of $A$,

$$\operatorname{qint} A^+ := \{x' \in X' \mid \forall a \in A \setminus \{0\} \colon \langle a, x' \rangle > 0\}$$

is called the **quasi-interior** of $A^+$.

The **bipolar cone** of $A$ is defined by

$$A^{++} := \{x \in X \mid \forall x' \in A^+ \colon \langle x, x' \rangle \geq 0\}.$$

The dual cone $A^+$ is a nonempty weak* closed convex cone with

$$A^+ = (\operatorname{cl}_{w^*} A)^+ = (\operatorname{cl}_{w^*} \operatorname{conv}(A \cup \{0\}))^+ = (\operatorname{cone} A)^+ = (\operatorname{cl}_{w^*} \operatorname{cone} A)^+.$$

Furthermore, $\operatorname{qint} A^+ \cup \{0\}$ is a convex cone,

$$\begin{aligned}\operatorname{qint} A^+ &= \operatorname{qint}(\operatorname{cone} A)^+,\\ (-A)^+ &= -A^+,\\ \operatorname{qint}(-A)^+ &= -\operatorname{qint} A^+.\end{aligned}$$

Obviously,

$$A \subseteq A^{++}. \tag{2.2.5}$$

The famous bipolar theorem (see e.g. [3, Theorem 5.103]) yields:

**Proposition 2.2.18** *Assume that* $(X, X', \langle ., . \rangle)$ *is a dual pair and* $A \subseteq X$ *is a nonempty cone. Then*

$$A^{++} = \operatorname{cl}_w \operatorname{conv} A.$$

*If* $A$ *is convex and* $\sigma(X, X')$*-closed, then* $A^{++} = A$.

If $A^+$ or its quasi-interior will be used without mentioning some dual pair, then they refer to the dual pair $(X, X^*, \langle ., . \rangle)$. In this case, $A^+$ is called the **continuous dual cone** of $A$.

The existence of a base is closely related to the dual cone [192, Theorem 2.2.12]:

**Proposition 2.2.19** *Let* $X$ *be a separated locally convex space and* $C \subset X$ *be a nontrivial convex cone. Then* $C$ *has a base* $B$ *with* $0 \notin \operatorname{cl} B$ *if and only if* $\operatorname{qint} C^+ \neq \emptyset$.

We will now introduce the recession cone, which is of basic importance for translation invariant functions.

**Definition 2.2.20** Let $X$ be a linear space. The **recession cone** of $A \subseteq X$ is the set

$$0^+A := \{u \in X \mid A + \mathbb{R}_+ u \subseteq A\}.$$

Note that the recession cone of the empty set is $X$.

A characterization of recession cones in the context of variational analysis will be given in Theorem 3.6.15.

**Lemma 2.2.21** *Assume that* $A$ *is a nonempty subset of a linear space* $X$.

(a) $0^+A$ *is a nonempty convex cone.*
(b) $A = 0^+A \iff A$ *is a convex cone.*
(c) $0^+(0^+A) = 0^+A$.
(d) $-0^+A = 0^+(-A)$.
(e) $0^+A \subseteq 0^+(\operatorname{conv} A)$.
(f) $0^+A \subseteq A \iff 0 \in A$.
(g) $0^+(X \setminus A) = -0^+A$.

*If* $X$ *is a topological vector space, then we have:*

(h) $0^+A \subseteq 0^+(\operatorname{cl} A)$.
(i) $0^+A \subseteq 0^+(\operatorname{int} A)$ *if* $\operatorname{int} A \neq \emptyset$.

(j) $0^+A$ *is closed if* $A$ *is closed.*
(k) $0^+(\text{int } A) = 0^+(\text{cl } A)$ *if* $\text{cl int } A = \text{cl } A$ *and* $\text{int cl } A = \text{int } A$.

***Proof*** (a)–(f) can easily be checked.

(g) Take any $k \in -0^+A$ and suppose, contrary to our claim, $k \notin 0^+(X \setminus A)$. Then there exist $x \in X \setminus A$ and $t \in \mathbb{R}_>$ with $x + tk \notin X \setminus A$, i.e., $x + tk \in A$. This implies $x \in A - \mathbb{R}_> k \subseteq A$, which contradicts $x \in X \setminus A$. Hence, $-0^+A \subseteq 0^+(X \setminus A)$ for each subset $A$ of $X$. Thus, $-0^+(X \setminus A) \subseteq 0^+A$. This yields the assertion.
(h) Take any $u \in 0^+A$. Then $\text{cl } A + \mathbb{R}_+ u \subseteq \text{cl}(A + \mathbb{R}_+ u) \subseteq \text{cl } A$. Hence, $u \in 0^+(\text{cl } A)$.
(i) Take any $u \in 0^+A$. Then $\text{int } A + \mathbb{R}_+ u \subseteq \text{int}(A + \mathbb{R}_+ u) \subseteq \text{int } A$. Therefore, $u \in 0^+(\text{int } A)$.
(j) Take any $a \in A$, $x \in \text{cl}(0^+A)$ and $\lambda \in \mathbb{R}_+$ as well as some arbitrary neighborhood $V$ of $a + \lambda x$. Then $U := V - a$ is a neighborhood of $\lambda x$. Since $0^+A$ is a cone, $\text{cl}(0^+A)$ is a cone as well, and $\lambda x \in \text{cl}(0^+A)$. Hence, there exists $u \in U \cap 0^+A$. Then $a + u \in V \cap A$, and $a + \lambda x \in \text{cl } A = A$. Thus, $x \in 0^+A$. Consequently, $0^+A$ is closed.
(k) $0^+(\text{int } A) \subseteq 0^+(\text{cl int } A) = 0^+(\text{cl } A)$ by (h). (i) implies $0^+(\text{cl } A) \subseteq 0^+(\text{int cl } A) = 0^+(\text{int } A)$. □

The assumptions of (k) are fulfilled by each convex set $A$ with nonempty interior. Even in this case, $0^+A = 0^+(\text{cl } A)$ does not necessarily hold.

**Example 2.2.22** Consider $A := \{(x_1, x_2)^T \in \mathbb{R}^2 \mid x_2 < 0\} \cup \{(0, 0)^T\}$. Then $(1, 0)^T \in 0^+(\text{cl } A) \setminus 0^+A$, although $A$ is convex.

Further sufficient conditions for the assumptions made in Lemma 2.2.21(k) are given in the following lemma.

**Lemma 2.2.23** *Assume* $X$ *to be a topological vector space,* $A, D \subseteq X$ *with* $0 \in \text{cl}(D \setminus \{0\})$.

(a) *If* $\text{cl } A + (D \setminus \{0\}) \subseteq A$, *then* $\text{int cl } A = \text{int } A$ *and* $\text{bd cl } A = \text{bd } A$.
(b) *If* $\text{cl } A + (D \setminus \{0\}) \subseteq \text{int } A$, *then* $\text{cl int } A = \text{cl } A$ *and* $0^+(\text{int } A) = 0^+(\text{cl } A)$.

***Proof***

(a) Take any $x \in \text{int cl } A$. Then there is an open set $U$ with $x \in U \subseteq \text{cl } A$. Since $x - U$ is a neighborhood of 0, there exists $d \in D \setminus \{0\}$ with $d \in x - U$. Hence, $x \in U + d \subseteq \text{cl } A + (D \setminus \{0\}) \subseteq A$. Thus, $x \in \text{int } A$. This implies statement (a).
(b) The first equation results from the assumption, the second from Lemma 2.2.21(k). □

**Proposition 2.2.24** *Assume that* $A$ *is a subset of a topological vector space* $X$ *with* $\text{int } 0^+A \neq \emptyset$. *Then we have:*

(a) *If* $A \neq \emptyset$, *then* $\text{cl}\, A + \text{int}\, 0^+A \subseteq \text{int}\, A$.
(b) $0^+(\text{int}\, A) = 0^+(\text{cl}\, A)$.
(c) $\text{core}\, A = \text{int}\, A = A + \mathbb{R}_> k$ *for each* $k \in \text{int}\, 0^+A$.
(d) $\text{int}\,\text{cl}\, A = \text{int}\, A$, $\text{bd}\,\text{cl}\, A = \text{bd}\, A$, *and* $\text{cl}\,\text{int}\, A = \text{cl}\, A$.
(e) *If* $A$ *is a proper subset of* $X$, *then* $\text{cl}\, A$ *is a proper subset of* $X$.
(f) *For each* $x \in X$ *and* $k \in \text{int}\, 0^+A$, *the inclusion* $x + \mathbb{R}k \subseteq A$ *holds if and only if* $x + \mathbb{R}k \subseteq \text{cl}\, A$.

***Proof*** The statements are obvious if $A$ is empty or $A = X$. Assume now that $A$ is a proper subset of $X$. For each $k \in -\text{int}\, 0^+A$, $\varphi_{A,k}$ is lower semicontinuous because of Corollary 4.12.4 and Proposition 3.3.14. We obtain $\text{cl}\, A + \text{int}\, 0^+A \subseteq A$ from Theorem 4.4.2. This yields (a). Theorem 4.6.6 implies (c). The other statements follow from Lemma 2.3.16. □

**Proposition 2.2.25** *Assume that A is a nonempty closed convex set in a topological vector space* $X$. *Then*

$$\forall a^0 \in A\colon\ 0^+A = \{u \in X \mid a^0 + \mathbb{R}_+ u \subseteq A\}.$$

***Proof*** Take any $a^0 \in A$. Obviously, $0^+A \subseteq \{u \in X \mid a^0 + \mathbb{R}_+ u \subseteq A\}$. Consider now any $u \in X$ with $a^0 + \mathbb{R}_+ u \subseteq A$. Take any $a \in A$ and $t \in \mathbb{R}_>$. Define $a^1 := a + tu$. Let $U$ be an arbitrary neighborhood of $a^1$. Then there exists $n \in \mathbb{N}_>$ such that $x := a^1 + \frac{1}{n}(a^0 - a) \in U$. We get $x = a + tu + \frac{1}{n}(a^0 - a) = (1 - \frac{1}{n})a + \frac{1}{n}(a^0 + ntu) \in A$ since $A$ is convex. Thus, $a^1 \in \text{cl}\, A = A$. Therefore, $u \in 0^+A$. □

The closedness of $A$ is an essential assumption in this proposition.

**Example 2.2.26** In $X = \mathbb{R}^2$, consider the convex cone $A := \{(x_1, x_2)^T \in X \mid x_1 > 0, x_2 \geq 0\} \cup \{(0, 0)^T\}$. Then $(1, 0)^T \in A$ and $(1, 0)^T + \mathbb{R}_+ \cdot (0, 1)^T \subseteq A$, but $(0, 1)^T \notin 0^+A$ since $(0, 0)^T + (0, 1)^T \notin A$.

Conditions for algebraic closedness as well as for directional closedness of a recession cone will be proved in Propositions 2.3.9 and 2.3.39 and in Corollary 2.3.38.

In many applications, cones are generated by finite sets.

**Definition 2.2.27** Suppose $S$ to be a nonempty subset of a linear space $X$. Then the **cone generated** by $S$, denoted by $\text{cone}\, S$, is the smallest cone that contains $S$. The **convex cone generated** by $S$ is the smallest convex cone that contains $S$. A convex cone generated by a finite subset of $X$ is called a **polyhedral cone**.

Obviously, $\text{cone}\, S = \mathbb{R}_+ S$, and the convex cone generated by $S$ consists of all non-negative linear combinations from $S$. The polyhedral cone C generated by the set $\{b^i \mid i \in \{1, \ldots, n\}\} \subset X$ is

$$C = \left\{ \sum_{i=1}^{n} \lambda_i b^i \mid \forall i \in \{1, \ldots, n\}\colon \lambda_i \in \mathbb{R}_+ \right\}.$$

We get the following from [3, Corollary 5.25].

**Lemma 2.2.28** *In a Hausdorff topological vector space, each polyhedral cone is closed.*

In the Euclidean space, each polyhedral cone can be represented by a matrix.

**Proposition 2.2.29** *A set $C \subseteq \mathbb{R}^\ell$ is a polyhedral cone if and only if there exists a matrix $W \in \mathbb{R}^{r,\ell}$ with $r \in \mathbb{N}_>$ such that*

$$C = \{x \in \mathbb{R}^\ell \mid Wx \geq 0\}.$$

*$C$ is pointed if and only if* $\operatorname{rank} W = \ell$.

***Proof*** The first statement was proved in [179, Theorem 12]. $C$ is pointed if and only if $Wx = 0$ has the unique solution $x = 0$. This is fulfilled if and only if $\operatorname{rank} W = \ell$. □

Each nonempty closed pointed convex cone in $\mathbb{R}^2$ is a polyhedral cone [136, Lemma 1.20]. The ice cream cone [4] is a nonempty closed pointed convex cone in $\mathbb{R}^3$ which is not polyhedral.

In $\mathbb{R}^\ell$, a **polyhedral set** or **polyhedron** is defined to be the intersection of a finite number of halfspaces. Hence, a polyhedral set $A$ can be given as

$$A = \{x \in \mathbb{R}^\ell \mid Wx \geq b\}$$

with $W \in \mathbb{R}^{r,\ell}$, $r \in \mathbb{N}_>$, and $b \in \mathbb{R}^r$. A bounded polyhedral set is called a **polytope**.

Obviously, polyhedral sets are convex, and each polyhedral cone is a polyhedral set. For details of polyhedral sets and cones, see [64, 167, 242, 389].

**Proposition 2.2.30** *Consider the polyhedral set*

$$A = \{x \in \mathbb{R}^\ell \mid Wx \geq b\}$$

*with $W \in \mathbb{R}^{r,\ell}$, $r \in \mathbb{N}_>$, and $b \in \mathbb{R}^r$. We have:*

(a) $\{x \in \mathbb{R}^\ell \mid Wx > b\} \subseteq \operatorname{int} A$.
(b) *If $W$ does not contain a row which consists of zeros only, then*

$$\operatorname{int} A = \{x \in \mathbb{R}^\ell \mid Wx > b\}.$$

(c) *If* $\operatorname{rank} W = r$, *then* $\operatorname{int} A = \{x \in \mathbb{R}^\ell \mid Wx > b\} \neq \emptyset$.

***Proof***

(a) Since the linear mapping given by $W$ is continuous, we get the assertion.
(b) To obtain a contradiction, suppose that $\{x \in \mathbb{R}^\ell \mid Wx > b\} \neq \operatorname{int} A$. Then there exists $\bar{x} \in \mathbb{R}^\ell$ such that $W\bar{x} \not> b$ and that $Wx \geq b$ holds for all $x$ in some neighborhood $U$ of $\bar{x}$. Since $W\bar{x} \not> b$, there is a row $W_p$ of $W$ with $W_p\bar{x} = b_p$. There

exists $t_0 \in \mathbb{R}_>$ such that $\bar{x} - tW_p^{\mathrm{T}} \in U$ holds for all $t \in \mathbb{R}$ with $|t| < t_0$. For all $t \in \mathbb{R}$ with $|t| < t_0$, we get $W_p(\bar{x} - tW_p^{\mathrm{T}}) \geq b_p$ and thus $W_p\bar{x} - tW_pW_p^{\mathrm{T}} \geq b_p$. Hence, $-tW_pW_p^{\mathrm{T}} \geq 0$ for all $t \in \mathbb{R}$ with $|t| < t_0$. We conclude that $W_pW_p^{\mathrm{T}} = 0$, which yields $W_p = 0$. This contradicts the assumption.

(c) rank $W = r$ implies that the system $Wx = b + (1, \ldots, 1)^{\mathrm{T}}$ has a solution. Hence, $\{x \in \mathbb{R}^\ell \mid Wx > b\} \neq \emptyset$. □

Later on, we will need the following lemmata from [475].

**Lemma 2.2.31** *Assume that $A$ and $D$ are closed convex sets in $\mathbb{R}^\ell$. If there exists $a \in \mathbb{R}^\ell$ such that $A \cap (a - D)$ is not bounded, then $0^+(A \cap (b - D)) \neq \{0\}$ for each $b \in \mathbb{R}^\ell$ with $A \cap (b - D) \neq \emptyset$.*

***Proof*** Since $A \cap (a - D)$ is closed, convex and not bounded, we get $0^+(A \cap (a - D)) \neq \{0\}$ from [389, Theorem 8.4]. By [389, Corollary 8.3.3], $0^+(A \cap (a - D)) = 0^+A \cap 0^+(a - D) = 0^+A \cap 0^+(b - D) = 0^+(A \cap (b - D))$ for each $b \in \mathbb{R}^\ell$ with $A \cap (b - D) \neq \emptyset$. □

**Lemma 2.2.32** *Assume that $D \subseteq \mathbb{R}^\ell$ is a closed convex set with $0 \in D$ and that $A \subset \mathbb{R}^\ell$ is a set for which there exist a polyhedral cone $H \subset \mathbb{R}^\ell$ and $u \in \mathbb{R}^\ell$ such that $-D \setminus \{0\} \subset \operatorname{int} H$ and $(A - u) \cap \operatorname{int} H = \emptyset$. Then each set $A \cap (b - D)$ with $b \in \mathbb{R}^\ell$ is bounded.*

***Proof*** The polyhedral cone $H$ has a representation $H = \{x \in \mathbb{R}^\ell \mid \forall i \in \{1, \ldots, m\}\colon (w^i)^T x \geq 0\}$ with $m \in \mathbb{N}_>$, $w^i \in \mathbb{R}^\ell \setminus \{0\}$ for each $i \in \{1, \ldots, m\}$. Then $\mathbb{R}^\ell \setminus \operatorname{int} H = \bigcup_{i=1}^{m} W_i$ with $W_i := \{x \in \mathbb{R}^\ell \mid (w^i)^T x \leq 0\}$ for all $i \in \{1, \ldots, m\}$. We get from $-D \setminus \{0\} \subset \operatorname{int} H$ and $0 \in W_i \subset \mathbb{R}^\ell \setminus \operatorname{int} H$ that $W_i \cap (-D) = \{0\}$ holds for all $i \in \{1, \ldots, m\}$. Therefore, $0^+(W_i \cap (-D)) = \{0\}$ for all $i \in \{1, \ldots, m\}$. By Lemma 2.2.31, $W_i \cap (a - D)$ is bounded for all $i \in \{1, \ldots, m\}$ and $a \in \mathbb{R}^\ell$. Hence, the set $(A - u) \cap (b - u - D) \subseteq (\mathbb{R}^\ell \setminus \operatorname{int} H) \cap (b - u - D) = \bigcup_{i=1}^{m}(W_i \cap (b - u - D))$ is bounded for each $b \in \mathbb{R}^\ell$. Consequently, $A \cap (b - D)$ is bounded for each $b \in \mathbb{R}^\ell$. □

**Corollary 2.2.33** *Assume that $D \subset \mathbb{R}^\ell$ is a closed pointed convex cone with $\mathbb{R}^\ell_+ \subseteq D$ and that $A \subset \mathbb{R}^\ell$ is a set for which there exists $u \in \mathbb{R}^\ell$ such that $A \subseteq u + D$. Then each set $A \cap (b - D)$ with $b \in \mathbb{R}^\ell$ is bounded.*

***Proof*** By [392, 6.48], there exist complementary open halfspaces $H_1$, $H_2$ with $D \setminus \{0\} \subset H_1$ and $-D \setminus \{0\} \subset H_2$. Choose $H$ as the closure of $H_2$. Then the assertion follows from Lemma 2.2.32. □

## 2.3 Algebraic Properties of Sets

### 2.3.1 *The Algebraic Interior and the Algebraic Closedness of Sets*

The algebraic interior and the algebraic closedness of sets were introduced in Definition 2.1.6. In Theorem 2.2.10, the algebraic interior was used for the construction of seminorms and norms. According to Corollary 2.2.11, the core of each nontrivial closed pointed convex cone in a topological vector space is the interior of this cone in some norm topology.

We will now discuss further properties of the algebraic interior and of algebraic closedness, which will be needed in the subsequent chapters.

**Lemma 2.3.1** *Suppose that* $X$ *is a linear space,* $A, B \subseteq X$, $x \in X$. *Then:*

(a) $A \subseteq B \implies \text{core}\, A \subseteq \text{core}\, B$.
(b) $A + \text{core}\, B \subseteq \text{core}(A + B)$ *if* $\text{core}\, B \neq \emptyset$.
(c) $x + \text{core}\, A = \text{core}(x + A)$ *if* $\text{core}\, A \neq \emptyset$.
(d) $\text{core}(-A) = -\,\text{core}\, A$.
(e) $x + \text{lin}\, A = \text{lin}(x + A)$.
(f) $\text{lin}(-A) = -\,\text{lin}\, A$.

Parts (a) and (d)–(f) of Lemma 2.3.1 are obvious. The other parts as well as further properties of the core can be found in [508].

Example 4.4.6 will illustrate that

$$\text{core}\, A \subseteq \text{core}(\text{core}\, A) \text{ is } \textbf{not} \text{ satisfied}$$

for an arbitrary set $A$ in $\mathbb{R}^2$. Thus, the algebraic interior of a set is not always algebraically open. Furthermore, we get with $B = \text{core}\, A$:

$$\text{core}\, A \subseteq B \text{ does } \textbf{not} \text{ imply } \text{core}\, A \subseteq \text{core}\, B.$$

In each topological vector space, the condition $\text{int}(\text{int}\, A) = \text{int}\, A$ is satisfied. Consequently, $\{\text{core}\, A \mid A \subseteq X\}$ is generally **not** a topology on the linear space $X$. However, the family of all algebraically open subsets of $X$ is a topology, and $X$ equipped with this topology is a separated topological vector space [331].

**Lemma 2.3.2** *Suppose that* $A$ *is a convex subset of a linear space* $X$.

(a) $\forall a \in \text{core}\, A, x \in A, \lambda \in [0, 1): \quad a + \lambda(x - a) \in \text{core}\, A$.
(b) $\text{core}\, A$ *is convex.*
(c) $\text{core}(\text{core}\, A) = \text{core}\, A$.
(d) *If* $A$ *is a nonempty set and the affine hull of* $A$ *is finite-dimensional, then* $\text{icr}\, A \neq \emptyset$.
(e) *If* $A$ *is algebraically closed, then* $\text{core}(X \setminus A) = X \setminus A$.

***Proof*** Parts (a)–(c) can be found in [289, p. 15], part (d) in [74, Theorem 2.3.5]. (e) results from the arguments given in [272, p. 16]. □

In each infinite-dimensional linear space, there exists a nonempty convex set $A$ with $\operatorname{icr} A = \emptyset$. This will be shown in Example 2.3.5.

**Proposition 2.3.3** *Suppose that $C$ is a cone in a linear space $X$. Then:*

(a) $\operatorname{core} C \cup \{0\}$ *is a cone.*
(b) $\operatorname{core} C = C + \operatorname{core} C$ *if $C$ is convex.*

This proposition results from the proof of [272, Lemma 1.12].

The relationship between reproducing cones and the algebraic interior is described by the following lemma.

**Lemma 2.3.4**

(a) *A cone $C$ in a linear space $X$ is reproducing if* $\operatorname{core} C \neq \emptyset$.
(b) *Each convex reproducing cone in a finite-dimensional linear space has a nonempty core.*

For the proof of (a), see [272, Lemma 1.13]. Part (b) follows from Lemma 2.3.2.

Ordering cones with an empty relative algebraic interior were mentioned in Sect. 2.2. For each infinite-dimensional linear space, the convex cone generated by an arbitrary basis of this space is reproducing, but has an empty core and an empty relative algebraic interior. The idea for the following example can be found in [74, p. 66].

**Example 2.3.5** Suppose $X$ to be an infinite-dimensional linear space and $B$ to be a basis of $X$. Define $C := \{\sum_{i=1}^{k} \alpha_i b^i \mid k \in \mathbb{N}_>,\ b^i \in B \text{ and } \alpha_i \in \mathbb{R}_+ \text{ for all } i \in \{1, \ldots, k\}\}$. $C$ is a nontrivial pointed convex cone with $\operatorname{core} C = \emptyset$. Since $C$ is reproducing, we have $\operatorname{aff} C = X$ and hence $\operatorname{icr} C = \operatorname{core} C = \emptyset$.

In order to show $\operatorname{core} C = \emptyset$, we assume that there exists a $c \in \operatorname{core} C$. Then there exist $n \in \mathbb{N}_>$, $b^i \in B$ and $\alpha_i \in \mathbb{R}_+$ for each $i \in \{1, \ldots, n\}$ with $c = \sum_{i=1}^{n} \alpha_i b^i$ and some $b \in B$ with $b \neq b^i$ for all $i \in \{1, \ldots, n\}$. $c \in \operatorname{core} C$ implies the existence of $r \in \mathbb{R}_>$ with $c - r(b - c) \in C$. Hence, $(1 + r)c - rb \in C$, which contradicts the fact that $C$ consists of linear combinations of the basis with nonnegative factors only.

In Sect. 2.3.2, sets will be investigated which can work as a substitute for the algebraic interior if the algebraic interior is empty.

**Proposition 2.3.6** *Suppose that $X$ is a topological vector space and $A \subseteq X$.*

(I) $\operatorname{int} A \subseteq \operatorname{core} A$, *and* $\operatorname{r-int} A \subseteq \operatorname{icr} A$.
(II) *In the case where $A$ is a convex set,*

$$\operatorname{int} A = \operatorname{core} A$$

*holds if one of the following conditions is fulfilled:*

(a) $X = \mathbb{R}^n$ *(equipped with the Euclidean topology),*
(b) $\text{int}\, A \neq \emptyset$,
(c) *$X$ is a Banach space and $A$ is closed.*

***Proof*** Lemma 2.1.22 yields part (I). The first assertions of part (II) are stated in [244, pp. 8 and 59]. (c) is given in [55, Remark 2.73]. □

Breckner and Trif [74, Example 2.3.9] gave an example of a convex set $A$ with $0 \in \text{core}\, A \setminus \text{int}\, A$ in the normed space $(l_0, \|\cdot\|_\infty)$ of all sequences $(x_n)_{n\in\mathbb{N}}$ of real numbers for which $\{n \in \mathbb{N} \mid x_n \neq 0\}$ is finite.

Example 4.4.6 will illustrate that $\text{int}\, A = \text{core}\, A$ is not satisfied for an arbitrary set $A$ in $\mathbb{R}^2$ even if $A$ has a nonempty interior and a nontrivial recession cone.

We now turn to the algebraic closedness of sets. In a topological vector space, each closed set is algebraically closed. The reverse statement does not hold even if the set is a cone.

**Example 2.3.7** In $X = \mathbb{R}^3$, the cone defined by

$$\left\{(x_1, x_2, x_3)^T \mid x_3 = \sqrt{x_1^2 + x_2^2}\right\} \setminus \{t \cdot (1, 1, \sqrt{2})^T \mid t \in \mathbb{R}_>\}$$

is algebraically closed, but not closed in the Euclidean topology.

We get the following from [244, p. 59].

**Proposition 2.3.8** *Assume that $A$ is a convex set in a topological vector space with* $\text{int}\, A \neq \emptyset$. *Then*

$$\text{lin}\, A = \text{cl}\, A\,.$$

**Proposition 2.3.9** *If $A$ is an algebraically closed set in a linear space $X$, then $0^+A$ is algebraically closed.*

***Proof*** Take any $a^0 \in 0^+A$ and $x \in X$ for which $a^0 + t(x - a^0) \in 0^+A$ holds for all $t \in (0, 1)$. Consider arbitrary elements $a \in A$ and $\lambda \in \mathbb{R}_>$. Since $0^+A$ is a cone, we get $\lambda a^0 + t(\lambda x - \lambda a^0) \in 0^+A$ for all $t \in (0, 1)$. Thus, $(a + \lambda a^0) + t((a + \lambda x) - (a + \lambda a^0)) = a + \lambda a^0 + t(\lambda x - \lambda a^0) \in A + 0^+A \subseteq A$ for all $t \in (0, 1)$. Since $a + \lambda a^0 \in A$ and $A$ is algebraically closed, we obtain $a + \lambda x \in A$. Therefore, $x \in 0^+A$. Hence, $0^+A$ is algebraically closed. □

Finally, it should be mentioned that further possibilities to work with other sets than with the topological interior or closure exist, especially if a topology is not assumed or if the interior of a set is empty.

Borwein and Lewis [62] defined the **quasi relative interior** of a convex set $A$ in a Hausdorff topological vector space $X$ as the set of all $a \in A$ for which $\text{cl cone}(A - a)$ is a linear subspace of $X$. They proved that this set coincides with the relative interior of $A$ if $A$ is a convex finite-dimensional set and $X$ a normed space. Their results include the statement that the quasi relative interior of each cone $L^p(\Omega)_+$ with $1 \leq p < +\infty$ is

$$\{f \in L^p(\Omega) \mid f(x) > 0 \text{ almost everywhere on } \Omega\}.$$

Adan and Novo [1] introduced the **vectorial closure** $\mathrm{vcl}(A)$ for an arbitrary set $A$ in a linear space $X$ as

$$\mathrm{vcl}(A) := \{x \in X \mid \exists k \in X\ \forall \lambda \in \mathbb{R}_>\ \exists t \in \mathbb{R}_> \text{ with } t \leq \lambda\colon x + tk \in A\}.$$

$\mathrm{vcl}(A)$ contains the algebraic closure of $A$ [1] and coincides with it if $A$ is convex [2]. In a topological vector space $X$, $\mathrm{vcl}(A) \subseteq \mathrm{cl}\, A$ [1] and $\mathrm{vcl}(A) = \mathrm{cl}\, A$ if $A$ is a convex set with nonempty interior [2]. However, the following example [1, Example 2] illustrates that $\mathrm{vcl}(\mathrm{vcl}(A)) = \mathrm{vcl}(A)$ is not necessarily fulfilled.

**Example 2.3.10** In $X = \mathbb{R}^2$, choose $A := \{(x_1, x_2)^{\mathrm{T}} \in X \mid x_1 > 0,\ 0 < x_2 \leq x_1^2\}$. Then $\mathrm{vcl}(A) = \{(x_1, x_2)^{\mathrm{T}} \in X \mid x_1 > 0,\ 0 \leq x_2 \leq x_1^2\}$, but $\mathrm{vcl}(\mathrm{vcl}(A)) = \mathrm{vcl}(A) \cup \{(0, 0)^{\mathrm{T}}\} = \mathrm{cl}\, A$.

### 2.3.2 Direction-Depending Properties of Sets

In this section, we will introduce direction-depending closures, interiors and boundaries of sets and study their properties. At the beginning, some further direction-depending statements are presented, which will be applied in the subsequent chapters. Especially the set $A - \mathbb{R}_> k$ and its relationship to the core of $A$ will be investigated, where $A$ is a subset of a linear space $Y$ and $k \in Y \setminus \{0\}$. The set $A - \mathbb{R}_> k$ can be used as a substitute for the algebraic interior of $A$ if $\mathrm{core}\, A = \emptyset$. Bibliographical notes related to Sect. 2.3.2 will be summarized in Remarks 2.3.25 and 2.3.57.

In Chap. 4, we will use conditions under which a cone together with some vector spans a linear space.

**Lemma 2.3.11** *Assume*

$$Y \text{ is a linear space, } A \subseteq Y \text{ and } k \in Y \setminus \{0\}, \qquad (\mathrm{Sp0}_{A,k})$$

*and let $A$ be a convex cone. Then $Y = A + \mathbb{R}k$ holds if and only if*

(a) *$A$ is a linear subspace of $Y$ of codimension 1 and $k \notin A$, or*
(b) $\{k, -k\} \cap \mathrm{core}\, A \neq \emptyset$.

**Corollary 2.3.12** *Assume that $Y$ is a linear space, $A \subseteq Y$ is a convex cone and $k \in A \setminus \{0\}$. Then $Y = A + \mathbb{R}k$ holds if and only if $k \in \mathrm{core}\, A$.*

Lemma 2.3.11 and Corollary 2.3.12 were proved by Zălinescu [506].

**Lemma 2.3.13** *Assume that* $Y$ *is a linear space and* $A \subseteq Y$ *is a cone with* $k \in -\operatorname{core} A$. *Then*

$$Y = A + \mathbb{R}_{>}k.$$

***Proof*** Take any $y \in Y$. Since $k \in -\operatorname{core} A$, there exists $t \in \mathbb{R}_{>}$ with $k + t(-y) \in -A$. Thus, $y \in A + \mathbb{R}_{>}k$ since $A$ is a cone. □

**Lemma 2.3.14** *Assume that* $Y$ *is a topological vector space,* $A$ *is a cone with* $\operatorname{int} A \neq \emptyset$ *and* $k \in \operatorname{int} A$. *Then*

$$Y = \operatorname{int} A - \mathbb{R}_{>}k. \tag{2.3.1}$$

***Proof*** The set $V := \operatorname{int} A - k$ is a neighborhood of 0. For each $t \in \mathbb{R}_{>}$, we get $tV = t(\operatorname{int} A - k) \subseteq \operatorname{int} A - tk$ since $A$ is a cone. Thus, $\mathbb{R}_{>}V \subseteq \operatorname{int} A - \mathbb{R}_{>}k$, which implies $Y = \mathbb{R}_{+}V \subseteq \operatorname{int} A - \mathbb{R}_{>}k$ because of $k \in \operatorname{int} A$. □

**Lemma 2.3.15** *Assume that* $Y$ *is a topological vector space,* $H$ *is a convex subset of* $Y$ *with* $\operatorname{int} H \neq \emptyset$, $k \in 0^{+}H \setminus \{0\}$ *and* $Y = H + \mathbb{R}k$. *Then*

$$H + \mathbb{R}_{>}k \subseteq \operatorname{int} H.$$

***Proof*** Take any $h^1 \in H$ and $t_1 \in \mathbb{R}_{>}$. Then $h^2 := h^1 + t_1 k \in H$ since $k \in 0^{+}H$.

Take any $y^1 \in Y$, and consider $y^2 := y^1 + h^2$. We get $Y = H - \mathbb{R}_{+}k$ from $Y = H + \mathbb{R}k$ and $k \in 0^{+}H$. Hence, there exist $h^3 \in H$ and $r \in \mathbb{R}_{+}$ with $y^2 = h^3 - rk$.

Assume first $r > 0$. Take $t_2 := \frac{r}{t_1 + r} \in (0, 1)$. Then $h^2 + (1 - t_2)y^1 = h^2 + (1 - t_2)(y^2 - h^2) = t_2 h^2 + (1 - t_2)y^2 = t_2(h^1 + t_1 k) + (1 - t_2)(h^3 - rk) = t_2 h^1 + (1 - t_2)h^3 \in H$ since $H$ is convex.

If $r = 0$, then $h^2 + y^1 = y^2 = h^3 \in H$.

Consequently, there exists $t_3 \in \mathbb{R}_{>}$ with $h^2 + t_3 y^1 \in H$. Since $H$ is convex, we obtain $h^2 + ty^1 \in H$ for each $t \in [0, t_3]$. This yields $h^2 \in \operatorname{core} H$. By Proposition 2.3.6, $h^2 \in \operatorname{int} H$. □

The next lemma will be used when studying lower semicontinuous translation invariant functions in Chap. 4.

**Lemma 2.3.16** *Assume that* $Y$ *is a topological vector space,* $A \subseteq Y$, $k \in Y \setminus \{0\}$ *and* $\operatorname{cl} A + \mathbb{R}_{>}k \subseteq A$. *Then the following statements are true.*

(a) $k \in 0^{+}A \cap 0^{+}(\operatorname{cl} A)$,
(b) $\operatorname{core} \operatorname{cl} A = \operatorname{core} A$,
(c) $\operatorname{int} \operatorname{cl} A = \operatorname{int} A$,
(d) $\operatorname{bd} \operatorname{cl} A = \operatorname{bd} A$.
(e) *If* $\operatorname{cl} A + \mathbb{R}_{>}k \subseteq \operatorname{int} A$, *then* $\operatorname{cl} \operatorname{int} A = \operatorname{cl} A$ *and* $0^{+}(\operatorname{int} A) = 0^{+}(\operatorname{cl} A)$.
(f) *If* $A$ *is a proper subset of* $Y$, *then* $\operatorname{cl} A$ *is a proper subset of* $Y$.
(g) *If* $k \in -0^{+}A$, *then* $A$ *is closed.*
$k \in -0^{+}A$ *holds if and only if* $k \in -0^{+}(\operatorname{cl} A)$.

(h) *For each* $y \in Y$, $y + \mathbb{R}k \subseteq A$ *holds if and only if* $y + \mathbb{R}k \subseteq \operatorname{cl} A$.

***Proof***

(a) is obvious.
(b) We have $\operatorname{core} \operatorname{cl} A \supseteq \operatorname{core} A$ since $A \subseteq \operatorname{cl} A$. Take now any $y^0 \in \operatorname{core} \operatorname{cl} A$ and $y \in Y$. There exists $\varepsilon \in \mathbb{R}_>$ such that $y^0 + t(y - k) \in \operatorname{cl} A$ for all $t \in (0, \varepsilon]$. Thus, $y^0 + ty \in \operatorname{cl} A + tk \subseteq A$. Hence, $y^0 \in \operatorname{core} A$.
(c) Take any $y \in \operatorname{int} \operatorname{cl} A$. Then there is an open set $U$ with $y \in U \subseteq \operatorname{cl} A$. Since $y - U$ is a neighborhood of 0, there exists $t \in \mathbb{R}_>$ with $tk \in y - U$. Hence, $y \in U + tk \subseteq \operatorname{cl} A + \mathbb{R}_> k \subseteq A$. Thus, $y \in \operatorname{int} A$.
(d) results from (c).
(e) $\operatorname{cl} \operatorname{int} A = \operatorname{cl} A$ is obvious. Lemma 2.2.21(k) implies $0^+(\operatorname{int} A) = 0^+(\operatorname{cl} A)$.
(f) follows from $\operatorname{cl} A + \mathbb{R}_> k \subseteq A$.
(g) Assume first $k \in -0^+ A$ and $y \in \operatorname{cl} A$. Then $a^0 := y + k \in A$ and $y = a^0 - k \in A$. Thus, $A$ is closed.
Assume now $k \in -0^+(\operatorname{cl} A), a \in A, t \in \mathbb{R}_>$. Then $a - tk = a - (t+1)k + k \in \operatorname{cl} A + k \subseteq A$. Hence, $k \in -0^+ A$.
(h) Assume $y \in Y$, $y + \mathbb{R}k \subseteq \operatorname{cl} A$ and $t \in \mathbb{R}$. Then $y + tk = y + (t-1)k + k \in \operatorname{cl} A + k \subseteq A$. □

We will now investigate the relationship between $\operatorname{core} A$ and sets $A - \mathbb{R}_> k$ with $k \in Y \setminus \{0\}$. The algebraic interior of $A$ is a subset of $A - \mathbb{R}_> k$ and of $A + \mathbb{R}_> k$.

**Lemma 2.3.17** *Assume that* $Y$ *is a linear space and* $A \subseteq Y$. *Then*

$$\operatorname{core} A \subseteq A + \mathbb{R}_> y \quad \textit{for each } y \in Y.$$

***Proof*** Assume $\operatorname{core} A \neq \emptyset$. Take any $a^0 \in \operatorname{core} A$. By the definition of the algebraic interior, there exists $t \in \mathbb{R}_>$ such that $a^0 + t(-y) \in A$. Hence, $a^0 \in A + \mathbb{R}_> y$. □

**Lemma 2.3.18** *Assume* $(\mathrm{Sp0}_{A,k})$.

(a) $A - \mathbb{R}_> k \subseteq A$ *holds if and only if* $k \in -0^+ A$.
(b) *If* $k \in -\operatorname{core} 0^+ A \setminus \{0\}$, *then* $\operatorname{core} A = A - \mathbb{R}_> k$.
(c) $(A - \mathbb{R}_> k) - \mathbb{R}_> k = A - \mathbb{R}_> k$.
(d) *If* $A \cap (-A) \subseteq \{0\}$ *and* $k \in -0^+ A$, *then:*
$0 \in A - \mathbb{R}_> k$ *or* $(A - \mathbb{R}_> k) \cap (-(A - \mathbb{R}_> k)) = \emptyset$.
(e) *If* $A$ *is convex, then* $A - \mathbb{R}_> k$ *is convex.*
(f) *If* $A$ *is a cone, then* $(A - \mathbb{R}_> k) \cup \{0\}$ *is a cone.*
(g) *If* $A + A \subseteq A$, *then* $(A - \mathbb{R}_> k) + (A - \mathbb{R}_> k) \subseteq A - \mathbb{R}_> k$.

***Proof***

(a), (c) and (e)–(g) are easy to verify.
(b) Proposition 2.3.3(a) implies $-\mathbb{R}_> k \subseteq \operatorname{core} 0^+ A$ because of $-k \in \operatorname{core} 0^+ A \setminus \{0\}$. Lemma 2.3.1(b) yields $A + \operatorname{core} 0^+ A \subseteq \operatorname{core}(A + 0^+ A) \subseteq \operatorname{core} A$. We obtain $A - \mathbb{R}_> k \subseteq \operatorname{core} A$. (b) follows from Lemma 2.3.17.

(d) Assume that the assumptions are fulfilled and that there exists $y \in (A - \mathbb{R}_{>}k) \cap (-(A - \mathbb{R}_{>}k))$. Then there are $a^1, a^2 \in A$ and $t_1, t_2 \in \mathbb{R}_{>}$ with $y = a^1 - t_1 k = -(a^2 - t_2 k)$. We get $-a^2 = a^1 - (t_1 + t_2)k \in A$ since $k \in -0^+A$. This implies $a^2 = 0$ because of $A \cap (-A) \subseteq \{0\}$. Thus, $0 = a^1 - (t_1 + t_2)k \in A - \mathbb{R}_{>}k$.

□

However, $\operatorname{core} A = A - \mathbb{R}_{>}k$ does not imply $k \in -\operatorname{core} 0^+A$. The core of $0^+A$ may be empty even if $A$ is convex and $\operatorname{int} A \neq \emptyset$.

**Example 2.3.19** In $Y = \mathbb{R}^2$, take $k := (0, -1)^{\mathrm{T}}$ and $A := \{(y_1, y_2)^{\mathrm{T}} \in Y \mid y_2 \geq y_1^2\}$. The set $A$ is convex, and $\operatorname{core} A = \operatorname{int} A = A - \mathbb{R}_{>}k$, but $0^+A = \operatorname{cone}\{-k\}$ has an empty algebraic interior.

Moreover, $k \in -0^+A \setminus \{0\}$ does not imply $\operatorname{core} A = A - \mathbb{R}_{>}k$. This becomes obvious by considering some closed cone $A$ with $k \in -\operatorname{bd} A \setminus \{0\}$. Nevertheless, we get the following statements.

**Proposition 2.3.20** *Assume that $Y$ is a linear space, that $A \subset Y$ is a proper convex subset of $Y$ with nonempty core, $k \in -0^+A$ and $Y = A + \mathbb{R}k$. Then*

$$\operatorname{core} A = A - \mathbb{R}_{>}k.$$

***Proof*** We have $\operatorname{core} A \subseteq A - \mathbb{R}_{>}k$ because of Lemma 2.3.17. Take now any $a^0 \in A$ and $t_0 \in \mathbb{R}_{>}$. To obtain a contradiction, suppose that $a^0 - t_0 k \notin \operatorname{core} A$. By [272, Theorem 3.14], there exists a linear functional $p \in Y^{\#} \setminus \{0\}$ such that $p(a) \leq p(a^0 - t_0 k)$ for each $a \in A$. Since $k \in -0^+A$, we get $p(a - tk) \leq p(a^0 - t_0 k)$ and hence $p(a^0) - tp(k) \leq p(a^0) - t_0 p(k)$ for each $a \in A$ and $t \in \mathbb{R}_+$. This yields $p(k) = 0$, and $p(a + tk) = p(a) \leq p(a^0)$ for each $a \in A$ and $t \in \mathbb{R}$. This contradicts $Y = A + \mathbb{R}k$ since $p \neq 0$ attains each real value. □

**Proposition 2.3.21** *Assume that $Y$ is a topological vector space, that $A \subseteq Y$ is a strictly convex set and $k \in -0^+A \setminus \{0\}$. Then*

$$\operatorname{core} A = \operatorname{int} A = A - \mathbb{R}_{>}k.$$

***Proof*** The statement is obvious if $A$ is empty. Assume now $A \neq \emptyset$. Take any $a \in A$ and $t \in \mathbb{R}_{>}$. Then $a - 2tk \in A \setminus \{a\}$, and $a - tk = \frac{1}{2}a + \frac{1}{2}(a - 2tk) \in \operatorname{int} A$ since $A$ is strictly convex. Hence, $A - \mathbb{R}_{>}k \subseteq \operatorname{int} A \neq \emptyset$. We obtain $\operatorname{int} A = \operatorname{core} A \subseteq A - \mathbb{R}_{>}k$ from Proposition 2.3.6 and Lemma 2.3.17. □

**Proposition 2.3.22** *Assume that $Y$ is a linear space and $A \subseteq Y$.*

(a) *We have*

$$\bigcap_{k \in -0^+A} (A - \mathbb{R}_{>}k) \subseteq A.$$

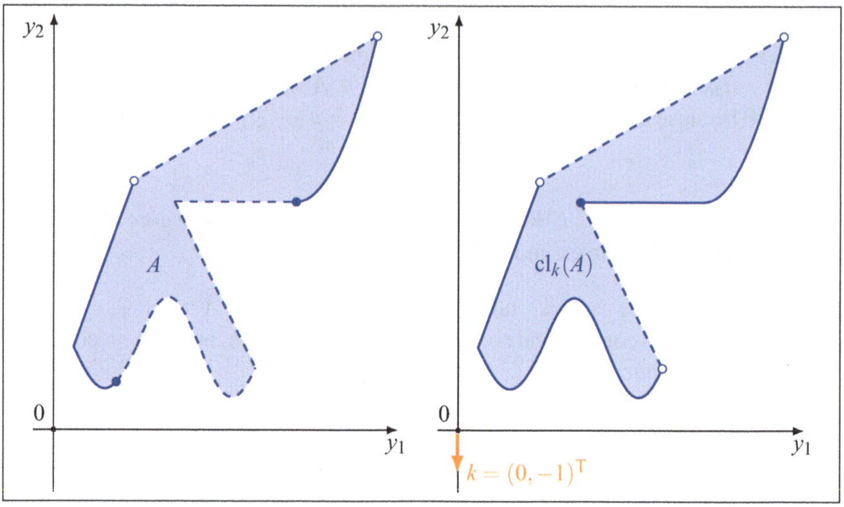


**Fig. 2.3** The $k$-directional closure of $A$

(b) *If* $\operatorname{core} 0^+A \neq \emptyset$, *then*

$$\operatorname{core} A = \bigcap_{k \in -\operatorname{core} 0^+A} (A - \mathbb{R}_> k) = \bigcap_{k \in -0^+A} (A - \mathbb{R}_> k) = \bigcap_{k \in Y} (A - \mathbb{R}_> k).$$

***Proof***

(a) is easy to verify.
(b) For $A = Y$, the statement is obvious. Assume now $A \neq Y$. This implies $0 \notin \operatorname{core} 0^+A$. The first equation results from Lemma 2.3.18(b). The other equations follow from $\bigcap_{k \in Y}(A - \mathbb{R}_> k) \subseteq \bigcap_{k \in -0^+A}(A - \mathbb{R}_> k) \subseteq \bigcap_{k \in -\operatorname{core} 0^+A}(A - \mathbb{R}_> k)$ and $\operatorname{core} A \subseteq \bigcap_{k \in Y}(A - \mathbb{R}_> k)$ because of Lemma 2.3.17.

□

Further properties of sets $A - \mathbb{R}_> k$ will be proved in Propositions 2.3.49, 2.3.53 and 2.3.55. We will now define the directional closure of a set (see Fig. 2.3).

**Definition 2.3.23** Assume $(\mathrm{Sp}0_{A,k})$.

The **k-directional closure** $\mathrm{cl}_k(A)$ of $A$ consists of all elements $y \in Y$ with the following property: For each $\lambda \in \mathbb{R}_>$, there exists $t \in \mathbb{R}_+$ with $t < \lambda$ such that $y - tk \in A$.

$A$ is said to be **k-directionally closed** if $A = \mathrm{cl}_k(A)$.

Obviously, $\mathrm{cl}_k(Y) = Y$ and $\mathrm{cl}_k(\emptyset) = \emptyset$.

It is easy to verify that directional closedness can also be characterized in the following way.

**Lemma 2.3.24** *Assume* ($\mathrm{Sp0}_{A,k}$).

(a) *$A$ is $k$-directionally closed if and only if, for each $y \in Y$, we have $y \in A$ whenever there exists a sequence $(t_n)_{n\in\mathbb{N}}$ of real numbers with $t_n \searrow 0$ and $y - t_n k \in A$.*
(b) $\mathrm{cl}_k(A) = \{y \in Y \mid \exists (t_n)_{n\in\mathbb{N}}$ *with* $t_n \in \mathbb{R}_+$ *and* $t_n \to 0\ \forall n \in \mathbb{N}\colon\ y - t_n k \in A\}$.

**Remark 2.3.25** The $k$-directional closure was introduced in Definition 2.3.23 in such a way that it is the closure in direction $k$. This definition was given in [485]. The property in Lemma 2.3.24(a) was used for the definition of $k$-directional closedness in [483].

Qiu and He [385] defined the $k$-vector closure of $A$ as

$$\mathrm{vcl}_k(A) := \{y \in Y \mid \exists (t_n)_{n\in\mathbb{N}} \text{ with } t_n \in \mathbb{R}_+ \text{ and } t_n \to 0\ \forall n \in \mathbb{N}\colon\ y + t_n k \in A\}.$$

Obviously, $\mathrm{vcl}_k(A) = \mathrm{cl}_{-k}(A)$. This $k$-vector closure was used for the investigation of the functionals $\varphi_{A,k}$, which we study in this book, in [385] under the assumption that $A$ is a convex cone with $k \in A \setminus (-A)$ and in [213] without this restrictive assumption. The vectorial closure of $A$ is the union of all sets $\mathrm{vcl}_k(A)$ with $k \in Y \setminus \{0\}$.

A property of $A$ which is equivalent to $k$-directional closedness in the case $k \in -0^+A$ was introduced by Hamel [219]. He defined $A$ to be **k-radially closed** if, for each $y \in A$, we have $y + tk \in A$ whenever there exists a sequence $(t_n)_{n\in\mathbb{N}}$ of real numbers with $t_n \to t$ and $y + t_n k \in A$. This definition was initiated by Martinez-Legaz, Rubinov and Singer [347, Definition 5], who called the same property **closedness along diagonal lines** for $Y = \mathbb{R}^\ell$ and $k = \mathbf{1}$. Obviously, a set is $k$-radially closed if it is $k$- and $(-k)$-directionally closed according to our definition.

**Lemma 2.3.26** *Assume* ($\mathrm{Sp0}_{A,k}$).

(a) $A \subseteq \mathrm{cl}_k(A) \subseteq A + \mathbb{R}_+ k$.
(b) $\mathrm{cl}_k(A) \subseteq \mathrm{cl}_k(B)$ *if* $A \subseteq B \subseteq Y$.
(c) $\mathrm{cl}_k(\mathrm{cl}_k(A)) = \mathrm{cl}_k(A)$.
(d) $\mathrm{cl}_k(-A) = -\mathrm{cl}_{-k}(A)$.
(e) $\mathrm{cl}_k(A + y) = \mathrm{cl}_k(A) + y$ *for each* $y \in Y$.
(f) $\mathrm{cl}_k(A) + \mathrm{cl}_k(B) \subseteq \mathrm{cl}_k(A + B)$ *for each* $B \subseteq Y$.
(g) $\mathrm{cl}_k(A + B) = \mathrm{cl}_k(\mathrm{cl}_k(A) + B) = \mathrm{cl}_k(\mathrm{cl}_k(A) + \mathrm{cl}_k(B))$ *for each* $B \subseteq Y$.
(h) $\mathrm{conv}\,\mathrm{cl}_k(A) \subseteq \mathrm{cl}_k(\mathrm{conv}\,A)$.
(i) *We have*

$$\mathrm{vcl}(A) = \bigcup_{k \in Y\setminus\{0\}} \mathrm{cl}_k(A).$$

***Proof***

(a), (b) and (e) follow immediately from Definition 2.3.23.
(c) For $A = \emptyset$, the assertion is true. Assume now $A \neq \emptyset$.

(i) The inclusion $A \subseteq \mathrm{cl}_k(A)$ implies $\mathrm{cl}_k(A) \subseteq \mathrm{cl}_k(\mathrm{cl}_k(A))$ because of (b).

(ii) Take any $y \in \mathrm{cl}_k(\mathrm{cl}_k(A))$ and $\lambda \in \mathbb{R}_>$. By Definition 2.3.23, there exists $t_0 \in \mathbb{R}_+$ with $t_0 < \frac{\lambda}{2}$ and $y - t_0 k \in \mathrm{cl}_k(A)$. If $t_0 = 0$, then $y \in \mathrm{cl}_k(A)$. Suppose now $t_0 \in \mathbb{R}_>$. By Definition 2.3.23, there is $t \in \mathbb{R}_+$ with $t < t_0$ and $(y - t_0 k) - tk \in A$, i.e., $y - (t_0 + t)k \in A$, where $0 < t + t_0 < 2t_0 < \lambda$. Hence, $y \in \mathrm{cl}_k(A)$.

(d) By Lemma 2.3.24(b), $y \in \mathrm{cl}_k(-A)$ holds if and only if there exists a sequence $(t_n)_{n\in\mathbb{N}}$ with $t_n \in \mathbb{R}_+$ and $t_n \to 0$ such that $y - t_n k \in -A$ for all $n \in \mathbb{N}$. This is equivalent to $-y \in \mathrm{cl}_{-k}(A)$ since $y - t_n k \in -A$ is equivalent to $-y - t_n(-k) \in A$.

(f) Take any $y^0 \in \mathrm{cl}_k(A) + \mathrm{cl}_k(B)$. Then there exist $y^1 \in \mathrm{cl}_k(A)$ and $y^2 \in \mathrm{cl}_k(B)$ with $y^0 = y^1 + y^2$. Because of Lemma 2.3.24(b), there is a sequence $(t_n)_{n\in\mathbb{N}}$ with $t_n \in \mathbb{R}_+$ and $t_n \to 0$ such that $y^1 - t_n k \in A$ for all $n \in \mathbb{N}$ and a sequence $(\tilde{t}_n)_{n\in\mathbb{N}}$ with $\tilde{t}_n \in \mathbb{R}_+$ and $\tilde{t}_n \to 0$ such that $y^2 - \tilde{t}_n k \in B$ for all $n \in \mathbb{N}$. For each $n \in \mathbb{N}$, we obtain $(y^1 + y^2) - (t_n + \tilde{t}_n)k \in A + B$. Hence, $y^0 \in \mathrm{cl}_k(A + B)$.

(g) $A + B \subseteq \mathrm{cl}_k(A) + B \subseteq \mathrm{cl}_k(A) + \mathrm{cl}_k(B)$ implies $\mathrm{cl}_k(A + B) \subseteq \mathrm{cl}_k(\mathrm{cl}_k(A) + B) \subseteq \mathrm{cl}_k(\mathrm{cl}_k(A) + \mathrm{cl}_k(B))$ by (b). Because of (f) and (c), we get $\mathrm{cl}_k(\mathrm{cl}_k(A) + \mathrm{cl}_k(B)) \subseteq \mathrm{cl}_k(A + B)$. This yields the assertion.

(h) Take any $y \in \mathrm{conv}\,\mathrm{cl}_k(A)$. Then there exist a finite index set $I$, numbers $\lambda_i \in \mathbb{R}_+$ with $\sum_{i\in I} \lambda_i = 1$ and elements $b_i \in \mathrm{cl}_k(A)$ such that $y = \sum_{i\in I} \lambda_i b_i$. For each $i \in I$, Lemma 2.3.24(b) implies the existence of sequences $(a_{in})_{n\in\mathbb{N}}$ with $a_{in} \in A$ and $(t_{in})_{n\in\mathbb{N}}$ with $t_{in} \in \mathbb{R}_+$ and $t_{in} \to 0$ such that $b_i - t_{in}k = a_{in}$ for all $n \in \mathbb{N}$. For each $n \in \mathbb{N}$, we get $y = \sum_{i\in I} \lambda_i (a_{in} + t_{in}k) = \sum_{i\in I} \lambda_i a_{in} + (\sum_{i\in I} \lambda_i t_{in})k$, where $\sum_{i\in I} \lambda_i a_{in} \in \mathrm{conv}\,A$. Moreover, $\sum_{i\in I} \lambda_i t_{in} \to 0$ for $n \to \infty$. Thus, $y \in \mathrm{cl}_k(\mathrm{conv}\,A)$ because of Lemma 2.3.24(b).

(i) results from Remark 2.3.25. □

In general, $\mathrm{cl}_k(A) + \mathrm{cl}_k(B) \neq \mathrm{cl}_k(A + B)$. The sum of two $k$-directionally closed sets is not necessarily $k$-directionally closed.

**Example 2.3.27** In $Y = \mathbb{R}^2$, the sets $A := \{(y_1, y_2)^T \in \mathbb{R}^2 \mid y_1 > 0,\, y_2 \geq \frac{1}{y_1}\}$ and $B := \{(y_1, y_2)^T \in \mathbb{R}^2 \mid y_1 \leq 0,\, y_2 = 0\}$ are closed convex sets and thus $k$-directionally closed for each $k \in Y \setminus \{0\}$ by Proposition 2.3.54. However, $A + B = \{(y_1, y_2)^T \in \mathbb{R}^2 \mid y_2 > 0\}$ is not closed. For each $k \in -(A + B)$, the set $A + B$ is not $k$-directionally closed.

$\mathrm{conv}\,\mathrm{cl}_k(A) = \mathrm{cl}_k(\mathrm{conv}\,A)$ is not necessarily fulfilled.

**Example 2.3.28** In $Y = \mathbb{R}^2$, the set $A := \{(y_1, y_2)^T \in \mathbb{R}^2 \mid y_1 \geq 0, y_2 = 0\} \cup \{(y_1, y_2)^T \in \mathbb{R}^2 \mid y_1 = 0,\ 0 \leq y_2 \leq 1\}$ is a closed set and thus $k$-directionally closed for $k := (0, 1)^T$ by Proposition 2.3.54. However, $\mathrm{conv}\,\mathrm{cl}_k(A) = \mathrm{conv}\,A = \{(y_1, y_2)^T \in \mathbb{R}^2 \mid y_1 \geq 0,\ 0 \leq y_2 < 1\} \cup \{(0, 1)^T\}$ is not $k$-directionally closed.

In applications, we often have to deal with the case $k \in -0^+ A$.

**Proposition 2.3.29** *Assume* $(\mathrm{Sp0}_{A,k})$ *and* $k \in -0^+A$. *Then*

(a) $\mathrm{cl}_k(A) = \{y \in Y \mid y - \mathbb{R}_> k \subseteq A\}$,
(b) $k \in -0^+(\mathrm{cl}_k(A))$.
(c) *If* $k \in -0^+A \setminus 0^+A$, *then* $k \in -0^+(\mathrm{cl}_k(A)) \setminus 0^+(\mathrm{cl}_k(A))$.

***Proof***

(a) (i) Assume first $y \in \mathrm{cl}_k(A)$. If $y \in A$, then $y - tk \in A$ holds for all $t \in \mathbb{R}_>$. Suppose now $y \notin A$. Take any $t \in \mathbb{R}_>$. By Definition 2.3.23, there exists $t_0 \in \mathbb{R}_>$ with $t_0 < t$ such that $y - t_0k \in A$. Then $y - tk = y - t_0k + (t_0 - t)k \in A + 0^+A \subseteq A$.
(ii) Take any $y \in Y$ for which $y - tk \in A$ holds for all $t \in \mathbb{R}_>$. Then $y \in \mathrm{cl}_k(A)$ by Definition 2.3.23.
(b) follows from (a).
(c) Consider any $k \in -0^+A \setminus 0^+A$. Then $k \in -0^+(\mathrm{cl}_k(A))$ because of (b).
To obtain a contradiction, suppose $k \in 0^+(\mathrm{cl}_k(A))$. Take any $a \in A$ and $\lambda \in \mathbb{R}_+$. Then $a + (\lambda + 1)k \in \mathrm{cl}_k(A)$ since $k \in 0^+(\mathrm{cl}_k(A))$. By Lemma 2.3.24, there exists a sequence $(t_n)_{n\in\mathbb{N}}$ with $t_n \in \mathbb{R}_+$ and $t_n \to 0$ such that $a + (\lambda + 1)k - t_nk \in A$ for all $n \in \mathbb{N}$. There is $n_0 \in \mathbb{N}$ with $t_{n_0} < 1$. We get $a + \lambda k \in A - (1 - t_{n_0})k \subseteq A - \mathbb{R}_+k \subseteq A$ since $k \in -0^+A$. This implies $A + \mathbb{R}_+k \subseteq A$, a contradiction to $k \notin 0^+A$. □

Example 2.3.47 illustrates that $k \in -0^+(\mathrm{cl}_k(A))$ does not result in $k \in -\mathrm{cl}_k(0^+A)$ and especially not in $k \in -0^+A$.

The following proposition can easily be verified using Lemma 2.3.24(b).

**Proposition 2.3.30** *Assume* $(\mathrm{Sp0}_{A,k})$.

(a) *If* $A$ *is a cone, then* $\mathrm{cl}_k(A)$ *is a cone.*
(b) *If* $A$ *is convex, then* $\mathrm{cl}_k(A)$ *is convex.*
(c) *If* $A + A \subseteq A$, *then* $\mathrm{cl}_k(A) + \mathrm{cl}_k(A) \subseteq \mathrm{cl}_k(A)$.

This implies together with Proposition 2.3.29(c):

**Corollary 2.3.31** *Assume that* $Y$ *is a linear space,* $A \subseteq Y$ *is a nontrivial convex cone and* $k \in -A \setminus A$. *Then* $\mathrm{cl}_k(A)$ *is a convex cone with* $k \in -\mathrm{cl}_k(A) \setminus \mathrm{cl}_k(A)$.

**Proposition 2.3.32** *Assume that* $Y$ *is a linear space,* $\{A_i \mid i \in I\}$ *is a family of subsets of* $Y$ *and* $k \in Y \setminus \{0\}$.

(a) *For each* $j \in I$,

$$\mathrm{cl}_k\left(\cap_{i\in I} A_i\right) \subseteq \mathrm{cl}_k(A_j) \subseteq \mathrm{cl}_k\left(\cup_{i\in I} A_i\right).$$

(b) *If* $I$ *is a finite set, then*

$$\cup_{i\in I}\, \mathrm{cl}_k(A_i) = \mathrm{cl}_k\left(\cup_{i\in I} A_i\right).$$

*Hence, the union of a finite number of* $k$*-directionally closed sets is* $k$*-directionally closed.*

(c) *If $A_i$ is $k$-directionally closed for each $i \in I$, then $\bigcap_{i\in I} A_i$ is $k$-directionally closed.*
(d) *If $k \in -0^+A_i$ for each $i \in I$, then*

$$\text{cl}_k\left(\bigcap_{i\in I} A_i\right) = \bigcap_{i\in I} \text{cl}_k(A_i).$$

***Proof*** (a) follows from Lemma 2.3.26(b), (b) from Definition 2.3.23. (c) results from Lemma 2.3.24(a).

(a) implies $\text{cl}_k\left(\bigcap_{i\in I} A_i\right) \subseteq \bigcap_{i\in I}\text{cl}_k(A_i)$. Assume now $k \in -0^+A_i$ for each $i \in I$. Then $k \in -0^+(\bigcap_{i\in I} A_i)$. Take any $y \in \bigcap_{i\in I}\text{cl}_k(A_i)$. From Proposition 2.3.29(a), we get $y - \mathbb{R}_> k \subseteq A_i$ for each $i \in I$. Therefore, $y - \mathbb{R}_> k \subseteq \bigcap_{i\in I} A_i$ and $y \in \text{cl}_k\left(\bigcap_{i\in I} A_i\right)$. This proves (d). □

The inequality $\bigcup_{i\in I}\text{cl}_k(A_i) \neq \text{cl}_k\left(\bigcup_{i\in I} A_i\right)$ can hold for an infinite set $I$ even if $k \in -0^+A_i$ for each $i \in I$.

**Example 2.3.33** In $Y = \mathbb{R}$, choose $k := 1$, $I := \mathbb{N}_>$, $A_i := \{y \in \mathbb{R} \mid y \leq 1 - \frac{1}{i}\}$ for each $i \in I$. Then $\text{cl}_k(A_i) = A_i$ for each $i \in I$. We get $\bigcup_{i\in I}\text{cl}_k(A_i) = (-\infty, 1)$, but $\text{cl}_k\left(\bigcup_{i\in I} A_i\right) = (-\infty, 1]$.

This example illustrates that the union of $k$-directionally closed sets is not necessarily $k$-directionally closed.

In general, $\text{cl}_k\left(\bigcap_{i\in I} A_i\right) \neq \bigcap_{i\in I}\text{cl}_k(A_i)$.

**Example 2.3.34** In $Y = \mathbb{R}$, take $k := 1$, $I := \{1, 2\}$, $A_1 := \{y \in \mathbb{Q} \mid y < 1\}$, and $A_2 := \{y \in \mathbb{R} \setminus \mathbb{Q} \mid y < 1\}$. Then $\text{cl}_k(A_1) = \text{cl}_k(A_2) = (-\infty, 1]$ and $A_1 \cap A_2 = \emptyset$ imply $\bigcap_{i\in I}\text{cl}_k(A_i) = (-\infty, 1]$, but $\text{cl}_k\left(\bigcap_{i\in I} A_i\right) = \emptyset$.

The next two propositions connect algebraic and directional closedness.

**Proposition 2.3.35** *Assume that $Y$ is a linear space, $A \subseteq Y$ is algebraically closed and $k \in -0^+A \setminus \{0\}$. Then $A$ is $k$-directionally closed.*

***Proof*** Suppose, contrary to our claim, that $A$ is not $k$-directionally closed. Then there exist $y \in Y \setminus A$ and a sequence $(t_n)_{n\in\mathbb{N}}$ with $t_n \in \mathbb{R}$, $t_n \searrow 0$ and $y - t_n k \in A$. Since $A$ is algebraically closed, there is a real number $\lambda \in (0, 1)$ with $y - t_1 k + \lambda(y - (y - t_1 k)) \notin A$, i.e., $y - t_1 k + \lambda t_1 k = y - (1 - \lambda)t_1 k \notin A$. There exists $j \in \mathbb{N}$ with $t_j < (1 - \lambda)t_1$. Since $y - t_j k \in A$, we get $y - (1 - \lambda)t_1 k = y - t_j k + (t_j - (1 - \lambda)t_1)k \in A$ from $k \in -0^+A$. This contradicts $y - (1 - \lambda)t_1 k \notin A$. □

The condition $k \in -0^+A \setminus \{0\}$ in Proposition 2.3.35 is not superfluous.

**Example 2.3.36** In $Y = \mathbb{R}^2$, consider $A := \{(y_1, y_2)^T \in Y \mid \exists n \in \mathbb{N}_> : y_2 = \frac{1}{n}\}$ and $k := (0, -1)^T$. $A$ is algebraically closed. It is not $k$-directionally closed since $\{(y_1, y_2)^T \in Y \mid y_2 = 0\} \subseteq \text{cl}_k(A) \setminus A$.

**Proposition 2.3.37** *Let $A$ be a nontrivial convex cone in a linear space $Y$ with $k \in -\,$core $A$. Then $A$ is $k$-directionally closed if and only if $A$ is algebraically closed.*

***Proof*** Suppose that $A$ is $k$-directionally closed. Take any $a \in A$ and $y \in Y$ with $a + \lambda(y - a) \in A$ for all $\lambda \in (0, 1)$. Since $Y = A + \mathbb{R}_{>}k$ by Lemma 2.3.13, there exist $a^0 \in A$ and $t \in \mathbb{R}_{>}$ with $y - a = a^0 + tk$. For each $n \in \mathbb{N}_{>}$, we get $y - \frac{1}{n}tk = y - \frac{1}{n}(y - a - a^0) = a + (1 - \frac{1}{n})(y - a) + \frac{1}{n}a^0 \in A + \frac{1}{n}a^0 \subseteq A$. Hence, $y \in A$ since $A$ is $k$-directionally closed. Thus, $A$ is algebraically closed. Proposition 2.3.35 yields the assertion. □

In Proposition 2.3.37, the assumption $k \in -\operatorname{core} A$ cannot be replaced by the condition $k \in (-A) \setminus A$. This is illustrated by Example 2.3.56. There $A$ is a nontrivial convex cone which is $k$-directionally closed for some $k \in (-A) \setminus A$ even though $A$ is not algebraically closed.

**Corollary 2.3.38** *Assume* $(\text{Sp}0_{A,k})$ *and* $k \in -\operatorname{core} 0^+A$. *Then* $0^+A$ *is* $k$*-directionally closed if and only if* $0^+A$ *is algebraically closed.*

**Proposition 2.3.39** *Assume* $(\text{Sp}0_{A,k})$. *If* $A$ *is* $k$*-directionally closed, then* $0^+A$ *is* $k$*-directionally closed.*

***Proof*** Assume that $A$ is $k$-directionally closed and that we have some $y \in Y$ for which there exists a sequence $(t_n)_{n \in \mathbb{N}}$ of real numbers with $t_n \searrow 0$ and $y - t_n k \in 0^+A$ for all $n \in \mathbb{N}$. Take any $a \in A$ and $\lambda \in \mathbb{R}_{>}$. Since $0^+A$ is a cone, we get $\lambda y - t_n \lambda k \in 0^+A$ for all $n \in \mathbb{N}$. Hence, $a + \lambda y - t_n \lambda k \in A$ for all $n \in \mathbb{N}$. Since $A$ is $k$-directionally closed and $\lambda t_n \searrow 0$, we obtain $a + \lambda y \in A$. Consequently, $y \in 0^+A$. Hence, $0^+A$ is $k$-directionally closed. □

Proposition 2.3.39 yields together with Corollary 2.3.38:

**Corollary 2.3.40** *Assume that* $Y$ *is a linear space and that* $A \subseteq Y$ *is* $k$*-directionally closed for some* $k \in -\operatorname{core} 0^+A \setminus \{0\}$. *Then* $0^+A$ *is algebraically closed.*

We will now introduce the directional boundary and the directional interior of a set (see Fig. 2.4).

**Definition 2.3.41** Assume $(\text{Sp}0_{A,k})$.

$a \in Y$ is said to be a **k-directional interior point** of $A$ if there exists $\lambda \in \mathbb{R}_{>}$ such that $a + tk \in A$ holds for all $t \in [0, \lambda]$. The set of all $k$-directional interior points of $A$ is the **k-directional interior** $\operatorname{int}_k(A)$ of $A$. The **k-directional boundary** of $A$ is given by $\operatorname{bd}_k(A) := \operatorname{cl}_k(A) \setminus \operatorname{int}_k(A)$.

Under the additional assumption $k \in -0^+A$, the $k$-directional interior and the $k$-directional boundary of $A$ are defined in [222] as $\{y \in A \mid \exists t \in \mathbb{R}_{>} : y + tk \in A\}$ and as $\{y \in Y \mid \forall t \in \mathbb{R}_{>} : (y + tk \notin A \wedge y - tk \in A)\}$, respectively. Propositions 2.3.53(a) and 2.3.29(a) underscore that these definitions are equivalent to ours if $k \in -0^+A$.

**Lemma 2.3.42** *Assume* $(\text{Sp}0_{A,k})$. *Then*

(a) $\operatorname{core} A \subseteq \operatorname{int}_k(A) \subseteq A \cap (A - \mathbb{R}_{>}k)$.

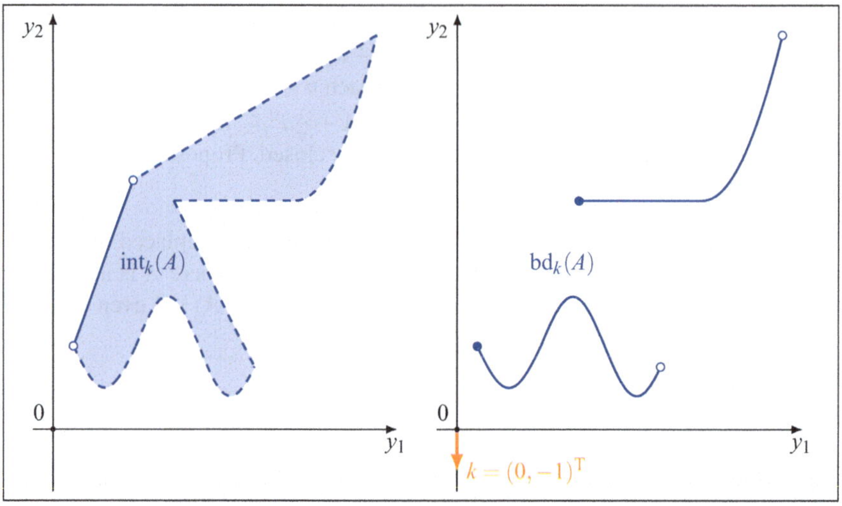


**Fig. 2.4** The $k$-directional interior and boundary of set $A$ from Fig. 2.3

(b) $\text{int}_k(A) = A - \mathbb{R}_> k$ *if and only if* $k \in -0^+ A$.
(c) $\text{int}_k(A) \subseteq \text{int}_k(B)$ *if* $A \subseteq B \subseteq Y$.
(d) $\text{int}_k(\text{int}_k(A)) = \text{int}_k(A)$.
(e) $\text{int}_k(-A) = -\text{int}_{-k}(A)$.
(f) $\text{int}_k(A + y) = \text{int}_k(A) + y$ *for each* $y \in Y$.
(g) $\text{int}_k(A) + B \subseteq \text{int}_k(A + B)$ *for each* $B \subseteq Y$.
(h) $\text{int}_k(A) = Y \setminus \text{cl}_{-k}(Y \setminus A)$.
(i) $\text{cl}_k(Y \setminus A) = Y \setminus \text{int}_{-k}(A)$.
(j) $\text{bd}_k(A) = \text{bd}_{-k}(Y \setminus A) = \text{cl}_k(A) \cap \text{cl}_{-k}(Y \setminus A)$.
(k) $\text{bd}_k(A + y) = \text{bd}_k(A) + y$ *for each* $y \in Y$.
(l) $\text{bd}_k(-A) = -\text{bd}_{-k}(A)$.

***Proof***

(a) follows from Definitions 2.3.41 and 2.1.6.
(b) $\text{int}_k(A) = A - \mathbb{R}_> k$ implies $A - \mathbb{R}_> k \subseteq A$ and thus $k \in -0^+ A$.
Assume now $k \in -0^+ A$. Take any $y \in A - \mathbb{R}_> k$. Then there exist $a \in A$ and $\lambda \in \mathbb{R}_>$ with $y = a - \lambda k$. For each $t \in [0, \lambda]$, we get $y + tk = a - (\lambda - t)k \in A$ since $k \in -0^+ A$. Hence, $y \in \text{int}_k(A)$. The assertion results from (a).
(c) follows from Definition 2.3.41.
(d) $\text{int}_k(\text{int}_k(A)) \subseteq \text{int}_k(A)$ because of (c).
Take now any $a \in \text{int}_k(A)$. Then there exists $\lambda \in \mathbb{R}_>$ such that $a + tk \in A$ holds for all $t \in [0, \lambda]$. This yields $a + tk \in \text{int}_k A$ for all $t \in [0, \frac{\lambda}{2}]$. Hence, $a \in \text{int}_k(\text{int}_k(A))$.
(e) Definition 2.3.41 implies:

$$\begin{aligned} y \in \operatorname{int}_k(-A) &\iff \exists \lambda \in \mathbb{R}_> \, \forall t \in [0, \lambda] : y + tk \in -A, \\ &\iff \exists \lambda \in \mathbb{R}_> \, \forall t \in [0, \lambda] : -y + t(-k) \in A, \\ &\iff -y \in \operatorname{int}_{-k}(A). \end{aligned}$$

(f) Definition 2.3.41 yields:

$$\begin{aligned} y^0 \in \operatorname{int}_k(A + y) &\iff \exists \lambda \in \mathbb{R}_> \, \forall t \in [0, \lambda] : y^0 + tk \in A + y, \\ &\iff \exists \lambda \in \mathbb{R}_> \, \forall t \in [0, \lambda] : y^0 - y + tk \in A, \\ &\iff y^0 - y \in \operatorname{int}_k(A). \end{aligned}$$

(g) For each $b \in B$, we have $\operatorname{int}_k(A) + b = \operatorname{int}_k(A + b) \subseteq \operatorname{int}_k(A + B)$ because of (f) and (c).

(h) We get from Definition 2.3.23:

$$\begin{aligned} y \in \operatorname{cl}_{-k}(Y \setminus A) &\iff \forall \lambda \in \mathbb{R}_> \, \exists t \in [0, \lambda) : y - t(-k) \in Y \setminus A, \\ &\iff \forall \lambda \in \mathbb{R}_> \, \exists t \in [0, \lambda) : y + tk \notin A, \\ &\iff y \notin \operatorname{int}_k(A). \end{aligned}$$

(i) follows from (h).

(j) We obtain from (h) that $\operatorname{bd}_k(A) = \operatorname{cl}_k(A) \setminus \operatorname{int}_k(A) = \operatorname{cl}_k(A) \cap \operatorname{cl}_{-k}(Y \setminus A)$ and $\operatorname{bd}_{-k}(Y \setminus A) = \operatorname{cl}_{-k}(Y \setminus A) \setminus \operatorname{int}_{-k}(Y \setminus A) = \operatorname{cl}_{-k}(Y \setminus A) \cap \operatorname{cl}_k(A)$.

(k) follows from (f) and Lemma 2.3.26(e).

(l) results from (e) and Lemma 2.3.26(d). □

For $A$ being a convex cone and $k \in A \setminus (-A)$, Qiu and He [385] denoted the set $A + \mathbb{R}_> k$ by $\operatorname{vint}_k(A)$.

In general, $\operatorname{int}_k(A) \cap \operatorname{cl}_k(Y \setminus A) \neq \emptyset$.

**Example 2.3.43** In $Y = \mathbb{R}$, consider $A := [0, 1]$ and $k := 1$. Then $0 \in \operatorname{int}_k(A) \cap \operatorname{cl}_k(Y \setminus A)$.

**Lemma 2.3.44** *Assume that $Y$ is a linear space and $A \subseteq Y$. Then*

$$\operatorname{core} A = \bigcap_{k \in Y \setminus \{0\}} \operatorname{int}_k(A).$$

*If* $k \in -\operatorname{core} 0^+ A \setminus \{0\}$, *then* $\operatorname{int}_k(A) = \operatorname{core} A$.

***Proof*** Definitions 2.3.41 and 2.1.6 imply the first equation. The second assertion follows from Lemmata 2.3.42(b) and 2.3.18(b). □

**Proposition 2.3.45** *Assume* $(\mathrm{Sp0}_{A,k})$.

(a) *If $A$ is convex, then* $\operatorname{int}_k(A)$ *is convex.*

(b) *If* $A$ *is a cone, then* $\text{int}_k(A) \cup \{0\}$ *is a cone.*
(c) *If* $A + A \subseteq A$*, then* $\text{int}_k(A) + \text{int}_k(A) \subseteq \text{int}_k(A)$.
(d) *If* $A$ *is a convex cone, then* $\text{int}_k(A) \cup \{0\}$ *is a convex cone.*
(e) *If* $A \cap (-A) \subseteq \{0\}$*, then* $0 \in \text{int}_k(A)$ *or* $\text{int}_k(A) \cap (-\text{int}_k(A)) = \emptyset$.

***Proof***

(a) Take any $a^1, a^2 \in \text{int}_k(A)$. Then there exists $\lambda_0 \in \mathbb{R}_>$ such that $a^1 + tk \in A$ and $a^2 + tk \in A$ for all $t \in [0, \lambda_0]$. Take any $\lambda \in (0, 1)$. For each $t \in [0, \lambda_0]$, we have $\lambda a^1 + (1 - \lambda)a^2 + tk = \lambda(a^1 + tk) + (1 - \lambda)(a^2 + tk) \in \lambda A + (1 - \lambda)A \subseteq A$. Hence, $\lambda a^1 + (1 - \lambda)a^2 \in \text{int}_k(A)$.
(b) Take any $a \in \text{int}_k(A)$. Then there exists $\lambda_0 \in \mathbb{R}_>$ such that $a + tk \in A$ for all $t \in [0, \lambda_0]$. Take any $\lambda \in \mathbb{R}_>$. For each $t \in [0, \lambda_0]$, we have $\lambda a + (\lambda t)k = \lambda(a + tk) \in \lambda A \subseteq A$. Hence, $\lambda a \in \text{int}_k(A)$.
(c) Take any $a^1, a^2 \in \text{int}_k(A)$. Then there exists $\lambda \in \mathbb{R}_>$ such that $a^1 + tk \in A$ and $a^2 + tk \in A$ for all $t \in [0, \lambda]$. For each $t \in [0, \lambda]$, we have $a^1 + a^2 + (2t)k = (a^1 + tk) + (a^2 + tk) \in A + A \subseteq A$. Hence, $a^1 + a^2 \in \text{int}_k(A)$.
(d) follows from (b) and (c).
(e) If $a \in \text{int}_k(A) \cap (-\text{int}_k(A))$, then $a \in A \cap (-A) \subseteq \{0\}$ and thus $0 \in \text{int}_k(A)$.

□

**Proposition 2.3.46** *Assume* $(\text{Sp}0_{A,k})$ *and* $k \in 0^+A \setminus \{0\}$. *Then* $A$ *is* $k$*-directionally closed,* $\text{int}_k(A) = A$ *and* $\text{bd}_k(A) = \emptyset$.

***Proof*** We obtain the first statement from Lemma 2.3.24. The second statement results from the definition of $\text{int}_k(A)$. This yields the third statement. □

Because of Proposition 2.3.46, $k \in 0^+A$ implies $k \in 0^+(\text{cl}_k(A))$. In general, $k \in 0^+(\text{cl}_k(A))$ does not imply $k \in \text{cl}_k(0^+A)$.

**Example 2.3.47** In $Y = \mathbb{R}$, consider the set $A := \mathbb{Q}$ and $k := 1$. Then $\text{cl}_k(A) = \mathbb{R}$ and $k \in 0^+(\text{cl}_k(A))$, but $\text{cl}_k(0^+A) = 0^+A = \{0\}$ and $k \notin \text{cl}_k(0^+A)$.

The extension of a set by rays in direction $k$ or $-k$ plays a fundamental role when investigating translation invariant functions in Chap. 4. This refers especially to the set $\text{cl}_k(A - \mathbb{R}_+k)$ (see Fig. 2.5).

**Proposition 2.3.48** *Assume* $(\text{Sp}0_{A,k})$.

(a) $\text{cl}_k(A) \subseteq A + \mathbb{R}_+k$.
(b) $A + \mathbb{R}_+k = \text{cl}_k(A) + \mathbb{R}_+k = \text{cl}_k(A + \mathbb{R}_+k) = \text{int}_k(A + \mathbb{R}_+k)$.
(c) $A + \mathbb{R}_>k = \text{cl}_k(A) + \mathbb{R}_>k = \text{cl}_k(A + \mathbb{R}_>k) = \text{int}_k(A + \mathbb{R}_>k)$.
(d) $A + \mathbb{R}k = \text{cl}_k(A) + \mathbb{R}k = \text{cl}_k(A + \mathbb{R}k) = \text{int}_k(A + \mathbb{R}k)$.
(e) $\text{bd}_k(A + \mathbb{R}_+k) = \text{bd}_k(A + \mathbb{R}_>k) = \text{bd}_k(A + \mathbb{R}k) = \emptyset$.

***Proof*** Definition 2.3.23 yields (a). The first equation of (b) follows from (a).

Since $k \in 0^+(A + \mathbb{R}_+k)$, Proposition 2.3.46 yields $A + \mathbb{R}_+k = \text{cl}_k(A + \mathbb{R}_+k) = \text{int}_k(A + \mathbb{R}_+k)$.

(c) and (d) can be proved analogously. (b)–(d) imply (e). □

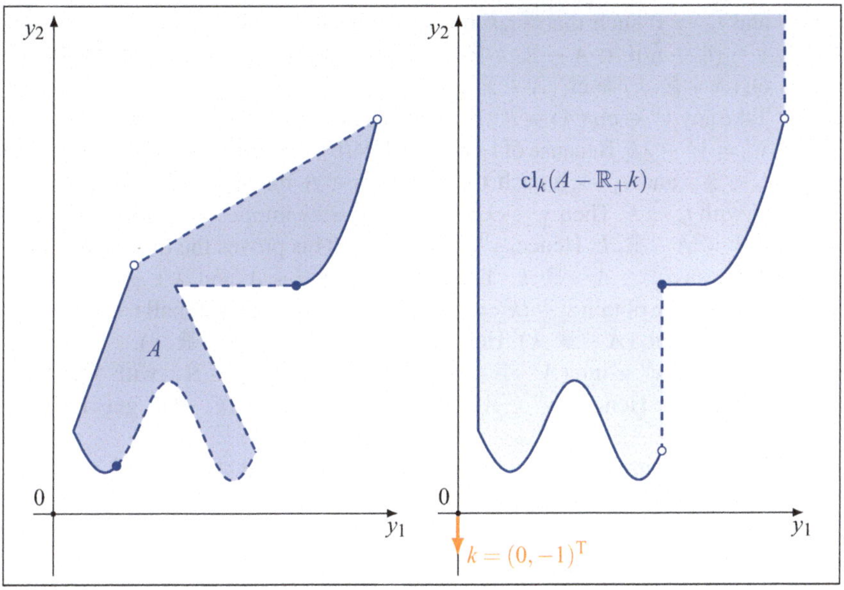


**Fig. 2.5** The set $\mathrm{cl}_k(A-\mathbb{R}_+k)$

**Proposition 2.3.49** *Assume* $(\mathrm{Sp0}_{A,k})$.

(1) $\mathrm{cl}_k(A) - \mathbb{R}_+k = \mathrm{cl}_k(A - \mathbb{R}_+k) = \mathrm{cl}_k(A - \mathbb{R}_>k)$.
(2) $A - \mathbb{R}_>k = \mathrm{cl}_k(A) - \mathbb{R}_>k = \mathrm{int}_k(A - \mathbb{R}_>k) = \mathrm{int}_k(A - \mathbb{R}_+k)$.
(3) $\mathrm{bd}_k(A - \mathbb{R}_>k) = \mathrm{bd}_k(A - \mathbb{R}_+k)$.
(4) $\mathrm{int}_k(A) - \mathbb{R}_+k \subseteq \mathrm{int}_k(A - \mathbb{R}_+k)$.

***Proof***

(1) (a) Since $\mathrm{cl}_k(-\mathbb{R}_+k) = -\mathbb{R}_+k$, Lemma 2.3.26(f) implies $\mathrm{cl}_k(A) - \mathbb{R}_+k = \mathrm{cl}_k(A) + \mathrm{cl}_k(-\mathbb{R}_+k) \subseteq \mathrm{cl}_k(A - \mathbb{R}_+k)$.
Take any $y \in Y$ for which there exists a sequence $(t_n)_{n\in\mathbb{N}}$ of real numbers with $t_n \searrow 0$ and $y - t_nk \in \mathrm{cl}_k(A) - \mathbb{R}_+k$ for all $n \in \mathbb{N}$. Hence, for each $n \in \mathbb{N}$, there is $\lambda_n \in \mathbb{R}_+$ with $y - t_nk \in \mathrm{cl}_k(A) - \lambda_nk$. If there exists $n \in \mathbb{N}$ with $t_n < \lambda_n$, then $y \in \mathrm{cl}_k(A) - (\lambda_n - t_n)k \in \mathrm{cl}_k(A) - \mathbb{R}_+k$. Assume now $t_n \geq \lambda_n$ for all $n \in \mathbb{N}$. Then $t_n - \lambda_n \to 0$ if $n \to \infty$, and $y - (t_n - \lambda_n)k \in \mathrm{cl}_k(A)$ for all $n \in \mathbb{N}$. Therefore, $y \in \mathrm{cl}_k(\mathrm{cl}_k(A)) = \mathrm{cl}_k(A)$. We conclude that $\mathrm{cl}_k(A) - \mathbb{R}_+k$ is $k$-directionally closed.
$A - \mathbb{R}_+k \subseteq \mathrm{cl}_k(A) - \mathbb{R}_+k \subseteq \mathrm{cl}_k(A - \mathbb{R}_+k)$ implies $\mathrm{cl}_k(A - \mathbb{R}_+k) \subseteq \mathrm{cl}_k(\mathrm{cl}_k(A) - \mathbb{R}_+k) \subseteq \mathrm{cl}_k(\mathrm{cl}_k(A - \mathbb{R}_+k))$, i.e., $\mathrm{cl}_k(A - \mathbb{R}_+k) \subseteq \mathrm{cl}_k(A) - \mathbb{R}_+k \subseteq \mathrm{cl}_k(A - \mathbb{R}_+k)$. This yields $\mathrm{cl}_k(A) - \mathbb{R}_+k = \mathrm{cl}_k(A - \mathbb{R}_+k)$.

(b) $\mathrm{cl}_k(A - \mathbb{R}_>k) \subseteq \mathrm{cl}_k(A - \mathbb{R}_+k)$ since $A - \mathbb{R}_>k \subseteq A - \mathbb{R}_+k$.
Take now any $y \in \mathrm{cl}_k(A - \mathbb{R}_+k)$. By Lemma 2.3.24(b), there exists a sequence $(t_n)_{n\in\mathbb{N}}$ with $t_n \in \mathbb{R}_+$ and $t_n \to 0$ such that $y - t_nk \in A - \mathbb{R}_+k$ for all $n \in \mathbb{N}$. Moreover, there exists a sequence $(\tilde{t}_n)_{n\in\mathbb{N}}$ with $\tilde{t}_n \in \mathbb{R}_+$

and $\tilde{t}_n \to 0$ such that $-\tilde{t}_n k \in -\mathbb{R}_> k$ for all $n \in \mathbb{N}$. Then $t_n + \tilde{t}_n \to 0$ and $y - (t_n + \tilde{t}_n)k \in A - \mathbb{R}_> k$ for all $n \in \mathbb{N}$. Thus, $y \in \mathrm{cl}_k(A - \mathbb{R}_> k)$. Hence, $\mathrm{cl}_k(A - \mathbb{R}_> k) = \mathrm{cl}_k(A - \mathbb{R}_+ k)$.

(2) (a) Take any $y^0 \in \mathrm{cl}_k(A) - \mathbb{R}_> k$. Then there exist $y^1 \in \mathrm{cl}_k(A)$ and $\lambda \in \mathbb{R}_>$ with $y^0 = y^1 - \lambda k$. Because of Lemma 2.3.24(b), there is a sequence $(t_n)_{n \in \mathbb{N}}$ with $t_n \in \mathbb{R}_+$ and $t_n \to 0$ such that $y^1 - t_n k \in A$ for all $n \in \mathbb{N}$. Take any $n \in \mathbb{N}$ with $t_n < \lambda$. Then $y^1 - \lambda k - t_n k \in A - \lambda k$ implies $y^1 - \lambda k \in A - (\lambda - t_n)k \subseteq A - \mathbb{R}_> k$. Hence, $y^0 \in A - \mathbb{R}_> k$. This proves the first equation.

(b) Take any $y \in A - \mathbb{R}_> k$. Then there exist $a \in A$ and $\lambda \in \mathbb{R}_>$ with $y = a - \lambda k$. We obtain $y + tk = a + (t - \lambda)k \in A - \mathbb{R}_> k$ for all $t \in [0, \lambda)$ and thus $y \in \mathrm{int}_k(A - \mathbb{R}_> k)$. Hence, $A - \mathbb{R}_> k = \mathrm{int}_k(A - \mathbb{R}_> k)$.

(c) Take any $y^0 \in \mathrm{int}_k(A - \mathbb{R}_+ k)$. Then there exists $\lambda \in \mathbb{R}_>$ with $y^0 + \lambda k \in A - \mathbb{R}_+ k$. Hence, $y^0 \in A - \mathbb{R}_+ k - \lambda k \subseteq A - \mathbb{R}_> k$. We get $\mathrm{int}_k(A - \mathbb{R}_+ k) \subseteq A - \mathbb{R}_> k$.
Moreover, $\mathrm{int}_k(A - \mathbb{R}_> k) \subseteq \mathrm{int}_k(A - \mathbb{R}_+ k)$ since $A - \mathbb{R}_> k \subseteq A - \mathbb{R}_+ k$. This yields the last equation.

(3) results from (1) and (2).

(4) follows from Lemma 2.3.42(g). □

The following examples illustrate the previous statements and show that

$$\mathrm{int}_k(A - \mathbb{R}_> k) = \mathrm{int}_k(A) - \mathbb{R}_> k \quad \text{and}$$
$$\mathrm{int}_k(A - \mathbb{R}_+ k) = \mathrm{int}_k(A) - \mathbb{R}_+ k$$

are not necessarily satisfied.

**Example 2.3.50** In $Y = \mathbb{R}$, choose $k := 1$ and $A := \{y \in \mathbb{Q} \mid y < 0\}$. Then
$\mathrm{cl}_k(A) = (-\infty, 0]$, $\mathrm{int}_k(A) = \emptyset$, $\mathrm{bd}_k(A) = (-\infty, 0]$,
$B := A - \mathbb{R}_+ k = A - \mathbb{R}_> k = (-\infty, 0)$, $\mathrm{cl}_k(B) = (-\infty, 0]$,
$\mathrm{int}_k(B) = (-\infty, 0)$, $\mathrm{bd}_k(B) = \{0\}$,
$\mathrm{cl}_k(A) - \mathbb{R}_+ k = (-\infty, 0]$, $\mathrm{int}_k(A) - \mathbb{R}_+ k = \emptyset$, $\mathrm{bd}_k(A) - \mathbb{R}_+ k = (-\infty, 0]$,
$\mathrm{cl}_k(A) - \mathbb{R}_> k = (-\infty, 0)$, $\mathrm{int}_k(A) - \mathbb{R}_> k = \emptyset$, $\mathrm{bd}_k(A) - \mathbb{R}_> k = (-\infty, 0)$.

**Example 2.3.51** In $Y = \mathbb{R}$, consider $k := 1$ and $A := [1, 2) \cup (3, 4]$. Then
$\mathrm{cl}_k(A) = [1, 2] \cup (3, 4]$, $\mathrm{int}_k(A) = [1, 2) \cup (3, 4)$, $\mathrm{bd}_k(A) = \{2, 4\}$,
$A - \mathbb{R}_+ k = (-\infty, 4]$, $\mathrm{cl}_k(A - \mathbb{R}_+ k) = (-\infty, 4]$, $\mathrm{int}_k(A - \mathbb{R}_+ k) = (-\infty, 4)$,
$A - \mathbb{R}_> k = (-\infty, 4)$, $\mathrm{cl}_k(A - \mathbb{R}_> k) = (-\infty, 4]$, $\mathrm{int}_k(A - \mathbb{R}_> k) = (-\infty, 4)$,
$\mathrm{bd}_k(A - \mathbb{R}_+ k) = \mathrm{bd}_k(A - \mathbb{R}_> k) = \{4\}$,
$\mathrm{cl}_k(A) - \mathbb{R}_+ k = (-\infty, 4]$, $\mathrm{int}_k(A) - \mathbb{R}_+ k = (-\infty, 4)$, $\mathrm{bd}_k(A) - \mathbb{R}_+ k = (-\infty, 4]$,
$\mathrm{cl}_k(A) - \mathbb{R}_> k = (-\infty, 4)$, $\mathrm{int}_k(A) - \mathbb{R}_> k = (-\infty, 4)$, $\mathrm{bd}_k(A) - \mathbb{R}_> k = (-\infty, 4)$.

Furthermore,

$$\mathrm{cl}_k(A) - \mathbb{R}_+ k = \mathrm{bd}_k(A) - \mathbb{R}_+ k \quad \text{and}$$
$$A - \mathbb{R}_> k = \mathrm{bd}_k(A) - \mathbb{R}_> k$$

do not necessarily hold.

**Example 2.3.52** In $Y = \mathbb{R}^2$, choose $k := (-1, 0)^T$ and $A := \mathbb{R}^2_+ \cup \{(y_1, y_2)^T \in Y \mid y_2 = 0\}$. We get
$\mathrm{cl}_k(A) = A$, $\mathrm{bd}_k(A) = \{(y_1, y_2)^T \in Y \mid y_1 = 0, y_2 > 0\}$,
$A - \mathbb{R}_+ k = \mathrm{cl}_k(A) - \mathbb{R}_+ k = A$,
$A - \mathbb{R}_> k = \mathrm{cl}_k(A) - \mathbb{R}_> k = A \setminus \{(y_1, y_2)^T \in Y \mid y_1 = 0, y_2 > 0\}$,
$\mathrm{bd}_k(A) - \mathbb{R}_+ k = A \setminus \{(y_1, y_2)^T \in Y \mid y_2 = 0\}$, $\mathrm{bd}_k(A) - \mathbb{R}_> k = \mathrm{int}\, \mathbb{R}^2_+$.

For the next proposition, recall that $k \in -0^+A$ implies $k \in -0^+(\mathrm{cl}_k(A))$ by Proposition 2.3.29.

**Proposition 2.3.53** *Assume* $(\mathrm{Sp0}_{A,k})$ *and* $k \in -0^+A \setminus \{0\}$. *Then*

(a) $\mathrm{int}_k(A) = \mathrm{int}_k(\mathrm{cl}_k(A)) = A - \mathbb{R}_> k$.
(b) $k \in -0^+(\mathrm{int}_k(A))$.
(c) $\mathrm{cl}_k(\mathrm{int}_k(A)) = \mathrm{cl}_k(A)$.
(d) $\mathrm{bd}_k(A) = \mathrm{bd}_k(\mathrm{cl}_k(A)) = \mathrm{bd}_k(\mathrm{int}_k(A))$.

***Proof*** We have $A - \mathbb{R}_+ k = A$ because of $k \in -0^+A \setminus \{0\}$.

(a) Lemma 2.3.42 yields $\mathrm{int}_k(A) = A - \mathbb{R}_> k$. By Proposition 2.3.49, $\mathrm{int}_k(\mathrm{cl}_k(A)) = \mathrm{int}_k(\mathrm{cl}_k(A - \mathbb{R}_+ k)) = \mathrm{int}_k(\mathrm{cl}_k(A) - \mathbb{R}_+ k) = \mathrm{cl}_k(A) - \mathbb{R}_> k = A - \mathbb{R}_> k$.
(b) $\mathrm{int}_k(A) = A - \mathbb{R}_> k$ implies $\mathrm{int}_k(A) - \mathbb{R}_+ k = A - \mathbb{R}_> k - \mathbb{R}_+ k = A - \mathbb{R}_> k = \mathrm{int}_k(A)$.
(c) By Proposition 2.3.49, $\mathrm{cl}_k(\mathrm{int}_k(A)) = \mathrm{cl}_k(A - \mathbb{R}_> k) = \mathrm{cl}_k(A - \mathbb{R}_+ k) = \mathrm{cl}_k(A)$.
(d) $\mathrm{bd}_k(\mathrm{cl}_k(A)) = \mathrm{cl}_k(\mathrm{cl}_k(A)) \setminus \mathrm{int}_k(\mathrm{cl}_k(A)) = \mathrm{cl}_k(A) \setminus \mathrm{int}_k(A) = \mathrm{bd}_k(A)$.
$\mathrm{bd}_k(\mathrm{int}_k(A)) = \mathrm{cl}_k(\mathrm{int}_k(A)) \setminus \mathrm{int}_k(\mathrm{int}_k(A)) = \mathrm{cl}_k(A) \setminus \mathrm{int}_k(A) = \mathrm{bd}_k(A)$.

□

The next propositions refer to the relationship between the introduced direction-depending properties and the topology in topological vector spaces.

**Proposition 2.3.54** *Assume that* $Y$ *is a topological vector space,* $A \subseteq Y$ *and* $k \in Y \setminus \{0\}$.

(a) $\mathrm{cl}\, A$ *is* $k$*-directionally closed, i.e.,* $\mathrm{cl}_k(\mathrm{cl}\, A) = \mathrm{cl}\, A$.
(b) $\mathrm{cl}_k(A) \subseteq \mathrm{cl}\, A$.
(c) $\mathrm{cl}\, \mathrm{cl}_k(A) = \mathrm{cl}\, A$.
(d) *If* $A$ *is closed, then it is* $k$*-directionally closed, i.e.,* $\mathrm{cl}_k(A) = A$.
(e) $\mathrm{cl}_k(A)$ *is closed if and only if* $\mathrm{cl}_k(A) = \mathrm{cl}\, A$.
(f) $\mathrm{cl}\, A - \mathbb{R}_> k \subseteq A$ *holds if and only if* $k \in -0^+A$ *and* $\mathrm{cl}_k(A) = \mathrm{cl}\, A$.

***Proof***

(a) Take any $y \in Y$ for which there exists a sequence $(t_n)_{n \in \mathbb{N}}$ of real numbers with $t_n \searrow 0$ and $y - t_n k \in \mathrm{cl}\, A$. Since each neighborhood of $y$ contains some $y - t_n k$, we get $y \in \mathrm{cl}(\mathrm{cl}\, A) = \mathrm{cl}\, A$. Hence, $\mathrm{cl}\, A$ is $k$-directionally closed.
(b) $A \subseteq \mathrm{cl}\, A$ yields $\mathrm{cl}_k(A) \subseteq \mathrm{cl}_k(\mathrm{cl}\, A) = \mathrm{cl}\, A$ by (a).

(c) (b) implies $\text{cl}\,\text{cl}_k(A) \subseteq \text{cl}\,A$. By Lemma 2.3.26(a), $A \subseteq \text{cl}_k(A)$. This results in $\text{cl}\,A \subseteq \text{cl}\,\text{cl}_k(A)$.
(d) follows from (a).
(e) Assume first that $\text{cl}_k(A)$ is closed. $A \subseteq \text{cl}_k(A)$ yields $\text{cl}\,A \subseteq \text{cl}\,\text{cl}_k(A) = \text{cl}_k(A)$. We obtain $\text{cl}\,A = \text{cl}_k(A)$ from (b). The reverse direction of the assertion is obvious.
(f) $\text{cl}\,A - \mathbb{R}_> k \subseteq A$ implies $k \in -0^+ A$ and $\text{cl}\,A \subseteq \text{cl}_k(A)$ by Proposition 2.3.29(a); hence $\text{cl}_k(A) = \text{cl}\,A$ because of (b).
If $k \in -0^+ A$ and $\text{cl}_k(A) = \text{cl}\,A$, then Proposition 2.3.29(a) yields $\text{cl}_k(A) - \mathbb{R}_> k \subseteq A$ and thus $\text{cl}\,A - \mathbb{R}_> k \subseteq A$. □

**Proposition 2.3.55** *Assume that $Y$ is a topological vector space, $A \subseteq Y$ and $k \in Y \setminus \{0\}$.*

(a) $\text{int}\,A \subseteq \text{int}_k(A)$.
(b) $A - \mathbb{R}_> k \subseteq \text{int}\,A$ *holds if and only if* $k \in -0^+ A$ *and* $\text{int}_k(A) = \text{int}\,A$.
(c) $\text{bd}_k(A) \subseteq \text{bd}\,A$.
(d) *If* $\text{cl}\,A - \mathbb{R}_> k \subseteq \text{int}\,A$, *then* $\text{bd}_k(A) = \text{bd}\,A$.

***Proof***

(a) $\text{int}\,A \subseteq \text{core}\,A \subseteq \text{int}_k(A)$ by Proposition 2.3.6(I) and Lemma 2.3.42(a).
(b) Assume first $A - \mathbb{R}_> k \subseteq \text{int}\,A$. Take any $y \in \text{int}_k(A)$. Then there exists $t \in \mathbb{R}_>$ with $y + tk \in A$. Hence, $y \in A - tk \subseteq \text{int}\,A$. This and (a) provide the first direction of the assertion.
Assume now $k \in -0^+ A$ and $\text{int}_k(A) = \text{int}\,A$. By Lemma 2.3.42(b), $k \in -0^+ A$ yields $\text{int}_k(A) = A - \mathbb{R}_> k$. Hence, $A - \mathbb{R}_> k = \text{int}\,A$.
(c) $\text{bd}_k(A) = \text{cl}_k(A) \setminus \text{int}_k(A) \subseteq \text{cl}\,A \setminus \text{int}\,A = \text{bd}\,A$ because of (a) and Proposition 2.3.54(b).
(d) Proposition 2.3.54(f) implies $\text{cl}_k(A) = \text{cl}\,A$. This and (b) yield the assertion.

□

Finally, we will illustrate the introduced notions and related statements for an ordering cone.

**Example 2.3.56** The lexicographic order in $\mathbb{R}^2$ is represented by the pointed convex cone $A = \{(y_1, y_2)^T \in \mathbb{R}^2 \mid y_1 > 0\} \cup \{(y_1, y_2)^T \in \mathbb{R}^2 \mid y_1{=}0,\ y_2{\geq}0\}$ (see Fig. 2.6). $A$ is not algebraically closed and thus not closed. Obviously, $0^+ A = A$ and $-0^+ A \setminus \{0\} = -A \setminus \{0\} = Y \setminus A = -A \setminus A$.

(a) For each $k \in A \setminus \{0\}$, the cone $A$ is $k$-directionally closed, $\text{int}_k(A) = A$, and $\text{bd}_k(A) = \emptyset$ by Proposition 2.3.46.
(b) Take any $k \in -\,\text{core}\,A = \{(y_1, y_2)^T \in \mathbb{R}^2 \mid y_1 < 0\}$. Then $\text{cl}\,A - \mathbb{R}_> k \subseteq \text{int}\,A$, and $\text{cl}_k(A) = \text{cl}\,A = \{(y_1, y_2)^T \in \mathbb{R}^2 \mid y_1 \geq 0\}$. Moreover, $\text{int}_k(A) = \text{int}\,A$ and $\text{bd}_k(A) = \text{bd}\,A = \{(y_1, y_2)^T \in \mathbb{R}^2 \mid y_1 = 0\}$.
(c) Take any $k \in -(A \setminus \text{core}\,A) \setminus \{0\} = \{(y_1, y_2)^T \in \mathbb{R}^2 \mid y_1 = 0,\ y_2 < 0\}$. Then $A$ is $k$-directionally closed and $\text{cl}\,A - \mathbb{R}_> k \not\subseteq A$. Furthermore, $\text{int}_k(A) = A \setminus \{0\} = \text{int}\,A \cup \{(y_1, y_2)^T \in \mathbb{R}^2 \mid y_1 = 0,\ y_2 > 0\}$ and $\text{bd}_k(A) = \{0\}$.

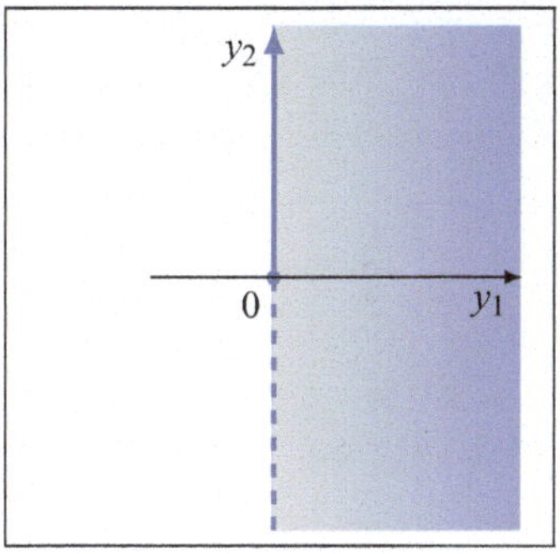


**Fig. 2.6** Lexicographic ordering cone in $\mathbb{R}^2$

**Remark 2.3.57** (*Bibliographical notes related to Sect.* 2.3.2)

Lemma 2.3.13 was proved in [483, Lemma 5.1], Lemma 2.3.14 in [472]. A statement which is analogous to Lemma 2.3.15 was shown in [472, Lemma 3.2.3]. Lemma 2.3.16 coincides with [484, Lemma 2.8].

The idea for the proof of $A - \mathbb{R}_{>}k \subseteq \operatorname{core} A$ in Proposition 2.3.20 is contained in [407, Lemma 3.1.11].

Lemma 2.3.24(a) and (b) were stated in [485, Lemma 2.2] and in [489, Lemma 2.1], respectively. The first inclusion of Lemma 2.3.26(a) as well as parts (c) and (e) of this lemma are mentioned for $\operatorname{cl}_{-k}(A)$ in [213, p. 2678f]. [485, Lemma 2.1] contains Lemma 2.3.26(b). Proposition 2.3.29(a) and (b) are taken from [485, Lemma 2.3] and [489, Lemma 2.2]. Proposition 2.3.30 is given in [489, Lemma 2.3]. Proposition 2.3.30(b) is mentioned for $\operatorname{cl}_{-k}(A)$ in [213, p. 2678f]. Corollary 2.3.31 can be derived from [385, Proposition 2.4]. Propositions 2.3.35 and 2.3.37 were given in [483, Lemmata 2.1 and 5.2]. Proposition 2.3.39 was proved in [483, Proposition 2.4(c)]. Proposition 2.3.54 coincides with [489, Lemma 2.4].

# Chapter 3
# Extended Real-Valued Functions

In this chapter, we present basic notations and properties for functions which attain values in $\overline{\mathbb{R}} := \mathbb{R} \cup \{-\infty, +\infty\}$, the **extended set of real numbers**. Of course, such functions may also be real-valued. But infima, suprema, limits, improper integrals or some measure can result in function values $-\infty$ or $+\infty$. Such values can also result from the extension of a function to the entire space or from the addition of an indicator function expressing certain restrictions in convex analysis.

In the first two sections, we will give a short overview about the extended set of real numbers and explain the way in which extended real-valued functions are handled in convex analysis. Here we will also discuss problems that arise by admitting values $\pm\infty$. In [480, 486], a unified approach to extended real-valued functions was presented which overcomes many drawbacks of the unilateral approach. But for the reader's convenience, we follow in this book the usual definitions from convex analysis.

Section 3.3 summarizes statements about the continuity and the semicontinuity of extended real-valued functions. The same is done for convexity and quasiconvexity in Sect. 3.4, for linearity and related algebraic properties in Sect. 3.5 and for monotonicity in Sect. 3.7. Basic notions for the variational analysis of such functions are collected in Sect. 3.6.

Since extended real-valued functions belong to the standard in convex analysis, most of the results in this chapter are well known. We add several new statements in order to point out that some properties of the functions which we focus on in this book turn out to be properties of each extended real-valued function.

## 3.1 The Extended Set of Real Numbers

The extended set of real numbers, $\overline{\mathbb{R}} = \mathbb{R} \cup \{-\infty, +\infty\}$, is often used in measure theory and in convex analysis. In measure theory, a set may have the measure $+\infty$. In convex analysis, the following properties of $\overline{\mathbb{R}}$ are relevant:



C. Tammer and P. Weidner, *Scalarization and Separation by Translation Invariant Functions*, Vector Optimization, https://doi.org/10.1007/978-3-030-44723-6_3

- $(\overline{\mathbb{R}}, \leq)$ is a totally ordered set and each subset of $\overline{\mathbb{R}}$ has an infimum and a supremum,
- $\overline{\mathbb{R}}$ is a compact Hausdorff space.

Recall first the rules which are defined for the calculation with $\pm\infty$ in $\overline{\mathbb{R}}$:

$$\begin{aligned}
&\forall y \in \mathbb{R}: \quad -\infty < y < +\infty,\\
&-(+\infty) = -\infty, \quad -(-\infty) = +\infty,\\
&(+\infty)\cdot(+\infty) = (-\infty)\cdot(-\infty) = +\infty,\\
&(+\infty)\cdot(-\infty) = (-\infty)\cdot(+\infty) = -\infty,
\end{aligned}$$

$$\begin{aligned}
\forall y \in \mathbb{R}_>: \quad & y\cdot(+\infty) = (+\infty)\cdot y = +\infty,\\
& y\cdot(-\infty) = (-\infty)\cdot y = -\infty,\\
\forall y \in -\mathbb{R}_>: \quad & y\cdot(+\infty) = (+\infty)\cdot y = -\infty,\\
& y\cdot(-\infty) = (-\infty)\cdot y = +\infty,\\
\forall y \in \mathbb{R}\cup\{+\infty\}: \quad & y + (+\infty) = +\infty + y = +\infty,\\
\forall y \in \mathbb{R}\cup\{-\infty\}: \quad & y + (-\infty) = -\infty + y = -\infty.
\end{aligned}$$

Moreover, we define

$$0\cdot(+\infty) := (+\infty)\cdot 0 := 0\cdot(-\infty) := (-\infty)\cdot 0 := 0.$$

$\overline{\mathbb{R}}$ is not a real linear space since $+\infty$ and $-\infty$ do not have an inverse element w.r.t. addition and since the calculation within $\overline{\mathbb{R}}$ is not possible in an analogous way as in $\mathbb{R}$.

**Example 3.1.1** The rule $(\lambda + \mu)x = \lambda x + \mu x$ does not hold for arbitrary $\lambda, \mu \in \mathbb{R}$ and $x \in \{+\infty, -\infty\}$ since otherwise, e.g.,

$$\begin{aligned}
+\infty &= (3-2)\cdot(+\infty) = 3\cdot(+\infty) + (-2)\cdot(+\infty)\\
&= +\infty + (-\infty) = 2\cdot(+\infty) + (-3)\cdot(+\infty) = (2-3)\cdot(+\infty)\\
&= -\infty.
\end{aligned}$$

Although $\overline{\mathbb{R}}$ is totally ordered, equivalent transformations of equations and inequalities in $\mathbb{R}$ do not work in the same way in $\overline{\mathbb{R}}$.

**Example 3.1.2** In $\overline{\mathbb{R}}$, $a + b = a + c$ does not imply $b = c$, and $a + b \leq a + c$ does not imply $b \leq c$. Choose, e.g., $a = -\infty$, $b = 0$ and $c = -3$.

Inequalities which are obvious in $\mathbb{R}$ can be wrong in $\overline{\mathbb{R}}$, e.g.,

$$a - t < a \text{ holds for } a \in \overline{\mathbb{R}} \text{ and } t \in \mathbb{R}_> \text{ if and only if } a \in \mathbb{R}.$$

But we can equip $\overline{\mathbb{R}}$ with the topology that is generated by a neighborhood base of the Euclidean topology on $\mathbb{R}$ together with the neighborhood bases $\mathscr{B}(+\infty) := \{\{x \in \overline{\mathbb{R}} \mid x > a\} \mid a \in \mathbb{R}\}$ and $\mathscr{B}(-\infty) := \{\{x \in \overline{\mathbb{R}} \mid x < a\} \mid a \in \mathbb{R}\}$. Whenever we will consider $\overline{\mathbb{R}}$ as a topological space, this topology will be supposed. Note that $\overline{\mathbb{R}}$ is a compact Hausdorff space, but not a topological vector space. Nevertheless, the following rules are valid in its topology $\tau$:

(a) $\forall O \in \tau \setminus \{\emptyset\}\ \forall \lambda \in \mathbb{R} \setminus \{0\} : \ \lambda O \in \tau$.
(b) $\forall O \in \tau \setminus \{\emptyset\}\ \forall x \in \mathbb{R} : \ x + O \in \tau$.

## 3.2 Definition of Extended Real-Valued Functions in Convex Analysis

The approach in this section is part of the basics of convex analysis as presented by Moreau [362] and Rockafellar [389]. It has become standard when studying minimization problems (see [12, 13, 64, 242, 508]), especially in the variational analysis for such problems [359, 392]. Originally, the concept focused on the minimization of convex real-valued functions which are extended to the entire space by the value $+\infty$ and/or for which the barrier $+\infty$ indicates the violation of restrictions. When minimizing a function in the case where feasible solutions exist, the function value $+\infty$ does not alter the minimum.

The consideration of extended real-valued functions in convex analysis is unilateral. The handling of the function depends on whether it is to be minimized or maximized. The usual definitions are given for minimization problems and can be applied to maximization by working with $-\infty$ instead of $+\infty$, where the rule for the addition of $+\infty$ and $-\infty$ has to be changed. The definition of an indicator function also refers to the minimization of the function to which the indicator function is added.

We will now explain the usual approach for a function $f$ which has to be minimized. Even if $f$ were real-valued, it could be extended to a function that is not real-valued any more for the following two reasons.

$f : C \to \overline{\mathbb{R}}$ with $C$ being a subset of some space $X$ can be extended to a function $g : X \to \overline{\mathbb{R}}$ by defining

$$g(x) := \begin{cases} f(x) & \text{if } x \in C, \\ +\infty & \text{if } x \in X \setminus C. \end{cases}$$

Furthermore, functions with values in $\overline{\mathbb{R}}$ play an important role in optimization theory when the problem

$$\inf\{f(x) \mid x \in C\}$$

with $f : X \to \overline{\mathbb{R}}$ is replaced by

$$\inf\{g(x) \mid x \in X\}$$

with $g := f + \iota_C$, where $\iota_C : X \to \overline{\mathbb{R}}$ is the indicator function of $C$.

**Definition 3.2.1** On a space $X$, the **indicator function** $\iota_C : X \to \overline{\mathbb{R}}$ of $C \subseteq X$ is defined by

$$\iota_C(x) := \begin{cases} 0 & \text{if } x \in C, \\ +\infty & \text{if } x \in X \setminus C. \end{cases}$$

Here the rule $+\infty + (-\infty) := +\infty$ has to be applied. Working with $g$ instead of $f$ offers an elegant way to avoid the necessity for investigating technical details of $C$ in advance. E.g., when studying the functional $\varphi_{A,k}$ given by

$$\varphi_{A,k}(x) := \inf\{t \in \mathbb{R} \mid x \in A + tk\}$$

for some set $A \subseteq X$ and some $k \in X \setminus \{0\}$ in a linear space $X$, we can simply write $\varphi_{A,k} : X \to \overline{\mathbb{R}}$ using the definition $\inf \emptyset := +\infty$.

Obviously, the value $+\infty$ serves as a symbol for a function value at points which are not feasible for the problem, and if a function value $+\infty$ exists since, e.g., the function is defined via a supremum, the related argument of the function is handled as if it did not belong to the domain of interest.

**Definition 3.2.2** Consider an extended real-valued function $\varphi \colon X \to \overline{\mathbb{R}}$ on some nonempty set $X$. We define its **(effective) domain** as

$$\operatorname{dom} \varphi := \{x \in X \mid \varphi(x) \neq +\infty\}.$$

$\varphi$ is called **finite-valued** on $X_0 \subseteq X$ if $\varphi(x) \in \mathbb{R}$ for all $x \in X_0$. $\varphi$ is said to be **finite-valued** if $\varphi$ is finite-valued on $X$. $\varphi$ is called **proper** if $\operatorname{dom} \varphi \neq \emptyset$ and $\varphi$ is finite-valued on $\operatorname{dom} \varphi$. Otherwise, it is called **improper**.

We will refer to extended real-valued functions also as functionals.

The finite-valued functionals are the functions $\varphi \colon X \to \mathbb{R}$ and are the proper functionals with $\operatorname{dom} \varphi = X$.

For two functions $f, g \colon X \to \overline{\mathbb{R}}$, we write $f \leq g$ if $f(x) \leq g(x)$ holds for all $x \in X$. If not mentioned otherwise, then the notions of greater and smaller—related to functions—refer to this relation. We say that $g$ **majorizes** $f$ if $f \leq g$. If $f$ attains the value $c \in \overline{\mathbb{R}}$ only, then we write $f = c$.

Take into consideration that each extended real-valued function has an upper bound and a lower bound, but that it is said to be **bounded above** or **bounded below** if there exists a real upper or lower bound, respectively. For $\varphi : M \to \overline{\mathbb{R}}$, $M$ being an arbitrary set, $t \in \overline{\mathbb{R}}$ is called a **minimum** or **minimal value** of $\varphi$ on $M$ if $t \in \mathbb{R}$, $t = \varphi(m^0)$ for some $m^0 \in M$ and $t \leq \varphi(m)$ for all $m \in M$. $t \in \overline{\mathbb{R}}$ is a **maximum** or **maximal value** of $\varphi$ on $M$ if $t \in \mathbb{R}$, $t = \varphi(m^0)$ for some $m^0 \in M$ and $t \geq \varphi(m)$ for all $m \in M$. We denote the maximum of $\varphi$ on $M$ by $\max_{x\in M} \varphi(x)$, the minimum of $\varphi$ on $M$ by $\min_{x\in M} \varphi(x)$ and define the **set of minimal solutions** or **minimizers** of $\varphi$ on $M$ as

$$\operatorname*{argmin}_{x\in M} \varphi(x) := \begin{cases} \{\bar{x} \in M \mid \varphi(\bar{x}) = \min_{x\in M} \varphi(x)\} & \text{if } \varphi \text{ attains a minimum on } M, \\ \emptyset & \text{otherwise.} \end{cases}$$

It is also denoted by $\operatorname{argmin}_M \varphi$.

Based on extended real-valued functions, useful tools for convex optimization have been developed, but we have to point out that such functions have to be handled with care. When working with extended real-valued functions, one should be aware of the following details.

(a) The way in which an extended real-valued function is handled in convex analysis depends on whether it is to be minimized or maximized. Supremum problems are studied analogously to the infimum problems above, but with $-\infty$ replacing $+\infty$ and vice versa. One consequence is the different definition of $+\infty + (-\infty)$ in both frameworks. The so-called **inf-addition**

$$+\infty + (-\infty) := (-\infty) + (+\infty) := +\infty$$

is applied for minimization problems, whereas the **sup-addition** for maximization problems is given by

$$+\infty + (-\infty) := (-\infty) + (+\infty) := -\infty.$$

Moreau [362] introduced both kinds of addition and, in connection with this, two different addition operators on $\overline{\mathbb{R}}$.
The indicator function defined above should not be added to functions that have to be maximized.

(b) The extension of a convex function $f : C \to \overline{\mathbb{R}}$, $C$ being a proper convex subset of some topological vector space $X$, to $g : X \to \overline{\mathbb{R}}$ by adding the function value $+\infty$ outside $C$ can destroy the continuity and even the lower semicontinuity of $f$.

**Example 3.2.3** Consider the functional $\varphi : \mathbb{R} \to \overline{\mathbb{R}}$ given by

$$\varphi(x) := \begin{cases} x & \text{if } x > 0, \\ +\infty & \text{if } x \le 0. \end{cases}$$

$\varphi$ is continuous on $\operatorname{dom} \varphi$, but not lower semicontinuous on $\mathbb{R}$.

We will see later—in Proposition 3.3.19—that an extended real-valued functional cannot be continuous if the effective domain is not open.

(c) Multiplication of functions with real numbers and subtraction of functions are only possible under certain assumptions (see [242, p. 389]).

(d) In connection with extended real-valued functions in convex analysis, the definitions $\inf \emptyset := +\infty$ and $\sup \emptyset := -\infty$ are given. Hence, $\sup \emptyset < \inf \emptyset$.

(e) If a function has been extended to the entire space by $+\infty$ or $-\infty$, then it cannot be handled like the original function any more. This is obviously the case since the way of extension depends on the purpose. The function extended by $+\infty$ does not have a finite Chebyshev norm, and the Chebyshev norm approximation

$$\inf_{y \in Y} \sup_{x \in X} |f(x) - g(x, y)|$$

cannot be applied. Here $X$ is the space to which the function has been extended and $Y$ is any nonempty set of parameters.

(f) In some optimization problems, a function cannot simply be extended to the entire space using only one of the values $+\infty$ or $-\infty$. Consider the problem

$$\inf_{x \in X_0} \sup_{y \in Y_0} L(x, y), \tag{3.2.1}$$

where $Y_0$ is a proper subset of some space $Y$, $X_0$ is a proper subset of some space $X$ and $L$ is real-valued. In such a case, $L$ is extended to a function $f$ which is defined on $X \times Y$ in the following way [392, Example 11.52]:

$$f(x, y) := \begin{cases} L(x, y) & \text{if } x \in X_0 \text{ and } y \in Y_0, \\ -\infty & \text{if } x \in X_0 \text{ and } y \in Y \setminus Y_0, \\ +\infty & \text{if } x \in X \setminus X_0. \end{cases}$$

The problem

$$\inf_{x \in X} \sup_{y \in Y} f(x, y)$$

has a real optimal value if and only if problem (3.2.1) has a real optimal value. In this case, the optimal values as well as the optimal solutions of both problems coincide.

(g) The previous item illustrates that the unilateral approach in convex analysis becomes complicated when infimum problems and supremum problems are combined. Duality theory usually works with such combined problems. Then one has to be very careful in using the notion of the effective domain or adding $+\infty$ and $-\infty$.

From now on, we define

$$\inf \emptyset := +\infty \text{ and } \sup \emptyset := -\infty$$

and apply the inf-addition

$$+\infty + (-\infty) := (-\infty) + (+\infty) := +\infty.$$

In this chapter, we will point out when a statement depends on this rule.

Note the following drawbacks of the inf-addition for $a, b, c \in \overline{\mathbb{R}}$:

$$-(b + c) \neq -b - c \text{ if } b = +\infty \text{ and } c = -\infty, \tag{3.2.2}$$

hence,

$$a \leq b + c \text{ implies } -a \geq -(b + c), \text{ but not } -a \geq -b - c$$

since, e.g., $3 \leq (+\infty) + (-\infty)$, but $-3 \not\geq (-\infty) + (+\infty)$.
But if $\{-\infty, +\infty\} \neq \{b, c\}$, then

$$-(b + c) = -b - c, \text{ and } \quad (a \leq b + c \implies -a \geq -b - c).$$

We have to mention another consequence of (3.2.2). If there exist elements $a \in A$ and $b \in B$ with $\{a, b\} = \{-\infty, +\infty\}$ for sets $A, B \subseteq \overline{\mathbb{R}}$, then

$$-\inf\{-a - b \mid a \in A, b \in B\} \tag{3.2.3}$$

should be written instead of

$$\sup\{a + b \mid a \in A, b \in B\}$$

since the sup-addition had to be applied in the last formula.

With the introduced rules, we can extend Definition 2.1.30 of the Minkowski functional.

**Definition 3.2.4** For each subset $A$ of a linear space $X$ with $0 \in A$, the **Minkowski functional** $p_A : X \to \overline{\mathbb{R}}$ of $A$ is defined by

$$p_A(x) := \inf\{\lambda \in \mathbb{R}_> \mid x \in \lambda A\}.$$

Then the Minkowski functional of a cone is its indicator function [362].

On $\overline{\mathbb{R}}$, one has to take into consideration that the familiar laws of arithmetic are not valid in full extent (see Example 3.1.1) and that equations as well as inequalities cannot be handled in the same way as for real numbers (see Example 3.1.2). Example 3.1.1 demonstrates that linear combinations of improper functionals have to be used carefully. This requires alternative definitions of properties like the convexity of functions.

Recall, e.g., that a function $f : C \to \mathbb{R}$, $C$ being a nonempty convex subset of a linear space $X$, is said to be convex on $C$ if

$$f(\lambda x^1 + (1 - \lambda)x^2) \leq \lambda f(x^1) + (1 - \lambda) f(x^2)$$

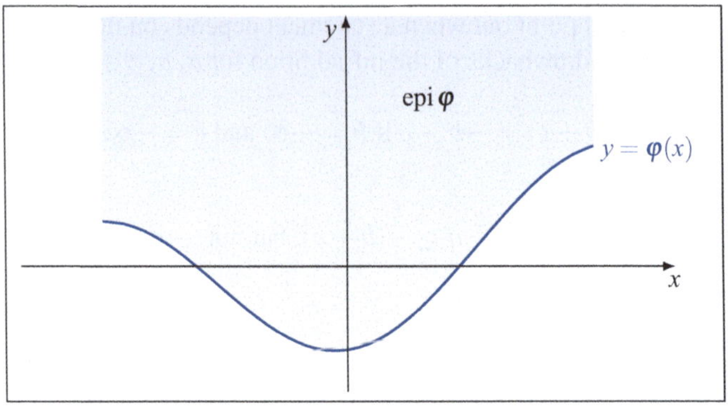


**Fig. 3.1** Epigraph of $\varphi$

holds for all $x^1, x^2 \in C$ and $\lambda \in (0, 1)$. The convexity of functions with values in $\overline{\mathbb{R}}$ has to be defined in another way or an appropriate rule for the addition of $+\infty$ and $-\infty$ has to be fixed.

We will describe essential properties of extended real-valued functions by the epigraph or the hypograph of the function. The epigraph (see Fig. 3.1) connects properties of a function with properties of a set.

**Definition 3.2.5** Assume that $X$ is a nonempty set and $\varphi : X \to \overline{\mathbb{R}}$. The **epigraph** of $\varphi$ is defined by

$$\operatorname{epi}\varphi := \{(x, t) \in X \times \mathbb{R} \mid \varphi(x) \leq t\}. \tag{3.2.4}$$

The **hypograph** or **subgraph** of $\varphi$ is the set

$$\operatorname{hypo}\varphi := \{(x, t) \in X \times \mathbb{R} \mid \varphi(x) \geq t\}.$$

Whenever $X$ is a topological space, the investigation of topological properties of the epigraph or of the hypograph refers to the space $X \times \mathbb{R}$.

The effective domain $\operatorname{dom}\varphi$ is the projection of $\operatorname{epi}\varphi$ onto $X$.

The next lemma contains some immediate implications of the definition.

**Lemma 3.2.6** *Assume that $X$ is a nonempty set and $\varphi : X \to \overline{\mathbb{R}}$.*

(a) *For each $x \in X$ and $t \in \mathbb{R}$, we have $(x, t) \in \operatorname{epi}\varphi$ if and only if $(x, -t) \in \operatorname{hypo}(-\varphi)$.*
(b) *For each $x \in X$, $\varphi(x) = -\infty$ if and only if $\{x\} \times \mathbb{R} \subseteq \operatorname{epi}\varphi$.*
*In this case, $(\{x\} \times \mathbb{R}) \cap \operatorname{hypo}\varphi = \emptyset$.*
(c) *For each $x \in X$, $\varphi(x) = +\infty$ if and only if $\{x\} \times \mathbb{R} \subseteq \operatorname{hypo}\varphi$.*
*In this case, $(\{x\} \times \mathbb{R}) \cap \operatorname{epi}\varphi = \emptyset$.*

Part (a) of this lemma yields two corollaries.

**Corollary 3.2.7** *Let $X$ be a linear space and $\varphi : X \to \overline{\mathbb{R}}$.*

(a) $\operatorname{hypo}\varphi + \operatorname{hypo}\varphi \subseteq \operatorname{hypo}\varphi \iff \operatorname{epi}(-\varphi) + \operatorname{epi}(-\varphi) \subseteq \operatorname{epi}(-\varphi)$.
(b) $\operatorname{hypo}\varphi$ *is convex* $\iff \operatorname{epi}(-\varphi)$ *is convex.*
(c) $\operatorname{hypo}\varphi$ *is a cone* $\iff \operatorname{epi}(-\varphi)$ *is a cone.*

**Corollary 3.2.8** *Let $X$ be a topological space and $\varphi : X \to \overline{\mathbb{R}}$. Then:*
$\operatorname{hypo}\varphi$ *is closed* $\iff \operatorname{epi}(-\varphi)$ *is closed.*

In Sect. 4.14, we will make use of the following statement [489, Lemma 3.1]:

**Lemma 3.2.9** *Assume that $X$ is a linear space, $\varphi\colon X \to \overline{\mathbb{R}}$ and $k := (0_X, -1) \in X \times \mathbb{R}$. Then:*

(a) $k \in -0^+(\operatorname{epi}\varphi) \setminus \{0\}$.
(b) $\operatorname{epi}\varphi$ *is $k$-directionally closed.*
(c) $\operatorname{epi}\varphi$ *is a proper subset of $X \times \mathbb{R}$ if and only if $\varphi$ is not a constant function with the function value $+\infty$ or $-\infty$.*

***Proof*** Define $A := \operatorname{epi}\varphi$.

(a) Take any $(x, t) \in A$, $\lambda \in \mathbb{R}_+$. Then $\varphi(x) \leq t \leq t + \lambda$, which implies $(x, t) - \lambda k = (x, t + \lambda) \in A$. Thus, $k \in -0^+A \setminus \{0\}$.
(b) Take any $(x, t) \in X \times \mathbb{R}$ for which there exists a sequence $(t_n)_{n\in\mathbb{N}}$ of real numbers with $t_n \searrow 0$ and $(x, t) - t_n k \in A$ for all $n \in \mathbb{N}$. Then $(x, t + t_n) = (x, t) - t_n k \in A$ and $\varphi(x) \leq t + t_n$ for all $n \in \mathbb{N}$. Thus, $\varphi(x) \leq t$. We obtain $(x, t) \in A$. Hence, $A$ is $k$-directionally closed.
(c) $A = \emptyset$ if and only if $\operatorname{dom}\varphi = \emptyset$.
$A = X \times \mathbb{R}$ if and only if $\varphi(x) = -\infty$ for each $x \in X$. □

**Proposition 3.2.10** *Assume that $X$ is a linear space and $\{\varphi_i \mid i \in I\}$ is a family of functions $\varphi_i : X \to \overline{\mathbb{R}}$.*

(a) *The intersection of epigraphs is the epigraph of some function. In detail:*

$$\operatorname{epi}(\sup_{i\in I}\varphi_i) = \bigcap_{i\in I}\operatorname{epi}(\varphi_i).$$

(b) *With $k := (0_X, -1) \in X \times \mathbb{R}$, we get*

$$\operatorname{epi}(\inf_{i\in I}\varphi_i) = \operatorname{cl}_k\left(\bigcup_{i\in I}\operatorname{epi}(\varphi_i)\right).$$

(c) *If $I$ is a finite set, then*

$$\operatorname{epi}(\inf_{i\in I}\varphi_i) = \bigcup_{i\in I}\operatorname{epi}(\varphi_i).$$

***Proof*** Definition 3.2.5 yields (a) and $\cup_{i\in I}\operatorname{epi}(\varphi_i) \subseteq \operatorname{epi}(\inf_{i\in I}\varphi_i)$.

Lemmata 2.3.26(b) and 3.2.9(b) imply $\operatorname{cl}_k(\cup_{i\in I}\operatorname{epi}(\varphi_i)) \subseteq \operatorname{cl}_k(\operatorname{epi}(\inf_{i\in I}\varphi_i)) = \operatorname{epi}(\inf_{i\in I}\varphi_i)$.

Take any $(x, s) \in (X \times \mathbb{R}) \setminus \mathrm{cl}_k(\bigcup_{i\in I}\mathrm{epi}(\varphi_i))$. By Definition 2.3.23, there exists $\lambda \in \mathbb{R}_>$ such that $(x, s) - tk \notin \bigcup_{i\in I}\mathrm{epi}(\varphi_i)$ holds for all $t \in \mathbb{R}_+$ with $t < \lambda$ . Hence, $(x, s + t) = (x, s) - tk \notin \mathrm{epi}(\varphi_i)$ for each $i \in I$ and each $t \in \mathbb{R}_+$ with $t < \lambda$, i.e., $\varphi_i(x) > s + t$ for each $i \in I$ and each $t \in \mathbb{R}_+$ with $t < \lambda$. Therefore, $\varphi_i(x) \geq s + \lambda$ for each $i \in I$. We get $\inf_{i\in I}\varphi_i(x) \geq s + \lambda$ and thus $(x, s) \notin \mathrm{epi}(\inf_{i\in I}\varphi_i)$. This results in (b).

(b) yields (c) because of Lemma 3.2.9(b) and Proposition 2.3.32(b). □

In general, $\mathrm{epi}(\inf_{i\in I}\varphi_i) \neq \bigcup_{i\in I}\mathrm{epi}(\varphi_i)$.

**Example 3.2.11** On $X = \mathbb{R}$, define $\varphi_i : X \to \overline{\mathbb{R}}$ by $\varphi_i(x) := \frac{1}{i}$ for all $x \in X$ and $i \in I := \mathbb{N}_>$. Then $\inf_{i\in I}\varphi_i(x) = 0$ and thus $(x, 0)^{\mathrm{T}} \in \mathrm{epi}(\inf_{i\in I}\varphi_i)$ for each $x \in X$. But we have $\varphi_i(x) > 0$ for all $x \in X$ and $i \in I$. Hence, $(x, 0)^{\mathrm{T}} \notin \mathrm{epi}(\varphi_i)$ for each $i \in I$. We get $\bigcup_{i\in I}\mathrm{epi}(\varphi_i) = X \times (0, +\infty)$, whereas $\mathrm{epi}(\inf_{i\in I}\varphi_i) = X \times [0, +\infty)$.

The example illustrates that the union of epigraphs is not necessarily the epigraph of some function.

Several regularization methods in optimization and allocation problems in mathematical economics deal with the inf-convolution, where the sum of epigraphs is of interest. However, the sum of two epigraphs is not necessarily an epigraph.

**Example 3.2.12** On $X = \mathbb{R}$, define $f_1 : X \to \overline{\mathbb{R}}$ by

$$f_1(x) := \begin{cases} 0 & \text{if } x \leq 0, \\ x & \text{if } x > 0, \end{cases}$$

and $f_2 : X \to \overline{\mathbb{R}}$ by $f_2(x) := e^{-x}$ for all $x \in X$. Then $\mathrm{epi} f_1 + \mathrm{epi} f_2 = \{(x, t) \in X \times \mathbb{R} \mid t > 0\}$. This set is not $k$-directionally closed for $k := (0, -1)$ and thus not the epigraph of some extended real-valued function on X. Note that $f_1$ and $f_2$ are finite-valued, continuous and convex.

The inf-convolution was studied by Moreau [362], who proved that it is a commutative and associative operation.

**Definition 3.2.13** Assume that $X$ is a linear space and $f_i : X \to \overline{\mathbb{R}}$ for each $i \in \{1, \ldots, m\}$, where $m \in \mathbb{N}$ with $m \geq 2$.

The **inf-convolution** of the functions $f_i$, $i \in \{1, \ldots, m\}$, is the function

$$f_1 \# f_2 \# \ldots \# f_m : X \to \overline{\mathbb{R}}$$

which is, for each $x \in X$, given by

$$(f_1 \# f_2 \# \ldots \# f_m)(x) := \inf \left\{ \sum_{i=1}^{m} f_i(x_i) \mid x_i \in X \text{ for each } i \in \{1, \ldots, m\}, x = \sum_{i=1}^{m} x_i \right\}.$$

In this definition, the rule of inf-addition is applied. An alternative way to define the inf-convolution without assuming inf-addition will be proved in Sect. 4.14.

The inf-convolution of a function $f$ with any other function preserves several properties of $f$ (see Lemma 3.7.11, Proposition 4.3.1(g)).

The following lemmata will be used in Sect. 4.9 to construct functions which are majorized by a given function and satisfy certain conditions.

**Lemma 3.2.14** *Under the assumptions of Definition 3.2.13, we have*

$$f_1 \# f_2 \# \ldots \# f_m \leq f_1 + \sum_{i=2}^{m} f_i(0), \tag{3.2.5}$$

$$\sum_{i=2}^{m} f_i(0) \leq 0 \implies f_1 \# f_2 \# \ldots \# f_m \leq f_1, \tag{3.2.6}$$

$$f_1 \leq f_2 \implies f_1 \# f_3 \leq f_2 \# f_3. \tag{3.2.7}$$

**Lemma 3.2.15** *Assume that $X$ is a linear space, $f : X \to \overline{\mathbb{R}}$, and $B \subseteq Y$. Then*

$$(f \# \iota_{-B})(x) = \inf_{b \in B} f(x+b) \qquad \textit{for all } x \in X.$$

***Proof*** For each $x \in X$,

$$\begin{aligned}(f \# \iota_{-B})(x) &= \inf\{f(x - x^1) + \iota_{-B}(x^1) \mid x^1 \in X\} \\ &= \inf\{f(x - x^1) + \iota_{-B}(x^1) \mid x^1 \in -B\} \\ &= \inf_{b \in B} f(x+b).\end{aligned}$$

□

Further properties of the inf-convolution will be investigated in Sect. 4.14.

## 3.3 Continuity and Semicontinuity

The continuity of functions $\varphi : D \to \overline{\mathbb{R}}$ on a subset $D$ of a topological space $X$ is defined by Definition 2.1.16. Its characterization by properties of the epigraph will be presented in Theorem 4.14.10, where the proof is based on results for translation invariant functions.

We start this section with an example for a continuous extended real-valued function which is not finite-valued.

**Example 3.3.1** The function $\varphi : \mathbb{R} \to \overline{\mathbb{R}}$ given by

$$\varphi(x) := \begin{cases} -\infty & \text{if } x \leq -\frac{\pi}{2}, \\ \tan(x) & \text{if } -\frac{\pi}{2} < x < \frac{\pi}{2}, \\ +\infty & \text{if } x \geq \frac{\pi}{2}, \end{cases}$$

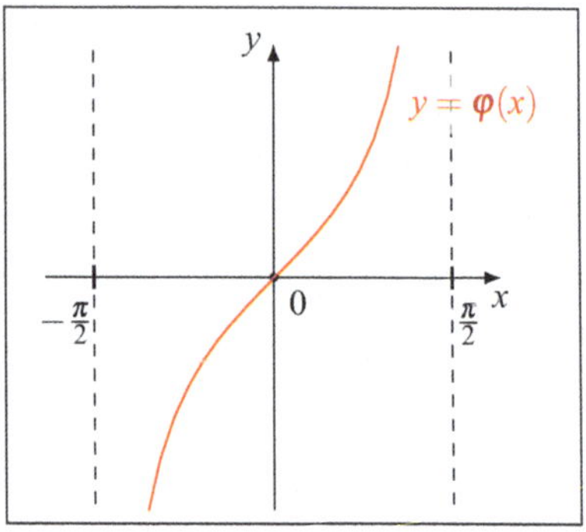


**Fig. 3.2** Function $\varphi$ in Example 3.3.1

is continuous (see Fig. 3.2).

The previous example also illustrates that there is a one-to-one-correspondence between the closed interval $[-\frac{\pi}{2}, \frac{\pi}{2}]$ and $\overline{\mathbb{R}}$ and also a one-to-one-correspondence between the open sets induced on $[-\frac{\pi}{2}, \frac{\pi}{2}]$ by the Euclidean topology and the open sets in $\overline{\mathbb{R}}$. Indeed, it is well known that $\overline{\mathbb{R}}$ can be considered as a two-point compactification of $\mathbb{R}$ since—as previously mentioned—$\overline{\mathbb{R}}$ is a compact topological space.

The definition of continuity implies:

**Lemma 3.3.2** *Let $X$ be a topological space and $\varphi : X \to \overline{\mathbb{R}}$.*

(a) *If $\varphi$ is continuous at $x^0 \in X$, then the functions $\lambda\varphi$ and $\varphi + \lambda$ with $\lambda \in \mathbb{R}$ are continuous at $x^0$.*
(b) *If $X$ is a topological vector space, $x^0, \overline{x} \in X$ and $\varphi$ is continuous at $x^0 - \overline{x}$, then $g : X \to \overline{\mathbb{R}}$ defined by $g(x) := \varphi(x - \overline{x})$ is continuous at $x^0$.*
(c) *If $X$ is a topological vector space, $x^0 \in X$, $\lambda \in \mathbb{R}$ and $\varphi$ is continuous at $\lambda x^0$, then $g : X \to \overline{\mathbb{R}}$ defined by $g(x) := \varphi(\lambda x)$ is continuous at $x^0$.*

The product of continuous extended real-valued functions is not necessarily continuous.

**Example 3.3.3** Define $f, g : \mathbb{R} \to \overline{\mathbb{R}}$ by

$$f(x) := \begin{cases} \frac{1}{|x|} & \text{if } x \neq 0, \\ +\infty & \text{if } x = 0, \end{cases}$$

$$g(x) := |x|.$$

Then

$$h(x) := g(x) \cdot f(x) = \begin{cases} 1 & \text{if } x \neq 0, \\ 0 \cdot (+\infty) = 0 & \text{if } x = 0. \end{cases}$$

$f$ and $g$ are continuous proper functions, but $h$ is not continuous at $x = 0$.

Function $f$ in the previous example illustrates that a continuous real-valued function which is not defined on the entire topological space can possibly be extended to a continuous extended real-valued function.

We now define the Lipschitz continuity of extended real-valued functions, which is of importance for fixed point theorems and differentiability.

**Definition 3.3.4** Let $X$ be a normed space, $\varphi : X \to \overline{\mathbb{R}}$ and $X_0 \subseteq X$ with $X_0 \neq \emptyset$.

$\varphi$ is called **Lipschitz continuous** or **Lipschitzian** on $X_0$ if $\varphi$ is finite-valued on $X_0$ and there exists $L \in \mathbb{R}_+$ such that $|\varphi(x^1) - \varphi(x^2)| \leq L\|x^1 - x^2\|$ for all $x^1, x^2 \in X_0$. In this case, $L$ is called a **Lipschitz constant** of $\varphi$ on $X_0$. $\varphi$ is **Lipschitz continuous** or **Lipschitzian** if $\varphi$ is Lipschitzian on $X$.

$\varphi$ is said to be **locally Lipschitz continuous** or **locally Lipschitzian** on $X_0$ if, for each $x \in X_0$, there exists a neighborhood $U$ of $x$ such that $\varphi$ is Lipschitzian on $U \cap X_0$. $\varphi$ is **locally Lipschitz continuous** or **locally Lipschitzian** if $\varphi$ is a finite-valued functional that is locally Lipschitzian on $X$.

If a function is locally Lipschitzian on some neighborhood of $x$, then it is continuous at $x$. Lipschitz continuity implies local Lipschitz continuity.

The Lipschitz continuity of extended real-valued functions will also be investigated in Sect. 4.14, where its characterization by properties of the epigraph is presented in Theorem 4.14.13.

We will now introduce the semicontinuity of functionals (see Fig. 3.3).

**Definition 3.3.5** Consider a topological space $(X, \tau)$ and a function $\varphi\colon X \to \overline{\mathbb{R}}$.

$\varphi$ is called **lower semicontinuous (l.s.c.)** at $x^0 \in X$ if $\varphi(x^0) = -\infty$ or, for each $h \in \mathbb{R}$ with $h < \varphi(x^0)$, there exists $U \in \mathcal{N}(x^0)$ such that $\varphi(x) > h$ for all $x \in U$.

$\varphi$ is called **upper semicontinuous (u.s.c.)** at $x^0 \in X$ if $\varphi(x^0) = +\infty$ or, for each $h \in \mathbb{R}$ with $h > \varphi(x^0)$, there exists $U \in \mathcal{N}(x^0)$ such that $\varphi(x) < h$ for all $x \in U$.

$\varphi$ is called a **lower semicontinuous function** or an **upper semicontinuous function** if $f$ is lower semicontinuous or upper semicontinuous, respectively, at each $x^0 \in X$. $\varphi$ is said to be **lower semicontinuous** or **upper semicontinuous on** a nonempty set $D \subset X$ if it is lower semicontinuous or upper semicontinuous, respectively, at each $x^0 \in D$ w.r.t. the topology induced by $\tau$ on $D$.

Semicontinuity is related to the convergence of the function.

**Proposition 3.3.6** *Let $X$ be a topological space, $\varphi : X \to \overline{\mathbb{R}}$ and $x \in X$.*

*Then $\varphi$ is lower semicontinuous at $x$ if and only if*

$$\varphi(x) \leq \liminf_{i \in I} \varphi(x_i)$$

*for each net $\{x_i\}_{i \in I}$ in $X$ with $x_i \to x$.*

For the proof, see [3, Lemma 2.42]. If $X$ is first countable, then nets in Proposition 3.3.6 can be replaced by sequences.

Note that each constant function is lower semicontinuous, upper semicontinuous and continuous.

Immediately from the definitions, we get the following statements.

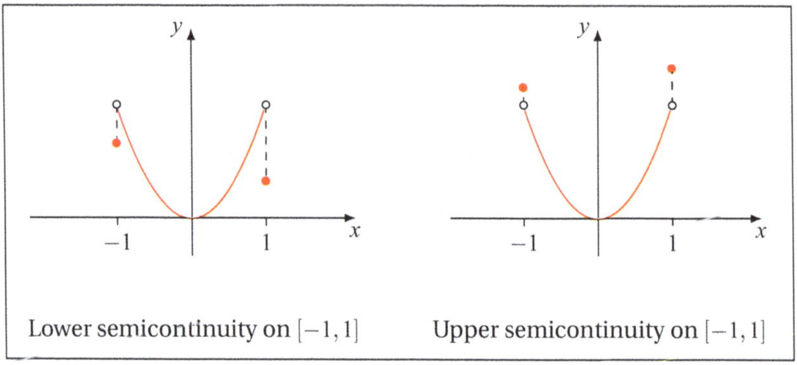


**Fig. 3.3** Examples of semicontinuous functions

**Lemma 3.3.7** *Let $X$ be a topological space, $\varphi : X \to \overline{\mathbb{R}}$, $x \in X$.*

(a) *$\varphi$ is upper semicontinuous at $x$ if and only if $-\varphi$ is lower semicontinuous at $x$.*
(b) *$\varphi$ is continuous at $x$ if and only if $\varphi$ is lower semicontinuous at $x$ and upper semicontinuous at $x$.*

**Lemma 3.3.8** *Let $X$ be a topological space and $\varphi : X \to \overline{\mathbb{R}}$.*

(a) *If $\varphi$ is lower semicontinuous at $x^0 \in X$, then the functions $\lambda\varphi$ with $\lambda \in \mathbb{R}_+$ and $\varphi + c$ with $c \in \mathbb{R}$ are lower semicontinuous at $x^0$.*
(b) *If $X$ is a topological vector space, $x^0, \overline{x} \in X$ and $\varphi$ is lower semicontinuous at $x^0 - \overline{x}$, then $g : X \to \overline{\mathbb{R}}$ defined by $g(x) := \varphi(x - \overline{x})$ is lower semicontinuous at $x^0$.*
(c) *If $X$ is a topological vector space, $x^0 \in X$, $\lambda \in \mathbb{R}$ and $\varphi$ is lower semicontinuous at $\lambda x^0$, then $g : X \to \overline{\mathbb{R}}$ defined by $g(x) := \varphi(\lambda x)$ is lower semicontinuous at $x^0$.*

*Analogous statements hold for upper semicontinuity.*

**Corollary 3.3.9** *Let $(X, \tau)$ be a topological space, $\varphi : X \to \overline{\mathbb{R}}$, $D \subseteq X$ and $D \neq \emptyset$. Then the following properties of $\varphi$ are equivalent to each other:*

(a) *$\varphi$ is lower semicontinuous on $D$.*
(b) *Each set $\{x \in D \mid \varphi(x) > t\}$ with $t \in \mathbb{R}$ is open w.r.t. the topology induced by $\tau$ on $D$.*
(c) *Each set $\{x \in D \mid \varphi(x) \leq t\}$ with $t \in \mathbb{R}$ is closed w.r.t. the topology induced by $\tau$ on $D$.*

We will use the following notation.

**Definition 3.3.10** Let $\mathscr{R}$ denote some binary relation on $\overline{\mathbb{R}}$, $X$ be a nonempty set, $\varphi : X \to \overline{\mathbb{R}}$. For $t \in \overline{\mathbb{R}}$, we define

$$\text{lev}_{\varphi,\mathscr{R}}(t) := \{x \in X \mid \varphi(x)\mathscr{R}t\}.$$

**Lemma 3.3.11** *Let $X$ be a topological space, $\varphi : X \to \overline{\mathbb{R}}$. Then* $\operatorname{epi}\varphi$ *is closed in* $X \times \mathbb{R}$ *if and only if the sublevel sets* $\operatorname{lev}_{\varphi,\le}(t) = \{x \in X \mid \varphi(x) \le t\}$ *are closed for all* $t \in \mathbb{R}$.

***Proof***

(i) $\operatorname{epi}\varphi = \{(x,t) \in X \times \mathbb{R} \mid \varphi(x) \le t\}$ is closed if and only if $(X \times \mathbb{R}) \setminus \operatorname{epi}\varphi = \{(x,t) \in X \times \mathbb{R} \mid \varphi(x) > t\} \cup ((X \setminus \operatorname{dom}\varphi) \times \mathbb{R})$ is open. Then for each $t \in \mathbb{R}$, the set $\{x \in X \mid \varphi(x) > t\} \cup (X \setminus \operatorname{dom}\varphi)$ is open and thus $\{x \in X \mid \varphi(x) \le t\}$ is closed.

(ii) Assume that $\operatorname{lev}_{\varphi,\le}(t)$ is closed for each $t \in \mathbb{R}$. Suppose, contrary to our claim, that $\operatorname{epi}\varphi$ is not closed. Then $(X \times \mathbb{R}) \setminus \operatorname{epi}\varphi$ is not open. Thus, there exists $(x,\lambda) \in (X \times \mathbb{R}) \setminus \operatorname{epi}\varphi$ such that for each $V \in \mathcal{N}((x,\lambda))$ there is $(x^V, t_V) \in V$ with $(x^V, t_V) \in \operatorname{epi}\varphi$, i.e., $\varphi(x^V) \le t_V$. Take any $\varepsilon > 0$, $U_\varepsilon(\lambda) := \{r \in \mathbb{R} \mid \lambda - \varepsilon < r < \lambda + \varepsilon\}$ and any $U \in \mathcal{N}(x)$. In $V_\varepsilon := U \times U_\varepsilon(\lambda) \in \mathcal{N}((x,\lambda))$, there exists $(x^\varepsilon, t_\varepsilon) \in V_\varepsilon$ with $\varphi(x^\varepsilon) \le t_\varepsilon < \lambda + \varepsilon$. For each $U \in \mathcal{N}(x)$, there is $x^\varepsilon \in U$ with $x^\varepsilon \in \operatorname{lev}_{\varphi,\le}(\lambda + \varepsilon)$. Therefore, $x \in \operatorname{cl}(\operatorname{lev}_{\varphi,\le}(\lambda+\varepsilon)) = \operatorname{lev}_{\varphi,\le}(\lambda+\varepsilon)$. We get $\varphi(x) \le \lambda + \varepsilon$ for all $\varepsilon > 0$. Hence, $\varphi(x) \le \lambda$, which contradicts $(x,\lambda) \notin \operatorname{epi}\varphi$. Thus, $\operatorname{epi}\varphi$ is closed. □

Corollary 3.3.9 and Lemma 3.3.11 imply:

**Proposition 3.3.12** *Let $X$ be a topological space, $\varphi : X \to \overline{\mathbb{R}}$.*
*Then the following statements are equivalent to each other:*

(a) *$\varphi$ is lower semicontinuous.*
(b) *The sets* $\operatorname{lev}_{\varphi,>}(t)$ *are open for all* $t \in \mathbb{R}$.
(c) *The sublevel sets* $\operatorname{lev}_{\varphi,\le}(t)$ *are closed for all* $t \in \mathbb{R}$.
(d) $\operatorname{epi}\varphi$ *is closed in* $X \times \mathbb{R}$.

**Remark 3.3.13** The equivalence between (a), (c) and (d) in Proposition 3.3.12 was shown in [449]. Rockafellar [389, Theorem 7.1] proved the equivalence between the conditions in Proposition 3.3.12 for $\varphi : \mathbb{R}^n \to \overline{\mathbb{R}}$. Moreau [362] defined lower semicontinuity by property (c) and stated the equivalence to property (d).

For each function $\varphi\colon X \to \overline{\mathbb{R}}$ on a topological space $X$, we will denote by $\operatorname{cl}\varphi$ the function $\operatorname{cl}\varphi\colon X \to \overline{\mathbb{R}}$ given by $\operatorname{epi}(\operatorname{cl}\varphi) = \operatorname{cl}(\operatorname{epi}\varphi)$, i.e., the greatest lower semicontinuous function majorized by $\varphi$.

For upper semicontinuity, statements analogous to Proposition 3.3.12 hold, with inverse relations and the epigraph being replaced by the hypograph. This follows from Lemma 3.3.7 and Corollary 3.2.8.

Using Proposition 3.3.12, we can prove a well-known fact about the pointwise supremum of lower semicontinuous functions.

**Proposition 3.3.14** *Let $X$ be a topological space and $\Gamma$ a collection of lower semicontinuous functions $f\colon X \to \overline{\mathbb{R}}$.*
*Then the function $\varphi\colon X \to \overline{\mathbb{R}}$ defined by*

$$\varphi(x) := \sup\{f(x) \mid f \in \Gamma\} \qquad \textit{for all } x \in X$$

*is lower semicontinuous.*

***Proof*** Take any $t \in \mathbb{R}$ and $x^0 \in \operatorname{lev}_{\varphi,>}(t)$. Then there exists $f^0 \in \Gamma$ with $x^0 \in \operatorname{lev}_{f^0,>}(t)$. Since $f^0$ is lower semicontinuous, there is a neighbourhood $U$ of $x^0$ such that $x \in \operatorname{lev}_{f^0,>}(t)$ for each $x \in U$. Hence, $x \in \operatorname{lev}_{\varphi,>}(t)$ for each $x \in U$. Consequently, $\varphi$ is lower semicontinuous. □

**Corollary 3.3.15** *Let $X$ be a topological space and $\Gamma$ a collection of upper semicontinuous functions $f : X \to \overline{\mathbb{R}}$.*

*Then the function $\varphi: X \to \overline{\mathbb{R}}$ defined by*

$$\varphi(x) := \inf\{f(x) \mid f \in \Gamma\} \qquad \textit{for all } x \in X$$

*is upper semicontinuous.*

***Proof*** The assertion follows from $\inf\{f(x) \mid f \in \Gamma\} = -\sup\{(-f)(x) \mid f \in \Gamma\}$ and Proposition 3.3.14. □

We get the following from [3, Theorem 2.43].

**Proposition 3.3.16** *A lower semicontinuous real-valued function on a nonempty compact subset of a topological space attains a minimum, and the nonempty set of minimizers is compact. An upper semicontinuous real-valued function on a nonempty compact subset of a topological space attains a maximum, and the nonempty set of maximizers is compact.*

The statement related to lower semicontinuous functions turns out to be a special case of the following proposition.

**Proposition 3.3.17** *Assume that the following conditions hold:*

(a) *$X$ is a topological space,*
(b) *$\varphi : X \to \overline{\mathbb{R}}$ is lower semicontinuous,*
(c) *$M \subseteq X$ is a nonempty compact set, and $\varphi$ is finite-valued on $M$.*

*Then the set of minimizers of $\varphi$ on $M$ is nonempty and compact.*

***Proof*** Consider the family $A$ of sets $M_t := \{x \in M \mid \varphi(x) \le t\}$ with $t \in \varphi(M)$. Each of these sets is closed in the topology induced on $M$, and the intersection of a finite number of sets from $A$ has a nonempty intersection. The compactness of $M$ implies that the intersection $B$ of all sets of $A$ is nonempty [3, Theorem 2.31]. Since $\varphi$ is finite-valued on $M$, $B$ is the set of minimizers of $\varphi$ on $M$. In the topology induced on $M$, $B$ is a closed subset of the compact set $M$ and thus compact. □

The relationship between the (semi)continuity of a function on its effective domain and the (semi)continuity on the entire space is investigated in the next proposition.

**Proposition 3.3.18** *Let $X$ be a topological space, $\varphi : X \to \overline{\mathbb{R}}$.*

(a) *If $\varphi$ is upper semicontinuous at $x \in \operatorname{dom} \varphi$, then $x \in \operatorname{int} \operatorname{dom} \varphi$.*
(b) *If $\varphi$ is lower semicontinuous on $\operatorname{dom} \varphi$, then it is lower semicontinuous at each $x \in \operatorname{dom} \varphi$ (w.r.t. the topology on $X$).*
(c) *$\varphi$ is upper semicontinuous if and only if $\varphi$ is upper semicontinuous on $\operatorname{dom} \varphi$ and $\operatorname{dom} \varphi$ is open.*
(d) *If $\varphi$ is lower semicontinuous on $\operatorname{dom} \varphi$ and $\operatorname{dom} \varphi$ is closed, then $\varphi$ is lower semicontinuous.*
(e) *$\varphi$ is continuous at each $x \in \operatorname{int}(X \setminus \operatorname{dom} \varphi)$ and upper semicontinuous at each $x \in X \setminus \operatorname{dom} \varphi$.*

***Proof*** Let $\tau$ denote the topology on $X$.

(a) and (b) follow from Definition 3.3.5.
(c) Assume first that $\varphi$ is upper semicontinuous. Then it is upper semicontinuous on $\operatorname{dom} \varphi$, and each set $\{x \in X \mid \varphi(x) < t\}$ with $t \in \mathbb{R}$ is open. Hence, $\operatorname{dom} \varphi = \bigcup_{t \in \mathbb{R}} \{x \in X \mid \varphi(x) < t\}$ is open.
Assume now that $\varphi$ is upper semicontinuous on $\operatorname{dom} \varphi$ and $\operatorname{dom} \varphi$ is open. Take any $t \in \mathbb{R}$. By Corollary 3.3.9 and Lemma 3.3.7, the set $D := \{x \in \operatorname{dom} \varphi \mid \varphi(x) < t\}$ is open in the topology induced by $\tau$ on $\operatorname{dom} \varphi$. Since $\operatorname{dom} \varphi$ is open, $D$ is open in the topology $\tau$. We conclude from $D = \{x \in X \mid \varphi(x) < t\}$ that $\varphi$ is upper semicontinuous.
(d) Take any $t \in \mathbb{R}$. By Corollary 3.3.9, the set $B := \{x \in \operatorname{dom} \varphi \mid \varphi(x) > t\}$ is open in the topology induced by $\tau$ on $\operatorname{dom} \varphi$. Moreover, $B$ is open in the topology $\tau$ since the function value of $\varphi$ is $+\infty$ outside $\operatorname{dom} \varphi$. Hence, $\{x \in X \mid \varphi(x) > t\} = B \cup (X \setminus \operatorname{dom} \varphi)$ is open. Consequently, $\varphi$ is lower semicontinuous.
(e) follows immediately from Definition 3.3.5 and Lemma 3.3.7. □

An extended real-valued function can only be continuous if its effective domain is open. This is caused by the function value $+\infty$ outside the effective domain. Proposition 3.3.18 implies:

**Proposition 3.3.19** *Let $X$ be a topological space, $\varphi : X \to \overline{\mathbb{R}}$.*
*$\varphi$ is continuous if and only if the following conditions are satisfied:*

(a) *$\varphi$ is lower semicontinuous,*
(b) *$\varphi$ is continuous on its effective domain,*
(c) *$\operatorname{dom} \varphi$ is open.*

## 3.4 Convexity and Quasiconvexity

In this section, we study the convexity of extended real-valued functions together with some related properties.

**Definition 3.4.1** Let $X$ be a linear space, $\varphi : X \to \overline{\mathbb{R}}$ and $X_0 \subseteq X$.

$\varphi$ is said to be **convex** on $X_0$ if $\operatorname{epi} \varphi \cap (X_0 \times \mathbb{R})$ is a convex set.

$\varphi$ is **concave** on $X_0$ if $\operatorname{hypo} \varphi \cap (X_0 \times \mathbb{R})$ is a convex set.

$\varphi$ is said to be a convex or concave function if it is convex or concave, respectively, on $X$.

Consequently, $\varphi$ is convex if and only if $\operatorname{epi} \varphi$ is convex, and $\varphi$ is concave if and only if $\operatorname{hypo} \varphi$ is convex. Recall that the empty set is convex.

**Remark 3.4.2** The characterization of a convex real-valued function on some finite-dimensional vector space by the epigraph of the function was given by Fenchel [161, p. 57].

For each function $\varphi : X \to \overline{\mathbb{R}}$ on a linear space $X$, we will denote by $\operatorname{conv} \varphi$ the function $\operatorname{conv} \varphi : X \to \overline{\mathbb{R}}$ given by $\operatorname{epi}(\operatorname{conv} \varphi) = \operatorname{conv}(\operatorname{epi} \varphi)$, i.e., the greatest convex function majorized by $\varphi$. On a topological vector space $X$, $\operatorname{cl} \operatorname{conv} \varphi$ is the greatest lower semicontinuous convex function majorized by $\varphi$.

We get from Lemma 3.2.6:

**Lemma 3.4.3** *Let $X$ be a linear space, $\varphi : X \to \overline{\mathbb{R}}$ and $X_0 \subseteq X$.*

*Then $\varphi$ is concave on $X_0$ if and only if $-\varphi$ is convex on $X_0$.*

Definition 3.4.1 is compatible with the usual definition of convexity for real-valued functions.

**Theorem 3.4.4** *Assume that $X$ is a linear space, $\varphi : X \to \overline{\mathbb{R}}$ and that $X_0 \subseteq X$ is convex.*

(a) *$\varphi$ is convex on $X_0$ if and only if*

$$\varphi(\lambda x^1 + (1-\lambda)x^2) \leq \lambda\varphi(x^1) + (1-\lambda)\varphi(x^2) \tag{3.4.1}$$

*holds for all $x^1, x^2 \in X_0$ and all $\lambda \in (0, 1)$.*

(b) *Assume that $\varphi$ attains only one or none of the values $+\infty$ and $-\infty$ on $X_0$. Then $\varphi$ is concave on $X_0$ if and only if*

$$\varphi(\lambda x^1 + (1-\lambda)x^2) \geq \lambda\varphi(x^1) + (1-\lambda)\varphi(x^2)$$

*holds for all $x^1, x^2 \in X_0$ and all $\lambda \in (0, 1)$.*

***Proof***

(i) Assume first that $\varphi$ is convex on $X_0$. Take any $x^1, x^2 \in X_0$ and $\lambda \in (0, 1)$. Obviously, (3.4.1) holds if $\varphi(x^1) = +\infty$ or $\varphi(x^2) = +\infty$. Here the inf-addition is applied. Assume now $\varphi(x^1) \neq +\infty$ and $\varphi(x^2) \neq +\infty$. Take any $t_1, t_2 \in \mathbb{R}$ with $\varphi(x^1) \leq t_1$ and $\varphi(x^2) \leq t_2$. Since $\operatorname{epi} \varphi \cap (X_0 \times \mathbb{R})$ is a convex set and $(x^1, t_1), (x^2, t_2) \in \operatorname{epi} \varphi \cap (X_0 \times \mathbb{R})$, we get $\lambda \cdot (x^1, t_1) + (1-\lambda) \cdot (x^2, t_2) \in \operatorname{epi} \varphi$. Hence, $\varphi(\lambda x^1 + (1-\lambda)x^2) \leq \lambda t_1 + (1-\lambda)t_2$. This implies $\varphi(\lambda x^1 + (1-\lambda)x^2) = -\infty$ if $\varphi(x^1) = -\infty$ or $\varphi(x^2) = -\infty$. Otherwise, it implies (3.4.1) when choosing $t_1 = \varphi(x^1)$ and $t_2 = \varphi(x^2)$.

(ii) Assume that (3.4.1) holds for all $x^1, x^2 \in X_0$, $\lambda \in (0, 1)$. Consider $(x^1, t_1)$, $(x^2, t_2) \in \operatorname{epi} \varphi \cap (X_0 \times \mathbb{R})$, i.e., $x^1, x^2 \in X_0$ and $t_1, t_2 \in \mathbb{R}$ with $\varphi(x^1) \leq t_1$, $\varphi(x^2) \leq t_2$. Then $\varphi(\lambda x^1 + (1-\lambda)x^2) \leq \lambda\varphi(x^1) + (1-\lambda)\varphi(x^2) \leq \lambda t_1 + (1-\lambda)t_2$ for all $\lambda \in (0, 1)$. Hence, $\lambda \cdot (x^1, t_1) + (1-\lambda) \cdot (x^2, t_2) \in \operatorname{epi} \varphi \cap (X_0 \times \mathbb{R})$ for all $\lambda \in (0, 1)$, i.e., $\operatorname{epi} \varphi \cap (X_0 \times \mathbb{R})$ is convex.

(b) follows from (a) by Lemma 3.4.3 since the assumptions imply $\{\varphi(x^1), \varphi(x^2)\} \neq \{-\infty, +\infty\}$ for all $x^1, x^2 \in X_0$. □

Inequality (3.4.1) holds for convex functions since we applied the inf-addition. Without this rule, the same assumption as in (b) would have to be made in (a).

We get the following from Definition 3.4.1.

**Lemma 3.4.5** *Let $X$ be a linear space and $\Gamma$ a collection of convex functions $f : X \to \overline{\mathbb{R}}$. Then the function $\varphi : X \to \overline{\mathbb{R}}$ defined by*

$$\varphi(x) := \sup\{f(x) \mid f \in \Gamma\} \qquad \text{for all } x \in X$$

*is convex.*

A continuous convex functional can attain the value $+\infty$.

**Example 3.4.6** Define $f : \mathbb{R} \to \overline{\mathbb{R}}$ by

$$f(x) := \begin{cases} +\infty & \text{if } x \leq 0, \\ \frac{1}{x} & \text{if } x > 0. \end{cases}$$

$f$ is a continuous convex proper function.

**Proposition 3.4.7** *Let $X$ be a linear space and $\varphi : X \to \overline{\mathbb{R}}$ be a convex function. Then:*

(a) $\operatorname{dom} \varphi$ *is convex.*
(b) *If $\varphi(x^0) = -\infty$ for some $x^0 \in X$, then $\varphi(\lambda x^0 + (1-\lambda)x) = -\infty$ for all $x \in \operatorname{dom} \varphi$ and all $\lambda \in (0, 1)$.*
(c) *If there exists $x^0 \in X$ such that $\varphi(x^0) = -\infty$, then $\varphi(x) = -\infty$ holds for all $x \in \operatorname{icr}(\operatorname{dom} \varphi)$.*
(d) *If $x \in X$ with $\varphi(x) = -\infty$ and $\varphi(0) \neq -\infty$, then $\varphi(-\lambda x) = +\infty$ for all $\lambda \in \mathbb{R}_>$.*
(e) *Assume that $X$ is a topological vector space, that $\varphi$ is lower semicontinuous on $\operatorname{dom} \varphi$ and attains the value $-\infty$. Then $\varphi$ does not attain any real value.*

***Proof***

(a) Consider $x^1, x^2 \in \operatorname{dom} \varphi$, $\lambda \in (0, 1)$, $x := \lambda x^1 + (1-\lambda)x^2$. There exist $t_1, t_2 \in \mathbb{R}$ such that $(x^1, t_1)$, $(x^2, t_2) \in \operatorname{epi} \varphi$. Since $\operatorname{epi} \varphi$ is convex, we conclude for $x$ and $t := \lambda t_1 + (1-\lambda)t_2 \in \mathbb{R}$ that $(x, t) \in \operatorname{epi} \varphi$. Hence, $\varphi(x) \leq t$, which yields $x \in \operatorname{dom} \varphi$.

(b) Assume $x^0 \in X$ with $\varphi(x^0) = -\infty$ and $x^1 \in \operatorname{dom}\varphi$. Since $x^0 \in \operatorname{dom}\varphi$, (a) implies:

$$\forall\lambda \in (0,1):\ \ x^\lambda := \lambda x^0 + (1-\lambda)x^1 \in \operatorname{dom}\varphi.$$

Moreover, there exists $t_1 \in \mathbb{R}$ with $(x^1, t_1) \in \operatorname{epi}\varphi$.
Take any $\lambda \in (0,1)$, $t \in \mathbb{R}$. For $t_2 := \frac{1}{\lambda}t - \frac{1-\lambda}{\lambda}t_1$, we have $t_2 \in \mathbb{R}$ and $t = \lambda t_2 + (1-\lambda)t_1$. Then $(x^\lambda, t) \in \operatorname{epi}\varphi$ because $(x^0, t_2), (x^1, t_1) \in \operatorname{epi}\varphi$ and $\operatorname{epi}\varphi$ is convex. Since $(x^\lambda, t) \in \operatorname{epi}\varphi$ for each $t \in \mathbb{R}$, we get $\{x^\lambda\} \times \mathbb{R} \subseteq \operatorname{epi}\varphi$ and $\varphi(x^\lambda) = -\infty$.

(c) Assume $x^0 \in X$ with $\varphi(x^0) = -\infty$ and $x^1 \in \operatorname{icr}(\operatorname{dom}\varphi)$. Then there exists $\varepsilon \in \mathbb{R}_>$ with $x^2 := x^1 + \varepsilon(x^1 - x^0) \in \operatorname{dom}\varphi$. We get $x^1 = \lambda x^2 + (1-\lambda)x^0$ for $\lambda := \frac{1}{1+\varepsilon} \in (0,1)$. (b) yields $\varphi(x^1) = -\infty$.

(d) Assume, under the assumptions of (d), that there exists $\lambda \in \mathbb{R}_>$ with $\varphi(-\lambda x) \neq +\infty$. Then $-\lambda x \in \operatorname{dom}\varphi$. For $t := \frac{\lambda}{1+\lambda} \in (0,1)$, we get $0 = tx + (1-t)(-\lambda x)$ and $\varphi(0) = -\infty$ because of (b). This contradicts the assumption of (d).

(e) Assume $x^0 \in X$ with $\varphi(x^0) = -\infty$ and $x \in \operatorname{dom}\varphi$. For each $\lambda \in (0,1)$, we have $x^\lambda := \lambda x^0 + (1-\lambda)x \in \operatorname{dom}\varphi$ and $\varphi(x^\lambda) = -\infty$ because of (b). For each $U \in \mathcal{N}(x)$, there exists $\lambda \in (0,1)$ with $x^\lambda \in U$. This results in $\varphi(x) = -\infty$ because of the definition of lower semicontinuity. □

**Remark 3.4.8** Part (c) of the proposition can be found in [508], parts (c) and (e) for $X = \mathbb{R}^n$ in [389].

Any discontinuous linear functional on some infinite-dimensional topological vector space is an example of a convex function which is not continuous on the interior of its domain. However, we can derive the following result from [3, Theorem 5.43] and Proposition 3.4.7(a).

**Proposition 3.4.9** *Let $X$ be a topological vector space and $\varphi : X \to \overline{\mathbb{R}}$ a convex proper function with* $\operatorname{int}\operatorname{dom}\varphi \neq \emptyset$. *Then the following statements are equivalent to each other:*

(a) *$\varphi$ is continuous on* $\operatorname{int}\operatorname{dom}\varphi$.
(b) *$\varphi$ is continuous at some point of* $\operatorname{int}\operatorname{dom}\varphi$.
(c) *$\varphi$ is upper semicontinuous on* $\operatorname{int}\operatorname{dom}\varphi$.
(d) *$\varphi$ is bounded above on a neighborhood of some point in* $\operatorname{int}\operatorname{dom}\varphi$.
(e) *$\varphi$ is bounded above on a neighborhood of each point in* $\operatorname{int}\operatorname{dom}\varphi$.

By Proposition 3.3.19, $\varphi$ cannot be continuous if $\operatorname{dom}\varphi$ is not open.

Propositions 4.1.4 and 4.1.5 from [64] result in the following lemma.

**Lemma 3.4.10** *Suppose that $X$ is a Banach space and $\varphi : X \to \overline{\mathbb{R}}$ is lower semicontinuous, convex, and proper. Then $\varphi$ is locally Lipschitzian on* $\operatorname{int}\operatorname{dom}\varphi$.

An example in [64, p. 129] illustrates that a lower semicontinuous convex proper function $\varphi : \mathbb{R}^2 \to \overline{\mathbb{R}}$ with nonempty compact domain is not necessarily continuous on $\operatorname{dom}\varphi$ and thus not necessarily locally Lipschitzian on $\operatorname{dom}\varphi$.

The previous lemma implies:

**Proposition 3.4.11** *Assume that $X$ is a Banach space and $\varphi : X \to \mathbb{R}$ is convex. Then the following properties are equivalent to each other:*

(a) *$\varphi$ is continuous.*
(b) *$\varphi$ is lower semicontinuous.*
(c) *$\varphi$ is locally Lipschitz continuous.*

**Definition 3.4.12** Assume that $X$ is a linear space, $\varphi : X \to \overline{\mathbb{R}}$ and that $X_0 \subseteq X$ is convex.

$\varphi$ is said to be

(a) **strictly convex** on $X_0$ if

$$\varphi(\lambda x^1 + (1-\lambda)x^2) < \lambda\varphi(x^1) + (1-\lambda)\varphi(x^2)$$

for all $x^1, x^2 \in X_0$ with $x^1 \neq x^2$ and all $\lambda \in (0, 1)$,

(b) **quasiconvex** on $X_0$ if

$$\varphi(\lambda x^1 + (1-\lambda)x^2) \leq \sup(\varphi(x^1), \varphi(x^2))$$

for all $x^1, x^2 \in X_0$ and all $\lambda \in (0, 1)$,

(c) **strictly quasiconvex** on $X_0$ if

$$\varphi(\lambda x^1 + (1-\lambda)x^2) < \sup(\varphi(x^1), \varphi(x^2))$$

for all $x^1, x^2 \in X_0$ with $x^1 \neq x^2$ and all $\lambda \in (0, 1)$,

(d) **quasiconcave** or **strictly quasiconcave** on $X_0$ if $-\varphi$ is quasiconvex or strictly quasiconvex, respectively, on $X_0$,

(e) strictly convex, quasiconvex, strictly quasiconvex, quasiconcave or strictly quasiconcave if it has this property on $X$.

The definition and the use of strict convexity depend on the fact that we assumed inf-addition. Otherwise, we would have to assume that $\varphi$ is proper or that $\operatorname{dom} \varphi = X$ holds.

We get from Theorem 3.4.4 and Definition 3.4.12:

**Lemma 3.4.13** *Let $X$ be a linear space, $\varphi : X \to \overline{\mathbb{R}}$ and $X_0 \subseteq X$ be convex.*

*If $\varphi$ is strictly convex on $X_0$, then it is convex on $X_0$ and strictly quasiconvex on $X_0$.*

*If $\varphi$ is convex on $X_0$ or strictly quasiconvex on $X_0$, then it is quasiconvex on $X_0$.*

**Proposition 3.4.14** *Let $X$ be a topological vector space and $\varphi : X \to \overline{\mathbb{R}}$ be finite-valued. If* $\operatorname{epi} \varphi$ *is strictly convex, then $\varphi$ is strictly convex.*

***Proof*** Take any $x^1, x^2 \in X$ with $x^1 \neq x^2$ and $\lambda \in (0, 1)$. Then $(x^1, \varphi(x^1)), (x^2, \varphi(x^2)) \in \operatorname{epi} \varphi$. We get $(\lambda x^1 + (1-\lambda)x^2, \lambda\varphi(x^1) + (1-\lambda)\varphi(x^2)) \in \operatorname{int}(\operatorname{epi} \varphi)$ since $\operatorname{epi} \varphi$ is strictly convex. Hence, there exists $t \in \mathbb{R}$ with $t < \lambda\varphi(x^1) + (1-\lambda)\varphi(x^2)$ and $(\lambda x^1 + (1-\lambda)x^2, t) \in \operatorname{epi} \varphi$. Thus, $\varphi(\lambda x^1 + (1-\lambda)x^2) \leq t < \lambda\varphi(x^1) + (1-\lambda)\varphi(x^2)$. □

**Proposition 3.4.15** *Let $X$ be a linear space, $\varphi : X \to \overline{\mathbb{R}}$ and $X_0 \subseteq X$ be convex.*

*$\varphi$ is quasiconvex on $X_0$ if and only if the sets $X_0 \cap \mathrm{lev}_{\varphi,\leq}(t)$ are convex for all $t \in \mathbb{R}$.*

***Proof***

(i) Assume that $\varphi$ is quasiconvex on $X_0$. Take any $t \in \mathbb{R}$, $x^1, x^2 \in X_0 \cap \mathrm{lev}_{\varphi,\leq}(t)$, $\lambda \in (0,1)$. Since $\varphi(x^1) \leq t$ and $\varphi(x^2) \leq t$, we obtain $\varphi(\lambda x^1 + (1-\lambda)x^2) \leq \sup(\varphi(x^1), \varphi(x^2)) \leq t$. Hence, $\lambda x^1 + (1-\lambda)x^2 \in \mathrm{lev}_{\varphi,\leq}(t)$.
Therefore, $X_0 \cap \mathrm{lev}_{\varphi,\leq}(t)$ is convex.

(ii) Assume now that the sets $X_0 \cap \mathrm{lev}_{\varphi,\leq}(t)$ are convex for all $t \in \mathbb{R}$. Take any $x^1, x^2 \in X_0$ and $\lambda \in (0,1)$. Define $t := \sup(\varphi(x^1), \varphi(x^2))$.
Assume first $t \in \mathbb{R}$. Since $x^1, x^2 \in \mathrm{lev}_{\varphi,\leq}(t)$, the convexity of $X_0 \cap \mathrm{lev}_{\varphi,\leq}(t)$ yields $\lambda x^1 + (1-\lambda)x^2 \in \mathrm{lev}_{\varphi,\leq}(t)$. Therefore, $\varphi(\lambda x^1 + (1-\lambda)x^2) \leq t = \sup(\varphi(x^1), \varphi(x^2))$.
Assume now $t = -\infty$. Then $x^1, x^2 \in X_0 \cap \mathrm{lev}_{\varphi,\leq}(w)$ for all $w \in \mathbb{R}$. Thus, the convexity of $X_0 \cap \mathrm{lev}_{\varphi,\leq}(w)$ yields $\lambda x^1 + (1-\lambda)x^2 \in X_0 \cap \mathrm{lev}_{\varphi,\leq}(w)$ for all $w \in \mathbb{R}$. Consequently, $\varphi(\lambda x^1 + (1-\lambda)x^2) = -\infty = \sup(\varphi(x^1), \varphi(x^2))$.
Assume now $t = +\infty$. Then $\varphi(\lambda x^1 + (1-\lambda)x^2) \leq \sup(\varphi(x^1), \varphi(x^2))$. □

Convex and strictly quasiconvex functions have well-known properties which are essential in optimization procedures. This is indicated by the following propositions.

**Proposition 3.4.16** *Let $X$ be a topological vector space, $\varphi : X \to \overline{\mathbb{R}}$ and $X_0 \subseteq X$ be convex. If $\varphi$ is convex on $X_0$ or strictly quasiconvex on $X_0$, then each local minimum of $\varphi$ on $X_0$ is also a global minimum of $\varphi$ on $X_0$.*

***Proof*** To obtain a contradiction, suppose that $x^0$ is a local minimizer of $\varphi$ on $X_0$ and that $\varphi(x^0)$ is not a global minimum of $\varphi$ on $X_0$. Then there exists $x^1 \in X_0$ with $\varphi(x^1) < \varphi(x^0)$. Thus, there is $\varepsilon \in \mathbb{R}_>$ with $\varphi(x^1) \leq \varphi(x^0) - \varepsilon$. Moreover, there exists a neighborhood $U$ of $x^0$ such that $\varphi(x^0)$ is a minimum of $\varphi$ on $X_0 \cap U$ and some $\lambda \in \mathbb{R}$ with $0 < \lambda < 1$ and $\lambda x^0 + (1-\lambda)x^1 \in U$.

If $\varphi$ is convex on $X_0$, then we get $\varphi(\lambda x^0 + (1-\lambda)x^1) \leq \lambda\varphi(x^0) + (1-\lambda) \cdot (\varphi(x^0) - \varepsilon) < \varphi(x^0)$. If $\varphi$ is strictly quasiconvex on $X_0$, we obtain $\varphi(\lambda x^0 + (1-\lambda)x^1) < \varphi(x^0)$. In both cases, this contradicts the statement that $x^0$ is a minimizer of $\varphi$ on $X_0 \cap U$. □

**Proposition 3.4.17** *Let $X$ be a linear space, $M \subseteq X$ be a convex set and $\varphi : X \to \overline{\mathbb{R}}$ be quasiconvex on $M$. Assume that $\varphi$ attains a minimum on $M$.*

(a) $\mathrm{argmin}_M\, \varphi$ *is a convex set.*
(b) *If $\varphi$ is strictly quasiconvex on $M$, then* $\mathrm{argmin}_M\, \varphi$ *contains one element only.*

***Proof*** Denote the minimum of $\varphi$ on $M$ by $t$. Since this minimum exists, we have $\varphi(x) \neq -\infty$ for all $x \in M$. Take any $x^1, x^2 \in \mathrm{argmin}_M\, \varphi$, $\lambda \in (0,1)$. Then $\varphi(x^1) = \varphi(x^2) = t \in \mathbb{R}$.

(a) Since $\varphi(\lambda x^1 + (1-\lambda)x^2) \leq \max(\varphi(x^1), \varphi(x^2)) = t$, we obtain $\lambda x^1 + (1-\lambda) \cdot x^2 \in \mathrm{argmin}_M\, \varphi$.

(b) Supposing $x^1 \neq x^2$, we get $\varphi(\lambda x^1 + (1-\lambda)x^2) < \max(\varphi(x^1), \varphi(x^2)) = t$, which contradicts the definition of $t$. □

**Lemma 3.4.18** *Let $X$ be a linear space and $\varphi : X \to \overline{\mathbb{R}}$.*

*If $\varphi$ is convex, concave, strictly convex, quasiconvex or strictly quasiconvex, then each of the following functions has the same property:*

(a) *$\lambda\varphi$ with $\lambda \in \mathbb{R}_>$,*
(b) *$\varphi + c$ with $c \in \mathbb{R}$,*
(c) *$g : X \to \overline{\mathbb{R}}$ defined by $g(x) := \varphi(x - \overline{x})$ with $\overline{x} \in X$,*
(d) *$g : X \to \overline{\mathbb{R}}$ defined by $g(x) := \varphi(\lambda x)$ with $\lambda \in \mathbb{R} \setminus \{0\}$.*

## 3.5 Linearity and Related Algebraic Properties

Usually, a linear function maps into a vector space. This is not the case for an extended real-valued function.

**Definition 3.5.1** Let $X$ be a linear space, $X_0 \subseteq X$.

$\varphi : X \to \overline{\mathbb{R}}$ is **affine** on $X_0$ if $\varphi$ is convex on $X_0$ and concave on $X_0$.

$\varphi : X \to \overline{\mathbb{R}}$ is an **affine** function if it is affine on $X$.

$\varphi : X \to \overline{\mathbb{R}}$ is a **linear** function if it is affine and $\varphi(0) = 0$ holds.

**Remark 3.5.2** Fenchel [161, p. 59] proved that the affine real-valued functions on finite-dimensional vector spaces defined in the traditional way by an equation (see Lemma 3.5.3 in combination with Corollary 3.5.19) are the functionals which are both convex and concave.

Obviously, each constant functional is affine. The constant functional with the function value $c = 0$ is linear.

We get the following from Definition 3.5.1.

**Lemma 3.5.3** *Let $X$ be a linear space and $\varphi : X \to \overline{\mathbb{R}}$.*

*$\varphi$ is an affine functional with $\varphi(0) \in \mathbb{R}$ if and only if it is the sum of some linear function $\varphi_l : X \to \overline{\mathbb{R}}$ and some real value.*

Definition 3.5.1 is compatible with the usual definition of linear and affine real-valued functions. This will be verified by Corollary 3.5.19.

Lemma 3.4.18 implies:

**Lemma 3.5.4** *Let $X$ be a linear space.*

*If $\varphi : X \to \overline{\mathbb{R}}$ is affine, then each of the following functions is also affine:*

(a) *$\lambda\varphi$ with $\lambda \in \mathbb{R}$,*
(b) *$\varphi + c$ with $c \in \mathbb{R}$,*
(c) *$g : X \to \overline{\mathbb{R}}$ defined by $g(x) := \varphi(x - \overline{x})$ with $\overline{x} \in X$,*
(d) *$g : X \to \overline{\mathbb{R}}$ defined by $g(x) := \varphi(\lambda x)$ with $\lambda \in \mathbb{R}$.*

*If* $\varphi : X \to \overline{\mathbb{R}}$ *is linear, then the functions in* (a) *and* (d) *are linear as well.*

An affine function $\varphi$ can have a value $\varphi(0) \notin \mathbb{R}$.

**Example 3.5.5** Define $\varphi \colon \mathbb{R} \to \overline{\mathbb{R}}$ by

$$\varphi(x) := \begin{cases} -\infty & \text{if } x \leq 0, \\ +\infty & \text{if } x > 0. \end{cases}$$

$\operatorname{epi}\varphi = (-\infty, 0] \times \mathbb{R}$ is convex and closed. $\operatorname{hypo}\varphi = (0, +\infty) \times \mathbb{R}$ is a convex set. $\varphi$ is a lower semicontinuous affine function.

The example illustrates that a lower semicontinuous affine functional is not necessarily continuous.

**Proposition 3.5.6** *Let* $(X, \tau)$ *be a topological vector space and* $\varphi : X \to \overline{\mathbb{R}}$.
*If* $\varphi$ *is affine and continuous, then it is constant or finite-valued.*

***Proof*** We consider a function $\varphi : X \to \overline{\mathbb{R}}$ which is convex, concave and continuous, but not finite-valued. Applying Proposition 3.4.7(e) to $\varphi$ and $-\varphi$ implies that $\varphi$ does not attain any real value. Suppose, contrary to our claim, that $\varphi$ is not constant. Then there exist $x^1, x^2 \in X$ with $\varphi(x^1) = -\infty$ and $\varphi(x^2) = +\infty$. For $M := \{tx^1 + (1 - t)x^2 \mid t \in [0, 1]\}$, consider $A := M \cap \operatorname{dom}\varphi$. Since $\varphi$ is continuous and $\operatorname{dom}\varphi = \{x \in X \mid \varphi(x) = -\infty\}$, there exists $U \in \mathcal{N}(x^2)$ with $U \cap \operatorname{dom}\varphi = \emptyset$. Thus, $x^2 \notin \operatorname{cl}\operatorname{dom}\varphi$. Because of $x^1 \in A$ and $x^2 \notin \operatorname{cl} A$, Lemma 2.1.26 implies the existence of $t_0 \in (0, 1]$ with $a := t_0 x^1 + (1 - t_0)x^2 \in \operatorname{bd}_M A$, where $\operatorname{bd}_M A$ denotes the boundary of $A$ w.r.t. the topology induced by $\tau$ on $M$. The continuity of $\varphi$ yields $\varphi(a) = -\infty$ by the definition of $A$ and the existence of a neighborhood $V$ of $a$ such that $\varphi(x) = -\infty$ for all $x \in V$. Since $a \in \operatorname{bd}_M A$, there exists $x^0 \in V \cap M$ with $x^0 \notin A$. The definition of $A$ implies $x^0 \in M \setminus \operatorname{dom}\varphi$, which contradicts $x^0 \in V \subseteq \operatorname{dom}\varphi$. Consequently, $\varphi$ is constant. □

**Corollary 3.5.7** *Assume that* $X$ *is a topological vector space and* $\varphi : X \to \overline{\mathbb{R}}$ *is continuous and linear. Then* $\varphi$ *is finite-valued.*

A linear functional may be improper without being constant.

**Example 3.5.8** Define $\varphi \colon \mathbb{R} \to \overline{\mathbb{R}}$ by

$$\varphi(x) := \begin{cases} -\infty & \text{if } x < 0, \\ 0 & \text{if } x = 0, \\ +\infty & \text{if } x > 0. \end{cases}$$

Then

$$\begin{aligned} \operatorname{epi}\varphi &= ((-\infty, 0) \times \mathbb{R}) \cup (\{0\} \times [0, +\infty)) \text{ and} \\ \operatorname{hypo}\varphi &= ((0, +\infty) \times \mathbb{R}) \cup (\{0\} \times (-\infty, 0]) \end{aligned}$$

are convex, but not closed. $\varphi$ is an improper linear functional.

We will now introduce further algebraic properties of extended real-valued functions.

**Definition 3.5.9** Assume that $X$ is a linear space and $\varphi : X \to \overline{\mathbb{R}}$.
$\varphi$ is said to be

(a) **positively homogeneous** if $\operatorname{epi}\varphi$ is a nonempty cone,
(b) **subadditive** if $\operatorname{epi}\varphi + \operatorname{epi}\varphi \subseteq \operatorname{epi}\varphi$ holds,
(c) **superadditive** if $\operatorname{hypo}\varphi + \operatorname{hypo}\varphi \subseteq \operatorname{hypo}\varphi$ holds,
(d) **additive** if it is subadditive and superadditive,
(e) **sublinear** if $\operatorname{epi}\varphi$ is a nonempty convex cone,
(f) **odd** if $\varphi(-x) = -\varphi(x)$ is satisfied for all $x \in X$,
(g) **homogeneous** if $\varphi$ is positively homogeneous and odd.

**Lemma 3.5.10** *Let $X$ be a linear space and $\varphi : X \to \overline{\mathbb{R}}$.*

(a) *$\varphi$ is superadditive if and only if $-\varphi$ is subadditive.*
(b) *$\varphi(0) \in \{0, -\infty\}$ holds for each positively homogeneous and for each sublinear functional.*

***Proof*** (a) results from Corollary 3.2.7(a). (b) follows from the fact that 0 belongs to each nonempty cone. □

For real-valued functions, the properties defined in Definition 3.5.9 coincide with the usual definitions.

**Theorem 3.5.11** *Let $X$ be a linear space and $\varphi : X \to \overline{\mathbb{R}}$.*

(a) *$\varphi$ is positively homogeneous with $\varphi(0) = 0$ if and only if*
$$\varphi(\lambda x) = \lambda\varphi(x)$$
*is satisfied for all $\lambda \in \mathbb{R}_+$ and all $x \in X$.*
(b) *$\varphi$ is homogeneous if and only if*
$$\varphi(\lambda x) = \lambda\varphi(x)$$
*holds for all $\lambda \in \mathbb{R}$ and all $x \in X$.*
(c) *$\varphi$ is subadditive if and only if*
$$\varphi(x^1 + x^2) \le \varphi(x^1) + \varphi(x^2) \tag{3.5.1}$$
*holds for all $x^1, x^2 \in X$.*
(d) *Assume that $\varphi$ is proper or that* $\operatorname{dom}\varphi \in \{\emptyset, X\}$. *Then $\varphi$ is superadditive if and only if*
$$\varphi(x^1 + x^2) \ge \varphi(x^1) + \varphi(x^2)$$
*holds for all $x^1, x^2 \in X$.*

***Proof***

(a) (i) Assume first $\varphi(0) = 0$ and that $\varphi$ is positively homogeneous, i.e., that $\operatorname{epi}\varphi$ is a nonempty cone. Take any $\lambda \in \mathbb{R}_>$.
Take first any $x \in X$ with $t := \varphi(x) \in \mathbb{R}$. Then $\lambda \cdot (x, t) \in \operatorname{epi}\varphi$ since $(x, t) \in \operatorname{epi}\varphi$. Hence, $\varphi(\lambda x) \leq \lambda t$. Suppose, contrary to our claim, $\varphi(\lambda x) < \lambda t$. Then there exists $\lambda_1 \in \mathbb{R}$ with $\varphi(\lambda x) < \lambda_1 < \lambda t$. We get $(\lambda x, \lambda_1) \in \operatorname{epi}\varphi$, and $\frac{1}{\lambda} \cdot (\lambda x, \lambda_1) \in \operatorname{epi}\varphi$, i.e., $(x, \frac{\lambda_1}{\lambda}) \in \operatorname{epi}\varphi$. Therefore, $t = \varphi(x) \leq \frac{\lambda_1}{\lambda}$, which contradicts $\lambda t > \lambda_1$. Thus, $\varphi(\lambda x) = \lambda t = \lambda\varphi(x)$.
Take now any $x \in X$ with $\varphi(x) = -\infty$. Then $(x, t) \in \operatorname{epi}\varphi$ for all $t \in \mathbb{R}$. We obtain $(\lambda x, \lambda t) \in \operatorname{epi}\varphi$ for all $t \in \mathbb{R}$ since $\operatorname{epi}\varphi$ is a cone. Hence, $\varphi(\lambda x) = -\infty = \lambda\varphi(x)$.
Take finally any $x \in X$ with $\varphi(x) = +\infty$. If $\varphi(\lambda x) \neq +\infty$, then $\varphi(x) = \varphi(\frac{1}{\lambda}\lambda x) = \frac{1}{\lambda}\varphi(\lambda x) \neq +\infty$ by the above statements, which contradicts $\varphi(x) = +\infty$.
(ii) Assume now that $\varphi(\lambda x) = \lambda\varphi(x)$ is satisfied for all $\lambda \in \mathbb{R}_+$ and $x \in X$. Take any $\lambda_1 \in \mathbb{R}_+$. If $(x^1, t_1) \in \operatorname{epi}\varphi$, then $\varphi(\lambda_1 x^1) = \lambda_1\varphi(x^1) \leq \lambda_1 t_1$. Hence, $(\lambda_1 x^1, \lambda_1 t_1) \in \operatorname{epi}\varphi$. Thus, $\operatorname{epi}\varphi$ is a cone. It is nonempty since $\varphi(0) = 0$.

(b) (i) Assume first that $\varphi$ is homogeneous, i.e., odd and positively homogeneous. Then $\varphi(-x) = -\varphi(x)$ and $\varphi(\lambda x) = \lambda\varphi(x)$ hold for all $\lambda \in \mathbb{R}_+$ and $x \in X$. Take any $x \in X$ and $\lambda_1 \in \mathbb{R}$ with $\lambda_1 < 0$. Then $\varphi(\lambda_1 x) = \varphi((-\lambda_1) \cdot (-x)) = (-\lambda_1)\varphi(-x) = (-\lambda_1) \cdot (-\varphi(x)) = \lambda_1\varphi(x)$. Consequently, $\varphi(\lambda x) = \lambda\varphi(x)$ holds for all $\lambda \in \mathbb{R}$ and $x \in X$.
(ii) The reverse direction of the equivalence is obvious because of (a).

(c) (i) Suppose first that $\varphi$ is subadditive, i.e., $\operatorname{epi}\varphi + \operatorname{epi}\varphi \subseteq \operatorname{epi}\varphi$.
If $x^1, x^2 \in X$ with $\varphi(x^1) = +\infty$ or $\varphi(x^2) = +\infty$, then (3.5.1) is satisfied.
Take now any $x^1, x^2 \in X$ with $\varphi(x^1) = -\infty$ and $\varphi(x^2) \neq +\infty$. Then $(x^1, t) \in \operatorname{epi}\varphi$ for all $t \in \mathbb{R}$. Moreover, there exists $\lambda \in \mathbb{R}$ with $(x^2, \lambda) \in \operatorname{epi}\varphi$. This implies $(x^1 + x^2, t + \lambda) \in \operatorname{epi}\varphi$ for all $t \in \mathbb{R}$. We get $\varphi(x^1 + x^2) = -\infty$. Hence, (3.5.1) holds for $x^1$ and $x^2$.
Take now any $x^1, x^2 \in X$ with $\varphi(x^1), \varphi(x^2) \in \mathbb{R}$. Since $(x^1, \varphi(x^1))$, $(x^2, \varphi(x^2)) \in \operatorname{epi}\varphi$, we obtain $(x^1, \varphi(x^1)) + (x^2, \varphi(x^2)) \in \operatorname{epi}\varphi$. Then $\varphi(x^1 + x^2) \leq \varphi(x^1) + \varphi(x^2)$.
(ii) Assume that $\varphi(x^1 + x^2) \leq \varphi(x^1) + \varphi(x^2)$ holds for all $x^1, x^2 \in X$. If $(x^3, t_3)$, $(x^4, t_4) \in \operatorname{epi}\varphi$, then the assumption yields $\varphi(x^3 + x^4) \leq \varphi(x^3) + \varphi(x^4) \leq t_3 + t_4$. We obtain $(x^3 + x^4, t_3 + t_4) \in \operatorname{epi}\varphi$.

(d) follows from (c) by Lemma 3.5.10 since the assumptions imply $\{\varphi(x^1), \varphi(x^2)\} \neq \{-\infty, +\infty\}$ for all $x^1, x^2 \in X$. □

Inequality (3.5.1) holds for subadditive functions since we assumed inf-addition. Without this assumption, (c) would be valid under the conditions assumed in (d).

The statement of Theorem 3.5.11(d) cannot be extended to functions which attain both of the values $+\infty$ and $-\infty$. This is illustrated by Example 3.5.8. There, the function $\varphi$ is additive, but

$$\varphi(-1 + 1) = 0 \neq +\infty = \varphi(-1) + \varphi(+1).$$

We get the following from Definition 3.5.9 by Lemma 2.2.6.

**Lemma 3.5.12** *Let $X$ be a linear space and $\varphi : X \to \overline{\mathbb{R}}$.*

$$\begin{aligned} \varphi \text{ is sublinear} \iff & \varphi \text{ is convex and positively homogeneous,} \\ \iff & \varphi \text{ is subadditive and positively homogeneous,} \\ \iff & \varphi \text{ is convex and subadditive and } \varphi(0) \leq 0. \end{aligned}$$

By Proposition 4.14.11, each continuous positively homogeneous function $\varphi: X \to \overline{\mathbb{R}}$ is finite-valued or the constant function with the function value $-\infty$.

**Proposition 3.5.13** *Assume that $X$ is a normed space and $\varphi : X \to \mathbb{R}$ is Lipschitz continuous on some neighborhood of 0 and sublinear. Then $\varphi$ is Lipschitz continuous.*

***Proof*** Since $\varphi$ is sublinear, it is subadditive and positively homogeneous. We have $\varphi(0) = 0$ since $\varphi$ is sublinear and finite-valued. $\varphi$ is Lipschitzian on some neighborhood $U$ of 0. Thus, there exists $L \in \mathbb{R}_+$ with $|\varphi(x)| \leq L\|x\|$ for all $x \in U$. Take any $x^0 \in X$. Then there are $\lambda \in \mathbb{R}_+$ and $x^1 \in U$ with $x^0 = \lambda x^1$. We obtain $|\varphi(x^0)| = |\varphi(\lambda x^1)| = \lambda|\varphi(x^1)| \leq \lambda L\|x^1\| = L\|\lambda x^1\| = L\|x^0\|$. Assume now $x^2, x^3 \in X$, and choose the notation such that $\varphi(x^2) \geq \varphi(x^3)$. Then $|\varphi(x^2 - x^3)| \leq L\|x^2 - x^3\|$. Since $\varphi(x^2) = \varphi(x^2 - x^3 + x^3) \leq \varphi(x^2 - x^3) + \varphi(x^3)$, we get $0 \leq \varphi(x^2) - \varphi(x^3) \leq \varphi(x^2 - x^3)$. Hence, $|\varphi(x^2) - \varphi(x^3)| \leq |\varphi(x^2 - x^3)| \leq L \cdot \|x^2 - x^3\|$. Thus, $\varphi$ is Lipschitz continuous. □

For finite-valued sublinear functionals on Banach spaces, continuity implies Lipschitz continuity. This results from Propositions 3.4.11 and 3.5.13.

**Corollary 3.5.14** *Assume that $X$ is a Banach space and $\varphi : X \to \mathbb{R}$ is sublinear. Then the following properties are equivalent to each other:*

(a) *$\varphi$ is lower semicontinuous.*
(b) *$\varphi$ is Lipschitz continuous.*

**Lemma 3.5.15** *Let $X$ be a linear space and $\varphi : X \to \overline{\mathbb{R}}$.*

(a) *If $\varphi$ is positively homogeneous, additive, odd or homogeneous, then $\lambda\varphi$ with $\lambda \in \mathbb{R}$ has the same property.*
(b) *If $\varphi$ is subadditive, superadditive or sublinear, then $\lambda\varphi$ with $\lambda \in \mathbb{R}_>$ has the same property.*
(c) *If $\varphi$ is subadditive, then $\varphi + c$ with $c \in \mathbb{R}_>$ is subadditive.*
(d) *If $\varphi$ is positively homogeneous, subadditive, superadditve, additive, sublinear, odd or homogeneous, then $g : X \to \overline{\mathbb{R}}$ defined by $g(x) = \varphi(\lambda x)$ with $\lambda \in \mathbb{R} \setminus \{0\}$ has the same property.*

Now, the relationship between additive and linear functions will be investigated.

**Lemma 3.5.16** *Let $X$ be a linear space and $\varphi: X \to \overline{\mathbb{R}}$ be a subadditive function.*

(a) *If $\varphi(0) = -\infty$, then $\varphi$ does not attain any real value.*

(b) *If* $\varphi(x) = -\infty$ *for some* $x \in X$ *and* $\varphi(0) \neq -\infty$, *then* $\varphi(-x) = +\infty$.
(c) $\operatorname{dom}\varphi = X$ *holds if and only if* $\varphi$ *is finite-valued or* $\varphi = -\infty$.
(d) *Assume* $\varphi(0) \in \mathbb{R}$.
*Then* $\operatorname{dom}\varphi = X$ *holds if and only if* $\varphi$ *is finite-valued.*
*Furthermore,* $\varphi(0) \geq 0$ *and:*
$\varphi(nx) = -\infty$ *for all* $n \in \mathbb{N}_>$ *and all* $x \in X$ *with* $\varphi(x) = -\infty$.

***Proof***

(a) Contrary to our claim, suppose $\varphi(0) = -\infty$ and $\lambda := \varphi(x) \in \mathbb{R}$ for some $x \in X$. Then $(0, t) \in \operatorname{epi}\varphi$ for all $t \in \mathbb{R}$, and $(x, \lambda) \in \operatorname{epi}\varphi$. We get $(0, t) + (x, \lambda) = (x, t + \lambda) \in \operatorname{epi}\varphi$ for all $t \in \mathbb{R}$. Thus, $\varphi(x) = -\infty$, which contradicts $\varphi(x) \in \mathbb{R}$.
(b) Suppose $\varphi(x) = -\infty$ as well as $\varphi(-x) \neq +\infty$ for some $x \in X$. Then there exists $\lambda \in \mathbb{R}$ with $(-x, \lambda) \in \operatorname{epi}\varphi$. Since $(x, t) \in \operatorname{epi}\varphi$ for all $t \in \mathbb{R}$ and $\varphi$ is subadditive, we obtain $(x - x, t + \lambda) \in \operatorname{epi}\varphi$ for all $t \in \mathbb{R}$. Therefore, $\varphi(0) = -\infty$.
(c) results from (a) and (b).
(d) The first statement follows from (c).
$\varphi(0) \in \mathbb{R}$ implies $(0, \varphi(0)) \in \operatorname{epi}\varphi$. We conclude that $(0, \varphi(0)) + (0, \varphi(0)) = (0, 2\varphi(0)) \in \operatorname{epi}\varphi$. Thus, $\varphi(0) \leq 2\varphi(0)$. We obtain $\varphi(0) \geq 0$.
Take any $x \in X$ with $\varphi(x) = -\infty$. Then $(x, t) \in \operatorname{epi}\varphi$ for all $t \in \mathbb{R}$. This results in $(nx, nt) \in \operatorname{epi}\varphi$ for all $t \in \mathbb{R}$, $n \in \mathbb{N}_>$. Therefore, $\varphi(nx) = -\infty$ for all $n \in \mathbb{N}_>$. □

**Proposition 3.5.17** *Let* $X$ *be a linear space and* $\varphi : X \to \overline{\mathbb{R}}$ *be an additive function.*

(a) *If* $\varphi(0) \neq 0$, *then* $\varphi$ *does not attain any real value.*
(b) *If* $\varphi(0) = 0$, *then* $\varphi$ *is odd and*

$$\forall t \in \mathbb{Q}\ \forall x \in X : \varphi(tx) = t\varphi(x).$$

***Proof*** Since $\varphi$ and $-\varphi$ are subadditive, we obtain from Lemma 3.5.16:
If $\varphi(0) \notin \mathbb{R}$, then $\varphi$ does not attain any real value. If $\varphi(0) \in \mathbb{R}$, then $\varphi(0) = 0$. This yields (a).

Assume now $\varphi(0) = 0$. For each $x \in X$, Lemma 3.5.16 implies that $\varphi(x) = -\infty$ holds if and only if $\varphi(-x) = +\infty$. Take now any $x \in X$ with $\varphi(x) \in \mathbb{R}$. Then $(x, \varphi(x))$, $(-x, \varphi(-x)) \in \operatorname{epi}\varphi \cap \operatorname{hypo}\varphi$. We conclude that $(x, \varphi(x)) + (-x, \varphi(-x)) = (0, \varphi(x) + \varphi(-x)) \in \operatorname{epi}\varphi \cap \operatorname{hypo}\varphi$. Hence, $0 = \varphi(0) = \varphi(x) + \varphi(-x)$. Thus, $\varphi(-x) = -\varphi(x)$. Consequently, $\varphi$ is odd.

Lemma 3.5.16 implies $\varphi(nx) = n\varphi(x)$ for all $n \in \mathbb{N}$ and $x \in X$ with $\varphi(x) \notin \mathbb{R}$. Take any $x \in X$ with $\varphi(x) \in \mathbb{R}$. Then $(x, \varphi(x)) \in \operatorname{epi}\varphi \cap \operatorname{hypo}\varphi$. We get $(nx, n\varphi(x)) \in \operatorname{epi}\varphi \cap \operatorname{hypo}\varphi$ for all $n \in \mathbb{N}$. Hence, $\varphi(nx) = n\varphi(x)$ for all $n \in \mathbb{N}$. Thus, $\varphi(nx) = n\varphi(x)$ for all $n \in \mathbb{N}$, $x \in X$. For $t \in -\mathbb{N}$, $x \in X$, we get $\varphi(tx) = \varphi((-t)(-x)) = -t\varphi(-x) = -t \cdot (-\varphi(x)) = t\varphi(x)$. Take now any $q \in \mathbb{Z} \setminus \{0\}$, $x \in X$. Then $\varphi(x) = \varphi(q\frac{1}{q}x) = q\varphi(\frac{1}{q}x)$, which yields $\varphi(\frac{1}{q}x) = \frac{1}{q}\varphi(x)$. Consequently, $\varphi(\frac{p}{q}x) = \frac{p}{q}\varphi(x)$ for all $x \in X$, $p \in \mathbb{N}$, $q \in \mathbb{Z} \setminus \{0\}$. □

**Theorem 3.5.18** *Let $X$ be a linear space and $\varphi : X \to \overline{\mathbb{R}}$.*

(1) *$\varphi$ is linear if and only if $\varphi$ is homogeneous and additive.*
(2) *$\varphi$ is linear if and only if $\varphi$ and $-\varphi$ are sublinear.*

***Proof***

(a) First, we prove that each linear functional is homogeneous.
Assume that $\varphi$ is linear.
Suppose, contrary to our claim, that $\varphi$ is not odd. Then there exists $x \in X$ with $\varphi(-x) \neq -\varphi(x)$. Proposition 3.4.7(d) implies $\varphi(x) \neq -\infty$ and $\varphi(-x) \neq -\infty$. Applying Proposition 3.4.7(d) to the convex function $-\varphi$, we get $\varphi(x) \neq +\infty$ and $\varphi(-x) \neq +\infty$. Thus, $t_1 := \varphi(x) \in \mathbb{R}$ and $t_2 := \varphi(-x) \in \mathbb{R}$. We get $(x, t_1), (-x, t_2) \in \operatorname{epi}\varphi \cap \operatorname{hypo}\varphi$. Therefore, $(\frac{1}{2}(x - x), \frac{1}{2}(t_1 + t_2)) \in \operatorname{epi}\varphi \cap \operatorname{hypo}\varphi$. We obtain $0 = \varphi(0) = \frac{1}{2}(t_1 + t_2)$. This yields $t_1 = -t_2$, which contradicts $\varphi(-x) \neq -\varphi(x)$. Hence, $\varphi$ is odd.
If $\varphi(x) = -\infty$, then Proposition 3.4.7(b) implies $\varphi(\lambda x) = -\infty$ for all $\lambda \in (0, 1)$. If $\varphi(x) = +\infty$ and we apply Proposition 3.4.7(b) to $-\varphi$, then we get $\varphi(\lambda x) = +\infty$ for all $\lambda \in (0, 1)$. Take now any $x \in X$ with $t := \varphi(x) \in \mathbb{R}$. Then $(0, 0), (x, t) \in \operatorname{epi}\varphi \cap \operatorname{hypo}\varphi$. We obtain $((1 - \lambda)\cdot 0 + \lambda x, (1 - \lambda)\cdot 0 + \lambda t) \in \operatorname{epi}\varphi \cap \operatorname{hypo}\varphi$ for all $\lambda \in [0, 1]$. Hence, $(\lambda x, \lambda t) \in \operatorname{epi}\varphi \cap \operatorname{hypo}\varphi$ for all $\lambda \in [0, 1]$. We conclude that $\varphi(\lambda x) = \lambda t$ for all $\lambda \in [0, 1]$. Thus, $\varphi(\lambda x) = \lambda\varphi(x)$ for all $x \in X,\ \lambda \in [0, 1]$.
Take any $x \in X$ and $\lambda > 1$. Then $\frac{1}{\lambda} \in (0, 1)$. Hence, $\varphi(x) = \varphi(\frac{1}{\lambda}(\lambda x)) = \frac{1}{\lambda}\varphi(\lambda x)$. We get $\varphi(\lambda x) = \lambda\varphi(x)$. Consequently, $\varphi(\lambda x) = \lambda\varphi(x)$ for all $\lambda \in \mathbb{R}_+$. Since $\varphi$ is odd, we obtain $\varphi(\lambda x) = \lambda\varphi(x)$ for all $\lambda \in \mathbb{R}$. Thus, $\varphi$ is homogeneous by Theorem 3.5.11.
(b) (2) follows from (a) because of Lemma 3.5.12.
(c) Applying Lemma 3.5.12 to $\varphi$ and $-\varphi$ implies the additivity of each linear function $\varphi$.
(d) If $\varphi$ is homogeneous and additive, then $\varphi$ and $-\varphi$ are sublinear by Lemma 3.5.12, and $\varphi$ is linear because of (2). □

For real-valued functions, our definition of linearity coincides with the usual one.

**Corollary 3.5.19** *Let $X$ be a linear space and $\varphi : X \to \mathbb{R}$.*
*Then $\varphi$ is linear if and only if*

$$\varphi(\lambda_1 x^1 + \lambda_2 x^2) = \lambda_1\varphi(x^1) + \lambda_2\varphi(x^2)$$

*holds for all $x^1, x^2 \in X$ and all $\lambda_1, \lambda_2 \in \mathbb{R}$.*

This results from Theorems 3.5.18 and 3.5.11.

**Corollary 3.5.20** *Assume that $X$ is a linear space and that $\varphi : X \to \overline{\mathbb{R}}$ is linear. If $\varphi$ is proper or* $\operatorname{dom}\varphi = X$, *then $\varphi$ is finite-valued.*

***Proof*** The assumptions imply that $\varphi$ cannot attain both of the values $+\infty$ and $-\infty$. $\varphi$ is homogeneous and hence odd by Theorem 3.5.18. Thus, $\varphi$ is finite-valued. □

**Theorem 3.5.21** *Let $X$ be a topological vector space and $\varphi : X \to \overline{\mathbb{R}}$ be continuous. Then $\varphi$ is linear if and only if $\varphi$ is additive and $\varphi(0) = 0$.*

***Proof*** The first direction of the assertion results from Definition 3.5.1 and Theorem 3.5.18(1).

Assume now that $\varphi$ is additive and $\varphi(0) = 0$. Contrary to our claim, suppose that $\varphi$ is not linear. Then $\varphi$ is not homogeneous by Theorem 3.5.18. Hence, there are $x^0 \in X$ and $\lambda \in \mathbb{R} \setminus \{0\}$ with $\varphi(\lambda x^0) \neq \lambda\varphi(x^0)$. There exist neighborhoods $V_1$ of $\varphi(\lambda x^0)$ and $V_2$ of $\lambda\varphi(x^0)$ such that $V_1 \cap V_2 = \emptyset$. Since $\varphi$ is continuous, the function $\varphi(\lambda x)$ is a continuous function of $x$. Thus, there exists a neighborhood $U$ of $x^0$ such that $\varphi(\lambda x) \in V_1$ for each $x \in U$. There is $p \in (0, 1)$ such that $tx^0 \in U$ for all $t \in (p, 1]$. Take any sequence $(q_n)_{n\in\mathbb{N}}$ of rational numbers which converges to $\lambda$ and fulfills the condition $q_n < \lambda$ for all $n \in \mathbb{N}$. Then there is $n_0 \in \mathbb{N}$ such that $\frac{q_n}{\lambda} \in (p, 1]$ for all $n > n_0$. We get $\frac{q_n}{\lambda}x^0 \in U$ for each $n > n_0$. For each $n > n_0$, Proposition 3.5.17 implies $q_n\varphi(x^0) = \varphi(q_n x^0) = \varphi(\lambda \frac{q_n}{\lambda} x^0) \in V_1$. Therefore, $q_n\varphi(x^0) \notin V_2$ for all $n > n_0$. This contradicts the fact that $q_n\varphi(x^0)$ converges to $\lambda\varphi(x^0)$ for $n \to \infty$. □

## 3.6 Variational Analysis

Mathematical algorithms are often based on linearization. Convex mappings are approximated by derivatives; convex sets are approximated by tangent spaces. In nonconvex analysis, conical approximations of sets are used instead of tangent spaces. Variational analysis deals with the different concepts of linearization.

In this section, we restrict ourselves to definitions and statements from the literature which are essential for the subsequent sections. Concepts of generalized subgradients and for vector- and set-valued functions will be presented in Chap. 8.

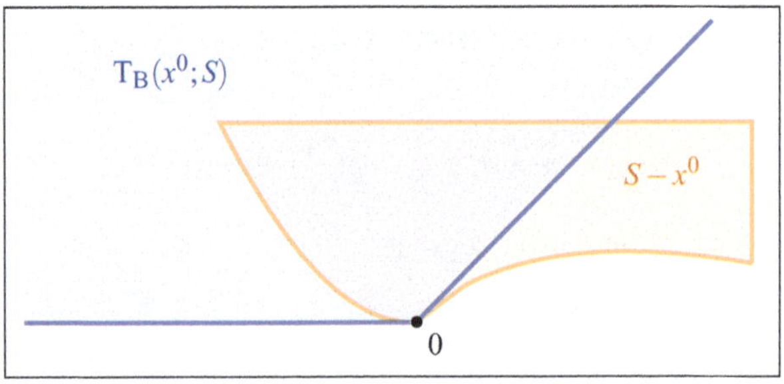


**Fig. 3.4** The Bouligand tangent cone to $S$ at $x^0$

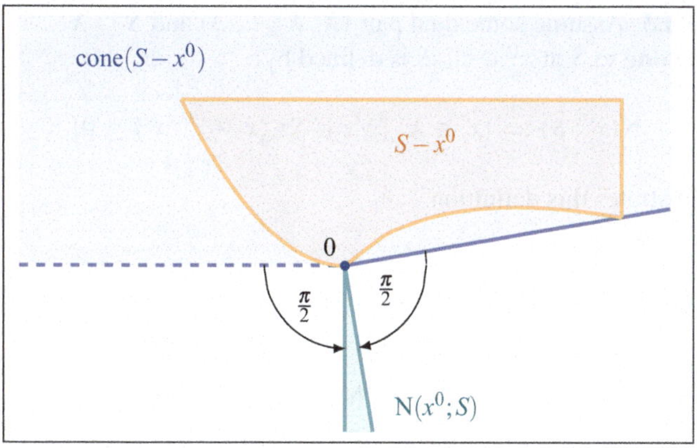


**Fig. 3.5** The normal cone to $S$ at $x^0$

We start with the classical conical approximations of sets in variational analysis. Remember that $X^*$ denotes the topological dual space of the space $X$ and $\iota_A$ stands for the indicator function of the set $A$. The following definition is illustrated by Fig. 3.4.

**Definition 3.6.1** Let $S$ be a nonempty subset of a normed space $X$. The **Bouligand tangent cone** or **contingent cone** to $S$ at $x^0 \in X$ is defined by

$$\mathrm{T_B}(x^0; S) := \{u \in X \mid \exists (t_n) \text{ with } t_n \in \mathbb{R}_+ \ \exists (x_n) \xrightarrow{S} x^0 \colon u = \lim_{n\to\infty} t_n(x_n - x^0)\}.$$

We get $\mathrm{T_B}(x^0; S) = \emptyset$ if $x^0 \notin \mathrm{cl}\, S$, and $\mathrm{T_B}(x^0; S) = X$ if $x^0 \in \mathrm{int} S$. For $x^0 \in \mathrm{cl}\, S$, the Bouligand tangent cone $\mathrm{T_B}(x^0; S)$ is a nonempty closed cone, which can also be described by

$$\mathrm{T_B}(x^0; S) = \{u \in X \mid \exists (t_n) \to +0 \ \exists (x_n) \to u \ \ \forall n \in \mathbb{N} \colon x^0 + t_n x_n \in S\}.$$

**Lemma 3.6.2** *Let $S$ be a nonempty subset of a normed space $X$. For each $x^0 \in \mathrm{cl}\, S$,*

$$\begin{aligned} \mathrm{T_B}(x^0; S) &= \mathrm{T_B}(0; S - x^0) \qquad \textit{and} \\ 0^+ S \subseteq \mathrm{T_B}(x^0; S) &\subseteq \mathrm{cl}\,\mathrm{cone}(S - x^0). \end{aligned} \tag{3.6.1}$$

*If $S$ is starshaped at $x^0 \in S$, then*

$$\mathrm{T_B}(x^0; S) = \mathrm{cl}\,\mathrm{cone}(S - x^0).$$

*If $S$ is convex, then $\mathrm{T_B}(x^0; S)$ is convex for each $x^0 \in S$.*

The mentioned properties of contingent cones are obvious or were proved in [272].

**Definition 3.6.3** Assume some dual pair $(X, X', \langle .,. \rangle)$ and $S \subseteq X$.
The **normal cone** to $S$ at $x^0 \in \mathrm{cl}_w\, S$ is defined by

$$\mathrm{N}(x^0; S) := \{x' \in X' \mid \forall x \in S\colon \langle x - x^0, x' \rangle \leq 0\}. \tag{3.6.2}$$

Figure 3.5 illustrates this definition.

For all $x^0 \in \mathrm{cl}_w\, S$,

$$\begin{aligned}
\mathrm{N}(x^0; S) &= -(S - x^0)^+ = -(\mathrm{cl}_w(S - x^0))^+ \\
&= -(\mathrm{cone}(S - x^0))^+ = -(\mathrm{cl}_w\,\mathrm{cone}(S - x^0))^+ \\
&= \mathrm{N}(0; S - x^0) = \mathrm{N}(0; \mathrm{cl}_w(S - x^0)) \\
&= \mathrm{N}(0; \mathrm{cone}(S - x^0)) = \mathrm{N}(0; \mathrm{cl}_w\,\mathrm{cone}(S - x^0)).
\end{aligned}$$

Consequently, the normal cone to $S$ is a nonempty weak* closed convex cone at each $x^0 \in \mathrm{cl}_w\, S$.

The properties of contingent cones imply

$$\mathrm{N}(x^0; S) = -(\mathrm{T_B}(x^0; S))^+ \tag{3.6.3}$$

for each set $S$ in a normed space which is starshaped at $x^0 \in S$.

**Definition 3.6.4** Assume some dual pair $(X, X', \langle .,. \rangle)$ and $\varphi\colon X \to \overline{\mathbb{R}}$.
The **(Fenchel) subdifferential** of $\varphi$ at $x^0 \in X$ is the set

$$\partial\varphi(x^0) := \begin{cases} \{x' \in X' \mid \forall x \in X\colon \langle x - x^0, x' \rangle \leq \varphi(x) - \varphi(x^0)\} & \text{if } \varphi(x^0) \in \mathbb{R}, \\ \emptyset & \text{if } \varphi(x^0) \notin \mathbb{R}. \end{cases} \tag{3.6.4}$$

Each element of $\partial\varphi(x^0)$ is called a **subgradient** of $\varphi$ at $x^0$. $\varphi$ is said to be **subdifferentiable** at $x^0 \in X$ if $\partial\varphi(x^0) \neq \emptyset$.

The **(Fenchel) conjugate** of $\varphi$ is the function $\varphi^*\colon X' \to \overline{\mathbb{R}}$ defined by

$$\varphi^*(x') := \sup\left\{\langle x, x' \rangle - \varphi(x) \mid x \in X\right\}.$$

The **support function** $\sigma_A : X' \to \overline{\mathbb{R}}$ of a set $A \subseteq X$ is given by

$$\sigma_A(x') := \sup\left\{\langle a, x' \rangle \mid a \in A\right\}.$$

The set $\mathrm{bar}\, A := \{x' \in X' \mid \sigma_A(x') \in \mathbb{R}\}$ is called the **barrier cone** of $A$.

The **(Fenchel) biconjugate** of $\varphi$ is the function $\varphi^{**}\colon X \to \overline{\mathbb{R}}$ defined by

$$\varphi^{**}(x) := \sup\left\{\langle x, x' \rangle - \varphi^*(x') \mid x' \in X'\right\}.$$

The subdifferential is defined in such a way that, for each $x^0 \in X$,

$$\partial\varphi(x^0) = \{x' \in X' \mid (x', -1) \in \mathrm{N}((x^0, \varphi(x^0)); \operatorname{epi}\varphi)\} \text{ if } \varphi(x^0) \in \mathbb{R}. \tag{3.6.5}$$

An immediate implication of (3.6.4) underscores the role of subdifferentials in optimization even for nonconvex problems.

**Theorem 3.6.5** *Suppose a dual pair $(X, X', \langle .,. \rangle)$ and that $\varphi\colon X \to \overline{\mathbb{R}}$ is a proper function. Then $\varphi$ attains its minimum on $X$ at $x^0 \in X$ if and only if $0 \in \partial\varphi(x^0)$.*

Definition 3.6.4 yields the following lemma, where the inf-addition is applied for the proof of the second inequality in (3.6.9).

**Lemma 3.6.6** *Suppose a dual pair $(X, X', \langle .,. \rangle)$ and $\varphi\colon X \to \overline{\mathbb{R}}$.*

(a) *If $\varphi$ attains the value $-\infty$ in some $x^0 \in X$, then* $\operatorname{dom}\varphi^* = \emptyset$.
*If* $\operatorname{dom}\varphi^* = \emptyset$, *then* $\inf\{\varphi(x) \mid x \in X\} = -\infty$.
(b) $\partial\varphi(x) \subseteq \operatorname{dom}\varphi^*$ *for all $x \in X$.*
(c) *If $x^0 \in X$ with $\varphi(x^0) \in \mathbb{R}$, then*

$$\partial\varphi(x^0) = \{x' \in X' \mid \varphi^*(x') \le \langle x^0, x' \rangle - \varphi(x^0)\}. \tag{3.6.6}$$

(d) *$\varphi^*$ is convex and lower semicontinuous w.r.t. the weak* topology $\sigma(X', X)$, where*

$$\varphi^*(0) = -\inf\{\varphi(x) \mid x \in X\}, \tag{3.6.7}$$
$$\varphi^*(x') \ge -\varphi(0) \quad \text{for all } x' \in X'. \tag{3.6.8}$$

(e) *$\varphi^{**}$ is convex and lower semicontinuous w.r.t. the weak topology $\sigma(X, X')$.*
(f) $\inf\{\varphi(x) \mid x \in X\} = \inf\{\varphi^{**}(x) \mid x \in X\}$, *more in detail,*

$$\inf\{\varphi(x) \mid x \in X\} \le \varphi^{**}(x^0) \le \varphi(x^0) \quad \text{for each } x^0 \in X. \tag{3.6.9}$$

(g) *For each function $g\colon X \to \overline{\mathbb{R}}$,*

$$(\varphi \,\#\, g)^* = \varphi^* + g^*,$$

*where the rule of sup-addition has to be applied, i.e.,*
$(+\infty) + (-\infty) := (-\infty) + (+\infty) := -\infty$.
(h) $(\varphi^{**})^* = \varphi^*$.

***Proof*** The first statement of (a), (3.6.7) and (3.6.8) are easy to verify. The second statement in (a) follows from (3.6.7).

Take any $x^0 \in X$ with $\varphi(x^0) \in \mathbb{R}$.

$$\begin{aligned}\partial\varphi(x^0) &= \{x' \in X' \mid \forall x \in X\colon \langle x - x^0, x'\rangle \le \varphi(x) - \varphi(x^0)\,\}\\ &= \{x' \in X' \mid \forall x \in X\colon \langle x, x'\rangle - \varphi(x) \le \langle x^0, x'\rangle - \varphi(x^0)\,\}\\ &= \{x' \in X' \mid \sup\left\{\langle x, x'\rangle - \varphi(x) \mid x \in X\right\} \le \langle x^0, x'\rangle - \varphi(x^0)\,\}\\ &= \{x' \in X' \mid \varphi^*(x') \le \langle x^0, x'\rangle - \varphi(x^0)\},\end{aligned}$$

which proves (c). (c) and (3.6.4) yield (b).

The other properties in (d) and (e) result from Proposition 3.3.14 and Lemma 3.4.5.

The definition of $\varphi^{**}$ implies $\varphi^{**}(x^0) \ge -\varphi^*(0)$ for all $x^0 \in X$ and thus together with (3.6.7) the first inequality in (3.6.9). Take now any $x^0 \in X$ and $x' \in X'$. We have $\varphi^*(x') \ge \langle x^0, x'\rangle - \varphi(x^0)$. This implies $\varphi^*(x') + \varphi(x^0) \ge \langle x^0, x'\rangle$, where the inf-addition has to be applied in the case $\{\varphi(x^0), \varphi^*(x')\} = \{-\infty, \infty\}$. We get $\varphi(x^0) \ge \langle x^0, x'\rangle - \varphi^*(x')$. This and the definition of the biconjugate result in the second inequality of (3.6.9) and thus in (f).

(g) and (h) were proved in [362, pp. 38 and 35]. □

We get the following from [190, p. 102].

**Lemma 3.6.7** *Suppose a dual pair $(X, X', \langle .,.\rangle)$, that $X$ is equipped with the weak topology $\sigma(X, X')$, that $X'$ is equipped with the weak* topology $\sigma(X', X)$, and that $\varphi\colon X \to \overline{\mathbb{R}}$ is a lower semicontinuous convex proper function. Then $\varphi^*$ is a proper function and*

$$\varphi = \varphi^{**}.$$

A more detailed statement is contained in the next lemma.

**Lemma 3.6.8** *Suppose a dual pair $(X, X', \langle .,.\rangle)$, that $X$ is equipped with the weak topology $\sigma(X, X')$, that $X'$ is equipped with the weak* topology $\sigma(X', X)$, and $\varphi\colon X \to \overline{\mathbb{R}}$ with* $\operatorname{dom}\varphi \neq \emptyset$. *Then*

$$\varphi^{**} \textit{ is proper if and only if } \operatorname{cl}\operatorname{conv}\varphi \textit{ is proper.}$$

*In this case,*

$$\varphi^{**} = \operatorname{cl}\operatorname{conv}\varphi.$$

*If* $\operatorname{cl}\operatorname{conv}\varphi$ *is not proper, then* $\varphi^{**} = -\infty$.

This statement was proved in [508, Theorem 2.3.4] for the dual space $(X, X^*, \langle .,.\rangle)$ and can be transferred to the general case because of (2.1.2) and (2.1.3).

Properties of subdifferentials and conjugate functions are broadly discussed in the literature, especially in convex analysis (e.g. in [3, 64, 508]). We obtain the following from [3, Theorems 7.12 and 7.13].

**Lemma 3.6.9** *Suppose a dual pair $(X, X', \langle .,.\rangle)$ and that $\varphi\colon X \to \overline{\mathbb{R}}$ is a convex proper function.*

(a) *If* $x \in \operatorname{dom}\varphi$, *then* $\partial\varphi(x)$ *is a weak* compact convex subset of* $X'$.
(b) *If* $\varphi$ *is* $\sigma(X, X')$*-continuous at* $x \in \operatorname{int}_{\mathrm{w}}\operatorname{dom}\varphi$, *then* $\partial\varphi(x) \neq \emptyset$.

The use of subdifferentials in convex optimization is pointed out by the well-known next statement (see e.g. [509, Theorem 45.2]). Note that $\partial\iota_S(x^0) = \mathrm{N}(x^0; S)$ for all $x^0 \in S$.

**Proposition 3.6.10** *Assume that* $X$ *is a separated locally convex space,* $\varphi\colon X \to \overline{\mathbb{R}}$ *is a convex finite-valued function,* $M \subseteq X$ *is a nonempty convex set with nonempty interior and that* $\varphi$ *is continuous at some* $x^0 \in M$. *Then* $x^0$ *solves the problem*

$$\inf_{x\in M} \varphi(x)$$

*if and only if*

$$0 \in \partial\varphi(x^0) + \partial\iota_M(x^0).$$

The following sum rule for subdifferentials (see e.g. [509, Satz 45.5]) is useful in this context.

**Proposition 3.6.11** *Suppose* $X$ *is a separated locally convex space. Let* $\varphi_i : X \to \overline{\mathbb{R}}$, $i \in \{1, \dots, n\}$, *be convex proper functionals,* $n \geq 2$. *Assume that there exists an element* $\tilde{x} \in \bigcap_{i=1}^n \operatorname{int}\operatorname{dom}\varphi_i$ *such that* $\varphi_1, \dots, \varphi_{n-1}$ *are continuous at* $\tilde{x}$. *Then*

$$\partial\left(\sum_{i=1}^n \varphi_i(x)\right) = \sum_{i=1}^n \partial\varphi_i(x) \quad \text{for all } x \in X.$$

Some basic facts about support functions should be mentioned.

**Lemma 3.6.12** *Suppose a dual pair* $(X, X', \langle .,. \rangle)$ *and that* $A$ *and* $B$ *are nonempty subsets of* $X$.

(a) $\sigma_A$ *is lower semicontinuous w.r.t. the weak* topology* $\sigma(X', X)$, *sublinear, and proper.*
(b) $\sigma_A = \sigma_{\operatorname{cl}_{\mathrm{w}}\operatorname{conv}A}$.
(c) $\sigma_A = (\iota_A)^*$.
(d) $\operatorname{bar} A$ *is a nonempty convex cone with* $-A^+ \subseteq \operatorname{bar} A$.
(e) *If* $A$ *is a nonempty cone, then* $\operatorname{bar} A = -A^+$ *and* $\sigma_A = \iota_{-A^+}$.
(f) $\operatorname{cl}_{\mathrm{w}}(\operatorname{bar} A) = -(0^+A)^+$.

***Proof*** $\sigma_A$ is proper since $0 \in \operatorname{dom}\sigma_A$ and $A \neq \emptyset$. $\sigma_A$ is lower semicontinuous w.r.t. $\sigma(X', X)$ because of Proposition 3.3.14. It is sublinear since it is the pointwise supremum of linear functions. (b) and (c) were proved in [508, p. 79]. (d) and (e) are easy to verify. (f) was proved in [441]. □

Since the Bouligand tangent cone is not convex in each case, other concepts of tangent cones have been developed, e.g. Clarke's tangent cone [98]. In each concept, the relationship between the generalized tangent cone, the generalized normal cone and the generalized subdifferential is based on the ideas presented by condition (3.6.1) together with the Eqs. (3.6.3) and (3.6.5).

**Definition 3.6.13** Let $S$ be a proper subset of a Banach space $X$. The **Clarke tangent cone** to $S$ at $x^0 \in \operatorname{cl} S$ is defined by

$$\mathrm{T_C}(x^0; S) := \{u \in X \mid \forall (t_n) \to +0 \ \forall (x_n) \xrightarrow{S} x^0 \ \exists (u_n) \to u \ \forall n \in \mathbb{N} : \ x_n + t_n u_n \in S\}.$$

The **Clarke normal cone** to $S$ at $x^0 \in S$ is

$$\mathrm{N_C}(x^0; S) := -(\mathrm{T_C}(x^0; S))^+. \tag{3.6.10}$$

For each $x^0 \in \operatorname{cl} S$, the set $\mathrm{T_C}(x^0; S)$ is a nonempty closed convex cone with

$$\mathrm{T_C}(x^0; S) \subseteq \mathrm{T_B}(x^0; S).$$

The following statement was proved by Jofré and Jourani [281, Cor. 3].

**Lemma 3.6.14** *Assume that $A$ is a proper closed subset of a Banach space $X$ and that $Z \subseteq X$ is a nonempty closed convex cone. Then*

$$A - Z \subseteq A \iff (\forall a \in A: \ Z \subseteq -\mathrm{T_C}(a; A)) \iff (\forall a \in A: \ \mathrm{N_C}(a; A) \subseteq Z^+).$$

We obtain from Lemma 3.6.14:

**Theorem 3.6.15** *Assume that $A$ is a proper closed subset of a Banach space. Then*

$$0^+ A = \bigcap_{a \in A} \mathrm{T_C}(a; A),$$

$$\bigcup_{a \in A} \mathrm{N_C}(a; A) \subseteq -(0^+ A)^+.$$

***Proof*** For $Z := -0^+ A$, Lemma 3.6.14 yields $0^+ A \subseteq \bigcap_{a \in A} \mathrm{T_C}(a; A)$ and $\bigcup_{a \in A} \mathrm{N_C}(a; A) \subseteq -(0^+ A)^+$. Since $\mathrm{T_C}(a; A)$ is a closed convex cone for each $a \in A$, Lemma 3.6.14 implies for $Z := -\bigcap_{a \in A} \mathrm{T_C}(a; A)$ that $A + \bigcap_{a \in A} \mathrm{T_C}(a; A) \subseteq A$ and hence $\bigcap_{a \in A} \mathrm{T_C}(a; A) \subseteq 0^+ A$. □

Clarke's generalized subdifferential will be introduced in Chap. 8.

## 3.7 Monotonicity

Functionals which are used for scalarization in vector optimization have to fulfill certain monotonicity conditions. The following definition is illustrated by Fig. 3.6.

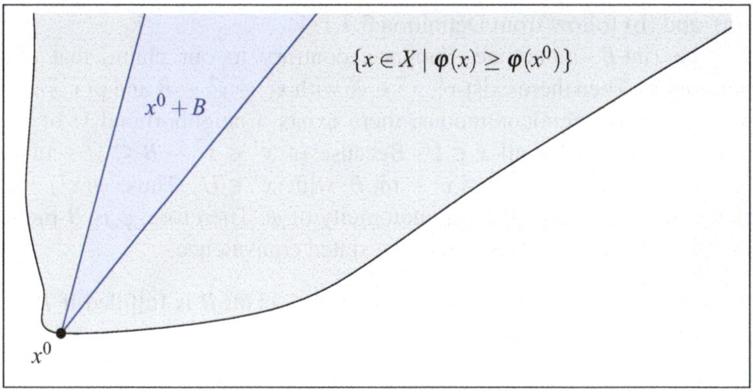


**Fig. 3.6** B-monotonicity on $X$

**Definition 3.7.1** Let $X$ be a linear space, $B,\ X_0 \subseteq X$ and $\varphi : X \to \overline{\mathbb{R}}$.
$\varphi$ is said to be

(a) **B-monotone** on $X_0$ if $x^1, x^2 \in X_0$ and $x^2 - x^1 \in B$ imply $\varphi(x^1) \leq \varphi(x^2)$,
(b) **strictly B-monotone** on $X_0$ if $x^1, x^2 \in X_0$ and $x^2 - x^1 \in B \setminus \{0\}$ imply $\varphi(x^1) < \varphi(x^2)$.

$\varphi$ is said to be **B-monotone** or **strictly B-monotone** if it is $B$-monotone or strictly $B$-monotone, respectively, on $X$.

We admit $B \subseteq \{0\}$ and that $X_0$ contains at most one element. In both cases, $\varphi$ is $B$-monotone and strictly $B$-monotone on $X_0$.

For a nonempty set $A$ in a separated locally convex space $X$, the dual cone $A^+$ consists of the continuous linear $A$-monotone real-valued functions on $X$, and its quasi-interior qint$A^+$ consists of the continuous linear strictly $A$-monotone real-valued functions on $X$. On $X = \mathbb{R}^\ell$, the $\mathbb{R}^\ell_+$-monotone functionals are the functionals that are monotonically increasing in each argument, and the linear $\mathbb{R}^\ell_+$-monotone functionals are the functions $\sum_{i=1}^{\ell} \lambda_i x_i$ with nonnegative parameters $\lambda_i$. The functionals which are strictly monotonically increasing in each argument are the strictly $\mathbb{R}^\ell_+$-monotone functionals on $\mathbb{R}^\ell$.

**Lemma 3.7.2** *Let $X$ be a linear space, $B,\ X_0 \subseteq X$ and $\varphi\colon X \to \overline{\mathbb{R}}$.*

(a) *If $\varphi$ is $B$-monotone on $X_0$ or strictly $B$-monotone on $X_0$, then it is $A$-monotone or strictly $A$-monotone, respectively, on $X_0$ for each subset $A$ of $B$.*
(b) *If $\varphi$ is strictly $B$-monotone on $X_0$, then it is $B$-monotone on $X_0$.*
(c) *Assume that $X$ is a topological vector space, $B \subseteq \operatorname{cl}\operatorname{int} B$ and that $\varphi$ is finite-valued and lower semicontinuous. Then $\varphi$ is* (int $B$)*-monotone if and only if it is $B$-monotone.*

***Proof*** (a) and (b) follow from Definition 3.7.1.

(c) Let $\varphi$ be (int $B$)-monotone. Suppose, contrary to our claim, that $\varphi$ is not $B$-monotone. Then there exist $x^1, x^2 \in X$ with $x^2 - x^1 \in B$ and $\varphi(x^1) > \varphi(x^2)$. Since $\varphi$ is lower semicontinuous, there exists a neighborhood U of $x^1$ such that $\varphi(x) > \varphi(x^2)$ for all $x \in U$. Because of $x^1 \in x^2 - B \subseteq x^2 - \operatorname{cl}\operatorname{int} B = \operatorname{cl}(x^2 - \operatorname{int} B)$, there is $x^3 \in x^2 - \operatorname{int} B$ with $x^3 \in U$. Thus, $\varphi(x^3) > \varphi(x^2)$, which contradicts the (int $B$)-monotonicity of $\varphi$. Therefore, $\varphi$ is $B$-monotone. (a) implies the reverse direction of the stated equivalence. □

Part (c) was proved in [472]. The assumption $B \subseteq \operatorname{cl}\operatorname{int} B$ is fulfilled if $B$ is some convex set with nonempty interior.

**Lemma 3.7.3** *Let $X$ be a linear space, $B \subseteq X$ and $\varphi\colon X \to \overline{\mathbb{R}}$.*
*The following properties are equivalent to each other.*

(a) *$\varphi$ is $B$-monotone.*
(b) $\operatorname{lev}_{\varphi,\leq}(t) - B \subseteq \operatorname{lev}_{\varphi,\leq}(t)$ *for all $t \in \mathbb{R}$.*
(c) $\operatorname{lev}_{\varphi,\geq}(t) + B \subseteq \operatorname{lev}_{\varphi,\geq}(t)$ *for all $t \in \mathbb{R}$.*
(d) $\operatorname{epi}\varphi + ((-B) \times \mathbb{R}_+) \subseteq \operatorname{epi}\varphi$.

***Proof*** The equivalence between (a) and (b) as well as between (a) and (c) results immediately from the definition of the $B$-monotonicity.

$$\begin{aligned}
\text{(d) holds} &\iff \forall t \in \mathbb{R}\ \forall x \in \operatorname{lev}_{\varphi,\leq}(t)\ \forall b \in B\ \forall \lambda \in \mathbb{R}_+ :\ (x,t) + (-b,\lambda) \in \operatorname{epi}\varphi,\\
&\iff \forall t \in \mathbb{R}\ \forall x \in \operatorname{lev}_{\varphi,\leq}(t)\ \forall b \in B\ \forall \lambda \in \mathbb{R}_+ :\ \varphi(x-b) \leq t + \lambda,\\
&\iff \forall t \in \mathbb{R}\ \forall x \in \operatorname{lev}_{\varphi,\leq}(t)\ \forall b \in B :\ \varphi(x-b) \leq t,\\
&\iff \text{(b) holds.}
\end{aligned}$$

□

Lemma 3.7.3 yields:

**Lemma 3.7.4** *Let $X$ be a linear space, $B \subseteq X$ and $\varphi\colon X \to \overline{\mathbb{R}}$ be $B$-monotone. Then*

$$\begin{aligned}
\operatorname{lev}_{\varphi,=}(-\infty) - B &\subseteq \operatorname{lev}_{\varphi,=}(-\infty),\\
\operatorname{lev}_{\varphi,=}(+\infty) + B &\subseteq \operatorname{lev}_{\varphi,=}(+\infty).
\end{aligned}$$

**Corollary 3.7.5** *Let $X$ be a linear space, $B \subseteq X$ with $B \setminus \{0\} \neq \emptyset$.*
*If $\varphi\colon X \to \overline{\mathbb{R}}$ is strictly $B$-monotone, then it is finite-valued.*

***Proof*** There exists $b \in B \setminus \{0\}$. Assume, contrary to our claim, that $\varphi$ is not finite-valued. Then there is $x \in X$ with $\varphi(x) \notin \mathbb{R}$. Suppose $\varphi(x) = -\infty$. Then $x - (x - b) = b \in B$. The strict $B$-monotonicity implies $\varphi(x - b) < \varphi(x)$, which contradicts $\varphi(x) = -\infty$. Hence, $\varphi(x) = +\infty$. Then $(x + b) - x = b \in B$. The strict $B$-monotonicity implies $\varphi(x) < \varphi(x + b)$, which contradicts $\varphi(x) = +\infty$. Consequently, $\varphi$ is finite-valued. □

The next lemma follows immediately from the definition of strict monotonicity.

**Lemma 3.7.6** *Let $X$ be a linear space, $B \subseteq X$ with $B \setminus \{0\} \neq \emptyset$ and $\varphi : X \to \mathbb{R}$. The following properties are equivalent to each other.*

(a) *$\varphi$ is strictly $B$-monotone.*
(b) $\mathrm{lev}_{\varphi,\leq}(t) - (B \setminus \{0\}) \subseteq \mathrm{lev}_{\varphi,<}(t)$ *for all $t \in \mathbb{R}$.*
(c) $\mathrm{lev}_{\varphi,\geq}(t) + (B \setminus \{0\}) \subseteq \mathrm{lev}_{\varphi,>}(t)$ *for all $t \in \mathbb{R}$.*

We are mainly interested in monotonicity on the effective domain.

**Proposition 3.7.7** *Let $X$ be a linear space, $B \subseteq X$ and $\varphi : X \to \overline{\mathbb{R}}$. $\varphi$ is $B$-monotone if and only if the following conditions hold:*

(a) *$\varphi$ is $B$-monotone on* $\mathrm{dom}\,\varphi$,
(b) $\mathrm{dom}\,\varphi - B \subseteq \mathrm{dom}\,\varphi$.

***Proof*** Assume first that $\varphi$ is B-monotone. Take any $x \in \mathrm{dom}\,\varphi$ and $b \in B$. Then $x - (x - b) = b \in B$. The $B$-monotonicity implies $\varphi(x - b) \leq \varphi(x) < +\infty$. Thus, $x - b \in \mathrm{dom}\,\varphi$. This yields (b).

Assume now that (a) and (b) hold. Suppose, contrary to the claim, that $\varphi$ is not $B$-monotone. Then there exist $x^1, x^2 \in X$ with $x^2 - x^1 \in B$ and $\varphi(x^1) > \varphi(x^2)$. Hence, $x^2 \in \mathrm{dom}\,\varphi$ and $x^1 \in x^2 - B \subseteq \mathrm{dom}\,\varphi - B \subseteq \mathrm{dom}\,\varphi$ because of (b). This contradicts (a). □

**Lemma 3.7.8** *Let $X$ be a linear space, $X_0, B \subseteq X$ and $\varphi : X \to \overline{\mathbb{R}}$. If $\varphi$ is $B$-monotone or strictly $B$-monotone on $X_0$, then*

(a) *$\lambda\varphi$ with $\lambda \in \mathbb{R}_>$ has the same property on $X_0$,*
(b) *$\varphi + c$ with $c \in \mathbb{R}$ has the same property on $X_0$,*
(c) *$g : X \to \overline{\mathbb{R}}$ defined by $g(x) := \varphi(x - x^0)$ with $x^0 \in X$ has the same property on $X_0 + x^0$,*
(d) *$g : X \to \overline{\mathbb{R}}$ defined by $g(x) := \varphi(\lambda x)$ with $\lambda \in \mathbb{R} \setminus \{0\}$ has the same property on $\frac{1}{\lambda}X_0$ if $\lambda B \subseteq B$,*
(e) *$g : X \to \overline{\mathbb{R}}$ defined by $g(x) := -\varphi(-x)$ has the same property on $-X_0$.*

Monotonicity is related to the conjugate and to the subdifferential of the function.

**Proposition 3.7.9** *Suppose a dual pair $(X, X', \langle .,. \rangle)$, $\varphi : X \to \overline{\mathbb{R}}$ and that $B$ is a nonempty subset of $X$.*

(a) *If $\varphi$ is $B$-monotone, then $\partial\varphi(x) \subseteq B^+$ for all $x \in X$.*
(b) *If $\varphi$ is $B$-monotone, $B$ is a cone and* $\mathrm{dom}\,\varphi \neq \emptyset$, *then* $\mathrm{dom}\,\varphi^* \subseteq B^+$.
(c) *If $X$ is equipped with the weak topology $\sigma(X, X')$, $X'$ is equipped with the weak* topology $\sigma(X', X)$ and $\varphi$ is lower semicontinuous, convex and proper with* $\mathrm{dom}\,\varphi^* \subseteq B^+$, *then $\varphi$ is $B$-monotone.*
(d) *If $X$ is equipped with the weak topology $\sigma(X, X')$, $X'$ is equipped with the weak* topology $\sigma(X', X)$, $B$ is a cone and $\varphi$ is $B$-monotone with* $\mathrm{dom}\,\varphi \neq \emptyset$, *then $\varphi^{**}$ is $B$-monotone.*

(e) *If* $\varphi$ *is strictly* $B$*-monotone, then* $\partial\varphi(x) \subseteq \operatorname{qint} B^+$ *for all* $x \in X$.
(f) *If* $\varphi$ *is strictly* $B$*-monotone on* $\operatorname{dom}\varphi$ *and* $\operatorname{dom}\varphi - B \subseteq \operatorname{dom}\varphi$*, then* $\partial\varphi(x) \subseteq \operatorname{qint} B^+$ *for all* $x \in X$.

***Proof***

(a) Take any $x \in X$. If $\partial\varphi(x) = \emptyset$, then the assertion is fulfilled. Assume now $\partial\varphi(x) \neq \emptyset$. Take any $x' \in \partial\varphi(x)$. For each $b \in B$, the $B$-monotonicity of $\varphi$ implies $\varphi(x - b) \leq \varphi(x)$, and $\langle -b, x'\rangle \leq \varphi(x - b) - \varphi(x) \leq 0$ by (3.6.4). Hence, $x' \in B^+$.
(b) If $B^+ = X'$, then $\operatorname{dom}\varphi^* \subseteq B^+$. Assume now $B^+ \neq X'$. Take any $x' \in X' \setminus B^+$. Then there exists $b \in B$ with $\langle b, x'\rangle < 0$. Choose any $x^0 \in \operatorname{dom}\varphi$ and $t \in \mathbb{R}_>$. The $B$-monotonicity of $\varphi$ implies $\varphi(x^0) \geq \varphi(x^0 - tb)$. Hence, $\varphi^*(x') \geq \langle x^0 - tb, x'\rangle - \varphi(x^0 - tb) \geq \langle x^0, x'\rangle - t\langle b, x'\rangle - \varphi(x^0)$. Since this inequality holds for each $t \in \mathbb{R}_>$, we obtain $\varphi^*(x') = +\infty$. Consequently, $\operatorname{dom}\varphi^* \subseteq B^+$.
(c) For all $x \in X$, we get from Lemma 3.6.7:

$$\varphi(x) = \sup\left\{\langle x, x'\rangle - \varphi^*(x') \mid x' \in X'\right\} = \sup\left\{\langle x, x'\rangle - \varphi^*(x') \mid x' \in \operatorname{dom}\varphi^*\right\}.$$

$\operatorname{dom}\varphi^* \subseteq B^+ \subseteq X'$ implies $\varphi(x) = \sup\left\{\langle x, x'\rangle - \varphi^*(x') \mid x' \in B^+\right\}$ for all $x \in X$. Take any $x^1, x^2 \in X$ with $x^2 - x^1 \in B$. For each $x' \in B^+$, we get $\langle x^2, x'\rangle \geq \langle x^1, x'\rangle$ and hence $\varphi(x^2) \geq \varphi(x^1)$.
(d) If $\varphi^{**}$ is not proper, then $\varphi^{**} = -\infty$ because of Lemma 3.6.8, and $\varphi^{**}$ is $B$-monotone. Assume now that $\varphi^{**}$ is proper. Because of (b) and Lemma 3.6.6(h), $\operatorname{dom}(\varphi^{**})^* = \operatorname{dom}\varphi^* \subseteq B^+$. Apply (c) to $\varphi^{**}$ instead of $\varphi$.
(e) Take any $x \in X$. If $\partial\varphi(x) = \emptyset$, then the assertion is fulfilled. Assume now $\partial\varphi(x) \neq \emptyset$. Take any $x' \in \partial\varphi(x)$. Then $\varphi(x) \in \mathbb{R}$. For each $b \in B \setminus \{0\}$, the strict $B$-monotonicity of $\varphi$ implies $\varphi(x - b) < \varphi(x)$, and $\langle -b, x'\rangle \leq \varphi(x - b) - \varphi(x) < 0$. Hence, $x' \in \operatorname{qint} B^+$.
(f) can be proved analogously to (e). □

**Remark 3.7.10** For the dual pair $(X, X^*, \langle ., .\rangle)$ and weakly closed convex cones $B$, Proposition 3.7.9(a)–(c) were proved by Filipović and Kupper [162], where the statement for the subdifferential was given under the additional assumption $\operatorname{dom}\varphi \neq \emptyset$ for $x \in \operatorname{dom}(-\varphi)$ only.

In Sect. 4.9, we will construct monotone envelopes of arbitrary extended real-valued functions. The inf-convolution will turn out to be a useful tool for this investigation.

**Lemma 3.7.11** *Let* $X$ *be a linear space,* $B$ *a nonempty subset of* $X$ *and* $\varphi, g\colon X \to \overline{\mathbb{R}}$.

*If* $g$ *is* $B$*-monotone, then* $\varphi \# g$ *is* $B$*-monotone.*

***Proof*** For each $x^0 \in X$ and $b \in B$,

$$\begin{aligned}(\varphi \# g)(x^0 + b) &= \inf\{\varphi(x^1) + g(x^0 + b - x^1) \mid x^1 \in X\}\\ &\geq \inf\{\varphi(x^1) + g(x^0 - x^1) \mid x^1 \in X\} = (\varphi \# g)(x^0).\end{aligned}$$ □

# Chapter 4
# Translation Invariant Functions

In this chapter, we introduce the functions on which our book focuses, discuss their properties and prove separation theorems for not necessarily convex sets in linear spaces. Ways to construct translation invariant functions in general as well as with desired properties will be pointed out and ways to approximate translation invariant functions by convex or concave functions. Relationships will be proved between translation invariant functions on the one hand, and Minkowski functionals and norms on the other hand. Moreover, we will derive formulas for translation invariant extensions as well as for translation invariant envelopes of arbitrary extended real-valued functions.

The results from this chapter are widely used in optimization and functional analysis, inside proofs or immediately in applications. This will be underscored by the subsequent chapters. Assumptions, definitions and statements related to translation invariant functions have immediate practical interpretations in mathematical finance as well as in consumer and production theory. Such interpretations are given throughout the chapter. Furthermore, applications in functional analysis and vector optimization will be indicated.

We investigate functions $\varphi_{A,k}$ on a real vector space $Y$ based on the formula

$$\varphi_{A,k}(y) := \inf\{t \in \mathbb{R} \mid y \in A + tk\},$$

where $A$ is a subset of $Y$ and $k \in Y \setminus \{0\}$. The class of these functions coincides with the class of translation invariant functions.

Researchers from different fields in mathematics and economic theory have applied functions of type $\varphi_{A,k}$ since these functions are appropriate for separating nonconvex sets and for the scalarization of relations like partial orders or preference relations. For detailed bibliographical notes related to the history of the functions under study, see Sect. 4.15.

Depending on the choice of $A$ and $k$, $\varphi_{A,k}$ can be real-valued, can attain the value $-\infty$ or can have an effective domain which is not the entire space. Details of extended



C. Tammer and P. Weidner, *Scalarization and Separation by Translation Invariant Functions*, Vector Optimization, https://doi.org/10.1007/978-3-030-44723-6_4

real-valued functions were explained in Chap. 3. For the application of this approach to $\varphi_{A,k}$, we have to keep in mind that $\inf \emptyset := +\infty$ and that we apply the inf-addition $(+\infty) + (-\infty) := (-\infty) + (+\infty) := +\infty$.

We will start our investigation in Sect. 4.1 with functions for which the sublevel sets are linear shifts of a set $A$ in the directions $k$ and $-k$. The class of these functions coincides with the class of translation invariant functions and with the class of functions $\varphi_{A,k}$ (Theorem 4.2.3). In Sect. 4.1, this relationship is discussed, further equivalent conditions for translation invariance are proved, and some examples of translation invariant functions are given.

In the sections to follow, translation invariant functions will be studied using their representation $\varphi_{A,k}$. Section 4.2 contains the fundamental theorems on arbitrary linear spaces. We will always try to find conditions that are sufficient and necessary for certain properties of $\varphi_{A,k}$, e.g. for continuity, convexity, sublinearity, strict quasiconcavity or monotonicity. Assumptions which are formulated using the recession cone $0^+A$ of $A$ often turn out to be equivalent to usual assumptions in production theory like the free disposal assumption and the strong free disposal assumption. Several statements connect $\varphi_{A,k}$ with the sublinear function $\varphi_{0^+A,k}$ (Corollary 4.2.8, Theorems 4.2.9 and 4.6.14, Proposition 4.10.10). Propositions 4.3.10 and 4.3.11 will give us a tool for transferring results which are related to the minimization of translation invariant functions to the maximization of such functions. This is of interest, e.g., for strictly quasiconcave utility functions.

Calculation rules for translation invariant functions will be proved in Sect. 4.3. This includes the investigation of functions $\varphi_{A,k}$ with varying $A$ and $k$. Interdependencies between $\varphi_{A,k}$ and the functions $\varphi_{A,\lambda k}$, $\varphi_{A+ck,k}$, $\varphi_{y^0+A,k}$ and $\varphi_{-A,-k}$ are often used in proofs and applications.

Section 4.5 deals with the monotonicity of $\varphi_{A,k}$ in the framework of scalarizing binary relations. Lower semicontinuous and continuous translation invariant functions on topological vector spaces are investigated in Sect. 4.4 and 4.6, respectively. Section 4.7 focuses on convex translation invariant functions, including statements for sublinear functionals.

Section 4.8 contains formulas for functions $\varphi_{A,k}$ with a set $A$ that is defined by linear inequalities. There, $A$ can be defined by a linear inequality system or by the union of sets satisfying some linear inequality.

Formulas for translation invariant and for monotone envelopes of arbitrary extended real-valued functions are derived in Sect. 4.9. The lower semicontinuous convex envelope of a translation invariant function is described in Theorem 4.10.7.

In Sect. 4.10, the Fenchel conjugate and the Fenchel subdifferential of $\varphi_{A,k}$ are given and further properties of translation invariant functions are proved which use dual cones and the mentioned conjugate and subdifferential.

In Sect. 4.11, we formulate separation theorems for use in the subsequent chapters. In Sect. 4.12, $\varphi_{A,k}$ is represented as the maximum or minimum of functions which turn out to be continuous, convex or concave under appropriate assumptions.

Section 4.13 contains relationships of $\varphi_{A,k}$ to Minkowski functionals and norms for the case where $A$ is a convex cone. In Proposition 4.13.1, we describe the Minkowski functional of $A + k$ by $\varphi_{A,k}$. The functional $\varphi_{A,k}$ coincides with the norm generated by an order interval on a part of the space. Another norm will be defined using $\varphi_{A,k}$ in Proposition 4.13.5.

In Sect. 4.14, each extended real-valued functional is characterized as the restriction of some translation invariant function. In this way, properties of arbitrary functionals are closely related to properties of translation invariant functions. This fact will be illustrated for the inf-convolution.

Detailed references are given at the end of each section. Section 4.15 contains bibliographical notes which refer to the history of translation invariant functions.

## 4.1 Translation Invariance, Uniform Sublevel Sets, and Functions $\varphi_{A,k}$

The purpose of scalarization is separation by values. A functional $\varphi$ separates two sets $V$ and $W$ in a space $Y$ if there exists $t \in \mathbb{R}$ such that one of the sets is contained in $M := \{y \in Y \mid \varphi(y) \leq t\}$, the other is contained in $\{y \in Y \mid \varphi(y) \geq t\}$ and $V \cup W \not\subseteq \{y \in Y \mid \varphi(y) = t\}$. Disjoint convex sets in a finite-dimensional vector space can be separated by some linear functional $\varphi$. In this case, $M = \tilde{A} + tk$ for some halfspace $\tilde{A}$ and some $k \in Y$. Being interested in nonconvex sets, we use this idea and investigate functionals $\varphi$ on a linear space $Y$ which fulfill the condition

$$\varphi(y) \leq t \iff y \in \tilde{A} + tk.$$

Here $\tilde{A}$ is assumed to be some subset of $Y$. Obviously, $\tilde{A} = \mathrm{lev}_{\varphi,\leq}(0)$.

**Proposition 4.1.1** *Consider a function $\varphi\colon Y \to \overline{\mathbb{R}}$ on some linear space $Y$, $k \in Y \setminus \{0\}$, and define $\tilde{A} := \mathrm{lev}_{\varphi,\leq}(0)$.*

*The following conditions are equivalent to each other:*

$$\mathrm{lev}_{\varphi,\leq}(t) = \tilde{A} + tk \quad \text{for all } t \in \mathbb{R}, \tag{4.1.1}$$

$$\mathrm{epi}\,\varphi = \{(y,t) \in Y \times \mathbb{R} \mid y \in \tilde{A} + tk\}, \tag{4.1.2}$$

$$\varphi(y + tk) = \varphi(y) + t \text{ for all } y \in Y,\ t \in \mathbb{R}, \tag{4.1.3}$$

$$\varphi(y) = \inf\{t \in \mathbb{R} \mid y \in \tilde{A} + tk\} \quad \text{for all } y \in Y. \tag{4.1.4}$$

*Each of them implies*

(a) $\mathrm{dom}\,\varphi = \tilde{A} + \mathbb{R}k$,
(b) $k \in -0^+(\mathrm{lev}_{\varphi,\leq}(t))$ *for each* $t \in \mathbb{R}$,
(c) $\mathrm{lev}_{\varphi,\leq}(t)$ *is $k$-directionally closed for each* $t \in \mathbb{R}$.

***Proof*** Obviously, each of the conditions implies $\mathrm{dom}\,\varphi = \tilde{A} + \mathbb{R}k$.

The equivalence between (4.1.1) and (4.1.2) is straightforward.

Assume now first that (4.1.1) is satisfied. Take any $y \in Y$. If $\varphi(y) = +\infty$, then $y \notin \tilde{A} + \mathbb{R}k$. Hence, $\varphi(y) = \inf\{t \in \mathbb{R} \mid y \in \tilde{A} + tk\}$ since $\inf \emptyset = +\infty$. If $\varphi(y) \in$

$\mathbb{R} \cup \{-\infty\}$, then $\varphi(y) = \inf\{t \in \mathbb{R} \mid \varphi(y) \le t\}$, hence $\varphi(y) = \inf\{t \in \mathbb{R} \mid y \in \tilde{A} + tk\}$ by (4.1.1). Thus, (4.1.4) holds.

Equation (4.1.4) implies, for each $\lambda \in \mathbb{R}$, $\varphi(y + \lambda k) = \inf\{t \in \mathbb{R} \mid y + \lambda k \in \tilde{A} + tk\} = \inf\{t \in \mathbb{R} \mid y \in \tilde{A} + (t - \lambda)k\} = \varphi(y) + \lambda$, and thus (4.1.3).

Assume now that (4.1.3) holds. For all $t \in \mathbb{R}$, we have:

$$\begin{aligned} \tilde{A} + tk &= \{y + tk \mid \varphi(y) \le 0,\ y \in Y\} \\ &= \{y \in Y \mid \varphi(y - tk) \le 0\} \\ &= \{y \in Y \mid \varphi(y) - t \le 0\} \\ &= \{y \in Y \mid \varphi(y) \le t\}. \end{aligned}$$

Hence, (4.1.1) is fulfilled.

Assume now (4.1.1).

For any $t \in \mathbb{R}_+$ and $y \in \tilde{A} - tk$, we get $\varphi(y) \le -t \le 0$ and thus $y \in \tilde{A}$. Hence, $k \in -0^+\tilde{A}$ and $k \in -0^+(\mathrm{lev}_{\varphi,\le}(t))$ for each $t \in \mathbb{R}$.

If $y \in Y$ and $y - t_n k \in \tilde{A}$ for some sequence $(t_n)_{n\in\mathbb{N}}$ with $t_n \in \mathbb{R}$ and $t_n \searrow 0$, then $\varphi(y) \le t_n$ for all $n \in \mathbb{N}$, and thus $\varphi(y) \le 0$, which implies $y \in \tilde{A}$. Hence, $\tilde{A}$ is $k$-directionally closed, and $\mathrm{lev}_{\varphi,\le}(t)$ is $k$-directionally closed for each $t \in \mathbb{R}$. □

Note that $k \in -0^+\tilde{A}$ is equivalent to $\tilde{A} = \tilde{A} - \mathbb{R}_+k$ and implies $\mathrm{int}\tilde{A} = \mathrm{int}\tilde{A} - \mathbb{R}_+k$.

$\varphi$ is said to be **translation invariant along** $k$ if property (4.1.3) is satisfied. It is said to be **translation invariant** if it is translation invariant along some $k \in Y \setminus \{0\}$. By Proposition 4.1.1, translation invariant functions are functions with uniform sublevel sets which can be generated by a linear shift of some set in a specified direction. In our book, we focus on this type of functions. Arbitrary functions with uniform sublevel sets were studied in [488]. An important subclass of the class of translation invariant functions consists of the **topical** functions, which have many applications in applied mathematics, especially when modeling discrete event systems (see [201, 202, 397]). A function $\varphi \colon \mathbb{R}^n \to \mathbb{R}^\ell$ is said to be topical if it is $\mathbb{R}^\ell_+$-monotone and translation invariant along $k = \mathbf{1}$.

The relationship between translation invariant functions $\varphi$ and sets $\mathrm{lev}_{\varphi,\le}(0)$ is of interest in mathematical economics, where risk measures are translation invariant functions and $\mathrm{lev}_{\varphi,\le}(0)$ turns out to be the acceptance set related to a risk measure $\varphi$ [11]. It is studied more in detail in Theorem 4.2.4.

For each function $\varphi \colon Y \to \overline{\mathbb{R}}$ on a nonempty set $Y$,

$$\varphi(y) = \inf\{t \in \mathbb{R} \mid y \in \mathrm{lev}_{\varphi,\le}(t)\}.$$

Based on this fact, Crouzeix [105] investigated functions on an arbitrary nonempty set $Y$ which are defined via a family $\{S_t \subseteq Y \mid t \in \mathbb{R}\}$ of sets by

$$\varphi(y) := \inf\{t \in \mathbb{R} \mid y \in S_t\}.$$

Martinez-Legaz [346] proved results for such functions on $\mathbb{R}^\ell$ and applied them to a linear control process. Drapeau and Kupper [119] defined risk measures in this way.

Further equivalent conditions for translation invariance are proved in the next proposition.

**Proposition 4.1.2** *Assume that $Y$ is a linear space, $k \in Y \setminus \{0\}$ and $\varphi\colon Y \to \overline{\mathbb{R}}$. Then the following conditions are equivalent to each other:*

$$\forall y \in Y,\ t \in \mathbb{R}\colon\ \varphi(y + tk) = \varphi(y) + t, \tag{4.1.5}$$

$$\forall y \in Y,\ t \in \mathbb{R}\colon\ \varphi(y + tk) \leq \varphi(y) + t, \tag{4.1.6}$$

$$\forall y \in Y,\ t \in \mathbb{R}\colon\ \varphi(y + tk) \geq \varphi(y) + t, \tag{4.1.7}$$

$$\{\lambda(k, 1) \in Y \times \mathbb{R} \mid \lambda \in \mathbb{R}\} \subseteq 0^+(\operatorname{epi}\varphi). \tag{4.1.8}$$

***Proof*** Obviously, (4.1.5) implies (4.1.6). Assume now that (4.1.6) holds. Take any $y \in Y$ and $t \in \mathbb{R}$. Then $\varphi(y) = \varphi((y + tk) - tk) \leq \varphi(y + tk) - t$. Thus, $\varphi(y + tk) \geq \varphi(y) + t$. By (4.1.6), $\varphi(y + tk) = \varphi(y) + t$. Hence, (4.1.5) is fulfilled.

The equivalence between (4.1.5) and (4.1.7) can be proved analogously.

Assume now that (4.1.5) holds. Take any $(y, t) \in \operatorname{epi}\varphi$ and $\lambda \in \mathbb{R}$. Then $\varphi(y) \leq t$ implies $\varphi(y + \lambda k) = \varphi(y) + \lambda \leq t + \lambda$. Hence, $(y, t) + \lambda(k, 1) \in \operatorname{epi}\varphi$. Thus, (4.1.8) is satisfied.

Assume now (4.1.8). Take first any $y \in Y$ with $\varphi(y) \in \mathbb{R}$. Then $(y, \varphi(y)) \in \operatorname{epi}\varphi$. This implies $(y + tk, \varphi(y) + t) \in \operatorname{epi}\varphi$ for each $t \in \mathbb{R}$ by (4.1.8). Hence, $\varphi(y + tk) \leq \varphi(y) + t$ holds for all $t \in \mathbb{R}$. Take now any $y \in Y$ with $\varphi(y) = -\infty$. Then $(y, \lambda) \in \operatorname{epi}\varphi$ for each $\lambda \in \mathbb{R}$. This yields $(y + tk, \lambda + t) \in \operatorname{epi}\varphi$ for each $t, \lambda \in \mathbb{R}$. Thus, $\varphi(y + tk) \leq \varphi(y) + t$ holds for all $t \in \mathbb{R}$, where $\varphi(y + tk) = -\infty$ for each $t \in \mathbb{R}$. For each $y \in Y$ with $\varphi(y) = +\infty$, we have $\varphi(y + tk) \leq \varphi(y) + t$ for all $t \in \mathbb{R}$. Hence, (4.1.6) is fulfilled. □

Replacing the right-hand side in (4.1.5) by $\varphi(y) + \xi(t)$ with some continuous function $\xi\colon \mathbb{R} \to \mathbb{R}$ does not extend the class of translation invariant functions which attain at least one real value. This was proved in [488, Prop. 5.6.1].

We throw a glance to the translation invariance of subadditive functions.

**Proposition 4.1.3** *Assume that $Y$ is a linear space with $k \in Y \setminus \{0\}$ and that $\varphi\colon Y \to \overline{\mathbb{R}}$ is a subadditive function with $\varphi(0) = 0$. Then $\varphi$ is translation invariant along $k$ if and only if*

$$\varphi(tk) = t \quad \textit{for each } t \in \mathbb{R}. \tag{4.1.9}$$

*If $\varphi$ is sublinear, then* (4.1.9) *is equivalent to*

$$\varphi(k) = 1 \textit{ and } \varphi(-k) = -1. \tag{4.1.10}$$

***Proof*** Equation (4.1.9) follows from $\varphi(0) = 0$ and the translation invariance of $\varphi$ along $k$.

Assume now (4.1.9). Since $\varphi$ is subadditive, Theorem 3.5.11 yields $\varphi(y+tk) \leq \varphi(y)+\varphi(tk) = \varphi(y)+t$ for any $y \in Y$ and $t \in \mathbb{R}$. Then $\varphi$ is translation invariant along $k$ because of Proposition 4.1.2.

If $\varphi$ is sublinear, then it is positively homogeneous. In this case, (4.1.10) implies $\varphi(tk) = t\varphi(k) = t$ and $\varphi(-tk) = t\varphi(-k) = -t$ for each $t \in \mathbb{R}_+$. This yields the assertion. □

In Proposition 4.10.2, we will prove that condition (4.1.9) is equivalent to translation invariance along $k$ for each lower semicontinuous convex proper function $\varphi$ with $\varphi(0) = 0$.

Because of Proposition 4.1.1, each function with uniform sublevel sets given by (4.1.1) and each translation invariant function can be constructed by the following formula.

**Definition 4.1.4** Assume

$$Y \text{ is a linear space, } A \subseteq Y \text{ and } k \in Y \setminus \{0\}. \qquad (\mathrm{Sp0}_{A,k})$$

The function $\varphi_{A,k}\colon Y \to \overline{\mathbb{R}}$ is defined by

$$\varphi_{A,k}(y) := \inf\{t \in \mathbb{R} \mid y \in A+tk\}. \qquad (4.1.11)$$

We get an immediate geometric interpretation of $\varphi_{A,k}$ since $A+tk$ is the set $A$ shifted by $tk$.

Note that $\mathrm{lev}_{\varphi_{A,k},\leq}(0)$ can be different from $A$ even if $k \in -0^+A$.

**Example 4.1.5** In $Y = \mathbb{R}^2$, take $k := (1,0)^{\mathrm{T}}$ and $A := \{(y_1, y_2)^{\mathrm{T}} \in \mathbb{R}^2 \mid y_1 < 0\}$.

Then $k \in -0^+A$ is satisfied. We have $\varphi_{A,k}((0,0)^{\mathrm{T}}) = 0$ even though $(0,0)^{\mathrm{T}} \notin A$. Thus, $\mathrm{lev}_{\varphi_{A,k},\leq}(0) \neq A$.

Moreover, $A$ and k cannot be used to define a functional $\varphi$ by the condition

$$\mathrm{lev}_{\varphi,\leq}(t) = A+tk \quad \text{for all } t \in \mathbb{R},$$

since this would imply $\varphi((0,0)^{\mathrm{T}}) = 0$ because of $\varphi((0,0)^{\mathrm{T}}) \leq t$ for all $t > 0$, however $(0,0)^{\mathrm{T}} \notin A$.

Nevertheless, we will prove in Theorem 4.2.3 that the class of functionals $\varphi_{A,k}$ coincides with the class of translation invariant functions and with the class of functions having uniform sublevel sets described by (4.1.1). In order to underscore the importance of this function class, we conclude this section with some examples.

**Example 4.1.6** Assume that $Y$ is the vector space of functions $f\colon X \to \mathbb{R}$, where $X$ is some nonempty set. Choose some positive function $k \in Y$ and

$$A := \{f \in Y \mid \forall x \in X\colon\ f(x) \leq 0\}.$$

For each $f \in Y$, we get

$$\begin{aligned}
\varphi_{A,k}(f) &= \inf\{t \in \mathbb{R} \mid f \in A + tk\} \\
&= \inf\{t \in \mathbb{R} \mid f - tk \in A\} \\
&= \inf\{t \in \mathbb{R} \mid \forall x \in X\colon\ f(x) - tk(x) \le 0\} \\
&= \inf\{t \in \mathbb{R} \mid \forall x \in X\colon\ \frac{f(x)}{k(x)} \le t\} \\
&= \sup_{x \in X} \frac{f(x)}{k(x)}, \text{ where} \\
\operatorname{dom} \varphi_{A,k} &= \{f \in Y \mid \sup_{x \in X} \frac{f(x)}{k(x)} \in \mathbb{R}\}.
\end{aligned}$$

If $k \in Y$ is the function with the constant value 1 at each $x \in X$, then we have

$$\begin{aligned}
\varphi_{A,k}(f) &= \sup_{x \in X} f(x) \text{ for each } f \in Y, \text{ where} \\
\operatorname{dom} \varphi_{A,k} &= \{f \in Y \mid \sup_{x \in X} f(x) \in \mathbb{R}\}.
\end{aligned}$$

If $Y$ is defined as the set of continuous functions $f\colon X \to \mathbb{R}$ on a compact topological space $X$, then inf and sup in Example 4.1.6 can be replaced by min and max, respectively. The formulas for this case were given by Rubinov and Gasimov [394, Example 4.1].

Interpreting $A$ as a set of acceptable vectors, a positive real value $\varphi_{A,k}(y)$ expresses the smallest number of units $-k$ which have to be added to $y$ in order to reach $A$, while the absolute value of a negative real value $\varphi_{A,k}(y)$ expresses the largest number of units $-k$ which can be subtracted from $y$ without leaving $A$. This interpretation was given by Artzner et al. [11], who considered functions $\varphi_{A,k}$ as risk measures. There, $Y$ is the set of random future values of investments, given as real-valued functions. The set $A$ consists of the final net worths which are accepted by all supervisors. $-k$ stands for a commonly accepted unit of payment. If $y$ is not an accepted future value, then $\varphi_{A,k}(y)$ is the number of units $-k$ which have to be paid extra in order to make the investment acceptable. If $y$ is an acceptable future value, then $-\varphi_{A,k}(y)$ units of $-k$ can be withdrawn from the position in the future.

The next example enumerates some of the risk measures which were investigated in [391].

**Example 4.1.7** Consider $Y = L^2(\Omega, M, P)$ with some probability measure $P$. Then the functions $\varphi\colon Y \to \overline{\mathbb{R}}$ defined in the following items are translation invariant along $k := -\mathbf{1} \in Y$. Here $E(X)$ and $\sigma(X)$ denote the expected value and the standard deviation of $X$, respectively.

(1) $\varphi(X) := \sup(-X)$ for each $X \in Y$.
Then $\varphi = \varphi_{A,k}$ for $A := \{X \in Y \mid X \ge 0\}$.

(2) $\varphi(X) := \lambda\sigma(X) - E(X)$ for each $X \in Y$, where $\lambda \in \mathbb{R}_+$.
Then $\varphi = \varphi_{A,k}$ for $A := \{X \in Y \mid E(X) \ge \lambda\sigma(X)\}$.

(3) $\varphi(X) := -\inf\{z \mid P(X \leq z) > \alpha\}$ for each $X \in Y$, where $\alpha \in (0, 1)$. Then $\varphi = \varphi_{A,k}$ for $A := \{X \in Y \mid P(X < 0) \leq \alpha\}$.

The risk measure in (3) calculates the quantiles of the functions. Artzner et al. [11, p. 216] proved that it is not subadditive and not convex.

Some applications of translation invariant functions in economics are contained in Sect. 15.1.

**Remark 4.1.8** (*Bibliographical notes related to Sect.* 4.1)

Jaschke and Küchler [280, p. 186] stated that (4.1.3) in Proposition 4.1.1 implies (4.1.4). References for the statement that (4.1.4) implies (4.1.3) will be given—related to (4.2.2)—in Remark 4.2.13. Hamel [221, Theorem 1] proved the equivalence of (4.1.1)–(4.1.3) in Proposition 4.1.1. For $Y = \mathbb{R}^\ell$ and $k = \mathbf{1}$, Rubinov and Singer [397, Lemma 4.1] had previously stated that (4.1.3) implies (4.1.1). Schrage [407, Satz 3.1.3] showed that (4.1.3) implies Proposition 4.1.1(b) and (c). In topological vector spaces, the relationship between functions introduced by (4.1.1) with some closed set $\tilde{A}$, functions of type (4.1.4) and functions with the property $\text{lev}_{\varphi,=}(t) = C + tk$ for some set $C$ and all $t \in \mathbb{R}$ was investigated by Weidner [470, 472].

For $Y = \mathbb{R}^\ell$ and $k = \mathbf{1}$, the equivalence between (4.1.5), (4.1.6) and (4.1.8) in Proposition 4.1.2 was proved in [397, Lemma 2.1, Prop. 3.1]. The equivalence between (4.1.5) and (4.1.6) was noted in [219] without proof, the equivalence between (4.1.5) and (4.1.7) in [222]. A proof of the equivalence between (4.1.5) and (4.1.8) was given in [221, Theorem 1].

Proposition 4.1.3 is based on an idea by Frittelli and Rosazza Gianin [170, p. 1485]. The history of Definition 4.1.4 will be described in Sect. 4.15.

## 4.2 Fundamental Theorems for Functions $\varphi_{A,k}$ on Linear Spaces

The direction-depending properties of sets which we studied in Sect. 2.3.2 turn out to be a helpful tool for the characterization of functions $\varphi_{A,k}$. When describing properties of a functional $\varphi_{A,k}$, the set $A - \mathbb{R}_+ k$ turns out to be of basic importance. This set can be considered to be the $(-k)$-directional extension of $A$. It coincides with $A$ if and only if $k \in -0^+ A$.

**Theorem 4.2.1** *Assume* $(\text{Sp0}_{A,k})$.

$$\text{dom}\,\varphi_{A,k} = A + \mathbb{R}k, \tag{4.2.1}$$

$$\varphi_{A,k}(y + tk) = \varphi_{A,k}(y) + t \quad \textit{for all } y \in Y,\ t \in \mathbb{R}, \tag{4.2.2}$$

$$\text{lev}_{\varphi_{A,k},<}(t) = \text{int}_k(A - \mathbb{R}_+ k) + tk = A - \mathbb{R}_> k + tk \quad \textit{for all } t \in \mathbb{R}, \tag{4.2.3}$$

$$\text{lev}_{\varphi_{A,k},\leq}(t) = \text{cl}_k(A - \mathbb{R}_+ k) + tk = \text{cl}_k(A) - \mathbb{R}_+ k + tk \quad \textit{for all } t \in \mathbb{R}, \tag{4.2.4}$$

$$\mathrm{lev}_{\varphi_{A,k},=}(t) = \mathrm{bd}_k(A - \mathbb{R}_+k) + tk \subseteq \mathrm{cl}_k(A) + tk \;\; \textit{for all } t \in \mathbb{R}, \tag{4.2.5}$$

$$\mathrm{epi}\, \varphi_{A,k} = \{(y,t) \in Y \times \mathbb{R} \mid y \in \mathrm{cl}_k(A - \mathbb{R}_+k) + tk\}. \tag{4.2.6}$$

$$\varphi_{A,k} = \varphi_{M,k} \;\; \textit{for each set } M \subseteq Y \textit{ with } A \subseteq M \subseteq \mathrm{cl}_k(A - \mathbb{R}_+k). \tag{4.2.7}$$

*Especially* $\varphi_{A,k} = \varphi_{\mathrm{cl}_k(A),k} = \varphi_{\mathrm{cl}_k(A-\mathbb{R}_+k),k} = \varphi_{A-\mathbb{R}_+k,k}$, (4.2.8)

$$\varphi_{A,k} = \varphi_{A-\mathbb{R}_>k,k}. \tag{4.2.9}$$

(a) *The condition*

$$\mathrm{lev}_{\varphi_{A,k},<}(t) \subseteq A + tk \;\; \textit{for all } t \in \mathbb{R} \tag{4.2.10}$$

*holds if and only if* $k \in -0^+A$.

(b) *We have*

$$A + tk \subseteq \mathrm{lev}_{\varphi_{A,k},\leq}(t) \;\; \textit{for all } t \in \mathbb{R}. \tag{4.2.11}$$

*The condition*

$$\mathrm{lev}_{\varphi_{A,k},\leq}(t) = A + tk \;\; \textit{for all } t \in \mathbb{R} \tag{4.2.12}$$

*is fulfilled if and only if* $k \in -0^+A$ *holds and* $A$ *is* $k$*-directionally closed.*

(c) *We have*

$$\mathrm{core}\, A + tk \subseteq \mathrm{lev}_{\varphi_{A,k},<}(t) \;\; \textit{for all } t \in \mathbb{R}. \tag{4.2.13}$$

*The following conditions are equivalent to each other:*

$$\mathrm{lev}_{\varphi_{A,k},<}(t) = \mathrm{core}\, A + tk \;\; \textit{for all } t \in \mathbb{R}, \tag{4.2.14}$$

$$A - \mathbb{R}_>k \subseteq \mathrm{core}\;\; A, \tag{4.2.15}$$

$$\mathrm{core}\, A = A - \mathbb{R}_>k. \tag{4.2.16}$$

*Condition* (4.2.15) *implies* $\mathrm{dom}\, \varphi_{A,k} = \mathrm{core}\, \mathrm{dom}\, \varphi_{A,k}$.

(d) *If* $k \in 0^+A$, *then* $\varphi_{A,k}(y) = -\infty$ *for all* $y \in \mathrm{dom}\, \varphi_{A,k}$.

(e) $\mathrm{bd}_k(A - \mathbb{R}_+k) + \mathbb{R}k = \{y \in Y \mid \varphi_{A,k}(y) \in \mathbb{R}\}$.

(f) *We have*

$$\varphi_{A,k}(y) = \min\{t \in \mathbb{R} \mid y \in \mathrm{cl}_k(A) + tk\}$$

*for each* $y \in Y$ *with* $\varphi_{A,k}(y) \in \mathbb{R}$, *i.e., for each* $y \in \mathrm{bd}_k(A - \mathbb{R}_+k) + \mathbb{R}k$.

(g) $\varphi_{Y\setminus A,-k} = \varphi_{Y\setminus \mathrm{int}_k(A),-k}$.

***Proof*** (4.2.1) is easy to verify.

(1) First, we prove (4.2.2). For each $y \in Y$ and $\lambda \in \mathbb{R}$, we obtain $\varphi_{A,k}(y + \lambda k) = \inf\{t \in \mathbb{R} \mid y + \lambda k \in A + tk\} = \inf\{t \in \mathbb{R} \mid y \in A + (t - \lambda)k\} = \inf\{u + \lambda \mid u \in \mathbb{R},\, y \in A + uk\} = \varphi_{A,k}(y) + \lambda$.

(2) We prove (4.2.3). Obviously, $\mathrm{lev}_{\varphi_{A,k},<}(t) = A - \mathbb{R}_>k + tk$ for all $t \in \mathbb{R}$. By Proposition 2.3.49, $\mathrm{int}_k(A - \mathbb{R}_+k) = A - \mathbb{R}_>k$.

(3) We will now prove (4.2.4), which implies (4.2.6). The second equation in (4.2.4) results from Proposition 2.3.49. Take any $t \in \mathbb{R}$.
Take first any $y \in Y$ with $\varphi_{A,k}(y) = t$ and $y \notin A + tk$. Then the definition of $\varphi_{A,k}$ implies the existence of a sequence $(t_n)_{n\in\mathbb{N}}$ of real numbers with $t_n \searrow t$ and $y \in A + t_n k$ for all $n \in \mathbb{N}$. This yields $y - tk - (t_n - t)k \in A$ for all $n \in \mathbb{N}$. Since $t_n - t \searrow 0$, we get $y - tk \in \operatorname{cl}_k(A)$ from Lemma 2.3.24. This implies $\operatorname{lev}_{\varphi_{A,k},\leq}(t) \subseteq \operatorname{cl}_k(A) - \mathbb{R}_+ k + tk$ by (4.2.3).
Take now any $y \in \operatorname{cl}_k(A) + tk$. Because of Lemma 2.3.24, there exists a sequence $(t_n)_{n\in\mathbb{N}}$ of nonnegative real numbers with $t_n \to 0$ and $y - tk - t_n k \in A$ for all $n \in \mathbb{N}$. Hence, $y \in A + (t + t_n)k$ for all $n \in \mathbb{N}$. This yields $\varphi_{A,k}(y) \leq t$, and $\operatorname{cl}_k(A) + tk \subseteq \operatorname{lev}_{\varphi_{A,k},\leq}(t)$. By Proposition 2.3.49, $\operatorname{cl}_k(A) - \mathbb{R}_> k = \operatorname{int}_k(A - \mathbb{R}_+ k)$. We obtain $\operatorname{cl}_k(A) - \mathbb{R}_+ k + tk \subseteq \operatorname{lev}_{\varphi_{A,k},\leq}(t)$ from (4.2.3).

(4) We prove (4.2.5). The first equation follows from (4.2.3) and (4.2.4).
Take any $y \in Y$ and $t \in \mathbb{R}$ with $\varphi_{A,k}(y) = t$. The definition of $\varphi_{A,k}$ implies the existence of a sequence $(t_n)_{n\in\mathbb{N}}$ with $t_n \in \mathbb{R}_+$, $t_n \to 0$ and $y \in A + (t + t_n)k$, i.e., $y - tk - t_n k \in A$. Hence, $y - tk \in \operatorname{cl}_k(A)$.

(5) We will now prove (4.2.7) and (4.2.8). Take any set $M \subseteq Y$ with $A \subseteq M \subseteq \operatorname{cl}_k(A - \mathbb{R}_+ k)$. We have $\operatorname{cl}_k(A - \mathbb{R}_+ k) \subseteq \operatorname{cl}_k(M - \mathbb{R}_+ k) \subseteq \operatorname{cl}_k(\operatorname{cl}_k(A - \mathbb{R}_+ k) - \mathbb{R}_+ k) = \operatorname{cl}_k(A - \mathbb{R}_+ k)$ by Lemma 2.3.26(b), Proposition 2.3.49(1) and Lemma 2.3.26(c). Hence, $\varphi_{A,k} = \varphi_{M,k}$ because of (4.2.4).

(6) By Proposition 2.3.49(a), $\operatorname{cl}_k(A - \mathbb{R}_> k) = \operatorname{cl}_k(A - \mathbb{R}_+ k)$. Equation (4.2.8) yields $\varphi_{A-\mathbb{R}_> k,k} = \varphi_{\operatorname{cl}_k(A-\mathbb{R}_> k),k} = \varphi_{\operatorname{cl}_k(A-\mathbb{R}_+ k),k} = \varphi_{A,k}$ and thus (4.2.9).

(a) First, assume that (4.2.10) is fulfilled. For each $y \in A - tk$ with $t \in \mathbb{R}_>$, we get $\varphi_{A,k}(y) \leq -t < 0$ and thus $y \in A$. Hence, $k \in -0^+ A$.
Conversely, assume $k \in -0^+ A$. Take any $y \in Y$ and $t \in \mathbb{R}$ with $\varphi_{A,k}(y) < t$. Then there exists $\lambda \in \mathbb{R}_>$ with $y \in A + (t - \lambda)k = A - \lambda k + tk$. Hence, $y \in A + tk$ because of $k \in -0^+ A$.

(b) Equation (4.2.11) results from (4.2.4).
We will now prove the second statement of (b). The reverse direction of the equivalence follows from (4.2.11), (a) and (4.2.5).
Assume now (4.2.12). Then $k \in -0^+ A$ by (a). If $y \in Y$ and $y - t_n k \in A$ for some sequence $(t_n)_{n\in\mathbb{N}}$ with $t_n \in \mathbb{R}$ and $t_n \searrow 0$, then $\varphi_{A,k}(y) \leq t_n$ for all $n \in \mathbb{N}$, and thus $\varphi_{A,k}(y) \leq 0$, which results in $y \in A$ by (4.2.12). Hence, $A$ is $k$-directionally closed.

(c) Inclusion (4.2.13) is easy to verify. The other statements of (c) are obvious if $\operatorname{core} A = \emptyset$. Suppose now $\operatorname{core} A \neq \emptyset$.
First, assume that (4.2.14) is satisfied. Take any $b \in A - \mathbb{R}_> k$. Then $\varphi_{A,k}(b) < 0$. We obtain $b \in \operatorname{core} A$ because of (4.2.14). Thus, (4.2.15) is fulfilled.
Conversely, assume that (4.2.15) holds. Then $A + \mathbb{R}k \subseteq \operatorname{core} A + \mathbb{R}k \subseteq \operatorname{core}(A + \mathbb{R}k)$ because of Lemma 2.3.1(b), i.e., $\operatorname{dom} \varphi_{A,k} \subseteq \operatorname{core} \operatorname{dom} \varphi_{A,k}$. Take any $t \in \mathbb{R}$ and $y \in Y$ with $\varphi_{A,k}(y) < t$. Then there exists $\lambda \in \mathbb{R}$ with $\lambda < t$ and $y \in A + \lambda k$. It follows that $y \in A + \lambda k = (A - (t - \lambda)k) + tk \subseteq \operatorname{core} A + tk$ by (4.2.15). This results in (4.2.14) by (4.2.13).
(4.2.15) is equivalent to (4.2.16) because of Lemma 2.3.17.

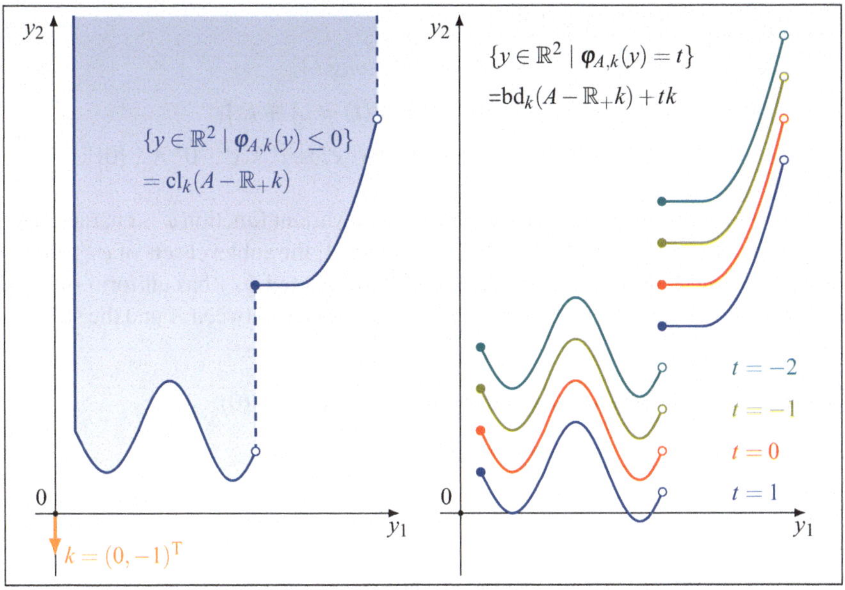


**Fig. 4.1** Sublevel sets and level sets of $\varphi_{A,k}$ for set $A$ from Fig. 2.5

(d) If $k \in 0^+A$, then $A + tk = A + (t - \lambda)k + \lambda k \subseteq A + (t - \lambda)k$ for all $t \in \mathbb{R}$, $\lambda \in \mathbb{R}_+$, and thus $\varphi_{A,k}(y) = -\infty$ for all $y \in \mathrm{dom}\, \varphi_{A,k}$.
(e) follows from (4.2.5).
(f) $\varphi_{A,k}(y) \in \mathbb{R}$ implies $\varphi_{A,k}(y) = \min\{t \in \mathbb{R} \mid \varphi_{A,k}(y) \leq t\} = \min\{t \in \mathbb{R} \mid y \in \mathrm{cl}_k(A) - \mathbb{R}_+k + tk\}$ by (4.2.4). Hence, $\varphi_{A,k}(y) = \min\{t \in \mathbb{R} \mid y \in \mathrm{cl}_k(A) + tk\}$.
(g) $\varphi_{Y \setminus A,-k} = \varphi_{\mathrm{cl}_{-k}(Y \setminus A),-k} = \varphi_{Y \setminus \mathrm{int}_k(A),-k}$ by (4.2.8) and Lemma 2.3.42(h). □

The sublevel sets and level sets of $\varphi_{A,k}$ are illustrated in Fig. 4.1.

Another essential property of $\varphi_{A,k}$, the representation as the inf-convolution of $A$'s indicator function with $\varphi_{\{0\},k}$, will be proved in Sect. 4.3 (Corollary 4.3.8).

By (4.2.5), $\varphi_{A,k}$ can be considered to be an oriented distance to $\mathrm{bd}_k(A - \mathbb{R}_+k)$.

Theorem 4.2.1(f) implies the following.

**Corollary 4.2.2** *Assume* ($\mathrm{Sp}0_{A,k}$) *and that* $\varphi_{A,k}$ *attains a minimal value* $t_0$ *on* $F \subset Y$. *Then*

$$\mathrm{argmin}_F\, \varphi_{A,k} = F \cap (\mathrm{cl}_k(A) + t_0k).$$

The class of functionals $\varphi_{A,k}$ turns out to be equivalent to the class of translation invariant functions. Proposition 4.1.1 results together with (4.2.2) and (4.2.8) in the following theorem.

**Theorem 4.2.3** *Assume that* $Y$ *is a linear space. For each* $k \in Y \setminus \{0\}$, *the following function classes coincide with each other:*

$$\{\varphi_{A,k}\colon Y \to \overline{\mathbb{R}} \mid A \subseteq Y\},$$
$$\{\varphi\colon Y \to \overline{\mathbb{R}} \mid \varphi \text{ translation invariant along } k\},$$
$$\{\varphi\colon Y \to \overline{\mathbb{R}} \mid \exists A \subseteq Y\ \forall t \in \mathbb{R}\colon\ \operatorname{lev}_{\varphi,\leq}(t) = A + tk\},$$
$$\{\varphi_{A,k}\colon Y \to \overline{\mathbb{R}} \mid A \subseteq Y \text{ is } k\text{-directionally closed}, k \in -0^+A \setminus \{0\}\}.$$

According to Proposition 4.1.1, each translation invariant function $\varphi$ is characterized by its zero-sublevel set $\operatorname{lev}_{\varphi,\leq}(0)$. By Theorem 4.2.1, the sublevel sets of $\varphi_{A,k}$ do not always have the shape of $A$. Nevertheless, each functional $\varphi_{A,k}$ has uniform sublevel sets. The following theorem studies the interdependence between $A$ and the sublevel sets of $\varphi_{A,k}$.

**Theorem 4.2.4** *Assume* $(\mathrm{Sp0}_{A,k})$. *Consider* $\tilde{A} := \operatorname{lev}_{\varphi_{A,k},\leq}(0)$.

(a) *$\tilde{A}$ is the unique set for which*

$$\operatorname{lev}_{\varphi_{A,k},\leq}(t) = \tilde{A} + tk \text{ for all } t \in \mathbb{R}. \tag{4.2.17}$$

(b) *$\tilde{A}$ is the unique set with the following properties:*

(i) *$\tilde{A}$ is $k$-directionally closed,*
(ii) *$k \in -0^+\tilde{A} \setminus \{0\}$, and*
(iii) *$\varphi_{A,k}$ coincides with $\varphi_{\tilde{A},k}$ on $Y$.*

(c) $\tilde{A} = \operatorname{cl}_k(A - \mathbb{R}_+k) = \{y \in Y \mid y - \mathbb{R}_>k \subseteq A - \mathbb{R}_+k\}$.

***Proof*** Consider first $\tilde{A} := \operatorname{lev}_{\varphi_{A,k},\leq}(0)$. Because of (4.2.2), Proposition 4.1.1 implies (4.2.17) and (i)–(iii). By (4.2.4), $\tilde{A} = \operatorname{cl}_k(A - \mathbb{R}_+k)$. Proposition 2.3.29 yields the second equation in (c). Obviously, $\tilde{A}$ is the unique set which can satisfy (4.2.17).

Assume now that (i)–(iii) hold for some set $\tilde{A} \subseteq Y$. (i) and (ii) imply $\tilde{A} = \operatorname{lev}_{\varphi_{\tilde{A},k},\leq}(0)$ by Theorem 4.2.1(b). Hence, $\tilde{A} = \operatorname{lev}_{\varphi_{A,k},\leq}(0)$ because of (iii). □

**Theorem 4.2.5** *Assume* $(\mathrm{Sp0}_{A,k})$ *and* $k \in -0^+(\operatorname{cl}_k(A))$. *Then*

$$\operatorname{lev}_{\varphi_{A,k},\leq}(t) = \operatorname{cl}_k(A) + tk \quad \text{for all } t \in \mathbb{R}, \tag{4.2.18}$$
$$\operatorname{epi}\varphi_{A,k} = \{(y,t) \in Y \times \mathbb{R} \mid y \in \operatorname{cl}_k(A) + tk\}, \tag{4.2.19}$$
$$\operatorname{dom}\varphi_{A,k} = A + \mathbb{R}_>k.$$

*Assume now* $k \in -0^+A$.

(a) $\forall t \in \mathbb{R}\colon\ \operatorname{lev}_{\varphi_{A,k},<}(t) = \operatorname{int}_k(A) + tk$.
(b) $\forall t \in \mathbb{R}\colon\ \operatorname{lev}_{\varphi_{A,k},=}(t) = \operatorname{bd}_k(A) + tk$.
(c) $\{y \in Y \mid \varphi_{A,k}(y) \in \mathbb{R}\} = \operatorname{bd}_k(A) + \mathbb{R}k$.
(d) $\forall y \in Y\colon\ (\varphi_{A,k}(y) = -\infty \iff y + \mathbb{R}k \subseteq A)$.
(e) $\varphi_{A,k}$ *is finite-valued on* $\operatorname{dom}\varphi_{A,k} \setminus A$.

(f) *$\varphi_{A,k}$ is proper if and only if $A$ is nonempty and $A$ does not contain lines parallel to $k$, i.e.,*

$$\forall a \in A\colon\ a + \mathbb{R}k \not\subseteq A. \tag{4.2.20}$$

*Condition* (4.2.20) *is equivalent to*

$$\forall y \in Y\colon\ y + \mathbb{R}k \not\subseteq A. \tag{4.2.21}$$

(g) *We have*

$$\varphi_{A,k}(y) = \min\{t \in \mathbb{R} \mid y \in \mathrm{bd}_k(A) + tk\}$$

*for each $y \in Y$ with $\varphi_{A,k}(y) \in \mathbb{R}$, i.e., for each $y \in \mathrm{bd}_k(A) + \mathbb{R}k$.*

***Proof*** $k \in -0^+(\mathrm{cl}_k(A))$ yields $\mathrm{cl}_k(A) - \mathbb{R}_+k = \mathrm{cl}_k(A)$. Thus, (4.2.18) and (4.2.19) follow from (4.2.4).

(4.2.18) implies $\mathrm{dom}\,\varphi_{A,k} = \mathrm{cl}_k(A) + \mathbb{R}k$. For each $t \in \mathbb{R}$ with $t \leq 0$, we get $\mathrm{cl}_k(A) + tk = \mathrm{cl}_k(A) + k + (t-1)k \subseteq \mathrm{cl}_k(A) + k$ from $k \in -0^+(\mathrm{cl}_k(A))$. Hence, $\mathrm{cl}_k(A) - \mathbb{R}_+k \subseteq \mathrm{cl}_k(A) + \mathbb{R}_>k$, and $\mathrm{dom}\,\varphi_{A,k} = \mathrm{cl}_k(A) + \mathbb{R}_>k = A + \mathbb{R}_>k$ because of Proposition 2.3.48.

(a) follows from (4.2.3).
(b) results from (a) and (4.2.18). It implies (c).
(d) Take first any $y \in Y$ with $\varphi_{A,k}(y) = -\infty$ and any $t \in \mathbb{R}$. Then there exists $\lambda < -t$ with $y \in A + \lambda k$. Hence, $\lambda + t < 0$ and $y - \lambda k \in A$, which results in $y + tk = (y - \lambda k) + (\lambda + t)k \in A + 0^+A \subseteq A$. This yields the forward direction of the equivalence.
For every $y \in Y$, the inclusion $y + \mathbb{R}k \subseteq A$ implies $y \in A - tk$ for any $t \in \mathbb{R}$. Thus, we get the reverse direction of the equivalence from (4.2.18).
(e) If $y \in Y$ with $\varphi_{A,k}(y) = -\infty$, then $y = y + 0 \cdot k \in A$ because of (d).
(f) Because of (d), $\varphi_{A,k}$ is proper if and only if $A \neq \emptyset$ and (4.2.21) hold. Condition (4.2.21) is equivalent to (4.2.20) since we obtain $y = y + 0 \cdot k \in A$ from $y + \mathbb{R}k \subseteq A$.
(g) follows from Theorem 4.2.1(f) and (b). □

The condition $k \in -0^+A$ implies $k \in -0^+(\mathrm{cl}_k(A))$ because of Proposition 2.3.29. It can be formulated in different ways.

**Proposition 4.2.6** *Suppose that $A$ is a proper subset of the linear space $Y$.*
*The following conditions are equivalent to each other for $A$ and $k \in Y \setminus \{0\}$.*

(a) *$k \in -0^+A$.*
(b) *$A = H - C$ for some proper subset $H$ of $Y$ and some convex cone $C \subset Y$ with $k \in C$.*
(c) *$A = A - C$ for some nontrivial convex cone $C \subset Y$ with $k \in C$.*
(d) *$A = A - C$ for some nontrivial cone $C \subset Y$ with $k \in C$.*

*If $Y$ is a topological vector space and $A$ is closed, then each of these conditions is equivalent to*

(e) $A = A - C$ *for some nontrivial closed convex cone* $C \subset Y$ *with* $k \in C$.

*If $Y$ is a Banach space and $A$ is closed, then* (a) *is equivalent to*

(f) $k \in -\mathrm{T_C}(a; A)$ *for each* $a \in A$.

***Proof*** (a) implies (b) with $H = A$ and $C = -0^+A$. (b) implies (c) since $A - C = H - C - C = H - C = A$. (c) yields (d). (d) implies (a) because of $C \subseteq -0^+A$.

Under the additional assumptions, (a) implies (e) with $C = -0^+A$. (e) yields (a).

(a) and (f) are equivalent because of Theorem 3.6.15. □

One of the basic assumptions for a set $A$ in production theory is the **free disposal assumption** $A = A - C$, where $-C$ is the ordering cone. Then $A$ is called a **free disposal set**. This kind of sets is frequently used in mathematical economics and optimization, e.g. in the well-known book by Debreu [110]. Further references related to this condition are given in [281], where the free disposal assumption is discussed for $C$ being an arbitrary closed set containing zero. In $Y = \mathbb{R}^\ell$, sets $A$ with the property $A = A - \mathbb{R}^\ell_+$ have been intensively investigated under the notion of **downward set**. Downward sets arise in the study of some problems in mathematical economics and game theory and in the study of inequality systems involving increasing functions [347].

**Proposition 4.2.7** *Assume* $(\mathrm{Sp0}_{A,k})$ *and* $A + C \subseteq A$ *for some nonempty set* $C \subseteq Y$. *For all* $y^0, y^1 \in Y$,

$$\varphi_{A,k}(y^0 + y^1) \leq \varphi_{A,k}(y^0) + \varphi_{C,k}(y^1). \tag{4.2.22}$$

***Proof*** Take any $y^0, y^1 \in Y$. Inequality (4.2.22) holds if $y^0 \notin A + \mathbb{R}k$ or $y^1 \notin C + \mathbb{R}k$. Assume now $y^0 \in A + \mathbb{R}k$ and $y^1 \in C + \mathbb{R}k$. There exists a monotonically decreasing sequence $(t_n)_{n \in \mathbb{N}}$ of real numbers with $y^0 \in A + t_n k$ which converges to $\varphi_{A,k}(y^0)$. Furthermore, there exists a monotonically decreasing sequence $(s_n)_{n \in \mathbb{N}}$ of real numbers with $y^1 \in C + s_n k$ which converges to $\varphi_{C,k}(y^1)$. For all $n \in \mathbb{N}$, we obtain $y^0 + y^1 \in (A + t_n k) + (C + s_n k) \subseteq A + (t_n + s_n)k$ and thus $\varphi_{A,k}(y^0 + y^1) \leq t_n + s_n$. Hence, $\varphi_{A,k}(y^0 + y^1) \leq \lim_{n \to +\infty}(t_n + s_n) = \varphi_{A,k}(y^0) + \varphi_{C,k}(y^1)$. □

The values of $\varphi_{A,k}$ are connected with the values of the sublinear function $\varphi_{0^+A,k}$. Proposition 4.2.7 implies:

**Corollary 4.2.8** *Assume* $(\mathrm{Sp0}_{A,k})$. *For all* $y^0, y^1 \in Y$,

$$\varphi_{A,k}(y^0 + y^1) \leq \varphi_{A,k}(y^0) + \varphi_{0^+A,k}(y^1).$$

**Theorem 4.2.9** *Assume that $A$ is a proper subset of the linear space $Y$ and* $k \in -\operatorname{core} 0^+A$.

(a) $\varphi_{A,k}$ *is finite-valued, and* $\operatorname{lev}_{\varphi_{A,k},<}(t) = \operatorname{core} A + tk$ *for all* $t \in \mathbb{R}$.
(b) $\varphi_{0^+A,k}$ *is finite-valued, and*

$$\varphi_{A,k}(y^0) - \varphi_{A,k}(y^1) \leq \varphi_{0^+A,k}(y^0 - y^1) \text{ for all } y^0, y^1 \in Y. \tag{4.2.23}$$

***Proof***

(a) By Lemma 2.3.11, $Y = 0^+A + \mathbb{R}k$. Thus, $Y = A + 0^+A + \mathbb{R}k$. Hence, $Y \subseteq A + \mathbb{R}k = \operatorname{dom} \varphi_{A,k}$. Suppose now, contrary to the claim, that $\varphi_{A,k}$ is not finite-valued. Then there exists $y \in Y$ with $\varphi_{A,k}(y) = -\infty$. From Theorem 4.2.5(d), we get $y + \mathbb{R}k \subseteq A$. This implies $Y = 0^+A + \mathbb{R}k + y \subseteq 0^+A + A \subseteq A$, which contradicts the assumption that $A$ is a proper subset of $Y$.
By Theorem 4.2.5, $\operatorname{lev}_{\varphi_{A,k},<}(t) = \operatorname{int}_k(A) + tk$ for all $t \in \mathbb{R}$. Lemma 2.3.44 yields $\operatorname{int}_k(A) = \operatorname{core} A$.
(b) Because of the assumptions, $0^+A$ is a proper subset of $Y$. $\varphi_{0^+A,k}$ is finite-valued by (a). Corollary 4.2.8 implies inequality (4.2.23). □

In the next theorem, we will characterize convexity of $\varphi_{A,k}$ and related properties.

**Theorem 4.2.10** *Assume $A$ is a proper subset of the linear space $Y$ and $k \in Y \setminus \{0\}$.*

(a) $\varphi_{A,k}$ *is convex if and only if* $\operatorname{cl}_k(A - \mathbb{R}_+k)$ *is convex.*
*Moreover,* $\varphi_{A,k}$ *is quasiconvex if and only if* $\varphi_{A,k}$ *is convex.*
(b) $\varphi_{A,k}$ *is positively homogeneous if and only if* $\operatorname{cl}_k(A - \mathbb{R}_+k)$ *is a cone.*
(c) $\varphi_{A,k}$ *is subadditive if and only if* $\operatorname{cl}_k(A - \mathbb{R}_+k) + \operatorname{cl}_k(A - \mathbb{R}_+k) \subseteq \operatorname{cl}_k(A - \mathbb{R}_+k)$.
(d) $\varphi_{A,k}$ *is sublinear if and only if* $\operatorname{cl}_k(A - \mathbb{R}_+k)$ *is a convex cone.*

*Each condition for* $\operatorname{cl}_k(A - \mathbb{R}_+k)$ *on the right-hand side of the statements* (a)–(d) *is fulfilled if it is fulfilled for $A$ or* $\operatorname{cl}_k(A)$.

***Proof***

(a) Assume first that $\varphi_{A,k}$ is convex. Then $\operatorname{epi} \varphi_{A,k}$ is convex by Definition 3.4.1. Hence, (4.2.6) implies that $\{(y, 0) \in Y \times \mathbb{R} \mid y \in \operatorname{cl}_k(A - \mathbb{R}_+k)\}$ is convex, i.e., $\operatorname{cl}_k(A - \mathbb{R}_+k)$ is convex.
Assume now that $\operatorname{cl}_k(A - \mathbb{R}_+k)$ is convex. Take any $(y^1, t_1), (y^2, t_2) \in \operatorname{epi} \varphi_{A,k}$, $\lambda \in [0, 1]$. For each $i \in \{1, 2\}$, we have $y^i \in \operatorname{cl}_k(A - \mathbb{R}_+k) + t_ik$ because of (4.2.6). Hence, $\lambda y^1 + (1 - \lambda)y^2 \in \operatorname{cl}_k(A - \mathbb{R}_+k) + (\lambda t_1 + (1 - \lambda)t_2)k$, i.e., $(\lambda y^1 + (1 - \lambda)y^2, \lambda t_1 + (1 - \lambda)t_2) \in \operatorname{epi} \varphi_{A,k}$. Thus, $\operatorname{epi} \varphi_{A,k}$ is convex. $\varphi_{A,k}$ is convex by Definition 3.4.1.
By (4.2.4), $\operatorname{cl}_k(A - \mathbb{R}_+k)$ is convex if and only if $\operatorname{lev}_{\varphi_{A,k},\leq}(t)$ is convex for each $t \in \mathbb{R}$. This condition is equivalent to the quasiconvexity of $\varphi_{A,k}$ because of Proposition 3.4.15.
(b) and (c) can be proved analogously to (a) using Definition 3.5.9.
(d) follows from (a) and (b) since—by Lemma 3.5.12—a functional is sublinear if and only if it is convex and positively homogeneous.

The last statement results from Proposition 2.3.30 since $\mathrm{cl}_k(A - \mathbb{R}_+k) = \mathrm{cl}_k(A) - \mathbb{R}_+k$. □

Theorems 4.4.4 and 4.6.6 contain statements related to the properties in Theorem 4.2.10 on topological vector spaces.

**Corollary 4.2.11** *Assume $A$ is a proper subset of the linear space $Y$ and $k \in -0^+A \setminus \{0\}$.*

(a) *$\varphi_{A,k}$ is concave if and only if $\mathrm{cl}_{-k}(Y \setminus A)$ is convex.*
*Moreover, $\varphi_{A,k}$ is quasiconcave if and only if $\varphi_{A,k}$ is concave.*
(b) *$\varphi_{A,k}$ is superadditive if and only if $\mathrm{cl}_{-k}(Y \setminus A) + \mathrm{cl}_{-k}(Y \setminus A) \subseteq \mathrm{cl}_{-k}(Y \setminus A)$.*

***Proof*** $\varphi_{A,k}$ is concave if and only if $-\varphi_{A,k} = \varphi_{Y\setminus A,-k}$ is convex because of Lemma 3.4.3 and Proposition 4.3.10. By Theorem 4.2.10, this is the case if and only if $\mathrm{cl}_{-k}((Y \setminus A) + \mathbb{R}_+k)$ is convex. According to Lemma 2.2.21(g), $k \in -0^+A \setminus \{0\}$ implies $k \in 0^+(Y \setminus A) \setminus \{0\}$. Hence, $(Y \setminus A) + \mathbb{R}_+k = Y \setminus A$, and $\mathrm{cl}_{-k}((Y \setminus A) + \mathbb{R}_+k) = \mathrm{cl}_{-k}(Y \setminus A)$.

The second statement of (a) as well as statement (b) can be proved using analogous arguments. □

Because of Lemma 2.3.42(i) and (b), we have

$$\mathrm{cl}_{-k}(Y \setminus A) = Y \setminus \mathrm{int}_k(A) = Y \setminus (A - \mathbb{R}_>k) \quad \text{if } k \in -0^+A \setminus \{0\}.$$

**Corollary 4.2.12** *Assume $A$ is a proper subset of the topological vector space $Y$, $k \in Y \setminus \{0\}$ and $A - \mathbb{R}_>k \subseteq \mathrm{int}\, A$.*

(a) *$\varphi_{A,k}$ is concave if and only if $Y \setminus \mathrm{int}\, A$ is convex.*
(b) *$\varphi_{A,k}$ is superadditive if and only if $(Y \setminus \mathrm{int}\, A) + (Y \setminus \mathrm{int}\, A) \subseteq Y \setminus \mathrm{int}\, A$.*

***Proof*** Proposition 2.3.55(b) implies $\mathrm{int}_k(A) = \mathrm{int}\, A$. Since $\mathrm{cl}_{-k}(Y \setminus A) = Y \setminus \mathrm{int}_k(A)$ by Lemma 2.3.42(i), we get $\mathrm{cl}_{-k}(Y \setminus A) = Y \setminus \mathrm{int}\, A$. Hence, the assertions follow from Corollary 4.2.11. □

Note that $Y \setminus \mathrm{int}\, A = \mathrm{cl}(Y \setminus A)$.

**Remark 4.2.13** (*Bibliographical notes related to Sect.* 4.2)

Property (4.2.2), the translation invariance of $\varphi_{A,k}$, was proved for $\varphi_{A,k}$ under special assumptions in [195] and [11, Prop. 2.1], but the proofs given there also work in the general case.

Theorem 4.2.1(a)–(d) without (4.2.16) were proved in [483, Theorem 2.1]. The equations $\varphi_{A,k} = \varphi_{\mathrm{cl}_k(A),k}$, $\mathrm{lev}_{\varphi_{A,k},<}(t) = \mathrm{cl}_k(A) - \mathbb{R}_>k + tk$ and $\mathrm{lev}_{\varphi_{A,k},\leq}(t) = \mathrm{cl}_k(A) - \mathbb{R}_+k + tk$ in Theorem 4.2.1 were also stated in [213]. The left equations in (4.2.3)–(4.2.5) correspond to results in [222]. Inclusion (4.2.11) was stated for $t = 0$ in [280]. There, the reverse direction of the second statement in Theorem 4.2.1(b) for $t = 0$ was mentioned for the case where $A$ is algebraically closed. Hamel [221, Prop. 3] and Schrage [407, Satz 3.1.3] showed the reverse direction of the second statement in Theorem 4.2.1(b) assuming radial closedness instead of directional closedness. The proof of [407, Satz 3.1.3] contains the statement that $\text{lev}_{\varphi_{A,k},\leq}(0)$ is the $k$-radial closure of $A - \mathbb{R}_+ k$. Statements which are equivalent to $\varphi_{A,k} = \varphi_{\text{cl}_k(A-\mathbb{R}_+k),k}$ and to (4.2.6) are contained in [221, Prop. 4] and [221, Cor. 1], respectively. For the case where $A$ coincides with the $k$-radial closure of $A - \mathbb{R}_+ k$, (4.2.12) was given in [407, Folgerung 3.6.1].

The idea to derive necessary and sufficient conditions for $\text{lev}_{\varphi_{A,k},\leq}(0) = A$ using directional properties of $A$ was introduced by Rubinov and Singer. They proved [397, Prop. 4.4] in $Y = \mathbb{R}^\ell$ for $k = \mathbf{1}$ that $\varphi_{A,k}$ is translation invariant along $k$ with $\text{lev}_{\varphi_{A,k},\leq}(0) = A$ if and only if $A$ is closed along diagonal lines and $k \in -0^+ A$.

Theorem 4.2.3 can be deduced from [407], taking into account that—in the case $k \in -0^+ A$—the set $A$ is $k$-directionally closed if and only if it is $k$-radially closed. The close relationship between functionals $\varphi_{A,k}$ and translation invariant functions via the zero-sublevel set of the functions was first pointed out by Artzner et al. [11] in the context of risk measures and acceptance sets. For $k \in -\operatorname{int}\mathbb{R}^\ell_+$, they proved the correspondence between sublinear $(-\mathbb{R}^\ell_+)$-monotone real-valued functions $f$ on $\mathbb{R}^\ell$ which are translation invariant along $k$ and functions $\varphi_{A,k}$ if $A \subset \mathbb{R}^\ell$ is a convex cone with $\mathbb{R}^\ell_+ \subseteq A$ and $A \cap (-\operatorname{int}\mathbb{R}^\ell_+) = \emptyset$ [11, Prop. 2.1 and 2.2]. Assuming $k = \mathbf{1}$ in $Y = \mathbb{R}^\ell$, Rubinov and Singer [397, Prop. 4.7] proved that $f\colon Y \to \overline{\mathbb{R}}$ is translation invariant along $k$ if and only if $k \in -0^+ \tilde{A}$ and $f = \varphi_{\tilde{A},k}$ for $\tilde{A} := \text{lev}_{f,\leq}(0)$.

Theorem 4.2.4 is taken from [485, Theorem 3.1]. Without the last equation in (c), Theorem 4.2.4 can also be deduced from [221, 407].

Under the assumption $k \in -0^+ A$, (4.2.18) and (4.2.19) were proved in [489, Lemma 2.5], $\text{dom}\,\varphi_{A,k} = A + \mathbb{R}_> k$ as well as Theorem 4.2.5(d)–(e) were proved in [483, Prop. 2.3].

The equivalence between the conditions (a)–(d) in Proposition 4.2.6 was given in [483, Prop. 2.2].

Inequality (4.2.22) was proved in [441] under the assumptions that $Y$ is a separated locally convex space, $A$ is a proper closed subset of $Y$, $C$ is a nontrivial closed convex cone with $A + C = A$ and $k \in -C \setminus C$.

Theorem 4.2.9 can be found in [483, Theorem 2.2 and Prop. 2.4].

Theorem 4.2.10 generalizes equivalences which were proved in [472] under certain assumptions in topological vector spaces. For the case where $A$ coincides with the $k$-radial closure of $A - \mathbb{R}_+ k$, Theorem 4.2.10(b)–(d) and the first statement of Theorem 4.2.10(a) are contained in [407, Satz 3.1.9]. Jaschke and Küchler [280, p. 187] stated that $\varphi_{A,k}$ is sublinear if $A$ is an algebraically closed convex cone with $k \in -A$ and that the sublevel set $\text{lev}_{\varphi_{A,k},\leq}(0)$ is a convex cone if $\varphi_{A,k}$ is sublinear.

Theorem 4.2.10(b) and the first statement of Theorem 4.2.10(a) were given under the assumption $k \in -0^+(\text{cl}_k(A))$ in [213, Theorem 14], with $\text{cl}_k(A)$ instead of $\text{cl}_k(A - \mathbb{R}_+ k)$. Because of $\text{cl}_k(A - \mathbb{R}_+ k) = \text{cl}_k(A) - \mathbb{R}_+ k$, the sets $\text{cl}_k(A - \mathbb{R}_+ k)$ and $\text{cl}_k(A)$ coincide if and only if $k \in -0^+(\text{cl}_k(A))$.

Corollary 4.2.12 was proved under the assumption $\text{cl}\,A - \mathbb{R}_> k \subseteq \operatorname{int} A$ in [484, Prop. 3.9].

## 4.3 Calculation Rules for Translation Invariant Functions

Some calculation rules for translation invariant functions are collected in the following proposition.

**Proposition 4.3.1** *Assume* $Y$ *is a linear space,* $k \in Y \setminus \{0\}$ *and that* $f : Y \to \overline{\mathbb{R}}$ *is translation invariant along* $k$.

(a) *For each* $c \in \mathbb{R}$, $f + c$ *is translation invariant along* $k$.
(b) *For each* $c \in \mathbb{R} \setminus \{0\}$, $cf$ *is translation invariant along* $\frac{1}{c}k$.
(c) *For each* $c \in \mathbb{R} \setminus \{0\}$, $g : Y \to \overline{\mathbb{R}}$ *defined by*

$$g(y) := f(cy)$$

*is translation invariant along* $\frac{1}{c}k$.
(d) *The function* $g : Y \to \overline{\mathbb{R}}$ *defined by*

$$g(y) := -f(-y)$$

*is translation invariant along* $k$.
(e) *For each* $y^0 \in Y$, $g : Y \to \overline{\mathbb{R}}$ *defined by*

$$g(y) := f(y + y^0)$$

*is translation invariant along* $k$.
(f) *If* $\{f_i : Y \to \overline{\mathbb{R}} \mid i \in I\}$ *is a family of functions which are translation invariant along* $k$, *then* $\sum_{i\in I} \lambda_i f_i$ *is translation invariant along* $k$ *for all* $\lambda_i \in \mathbb{R}$ *with* $\sum_{i\in I} \lambda_i = 1$.
(g) *For each function* $g : Y \to \overline{\mathbb{R}}$, $f \# g$ *is translation invariant along* $k$.

***Proof***

(a) For each $y \in Y$ and $t \in \mathbb{R}$, $(f + c)(y + tk) = f(y + tk) + c = f(y) + t + c = (f + c)(y) + t$.
(b) For each $y \in Y$ and $t \in \mathbb{R}$,

$$(cf)(y + t \cdot \frac{1}{c}k) = cf(y + \frac{t}{c}k) = c(f(y) + \frac{t}{c}) = (cf)(y) + t.$$

(c) For each $y \in Y$ and $t \in \mathbb{R}$,

$$g(y + t \cdot \frac{1}{c}k) = f(c(y + t \cdot \frac{1}{c}k)) = f(cy + tk) = f(cy) + t = g(y) + t.$$

(d) follows from (b) and (c).
(e) For each $y \in Y$ and $t \in \mathbb{R}$, $g(y + tk) = f(y + tk + y^0) = f(y + y^0) + t = g(y) + t$.

(f) For each $y \in Y$ and $t \in \mathbb{R}$,

$$\sum_{i\in I} \lambda_i f_i(y+tk) = \sum_{i\in I} \lambda_i (f_i(y)+t) = \sum_{i\in I} \lambda_i f_i(y) + t \sum_{i\in I} \lambda_i = \sum_{i\in I} \lambda_i f_i(y) + t.$$

(g) For each $y \in Y$ and $t \in \mathbb{R}$,

$$\begin{aligned}(f \# g)(y+tk) &= \inf\{f(y+tk-y^2)+g(y^2) \mid y^2 \in Y\}\\ &= \inf\{f(y-y^2)+g(y^2)+t \mid y^2 \in Y\}\\ &= \inf\{f(y-y^2)+g(y^2) \mid y^2 \in Y\}+t\\ &= (f \# g)(y)+t.\end{aligned}$$

□

The function in Proposition 4.3.1(d) is applied for the construction of upper function envelopes (see Sect. 4.9).

**Proposition 4.3.2** *Let $Y$ be a linear space, $k \in Y \setminus \{0\}$ and $\{\varphi_i : Y \to \overline{\mathbb{R}} \mid i \in I\}$ a collection of functions which are translation invariant along $k$.*

*Then the functions $f, g : Y \to \overline{\mathbb{R}}$ defined by*

$$f(y) := \sup\{\varphi_i(y) \mid i \in I\} \quad \textit{and} \quad g(y) := \inf\{\varphi_i(y) \mid i \in I\} \quad \textit{for all } y \in Y$$

*are translation invariant along $k$.*

*Using the representation $\varphi_i = \varphi_{A_i,k}$ for each $i \in I$, we get*

$$\begin{aligned} f &= \varphi_{A,k} \quad \textit{for } A := \cap_{i\in I}(A_i - \mathbb{R}_+ k),\\ g &= \varphi_{B,k} \quad \textit{for } B := \cup_{i\in I} A_i. \end{aligned}$$

***Proof*** Proposition 4.1.2 implies $\{\lambda(k,1) \in Y \times \mathbb{R} \mid \lambda \in \mathbb{R}\} \subseteq 0^+(\operatorname{epi} \varphi_i)$ for each $i \in I$. Hence, $\{\lambda(k,1) \in Y \times \mathbb{R} \mid \lambda \in \mathbb{R}\} \subseteq 0^+(\cap_{i\in I} \operatorname{epi} \varphi_i)$. Because of Proposition 3.2.10(a), we have $\cap_{i\in I} \operatorname{epi} \varphi_i = \operatorname{epi} f$. This and Proposition 4.1.2 yield the translation invariance of $f$ along $k$.

Applying the translation invariance of $f$ and Proposition 4.3.1(b) for $c = -1$, we obtain the translation invariance of $g$ since $g(y) = -\sup\{-\varphi_i(y) \mid i \in I\}$ for each $y \in Y$.

Since $f$ and $g$ are translation invariant along $k$, there exist sets $\tilde{A}, \tilde{B} \subseteq Y$ with $f = \varphi_{\tilde{A},k}$ and $g = \varphi_{\tilde{B},k}$. The definition of $f$ implies $\operatorname{lev}_{f,\leq}(0) = \cap_{i\in I} \operatorname{lev}_{\varphi_i,\leq}(0)$ and $\operatorname{cl}_k(\tilde{A} - \mathbb{R}_+ k) = \cap_{i\in I} \operatorname{cl}_k(A_i - \mathbb{R}_+ k)$ by (4.2.4). By (4.2.8), we have $\varphi_{\tilde{A},k} = \varphi_{\operatorname{cl}_k(\tilde{A}-\mathbb{R}_+k),k}$. This yields $f = \varphi_{\bar{A},k}$ for $\bar{A} := \cap_{i\in I} \operatorname{cl}_k(A_i - \mathbb{R}_+ k)$. By Lemma 2.3.32(d), $\bar{A} = \operatorname{cl}_k(\cap_{i\in I}(A_i - \mathbb{R}_+ k))$, which proves the statement for $f$ because of (4.2.8).

The definition of $g$ implies $\operatorname{lev}_{g,<}(0) = \cup_{i\in I} \operatorname{lev}_{\varphi_i,<}(0)$ and $\tilde{B} - \mathbb{R}_> k = \cup_{i\in I}(A_i - \mathbb{R}_> k)$ by (4.2.3). By (4.2.9), we have $\varphi_{\tilde{B},k} = \varphi_{\tilde{B}-\mathbb{R}_>k,k}$. This yields $g = \varphi_{\bar{B},k}$ for $\bar{B} := \cup_{i\in I}(A_i - \mathbb{R}_> k) = (\cup_{i\in I} A_i) - \mathbb{R}_> k$. The assertion for $g$ results from (4.2.9). □

Suprema and infima of functions are of interest in duality theory. Suprema of linear translation invariant functions $\varphi_i$ were considered as risk measures by Artzner et al. [11, Prop. 3.1] and by Jaschke and Küchler [280, Prop. 6], who interpreted the functions $\varphi_i$ as normalized price systems.

We continue using the representation $\varphi_{A,k}$ of translation invariant functions and study the influence of varying sets $A$ and vectors $k$ on the functionals.

**Proposition 4.3.3** *Assume* $(\mathrm{Sp0}_{A,k})$ *and* $\lambda \in \mathbb{R}_>$. *Then* $\operatorname{dom} \varphi_{A,\lambda k} = \operatorname{dom} \varphi_{A,k}$ *and*

$$\varphi_{A,\lambda k}(y) = \frac{1}{\lambda}\varphi_{A,k}(y) \text{ for all } y \in Y.$$

$\varphi_{A,\lambda k}$ *is proper, finite-valued, convex, concave, strictly quasiconvex, subadditive, superadditive, sublinear, positively homogeneous, homogeneous, odd, affine or linear if and only if* $\varphi_{A,k}$ *has the same property.*

*For* $B,\ F \subseteq Y$, *the functional* $\varphi_{A,\lambda k}$ *is* $B$*-monotone on* $F$ *or strictly* $B$*-monotone on* $F$ *if and only if* $\varphi_{A,k}$ *has the same property on* $F$.

*If* $Y$ *is a topological vector space, then* $\varphi_{A,\lambda k}$ *is continuous, lower semicontinuous or upper semicontinuous at* $y \in Y$ *if and only if this is the case for* $\varphi_{A,k}$.

***Proof*** For all $y \in Y$,

$$\begin{aligned}\varphi_{A,\lambda k}(y) &= \inf\{t \in \mathbb{R} \mid y \in A + t(\lambda k)\} = \inf\{t \in \mathbb{R} \mid y \in A + (\lambda t)k\}\\ &= \inf\{\frac{1}{\lambda}u \mid u \in \mathbb{R},\ y \in A + uk\} = \frac{1}{\lambda}\inf\{u \in \mathbb{R} \mid y \in A + uk\}\\ &= \frac{1}{\lambda}\varphi_{A,k}(y).\end{aligned}$$

The statements in a topological vector space result from Lemmata 3.3.2 and 3.3.8. The monotonicity statements follow from Lemma 3.7.8. Lemma 3.4.18 implies the statements for convex, concave and strictly quasiconvex functions. Lemma 3.5.4 yields the statements for affine and linear functions. The other properties follow from Lemma 3.5.15 or are obvious. □

The proposition underscores that replacing $k$ by another vector in the same direction scales the functional. Consequently, $\varphi_{A,k}$ and $\varphi_{A,\lambda k}$ with $\lambda \in \mathbb{R}_>$ attain optimal values on some set $F \subseteq Y$ at the same elements of $F$. Hence, it is sufficient to consider only one vector $k$ per direction in optimization problems, e.g., to restrict $k$ to unit vectors if $Y$ is a normed space.

Note that the sum of two functions $\varphi_{A,k^1}$ and $\varphi_{A,k^2}$ with $k^1, k^2 \in Y \setminus \{0\}$ is not necessarily a function $\varphi_{A,k}$ with $k \in Y \setminus \{0\}$. This will be shown in Example 4.7.4.

If $\varphi_{A,k}(0) \in \mathbb{R}$, then there exists a translation invariant functional for which the sublevel sets have the same shape as those of $\varphi_{A,k}$ and which attains the function value 0 at 0. This functional is $\varphi_{A+ck,k}$ with $c = \varphi_{A,k}(0)$.

**Proposition 4.3.4** *Assume* $(\mathrm{Sp0}_{A,k})$ *and* $c \in \mathbb{R}$. *Then* $\operatorname{dom} \varphi_{A+ck,k} = \operatorname{dom} \varphi_{A,k}$ *and*

$$\varphi_{A+ck,k}(y) = \varphi_{A,k}(y) - c \text{ for all } y \in Y.$$

*$\varphi_{A+ck,k}$ is proper, finite-valued, convex, concave, strictly quasiconvex or affine if and only if $\varphi_{A,k}$ has the same property. If $\varphi_{A,k}$ is subadditive and $c < 0$, then $\varphi_{A+ck,k}$ is subadditive.*

*For $B,\ F \subseteq Y$, the functional $\varphi_{A+ck,k}$ is $B$-monotone on $F$ or strictly $B$-monotone on $F$ if and only if $\varphi_{A,k}$ has the same property on $F$.*

*If $Y$ is a topological vector space, then $\varphi_{A+ck,k}$ is continuous, lower semicontinuous or upper semicontinuous at $y \in Y$ if and only if this is the case for $\varphi_{A,k}$.*

***Proof*** For all $y \in Y$,

$$\begin{aligned}\varphi_{A+ck,k}(y) &= \inf\{t \in \mathbb{R} \mid y \in A + ck + tk\} = \inf\{t \in \mathbb{R} \mid y \in A + (c+t)k\} \\ &= \inf\{u - c \mid u \in \mathbb{R},\ y \in A + uk\} = \varphi_{A,k}(y) - c.\end{aligned}$$

The statements in a topological vector space result from Lemmata 3.3.2 and 3.3.8. The monotonicity statements follow from Lemma 3.7.8. Lemma 3.4.18 implies the statements for convex, concave and strictly quasiconvex functions. Lemma 3.5.4 yields the statement for affine functions. The other properties follow from Lemma 3.5.15 or are obvious. □

In vector optimization, scalarizing functionals often depend on some given point $y^0$.

**Proposition 4.3.5** *Assume* (Sp0$_{A,k}$) *and* $y^0 \in Y$*. Then* $\text{dom}\,\varphi_{y^0+A,k} = y^0 + \text{dom}\,\varphi_{A,k}$ *and*

$$\varphi_{y^0+A,k}(y) = \varphi_{A,k}(y - y^0) \text{ for all } y \in Y. \tag{4.3.1}$$

*$\varphi_{y^0+A,k}$ is proper, finite-valued, convex, concave, strictly quasiconvex or affine if and only if $\varphi_{A,k}$ has the same property.*

*For $B,\ F \subseteq Y$, the functional $\varphi_{y^0+A,k}$ is $B$-monotone on $F$ or strictly $B$-monotone on $F$ if and only if $\varphi_{A,k}$ has the same property on $F + y^0$.*

*If $Y$ is a topological vector space, then $\varphi_{y^0+A,k}$ is continuous, lower semicontinuous or upper semicontinuous at $y \in Y$ if and only if this is the case for $\varphi_{A,k}$ at $y - y^0$.*

***Proof*** For $y \in Y$, we get $\varphi_{y^0+A,k}(y) = \inf\{t \in \mathbb{R} \mid y \in y^0 + A + tk\} = \inf\{t \in \mathbb{R} \mid y - y^0 \in A + tk\} = \varphi_{A,k}(y - y^0)$. The statements in a topological vector space result from Lemmata 3.3.2 and 3.3.8. The monotonicity statements follow from Lemma 3.7.8. Lemma 3.4.18 implies the statements for convex, concave and strictly quasiconvex functions. Lemma 3.5.4 yields the statement for affine functions. The other properties are obvious. □

The values of each function $\varphi_{A,k}$ can be calculated by means of some function $\varphi_{\tilde{A},k}$ with $0 \in \tilde{A}$.

**Corollary 4.3.6** *Assume* $(\mathrm{Sp0}_{A,k})$.

(a) *For any* $a^0 \in A$, *we have* $0 \in A - a^0$ *and* $\varphi_{A,k}(y) = \varphi_{A-a^0,k}(y - a^0)$ *for all* $y \in Y$.
(b) *If* $0 \in \operatorname{dom} \varphi_{A,k}$, *then there exists* $\lambda \in \mathbb{R}$ *with* $0 \in A + \lambda k$ *and* $\varphi_{A,k} = \varphi_{A+\lambda k,k} + \lambda$.

***Proof***

(a) follows from Propositions 4.3.5.
(b) results from Propositions 4.3.4 since $\operatorname{dom} \varphi_{A,k} = A + \mathbb{R}k$. □

The infimum of a translation invariant function on a given set can be described as the function value of some translation invariant function.

**Proposition 4.3.7** *Assume* $(\mathrm{Sp0}_{A,k})$ *and* $B \subseteq Y$. *Then*

$$\varphi_{A-B,k} = \varphi_{A,k} \,\#\, \iota_{-B},$$
$$\varphi_{A-B,k}(y) = \inf_{b\in B} \varphi_{A,k}(y+b) \quad \text{for all } y \in Y,$$
$$\varphi_{A-B,k}(0) = \inf_{b\in B} \varphi_{A,k}(b).$$

***Proof*** Take any $y \in Y$. Lemma 3.2.15 yields $(\varphi_{A,k} \,\#\, \iota_{-B})(y) = \inf_{b\in B} \varphi_{A,k}(y + b)$. Propositions 4.3.5 and 4.3.2 imply $\inf_{b\in B} \varphi_{A,k}(y + b) = \inf_{b\in B} \varphi_{A-b,k}(y) = \varphi_{C,k}(y)$ with $C := \bigcup_{b\in B}(A - b) = A - B$. □

**Corollary 4.3.8** *Assume* $(\mathrm{Sp0}_{A,k})$. *Then*

$$\varphi_{A,k} = \iota_A \,\#\, \varphi_{\{0_Y\},k},$$

*i.e.,* $\varphi_{A,k}$ *is the inf-convolution of* $A$*'s indicator function with* $\varphi_{\{0_Y\},k}$.

This representation of $\varphi_{A,k}$ will be used in Sect. 4.9 when proving formulas for translation invariant function envelopes and in the variational analysis of translation invariant functions in Sect. 4.10. Properties of $\varphi_{\{0\},k}$ will be proved in Lemma 4.7.6.

**Proposition 4.3.9** *Assume* $(\mathrm{Sp0}_{A,k})$.

(a) *Then* $\operatorname{dom} \varphi_{-A,-k} = -\operatorname{dom} \varphi_{A,k}$, *and*

$$\varphi_{-A,-k}(y) = \varphi_{A,k}(-y) \quad \text{for all } y \in Y.$$

(b) *Furthermore,*

$$\varphi_{A,-k}(y) = -\sup\{t \in \mathbb{R} \mid y \in A + tk\} \quad \text{for all } y \in Y.$$

***Proof*** Take any $y \in Y$.

$$\begin{aligned}\varphi_{-A,-k}(y) &= \inf\{t \in \mathbb{R} \mid y \in -A + t(-k)\}\\ &= \inf\{t \in \mathbb{R} \mid -y \in A + tk\}\\ &= \varphi_{A,k}(-y).\\ \sup\{t \in \mathbb{R} \mid y \in A + tk\} &= -\inf\{-t \in \mathbb{R} \mid y \in A + tk\}\\ &= -\inf\{t \in \mathbb{R} \mid y \in A - tk\}\\ &= -\varphi_{A,-k}(y).\end{aligned}$$

□

In many cases, properties of a function $\varphi$ become obvious by certain properties of $-\varphi$.

**Proposition 4.3.10** *Assume* $(\mathrm{Sp0}_{A,k})$ *and* $k \in -0^+A \setminus \{0\}$.
*Then* $-k \in -0^+(Y \setminus A) \setminus \{0\}$, *and*

$$\varphi_{A,k} = -\varphi_{Y\setminus A,-k} = -\varphi_{Y\setminus \mathrm{int}_k(A),-k}, \tag{4.3.2}$$

*where*

(a) $\mathrm{bd}_k(A) + \mathbb{R}k = \{y \in Y \mid \varphi_{A,k}(y) \in \mathbb{R}\} = \{y \in Y \mid \varphi_{Y\setminus A,-k}(y) \in \mathbb{R}\}$,
(b) $\mathrm{dom}\, \varphi_{A,k} \cap \mathrm{dom}\, \varphi_{Y\setminus A,-k} = \mathrm{bd}_k(A) + \mathbb{R}k$.

*Moreover,* $k \in -0^+(Y \setminus (-A))$, *and*

$$\varphi_{Y\setminus(-A),k}(y) = -\varphi_{A,k}(-y) \quad \textit{for all } y \in Y.$$

***Proof*** Lemma 2.2.21(g) and (d) yield $-k \in -0^+(Y \setminus A) \setminus \{0\}$ and $k \in -0^+(Y \setminus (-A))$.
Take any $t \in \mathbb{R}$.

$$\begin{aligned}\varphi_{A,k}(y) = t &\iff y \in \mathrm{bd}_k(A) + tk \text{ by Theorem 4.2.5(b)},\\ &\iff y \in \mathrm{bd}_{-k}(Y \setminus A) + tk \text{ because of Lemma 2.3.42(j)},\\ &\iff y \in \mathrm{bd}_{-k}(Y \setminus A) + (-t)\cdot(-k),\\ &\iff \varphi_{Y\setminus A,-k}(y) = -t \text{ by Theorem 4.2.5(b)}.\end{aligned}$$

We obtain (a) and $\varphi_{A,k}(y) = -\varphi_{Y\setminus A,-k}(y)$ for all $y \in \mathrm{bd}_k(A) + \mathbb{R}k$.

$y \in \mathrm{dom}\, \varphi_{A,k} \setminus (\mathrm{bd}_k(A) + \mathbb{R}k)$ is equivalent to $\varphi_{A,k}(y) = -\infty$ and to $y + \mathbb{R}k \subseteq A$ by Theorem 4.2.5(d). The previous inclusion implies $y \notin (Y \setminus A) - \mathbb{R}k = \mathrm{dom}\, \varphi_{Y\setminus A,-k}$. Hence, (b) is fulfilled.

Since $\mathrm{dom}\, \varphi_{A,k} \cup \mathrm{dom}\, \varphi_{Y\setminus A,-k} = Y$, we get $\varphi_{A,k} = -\varphi_{Y\setminus A,-k}$.

The second equation in (4.3.2) was proved in Theorem 4.2.1(g). For each $y \in Y$,

$$\begin{aligned}\varphi_{Y\setminus(-A),k}(y) &= -\varphi_{-A,-k}(y) && \text{because of (4.3.2)}\\ &= -\varphi_{A,k}(-y) && \text{because of Proposition 4.3.9.}\end{aligned}$$

□

**Proposition 4.3.11** *Assume*

$$(\mathrm{Sp2}_{A,k}):\ Y \text{ is a topological vector space, } A \text{ is a proper subset of } Y,\ k \in Y \setminus \{0\} \text{ and } \operatorname{cl} A - \mathbb{R}_{>}k \subseteq \operatorname{int} A.$$

*Then* $(\mathrm{Sp2}_{Y\setminus A,-k})$ *holds and*

$$\varphi_{A,k} = -\varphi_{Y\setminus A,-k},$$

*where*

(a) $\operatorname{bd} A + \mathbb{R}k = \{y \in Y \mid \varphi_{A,k}(y) \in \mathbb{R}\} = \{y \in Y \mid \varphi_{Y\setminus A,-k}(y) \in \mathbb{R}\}$,
(b) $\operatorname{dom} \varphi_{A,k} \cap \operatorname{dom} \varphi_{Y\setminus A,-k} = \operatorname{bd} A + \mathbb{R}k$.

*Moreover,* $\varphi_{Y\setminus A,-k} = \varphi_{Y\setminus \operatorname{int} A,-k}$.

***Proof*** Take any $y \in \operatorname{cl}(Y \setminus A)$ and $t \in \mathbb{R}_{>}$. Then $y \notin \operatorname{int} A$ because of Lemma 2.1.11(n). If $y + tk \in \operatorname{cl} A$, then $y \in \operatorname{cl} A - tk \subseteq \operatorname{int} A$ by $(\mathrm{Sp2}_{A,k})$, which contradicts $y \notin \operatorname{int} A$. Hence, $\operatorname{cl}(Y \setminus A) - \mathbb{R}_{>} \cdot (-k) \subseteq Y \setminus \operatorname{cl} A = \operatorname{int}(Y \setminus A)$, i.e., $(\mathrm{Sp2}_{Y\setminus A,-k})$ holds.

Proposition 2.3.55 implies $\operatorname{bd}_k A = \operatorname{bd} A$ and $\operatorname{int}_k A = \operatorname{int} A$. Thus, Proposition 4.3.10 yields the assertion. □

The assumptions of Proposition 4.3.11 will be discussed in Sect. 4.6. We illustrate the above proposition.

**Example 4.3.12** In $Y = \mathbb{R}^2$, consider $k := (1, 0)^T$ and $A := \{(y_1, y_2)^T \in \mathbb{R}^2 \mid y_2 \geq e^{y_1}\}$. Then $(\mathrm{Sp2}_{A,k})$ is fulfilled. $\varphi_{A,k}$ is a proper functional with $\operatorname{dom} \varphi_{A,k} = A + \mathbb{R}k = \{(y_1, y_2)^T \in \mathbb{R}^2 \mid y_2 > 0\}$, but $\operatorname{dom} \varphi_{Y\setminus A,-k} = Y$. We have $\varphi_{Y\setminus A,-k} = -\varphi_{A,k}$ and $\operatorname{dom} \varphi_{A,k} = \{y \in \mathbb{R}^2 \mid \varphi_{Y\setminus A,-k}(y) \in \mathbb{R}\}$.

In our book, we will investigate and apply minimization problems with translation invariant objective functions. Propositions 4.3.10 and 4.3.11 give us a tool to make statements about the maximization of $\varphi_{A,k}$ using results for the minimization of $\varphi_{Y\setminus A,-k}$.

**Proposition 4.3.13** *Assume* $(\mathrm{Sp0}_{A,k})$ *and* $A_0 \subseteq Y$.

(a) *Then*

$$\varphi_{A_0,k} \leq \varphi_{A,k} \tag{4.3.3}$$

*if and only if* $A \subseteq \operatorname{cl}_k(A_0 - \mathbb{R}_+k)$.
(4.3.3) *implies* $\operatorname{dom} \varphi_{A,k} \subseteq \operatorname{dom} \varphi_{A_0,k}$.

(b) *If* $\varphi_{A,k}$ *is proper, then*

$$\varphi_{A_0,k}(y) < \varphi_{A,k}(y) \ \textit{for all}\ y \in \operatorname{dom} \varphi_{A,k} \tag{4.3.4}$$

*if and only if* $\operatorname{cl}_k(A) \subseteq A_0 - \mathbb{R}_{>}k$.

***Proof***

(a) If (4.3.3) holds, then $y^0 \in A$ implies $\varphi_{A_0,k}(y^0) \le \varphi_{A,k}(y^0) \le 0$, which yields $y^0 \in \mathrm{cl}_k(A_0 - \mathbb{R}_+k)$ because of (4.2.4).
Assume now $A \subseteq \mathrm{cl}_k(A_0 - \mathbb{R}_+k)$. We obtain $A - \mathbb{R}_+k \subseteq \mathrm{cl}_k(A_0 - \mathbb{R}_+k) - \mathbb{R}_+k = \mathrm{cl}_k(A_0) - \mathbb{R}_+k = \mathrm{cl}_k(A_0 - \mathbb{R}_+k)$ because of Proposition 2.3.49(1), and $\mathrm{cl}_k(A - \mathbb{R}_+k) \subseteq \mathrm{cl}_k(A_0 - \mathbb{R}_+k)$ by Lemma 2.3.26(c) and (b). Take any $y \in Y$. If $\varphi_{A,k}(y) = +\infty$, then $\varphi_{A_0,k}(y) \le \varphi_{A,k}(y)$. Assume now $\varphi_{A,k}(y) \ne +\infty$. Take any $t \in \mathbb{R}$ with $\varphi_{A,k}(y) \le t$. Because of (4.2.4), $y \in \mathrm{cl}_k(A - \mathbb{R}_+k) + tk \subseteq \mathrm{cl}_k(A_0 - \mathbb{R}_+k) + tk$ and hence $\varphi_{A_0,k}(y) \le t$. This yields (4.3.3).
The last statement of (a) is obvious.

(b) If (4.3.4) holds, then $y^0 \in \mathrm{cl}_k(A)$ implies $\varphi_{A_0,k}(y^0) < \varphi_{A,k}(y^0) \le 0$, which yields $y^0 \in A_0 - \mathbb{R}_> k$ because of (4.2.3).
Assume now $\mathrm{cl}_k(A) \subseteq A_0 - \mathbb{R}_> k$. Take any $y \in \mathrm{dom}\, \varphi_{A,k}$. Then $t := \varphi_{A,k}(y) \in \mathbb{R}$. Because of (4.2.4), $y \in \mathrm{cl}_k(A - \mathbb{R}_+k) + tk = \mathrm{cl}_k(A) - \mathbb{R}_+k + tk \subseteq A_0 - \mathbb{R}_> k + tk$ and hence $\varphi_{A_0,k}(y) < t$. This yields (4.3.4). □

Proposition 4.3.13 implies:

**Corollary 4.3.14** *Assume* (Sp0$_{A,k}$). *If* $A \subseteq A_0 \subseteq Y$, *then* $\mathrm{dom}\, \varphi_{A,k} \subseteq \mathrm{dom}\, \varphi_{A_0,k}$ *and*

$$\varphi_{A_0,k} \le \varphi_{A,k}.$$

The relationship between $\varphi_{A,k}$ and $\varphi_{\mathrm{cl}\, A,k}$ on a topological vector space will be discussed in Sect. 4.4.

**Remark 4.3.15** (*Bibliographical notes related to Sect.* 4.3)
Proposition 4.3.1(b)–(c) with $c = -1$, (d) and (f)–(g) were mentioned in [222]. Hamel [221, Cor. 3] proved the translation invariance of the functions $f$ and $g$ in Proposition 4.3.2. Propositions 4.3.3, 4.3.4 and 4.3.5 are taken from [483, Sect. 4]. Equation (4.3.1) has often been used inside proofs in the literature and was mentioned for the general case, e.g., in [407].
Corollary 4.3.8 is based on an idea from [225, Sect. 5.5], where it was stated without proof for the case where $A$ is the zero-sublevel set of $\varphi_{A,k}$. There, $\varphi_k$ introduced by the right-hand side of (4.7.5) was used instead of $\varphi_{\{0\},k}$.
Proposition 4.3.9(a) was proved in [487, Lemma 2.9]. Proposition 4.3.11 and Example 4.3.12 were given for the function $\varphi_{Y \setminus \mathrm{int}\, A,-k}$ in [484, Prop. 3.7, Example 3.8]. Corollary 4.3.14 coincides with [483, Theorem 2.1(e)].

## 4.4 Lower Semicontinuous Translation Invariant Functions

We will now investigate properties of lower semicontinuous translation invariant functions. Furthermore, the relationship between $\varphi_{A,k}$ and $\varphi_{\mathrm{cl}\, A,k}$ will be discussed.
(4.2.4) implies because of Proposition 3.3.12:

**Proposition 4.4.1** *Assume that $Y$ is a topological vector space, $A \subseteq Y$ and $k \in Y \setminus \{0\}$. Then $\varphi_{A,k}$ is lower semicontinuous if and only if $\mathrm{cl}_k(A - \mathbb{R}_+ k)$ is a closed set.*

For the next theorem, recall that $0^+A \subseteq 0^+(\mathrm{cl}_k(A))$ because of Proposition 2.3.29.

**Theorem 4.4.2** *Assume that $Y$ is a topological vector space, $A \subseteq Y$ and*

$$k \in -0^+(\mathrm{cl}_k(A)) \setminus \{0\}.$$

*The following conditions are equivalent to each other:*

(a) $\varphi_{A,k}$ *is lower semicontinuous.*
(b) $\mathrm{cl}_k(A) = \mathrm{cl}\, A$.
(c) $\mathrm{lev}_{\varphi_{A,k},\leq}(t) = \mathrm{cl}\, A + tk$ *holds for all* $t \in \mathbb{R}$.

*If $k \in -0^+A$, then each of these conditions is equivalent to:*

(d) $\mathrm{cl}\, A - \mathbb{R}_> k \subseteq A$.

***Proof*** Because of (4.2.18) and Proposition 3.3.12, $\varphi_{A,k}$ is lower semicontinuous if and only if $\mathrm{cl}_k(A)$ is a closed set. This is equivalent to $\mathrm{cl}_k(A) = \mathrm{cl}\, A$ by Proposition 2.3.54(e).

(b) is equivalent to (c) because of (4.2.18).

(b) is equivalent to (d) for $k \in -0^+A$ by Proposition 2.3.54(f). □

We can restrict the class of sets $A$ and vectors $k$ which have to be used in order to construct all lower semicontinuous translation invariant functions.

**Theorem 4.4.3** *Assume $Y$ to be a topological vector space, $k \in Y \setminus \{0\}$. The class*

$$\{\varphi_{A,k} \colon Y \to \overline{\mathbb{R}} \mid A \subseteq Y,\ \varphi_{A,k} \text{ lower semicontinuous}\} \tag{4.4.1}$$

*of all lower semicontinuous functions on $Y$ which are translation invariant along $k$ coincides with*

$$\{\varphi_{A,k} \colon Y \to \overline{\mathbb{R}} \mid A \subseteq Y,\ \mathrm{cl}\, A - \mathbb{R}_> k \subseteq A\} \tag{4.4.2}$$

*and with*

$$\{\varphi_{A,k} \colon Y \to \overline{\mathbb{R}} \mid A \subseteq Y \text{ closed},\ k \in -0^+A \setminus \{0\}\}. \tag{4.4.3}$$

***Proof***

(a) Take first any $A \subseteq Y$ for which $\varphi_{A,k}$ is lower semicontinuous. Because of Theorem 4.2.3, there exists $\tilde{A} \subseteq Y$ with $k \in -0^+\tilde{A} \setminus \{0\}$ and $\varphi_{A,k} = \varphi_{\tilde{A},k}$. We get $\mathrm{cl}\, \tilde{A} - \mathbb{R}_> k \subseteq \tilde{A}$ because of Theorem 4.4.2. Hence, set (4.4.1) is contained in set (4.4.2).

(b) Take now any $A \subseteq Y$ with $\operatorname{cl} A - \mathbb{R}_{>}k \subseteq A$. Consider $\varphi := \varphi_{A,k}$. From $\operatorname{cl} A - \mathbb{R}_{>}k \subseteq A$, we obtain $k \in -0^{+}A \setminus \{0\}$ and $k \in -0^{+}(\operatorname{cl} A) \setminus \{0\}$. Theorem 4.4.2 results in $\operatorname{lev}_{\varphi,\leq}(t) = \operatorname{cl} A + tk$ for all $t \in \mathbb{R}$. Proposition 4.1.1 yields $\varphi = \varphi_{\operatorname{cl} A,k}$. Hence, set (4.4.2) is contained in set (4.4.3).

(c) Take any closed set $A \subseteq Y$ with $k \in -0^{+}A \setminus \{0\}$. Then $\operatorname{cl} A - \mathbb{R}_{>}k = A - \mathbb{R}_{>}k \subseteq A$. $\varphi_{A,k}$ is lower semicontinuous because of Theorem 4.4.2. Hence, set (4.4.3) is contained in set (4.4.1). □

**Theorem 4.4.4** *Assume*

$$(\mathrm{Sp1}_{A,k}):\ Y \text{ is a topological vector space, } A \text{ is a proper subset of } Y,\ k \in Y \setminus \{0\} \text{ and } \operatorname{cl} A - \mathbb{R}_{>}k \subseteq A.$$

*Then* $\varphi_{A,k}$ *is lower semicontinuous,*

$$\operatorname{dom} \varphi_{A,k} = A + \mathbb{R}k = A + \mathbb{R}_{>}k \neq \emptyset, \tag{4.4.4}$$

$$\operatorname{lev}_{\varphi_{A,k},\leq}(t) = \operatorname{cl} A + tk \ \text{ for all } t \in \mathbb{R}, \tag{4.4.5}$$

$$\operatorname{int} A + tk \subseteq \operatorname{lev}_{\varphi_{A,k},<}(t) \ \text{ for all } t \in \mathbb{R}, \tag{4.4.6}$$

$$\operatorname{lev}_{\varphi_{A,k},=}(t) \subseteq \operatorname{bd} A + tk \ \text{ for all } t \in \mathbb{R}, \tag{4.4.7}$$

$$\operatorname{epi} \varphi_{A,k} = \{(y,t) \in Y \times \mathbb{R} \mid y \in \operatorname{cl} A + tk\}, \tag{4.4.8}$$

$$\varphi_{A,k} = \varphi_{\operatorname{cl} A,k}. \tag{4.4.9}$$

*Moreover:*

(a) $\forall y \in Y:\ (\varphi_{A,k}(y) = -\infty \iff y + \mathbb{R}k \subseteq A)$.
(b) *If* $k \in 0^{+}A$, *then* $\operatorname{dom} \varphi_{A,k} = A$ *and* $\varphi_{A,k}(y) = -\infty$ *for all* $y \in A$.
(c) $\varphi_{A,k}$ *is finite-valued on* $\operatorname{dom} \varphi_{A,k} \setminus A$.
(d) *If* $k \in -\operatorname{core} 0^{+}(\operatorname{cl} A)$, *then* $\varphi_{A,k}$ *is finite-valued and*

$$\operatorname{lev}_{\varphi_{A,k},<}(t) = \operatorname{core} A + tk \ \text{ for all } t \in \mathbb{R}.$$

(e) *If* $\varphi_{A,k}$ *is proper, then*

$$\operatorname{cl} A = \operatorname{bd} A - \mathbb{R}_{+}k \tag{4.4.10}$$

*and*

$$\operatorname{dom} \varphi_{A,k} = \operatorname{bd} A + \mathbb{R}k. \tag{4.4.11}$$

(f) $\varphi_{A,k}$ *is quasiconvex* $\iff$ $\varphi_{A,k}$ *is convex* $\iff$ $\operatorname{cl} A$ *is convex.*
(g) $\varphi_{A,k}$ *is positively homogeneous* $\iff$ $\operatorname{cl} A$ *is a cone.*
(h) $\varphi_{A,k}$ *is subadditive* $\iff$ $\operatorname{cl} A + \operatorname{cl} A \subseteq \operatorname{cl} A$.
(i) $\varphi_{A,k}$ *is sublinear* $\iff$ $\operatorname{cl} A$ *is a convex cone.*
(j) *We have*

$$\begin{aligned}\varphi_{A,k}(y) &= \min\{t \in \mathbb{R} \mid y \in \operatorname{cl} A + tk\} \\ &= \min\{t \in \mathbb{R} \mid y \in \operatorname{bd} A + tk\}\end{aligned} \tag{4.4.12}$$

*for each* $y \in Y$ *with* $\varphi_{A,k}(y) \in \mathbb{R}$.

***Proof*** Theorem 4.4.2 implies the lower semicontinuity of $\varphi_{A,k}$ and (4.4.5). Inclusion (4.4.6) results from (4.2.13) since $\operatorname{int} A \subseteq \operatorname{core} A$. Inclusion (4.4.7) follows immediately from (4.4.5) and (4.4.6). Equation (4.4.8) is equivalent to (4.4.5). We obtain (4.4.9) from (4.2.8) since Proposition 2.3.54(f) yields $\operatorname{cl} A = \operatorname{cl}_k(A)$.

(a) , (c) and (4.4.4) follow from Theorem 4.2.5.
(b) is implied by (4.4.4) and (a).
(d) Apply Theorem 4.2.9 to $\varphi_{\operatorname{cl} A,k}$. The assertion results from (4.4.9) and Lemma 2.3.16(b).
(e) Assume first that $A$ is closed.
Take any $a \in A$. Because of (a), there exists $t \in \mathbb{R}$ with $a + tk \notin A$. ($\mathrm{Sp}1_{A,k}$) implies $t > 0$. Because of Lemma 2.1.26, there exists $\lambda \in (0, 1]$ with $\lambda a + (1 - \lambda)(a + tk) \in \operatorname{bd} A$, i.e., $a + (1 - \lambda)tk \in \operatorname{bd} A$. Thus, $a \in \operatorname{bd} A - \mathbb{R}_+k$. Consequently, $A \subseteq \operatorname{bd} A - \mathbb{R}_+k$. This and ($\mathrm{Sp}1_{A,k}$) yield (4.4.10). We get (4.4.11) from (4.4.4).
Assume now that $A$ is an arbitrary set fulfilling ($\mathrm{Sp}1_{A,k}$). Then ($\mathrm{Sp}1_{\operatorname{cl} A,k}$) holds. Since we proved (e) for closed sets $A$, we can apply it to $\varphi_{\operatorname{cl} A,k}$. Together with (4.4.9) and Lemma 2.3.16(d), this yields the assertion.
(f) –(i) The assumptions imply $k \in -0^+(\operatorname{cl} A)$ and $\operatorname{cl} A = \operatorname{cl}_k(A)$ by Proposition 2.3.54(f). Hence, $\operatorname{cl}_k(A - \mathbb{R}_+k) = \operatorname{cl}_k(A) - \mathbb{R}_+k = \operatorname{cl} A - \mathbb{R}_+k = \operatorname{cl} A$. The assertions result from Theorem 4.2.10.
(j) $\varphi_{A,k}(y) \in \mathbb{R}$ yields $\varphi_{A,k}(y) = \min\{t \in \mathbb{R} \mid \varphi_{A,k}(y) \le t\} = \min\{t \in \mathbb{R} \mid y \in \operatorname{cl} A + tk\}$ because of (4.4.5). Thus, we get the first equation. The second equation results from (4.4.7). □

($\mathrm{Sp}1_{A,k}$) implies further properties of $A$ and $k$, which were proved in Lemma 2.3.16.

Theorem 4.4.4 is often applied to closed sets $A$. For such sets, assumption ($\mathrm{Sp}1_{A,k}$) coincides with:

$$Y \text{ is a topological vector space, } A \text{ is a proper closed subset of } Y, \text{ and } k \in -0^+A \setminus \{0\}. \qquad (\mathrm{H}1_{A,k})$$

The sublevel sets of $\varphi_{A,k}$ under assumption ($\mathrm{H}1_{A,k}$) are illustrated in Fig. 4.2.

Alternative formulations of assumption ($\mathrm{H}1_{A,k}$) result immediately from Proposition 4.2.6.

In general, the inclusion in (4.4.7) is strict. The following example illustrates that $\varphi_{A,k}$ in Theorem 4.4.4 can take values other than zero on the boundary of $A$ (finite ones as well as $-\infty$) and that $\{y \in Y \mid \varphi_{A,k}(y) < t\} = \operatorname{int} A + tk$ does not necessarily hold.

**Example 4.4.5** In $Y = \mathbb{R}^2$, consider $k := (1, 0)^T$ and

$$A := (-\mathbb{R}^2_+) \cup \{(y_1, y_2)^T \in \mathbb{R}^2 \mid y_1 \le -1\} \cup \{(y_1, y_2)^T \in \mathbb{R}^2 \mid y_2 \le -1\}$$

(see Fig. 4.3). Then $(\mathrm{H}1_{A,k})$ holds, and

$$\varphi_{A,k}((y_1,y_2)^{\mathrm{T}}) = \begin{cases} -\infty & \text{if } y_2 \le -1, \\ y_1 & \text{if } -1 < y_2 \le 0, \\ y_1+1 & \text{if } y_2 > 0. \end{cases}$$

In particular, $\varphi_{A,k}((0,-1)^{\mathrm{T}}) = -\infty$ and $\varphi_{A,k}((-1,0)^{\mathrm{T}}) = -1$, although $(0,-1)^{\mathrm{T}} \in \operatorname{bd} A$ and $(-1,0)^{\mathrm{T}} \in \operatorname{bd} A$. Consequently, $\operatorname{lev}_{\varphi_{A,k},=}(0) \ne \operatorname{bd} A$, and $\operatorname{lev}_{\varphi_{A,k},<}(0) \ne \operatorname{int} A$.

core $A$ and int $A$ may differ under assumption $(\mathrm{H}1_{A,k})$ and thus also under assumption $(\mathrm{Sp}1_{A,k})$.

**Example 4.4.6** In $Y = \mathbb{R}^2$, take $k := (1,0)^{\mathrm{T}}$ and

$$A := \mathbb{R}^2 \setminus \{(y_1,y_2)^{\mathrm{T}} \in \mathbb{R}^2 \mid y_2 \ne 0,\, y_1 > 0,\, -y_1^2 < y_2 < y_1^2\}$$

(see Fig. 4.4). Then $(\mathrm{H}1_{A,k})$ holds, but $(0,0)^{\mathrm{T}} \in \operatorname{core} A \setminus \operatorname{int} A$.

Moreover, $(0,0)^{\mathrm{T}} \notin \operatorname{core}(\operatorname{core} A)$.

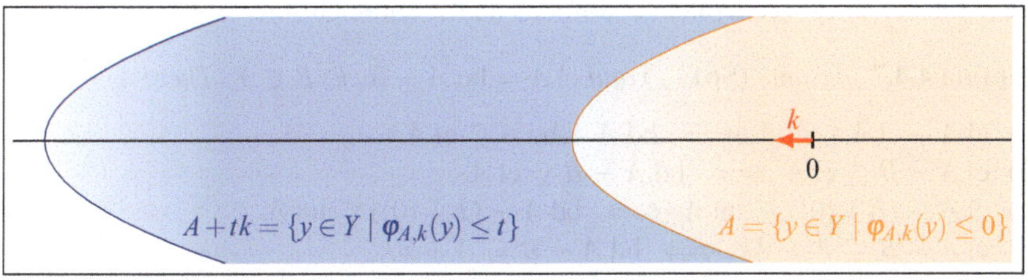


**Fig. 4.2** Sublevel sets of $\varphi_{A,k}$ under assumption $(\mathrm{H}1_{A,k})$

**Fig. 4.3** Set $A$ and vector $k$ in Example 4.4.5

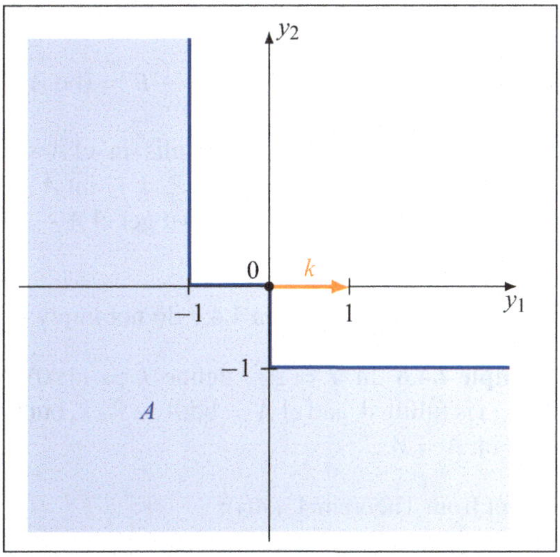

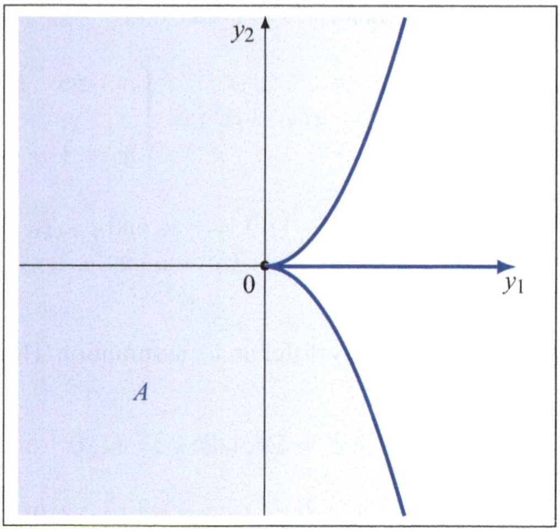


**Fig. 4.4** Set $A$ in Example 4.4.6

In parts of Theorem 4.4.4 and when investigating the monotonicity of $\varphi_{A,k}$, assertions can be proved which make assumptions on the boundary of $A$ only. Such assumptions are not necessarily weaker than conditions on the entire closure of $A$.

**Lemma 4.4.7** *Assume* $(\mathrm{Sp1}_{A,k})$ *and* $\operatorname{cl} A = \operatorname{bd} A - \mathbb{R}_+k$, $B \subset Y$. *Then:*

(a) $\operatorname{cl} A + \operatorname{cl} A \subseteq \operatorname{cl} A \iff \operatorname{bd} A + \operatorname{bd} A \subseteq \operatorname{cl} A$.
(b) $\operatorname{cl} A - B \subseteq \operatorname{cl} A \iff \operatorname{bd} A - B \subseteq \operatorname{cl} A$.
(c) $\operatorname{cl} A - (B \setminus \{0\}) \subseteq \operatorname{int} A \iff \operatorname{bd} A - (B \setminus \{0\}) \subseteq \operatorname{int} A$.
(d) $\operatorname{cl} A - B \subseteq A + \mathbb{R}k \iff \operatorname{bd} A - B \subseteq A + \mathbb{R}k$.

***Proof***

(a) $\operatorname{bd} A + \operatorname{bd} A \subseteq \operatorname{cl} A$ yields $\operatorname{cl} A + \operatorname{cl} A = (\operatorname{bd} A - \mathbb{R}_+k) + (\operatorname{bd} A - \mathbb{R}_+k) \subseteq \operatorname{cl} A - \mathbb{R}_+k \subseteq \operatorname{cl} A$.
(b) $\operatorname{bd} A - B \subseteq \operatorname{cl} A$ implies $\operatorname{cl} A - B = (\operatorname{bd} A - \mathbb{R}_+k) - B = (\operatorname{bd} A - B) - \mathbb{R}_+k \subseteq \operatorname{cl} A - \mathbb{R}_+k \subseteq \operatorname{cl} A$.
(c) $\operatorname{bd} A - (B \setminus \{0\}) \subseteq \operatorname{int} A$ results in $\operatorname{cl} A - (B \setminus \{0\}) = (\operatorname{bd} A - \mathbb{R}_+k) - (B \setminus \{0\}) = (\operatorname{bd} A - (B \setminus \{0\})) - \mathbb{R}_+k \subseteq \operatorname{int} A - \mathbb{R}_+k \subseteq \operatorname{int} A$.
(d) From $\operatorname{bd} A - B \subseteq A + \mathbb{R}k$, we get $\operatorname{cl} A - B = (\operatorname{bd} A - \mathbb{R}_+k) - B = (\operatorname{bd} A - B) - \mathbb{R}_+k \subseteq A + \mathbb{R}k$. □

The assumptions of Lemma 4.4.7 do not imply $\operatorname{bd} A - \mathbb{R}_>k \subseteq \operatorname{int} A$.

**Example 4.4.8** In $Y = \mathbb{R}^2$, define $k := (1,0)^T$, $A := \{(y_1, y_2)^T \mid y_2 = 0\}$. Then $(\mathrm{H1}_{A,k})$ is fulfilled, and $\operatorname{cl} A = \operatorname{bd} A - \mathbb{R}_+k$, but $\operatorname{bd} A - \mathbb{R}_>k \subseteq \operatorname{int} A$ is not satisfied since $\operatorname{int} A = \emptyset$.

We get from Theorem 4.4.4(j):

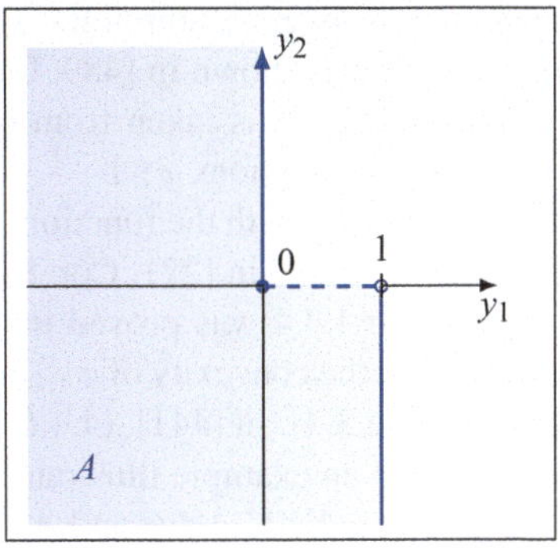


**Fig. 4.5** Set $A$ in Example 4.4.11

**Corollary 4.4.9** *Assume* $(\mathrm{Sp1}_{A,k})$ *and that* $\varphi_{A,k}$ *attains a minimal value* $t_0$ *on* $F \subset Y$. *Then*

$$\operatorname{argmin}_F \varphi_{A,k} = F \cap (\operatorname{cl} A + t_0 k) = F \cap (\operatorname{bd} A + t_0 k).$$

Theorem 4.4.4 contains a sufficient condition for $\varphi_{A,k} = \varphi_{\operatorname{cl} A,k}$. We will now discuss the relationship between these two functions.

The following is due to Corollary 4.3.14.

**Proposition 4.4.10** *Assume that* $Y$ *is a topological vector space,* $A \subseteq Y$ *and* $k \in Y \setminus \{0\}$.

*Then* $\operatorname{dom} \varphi_{A,k} \subseteq \operatorname{dom} \varphi_{\operatorname{cl} A,k}$, *and* $\varphi_{\operatorname{cl} A,k} \leq \varphi_{A,k}$.

Nevertheless, $\varphi_{A,k}$ can differ from $\varphi_{\operatorname{cl} A,k}$ on $\operatorname{dom} \varphi_{A,k}$ even for $k \in -0^+A$ . In the first example, we will illustrate this for finite-valued functionals, in the second and third for a nontrivial pointed convex cone $A$, where both functions will be proper in Example 4.4.12, whereas $\varphi_{\operatorname{cl} A,k}$ will not be proper in Example 4.4.13.

**Example 4.4.11** In $Y = \mathbb{R}^2$, consider $A := \{(y_1, y_2)^T \in \mathbb{R}^2 \mid y_1 \leq 0\} \cup \{(y_1, y_2)^T \in \mathbb{R}^2 \mid y_1 \leq 1, y_2 < 0\}$ (see Fig. 4.5) and $k := (1, 0)^T$. The functions $\varphi_{A,k}$ and $\varphi_{\operatorname{cl} A,k}$ are finite-valued. We get $\varphi_{A,k}((0,0)^T) = 0$, but $\varphi_{\operatorname{cl} A,k}((0,0)^T) = -1$.

**Example 4.4.12** In $Y = \mathbb{R}^3$, take $A := \mathbb{R}^3_+ \setminus \{(y_1, y_2, y_3)^T \in \mathbb{R}^3 \mid y_1 > 0,\ y_2 < y_1,\ y_3 = 0\}$ and $k := (0, -1, 0)^T \in -0^+A$. Then $\varphi_{A,k}$ and $\varphi_{\operatorname{cl} A,k}$ are proper and have the same effective domain. We have $\varphi_{A,k}((1,0,0)^T) = 1$, however $\varphi_{\operatorname{cl} A,k}((1,0,0)^T) = 0$. Nevertheless, $\operatorname{lev}_{\varphi_{A,k},\leq}(t) = \operatorname{cl} A + tk$ holds for all $t \in \mathbb{R}$.

**Example 4.4.13** Let $A$ be the lexicographic ordering cone in $Y = \mathbb{R}^2$ (see Example 2.3.56), and $k := (0, -1)^T$. $A$ is a nontrivial pointed convex cone. Then $\varphi_{A,k}((0,0)^T) = 0$, but $\varphi_{\operatorname{cl} A,k}((0,0)^T) = -\infty$.

**Remark 4.4.14** (*Bibliographical notes related to Sect.* 4.4)

A statement which is equivalent to Proposition 4.4.1 can be deduced from [221]. The equivalence between parts (a) and (b) of Theorem 4.4.2 was also stated in [222] under the stronger assumption $k \in -0^+A \setminus \{0\}$. The equivalence of (a) and (b) to (d) in this case was shown in [489, Lemma 2.6].

Theorem 4.4.3 is taken from [488, Theorem 5.5.13]. The equivalence between the family of functions $\varphi\colon Y \to \overline{\mathbb{R}}$ which are translation invariant along $k$ and lower semicontinuous with the function class $\{\varphi_{A,k} \mid A \subseteq Y \text{ closed},\ k \in -0^+A \setminus \{0\}\}$ was previously proved in [221, Cor. 8].

Theorem 4.4.4 was proved in [484, Theorem 2.9, Prop. 2.7] without (j). Equation (4.4.5), the convexity of $\varphi_{A,k}$ for convex cl $A$ and (4.4.9) were also stated in [164, Cor. 3.3 and 3.4]. In [441], (4.4.6) and (4.4.7) are also mentioned under assumption $(\mathrm{H1}_{A,k})$ and an example illustrates that the inclusion in (4.4.7) can be strict.

Examples 4.4.5, 4.4.6 and 4.4.8 can be found in [484, Sect. 2]. Lemma 4.4.7 coincides with [484, Lemma 2.10]. The equivalence of Lemma 4.4.7(b) was proved in [472, Satz 3.1.11] for any function which fulfills (4.4.5) with a closed set $A$ and in [476, Prop. 3] for the case that $\varphi_{A,k}$ is finite-valued. [192] contains Lemma 4.4.7 (a)–(c) assuming the stronger condition $(\mathrm{H2}_{A,k})$ and some additional assumptions.

Proposition 4.4.10 was proved in [484, Prop. 2.7]. Equation (4.4.9) was also stated in [164, Cor. 3.3]. Example 4.4.12 coincides with [484, Example 2.6].

## 4.5 Representation of Binary Relations by Translation Invariant Functions and Monotonicity

Binary relations, especially partial orders, can structure a space or express preferences in decision making and optimization. Thus, the representation of such relations by real-valued functions serves as a useful tool in proofs, e.g. in operator theory [310], but also as a basis for scalarization methods in vector optimization [472] and for the development of risk measures in mathematical finance [11].

If $C$ is an algebraically closed ordering cone in a linear space $Y$, then the corresponding order $\leq_C$ can be represented by $\varphi_{-C,k}$ with an arbitrary $k \in C \setminus \{0\}$ since

$$\forall y^1, y^2 \in Y\colon\ (y^1 \leq_C y^2 \iff \varphi_{-C,k}(y^1 - y^2) \leq 0).$$

This follows from (4.2.18) together with Proposition 2.3.35.

For this case, we will show in Proposition 4.7.1(b) that $\varphi_{-C,k}$ is $C$-monotone. Thus, we get

$$\forall y^1, y^2 \in Y\colon\ (y^1 \leq_C y^2 \implies \varphi_{-C,k}(y^1) \leq \varphi_{-C,k}(y^2)). \tag{4.5.1}$$

More generally, if a binary relation can be described by some proper algebraically closed subset $A$ of $Y$ with $0^+A \neq \{0\}$ as

$$\mathscr{R}_A = \{(y^1, y^2) \in Y \times Y \mid y^2 - y^1 \in A\},$$

then for each $k \in 0^+A \setminus \{0\}$,

$$\forall y^1, y^2 \in Y: \; (y^1 \mathscr{R}_A y^2 \iff \varphi_{-A,k}(y^1 - y^2) \leq 0).$$

If the function $\varphi_{-A,k}$ is additionally $A$-monotone, then

$$\forall y^1, y^2 \in Y: \; (y^1 \mathscr{R}_A y^2 \implies \varphi_{-A,k}(y^1) \leq \varphi_{-A,k}(y^2)).$$

The reverse implication is not true since it is not true for (4.5.1).

**Example 4.5.1** In $Y = \mathbb{R}^2$, consider $C := \mathbb{R}^2_+$ and $k := (1, 1)^T$. Then $\operatorname{dom} \varphi_{-C,k} = Y$. For $y^1 := (-1, -1)^T$ and $y^2 := (-2, 0)^T$, we have $\varphi_{-C,k}(y^1) = -1 \leq 0 = \varphi_{-C,k}(y^2)$, but $y^1 \leq_C y^2$ does not hold since $y^2 - y^1 = (-1, 1)^T \notin C$.

However the following local representation of $\mathscr{R}_A$ results from Proposition 4.3.5. For all $y^1, y^2 \in Y$,

$$y^1 \mathscr{R}_A y^2 \text{ is equivalent to } \varphi_{y^2 - A,k}(y^1) \leq 0.$$

Now, we will characterize monotonicity of $\varphi_{A,k}$, which will especially be applied to scalarization in vector optimization (see Chaps. 6 and 11).

**Theorem 4.5.2** *Assume* $(\mathrm{Sp0}_{A,k})$ *and* $B \subseteq Y$.

(a) $\varphi_{A,k}$ *is* $B$*-monotone* $\iff A - B \subseteq \mathrm{cl}_k(A) - \mathbb{R}_+ k$.
(b) $\varphi_{A,k}$ *is strictly* $B$*-monotone* $\implies \mathrm{cl}_k(A) - (B \setminus \{0\}) \subseteq A - \mathbb{R}_> k$.
*If* $\varphi_{A,k}$ *is strictly* $B$*-monotone and* $B \setminus \{0\} \neq \emptyset$, *then* $\varphi_{A,k}$ *is finite-valued.*

*Assume additionally that* $\varphi_{A,k}$ *is finite-valued on* $F \subseteq Y$.

(c) $A - B \subseteq \mathrm{cl}_k(A) - \mathbb{R}_+ k \implies \varphi_{A,k}$ *is strictly* $(B + \mathbb{R}_> k)$*-monotone on* $F$.
*Especially:*
$A - B \subseteq \mathrm{cl}_k(A) - \mathbb{R}_+ k \implies \varphi_{A,k}$ *is strictly* $(\mathrm{core}\, B)$*-monotone on* $F$.
(d) $\mathrm{cl}_k(A) - (B \setminus \{0\}) \subseteq A - \mathbb{R}_> k \implies \varphi_{A,k}$ *is strictly* $B$*-monotone on* $F$.
*Especially:*
$\mathrm{cl}_k(A) - (B \setminus \{0\}) \subseteq \mathrm{core}\, A \implies \varphi_{A,k}$ *is strictly* $B$*-monotone on* $F$.

***Proof*** The assertions are true if $A = \emptyset$ or $B = \emptyset$. Assume now $A \neq \emptyset$ and $B \neq \emptyset$.

(a) Assume first that $\varphi_{A,k}$ is $B$-monotone. Take any $a \in A$, $b \in B$. Since $a - (a - b) = b \in B$, we obtain $\varphi_{A,k}(a - b) \leq \varphi_{A,k}(a) \leq 0$, thus $a - b \in \mathrm{cl}_k(A) - \mathbb{R}_+ k$ by (4.2.4). Consequently, $A - B \subseteq \mathrm{cl}_k(A) - \mathbb{R}_+ k$.
Assume now $A - B \subseteq \mathrm{cl}_k(A) - \mathbb{R}_+ k$. Take any $y^1, y^2 \in Y$ with $y^2 - y^1 \in B$. If $\varphi_{A,k}(y^2) = +\infty$, then $\varphi_{A,k}(y^1) \leq \varphi_{A,k}(y^2)$. Assume $\varphi_{A,k}(y^2) \neq +\infty$. Then $y^2 \in \operatorname{dom} \varphi_{A,k} = A + \mathbb{R}k$. There exists a sequence $(t_n)_{n \in \mathbb{N}}$ of real numbers $t_n \geq \varphi_{A,k}(y^2)$ which converges to $\varphi_{A,k}(y^2)$ with $y^2 \in A + t_n k$ for all $n \in \mathbb{N}$. We get $y^1 \in y^2 - B \subseteq (A - B) + t_n k$ and thus $y^1 \in \mathrm{cl}_k(A) - \mathbb{R}_+ k + t_n k$ for all $n \in \mathbb{N}$. By (4.2.4), $\varphi_{A,k}(y^1) \leq t_n$ for all $n \in \mathbb{N}$. Thus, $\varphi_{A,k}(y^1) \leq \varphi_{A,k}(y^2)$. Hence, $\varphi_{A,k}$ is $B$-monotone.

(b) The first statement is satisfied if $B = \{0\}$. Assume now $B \neq \{0\}$. Take any $a \in \mathrm{cl}_k(A)$ and $b \in B \setminus \{0\}$. From (4.2.4), we deduce that $\varphi_{A,k}(a) \leq 0$. Since $a - (a-b) = b \in B \setminus \{0\}$, we obtain $\varphi_{A,k}(a-b) < \varphi_{A,k}(a) \leq 0$. Thus, $a - b \in A - \mathbb{R}_> k$ by (4.2.3). Consequently, $\mathrm{cl}_k(A) - (B \setminus \{0\}) \subseteq A - \mathbb{R}_> k$. The second statement results from Corollary 3.7.5.

(c) and (d) hold if $F = \emptyset$. Assume from now on $F \neq \emptyset$.
Suppose $A - B \subseteq \mathrm{cl}_k(A) - \mathbb{R}_+ k$, and take any $y^1, y^2 \in F$ with $y^2 - y^1 \in B + \mathbb{R}_> k$. Then there exists $\lambda \in \mathbb{R}_>$ with $y^2 - y^1 \in B + \lambda k$. Hence, $(y^2 - \lambda k) - y^1 \in B$. Since $\varphi_{A,k}$ is $B$-monotone by (a), we get $\varphi_{A,k}(y^1) \leq \varphi_{A,k}(y^2 - \lambda k) = \varphi_{A,k}(y^2) - \lambda < \varphi_{A,k}(y^2)$ because of $\varphi_{A,k}(y^2) \in \mathbb{R}$. Thus, $\varphi_{A,k}$ is strictly $(B + \mathbb{R}_> k)$-monotone on $F$. The second statement results from Lemmata 2.3.17 and 3.7.2(a).

(d) Assume $\mathrm{cl}_k(A) - (B \setminus \{0\}) \subseteq A - \mathbb{R}_> k$. Take any $y^1, y^2 \in F$ with $y^2 - y^1 \in B \setminus \{0\}$. Then $t := \varphi_{A,k}(y^2) \in \mathbb{R}$. We get $y^2 \in \mathrm{cl}_k(A) + tk$ by (4.2.5). Hence, $y^1 \in y^2 - (B \setminus \{0\}) \subseteq (\mathrm{cl}_k(A) - (B \setminus \{0\})) + tk \subseteq A - \mathbb{R}_> k + tk$. This implies $\varphi_{A,k}(y^1) < t = \varphi_{A,k}(y^2)$. Consequently, $\varphi_{A,k}$ is strictly $B$-monotone on $F$. By Lemma 2.3.17, $A - (B \setminus \{0\}) \subseteq A - \mathbb{R}_> k$ is satisfied if $A - (B \setminus \{0\}) \subseteq \mathrm{core}\, A$. □

Theorem 4.5.2 implies:

**Corollary 4.5.3** *Assume that $Y$ is a linear space, that $A \subseteq Y$ is a $k$-directionally closed set with $k \in -0^+ A \setminus \{0\}$ and $F, B \subseteq Y$.*

(a) *$\varphi_{A,k}$ is $B$-monotone $\iff A - B \subseteq A$.*
(b) *Assume additionally that $\varphi_{A,k}$ is finite-valued on $F$. Then:*
*$A - B \subseteq A \implies \varphi_{A,k}$ is strictly $(B + \mathbb{R}_> k)$-monotone on $F$.*
*Especially: $A - B \subseteq A \implies \varphi_{A,k}$ is strictly* (core $B$)*-monotone on $F$.*

***Proof*** Clearly, $A - B \subseteq A$ implies $A - B \subseteq \mathrm{cl}_k(A) - \mathbb{R}_+ k$. For a $k$-directionally closed set $A$ with $k \in -0^+ A$, the inclusion $A - B \subseteq \mathrm{cl}_k(A) - \mathbb{R}_+ k$ yields $A - B \subseteq A$. □

Note that the condition $A - B \subseteq A$ is fulfilled if $B \subseteq -0^+ A$. In general, Corollary 4.5.3(a) is not satisfied without the $k$-directional closedness of $A$.

**Example 4.5.4** In $Y = \mathbb{R}^2$, take $A := -\,\mathrm{int}\, \mathbb{R}^2_+ \cup \{(0,0)^T\}$, $B := \mathbb{R}^2_+$, and $k := (1,1)^T$. Then $k \in -0^+ A$ and $A - B \subseteq \mathrm{cl}_k(A) - \mathbb{R}_+ k = -\mathbb{R}^2_+$. The function $\varphi_{A,k}$ is $B$-monotone because of Theorem 4.5.2, but $A - B = -\mathbb{R}^2_+ \not\subseteq A$.

**Theorem 4.5.5** *Assume* $(\mathrm{Sp1}_{A,k})$ *and $B \subseteq Y$.*

(a) *$\varphi_{A,k}$ is $B$-monotone $\iff A - B \subseteq \mathrm{cl}\, A$.*
(b) *$\varphi_{A,k}$ is strictly $B$-monotone $\implies \mathrm{cl}\, A - (B \setminus \{0\}) \subseteq A - \mathbb{R}_> k$.*

*Assume additionally that $\varphi_{A,k}$ is finite-valued on $F \subseteq Y$.*

(c) *$A - B \subseteq \mathrm{cl}\, A \implies \varphi_{A,k}$ is strictly $(B + \mathbb{R}_> k)$-monotone on $F$.*
*Especially: $A - B \subseteq \mathrm{cl}\, A \implies \varphi_{A,k}$ is strictly* (core $B$)*-monotone on $F$.*

(d) $\operatorname{cl} A-(B\setminus\{0\})\subseteq A-\mathbb{R}_{>}k \implies \varphi_{A,k}$ *is strictly* $B$*-monotone on* $F$.
*Especially:* $\operatorname{cl} A-(B\setminus\{0\})\subseteq \operatorname{core} A \implies \varphi_{A,k}$ *is strictly* $B$*-monotone on* $F$.

*Assume now additionally that* $\varphi_{A,k}$ *is finite-valued.*

(e) *If* $A-\mathbb{R}_{>}k\subseteq \operatorname{int} A$, *then:*
$\varphi_{A,k}$ *is strictly* $B$*-monotone* $\iff \operatorname{cl} A-(B\setminus\{0\})\subseteq \operatorname{int} A$.
(f) $\varphi_{A,k}$ *is strictly* $(\operatorname{int} B)$*-monotone* $\iff \varphi_{A,k}$ *is* $(\operatorname{int} B)$*-monotone.*
*If* $B\subseteq \operatorname{cl}\operatorname{int} B$, *then:*
$\varphi_{A,k}$ *is strictly* $(\operatorname{int} B)$*-monotone* $\iff \varphi_{A,k}$ *is* $B$*-monotone.*

***Proof*** The assumption implies $k\in -0^+A$ and $k\in -0^+(\operatorname{cl} A)$. We get $\operatorname{cl}_k(A)=\operatorname{cl} A$ from Theorem 4.4.2. Hence, $\operatorname{cl} A-\mathbb{R}_+k=\operatorname{cl} A$, and $A-B\subseteq \operatorname{cl}_k(A)-\mathbb{R}_+k$ is equivalent to $A-B\subseteq \operatorname{cl} A$. Theorem 4.5.2 yields (a)–(d).

(e) We obtain $\operatorname{int} A\subseteq \operatorname{core} A\subseteq A-\mathbb{R}_{>}k$ by Lemma 2.3.17. Hence, $\operatorname{int} A=A-\mathbb{R}_{>}k$. The assertion follows from (b) and (d).
(f) $(\mathrm{Sp}1_{A,k})$ implies $\operatorname{int}\operatorname{cl} A=\operatorname{int} A$ by Lemma 2.3.16(c). Assume that $\varphi_{A,k}$ is $(\operatorname{int} B)$-monotone. Then $A-\operatorname{int} B\subseteq \operatorname{cl} A$ because of (a). Thus, $\operatorname{cl} A-\operatorname{int} B\subseteq \operatorname{cl} A$, and $\operatorname{cl} A-\operatorname{int} B\subseteq \operatorname{int}\operatorname{cl} A=\operatorname{int} A\subseteq \operatorname{core} A$. Hence, $\varphi_{A,k}$ is strictly $(\operatorname{int} B)$-monotone because of (d). The assertion follows by Lemma 3.7.2 since $\varphi_{A,k}$ is lower semicontinuous because of Theorem 4.4.4. □

In parts of Theorem 4.5.5, assertions can be proved which make assumptions on the boundary of $A$ only instead on the whole set $A$. According to Lemma 4.4.7, such assumptions are not necessarily weaker than conditions for the entire set $A$.

Since monotonicity is often required for finite-valued functions $\varphi_{A,k}$ which are defined using a closed set $A$, we summarize some results of Theorem 4.5.5 for this case.

**Corollary 4.5.6** *Assume* $(\mathrm{H}1_{A,k})$, *that* $\varphi_{A,k}$ *is finite-valued and* $B\subseteq Y$.

(a) $\varphi_{A,k}$ *is* $B$*-monotone* $\iff A-B\subseteq A$.
(b) $A-(B\setminus\{0\})\subseteq \operatorname{core} A \implies \varphi_{A,k}$ *is strictly* $B$*-monotone.*
(c) *If* $A-\mathbb{R}_{>}k\subseteq \operatorname{core} A$, *then:*
$\varphi_{A,k}$ *is strictly* $B$*-monotone* $\iff A-(B\setminus\{0\})\subseteq \operatorname{core} A$.
(d) *If* $A-\mathbb{R}_{>}k\subseteq \operatorname{int} A$, *then:*
$\varphi_{A,k}$ *is strictly* $B$*-monotone* $\iff A-(B\setminus\{0\})\subseteq \operatorname{int} A$.

There is a relationship between the monotonicity of $\varphi_{A,k}$ w.r.t. cones on the one hand and the Clarke tangent cones to $A$ on the other hand.

**Proposition 4.5.7** *Assume* $(\mathrm{H}1_{A,k})$ *and that* $B\subseteq Y$ *is a cone. Then:*

$$\varphi_{A,k} \text{ is } B\text{-monotone} \iff B\subseteq -0^+A.$$

*If* $Y$ *is a Banach space, then:*

$$\varphi_{A,k} \text{ is } B\text{-monotone} \iff \forall a\in A:\; B\subseteq -\mathrm{T_C}(a;A).$$

***Proof*** By Theorem 4.5.5(a), $\varphi_{A,k}$ is $B$-monotone if and only if $A - B \subseteq A$. Since $B$ is a cone, $A - B \subseteq A$ is equivalent to $B \subseteq -0^+ A$. The last assertion of the proposition follows from Theorem 3.6.15. □

Theorem 4.5.2 contains some special cases which are of importance in vector optimization.

**Corollary 4.5.8** *Assume* $(\mathrm{Sp0}_{A,k})$.

(a) $\varphi_{A,k}$ *is* $(-0^+ A)$*-monotone.*
(b) *If* $\varphi_{A,k}$ *is finite-valued on* $F \subseteq Y$, *then* $\varphi_{A,k}$ *is strictly* $(-\operatorname{core} 0^+ A)$*-monotone on* $F$.
(c) *If* $A$ *is a proper subset of* $Y$ *and* $k \in -\operatorname{core} 0^+ A$, *then* $\varphi_{A,k}$ *is finite-valued and strictly* $(-\operatorname{core} 0^+ A)$*-monotone.*

Part (c) of Corollary 4.5.8 results from (b) because of Theorem 4.2.9.

We get the following statement from Theorem 4.5.5(a) and (c).

**Corollary 4.5.9** *Assume* $(\mathrm{Sp1}_{A,k})$.

(a) $\varphi_{A,k}$ *is* $(-0^+(\operatorname{cl} A))$*-monotone.*
(b) *If* $\varphi_{A,k}$ *is finite-valued on* $F \subseteq Y$, *then* $\varphi_{A,k}$ *is strictly* $(-\operatorname{core} 0^+(\operatorname{cl} A))$*-monotone on* $F$.

Furthermore, Theorem 4.5.2 yields the next corollary.

**Corollary 4.5.10** *Assume* $(\mathrm{Sp0}_{A,k})$ *and* $A + A \subseteq A$.

(a) $\varphi_{A,k}$ *is* $(-A)$*-monotone.*
(b) *If* $\varphi_{A,k}$ *is finite-valued on* $F \subseteq Y$, *then* $\varphi_{A,k}$ *is strictly* $(-A + \mathbb{R}_{>}k)$*-monotone on* $F$, *and thus strictly* $(-\operatorname{core} A)$*-monotone on* $F$.

The condition in Corollary 4.5.10(b) cannot be omitted, which will be illustrated in Example 4.7.4(b).

The assumptions of all parts of Corollary 4.5.10 do not imply that $A$ is a cone or a shifted cone even if $A$ is a closed convex set and $\varphi_{A,k}$ is proper.

**Example 4.5.11** In $Y = \mathbb{R}^2$, consider $A := \{(y_1, y_2)^T \in \mathbb{R}^2 \mid y_2 \geq \frac{1}{y_1}, y_1 > 0\}$ and $k := (-1, 0)^T$. Then $k \in -0^+ A$, and $A$ is a proper closed convex subset of $Y$ for which $A + A \subseteq A$ holds. $\varphi_{A,k}$ is finite-valued on $\operatorname{dom} \varphi_{A,k} = \{(y_1, y_2)^T \in \mathbb{R}^2 \mid y_2 > 0\}$. However, $A$ is not a cone or a shifted cone.

Proposition 3.7.7 results in a statement about monotonicity on the effective domain. Recall that $\operatorname{dom} \varphi_{A,k} = A + \mathbb{R}k$.

**Proposition 4.5.12** *Assume* $(\mathrm{Sp0}_{A,k})$ *and* $B \subseteq Y$.

$\varphi_{A,k}$ *is* $B$*-monotone if and only if the following conditions hold:*

(a) $\varphi_{A,k}$ *is* $B$*-monotone on* $\operatorname{dom} \varphi_{A,k}$,
(b) $A - B \subseteq A + \mathbb{R}k$.

Furthermore, we have:

**Proposition 4.5.13** *Assume* $(\text{Sp0}_{A,k})$ *and that* $\varphi_{A,k}$ *is* $B$*-monotone for some set* $B \subseteq Y$ *with* $Y = B + \mathbb{R}k$*. Then* $\varphi_{A,k}$ *is finite-valued or* $\varphi_{A,k} = +\infty$ *or* $\varphi_{A,k} = -\infty$*.*

***Proof*** Take any $y \in Y$.

Assume first $\varphi_{A,k}(y^0) = +\infty$ for some $y^0 \in Y$. Since $Y = B + \mathbb{R}k$, there exists $t \in \mathbb{R}$ with $y - y^0 \in B + tk$. The $B$-monotonicity yields $\varphi_{A,k}(y^0) \leq \varphi_{A,k}(y - tk) = \varphi_{A,k}(y) - t$. Hence, $\varphi_{A,k}(y) = +\infty$.

Assume now $\varphi_{A,k}(y^0) = -\infty$ for some $y^0 \in Y$. Since $Y = B + \mathbb{R}k$, there exists $t \in \mathbb{R}$ with $y^0 - y \in B + tk$. The $B$-monotonicity implies $\varphi_{A,k}(y^0) \geq \varphi_{A,k}(y + tk) = \varphi_{A,k}(y) + t$. Hence, $\varphi_{A,k}(y) = -\infty$. □

**Remark 4.5.14** (*Bibliographical notes related to Sect.* 4.5)

Theorem 4.5.2 was proved in [490, Theorem 2.5]. [213, Theorem 8] contains an equivalent statement to Theorem 4.5.2(a). For finite-valued functions, (b) and the first statement of (d) were also given in [213, p. 2684].

For the case where $A$ coincides with the $k$-radial closure of $A - \mathbb{R}_+ k$, Corollary 4.5.3(a) was mentioned in [407, Satz 3.1.9]. Corollary 4.5.9 was proved in [484, Cor. 2.17]. Corollaries 4.5.8 and 4.5.10 without the strict $(-\operatorname{core} A)$-monotonicity as well as Example 4.5.11 are taken from [483, Cor. 3.1 and 3.2, Example 3.3].

## 4.6 Continuous Translation Invariant Functions and Lipschitz Continuity

We will now investigate continuous translation invariant functions. By Theorem 4.4.3, the lower semicontinuous translation invariant functions are the functions which can be represented as $\varphi_{A,k}$ with $A$ and $k$ fulfilling the assumptions of the following lemma.

**Lemma 4.6.1** *Assume that* $Y$ *is a topological vector space,* $A \subseteq Y$*,* $k \in Y \setminus \{0\}$ *and*

$$\operatorname{cl} A - \mathbb{R}_> k \subseteq A.$$

*The following conditions are equivalent to each other:*

$$\varphi_{A,k} \text{ is continuous,}$$

$$\operatorname{cl} A - \mathbb{R}_> k \subseteq \operatorname{int} A, \tag{4.6.1}$$

$$\operatorname{bd} A - \mathbb{R}_> k \subseteq \operatorname{int} A, \tag{4.6.2}$$

$$\operatorname{lev}_{\varphi_{A,k},<}(t) = \operatorname{int} A + tk \;\; \text{for all } t \in \mathbb{R}, \tag{4.6.3}$$

$$\mathrm{lev}_{\varphi_{A,k},=}(t) = \mathrm{bd}\, A + tk \quad \textit{for all } t \in \mathbb{R}. \tag{4.6.4}$$

***Proof*** For $A = \emptyset$ and $A = Y$, the statements are obvious. Assume now that $A$ is a proper subset of $Y$. Then $(\mathrm{Sp}1_{A,k})$ is fulfilled.

(a) In this part of the proof, let $A$ be closed.
Assume first that (4.6.1) holds. Let $t \in \mathbb{R}$ and $y \in Y$ be such that $\varphi_{A,k}(y) < t$. Then there exists $\lambda \in \mathbb{R}$ with $\lambda < t$ and $y \in A + \lambda k$. It follows that $y \in A + \lambda k = (A - (t - \lambda)k) + tk \subseteq \mathrm{int} A + tk$. This results in (4.6.3) because of (4.4.6).
Assume now that (4.6.3) is satisfied. Take any $y \in A - \mathbb{R}_{>}k$. Then $\varphi_{A,k}(y) < 0$. We obtain $y \in \mathrm{int} A$ by (4.6.3). Thus, (4.6.1) is fulfilled.
(4.6.1) is equivalent to (4.6.2) because of $A - \mathbb{R}_{>}k \subseteq A$. Equation (4.6.3) is equivalent to (4.6.4) because of (4.4.5).
Consequently, (4.6.1)–(4.6.4) are equivalent to each other.
$\varphi_{A,k}$ is lower semicontinuous by Theorem 4.4.4. Assume first that (4.6.3) holds. Together with Lemma 3.3.7 and Proposition 3.3.12, this implies the upper semicontinuity and the continuity of $\varphi_{A,k}$.
Assume now that $\varphi_{A,k}$ is continuous. Then $\varphi_{A,k}$ is upper semicontinuous, and all sets $\mathrm{lev}_{\varphi_{A,k},<}(t)$ with $t \in \mathbb{R}$ are open. For each $t \in \mathbb{R}$, we obtain $\mathrm{int}\, A + tk \subseteq \mathrm{lev}_{\varphi_{A,k},<}(t) \subseteq A + tk$ by Theorem 4.4.4. This implies (4.6.3).

(b) Consider now an arbitrary set $A$ which fulfills the assumptions.
Then $\varphi_{A,k} = \varphi_{\mathrm{cl}\, A,k}$ by Theorem 4.4.4. Since we proved the lemma for closed sets $A$, we can apply it for $\varphi_{\mathrm{cl}\, A,k}$. Together with Lemma 2.3.16, this yields the assertion. □

A functional which fulfills condition (4.6.4), but is not semicontinuous and does not satisfy (4.6.3), will be given in Example 4.6.9.

A necessary and sufficient condition for continuity under weaker assumptions can be derived.

**Theorem 4.6.2** *Assume that $Y$ is a topological vector space, $A \subseteq Y$ and $k \in -0^+A \setminus \{0\}$. Then $\varphi_{A,k}$ is continuous if and only if* $\mathrm{cl}\, A - \mathbb{R}_{>}k \subseteq \mathrm{int}\, A$.

***Proof*** For $A = \emptyset$ and $A = Y$, the statement is obvious. Assume now that $A$ is a proper subset of $Y$. The inclusion $\mathrm{cl}\, A - \mathbb{R}_{>}k \subseteq \mathrm{int}\, A$ implies the continuity of $\varphi_{A,k}$ by Lemma 4.6.1. Assume now that $\varphi_{A,k}$ is continuous. Then $\varphi_{A,k}$ is lower semicontinuous. We get $\mathrm{cl}\, A - \mathbb{R}_{>}k \subseteq A$ from Theorem 4.4.2. Therefore, $\mathrm{cl}\, A - \mathbb{R}_{>}k \subseteq \mathrm{int}\, A$ follows from Lemma 4.6.1. □

We can restrict the class of sets $A$ and vectors $k$ which have to be used in order to construct all continuous translation invariant functions.

**Theorem 4.6.3** *Assume $Y$ to be a topological vector space, $k \in Y \setminus \{0\}$. The class*

$$\{\varphi_{A,k} \colon Y \to \overline{\mathbb{R}} \mid A \subseteq Y,\ \varphi_{A,k} \textit{ continuous}\} \tag{4.6.5}$$

*of all continuous functions on Y which are translation invariant along k coincides with*

$$\{\varphi_{A,k}\colon Y \to \overline{\mathbb{R}} \mid A \subseteq Y,\ \operatorname{cl} A - \mathbb{R}_{>}k \subseteq \operatorname{int} A\} \tag{4.6.6}$$

*and with*

$$\{\varphi_{A,k}\colon Y \to \overline{\mathbb{R}} \mid A \subseteq Y \text{ closed},\ A - \mathbb{R}_{>}k \subseteq \operatorname{int} A\}. \tag{4.6.7}$$

***Proof***

(a) Take first any $A \subseteq Y$ for which $\varphi_{A,k}$ is continuous. Because of Theorem 4.2.3, there exists $\tilde{A} \subseteq Y$ with $k \in -0^+\tilde{A} \setminus \{0\}$ and $\varphi_{A,k} = \varphi_{\tilde{A},k}$. Theorem 4.6.2 implies $\operatorname{cl} \tilde{A} - \mathbb{R}_{>}k \subseteq \operatorname{int} \tilde{A}$. Hence, set (4.6.5) is contained in set (4.6.6).

(b) Take now any $A \subseteq Y$ with $\operatorname{cl} A - \mathbb{R}_{>}k \subseteq \operatorname{int} A$. Consider $\varphi := \varphi_{A,k}$. We get $k \in -0^+A \setminus \{0\}$ from $\operatorname{cl} A - \mathbb{R}_{>}k \subseteq \operatorname{int} A$. Theorem 4.4.2 yields $\operatorname{lev}_{\varphi_{A,k},\leq}(t) = \operatorname{cl} A + tk$ for all $t \in \mathbb{R}$. Hence, $\varphi = \varphi_{\operatorname{cl} A,k}$ by Proposition 4.1.1. Because of $\operatorname{cl} A - \mathbb{R}_{>}k \subseteq \operatorname{int} A \subseteq \operatorname{int} \operatorname{cl} A$, set (4.6.6) is contained in set (4.6.7).

(c) Take any closed set $A \subseteq Y$ with $A - \mathbb{R}_{>}k \subseteq \operatorname{int} A$. Then $\varphi_{A,k}$ is continuous by Theorem 4.6.2. Hence, set (4.6.7) is contained in set (4.6.5). □

By Proposition 3.3.19, $\operatorname{dom} \varphi_{A,k}$ has to be open if $\varphi_{A,k}$ is continuous.

In Example 4.4.5, $\operatorname{dom} \varphi_{A,k} = \mathbb{R}^2$ is open, but $\operatorname{cl} A - \mathbb{R}_{>}k \not\subseteq \operatorname{int} A$ and $\varphi_{A,k}$ is not continuous at $(0,-1)^{\mathrm{T}}$.

A function $\varphi_{A,k}$ with an open effective domain can be continuous on its effective domain without being lower semicontinuous.

**Example 4.6.4** In $Y = \mathbb{R}^2$, take $A := \{(y_1, y_2)^{\mathrm{T}} \in \mathbb{R}^2 \mid y_1 > 0,\ y_2 \geq y_1\}$ and $k := (0,-1)^{\mathrm{T}} \in -0^+A$. Then $\varphi_{A,k}$ is continuous on the open effective domain $\operatorname{dom} \varphi_{A,k} = \{(y_1, y_2)^{\mathrm{T}} \in \mathbb{R}^2 \mid y_1 > 0\}$. However, $\varphi_{A,k}$ is not lower semicontinuous since $\operatorname{cl} A - \mathbb{R}_{>}k \not\subseteq A$.

We will now give an example of a lower semicontinuous function $\varphi_{A,k}$ which is continuous on its effective domain, but not continuous on $Y$.

**Example 4.6.5** In $Y = \mathbb{R}^2$, consider $A := -\mathbb{R}^2_+$ and $k := (1,0)^{\mathrm{T}} \in -0^+A$. Then $A$ and $\operatorname{dom} \varphi_{A,k}$ are closed. $\varphi_{A,k}$ is continuous on $\operatorname{dom} \varphi_{A,k}$. It is lower semicontinuous since $\operatorname{cl} A - \mathbb{R}_{>}k \subseteq A$, but not continuous since $\operatorname{cl} A - \mathbb{R}_{>}k \not\subseteq \operatorname{int} A$.

If $\varphi_{A,k}$ is continuous and $k \in -0^+A$, then the level sets $\operatorname{lev}_{\varphi_{A,k},=}(t)$ are given by the shifted boundary of $A$. The next theorem points out this structural dependence and summarizes additional properties of continuous functionals $\varphi_{A,k}$.

**Theorem 4.6.6** *Assume* (Sp2$_{A,k}$), *i.e., Y is a topological vector space, A is a proper subset of Y, $k \in Y \setminus \{0\}$ and*

$$\operatorname{cl} A - \mathbb{R}_{>}k \subseteq \operatorname{int} A. \tag{4.6.8}$$

*Then the following statements hold beside the properties stated in Theorem 4.4.4.*

$$\operatorname{cl} A = \operatorname{cl} \operatorname{int} A, \tag{4.6.9}$$

$$\operatorname{core} A = \operatorname{int} A = A - \mathbb{R}_{>}k, \tag{4.6.10}$$

$$\varphi_{A,k} \textit{ is continuous},$$

$$\operatorname{dom} \varphi_{A,k} = A + \mathbb{R}k = \operatorname{int} A + \mathbb{R}k \neq \emptyset \textit{ is an open set}, \tag{4.6.11}$$

$$\operatorname{lev}_{\varphi_{A,k},\leq}(t) = \operatorname{cl} A + tk \quad \textit{for all } t \in \mathbb{R}, \tag{4.6.12}$$

$$\operatorname{lev}_{\varphi_{A,k},<}(t) = \operatorname{int} A + tk \quad \textit{for all } t \in \mathbb{R}, \tag{4.6.13}$$

$$\operatorname{lev}_{\varphi_{A,k},=}(t) = \operatorname{bd} A + tk \quad \textit{for all } t \in \mathbb{R}. \tag{4.6.14}$$

*Moreover,*

(a) $\forall y \in Y\colon\ (\varphi_{A,k}(y) = -\infty \iff y + \mathbb{R}k \subseteq \operatorname{int} A)$.
(b) $\{y \in Y \mid \varphi_{A,k}(y) \in \mathbb{R}\} = \operatorname{bd} A + \mathbb{R}k$, *and* $\operatorname{dom} \varphi_{A,k} \setminus \operatorname{int} A \subseteq \operatorname{bd} A + \mathbb{R}_{+}k$.
(c) *The following conditions are equivalent to each other:*

$$\varphi_{A,k} \textit{ is proper},$$

$$\operatorname{int} A = \operatorname{bd} A - \mathbb{R}_{>}k, \tag{4.6.15}$$

$$\operatorname{cl} A = \operatorname{bd} A - \mathbb{R}_{+}k, \tag{4.6.16}$$

$$\operatorname{dom} \varphi_{A,k} = \operatorname{bd} A + \mathbb{R}k. \tag{4.6.17}$$

(d) $\varphi_{A,k}$ *is finite-valued* $\iff Y = \operatorname{bd} A + \mathbb{R}k$.
(e) $\varphi_{A,k}$ *is strictly quasiconvex on* $\operatorname{dom} \varphi_{A,k} \iff \operatorname{cl} A$ *is a strictly convex set.*
*In this case,* $\varphi_{A,k}$ *is proper and strictly* $(-\operatorname{core} 0^{+}(\operatorname{cl} A))$*-monotone on* $\operatorname{dom} \varphi_{A,k}$.
(f) *If* $\operatorname{cl}(Y \setminus A)$ *is a strictly convex set, then* $\varphi_{A,k}$ *is a function with* $\operatorname{dom} \varphi_{A,k} = Y$, *which is strictly quasiconcave on* $\operatorname{bd} A + \mathbb{R}k$ *and strictly* $(\operatorname{core} 0^{+}(\operatorname{cl}(Y \setminus A)))$*-monotone on* $\operatorname{bd} A + \mathbb{R}k$.
*If* $\varphi_{A,k}$ *is finite-valued, then:*
$\varphi_{A,k}$ *is strictly quasiconcave* $\iff \operatorname{cl}(Y \setminus A)$ *is a strictly convex set.*
(g) $\varphi_{A,k}$ *is linear* $\iff \operatorname{bd} A$ *is a linear subspace of* $Y$.
*In this case,* $\varphi_{A,k}$ *is finite-valued and* $\operatorname{bd} A$ *is a linear subspace of codimension* 1.
(h) $\varphi_{A,k}$ *is affine* $\iff \operatorname{bd} A$ *is an affine set.*
*In this case,* $\varphi_{A,k}$ *is finite-valued and* $\operatorname{bd} A$ *is the sum of some* $y^0 \in Y$ *and a linear subspace of codimension* 1.

***Proof*** Assume first that $A$ is closed and $(\mathrm{Sp2}_{A,k})$ holds. Since $(\mathrm{Sp2}_{A,k})$ implies $(\mathrm{Sp1}_{A,k})$, Theorem 4.4.4 can be applied.

If $y \in A$, then $y - \frac{1}{n}k \in \operatorname{int} A$ for any $n \in \mathbb{N}_{>}$, and thus $y \in \operatorname{cl} \operatorname{int} A$. Hence, $A = \operatorname{cl} \operatorname{int} A$.

$\operatorname{int} A \subseteq \operatorname{core} A \subseteq A - \mathbb{R}_{>}k \subseteq \operatorname{int} A$ yields (4.6.10).

$\varphi_{A,k}$ is continuous because of Theorem 4.6.2. By Proposition 3.3.19, $\operatorname{dom} \varphi_{A,k}$ is open. Lemma 4.6.1 implies (4.6.13) and (4.6.14) and thus (4.6.12).

If $y \in A + \mathbb{R}k$, then there exist $a \in A$ and $t \in \mathbb{R}$ with $y = a + tk = a - k + (t + 1)k \in \operatorname{int} A + \mathbb{R}k$. Hence, $A + \mathbb{R}k = \operatorname{int} A + \mathbb{R}k$.

(a) For each $y \in Y$,

$$\begin{aligned}\varphi_{A,k}(y) = -\infty &\iff \forall t \in \mathbb{R}\colon\ \varphi_{A,k}(y) < t,\\ &\iff \forall t \in \mathbb{R}\colon\ y \in \operatorname{int} A + tk,\\ &\iff y + \mathbb{R}k \subseteq \operatorname{int} A.\end{aligned}$$

(b) The first statement results from (4.6.14). If $y \in Y$ with $\varphi_{A,k}(y) = -\infty$, then $y = y + 0 \cdot k \in \operatorname{int} A$ because of (a). This yields the second statement of (b).

(c) Assume first that $\varphi_{A,k}$ is proper. Take any $a \in \operatorname{int} A$. Equation (4.6.13) implies $t := \varphi_{A,k}(a) < 0$ with $t \in \mathbb{R}$, and $a \in \operatorname{bd} A + tk \subseteq \operatorname{bd} A - \mathbb{R}_{>}k$ by (4.6.14). Thus, (4.6.15) holds because of (4.6.8).
Equation (4.6.16) follows from (4.6.15) since $A = \operatorname{bd} A \cup \operatorname{int} A$.
If $A = \operatorname{bd} A - \mathbb{R}_{+}k$, then $\operatorname{dom} \varphi_{A,k} = A + \mathbb{R}k = \operatorname{bd} A - \mathbb{R}_{+}k + \mathbb{R}k = \operatorname{bd} A + \mathbb{R}k$.
If $\operatorname{dom} \varphi_{A,k} = \operatorname{bd} A + \mathbb{R}k$, then $\varphi_{A,k}$ is proper because of (4.6.14).

(d) results from (b).

(e) Take any $\lambda \in (0, 1)$.
Assume first that $\varphi_{A,k}$ is strictly quasiconvex on $\operatorname{dom} \varphi_{A,k}$. If there exist $a^1, a^2 \in A$ with $a^1 \neq a^2$, then we have $\varphi_{A,k}(a^1) \leq 0$, $\varphi_{A,k}(a^2) \leq 0$, and $\varphi_{A,k}(\lambda a^1 + (1-\lambda)a^2) < \sup(\varphi_{A,k}(a^1), \varphi_{A,k}(a^2)) \leq 0$, which yields $\lambda a^1 + (1-\lambda)a^2 \in \operatorname{int} A$. Thus, $A$ is a strictly convex set.
Assume now that $A$ is a strictly convex set. Since $A$ is a proper subset of $Y$, there exists $a^0 \in \operatorname{bd} A$. Thus, $k \in (-0^+A) \setminus 0^+A$ since otherwise $a^0 + \mathbb{R}k \subseteq A$ and $a^0 \in \operatorname{int} A$ by the strict convexity of $A$. By Proposition 4.7.7, $\varphi_{A,k}$ is proper and strictly $(-\operatorname{core} 0^+A)$-monotone on $\operatorname{dom} \varphi_{A,k}$.
Since $A$ is a strictly convex set, $\operatorname{dom} \varphi_{A,k}$ is convex. Take any $y^1, y^2 \in \operatorname{dom} \varphi_{A,k}$ with $y^1 \neq y^2$. Then $t_i := \varphi_{A,k}(y^i) \in \mathbb{R}$ for each $i \in \{1, 2\}$. Hence, there exist $a^1, a^2 \in A$ with $y^1 = a^1 + t_1 k$, $y^2 = a^2 + t_2 k$ because of (4.4.12). If $a^1 = a^2$, then $t_1 \neq t_2$ and $\lambda y^1 + (1-\lambda)y^2 = a^1 + (\lambda t_1 + (1-\lambda)t_2)k$, which implies $\varphi_{A,k}(\lambda y^1 + (1-\lambda)y^2) \leq \lambda t_1 + (1-\lambda)t_2 < \max(t_1, t_2)$. If $a^1 \neq a^2$, then $\lambda y^1 + (1-\lambda)y^2 = \lambda a^1 + (1-\lambda)a^2 + (\lambda t_1 + (1-\lambda)t_2)k$, which results in $\varphi_{A,k}(\lambda y^1 + (1-\lambda)y^2) < \lambda t_1 + (1-\lambda)t_2 \leq \max(t_1, t_2)$ because of $\lambda a^1 + (1-\lambda)a^2 \in \operatorname{int} A$. Thus, $\varphi_{A,k}$ is strictly quasiconvex on $\operatorname{dom} \varphi_{A,k}$.

(g) If $\varphi_{A,k}$ is linear, then it is finite-valued by Corollary 3.5.7 since it is continuous. By applying (4.6.14) to $t = 0$, it follows that $\operatorname{bd} A$ is a linear subspace of $Y$.
Assume now that $\operatorname{bd} A$ is a linear subspace of $Y$. Take any $y \in Y \setminus A$ and $a^1 \in \operatorname{bd} A$. Then $a^2 := a^1 - k \in \operatorname{int} A$, and there exist $a^3 \in \operatorname{bd} A$ and $\lambda \in (0, 1)$ with $a^3 = y + \lambda(a^2 - y)$ because of Lemma 2.1.26. Define $t := \frac{1}{1-\lambda}$. Hence, $t - t\lambda = 1$, $1 + t\lambda - t = 0$, and $y - t\lambda k = a^1(1 + t\lambda - t) + y(t - t\lambda) - t\lambda k = a^1 + t(a^3 - a^1) \in \operatorname{bd} A$. Therefore, $y \in \operatorname{bd} A + \mathbb{R}_{>}k$. This yields $Y \setminus A \subseteq \operatorname{bd} A + \mathbb{R}_{>}k$. Equation (4.6.14) implies $\operatorname{bd} A + \mathbb{R}_{>}k \subseteq Y \setminus A$ and thus $Y \setminus A = \operatorname{bd} A + \mathbb{R}_{>}k$.

Take any $a^4 \in \operatorname{int} A$ and $y \in Y \setminus A$. There exist $a^5 \in \operatorname{bd} A$ and $\lambda_1 \in (0, 1)$ with $a^5 = a^4 + \lambda_1(y - a^4)$. Furthermore, there exist $a^6 \in \operatorname{bd} A$ and $\lambda_2 \in \mathbb{R}_>$ with $y = a^6 + \lambda_2 k$. Then $a^5 = a^4 + \lambda_1((a^6 + \lambda_2 k) - a^4)$ implies $a^4 = \frac{1}{1-\lambda_1} a^5 - \frac{\lambda_1}{1-\lambda_1} a^6 - \frac{\lambda_1\lambda_2}{1-\lambda_1} k \in \operatorname{bd} A - \mathbb{R}_> k$. Thus, $\operatorname{int} A = \operatorname{bd} A - \mathbb{R}_> k$.
Hence, $Y = \operatorname{bd} A + \mathbb{R}k$, and $\operatorname{bd} A$ is a linear subspace of codimension 1. Because of $Y = \operatorname{bd} A + \mathbb{R}k$, $\varphi_{A,k}$ is finite-valued. Equation (4.6.14) yields the linearity of $\varphi_{A,k}$.

(h) Assume first that $\varphi_{A,k}$ is affine. Since it is continuous and $A$ is a proper subset of $Y$, $\varphi_{A,k}$ is finite-valued by Proposition 3.5.6. Because of Lemma 3.5.3, there exist a linear function $f\colon Y \to \overline{\mathbb{R}}$ and $c \in \mathbb{R}$ with $\varphi_{A,k} = f + c$. By Proposition 4.3.1(a), $f$ is translation invariant along $k$. The function $f = \varphi_{A,k} - c$ is continuous. Because of Theorem 4.6.3, there exists a closed set $B$ which fulfills ($\text{Sp2}_{B,k}$) such that $f = \varphi_{B,k}$. Then $\operatorname{bd} B$ is a linear subspace of $Y$ of codimension 1 because of (g). We get $\varphi_{A,k} = \varphi_{B-ck,k}$ by Proposition 4.3.4. Equation (4.6.14) yields $\operatorname{bd} A = \operatorname{bd}(B - ck) = \operatorname{bd} B - ck$. Thus, the forward direction of the first statement as well as the additional statements in (h) are satisfied.
Assume now that $\operatorname{bd} A$ is an affine set. Since it is nonempty, there exist a linear subspace $C \subset Y$ and some $y^0 \in Y$ with $\operatorname{bd} A = C + y^0$. ($\text{Sp2}_{A,k}$) implies ($\text{Sp2}_{A-y^0,k}$). Since $C = \operatorname{bd}(A - y^0)$ is a linear subspace, we get $Y = C + \mathbb{R}k$ from (g) and (d). Thus, there exists $t \in \mathbb{R}$ with $y^0 \in C + tk$, and $\operatorname{bd} A = C + tk$. Hence, $Y = \operatorname{bd} A + \mathbb{R}k$, and $\varphi_{A,k}$ is finite-valued. Define $g\colon Y \to \mathbb{R}$ by $g := \varphi_{A,k} + t$. Take any $y^1, y^2 \in Y$ and $\mu, \lambda \in \mathbb{R}$. Consider $t_1 := g(y^1)$, $t_2 := g(y^2)$. For each $i \in \{1, 2\}$, $\varphi_{A,k}(y^i) = t_i - t$ implies $y^i \in \operatorname{bd} A + (t_i - t)k = C + t_i k$. Hence, $\mu y^1 + \lambda y^2 \in C + (\mu t_1 + \lambda t_2)k = \operatorname{bd} A - tk + (\mu t_1 + \lambda t_2)k$, which yields $\varphi_{A,k}(\mu y^1 + \lambda y^2) = \mu t_1 + \lambda t_2 - t$, and $g(\mu y^1 + \lambda y^2) = \mu t_1 + \lambda t_2$. Hence, $g$ is a linear function, and $\varphi_{A,k}$ is affin.

Thus, we have proved all statements but (f) for the case where $A$ is a closed set. Take now any set $A$ which fulfills ($\text{Sp2}_{A,k}$). By Lemma 2.3.16(c), ($\text{Sp2}_{\operatorname{cl} A,k}$) is fulfilled. We get $\varphi_{\operatorname{cl} A,k} = \varphi_{A,k}$ from Theorem 4.4.4. Applying the proved statements to the closed set $\operatorname{cl} A$ and taking into consideration Lemma 2.3.16 implies these statements for $A$.

We will now prove (f). Because of Proposition 4.3.11, ($\text{Sp2}_{Y\setminus A,-k}$) holds, we have $\varphi_{A,k} = -\varphi_{Y\setminus A,-k}$, and $\operatorname{bd} A + \mathbb{R}k$ is the set on which $\varphi_{Y\setminus A,-k}$ is finite-valued.
Assume first that $\operatorname{cl}(Y \setminus A)$ is a strictly convex set. Because of (e), $\varphi_{Y\setminus A,-k}$ is strictly quasiconvex on $\operatorname{dom} \varphi_{Y\setminus A,-k}$ and proper. Hence, $\operatorname{dom} \varphi_{Y\setminus A,-k} = \operatorname{bd} A + \mathbb{R}k$. Then $\varphi_{A,k} = -\varphi_{Y\setminus A,-k}$ is strictly quasiconcave on $\operatorname{bd} A + \mathbb{R}k$. Since $\varphi_{Y\setminus A,-k}$ is proper, we obtain $\operatorname{dom} \varphi_{A,k} = Y$. Because of (e), $\varphi_{Y\setminus A,-k}$ is strictly $(-\operatorname{core} 0^+(\operatorname{cl}(Y \setminus A)))$-monotone on $\operatorname{bd} A + \mathbb{R}k$. Thus, $\varphi_{A,k}$ is strictly $(\operatorname{core} 0^+(\operatorname{cl}(Y \setminus A)))$-monotone on $\operatorname{bd} A + \mathbb{R}k$. If $\varphi_{A,k}$ is finite-valued, then (d) implies the reverse direction of the last equivalence in (f).
Assume now that $\varphi_{A,k}$ is finite-valued and strictly quasiconcave. Then $\varphi_{Y\setminus A,-k}$ is strictly quasiconvex, we have $\operatorname{dom} \varphi_{Y\setminus A,-k} = \operatorname{dom} \varphi_{A,k} = Y$, and $\operatorname{cl}(Y \setminus A)$ is a strictly convex set because of (e). □

The level sets of $\varphi_{A,k}$ under assumption ($\text{Sp2}_{A,k}$) are illustrated in Fig. 4.6.

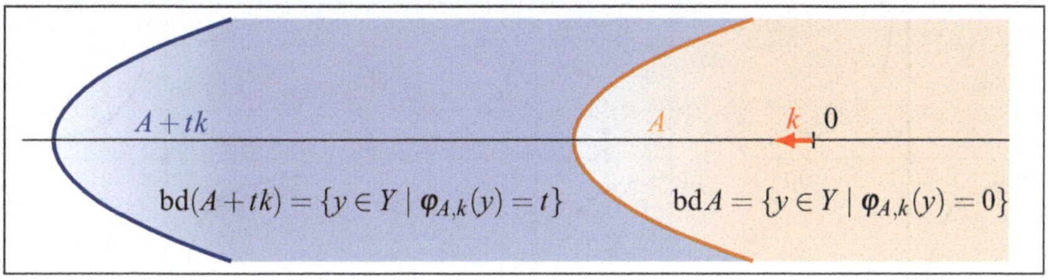


**Fig. 4.6** The level sets of $\varphi_{A,k}$ under assumption (Sp2$_{A,k}$)

**Remark 4.6.7** In [472, Satz 3.1.10], it was proved that a function which is defined by (4.6.12) for a closed set $A$ and $k \in Y \setminus \{0\}$ is proper if and only if (4.2.20) and (H1$_{A,k}$) hold, and that these conditions are fulfilled for each closed pointed convex cone $A$ with $k \in -A \setminus \{0\}$. The equivalence of Theorem 4.5.5(a) was shown in [472, Satz 3.1.11] for any function which fulfills (4.6.12) for a closed set $A$. Moreover, lower semicontinuity, monotonicity, convexity and quasiconvexity of such a function were studied in [472]. According to [472, Satz 3.2.1], the functionals given by one of the conditions (4.6.12)–(4.6.14) and $\varphi_{A,k}$ coincide for a closed set $A$ on $A + \mathbb{R}k$ and are well defined if and only if $\operatorname{int} A = \operatorname{bd} A - \mathbb{R}_{>}k$. In [472], sets with this property were investigated and it was proved that $\operatorname{cl} A = \operatorname{cl} \operatorname{int} A$ holds for these sets.

In decision making, utility functions are often assumed to be strictly quasiconcave. A functional of type $\varphi_{A,k}$ cannot be strictly convex or strictly concave since it is affine on each line in direction $k$.

Note that even for a convex set $A$, the inclusion $A - \mathbb{R}_{>}k \subseteq \operatorname{int} A$ does not imply $\operatorname{cl} A - \mathbb{R}_{>}k \subseteq \operatorname{int} A$.

**Example 4.6.8** For $A := \operatorname{int} \mathbb{R}^2_+$ and $k := (-1, 0)^T$, $A$ is convex, $A - \mathbb{R}_{>}k \subseteq \operatorname{int} A$, but $\operatorname{cl} A - \mathbb{R}_{>}k \not\subseteq \operatorname{int} A$.

Consider a function $\varphi\colon \mathbb{R}^\ell \to \mathbb{R}$ for which there exists a set $A \subset \mathbb{R}^\ell$ such that

$$\operatorname{lev}_{\varphi,=}(t) = \operatorname{bd} A + tk \text{ for all } t \in \mathbb{R}.$$

This condition does not imply that $\varphi$ is semicontinuous, especially not that

$$\operatorname{lev}_{\varphi,<}(t) = \operatorname{int} \tilde{A} + tk \qquad \text{or} \qquad \operatorname{lev}_{\varphi,\leq}(t) = \operatorname{cl} \tilde{A} + tk$$

holds for some set $\tilde{A} \subset \mathbb{R}^\ell$ and all $t \in \mathbb{R}$. We underscore this fact by the following example.

**Example 4.6.9** Define $A := \{(y_1, y_2)^T \in \mathbb{R}^2 \mid y_1 > 0,\ y_2 \geq \frac{1}{y_1}\} \cup \{(y_1, y_2)^T \in \mathbb{R}^2 \mid y_1 < 0,\ y_2 \leq \frac{1}{y_1}\} \cup \{(0, 0)^T\}$, $k := (1, 0)^T$ and $\varphi\colon \mathbb{R}^2 \to \mathbb{R}$ by

$$\operatorname{lev}_{\varphi,=}(t) = \operatorname{bd} A + tk \text{ for all } t \in \mathbb{R}.$$

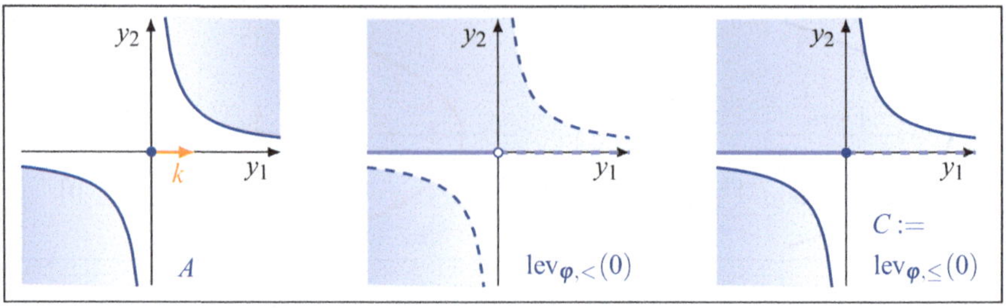


**Fig. 4.7** Illustration of Example 4.6.9

Then $\mathrm{lev}_{\varphi,<}(0) = \{(y_1, y_2)^T \in \mathbb{R}^2 \mid y_1 > 0,\ 0 < y_2 < \frac{1}{y_1}\} \cup \{(y_1, y_2)^T \in \mathbb{R}^2 \mid y_1 < 0,\ y_2 < \frac{1}{y_1}\} \cup \{(y_1, y_2)^T \in \mathbb{R}^2 \setminus \{0\} \mid y_1 \leq 0,\ y_2 \geq 0\}$ is not open and $C := \mathrm{lev}_{\varphi,\leq}(0) = \{(y_1, y_2)^T \in \mathbb{R}^2 \mid y_1 > 0,\ 0 < y_2 \leq \frac{1}{y_1}\} \cup \{(y_1, y_2)^T \in \mathbb{R}^2 \mid y_1 < 0,\ y_2 \leq \frac{1}{y_1}\} \cup \{(y_1, y_2)^T \in \mathbb{R}^2 \mid y_1 \leq 0,\ y_2 \geq 0\}$ is not closed. Consequently, $\varphi$ is neither lower semicontinuous nor upper semicontinuous.

Note that $\varphi = \varphi_{C,k}$, where $\mathrm{cl}\, C - \mathbb{R}_{>}k \subseteq C$ is not satisfied and the condition $\mathrm{lev}_{\varphi,=}(t) = \mathrm{bd}\, C + tk$ is not fulfilled on the $y_1$-axis since $(y_1, 0)^T \in \mathrm{bd}\, C + tk$ holds for all $t \in \mathbb{R}$.

The sets in this example are illustrated by Fig. 4.7.

The proposition to come will show alternative formulations of assumption (Sp2$_{A,k}$).

**Lemma 4.6.10** *Assume that $Y$ is a topological vector space, $A \subseteq Y$ and that $C \subset Y$ is a nontrivial cone. Then*

$$\mathrm{int}\, A \subseteq A - (C \setminus \{0\}),$$

*and the following conditions are equivalent to each other:*

(a) $A - (C \setminus \{0\}) \subseteq \mathrm{int}\, A$.
(b) $A - (C \setminus \{0\}) = \mathrm{int}\, A$.

***Proof*** Take any $a \in \mathrm{int}\, A$ and $c \in C \setminus \{0\}$. Then there exists $\lambda \in \mathbb{R}_{>}$ with $a^1 := a + \lambda c \in A$. This implies $\lambda c \in C \setminus \{0\}$ and $a = a^1 - \lambda c$. Thus, $\mathrm{int}\, A \subseteq A - (C \setminus \{0\})$. Hence, (a) is equivalent to (b). □

**Proposition 4.6.11** *Assume that $Y$ is a topological vector space and $A$ is a proper subset of $Y$. The following conditions are equivalent to each other for $A$ and $k \in Y \setminus \{0\}$.*

(a) $\mathrm{cl}\, A - \mathbb{R}_{>}k \subseteq \mathrm{int}\, A$.
(b) $\mathrm{cl}\, A - \mathbb{R}_{>}k = \mathrm{int}\, A$.
(c) $\mathrm{cl}\, A - (C \setminus \{0\}) = \mathrm{int}\, A$ *for some nontrivial convex cone $C \subset Y$ with $k \in C$.*
(d) $\mathrm{cl}\, A - (C \setminus \{0\}) = \mathrm{int}\, A$ *for some nontrivial cone $C \subset Y$ with $k \in C$.*

*If $Y$ is a Hausdorff space, then these conditions are equivalent to*

(e) $\operatorname{cl} A - (C \setminus \{0\}) = \operatorname{int} A$ *for some nontrivial closed convex cone* $C \subset Y$ *with* $k \in C$.

***Proof*** If (a) holds, then $\operatorname{int} \operatorname{cl} A = \operatorname{int} A$ by Lemma 2.3.16. The application of Lemma 4.6.10 to $\operatorname{cl} A$ proves the equivalence of (a) and (b).

(b) implies (c) with $C = \mathbb{R}_+ k$. (c) implies (d). (a) follows from (d).

(e) implies (a), and if $Y$ is a Hausdorff space, then (b) implies (e) with $C = \mathbb{R}_+ k$. □

The property $A - (C \setminus \{0\}) = \operatorname{int} A$ of the set $A$ with $C$ being an ordering cone is known in production theory as the **strong free disposal assumption**.

Theorem 4.6.6 is often applied to closed sets $A$. For such sets, assumption $(\mathrm{Sp2}_{A,k})$ coincides with:

$$Y \text{ is a topological vector space, } A \text{ is a proper closed subset of } Y,\ k \in Y \setminus \{0\} \text{ and } A - \mathbb{R}_> k \subseteq \operatorname{int} A. \qquad (\mathrm{H2}_{A,k})$$

Even under condition $(\mathrm{H2}_{A,k})$, $A$ is not necessarily convex and $\varphi_{A,k}$ is not necessarily proper.

**Example 4.6.12** In $Y = \mathbb{R}^2$, define $k := (-1, 0)^T$ and $A := \{(y_1, y_2)^T \in \mathbb{R}^2 \mid -e^{y_1} \le y_2 \le e^{y_1}\}$. $(\mathrm{H2}_{A,k})$ is fulfilled. We obtain $\operatorname{dom} \varphi_{A,k} = Y$, but $\varphi_{A,k}$ is not proper since $\varphi_{A,k}(y) = -\infty$ for all $y \in Y$ with $y_2 = 0$.

The assumptions of Theorem 4.6.6 do not result in $\operatorname{dom} \varphi_{A,k} = Y$, which is illustrated by Example 4.3.12.

We now turn to sets $A$ for which the recession cone has a nonempty interior and $-k$ is an element of this interior. Properties of such sets were proved in Proposition 2.2.24. If $A$ is a proper closed subset of a Banach space, then $0^+A = \bigcap_{a \in A} T_C(a; A)$ by Theorem 3.6.15. In this case, $\operatorname{int} 0^+A$ is contained in the interior of each Clarke tangent cone to $A$ at some $a \in A$.

For each nontrivial convex cone $A$, $(\mathrm{Sp2}_{A,k})$ implies $k \in -\operatorname{int} 0^+A$.

**Proposition 4.6.13** *Assume that A is a proper subset of the topological vector space Y and*

$$k \in -\operatorname{int} 0^+A.$$

*Then* $\varphi_{A,k}$ *is continuous and finite-valued.*

***Proof*** Proposition 2.2.24(a) implies $\operatorname{cl} A + \operatorname{int} 0^+A \subseteq \operatorname{int} A$. By Theorem 4.6.2, $\varphi_{A,k}$ is continuous. It is finite-valued because of Theorem 4.2.9. □

Now, the Lipschitz continuity of $\varphi_{A,k}$ will be investigated.

**Theorem 4.6.14** *Assume that Y is a Banach space and A is a proper closed subset of Y for which there exists* $k \in -0^+A \setminus \{0\}$. *Then:*

$$\varphi_{A,k} \textit{ is Lipschitz continuous} \iff \varphi_{0^+A,k} \textit{ is Lipschitz continuous}$$
$$\iff k \in -\operatorname{int} 0^+A.$$

*If* $k \in -\operatorname{int} 0^+A$, *then each Lipschitz constant of* $\varphi_{0^+A,k}$ *is also a Lipschitz constant of* $\varphi_{A,k}$.

***Proof*** The assumption implies $(\mathrm{Sp1}_{A,k})$ and $(\mathrm{Sp1}_{0^+A,k})$.

(a) Assume first $k \in -\operatorname{int} 0^+A$. By Proposition 4.6.13, $\varphi_{A,k}$ and $\varphi_{0^+A,k}$ are continuous and finite-valued. Since $0^+A$ is a convex cone, $\varphi_{0^+A,k}$ is sublinear by Theorem 4.4.4(i) and thus Lipschitzian with some Lipschitz constant $L$ because of Corollary 3.5.14. By Lemma 3.5.10(b), $\varphi_{0^+A,k}(0) = 0$. Take any $y^0, y^1 \in Y$, and choose the notation such that $\varphi_{A,k}(y^0) \geq \varphi_{A,k}(y^1)$. Then $0 \leq \varphi_{A,k}(y^0) - \varphi_{A,k}(y^1) \leq \varphi_{0^+A,k}(y^0 - y^1)$ by (4.2.23), and $|\varphi_{A,k}(y^0) - \varphi_{A,k}(y^1)| \leq |\varphi_{0^+A,k}(y^0 - y^1)| = |\varphi_{0^+A,k}(y^0 - y^1) - \varphi_{0^+A,k}(0)| \leq L\|y^0 - y^1\|$. Hence, $\varphi_{A,k}$ is Lipschitzian with Lipschitz constant $L$.

(b) Assume now that $\varphi_{A,k}$ is Lipschitz continuous with some Lipschitz constant $L$. Then it is finite-valued. The set $U := \{y \in Y \mid \|y\| \leq \frac{1}{L}\}$ is a neighborhood of 0. Take any $u \in U$, $a \in A$ and $\lambda \in \mathbb{R}_+$. We get $|\varphi_{A,k}(a + \lambda u) - \varphi_{A,k}(a)| \leq L\|\lambda u\| = L\lambda\|u\| \leq \lambda$. Then $\varphi_{A,k}(a + \lambda u) \leq \varphi_{A,k}(a) + \lambda$, and thus $\varphi_{A,k}(a + \lambda(u - k)) = \varphi_{A,k}(a + \lambda u) - \lambda \leq \varphi_{A,k}(a) \leq 0$. We obtain $a + \lambda(u - k) \in A$ because of (4.4.5). Hence, $u - k \in 0^+A$ for all $u \in U$. Consequently, $k \in -\operatorname{int} 0^+A$.

(c) Assume that $\varphi_{0^+A,k}$ is Lipschitz continuous. Apply (b) to $0^+A$ instead of $A$. This yields $k \in -\operatorname{int} 0^+(0^+A) = -\operatorname{int} 0^+A$. □

**Theorem 4.6.15** *Assume* $Y$ *to be a Banach space. The class*

$$\{\varphi_{A,k}\colon Y \to \overline{\mathbb{R}} \mid A \subseteq Y,\ k \in Y \setminus \{0\},\ \varphi_{A,k} \textit{ Lipschitz continuous}\} \tag{4.6.18}$$

*of all Lipschitz continuous translation invariant functions on* $Y$ *coincides with*

$$\{\varphi_{A,k}\colon Y \to \overline{\mathbb{R}} \mid A \textit{ is a proper closed subset of } Y,\ k \in -\operatorname{int} 0^+A\}. \tag{4.6.19}$$

***Proof*** Take first any $A \subseteq Y$ and $k \in Y \setminus \{0\}$ for which $\varphi := \varphi_{A,k}$ is Lipschitz continuous. For $\tilde{A} := \mathrm{cl}_k(A - \mathbb{R}_+k)$, we obtain $\varphi = \varphi_{\tilde{A},k}$ by (4.2.8). $\tilde{A}$ is a proper subset of $Y$ since $\varphi$ is finite-valued. We have $k \in -0^+(A - \mathbb{R}_+k)$. Proposition 2.3.29 implies $k \in -0^+\tilde{A}$. Since $\varphi$ is lower semicontinuous, $\tilde{A}$ is closed by Theorem 4.4.2. Hence, $k \in -\operatorname{int} 0^+\tilde{A}$ because of Theorem 4.6.14. Therefore, set (4.6.18) is contained in set (4.6.19). The reverse inclusion follows from Theorem 4.6.14. □

Durea [121, Lemma 2.1] proved:

**Lemma 4.6.16** *Assume that* $Y$ *is a Banach space,* $C \subset Y$ *is a closed convex cone with nonempty interior, and* $k \in -\operatorname{int} C$. *Then* $\varphi_{C,k}$ *is Lipschitz continuous with the Lipschitz constant* $\frac{1}{d(-k,\mathrm{bd}\, C)}$.

Here $d$ stands for the distance function defined in Definition 2.1.28. If the assumptions of Theorem 4.6.14 are satisfied and $\varphi_{A,k}$ is Lipschitz continuous, then $C := 0^+A$ fulfills the assumptions of Lemma 4.6.16. Consequently, the reciprocal of $d(-k, \operatorname{bd} 0^+A)$ is a Lipschitz constant of $\varphi_{0^+A,k}$ and thus a Lipschitz constant of $\varphi_{A,k}$.

According to Theorem 4.6.14, the following condition guarantees the Lipschitz continuity of $\varphi_{A,k}$ if $Y$ is a Banach space:

$$\begin{array}{c} Y \text{ is a topological vector space, } A \text{ is a proper closed subset of } Y, \\ \text{and } k \in -\operatorname{int} 0^+A. \end{array} \qquad (\mathrm{H3}_{A,k})$$

If $A$ is a convex cone, then $(\mathrm{H3}_{A,k})$ is equivalent to $(\mathrm{H2}_{A,k})$. However, $(\mathrm{H2}_{A,k})$ does not imply $k \in -\operatorname{int} 0^+A$ for an arbitrary set $A$. The interior of $0^+A$ may be empty even if $A$ is convex, has a nonempty interior and $\varphi_{A,k}$ is finite-valued.

**Example 4.6.17** In $Y = \mathbb{R}^2$, define $k := (0, -1)^T$ and $A := \{(y_1, y_2)^T \in \mathbb{R}^2 \mid y_2 \geq y_1^2\}$. Then assumption $(\mathrm{H2}_{A,k})$ is fulfilled. $\varphi_{A,k}$ is finite-valued and $A$ is convex, but $0^+A = -\mathbb{R}_+k$.

Condition $(\mathrm{H3}_{A,k})$ could also be formulated in alternative ways.

**Proposition 4.6.18** *Assume that $Y$ is a topological vector space and $A$ is a proper closed subset of $Y$. The following conditions are equivalent to each other for $A$ and $k \in Y$.*

(a) $k \in -\operatorname{int} 0^+A$.
(b) $A = H - C$ *for some proper subset $H$ of $Y$ and some nontrivial convex cone $C \subset Y$ with $k \in \operatorname{int} C$.*
(c) $A = A - C$ *for some nontrivial closed convex cone $C \subset Y$ with $k \in \operatorname{int} C$.*
(d) $A = A - C$ *for some nontrivial cone $C \subset Y$ with $k \in \operatorname{int} C$.*
(e) $A - \operatorname{int} C \subseteq A$ *for some closed convex cone $C \subset Y$ with $k \in \operatorname{int} C$.*
(f) $A - \operatorname{int} C \subseteq A$ *for some cone $C \subset Y$ with $k \in \operatorname{int} C$.*

***Proof*** (a) implies (b) with $H = A$ and $C = -0^+A$. (b) implies (a) since $C \subseteq -0^+A$.

(a) implies (c) with $C = -0^+A$. (c) yields (d). (d) implies (a) because of $C \subseteq -0^+A$.

(c) yields (e). (e) implies (f). (f) implies (a). □

$(\mathrm{H3}_{A,k})$ can hold even though $A$ is not convex and $A + A \not\subseteq A$.

**Example 4.6.19** In $Y = \mathbb{R}^2$, define $k := (-1, -1)^T$ and $A := (\mathbb{R}^2_+ + (-1, 0)^T) \cup (\mathbb{R}^2_+ + (0, -1)^T)$ (see Fig. 4.8). Then $(\mathrm{H3}_{A,k})$ is fulfilled, but $A$ is not convex and $A + A \not\subseteq A$ since $(-1, 0)^T + (0, -1)^T \notin A$.

The function $\varphi_{A,k}$ given in Example 4.6.5 is Lipschitzian on $\operatorname{dom} \varphi_{A,k}$, but not Lipschitzian since it is not finite-valued. Of course, $k \notin -\operatorname{int} 0^+A$ in this example.

In Proposition 4.7.9, we will show that $\varphi_{A,k}$ is locally Lipschitzian on $\operatorname{int} \operatorname{dom} \varphi_{A,k}$ if $A$ is a proper closed convex subset of a Banach space $Y$ and $k \in (-0^+A) \setminus 0^+A$. In the following example, the function $\varphi_{A,k}$ is not locally Lipschitzian on $\operatorname{int} \operatorname{dom} \varphi_{A,k}$.

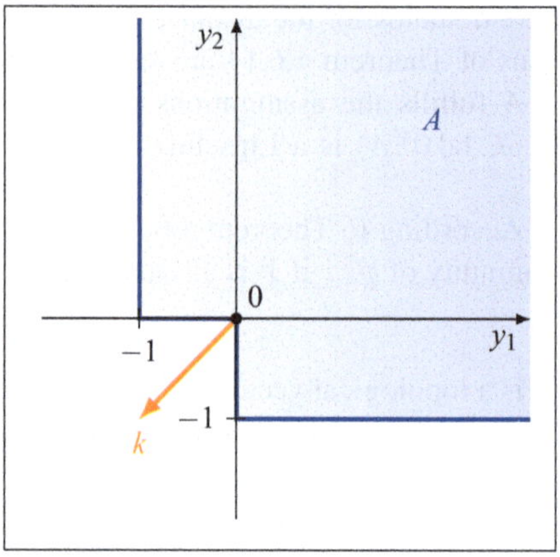


**Fig. 4.8** Set $A$ and vector $k$ in Example 4.6.19

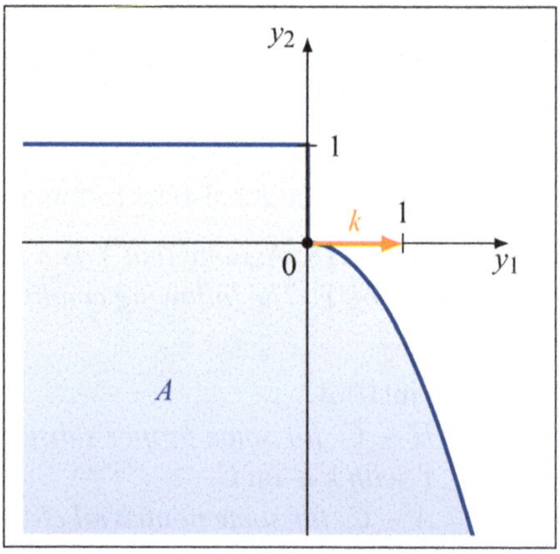


**Fig. 4.9** Set $A$ and vector $k$ in Example 4.6.20

**Example 4.6.20** In $Y = \mathbb{R}^2$, choose $A := (-\mathbb{R}^2_+ + (0,1)^T) \cup \{(y_1, y_2)^T \in \mathbb{R}^2 \mid y_2 \leq -y_1^2\}$ (see Fig. 4.9) and $k := (1,0)^T$. Then $A$ is not convex, and

$$\varphi_{A,k}((y_1, y_2)^T) = \begin{cases} y_1 - \sqrt{-y_2} & \text{if } y_2 \leq 0, \\ y_1 & \text{if } 0 < y_2 \leq 1, \\ +\infty & \text{if } y_2 > 1. \end{cases}$$

$\varphi_{A,k}$ is continuous at $(0,0)^T \in \operatorname{int} \operatorname{dom} \varphi_{A,k}$, but there does not exist any neighborhood of $(0,0)^T$ on which $\varphi_{A,k}$ is Lipschitzian. This can be proved by assuming that $\varphi_{A,k}$ is Lipschitzian on some neighborhood $\{y \in Y \mid \|y\|_2 < \varepsilon\}$, $\varepsilon \in \mathbb{R}_>$, of $(0,0)^T$ with some Lipschitz constant $L$ and showing that the Lipschitz condition is not fulfilled for $(0, y_2)^T$ and $(0, -y_2)^T$ if $y_2 := \frac{1}{2}\max\{1, \varepsilon, \frac{1}{4L^2}\}$.

Conditions under which $\varphi_{A,k}$ is continuous at some point or Lipschitzian on some neighborhood of a point were given by Tammer and Zălinescu [441]. These conditions depend on the definition $\inf \emptyset = +\infty$.

**Remark 4.6.21** (*Bibliographical notes related to Sect.* 4.6)
Lemma 4.6.1 is contained in [484, Theorem 2.9]. Theorem 4.6.2 was proved in [489, Lemma 2.6]. Theorem 4.6.3 is taken from [488, Theorem 5.5.13]. For topological vector spaces $Y$ and $k \in Y \setminus \{0\}$, the equivalence between

$$\{\varphi\colon Y \to \overline{\mathbb{R}} \mid \varphi \text{ translation invariant along } k \text{ and continuous}\}$$

and $\{\varphi_{A,k} \mid A \subseteq Y \text{ closed},\ A - \mathbb{R}_{>}k \subseteq \operatorname{int} A\}$ can be deduced from [221, Cor. 9]. The characterization of all continuous positively homogeneous translation invariant functions on a Banach space as a class of functions $\varphi_{A,k}$ with $A$ and $k$ belonging to special sets was given in [394, Sect. 3].

Theorem 4.6.6 was proved without (f) and (h) in [484, Theorem 3.1]. A statement which is equivalent to the last equivalence in (f) was given in [484, Prop. 3.9]. Equation (4.6.12), (4.6.13) and (4.6.14) were also stated in [164, Cor. 3.4].

Example 4.6.8 and Example 4.6.9 coincide with [484, Example 3.2] and an example in [470], respectively. Lemma 4.6.10, Proposition 4.6.11 and Example 4.6.12 are taken from [484, Lemma 3.3, Prop. 3.4, Example 3.5]. An example in [441] illustrated that $(\mathrm{H2}_{A,k})$ does not imply properness of $\varphi_{A,k}$.

Tammer and Zălinescu [441] proved—in more general spaces than those in Theorem 4.6.14, but under the assumptions given there for $A$ and $k$—that $\varphi_{A,k}$ is finite-valued and Lipschitzian if and only if $k \in -\operatorname{int} 0^+A$. They noted that the function $\varphi_{A,k}$ in Example 4.6.5 is Lipschitzian on $\operatorname{dom} \varphi_{A,k}$ without being finite-valued and gave Example 4.6.20.

Example 4.6.17, Proposition 4.6.18 and Example 4.6.19 can be found in [484, Example 3.11, Prop. 3.13, Example 3.14].

## 4.7 Convex Translation Invariant Functions

The previous sections contain many properties of translation invariant functions that are also essential if these functions are convex. Now, we will prove additional results for this special case. Sets $A$ for which $\varphi_{A,k}$ is convex were characterized in Theorem 4.2.10. We first turn to functions $\varphi_{A,k}$ which are generated by convex cones $A$.

In many applications, the set $A$ in the definition of $\varphi_{A,k}$ is a nontrivial convex cone in order to obtain a scalarization that is related to a cone order (see Sect. 4.5). In the representation $\varphi_{A,k}$ of a coherent risk measure, the so-called acceptance set $A$ can coincide with the ordering cone in a function space $L^p(\Omega, M, P)$ (see e.g. [391]). This cone has an empty interior if $1 \leq p < \infty$. There are notable links between coherent risk measures, convex games, distorted probabilities, the Choquet integral representation and insurance prices (see the references in [170, p. 1481]).

Several properties of $\varphi_{A,k}$ for convex cones $A$ follow immediately from the previous sections, taking into consideration that the recession cone of a convex cone $A$ is $A$.

**Proposition 4.7.1** *Assume* $(\mathrm{Sp0}_{A,k})$ *and that $A$ is a nontrivial convex cone. Then:*

(a) $\operatorname{dom}\varphi_{A,k}$ *is convex.*
(b) $\varphi_{A,k}$ *is sublinear and* $(-A)$*-monotone.*
(c) *On each set $F \subseteq Y$ on which $\varphi_{A,k}$ is finite-valued, $\varphi_{A,k}$ is strictly* $(-\operatorname{core} A)$*-monotone and strictly* $(-A+\mathbb{R}_{>}k)$*-monotone.*
(d) *If $k \in A$, then $\varphi_{A,k}(y) = -\infty$ for all $y \in \operatorname{dom}\varphi_{A,k}$.*
*If $k \in (-A) \cap A$, then* $\operatorname{dom}\varphi_{A,k} = A$.
(e) *If $k \in -A$, then* $\operatorname{dom}\varphi_{A,k} = A + \mathbb{R}_{>}k$ *and*

$$\operatorname{lev}_{\varphi_{A,k},\leq}(t) = \operatorname{cl}_k(A) + tk \text{ for all } t \in \mathbb{R}. \tag{4.7.1}$$

(f) $\varphi_{A,k}$ *is finite-valued if and only if*

(i) $k \in -\operatorname{core} A$ *or*
(ii) *$A$ is a linear subspace of $Y$ of codimension* 1 *and $k \notin A$.*

*In case* (ii), $\varphi_{A,k}$ *is a linear function.*
(g) $\operatorname{dom}\varphi_{A,k} = Y$ *holds if and only if $\varphi_{A,k}$ is finite-valued or $k \in \operatorname{core} A$.*
(h) $\varphi_{A,k} = -\infty$ *if and only if $k \in \operatorname{core} A$.*

***Proof***

(a) follows from (4.2.1).
(b) $\operatorname{cl}_k(A)$ is a convex cone by Proposition 2.3.30. Thus, $\operatorname{cl}_k(A - \mathbb{R}_+k) = \operatorname{cl}_k(A) - \mathbb{R}_+k$ is a convex cone, and $\varphi_{A,k}$ is sublinear because of Theorem 4.2.10. The monotonicity statements in (b) and (c) follow from Corollary 4.5.10.
(d) The first statement results from Theorem 4.2.1(d). Since $k \in (-A) \cap A$, we get $a + \mathbb{R}k \subseteq A$ for all $a \in A$, and thus $\operatorname{dom}\varphi_{A,k} = A + \mathbb{R}k = A$.
(e) follows from Theorem 4.2.5.
(f) If $\varphi_{A,k}$ is finite-valued, then $k \notin A$ by (d), which together with Lemma 2.3.11 results in the cases (i) and (ii).
If (i) holds, then $\varphi_{A,k}$ is finite-valued because of Theorem 4.2.9.
Assume now (ii). Since each $y \in Y$ has a unique representation $y = a + tk$ with $a \in A$ and $t \in \mathbb{R}$, the function $\varphi_{A,k}$ is finite-valued and linear.
(g) follows from (f) and Lemma 2.3.11.
(h) results from (g) and (d). □

We summarize statements for a convex cone $A$ with $k \in -\operatorname{core} A$. Proposition 4.7.1 and Theorem 4.2.9 imply the following.

**Corollary 4.7.2** *Let $A$ be a nontrivial convex cone in a linear space $Y$ with $k \in -\operatorname{core} A$. Then $\varphi_{A,k}$ is finite-valued, sublinear,* $(-A)$*-monotone, strictly* $(-\operatorname{core} A)$*-monotone and strictly* $(-A+\mathbb{R}_{>}k)$*-monotone,*

$$\operatorname{lev}_{\varphi_{A,k},\leq}(t) = \operatorname{cl}_k(A) + tk \quad \textit{for all } t \in \mathbb{R}, \tag{4.7.2}$$

$$\operatorname{lev}_{\varphi_{A,k},<}(t) = \operatorname{core} A + tk \quad \textit{for all } t \in \mathbb{R}. \tag{4.7.3}$$

Now, we turn to functionals on topological vector spaces.

**Proposition 4.7.3** *Assume that $Y$ is a topological vector space, $A \subset Y$ is a nontrivial convex cone and $k \in Y \setminus \{0\}$.*

(a) *If $k \in (-A) \setminus \operatorname{cl} A$, then $\varphi_{A,k}$ is proper, strictly $(-\operatorname{core} A)$-monotone on $\operatorname{dom} \varphi_{A,k}$ and strictly $(-A + \mathbb{R}_> k)$-monotone on $\operatorname{dom} \varphi_{A,k}$.*
(b) *If $k \in \operatorname{cl} A$, then $\varphi_{A,k}(y) = -\infty$ for all $y \in \operatorname{int} A$.*
(c) *$(\mathrm{Sp2}_{A,k})$ holds if and only if $k \in -\operatorname{int} A$.*
(d) *If $k \in -\operatorname{int} A$, then $\varphi_{A,k}$ is continuous, finite-valued, and strictly $(-\operatorname{core} A)$-monotone.*
(e) *If $A$ is closed and $k \in -A$, then $\varphi_{A,k}$ is lower semicontinuous.*

***Proof***

(a) Take any $a \in A$. Since $k \notin \operatorname{cl} A$, there exists a neighborhood $U$ of $k$ such that $U \subseteq Y \setminus A$. There is $n \in \mathbb{N}_>$ with $y := k + \frac{1}{n}(a + k) \in U$. Thus, $y \notin A$. This implies $ny = a + (n+1)k \notin A$ since $A$ is a cone. Hence, $\varphi_{A,k}$ is proper by Theorem 4.2.5(d). This yields the monotonicity because of Proposition 4.7.1(c).
(b) Take any $a \in \operatorname{int} A$. Then $a + \mathbb{R}_> k \subseteq \operatorname{int} \operatorname{cl} A = \operatorname{int} A$ by Lemma 2.1.24. Hence, $\varphi_{A,k}(a) = -\infty$.
(c) The forward direction follows from $0 \in A$, the other direction with $\operatorname{cl} A - \mathbb{R}_> k \subseteq \operatorname{int} \operatorname{cl} A = \operatorname{int} A$.
(d) Because of (c), Theorem 4.6.6 and Corollary 4.7.2 yield the assertion.
(e) results from Theorem 4.4.2. □

Further sufficient conditions for the continuity of $\varphi_{A,k}$ in the case where $A$ is convex will be given in Proposition 4.7.9.

Proposition 4.7.3(a) does not hold for an arbitrary $k \in (-A) \setminus A$, and the assumption in (b) does not imply $\varphi_{A,k}(y) = -\infty$ for all $y \in A$. This will be shown in part (b) of the following example. Example 4.7.4 illustrates several statements of Propositions 4.7.1 and 4.7.3.

**Example 4.7.4** Let $A$ be the lexicographic ordering cone in $\mathbb{R}^2$ (see Example 2.3.56). For each $k \in \mathbb{R}^2 \setminus \{0\}$, $\varphi_{A,k}$ is sublinear and $(-A)$-monotone.

(a) For each $k \in -\operatorname{int} A$, $\varphi_{A,k}$ is continuous, finite-valued, and strictly $(-\operatorname{int} A)$-monotone. For $k := (-1, 0)^T$ and all $y = (y_1, y_2)^T \in \mathbb{R}^2$,

$$\varphi_{A,k}(y) = -y_1.$$

(b) Consider now $k := (0, -1)^T$. Then $k \in (-A) \setminus A$, but $k \notin -\operatorname{int} A$ and $k \in \operatorname{cl} A$. We get $\operatorname{dom} \varphi_{A,k} = \{(y_1, y_2)^T \in \mathbb{R}^2 \mid y_1 \geq 0\}$, and

$$\varphi_{A,k}((y_1, y_2)^T) = \begin{cases} +\infty & \text{if } y_1 < 0, \\ -y_2 & \text{if } y_1 = 0, \\ -\infty & \text{if } y_1 > 0. \end{cases}$$

Thus, $\varphi_{A,k}(y) = -\infty$ for all $y \in \operatorname{int} A$, but $\varphi_{A,k}(y) \in \mathbb{R}$ for each $y \in \operatorname{bd} A$. Here $A$ coincides with $\operatorname{lev}_{\varphi_{A,k},\leq}(0)$, and $\operatorname{lev}_{\varphi_{A,k},=}(t) = \{(0, -t)^T\} \subset A + tk$ for all $t \in \mathbb{R}$.

$A - \mathbb{R}_> k \not\subseteq \operatorname{int} A$, and $\operatorname{lev}_{\varphi_{A,k},<}(0) = A \setminus \{0\}$ is not contained in $\operatorname{int} A$.

$\varphi_{A,k}$ is not proper and not strictly $(-\operatorname{int} A)$-monotone. For example, $y^2 - y^1 \in -\operatorname{int} A$ holds for $y^1 := (2, 0)^T$ and $y^2 := (1, 0)^T$, but $\varphi_{A,k}(y^1) \not< \varphi_{A,k}(y^2)$ since $\varphi_{A,k}(y^1) = \varphi_{A,k}(y^2) = -\infty$.

The lexicographic order $\leq_A$ can be completely described by the function $\varphi_{A,k}$. For all $y^1, y^2 \in Y$,

$$y^1 \leq_A y^2 \iff (y^1 = y^2 \text{ or } \varphi_{A,k}(y^2 - y^1) < 0).$$

(c) Take any $k \in (-A) \setminus A$. Then $k = (0, -t)^T$ for some $t \in \mathbb{R}_>$. Proposition 4.3.3 and (b) imply

$$\varphi_{A,k}((y_1, y_2)^T) = \begin{cases} +\infty & \text{if } y_1 < 0, \\ -\frac{y_2}{t} & \text{if } y_1 = 0, \\ -\infty & \text{if } y_1 > 0. \end{cases}$$

(d) For each $k \in \operatorname{int} A$, we have $\varphi_{A,k} = -\infty$.

(e) For each $k \in A \setminus \operatorname{int} A$, we get $\operatorname{dom} \varphi_{A,k} = \{(y_1, y_2)^T \in \mathbb{R}^2 \mid y_1 \geq 0\}$, and

$$\varphi_{A,k}((y_1, y_2)^T) = \begin{cases} +\infty & \text{if } y_1 < 0, \\ -\infty & \text{if } y_1 \geq 0. \end{cases}$$

The sum of two functions $\varphi_{A,k^1}$ and $\varphi_{A,k^2}$ with $k^1, k^2 \in \mathbb{R}^2 \setminus \{0\}$ is not necessarily a function $\varphi_{A,k}$ with $k \in \mathbb{R}^2 \setminus \{0\}$. This becomes obvious by choosing $k^1 := (-1, 0)^T$ and $k^2 := (0, -1)^T$.

The functional $\varphi_{A,k}$ is often used locally in such a way that we attach some set to a point $y^0$ and use this set as $A$. Properties of $\varphi_{y^0-C,k}$ for closed convex cones $C$ are summarized in the next corollary.

**Corollary 4.7.5** *Assume that $Y$ is a topological vector space, $C \subset Y$ is a nontrivial closed convex cone, $k \in C \setminus \{0\}$ and $y^0 \in Y$. Then $\varphi_{y^0-C,k}$ is a convex $C$-monotone functional, which is lower semicontinuous.*

(a) *If $k \in C \cap (-C)$, then* $\operatorname{dom} \varphi_{y^0-C,k} = y^0 - C$ *and $\varphi_{y^0-C,k}$ does not attain any real value.*

(b) *Assume $k \in C \setminus (-C)$. Then $\varphi_{y^0-C,k}$ is proper and strictly* (core $C$)*-monotone on* $\operatorname{dom} \varphi_{y^0-C,k}$. *Moreover, it is strictly $(C + \mathbb{R}_> k)$-monotone on* $\operatorname{dom} \varphi_{y^0-C,k}$.

(c) *If $k \in$* core $C$, *then $\varphi_{y^0-C,k}$ is finite-valued.*

(d) *If* $k \in \operatorname{int} C$*, then* $\varphi_{y^0-C,k}$ *is continuous, finite-valued and strictly* $(\operatorname{int} C)$*-monotone.*
(e) $\varphi_{y^0-C,k}$ *is subadditive if and only if* $y^0 \in -C$.
(f) $\varphi_{y^0-C,k}$ *is sublinear if and only if* $y^0 \in C \cap (-C)$.

***Proof*** Apply Propositions 4.7.1 and 4.7.3 to $A = -C$. Then (a)–(d) and the statements in front of (a) follow from Proposition 4.3.5.

Since $C + C = C$, the condition $(y^0 - C) + (y^0 - C) \subseteq y^0 - C$ is equivalent to $y^0 + (y^0 - C) \subseteq y^0 - C$ and thus equivalent to $y^0 - C \subseteq -C$. Because of $0 \in C$, we get

$$(y^0 - C) + (y^0 - C) \subseteq y^0 - C \iff y^0 \in -C. \tag{4.7.4}$$

(e) results from Theorem 4.4.4(h) and (4.7.4).
(f) Since $y^0 - C$ is convex, Lemma 2.2.6 and (4.7.4) imply that $y^0 - C$ is a convex cone if and only if $0 \in y^0 - C$ and $(y^0 - C) + (y^0 - C) \subseteq y^0 - C$, i.e., if and only if $y^0 \in C$ and $y^0 \in -C$. Theorem 4.4.4(i) yields the assertion. □

By Corollary 4.3.8, we have $\varphi_{A,k} = \iota_A \# \varphi_{\{0\},k}$ for each function $\varphi_{A,k}$. The following lemma contains properties of $\varphi_{\{0\},k}$, which will be used in the proofs of the subsequent sections.

**Lemma 4.7.6** *Assume* $Y$ *is a linear space and* $k \in Y \setminus \{0\}$.

(a) $\varphi_{\{0_Y\},k}$ *is a sublinear functional with*

$$\varphi_{\{0_Y\},k}(y) = \begin{cases} t & \text{if } y = tk \text{ with } t \in \mathbb{R}, \\ +\infty & \text{otherwise.} \end{cases} \tag{4.7.5}$$

*It is lower semicontinuous if* $Y$ *is a separated locally convex space.*
(b) $\varphi_{\{0_Y\},k}$ *is the greatest functional* $g\colon Y \to \overline{\mathbb{R}}$ *with* $g(0) \le 0$ *which is translation invariant along* $k$.
(c) *Suppose that* $Y$ *belongs to a dual pair* $(Y, Y', \langle .,. \rangle)$*. For each* $y' \in Y'$,

$$\varphi^*_{\{0_Y\},k}(y') = \begin{cases} 0 & \text{if } \langle k, y' \rangle = 1, \\ +\infty & \text{otherwise.} \end{cases} \tag{4.7.6}$$

*Hence,*

$$\varphi^*_{\{0_Y\},k} = \iota_D \quad \text{for } D := \{y' \in Y' \mid \langle k, y' \rangle = 1\}. \tag{4.7.7}$$

***Proof*** Define $\varphi_k := \varphi_{\{0_Y\},k}$.

(a) Since $0_Y - \mathbb{R}_+ k$ is a convex cone, the set $\operatorname{cl}_k(0_Y - \mathbb{R}_+ k)$ is a convex cone by Lemma 2.3.30. The sublinearity of $\varphi_k$ follows from Theorem 4.2.10(d). Equation (4.7.5) is easy to verify. If $Y$ is a separated locally convex space, then the set $\mathbb{R}_+ k$ is closed, and $\varphi_k$ is lower semicontinuous because of Theorem 4.4.4.
(b) Take any $g\colon Y \to \overline{\mathbb{R}}$ with $g(0) \le 0$ which is translation invariant along $k$. Then $g(tk) = g(0 + tk) = g(0) + t \le t$ for each $t \in \mathbb{R}$. Thus, $g \le \varphi_k$.

(c) For each $y' \in Y'$,

$$\begin{aligned}\varphi_k^*(y') &= \sup\{\langle y, y'\rangle - \varphi_k(y) \mid y \in Y\}\\ &= \sup\{\langle tk, y'\rangle - t \mid t \in \mathbb{R}\}\\ &= \sup\{t(\langle k, y'\rangle - 1) \mid t \in \mathbb{R}\}.\end{aligned}$$

This implies (4.7.6). □

We will now investigate $\varphi_{A,k}$ for arbitrary convex sets $A$.

**Proposition 4.7.7** *Assume that $Y$ is a topological vector space, that $A \subseteq Y$ is a convex set, and $k \in (-0^+A) \setminus 0^+(\operatorname{cl} A)$. Then $\varphi_{A,k}$ is proper and strictly $(-\operatorname{core} 0^+A)$-monotone on $\operatorname{dom} \varphi_{A,k}$.*

***Proof*** Because of $k \in (-0^+A) \setminus 0^+(\operatorname{cl} A)$, $A$ is a proper subset of $Y$. To obtain a contradiction, suppose that $\varphi_{A,k}$ is not proper. Theorem 4.2.5(f) implies the existence of $a^0 \in A$ with $a^0 + \mathbb{R}k \subseteq A$. Take any $a \in \operatorname{cl} A$ and $t \in \mathbb{R}_>$. Define $a^1 := a + tk$. Let $U$ be an arbitrary neighborhood of $a^1$. Then there exists $n \in \mathbb{N}_>$ with $y := a^1 + \frac{1}{n}(a^0 - a) \in U$. We get $y = a + tk + \frac{1}{n}(a^0 - a) = (1 - \frac{1}{n})a + \frac{1}{n}(a^0 + ntk) \in \operatorname{cl} A$ since $A$ is convex. Thus, $a^1 \in \operatorname{cl} A$, and $a + \mathbb{R}_+k \subseteq \operatorname{cl} A$ for all $a \in \operatorname{cl} A$. This contradicts $k \notin 0^+(\operatorname{cl} A)$. Since $\varphi_{A,k}$ is proper, the monotonicity results from Corollary 4.5.8(b). □

Note that the recession cone of each unbounded closed convex set in a finite-dimensional separated topological vector space does not contain zero only [508]. However, Zălinescu [508, Example 1.1.1] gave an example of an unbounded closed convex set $A$ in $\ell^p$, $p \in [1, \infty]$, with $0^+A = \{0\}$.

Example 4.7.4(b) illustrates that $\varphi_{A,k}$ in Proposition 4.7.7 is not necessarily proper for $k \in -0^+A \setminus 0^+A$. In Theorem 4.2.1(d), we proved that $\varphi_{A,k}$ does not attain any real value if $k \in 0^+A$.

**Corollary 4.7.8** *Assume* $(\mathrm{H1}_{A,k})$. *If $\varphi_{A,k}$ is convex, then it is proper or does not attain any real value.*

***Proof*** By Theorem 4.4.4(f), $A$ is convex. If $k \in 0^+A$, then $\varphi_{A,k}$ does not attain any real value. Otherwise, $k \in (-0^+A) \setminus 0^+A$, which implies that $\varphi_{A,k}$ is proper because of Proposition 4.7.7. □

**Proposition 4.7.9** *Assume that $Y$ is a topological vector space, $A$ is a proper subset of $Y$ for which* $\operatorname{cl} A$ *is convex and $k \in (-0^+A) \setminus 0^+A$ such that*

$$\operatorname{cl} A - \mathbb{R}_>k \subseteq A.$$

(a) *$\varphi_{A,k}$ is convex, proper and lower semicontinuous on $Y$ and strictly $(-\operatorname{core} 0^+(\operatorname{cl} A))$-monotone on $\operatorname{dom} \varphi_{A,k}$.*
(b) $\operatorname{dom} \varphi_{A,k} = A + \mathbb{R}k$ *is convex.*

(c) $\operatorname{dom} \varphi_{A,k} = \operatorname{core} \operatorname{dom} \varphi_{A,k}$ *if and only if* $\operatorname{cl} A - \mathbb{R}_> k \subseteq \operatorname{core} A$.
(d) *If* $\operatorname{int} A \neq \emptyset$ *or* $Y = \mathbb{R}^n$ *(equipped with the Euclidean topology), then* $\varphi_{A,k}$ *is continuous if and only if* $\operatorname{dom} \varphi_{A,k}$ *is open.*
(e) *If* $\operatorname{int} A \neq \emptyset$*, then* $\varphi_{A,k}$ *is continuous on* $\operatorname{int} \operatorname{dom} \varphi_{A,k}$.
(f) *If* $Y$ *is a Banach space, then* $\varphi_{A,k}$ *is locally Lipschitzian on* $\operatorname{int} \operatorname{dom} \varphi_{A,k}$.

***Proof*** ($\mathrm{Sp1}_{A,k}$) is fulfilled, and $\varphi_{A,k} = \varphi_{\operatorname{cl} A,k}$ by (4.4.9). Lemma 2.3.16 implies $\operatorname{core} \operatorname{cl} A = \operatorname{core} A$ and $\operatorname{int} \operatorname{cl} A = \operatorname{int} A$.

(a) We get $k \notin 0^+(\operatorname{cl} A)$ from Lemma 2.3.16(g). $\varphi_{A,k}$ is convex and lower semicontinuous by Theorem 4.4.4, proper and strictly $(-\operatorname{core} 0^+(\operatorname{cl} A))$-monotone on $\operatorname{dom} \varphi_{A,k}$ by Proposition 4.7.7 (applied to $\operatorname{cl} A$).
(b) $\operatorname{dom} \varphi_{A,k} = \operatorname{cl} A + \mathbb{R}k$ is convex since $\operatorname{cl} A$ is convex.
(c) Assume $\operatorname{dom} \varphi_{A,k} = \operatorname{core} \operatorname{dom} \varphi_{A,k}$, and take any $y^1 \in \operatorname{cl} A - \mathbb{R}_> k$ and $y \in Y$. Hence, $y^1 \in A$. There exist $a^1 \in \operatorname{cl} A$ and $t_1 \in \mathbb{R}_>$ with $y^1 = a^1 - t_1 k$. Then $y^1 \in \operatorname{cl} A + \mathbb{R}k = \operatorname{core}(\operatorname{cl} A + \mathbb{R}k)$. There is $t_2 \in \mathbb{R}_>$ with $y^2 := y^1 + t_2(y - y^1) \in \operatorname{cl} A + \mathbb{R}k$. If $y^2 \in \operatorname{cl} A$, then the convexity of $\operatorname{cl} A$ implies $y^1 + t(y - y^1) \in \operatorname{cl} A$ for each $t \in \mathbb{R}_+$ with $t \leq t_2$. Suppose now $y^2 \notin \operatorname{cl} A$. By Lemma 2.3.16(a), $k \in -0^+(\operatorname{cl} A)$. There exist $a^2 \in \operatorname{cl} A$ and $t_3 \in \mathbb{R}_>$ with $y^2 = a^2 + t_3 k$. We obtain $y^1 + \frac{t_1 t_2}{t_1+t_3}(y - y^1) = y^1 + \frac{t_1}{t_1+t_3}(y^2 - y^1) = a^1 - t_1 k + \frac{t_1}{t_1+t_3}(a^2 + t_3 k - (a^1 - t_1 k)) = a^1 + \frac{t_1}{t_1+t_3}(a^2 - a^1) \in \operatorname{cl} A$ since $\operatorname{cl} A$ is convex. Because of $y^1 \in A$, we get $y^1 + t(y - y^1) \in \operatorname{cl} A$ for each $t \in \mathbb{R}_+$ with $t \leq \frac{t_1 t_2}{t_1+t_3}$. Thus, $y^1 \in \operatorname{core} \operatorname{cl} A = \operatorname{core} A$.
The reverse implication in (c) follows from Theorem 4.2.1(c) by applying the theorem to $\operatorname{cl} A$ instead of $A$.
(d) The assumptions of (d) imply $\operatorname{int} A = \operatorname{core} A$ and $\operatorname{int} \operatorname{dom} \varphi_{A,k} = \operatorname{core} \operatorname{dom} \varphi_{A,k}$ when applying Proposition 2.3.6 to $\operatorname{cl} A$ and to $\operatorname{dom} \varphi_{A,k}$. Hence, (c) implies that $\operatorname{dom} \varphi_{A,k}$ is open if and only if $\operatorname{cl} A - \mathbb{R}_> k \subseteq \operatorname{int} A$. The assertion follows from Theorem 4.6.2.
(e) If $\operatorname{int} A \neq \emptyset$, then $\operatorname{int} \operatorname{dom} \varphi_{A,k} \neq \emptyset$, and $\varphi_{A,k}$ is bounded above by zero on $\operatorname{int} A$ because of (4.4.6). Thus, $\varphi_{A,k}$ is continuous on $\operatorname{int} \operatorname{dom} \varphi_{A,k}$ by Proposition 3.4.9.
(f) follows from Lemma 3.4.10. □

Tammer and Zălinescu [441] gave an example of a nontrivial closed convex cone $A$ in $\mathbb{R}^3$ and $k \in (-A) \setminus A$ for which $\varphi_{A,k}$ is not Lipschitzian on $\operatorname{int} \operatorname{dom} \varphi_{A,k}$. Example 4.4.5 illustrates that the statements (c)–(f) of Proposition 4.7.9 are not valid for an arbitrary closed nonconvex set $A$ which fulfills ($\mathrm{Sp1}_{A,k}$).

We get the following from Proposition 4.7.9.

**Corollary 4.7.10** *Assume that* $Y$ *is a topological vector space,* $A$ *is a proper closed convex subset of* $Y$, $\operatorname{int} A \neq \emptyset$, $k \in -0^+ A$ *and* $Y = A + \mathbb{R}k$*. Then* $\varphi_{A,k}$ *is convex, continuous and finite-valued.*

The assumptions of Corollary 4.7.10 imply $k \in (-0^+ A) \setminus 0^+ A$ since otherwise $Y = A$.

Further statements for special convex and sublinear translation invariant functions are contained in the next section.

**Remark 4.7.11** (*Bibliographical notes related to Sect.* 4.7)

Proposition 4.7.1 was shown without the strict $(-A + \mathbb{R}_{>}k)$-monotonicity in (c), without (4.7.1) and without (g)–(h) in [483, Prop. 5.1]. Equation (4.7.3) in Corollary 4.7.2 was proved in [483, Prop. 5.1]. For an algebraically closed cone, the other statements of Corollary 4.7.2 without the strict $(-A + \mathbb{R}_{>}k)$-monotonicity were shown in [483, Cor. 5.1]. Under the additional assumptions that $Y$ is a separated locally convex space and $A$ is closed, the statements of Corollary 4.7.2 without the last monotonicity assertion were given in [195, Lemma 7].

Proposition 4.7.3 was proved without the last statement in (a) in [484]. Under the assumptions that $Y$ is a separated locally convex space, $A$ is a closed convex cone and $k \in -A \setminus A$, it was shown in [195, Lemma 7] that $\varphi_{A,k}$ is a lower semicontinuous sublinear proper $(-A)$-monotone function for which $\text{lev}_{\varphi_{A,k},\leq}(t) = A + tk$ holds for all $t \in \mathbb{R}$.

Example 4.7.4 extends an example from [485, Sect. 3].

Filipović and Kupper [162] proved in a separated locally convex space $Y$ that $\iota_D^*$ with $D$ given by (4.7.7) in the dual system $(Y, Y^*, \langle ., . \rangle)$ is the greatest lower semicontinuous convex function which is translation invariant along $k$ and zero at $y = 0$. Under their assumptions, $\iota_D^*$ coincides with $\varphi_{\{0_Y\},k}$ because of (4.7.7).

Proposition 4.7.7, Corollary 4.7.8 and Proposition 4.7.9 are taken from [484, Prop. 4.3, Cor. 4.4, Prop. 4.5]. Proposition 4.7.9(f) was proved under the additional assumptions that $A$ is closed and $\text{int}\, A \neq \emptyset$ in [441].

Corollary 4.7.10 is contained in [186, Theorem 2.2].

## 4.8 Translation Invariant Functions Generated by Linear Inequalities

In this section, we start with sets which are given by a linear inequality system and turn afterwards to sets in which some inequalities can be violated.

First, we will work with sets $A$ fulfilling the following assumption:

(HT1$_A$): $(Y, Y', \langle ., . \rangle)$ is a dual pair, where $Y$ is equipped with the weak topology $\sigma(Y, Y')$, $\Gamma \subset Y' \setminus \{0\}$, $A$ is a proper subset of $Y$ given by $A = \{y \in Y \mid \forall g \in \Gamma\colon \langle y, g \rangle \leq b_g\}$ with $b_g \in \mathbb{R}$ for each $g \in \Gamma$ such that $0^+A \neq \{0\}$.

Consider any set $A$ which fulfills (HT1$_A$). It is easy to check that $A$ is a closed convex subset of $Y$ with

$$0^+A = \{y \in Y \mid \forall g \in \Gamma\colon \langle y, g \rangle \leq 0\}.$$

$k \in -0^+A$ is equivalent to the condition that $\langle k, g \rangle \geq 0$ holds for all $g \in \Gamma$. The set $A$ fulfills (H1$_{A,k}$) for each $k \in -0^+A \setminus \{0\}$.

**Proposition 4.8.1** *Assume* (HT1$_A$) *and* $k \in Y \setminus \{0\}$ *with* $\langle k, g\rangle \geq 0$ *for all* $g \in \Gamma$.

(a) $\varphi_{A,k}$ *is convex and lower semicontinuous.*
(b) $\varphi_{A,k}$ *is subadditive if and only if* $b_g \leq 0$ *for all* $g \in \Gamma$.
(c) $\varphi_{A,k}$ *is sublinear if and only if* $b_g = 0$ *for all* $g \in \Gamma$.
(d) $\varphi_{A,k}$ *is* $C$*-monotone for* $C := \{y \in Y \mid \forall g \in \Gamma\colon \langle y, g\rangle \geq 0\}$.

*Assume now additionally that there exists* $g \in \Gamma$ *with* $\langle k, g\rangle > 0$.

(e) $\varphi_{A,k}$ *is proper.*
(f) *If* $\operatorname{int} A \neq \emptyset$, *then* $\varphi_{A,k}$ *is continuous on* $\operatorname{int}\operatorname{dom}\varphi_{A,k}$.
(g) *If* $Y$ *is a Banach space, then* $\varphi_{A,k}$ *is locally Lipschitzian on* $\operatorname{int}\operatorname{dom}\varphi_{A,k}$.
(h) $\varphi_{A,k}$ *is strictly* (core $C$)*-monotone on* $\operatorname{dom}\varphi_{A,k}$.

***Proof*** (a)–(c) result from Theorem 4.4.4. (d) follows from Corollary 4.5.8.
(e)–(h) result from Proposition 4.7.9. □

The formula for the considered functional $\varphi_{A,k}$ is given in the following proposition.

**Proposition 4.8.2** *Assume* (HT1$_A$), *and take any* $k \in Y \setminus \{0\}$ *with* $\langle k, g\rangle \geq 0$ *for all* $g \in \Gamma$.

(a) *If* $\langle k, g\rangle = 0$ *for all* $g \in \Gamma$, *then we have* $\operatorname{dom}\varphi_{A,k} = A$ *and*

$$\varphi_{A,k}(y) = -\infty \quad \textit{for each } y \in A.$$

(b) *If there exists* $g_1 \in \Gamma$ *such that* $\langle k, g_1\rangle > 0$, *then we get*

$$\operatorname{dom}\varphi_{A,k} = \left\{ y \in Y \mid \langle y, g\rangle \leq b_g \textit{ for all } g \in \Gamma^{\mathrm{act}}, \textit{ and } \sup_{g \notin \Gamma^{\mathrm{act}}} \frac{\langle y, g\rangle - b_g}{\langle k, g\rangle} \in \mathbb{R} \right\},$$

$$\varphi_{A,k}(y) = \sup_{g \notin \Gamma^{\mathrm{act}}} \frac{\langle y, g\rangle - b_g}{\langle k, g\rangle} \quad \textit{for each } y \in \operatorname{dom}\varphi_{A,k},$$

*where* $\Gamma^{\mathrm{act}} := \{g \in \Gamma \mid \langle k, g\rangle = 0\}$.
(c) *If* $\Gamma$ *consists of a finite number of functions and* $\langle k, g\rangle > 0$ *for all* $g \in \Gamma$, *then* (H2$_{A,k}$) *holds and* $\varphi_{A,k}$ *is finite-valued with*

$$\varphi_{A,k}(y) = \max_{g \in \Gamma} \frac{\langle y, g\rangle - b_g}{\langle k, g\rangle} \quad \textit{for all } y \in Y.$$

***Proof*** For all $y \in Y$ and $t \in \mathbb{R}$,

$$\begin{aligned} y \in A + tk &\iff y - tk \in A \iff \forall g \in \Gamma\colon \langle y - tk, g\rangle \leq b_g, \\ &\iff \forall g \in \Gamma\colon \langle y, g\rangle - t\langle k, g\rangle \leq b_g. \end{aligned}$$

(a) For all $y \in Y$ and $t \in \mathbb{R}$, the assumption implies that $y \in A + tk$ is equivalent to $y \in A$. By the definition of $\varphi_{A,k}$, this results in the assertion since $\operatorname{dom}\varphi_{A,k} = A + \mathbb{R}k$.

(b) If there exists $g_1 \in \Gamma$ such that $\langle k, g_1\rangle > 0$, then for all $y \in Y$ and all $t \in \mathbb{R}$,

$$y \in A + tk \iff \left( \langle y, g\rangle \leq b_g \text{ for all } g \in \Gamma^{\text{act}}, \text{ and } t \geq \frac{\langle y, g\rangle - b_g}{\langle k, g\rangle} \text{ for all } g \notin \Gamma^{\text{act}} \right).$$

This yields (b).

(c) We have $A - \mathbb{R}_{>}k \subseteq \{y \in Y \mid \forall g \in \Gamma\colon \langle y, g\rangle < b_g\}$. Then $\{y \in Y \mid \forall g \in \Gamma\colon \langle y, g\rangle < b_g\} \subseteq \operatorname{int} A$ results from the continuity of the functions $\langle \cdot, g\rangle$. Hence, $(\text{H2}_{A,k})$ holds. The other statements follow from (b). □

In the next proposition, we describe in which way shifts in the constraints of $A$ influence the values of $\varphi_{A,k}$ and how perturbations in the function value of $\varphi_{A,k}$ can be interpreted as perturbations in the constraints of $A$.

**Proposition 4.8.3** *Assume* $(\text{HT1}_A)$. *Take any* $k \in Y \setminus \{0\}$ *with* $\langle k, g\rangle \geq 0$ *for all* $g \in \Gamma$.

(a) *For* $\tilde{A} := \{y \in Y \mid \forall g \in \Gamma\colon \langle y, g\rangle \leq b_g + \varepsilon\langle k, g\rangle\}$ *with* $\varepsilon \in \mathbb{R}$, *we get*

$$\varphi_{\tilde{A},k}(y) = \varphi_{A,k}(y) - \varepsilon \text{ for all } y \in Y.$$

(b) *For* $\tilde{A} := \{y \in Y \mid \forall g \in \Gamma\colon \langle y, g\rangle \leq b_g + \langle y^0, g\rangle\}$ *with* $y^0 \in Y$, *we have*

$$\varphi_{\tilde{A},k}(y) = \varphi_{A,k}(y - y^0) \text{ for all } y \in Y.$$

***Proof***

(a) results from Proposition 4.3.4 since

$$\begin{aligned} A + \varepsilon k &= \{y + \varepsilon k \in Y \mid \forall g \in \Gamma\colon \langle y, g\rangle \leq b_g\} \\ &= \{y \in Y \mid \forall g \in \Gamma\colon \langle y - \varepsilon k, g\rangle \leq b_g\} \\ &= \{y \in Y \mid \forall g \in \Gamma\colon \langle y, g\rangle - \varepsilon\langle k, g\rangle \leq b_g\} \\ &= \{y \in Y \mid \forall g \in \Gamma\colon \langle y, g\rangle \leq b_g + \varepsilon\langle k, g\rangle\}. \end{aligned}$$

(b) follows from Proposition 4.3.5 since

$$\begin{aligned} A + y^0 &= \{y + y^0 \in Y \mid \forall g \in \Gamma\colon \langle y, g\rangle \leq b_g\} \\ &= \{y \in Y \mid \forall g \in \Gamma\colon \langle y - y^0, g\rangle \leq b_g\} \\ &= \{y \in Y \mid \forall g \in \Gamma\colon \langle y, g\rangle \leq b_g + \langle y^0, g\rangle\}. \end{aligned}$$ □

Consider now any polyhedral set $A$ in a Euclidean space which is given by the assumption

$(\text{H1}_A)$: $A = \{y \in \mathbb{R}^\ell \mid Wy \leq b\}$ is a proper unbounded subset of $Y = \mathbb{R}^\ell$ with $W \in \mathbb{R}^{r,\ell}$ and $b \in \mathbb{R}^r$, where $W = (w^1 \cdots w^r)^T$ with $w^i \in \mathbb{R}^\ell \setminus \{0\}$ for all $i \in \{1, \ldots, r\}$.

Then $w^i$ stands for the i-th row vector of $W$. It is easy to check that $0^+A = \{y \in \mathbb{R}^\ell \mid Wy \leq 0\}$. Since $A$ is an unbounded closed convex set, we have $0^+A \neq \{0\}$. The condition $k \in -0^+A$ is equivalent to $Wk \geq 0$.

**Proposition 4.8.4** *Assume* (H1$_A$). *Take any* $k \in \mathbb{R}^\ell \setminus \{0\}$ *with* $Wk \geq 0$.

(a) $\varphi_{A,k}$ *is convex and lower semicontinuous.*
(b) $\varphi_{A,k}$ *is subadditive if and only if* $b \leq 0$.
(c) $\varphi_{A,k}$ *is sublinear if and only if* $b = 0$.
(d) $\varphi_{A,k}$ *is* $C$*-monotone for* $C := \{y \in \mathbb{R}^\ell \mid Wy \geq 0\}$.

*Assume now additionally* $Wk \neq 0$.

(e) $\varphi_{A,k}$ *is proper.*
(f) $\varphi_{A,k}$ *is locally Lipschitzian on* int dom $\varphi_{A,k}$ *and hence continuous on* int dom $\varphi_{A,k}$.
(g) $\varphi_{A,k}$ *is strictly* (int $C$)*-monotone on* dom $\varphi_{A,k}$, *where* int $C = \{y \in \mathbb{R}^\ell \mid Wy > 0\}$.

***Proof*** Apply Propositions 4.8.1, 2.2.30(b) and 2.3.6. □

The formula for functionals $\varphi_{A,k}$ with $A$ being a polyhedral set is given in the following proposition.

**Proposition 4.8.5** *Assume* (H1$_A$). *Take any* $k \in \mathbb{R}^\ell \setminus \{0\}$ *with* $Wk \geq 0$.
*Then* (H1$_{A,k}$) *holds, and*

$$\operatorname{dom} \varphi_{A,k} = \{y \in \mathbb{R}^\ell \mid \forall i \in I^{\text{act}} \colon (w^i)^T y \leq b_i\},$$

$$\varphi_{A,k}(y) = \sup_{i \notin I^{\text{act}}} \frac{(w^i)^T y - b_i}{(w^i)^T k} \quad \text{for each } y \in \operatorname{dom} \varphi_{A,k},$$

*where* $I^{\text{act}} := \{i \in \{1, \ldots, r\} \mid (w^i)^T k = 0\}$.

(a) *If* $Wk = 0$, *then* dom $\varphi_{A,k} = A$ *and*

$$\varphi_{A,k}(y) = -\infty \quad \text{for each } y \in A.$$

(b) *In the case* $Wk > 0$, *the function* $\varphi_{A,k}$ *is finite-valued and*

$$\varphi_{A,k}(y) = \max_{i=1,\ldots,r} \frac{(w^i)^T y - b_i}{(w^i)^T k} \quad \text{for all } y \in \mathbb{R}^\ell.$$

(H2$_{A,k}$) *holds if and only if* $Wk > 0$.

***Proof*** int $A = \{y \in \mathbb{R}^\ell \mid Wy < b\}$ by Proposition 2.2.30. Hence, (H2$_{A,k}$) implies $Wk > 0$. The other statements result from Proposition 4.8.2. □

Günther and Popovici [204] used functions of the type described in Proposition 4.8.5(b) as sorting functions in Graef-Younes algorithms. Such algorithms eliminate dominated elements of finite sets in vector optimization (see [272, Sect. 12.4]).

Choose now the identity matrix as matrix $W$.

**Corollary 4.8.6** *Assume $A = b - \mathbb{R}^\ell_+$ with $b \in \mathbb{R}^\ell$ and $k \in \mathbb{R}^\ell_+ \setminus \{0\}$. Then*

$$\operatorname{dom} \varphi_{A,k} = \{y \in \mathbb{R}^\ell \mid \forall i \in I^{\text{act}} \colon y_i \le b_i\}, \text{ and}$$

$$\varphi_{A,k}(y) = \max_{i \notin I^{\text{act}}} \frac{y_i - b_i}{k_i} \quad \text{for each } y \in \operatorname{dom} \varphi_{A,k},$$

*where $I^{\text{act}} := \{i \in \{1, \dots, \ell\} \mid k_i = 0\}$.*

*In the case $k > 0$, the function $\varphi_{A,k}$ is finite-valued and*

$$\varphi_{A,k}(y) = \max_{i=1,\dots,\ell} \frac{y_i - b_i}{k_i} \quad \text{for all } y \in \mathbb{R}^\ell. \tag{4.8.1}$$

Functions of type (4.8.1) have been applied with $k = \mathbf{1}$ as scalarizing functionals in vector optimization [393] and with $b = 0$ as production functions in microeconomic theory [349].

Considering Proposition 4.8.5 with $r = 1$, we get the following statement for halfspaces $A$.

**Corollary 4.8.7** *Assume $A = \{y \in \mathbb{R}^\ell \mid w^T y \le b\}$ with $w \in \mathbb{R}^\ell \setminus \{0\}$, $b \in \mathbb{R}$.*

*For $k \in \mathbb{R}^\ell \setminus \{0\}$ with $w^T k = 0$, we have $\operatorname{dom} \varphi_{A,k} = A$ and*

$$\varphi_{A,k}(y) = -\infty \quad \text{for each } y \in A.$$

*For $k \in \mathbb{R}^\ell \setminus \{0\}$ with $w^T k > 0$, the function $\varphi_{A,k}$ is finite-valued and*

$$\varphi_{A,k}(y) = \frac{w^T y - b}{w^T k} \quad \text{for all } y \in \mathbb{R}^\ell. \tag{4.8.2}$$

Proposition 4.8.3 implies:

**Proposition 4.8.8** *Assume* (H1$_A$). *Take any $k \in \mathbb{R}^\ell \setminus \{0\}$ with $Wk \ge 0$.*

(a) *For $\tilde{A} := \{y \in \mathbb{R}^\ell \mid Wy \le b + \varepsilon Wk\}$ with $\varepsilon \in \mathbb{R}$, we get $\varphi_{\tilde{A},k}(y) = \varphi_{A,k}(y) - \varepsilon$ for all $y \in \mathbb{R}^\ell$.*
(b) *For $\tilde{A} := \{y \in \mathbb{R}^\ell \mid Wy \le b + Wy^0\}$ with $y^0 \in \mathbb{R}^\ell$, we have $\varphi_{\tilde{A},k}(y) = \varphi_{A,k}(y - y^0)$ for all $y \in \mathbb{R}^\ell$.*

**Corollary 4.8.9** *Assume that $A = \{y \in \mathbb{R}^\ell \mid Wy \le b\}$ holds for some $b \in \mathbb{R}^\ell$ and some regular matrix $W \in \mathbb{R}^{\ell,\ell}$. Take any $k \in \mathbb{R}^\ell \setminus \{0\}$ with $Wk \ge 0$. For*

$$\tilde{A} := \{y \in \mathbb{R}^\ell \mid Wy \le b + s\} \text{ with } s \in \mathbb{R}^\ell,$$

*we have* $\varphi_{\tilde{A},k}(y) = \varphi_{A,k}(y - W^{-1}s)$ *for all* $y \in \mathbb{R}^\ell$.

In combinatorial optimization, a polyhedral set $A \subseteq \mathbb{R}^\ell$ is usually given implicitly via $A = P + C$ with a polytope $P$ and a polyhedral cone $C$. It is well known that the sums of polytopes and polyhedral cones are the polyhedral sets.

**Proposition 4.8.10** *Assume*

$$A = P + C,$$

*where* $P \subseteq \mathbb{R}^\ell$ *is a nonempty polytope and* $C = \{y \in \mathbb{R}^\ell \mid Wy \leq 0\}$ *is a nontrivial cone with* $W \in \mathbb{R}^{r,\ell}$, *where* $W = (w^1 \cdots w^r)^{\mathrm{T}}$ *with* $w^i \in \mathbb{R}^\ell \setminus \{0\}$ *for all* $i \in \{1, \dots, r\}$.

*Take any* $k \in \mathbb{R}^\ell \setminus \{0\}$ *with* $Wk \geq 0$. *Then*

$$\operatorname{dom} \varphi_{A,k} = P + \{y \in \mathbb{R}^\ell \mid \forall i \in I^{\mathrm{act}}: (w^i)^T y \leq 0\}, \text{ and} \tag{4.8.3}$$

$$\varphi_{A,k}(y^0) = \inf_{p \in P \cap K(y^0)} \sup_{i \notin I^{\mathrm{act}}} \frac{(w^i)^{\mathrm{T}}(y^0 - p)}{(w^i)^T k} \quad \text{for all } y^0 \in \operatorname{dom} \varphi_{A,k}, \tag{4.8.4}$$

*where*

$$I^{\mathrm{act}} := \{i \in \{1, \dots, r\} \mid (w^i)^T k = 0\}, \quad \text{and}$$
$$K(y^0) := \{y \in \mathbb{R}^\ell \mid \forall i \in I^{\mathrm{act}}: (w^i)^{\mathrm{T}}(y^0) \leq (w^i)^T y\} \quad \text{for each } y^0 \in \mathbb{R}^\ell.$$

(a) *If* $Wk = 0$, *then* $\operatorname{dom} \varphi_{A,k} = A$ *and*

$$\varphi_{A,k}(y) = -\infty \quad \text{for each } y \in A.$$

(b) *In the case* $Wk > 0$, *the function* $\varphi_{A,k}$ *is finite-valued and*

$$\varphi_{A,k}(y) = \min_{p \in P} \max_{i=1,\dots,r} \frac{(w^i)^{\mathrm{T}}(y - p)}{(w^i)^T k} \quad \text{for all } y \in \mathbb{R}^\ell.$$

***Proof*** For each $y \in \mathbb{R}^\ell$, Proposition 4.3.7 implies

$$\varphi_{A,k}(y) = \varphi_{P+C,k}(y) = \inf_{p \in -P} \varphi_{C,k}(y + p) = \inf_{p \in P} \varphi_{C,k}(y - p). \tag{4.8.5}$$

For each $y \in \mathbb{R}^\ell$, $\varphi_{A,k}(y) = +\infty$ holds if and only if $\varphi_{C,k}(y - p) = +\infty$ for each $p \in P$, i.e., if and only if $y - p \notin \operatorname{dom} \varphi_{C,k}$ for each $p \in P$, which is equivalent to $y \notin P + \operatorname{dom} \varphi_{C,k}$. Hence, $\operatorname{dom} \varphi_{A,k} = P + \operatorname{dom} \varphi_{C,k}$.

Applying Proposition 4.8.5, we get $\operatorname{dom} \varphi_{C,k} = \{y \in \mathbb{R}^\ell \mid \forall i \in I^{\mathrm{act}}: (w^i)^T y \leq 0\}$, which yields (4.8.3). Furthermore, we obtain from Proposition 4.8.5:

$$\varphi_{C,k}(y - p) = \sup_{i \notin I^{\mathrm{act}}} \frac{(w^i)^{\mathrm{T}}(y - p)}{(w^i)^T k}$$

for each $y \in \mathbb{R}^\ell$ and $p \in P$ with $y - p \in \operatorname{dom} \varphi_{C,k}$.

For each $y^0 \in \operatorname{dom}\varphi_{A,k}$, the last statement implies together with (4.8.5):

$$\begin{aligned}\varphi_{A,k}(y^0) &= \inf_{p\in P}\varphi_{C,k}(y^0-p) = \inf\{\varphi_{C,k}(y^0-p) \mid p\in P,\ y^0-p\in\operatorname{dom}\varphi_{C,k}\}\\ &= \inf\{\sup_{i\notin I^{\text{act}}}\frac{(w^i)^{\mathrm T}(y^0-p)}{(w^i)^T k} \mid p\in P,\ y^0-p\in\operatorname{dom}\varphi_{C,k}\},\end{aligned}$$

where, for each $p \in P$,

$$\begin{aligned}y^0-p\in\operatorname{dom}\varphi_{C,k} &\iff p\in y^0-\operatorname{dom}\varphi_{C,k},\\ &\iff p\in\{y^0-y \mid y\in\mathbb{R}^\ell,\ (w^i)^T y\le 0 \text{ for all } i\in I^{\text{act}}\},\\ &\iff p\in\{y\in\mathbb{R}^\ell \mid \forall i\in I^{\text{act}}: (w^i)^{\mathrm T}(y^0-y)\le 0\},\\ &\iff p\in\{y\in\mathbb{R}^\ell \mid \forall i\in I^{\text{act}}: (w^i)^{\mathrm T}(y^0)\le(w^i)^T y\}=K(y^0).\end{aligned}$$

This results in (4.8.4).

(a) follows straightforward.
(b) $\varphi_{C,k}$ is finite-valued because of Proposition 4.8.5 and lower semicontinuous by Proposition 4.8.4. Since $P$ is nonempty and compact, Proposition 3.3.17 implies $\inf_{p\in P}\varphi_{C,k}(y-p) = \min_{p\in P}\varphi_{C,k}(y-p)$. (b) results from (4.8.5) and the application of Proposition 4.8.5(b) to $\varphi_{C,k}(y-p)$. □

Up to now, we have considered sets $A$ which are given by an inequality system. Now, we turn to sets of points in which some inequalities can be violated.

First, we will work with sets $A$ fulfilling the following assumption:

(HT2$_A$): $(Y, Y', \langle .,. \rangle)$ is a dual pair, where $Y$ is equipped with the weak topology $\sigma(Y, Y')$, $\Gamma \subset Y' \setminus \{0\}$,
$A = \{y \in Y \mid \exists g \in \Gamma: \langle y, g\rangle \le b_g\}$ with $b_g \in \mathbb{R}$ for each $g \in \Gamma$ such that $A_0 := \{y \in Y \mid \forall g \in \Gamma: \langle y, g\rangle \le -b_g\}$ is a proper subset of $Y$ with $0^+A_0 \neq \{0\}$ and $\operatorname{int} A_0 = \{y \in Y \mid \forall g \in \Gamma: \langle y, g\rangle < -b_g\}$.

In general, the set $A$ in (HT2$_A$) is nonconvex.

(HT2$_A$) implies $A = Y \setminus \operatorname{int}(-A_0)$ since

$$\begin{aligned}y\in\operatorname{int}(-A_0) &\iff -y\in\operatorname{int}A_0,\\ &\iff \forall g\in\Gamma:\ \langle -y,g\rangle < -b_g,\\ &\iff \forall g\in\Gamma:\ -\langle y,g\rangle < -b_g,\\ &\iff \forall g\in\Gamma:\ \langle y,g\rangle > b_g,\\ &\iff y\notin A.\end{aligned}$$

Assume (HT2$_A$), and take any $k \in Y \setminus \{0\}$ with $\langle k, g\rangle > 0$ for all $g \in \Gamma$. For each $y \in Y$,

$$\begin{aligned}\varphi_{A,k}(y) &= \varphi_{Y\setminus \operatorname{int}(-A_0),k}(y)\\ &= -\varphi_{-A_0,-k}(y) \quad \text{by Proposition 4.3.11, and}\\ \varphi_{A,k}(y) &= -\varphi_{A_0,k}(-y) \quad \text{by Proposition 4.3.9.}\end{aligned}$$

Hence, Proposition 4.8.1 results in the following statements.

**Proposition 4.8.11** *Assume* (HT2$_A$) *and* $k \in Y \setminus \{0\}$ *with* $\langle k, g\rangle > 0$ *for all* $g \in \Gamma$.

(a) $\varphi_{A,k}$ *is concave and upper semicontinuous.*
(b) $\varphi_{A,k}$ *is superadditive if and only if* $b_g \geq 0$ *for all* $g \in \Gamma$.
(c) $-\varphi_{A,k}$ *is sublinear if and only if* $b_g = 0$ *for all* $g \in \Gamma$.
(d) $\varphi_{A,k}$ *is* $C$*-monotone for* $C := \{y \in Y \mid \forall g \in \Gamma\colon \langle y, g\rangle \geq 0\}$.
(e) *If* $\operatorname{int} A_0 \neq \emptyset$, *then* $\varphi_{A,k}$ *is continuous on* $\operatorname{int} B$ *with* $B := \{y \in Y \mid \varphi_{A,k}(y) \neq -\infty\}$.
(f) *If* $Y$ *is a Banach space, then* $\varphi_{A,k}$ *is locally Lipschitzian on* $\operatorname{int} B$.
(g) $\varphi_{A,k}$ *is strictly* (core $C$)*-monotone on* $B$.

Furthermore, we obtain the following from Proposition 4.8.2.

**Proposition 4.8.12** *Assume* (HT2$_A$), *and take any* $k \in Y \setminus \{0\}$ *with* $\langle k, g\rangle > 0$ *for all* $g \in \Gamma$. *Then* $\operatorname{dom} \varphi_{A,k} = Y$, *and*

$$\varphi_{A,k}(y) = \inf_{g\in\Gamma} \frac{\langle y, g\rangle - b_g}{\langle k, g\rangle} \textit{ for all } y \in Y.$$

The next proposition can be proved analogously to Proposition 4.8.3.

**Proposition 4.8.13** *Assume* (HT2$_A$). *Take any* $k \in Y \setminus \{0\}$ *with* $\langle k, g\rangle > 0$ *for all* $g \in \Gamma$.

(a) *For* $\tilde{A} := \{y \in Y \mid \exists g \in \Gamma\colon \langle y, g\rangle \leq b_g + \varepsilon\langle k, g\rangle\}$ *with* $\varepsilon \in \mathbb{R}$, *we get*

$$\varphi_{\tilde{A},k}(y) = \varphi_{A,k}(y) - \varepsilon \textit{ for all } y \in Y.$$

(b) *For* $\tilde{A} := \{y \in Y \mid \exists g \in \Gamma\colon \langle y, g\rangle \leq b_g + \langle y^0, g\rangle\}$ *with* $y^0 \in Y$, *we have*

$$\varphi_{\tilde{A},k}(y) = \varphi_{A,k}(y - y^0) \textit{ for all } y \in Y.$$

From now on, we consider sets $A$ in a Euclidean space which are given by the assumption

(H2$_A$): $A = \{y \in \mathbb{R}^\ell \mid \exists i \in \{1, \dots, r\}\colon (w^i)^T y \leq b_i\}$ with $b \in \mathbb{R}^r$ and $W \in \mathbb{R}^{r,\ell}$, where $W = (w^1 \cdots w^r)^T$ with $w^i \in \mathbb{R}$ for all $i \in \{1, \dots, r\}$ and rank $W = r$ such that
$A_0 := \{y \in \mathbb{R}^\ell \mid Wy \leq -b\}$ is a proper unbounded subset of $Y = \mathbb{R}^\ell$.

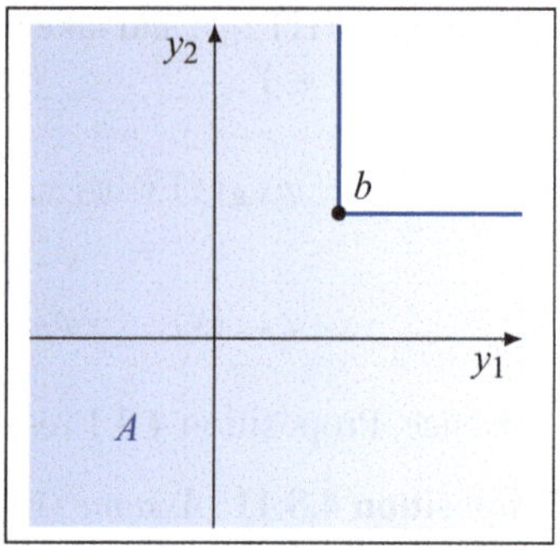


**Fig. 4.10** Set $A$ in Corollary 4.8.16 for $\ell = 2$

Again, $w^i$ stands for the i-th row vector of $W$.

**Proposition 4.8.14** *Assume* (H2$_A$) *and* $k \in \mathbb{R}^\ell \setminus \{0\}$ *with* $Wk > 0$.

(a) $\varphi_{A,k}$ *is concave and upper semicontinuous.*
(b) $\varphi_{A,k}$ *is superadditive if and only if* $b \geq 0$.
(c) $-\varphi_{A,k}$ *is sublinear if and only if* $b = 0$.
(d) $\varphi_{A,k}$ *is* $C$*-monotone for* $C := \{y \in Y \mid Wy \geq 0\}$.
(e) $\varphi_{A,k}$ *is locally Lipschitzian on* $\operatorname{int} B$ *and hence continuous on* $\operatorname{int} B$ *for* $B := \{y \in Y \mid \varphi_{A,k}(y) \neq -\infty\}$.
(f) $\varphi_{A,k}$ *is strictly* $(\operatorname{int} C)$*-monotone on* $B$.

***Proof*** Proposition 2.2.30 yields $\operatorname{int} A_0 \neq \emptyset$. The assertion follows from Proposition 4.8.11. □

Proposition 4.8.12 implies the next proposition.

**Proposition 4.8.15** *Assume* (H2$_A$), *and take any* $k \in \mathbb{R}^\ell \setminus \{0\}$ *with* $Wk > 0$. *Then* $\varphi_{A,k}$ *is finite-valued, and*

$$\varphi_{A,k}(y) = \min_{i=1,\ldots,r} \frac{(w^i)^T y - b_i}{(w^i)^T k} \quad \textit{for all } y \in \mathbb{R}^\ell.$$

Choosing $W$ as the identity matrix, we get the following from Proposition 4.8.15. Note that the assumption is illustrated by Fig. 4.10.

**Corollary 4.8.16** *Assume* $A = \{y \in \mathbb{R}^\ell \mid \exists i \in \{1, \ldots, \ell\} \colon y_i \leq b_i\}$ *with* $b \in \mathbb{R}^\ell$ *and* $k \in \operatorname{int} \mathbb{R}^\ell_+$. *Then* $\varphi_{A,k}$ *is finite-valued, and*

$$\varphi_{A,k}(y) = \min_{i=1,\ldots,\ell} \frac{y_i - b_i}{k_i} \quad \textit{for all } y \in \mathbb{R}^\ell. \tag{4.8.6}$$

The set $A$ in Corollary 4.8.16 can also be described by

$$A = \mathbb{R}^\ell \setminus (b + \operatorname{int} \mathbb{R}^\ell_+)$$

and by

$$A = \{y \in \mathbb{R}^\ell \mid \min_{i=1,\dots,\ell} (y_i - b_i) \leq 0\}.$$

Rubinov and Gasimov [394, Example 4.3] pointed out that

$$\varphi_{-A,k}(y) = \min_{i=1,\dots,\ell} \frac{y_i}{k_i} \quad \text{for all } y \in \mathbb{R}^\ell \tag{4.8.7}$$

if $A = \{y \in \mathbb{R}^\ell \mid \max_{i=1,\dots,\ell} y_i \geq 0\}$ and $k \in \operatorname{int} \mathbb{R}^\ell_+$. This statement results immediately from Corollary 4.8.16. Leontief [322] used utility functions of type (4.8.7) for consumer preferences on complementary goods.

Functions $\varphi\colon \mathbb{R}^\ell \times \mathbb{R}^\ell \to \mathbb{R}$ of the type

$$\varphi(y, w) = \min_{i=1,\dots,\ell} (y_i + w_i)$$

turn out to be relevant in many applications [201, 397]. Their separation properties were investigated in [347, Sect. 8] and [397, Prop. 2.1].

Functions $\varphi\colon \mathbb{R}^\ell \times (\mathbb{R}^\ell_+ \setminus \{0\}) \to \mathbb{R}$ of the type

$$\varphi(y, p) = \min_{i \notin I^{\text{act}}} p_i y_i,$$

where $I^{\text{act}} := \{i \in \{1, \dots, \ell\} \mid p_i = 0\}$, were used as separating functionals in [396].

Proposition 4.8.13 implies:

**Proposition 4.8.17** *Assume* (H2$_A$). *Take any* $k \in \mathbb{R}^\ell \setminus \{0\}$ *with* $Wk > 0$.

(a) *For* $\tilde{A} := \{y \in \mathbb{R}^\ell \mid \exists i \in \{1, \dots, r\}\colon (w^i)^T y \leq b_i + \varepsilon (w^i)^T k\}$ *with* $\varepsilon \in \mathbb{R}$, *we get*

$$\varphi_{\tilde{A},k}(y) = \varphi_{A,k}(y) - \varepsilon \quad \textit{for all } y \in \mathbb{R}^\ell.$$

(b) *For* $\tilde{A} := \{y \in \mathbb{R}^\ell \mid \exists i \in \{1, \dots, r\}\colon (w^i)^T y \leq b_i + (w^i)^T y^0\}$ *with* $y^0 \in \mathbb{R}^\ell$, *we have*

$$\varphi_{\tilde{A},k}(y) = \varphi_{A,k}(y - y^0) \quad \textit{for all } y \in \mathbb{R}^\ell.$$

**Remark 4.8.18** (*Bibliographical notes related to Sect.* 4.8)

Tammer and Winkler [438] studied $\varphi_{A,k}$ for polyhedral sets $A$ which are constructed via the unit ball of some block norm (see Sect. 9.1). These sets $A$ fulfill the property $\operatorname{bd} A - \mathbb{R}^\ell_+ \subseteq A$. For this special case, Tammer and Winkler proved the formula from Proposition 4.8.5(b) for $k \in \operatorname{int} \mathbb{R}^\ell_+$ and that $\varphi_{A,k}$ is finite-valued, continuous, convex, $\mathbb{R}^\ell_+$-monotone and, under an additional assumption, also $(\operatorname{int} \mathbb{R}^\ell_+)$-monotone. The formula from Proposition 4.8.5(b) was also given in [297, Cor. 3.14] for the special case where $A$ is a polyhedral cone with $W \geq 0$. Equation (4.8.1) was shown for $A = b - \mathbb{R}^\ell_+$ and $k > 0$ in [93, p. 14]. The special case of (4.8.2) under the additional assumptions $b = 0$ and $w > 0$ can be found in [301, Example 5]. Equation (4.8.1) with $b \in \mathbb{R}^\ell_+$ and (4.8.6) were given for $k = \mathbf{1}$ in [347, p. 118]. Proposition 4.8.10(b) is equivalent to a statement by Günther and Popovici [204].

## 4.9 Monotone and Translation Invariant Function Envelopes

In this section, monotone as well as translation invariant functions are constructed which majorize a given extended real-valued function or are majorized by it.

The greatest function of a given type which is majorized by a function $f$ is often called the **lower envelope** of $f$, whereas the smallest function of a given type which majorizes $f$ is called the **upper envelope** of $f$. In this section, results for lower envelopes are studied using the inf-convolution. The results for the upper envelopes can be formulated by means of the so-called sup-convolution, where the sup-addition has to be applied and where we would have to deal with two different operators for addition (see [362] for details of the sup-convolution). For this reason, we avoid the approach via the sup-convolution. Note that this problem would not arise if extended real-valued functions were handled in the way proposed in [486].

We will first construct monotone envelopes of extended real-valued functions. Lemma 3.7.3(d) implies:

**Lemma 4.9.1** *Let $Y$ be a linear space, $B \subseteq Y$ and $\varphi\colon Y \to \overline{\mathbb{R}}$ be $B$-monotone. Then* $\operatorname{conv} \varphi$ *is $B$-monotone. If $Y$ is a topological vector space, then* $\operatorname{cl} \varphi$ *is $B$-monotone.*

In the next lemma, recall that $\iota_A$ denotes the indicator function of the set $A$.

**Lemma 4.9.2** *Let $Y$ be a linear space and $B \subseteq Y$.*

*For each $B$-monotone function $\varphi\colon Y \to \overline{\mathbb{R}}$ with $\varphi(0) \le 0$, we have*

$$\varphi \le \iota_{-B}.$$

*Assume now $0 \in B$ and $B + B \subseteq B$.*

(a) *$\iota_{-B}$ is the greatest $B$-monotone function $g\colon Y \to \overline{\mathbb{R}}$ with $g(0) \le 0$.*
(b) *$\iota_{-\operatorname{conv} B}$ is the greatest convex $B$-monotone function $g\colon Y \to \overline{\mathbb{R}}$ with $g(0) \le 0$. Moreover, $\iota_{-\operatorname{conv} B}$ is sublinear.*
(c) *In a topological vector space $Y$, the function $\iota_{-\operatorname{cl} B}$ is the greatest lower semicontinuous $B$-monotone function $g\colon Y \to \overline{\mathbb{R}}$ with $g(0) \le 0$, and $\iota_{-\operatorname{cl}\operatorname{conv} B}$ is the greatest lower semicontinuous convex $B$-monotone function $g\colon Y \to \overline{\mathbb{R}}$ with $g(0) \le 0$. Moreover, $\iota_{-\operatorname{cl}\operatorname{conv} B}$ is sublinear.*

***Proof*** $\varphi(0) \le 0$ and Lemma 3.7.3(b) imply $\varphi(-b) \le 0$ for all $b \in B$. Hence, $\varphi \le \iota_{-B}$.

Assume now $0 \in B$ and $B + B \subseteq B$.

(a) Contrary to the claim, suppose that $\iota_{-B}$ is not $B$-monotone. Then there exist $y^1, y^2 \in Y$ with $y^2 - y^1 \in B$ and $\iota_{-B}(y^1) > \iota_{-B}(y^2)$. This implies $y^2 \in -B$ and $y^1 \notin -B$. We get $y^1 \in y^2 - B \subseteq -B - B \subseteq -B$, which contradicts $y^1 \notin -B$. Hence, $\iota_{-B}$ is $B$-monotone.

(b) $\operatorname{conv} \iota_{-B}$ is $B$-monotone because of Lemma 4.9.1. The statement for convex $B$-monotone functions follows from $\operatorname{conv} \iota_{-B} = \iota_{-\operatorname{conv} B}$. By Lemma 2.2.6, $\operatorname{conv} B$ is a convex cone. Hence, $\iota_{-\operatorname{conv} B}$ is sublinear.

(c) In a topological vector space $Y$, we have $\operatorname{cl} \iota_{-B} = \iota_{-\operatorname{cl} B}$ since $\operatorname{cl} \operatorname{epi} \iota_{-B} = \operatorname{cl}((-B) \times \mathbb{R}_+) = \operatorname{cl}(-B) \times \operatorname{cl}(\mathbb{R}_+) = \operatorname{cl}(-B) \times \mathbb{R}_+ = \operatorname{epi} \iota_{-\operatorname{cl} B}$. Because of Lemma 4.9.1, $\operatorname{cl} \iota_{-B}$ is $B$-monotone. This implies the statements for $\operatorname{cl} \iota_{-B}$ and for $\operatorname{cl} \operatorname{conv} \iota_{-B}$. □

**Theorem 4.9.3** *Let $Y$ be a linear space, $B$ a proper subset of $Y$ and $f\colon Y \to \overline{\mathbb{R}}$. Define*

$$f_B := f \,\#\, \iota_{-B}.$$

*Then $f_B(y) = \inf\{f(y + b) \mid b \in B\}$ for each $y \in Y$.*

(a) *The function $f$ is $B$-monotone if and only if $f \leq f_B$.*
*If $0 \in B$, then $f \geq f_B$.*

(b) *If $B + B \subseteq B$, then $f_B$ is a $B$-monotone function.*

(c) *If $0 \in B$ and $B + B \subseteq B$, then $f_B$ is the greatest $B$-monotone function majorized by $f$.*

(d) *If $0 \in B$ and $B + B \subseteq B$, then $\operatorname{cl} \operatorname{conv} f_B$ is the greatest lower semicontinuous convex $B$-monotone function majorized by $f$.*

***Proof*** The first statement is due to Lemma 3.2.15.

(a) $f$ is $B$-monotone if and only if $f(y) \leq f(y + b)$ for all $y \in Y$ and $b \in B$, which is equivalent to $f \leq f_B$. The second statement in (a) is obvious.

(b) Take any $y^1, y^2 \in Y$ with $y^2 - y^1 \in B$. Then $y^2 \in y^1 + B$ and thus $y^2 + B \subseteq y^1 + B + B \subseteq y^1 + B$. Consequently, $f_B(y^1) \leq f_B(y^2)$. Hence, $f_B$ is $B$-monotone.

(c) The assumptions imply that $f_B$ is $B$-monotone and majorized by $f$. Take any $B$-monotone function $g\colon Y \to \overline{\mathbb{R}}$ which is majorized by $f$. Then $g = g_B = g \,\#\, \iota_{-B} \leq f \,\#\, \iota_{-B} = f_B$ because of (a) and (3.2.7).

(d) $\operatorname{cl} \operatorname{conv} f_B$ is $B$-monotone because of Lemma 4.9.1. This and (c) result in the assertion. □

In Theorem 4.9.3(d), $\operatorname{cl} \operatorname{conv} f_B$ cannot be replaced by $f_B^{**}$ if $\operatorname{cl} \operatorname{conv} f_B$ is not proper.

**Example 4.9.4** In $Y = \mathbb{R}^2$, consider $B := \{(y_1, 0)^T \in Y \mid y_1 \geq 0\}$ and $f\colon Y \to \overline{\mathbb{R}}$ given by

$$f((y_1, y_2)^T) := \begin{cases} -y_1 & \text{if } y_1 \geq 0 \text{ and } y_2 = 0, \\ +\infty & \text{otherwise.} \end{cases}$$

Then

$$f_B((y_1, y_2)^T) = \begin{cases} -\infty & \text{if } y_2 = 0, \\ +\infty & \text{otherwise.} \end{cases}$$

Consider the dual pair $(Y, Y^*, \langle .,. \rangle)$ or any other dual pair $(Y, Y', \langle .,. \rangle)$ which satisfies the assumptions of Lemma 3.6.8. Then $f_B^{**} = -\infty$ by Lemma 3.6.8. Because of Theorem 4.9.3(b), $f_B$ is $B$-monotone. Furthermore, it is lower semicontinuous, convex, and majorized by $f$, but $f_B \leq f_B^{**}$ is not satisfied.

**Lemma 4.9.5** *Let $Y$ be a linear space and $B \subseteq Y$.*
*For each $B$-monotone function $\varphi \colon Y \to \overline{\mathbb{R}}$ with $\varphi(0) \geq 0$, we have*

$$\varphi \geq -\iota_B.$$

*If $0 \in B$ and $B + B \subseteq B$, then $-\iota_B$ is the smallest $B$-monotone function $g \colon Y \to \overline{\mathbb{R}}$ with $g(0) \geq 0$.*

***Proof*** $\varphi(0) \geq 0$ and Lemma 3.7.3(c) yield $\varphi(b) \geq 0$ for all $b \in B$. Hence, $\varphi \geq -\iota_B$.
Assume now $0 \in B$ and $B + B \subseteq B$. By Lemma 4.9.2, $\iota_{-B}$ is $B$-monotone. Lemma 3.7.8(e) implies the $B$-monotonicity of $-\iota_B$ since $-\iota_B(y) = -\iota_{-B}(-y)$ for each $y \in Y$. □

**Theorem 4.9.6** *Let $Y$ be a linear space, $B$ a proper subset of $Y$ and $f \colon Y \to \overline{\mathbb{R}}$. Define $\tilde{f}_B \colon Y \to \overline{\mathbb{R}}$ by*

$$\tilde{f}_B(y) := \sup\{f(y - b) \mid b \in B\} \quad \textit{for each } y \in Y.$$

(a) *The function $f$ is $B$-monotone if and only if $f \geq \tilde{f}_B$.*
*If $0 \in B$, then $f \leq \tilde{f}_B$.*
(b) *If $B + B \subseteq B$, then $\tilde{f}_B$ is a $B$-monotone function.*
(c) *If $0 \in B$ and $B + B \subseteq B$, then $\tilde{f}_B$ is the smallest $B$-monotone function which majorizes $f$.*

***Proof***

(a) $f$ is $B$-monotone if and only if $f(y) \geq f(y - b)$ for all $y \in Y$ and $b \in B$, which is equivalent to $f \geq \tilde{f}_B$. The second statement in (a) is obvious.
(b) Take any $y^1, y^2 \in Y$ with $y^2 - y^1 \in B$. Then $y^1 \in y^2 - B$ and thus $y^1 - B \subseteq y^2 - B - B \subseteq y^2 - B$. Consequently, $\tilde{f}_B(y^1) \leq \tilde{f}_B(y^2)$. Hence, $\tilde{f}_B$ is $B$-monotone.
(c) $\tilde{f}_B$ is $B$-monotone by (b). Take any $B$-monotone function $g \colon Y \to \overline{\mathbb{R}}$ which majorizes $f$. Because of (a), $g = \tilde{g}_B \geq \tilde{f}_B \geq f$. □

**Corollary 4.9.7** *Let $Y$ be a linear space, $B$ a proper subset of $Y$ with $0 \in B$ and $B + B \subseteq B$, $S$ a nonempty subset of $Y$, and $f \colon Y \to \overline{\mathbb{R}}$.*

(1) *$f_B^{S,1} \colon Y \to \overline{\mathbb{R}}$ defined by*

$$f_B^{S,1}(y) := \inf\{f(s) \mid s \in S \cap (y + B)\} \quad \textit{for each } y \in Y$$

*is the greatest $B$-monotone function majorized by $f$ on $S$.*

(2) $f_B^{S,2}\colon Y \to \overline{\mathbb{R}}$ *defined by*

$$f_B^{S,2}(y) := \sup\{f(s) \mid s \in S \cap (y - B)\} \quad \textit{for each } y \in Y$$

*is the smallest $B$-monotone function which majorizes $f$ on $S$.*

(3) *The following conditions are equivalent to each other:*

(a) *$f$ is $B$-monotone on $S$.*
(b) *$f(y) = f_B^{S,1}(y)$ for each $y \in S$.*
(c) *$f(y) = f_B^{S,2}(y)$ for each $y \in S$.*
(d) *$f_B^{S,1}(y) = f_B^{S,2}(y)$ for each $y \in S$.*

***Proof***

(1) The greatest $B$-monotone function majorized by $f$ on $S$ is the greatest $B$-monotone function majorized by $f^0\colon Y \to \overline{\mathbb{R}}$ defined by

$$f^0(y) := \begin{cases} f(y) & \text{if } y \in S, \\ +\infty & \text{otherwise.} \end{cases}$$

By Theorem 4.9.3, $f_B^0\colon Y \to \overline{\mathbb{R}}$ given by

$$f_B^0(y) := \inf\{f^0(y + b) \mid b \in B\} \quad \text{for each } y \in Y$$

is the greatest $B$-monotone function majorized by $f^0$. For each $y \in Y$,

$$f_B^0(y) = \inf\{f(y + b) \mid b \in B,\, y + b \in S\} = \inf\{f(s) \mid s \in S \cap (y + B)\}.$$

(2) The smallest $B$-monotone function which majorizes $f$ on $S$ is the smallest $B$-monotone function which majorizes $f^1\colon Y \to \overline{\mathbb{R}}$ defined by

$$f^1(y) := \begin{cases} f(y) & \text{if } y \in S, \\ -\infty & \text{otherwise.} \end{cases}$$

By Theorem 4.9.6, $f_B^1\colon Y \to \overline{\mathbb{R}}$ given by

$$f_B^1(y) := \sup\{f^1(y - b) \mid b \in B\} \quad \text{for each } y \in Y$$

is the smallest $B$-monotone function which majorizes $f^1$. For each $y \in Y$,

$$f_B^1(y) = \sup\{f(y - b) \mid b \in B,\, y - b \in S\} = \inf\{f(s) \mid s \in S \cap (y - B)\}.$$

(3) Assume first that (a) holds. Take any $y \in S$. Then $f(y + b) \geq f(y)$ for each $b \in B$ with $y + b \in S$, and $f(y - b) \leq f(y)$ for each $b \in B$ with $y - b \in S$. We obtain (b) and (c) and thus (d).

Conversely, (a) follows from each of the conditions (b) and (c) since $B$-monotonicity on $Y$ contains $B$-monotonicity on $S$.
If (d) holds, then (1) and (2) imply $f_B^{S,1}(y) = f_B^{S,2}(y) = f(y)$ for each $y \in S$, which yields (a). □

We will now investigate translation invariant function envelopes.

**Theorem 4.9.8** *Assume $Y$ is a linear space, $k \in Y \setminus \{0\}$, and $f : Y \to \overline{\mathbb{R}}$.*

(a) *For all $y \in Y$,*

$$(f \# \varphi_{\{0_Y\},k})(y) = \inf\{f(y - tk) + t \mid t \in \mathbb{R}\}. \tag{4.9.1}$$

(b) *If $f$ is translation invariant along $k$, then $f = f \# \varphi_{\{0_Y\},k}$.*
(c) *The inf-convolution $f \# \varphi_{\{0_Y\},k}$ is the greatest function which is majorized by $f$ and translation invariant along $k$.*

***Proof*** Define $\varphi_k := \varphi_{\{0_Y\},k}$.

(a) (4.7.5) implies, for each $y \in Y$,

$$\begin{aligned}(f \# \varphi_k)(y) &= \inf\{f(y^1) + \varphi_k(y^2) \mid y = y^1 + y^2, y^1 \in Y, y^2 \in Y\}\\ &= \inf\{f(y^1) + t \mid y = y^1 + tk, y^1 \in Y, t \in \mathbb{R}\}\\ &= \inf\{f(y - tk) + t \mid t \in \mathbb{R}\}.\end{aligned}$$

(b) results from (a).
(c) By (3.2.5), $\varphi_k(0) = 0$ implies $f \# \varphi_k \leq f$. The function $\varphi_k$ is translation invariant along $k$ because of Lemma 4.7.6(b). Thus, $f \# \varphi_k$ is translation invariant along $k$ by Proposition 4.3.1(g).
Take any function $g : Y \to \overline{\mathbb{R}}$ which is majorized by $f$ and translation invariant along $k$. Then (b) and (3.2.7) imply $g = g \# \varphi_k \leq f \# \varphi_k \leq f$ since $g \leq f$. □

**Theorem 4.9.9** *Assume* $(\mathrm{Sp0}_{A,k})$.

(a) *If $A$ is a proper subset of $Y$ with $0 \in A$ and $A + A \subseteq A$, then $\varphi_{A,k}$ is the greatest $(-A)$-monotone function $g$ with $g(0) \leq 0$ which is translation invariant along $k$.*
(b) *$\varphi_{\operatorname{conv} A,k}$ is the greatest convex function which is majorized by $\varphi_{A,k}$ and translation invariant along $k$.*
(c) *If $Y$ is a topological vector space, then $\varphi_{\operatorname{cl}(A-\mathbb{R}_+k),k}$ is the greatest lower semicontinuous function which is majorized by $\varphi_{A,k}$, i.e.,*

$$\operatorname{cl} \varphi_{A,k} = \varphi_{\operatorname{cl}(A-\mathbb{R}_+k),k}.$$

(d) *If $Y$ is a topological vector space, then $\varphi_{\operatorname{cl}(\operatorname{conv} A-\mathbb{R}_+k),k}$ is the greatest lower semicontinuous convex function which is majorized by $\varphi_{A,k}$ and translation invariant along $k$.*

***Proof*** Define $\varphi_k := \varphi_{\{0_Y\},k}$.

(a) $\varphi_{A,k}(0) \leq 0$ because of $0 \in A$. By Lemma 4.7.6(b), $\varphi_k$ is the greatest function $\tilde{g}$ with $\tilde{g}(0) \leq 0$ which is translation invariant along $k$. Because of Corollary 4.3.8 and Theorem 4.9.3(c), $\varphi_{A,k} = \varphi_k \,\#\, \iota_A$ is the greatest $(-A)$-monotone function majorized by $\varphi_k$.

(b) Take any convex function $g$ which is majorized by $\varphi_{A,k}$ and translation invariant along $k$. Then $g = \varphi_{B,k}$ for some $B \subseteq Y$. Because of Proposition 4.3.13(a), $g \leq \varphi_{A,k}$ implies $A \subseteq \mathrm{cl}_k(B - \mathbb{R}_+k)$. By Theorem 4.2.10(a), $\mathrm{cl}_k(B - \mathbb{R}_+k)$ is convex. Hence, $\mathrm{conv}\, A \subseteq \mathrm{cl}_k(B - \mathbb{R}_+k)$. Proposition 4.3.13 yields $g \leq \varphi_{\mathrm{conv}\, A,k} \leq \varphi_{A,k}$. Because of Theorem 4.2.10(a), $\varphi_{\mathrm{conv}\, A,k}$ is convex.

(c) By Proposition 2.3.54(a), $\mathrm{cl}_k(\mathrm{cl}(A - \mathbb{R}_+k)) = \mathrm{cl}(A - \mathbb{R}_+k)$. Furthermore, $k \in -0^+(A - \mathbb{R}_+k)$. Thus, $f := \varphi_{\mathrm{cl}(A-\mathbb{R}_+k),k}$ is lower semicontinuous because of Theorem 4.4.2. By Proposition 4.3.13(a), $A \subseteq \mathrm{cl}(A - \mathbb{R}_+k)$ yields $f \leq \varphi_{A,k}$. Take any lower semicontinuous function $g$ which is majorized by $\varphi_{A,k}$ and any $t \in \mathbb{R}$. Then $g \leq \varphi_{A,k}$ implies $\mathrm{lev}_{\varphi_{A,k},\leq}(t) \subseteq \mathrm{lev}_{g,\leq}(t)$. By (4.2.4), $\mathrm{cl}_k(A - \mathbb{R}_+k) + tk \subseteq \mathrm{lev}_{g,\leq}(t)$. Since $g$ is lower semicontinuous, $\mathrm{lev}_{g,\leq}(t)$ is closed. Consequently, $\mathrm{cl}(\mathrm{cl}_k(A - \mathbb{R}_+k) + tk) \subseteq \mathrm{lev}_{g,\leq}(t)$.

$$\begin{aligned}\mathrm{lev}_{f,\leq}(t) &= \mathrm{cl}_k(\mathrm{cl}(A - \mathbb{R}_+k) - \mathbb{R}_+k) + tk\\ &\subseteq \mathrm{cl}_k(\mathrm{cl}(A - \mathbb{R}_+k - \mathbb{R}_+k)) + tk \quad \text{because of Lemma 2.3.26(b).}\end{aligned}$$

By parts (a) and (c) of Proposition 2.3.54,

$$\begin{aligned}\mathrm{cl}_k(\mathrm{cl}(A - \mathbb{R}_+k - \mathbb{R}_+k)) + tk = \mathrm{cl}(A - \mathbb{R}_+k) + tk &= \mathrm{cl}(\mathrm{cl}_k(A - \mathbb{R}_+k)) + tk\\ &\subseteq \mathrm{cl}(\mathrm{cl}_k(A - \mathbb{R}_+k) + tk)\\ &\subseteq \mathrm{lev}_{g,\leq}(t).\end{aligned}$$

Hence, $g \leq f$.

(d) The set $\mathrm{cl}(\mathrm{conv}\, A - \mathbb{R}_+k)$ is convex. Thus, $\varphi_{\mathrm{cl}(\mathrm{conv}\, A-\mathbb{R}_+k),k}$ is convex. (d) results from (b) and (c), where (c) has to be applied to $\mathrm{conv}\, A$ instead of $A$. □

In Theorem 4.10.7,

$$\mathrm{cl}\,\mathrm{conv}\, \varphi_{A,k} = \varphi_{\mathrm{cl}(\mathrm{conv}\, A-\mathbb{R}_+k),k}$$

will be proved for the case where $Y$ is equipped with the weak topology.

**Theorem 4.9.10** *Assume* $(\mathrm{Sp0}_{A,k})$, *that $A$ is a proper subset of $Y$ and $f: Y \to \overline{\mathbb{R}}$.*

(a) *For each $y \in Y$,*

$$(f \,\#\, \varphi_{A,k})(y) = \inf\{f(y - tk - a) + t \mid t \in \mathbb{R},\ a \in A\}.$$

(b) *Let $f$ be $(-A)$-monotone and translation invariant along $k$. Then $f \leq f \,\#\, \varphi_{A,k}$. Moreover, $f = f \,\#\, \varphi_{A,k}$ if $0 \in A$.*

(c) *If* $0 \in A$ *and* $A + A \subseteq A$*, then* $f \mathbin{\#} \varphi_{A,k}$ *is the greatest* $(-A)$*-monotone function which is majorized by* $f$ *and translation invariant along* $k$.

***Proof*** Define $\varphi_k := \varphi_{\{0_Y\},k}$.

(a) For each $y \in Y$,

$$\begin{aligned}(f \mathbin{\#} \varphi_{A,k})(y) &= ((f \mathbin{\#} \varphi_k) \mathbin{\#} \iota_A)(y) \quad \text{by Corollary 4.3.8} \\ &= \inf_{a\in A} (f \mathbin{\#} \varphi_k)(y-a) \quad \text{by Lemma 3.2.15} \\ &= \inf_{a\in A} \inf\{f(y-a-tk)+t \mid t \in \mathbb{R}\} \quad \text{by Theorem 4.9.8.}\end{aligned}$$

(b) Since $f$ is $(-A)$-monotone, we obtain $f \le f \mathbin{\#} \iota_A$ from Theorem 4.9.3(a). Since $f$ is translation invariant along $k$, Theorem 4.9.8(b) implies $f = f \mathbin{\#} \varphi_k$. We get $f = f \mathbin{\#} \varphi_k \le (f \mathbin{\#} \iota_A) \mathbin{\#} \varphi_k$ from (3.2.7). Hence, $f \le f \mathbin{\#} \varphi_{A,k}$ because of Corollary 4.3.8. If $0 \in A$, we can replace "$\le$" by "$=$" in the previous lines of the proof.

(c) By Theorem 4.9.8, $f \mathbin{\#} \varphi_k$ is the greatest function which is majorized by $f$ and translation invariant along $k$. By Theorem 4.9.3(c), $(f \mathbin{\#} \varphi_k) \mathbin{\#} \iota_A$ is the greatest $(-A)$-monotone function which is majorized by $f \mathbin{\#} \varphi_k$. Because of Corollary 4.3.8, $(f \mathbin{\#} \varphi_k) \mathbin{\#} \iota_A = f \mathbin{\#} \varphi_{A,k}$. Proposition 4.3.1(g) implies that $f \mathbin{\#} \varphi_{A,k}$ is translation invariant along $k$. □

Formulas for upper translation invariant function envelopes will be obtained by means of the following lemma.

**Lemma 4.9.11** *Assume* $Y$ *is a linear space and* $k \in Y \setminus \{0\}$*. Define* $\varphi_k := \varphi_{Y\setminus\mathbb{R}_+k,k}$.

(a) *Then*

$$\varphi_k(y) = \begin{cases} t & \text{if } y = tk \text{ with } t \in \mathbb{R}, \\ -\infty & \text{otherwise.} \end{cases}$$

*Moreover,* $\varphi_k(y) = -\varphi_{\{0_Y\},k}(-y)$ *for each* $y \in Y$.

(b) $\varphi_k$ *is the smallest function* $g\colon Y \to \overline{\mathbb{R}}$ *with* $g(0) \ge 0$ *which is translation invariant along* $k$.

***Proof***

(a) For each $y \in Y$,

$$\begin{aligned}\varphi_{\{0\},k}(y) = \varphi_{-\mathbb{R}_+k,k}(y) &= -\varphi_{Y\setminus(-\mathbb{R}_+k),-k}(y) \quad \text{by (4.2.8) and (4.3.2)} \\ &= -\varphi_{Y\setminus\mathbb{R}_+k,k}(-y) \quad \text{by Proposition 4.3.9.}\end{aligned}$$

This implies the second statement of (a) and the first one because of (4.7.5).

(b) Take any $g\colon Y \to \overline{\mathbb{R}}$ with $g(0) \ge 0$ which is translation invariant along $k$. Then $g(tk) = g(0+tk) = g(0) + t \ge t$ for each $t \in \mathbb{R}$. Thus, $g \ge \varphi_k$. □

**Theorem 4.9.12** *Assume $Y$ is a linear space, $k \in Y \setminus \{0\}$ and $f: Y \to \overline{\mathbb{R}}$. Define $f_k: Y \to \overline{\mathbb{R}}$ by*

$$f_k(y) := \sup\{f(y - tk) + t \mid t \in \mathbb{R}\} \quad \textit{for all } y \in Y.$$

(a) *If $f$ is translation invariant along $k$, then $f = f_k$.*
(b) *$f_k$ is the smallest function which majorizes $f$ and is translation invariant along $k$.*

***Proof***

(a) is obvious.
(b) For each $y \in Y$ and $\lambda \in \mathbb{R}$,

$$\begin{aligned} f_k(y + \lambda k) &= \sup\{f(y + (\lambda - t)k) + t \mid t \in \mathbb{R}\} \\ &= \sup\{f(y - t_0 k) + \lambda + t_0 \mid t_0 \in \mathbb{R}\} \\ &= f_k(y) + \lambda. \end{aligned}$$

Thus, $f_k$ is translation invariant along $k$. We get $f \le f_k$ since $f(y - 0 \cdot k) + 0 = f(y)$ for each $y \in Y$.
Take any function $g: Y \to \overline{\mathbb{R}}$ which majorizes $f$ and is translation invariant along $k$. (a) implies $g = g_k \ge f_k$ since $g \ge f$. □

**Proposition 4.9.13** *Assume* $(\mathrm{Sp0}_{A,k})$ *and that $A$ is a proper subset of $Y$ with $0 \in A$ and $A + A \subseteq A$. $\varphi: Y \to \overline{\mathbb{R}}$ defined by*

$$\varphi(y) := \sup\{t \in \mathbb{R} \mid y \in -A + tk\} \quad \textit{for each } y \in Y$$

*is the smallest $(-A)$-monotone function $g$ with $g(0) \ge 0$ which is translation invariant along $k$.*

*For each $y \in Y$,*

$$\varphi(y) = -\varphi_{A,k}(-y).$$

*Furthermore,*

$$\varphi = \varphi_{C,k} \ \textit{for} \ C := Y \setminus (-(A - \mathbb{R}_+ k)).$$

***Proof*** By Lemma 4.9.11, $\varphi_k := \varphi_{Y \setminus \mathbb{R}_+ k, k}$ is the smallest function $\tilde{g}: Y \to \overline{\mathbb{R}}$ with $\tilde{g}(0) \ge 0$ which is translation invariant along $k$. By Theorem 4.9.6(c), $f: Y \to \overline{\mathbb{R}}$ defined by

$$f(y) := \sup\{\varphi_k(y + a) \mid a \in A\} \quad \text{for each } y \in Y$$

is the smallest $(-A)$-monotone function which majorizes $\varphi_k$. Because of Lemma 4.9.11(a), $f(y) = \sup\{t \in \mathbb{R} \mid y + a = tk, a \in A\} = \varphi(y)$ for each $y \in Y$, i.e., $f = \varphi$. Thus, $\varphi(0) \ge 0$. Take any $y \in Y$.

$$
\begin{aligned}
\varphi(y) &= \sup\{-(-t) \mid -y \in A + (-t)k,\ t \in \mathbb{R}\} \\
&= \sup\{-t \mid -y \in A + tk,\ t \in \mathbb{R}\} \\
&= -\inf\{t \in \mathbb{R} \mid -y \in A + tk\} \\
&= -\varphi_{A,k}(-y). \\
\varphi_{A,k}(y) &= \varphi_{A-\mathbb{R}_+k,k}(y) = -\varphi_{Y\setminus(A-\mathbb{R}_+k),-k}(y) \quad \text{by (4.2.8) and (4.3.2)} \\
&= -\varphi_{Y\setminus(-(A-\mathbb{R}_+k)),k}(-y) \quad \text{by Proposition 4.3.9.}
\end{aligned}
$$

Thus, $\varphi_{C,k}(y) = -\varphi_{A,k}(-y) = \varphi(y)$. Hence, $\varphi = \varphi_{C,k}$ is translation invariant along $k$. □

**Theorem 4.9.14** *Assume* (Sp0$_{A,k}$)*, that $A$ is a proper subset of $Y$ and $f\colon Y \to \overline{\mathbb{R}}$. Define $f^{A,k}\colon Y \to \overline{\mathbb{R}}$ by*

$$
f^{A,k}(y) := \sup\{f(y - tk + a) + t \mid t \in \mathbb{R}, a \in A\}
$$

*for each $y \in Y$.*

(a) *Let $f$ be $(-A)$-monotone and translation invariant along $k$. Then $f \geq f^{A,k}$. Moreover, $f = f^{A,k}$ if $0 \in A$.*
(b) *If $0 \in A$ and $A + A \subseteq A$, then $f^{A,k}$ is the smallest $(-A)$-monotone function which majorizes $f$ and is translation invariant along $k$.*

***Proof*** Define $f_k$ as in Theorem 4.9.12.

(a) Since $f$ is $(-A)$-monotone, we obtain $f(y) \geq \sup\{f(y + a) \mid a \in A\}$ for each $y \in Y$ by Theorem 4.9.6(a). Since $f$ is translation invariant along $k$, Theorem 4.9.12 implies, for each $y \in Y$,

$$
\begin{aligned}
f(y) &= f_k(y) = \sup\{f(y - tk) + t \mid t \in \mathbb{R}\} \\
&\geq \sup\{\sup\{f(y - tk + a) \mid a \in A\} + t \mid t \in \mathbb{R}\} = f^{A,k}(y).
\end{aligned}
$$

If $0 \in A$, we can replace "$\geq$" by "$=$" in the previous lines of the proof.
(b) By Theorem 4.9.12, $f_k$ is the smallest function which majorizes $f$ and is translation invariant along $k$. By Theorem 4.9.6, the smallest $(-A)$-monotone function which majorizes $f_k$ is $\varphi\colon Y \to \overline{\mathbb{R}}$ defined by

$$
\varphi(y) := \sup\{f_k(y + a) \mid a \in A\} \quad \text{for each } y \in Y.
$$

$\varphi$ coincides with $f^{A,k}$. For each $y \in Y$ and $\lambda \in \mathbb{R}$,

$$
\begin{aligned}
f^{A,k}(y + \lambda k) &= \sup\{f(y + (\lambda - t)k + a) + t \mid t \in \mathbb{R}, a \in A\} \\
&= \sup\{f(y - t_0 k + a) + t_0 + \lambda \mid t_0 \in \mathbb{R}, a \in A\} \\
&= f^{A,k}(y) + \lambda.
\end{aligned}
$$

Thus, $f^{A,k}$ is translation invariant along $k$. □

Under the assumptions that $Y$ is the space of extended real-valued functions on some nonempty set $X$ and $B$ is the standard ordering cone in this space, function envelopes which are $B$-monotone, translation invariant along $k = \mathbf{1}$ or have both properties were intensively studied by Dolecki and Greco [117]. They made use of the following idea for constructing lower function envelopes by means of upper function envelopes: Define the function $f^C\colon Y \to \overline{\mathbb{R}}$ for a given function $f\colon Y \to \overline{\mathbb{R}}$ by $f^C(y) := -f(-y)$. If $g \geq f^C$ for some function $g\colon Y \to \overline{\mathbb{R}}$, then $g^C \leq f$. Dolecki and Greco pointed out that the translation invariant functions which they investigated are applied for the regularization of functions, in the calculus of variations and in the theory of general convergence.

**Remark 4.9.15** (*Bibliographical notes related to Sect.* 4.9)
Filipović and Kupper [162] investigated function envelopes under the assumptions that $A$ and $B$ are weakly closed convex cones in a separated locally convex space $Y$. The following references to their paper refer to these restrictive assumptions.

They proved that $\iota_{-B}$ is the greatest lower semicontinuous convex $B$-monotone function which is zero at $y = 0$, Theorem 4.9.3(a)–(b) and Theorem 4.9.3(d) with $f_B^{**}$ in the dual system $(Y, Y^*, \langle .,. \rangle)$ instead of cl conv $f_B$. Note that $f_B^{**} = \operatorname{cl}\operatorname{conv} f_B$ is not necessarily satisfied if cl conv $f_B$ is not proper (see Example 4.9.4).

Filipović and Kupper stated that the function $g\colon Y \to \overline{\mathbb{R}}$ given by $g(y) := \sup\{y^*(y) \mid y^* \in -A^+, y^*(k) = 1\}$ is the greatest lower semicontinuous convex $(-A)$-monotone function which is translation invariant along $k$ and zero at $y = 0$. By Theorem 4.10.8, $g = \varphi_{A,k}^{**}$ if $k \in -A \setminus A$. Under the mentioned assumptions, $\varphi_{A,k}^{**}$ in the dual system $(Y, Y^*, \langle .,. \rangle)$ coincides with $\varphi_{A,k}$ if $\varphi_{A,k}$ is proper (see Lemma 3.6.7).

Furthermore, Filipović and Kupper proved Theorem 4.9.10(a) and (b) with $f \# \iota_D^* \# \iota_A$ instead of $f \# \varphi_{A,k}$ and that $(f \# \iota_D^* \# \iota_A)^{**}$ is the greatest lower semicontinuous convex $(-A)$-monotone function which is majorized by $f$ and translation invariant along $k$, where $D$ was given by (4.7.7) in the dual system $(Y, Y^*, \langle .,. \rangle)$. Under their assumptions, $\iota_D^*$ coincides with $\varphi_{\{0_Y\},k}$ because of (4.7.7).

Theorem 4.9.8(a) and the translation invariance of $f \# \varphi_k$ were mentioned under special assumptions by Hamel [224, p. 7], who introduced $\varphi_k$ by the right-hand side of (4.7.5). Theorem 4.9.8(c) was proved for the function given by the right-hand side of (4.9.1) in [222]. Theorem 4.9.12(b) was stated in [222] without proof.

## 4.10 Variational Analysis

In this section, we use the notation introduced in Sect. 3.6.

Translation invariance is related to the conjugate and to the subdifferential of the function.

**Proposition 4.10.1** *Suppose a dual pair* $(Y, Y', \langle .,. \rangle)$, $k \in Y \setminus \{0\}$ *and* $f\colon Y \to \overline{\mathbb{R}}$.
*Define* $D := \{y' \in Y' \mid \langle k, y' \rangle = 1\}$.

(a) *If $f$ is translation invariant along $k$ and* $\operatorname{dom} f \neq \emptyset$, *then for all* $y \in Y$,

$$\operatorname{dom} f^* \subseteq D \quad \textit{and} \quad \partial f(y) \subseteq D.$$

(b) *If $Y$ is equipped with the weak topology $\sigma(Y, Y')$, $Y'$ is equipped with the weak* topology $\sigma(Y', Y)$ and $f$ is lower semicontinuous, convex and proper, then*

$$f \textit{ is translation invariant along } k \textit{ if and only if } \operatorname{dom} f^* \subseteq D.$$

***Proof***

(a) If there exists $y \in Y$ with $f(y) = -\infty$, then $\operatorname{dom} f^* = \emptyset$ by Lemma 3.6.6(a) and thus $\operatorname{dom} f^* \subseteq D$. Assume now that $f$ is proper. Then there exists $y^0 \in Y$ with $f(y^0) \in \mathbb{R}$. Take any $y' \in Y'$ and $t \in \mathbb{R}$. The definition of $f^*$ yields

$$f^*(y') \geq \langle y^0 + tk, y' \rangle - f(y^0 + tk) = \langle y^0, y' \rangle - f(y^0) + t(\langle k, y' \rangle - 1).$$

Since this inequality is satisfied for each $t \in \mathbb{R}$, we get from $f^*(y') \neq +\infty$ that $\langle k, y' \rangle = 1$ holds. Consequently, $\operatorname{dom} f^* \subseteq D$. The second statement of the assertion in (a) results from Lemma 3.6.6(b).

(b) Assume $\operatorname{dom} f^* \subseteq D$. For all $y \in Y$,

$$f(y) = \sup \left\{ \langle y, y' \rangle - f^*(y') \mid y' \in Y' \right\} = \sup \left\{ \langle y, y' \rangle - f^*(y') \mid y' \in \operatorname{dom} f^* \right\}$$

by Lemma 3.6.7.
$\operatorname{dom} f^* \subseteq D \subseteq Y'$ implies $f(y) = \sup \left\{ \langle y, y' \rangle - f^*(y') \mid y' \in D \right\}$ for all $y \in Y$.
For all $y \in Y$ and $t \in \mathbb{R}$,

$$\begin{aligned} f(y + tk) &= \sup \left\{ \langle y + tk, y' \rangle - f^*(y') \mid y' \in D \right\} \\ &= \sup \left\{ \langle y, y' \rangle + t - f^*(y') \mid y' \in D \right\} \\ &= \sup \left\{ \langle y, y' \rangle - f^*(y') \mid y' \in D \right\} + t \\ &= f(y) + t. \end{aligned}$$

□

Proposition 4.10.1 implies a necessary and sufficient condition for translation invariance of certain lower semicontinuous convex functions.

**Proposition 4.10.2** *Suppose a dual pair $(Y, Y', \langle .,. \rangle)$, that $Y$ is equipped with the weak topology $\sigma(Y, Y')$, that $Y'$ is equipped with the weak* topology $\sigma(Y', Y)$, and that $f : Y \to \overline{\mathbb{R}}$ is a lower semicontinuous convex proper function with $f(0) = 0$. Then $f$ is translation invariant along $k$ if and only if*

$$f(tk) = t \quad \textit{for all } t \in \mathbb{R}. \tag{4.10.1}$$

***Proof*** Because of $f(0) = 0$, the translation invariance of $f$ along $k$ implies (4.10.1).

Assume now (4.10.1). Take any $y' \in Y'$. Since $f^*(y') \geq \langle tk, y'\rangle - f(tk) = t(\langle k, y'\rangle - 1)$ is satisfied for each $t \in \mathbb{R}$, we get from $f^*(y') \neq +\infty$ that $\langle k, y'\rangle = 1$ holds. Consequently, $\operatorname{dom} f^* \subseteq \{y' \in Y' \mid \langle k, y'\rangle = 1\}$. Because of Proposition 4.10.1, $f$ is translation invariant along $k$. □

In order to prove duality assertions (see Chap. 12) as well as optimality conditions, it is important to know the structure of the conjugate and of the subdifferential of $\varphi_{A,k}$.

**Theorem 4.10.3** *Suppose a dual pair $(Y, Y', \langle .,. \rangle)$, a nonempty subset $A$ of $Y$ and $k \in Y \setminus \{0\}$. Then*

$$\varphi^*_{A,k}(y') = \begin{cases} \sigma_A(y') & \text{if } y' \in \operatorname{bar} A \text{ with } \langle k, y'\rangle = 1, \\ +\infty & \text{otherwise.} \end{cases} \tag{4.10.2}$$

$$\partial\varphi_{A,k}(y) = \begin{cases} \{y' \in Y' \mid \langle k, y'\rangle = 1,\ \sigma_A(y') \leq \langle y, y'\rangle - \varphi_{A,k}(y)\} & \text{if } \varphi_{A,k}(y) \in \mathbb{R}, \\ \emptyset & \text{if } \varphi_{A,k}(y) \notin \mathbb{R}. \end{cases} \tag{4.10.3}$$

*For each $y \in Y$,*

$$\partial\varphi_{A,k}(y) \subseteq \{y' \in \operatorname{bar} A \mid \langle k, y'\rangle = 1\} \subseteq \{y' \in -(0^+A)^+ \mid \langle k, y'\rangle = 1\}. \tag{4.10.4}$$

*Moreover, for each $y \in Y$,*

$$\varphi_{A,k}(y) \geq \varphi^{**}_{A,k}(y) = \sup\{\langle y, y'\rangle - \sigma_A(y') \mid y' \in \operatorname{bar} A,\ \langle k, y'\rangle = 1\}. \tag{4.10.5}$$

*$\varphi^{**}_{A,k}$ is translation invariant along $k$.*

*If $Y$ is equipped with the weak topology $\sigma(Y, Y')$, $Y'$ is equipped with the weak* topology $\sigma(Y', Y)$ and $A$ is a proper closed convex subset of $Y$ with $k \in -0^+A \setminus 0^+A$, then for each $y \in Y$,*

$$\varphi_{A,k}(y) = \sup\{\langle y, y'\rangle - \sigma_A(y') \mid y' \in \operatorname{bar} A,\ \langle k, y'\rangle = 1\}. \tag{4.10.6}$$

*If $A$ is a proper $\sigma(Y, Y')$-closed convex set and* $\operatorname{int}_w A = \operatorname{bd}_w A - \mathbb{R}_> k$, *then $\varphi_{A,k}$ is a $\sigma(Y, Y')$-continuous convex proper functional, and*

$$\partial\varphi_{A,k}(y) \neq \emptyset \text{ for each } y \in \operatorname{dom}\varphi_{A,k}. \tag{4.10.7}$$

***Proof*** Define $\varphi_k := \varphi_{\{0_Y\},k}$. By Corollary 4.3.8, $\varphi_{A,k} = \iota_A \ \# \ \varphi_k$. Lemma 3.6.6(g) and 3.6.12(c) imply $\varphi^*_{A,k} = \iota^*_A + \varphi^*_k = \sigma_A + \varphi^*_k$, where $\sigma_A$ and $\varphi^*_k$ do not attain the value $-\infty$. Equation (4.10.2) follows from (4.7.6).

For each $y \in Y$ with $\varphi_{A,k}(y) \in \mathbb{R}$, (3.6.6) yields $\partial\varphi_{A,k}(y) = \{y' \in Y' \mid \varphi^*_{A,k}(y') \leq \langle y, y'\rangle - \varphi_{A,k}(y)\}$ and thus $\partial\varphi_{A,k}(y) = \{y' \in \operatorname{dom}\varphi^*_{A,k} \mid \varphi^*_{A,k}(y') \leq \langle y, y'\rangle - \varphi_{A,k}(y)\}$. Consequently, (4.10.3) and the first inclusion in (4.10.4) hold. Lemma 3.6.12(f) implies $\operatorname{bar} A \subseteq -(0^+A)^+$ and thus (4.10.4).

For each $y \in Y$, Lemma 3.6.6(f) and the definition of the biconjugate yield

$$\varphi_{A,k}(y) \geq \varphi^{**}_{A,k}(y) = \sup\{\langle y, y' \rangle - \varphi^{*}_{A,k}(y') \mid y' \in \operatorname{dom} \varphi^{*}_{A,k}\}$$

and hence (4.10.5).

The translation invariance of $\varphi^{**}_{A,k}$ along $k$ is straightforward because of (4.10.5).

The assumptions in front of (4.10.6) imply by Proposition 4.7.7 and Theorem 4.4.4 that $\varphi_{A,k}$ is a lower semicontinuous convex proper function. Lemma 3.6.7 yields (4.10.6).

Together with Theorems 4.6.6 and 4.4.4(f), the assumptions in front of (4.10.7) imply the mentioned properties of $\varphi_{A,k}$ and $\operatorname{int} \operatorname{dom} \varphi_{A,k} = \operatorname{dom} \varphi_{A,k}$. Thus, (4.10.7) follows from Lemma 3.6.9. □

By (4.10.4),

$$0 \notin \partial \varphi_{A,k}(y) \quad \text{for each } y \in Y.$$

This corresponds to Theorem 3.6.5 since $\varphi_{A,k}$ attains each real value if it is proper.

The conjugate and the biconjugate of a proper translation invariant function can be improper.

**Proposition 4.10.4** *Suppose a dual pair $(Y, Y', \langle ., . \rangle)$, a nonempty subset $A$ of $Y$, $k \in Y \setminus \{0\}$, and that $\varphi_{\mathrm{cl_w} \operatorname{conv} A,k}$ is not proper. Then*

$$\{y' \in \operatorname{bar} A \mid \langle k, y' \rangle = 1\} = \emptyset,$$
$$\varphi^{*}_{A,k} = +\infty, \ \varphi^{**}_{A,k} = -\infty,$$
$$\partial \varphi_{A,k}(y) = \emptyset \textit{ for each } y \in Y.$$

***Proof*** Since $\varphi_{\mathrm{cl_w} \operatorname{conv} A,k}$ is not proper, there exist $y \in Y$ and sequences $(a_n)_{n \in \mathbb{N}}$ and $(t_n)_{n \in \mathbb{N}}$ with $y = a_n + t_n k$, $a_n \in \mathrm{cl_w} \operatorname{conv} A$, $t_n \in \mathbb{R}$ for all $n \in \mathbb{N}$ and $\lim\limits_{n \to \infty} t_n = -\infty$. Assume that there is $y' \in Y'$ with $\langle k, y' \rangle = 1$. For each $n \in \mathbb{N}$, $\langle y, y' \rangle = \langle a_n, y' \rangle + t_n$ and thus $\langle a_n, y' \rangle = \langle y, y' \rangle - t_n$. Since $\lim\limits_{n \to \infty} t_n = -\infty$, we have $y' \notin \operatorname{bar}(\mathrm{cl_w} \operatorname{conv} A)$. Because of $\operatorname{bar}(\mathrm{cl_w} \operatorname{conv} A) = \operatorname{bar} A$, we obtain $\{y' \in \operatorname{bar} A \mid \langle k, y' \rangle = 1\} = \emptyset$. Theorem 4.10.3 implies the other statements of the proposition. □

$\varphi_{\mathrm{cl} \operatorname{conv} A,k}$ can be improper even though $\varphi_{A,k}$ is proper.

**Example 4.10.5** Consider $A := \{(y_1, y_2)^{\mathrm{T}} \in \mathbb{R}^2 \mid y_1 \geq 0, y_2 = 0\} \cup \{(y_1, y_2)^{\mathrm{T}} \in \mathbb{R}^2 \mid y_2 = e^{y_1}\}$ and $k := (-1, 0)^{\mathrm{T}}$. Then $\varphi_{A,k}$ is proper, but $\varphi_{\mathrm{cl} \operatorname{conv} A,k}$ is improper since $\{(y_1, y_2)^{\mathrm{T}} \in \mathbb{R}^2 \mid y_2 = 0\} \subseteq \mathrm{cl} \operatorname{conv} A$.

The following statement underscores that the Fenchel conjugate was designed for convex problems. The formula for the so-called limiting subdifferential of $\varphi_{A,k}$, which was introduced to tackle nonconvex problems, will be given in Theorem 12.1.8.

**Corollary 4.10.6** *Suppose a dual pair $(Y, Y', \langle ., . \rangle)$, a nonempty subset $A$ of $Y$ and $k \in Y \setminus \{0\}$. Then*

$$\varphi^*_{A,k} = \varphi^*_{\mathrm{cl_w\,conv}\,A,k} \quad \textit{and} \quad \varphi^{**}_{A,k} = \varphi^{**}_{\mathrm{cl_w\,conv}\,A,k}.$$

***Proof*** By Lemma 3.6.12, $\sigma_A = \sigma_{\mathrm{cl_w\,conv}\,A}$ and hence bar $A =$ bar$(\mathrm{cl_w\,conv}\,A)$. (4.10.2) yields the first statement of the corollary, which implies the second statement by the definition of the biconjugate. □

Using Corollary 4.10.6, we can prove a formula for the lower semicontinuous convex function envelope of $\varphi_{A,k}$.

**Theorem 4.10.7** *Suppose a dual pair $(Y, Y', \langle .,. \rangle)$, $A \subseteq Y$, $k \in Y \setminus \{0\}$ and that $Y$ is equipped with the weak topology $\sigma(Y, Y')$. Then*

$$\mathrm{cl\,conv}\,\varphi_{A,k} = \varphi_{\mathrm{cl}(\mathrm{conv}\,A - \mathbb{R}_+ k),k}.$$

***Proof*** If $A = \emptyset$ or $A = Y$, then the statement is obvious. Assume now that $A$ is a proper subset of $Y$. Corollary 4.10.6 and (4.2.8) yield

$$\varphi^{**}_{\mathrm{cl}(\mathrm{conv}\,A - \mathbb{R}_+ k),k} = \varphi^{**}_{\mathrm{cl\,conv}(A - \mathbb{R}_+ k),k} = \varphi^{**}_{A - \mathbb{R}_+ k,k} = \varphi^{**}_{A,k}. \tag{4.10.8}$$

By Theorem 4.4.4, $\varphi_{\mathrm{cl}(\mathrm{conv}\,A - \mathbb{R}_+ k),k}$ is convex and lower semicontinuous. Lemma 3.6.8 implies that $\varphi_{\mathrm{cl}(\mathrm{conv}\,A - \mathbb{R}_+ k),k}$ is proper if and only if the function $\varphi^{**}_{\mathrm{cl}(\mathrm{conv}\,A - \mathbb{R}_+ k),k}$ is proper, and that $\mathrm{cl\,conv}\,\varphi_{A,k}$ is proper if and only if $\varphi^{**}_{A,k}$ is proper. Thus, we obtain from (4.10.8) that $\varphi_{\mathrm{cl}(\mathrm{conv}\,A - \mathbb{R}_+ k),k}$ is proper if and only if $\mathrm{cl\,conv}\,\varphi_{A,k}$ is proper. In this case, both functions coincide because of Lemma 3.6.8 and (4.10.8). Consider now the case where $\varphi_{\mathrm{cl}(\mathrm{conv}\,A - \mathbb{R}_+ k),k}$ is not proper. Then $\mathrm{cl\,conv}\,\varphi_{A,k}$ is not proper. Both functions do not attain any real value because of Lemma 3.4.7(e). We have $\varphi_{\mathrm{cl}(\mathrm{conv}\,A - \mathbb{R}_+ k),k} \leq \mathrm{cl\,conv}\,\varphi_{A,k}$ since $\varphi_{\mathrm{cl}(\mathrm{conv}\,A - \mathbb{R}_+ k),k}$ is convex and lower semicontinuous. Hence,

$$\mathrm{dom\,cl\,conv}\,\varphi_{A,k} \subseteq \mathrm{dom}\,\varphi_{\mathrm{cl}(\mathrm{conv}\,A - \mathbb{R}_+ k),k}.$$

Since $\mathrm{cl\,conv}\,\varphi_{A,k}$ is lower semicontinuous, its sublevel sets are closed. Thus, $\mathrm{dom\,cl\,conv}\,\varphi_{A,k}$ is closed since $\mathrm{cl\,conv}\,\varphi_{A,k}$ does not attain any real value. Moreover, $\mathrm{dom\,cl\,conv}\,\varphi_{A,k}$ is convex since $\mathrm{cl\,conv}\,\varphi_{A,k}$ is convex. From $\mathrm{cl\,conv}\,\varphi_{A,k} \leq \varphi_{A,k}$, we obtain $\mathrm{dom}\,\varphi_{A,k} \subseteq \mathrm{dom\,cl\,conv}\,\varphi_{A,k}$, i.e., $A + \mathbb{R}k \subseteq \mathrm{dom\,cl\,conv}\,\varphi_{A,k}$. Therefore, $\mathrm{conv}\,A - \mathbb{R}_+ k + \mathbb{R}k \subseteq \mathrm{dom\,cl\,conv}\,\varphi_{A,k}$ since $\mathrm{dom\,cl\,conv}\,\varphi_{A,k}$ is convex. We get $\mathrm{cl}(\mathrm{conv}\,A - \mathbb{R}_+ k + \mathbb{R}k) \subseteq \mathrm{dom\,cl\,conv}\,\varphi_{A,k}$ since $\mathrm{dom\,cl\,conv}\,\varphi_{A,k}$ is closed. Hence,

$$\mathrm{dom}\,\varphi_{\mathrm{cl}(\mathrm{conv}\,A - \mathbb{R}_+ k),k} = \mathrm{cl}(\mathrm{conv}\,A - \mathbb{R}_+ k) + \mathbb{R}k \subseteq \mathrm{dom\,cl\,conv}\,\varphi_{A,k}.$$

Consequently, $\mathrm{cl\,conv}\,\varphi_{A,k} = \varphi_{\mathrm{cl}(\mathrm{conv}\,A - \mathbb{R}_+ k),k}$. □

We will now specify the formulas for the subdifferential and the conjugate of $\varphi_{A,k}$ in the case where $A$ is a convex cone.

**Theorem 4.10.8** *Suppose a topological vector space $Y$ with a dual pair $(Y, Y', \langle.,.\rangle)$, a nontrivial convex cone $A \subset Y$ and $k \in -A \setminus \operatorname{cl} A$. Then $\varphi_{A,k}$ is sublinear and proper. For each $y' \in Y'$, we have*

$$\varphi^*_{A,k}(y') = \begin{cases} 0 & \text{if } y' \in -A^+ \text{ with } \langle k, y'\rangle = 1, \\ +\infty & \text{otherwise.} \end{cases} \tag{4.10.9}$$

*For each $y \in Y$,*

$$\partial\varphi_{A,k}(y) = \{y' \in -A^+ \mid \langle k, y'\rangle = 1, \langle y, y'\rangle = \varphi_{A,k}(y)\}. \tag{4.10.10}$$

$$\partial\varphi_{A,k}(0) = \{y' \in -A^+ \mid \langle k, y'\rangle = 1\}. \tag{4.10.11}$$

*Moreover, for each $y \in Y$,*

$$\varphi_{A,k}(y) \geq \varphi^{**}_{A,k}(y) = \sup\{\langle y, y'\rangle \mid y' \in -A^+ \text{ with } \langle k, y'\rangle = 1\}. \tag{4.10.12}$$

*$\varphi^{**}_{A,k}$ is $(-A)$-monotone.*

*If $Y$ is equipped with the weak topology $\sigma(Y, Y')$, $Y'$ is equipped with the weak* topology $\sigma(Y', Y)$ and $A$ is closed, then*

$$\varphi_{A,k}(y) = \sup\{\langle y, y'\rangle \mid y' \in -A^+ \text{ with } \langle k, y'\rangle = 1\} \text{ for each } y \in Y. \tag{4.10.13}$$

*If $k \in -\operatorname{int} A$, then $\varphi_{A,k}$ is a $\sigma(Y, Y')$-continuous finite-valued function, and*

$$\partial\varphi_{A,k}(y) \neq \emptyset \text{ for each } y \in Y. \tag{4.10.14}$$

***Proof*** Propositions 4.7.1(b) and 4.7.3(a) imply that $\varphi_{A,k}$ is sublinear and proper. Hence, $\varphi_{A,k}(0) = 0$ by Lemma 3.5.10. Because of Lemma 3.6.12, bar $A = -A^+$ and $\sigma_A = \iota_{-A^+}$. Theorem 4.10.3 implies (4.10.9), (4.10.12), (4.10.13) and, for each $y \in Y$,

$$\begin{aligned} \partial\varphi_{A,k}(y) &= \{y' \in Y' \mid \langle k, y'\rangle = 1, \sigma_A(y') \leq \langle y, y'\rangle - \varphi_{A,k}(y)\} \\ &= \{y' \in -A^+ \mid \langle k, y'\rangle = 1, \langle y, y'\rangle \geq \varphi_{A,k}(y)\} \quad \text{since } \sigma_A = \iota_{-A^+}. \end{aligned}$$

This yields (4.10.10) because of the inequality in (4.10.12). Equation (4.10.11) results from (4.10.10).

If $y^1, y^2 \in Y$ with $y^2 - y^1 \in -A$, then $\langle y^2 - y^1, y'\rangle \geq 0$ and thus $\langle y^2, y'\rangle \geq \langle y^1, y'\rangle$ holds for each $y' \in -A^+$, which implies $\varphi^{**}_{A,k}(y^2) \geq \varphi^{**}_{A,k}(y^1)$ by (4.10.12). Thus, $\varphi^{**}_{A,k}$ is $(-A)$-monotone.

For $k \in -\operatorname{int} A$, Proposition 4.7.3(d) results in the mentioned properties of $\varphi_{A,k}$ and (4.10.14) because of Lemma 3.6.9(b). □

If $A$ is a closed convex cone, then $\varphi_{A,k}$ can be described by the dual cone of $A$.

**Proposition 4.10.9** *Suppose a dual pair $(Y, Y', \langle ., . \rangle)$, a nonempty subset $A$ of $Y$ and $k \in Y \setminus \{0\}$.*

*For each $y \in Y$,*

$$\varphi_{A,k}(y) \geq \inf\{t \in \mathbb{R} \mid \forall y' \in A^+ : \langle y, y' \rangle \geq t \langle k, y' \rangle\}. \tag{4.10.15}$$

*If $A$ is a $\sigma(Y, Y')$-closed convex cone, then*

$$\varphi_{A,k}(y) = \inf\{t \in \mathbb{R} \mid \forall y' \in A^+ : \langle y, y' \rangle \geq t \langle k, y' \rangle\} \quad \text{for each } y \in Y. \tag{4.10.16}$$

***Proof*** Since $\varphi_{A,k}(y) = \inf\{t \in \mathbb{R} \mid y - tk \in A\}$, (4.10.15) and (4.10.16) result from (2.2.5) and from Proposition 2.2.18, respectively. □

Further properties of the subdifferential for translation invariant functions are collected in the subsequent propositions. The next proposition follows from Theorem 4.2.9 and (3.6.4) together with Proposition 4.7.1(b).

**Proposition 4.10.10** *Suppose a dual pair $(Y, Y', \langle ., . \rangle)$, a proper subset $A$ of $Y$ and $k \in -\operatorname{core} 0^+A$. Then $\varphi_{A,k}$ is a finite-valued function, and $\varphi_{0^+A,k}$ is a sublinear finite-valued function. For each $y^0 \in Y$, we get*

$$y' \in \partial \varphi_{A,k}(y^0) \implies \forall y \in Y : \langle y - y^0, y' \rangle \leq \varphi_{0^+A,k}(y - y^0).$$

In Proposition 3.7.9, monotonicity assumptions are given which guarantee that $\operatorname{dom} \varphi^*_{A,k}$ or the subdifferentials of $\varphi_{A,k}$ are contained in the dual cone of some set or in its quasi-interior. Section 4.5 contains necessary and sufficient conditions for the monotonicity of $\varphi_{A,k}$. This allows us to deduce statements like those in the following proposition.

**Proposition 4.10.11** *Suppose a dual pair $(Y, Y', \langle ., . \rangle)$, nonempty subsets $A$, $B$ of $Y$ and $k \in Y \setminus \{0\}$.*

(a) *If $A - B \subseteq \operatorname{cl}_k(A) - \mathbb{R}_+ k$, especially if $B \subseteq -0^+A$, then*

$$\partial \varphi_{A,k}(y) \subseteq B^+ \quad \text{for all } y \in Y. \tag{4.10.17}$$

(b) *Assume that $B$ is a cone.*
*If $A - B \subseteq \operatorname{cl}_k(A) - \mathbb{R}_+ k$, especially if $B \subseteq -0^+A$, then*

$$\operatorname{dom} \varphi^*_{A,k} \subseteq B^+. \tag{4.10.18}$$

(c) *If $k \in -\operatorname{core} 0^+A$ and $\operatorname{cl}_k(A) - (B \setminus \{0\}) \subseteq A - \mathbb{R}_> k$, then*

$$\partial \varphi_{A,k}(y) \subseteq \operatorname{qint} B^+ \quad \text{for all } y \in Y. \tag{4.10.19}$$

***Proof*** The assumptions in (a) and (b) imply by Theorem 4.5.2(a) that $\varphi_{A,k}$ is $B$-monotone. Under the assumptions of (c), $\varphi_{A,k}$ is finite-valued by Theorem

4.2.9(a) and strictly $B$-monotone by Theorem 4.5.2(d). Proposition 3.7.9 yields the assertions. □

For the special case where $A$ is convex and $Y$ is a space of real-valued functions on some set that is equipped with a probability measure, duality statements for functions $\varphi_{A,k}$ were proved by Föllmer and Schied [168].

**Remark 4.10.12** (*Bibliographical notes related to Sect.* 4.10)

Proposition 4.10.1 was proved for the dual pair $(Y, Y^*, \langle .,. \rangle)$ by Filipović and Kupper in [162, Lemma 3.2], where the statement for the subdifferential was given for $y \in \operatorname{dom}(-f)$ only. The proofs of Proposition 4.10.2 and 4.10.9 are based on ideas from [170] and [280, p. 189], respectively.

For the dual pair $(Y, Y^*, \langle .,. \rangle)$ and under the additional assumptions that $A$ is a proper closed convex set and that $\varphi_{A,k}$ is proper, (4.10.2) and (4.10.4) were established in [441], (4.10.3) was proved by Durea and Tammer [129] in normed spaces $Y$ and in [441] for separated locally convex spaces $Y$, (4.10.2) and (4.10.6) were shown in [221, Sect. 3.3], (4.10.19) was proved for a proper closed convex cone $B$ with $A - (B \setminus \{0\}) \subseteq \operatorname{int} A$ in [291, Cor. 5.2.24].

The idea to use Corollary 4.3.8 for deriving the conjugate's formula is contained in [225, Sect. 5.5].

Equation (4.10.12) can be deduced from [280, Theorem 2]. This is also the case for (4.10.13) if $k \in -\operatorname{core} A$.

The following citations refer to the dual space $(Y, Y^*, \langle .,. \rangle)$ and to the additional assumption that $A$ is weakly closed. (4.10.9), (4.10.11) and (4.10.13) were shown in [221, Sect. 3.3], where an additional sufficient condition for properness of $\varphi_{A,k}$ was used. For $k \in -\operatorname{int} A$, (4.10.11) can also be found in [395, p. 139f]. Equation (4.10.10) under the additional assumption $k \in -\operatorname{int} A$ and the statement in Theorem 4.10.8 related to the existence of subgradients were proved by Durea and Tammer [129] in a normed space $Y$.

## 4.11 Separation of Sets by Translation Invariant Functions

The functional $\varphi_{A,k}$ is constructed in such a way that it can be used for the separation of not necessarily convex sets. We will now prove some separation theorems, based on the results of the previous sections. These theorems are used explicitly or implicitly in most of the cases where translation invariant functions are applied, e.g., for the characterization of solutions in vector optimization, for the proof of Lagrangian multiplier rules, duality assertions and variational principles.

Each inequality can be considered to be a separation statement. Consequently, this section contains a choice of possible theorems which are evoked by the properties of $\varphi_{A,k}$. For the same reason, the investigation of properties of $\varphi_{A,k}$ has been accompanied by the formulation of separation statements (see [184, 186, 222, 407]).

The idea that separation by a function value corresponds to separation by a sublevel set and that a separating function can be constructed using a separating set does also work for other than translation invariant functions. For example, Kasimbeyli [287] proved scalarization statements for nonconvex vector optimization problems on the basis of cone separation theorems.

Theorem 4.2.1 implies a basic separation theorem in linear spaces.

**Theorem 4.11.1** *Assume* $(\mathrm{Sp0}_{A,k})$ *and* $D \subseteq Y$.

(a) $(\forall d \in D\colon \varphi_{A,k}(d) \geq 0) \implies \operatorname{core} A \cap D = \emptyset$.
(b) *If* $A - \mathbb{R}_{>}k \subseteq \operatorname{core} A$, *especially if* $k \in -\operatorname{core} 0^+A$, *then:*
$\operatorname{core} A \cap D = \emptyset \iff \forall d \in D\colon \varphi_{A,k}(d) \geq 0$.
(c) $(\forall d \in D\colon \varphi_{A,k}(d) > 0) \implies A \cap D = \emptyset$.
(d) *If* $k \in -0^+A \setminus \{0\}$ *and* $A$ *is* $k$*-directionally closed, then:*
$A \cap D = \emptyset \iff \forall d \in D\colon \varphi_{A,k}(d) > 0$.

***Proof*** (a) follows from (4.2.13). (b) results from (4.2.14) and Lemma 2.3.18(b). Inclusion (4.2.11) and (4.2.12) yield (c) and (d), respectively. □

The following separation theorems refer to topological vector spaces. On such spaces, we will use the next lemma.

**Lemma 4.11.2** *Consider two subsets* $S_1, S_2$ *of a topological space* $Y$.

(a) $\operatorname{int} S_1 \cap S_2 = \emptyset \iff \operatorname{int} S_1 \cap \operatorname{cl} S_2 = \emptyset$.
(b) *If* $S_1 \subseteq \operatorname{cl} \operatorname{int} S_1$, *then:*
$\operatorname{int} S_1 \cap S_2 = \emptyset \implies S_1 \cap \operatorname{int} S_2 = \emptyset$.

***Proof***

(a) Assume $\operatorname{int} S_1 \cap S_2 = \emptyset$. Contrary to the claim, suppose that there exists $y \in \operatorname{int} S_1 \cap \operatorname{cl} S_2$. Since $\operatorname{int} S_1$ is a neighborhood of $y$, there is $s \in \operatorname{int} S_1$ with $s \in S_2$, which contradicts the assumption.
(b) Assume $\operatorname{int} S_1 \cap S_2 = \emptyset$. Then $\operatorname{int} S_1 \cap \operatorname{int} S_2 = \emptyset$, which implies $\operatorname{cl} \operatorname{int} S_1 \cap \operatorname{int} S_2 = \emptyset$ by (a). We get $S_1 \cap \operatorname{int} S_2 = \emptyset$ since $S_1 \subseteq \operatorname{cl} \operatorname{int} S_1$. □

Separating functionals should have properties which are required in proofs or algorithms. For this reason, we will now formulate a theorem in which essential properties of the separating functional are summarized.

**Theorem 4.11.3** *Assume* $(\mathrm{Sp1}_{A,k})$.

(1) *Then* $\varphi_{A,k}$ *is lower semicontinuous and* $(-0^+A)$*-monotone. If* $\varphi_{A,k}$ *is finite-valued on* $F \subseteq Y$, *then it is strictly* $(-\operatorname{core} 0^+A)$*-monotone on* $F$. $\varphi_{A,k}$ *is convex if* $\operatorname{cl} A$ *is convex. It is sublinear if* $\operatorname{cl} A$ *is a convex cone.* $\varphi_{A,k}$ *is finite-valued if* $k \in -\operatorname{core} 0^+A$.
*We have* $\varphi_{A,k}(a) \leq 0$ *for all* $a \in \operatorname{cl} A$, *and* $\varphi_{A,k}(a) < 0$ *for all* $a \in \operatorname{core} A$.
*For each set* $D \subseteq Y$, *Theorem 4.11.1*(a)–(b) *hold and:*
   (a) $\operatorname{cl} A \cap D = \emptyset \iff \forall d \in D\colon \varphi_{A,k}(d) > 0$.
   (b) *If* $D \subseteq \operatorname{cl} \operatorname{int} D$, *then:*
   $(\forall d \in \operatorname{int} D\colon \varphi_{A,k}(d) > 0) \implies \operatorname{int} A \cap D = \emptyset$.

(2) *Assume* (Sp2$_{A,k}$).
*Then* $\varphi_{A,k}$ *is continuous. For each set* $D \subseteq Y$*, we have:*

(a) $\operatorname{int} A \cap D = \emptyset \iff \forall d \in D\colon \varphi_{A,k}(d) \geq 0.$
(b) $\operatorname{cl} A \cap \operatorname{int} D = \emptyset \iff \forall d \in \operatorname{int} D\colon \varphi_{A,k}(d) > 0.$
(c) $\operatorname{int} A \cap D = \emptyset \implies \operatorname{cl} A \cap \operatorname{int} D = \emptyset.$
(d) *If* $D \subseteq \operatorname{cl} \operatorname{int} D$*, then:*
$\operatorname{int} A \cap D = \emptyset \iff \operatorname{cl} A \cap \operatorname{int} D = \emptyset.$

***Proof***

(1) The statements in front of (a) result from Theorem 4.4.4, Corollary 4.5.8, (4.2.13), Theorems 4.2.10 and 4.2.9.

(a) follows from (4.4.5).
(b) If $\varphi_{A,k}(d) > 0$ for all $d \in \operatorname{int} D$, then $\operatorname{cl} A \cap \operatorname{int} D = \emptyset$. Because of Lemma 4.11.2(b), we have $\operatorname{int} \operatorname{cl} A \cap D = \emptyset$. This implies the assertion since $\operatorname{int} \operatorname{cl} A = \operatorname{int} A$ by Lemma 2.3.16(c).

(2) We apply Theorem 4.6.6.

(a) follows from (4.6.13).
(b) Apply (1)(a) to $\operatorname{int} D$ instead of $D$.
(c) By (4.6.9), $\operatorname{cl} A = \operatorname{cl} \operatorname{int} A$. For $C := \operatorname{cl} A$, we get $\operatorname{cl} \operatorname{int} C = \operatorname{cl} \operatorname{int} A = \operatorname{cl} A$, and $\operatorname{int} C = \operatorname{int} A$. Apply Lemma 4.11.2(b) to $S_1 = C$ and $S_2 = D$.
(d) For the reverse direction, apply Lemma 4.11.2(b) to $S_1 = D$ and $S_2 = \operatorname{cl} A$. □

The condition $D \subseteq \operatorname{cl} \operatorname{int} D$ is fulfilled if $D$ is a convex set with nonempty interior.

For the convenience in applications, we present the special results related to the separation of a closed convex cone $A$ and an arbitrary set $D$.

**Theorem 4.11.4** *Assume that* $Y$ *is a topological vector space and that* $A \subset Y$ *is a nontrivial closed convex cone with* $k \in -A \setminus A$.

(1) $\varphi_{A,k}$ *is a lower semicontinuous sublinear proper* $(-A)$*-monotone functional, which is strictly* $(-\operatorname{core} A)$*-monotone on* $\operatorname{dom} \varphi_{A,k}$. *It is locally Lipschitzian on* $\operatorname{int} \operatorname{dom} \varphi_{A,k}$ *if* $Y$ *is a Banach space.*
*Moreover,* $\varphi_{A,k}(a) \leq 0$ *for all* $a \in A$*, and* $\varphi_{A,k}(a) < 0$ *for all* $a \in \operatorname{core} A$.
*For each set* $D \subseteq Y$*, we have:*

(a) $A \cap D = \emptyset \iff \forall d \in D\colon \varphi_{A,k}(d) > 0.$
(b) *If* $D \subseteq \operatorname{cl} \operatorname{int} D$*, then:*
$(\forall d \in \operatorname{int} D\colon \varphi_{A,k}(d) > 0) \implies \operatorname{int} A \cap D = \emptyset.$
(c) $(\forall d \in D\colon \varphi_{A,k}(d) \geq 0) \implies \operatorname{core} A \cap D = \emptyset.$

(2) *If* $k \in -\operatorname{core} A$*, then* $\varphi_{A,k}$ *is finite-valued, and for each* $D \subseteq Y$*:*
$\operatorname{core} A \cap D = \emptyset \iff \forall d \in D\colon \varphi_{A,k}(d) \geq 0.$

(3) *Suppose* $k \in -\operatorname{int} A$.
*Then* $\varphi_{A,k}$ *is continuous and finite-valued.* $\varphi_{A,k}$ *is Lipschitzian if* $Y$ *is a Banach space.*
*For each set* $D \subseteq Y$, *we have:*

(a) $\operatorname{int} A \cap D = \emptyset \iff \forall d \in D\colon \varphi_{A,k}(d) \geq 0.$
(b) $A \cap \operatorname{int} D = \emptyset \iff \forall d \in \operatorname{int} D\colon \varphi_{A,k}(d) > 0.$
(c) $\operatorname{int} A \cap D = \emptyset \implies A \cap \operatorname{int} D = \emptyset.$
(d) *If* $D \subseteq \operatorname{cl} \operatorname{int} D$, *then:*
$\operatorname{int} A \cap D = \emptyset \iff A \cap \operatorname{int} D = \emptyset.$

***Proof*** The properties of $\varphi_{A,k}$ which are mentioned beside the separation statements result from Propositions 4.7.1(b), 4.7.3(a), (d) and (e), 4.7.9(f), (4.2.11), (4.2.13), and Theorem 4.6.14. The separation statements follow from Theorem 4.11.3. □

Finally, we prove a separation result for the sum of a set and some cone interior.

**Proposition 4.11.5** *Assume that* $Y$ *is a topological vector space, that* $B$ *is a nonempty subset of* $Y$ *and that* $K \subset Y$ *is a nontrivial convex cone with* $k \in \operatorname{int} K$ *such that* $B - \operatorname{int} K \neq Y$. *Then* $\varphi_{B-\operatorname{int} K,k}$ *is continuous, finite-valued,* $K$*-monotone and strictly* $(\operatorname{int} K)$*-monotone. Moreover,* $B \subseteq \operatorname{cl}(B - \operatorname{int} K)$ *and*

$$\varphi_{B-\operatorname{int} K,k}(y) \leq 0 \text{ for all } y \in \operatorname{cl}(B - \operatorname{int} K).$$

*For each set* $D \subseteq Y$, *we have:*

$$(B - \operatorname{int} K) \cap D = \emptyset \iff \forall d \in D\colon \varphi_{B-\operatorname{int} K,k}(d) \geq 0.$$

***Proof*** The set $A := B - \operatorname{int} K$ is a proper subset of $Y$. We have $K \subseteq -0^+A$ and thus $k \in -\operatorname{int} 0^+A$. $\varphi_{A,k}$ is continuous and finite-valued by Proposition 4.6.13. It is $K$-monotone and strictly $(\operatorname{int} K)$-monotone because of Corollary 4.5.8. Obviously, $B \subseteq \operatorname{cl} A$ since $\operatorname{int} K \cup \{0\}$ is a cone. $(\mathrm{Sp2}_{A,k})$ holds because of Theorem 4.6.2. Theorem 4.11.3(2a) implies the separation statement since $\operatorname{int} A = A$. □

**Remark 4.11.6** (*Bibliographical notes related to Sect.* 4.11)
Theorem 4.11.1 coincides with [483, Prop. 2.5]. Lemma 4.11.2 and the separation statements in Theorem 4.11.3 were proved in [484], where Lemma 4.11.2(a) is taken from [464, Satz 2.4.2]. Proposition 4.11.5 extends a necessary condition for the separation of sets which was proved by Luc [338, p. 83].

## 4.12 Approximation of Translation Invariant Functions

In this section, $\varphi_{A,k}$ is represented as the maximum or minimum of functions which turn out to be continuous, convex, sublinear or concave under appropriate assumptions.

The following theorem is illustrated by Fig. 4.11.

**Theorem 4.12.1** *Assume that $A$ and $C$ are proper subsets of the linear space $Y$ with $C \subseteq 0^+A$, $0 \in \mathrm{bd}_k(C)$ and $k \in -(0^+C \cap \mathrm{core}\, 0^+A)$.*
*Define $B(y) := C + y - \varphi_{A,k}(y) \cdot k$ for every $y \in Y$.*
*For each $y^0 \in Y$,*

$$\begin{aligned}\varphi_{A,k}(y^0) &= \min_{y\in Y} \varphi_{B(y),k}(y^0)\\ &= \min_{y\in \mathrm{bd}_k(A)} \varphi_{y+C,k}(y^0) = \min_{y\in \mathrm{bd}_k(A)} \varphi_{C,k}(y^0 - y),\end{aligned} \tag{4.12.1}$$

*where $y^0$ is a minimizer of the first optimization problem in (4.12.1).*
*The function $\varphi_{C,k}$ as well as each function $\varphi_{B(y),k}$ and $\varphi_{y+C,k}$ with $y \in Y$ has the following properties:*

(a) *It is finite-valued if $k \in -\mathrm{core}\, 0^+C$.*
(b) *It is convex if $C$ is convex.*
(c) *It is lower semicontinuous if $Y$ is a topological vector space and $C$ is closed.*
(d) *It is continuous if $Y$ is a topological vector space and $C$ is a convex cone with $k \in -\mathrm{int}\, C$.*

*Moreover, $\varphi_{C,k}$ is a sublinear function if $C$ is a convex cone.*

***Proof*** $\varphi_{A,k}$ and $\varphi_{0^+A,k}$ are finite-valued by Theorem 4.2.9. Take any $y^0, y \in Y$.

$$\begin{aligned}\varphi_{B(y),k}(y^0) &= \varphi_{C,k}(y^0 - y + \varphi_{A,k}(y)\cdot k) \qquad \text{because of Proposition 4.3.5}\\ &= \varphi_{C,k}(y^0 - y) + \varphi_{A,k}(y).\end{aligned}$$

By (4.2.5), $0 \in \mathrm{bd}_k(C) = \mathrm{bd}_k(C - \mathbb{R}_+k)$ yields $\varphi_{C,k}(0) = 0$ and thus $\varphi_{B(y^0),k}(y^0) = \varphi_{A,k}(y^0)$. Since $C \subseteq 0^+A$, Corollary 4.3.14 implies $\varphi_{0^+A,k}(y^1) \le \varphi_{C,k}(y^1)$ for each $y^1 \in Y$. Hence, $\varphi_{B(y),k}(y^0) \ge \varphi_{0^+A,k}(y^0 - y) + \varphi_{A,k}(y)$. Then $\varphi_{B(y),k}(y^0) \ge \varphi_{A,k}(y^0)$ because of (4.2.23). This yields the first equation in (4.12.1) and the related statement about a minimizer.

We have $B(y) = y^2 + C$ with $y^2 := y - \varphi_{A,k}(y)\cdot k$. Theorem 4.2.5(b) implies $y \in \mathrm{bd}_k(A) + \varphi_{A,k}(y)\cdot k$ and hence $y^2 \in \mathrm{bd}_k(A)$. Conversely, we get for each $y^3 \in \mathrm{bd}_k(A)$ that $\varphi_{A,k}(y^3) = 0$ and thus $B(y^3) = y^3 + C$ holds. This yields the second equation in (4.12.1). The last equation in (4.12.1) follows from Proposition 4.3.5.

(a) results from Theorem 4.2.9, (b) from Theorem 4.2.10, and (c) from Theorem 4.4.4. Proposition 4.7.1 yields the last statement of the theorem.

Assume now that $Y$ is a topological vector space. If $C$ is a convex cone with $k \in -\mathrm{int}\, C$, then we obtain $\mathrm{cl}\, C - \mathbb{R}_>k \subseteq \mathrm{int}\, C$ from Lemma 2.2.7(b). Hence, $\mathrm{cl}\, B(y) - \mathbb{R}_>k \subseteq \mathrm{int}\, B(y)$, and Theorem 4.6.6 implies (d). □

**Corollary 4.12.2** *Assume that $A$ is a proper subset of the linear space $Y$ with $0 \in \mathrm{bd}_k(0^+A)$ and $k \in -\mathrm{core}\, 0^+A$.*
*Define $B(y) := 0^+A + y - \varphi_{A,k}(y)\cdot k$ for every $y \in Y$.*
*For each $y^0 \in Y$,*

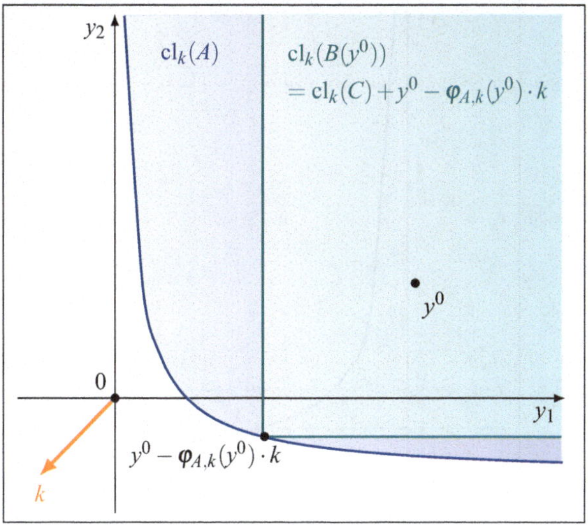


**Fig. 4.11** Illustration of Theorem 4.12.1: $\varphi_{A,k}(y) \leq \varphi_{B(y^0),k}(y)$ for each $y \in Y$, and $\varphi_{A,k}(y^0) = \varphi_{B(y^0),k}(y^0)$

$$\varphi_{A,k}(y^0) = \min_{y \in Y} \varphi_{B(y),k}(y^0) = \min_{y \in \mathrm{bd}_k(A)} \varphi_{y+0^+A,k}(y^0) = \min_{y \in \mathrm{bd}_k(A)} \varphi_{0^+A,k}(y^0 - y).$$

*The function $\varphi_{0^+A,k}$ as well as each function $\varphi_{B(y),k}$ and $\varphi_{y+0^+A,k}$ with $y \in Y$ has the following properties:*

(a) *It is convex and finite-valued.*
(b) *It is lower semicontinuous if $Y$ is a topological vector space and $A$ is closed.*
(c) *It is continuous if $Y$ is a topological vector space and $k \in -\mathrm{int}\, 0^+A$.*

*Moreover, $\varphi_{0^+A,k}$ is a sublinear function.*

Figure 4.12 illustrates the representation of $\varphi_{A,k}$ by the maxima of functions in the next theorem.

**Theorem 4.12.3** *Assume that $A$ and $C$ are proper subsets of the linear space $Y$ with $C \subseteq 0^+A$, $0 \in \mathrm{bd}_k(C)$ and $k \in -(0^+C \cap \mathrm{core}\, 0^+A)$.*
*Define $B(y) := (Y \setminus (-C)) + y - \varphi_{A,k}(y) \cdot k$ for every $y \in Y$.*
*For each $y^0 \in Y$,*

$$\begin{aligned}\varphi_{A,k}(y^0) &= \max_{y \in Y} \varphi_{B(y),k}(y^0) \\ &= \max_{y \in \mathrm{bd}_k(A)} \varphi_{y+(Y\setminus(-C)),k}(y^0) = \max_{y \in \mathrm{bd}_k(A)} \varphi_{Y\setminus(-C),k}(y^0 - y),\end{aligned} \tag{4.12.2}$$

*where $y^0$ is a maximizer of the first optimization problem in (4.12.2).*

*The function $\varphi_{Y\setminus(-C),k}$ as well as each of the functions $\varphi_{B(y),k}$ and $\varphi_{y+(Y\setminus(-C)),k}$ with $y \in Y$ has the following properties:*

(a) *It is finite-valued if $k \in -\mathrm{core}\, 0^+C$.*
(b) *It is concave if $C$ is convex.*

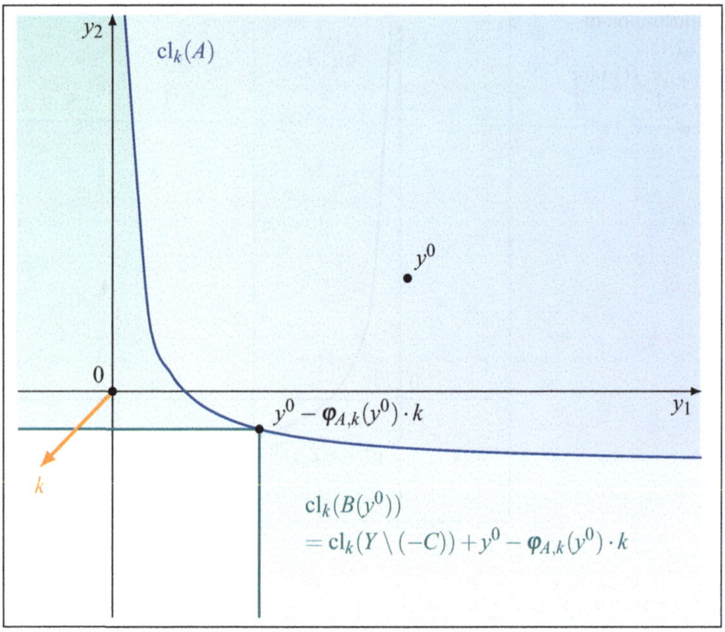


**Fig. 4.12** Illustration of Theorem 4.12.3: $\varphi_{A,k}(y) \geq \varphi_{B(y^0),k}(y)$ for each $y \in Y$, and $\varphi_{A,k}(y^0) = \varphi_{B(y^0),k}(y^0)$

(c) *It is continuous if $Y$ is a topological vector space and $C$ is a convex cone with $k \in -\operatorname{int} C$.*

*Moreover, $-\varphi_{Y \setminus (-C),k}$ is a sublinear function if $C$ is a convex cone.*

***Proof*** $\varphi_{A,k}$ and $\varphi_{0^+A,k}$ are finite-valued by Theorem 4.2.9. Take any $y^0, y \in Y$.

$$\begin{aligned}
\varphi_{B(y),k}(y^0) &= \varphi_{Y \setminus (-C),k}(y^0 - y + \varphi_{A,k}(y) \cdot k) && \text{because of Proposition 4.3.5}\\
&= \varphi_{Y \setminus (-C),k}(y^0 - y) + \varphi_{A,k}(y) &&\\
&= -\varphi_{-C,-k}(y^0 - y) + \varphi_{A,k}(y) && \text{because of Proposition 4.3.10}\\
&= -\varphi_{C,k}(y - y^0) + \varphi_{A,k}(y) && \text{because of Proposition 4.3.9.}
\end{aligned}$$

By (4.2.5), $0 \in \mathrm{bd}_k(C) = \mathrm{bd}_k(C - \mathbb{R}_+ k)$ yields $\varphi_{C,k}(0) = 0$ and thus $\varphi_{B(y^0),k}(y^0) = \varphi_{A,k}(y^0)$. Since $C \subseteq 0^+A$, Corollary 4.3.14 implies $\varphi_{0^+A,k}(y^1) \leq \varphi_{C,k}(y^1)$ for each $y^1 \in Y$. Hence, $\varphi_{B(y),k}(y^0) \leq -\varphi_{0^+A,k}(y - y^0) + \varphi_{A,k}(y)$. Then $\varphi_{B(y),k}(y^0) \leq \varphi_{A,k}(y^0)$ because of Theorem 4.2.9. This yields the first equation in (4.12.2) and the related statement about a maximizer.

We have $B(y) = (Y \setminus (-C)) + y^2$ with $y^2 := y - \varphi_{A,k}(y) \cdot k$. Theorem 4.2.5(b) implies $y \in \mathrm{bd}_k(A) + \varphi_{A,k}(y) \cdot k$ and hence $y^2 \in \mathrm{bd}_k(A)$. Conversely, we get for each $y^3 \in \mathrm{bd}_k(A)$ that $\varphi_{A,k}(y^3) = 0$ and thus $B(y^3) = y^3 + (Y \setminus (-C))$ holds. This yields the second equation in (4.12.2). The last equation in (4.12.2) results from Proposition 4.3.5.

We obtain $0^+(B(y)) = 0^+(Y \setminus (-C)) = 0^+C$ from Lemma 2.2.21(g). Thus, $k \in -0^+(B(y)) = -0^+(Y \setminus (-C))$. If $k \in -\operatorname{core} 0^+C$, then $k \in -\operatorname{core} 0^+(B(y)) = -\operatorname{core} 0^+(Y \setminus (-C))$, and (a) holds by Theorem 4.2.9.

If $C$ is convex, then $\operatorname{cl}_{-k}(Y \setminus B(y)) = \operatorname{cl}_{-k}(-C + y - \varphi_{A,k}(y) \cdot k)$, $\operatorname{cl}_{-k}(Y \setminus (Y \setminus (-C))) = \operatorname{cl}_{-k}(-C)$ and $\operatorname{cl}_{-k}(Y \setminus (y + (Y \setminus (-C)))) = \operatorname{cl}_{-k}(-C + y)$ are convex sets because of Proposition 2.3.30. Corollary 4.2.11 implies (b).

We get $-\varphi_{Y\setminus(-C),k} = \varphi_{-C,-k}$ from Proposition 4.3.10. By Proposition 4.7.1, $\varphi_{-C,-k}$ is sublinear if $C$ is a convex cone.

Assume now that $Y$ is a topological vector space, $C$ is a convex cone and $k \in -\operatorname{int} C$. Lemma 2.2.7(b) yields $\operatorname{cl}(Y \setminus (-C)) - \mathbb{R}_{>}k \subseteq \operatorname{int}(Y \setminus (-C))$. Hence, $\operatorname{cl} B(y) - \mathbb{R}_{>}k \subseteq \operatorname{int} B(y)$. (c) follows from Theorem 4.6.6. □

**Corollary 4.12.4** *Assume that $A$ is a proper subset of the linear space $Y$ with $0 \in \operatorname{bd}_k(0^+A)$ and $k \in -\operatorname{core} 0^+A$.*
*Define $B(y) := (Y \setminus (-0^+A)) + y - \varphi_{A,k}(y) \cdot k$ for every $y \in Y$.*
*For each $y^0 \in Y$,*

$$\begin{aligned}\varphi_{A,k}(y^0) &= \max_{y\in Y} \varphi_{B(y),k}(y^0)\\ &= \max_{y\in \operatorname{bd}_k(A)} \varphi_{y+(Y\setminus(-0^+A)),k}(y^0) = \max_{y\in \operatorname{bd}_k(A)} \varphi_{Y\setminus(-0^+A),k}(y^0 - y).\end{aligned}$$

*The function $\varphi_{Y\setminus(-0^+A),k}$ as well as each of the functions $\varphi_{B(y),k}$ and $\varphi_{y+(Y\setminus(-0^+A)),k}$ with $y \in Y$ has the following properties:*

(a) *It is finite-valued and concave.*
(b) *It is continuous if $Y$ is a topological vector space and $k \in -\operatorname{int} 0^+A$.*

*Moreover, $-\varphi_{Y\setminus(-0^+A),k}$ is a sublinear function.*

The previous statements will now be illustrated by a special case.

**Corollary 4.12.5** *Assume that $A$ is a proper subset of $Y = \mathbb{R}^\ell$ with $A - \mathbb{R}^\ell_+ \subseteq A$ and $k := \mathbf{1} \in Y$. For each $y^0 \in Y$, we have*

$$\varphi_{A,k}(y^0) = \max_{y\in Y}(\varphi_{A,k}(y) + \min_{i=1,\dots,\ell}(y_i^0 - y_i)) \tag{4.12.3}$$

$$= \min_{y\in Y}(\varphi_{A,k}(y) + \max_{i=1,\dots,\ell}(y_i^0 - y_i)) \tag{4.12.4}$$

*Moreover,*

$$\forall y^0 \in Y:\ \varphi_{A,k}(y^0) = \max_{y\in \operatorname{bd} A}\ \min_{i=1,\dots,\ell}(y_i^0 - y_i) = \min_{y\in \operatorname{bd} A}\ \max_{i=1,\dots,\ell}(y_i^0 - y_i). \tag{4.12.5}$$

***Proof*** We have $C := -\mathbb{R}^\ell_+ \subseteq 0^+A$ and $k \in -\operatorname{int} 0^+A$ since $A - \mathbb{R}^\ell_+ \subseteq A$. By Theorem 4.2.9, $\varphi_{A,k}$ is finite-valued. Take any $y^0 \in Y$.

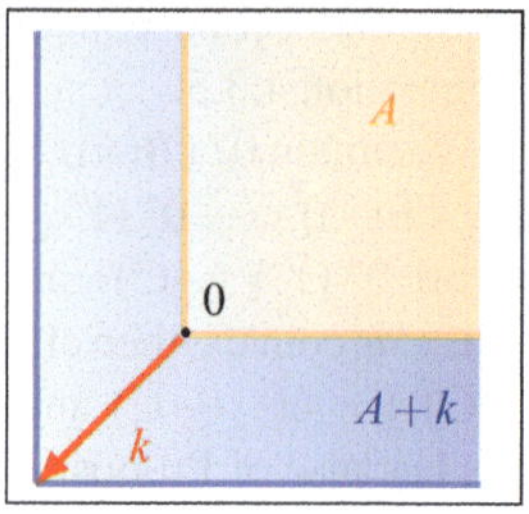


**Fig. 4.13** The sets $A$ and $A+k$ in Proposition 4.13.1

By Theorem 4.12.3, $\varphi_{A,k}(y^0) = \max_{y\in Y} \varphi_{B(y),k}(y^0)$, where $B(y) := Y \setminus (\mathbb{R}^\ell_+ + y - \varphi_{A,k}(y)\cdot k)$ for each $y \in Y$. Take any $y \in Y$. Because of Proposition 4.3.11, $\varphi_{B(y),k}(y^0) = \varphi_{\tilde{B}(y),k}(y^0)$ with $\tilde{B}(y) := Y \setminus (\operatorname{int}\mathbb{R}^\ell_+ + y - \varphi_{A,k}(y)\cdot k)$.
By Corollary 4.8.16, $\varphi_{\tilde{B}(y),k}(y^0) = \min_{i=1,\dots,\ell} (y_i^0 - (y_i - \varphi_{A,k}(y)) = \varphi_{A,k}(y) + \min_{i=1,\dots,\ell}(y_i^0 - y_i)$. This implies (4.12.3).

By Theorem 4.12.1, $\varphi_{A,k}(y^0) = \min_{y\in Y} \varphi_{D(y),k}(y^0)$, where $D(y) := -\mathbb{R}^\ell_+ + y - \varphi_{A,k}(y)\cdot k$ for every $y \in Y$. For each $y \in Y$, $\varphi_{D(y),k}(y^0) = \varphi_{A,k}(y) + \max_{i=1,\dots,\ell}(y_i^0 - y_i)$ by Corollary 4.8.6. This implies (4.12.4).

Proposition 2.2.24(a) yields $\operatorname{cl} A - \mathbb{R}_{>}k \subseteq \operatorname{int} A$. We get $\operatorname{bd}_k(A) = \operatorname{bd} A$ from Proposition 2.3.55(d).

By Theorem 4.12.3, $\varphi_{A,k}(y^0) = \max_{y\in \operatorname{bd}_k(A)} \varphi_{Y\setminus(-C),k}(y^0 - y)$. Take any $y \in Y$. Because of Proposition 4.3.11, $\varphi_{Y\setminus(-C),k}(y^0 - y) = \varphi_{Y\setminus(-\operatorname{int} C),k}(y^0 - y)$. We get $\varphi_{Y\setminus(-\operatorname{int} C),k}(y^0 - y) = \min_{i=1,\dots,\ell}(y_i^0 - y_i)$ from Corollary 4.8.16. This implies the first equation of (4.12.5).

By Theorem 4.12.1, $\varphi_{A,k}(y^0) = \min_{y\in \operatorname{bd}_k(A)} \varphi_{C,k}(y^0 - y)$. For each $y \in Y$, $\varphi_{C,k}(y^0 - y) = \max_{i=1,\dots,\ell}(y_i^0 - y_i)$ by Corollary 4.8.6. This implies the second equation of (4.12.5). □

**Remark 4.12.6** A statement which is equivalent to (4.12.3) was proved by Rubinov and Singer [397, Theorem 5.2].

## 4.13 Relationship to Minkowski Functionals and Norms

Translation invariant functions that are generated by cones often coincide with a Minkowski functional on a subset of the space. In the following, $p_A$ denotes the Minkowski functional of the set $A$.

The sets of the next proposition are illustrated in Fig. 4.13.

**Proposition 4.13.1** *Assume that $A$ is a nontrivial convex cone in a linear space $Y$ with $k \in -\operatorname{core} A$. For the Minkowski functional $p_{A+k}$, we get*

$$p_{A+k}(y) = \begin{cases} \varphi_{A,k}(y) & \text{if } y \in Y \setminus \operatorname{cl}_k(A), \\ 0 & \text{if } y \in \operatorname{cl}_k(A), \end{cases}$$

*i.e.,*

$$p_{A+k}(y) = \max\{\varphi_{A,k}(y), 0\} \quad \textit{for all } y \in Y.$$

$p_{A+k}$ *is finite-valued and sublinear. It is* $(-A)$*-monotone and strictly* $(-\operatorname{core} A)$*-monotone on* $Y \setminus \operatorname{cl}_k(A)$.

*If* $Y$ *is a topological vector space,* $A$ *is closed and* $k \in -\operatorname{int} A$*, then* $p_{A+k}$ *is continuous.*

***Proof*** By Corollary 4.7.2, $\varphi_{A,k}$ is finite-valued, $(-A)$-monotone and strictly $(-\operatorname{core} A)$-monotone.

For each $y \in Y$, we have $p_{A+k}(y) = \inf\{\lambda > 0 \mid y \in \lambda(A+k)\}$ and therefore $p_{A+k}(y) = \inf\{\lambda > 0 \mid y \in A + \lambda k\}$. Hence, $p_{A+k}(y) = \varphi_{A,k}(y)$ if $\varphi_{A,k}(y) > 0$. By (4.7.2), $\varphi_{A,k}(y) > 0$ holds if and only if $y \in Y \setminus \operatorname{cl}_k(A)$.

Take any $y \in \operatorname{cl}_k(A)$. Proposition 2.3.29 implies $y - \mathbb{R}_{>}k \subseteq A$ and thus $y \in A + \lambda k$ for each $\lambda \in \mathbb{R}_{>}$. We get $p_{A+k}(y) = 0$. Therefore, $p_{A+k}(y) = \max\{\varphi_{A,k}(y), 0\}$ for all $y \in Y$. Since $A + k$ is convex and absorbing, $p_{A+k}$ is sublinear by Lemma 2.1.31(a).

Assume now that $Y$ is a topological vector space, that $A$ is closed and $k \in -\operatorname{int} A$. $\varphi_{A,k}$ is continuous because of Proposition 4.7.3(d). Since $A = \operatorname{cl}_k(A)$ by Proposition 2.3.54(a), $Y \setminus \operatorname{cl}_k(A)$ is open and $\operatorname{bd} \operatorname{cl}_k(A) = \operatorname{bd} A$. By (4.6.14) in Theorem 4.6.6, $\operatorname{bd} A = \operatorname{lev}_{\varphi_{A,k},=}(0)$. This implies the continuity of $p_{A+k}$. □

Nam and Zălinescu [366] proved that, for all vectors $y$ of a normed space $Y$,

$$\max\{\varphi_{A,k}(y), 0\} = \inf\{t \geq 0 \mid y \in A + tk\} \tag{4.13.1}$$

if $A$ is a nonempty closed subset of $Y$ and $k \in -0^+A$. They studied the functional defined by the right-hand side of (4.13.1). For more details related to this functional, see Sect. 9.2.

Furthermore, the following statement was proved in [347, Theorem 1]:
If $A \subseteq \mathbb{R}^\ell$ is a closed set with $A - \mathbb{R}^\ell_+ \subseteq A$, then

$$\max\{\varphi_{A,k}(y), 0\} = \inf_{a \in A} \|y - a\|_\infty \quad \text{holds for } k = \mathbf{1} \text{ and all } y \in \mathbb{R}^\ell.$$

We obtain immediately from Lemma 2.1.29 and Corollary 4.7.2:

**Proposition 4.13.2** *Suppose that* $Y$ *is a linear space and* $A \subset Y$ *a nontrivial convex cone with* $k \in \operatorname{core} A$*. Then the function* $q\colon Y \to \overline{\mathbb{R}}$ *defined by*

$$q(y) := \max\{\varphi_{A,-k}(y), \varphi_{A,-k}(-y)\} \quad \textit{for each } y \in Y$$

*is a seminorm on* $Y$.

If $A$ is an ordering cone, then the function $q$ in Proposition 4.13.2 is the seminorm $p_{A,k}$ which we investigated in Theorem 2.2.10. This seminorm is the order unit norm if $A$ is an algebraically closed ordering cone. We turn to this norm in the next proposition.

**Proposition 4.13.3** *Suppose that $Y$ is a linear space and $A \subset Y$ is a nontrivial algebraically closed pointed convex cone with $k \in$ core $A$. Denote by $\|\cdot\|_{A,k}$ the norm which is given as the Minkowski functional of the order interval $[-k,k]_A$ (see Theorem 2.2.10). Then*

$$\|y\|_{A,k} = \max\{\varphi_{A,-k}(y), \varphi_{A,-k}(-y)\} \text{ for all } y \in Y. \tag{4.13.2}$$

*Hence,*

$$\|y\|_{A,k} = \begin{cases} \varphi_{-A,k}(y) & \text{if } y \in A, \\ \varphi_{A,-k}(y) & \text{if } y \in -A. \end{cases} \tag{4.13.3}$$

*On $A$, $\|\cdot\|_{A,k}$ is $A$-monotone and strictly (core $A$)-monotone.*
*On $-A$, $\|\cdot\|_{A,k}$ is $(-A)$-monotone and strictly $(-$ core $A)$-monotone.*

***Proof*** By Proposition 2.3.35, $\mathrm{cl}_{-k}(A) = A$ and $\mathrm{cl}_k(-A) = -A$.
$\varphi_{-A,k}(y) = \varphi_{A,-k}(-y)$ because of Proposition 4.3.9. Take any $y \in Y$.

$$\begin{aligned} \|y\|_{A,k} &= \inf\{\lambda > 0 \mid y \in \lambda((A-k) \cap (k-A))\} \\ &= \inf\{\lambda > 0 \mid y \in (A - \lambda k) \cap (\lambda k - A)\}. \end{aligned}$$

Thus, $\|y\|_{A,k} \geq \alpha := \max\{\varphi_{A,-k}(y), \varphi_{-A,k}(y)\}$. From now on, we apply the equations in Corollary 4.7.2 for $t = 0$.

Suppose $\alpha \leq 0$. Then $\varphi_{A,-k}(y) \leq 0$ and $\varphi_{-A,k}(y) \leq 0$ imply $y \in \mathrm{cl}_{-k}(A) = A$ and $y \in \mathrm{cl}_k(-A) = -A$. Thus, $y \in A \cap (-A) = \{0\}$. Consequently, $\alpha > 0$ if $y \in Y \setminus \{0\}$. Furthermore, $0 \in A \setminus (\text{core } A)$ implies $\varphi_{A,-k}(0) = 0$ and $\varphi_{-A,k}(0) = 0$. Hence, $\alpha = 0$ if $y = 0$.

Assume, without loss of generality, $\varphi_{-A,k}(y) \geq \varphi_{A,-k}(y)$. Then $\alpha = \varphi_{-A,k}(y)$, and $y \in -A + \alpha k$.

Since $a - \varphi_{A,-k}(y)k = a + (\alpha - \varphi_{A,-k}(y))k - \alpha k$ for all $a \in A$, we get $A - \varphi_{A,-k}(y)k \subseteq A - \alpha k$. Therefore, $y \in (A - \alpha k) \cap (\alpha k - A)$. This yields $\|y\|_{A,k} \leq \alpha$. Thus, (4.13.2) holds. Equation (4.13.3) and the monotonicity result from Corollary 4.7.2. □

**Example 4.13.4** Consider the space $Y = C(T)$ of continuous functions on a compact space $T$, its ordering cone $A = \{f \in C(T) \mid \forall x \in T\colon f(x) \geq 0\}$ and the constant function $k = \mathbf{1}$. Then $\varphi_{-A,k}(f) = \max_{x \in T} f(x)$ (see Example 4.1.6) and $\|f\|_{A,k} = \max_{x \in T} |f(x)|$ for all $f \in C(T)$.

**Proposition 4.13.5** *Assume that $Y$ is a linear space, $A \subset Y$ is an algebraically closed pointed convex cone with $k \in$ core $A$. Define*

$$\mu_{A,k}(y) := \max\{\varphi_{A,-k}(y), 0\} + \max\{\varphi_{A,-k}(-y), 0\} \quad \text{for all } y \in Y. \tag{4.13.4}$$

*Then for all $y \in Y$,*

$$\mu_{A,k}(y) = \max\{\varphi_{A,-k}(y), 0\} + \max\{\varphi_{-A,k}(y), 0\} \tag{4.13.5}$$
$$= p_{A-k}(y) + p_{k-A}(y). \tag{4.13.6}$$

*Moreover,*

$$\mu_{A,k}(y) = \mu_{-A,-k}(y) \quad \textit{for all } y \in Y, \tag{4.13.7}$$

$$\mu_{A,k}(y) = \begin{cases} \varphi_{-A,k}(y) & \textit{if } y \in A, \\ \varphi_{A,-k}(y) & \textit{if } y \in -A, \\ \varphi_{A,-k}(y) + \varphi_{-A,k}(y) & \textit{otherwise}, \end{cases} \tag{4.13.8}$$

*i.e.,*

$$\mu_{A,k}(y) = \|y\|_{A,k} \quad \textit{for all } y \in A \cup (-A).$$

*We have:*

(a) $\mu_{A,k}$ *is a norm on Y which is equivalent to* $\|\cdot\|_{A,k}$.
(b) $\{y \in Y \mid \mu_{A,k}(y) \leq 1\} \subseteq (A - k) \cap (k - A)$.
(c) $\{y \in Y \mid \mu_{A,k}(y) < 1\} \subseteq (\text{core } A - k) \cap (k - \text{core } A)$.
(d) $\mu_{A,k}$ *is A-monotone on A.*
(e) $\mu_{A,k}$ *is strictly* (core $A$)*-monotone on A.*
(f) *If Y is a normed space, A has a closed bounded base and* $k \in \text{int } A$, *then* $\mu_{A,k}$ *is a norm which is equivalent to the norm given on Y.*

***Proof*** Equations (4.13.5) and (4.13.6) result from Lemma 4.3.9 and Proposition 4.13.1. Equation (4.13.5) yields (4.13.7). From now on, we apply Corollary 4.7.2, where the contained equations are used for $t = 0$. Since $A$ is pointed, (4.13.5) implies (4.13.8) by the first equation in Corollary 4.7.2.

(a) $\varphi_{A,-k}$ and $\varphi_{-A,k}$ are finite-valued. Hence, $\mu_{A,k}$ is finite-valued.
Since $A$ is pointed, we get $\mu_{A,k}(y) \geq 0$ for all $y \in Y$ and that $\mu_{A,k}(y) = 0$ implies $y = 0$. The functions $\varphi_{A,-k}$ and $\varphi_{-A,k}$ are sublinear. This yields $\mu_{A,k}(\lambda y) = \lambda \mu_{A,k}(y)$ for all $\lambda \in \mathbb{R}_+$, $y \in Y$. Since $\mu_{A,k}(-y) = \mu_{A,k}(y)$ for each $y \in Y$, we obtain $\mu_{A,k}(\lambda y) = |\lambda| \mu_{A,k}(y)$ for all $\lambda \in \mathbb{R}$, $y \in Y$. The sublinearity of $\varphi_{A,-k}$ and $\varphi_{-A,k}$ yields $\mu_{A,k}(y^1 + y^2) \leq \mu_{A,k}(y^1) + \mu_{A,k}(y^2)$ for all $y^1, y^2 \in Y$. Hence, $\mu_{A,k}$ is a norm on $Y$.
For all $y \in Y$, $\|y\|_{A,k} \leq \mu_{A,k} \leq 2\|y\|_{A,k}$ by (4.13.2), (4.13.3) and (4.13.8).
(b) –(e) follow from Corollary 4.7.2.
(f) results from (a) and Theorem 2.2.10(g). □

Under the assumptions of Proposition 4.13.5, the function $g_{A,k}$ defined by

$$g_{A,k}(y) := \varphi_{A,-k}(y) + \varphi_{-A,k}(y) \quad \text{for all } y \in Y$$

is a seminorm. This follows from Lemma 2.1.29 since $g_{A,k}(y) = \varphi_{A,-k}(y) + \varphi_{A,-k}(-y)$ by Proposition 4.3.9 and $\varphi_{A,-k}$ is finite-valued and sublinear by Corollary 4.7.2. However, $g_{A,k}$ is not a norm since $g_{A,k}(tk) = 0$ for all $t \in \mathbb{R}$.

The relationship between functions $\varphi_{A,k}$ and block norms on the Euclidean space will be investigated in Sect. 9.1. Since $\varphi_{A,k}$ can be considered to be an oriented distance to $\mathrm{bd}_k(A - \mathbb{R}_+k)$, connections to different distance functions exist. These will be pointed out for the directional minimal time function and for the Hiriart-Urruty oriented distance in Chap. 9. Both types of distance functions can also be used for distances to sets which are not necessarily convex or cones. For this case, relationships between the Hiriart-Urruty oriented distance and functions $\varphi_{A,k}$ were proved in [325].

**Remark 4.13.6** (*Bibliographical notes related to Sect.* 4.13)
The statements in the first two sentences of Proposition 4.13.1 were proved in [483, Prop. 5.2] for cones which are additionally algebraically closed.

Equation (4.13.2) and (4.13.3) were proved in [487, Prop. 7.4]. Rubinov [395, p. 140] gave Example 4.13.4 and stated (4.13.2) under the additional assumptions that $Y$ is a locally convex space and $k \in \mathrm{int}\, A$.

Proposition 4.13.5 was proved without (b)–(f) in [487, Prop. 7.6]. The definition of $\mu_{A,k}$ in (4.13.4) was introduced by Dauer and Saleh [109, p. 232] for a pointed convex cone $A$ with $k \in \mathrm{int}\, A$ in a topological vector space $Y$. Under these assumptions, they proved that $\mu_{A,k}$ is a continuous seminorm, which is equivalent to $\|\cdot\|_{A,k}$. Furthermore, they proved—without a closedness assumption for $A$—that in Proposition 4.13.5 (b)–(d), (f) and a statement that is contained in (e) hold.

## 4.14 Translation Invariant Extensions of Arbitrary Functionals

In this section, each extended real-valued function will be shown to be the restriction of some translation invariant function to a hyperspace.

The following definition is illustrated by Fig. 4.14.

**Definition 4.14.1** Assume that $X$ is a linear space and $f: X \to \overline{\mathbb{R}}$.
The function $\varphi_{A,k}: X \times \mathbb{R} \to \overline{\mathbb{R}}$ with $A := \mathrm{epi}\, f$ and $k := (0_X, -1) \in X \times \mathbb{R}$ is called the **epigraphical extension** $f_{\mathrm{epi}}$ of $f$, i.e.,

$$f_{\mathrm{epi}}(x, s) := \inf\{t \in \mathbb{R} \mid (x, s) \in \mathrm{epi}\, f + t(0_X, -1)\} \quad \text{for all } (x, s) \in X \times \mathbb{R}.$$

**Proposition 4.14.2** *Assume that $X$ is a linear space and $f: X \to \overline{\mathbb{R}}$.*
*Then* $\mathrm{dom}\, f_{\mathrm{epi}} = \mathrm{dom}\, f \times \mathbb{R}$, *and*

$$f_{\mathrm{epi}}(x, s) = f(x) - s \quad \textit{for all } (x, s) \in X \times \mathbb{R}. \tag{4.14.1}$$

***Proof*** Take any $(x, s) \in X \times \mathbb{R}$.

$$\begin{aligned} f_{\rm epi}(x,s) &= \inf\{t\in\mathbb{R}\mid (x,s)\in \operatorname{epi} f + t(0_X,-1)\}\\ &= \inf\{t\in\mathbb{R}\mid (x,s+t)\in \operatorname{epi} f\}\\ &= \inf\{t\in\mathbb{R}\mid f(x)\le s+t\}\\ &= \inf\{t\in\mathbb{R}\mid f(x)-s\le t\}\\ &= f(x)-s. \end{aligned}$$

□

Proposition 4.14.2 implies:

**Theorem 4.14.3** *Each extended real-valued function $f$ on a linear space $X$ is the restriction of some translation invariant function to a hyperspace.*

*In detail,* $\operatorname{dom} f = \{x\in X\mid (x,0)\in \operatorname{dom} f_{\rm epi}\}$ *and*

$$f(x) = f_{\rm epi}(x,0) \quad \textit{for each } x\in X. \tag{4.14.2}$$

Conversely, we have:

**Lemma 4.14.4** *Assume that $X$ is a linear space, $k := (0_X,-1)\in X\times\mathbb{R}$ and that $\tilde{A}$ is a subset of $X\times\mathbb{R}$ which is $k$-directionally closed and for which $k\in -0^+\tilde{A}$ holds. Define $f\colon X\to\overline{\mathbb{R}}$ by*

$$f(x) := \varphi_{\tilde{A},k}(x,0) \quad \textit{for each } x\in X.$$

*Then* $\tilde{A} = \operatorname{epi} f$, $\varphi_{\tilde{A},k} = f_{\rm epi}$ *and* $\operatorname{dom} f = \{x\in X\mid (x,0)\in \operatorname{dom}\varphi_{\tilde{A},k}\}$.

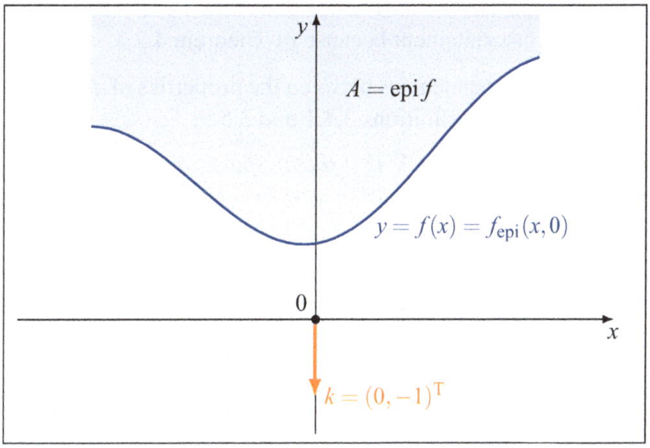


**Fig. 4.14** The epigraphical extension $f_{\rm epi} = \varphi_{A,k}$ of $f$

***Proof*** We get

$$\begin{aligned}\operatorname{epi} f &= \{(x,t) \in X \times \mathbb{R} \mid f(x) \le t\} = \{(x,t) \in X \times \mathbb{R} \mid \varphi_{\tilde{A},k}(x,0) \le t\}\\ &= \{(x,t) \in X \times \mathbb{R} \mid (x,0) \in \tilde{A} + tk\} \qquad \text{because of (4.2.12)}\\ &= \{(x,t) \in X \times \mathbb{R} \mid (x,0) \in \tilde{A} + t\cdot(0_X,-1)\} = \{(x,t) \in X \times \mathbb{R} \mid (x,t) \in \tilde{A}\} = \tilde{A}.\end{aligned}$$

Because of Definition 4.14.1, $f_{\text{epi}} = \varphi_{\tilde{A},k}$. This implies the last assertion of the lemma by Theorem 4.14.3. □

The set of epigraphs and the class of epigraphical extensions can be completely characterized.

**Theorem 4.14.5** *Assume that $X$ is a linear space. Consider $k := (0_X, -1) \in X \times \mathbb{R}$.*

*The set of all epigraphs related to extended real-valued functions on $X$ is*

$$\{A \subseteq X \times \mathbb{R} \mid A \text{ is } k\text{-directionally closed}, k \in -0^+A\}.$$

*The class of all epigraphical extensions related to extended real-valued functions on $X$ is the set*

$$\{\varphi_{A,k} \colon X \times \mathbb{R} \to \overline{\mathbb{R}} \mid A \subseteq X \times \mathbb{R} \text{ is } k\text{-directionally closed}, k \in -0^+A\}.$$

*This class coincides with the set*

$$\{\varphi_{A,k} \colon X \times \mathbb{R} \to \overline{\mathbb{R}} \mid A \subseteq X \times \mathbb{R}\}.$$

***Proof*** The first two statements follow from Lemmata 4.14.4 and 3.2.9. The second statement implies the last statement because of Theorem 4.2.3. □

We will now study interdependencies between the properties of $f$ and $f_{\text{epi}}$. Theorem 4.2.10 yields together with Definitions 3.4.1 and 3.5.9:

**Proposition 4.14.6** *Assume that $X$ is a linear space and that $f \colon X \to \overline{\mathbb{R}}$ is not a constant function with the function value $+\infty$ or $-\infty$. Then $f$ is convex, positively homogeneous, subadditive or sublinear if and only if $f_{\text{epi}}$ has this property.*

Under the assumptions of Proposition 4.14.6, $f_{\text{epi}}$ is quasiconvex if and only if it is convex. This was proved in Theorem 4.2.10. However, it is well known that $f$ can be quasiconvex without being convex.

Furthermore, we have:

**Proposition 4.14.7** *Assume that $X$ is a linear space and $f \colon X \to \overline{\mathbb{R}}$.*

*If $f_{\text{epi}}$ is strictly quasiconvex, then $f$ is strictly quasiconvex.*

***Proof*** Take any $x^1, x^2 \in X$ with $x^1 \neq x^2$ and $\lambda \in (0,1)$. By Theorem 4.14.3,

$$\begin{aligned}f(\lambda x^1 + (1-\lambda)x^2) &= f_{\text{epi}}(\lambda x^1 + (1-\lambda)x^2, 0)\\ &< \sup(f_{\text{epi}}(x^1,0), f_{\text{epi}}(x^2,0)) = \sup(f(x^1), f(x^2)).\end{aligned}$$ □

Note that $f_{\text{epi}}$ cannot be strictly convex since it is affine on each line in direction $(0_X, 1)$.

**Proposition 4.14.8** *Assume $X$ is a linear space, $F, B \subseteq X$ with $F \neq \emptyset$ and $f: X \to \overline{\mathbb{R}}$.*

(1) *$f$ is $B$-monotone on $F$ if and only if $f_{\text{epi}}$ is $(B \times (-\mathbb{R}_+))$-monotone on $F \times \mathbb{R}$.*
(2) *$f_{\text{epi}}$ is strictly $(B \times (-\mathbb{R}_+))$-monotone on $F \times \mathbb{R}$ if and only if*
 (a) *$f$ is strictly $B$-monotone on $F$, and*
 (b) *$f$ is finite-valued on $F$ or $0_X \notin B$.*

***Proof***

(1) Assume first that $f$ is $B$-monotone on $F$. Take any $(x^1, s_1), (x^2, s_2) \in F \times \mathbb{R}$ with $(x^2, s_2) - (x^1, s_1) \in B \times (-\mathbb{R}_+)$. Then $f(x^1) \leq f(x^2)$ since $f$ is $B$-monotone on $F$, and $s_1 \geq s_2$. Hence, $f_{\text{epi}}(x^1, s_1) \leq f_{\text{epi}}(x^2, s_2)$ by (4.14.1). Thus, $f_{\text{epi}}$ is $(B \times (-\mathbb{R}_+))$-monotone on $F \times \mathbb{R}$.
Assume now that $f_{\text{epi}}$ is $(B \times (-\mathbb{R}_+))$-monotone on $F \times \mathbb{R}$. Take any $x^1, x^2 \in F$ with $x^2 - x^1 \in B$. Then $f_{\text{epi}}(x^1, 0) \leq f_{\text{epi}}(x^2, 0)$ by the monotonicity. Hence, $f(x^1) \leq f(x^2)$ by (4.14.2). Thus, $f$ is $B$-monotone on $F$.
(2) (i) Assume first that $f_{\text{epi}}$ is strictly $(B \times (-\mathbb{R}_+))$-monotone on $F \times \mathbb{R}$. Take any $x^1, x^2 \in F$ with $x^2 - x^1 \in B \setminus \{0\}$. Then $f_{\text{epi}}(x^1, 0) < f_{\text{epi}}(x^2, 0)$ by the assumed monotonicity. Hence, $f(x^1) < f(x^2)$ by (4.14.2). Thus, $f$ is strictly $B$-monotone on $F$.
Suppose $0 \in B$. Take any $x \in F$. Then $(x, 0) - (x, 1) \in (B \times (-\mathbb{R}_+)) \setminus \{(0_X, 0)\}$ implies $f_{\text{epi}}(x, 1) < f_{\text{epi}}(x, 0)$. By (4.14.2), $f(x) - 1 < f(x) - 0$. Hence, $f(x) \in \mathbb{R}$. Thus, $f$ is finite-valued on $F$.
 (ii) Assume now that (a) and (b) hold. Take any $(x^1, s_1), (x^2, s_2) \in F \times \mathbb{R}$ with $(x^2, s_2) - (x^1, s_1) \in (B \times (-\mathbb{R}_+)) \setminus \{(0_X, 0)\}$. If $x^1 = x^2$, then $s_1 > s_2$. Otherwise, $f(x^1) < f(x^2)$ since $f$ is strictly $B$-monotone on $F$, and $s_1 \geq s_2$. If $f$ is finite-valued on $F$, then we get $f(x^1) - s_1 < f(x^2) - s_2$. If $0_X \notin B$, then $x^1 \neq x^2$, and we obtain $f(x^1) - s_1 < f(x^2) - s_2$. This implies $f_{\text{epi}}(x^1, s_1) < f_{\text{epi}}(x^2, s_2)$ by (4.14.1). Thus, $f_{\text{epi}}$ is strictly $(B \times (-\mathbb{R}_+))$-monotone on $F \times \mathbb{R}$. □

**Proposition 4.14.9** *Assume that $X$ is a topological vector space and $f: X \to \overline{\mathbb{R}}$.*

(a) *$f$ is lower semicontinuous if and only if $f_{\text{epi}}$ is lower semicontinuous.*
(b) *$f$ is continuous if and only if $f_{\text{epi}}$ is continuous.*

***Proof*** Consider $A := \text{epi}\, f$ and $k := (0_X, -1) \in X \times \mathbb{R}$. Then $f_{\text{epi}} = \varphi_{A,k}$.

(a) Apply Proposition 3.3.12. $f$ is lower semicontinuous if and only if $A$ is closed. $\varphi_{A,k}$ is lower semicontinuous if and only if $\text{lev}_{\varphi_{A,k},\leq}(t)$ is closed for each $t \in \mathbb{R}$. By Lemma 3.2.9, $A$ is $k$-directionally closed and $k \in -0^+A \setminus \{0\}$. We obtain $\text{lev}_{\varphi_{A,k},\leq}(t) = A + tk$ for each $t \in \mathbb{R}$ because of (4.2.12). Hence, $A$ is closed if and only if $\text{lev}_{\varphi_{A,k},\leq}(t)$ is closed for each $t \in \mathbb{R}$.

(b) results from (4.14.1) by the properties of continuous functions. □

This implies a characterization of any continuous function by its epigraph.

**Theorem 4.14.10** *Assume that $X$ is a topological vector space and $f: X \to \overline{\mathbb{R}}$. $f$ is continuous if and only if* $\operatorname{epi} f$ *is closed and*

$$\operatorname{epi} f + \mathbb{R}_> \cdot (0_X, 1) \subseteq \operatorname{int}(\operatorname{epi} f). \tag{4.14.3}$$

***Proof*** Define $A := \operatorname{epi} f$ and $k := (0_X, -1) \in X \times \mathbb{R}$. Then $f_{\text{epi}} = \varphi_{A,k}$. By Proposition 4.14.9, $f$ is continuous if and only if $\varphi_{A,k}$ is continuous. Because of Lemma 3.2.9, $k \in -0^+A \setminus \{0\}$ holds. We get from Theorem 4.6.2 that $\varphi_{A,k}$ is continuous if and only if $\operatorname{cl} A - \mathbb{R}_> \cdot k \subseteq \operatorname{int} A$. Hence, $f$ is continuous if and only if

$$\operatorname{cl} A + \mathbb{R}_> \cdot (0_X, 1) \subseteq \operatorname{int} A.$$

This yields the assertion since each continuous function is lower semicontinuous and each lower semicontinuous function has a closed epigraph. □

We apply this result to positively homogeneous functions.

**Proposition 4.14.11** *Assume that $X$ is a topological vector space. Each continuous positively homogeneous function $f: X \to \overline{\mathbb{R}}$ is finite-valued or the constant function with the function value $-\infty$.*

***Proof*** Since $f$ is positively homogeneous, $A := \operatorname{epi} f$ is a nonempty cone. By Theorem 4.14.10, $A - \mathbb{R}_> \cdot k \subseteq \operatorname{int} A$ for $k := (0_X, -1)$. Hence, $k \in -\operatorname{int} A = -\operatorname{int}(0^+A)$. If $A = X$, then $f$ is the constant function with the function value $-\infty$. Otherwise, $f_{\text{epi}} = \varphi_{A,k}$ is finite-valued because of Theorem 4.2.9. Thus, $f$ is finite-valued by (4.14.1). □

A sublinear extended real-valued function which is neither finite-valued nor constant is given by $\varphi : \mathbb{R} \to \overline{\mathbb{R}}$ with

$$\varphi(x) := \begin{cases} -\infty & \text{if } x < 0, \\ 0 & \text{if } x = 0, \\ +\infty & \text{if } x > 0. \end{cases}$$

**Proposition 4.14.12** *Assume that $X$ is a normed space and $f: X \to \overline{\mathbb{R}}$.*

(a) *If $f_{\text{epi}}$ is Lipschitz continuous with Lipschitz constant $L$, then $f$ is Lipschitz continuous with Lipschitz constant $L$.*
(b) *If $f$ is Lipschitz continuous with Lipschitz constant $L$, then $f_{\text{epi}}$ is Lipschitz continuous with Lipschitz constant $\sqrt{2} \cdot \max\{L, 1\}$.*

***Proof***

(a) results immediately from Theorem 4.14.3.

(b) Consider first any $a = (a_1, a_2)^T \geq 0$ and $b := (1, 1)^T$ in the Euclidean space $\mathbb{R}^2$. Cauchy's inequality $|\langle a, b\rangle| \leq \|a\| \cdot \|b\|$ yields $a_1 + a_2 \leq \sqrt{2} \cdot \sqrt{a_1^2 + a_2^2}$.
Define $L_1 := \sqrt{2} \cdot \max\{L, 1\}$. Take any $(x^1, s_1), (x^2, s_2) \in X \times \mathbb{R}$.
For $d := |f_{\text{epi}}(x^1, s_1) - f_{\text{epi}}(x^2, s_2)|$,

$$\begin{aligned}
d &= |f(x^1) - s_1 - (f(x^2) - s_2)| \qquad \text{by (4.14.1)},\\
&\leq |f(x^1) - f(x^2)| + |s_2 - s_1|\\
&\leq L\|x^1 - x^2\|_X + |s_2 - s_1|\\
&\leq \sqrt{2} \cdot \sqrt{(L\|x^1 - x^2\|_X)^2 + |s_2 - s_1|^2}\\
&\leq \sqrt{2} \cdot \sqrt{(\max\{L, 1\}\|x^1 - x^2\|_X)^2 + (\max\{L, 1\}|s_2 - s_1|)^2}, \text{ i.e.,}\\
d &\leq L_1 \cdot \sqrt{\|x^1 - x^2\|_X^2 + |s_2 - s_1|^2} = L_1 \cdot \|(x^1, s_1) - (x^2, s_2)\|_{X\times\mathbb{R}}.
\end{aligned}$$

□

**Theorem 4.14.13** *Assume that $X$ is a Banach space and $f: X \to \overline{\mathbb{R}}$.*
*$f$ is Lipschitz continuous if and only if* epi $f$ *is closed,* epi $f \neq X \times \mathbb{R}$ *and*

$$(0_X, 1) \in \operatorname{int} 0^+(\operatorname{epi} f). \tag{4.14.4}$$

***Proof*** Consider $Y := X \times \mathbb{R}$, $A := \operatorname{epi} f$ and $k := (0_X, -1) \in X \times \mathbb{R}$.

Assume first that $f$ is Lipschitz continuous. Then $f$ is lower semicontinuous, and hence epi $f$ is closed. $f_{\text{epi}} = \varphi_{A,k}$ is Lipschitz continuous by Proposition 4.14.12. Theorem 4.6.14 implies $k \in -\operatorname{int} 0^+A$ and thus (4.14.4). We have epi $f \neq Y$ since $f$ is finite-valued.

Assume now that epi $f$ is closed, epi $f \neq Y$ and (4.14.4) holds. Then $\varphi_{A,k}$ is Lipschitz continuous by Theorem 4.6.14. Hence, $f$ is Lipschitz continuous because of Proposition 4.14.12. □

**Proposition 4.14.14** *Assume that $X$ is a Banach space and $f: X \to \overline{\mathbb{R}}$ is Lipschitz continuous.*

(a) *There exists a Lipschitz continuous sublinear function $g: X \to \overline{\mathbb{R}}$ such that*

$$f(x^0) = \min_{x\in X}(f(x) + g(x^0 - x)) \quad \textit{for all } x^0 \in X. \tag{4.14.5}$$

*Such a function can be given by*

$$g(x) := \varphi_{\bar{A},k}(x, 0) \quad \textit{for all } x \in X$$

*with $\bar{A} := 0^+(\operatorname{epi} f)$ and $k := (0_X, -1)$, where $\varphi_{\bar{A},k} = g_{\text{epi}}$.*

(b) *There exists a Lipschitz continuous sublinear function $\tilde{g}: X \to \overline{\mathbb{R}}$ such that*

$$f(x^0) = \max_{x\in X}(f(x) - \tilde{g}(x^0 - x)) \quad \textit{for all } x^0 \in X.$$

*Such a function can be given by*

$$\tilde{g}(x) := \varphi_{\tilde{A},k}(x,0) \;\; \textit{for all } x \in X$$

*with* $\tilde{A} := 0^+(\operatorname{epi}(-f))$ *and* $k := (0_X, -1)$, *where* $\varphi_{\tilde{A},k} = \tilde{g}_{\mathrm{epi}}$.

***Proof***

(a) By Theorem 4.14.13, $A := \operatorname{epi} f$ is a proper closed subset of $Y := X \times \mathbb{R}$ with $k \in -\operatorname{int} 0^+ A$. Since epi $f$ is closed, its recession cone $\bar{A}$ is closed. We have $k \in -\operatorname{int} 0^+ A = -\operatorname{int} 0^+(0^+ A) = -\operatorname{int} 0^+ \bar{A}$. Hence, we get $\bar{A} = \operatorname{epi} g$ from Lemma 4.14.4. $\varphi_{\bar{A},k} = g_{\mathrm{epi}}$ is Lipschitzian by Theorem 4.6.14. Thus, $g$ is Lipschitzian because of Proposition 4.14.12. By Theorem 4.2.10, $\varphi_{\bar{A},k}$ is sublinear. Hence, $g$ is sublinear by Proposition 4.14.6. Theorem 4.2.9 implies

$$\varphi_{A,k}(x^0,0) - \varphi_{A,k}(x,0) \le \varphi_{\bar{A},k}((x^0,0)-(x,0)) \;\; \text{for all } x^0, x \in X$$

and thus

$$f(x^0) - f(x) \le g(x^0 - x) \;\; \text{for all } x^0, x \in X$$

because of Theorem 4.14.3. Since $g$ is sublinear and finite-valued, $g(x^0 - x^0) = g(0) = 0$ holds. This results in (4.14.5).

(b) The function $-f$ is Lipschitzian. Apply (a) to $-f$. Then

$$(-f)(x^0) = \min_{x \in X}(-f(x) + \tilde{g}(x^0 - x)) \;\; \text{for all } x^0 \in X, \text{ i.e.,}$$
$$f(x^0) = -\min_{x \in X}(-(f(x) - \tilde{g}(x^0 - x))) \;\; \text{for all } x^0 \in X,$$

which yields the assertion. □

Note that the definition of Lipschitz continuity implies

$$f(x^0) = \min_{x \in X}(f(x) + L\|x^0 - x\|) = \max_{x \in X}(f(x) - L\|x^0 - x\|) \;\; \text{for all } x^0 \in X$$

if $f : X \to \overline{\mathbb{R}}$ is Lipschitz continuous on a Banach space $X$ with Lipschitz constant $L$.

Example 3.2.12 illustrates that the sum of two epigraphs is not necessarily an epigraph. Nevertheless, our results yield a description of the inf-convolution's epigraph. In the following statements, we will assume $m \in \mathbb{N}$ with $m \ge 2$.

**Theorem 4.14.15** *Assume that* $X$ *is a linear space and* $f_i : X \to \overline{\mathbb{R}}$ *for each* $i \in \{1, \ldots, m\}$. *Then* $\varphi_{A,k}$ *with* $A := \sum_{i=1}^m \operatorname{epi} f_i$ *and* $k := (0_X, -1) \in X \times \mathbb{R}$ *is the epigraphical extension of* $f_1 \# f_2 \# \ldots \# f_m$. *Moreover, we have*

$$\operatorname{epi}(f_1 \# f_2 \# \ldots \# f_m) = \operatorname{cl}_k\left(\sum_{i=1}^m \operatorname{epi} f_i\right). \tag{4.14.6}$$

***Proof*** Take any $x \in X$ and $\lambda \in \mathbb{R}$.

$$\begin{aligned}\varphi_{A,k}(x,\lambda) &= \inf\{t \in \mathbb{R} \mid (x,\lambda) \in t(0_X,-1) + \sum_{i=1}^{m} \operatorname{epi} f_i\}\\ &= \inf\{t \in \mathbb{R} \mid x_i \in X, t_i \in \mathbb{R} \text{ and } f_i(x_i) \le t_i \text{ for each } i \in \{1,\dots,m\},\\ &\qquad x = \sum_{i=1}^{m} x_i,\ \lambda = -t + \sum_{i=1}^{m} t_i\}.\end{aligned}$$

$\lambda = -t + \sum_{i=1}^{m} t_i$ and $f_i(x_i) \le t_i$ imply $t = -\lambda + \sum_{i=1}^{m} t_i \ge -\lambda + \sum_{i=1}^{m} f_i(x_i)$. Hence,

$$\begin{aligned}\varphi_{A,k}(x,\lambda) &= \inf\{-\lambda + \sum_{i=1}^{m} f_i(x_i) \mid x_i \in X \text{ and } f_i(x_i) < +\infty \text{ for each } i \in \{1,\dots,m\},\\ &\qquad x = \sum_{i=1}^{m} x_i\}\\ &= \inf\{-\lambda + \sum_{i=1}^{m} f_i(x_i) \mid x_i \in X \text{ for each } i \in \{1,\dots,m\},\ x = \sum_{i=1}^{m} x_i\},\end{aligned}$$

and $\varphi_{A,k}(x,0) = (f_1 \# f_2 \# \dots \# f_m)(x)$.

Since $k \in -0^+A$, we have $k \in -0^+(\operatorname{cl}_k(A))$ by Proposition 2.3.29. Theorem 4.2.1 yields $\varphi_{A,k} = \varphi_{\operatorname{cl}_k(A),k}$. Lemma 4.14.4 implies that $\varphi_{\operatorname{cl}_k(A),k}$ is the epigraphical extension of $f_1 \# f_2 \# \dots \# f_m$ and that (4.14.6) holds. □

Note that (4.14.6) offers an alternative way to define the inf-convolution without assuming inf-addition.

**Corollary 4.14.16** *Assume that $X$ is a linear space and $f_i : X \to \overline{\mathbb{R}}$ for each $i \in \{1,\dots,m\}$.*

*$f_1 \# f_2 \# \dots \# f_m$ is*

(a) *convex if and only if* $\operatorname{cl}_k\left(\sum_{i=1}^{m} \operatorname{epi} f_i\right)$ *is convex,*

(b) *positively homogeneous if and only if* $\operatorname{cl}_k\left(\sum_{i=1}^{m} \operatorname{epi} f_i\right)$ *is a nonempty cone,*

(c) *subadditive if and only if* $\operatorname{cl}_k\left(\sum_{i=1}^{m} \operatorname{epi} f_i\right) + \operatorname{cl}_k\left(\sum_{i=1}^{m} \operatorname{epi} f_i\right) \subseteq \operatorname{cl}_k\left(\sum_{i=1}^{m} \operatorname{epi} f_i\right)$,

(d) *sublinear if and only if* $\operatorname{cl}_k\left(\sum_{i=1}^{m} \operatorname{epi} f_i\right)$ *is a nonempty convex cone.*

*If $X$ is a topological vector space, then $f_1 \# f_2 \# \dots \# f_m$ is*

(e) *lower semicontinuous if and only if* $\operatorname{cl}_k\left(\sum_{i=1}^{m} \operatorname{epi} f_i\right)$ *is closed,*

(f) *continuous if and only if it is lower semicontinuous and*

$$\operatorname{cl}\left(\sum_{i=1}^{m}\operatorname{epi} f_i\right) + \mathbb{R}_> \cdot (0_X, 1) \subseteq \operatorname{int}\left(\operatorname{cl}\left(\sum_{i=1}^{m}\operatorname{epi} f_i\right)\right).$$

*If $X$ is a Banach space, then $f_1 \# f_2 \# \dots \# f_m$ is*

(g) *Lipschitz continuous if and only if it is lower semicontinuous,* $\operatorname{cl}\left(\sum_{i=1}^{m}\operatorname{epi} f_i\right) \neq Y$ *and*

$$(0_X, 1) \in \operatorname{int} 0^+\left(\operatorname{cl}\left(\sum_{i=1}^{m}\operatorname{epi} f_i\right)\right).$$

***Proof***

(a) follows from Definition 3.4.1.
(b) –(d) follow from Definition 3.5.9.
(e) results from Proposition 3.3.12.

If $f_1 \# f_2 \# \dots \# f_m$ is lower semicontinuous, then $\operatorname{cl}_k\left(\sum_{i=1}^{m}\operatorname{epi} f_i\right)$ is closed and coincides with $\operatorname{cl}\left(\sum_{i=1}^{m}\operatorname{epi} f_i\right)$ because of Proposition 2.3.54. Hence, (f) follows from Theorem 4.14.10, and (g) from Theorem 4.14.13. □

Corollary 4.14.16 implies sufficient conditions for properties of the inf-convolution.

**Corollary 4.14.17** *Assume that $X$ is a linear space and $f_i \colon X \to \overline{\mathbb{R}}$ for each $i \in \{1, \dots, m\}$. Then $f_1 \# f_2 \# \dots \# f_m$ is convex, positively homogeneous, subadditive or sublinear if each function $f_i$ with $i \in \{1, \dots, m\}$ has this property.*

***Proof*** The assertions follow from Definitions 3.4.1 and 3.5.9 by Proposition 2.3.30. □

**Remark 4.14.18** (*Bibliographical notes related to Sect.* 4.14)
The results of Sect. 4.14 are contained in [489], with the exception of Propositions 4.14.7, 4.14.8 and 4.14.14(b). The results related to the inf-convolution were given in [489] for the inf-convolution of two functions only. Proposition 4.14.8 was proved in [488, Prop. 5.7.5].

## 4.15 Bibliographical Notes

This section presents a short history related to the investigation of properties for translation invariant functions, with a focus on their representation as $\varphi_{A,k}$.

Because of the strong connection to partial orders—pointed out at the beginning of Sect. 4.5—functions of type $\varphi_{A,k}$ have been used for the construction of sublinear functionals in different fields of mathematics, mainly as a tool within proofs. In these cases, $A$ is a pointed convex cone, usually the ordering cone of the considered space, $k \in -A$, and

$$\varphi_{A,k}(y) = \inf\{t \in \mathbb{R} \mid y \geq_A tk\}.$$

Among the earliest references listed in [223] are papers by Bonsall from 1954 [59], by Krasnosel'skiĭ from 1964 [310] and by Rubinov from 1977 [395], where the functional was applied in operator theory.

For convex sets $A$ which are not necessarily cones, Gerstewitz and Iwanow [184] used the functional given by

$$\varphi(y) := \sup\{t \in \mathbb{R} \mid y \in \operatorname{cl} A + tk\}$$

in separation theorems which they applied to the definition of a properly efficient point set and the investigation of dual problems in vector optimization. This functional was based on the idea by Gerstewitz (later Gerth, now Tammer) [182, 183] to define a functional $\varphi$ for the separation of an arbitrary set from a convex set $A$ in a topological vector space $Y$ with $k \in Y$ by

$$\varphi(y) := \sup\{t \in \mathbb{R} \mid y \in \operatorname{cl}(A + tk),\ t \leq r(y)\},$$

where the values $r(y)$ depend on special assumptions on $A$ and $k$ and were not given by a formula. Gerstewitz published first approaches to this idea in [180, 181]. Iwanow and Nehse discussed the results from [180] for $Y = \mathbb{R}^\ell$ in [263], where they altered the definition of the functional into

$$\varphi(y) := \sup\{t \in \mathbb{R} \mid y \in \operatorname{cl} A + t \cdot \mathbf{1}\},$$

under the assumption that $A$ is a nonempty open convex set with $A + (\mathbb{R}^\ell_+ \setminus \{0\}) \subseteq A$. They proved that this functional turns out to be a special case of the function $h$ in the following statement by Crouzeix [105].

**Lemma 4.15.1** *Assume that* $\{S_\lambda \subseteq \mathbb{R}^\ell \mid \lambda \in \mathbb{R}\}$ *is a family of sets with* $S_\lambda = \bigcap_{\mu<\lambda} S_\mu$ *for each* $\lambda \in \mathbb{R}$*. Define*

$$h(y) := \sup\{\lambda \in \mathbb{R} \mid y \in S_\lambda\} \text{ for each } y \in \mathbb{R}^\ell.$$

*Then*

(a) $S_\lambda = \{y \in \mathbb{R}^\ell \mid h(y) \geq \lambda\}$.
(b) *If* $S_\lambda$ *is convex for each* $\lambda \in \mathbb{R}$*, then* $h$ *is quasiconcave.*
(c) *If* $S_\lambda$ *is closed for each* $\lambda \in \mathbb{R}$*, then* $h$ *is upper semicontinuous.*

Lemma 4.15.1(a) points out the strong relationship between properties of functions $\varphi_{A,k}$ and properties of their sublevel sets, which has been intensively used in proofs since the middle of the 1980s.

In the mentioned paper by Gerstewitz and Iwanow [184], $A$ is supposed to be an open convex set with $Y = \mathrm{cl}\, A + \mathbb{R}k$, $K$ is the ordering cone of the space $Y$, $k \in K \setminus \{0\}$. Then the strict $K$-monotonicity of their functional is stated under the assumption $\mathrm{cl}\, A + (K \setminus \{0\}) \subset A$ in a Hausdorff topological vector space $Y$, and the $K$-monotonicity is stated under the assumption $\mathrm{cl}\, A + K \subseteq \mathrm{cl}\, A$ in a barrelled locally convex space $Y$. Moreover, in both cases concavity, continuity and a condition which is equivalent to (4.6.12) were proved.

Zălinescu [507] considered $\varphi_{A,k}$, given by formula (4.1.11), as a function which maps into $\overline{\mathbb{R}}$ and not necessarily into $\mathbb{R}$, under the assumption that $A$ is a proper closed convex subset of a Hausdorff topological vector space with $0 \in \mathrm{bd}\, A$ and $k \in (-0^+A) \cap (-\,\mathrm{int}\, A)$.

Properties of $\varphi_{A,k}$ on topological vector spaces for not necessarily convex sets $A$ were proved in [186, 472], later followed by [129, 192, 441]. In [186], Gerth and Weidner investigated functionals $\varphi_{A,k}$ under the following alternative assumptions on an arbitrary topological vector space $Y$:

(a) $A$ is the closure of some proper open convex subset of $Y$, $k \in -0^+A$ and $Y = A + \mathbb{R}k$;
(b) $A$ is the closure of some proper subset $C$ of $Y$ for which there exists a cone $K \subset Y$ with $\mathrm{int}\, K \neq \emptyset$ and $A - \mathrm{int}\, K \subseteq \mathrm{int}\, C$, $k \in \mathrm{int}\, K$.

($\mathrm{H3}_{A,k}$) implies (b) with $C = \mathrm{int}\, A$ because of Theorem 4.6.6. Consequently, (b) is equivalent to ($\mathrm{H3}_{A,k}$) for $A \neq Y$. Under each of the assumptions (a) and (b), Gerth and Weidner proved that $\varphi_{A,k}$ is continuous and finite-valued, that the statements (4.6.12)–(4.6.14) hold, and gave sufficient conditions for the monotonicity, the strict monotonicity, and the subadditivity of the functional, which referred to the boundary of $A$. They proved the strict ($\mathrm{int}\, K$)-monotonicity of $\varphi_{A,k}$ under assumption (b), the convexity of $\varphi_{A,k}$ under assumption (a) and showed that $\varphi_{A,k}$ is also convex under assumption (b) if $A$ is convex. The case where $\varphi_{A,k}$ becomes a sublinear function, i.e., $C = K$ is a cone in (b), was also investigated. Tammer and Weidner formulated separation theorems based on these results.

In [470, 472], Weidner compared functionals on topological vector spaces which are defined by one of the conditions (4.6.12)–(4.6.14), by condition (4.6.14) with $A$ instead of $\mathrm{bd}\, A$, as $\varphi_{A,k}$, or as $\varphi_{\mathrm{cl}\, A,k}$ without restrictive assumptions. She investigated under which conditions these definitions result in well-defined functionals as well as the properties of these functionals. Under the supposition that

$$\mathrm{int}\, A = \mathrm{bd}\, A - \mathbb{R}_{>}k \text{ and } Y = \mathrm{bd}\, A + \mathbb{R}k,$$

she proved that $\varphi_{A,k}$ is finite-valued and continuous and gave necessary and sufficient conditions for the following properties: monotonicity, strict monotonicity, convexity, quasiconvexity, strict quasiconvexity, subadditivity, superadditivity, concavity and its modifications, positive-homogeneity, sublinearity, oddness, homogeneity and linearity (see also [476]). It was shown in [472] that this supposition is fulfilled if assumption (a) from [186] mentioned above or ($H3_{A,k}$) is satisfied, and that ($H3_{A,k}$) is equivalent to the conditions stated in Proposition 4.6.18, which also includes the above assumption (b) for closed sets $C$.

In [192, Theorem 2.3.1], $\varphi_{A,k}$ is considered as an extended real-valued functional. Under assumption ($H1_{A,k}$), the lower semicontinuity, (4.6.12) and necessary and sufficient conditions for $\varphi_{A,k}$ to be convex, subadditive, positively homogeneous, proper or finite-valued were proved. Under assumption ($H2_{A,k}$), the continuity of $\varphi_{A,k}$, (4.6.13), (4.6.14) and (4.6.9) were shown. Necessary and sufficient conditions for the Lipschitz continuity of functions $\varphi_{A,k}$ were proved by Tammer and Zălinescu in [441].

For sets $A$ that are not necessarily closed, functions $\varphi_{A,k}$ on linear topological spaces were studied without using direction-depending properties of sets by Flores-Bazán and Hernández [164] and by Weidner [484].

The relationship between functions of type $\varphi_{A,k}$, translation invariant functions and the sublevel sets of these functions was investigated under special assumptions in the context of risk measures by Artzner et al. [11]. Although translation invariance had previously been used in mathematical finance and statistics (see [223] for references), the paper by Artzner et al. stimulated the research in this field. Jaschke and Küchler [280] investigated risk measures which are in fact functions of type $\varphi_{A,k}$ for arbitrary sets $A$ in a linear space $Y$ and $k \in Y \setminus \{0\}$. They proved that each translation invariant function has a representation $\varphi_{A,k}$.

Rubinov and Singer [397] studied the relationship between translation invariance, the epigraph of the function and the sublevel sets of the function in $Y = \mathbb{R}^\ell$ for $k = \mathbf{1}$. Under the same assumptions, Martinez-Legaz et al. [347] introduced closedness along lines. A notion of directional closedness in general linear spaces was defined by Qiu and He [385] and used for establishing properties of functions $\varphi_{A,k}$ in [213, 385]. Independently, directional closedness was introduced for the investigation of functions $\varphi_{A,k}$ on linear spaces by Weidner [482, 483, 485].

In [219, 221, 407], Hamel and Schrage considered functionals $\varphi_{A,k}$ for arbitrary sets $A$ in a linear space $Y$ and $k \in Y \setminus \{0\}$, using closedness along lines. One of their main results is a statement that is equivalent to $\varphi_{A,k} = \varphi_{\mathrm{cl}_k(A-\mathbb{R}_+k),k}$. They generalized results from Artzner et al. [11], Rubinov and Singer [397], Martinez-Legaz et al. [347], Tammer, Weidner and Zălinescu [186, 192, 472] and from various other authors who had studied translation invariant functions as risk measures.

The variational analysis for functions $\varphi_{A,k}$ started with contributions by Rubinov [395] and Jaschke and Küchler [280] for convex cones $A$. Results for the case of arbitrary convex sets $A$ were obtained by Föllmer and Schied [168] as well as by Tammer in collaboration with Durea [129] and Zălinescu [441], for the case of nonconvex sets $A$ by Tammer, Bao and Hillmann [19, 29]. Filipović and Kupper [162] also contributed to this field and derived formulas for monotone and for translation invariant envelopes of extended real-valued functions on separated locally convex spaces. The idea to represent functions $\varphi_{A,k}$ using the inf-convolution was given in [225].

Relationships between translation invariant functions and norms were proved by Rubinov [395], Dauer and Saleh [109] and Weidner [483, 487]. In [483], certain Minkowski functionals are described by translation invariant functions on a subset of the space.

The characterization of arbitrary extended real-valued functions as restrictions of translation invariant functions and related implications for the investigation of extended real-valued functions were presented by Weidner in [489].

# Chapter 5
# Minimizers of Translation Invariant Functions

When applying a translation invariant function for the separation of sets or the scalarization of problems, then it usually has to be minimized over a given set. In Chap. 6, we will characterize solutions to vector optimization problems by minimizers of translation invariant functions $\varphi_{a-H,k}$ on a set $F$ of feasible elements. Jaschke and Küchler [280, p. 193] studied good-deal bounds for a given cash stream $a$, which turn out to be of the type $\inf_{y \in F} \varphi_{a-H,k}(y)$ and $-\inf_{y \in F} \varphi_{-a-H,k}(y)$. In this chapter, we investigate properties of the scalar optimization problem

$$\min_{y \in F} \varphi_{a-H,k}(y) \tag{5.0.1}$$

under the assumption

$$(\text{Sp0--OP}_{F,a,H,k}):\; Y \text{ is a linear space, } F, H \subseteq Y, a \in Y, \text{ and } k \in Y \setminus \{0\}.$$

Recall that $\varphi_{a-H,k} \colon Y \to \overline{\mathbb{R}}$ is given by

$$\varphi_{a-H,k}(y) = \inf\{t \in \mathbb{R} \mid y \in a - H + tk\} \quad \text{for each } y \in Y.$$

Section 5.1 connects the minimization problem (5.0.1) with a generalization of an optimization problem introduced by Pascoletti and Serafini [376] and with other optimization problems which have the same set of minimizers. Formulas for the optimal value of (5.0.1) and for the minimizer set will be deduced. Moreover, relationships to problems with an altered feasible set are proved. In Sect. 5.2, the existence of optimal solutions and properties of the solution set are studied. Section 5.3 deals with the control of the parameters $a$ and $k$ in the case where problems (5.0.1) are applied with varying parameters. Interdependencies between properties of problems (5.0.1) with different parameters are discussed in Sect. 5.4. Implications for minimizers of certain Minkowski functionals and norms are shown in Sect. 5.5.



C. Tammer and P. Weidner, *Scalarization and Separation by Translation Invariant Functions*, Vector Optimization, https://doi.org/10.1007/978-3-030-44723-6_5

The content of this chapter is taken from [487], an extension of results in [472], but Theorems 5.1.3 and 5.1.4 are proved under weaker assumptions, and part (a) of both theorems as well as Example 5.1.5 were added.

## 5.1 Problem Formulation and Related Problems

In this chapter, we investigate the optimization problems (5.0.1) and

$$
\begin{aligned}
(\mathrm{P}_{F,a,H,k}): \ & t \to \min \\
& y \in a - H + tk \\
& y \in F,\ t \in \mathbb{R}
\end{aligned}
$$

under assumption (Sp0–OP$_{F,a,H,k}$).

Since $\varphi_{a-H,k}$ is defined as an extended real-valued functional, the set of feasible vectors $y$ in both problems is $F \cap \operatorname{dom} \varphi_{a-H,k} = F \cap (a - H + \mathbb{R}k)$.

Geometrical interpretation of (P$_{F,a,H,k}$):

Stick the set $-H$ to the point $a$ and shift the set $a - H$ along the line $\{a + tk \mid t \in \mathbb{R}\}$ until the smallest $t$ is reached for which the intersection of the set with $F$ is not empty, i.e., for which $F \cap (a - H + tk) \neq \emptyset$. Then $t$ is the optimal value of (P$_{F,a,H,k}$), and the set of points in which $F$ and $a - H + tk$ intersect is the set of optimal solutions $y$ to (P$_{F,a,H,k}$).

We illustrate this by an example (see Fig. 5.1).

**Example 5.1.1** In $Y = \mathbb{R}^2$, choose $F := \{(y_1, y_2)^T \in \mathbb{R}^2 \mid 0 \leq y_2 \leq y_1 \leq 1\}$, $H := \mathbb{R}^2_+$, $a := (-1, 0)^T$, and $k := (1, 1)^T$. (P$_{F,a,H,k}$) attains an optimal value $t_{\text{opt}}$. The vector $y^{\text{opt}} := (0, 0)^T$ is the unique optimal solution to (P$_{F,a,H,k}$).

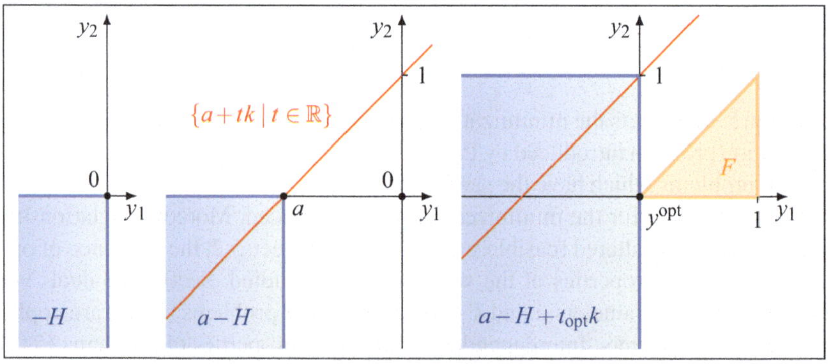


**Fig. 5.1** The optimal solution to (P$_{F,a,H,k}$) in Example 5.1.1. Adapted from [472, p. 79]

Problem $(\mathrm{P}_{F,a,H,k})$ generalizes a scalarizing problem by Pascoletti and Serafini [376] (for details, see Sect. 6.4).

The following definition of equivalent optimization problems will be used from now on.

**Definition 5.1.2** Two optimization problems $\min_{y\in F}\varphi_1(y)$ and $\min_{y\in F}\varphi_2(y)$ with $\varphi_1, \varphi_2 \colon Y \to \overline{\mathbb{R}}$, $F$ being an arbitrary subset of the nonempty set $Y$, are said to be **equivalent** to each other if the feasible sets as well as the optimal point sets of both problems coincide and if the optimal values of both problems coincide in the case that the optimal point set is nonempty. The problem $\min_{y\in F}\varphi_1(y)$ is said to be **equivalent** to problem $(\mathrm{P}_{F,a,H,k})$ under assumption (Sp0–OP$_{F,a,H,k}$) if both problems have the same set of feasible vectors $y$, if each problem has an optimal solution if and only if the other has an optimal solution, and if the sets of optimal solutions $y$ as well as the optimal values of both problems coincide in this case.

**Theorem 5.1.3** *Assume* (Sp0–OP$_{F,a,H,k}$).

*(a) If* (5.0.1) *has an optimal solution, then the optimal value of* (5.0.1) *is*

$$t_{\mathrm{opt}} := \varphi_{F+H,-k}(a), \qquad \textit{and}$$
$$\operatorname{argmin}_F \varphi_{a-H,k} = F \cap (a - \mathrm{cl}_{-k}(H) + t_{\mathrm{opt}}k).$$

*(b) Problem (5.0.1) is equivalent to the optimization problem* $(\mathrm{P}_{F,a,\mathrm{cl}_{-k}(H),k})$.

*(c) If* $k \in 0^+H$ *and* (5.0.1) *has a feasible solution* $y^0 \in F \cap (a - H + t_0k)$ *with* $t_0 \in \mathbb{R}$, *then* (5.0.1) *has an optimal solution if and only if the problem*

$$\min_{y\in F\cap(a-H+t_0k)} \varphi_{a-H,k}(y) \tag{5.1.1}$$

*has an optimal solution. In this case, the optimal point sets of both problems as well as their optimal values coincide.*

***Proof***

(a) By Propositions 4.3.7, 4.3.5 and 4.3.9,

$$\inf_{y\in F}\varphi_{a-H,k}(y) = \varphi_{a-(F+H),k}(0) = \varphi_{-(F+H),k}(-a) = \varphi_{F+H,-k}(a).$$

This implies the statement for the optimal value. By Corollary 4.2.2, $\operatorname{argmin}_F \varphi_{a-H,k} = F \cap (\mathrm{cl}_k(a - H) + t_{\mathrm{opt}}k)$. Because of Lemma 2.3.26(e) and (d), $\mathrm{cl}_k(a - H) = a + \mathrm{cl}_k(-H) = a - \mathrm{cl}_{-k}(H)$.

(b) The set of feasible vectors $y$ in $(\mathrm{P}_{F,a,\mathrm{cl}_{-k}(H),k})$ is $F \cap (a - \mathrm{cl}_{-k}(H) + \mathbb{R}k) = F \cap (\mathrm{cl}_k(a - H) + \mathbb{R}k) = F \cap (a - H + \mathbb{R}k)$ because of Proposition 2.3.48(d). Thus, it coincides with the feasible set $F \cap \operatorname{dom} \varphi_{a-H,k}$ of (5.0.1).
Because of Theorem 4.2.1(f), (4.2.3) and (4.2.4), the problem (5.0.1) has an optimal solution if and only if there exists $t_0 \in \mathbb{R}$ such that $F \cap (\mathrm{cl}_k(a - H) + t_0k) \neq \emptyset$ and $F \cap (a - H - \mathbb{R}_> k + t_0k) = \emptyset$. By Proposition 2.3.49(2),

$a - H - \mathbb{R}_{>}k = \mathrm{cl}_k(a - H) - \mathbb{R}_{>}k = a - \mathrm{cl}_{-k}(H) - \mathbb{R}_{>}k$. Thus, (5.0.1) has an optimal solution if and only if $(\mathrm{P}_{F,a,\mathrm{cl}_{-k}(H),k})$ has an optimal solution. This and (a) imply the assertion.

(c) Since $k \in 0^+H$, we get $a - H + tk = a - H + t_0k - (t_0 - t)k \subseteq a - H + t_0k$ for all $t \leq t_0$.

□

**Theorem 5.1.4** *Assume*

$$(\mathrm{Sp1{-}OP}_{F,a,H,k}):\quad \begin{array}{l} Y \text{ is a topological vector space, } F \subseteq Y, a \in Y, k \in Y \setminus \{0\}, \\ H \text{ is a proper subset of } Y \text{ with } \mathrm{cl}H + \mathbb{R}_{>}k \subseteq H. \end{array}$$

*(a) If* (5.0.1) *has the optimal value* $t_{\mathrm{opt}}$, *then*

$$\mathrm{argmin}_F \varphi_{a-H,k} = F \cap (a - \mathrm{cl}H + t_{\mathrm{opt}}k) = F \cap (a - \mathrm{bd}H + t_{\mathrm{opt}}k).$$

*(b) Problem* (5.0.1) *is equivalent to the optimization problem* $(\mathrm{P}_{F,a,\mathrm{cl}H,k})$.

*(c) If* (5.0.1) *has an optimal solution, then the optimization problem* $(\mathrm{P}_{F,a,\mathrm{bd}H,k})$ *has the same optimal value and the same set of optimal solutions* $y$.

*(d) If* $\varphi_{a-H,k}$ *is proper, then* (5.0.1) *is equivalent to the optimization problem* $(\mathrm{P}_{F,a,\mathrm{bd}H,k})$.

*(e) If* $F$ *is convex, then* (5.0.1) *has an optimal solution if and only if the optimization problem* $(\mathrm{P}_{F,a,\mathrm{bd}H,k})$ *has an optimal solution. In this case, the sets of optimal solutions* $y$ *as well as the optimal values of both problems coincide.*

***Proof***

(a) follows from Corollary 4.4.9.

(b) The assumptions imply $\mathrm{cl}_{-k}(H) = \mathrm{cl}H$ and $k \in 0^+H$ because of Proposition 2.3.54(f). Thus, Theorem 5.1.3(b) yields the equivalence of the problems (5.0.1) and $(\mathrm{P}_{F,a,\mathrm{cl}H,k})$.

(c) follows from (a) and (b).

(d) By (4.4.10), $a - \mathrm{cl}H = a - \mathrm{bd}H - \mathbb{R}_+k$. Hence, the sets of feasible vectors $y$ in $(\mathrm{P}_{F,a,\mathrm{cl}H,k})$ and in $(\mathrm{P}_{F,a,\mathrm{bd}H,k})$ coincide. If $(\mathrm{P}_{F,a,\mathrm{bd}H,k})$ has an optimal value $t_{\mathrm{opt}}$, then $t_{\mathrm{opt}}$ is also the optimal value of $(\mathrm{P}_{F,a,\mathrm{cl}H,k})$, and both problems have the same set of optimal solutions $y$. This, (b) and (c) yield the assertion.

(e) Assume that $(\mathrm{P}_{F,a,\mathrm{bd}H,k})$ has an optimal value $t_{\mathrm{opt}}$. Then there exists $y^0 \in F$ with $y^0 \in a - \mathrm{bd}H + t_{\mathrm{opt}}k$. If we had $a - y^0 + tk \in \mathrm{cl}H$ for some $t < t_{\mathrm{opt}}$, this would imply $a - y^0 + tk \in \mathrm{int}\,H$ and thus $a - y^0 + t_{\mathrm{opt}}k = a - y^0 + tk + (t_{\mathrm{opt}} - t)k \in \mathrm{int}\,H + \mathbb{R}_{>}k \subseteq \mathrm{int}\,H$, which contradicts $a - y^0 + t_{\mathrm{opt}}k \in \mathrm{bd}H$. Therefore, $a - y^0 + tk \notin \mathrm{cl}H$ for each $t < t_{\mathrm{opt}}$.

Contrary to the claim, suppose that $t_{\mathrm{opt}}$ is not the optimal value of (5.0.1). By (b), it is not the optimal value of $(\mathrm{P}_{F,a,\mathrm{cl}H,k})$. Then there exist $y^1 \in F$ and $t_1 < t_{\mathrm{opt}}$ with $y^1 \in a - \mathrm{int}\,H + t_1k$. Since $a - y^0 + t_1k \notin \mathrm{cl}H$, Lemma 2.1.26 implies the existence of $t_2 \in (0, 1)$ with $t_2(a - y^1 + t_1k) + (1 - t_2)(a - y^0 + t_1k) \in \mathrm{bd}H$,

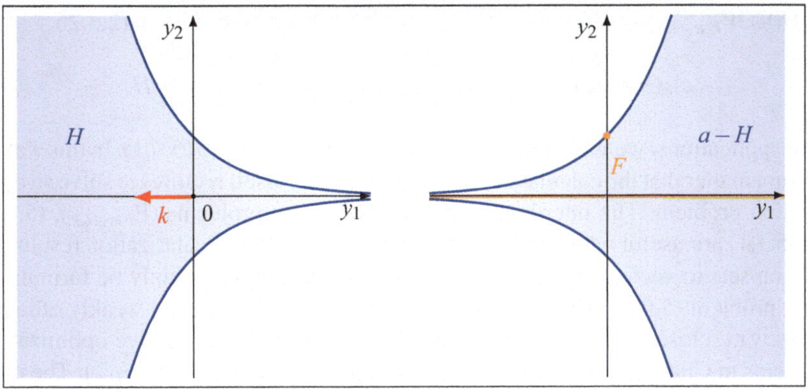


**Fig. 5.2** The sets in Example 5.1.5

i.e., $a - y^2 + t_1 k \in \operatorname{bd} H$ for $y^2 := t_2 y^1 + (1 - t_2) y^0 \in F$. This contradicts the fact that $t_{\text{opt}} > t_1$ is the optimal value of $(\text{P}_{F,a,\operatorname{bd}H,k})$. Consequently, $t_{\text{opt}}$ is the optimal value of $(\text{P}_{F,a,\operatorname{cl}H,k})$. The set of optimal solutions $y$ to $(\text{P}_{F,a,\operatorname{cl}H,k})$ is the same as to $(\text{P}_{F,a,\operatorname{bd}H,k})$ because of (a) and (b).

This and (c) prove the assertion. □

For the special case where $H$ is a closed convex cone in $Y = \mathbb{R}^\ell$, Pascoletti and Serafini [376] pointed out that the optimal solutions to $(\text{P}_{F,a,H,k})$ solve $(\text{P}_{F,a,\operatorname{bd}H,k})$ and studied local optimality.

The assumptions in Theorem 5.1.4(d) and (e) are not superfluous.

**Example 5.1.5** In $Y = \mathbb{R}^2$, consider $H := \{(y_1, y_2)^T \in Y \mid -e^{-y_1} \le y_2 \le e^{-y_1}\}$, $a := (0, 0)^T$, $k := (-1, 0)^T$, and $F := \{(0, 1)^T\} \cup \{(y_1, y_2)^T \in Y \mid y_2 = 0\}$ (see Fig. 5.2). Then $(\text{Sp1–OP}_{F,a,H,k})$ is fulfilled. The set of feasible vectors $y$ in $(\text{P}_{F,a,\operatorname{cl}H,k})$ is $F$, whereas the set of feasible vectors $y$ in $(\text{P}_{F,a,\operatorname{bd}H,k})$ is $\{(0, 1)^T\}$. $(\text{P}_{F,a,\operatorname{cl}H,k})$ does not attain a minimal value, but the optimal value of $(\text{P}_{F,a,\operatorname{bd}H,k})$ is zero.

We will now investigate such problems (5.0.1) which are equivalent to $(\text{P}_{F,a,H,k})$. This is the case under the assumption

$(\text{H0–OP}_{F,a,H,k})$: $Y$ is a linear space, $F \subseteq Y$, $a \in Y$, and
$H$ is a proper $(-k)$-directionally closed subset of $Y$
with $k \in 0^+H \setminus \{0\}$,

and under the stronger assumption

$(\text{H1–OP}_{F,a,H,k})$: $Y$ is a topological vector space, $F \subseteq Y$, $a \in Y$, and
$H$ is a proper closed subset of $Y$ with $k \in 0^+H \setminus \{0\}$.

(H0–OP$_{F,a,H,k}$) implies $k \in -0^+(a-H)$ and, because of Lemma 2.3.26,

$$\mathrm{cl}_k(a-H) = a + \mathrm{cl}_k(-H) = a - \mathrm{cl}_{-k}(H) = a - H.$$

In applications, we will prefer the problem type (P$_{F,a,H,k}$) to (5.0.1). In this way, it does not matter that the calculation of values for $\varphi_{a-H,k}$ itself requires to solve an optimization problem. The interdependencies between the problems (P$_{F,a,H,k}$), (5.0.1) and (5.1.1) are useful when proving properties of (P$_{F,a,H,k}$). Scalarization results for solution sets to vector optimization problems in Chap. 6 will mainly be formulated using problem (5.0.1). The scalarization methods for efficient and weakly efficient points w.r.t. closed convex cones in Sect. 6.2.5 and for multiobjective optimization problems in Chap. 7 will be described by solutions to problems (P$_{F,a,H,k}$). The relationship to (P$_{F,a,\mathrm{bd}H,k}$) can be used for restricting the attention to the boundary of $H$ in certain steps of algorithms.

From now on, let $\mathrm{M}_{F,a,H,k}$ denote the set of optimal solutions $y$ to problem (P$_{F,a,H,k}$). According to Theorem 5.1.3(b), this set coincides with $\mathrm{argmin}_F \varphi_{a-H,k}$ if $H$ is $(-k)$-directionally closed.

$F + (D \cup \{0\})$ may be closed or convex for some set $D \subset Y$ even though $F$ does not have this property. In this case, the following proposition can be useful.

**Proposition 5.1.6** *Assume* (H0–OP$_{F,a,H,k}$) *and that there exists a proper subset* $D \neq \{0\}$ *of* $Y$ *such that* $H + (D \setminus \{0\}) \subseteq H$. *Consider any set* $\tilde{F} \subseteq Y$ *with* $F \subseteq \tilde{F} \subseteq F + (D \cup \{0\})$. *Then* (P$_{F,a,H,k}$) *has an optimal solution if and only if* (P$_{\tilde{F},a,H,k}$) *has an optimal solution. In this case, both optimal values coincide. Moreover,*

$$\begin{aligned}\mathrm{M}_{F,a,H,k} &= \mathrm{M}_{\tilde{F},a,H,k} \cap F,\\ \mathrm{M}_{\tilde{F},a,H,k} &\subseteq \mathrm{M}_{F,a,H,k} + (D \cup \{0\}).\end{aligned}$$

***Proof*** First, assume that $\varphi_{a-H,k}$ attains on $F$ the minimal value $t$. Then $\varphi_{a-H,k}$ attains the function value $t$ on $\tilde{F}$ as well. To obtain a contradiction, suppose that $t$ is not the minimal value of $\varphi_{a-H,k}$ on $F + (D \cup \{0\})$. Then there exist $\bar{t} < t$, $y \in F$ and $d \in D \setminus \{0\}$ with $\varphi_{a-H,k}(y+d) = \bar{t}$. We get $y + d \in a - H + \bar{t}k$ from (4.2.18). Thus, $y \in a - d - H + \bar{t}k \subseteq a - H + \bar{t}k$. Then (4.2.18) implies $\varphi_{a-H,k}(y) \leq \bar{t}$, which contradicts the definition of $t$. Hence, $t$ is also the minimal value of $\varphi_{a-H,k}$ on $F + (D \cup \{0\})$ and thus on $\tilde{F}$. This results in $\mathrm{M}_{F,a,H,k} \subseteq \mathrm{M}_{\tilde{F},a,H,k}$.

Assume now that $t$ is the minimum of $\varphi_{a-H,k}$ on $\tilde{F}$. Then $\varphi_{a-H,k}$ does not attain any smaller value than $t$ on $F$. Take any $y \in F$ and $d \in D \cup \{0\}$ with $\varphi_{a-H,k}(y+d) = t$. Then $y + d \in a - H + tk$, which implies $y \in a - H + tk$. Therefore, $\varphi_{a-H,k}(y) = t$. Consequently, $y$ is a minimizer of $\varphi_{a-H,k}$ on $F$. This yields the assertion. □

We obtain for $D = 0^+H$:

**Corollary 5.1.7** *Assume* (H0–OP$_{F,a,H,k}$). *Consider any set* $\tilde{F} \subseteq Y$ *with* $F \subseteq \tilde{F} \subseteq F + 0^+H$. *Then* (P$_{F,a,H,k}$) *has an optimal solution if and only if* (P$_{\tilde{F},a,H,k}$) *has an optimal solution. In this case, both optimal values coincide. Moreover,*

$$\begin{aligned} \mathrm{M}_{F,a,H,k} &= \mathrm{M}_{\tilde{F},a,H,k} \cap F, \\ \mathrm{M}_{\tilde{F},a,H,k} &\subseteq \mathrm{M}_{F,a,H,k} + 0^+ H. \end{aligned}$$

## 5.2 Existence of Optimal Solutions and Properties of the Solution Set

First, we are interested in convexity and closedness of the solution set.

**Proposition 5.2.1** *Assume* (H0–OP$_{F,a,H,k}$).

*(a) If $F$ and $H$ are convex, then $\mathrm{M}_{F,a,H,k}$ as well as the set of feasible vectors $y$ in* (P$_{F,a,H,k}$) *are convex.*
*(b) If* (H1–OP$_{F,a,H,k}$) *holds and $F$ is closed, then $\mathrm{M}_{F,a,H,k}$ is closed.*

***Proof*** If $\mathrm{M}_{F,a,H,k} = \emptyset$, then the statement of the proposition is fulfilled. Otherwise, $\mathrm{M}_{F,a,H,k} = F \cap (a - H + t_0 k)$, where $t_0$ denotes the optimal value of (P$_{F,a,H,k}$). This yields the assertion. □

In the case where $F$ is not closed, the next proposition shows a relationship between the solution set to (P$_{F,a,H,k}$) and the closure of this solution set. Here we also consider the efficient elements of the minimizer set. These elements will be used for the scalarization of efficient solutions in vector optimization (see Chap. 6).

**Proposition 5.2.2** *Assume* (H1–OP$_{F,a,H,k}$) *and that there exists a proper subset $D \neq \{0\}$ of $Y$ such that $H + (D \setminus \{0\}) \subseteq H$ and that $F + (D \cup \{0\})$ is closed. Then*

$$\begin{aligned} \mathrm{cl}\, \mathrm{M}_{F,a,H,k} &\subseteq \mathrm{M}_{F,a,H,k} + (D \cup \{0\}) \text{ and} \\ \mathrm{Eff}(\mathrm{cl}\, \mathrm{M}_{F,a,H,k}, D) &\subseteq \mathrm{Eff}(\mathrm{M}_{F,a,H,k}, D). \end{aligned}$$

*If additionally $D + D \subseteq D$ and $D \cap (-D) \subseteq \{0\}$ hold, then*

$$\mathrm{Eff}(\mathrm{cl}\, \mathrm{M}_{F,a,H,k}, D) = \mathrm{Eff}(\mathrm{M}_{F,a,H,k}, D).$$

***Proof*** For $\mathrm{M}_{F,a,H,k} = \emptyset$, the statement is obvious. Assume now $\mathrm{M}_{F,a,H,k} \neq \emptyset$ and $t$ to be the minimum of $\varphi_{a-H,k}$ on $F$. Then $\mathrm{M}_{F,a,H,k} = F \cap (a - H + tk)$. Take any $\bar{y} \in \mathrm{cl}\, \mathrm{M}_{F,a,H,k}$. We have $\mathrm{cl}\, \mathrm{M}_{F,a,H,k} \subseteq \mathrm{cl} F \subseteq \mathrm{cl}(F + (D \cup \{0\})) = F + (D \cup \{0\})$. Hence, there exist $y \in F$ and $d \in D \cup \{0\}$ with $\bar{y} = y + d$. We get $\mathrm{cl}\, \mathrm{M}_{F,a,H,k} \subseteq a - H + tk$ since $H$ is closed. Therefore, $y + d \in a - H + tk$. This implies $y \in a - H + tk$. Consequently, $y \in \mathrm{M}_{F,a,H,k}$. Thus, $\bar{y} \in \mathrm{M}_{F,a,H,k} + (D \cup \{0\})$. We conclude that $\mathrm{cl}\, \mathrm{M}_{F,a,H,k} \subseteq \mathrm{M}_{F,a,H,k} + (D \cup \{0\})$. Lemma 6.2.5(d) yields the assertion. □

We will now investigate the existence of optimal solutions to the considered optimization problems and the compactness of the solution sets.

Immediately from (P$_{F,a,H,k}$), we get necessary conditions for the existence of optimal solutions.

**Proposition 5.2.3** *Assume* (H0–OP$_{F,a,H,k}$) *and* $\mathrm{M}_{F,a,H,k} \neq \emptyset$.
*Then the following conditions are fulfilled:*

$$F \cap (a - H + \mathbb{R}k) \neq \emptyset, \tag{5.2.1}$$

$$\exists t \in \mathbb{R} : \; F \cap (a - H + tk) = \emptyset. \tag{5.2.2}$$

Condition (5.2.1) expresses the existence of a feasible solution for (P$_{F,a,H,k}$) and is equivalent to $F \cap \operatorname{dom} \varphi_{a-H,k} \neq \emptyset$. It holds if $F$ is nonempty and $\varphi_{a-H,k}$ is finite-valued on $F$.

Condition (5.2.2) is fulfilled if and only if $\varphi_{a-H,k}$ is bounded below on $F$. In this case, $\varphi_{a-H,k}$ is finite-valued on $F \cap \operatorname{dom} \varphi_{a-H,k}$.

**Lemma 5.2.4** *Assume* (H0–OP$_{F,a,H,k}$). *The following conditions are equivalent to each other:*

*(a)* $\varphi_{a-H,k}$ *is finite-valued on* $F \cap \operatorname{dom} \varphi_{a-H,k}$.
*(b)* $\varphi_{a-H,k}$ *is finite-valued on* $F \cap (a - H + tk)$ *for some* $t \in \mathbb{R}$.

***Proof*** Obviously, (a) implies (b). Assume now that (a) is not fulfilled. Then there exists $y \in F \cap \operatorname{dom} \varphi_{a-H,k}$ with $\varphi_{a-H,k}(y) = -\infty$. This implies $y + \mathbb{R}k \subseteq a - H$ by Theorem 4.2.5(d). For each $t \in \mathbb{R}$, we get $y - tk \in a - H$. Hence, $y \in a - H + tk$ for each $t \in \mathbb{R}$. Then (b) is not fulfilled. □

We will now study sufficient conditions for the existence of optimal solutions, based on the following theorem.

**Theorem 5.2.5** *Assume* (H1–OP$_{F,a,H,k}$) *and*

$$\exists t_1 \in \mathbb{R} : \; F \cap (a - H + t_1 k) \text{ is nonempty and compact.} \tag{5.2.3}$$

*(a) If* (5.2.2) *holds, then* $\mathrm{M}_{F,a,H,k}$ *is nonempty and compact.*
*(b) Assumption* (5.2.2) *holds if and only if* $\varphi_{a-H,k}$ *is finite-valued on* $F \cap \operatorname{dom} \varphi_{a-H,k}$.

***Proof*** Equation (5.2.2) implies by (4.4.5) that $\varphi_{a-H,k}$ is finite-valued on $F \cap \operatorname{dom} \varphi_{a-H,k}$.

Assume now that $\varphi_{a-H,k}$ is finite-valued on $F \cap \operatorname{dom} \varphi_{a-H,k}$. The set $B := F \cap (a - H + t_1 k)$ is nonempty and compact. By Theorem 5.1.3(c), $\mathrm{M}_{F,a,H,k} = \mathrm{M}_{B,a,H,k}$. Proposition 3.3.17 yields that $\mathrm{M}_{B,a,H,k}$ is nonempty and compact. This proves (a) and, because of (4.4.5), also (b). □

Proposition 3.3.16 implies by Theorem 4.4.4:

**Proposition 5.2.6** *Assume* (H1–OP$_{F,a,H,k}$). *Then* $\mathrm{M}_{F,a,H,k}$ *is nonempty and compact if* $F$ *is a nonempty compact set and* $\varphi_{a-H,k}$ *is finite-valued.*

We obtain especially from Theorem 4.2.9:

**Corollary 5.2.7** *Assume* (H1–OP$_{F,a,H,k}$)*, that* $F$ *is a nonempty compact set and* $k \in \operatorname{core} 0^+H$*. Then* $\mathrm{M}_{F,a,H,k}$ *is nonempty and compact.*

We will now investigate the set of minimizers for the case where $\varphi_{a-H,k}$ is not necessarily finite-valued.

**Proposition 5.2.8** *Assume* (H1–OP$_{F,a,H,k}$)*, that* $F$ *is a compact set and that the problem* (P$_{F,a,H,k}$) *has a feasible solution.*

*Then* $\mathrm{M}_{F,a,H,k}$ *is nonempty and compact under each of the following conditions:*

*(a)* $F \cap (a - H) = \emptyset$,
*(b)* $F \cap (a - \operatorname{int} H) = \emptyset$ *and* $H + \mathbb{R}_{>}k \subset \operatorname{int} H$,
*(c)* $H$ *does not contain any line in direction* $k$,
*(d)* $H$ *is convex and* $k \notin -0^+H$.

***Proof*** (P$_{F,a,H,k}$) has a feasible solution. Then there exists $t_0 \in \mathbb{R}$ with $B := F \cap (a - H + t_0k) \neq \emptyset$. By Theorem 5.1.3(c), $\mathrm{M}_{F,a,H,k} = \mathrm{M}_{B,a,H,k}$. The set $B$ is compact since $F$ is compact and $H$ is closed. $\varphi_{a-H,k}$ is finite-valued on $B$ in the cases (a) and (c) by Theorem 4.4.4, in case (b) by Theorem 4.6.6, in case (d) by Proposition 4.7.7. Proposition 3.3.17 yields the assertion. □

For $Y = \mathbb{R}^\ell$, the existence of optimal solutions to (P$_{F,a,H,k}$) can be guaranteed without the assumption that $F$ is compact.

**Proposition 5.2.9** *Assume that the following conditions are fulfilled:*

*(1)* $a \in \mathbb{R}^\ell$, $F \subseteq \mathbb{R}^\ell$ *is a nonempty closed set, and there exists* $u \in \mathbb{R}^\ell$ *with* $F \subseteq u + \mathbb{R}^\ell_+$.
*(2)* $H$ *is a proper closed subset of* $\mathbb{R}^\ell$ *with* $\mathbb{R}^\ell_+ \subseteq 0^+H$ *and* $k \in \operatorname{core} 0^+H$ *that, for each* $j \in \{1, \ldots, \ell\}$*, does not contain any line in direction* $\mathfrak{e}^j$.

*Then* $\mathrm{M}_{F,a,H,k}$ *is nonempty and compact.*

***Proof*** By Theorem 4.2.9, $\varphi_{a-H,k}$ is finite-valued. Hence, (P$_{F,a,H,k}$) has a feasible solution. Then there exists $t_0 \in \mathbb{R}$ with $B := F \cap (a - H + t_0k) \neq \emptyset$.

For each $j \in \{1, \ldots, \ell\}$, $H$ does not contain any line in direction $\mathfrak{e}^j$.

Hence, for all $j \in \{1, \ldots, \ell\}$, there exists $s_j \in \mathbb{R}$ such that

$$(-u + a + t_0k) - s_j\mathfrak{e}^j \notin H. \tag{5.2.4}$$

Define $z_j := u_j + s_j$ for all $j \in \{1, \ldots, \ell\}$. Suppose that there exist $y \in B$ and $i \in \{1, \ldots, \ell\}$ with $y_i > z_i$. Then $y \in F \subseteq u + \mathbb{R}^\ell_+$ implies $u + s_i\mathfrak{e}^i \in y - \mathbb{R}^\ell_+ \subseteq (a - H + t_0k) - \mathbb{R}^\ell_+ \subseteq a - H + t_0k$, which contradicts (5.2.4). Thus, $B \subseteq z - \mathbb{R}^\ell_+$. Hence, $B$ is compact. By Corollary 5.2.7, $\mathrm{M}_{B,a,H,k}$ is nonempty and compact. Theorem 5.1.3(c) yields the assertion. □

Note that $\mathbb{R}^\ell_+ \subseteq 0^+H$ and $k \in \operatorname{int} \mathbb{R}^\ell_+$ imply $k \in \operatorname{core} 0^+H$.

The assumption in Proposition 5.2.9 that refers to lines in directions $\mathfrak{e}^j$ is not superfluous even for convex sets $H$.

**Example 5.2.10** In $Y = \mathbb{R}^2$, $F := \{(y_1, y_2)^T \in \mathbb{R}^2 \mid y_1 > 0,\ y_2 = \frac{1}{y_1}\}$, $a := (0, 0)^T$, $k := (1, 1)^T$, and $H := \{(y_1, y_2)^T \in \mathbb{R}^2 \mid y_1 \geq 0\}$ fulfill all assumptions of Proposition 5.2.9 but the condition that $H$ does not contain any line in direction $(0, 1)^T$. We have $F \cap (a - H + tk) = \emptyset$ for all $t \leq 0$, and $F \cap (a - H + tk) \neq \emptyset$ for all $t > 0$. Thus, $(\mathrm{P}_{F,a,H,k})$ does not have any optimal solution.

Furthermore, the statement of Proposition 5.2.9 is not true any more without the assumption $F \subseteq u + \mathbb{R}^\ell_+$.

**Example 5.2.11** In $Y = \mathbb{R}^2$, consider $F := \{(y_1, y_2)^T \in \mathbb{R}^2 \mid y_2 = 0\}$, $H := \{(y_1, y_2)^T \in \mathbb{R}^2 \mid y_1 > 0,\ y_2 \geq \frac{1}{y_1}\}$, and $k := (1, 1)^T$. For each $a \in \mathbb{R}^2$, $(\mathrm{P}_{F,a,H,k})$ does not have any optimal solution.

Together with the following lemma, Proposition 5.2.9 implies a special statement for convex cones $H$.

**Lemma 5.2.12** *Assume that $H \subset \mathbb{R}^\ell$ is a nontrivial closed pointed convex cone with $\mathbb{R}^\ell_+ \subseteq H$. Then for each $j \in \{1, \ldots, \ell\}$, $H$ does not contain any line in direction $\mathfrak{e}^j$.*

***Proof*** Take any $j \in \{1, \ldots, \ell\}$. We have $\mathfrak{e}^j \in H$ by $\mathbb{R}^\ell_+ \subseteq H$. Since $H$ is pointed, we get $\mathfrak{e}^j \in H \setminus (-H)$. By Proposition 4.7.7, $\varphi_{-H,\mathfrak{e}^j}$ is proper. Thus, Theorem 4.2.5 implies that $H$ does not contain any line in direction $\mathfrak{e}^j$. □

Thus, we obtain from Proposition 5.2.9:

**Corollary 5.2.13** *Assume that the following conditions are fulfilled:*

*(1) $a \in \mathbb{R}^\ell$, $F \subseteq \mathbb{R}^\ell$ is a nonempty closed set, and there exists $u \in \mathbb{R}^\ell$ with $F \subseteq u + \mathbb{R}^\ell_+$.*
*(2) $H \subseteq \mathbb{R}^\ell$ is a nontrivial closed pointed convex cone with $\mathbb{R}^\ell_+ \subseteq H$ and $k \in \operatorname{int} H$.*

*Then $\mathrm{M}_{F,a,H,k}$ is nonempty and compact.*

Consider now the optimization problems on $Y = \mathbb{R}^\ell$ for the case where $k$ is not necessarily an element of core $0^+H$.

**Proposition 5.2.14** *Assume that the following conditions hold:*

*(1) $a \in \mathbb{R}^\ell$, $F \subseteq \mathbb{R}^\ell$ is closed, and there exists $u \in \mathbb{R}^\ell$ with $F \subseteq u + \mathbb{R}^\ell_+$.*
*(2) $H$ is a proper closed subset of $\mathbb{R}^\ell$ with $\mathbb{R}^\ell_+ \subseteq 0^+H$ and $k \in 0^+H \setminus \{0\}$ that, for each $j \in \{1, \ldots, \ell\}$, does not contain any line in direction $\mathfrak{e}^j$.*
*(3) The problem $(\mathrm{P}_{F,a,H,k})$ has a feasible solution.*
*(4) One of the following assumptions is fulfilled:*

*(a) $F \cap (a - H) = \emptyset$,*
*(b) $F \cap (a - \operatorname{int} H) = \emptyset$ and $H + \mathbb{R}_> k \subset \operatorname{int} H$,*
*(c) $H$ does not contain any line in direction $k$,*
*(d) $H$ is convex and $k \notin -0^+H$.*

*Then $\mathrm{M}_{F,a,H,k}$ is nonempty and compact.*

***Proof*** $(\mathrm{P}_{F,a,H,k})$ has a feasible solution. Then there exists $t_0 \in \mathbb{R}$ with $B := F \cap (a - H + t_0 k) \neq \emptyset$. By the same arguments as in the proof of Proposition 5.2.9, $B$ is compact. By Proposition 5.2.8, $\mathrm{M}_{B,a,H,k}$ is nonempty and compact. Theorem 5.1.3(c) yields the assertion. □

For convex sets $H$, Proposition 5.2.14 implies a statement which can be proved by means of the following lemma.

**Lemma 5.2.15** *Assume that $H \subset \mathbb{R}^\ell$ is a proper closed convex subset of $\mathbb{R}^\ell$ with $H + (\mathbb{R}^\ell_+ \setminus \{0\}) \subseteq \operatorname{int} H$, $H + \mathbb{R}_> k \subseteq \operatorname{int} H$ and $k \notin -0^+ H$. For each $j \in \{1, \ldots, \ell\}$, $H$ does not contain any line in direction $\mathfrak{e}^j$.*

***Proof*** Contrary to the claim, suppose that there exists $i \in \{1, \ldots, \ell\}$ such that $H$ contains some line in direction $\mathfrak{e}^i$. Then there exists $u \in \mathbb{R}^\ell$ with $\{u + s\,\mathfrak{e}^i \mid s \in \mathbb{R}\} \subseteq H$. We obtain $\{u + s\,\mathfrak{e}^i \mid s \in \mathbb{R}\} = \{u + (s-1)\mathfrak{e}^i \mid s \in \mathbb{R}\} + \mathfrak{e}^i \subseteq H + (\mathbb{R}^\ell_+ \setminus \{0\}) \subseteq \operatorname{int} H$.

By Proposition 4.7.7, $\varphi_{-H,k}$ is proper. Hence, Theorem 4.6.6 yields $\operatorname{int} H = \operatorname{bd} H + \mathbb{R}_> k$. Since $u \in \operatorname{int} H$, there are $h \in \operatorname{bd} H$ and $t_0 > 0$ with $h = u - t_0 k$. We have $h + \mathfrak{e}^i \in \operatorname{int} H$ because of $H + (\mathbb{R}^\ell_+ \setminus \{0\}) \subseteq \operatorname{int} H$. Therefore, there exists $t_1 > 0$ with $h + \mathfrak{e}^i - t_1 k \in \operatorname{int} H$.

For $t_2 := \frac{t_0}{t_0 + t_1} \in (0, 1)$ and $s_0 := -\frac{t_2}{1 - t_2}$, we get $t_2 + (1 - t_2)s_0 = 0$, $t_2(t_0 + t_1) = t_0$. Hence, $h = u - t_0 k = u + (t_2 + (1 - t_2)s_0)e^i - t_2(t_0 + t_1)k = t_2(u + e^i - (t_0 + t_1)k) + (1 - t_2)(u + s_0\mathfrak{e}^i) = t_2(h + \mathfrak{e}^i - t_1 k) + (1 - t_2)(u + s_0\mathfrak{e}^i) \in \operatorname{int} H$ since $\operatorname{int} H$ is convex. This contradicts $h \in \operatorname{bd} H$. □

Thus, we get from Proposition 5.2.14:

**Corollary 5.2.16** *Assume that the following conditions hold:*

*(1) $a \in \mathbb{R}^\ell$, $F \subseteq \mathbb{R}^\ell$ is closed and there exists $u \in \mathbb{R}^\ell$ with $F \subseteq u + \mathbb{R}^\ell_+$.*
*(2) $H$ is a proper closed convex subset of $\mathbb{R}^\ell$ with $H + (\mathbb{R}^\ell_+ \setminus \{0\}) \subseteq \operatorname{int} H$, $H + \mathbb{R}_> k \subseteq \operatorname{int} H$ and $k \notin -0^+ H$.*
*(3) The problem $(\mathrm{P}_{F,a,H,k})$ has a feasible solution.*

*Then $\mathrm{M}_{F,a,H,k}$ is nonempty and compact.*

Without the assumption $F \subseteq u + \mathbb{R}^\ell_+$ in Corollary 5.2.16, the existence of optimal solutions to $(\mathrm{P}_{F,a,H,k})$ can depend on the choice of $a$ even if $H$ is a closed convex cone.

**Example 5.2.17** Assume $Y = \mathbb{R}^3$ and that $H$ is the convex cone generated by the vectors $(2, 0, -1)^{\mathrm{T}}$, $(0, 2, -1)^{\mathrm{T}}$ and $(-1, 0, 2)^{\mathrm{T}}$. Choose $k := (1, 1, 1)^{\mathrm{T}}$, $a := (0, 0, 0)^{\mathrm{T}}$, and $b := (1, -1, -\frac{1}{2})^{\mathrm{T}}$. The set $F := \{(y_1, y_2, y_3)^{\mathrm{T}} \in \mathbb{R}^3 \mid 2y_1 + y_2 + 2y_3 = 0,\ y_1 > 0,\ y_2 \leq -\frac{1}{y_1}\}$ is contained in the same hyperplane as the points $(0, 0, 0)^{\mathrm{T}}$,$(2, 0, -2)^{\mathrm{T}}$ und $(0, -2, 1)^{\mathrm{T}}$. It is closed and convex. Since $(a - H) \cap F = \emptyset$ and $(a - \operatorname{int} H + tk) \cap F \neq \emptyset$ for all $t > 0$, the problem $(\mathrm{P}_{F,a,H,k})$ does not have any optimal solution. However, there exist optimal solutions to the problem $(\mathrm{P}_{F,b,H,k})$. These are the elements of the set $(b - \operatorname{bd} H) \cap F = (b - H) \cap F = \{(y_1, y_2, y_3)^{\mathrm{T}} \in \mathbb{R}^3 \mid y_1 = 1,\ y_2 = -2\lambda - 1,\ y_3 = \lambda - \frac{1}{2},\ \lambda \in \mathbb{R}_+\}$.

Proposition 5.2.14 and Lemma 5.2.12 imply for convex cones $H$:

**Corollary 5.2.18** *Assume that the following conditions hold:*

*(1)* $a \in \mathbb{R}^\ell$*,* $F \subseteq \mathbb{R}^\ell$ *is closed, and there exists* $u \in \mathbb{R}^\ell$ *with* $F \subseteq u + \mathbb{R}^\ell_+$*.*
*(2)* $H \subset \mathbb{R}^\ell$ *is a nontrivial closed pointed convex cone with* $\mathbb{R}^\ell_+ \subseteq H$ *and* $k \in H \setminus (-H)$*.*
*(3) The problem* $(\mathrm{P}_{F,a,H,k})$ *has a feasible solution.*

*Then* $\mathrm{M}_{F,a,H,k}$ *is nonempty and compact.*

Without the assumption $F \subseteq u + \mathbb{R}^\ell_+$ in Corollary 5.2.18, the existence of optimal solutions to $(\mathrm{P}_{F,a,H,k})$ can depend on the choice of $a$ even if $H$ is the nonnegative orthant.

**Example 5.2.19** In $Y = \mathbb{R}^3$, consider $H := \mathbb{R}^3_+$, $k := (1,1,1)^T$, $a := (0,0,0)^T$, and $b := (1,-1,-1)^T$. The set $F := \{(y_1, y_2, y_3)^T \in \mathbb{R}^3 \mid y_1 + y_3 = 0,\ y_2 \leq -\frac{1}{y_1},\ y_1 > 0\}$ is closed, convex, and is contained in the hyperplane $\{(y_1, y_2, y_3)^T \in \mathbb{R}^3 \mid y_1 + y_3 = 0\}$. We have $(a - H) \cap F = \emptyset$, but $(a - \operatorname{int} H + tk) \cap F \neq \emptyset$ for all $t > 0$. Thus, $(\mathrm{P}_{F,a,H,k})$ does not have any optimal solution. However, $(b - \operatorname{bd} H) \cap F = (b - H) \cap F = \{(y_1, y_2, y_3)^T \in \mathbb{R}^3 \mid y_1 = 1,\ y_2 \leq -1,\ y_3 = -1\}$ is the set of minimizers $y$ of $(\mathrm{P}_{F,b,H,k})$.

We now turn to statements in $Y = \mathbb{R}^\ell$ without the assumption that $F \subseteq u + \mathbb{R}^\ell_+$ holds for some $u \in \mathbb{R}^\ell$.

**Proposition 5.2.20** *Assume that the following conditions are fulfilled:*

*(1)* $a \in \mathbb{R}^\ell$*,* $F \subseteq \mathbb{R}^\ell$ *is a nonempty closed set.*
*(2)* $H$ *is a proper closed subset of* $\mathbb{R}^\ell$ *with* $k \in \operatorname{core} 0^+ H$*.*
*(3) There exist a polyhedral cone* $C \subset \mathbb{R}^\ell$*,* $z \in \operatorname{cl}\operatorname{conv} H$ *and* $u \in \mathbb{R}^\ell$ *such that*

$$z - \operatorname{cl}\operatorname{conv} H \subset \operatorname{int} C \cup \{0\} \quad \textit{and}$$
$$(F - u) \cap \operatorname{int} C = \emptyset.$$

*Then* $\mathrm{M}_{F,a,H,k}$ *is nonempty and compact.*

***Proof*** By Theorem 4.2.9, $\varphi_{a-H,k}$ is finite-valued. Hence, $(\mathrm{P}_{F,a,H,k})$ has a feasible solution. Then there exists $t_0 \in \mathbb{R}$ with $B := F \cap (a - H + t_0 k) \neq \emptyset$. The set $B \subseteq F \cap (a + t_0 k - z + (z - \operatorname{cl}\operatorname{conv} H))$ is bounded by Lemma 2.2.32, hence compact. By Corollary 5.2.7, $\mathrm{M}_{B,a,H,k}$ is nonempty and compact. Theorem 5.1.3(c) yields the assertion. □

The assumptions of Proposition 5.2.20 do not imply that $F \subseteq u + \mathbb{R}^\ell_+$ holds for some $u \in \mathbb{R}^\ell$ or that $H$ is a cone.

**Example 5.2.21** The sets $H := \{(y_1, y_2)^T \in \mathbb{R}^2 \mid y_1 \geq -1, y_2 \geq -1, (y_1 + 1)(y_2 + 1) \geq 1\}$ and $F := \{(y_1, y_2)^T \in \mathbb{R}^2 \mid y_1 \geq 0, y_2 \geq -\frac{y_1}{2}\}$ fulfill the conditions in Proposition 5.2.20 with $C := \{(y_1, y_2)^T \in \mathbb{R}^2 \mid y_1 + y_2 \leq 0\}$ and $u := z := (0,0)^T$.

For the case where $k$ is not necessarily an element of core $0^+H$, we can prove the next statement.

**Proposition 5.2.22** *Assume that the following conditions are satisfied:*

*(1) $a \in \mathbb{R}^\ell$, $F \subseteq \mathbb{R}^\ell$ is a nonempty closed set.*
*(2) $H$ is a proper closed subset of $\mathbb{R}^\ell$ with $k \in 0^+H \setminus \{0\}$.*
*(3) There exist a polyhedral cone $C \subset \mathbb{R}^\ell$, $z \in \operatorname{cl}\operatorname{conv} H$ and $u \in \mathbb{R}^\ell$ such that*

$$z - \operatorname{cl}\operatorname{conv} H \subset \operatorname{int} C \cup \{0\} \quad \textit{and}$$
$$(F - u) \cap \operatorname{int} C = \emptyset.$$

*(4) The problem $(\mathrm{P}_{F,a,H,k})$ has a feasible solution.*
*(5) One of the following assumptions is fulfilled:*

*(a) $F \cap (a - H) = \emptyset$,*
*(b) $F \cap (a - \operatorname{int} H) = \emptyset$ and $H + \mathbb{R}_{>}k \subset \operatorname{int} H$,*
*(c) $H$ does not contain any line in direction $k$.*

*Then $\mathrm{M}_{F,a,H,k}$ is nonempty and compact.*

***Proof*** $(\mathrm{P}_{F,a,H,k})$ has a feasible solution. Thus, there exists $t_0 \in \mathbb{R}$ with $B := F \cap (a - H + t_0 k) \neq \emptyset$. The set $B \subseteq F \cap (a + t_0 k - z + (z - \operatorname{cl}\operatorname{conv} H))$ is bounded by Lemma 2.2.32, hence compact. By Proposition 5.2.8, $\mathrm{M}_{B,a,H,k}$ is nonempty and compact. Theorem 5.1.3(c) yields the assertion. □

Apply now the idea of the previous proposition to convex sets $H$.

**Proposition 5.2.23** *Assume that the following conditions are fulfilled:*

*(1) $a \in \mathbb{R}^\ell$, $F \subseteq \mathbb{R}^\ell$ is a nonempty closed set.*
*(2) $H$ is a proper closed convex subset of $\mathbb{R}^\ell$ with $k \in 0^+H \setminus (-0^+H)$.*
*(3) There exist a polyhedral cone $C \subset \mathbb{R}^\ell$, $h \in H$ and $u \in \mathbb{R}^\ell$ such that*

$$h - H \subset \operatorname{int} C \cup \{0\} \quad \textit{and}$$
$$(F - u) \cap \operatorname{int} C = \emptyset.$$

*(4) The problem $(\mathrm{P}_{F,a,H,k})$ has a feasible solution.*

*Then $\mathrm{M}_{F,a,H,k}$ is nonempty and compact.*

***Proof*** $(\mathrm{P}_{F,a,H,k})$ has a feasible solution. Thus, there exists $t_0 \in \mathbb{R}$ with $B := F \cap (a - H + t_0 k) \neq \emptyset$. The set $B = F \cap (a - h + t_0 k + (h - H))$ is bounded by Lemma 2.2.32, hence compact. By Proposition 5.2.8, $\mathrm{M}_{B,a,H,k}$ is nonempty and compact. Theorem 5.1.3(c) yields the assertion. □

Condition (4) in Proposition 5.2.23 is fulfilled under the assumptions (1)–(3) if $k \in$ core $0^+H$.

The connectedness of the solution sets to problems $(\mathrm{P}_{F,a,H,k})$ in $Y = \mathbb{R}^\ell$ was investigated for closed convex cones $H$ in [248].

## 5.3 Parameter Control

Problems $(\mathrm{P}_{F,a,H,k})$ with varying parameters $a$ and $k$ will be used for the scalarization of vector optimization problems in Chapters 6 and 7. We will now examine the parameter control for these scalarizing problems.

**Proposition 5.3.1** *Assume that $Y$ is a linear space, $H$ is a proper algebraically closed convex subset of $Y$ with $0^+H \neq \{0\}$ and that $F$ is a nonempty subset of $Y$ for which $F + 0^+H$ is convex.*

*Then $\{(a,k) \in Y \times (0^+H \setminus \{0\}) \mid (P_{F,a,H,k})$ has a feasible solution$\}$ is a convex set.*

***Proof*** Assume $(a^1,k^1),(a^2,k^2) \in Y \times (0^+H \setminus \{0\})$ such that $(\mathrm{P}_{F,a^1,H,k^1})$ and $(\mathrm{P}_{F,a^2,H,k^2})$ have feasible solutions. Then there exist $y^1, y^2 \in F$ and $t_1, t_2 \in \mathbb{R}$ such that $y^1 \in a^1 - H + t_1k^1$ and $y^2 \in a^2 - H + t_2k^2$. Take any $\lambda \in (0,1)$. We obtain $\lambda y^1 + (1-\lambda)y^2 \in \lambda a^1 + (1-\lambda)a^2 - H + \lambda t_1 k^1 + (1-\lambda)t_2k^2$ since $H$ is convex. For $t := \max(t_1,t_2)$, we get $h := \lambda(t-t_1)k^1 + (1-\lambda)(t-t_2)k^2 \in 0^+H$ since $0^+H$ is a convex cone. Then $\lambda t_1k^1 + (1-\lambda)t_2k^2 = t(\lambda k^1 + (1-\lambda)k^2) - h$ implies $\lambda y^1 + (1-\lambda)y^2 \in \lambda a^1 + (1-\lambda)a^2 - H + t(\lambda k^1 + (1-\lambda)k^2)$. Since $F + 0^+H$ is convex, there exist $y^3 \in F$ and $h^1 \in 0^+H$ such that $\lambda y^1 + (1-\lambda)y^2 = y^3 + h^1$. Hence, $y^3$ is a feasible solution $y$ to $(\mathrm{P}_{F,\lambda a^1+(1-\lambda)a^2,H,\lambda k^1+(1-\lambda)k^2})$. □

Note that $F + 0^+H$ is convex if $F$ is convex.

By Theorem 4.2.1(d) and Proposition 4.3.3, we can restrict the set of parameters $k$ which have to be considered.

**Proposition 5.3.2** *Assume* $(\text{H0–OP}_{F,a,H,k})$.

*(a) If $k \in -0^+H$, then $(\mathrm{P}_{F,a,H,k})$ does not have an optimal solution.*
*(b) For each $\lambda \in \mathbb{R}_>$, the problem $(\mathrm{P}_{F,a,H,\lambda k})$ has the same feasible vectors $y$ and the same optimal solutions $y$ as $(\mathrm{P}_{F,a,H,k})$.*

The proposition underscores that replacing $k$ by another vector in the same direction does not alter the optimal solutions. Hence, it is sufficient to consider only one vector $k$ per direction, e.g., to restrict $k$ to unit vectors if $Y$ is a normed space. In $Y = \mathbb{R}^\ell$, the range for $k$ can also be restricted by the following proposition.

**Proposition 5.3.3** *Assume $F \subseteq \mathbb{R}^\ell$, $a \in \mathbb{R}^\ell$ and $H$ is a proper closed subset of $\mathbb{R}^\ell$ with $0^+H \neq \{0\}$. We get*

$$\bigcup_{k\in\{y\in 0^+H \mid \sum_{i=1}^{\ell} y_i>0\}} \mathrm{M}_{F,a,H,k} = \bigcup_{k\in\{y\in 0^+H \mid \sum_{i=1}^{\ell} y_i=1\}} \mathrm{M}_{F,a,H,k} \quad \textit{and}$$

$$\bigcup_{k\in\{y\in 0^+H \mid \sum_{i=1}^{\ell} y_i<0\}} \mathrm{M}_{F,a,H,k} = \bigcup_{k\in\{y\in 0^+H \mid \sum_{i=1}^{\ell} y_i=-1\}} \mathrm{M}_{F,a,H,k}.$$

***Proof*** Proposition 5.3.2 implies $\mathrm{M}_{F,a,H,k} = \mathrm{M}_{F,a,H,\lambda k}$ for $\lambda := (\sum_{i=1}^{\ell} k_i)^{-1}$ in the first equation and $\lambda := -(\sum_{i=1}^{\ell} k_i)^{-1}$ in the second equation. □

We now investigate whether the set of parameters $a$ can be restricted. We obtain the following from Proposition 4.3.4.

**Proposition 5.3.4** *Assume* (H0–OP$_{F,a,H,k}$). *Then* $\mathrm{M}_{F,a,H,k} = \mathrm{M}_{F,a+ck,H,k}$ *for each* $c \in \mathbb{R}$.

This implies:

**Proposition 5.3.5** *Assume $Y$ is a linear space, $F \subseteq Y$, and $H$ is a proper $(-k)$-directionally closed subset of $Y$ with $k \in 0^+H \setminus \{0\}$.*
*If $\Lambda \subset Y$ with $Y = \{\Lambda + \beta k | \beta \in \mathbb{R}\}$, then*

$$\bigcup_{a \in Y} \mathrm{M}_{F,a,H,k} = \bigcup_{a \in \Lambda} \mathrm{M}_{F,a,H,k}.$$

$\Lambda$ can be chosen as the complementary space of the linear subspace $\{\beta k \mid \beta \in \mathbb{R}\}$. Thus, in $Y = \mathbb{R}^\ell$, the dimension of the parameter set for $a$ can be reduced by one.

**Proposition 5.3.6** *Assume $F \subseteq \mathbb{R}^\ell$ and $H$ is a proper $(-k)$-directionally closed subset of $\mathbb{R}^\ell$ with $k \in 0^+H \setminus \{0\}$.*

*(a) For each $j \in \{1, \ldots, \ell\}$ with $k_j \neq 0$, we have*

$$\bigcup_{a \in \mathbb{R}^\ell} \mathrm{M}_{F,a,H,k} = \bigcup_{a \in \{y \in \mathbb{R}^\ell | y_j = 0\}} \mathrm{M}_{F,a,H,k}.$$

*(b) If $k \in \operatorname{int} \mathbb{R}^\ell_+ \cup (-\operatorname{int} \mathbb{R}^\ell_+)$, then we have*

$$\bigcup_{a \in \mathbb{R}^\ell} \mathrm{M}_{F,a,H,k} = \bigcup_{a \in \mathbb{R}^\ell_+} \mathrm{M}_{F,a,H,k} = \bigcup_{a \in -\mathbb{R}^\ell_+} \mathrm{M}_{F,a,H,k}.$$

*(c) If $\sum_{i=1}^{\ell} k_i \neq 0$, then*

$$\bigcup_{a \in \mathbb{R}^\ell} \mathrm{M}_{F,a,H,k} = \bigcup_{a \in \{y \in \mathbb{R}^\ell | \sum_{i=1}^{\ell} y_i = 0\}} \mathrm{M}_{F,a,H,k}.$$

***Proof*** Apply Proposition 5.3.4.

(a) For $c = -\frac{a_j}{k_j}$, we get $(a + ck)_j = 0$.

(b) Choose $n \in \{1, \ldots, \ell\}$ with $\frac{a_n}{k_n} = \max_{i \in \{1,\ldots,\ell\}} \frac{a_i}{k_i}$, and $m \in \{1, \ldots, \ell\}$ with $\frac{a_m}{k_m} = \min_{i \in \{1,\ldots,\ell\}} \frac{a_i}{k_i}$.
If $k \in \operatorname{int} \mathbb{R}^\ell_+$, then we get $a + ck \in -\mathbb{R}^\ell_+$ for $c = -\frac{a_n}{k_n}$ and $a + ck \in \mathbb{R}^\ell_+$ for $c = -\frac{a_m}{k_m}$.
If $k \in -\operatorname{int} \mathbb{R}^\ell_+$, then the two vectors have the opposite sign.

(c) For $c = -\frac{\sum_{i=1}^{\ell} a_i}{\sum_{i=1}^{\ell} k_i}$, we obtain $\sum_{i=1}^{\ell} (a + ck)_i = 0$. □

Hence, in all statements to come, the existence of a parameter vector $a \in \mathbb{R}^\ell$ for which a given point $y^0$ solves the problem $(\mathrm{P}_{F,a,H,k})$ with $k_j \neq 0$ is equivalent to the existence of a corresponding parameter vector $a \in \mathbb{R}^\ell$ with $a_j = 0$, in the case $k \in \operatorname{int} \mathbb{R}^\ell_+ \cup (-\operatorname{int} \mathbb{R}^\ell_+)$ equivalent to the existence of a corresponding $a \in \mathbb{R}^\ell_+$.

## 5.4 Interdependencies Between Problems with Different Parameters

Solving a problem $(\mathrm{P}_{F,a,H,k})$, we can also derive some information about problems with altered parameters $a$ and $k$.

**Lemma 5.4.1** *Assume that $Y$ is a linear space, $F \subseteq Y$ and that $H$ is a proper subset of $Y$. Then:*

*(a) If $k \in \operatorname{core} 0^+H$, then each function $\varphi_{a-H,k}$ with $a \in Y$ is finite-valued.*
*(b) Suppose $Y$ is a topological vector space and $H$ is a closed set. If $\varphi_{a-H,k}$ is bounded below on $F$ for some $a \in Y$ and $k \in \operatorname{int} 0^+H$, then $\varphi_{b-H,k^0}$ is bounded below on $F$ for all $b \in Y$ and $k^0 \in \operatorname{int} 0^+H$.*

***Proof***

(a) results from Theorem 4.2.9.
(b) Assume that $\varphi_{a-H,k}$ is bounded below on $F$ for some $a \in Y$ and $k \in \operatorname{int} 0^+H$. Then there exists $s \in \mathbb{R}$ with $\varphi_{a-H,k}(y) \geq s$ for all $y \in F$. Theorem 4.6.6 implies $(a - \operatorname{int} H + sk) \cap F = \emptyset$. Take any $b \in Y$ and $k^0 \in \operatorname{int} 0^+H$. By Lemma 2.3.14, $Y = \bigcup_{\alpha \in \mathbb{R}} (\operatorname{int} 0^+H + \alpha k^0) = \bigcup_{\alpha \in \mathbb{R}} (a + sk - \operatorname{int} 0^+H - \alpha k^0)$. Hence, there exists $t \in \mathbb{R}$ with $b \in a + sk - \operatorname{int} 0^+H - tk^0$, which implies $b + tk^0 \in a + sk - \operatorname{int} 0^+H$. We get $b - \operatorname{int} H + tk^0 \subseteq a + sk - (\operatorname{int} H + \operatorname{int} 0^+H) \subseteq a + sk - \operatorname{int} H$. Therefore, $(b - \operatorname{int} H + tk^0) \cap F = \emptyset$. Theorem 4.6.6 yields $\varphi_{b-H,k^0}(y) \geq t$ for all $y \in F$, i.e., $\varphi_{b-H,k^0}$ is bounded below on $F$. □

In Lemma 5.4.1, it is essential that $k$ and $k^0$ belong to the interior of $0^+H$.

**Example 5.4.2** The set $H := \{(y_1, y_2)^T \in \mathbb{R}^2 \mid y_2 \geq y_1^2\}$ fulfills the assumptions of Lemma 5.4.1 without the condition that the interior of $0^+H$ is nonempty. We have $k := (0, 1)^T \in 0^+H$. Consider $F := \{(y_1, y_2)^T \in \mathbb{R}^2 \mid y_1 \geq 0,\ y_2 = -y_1^2\} \cup \{(-1, 0)^T\}$. Then $F \cap -\operatorname{int} H = \emptyset$. Hence, $\varphi_{-H,k}(y) \geq 0$ for all $y \in F$, i.e., $\varphi_{-H,k}$ is bounded below on $F$. Take $b := (1, 0)^T$ and any $t \in -\mathbb{R}_>$. Define $\tilde{y}_1 := \frac{3}{4} - \frac{t}{2} > 0$ and $\tilde{y}_2 := -\tilde{y}_1^2$. Then $\tilde{y} := (\tilde{y}_1, \tilde{y}_2)^T \in F$. We obtain $\bar{y} := \tilde{y} - b - tk = (-\frac{t}{2} - \frac{1}{4}, -\frac{t^2}{4} - \frac{t}{4} - \frac{9}{16})^T \in -\operatorname{int} H$ since $\bar{y}_2 < -\bar{y}_1^2$. Hence, $\tilde{y} \in b - \operatorname{int} H + tk$, and $\varphi_{b-H,k}(\tilde{y}) < t$. Consequently, $\varphi_{b-H,k}$ is not bounded below on $F$.

Lemma 5.4.1 implies the following by Theorem 5.1.3(b).

**Proposition 5.4.3** *Assume that $Y$ is a linear space, $F \subseteq Y$ and that $H$ is a proper subset of $Y$.*

(a) *If $H$ is $(-k)$-directionally closed and $k \in \operatorname{core} 0^+H$, then $F$ is the set of feasible solutions $y$ of each problem $(\mathrm{P}_{F,a,H,k})$ with $a \in Y$.*
(b) *Suppose $Y$ is a topological vector space and $H$ is a closed set. If the objective function of $(\mathrm{P}_{F,a,H,k})$ is not bounded below on $F$ for some $a \in Y$ and $k \in \operatorname{int} 0^+H$, then none of the problems $(\mathrm{P}_{F,b,H,k^0})$ with $b \in Y$ and $k^0 \in \operatorname{int} 0^+H$ has an optimal solution.*

**Proposition 5.4.4** *Assume $F \subseteq \mathbb{R}^\ell$ is a nonempty closed convex set, $a \in \mathbb{R}^\ell$, $H$ is a proper closed convex subset of $\mathbb{R}^\ell$ and $k \in 0^+H \setminus (-0^+H)$.*

(a) *If $F \cap (a - H + \mathbb{R}k) \neq \emptyset$ and $\mathrm{M}_{F,a,H,k} = \emptyset$ or if $\mathrm{M}_{F,a,H,k}$ is not bounded, then for each $b \in \mathbb{R}^\ell$ and each $k^0 \in 0^+H \setminus (-0^+H)$, we have $\mathrm{M}_{F,b,H,k^0} = \emptyset$ or $0^+\mathrm{M}_{F,b,H,k^0} \neq \{0\}$.*
(b) *If $\mathrm{M}_{F,a,H,k}$ is nonempty and bounded, then for each $b \in \mathbb{R}^\ell$ and for each $k^0 \in 0^+H \setminus (-0^+H)$ with $F \cap (b - H + \mathbb{R}k^0) \neq \emptyset$, the set $\mathrm{M}_{F,b,H,k^0}$ is nonempty and compact.*

***Proof*** By Proposition 4.7.7, each function $\varphi_{b-H,k^0}$ with $b \in Y$ and $k^0 \in 0^+H \setminus (-0^+H)$ is proper.

(a) Consider first the case where $\mathrm{M}_{F,a,H,k} = \emptyset$ and where exists $t \in \mathbb{R}$ with $F \cap (a + tk - H) \neq \emptyset$. Then $F \cap (a + tk - H)$ cannot be compact since otherwise $\varphi_{a-H,k}$ would attain a minimum on $F \cap (a + tk - H)$ und thus on $F$. Since $F \cap (a + tk - H)$ is closed, it is not bounded.
Consider now the case where $\mathrm{M}_{F,a,H,k}$ is not bounded. Let $t$ be the minimum of $\varphi_{a-H,k}$ on $F \cap \operatorname{dom} \varphi_{a-H,k}$. Then $F \cap (a + tk - H)$ is not bounded.
Take any $b \in \mathbb{R}^\ell$ and $k^0 \in 0^+H \setminus (-0^+H)$. If the minimum $\bar{t}$ of $\varphi_{b-H,k^0}$ on $F$ exists, then $\mathrm{M}_{F,b,H,k^0} = F \cap (b + \bar{t}k^0 - H)$, and $0^+\mathrm{M}_{F,b,H,k^0} \neq \{0\}$ by Lemma 2.2.31.
(b) follows from (a). □

Note that the condition $F \cap (a - H + \mathbb{R}k) \neq \emptyset$ is equivalent to the existence of feasible solutions to problem $(\mathrm{P}_{F,a,H,k})$.

**Proposition 5.4.5** *Assume $F \subseteq \mathbb{R}^\ell$ is a nonempty closed convex set, $a \in \mathbb{R}^\ell$, $H$ is a proper closed convex subset of $\mathbb{R}^\ell$ and $k \in \operatorname{int} 0^+H$.*

(a) *If* $\mathrm{M}_{F,a,H,k} = \emptyset$ *or if* $\mathrm{M}_{F,a,H,k}$ *is not bounded, then for each* $b \in \mathbb{R}^\ell$ *and each* $k^0 \in 0^+H \setminus (-0^+H)$, *we have* $\mathrm{M}_{F,b,H,k^0} = \emptyset$ *or* $0^+\mathrm{M}_{F,b,H,k^0} \neq \{0\}$.

(b) *If* $\mathrm{M}_{F,a,H,k}$ *is nonempty and bounded, then the set* $\mathrm{M}_{F,b,H,k^0}$ *is nonempty and compact for each* $b \in \mathbb{R}^\ell$ *and* $k^0 \in \operatorname{int} 0^+H$.

***Proof*** By Theorem 4.2.9, each function $\varphi_{b-H,k^0}$ with $b \in Y$ and $k^0 \in \operatorname{int} 0^+H$ is finite-valued. Apply Proposition 5.4.4. □

If $(\mathrm{P}_{F,a,H,k})$ in Proposition 5.4.5(b) has a unique minimal solution, then this is not necessarily the case for the considered problems $(\mathrm{P}_{F,b,H,k^0})$.

**Example 5.4.6** Choose $F := H := \mathbb{R}^2_+$, $a := (0,0)^T$, $b := (1,0)^T$, and $k := (1,1)^T$. Then $F \cap (a - H) = \{(0,0)^T\}$, and $F \cap (b - H) = F \cap (b - \operatorname{bd} H) = \{(y_1, y_2)^T \in \mathbb{R}^2 \mid 0 \leq y_1 \leq 1,\ y_2 = 0\}$. Thus, $\varphi_{a-H,k}$ attains its minimum on $F$ only at $(0,0)^T$, whereas each element of $F \cap (b - H)$ is a minimizer of $\varphi_{b-H,k}$ on $F$.

Further sensitivity results for the studied problems $(\mathrm{P}_{F,a,H,k})$ in $Y = \mathbb{R}^\ell$ were proved by Pascoletti and Serafini [376] for polyhedral cones $H$, later by Helbig for closed convex cones $H$ [234] as well as for closed convex sets [233] and by Eichfelder [136] for closed pointed convex cones $H$. The way in which the solution set to problems $(\mathrm{P}_{F,a,H,k})$ depends on alternations in $a$ and $k$ was also studied for convex cones $H$ in more general spaces $Y$ by Sterna-Karwat [423, 424]. In [136], further references related to sensitivity are given.

## 5.5 Minimization of Minkowski Functionals and Norms

Certain Minkowski functionals and norms coincide with some functional $\varphi_{A,k}$ on a subset of the space. In practical applications, the set on which such a norm is minimized is usually contained in this subset. We apply Sect. 4.13 in order to show that certain norm minimization problems are equivalent to problems $(\mathrm{P}_{F,a,H,k})$, which were investigated in the previous sections of this chapter.

**Proposition 5.5.1** *Assume that* $Y$ *is a linear space,* $A \subseteq Y$ *is a nontrivial algebraically closed convex cone with* $k \in -\operatorname{core} A$ *and* $a \in Y$. *Then*

$$p_{A+k}(y-a) = \max\{\varphi_{a+A,k}(y), 0\} \text{ for all } y \in Y, \text{ i.e.,}$$

$$p_{A+k}(y-a) = \begin{cases} \varphi_{a+A,k}(y) & \text{if } y \in Y \setminus (a+A), \\ 0 & \text{if } y \in a+A. \end{cases}$$

*The problem*

$$\min_{y \in F} p_{A+k}(y-a)$$

*is equivalent to the problem* $(\mathrm{P}_{F,a,-A,k})$ *for each set* $F \subseteq Y \setminus (a+A)$.

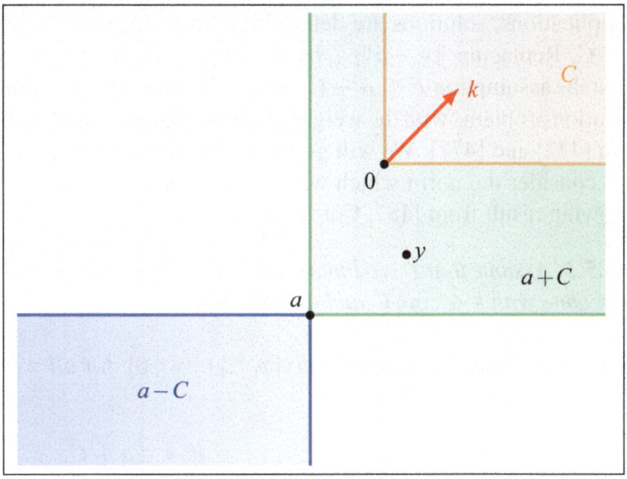


**Fig. 5.3** $\|y-a\|_{C,k}=\varphi_{a-C,k}(y)$ on $a+C$ in Proposition 5.5.2

***Proof*** By Proposition 2.3.37, $A$ is $k$-directionally closed. Apply Proposition 4.13.1 to $y-a$ instead of $y$. Lemma 4.3.5 yields the assertion. □

An essential statement of the next proposition is illustrated by Fig. 5.3.

**Proposition 5.5.2** *Suppose that $Y$ is a linear space, $C\subset Y$ is an algebraically closed pointed convex cone with $k\in\operatorname{core}C$, $a\in Y$. Denote by $\|\cdot\|_{C,k}$ the norm which is given as the Minkowski functional of the order interval $[-k,k]_C$ (see Theorem 2.2.10). Then*

$$\|y-a\|_{C,k}=\max\{\varphi_{a+C,-k}(y),\varphi_{a-C,k}(y)\}\ \textit{for all } y\in Y.$$

*Hence,*

$$\|y-a\|_{C,k}=\begin{cases}\varphi_{a-C,k}(y) & \textit{if } y\in a+C,\\ \varphi_{a+C,-k}(y) & \textit{if } y\in a-C.\end{cases}\tag{5.5.1}$$

*The problem*

$$\min_{y\in F}\|y-a\|_{C,k}$$

*is equivalent to the problem* $(\mathrm{P}_{F,a,C,k})$ *for each set* $F\subseteq a+C$
*and equivalent to the problem* $(\mathrm{P}_{F,a,-C,-k})$ *for each set* $F\subseteq a-C$.

***Proof*** Apply Proposition 4.13.3 to $A=C$ and to $y-a$ instead of $y$. Lemma 4.3.5 yields the assertion. □

Jahn [269, 270] characterized solutions to nonconvex vector optimization problems in linear spaces with a convex cone as domination set by minimizers of norms. He also used special Minkowski functionals for the construction of norms.

In many applications, solutions are determined by problems $\min_{y\in F}\|y-a\|_{C,k}$ with $F\subseteq a+C$. Replacing $\|y-a\|_{C,k}$ by $\varphi_{a-C,k}(y)$, this approach can often be applied without the assumption $F\subseteq a+C$. This is illustrated for the scalarization of vector optimization problems with the weighted Chebyshev norm and with extensions of this norm in [472] and [477]. We will go more into detail in Chap. 7.

Finally, we consider the norm which was investigated in Proposition 4.13.5 and prove the following result from [487, Cor. 7.7].

**Proposition 5.5.3** *Assume that* $Y$ *is a linear space,* $C\subset Y$ *is an algebraically closed pointed convex cone with* $k\in \operatorname{core} C$ *and* $a\in Y$. *Define*

$$\mu_{C,k}(y):=\max\{\varphi_{C,-k}(y),0\}+\max\{\varphi_{C,-k}(-y),0\}\quad \textit{for all } y\in Y.$$

*Then*

$$\mu_{C,k}(y-a)=\begin{cases}\varphi_{a-C,k}(y) & \textit{if } y\in a+C,\\ \varphi_{a+C,-k}(y) & \textit{if } y\in a-C,\\ \varphi_{a+C,-k}(y)+\varphi_{a-C,k}(y) & \textit{otherwise.}\end{cases}\qquad(5.5.2)$$

*The problem*

$$\min_{y\in F}\mu_{C,k}(y-a)$$

*is equivalent to the problem* $(\mathrm{P}_{F,a,C,k})$ *for each set* $F\subseteq a+C$ *and equivalent to the problem* $(\mathrm{P}_{F,a,-C,-k})$ *for each set* $F\subseteq a-C$.

***Proof*** Apply Proposition 4.13.5 to $A=C$ and to $y-a$ instead of $y$. Lemma 4.3.5 yields the assertion. □

# Chapter 6
# Vector Optimization in General Spaces

In this chapter, vector optimization problems in linear spaces are studied, where the notion of vector optimization will be used in a very general way. A vector optimization problem with a variable domination structure is equivalent to the problem of finding nondominated solutions in a set related to some arbitrary binary relation. Each decision problem can be shown to be of this type. Section 6.1 contains the basic concept for variable domination structures and connects it to decision problems. In Sects. 6.2 and 6.3, we investigate the case where the domination structure can be described by a single set—the domination set of the vector optimization problem.

In Sect. 6.1, we will start with a general decision problem, where the decision maker's preferences are composed in a domination structure as well as in some alternative structure. Optimal decisions are investigated as optima w.r.t. the preference relation, and the results can straightforwardly be transferred to optima w.r.t. arbitrary binary relations. It will be proved that—even for variable domination structures—efficient points w.r.t. some fixed reference set can be helpful in finding the best solutions. The following definition is illustrated by Fig. 6.1.

**Definition 6.0.1** Suppose $Y$ to be a linear space and $F, D \subseteq Y$. An element $y^0 \in F$ is called an **efficient point** of $F$ w.r.t. $D$ if

$$F \cap (y^0 - D) \subseteq \{y^0\}.$$

We denote the set of efficient points of $F$ w.r.t. $D$ by $\mathrm{Eff}(F, D)$ and call $D$ the **reference set** of the efficient point set $\mathrm{Eff}(F, D)$.

Here $D$ is not necessarily the domination set of the vector optimization problem and not necessarily a cone. This general approach is motivated by the following facts:



C. Tammer and P. Weidner, *Scalarization and Separation by Translation Invariant Functions*, Vector Optimization, https://doi.org/10.1007/978-3-030-44723-6_6

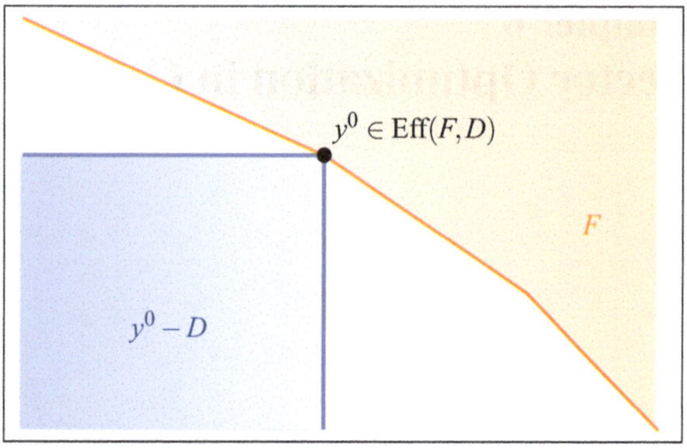


**Fig. 6.1** An efficient point $y^0$ of $F$ with respect to $D$

- Efficiency immediately corresponds to the separation of sets.
- The preference relation of a decision maker is not always a partial order.
- Weak efficiency can be handled as efficiency w.r.t. the algebraic or topological interior of the domination set.
- An efficient point set w.r.t. an appropriate reference set can act as a substitute for the weakly efficient point set in the case where the interior of the domination set is empty.
- Proper efficiency can be defined by efficient point sets w.r.t. special sets which contain the domination set.
- Data perturbations, uncertainties of the decision maker or criteria which cannot be expressed as functions can cause a domination set which is not a cone and does not contain zero in its boundary (see e.g. [208, 337]).
- In multiobjective optimization, different scalarization methods are applied. They can be interpreted as decision making with a special domination structure. Ester [159] introduced a scalarization based on the idea of a strictly convex domination set (see Sect. 7.3.10). In [479], this scalarization is used to calculate trade-off directions.
- For certain vector optimization problems, the domination set can be mapped from the space of outcomes into the decision space in such a way that efficient solutions can be determined immediately in the decision space. The image of the domination set in the decision space usually has other properties than the domination set.

Moreover, the general assumptions for efficiency result in more transparent proofs, new insights in the necessity of conditions, and help to uncover the reason for certain properties. Results which were proved under special assumptions can often be transferred to other applications.

Properties of efficient point sets will be investigated in Sect. 6.2. In this section, the vector optimization problem for arbitrary domination sets will be introduced, where we first focus on efficient and weakly efficient solutions, including surrogates for the weakly efficient point set and problems with uncertainties or perturbations. Scalarization results will be proved, with an emphasis on translation invariant functions and implications for norms. Beside this, some basic properties of efficient and weakly efficient points will be studied, especially their existence. Proper efficiency and related subsets of the efficient point set will be examined in Sect. 6.3, based on the results from Sect. 6.2. Detailed bibliographical notes are contained in Sect. 6.4.

## 6.1 Decision Making and Optimality Defined by Relations or Sets

Consider the following general decision problem:
A decision maker (DM) wants to make a decision by choosing an element from a set $S$ of feasible decisions, where the outcomes of the decisions are given by some function $f : S \to Y$, $Y$ being some arbitrary set.

What is a best decision depends on the DM's preferences in the set $F := f(S)$ of outcomes. A binary relation $\succ$ is called a **strict preference relation** on $Y$ if

$$\forall y^1, y^2 \in Y\colon (y^1 \succ y^2 \iff y^1 \text{ is strictly preferred to } y^2).$$

A binary relation $\succ$ is said to be a **weak preference relation** on $Y$ if

$$\forall y^1, y^2 \in Y\colon (y^1 \succ y^2 \iff y^2 \text{ is not preferred to } y^1).$$

Let $\succ$ denote the DM's strict or weak preference relation on $Y$. Then the set of decision outcomes that are optimal for the DM is

$$\mathrm{ND}(F, \succ) = \{y^0 \in F \mid \forall y \in F : (y \succ y^0 \Rightarrow y^0 \succ y)\},$$

the set of nondominated elements of $F$ w.r.t. $\succ$, i.e., optimal decisions are the elements of $\mathrm{ND}_f(S, \succ) := \{x \in S \mid f(x) \in \mathrm{ND}(f(S), \succ)\}$.

Note that $\succ$ consists of the preferences the DM is aware of. This relation is refined during the decision process, but fixed in each single step of the decision process, where information about $\mathrm{ND}(F, \succ)$ should support the DM in formulating further preferences. In the final phase of the decision process, the DM chooses one decision, but in the previous phases $\mathrm{ND}(F, \succ)$ contains more than one element.

Although the investigation in this section is motivated by the above decision problem, the results can be applied to nondominated elements with respect to an arbitrary binary relation.

If $\succ$ is asymmetric or antisymmetric, then

$$\mathrm{ND}(F, \succ) = \{y^0 \in F \mid \nexists y \in F \setminus \{y^0\}\colon y \succ y^0\}.$$

Each strict preference relation is asymmetric.

**Definition 6.1.1** Suppose $\succ$ to be a binary relation on the linear space $Y$.

$d \in Y$ is said to be a **domination factor** of $y \in Y$ if $y \succ y + d$. We define the **domination structure** of $\succ$ by $D_\succ : Y \to \mathcal{P}(Y)$ with $D_\succ(y) := \{d \in Y \mid y \succ y + d\}$ for each $y \in Y$. If there exists a set $D \subseteq Y$ with $D_\succ(y) = D$ for all $y \in Y$, then $D$ is called the **domination set** of $\succ$. We call $D \subseteq Y$ the **domination set** of $\succ$ on $F \subseteq Y$ if $D_\succ(y) = D$ for all $y \in F$.

The problem to find nondominated elements of a set $F$ w.r.t. some binary relation is also called a **vector optimization problem with a variable domination structure**. If we call it a **vector optimization problem** only, then we usually assume that a domination set exists.

A transitive relation can have a domination structure which consists of closed convex cones without having a domination set.

**Example 6.1.2** On $Y = \mathbb{R}^2$, consider the binary relation $\succ$ which is given by

$$D_\succ(y^0) := \begin{cases} \{(y_1, y_2)^T \in Y \mid y_2 = 0\} & \text{if } y_1^0 = 0, \\ \{(y_1, y_2)^T \in Y \mid y_1 \geq 0,\, y_2 = 0\} & \text{if } y_1^0 > 0, \\ \{(y_1, y_2)^T \in Y \mid y_1 \leq 0,\, y_2 = 0\} & \text{if } y_1^0 < 0, \end{cases}$$

for each $y^0 \in Y$.

Definition 6.1.1 implies:

**Proposition 6.1.3** *Suppose $\succ$ to be a binary relation on the linear space $Y$, $D \subseteq Y$.*
*There exists a domination set of $\succ$ if and only if:*

$$\forall y^1, y^2, y \in Y : \;\; (y^1 \succ y^2 \implies (y^1 + y) \succ (y^2 + y)). \tag{6.1.1}$$

*$D$ is a domination set of $\succ$ if and only if:*

$$\forall y^1, y^2 \in Y : \;\; (y^1 \succ y^2 \iff y^2 \in y^1 + D).$$

*If $D$ is a domination set of $\succ$, then we have:*

(a) *$\succ$ is reflexive $\iff 0 \in D$.*
(b) *$\succ$ is asymmetric $\iff D \cap (-D) = \emptyset$.*
(c) *$\succ$ is antisymmetric $\iff D \cap (-D) = \{0\}$.*
(d) *$\succ$ is transitive $\iff D + D \subseteq D$.*
(e) *$\succ$ fulfills the condition*

$$\forall\, y^1, y^2 \in Y \;\; \forall \lambda \in \mathbb{R}_> : \;\; y^1 \succ y^2 \implies (\lambda y^1) \succ (\lambda y^2) \tag{6.1.2}$$

*if and only if $D \cup \{0\}$ is a cone.*
(f) *$\succ$ is a transitive relation that fulfills condition (6.1.2) if and only if $D \cup \{0\}$ is a convex cone.*
(g) *$\succ$ is a partial order that fulfills condition (6.1.2) if and only if $D$ is a pointed convex cone.*

**Example 6.1.4** Not each preference relation fulfills the conditions (6.1.1) and (6.1.2). Consider $y$ to be the number of teaspoons full of sugar that a person puts into a cup of coffee. Two could be preferred to one, but probably not $2+2$ to $1+2$ or $2 \times 2$ to $2 \times 1$.

The domination set of the partial order $\leq_C$ is the ordering cone $C$.

A domination set can contain zero in its interior, but it is also possible that it does not contain zero in its closure.

**Example 6.1.5** Pirlot and Vincke [382] intensively studied relations of the following type.

A binary relation $\mathcal{R}$ is said to be a semiorder on the set $A$ if there exist a real-valued function $g$ and $\varepsilon \in \mathbb{R}_>$ such that

$$\forall a^1, a^2 \in A\colon (a^1 \mathcal{R} a^2 \iff g(a^1) \geq g(a^2) - \varepsilon).$$

The binary relation $\mathcal{R}^S$ is called the strict part of the given semiorder on $A$ if

$$\forall a^1, a^2 \in A\colon (a^1 \mathcal{R}^S a^2 \iff g(a^1) > g(a^2) + \varepsilon).$$

The idea behind these definitions is the indifference between two elements $a^1, a^2$ whenever $|g(a^1) - g(a^2)| \leq \varepsilon$. Pirlot and Vincke gave numerous examples of practical applications.

If $g$ is the mapping which assigns a computer number of a given datatype to a real number, then the "$\geq$"-comparison of two numbers via their computer numbers is a semiorder, and the "$>$"-comparison of these numbers via their computer numbers is the strict part of this semiorder.

In the image space $\mathbb{R}$, the domination set of the strict part of each semiorder is $-\varepsilon - \mathbb{R}_>$, whereas the domination set of the semiorder is $\varepsilon - \mathbb{R}_+$ and thus contains zero in its interior. The domination set $\varepsilon - \mathbb{R}_+$ does not fulfill any of the properties in Proposition 6.1.3(b)–(g).

A domination set may be bounded.

**Example 6.1.6** Consider the following decision situation: A location for a new plant has to be chosen. The location has to fulfill the following conditions:

- It has to be built in an existing settlement.
- Since north-northeast winds prevail in the area, the plant has to be situated more than four miles away from all other settlements to the south-southwest.

We put a coordinate system into the considered area such that one unit corresponds to one mile and each settlement has nonnegative coordinates. The set of these coordinate pairs is the set $F$ of possible locations. An element $y^0 \in F$ meets the decision maker's requirements if and only if there does not exist any other $y \in F$ with $y^0 \in y + D$, where $D := \{(y_1, y_2)^T \in \mathbb{R}^2 \mid y_1^2 + y_2^2 \leq 16,\ y_2 \geq y_1 \geq 0\}$. Hence, the solution set of the problem is $\mathrm{ND}(F, \succ) = \{y^0 \in F \mid \nexists y \in F \setminus \{y^0\}\colon y \succ y^0\}$, where $\succ$ is the binary relation on $Y$ given by the domination set $D$.

We will now discuss the transitivity condition $D + D \subseteq D$. Note that $D + D \subseteq D$ implies $nD \subseteq D$ for each $n \in \mathbb{N}_>$.

**Proposition 6.1.7** *Assume that $Y$ is a linear space. A nonempty subset $D$ of $Y$ fulfills the condition $D + D \subseteq D$ under each of the following assumptions:*

(a) *$D$ is a convex cone.*
(b) *$D = a + A$ for $A \subseteq Y$ with $A + A \subseteq A$ and $a \in A$.*
(c) *$D = A + H$ for two sets $A, H \subseteq Y$ with $A + A \subseteq A$ and $H \subseteq A$.*
(d) *$D = A \setminus B$ for some sets $A, B \subseteq Y$ for which $A \setminus B$ is convex, $A \cup \{0\}$ is a pointed convex cone, and*

$$\forall y \in Y \setminus \{0\}\ \exists t \in \mathbb{R}\colon\ (t > 1 \wedge ty \notin B).$$

***Proof***

(a) –(c) are easy to verify.
(d) Take any $d^1, d^2 \in D \setminus \{0\}$. Then $a := d^1 + d^2 \neq 0$. There exists $t \in \mathbb{R}$ with $t > 1$ and $ta \notin B$. Since $A \cup \{0\}$ is a convex cone, we have $ta \in D$. For $t_1 := \frac{t}{2t-1} \in (0, 1)$ and $t_2 := 1 - \frac{1}{t} \in (0, 1)$, we get $t_3 := 1 - t_1 \in (0, 1)$ and $d^3 := t_2 d^2 + (1 - t_2)ta \in D$ since $D$ is convex. Then

$$t_3 d^1 + (1 - t_3)d^3 \in D. \tag{6.1.3}$$

We obtain $t_3 d^1 + (1 - t_3)d^3 = t_3 d^1 + (1 - t_3)t_2 d^2 + (1 - t_3)(1 - t_2)t(d^1 + d^2)$. Since $(1 - t_3)(1 - t_2)t = t_1 \frac{1}{t} t = t_1$ and $(1 - t_3)t_2 = t_1 t_2 = \frac{t}{2t-1} - \frac{1}{2t-1} = 1 - \frac{t}{2t-1} = 1 - t_1$, we get $t_3 d^1 + (1 - t_3)d^3 = (1 - t_1)d^1 + (1 - t_1)d^2 + t_1(d^1 + d^2) = d^1 + d^2$. Equation (6.1.3) yields $d^1 + d^2 \in D$. □

The next lemma is related to Proposition 6.1.7(b).

**Lemma 6.1.8** *Assume that $Y$ is a topological vector space, $A \subset Y$ is a nontrivial cone, $y \in Y$ and $D := y + A$ fulfills the condition $D + D \subseteq D$. Then $y \in \operatorname{cl} A$.*

***Proof*** Take any $a \in A \setminus \{0\}$. Then $\mathbb{R}_> a \subseteq A \setminus \{0\}$, and $(y + \mathbb{R}_> a) + (y + \mathbb{R}_> a) \subseteq D + D \subseteq D = y + A$. Hence, $y + \mathbb{R}_> a = y + \mathbb{R}_> a + \mathbb{R}_> a \subseteq A$, which implies $y \in \operatorname{cl} A$. □

In Fig. 6.2, some examples of sets $D$ with $D + D \subseteq D$ are given. In Fig. 6.2c, $A$ is the interior of a convex cone, and the boundary of $D$ is one branch of a hyperbola. Furthermore, the set $D := \{y \in \mathbb{R} \mid \exists n \in \mathbb{N}_>\colon |y - n| < \frac{n}{10}\}$, i.e.

$$D = \left(\frac{9}{10}, \frac{11}{10}\right) \bigcup \left(\frac{18}{10}, \frac{22}{10}\right) \bigcup \left(\frac{27}{10}, \frac{33}{10}\right) \bigcup \left(\frac{36}{10}, \frac{44}{10}\right) \bigcup \left\{y \in \mathbb{R} \mid y > \frac{45}{10}\right\},$$

is an example of a nonconvex set which fulfills $D + D \subseteq D$, has a nonempty interior and is not connected.

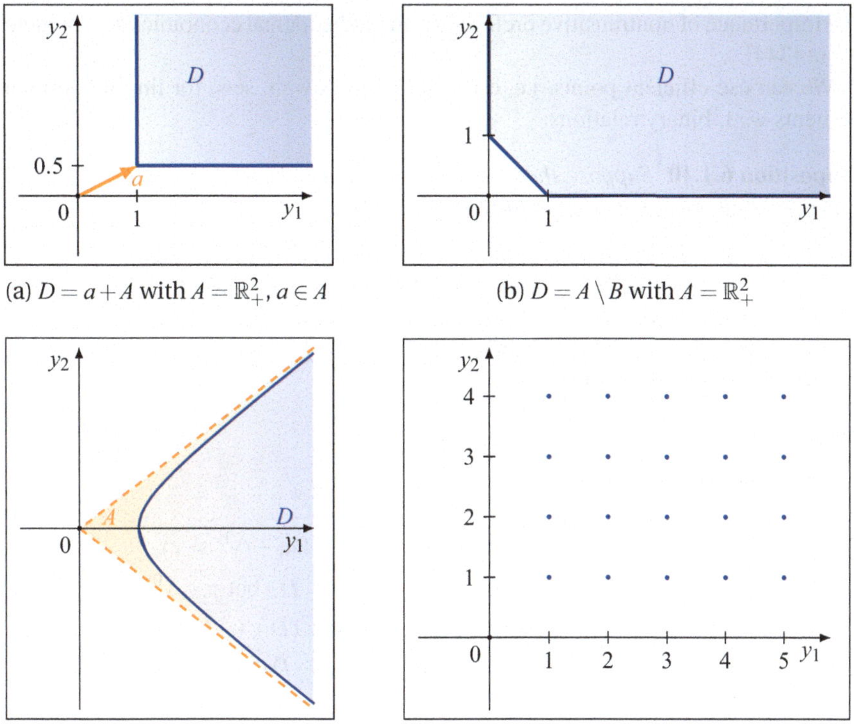


(a) $D = a + A$ with $A = \mathbb{R}^2_+$, $a \in A$

(b) $D = A \setminus B$ with $A = \mathbb{R}^2_+$

(c) $D = A \setminus B$

(d) $D = \mathbb{N}_> \times \mathbb{N}_>$

**Fig. 6.2** Sets $D$ fulfilling the condition $D + D \subseteq D$. Adapted from [464, p. 23]

In applications, the transitivity condition can be important for subsets or extensions of the domination set.

**Proposition 6.1.9** *Assume that $Y$ is a linear space, $D \subseteq Y$ and $D + D \subseteq D$.*

(a) $D^1 + D^1 \subseteq D^1$ *for* $D^1 := D \cap (-D)$.
(b) $D^2 + D^2 \subseteq D^2$ *for* $D^2 := D \setminus (-D)$.
(c) $D^3 + D^3 \subseteq D^3$ *for* $D^3 := D \cup \{0\}$.
(d) *If $Y$ is a topological vector space, then* $\operatorname{cl} D + \operatorname{cl} D \subseteq \operatorname{cl} D$.

***Proof***

(a) $D + D \subseteq D$ implies $(-D) + (-D) \subseteq -D$ and thus the assertion.
(b) Take any $d^1, d^2 \in D^2$. Consider $d := d^1 + d^2 \in D$. If $d \in -D$, then $-d^2 = d^1 - d \in D + D \subseteq D$, which contradicts $d^2 \in D^2$.
(c) is obvious.
(d) $\operatorname{cl} D + \operatorname{cl} D \subseteq \operatorname{cl}(D + D) \subseteq \operatorname{cl} D$.

□

The importance of nontransitive preferences in mathematical economics was pointed out in [344].

We can use efficient points, i.e. optimal elements w.r.t. sets, for finding optimal elements w.r.t. binary relations.

**Proposition 6.1.10** *Suppose that $Y$ is a linear space, $F, D \subseteq Y$, and that $\succ$ is a binary relation on $Y$ with domination structure $D_\succ$.*

(a) *If $D_\succ(y) = D$ for all $y \in F$, then* $\mathrm{ND}(F, \succ) = \mathrm{Eff}(F, D \setminus (-D))$.
*If $\succ$ is asymmetric or antisymmetric, then* $\mathrm{ND}(F, \succ) = \mathrm{Eff}(F, D)$.
(b) *If $D_\succ(y) \subseteq D$ for all $y \in F$, then* $\mathrm{Eff}(F, D) \subseteq \mathrm{ND}(F, \succ)$.
(c) *If $\succ$ is asymmetric or antisymmetric and $D \subseteq D_\succ(y)$ for all $y \in F$, then* $\mathrm{ND}(F, \succ) \subseteq \mathrm{Eff}(F, D)$.

***Proof***

(a) Take any $y^0 \in F$.

$$\begin{aligned} y^0 \notin \mathrm{ND}(F, \succ) &\iff \exists y \in F : y \succ y^0, \text{ but } \neg(y^0 \succ y), \\ &\iff \exists y \in F : y^0 \in y + D, \text{ but } y \notin y^0 + D, \\ &\iff \exists y \in F : y^0 \in y + (D \setminus (-D)), \\ &\iff y^0 \notin \mathrm{Eff}(F, D \setminus (-D)). \end{aligned}$$

If $\succ$ is asymmetric or antisymmetric, then $D \cap (-D) \subseteq \{0\}$. Thus, $D \setminus (-D) = D \setminus \{0\}$. This implies $\mathrm{Eff}(F, D \setminus (-D)) = \mathrm{Eff}(F, D)$.
(b) Take any $y^0 \in F \setminus \mathrm{ND}(F, \succ)$. Then there exists $y \in F$ for which $y \succ y^0$, but not $y^0 \succ y$ holds. Then $y^0 \in y + D_\succ(y)$, and $y \in y^0 - D_\succ(y) \subseteq y^0 - D$. Consequently, $y^0 \notin \mathrm{Eff}(F, D)$.
(c) Take any $y^0 \in F \setminus \mathrm{Eff}(F, D)$. Then there exists $y \in F \setminus \{y^0\}$ with $y \in y^0 - D \subseteq y^0 - D_\succ(y)$. We get $y^0 \in y + D_\succ(y)$ and thus $y \succ y^0$. Therefore, $y^0 \notin \mathrm{ND}(F, \succ)$ since $\succ$ is asymmetric or antisymmetric. □

The domination factors refer to elements that are dominated. Of course, a structure could also be built by dominating elements.

**Definition 6.1.11** Suppose $\succ$ to be a binary relation on the linear space $Y$.

$\tilde{d} \in Y$ is said to be a **pre-domination factor** of $y \in Y$ if $y - \tilde{d} \succ y$. We define the **pre-domination structure** of $\succ$ by $\tilde{D}_\succ : Y \to \mathcal{P}(Y)$ with $\tilde{D}_\succ(y) := \{\tilde{d} \in Y \mid y - \tilde{d} \succ y\}$ for each $y \in Y$.

A pre-domination structure is constant on the entire space if and only if this is the case for the domination structure of the same relation.

**Proposition 6.1.12** *Suppose $\succ$ to be a binary relation on the linear space $Y$ with pre-domination structure $\tilde{D}_\succ$. There exists a domination set $D \subseteq Y$ of $\succ$ if and only if $\tilde{D}_\succ(y) = D$ for all $y \in Y$.*

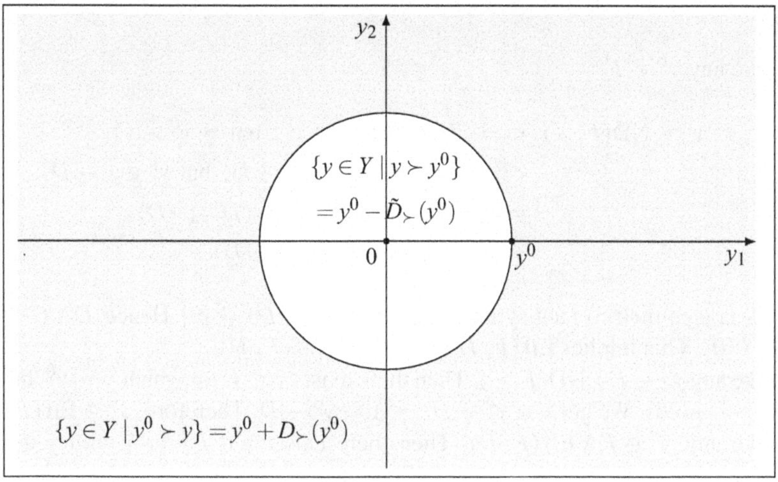


**Fig. 6.3** Domination structure $D_\succ$ and pre-domination structure $\tilde{D}_\succ$ in Example 6.1.13

***Proof*** Let $D_\succ$ denote the domination structure of $\succ$. Then

$$\begin{aligned}(\forall y \in Y\colon D_\succ(y) = D) &\iff \forall y \in Y\colon (y \succ y + d \iff d \in D),\\ &\iff \forall y \in Y\colon (y - d \succ y \iff d \in D),\\ &\iff \forall y \in Y\colon \tilde{D}_\succ(y) = D.\end{aligned}$$

□

The pre-domination structure may consist of convex sets even though this is not the case for the domination structure.

**Example 6.1.13** On $Y = \mathbb{R}^2$, define the binary relation $\succ$ by

$$y^1 \succ y^2 \iff \|y^1\|_2 \le \|y^2\|_2.$$

The pre-domination structure, but not the domination structure, consists of convex sets (see Fig. 6.3).

Analogously to Proposition 6.1.10, the following relationships between pre-domination structures and efficiency w.r.t. sets hold.

**Proposition 6.1.14** *Suppose that $Y$ is a linear space, $F$, $D \subseteq Y$, and that $\succ$ is a binary relation on $Y$ with pre-domination structure $\tilde{D}_\succ$.*

(a) *Assume $\tilde{D}_\succ(y) = D$ for all $y \in F$. Then* $\mathrm{ND}(F, \succ) = \mathrm{Eff}(F, D \setminus (-D))$.
*If $\succ$ is asymmetric or antisymmetric, then* $\mathrm{ND}(F, \succ) = \mathrm{Eff}(F, D)$.
(b) *If $\tilde{D}_\succ(y) \subseteq D$ for all $y \in F$, then* $\mathrm{Eff}(F, D) \subseteq \mathrm{ND}(F, \succ)$.
(c) *If $\succ$ is asymmetric or antisymmetric and $D \subseteq \tilde{D}_\succ(y)$ for all $y \in F$, then* $\mathrm{ND}(F, \succ) \subseteq \mathrm{Eff}(F, D)$.

***Proof***

(a) Take any $y^0 \in F$.

$$\begin{aligned} y^0 \notin \mathrm{ND}(F, \succ) &\iff \exists y \in F: \ y \succ y^0, \ \text{but} \ \neg(y^0 \succ y), \\ &\iff \exists y \in F: \ y \in y^0 - D, \ \text{but} \ y^0 \notin y - D, \\ &\iff \exists y \in F: \ y^0 \in y + (D \setminus (-D)), \\ &\iff y^0 \notin \mathrm{Eff}(F, D \setminus (-D)). \end{aligned}$$

If $\succ$ is asymmetric or antisymmetric, then $D \cap (-D) \subseteq \{0\}$. Hence, $D \setminus (-D) = D \setminus \{0\}$. This implies $\mathrm{Eff}(F, D \setminus (-D)) = \mathrm{Eff}(F, D)$.

(b) Take any $y^0 \in F \setminus \mathrm{ND}(F, \succ)$. Then there exists $y \in F$ for which $y \succ y^0$, but not $y^0 \succ y$ holds. We get $y \in y^0 - \tilde{D}_\succ(y^0) \subseteq y^0 - D$. Therefore, $y^0 \notin \mathrm{Eff}(F, D)$.

(c) Take any $y^0 \in F \setminus \mathrm{Eff}(F, D)$. Then there exists $y \in F \setminus \{y^0\}$ with $y \in y^0 - D \subseteq y^0 - \tilde{D}_\succ(y^0)$. Thus, $y \succ y^0$. We obtain $y^0 \notin \mathrm{ND}(F, \succ)$ since $\succ$ is asymmetric or antisymmetric.

□

Since the domination structure as well as the pre-domination structure completely characterize the binary relation, the nondominated point set w.r.t. the relation can be described via the domination structure or via the pre-domination structure.

**Lemma 6.1.15** *Suppose that $Y$ is a linear space, $F \subseteq Y$, and that $\succ$ is a binary relation on $Y$ with domination structure $D_\succ$ and pre-domination structure $\tilde{D}_\succ$. Then*

$$\begin{aligned} \mathrm{ND}(F, \succ) &= \{y^0 \in F \mid \forall y \in F: \ (y^0 \in y + D_\succ(y) \Rightarrow y \in y^0 + D_\succ(y^0))\} \\ &= \{y^0 \in F \mid \forall y \in F: \ (y \in y^0 - \tilde{D}_\succ(y^0) \Rightarrow y^0 \in y - \tilde{D}_\succ(y))\} \\ &= \{y^0 \in F \mid \forall y \in F: \ (y^0 \in y + \tilde{D}_\succ(y^0) \Rightarrow y \in y^0 + \tilde{D}_\succ(y))\}. \end{aligned}$$

*If $\succ$ is asymmetric or antisymmetric, we get*

$$\begin{aligned} \mathrm{ND}(F, \succ) &= \{y^0 \in F \mid \nexists y \in F \setminus \{y^0\}: \ y^0 \in y + D_\succ(y)\} \\ &= \{y^0 \in F \mid \nexists y \in F \setminus \{y^0\}: \ y \in y^0 - \tilde{D}_\succ(y^0)\} \\ &= \{y^0 \in F \mid \nexists y \in F \setminus \{y^0\}: \ y^0 \in y + \tilde{D}_\succ(y^0)\}. \end{aligned}$$

The investigation of optimality related to variable domination structures will be continued in Chap. 11. In the next sections, we turn to the case where the domination structure can be described by a domination set. Then the nondominated point set becomes an efficient point set.

## 6.2 Efficient and Weakly Efficient Points in Vector Optimization

In this section, we investigate properties of efficient point sets w.r.t. arbitrary reference sets and vector optimization problems with arbitrary domination sets. For these vector optimization problems, the efficient point set, the weakly efficient point set and surrogates for the weakly efficient point set will be studied. The investigation includes results for problems with uncertainties or perturbations. The considered type of vector optimization problems and solutions will be introduced in Sect. 6.2.1. Section 6.2.1 contains some basic properties of these solution sets. In this subsection, sufficient conditions for the solutions are examined which make use of scalarizing functionals. Scalarization results using translation invariant functions are proved in Sect. 6.2.2. Implications for the scalarization by norms will be presented in Sect. 6.2.3. In Sect. 6.2.4, statements for the existence of efficient and weakly efficient points will be deduced. Additional scalarization and existence results for efficient and weakly efficient points w.r.t. closed convex cones will be proved in Sect. 6.2.5.

### *6.2.1 Vector Optimization Problems and Their Solutions*

Consider the **vector optimization problem** given as

$$D\text{-}\operatorname*{Opt}_{s\in S} f(s) \tag{VOP}$$

under the assumption that $S_0$ is a nonempty set with $S \subseteq S_0$, $Y$ is a linear space, $f\colon S_0 \to Y$, and that $D$ is a nonempty subset of $Y$.

$S$ stands for the set of feasible elements. The value of $f$ expresses properties and/or evaluations of $s$ which are of interest in the specific context. $D$ is the set which defines the solution concept for (VOP). For each $y^0 \in F := f(S)$, imagine that the set of elements in $F$ which is preferred to $y^0$ is $F \cap (y^0 - (D \setminus \{0\}))$. Then we are interested in the set $\operatorname{Eff}(F, D) = \{y^0 \in F \mid F \cap (y^0 - D) \subseteq \{y^0\}\}$ of efficient points of $F$ w.r.t. $D$ and in the set $\operatorname{Eff}_f(S, D) := \{s^0 \in S \mid f(s^0) \in \operatorname{Eff}(f(S), D)\}$ of **efficient solutions** to (VOP). We will call $D$ the **domination set** of (VOP).

The author of this chapter studied vector optimization problems under such general assumptions motivated by decision theory [463, 464, 472]. In this chapter, only those results are presented which will be used in the context of our book. If $D$ is an ordering cone in $Y$, then $\operatorname{Eff}(F, D)$ is the set of elements of $F$ that are nondominated w.r.t. the cone order $\leq_D$. In the literature, vector optimization problems are usually defined with domination sets that are ordering cones. The sets $\operatorname{Eff}(F, D)$ and $\operatorname{Eff}_f(S, D)$ are denoted in the literature in different ways, e.g. as sets of minimal, efficient, nondominated or Pareto optimal points.

Efficiency in other than Euclidean spaces was applied by Peleg [378] to prices for optimal consumption plans, based on statements by Radner [387] in the space of all bounded sequences of real numbers. Section 15.2 deals with a vector-valued optimal control problem having outcomes in a Banach space.

Although the vector optimization problem gives rise to the investigation in this and in the following section, the results of both sections can independently be applied to efficient point sets of some arbitrary set $F$ w.r.t. an arbitrary reference set $D$.

First, we summarize equivalent conditions for efficiency.

**Lemma 6.2.1** *Assume that $Y$ is a linear space, $F, D \subseteq Y$, $y^0 \in F$. Then we have:*

$$\begin{aligned} y^0 \in \mathrm{Eff}(F, D) &\iff F \cap (y^0 - D) \subseteq \{y^0\}, \\ &\iff (F - y^0) \cap (-D) \subseteq \{0\}, \\ &\iff \nexists y \in F \setminus \{y^0\} : y \in y^0 - D, \\ &\iff \nexists y \in F \setminus \{y^0\} : y - y^0 \in -D, \\ &\iff F \setminus \{y^0\} \subseteq y^0 + (Y \setminus (-D)). \end{aligned}$$

The last equivalence states that efficient points of $F$ w.r.t. $D$ are so-called ideal elements of $F$ w.r.t. $Y \setminus (-D)$. This fact was pointed out in [209]. Moreover, it is shown in [209] that the main approximate solution concepts can be described by sets $\mathrm{Eff}(F, D)$ with an appropriate set $D$.

If $Y$ is a topological vector space, then it turns out that—in general—it is easier to determine solutions to vector optimization problems w.r.t. open domination sets.

**Definition 6.2.2** Suppose $Y$ to be a topological vector space and $F, D \subseteq Y$. We define

$$\mathrm{WEff}(F, D) := \mathrm{Eff}(F, \mathrm{int}\, D)$$

as the set of **weakly efficient points** of $F$ w.r.t. $D$ and

$$\mathrm{WEff}_f(S, D) := \mathrm{Eff}_f(S, \mathrm{int}\, D)$$

as the set of **weakly efficient solutions** to (VOP).

In linear spaces, it is also usual to define weak efficiency as efficiency w.r.t. the algebraic interior of the domination set [272]. Proposition 2.3.6 implies by Lemma 6.2.5(b):

**Proposition 6.2.3** *Assume that $Y$ is a topological vector space, $F, D \subseteq Y$. Then*

$$\mathrm{Eff}(F, \mathrm{core}\, D) \subseteq \mathrm{WEff}(F, D).$$

*Moreover,* $\mathrm{Eff}(F, \mathrm{core}\, D) = \mathrm{WEff}(F, D)$ *holds for a convex set $D$ under each of the following conditions:*

(a) $Y = \mathbb{R}^n$ *(equipped with the Euclidean topology),*
(b) $\mathrm{int}\, D \neq \emptyset$,

(c) *Y is a Banach space and D is closed.*

The sets $\mathrm{Eff}(F, D + \mathbb{R}_{>}k)$ show certain properties which usually only weakly efficient point sets have (see Theorem 6.2.29). In vector optimization, we can work with these sets instead of the weakly efficient point set if $D$ has an empty interior. By Lemmata 2.3.17 and 2.3.18, we get the following from Proposition 6.2.3.

**Proposition 6.2.4** *Assume that $Y$ is a linear space, $F, D \subseteq Y$.*

(a) *If $k \in 0^+D \setminus \{0\}$, then* $\mathrm{Eff}(F, D) \subseteq \mathrm{Eff}(F, D + \mathbb{R}_{>}k)$.
(b) $\mathrm{Eff}(F, D + \mathbb{R}_{>}k) \subseteq \mathrm{Eff}(F, \mathrm{core}\, D)$ *for each $k \in Y \setminus \{0\}$.*
(c) $\mathrm{Eff}(F, \mathrm{core}\, D) = \mathrm{Eff}(F, D + \mathbb{R}_{>}k)$ *if $D + \mathbb{R}_{>}k \subseteq \mathrm{core}\, D$, especially if $k \in \mathrm{core}\, 0^+ D$.*
(d) *If $Y$ is a topological vector space, then* $\mathrm{Eff}(F, D + \mathbb{R}_{>}k) \subseteq \mathrm{WEff}(F, D)$ *for each $k \in Y \setminus \{0\}$, and* $\mathrm{WEff}(F, D) = \mathrm{Eff}(F, D + \mathbb{R}_{>}k)$ *if $k \in \mathrm{int}\, 0^+ D$.*

We will now prove some basic properties of efficient and weakly efficient point sets. These statements include relationships between the efficient point set of $F$ and the efficient point set of $F + D$. In applications, $F + D$ may be convex or closed even though $F$ does not have this property. Extensions of the feasible set which leave invariant the efficient point set were discussed in [466].

**Lemma 6.2.5** *Assume that $Y$ is a linear space, $F, D \subseteq Y$. Then we have:*

(a) $\mathrm{Eff}(F, D) = \mathrm{Eff}(F, D \cup \{0\}) = \mathrm{Eff}(F, D \setminus \{0\})$.
(b) $D_1 \subseteq D \implies \mathrm{Eff}(F, D) \subseteq \mathrm{Eff}(F, D_1)$.
(c) $F_1 \subseteq F \implies \mathrm{Eff}(F, D) \cap F_1 \subseteq \mathrm{Eff}(F_1, D)$.
(d) *Suppose $F \subseteq A \subseteq F + (D \cup \{0\})$.*
*Then* $\mathrm{Eff}(A, D) \subseteq \mathrm{Eff}(F, D)$.
*If $D \cap (-D) \subseteq \{0\}$ and $D + D \subseteq D$, then* $\mathrm{Eff}(A, D) = \mathrm{Eff}(F, D)$.
(e) *If $D + D \subseteq D$, then* $\mathrm{Eff}(F \cap (y - D), D) = \mathrm{Eff}(F, D) \cap (y - D)$ *for all $y \in Y$.*

***Proof***

(a) –(c) follow immediately from the definition of efficient points.
(d) Take any $y^0 \in \mathrm{Eff}(A, D)$. Since $A \subseteq F + (D \cup \{0\})$, there exist $y^1 \in F$ and $d \in D \cup \{0\}$ with $y^0 = y^1 + d$. Then $y^1 \in y^0 - (D \cup \{0\})$. We get $y^1 = y^0$ because of $y^0 \in \mathrm{Eff}(A, D)$ and $F \subseteq A$. Therefore, $y^0 \in F$. Hence, $\mathrm{Eff}(A, D) \subseteq F$. We obtain $\mathrm{Eff}(A, D) \subseteq \mathrm{Eff}(F, D)$ from (c) since $F \subseteq A$.
Assume now $D \cap (-D) \subseteq \{0\}$ and $D + D \subseteq D$. Contrary to the claim, suppose that $\mathrm{Eff}(A, D) \neq \mathrm{Eff}(F, D)$. Then there exists $y^0 \in \mathrm{Eff}(F, D) \setminus \mathrm{Eff}(A, D)$. Hence, there is $a \in A$ with $a \in y^0 - (D \setminus \{0\})$. The inclusion $A \subseteq F + (D \cup \{0\})$ implies the existence of $y \in F$ and $d \in D \cup \{0\}$ with $y + d \in y^0 - (D \setminus \{0\})$. Therefore, $y^0 - y \in d + (D \setminus \{0\}) \subseteq D$. We obtain $y^0 = y$ since $y^0 \in \mathrm{Eff}(F, D)$. Then $y + d \in y^0 - (D \setminus \{0\})$ implies $d \in -(D \setminus \{0\})$, which contradicts $D \cap (-D) \subseteq \{0\}$.

(e) Take any $y \in Y$. From (c), we get $\operatorname{Eff}(F \cap (y - D), D) \supseteq \operatorname{Eff}(F, D) \cap (y - D)$. To obtain a contradiction, suppose that there exists $y^0 \in \operatorname{Eff}(F \cap (y - D), D) \setminus \operatorname{Eff}(F, D)$. Then $(y^0 - D) \cap (F \cap (y - D)) \subseteq \{y^0\}$, and there is $y^1 \neq y^0$ with $y^1 \in F \cap (y^0 - D)$. We have $y^1 \in y^0 - D \subseteq (y - D) - D \subseteq y - D$. Thus, $y^1 \in (y^0 - D) \cap (F \cap (y - D))$, which contradicts $(y^0 - D) \cap (F \cap (y - D)) \subseteq \{y^0\}$.

□

Yu [502, p. 20] gave an example of a convex cone that is not pointed and for which $\operatorname{Eff}(F, D)$ is not a subset of $\operatorname{Eff}(F + D, D)$.

We obtain the following for weakly efficient points.

**Lemma 6.2.6** *Assume that $Y$ is a topological vector space, $F, D \subseteq Y$. Then:*

(a) $\operatorname{Eff}(F, D) \subseteq \operatorname{WEff}(F, D)$.
(b) $\operatorname{int} D_1 \subseteq \operatorname{int} D \implies \operatorname{WEff}(F, D) \subseteq \operatorname{WEff}(F, D_1)$.
(c) $F_1 \subseteq F \implies \operatorname{WEff}(F, D) \cap F_1 \subseteq \operatorname{WEff}(F_1, D)$.
(d) *Suppose* $F \subseteq A \subseteq F \cup (F + \operatorname{int} D)$.
*Then* $\operatorname{WEff}(A, D) \subseteq \operatorname{WEff}(F, D)$.
*If* $\operatorname{int} D \cap (-\operatorname{int} D) \subseteq \{0\}$ *and* $\operatorname{int} D + \operatorname{int} D \subseteq D$, *then* $\operatorname{WEff}(A, D) = \operatorname{WEff}(F, D)$.
(e) *Suppose* $F \subseteq A \subseteq F + (D \cup \{0\})$ *and* $D + \operatorname{int} D \subseteq D \setminus \{0\}$.
*Then* $\operatorname{WEff}(A, D) \cap F = \operatorname{WEff}(F, D)$.
(f) *If* $D + \operatorname{int} D \subseteq D$, *then* $\operatorname{WEff}(F \cap (y - D), D) = \operatorname{WEff}(F, D) \cap (y - D)$ *for all* $y \in Y$.
(g) $\operatorname{WEff}(F, D) = F$ *if* $\operatorname{int} D = \emptyset$.
(h) $\operatorname{WEff}(F, D) = \operatorname{WEff}(F, \operatorname{cl} D)$ *if* $D$ *is convex and* $\operatorname{int} D \neq \emptyset$.

***Proof***

(a) follows from Lemma 6.2.5(b) with $D_1 = \operatorname{int} D$.
(b) –(d) result from Lemma 6.2.5(b)–(d) when replacing $D$ and $D_1$ by their interior.
(e) Take any $y^0 \in F$ with $y^0 \notin \operatorname{WEff}(A, D)$. Then there exists $a \in A \setminus \{y^0\}$ with $y^0 \in a + \operatorname{int} D$. Since $A \subseteq F + (D \cup \{0\})$, there are $y^1 \in F$ and $d \in D \cup \{0\}$ such that $y^0 \in y^1 + d + \operatorname{int} D$. The inclusion $D + \operatorname{int} D \subseteq D \setminus \{0\}$ implies $0 \notin \operatorname{int} D$ and $y^0 \in y^1 + (\operatorname{int} D \setminus \{0\})$. Consequently, $y^0 \notin \operatorname{WEff}(F, D)$.
Hence, $\operatorname{WEff}(F, D) \subseteq \operatorname{WEff}(A, D)$. The assertion follows by (c) since $F \subseteq A$.
(f) Take any $y \in Y$. From (c), we get $\operatorname{WEff}(F \cap (y - D), D) \supseteq \operatorname{WEff}(F, D) \cap (y - D)$. To obtain a contradiction, suppose that there exists $y^0 \in \operatorname{WEff}(F \cap (y - D), D) \setminus \operatorname{WEff}(F, D)$. Then $(y^0 - \operatorname{int} D) \cap (F \cap (y - D)) \subseteq \{y^0\}$, and there is $y^1 \neq y^0$ with $y^1 \in F \cap (y^0 - \operatorname{int} D)$. We get $y^1 \in y^0 - \operatorname{int} D \subseteq (y - D) - \operatorname{int} D \subseteq y - D$. Hence, $y^1 \in (y^0 - \operatorname{int} D) \cap (F \cap (y - D))$, which contradicts $(y^0 - \operatorname{int} D) \cap (F \cap (y - D)) \subseteq \{y^0\}$.
(g) is obvious.
(h) The assumptions imply $\operatorname{int} D = \operatorname{int} \operatorname{cl} D$. □

We can prove analogously:

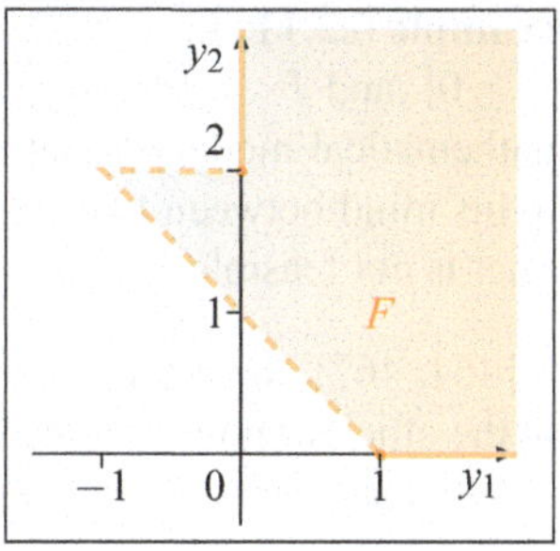


**Fig. 6.4** Set $F$ in Example 6.2.10

**Lemma 6.2.7** *Assume that $Y$ is a linear space, $F, D \subseteq Y$. Then:*

(a) $\mathrm{Eff}(F, D) \subseteq \mathrm{Eff}(F, \mathrm{core}\, D)$.
(b) $\mathrm{core}\, D_1 \subseteq \mathrm{core}\, D \implies \mathrm{Eff}(F, \mathrm{core}\, D) \subseteq \mathrm{Eff}(F, \mathrm{core}\, D_1)$.
(c) $F_1 \subseteq F \implies \mathrm{Eff}(F, \mathrm{core}\, D) \cap F_1 \subseteq \mathrm{Eff}(F_1, \mathrm{core}\, D)$.
(d) *Suppose* $F \subseteq A \subseteq F \cup (F + \mathrm{core}\, D)$.
*Then* $\mathrm{Eff}(A, \mathrm{core}\, D) \subseteq \mathrm{Eff}(F, \mathrm{core}\, D)$.
*If* $\mathrm{core}\, D \cap (-\mathrm{core}\, D) \subseteq \{0\}$ *and* $\mathrm{core}\, D + \mathrm{core}\, D \subseteq \mathrm{core}\, D$, *then* $\mathrm{Eff}(A, \mathrm{core}\, D) = \mathrm{Eff}(F, \mathrm{core}\, D)$.
(e) *Suppose* $F \subseteq A \subseteq F + (D \cup \{0\})$, $0 \notin \mathrm{core}\, D$ *and* $D + \mathrm{core}\, D \subseteq \mathrm{core}\, D$.
*Then* $\mathrm{Eff}(A, \mathrm{core}\, D) \cap F = \mathrm{Eff}(F, \mathrm{core}\, D)$.
(f) *If* $D + \mathrm{core}\, D \subseteq D$, *then* $\mathrm{Eff}(F \cap (y - D), \mathrm{core}\, D) = \mathrm{Eff}(F, \mathrm{core}\, D) \cap (y - D)$ *for all* $y \in Y$.
(g) $\mathrm{Eff}(F, \mathrm{core}\, D) = F$ *if* $\mathrm{core}\, D = \emptyset$.

**Corollary 6.2.8** *Assume that $Y$ is a linear space, $F \subseteq Y$, that $D \subseteq Y$ is a nontrivial pointed convex cone and $F \subseteq A \subseteq F + D$.*

(a) $\mathrm{Eff}(A, D) = \mathrm{Eff}(F, D)$.
(b) *If $Y$ is a linear topological space, then* $\mathrm{WEff}(A, D) \cap F = \mathrm{WEff}(F, D)$.

Podinovskij and Nogin [384] gave the following example that $\mathrm{WEff}(F + \mathbb{R}^\ell_+, \mathbb{R}^\ell_+) = \mathrm{WEff}(F, \mathbb{R}^\ell_+)$ is generally not fulfilled in $Y = \mathbb{R}^\ell$.

**Example 6.2.9** For $F := \{(y_1, y_2)^T \in \mathbb{R}^2 \mid y_2 > 0\} \cup \{0\}$, we have $F + \mathbb{R}^2_+ = F \cup \{(y_1, y_2)^T \in \mathbb{R}^2 \mid y_1 > 0, y_2 = 0\}$. Then $(1, 0)^T \in \mathrm{WEff}(F + \mathbb{R}^2_+, \mathbb{R}^2_+) \setminus \mathrm{WEff}(F, \mathbb{R}^2_+)$ since $(1, 0)^T \notin F$.

Note that the efficient point set can alter if feasible points which are not efficient are neglected.

**Example 6.2.10** In $Y = \mathbb{R}^2$, take $F := \{(y_1, y_2)^T \in Y \mid 0 \le y_2 < 2,\ y_1 > 1 - y_2\} \cup \{(y_1, y_2)^T \in Y \mid y_1 \ge 0,\ y_2 \ge 2\} \cup \{(1, 0)^T\}$ (see Fig. 6.4), $D := \mathbb{R}^2_+$, and $\tilde{F} := F \setminus \mathrm{int}\, F$. Then $\mathrm{Eff}(F, D) = \{(1, 0)^T\}$, but $\mathrm{Eff}(\tilde{F}, D) = \{(1, 0)^T, (0, 2)^T\}$.

**Example 6.2.11** In $Y = \mathbb{R}^2$, take $F := \{(0,0)^T, (0,1)^T\}$, $D := \{(y_1, y_2)^T \in Y \mid y_1 \geq 0\}$, and $\tilde{F} := \{(0,0)^T\}$. Then $\mathrm{Eff}(F, D) = \emptyset$, but $\mathrm{Eff}(\tilde{F}, D) = \tilde{F}$. This is the mathematical model of a situation in which the decision maker is not able to make up his mind between two alternatives, but is satisfied with the remaining one if the other is not feasible any more.

In [464, 467], the influence of neglecting feasible points and of varying preferences on the efficient point set and on the weakly efficient point set were investigated.

Efficient points are often located on the boundary of a set.

**Theorem 6.2.12** *Assume that* $Y$ *is a topological vector space,* $F, D \subseteq Y$ *and that* $0 \in \mathrm{bd}(D \setminus \{0\})$. *Then*

(a) $\mathrm{Eff}(F, D) \subseteq \mathrm{bd}\, F$.
(b) *If additionally* $D \cap (-D) \subseteq \{0\}$ *and* $D + D \subseteq D$, *then*

$$\mathrm{Eff}(F, D) \subseteq \mathrm{bd}\, A$$

*for each set* $A \subseteq Y$ *with* $F \subseteq A \subseteq F + (D \cup \{0\})$.

***Proof***

(a) Contrary to the claim, suppose that there exists $y^0 \in \mathrm{Eff}(F, D) \setminus \mathrm{bd}\, F$. Then $y^0 \in \mathrm{int}\, F$. Hence, there is a neighborhood $U$ of $y^0$ with $U \subseteq F$. Thus, $W := y^0 - U$ is a neighborhood of 0. Since $0 \in \mathrm{bd}(D \setminus \{0\})$, there is $d \in W \cap (D \setminus \{0\}) \subseteq (y^0 - F) \cap (D \setminus \{0\})$. Then there exists $y \in F$ with $d = y^0 - y \in D \setminus \{0\}$. Therefore, $y \in y^0 - (D \setminus \{0\})$, which contradicts $y^0 \in \mathrm{Eff}(F, D)$.
(b) results from Lemma 6.2.5. □

We get for weakly efficient points:

**Theorem 6.2.13** *Let* $Y$ *be a topological vector space,* $F, D \subseteq Y$. *Then*

$$F \cap \mathrm{bd}(F + D) \subseteq \mathrm{WEff}(F, D). \tag{6.2.1}$$

*Assume now* $0 \in \mathrm{bd}(\mathrm{int}\, D \setminus \{0\})$. *Then:*

(a) $\mathrm{WEff}(F, D) \subseteq \mathrm{bd}\, F$.
(b) *If* $D + \mathrm{int}\, D \subseteq D \setminus \{0\}$, *then*

$$\mathrm{WEff}(F, D) = F \cap \mathrm{bd}(F + D)$$

*and*

$$\mathrm{WEff}(F, D) \subseteq \mathrm{bd}\, A$$

*for each set* $A \subseteq Y$ *with* $F \subseteq A \subseteq F + (D \cup \{0\})$.

(c) *If* $\operatorname{int} D \cap (-\operatorname{int} D) \subseteq \{0\}$ *and* $\operatorname{int} D + \operatorname{int} D \subseteq D$, *then*

$$\operatorname{WEff}(F, D) \subseteq \operatorname{bd} A$$

*for each set* $A \subseteq Y$ *with* $F \subseteq A \subseteq F \cup (F + \operatorname{int} D)$.

***Proof*** If $y^0 \in F \setminus \operatorname{WEff}(F, D)$, then there exists $y \in F \cap (y^0 - \operatorname{int} D)$, which implies $y^0 \in \operatorname{int}(F + D)$. Thus, (6.2.1) holds.

(a) and (c) follow immediately from Theorem 6.2.12.
(b) The second statement results from (a) and Lemma 6.2.6(e) since $\operatorname{WEff}(F, D) \subseteq \operatorname{WEff}(A, D) \subseteq \operatorname{bd} A$, the first one from the second by (6.2.1). □

**Corollary 6.2.14** *Assume* $Y$ *is a topological vector space,* $F \subseteq Y$, *and that* $D \subset Y$ *is a nontrivial pointed convex cone. Then*

$$\operatorname{Eff}(F, D) \subseteq \operatorname{bd} F \quad \textit{and}$$
$$\operatorname{Eff}(F, D) \subseteq \operatorname{bd}(F + D).$$

*If* $D$ *has a nonempty interior, then*

$$\operatorname{WEff}(F, D) \subseteq \operatorname{bd} F \quad \textit{and}$$
$$\operatorname{WEff}(F, D) = F \cap \operatorname{bd}(F + D).$$

In many applications, efficient points in a linear space do not belong to the algebraic interior of the feasible set $F$.

**Theorem 6.2.15** *Let* $Y$ *be a linear space,* $F, D \subseteq Y$.

(a) *Assume that there exist* $d \in D \setminus \{0\}$ *and a sequence* $(t_n)_{n \in \mathbb{N}}$ *of real numbers with* $\lim_{n \to \infty} t_n = 0$ *and* $t_n d \in D$ *for all* $n \in \mathbb{N}$. *Then*

$$\operatorname{Eff}(F, D) \subseteq F \setminus \operatorname{core} F.$$

(b) $F \setminus \operatorname{core}(F + D) \subseteq \operatorname{Eff}(F, \operatorname{core} D)$.
(c) *If* $D$ *is a convex cone, then*

$$\operatorname{Eff}(F, \operatorname{core} D) = F \setminus \operatorname{core}(F + D).$$

***Proof***

(a) Take any $y^0 \in \operatorname{core} F$. There exists $n \in \mathbb{N}$ with $y^0 - t_n d \in F$. Thus, $y^0 \notin \operatorname{Eff}(F, D)$.
(b) If $y^0 \in F \setminus \operatorname{Eff}(F, \operatorname{core} D)$, then there exists $y \in F \cap (y^0 - \operatorname{core} D)$, which implies $y^0 \in \operatorname{core}(F + D)$.

(c) For $D = Y$, the statement is obvious. Assume now $D \neq Y$. Then $0 \notin \text{core } D$. Lemma 6.2.7(e) yields $\text{Eff}(F, \text{core } D) = \text{Eff}(F + D, \text{core } D) \cap F$. We obtain $\text{Eff}(F + D, \text{core } D) \subseteq (F + D) \setminus \text{core}(F + D)$ because of (a). □

The assumption in Theorem 6.2.15(a) is fulfilled if $D \setminus \{0\} \neq \emptyset$ and $D \cup \{0\}$ is starshaped about zero. This is the case if $D$ is a nontrivial cone or if $D \cup \{0\}$ is convex and $D \setminus \{0\} \neq \emptyset$.

Now, we prove sufficient conditions for solutions to vector optimization problems using minimizers of extended real-valued functions.

**Proposition 6.2.16** *Assume that $Y$ is a linear space, $F, D \subseteq Y$ and $\varphi : Y \to \overline{\mathbb{R}}$.*

(a) $\text{Eff}(F, D) \cap \text{argmin}_F\, \varphi \subseteq \text{Eff}(\text{argmin}_F\, \varphi, D)$.
(b) *Suppose that $\varphi$ is $D$-monotone on $F$.*
*Then* $\text{Eff}(F, D) \cap \text{argmin}_F\, \varphi = \text{Eff}(\text{argmin}_F\, \varphi, D)$.
*If* $\text{argmin}_F\, \varphi = \{y^0\}$, *then* $y^0 \in \text{Eff}(F, D)$.
(c) *If $\varphi$ is strictly $D$-monotone on $F$, then* $\text{argmin}_F\, \varphi \subseteq \text{Eff}(F, D)$.
(d) *If $\varphi$ is strictly $D$-monotone on some set $A \subseteq Y$ with $F \subset A \subseteq F + (D \cup \{0\})$, then* $\text{argmin}_F\, \varphi = \text{argmin}_A\, \varphi$.

***Proof***

(a) results from Lemma 6.2.5(c) since $\text{argmin}_F\, \varphi \subseteq F$.
(b) Take any $y^0 \in \text{Eff}(\text{argmin}_F\, \varphi, D)$. Contrary to the claim, suppose that $y^0 \notin \text{Eff}(F, D) \cap \text{argmin}_F\, \varphi$. Because of $y^0 \notin \text{Eff}(F, D)$, there exists $y \neq y^0$ with $y \in F \cap (y^0 - D)$. Then $y^0 - y \in D$, and $\varphi(y^0) \geq \varphi(y)$ since $\varphi$ is $D$-monotone on $F$. Therefore, $y \in \text{argmin}_F\, \varphi \cap (y^0 - D)$, which contradicts $y^0 \in \text{Eff}(\text{argmin}_F\, \varphi, D)$.
If $\text{argmin}_F\, \varphi = \{y^0\}$, then $\text{Eff}(\text{argmin}_F\, \varphi, D) = \{y^0\}$. This yields the assertion.
(c) Take any $y^0 \in \text{argmin}_F\, \varphi$. To obtain a contradiction, suppose $y^0 \notin \text{Eff}(F, D)$. Then there exists $y \neq y^0$ with $y \in F \cap (y^0 - D)$. Thus, $y^0 - y \in D \setminus \{0\}$, which implies $\varphi(y^0) > \varphi(y)$ since $\varphi$ is strictly $D$-monotone on $F$. This contradicts $y^0 \in \text{argmin}_F\, \varphi$.
(d) Take first any $y^0 \in \text{argmin}_F\, \varphi$ and $a \in A \setminus F$. Then there exists $y^1 \in F$ with $a \in y^1 + (D \setminus \{0\})$. Since $\varphi$ is strictly $D$-monotone on $A$, we get $\varphi(a) > \varphi(y^1)$. Thus, $y^0 \in \text{argmin}_A\, \varphi$.
Take now any $a^0 \in \text{argmin}_A\, \varphi$. To obtain a contradiction, suppose $a^0 \notin F$. The strict $D$-monotonicity of $\varphi$ implies the existence of $y^2 \in F$ with $a^0 \in y^2 + (D \setminus \{0\})$ and $\varphi(a^0) > \varphi(y^2)$. This contradicts $a^0 \in \text{argmin}_A\, \varphi$. □

Immediately from the previous proposition, we obtain the related statements for weakly efficient points.

**Proposition 6.2.17** *Assume that $Y$ is a topological vector space, $F, D \subseteq Y$ and $\varphi : Y \to \overline{\mathbb{R}}$.*

(a) $\text{WEff}(F, D) \cap \text{argmin}_F\, \varphi \subseteq \text{WEff}(\text{argmin}_F\, \varphi, D)$.

(b) *Suppose that* $\varphi$ *is* (int $D$)*-monotone on* $F$.
*Then* $\text{WEff}(F, D) \cap \text{argmin}_F\, \varphi = \text{WEff}(\text{argmin}_F\, \varphi, D)$.
*If* $\text{argmin}_F\, \varphi = \{y^0\}$, *then* $y^0 \in \text{WEff}(F, D)$.
(c) $\text{argmin}_F\, \varphi \subseteq \text{WEff}(F, D)$ *holds if* $\varphi$ *is strictly* (int $D$)*-monotone on* $F$.

The influence of perturbations in the domination set or uncertainties of the decision maker on the solution sets is investigated in the following proposition.

**Proposition 6.2.18** *Assume* $Y$ *is a linear space,* $F, D \subseteq Y$ *and* $d \in Y$ *with* $D + d \subseteq D$. *Then we have:*

(a) $\text{Eff}(F, D) \subseteq \text{Eff}(F, D + d)$.
(b) $\text{WEff}(F, D) \subseteq \text{WEff}(F, D + d)$.

*Moreover, we get for each function* $\varphi : Y \to \overline{\mathbb{R}}$:

(c) $\text{Eff}(F, D) \cap \text{argmin}_F\, \varphi \subseteq \text{Eff}(\text{argmin}_F\, \varphi, D + d)$.
(d) $\text{WEff}(F, D) \cap \text{argmin}_F\, \varphi \subseteq \text{WEff}(\text{argmin}_F\, \varphi, D + d)$.

***Proof***

(a) follows from Lemma 6.2.5(b),
(b) from Lemma 6.2.6(b).
(c) results from (a) and Proposition 6.2.16(a),
(d) from (b) and Proposition 6.2.17(a). □

Obviously, $D + d \subseteq D$ is fulfilled if $d \in 0^+D$.

If the domination set of the vector optimization problem can be described by the sublevel set of some functional, then a scalarizing function is available.

**Proposition 6.2.19** *Assume* $Y$ *is a linear space,* $F \subseteq Y$, $\varphi : Y \to \overline{\mathbb{R}}$ *and* $t \in \mathbb{R}$. *Then:*

(a) $\text{Eff}(F, -lev_{\varphi,\leq}(t)) = \{y^0 \in F \mid \forall y \in F : \varphi(y - y^0) > t\}$.
(b) $\text{Eff}(F, -lev_{\varphi,<}(t)) = \{y^0 \in F \mid \forall y \in F : \varphi(y - y^0) \geq t\}$.

*Assume now* $\varphi(0) \in \mathbb{R}$.

(c) $\text{Eff}(F, -lev_{\varphi,\leq}(\varphi(0))) = \{y^0 \in F \mid \forall y \in F : \varphi(y^0 - y^0) < \varphi(y - y^0)\}$.
(d) $\text{Eff}(F, -lev_{\varphi,<}(\varphi(0))) = \{y^0 \in F \mid \forall y \in F : \varphi(y^0 - y^0) \leq \varphi(y - y^0)\}$.

***Proof*** We have

$$\begin{aligned} y^0 \in \text{Eff}(F, -lev_{\varphi,\leq}(t)) &\iff \nexists y \in F \setminus \{y^0\} : y - y^0 \in lev_{\varphi,\leq}(t), \\ &\iff \nexists y \in F \setminus \{y^0\} : \varphi(y - y^0) \leq t, \\ &\iff \forall y \in F \setminus \{y^0\} : \varphi(y - y^0) > t. \end{aligned}$$

This implies (a). (b) can be proved analogously. (c) and (d) follow with $t = \varphi(0)$. □

### 6.2.2 Scalarization by Translation Invariant Functions

We will now derive conditions for efficient and weakly efficient points by translation invariant functions. The investigation includes surrogates for the weakly efficient point set. Here we use functionals $\varphi_{a-H,k}$, where $a \in Y$ can be considered to be some reference point and $H \subset Y$ is a set related to the domination set $D$ of a vector optimization problem or to the reference set $D$ of a general efficient point set.

Recall that $\varphi_{a-H,k}: Y \to \overline{\mathbb{R}}$ is given by

$$\varphi_{a-H,k}(y) = \inf\{t \in \mathbb{R} \mid y \in a - H + tk\} \quad \text{for each } y \in Y,$$

where $Y$ is a linear space, $F, H \subseteq Y$, $a \in Y$ and $k \in Y \setminus \{0\}$.

Properties of minimizer sets of such functionals on a given set were deduced in Chap. 5. Our scalarization extends the approach by Pascoletti and Serafini [376] in the way that is described in Sect. 6.4.

Applications of the results from this section in multiobjective optimization will be studied in Chap. 7. The importance of the investigated scalarization for more general vector optimization problems is underscored by the fact that a modification of the scalarizing problem by Pascoletti and Serafini is applied to a vector optimization problem with PDE constraints, a Banach space as objective space and a closed pointed convex reproducing cone with empty interior as domination set in [323] (see Sect. 15.2). There, efficiency w.r.t. $D + \mathbb{R}_{>}k$ is considered.

In statements about functionals $\varphi_{a-H,k}$, we will require conditions under which they were studied in Chap. 4. There, further properties of $\varphi_{a-H,k}$ were proved, especially the lower semicontinuity under assumption ($\text{Sp1}_{a-H,k}$) and the continuity under assumption ($\text{Sp2}_{a-H,k}$). In several statements, $\varphi_{a-H,k}$ has to be finite-valued. Sufficient conditions for this property were given in Theorems 4.2.9, 4.4.4 and 4.6.6 as well as in Proposition 4.7.1 and Corollary 4.7.10, but also in more special statements in Chap. 4.

We will use the supposition

($\text{H0–VOP}_{H,k}$): $Y$ is a linear space, $F, D \subset Y$, $H \neq \{0\}$ is a proper subset of $Y$, $k \in Y \setminus \{0\}$,

and the stronger assumption

($\text{H1–VOP}_{H,k}$): $Y$ is a topological vector space, $F, D \subset Y$, and $H \neq \{0\}$ is a proper closed subset of $Y$ with $k \in 0^+H \setminus \{0\}$.

Note that ($\text{H1–VOP}_{H,k}$) implies ($\text{H1}_{-H,k}$) and ($\text{H1}_{a-H,k}$) for each $a \in Y$. Hence, $\varphi_{a-H,k}$ is lower semicontinuous if ($\text{H1–VOP}_{H,k}$) holds. $\varphi_{a-H,k}$ is finite-valued under assumption ($\text{H0–VOP}_{H,k}$) if $k \in \operatorname{core} 0^+H$.

Even if a function $\varphi_{a-H,k}$ does not contain the entire set $F$ in its effective domain, it can be used to find efficient and weakly efficient points of $F$.

**Lemma 6.2.20** *Suppose* (H0–VOP$_{H,k}$), $a \in Y$.

(a) $\mathrm{Eff}(F, D) \cap \mathrm{dom}\, \varphi_{a-H,k} \subseteq \mathrm{Eff}(F \cap \mathrm{dom}\, \varphi_{a-H,k}, D)$.
(b) $H + D \subseteq H + \mathbb{R}k \implies \mathrm{Eff}(F, D) \cap \mathrm{dom}\, \varphi_{a-H,k} = \mathrm{Eff}(F \cap \mathrm{dom}\, \varphi_{a-H,k}, D)$.

*Assume now additionally that $Y$ is a topological vector space.*

(c) $\mathrm{WEff}(F, D) \cap \mathrm{dom}\, \varphi_{a-H,k} \subseteq \mathrm{WEff}(F \cap \mathrm{dom}\, \varphi_{a-H,k}, D)$.
(d) $H + \mathrm{int}\, D \subseteq H + \mathbb{R}k$ *implies:*
$\mathrm{WEff}(F, D) \cap \mathrm{dom}\, \varphi_{a-H,k} = \mathrm{WEff}(F \cap \mathrm{dom}\, \varphi_{a-H,k}, D)$.

***Proof***

(a) results from Lemma 6.2.5(c).
(b) Take any $y^0 \in F \cap \mathrm{dom}\, \varphi_{a-H,k}$. We have $y^0 \in a - H + tk$ for some $t \in \mathbb{R}$. Assume $y^0 \notin \mathrm{Eff}(F, D)$. Then there exists $y \in F \cap (y^0 - (D \setminus \{0\}))$. We get $y \in a - H + tk - D \subseteq a - H + \mathbb{R}k = \mathrm{dom}\, \varphi_{a-H,k}$. Consequently, $y^0 \notin \mathrm{Eff}(F \cap \mathrm{dom}\, \varphi_{a-H,k}, D)$.
(c) and (d) follow from (a) and (b) with $\mathrm{int}\, D$ instead of $D$. □

We will now give some sufficient conditions for efficient and weakly efficient points by minimizers of functions $\varphi_{a-H,k}$. Keep in mind that $F \cap \mathrm{dom}\, \varphi_{a-H,k}$ is the feasible set of $\min_{y \in F} \varphi_{a-H,k}(y)$ and

$$\mathrm{argmin}_F\, \varphi_{a-H,k} = \mathrm{argmin}_{F \cap \mathrm{dom}\, \varphi_{a-H,k}}\, \varphi_{a-H,k}.$$

In Chap. 5, we pointed out the relationship between problem $\min_{y \in F} \varphi_{a-H,k}(y)$ and problems of the type

$$\begin{aligned} (\mathrm{P}_{F,a,H,k}):\ & t \to \min \\ & y \in a - H + tk \\ & y \in F,\ t \in \mathbb{R}. \end{aligned}$$

According to Theorem 5.1.3(b), the solution sets of both problems coincide if $H$ is $(-k)$-directionally closed. Consequently, the statements in this chapter also connect efficient, weakly efficient and properly efficient points with solutions to problems $(\mathrm{P}_{F,a,H,k})$.

**Theorem 6.2.21** *Suppose* (H0–VOP$_{H,k}$), $a \in Y$. *Define*

$$\Psi := \mathrm{argmin}_F\, \varphi_{a-H,k}.$$

*Then:*

(a) $\mathrm{Eff}(F, D) \cap \Psi \subseteq \mathrm{Eff}(\Psi, D)$.
(b) $H + D \subseteq \mathrm{cl}_{-k}(H) + \mathbb{R}_+ k \implies \mathrm{Eff}(F, D) \cap \Psi = \mathrm{Eff}(\Psi, D)$.
(c) $H + D \subseteq \mathrm{cl}_{-k}(H) + \mathbb{R}_+ k$ *and* $\Psi = \{y^0\}$ *imply* $y^0 \in \mathrm{Eff}(F, D)$.

(d) *If* $\text{cl}_{-k}(H) + (D \setminus \{0\}) \subseteq H + \mathbb{R}_{>}k$, *especially if* $\text{cl}_{-k}(H) + (D \setminus \{0\}) \subseteq \text{core}\, H$, *then* $\Psi \subseteq \text{Eff}(F, D)$.

(e) *Assume* (H1–VOP$_{H,k}$), *that* $F \cap \text{dom}\, \varphi_{a-H,k}$ *is convex and that* $H$ *is a strictly convex set with* $H + D \subseteq H + \mathbb{R}_+ k$ *and* $H + \mathbb{R}_> k \subseteq \text{int}\, H$.
*Then* $\Psi \subseteq \text{Eff}(F, D)$.
*In this case,* $\Psi$ *contains at most one element, which is the only local minimizer of* $\varphi_{a-H,k}$ *on* $F$.

(f) *Assume* $H + D \subseteq \text{cl}_{-k}(H) + \mathbb{R}_+ k$ *or* $\text{cl}_{-k}(H) + \text{core}\, D \subseteq H + \mathbb{R}_> k$.
*Then* $\Psi \subseteq \text{Eff}(F, \text{core}\, D)$.

(g) *If* $H + D \subseteq \text{cl}_{-k}(H) + \mathbb{R}_+ k$ *and* $D + \mathbb{R}_> k \neq Y$, *then* $\Psi \subseteq \text{Eff}(F, D + \mathbb{R}_> k)$.

(h) *Assume that* $Y$ *is a topological vector space.*
*If* $H + D \subseteq \text{cl}_{-k}(H) + \mathbb{R}_+ k$ *or* $\text{cl}_{-k}(H) + (\text{int}\, D \setminus \{0\}) \subseteq H + \mathbb{R}_> k$, *then* $\Psi \subseteq \text{WEff}(F, D)$.
$\text{cl}_{-k}(H) + (\text{int}\, D \setminus \{0\}) \subseteq H + \mathbb{R}_> k$ *is satisfied if* $H$ *is* $(-k)$*-directionally closed and* $H + \text{int}\, D \subseteq H$.

***Proof***

(a) follows from Lemma 6.2.5(c).

(b) $H + D \subseteq \text{cl}_{-k} H + \mathbb{R}_+ k$ yields $(a - H) - D \subseteq a - \text{cl}_{-k}(H) - \mathbb{R}_+ k$. Since $a - \text{cl}_{-k}(H) - \mathbb{R}_+ k = a + \text{cl}_k(-H) - \mathbb{R}_+ k = \text{cl}_k(a - H) - \mathbb{R}_+ k$, the function $\varphi_{a-H,k}$ is $D$-monotone by Theorem 4.5.2. This results in the assertion by Proposition 6.2.16(b).

(c) follows immediately from (b).

(d) If $\varphi_{a-H,k}$ is not finite-valued on $F \cap \text{dom}\, \varphi_{a-H,k}$, then the assertion is fulfilled. Assume now that $\varphi_{a-H,k}$ is finite-valued on $F \cap \text{dom}\, \varphi_{a-H,k}$. Then $\varphi_{a-H,k}$ is strictly $D$-monotone on $F \cap \text{dom}\, \varphi_{a-H,k}$ by Theorem 4.5.2. We obtain $\Psi \subseteq \text{Eff}(F \cap \text{dom}\, \varphi_{a-H,k}, D)$ by Proposition 6.2.16(c). By Lemma 6.2.20(b), $\Psi \subseteq \text{Eff}(F, D)$.
The inclusion $\text{core}\, H \subseteq H + \mathbb{R}_> k$ implies the assertion.

(e) $\varphi_{a-H,k}$ is strictly quasiconvex on $\text{dom}\, \varphi_{a-H,k}$ by Theorem 4.6.6. Propositions 3.4.17 and 3.4.16 imply together with (c) the assertion.

(f) The statement under the second assumption follows from (d). The remaining statement will be proved in (g).

(g) If $\varphi_{a-H,k}$ is not finite-valued on $F \cap \text{dom}\, \varphi_{a-H,k}$, then the assertion is fulfilled. Assume now that $\varphi_{a-H,k}$ is finite-valued on $F \cap \text{dom}\, \varphi_{a-H,k}$. By Theorem 4.5.2, $\varphi_{a-H,k}$ is strictly $(D + \mathbb{R}_> k)$-monotone on $F \cap \text{dom}\, \varphi_{a-H,k}$ and strictly $(\text{core}\, D)$-monotone on $F \cap \text{dom}\, \varphi_{a-H,k}$. We obtain $\Psi \subseteq \text{Eff}(F \cap \text{dom}\, \varphi_{a-H,k}, D + \mathbb{R}_> k)$ and $\Psi \subseteq \text{Eff}(F \cap \text{dom}\, \varphi_{a-H,k}, \text{core}\, D)$ by Proposition 6.2.16(c). Lemma 6.2.20(b) and $H + (D + \mathbb{R}_> k) \subseteq \text{cl}_{-k}(H) + \mathbb{R}k = H + \mathbb{R}k$ imply the assertion.

(h) $\text{Eff}(F, \text{core}\, D) \subseteq \text{Eff}(F, \text{int}\, D)$ and (f) yield the first assertion. For the second assertion, apply (d) to $\text{int}\, D$ instead of $D$. If $H$ is $(-k)$-directionally closed and $H + \text{int}\, D \subseteq H$, then we have $\text{cl}_{-k}(H) + \text{int}\, D = H + \text{int}\, D \subseteq \text{int}\, H \subseteq \text{core}\, H \subseteq H + \mathbb{R}_> k$. □

The assumption $H + D \subseteq \mathrm{cl}_{-k}(H) + \mathbb{R}_+ k$ is fulfilled if $H + D \subseteq H$.

This condition and (H1–VOP$_{H,k}$) do not imply that $H$ or $D$ is a convex cone or that zero belongs to the boundary of $D$.

**Example 6.2.22** In $Y = \mathbb{R}^2$, $H := D := \mathbb{R}^2_+ + (1,1)^T$ and $k := (1,1)^T$ fulfill the assumptions for $H$ and $k$ in (H1–VOP$_{H,k}$) and $H + D \subseteq H$, although $D$ is not a convex cone and $0 \notin \mathrm{bd}\, D$.

**Corollary 6.2.23** *Suppose* (H0–VOP$_{D,k}$)*,* $a \in Y$ *and that* $D$ *is a nontrivial convex cone. Define*

$$\Psi := \operatorname{argmin}_F \varphi_{a-D,k}.$$

(a) $\Psi \subseteq \mathrm{Eff}(F, \mathrm{core}\, D)$. *Moreover,* $\Psi \subseteq \mathrm{Eff}(F, D + \mathbb{R}_> k)$ *if* $D + \mathbb{R}_> k \neq Y$. *If* $Y$ *is a topological vector space, then* $\Psi \subseteq \mathrm{WEff}(F, D)$.
(b) $\mathrm{Eff}(F, D) \cap \Psi = \mathrm{Eff}(\Psi, D)$.
(c) $\Psi = \{y^0\} \implies y^0 \in \mathrm{Eff}(F, D)$.

We will now characterize the efficient point set and the weakly efficient point set by minimizers of functions $\varphi_{a-D,k}$.

**Theorem 6.2.24** *Assume* $Y$ *is a linear space,* $F \subset Y$*, and* $D \neq \{0\}$ *is a proper* $(-k)$*-directionally closed subset of* $Y$ *with* $k \in 0^+ D \setminus \{0\}$*. Then*

$$\begin{aligned} \mathrm{Eff}(F, D) &= \{y^0 \in F \mid \forall y \in F \setminus \{y^0\} \colon \varphi_{y^0-D,k}(y) > 0\} \\ &= \{y^0 \in F \mid \forall y \in F \setminus \{y^0\} \colon \varphi_{-D,k}(y - y^0) > 0\}. \end{aligned} \tag{6.2.2}$$

*If* $0 \in D$*, then* $\varphi_{y^0-D,k}(y^0) = \varphi_{-D,k}(y^0 - y^0) \leq 0$ *for each* $y^0 \in Y$.

*If* $0 \in D \setminus (D + \mathbb{R}_> k)$*, then* $\varphi_{y^0-D,k}(y^0) = \varphi_{-D,k}(y^0 - y^0) = 0$ *for each* $y^0 \in Y$ *and*

$$\begin{aligned} \mathrm{Eff}(F, D) &= \{y^0 \in F \mid \forall y \in F \setminus \{y^0\} \colon \varphi_{y^0-D,k}(y^0) < \varphi_{y^0-D,k}(y)\} \\ &= \{y^0 \in F \mid \forall y \in F \setminus \{y^0\} \colon \varphi_{-D,k}(y^0 - y^0) < \varphi_{-D,k}(y - y^0)\}. \end{aligned}$$

***Proof*** For the relationship between $\varphi_{-D,k}$ and $\varphi_{y^0-D,k}$, see Proposition 4.3.5.

Take any $y^0 \in F$. Then $y^0 - D \subseteq \mathrm{dom}\, \varphi_{y^0-D,k}$. We have

$$\begin{aligned} y^0 \in \mathrm{Eff}(F, D) &\iff F \cap (y^0 - D) \subseteq \{y^0\}, \\ &\iff \forall y \in F \setminus \{y^0\} \colon \varphi_{y^0-D,k}(y) > 0, \text{ because of (4.2.12).} \end{aligned}$$

$0 \in D$ yields $y^0 \in y^0 - D$ and thus $\varphi_{y^0-D,k}(y^0) \leq 0$ because of (4.2.12).

$0 \in D \setminus (D + \mathbb{R}_> k)$ implies $\varphi_{y^0-D,k}(y^0) = 0$ by (4.2.3) and thus the assertion. □

Note that $\varphi_{-D,k}$ and each $\varphi_{y^0-D,k}$ are finite-valued if $k \in \mathrm{core}\, 0^+ D$.

Under assumption (H1–VOP$_{D,k}$), the following statements hold. $\varphi_{-D,k}$ and $\varphi_{y^0-D,k}$ in Theorem 6.2.24 are lower semicontinuous and even continuous in the case $D + \mathbb{R}_{>}k \subset \text{int } D$. Because of Theorem 4.4.4, the functionals $\varphi_{y^0-D,k}$ in Theorem 6.2.24 and in Theorem 6.2.31 are convex if and only if $D$ is a convex set, and the functionals $\varphi_{-D,k}$ in these theorems are sublinear if and only if $D$ is a convex cone.

**Corollary 6.2.25** *Suppose that $Y$ is a linear space, $F \subset Y$, and that $D$ is a nontrivial algebraically closed convex cone in $Y$ with $k \in D \setminus \{0\}$. If $D$ is pointed or $k \in$ core $D$, then*

$$\begin{aligned}\text{Eff}(F, D) &= \{y^0 \in F \mid \forall y \in F \setminus \{y^0\} \colon \varphi_{y^0-D,k}(y^0) < \varphi_{y^0-D,k}(y)\} \\ &= \{y^0 \in F \mid \forall y \in F \setminus \{y^0\} \colon \varphi_{-D,k}(y^0 - y^0) < \varphi_{-D,k}(y - y^0)\},\end{aligned}$$

*and $\varphi_{y^0-D,k}(y^0) = \varphi_{-D,k}(y^0 - y^0) = 0$ for each $y^0 \in Y$.*

***Proof*** The set $D$ is $(-k)$-directionally closed by Proposition 2.3.35. Since $D$ is a nontrivial cone, we have $0 \notin \text{core } D$. For $k \in \text{core } D$, we get $D + \mathbb{R}_{>}k = \text{core } D$ by Lemma 2.3.18(b) and thus $0 \notin D + \mathbb{R}_{>}k$. Furthermore, $0 \notin D + \mathbb{R}_{>}k$ if $D$ is pointed. Hence, the assertion follows immediately from Theorem 6.2.24. □

**Proposition 6.2.26** *Assume that $Y$ is a linear space, $F \subseteq Y$, and that $D$ is a proper $(-k)$-directionally closed subset of $Y$ with $0 \in D$, $k \in 0^+D \setminus \{0\}$ and $\mathbb{R}k \not\subseteq D$.*

(a) *For each $y^0 \in \text{Eff}(F, D)$, we have $\varphi_{y^0-D,k}(y^0) \leq 0$ and*

$$\{y^0\} = \text{argmin}_F\, \varphi_{y^0-D,k} = \underset{y \in F}{\text{argmin}}\, \varphi_{-D,k}(y - y^0).$$

(b) *If $D + D \subseteq D$ and $y^0 \in Y$, then*

$$y^0 \in \text{Eff}(F, D) \iff \text{argmin}_F\, \varphi_{y^0-D,k} = \{y^0\}.$$

***Proof***

(a) $\mathbb{R}k \not\subseteq D$ implies $y^0 + \mathbb{R}k \not\subseteq y^0 - D$. We get $\varphi_{y^0-D,k}(y^0) \in \mathbb{R}$ from Theorem 4.2.5(d). Theorem 6.2.24 yields the assertion.
(b) results from (a) and Theorem 6.2.21(c).

□

Proposition 6.2.26 implies the following.

**Corollary 6.2.27** *Assume $Y = \mathbb{R}^\ell$, $F \subseteq Y$, $D = \{y \in \mathbb{R}^\ell \mid W^T y \geq b\}$ with $b \in -\mathbb{R}^\ell_+$ and $W \in \mathbb{R}^{\ell,\ell}$, where rank $W = \ell$ and the columns $w^1, \ldots, w^\ell$ of $W$ fulfill the condition $(w^i)^T w^j = 0$ for all $i, j \in \{1, \ldots, \ell\}$ with $i \neq j$. Then*

$$y^0 \in \text{Eff}(F, D) \implies \forall i \in \{1, \ldots, \ell\} \colon \text{argmin}_F\, \varphi_{y^0-D,w^i} = \{y^0\} \text{ and } \varphi_{y^0-D,w^i}(y^0) \leq 0.$$

In Corollary 6.2.27, $W$ can be the identity matrix. In this case, $D = b + \mathbb{R}^\ell_+$.

**Proposition 6.2.28** *Suppose that the assumptions of Corollary 6.2.27 are fulfilled with $b = 0$.*

(a) $\left(\exists a \in \mathbb{R}^\ell \ \ \forall i \in \{1, \ldots, \ell\} \colon \ y^0 \in \operatorname{argmin}_F \varphi_{a-D,w^i}\right) \implies y^0 \in \operatorname{Eff}(F, D)$.
(b) *We have*

$$\begin{aligned} y^0 \in \operatorname{Eff}(F, D) &\iff \forall i \in \{1, \ldots, \ell\} \colon y^0 \in \operatorname{argmin}_F \varphi_{y^0-D,w^i}, \\ &\iff \forall i \in \{1, \ldots, \ell\} \colon \operatorname{argmin}_F \varphi_{y^0-D,w^i} = \{y^0\}, \\ &\iff \exists i \in \{1, \ldots, \ell\} \colon \operatorname{argmin}_F \varphi_{y^0-D,w^i} = \{y^0\}. \end{aligned}$$

***Proof***

(a) Take any $a \in \mathbb{R}^\ell$ and $y^0 \in F \setminus \operatorname{Eff}(F, D)$. Then $y^0 - y^1 \in D \setminus \{0\}$ for some $y^1 \in F$. Since rank $W = \ell$, we get $W^{\mathrm{T}}(y^0 - y^1) \neq 0$ and the existence of $i \in \{1, \ldots, \ell\}$ with $(w^i)^{\mathrm{T}}(y^0 - y^1) > 0$.
Thus, $W^{\mathrm{T}}(y^0 - y^1 - tw^i) \geq 0$ for some $t \in \mathbb{R}_>$. We obtain $y^0 - y^1 - tw^i \in D$, which implies $y^1 \in y^0 - D - tw^i$.
Contrary to the claim, suppose that $y^0 \in \operatorname{argmin}_F \varphi_{a-D,w^i}$.
Define $r := \min_F \varphi_{a-D,w^i}$. Then $y^0 \in a - D + rw^i$. Hence, $y^1 \in (a - D + rw^i) - D - tw^i \subset a - D + (r - t)w^i$. We obtain $\varphi_{a-D,w^i}(y^1) < r$, which contradicts the definition of $r$.
(b) results from (a) and Proposition 6.2.26(b). □

In the previous statements, efficient points $y^0$ w.r.t. $D$ were described as unique minimizers of $\varphi_{y^0-D,k}$. Without the uniqueness, we get efficient points w.r.t. some reference set that usually does not coincide with the domination set. Under appropriate assumptions, these are the weakly efficient points.

**Theorem 6.2.29** *Assume $Y$ is a linear space, $F, D \subset Y$, $D \neq \emptyset$ and $k \in Y \setminus \{0\}$ with $D + \mathbb{R}_> k \neq Y$. Then*

$$\begin{aligned} \operatorname{Eff}(F, D + \mathbb{R}_> k) &= \{y^0 \in F \mid \forall y \in F \setminus \{y^0\} \colon \varphi_{y^0-D,k}(y) \geq 0\} \\ &= \{y^0 \in F \mid \forall y \in F \setminus \{y^0\} \colon \varphi_{-D,k}(y - y^0) \geq 0\}. \end{aligned} \tag{6.2.3}$$

*If $0 \in D$, then $\varphi_{y^0-D,k}(y^0) = \varphi_{-D,k}(y^0 - y^0) \leq 0$ for each $y^0 \in Y$.*
*If $0 \in D \setminus (D + \mathbb{R}_> k)$, then $\varphi_{y^0-D,k}(y^0) = \varphi_{-D,k}(y^0 - y^0) = 0$ for each $y^0 \in Y$ and*

$$\begin{aligned} \operatorname{Eff}(F, D + \mathbb{R}_> k) &= \{y^0 \in F \mid \varphi_{y^0-D,k}(y^0) = \min_{y \in F} \varphi_{y^0-D,k}(y)\} \\ &= \{y^0 \in F \mid \varphi_{-D,k}(y^0 - y^0) = \min_{y \in F} \varphi_{-D,k}(y - y^0)\}. \end{aligned}$$

***Proof*** Take any $y^0 \in F$. Then $y^0 - D \subseteq \operatorname{dom} \varphi_{y^0-D,k}$. We have

$$\begin{aligned} y^0 \in \operatorname{Eff}(F, D + \mathbb{R}_> k) &\iff F \cap (y^0 - (D + \mathbb{R}_> k)) \subseteq \{y^0\}, \\ &\iff \forall y \in F \setminus \{y^0\} \colon \varphi_{y^0-D,k}(y) \geq 0, \text{ by (4.2.3)}. \end{aligned}$$

The other statements follow in the same way as in the proof of Theorem 6.2.24. □

We derive the following from Theorem 6.2.29 by Lemma 2.3.18(b) and Proposition 2.3.3.

**Corollary 6.2.30** *Suppose $Y$ is a linear space, $F, D \subset Y$, and $D$ is a nontrivial convex cone with $k \in \operatorname{core} D$. Then*

$$\begin{aligned}\operatorname{Eff}(F, \operatorname{core} D) &= \{y^0 \in F \mid \varphi_{y^0-D,k}(y^0) = \min_{y\in F} \varphi_{y^0-D,k}(y)\} \\ &= \{y^0 \in F \mid \varphi_{-D,k}(y^0 - y^0) = \min_{y\in F} \varphi_{-D,k}(y - y^0)\},\end{aligned}$$

*and $\varphi_{y^0-D,k}(y^0) = \varphi_{-D,k}(y^0 - y^0) = 0$ for each $y^0 \in Y$.*

Theorem 6.2.29 implies:

**Theorem 6.2.31** *Assume $Y$ is a topological vector space, $F \subset Y$, $k \in Y \setminus \{0\}$, and $D$ is a proper subset of $Y$ with $D + \mathbb{R}_{>}k \subseteq \operatorname{int} D$.*
*Then*

$$\begin{aligned}\operatorname{WEff}(F, D) &= \{y^0 \in F \mid \forall y \in F \setminus \{y^0\} \colon \varphi_{y^0-D,k}(y) \geq 0\} \\ &= \{y^0 \in F \mid \forall y \in F \setminus \{y^0\} \colon \varphi_{-D,k}(y - y^0) \geq 0\}.\end{aligned}$$

*If $0 \in D$, then $\varphi_{y^0-D,k}(y^0) = \varphi_{-D,k}(y^0 - y^0) \leq 0$ for each $y^0 \in Y$.*
*If $0 \in D \setminus \operatorname{int} D$, then $\varphi_{y^0-D,k}(y^0) = \varphi_{-D,k}(y^0 - y^0) = 0$ for each $y^0 \in Y$ and*

$$\begin{aligned}\operatorname{WEff}(F, D) &= \{y^0 \in F \mid \varphi_{y^0-D,k}(y^0) = \min_{y\in F} \varphi_{y^0-D,k}(y)\} \\ &= \{y^0 \in F \mid \varphi_{-D,k}(y^0 - y^0) = \min_{y\in F} \varphi_{-D,k}(y - y^0)\}.\end{aligned}$$

***Proof*** By Lemma 2.3.17, we have $\operatorname{int} D \subseteq D + \mathbb{R}_{>}k$. Hence, the assumptions imply $\operatorname{int} D = D + \mathbb{R}_{>}k$. Apply Theorem 6.2.29. □

The equation $\operatorname{WEff}(F, D) = \{y^0 \in F \mid \varphi_{y^0-D,k}(y^0) = \min_{y\in F} \varphi_{y^0-D,k}(y)\}$ was proved in [186, Theorem 3.1] under the stronger assumption that $D$ is a set with $0 \in D \setminus \operatorname{int} D$ for which there exists a cone $K$ with nonempty interior such that $\operatorname{cl} D + \operatorname{int} K \subseteq \operatorname{int} D$ and $k \in \operatorname{int} K$. It was also proved under assumptions which are equivalent to $D$ being a closed convex set with $0 \in \operatorname{bd} D$, $\operatorname{int} D \neq \emptyset$, $k \in -0^+D$ and $Y = D + \mathbb{R}k$ in [186, Theorem 3.2]. In both statements, $\varphi_{y^0-D,k}$ is continuous and finite-valued.

In Theorem 6.2.31, the assumption $D + \mathbb{R}_{>}k \subseteq \operatorname{int} D$ is not superfluous.

**Example 6.2.32** In $Y = \mathbb{R}^2$, consider $D := \mathbb{R}^2_+$, $k := (1, 0)^T \in D$, and $F := \{(y_1, y_2)^T \in Y \mid -2 \leq y_1 \leq 0, -1 \leq y_2 \leq 0\} \cup \{(y_1, y_2)^T \in Y \mid -1 \leq y_1 \leq 0, -2 \leq y_2 \leq -1\}$ (see Fig. 6.5). Then $y^0 := (-1, -1)^T \in \operatorname{WEff}(F, D)$, but $\varphi_{y^0-D,k}(y^0) = 0$ and $\varphi_{y^0-D,k}((-2, -1)^T) = -1 < 0$.

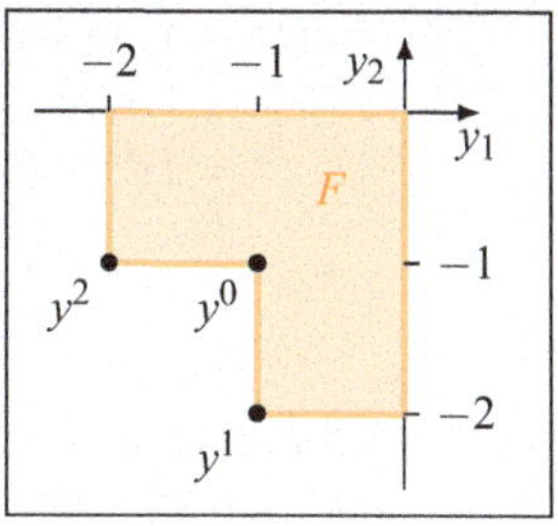


**Fig. 6.5** Set $F$ in Example 6.2.32. Adapted from [472, p. 139]

**Corollary 6.2.33** *Suppose that $Y$ is a topological vector space, $F \subset Y$, and that $D$ is a nontrivial convex cone in $Y$ with $k \in \operatorname{int} D$. Then*

$$\begin{aligned}\mathrm{WEff}(F, D) &= \{y^0 \in F \mid \varphi_{y^0-D,k}(y^0) = \min_{y\in F} \varphi_{y^0-D,k}(y)\}\\ &= \{y^0 \in F \mid \varphi_{-D,k}(y^0 - y^0) = \min_{y\in F} \varphi_{-D,k}(y - y^0)\},\end{aligned}$$

*and* $\varphi_{y^0-D,k}(y^0) = \varphi_{-D,k}(y^0 - y^0) = 0$ *for each* $y^0 \in Y$.

We can also characterize efficiency by means of Theorem 6.2.29.

**Proposition 6.2.34** *Suppose $Y$ is a linear space, $F, D \subset Y$, and $D$ is a nontrivial pointed convex cone. Then*

$$\begin{aligned}\mathrm{Eff}(F, D) &= \bigcap_{k\in D\setminus\{0\}} \mathrm{Eff}(F, D + \mathbb{R}_{>}k)\\ &= \{y^0 \in F \mid \forall k \in D \setminus \{0\}\colon \varphi_{y^0-D,k}(y^0) = \min_{y\in F} \varphi_{y^0-D,k}(y)\}\\ &= \{y^0 \in F \mid \forall k \in D \setminus \{0\}\colon \varphi_{-D,k}(y^0 - y^0) = \min_{y\in F} \varphi_{-D,k}(y - y^0)\}.\end{aligned}$$

***Proof*** Obviously, $\bigcup_{k\in D\setminus\{0\}}(D + \mathbb{R}_{>}k) \subseteq D \setminus \{0\}$. For each $d \in D \setminus \{0\}$, we have $d = \frac{1}{2}d + \frac{1}{2}d \in D + \frac{1}{2}k$ with $k = d$. Thus, $D \setminus \{0\} = \bigcup_{k\in D\setminus\{0\}}(D + \mathbb{R}_{>}k)$. This yields $\mathrm{Eff}(F, D) = \mathrm{Eff}(F, \bigcup_{k\in D\setminus\{0\}}(D + \mathbb{R}_{>}k)) = \bigcap_{k\in D\setminus\{0\}} \mathrm{Eff}(F, D + \mathbb{R}_{>}k)$. The last two equations of the proposition result from Theorem 6.2.29. □

The entire efficient point set can be generated by varying $k$ in a proper subset of $D \setminus \{0\}$. This follows from statements related to the parameter control of $k$ in Chap. 5 and in Sect. 4.3. One result is applied in the next corollary.

**Corollary 6.2.35** *Suppose that $Y$ is a normed space, $F, D \subset Y$, and that $D$ is a nontrivial pointed convex cone. Define $K := \{k \in D \mid \|k\| = 1\}$. Then*

$$\begin{aligned}\operatorname{Eff}(F, D) &= \bigcap_{k\in K} \operatorname{Eff}(F, D + \mathbb{R}_{>}k)\\ &= \{y^0 \in F \mid \forall k \in K : \varphi_{y^0-D,k}(y^0) = \min_{y\in F} \varphi_{y^0-D,k}(y)\}\\ &= \{y^0 \in F \mid \forall k \in K : \varphi_{-D,k}(y^0 - y^0) = \min_{y\in F} \varphi_{-D,k}(y - y^0)\}.\end{aligned}$$

***Proof*** By Proposition 4.3.3, $\varphi_{A,\lambda k} = \frac{1}{\lambda}\varphi_{A,k}$ for each nonempty subset $A$ of $Y$ and $\lambda \in \mathbb{R}_{>}$. For each $d \in D \setminus \{0\}$, there exist $k \in K$ and $\lambda \in \mathbb{R}_{>}$ such that $d = \lambda k$. This yields the assertion because of Proposition 6.2.34 and Theorem 6.2.29. □

Up to now, we have used functionals $\varphi_{y^0-D,k}$ for scalarizing the weakly efficient point set and the efficient point set, where $y^0$ was (weakly) efficient. We now turn to scalarization by functions $\varphi_{a-D,k}$, where $a$ is a fixed vector. In this case, $a$ can be a lower or an upper bound of $F$ and $D$ has to be a convex cone. For a vector optimization problem with the domination set $D$, we denote $a \in Y$ as a lower or upper bound of $F$ if $F \subseteq a + D$ or $F \subseteq a - D$, respectively. In applications, an upper bound can easily be added to the vector optimization problem without any influence on the solution set. Note that scalarizations that are based on norms require a lower bound of $F$ or a similar condition.

**Theorem 6.2.36** *Suppose $Y$ is a linear space, $F, D \subset Y$, $D$ is a nontrivial convex cone with* core $D \neq \emptyset$ *and there exists $a \in Y$ such that $F \subseteq a -$* core $D$ *or $F \subseteq a +$* core $D$. *Then*

$$\begin{aligned}\operatorname{Eff}(F, \operatorname{core} D) &= \{y^0 \in F \mid \exists k \in \operatorname{core} D : \ \varphi_{a-D,k}(y^0) = \min_{y\in F} \varphi_{a-D,k}(y)\}\\ &= \{y^0 \in F \mid \exists k \in \operatorname{core} D : \ \varphi_{-D,k}(y^0 - a) = \min_{y\in F} \varphi_{-D,k}(y - a)\}.\end{aligned}$$

*If $D$ is algebraically closed, then*

$$\begin{aligned}\operatorname{Eff}(F, D) &= \{y^0 \in F \mid \exists k \in \operatorname{core} D \ \forall y \in F \setminus \{y^0\} : \ \varphi_{a-D,k}(y^0) < \varphi_{a-D,k}(y)\}\\ &= \{y^0 \in F \mid \exists k \in \operatorname{core} D \ \forall y \in F \setminus \{y^0\} : \ \varphi_{-D,k}(y^0 - a) < \varphi_{-D,k}(y - a)\}.\end{aligned}$$

*In the case $F \subseteq a -$* core $D$, *we can choose $k = a - y^0$ for $y^0 \in$* Eff$(F,$ core $D)$, *which results in $\varphi_{a-D,k}(y^0) = -1$. In the case $F \subseteq a +$* core $D$, *we can choose $k = y^0 - a$ for $y^0 \in$* Eff$(F,$ core $D)$, *which results in $\varphi_{a-D,k}(y^0) = 1$.*
*The same choice works for $y^0 \in$* Eff$(F, D)$.

*All considered functions $\varphi_{a-D,k}$ and $\varphi_{-D,k}$ are finite-valued and convex. Moreover, each function $\varphi_{-D,k}$ is sublinear. If $Y$ is a topological vector space and $D$ is closed, then all functions $\varphi_{a-D,k}$ and $\varphi_{-D,k}$ are lower semicontinuous.*

***Proof*** Because of Corollary 6.2.23, we only have to show the inclusions $\subseteq$ of the equations.

Assume $F \subseteq a -$ core $D$, and take any $y^0 \in$ Eff$(F,$ core $D)$.

Then $y^0 \in F \subseteq a - \text{core } D$. We get $k := a - y^0 \in \text{core } D$, and $D + \mathbb{R}_{>}k = \text{core } D$. Apply Theorem 4.2.1.

Because of (4.2.11) and (4.2.3), $0 \in D \setminus \text{core } D$ implies $y^0 \in y^0 - (D \setminus \text{core } D) = a - (D \setminus \text{core } D) - k$ and $\varphi_{a-D,k}(y^0) = -1$. For each $y \in Y$,

$$\varphi_{a-D,k}(y) < -1 \iff y \in a - \text{core } D - k = y^0 - \text{core } D.$$

Since $F \cap (y^0 - \text{core } D) \subseteq \{y^0\}$, we obtain $\varphi_{a-D,k}(y) \nleq -1$ for all $y \in F \setminus \{y^0\}$. Thus, $\varphi_{a-D,k}(y^0) = \min_{y \in F} \varphi_{a-D,k}(y)$.

For $y^0 \in \text{Eff}(F, D)$, we have $\{y^0\} = F \cap (y^0 - D) = F \cap (a - D - k)$. Equation (4.2.12) implies $\varphi_{a-D,k}(y) \nleq -1$ for all $y \in F \setminus \{y^0\}$.

The case $F \subseteq a + \text{core } D$ can be handled analogously.

Each of the considered functions $\varphi_{a-D,k}$ is finite-valued because of Theorem 4.2.9. The convexity and sublinearity of the functions follow from Theorem 4.2.10. Theorem 4.4.2 yields the statement about the lower semicontinuity. □

**Theorem 6.2.37** *Suppose that $Y$ is a topological vector space, $F, D \subset Y$, $D$ is a nontrivial convex cone with* $\text{int } D \neq \emptyset$ *and there exists $a \in Y$ such that $F \subseteq a - \text{int } D$ or $F \subseteq a + \text{int } D$. Then*

$$\begin{aligned}\text{WEff}(F, D) &= \{y^0 \in F \mid \exists k \in \text{int } D : \ \varphi_{a-D,k}(y^0) = \min_{y \in F} \varphi_{a-D,k}(y)\} \\ &= \{y^0 \in F \mid \exists k \in \text{int } D : \ \varphi_{-D,k}(y^0 - a) = \min_{y \in F} \varphi_{-D,k}(y - a)\}.\end{aligned}$$

*If $D$ is closed, then*

$$\begin{aligned}\text{Eff}(F, D) &= \{y^0 \in F \mid \exists k \in \text{int } D \ \forall y \in F \setminus \{y^0\} : \ \varphi_{a-D,k}(y^0) < \varphi_{a-D,k}(y)\} \\ &= \{y^0 \in F \mid \exists k \in \text{int } D \ \forall y \in F \setminus \{y^0\} : \ \varphi_{-D,k}(y^0 - a) < \varphi_{-D,k}(y - a)\}.\end{aligned}$$

*In the case $F \subseteq a - \text{int } D$, we can choose $k = a - y^0$ for $y^0 \in \text{WEff}(F, D)$, which results in $\varphi_{a-D,k}(y^0) = -1$. In the case $F \subseteq a + \text{int } D$, we can choose $k = y^0 - a$ for $y^0 \in \text{WEff}(F, D)$, which results in $\varphi_{a-D,k}(y^0) = 1$.*
*The same choice works for $y^0 \in \text{Eff}(F, D)$.*

*All considered functions $\varphi_{a-D,k}$ and $\varphi_{-D,k}$ are finite-valued, convex and continuous. Moreover, each function $\varphi_{-D,k}$ is sublinear.*

***Proof*** Since $D$ is convex and $\text{int } D \neq \emptyset$, we have $\text{int } D = \text{core } D$. Apply Theorem 6.2.36. The continuity follows from Theorem 4.6.2. □

Each function $\varphi_{a-D,k}$ in Theorems 6.2.36 and 6.2.37 is finite-valued. This is not always the case in the following two theorems, which can be applied if the algebraic interior of $D$ is empty.

**Theorem 6.2.38** *Suppose that $Y$ is a linear space, that $D \subset Y$ is a nontrivial algebraically closed pointed convex cone, and that $F \subseteq a - D$ holds for some $a \in Y$ with $F \neq \{a\}$. Then*

$$\mathrm{Eff}(F, D) = \{y^0 \in F \mid \exists k \in D \setminus \{0\}\ \forall y \in F \setminus \{y^0\}\colon \varphi_{a-D,k}(y^0) = -1 < \varphi_{a-D,k}(y)\}.$$

*All considered functions $\varphi_{a-D,k}$ are convex. They are proper if $Y$ is a topological vector space and $D$ is closed.*

*For $y^0 \in \mathrm{Eff}(F, D)$, we can choose $k = a - y^0$, where $\varphi_{a-D,k}$ is finite-valued on $F$.*

***Proof*** Theorem 4.2.10 implies the convexity of the functions. The statement about the properness of the functionals follows from Corollary 4.7.8. The inclusion $\supseteq$ results from Corollary 6.2.23(c). The case $F = \emptyset$ is trivial. Assume now $y^0 \in \mathrm{Eff}(F, D)$, i.e., $F \cap (y^0 - D) = \{y^0\}$.

$F \neq \{a\}$ implies $y^0 \neq a$. Thus, $k := a - y^0 \in D \setminus \{0\}$. From (4.2.12), we conclude for each $y \in Y$ that

$$\varphi_{a-D,k}(y) \leq -1 \iff y \in a - D - k = y^0 - D.$$

Hence, $\varphi_{a-D,k}(y^0) \leq -1 < \varphi_{a-D,k}(y)$ for all $y \in F \setminus \{y^0\}$. Thus, $\varphi_{a-D,k}$ is finite-valued on $F \setminus \{y^0\}$ since $F \subseteq \mathrm{dom}\,\varphi_{a-D,k}$. Contrary to the claim, suppose that $\varphi_{a-D,k}(y^0) < -1$. Then $y^0 \in a - D - \mathbb{R}_{>}k - k$ by Theorem 4.2.1. Since $y^0 = a - k$, we get $0 \in D + \mathbb{R}_{>}k$. Therefore, $(\mathbb{R}_{>}k) \cap (-D) \neq \emptyset$, which contradicts the fact that $D$ is pointed. □

**Theorem 6.2.39** *Suppose that $Y$ is a linear space, that $D \subset Y$ is a nontrivial algebraically closed convex cone, $F \subset Y$, and that $F \subseteq a + D$ holds for some $a \in Y \setminus F$. Then*

$$\mathrm{Eff}(F, D) \subseteq \{y^0 \in F \mid \exists k \in D \setminus \{0\}\ \forall y \in F \setminus \{y^0\}\colon \varphi_{a-D,k}(y^0) < \varphi_{a-D,k}(y)\}.$$

*If $D$ is pointed, then*

$$\mathrm{Eff}(F, D) = \{y^0 \in F \mid \exists k \in D \setminus \{0\}\ \forall y \in F \setminus \{y^0\}\colon \varphi_{a-D,k}(y^0) = 1 < \varphi_{a-D,k}(y)\}.$$

*For $y^0 \in \mathrm{Eff}(F, D)$, we can choose $k = y^0 - a$, which results in $\varphi_{a-D,k}(y^0) \leq 1 < \varphi_{a-D,k}(y)$ for all $y \in F \setminus \{y^0\}$. In this case, $\varphi_{a-D,k}(y^0) = 1$ if $D$ is pointed.*

*All considered functions $\varphi_{a-D,k}$ are convex. They are proper if $Y$ is a topological vector space and $D$ is closed and pointed.*

***Proof*** Theorem 4.2.10 implies the convexity of the functions. The statement about the properness of the functionals follows from Proposition 4.7.7. If $D$ is pointed, then the inclusion $\supseteq$ results from Corollary 6.2.23(c).

The case $F = \emptyset$ is trivial. Assume now $F \neq \emptyset$ and $y^0 \in \mathrm{Eff}(F, D)$, i.e., $F \cap (y^0 - D) = \{y^0\}$.

$a \notin F$ implies $k := y^0 - a \in D \setminus \{0\}$. By (4.2.12), we conclude for each $y \in Y$ that

$$\varphi_{a-D,k}(y) \leq 1 \iff y \in a - D + k = y^0 - D.$$

Hence, $\varphi_{a-D,k}(y^0) \leq 1$ and $\varphi_{a-D,k}(y) > 1$ for all $y \in F \setminus \{y^0\}$.

Assume now $\varphi_{a-D,k}(y^0) < 1$. Then $y^0 \in a - D - \mathbb{R}_{>}k + k$ by Theorem 4.2.1. Since $k = y^0 - a$, we have $0 \in D + \mathbb{R}_{>}k$. Then $(\mathbb{R}_{>}k) \cap (-D) \neq \emptyset$. Thus, $D$ is not pointed. □

Note that $F \subseteq a + D$ and $a \in F$ would imply $\mathrm{Eff}(F, D) \subseteq \{a\}$.

If $D$ is not a linear subspace of $Y$ and $k \notin \mathrm{core}\, D$, then the functions $\varphi_{a-D,k}$ in Theorems 6.2.38 and 6.2.39 are not finite-valued. This follows from Proposition 4.7.1. In Theorem 6.2.39, the choice $k = y^0 - a$ could result in a function $\varphi_{a-D,k}$ that does not attain any real value if $D$ were not pointed.

**Example 6.2.40** In $Y = \mathbb{R}^2$, consider $F := \{(y_1, y_2)^T \in Y \mid 1 \leq y_2 \leq y_1 + 1\}$ and the closed convex cone $D := \{(y_1, y_2)^T \in Y \mid y_1 \geq 0\}$. Then $F \subseteq a + D$ and $a \notin F$ for $a := (0, 0)^T$. We get $\mathrm{Eff}(F, D) = \{(0, 1)^T\}$. For $k := (0, 1)^T - a = (0, 1)^T$, we have $k \in D \cap (-D)$. Hence, $\varphi_{a-D,k}$ does not attain real values. We obtain $-\infty = \varphi_{a-D,k}((0, 1)^T) < \varphi_{a-D,k}(y) = +\infty$ for all $y \in F \setminus \{(0, 1)^T\}$.

Under certain assumptions, each weakly efficient point set of $F$ w.r.t. a convex cone $D$ that has a nonempty interior can be represented as the minimizer set of one single continuous strictly (int $D$)-monotone functional. In the convex case, the weakly efficient point set turns out to be the maximizer set of one single convex function. These facts result from the following proposition.

**Proposition 6.2.41** *Assume that $Y$ is a topological vector space, $F \subset Y$, that $D \subset Y$ is a convex cone with $k \in \mathrm{int}\, D$ and that $F + D$ is a proper closed subset of $Y$. Then*

$$\mathrm{WEff}(F, D) = \{y^0 \in F \mid \varphi_{F+D,-k}(y^0) = \max_{y \in F} \varphi_{F+D,-k}(y)\}$$
$$= \{y^0 \in F \mid \varphi_{B,k}(y^0) = \min_{y \in F} \varphi_{B,k}(y)\}$$

*with $B := Y \setminus \mathrm{int}(F + D)$, where*

$$\max_{y \in F} \varphi_{F+D,-k}(y) = \min_{y \in F} \varphi_{B,k}(y) = 0.$$

*$\varphi_{F+D,-k}$ and $\varphi_{B,k}$ are finite-valued and continuous.*

*$\varphi_{B,k}$ is strictly (int $D$)-monotone, whereas $\varphi_{F+D,-k}$ is strictly (− int $D$)-monotone.*

*If $F + D$ is convex, then $\varphi_{F+D,-k}$ is convex and $\varphi_{B,k}$ is concave.*

***Proof***

(a) $(\mathrm{H2}_{F+D,-k})$ holds. Thus, $\varphi_{F+D,-k}$ is continuous. Since $k \in \mathrm{core}\, D$ and $D \subseteq 0^+(F + D)$, we get $k \in \mathrm{core}\, 0^+(F + D)$. Hence, $\varphi_{F+D,-k}$ is finite-valued by Theorem 4.2.9. Because of Proposition 4.3.11, $\varphi_{B,k} = -\varphi_{F+D,-k}$ is also finite-valued and continuous. If $F + D$ is convex, then $\varphi_{F+D,-k}$ is convex by Theorem 4.4.4; thus, $\varphi_{B,k}$ is concave.

(b) To obtain a contradiction, suppose that $B - \operatorname{int} D \subseteq B$ does not hold. Then there exist $y \in Y$ and $d \in \operatorname{int} D$ such that $y \notin \operatorname{int}(F + D)$ and $y - d \in \operatorname{int}(F + D)$. We obtain $y \in d + \operatorname{int}(F + D) \subseteq \operatorname{int}(F + D)$, which contradicts $y \notin \operatorname{int}(F + D)$. Thus, $B - \operatorname{int} D \subseteq B$.
The function $\varphi_{B,k}$ is strictly ($\operatorname{int} D$)-monotone by Corollary 4.5.6(a) and Theorem 4.5.5(f). Hence, $\varphi_{F+D,-k} = -\varphi_{B,k}$ is strictly ($-\operatorname{int} D$)-monotone. By Proposition 6.2.17, $\{y^0 \in F \mid \varphi_{B,k}(y^0) = \min_{y \in F} \varphi_{B,k}(y)\} \subseteq \operatorname{WEff}(F, D)$.
We have $\varphi_{F+D,-k}(y) \leq 0$ for all $y \in F + D$. Apply Theorem 4.6.6. We obtain:

$$\forall y \in Y\colon\ (\varphi_{F+D,-k}(y) = 0 \iff y \in \operatorname{bd}(F + D));$$
$$\forall y \in Y\colon\ (\varphi_{B,k}(y) = 0 \iff y \in \operatorname{bd} B = \operatorname{bd}(F + D)).$$

Consequently, $\max_{y \in F} \varphi_{F+D,-k}(y) = 0 = \min_{y \in F} \varphi_{B,k}(y)$ since $\varphi_{F+D,-k} = -\varphi_{B,k}$. Thus, $\{y^0 \in F \mid \varphi_{F+D,-k}(y^0) = \max_{y \in F} \varphi_{F+D,-k}(y)\} = \{y^0 \in F \mid \varphi_{B,k}(y^0) = \min_{y \in F} \varphi_{B,k}(y)\} \subseteq \operatorname{WEff}(F, D)$.

(c) By Theorem 6.2.13, $\operatorname{WEff}(F, D) \subseteq \operatorname{bd}(F + D)$. Therefore, $\varphi_{F+D,-k}(y) = 0$ for all $y \in \operatorname{WEff}(F, D)$. This yields the assertion. □

Further scalarization results for efficient and weakly efficient points w.r.t. closed convex cones will be proved in Sect. 6.2.5.

In numerical methods, we have to deal with slightly altered function values. Moreover, uncertainties of the decision maker can result in an altered domination set. Statements for these cases are contained in the following proposition.

**Proposition 6.2.42** *Suppose $Y$ is a linear space, $F, D \subset Y$ with $D \neq \emptyset$, $k \in Y \setminus \{0\}$ with $D + \mathbb{R}_> k \neq Y$, and $\varepsilon \in \mathbb{R}$.*
*We have*

$$\begin{aligned}\operatorname{Eff}(F, D + \mathbb{R}_> k + \varepsilon k) &= \{y^0 \in F \mid \forall y \in F \setminus \{y^0\}\colon \varphi_{y^0 - D,k}(y) \geq -\varepsilon\} \\ &= \{y^0 \in F \mid \forall y \in F \setminus \{y^0\}\colon \varphi_{-D,k}(y - y^0) \geq -\varepsilon\}.\end{aligned} \tag{6.2.4}$$

*If $D + \mathbb{R}_> k \subseteq \operatorname{core} D$, then*

$$\begin{aligned}\operatorname{Eff}(F, \operatorname{core} D + \varepsilon k) &= \{y^0 \in F \mid \forall y \in F \setminus \{y^0\}\colon \varphi_{y^0 - D,k}(y) \geq -\varepsilon\} \\ &= \{y^0 \in F \mid \forall y \in F \setminus \{y^0\}\colon \varphi_{-D,k}(y - y^0) \geq -\varepsilon\}.\end{aligned} \tag{6.2.5}$$

*If $k \in 0^+ D$ and $D$ is $(-k)$-directionally closed, then*

$$\begin{aligned}\operatorname{Eff}(F, D + \varepsilon k) &= \{y^0 \in F \mid \forall y \in F \setminus \{y^0\}\colon \varphi_{y^0 - D,k}(y) > -\varepsilon\} \\ &= \{y^0 \in F \mid \forall y \in F \setminus \{y^0\}\colon \varphi_{-D,k}(y - y^0) > -\varepsilon\}.\end{aligned} \tag{6.2.6}$$

*If $Y$ is a topological vector space and $D + \mathbb{R}_> k \subseteq \operatorname{int} D$, then we get*

$$\text{WEff}(F, D + \varepsilon k) = \{y^0 \in F \mid \forall y \in F \setminus \{y^0\}\colon \varphi_{y^0-D,k}(y) \geq -\varepsilon\} \quad (6.2.7)$$
$$= \{y^0 \in F \mid \forall y \in F \setminus \{y^0\}\colon \varphi_{-D,k}(y - y^0) \geq -\varepsilon\}.$$

*If* $0 \in D + \varepsilon k$*, then* $\varphi_{y^0-D,k}(y^0) \leq -\varepsilon$ *for each* $y^0 \in Y$.

***Proof*** By Proposition 4.3.4, $\varphi_{y^0-(D+\varepsilon k),k}(y) = \varphi_{y^0-D,k}(y) + \varepsilon$. We obtain (6.2.4) by applying Theorem 6.2.29 to $D + \varepsilon k$ instead of $D$. The inclusion $D + \mathbb{R}_{>}k \subseteq \operatorname{core} D$ implies $D + \mathbb{R}_{>}k = \operatorname{core} D$ and thus (6.2.5) because of (6.2.4). (6.2.7) can be proved analogously. Applying Theorem 6.2.24 to $D + \varepsilon k$ instead of $D$ yields (6.2.6). If $0 \in D + \varepsilon k$, then $\varphi_{y^0-(D+\varepsilon k),k}(y^0) \leq 0$ by (4.2.11). □

**Corollary 6.2.43** *Suppose* $Y$ *is a linear space,* $F, D \subset Y$ *with* $D \neq \emptyset$*,* $k \in Y \setminus \{0\}$ *with* $0 \in \operatorname{bd}_{-k}(D + \mathbb{R}_{+}k)$ *and* $D + \mathbb{R}_{>}k \neq Y$*, and* $\varepsilon \in \mathbb{R}$.

*Then* $\varphi_{y^0-D,k}(y^0) = 0$ *for each* $y^0 \in Y$*. We have*

$$\text{Eff}(F, D + \mathbb{R}_{>}k + \varepsilon k) = \{y^0 \in F \mid \forall y \in F \setminus \{y^0\}\colon \varphi_{y^0-D,k}(y^0) \leq \varphi_{y^0-D,k}(y) + \varepsilon\}.$$

*If* $D + \mathbb{R}_{>}k \subseteq \operatorname{core} D$*, then*

$$\text{Eff}(F, \operatorname{core} D + \varepsilon k) = \{y^0 \in F \mid \forall y \in F \setminus \{y^0\}\colon \varphi_{y^0-D,k}(y^0) \leq \varphi_{y^0-D,k}(y) + \varepsilon\}. \quad (6.2.8)$$

*If* $k \in 0^+D$ *and* $D$ *is* $(-k)$*-directionally closed, then*

$$\text{Eff}(F, D + \varepsilon k) = \{y^0 \in F \mid \forall y \in F \setminus \{y^0\}\colon \varphi_{y^0-D,k}(y^0) < \varphi_{y^0-D,k}(y) + \varepsilon\}. \quad (6.2.9)$$

*If* $Y$ *is a topological vector space and* $D + \mathbb{R}_{>}k \subseteq \operatorname{int} D$*, then we get*

$$\text{WEff}(F, D + \varepsilon k) = \{y^0 \in F \mid \forall y \in F \setminus \{y^0\}\colon \varphi_{y^0-D,k}(y^0) \leq \varphi_{y^0-D,k}(y) + \varepsilon\}.$$

***Proof*** For each $y^0 \in Y$, the assumption $0 \in \operatorname{bd}_{-k}(D + \mathbb{R}_{+}k)$ implies $y^0 \in \operatorname{bd}_k(y^0 - D - \mathbb{R}_{+}k)$ and thus $\varphi_{y^0-D,k}(y^0) = 0$ by (4.2.5). Hence, Proposition 6.2.42 yields the assertions. □

**Proposition 6.2.44** *Suppose* $Y$ *is a linear space,* $a \in Y$*,* $\varepsilon \in \mathbb{R}$*,* $F, D \subset Y$ *with* $D \neq \emptyset$ *and* $D + D \subseteq D$*,* $k \in \operatorname{core} 0^+D$*, and that* $D$ *is* $(-k)$*-directionally closed. Then*

$$\{y^0 \in F \mid \forall y \in F \setminus \{y^0\}\colon \varphi_{a-D,k}(y^0) \leq \varphi_{a-D,k}(y) + \varepsilon\} \subseteq \text{Eff}(F, D + \mathbb{R}_{>}k + \varepsilon k),$$
$$\{y^0 \in F \mid \forall y \in F \setminus \{y^0\}\colon \varphi_{a-D,k}(y^0) < \varphi_{a-D,k}(y) + \varepsilon\} \subseteq \text{Eff}(F, D + \varepsilon k). \quad (6.2.10)$$

*For each* $\bar{\varepsilon} > \varepsilon$*,*

$$\{y^0 \in F \mid \forall y \in F \setminus \{y^0\}\colon \varphi_{a-D,k}(y^0) \leq \varphi_{a-D,k}(y) + \varepsilon\} \subseteq \text{Eff}(F, D + \bar{\varepsilon}k). \quad (6.2.11)$$

***Proof*** Take any $y^0 \in F$. The assumption $k \in \operatorname{core} 0^+D$ yields $k \in -\operatorname{core} 0^+(a-D)$. By Theorem 4.2.9, $\varphi_{a-D,k}$ is finite-valued. Theorem 4.2.1(f) implies $y^0 \in \operatorname{cl}_k(a-D) + \varphi_{a-D,k}(y^0)k = a + \operatorname{cl}_k(-D) + \varphi_{a-D,k}(y^0)k = a - \operatorname{cl}_{-k}(D) + \varphi_{a-D,k}(y^0)k = a - D + \varphi_{a-D,k}(y^0)k$.

Assume first $y^0 \notin \operatorname{Eff}(F, D + \mathbb{R}_> k + \varepsilon k)$. Then there exists $y \in F \setminus \{y^0\}$ with $y \in y^0 - D - \mathbb{R}_> k - \varepsilon k$. Therefore, $y \in (a - D + \varphi_{a-D,k}(y^0)k) - D - \mathbb{R}_> k - \varepsilon k \subseteq (a-D) - \mathbb{R}_> k + (\varphi_{a-D,k}(y^0) - \varepsilon)k$ since $D + D \subseteq D$. Equation (4.2.3) implies $\varphi_{a-D,k}(y) < \varphi_{a-D,k}(y^0) - \varepsilon$ and thus $\varphi_{a-D,k}(y^0) > \varphi_{a-D,k}(y) + \varepsilon$. Hence, the first inclusion holds.

Assume now $y^0 \notin \operatorname{Eff}(F, D + \varepsilon k)$. Then there exists $y \in F \setminus \{y^0\}$ with $y \in y^0 - D - \varepsilon k$. Therefore, $y \in (a - D + \varphi_{a-D,k}(y^0)k) - D - \varepsilon k \subseteq (a-D) + (\varphi_{a-D,k}(y^0) - \varepsilon)k$. Inclusion (4.2.11) implies $\varphi_{a-D,k}(y) \le \varphi_{a-D,k}(y^0) - \varepsilon$ and thus $\varphi_{a-D,k}(y^0) \ge \varphi_{a-D,k}(y) + \varepsilon$. Hence, (6.2.10) holds.

For each $\bar{\varepsilon} > \varepsilon$, we obtain $D + \bar{\varepsilon}k = D + \varepsilon k + (\bar{\varepsilon} - \varepsilon)k \subseteq D + \varepsilon k$ since $k \in 0^+D$. This yields $\operatorname{Eff}(F, D + \varepsilon k) \subseteq \operatorname{Eff}(F, D + \bar{\varepsilon}k)$ and thus (6.2.11) because of (6.2.10). □

For each $\bar{\varepsilon} > \varepsilon$, we have $D + \bar{\varepsilon}k \subseteq D + \mathbb{R}_> k + \varepsilon k$ and thus $\operatorname{Eff}(F, D + \mathbb{R}_> k + \varepsilon k) \subseteq \operatorname{Eff}(F, D + \bar{\varepsilon}k)$.

In many statements of this section, values of some function $\varphi_{y^0-D,k}$ are compared. These statements would also hold if $\varphi_{y^0-D,k}$ were replaced by $\varphi_{(y^0+ck)-D,k}$ with some arbitrary $c \in \mathbb{R}$ since $\varphi_{(y^0+ck)-D,k} = \varphi_{y^0-D,k} - c$.

### 6.2.3 Scalarization by Norms

Section 6.2.2 implies—because of Sects. 4.13 and 5.5—scalarization results for efficient and weakly efficient points by norms.

In this section, $\|\cdot\|_{C,k}$ will denote the norm that is defined as the Minkowski functional of the order interval $[-k, k]_C$ (see Theorem 2.2.10).

First, we give some sufficient conditions for efficient and weakly efficient points by minimizers of norms. The next theorem results from Theorem 6.2.21 and Proposition 5.5.2.

**Theorem 6.2.45** *Suppose that $Y$ is a linear space, $D \subset Y$, $C$ is a nontrivial algebraically closed pointed convex cone in $Y$ with $k \in \operatorname{core} C$ and $F \subseteq a + C$ for some $a \in Y$. Define*

$$\Psi := \operatorname*{argmin}_{y \in F} \|y - a\|_{C,k}.$$

*Then:*

(a) $\operatorname{Eff}(F, D) \cap \Psi \subseteq \operatorname{Eff}(\Psi, D)$.
(b) $C + D \subseteq C \implies \operatorname{Eff}(F, D) \cap \Psi = \operatorname{Eff}(\Psi, D)$.
(c) $C + D \subseteq C$ *and* $\Psi = \{y^0\}$ *imply* $y^0 \in \operatorname{Eff}(F, D)$.

(d) $C + (D \setminus \{0\}) \subseteq \operatorname{core} C \implies \Psi \subseteq \operatorname{Eff}(F, D)$.
(e) $\Psi \subseteq \operatorname{Eff}(F, D + \mathbb{R}_{>}k)$ *if* $C + D \subseteq C$ *and* $D + \mathbb{R}_{>}k \neq Y$.
(f) $\Psi \subseteq \operatorname{Eff}(F, \operatorname{core} D)$ *if* $C + D \subseteq C$ *or* $C + \operatorname{core} D \subseteq \operatorname{core} C$.
(g) *Assume that* $Y$ *is a topological vector space. Then*
$\Psi \subseteq \operatorname{WEff}(F, D)$ *if* $C + D \subseteq C$ *or* $C + (\operatorname{int} D \setminus \{0\}) \subseteq \operatorname{core} C$.

**Corollary 6.2.46** *Suppose that* $Y$ *is a linear space,* $D$ *is a nontrivial algebraically closed pointed convex cone in* $Y$ *with* $k \in \operatorname{core} D$ *and* $F \subseteq a + D$ *for some* $a \in Y$. *Define*

$$\Psi := \operatorname*{argmin}_{y \in F} \|y - a\|_{D,k}.$$

*Then:*

(a) $\operatorname{Eff}(F, D) \cap \Psi = \operatorname{Eff}(\Psi, D)$.
(b) $\Psi = \{y^0\}$ *implies* $y^0 \in \operatorname{Eff}(F, D)$.
(c) $\Psi \subseteq \operatorname{Eff}(F, \operatorname{core} D)$.
(d) $\Psi \subseteq \operatorname{WEff}(F, D)$ *if* $Y$ *is a topological vector space.*

We will now characterize the efficient point set and the weakly efficient point set by norm minimizers. We get the following from Theorem 6.2.36 by Lemma 5.5.2.

**Theorem 6.2.47** *Suppose that* $Y$ *is a linear space,* $D \subset Y$ *is a nontrivial algebraically closed pointed convex cone with nonempty algebraic interior, and* $F \subseteq a + \operatorname{core} D$ *for some* $a \in Y$. *Then*

$$\operatorname{Eff}(F, \operatorname{core} D) = \{y^0 \in F \mid \exists k \in \operatorname{core} D : \ \|y^0 - a\|_{D,k} = \min_{y \in F} \|y - a\|_{D,k}\} \quad \textit{and}$$

$$\operatorname{Eff}(F, D) = \{y^0 \in F \mid \exists k \in \operatorname{core} D \ \forall y \in F \setminus \{y^0\} : \ \|y^0 - a\|_{D,k} < \|y - a\|_{D,k}\}.$$

*For* $y^0 \in \operatorname{Eff}(F, \operatorname{core} D)$, *we can choose* $k = y^0 - a$, *which yields* $\|y^0 - a\|_{D,k} = 1$. *The same choice works for* $y^0 \in \operatorname{Eff}(F, D)$.

Theorem 6.2.37 and Proposition 5.5.2 imply:

**Theorem 6.2.48** *Suppose that* $Y$ *is a topological vector space and* $D \subset Y$ *a nontrivial closed pointed convex cone with* $\operatorname{int} D \neq \emptyset$.

*If* $F \subseteq a + \operatorname{int} D$ *for some* $a \in Y$, *then*

$$\operatorname{WEff}(F, D) = \{y^0 \in F \mid \exists k \in \operatorname{int} D : \ \|y^0 - a\|_{D,k} = \min_{y \in F} \|y - a\|_{D,k}\} \quad \textit{and}$$

$$\operatorname{Eff}(F, D) = \{y^0 \in F \mid \exists k \in \operatorname{int} D \ \forall y \in F \setminus \{y^0\} : \ \|y^0 - a\|_{D,k} < \|y - a\|_{D,k}\}.$$

*For* $y^0 \in \operatorname{WEff}(F, D)$, *we can choose* $k = y^0 - a$, *which results in* $\|y^0 - a\|_{D,k} = 1$. *The same choice works for* $y^0 \in \operatorname{Eff}(F, D)$.

### 6.2.4 Existence of Efficient and Weakly Efficient Points

Theorem 6.2.13 yields the following statement about the existence of weakly efficient points.

**Proposition 6.2.49** *Assume $Y$ is a topological vector space, $F, D \subseteq Y$. Then*

$$\mathrm{WEff}(F, D) \neq \emptyset$$

*if $F \cap \mathrm{bd}(F + D) \neq \emptyset$.*

*This condition is fulfilled if $0 \in D$, $D + D = D$, $F \neq \emptyset$ and $F + D$ is not open.*

***Proof*** The first statement follows immediately from (6.2.1). Suppose now $0 \in D$, $D + D = D$, $F \neq \emptyset$ and that $F + D$ is not open. Since $F + D$ is not open, there exist $y \in F$ and $d \in D$ with $y + d \in \mathrm{bd}(F + D)$. If $y \in \mathrm{int}(F + D)$, then $y + d \in d + \mathrm{int}(F + D) \subseteq \mathrm{int}(F + D)$ because of $D + D = D$, which contradicts $y + d \in \mathrm{bd}(F + D)$. This and $y \in F \subseteq F + D$ imply $y \in F \cap \mathrm{bd}(F + D)$. □

**Corollary 6.2.50** *Assume $Y$ is a topological vector space, that $D \subseteq Y$ is a nontrivial convex cone, $F \subseteq Y$, and that $F + D$ is not open. Then*

$$\mathrm{WEff}(F, D) \neq \emptyset.$$

For efficient points, we can prove the following existence result.

**Theorem 6.2.51** *Assume that $Y$ is a topological vector space,*

(1) *$D \subset Y$ is a set for which there exists a proper closed subset $H$ of $Y$ with $\mathrm{core}\, 0^+H \neq \emptyset$ and $H - (D \setminus \{0\}) \subseteq \mathrm{core}\, H$, and*
(2) *one of the following conditions holds:*
    (i) *$F \subseteq Y$ is a nonempty compact set,*
    (ii) *$D + D \subseteq D$ and there exists $y \in Y$ such that $F \cap (y - D)$ is a nonempty compact set.*

*Then*

$$\mathrm{Eff}(F, D) \neq \emptyset.$$

*Each set $D$ for which $D \setminus \{0\}$ is contained in the core of a nontrivial closed convex cone $C$ fulfills assumption* (1). *In $Y = \mathbb{R}^\ell$, each nontrivial closed pointed convex cone $D$ satisfies condition* (1).

***Proof***

(a) First we assume that $F$ is nonempty and compact.
Choose some $k \in -\mathrm{core}\, 0^+H$. $(\mathrm{Sp1}_{H,k})$ is fulfilled, $\varphi_{H,k}$ is lower semicontinuous and finite-valued because of Theorem 4.4.4. The function $\varphi_{H,k}$ attains a minimum on $F$ by Proposition 3.3.16. Moreover, $\varphi_{H,k}$ is strictly $D$-monotone by

Theorem 4.5.5(d). Hence, the minimizers of $\varphi_{H,k}$ belong to $\mathrm{Eff}(F, D)$ because of Proposition 6.2.16(c).

(b) Assume now $D + D \subseteq D$ and that $F \cap (y - D)$ is nonempty and compact for some $y \in Y$. We obtain $\mathrm{Eff}(F \cap (y - D), D) \neq \emptyset$ from (a). By Lemma 6.2.5(e), $\mathrm{Eff}(F \cap (y - D), D) = \mathrm{Eff}(F, D) \cap (y - D)$, which yields $\mathrm{Eff}(F, D) \neq \emptyset$.

For the last but one statement, take $H = -C$. The last statement can be derived from the cone separation theorem in [272, Theorem 3.22]. □

**Proposition 6.2.52** *Assume $Y$ is a topological vector space, $F \subseteq Y$, and that $D$ is a proper closed subset of $Y$ with $\mathrm{int}\, 0^+D \neq \emptyset$. If $\mathrm{WEff}(F, D) \neq \emptyset$, then each function $\varphi_{a-D,k}$ with $a \in Y$ and $k \in \mathrm{int}\, 0^+D$ is bounded below on $F$, finite-valued and continuous.*

***Proof*** Take any $k \in \mathrm{int}\, 0^+D$. By Proposition 4.6.13, each function $\varphi_{a-D,k}$ with $a \in Y$ is finite-valued and continuous. There exists $y \in \mathrm{WEff}(F, D)$. Hence, $F \cap (y - \mathrm{int}\, D) \subseteq \{y\}$. Thus, $\varphi_{y-D,k}$ is bounded below on $F$ by (4.4.5). Because of Lemma 5.4.1, each function $\varphi_{a-D,k}$ with $a \in Y$ is bounded below on $F$. □

Example 5.4.2 illustrates that the functionals in Proposition 6.2.52 can fail to be bounded below on $F$ if the conditions $\mathrm{int}\, 0^+D \neq \emptyset$ and $k \in \mathrm{int}\, 0^+D$ are replaced by $D + \mathbb{R}_{>}k \subseteq \mathrm{int}\, D$.

Even in $Y = \mathbb{R}^3$ with $D = \mathbb{R}^3_+$, $k \in \mathrm{int}\, D$ and a closed convex set $F$ with $\mathrm{WEff}(F, D) \neq \emptyset$, it can depend on $a \in Y$ whether $\varphi_{a-D,k}$ attains a minimum on $F$. This is illustrated by Example 5.2.19.

Proposition 6.2.26(a) and Theorem 6.2.21(c) imply the following statement.

**Proposition 6.2.53** *Assume that $Y$ is a linear space, $F \subseteq Y$, and that $D$ is a proper $(-k)$-directionally closed subset of $Y$ with $0 \in D$, $D + D \subseteq D$, $k \in 0^+D \setminus \{0\}$ and $\mathbb{R}k \not\subseteq D$. Then $\mathrm{Eff}(F, D) \neq \emptyset$ if and only if there exists $a \in Y$ such that $\varphi_{a-D,k}$ has a unique minimizer on $F$.*

**Proposition 6.2.54** *Assume $Y = \mathbb{R}^\ell$, that $F \subseteq Y$ is a closed set and that $D$ is a proper closed subset of $Y$ with $0 \in D$, $k \in 0^+D \setminus \{0\}$ and $D + D \subseteq D$ for which there exists a nontrivial closed convex cone $C$ with $D \setminus \{0\} \subseteq \mathrm{int}\, C$. Then the following conditions are equivalent to each other:*

(a) $\mathrm{Eff}(F, D) \neq \emptyset$.
(b) *There exists $a \in Y$ such that $\mathrm{argmin}_F\, \varphi_{a-D,k}$ is nonempty and bounded.*
(c) *There exists $a \in Y$ such that $\mathrm{argmin}_F\, \varphi_{a-D,k}$ is nonempty and compact.*

***Proof*** (a) yields (b) by Proposition 6.2.53.
Assume now (b). The set $M := \mathrm{argmin}_F\, \varphi_{a-D,k}$ is compact by Proposition 5.2.1(b). Then $\mathrm{Eff}(M, D) \neq \emptyset$ because of Theorem 6.2.51, applied to $H = -C$. Theorem 6.2.21(b), applied to $H = D$, implies $\mathrm{Eff}(F, D) \neq \emptyset$. □

In $Y = \mathbb{R}^\ell$, each nontrivial closed pointed convex cone $D \subset Y$ fulfills the assumptions of Proposition 6.2.53 and of Proposition 6.2.54 with $k \in D \setminus \{0\}$.

**Proposition 6.2.55** *Assume that $F \subseteq \mathbb{R}^\ell$ is a nonempty closed convex set and $D \subset \mathbb{R}^\ell$ is a nontrivial closed pointed convex cone with $k \in D \setminus \{0\}$. Then* $\mathrm{Eff}(F, D) \neq \emptyset$ *holds if and only if* $\mathrm{argmin}_F\, \varphi_{a-D,k}$ *is nonempty and compact for each $a \in \mathbb{R}^\ell$ with $F \cap \mathrm{dom}\, \varphi_{a-D,k} \neq \emptyset$.*

***Proof*** Apply Propositions 6.2.54 and 5.4.4. □

In Proposition 6.2.55, $F \cap \mathrm{dom}\, \varphi_{a-D,k} \neq \emptyset$ is satisfied for each $a \in \mathbb{R}^\ell$ if $k \in \mathrm{int}\, D$.

If $\mathrm{Eff}(F, D)$ is empty, then the minimizer set $\mathrm{argmin}_F\, \varphi_{a-D,k}$ may be empty for some $a = a^1 \in \mathbb{R}^\ell$ and nonempty for some $a = a^2 \in \mathbb{R}^\ell$ even though the assumptions of Proposition 6.2.55 are fulfilled with $k \in \mathrm{int}\, D$. This is illustrated for $Y = \mathbb{R}^3$ and $D = \mathbb{R}^3_+$ by Example 5.2.19 and by an example from Helbig [226, Example III.4.], where $F$ is defined by an infinite number of linear restrictions.

More existence results for efficient and weakly efficient points w.r.t. closed convex cones will be proved in Sect. 6.2.5. Further statements about the existence of efficient points w.r.t. general reference sets are given in [464]. A comprehensive study of existence results for nondominated elements w.r.t. binary relations and for efficient points is contained in [192]. The existence of efficient and weakly efficient points w.r.t. convex cones was also studied by Luc [338, Chap. 2].

### 6.2.5 Additional Results Related to Closed Convex Cones

In this section, we investigate efficient and weakly efficient points in a topological vector space $Y$ w.r.t. some closed convex cone $D$ using scalarizing functions $\varphi_{a-D,k}$ with $a \in Y$ and $k \in D \setminus \{0\}$. The previous sections contain various results for this case. Now, we will add several details, two algorithms for the determination of efficient points w.r.t. $D$, and a scalarizing problem which makes use of the dual space. For finding solutions to some vector optimization problem (VOP), we will apply the problem

$$\min_{y \in F} \varphi_{a-D,k}(y),$$

which is equivalent to the problem

$$\begin{aligned} (\mathrm{P}_{F,a,D,k}):\ & t \to \min \\ & y \in a - D + tk \\ & y \in F,\ t \in \mathbb{R} \end{aligned}$$

in the sense of Definition 5.1.2 because of Theorem 5.1.3(b). Recall that $\mathrm{M}_{F,a,D,k}$ denotes the set of optimal solutions $y$ to problem $(\mathrm{P}_{F,a,D,k})$.

First, we summarize implications of the previous sections for closed convex cones.

**Proposition 6.2.56** *Assume that $Y$ is a topological vector space, that $D \subset Y$ is a nontrivial closed pointed convex cone, and $F \subseteq Y$.*

(a) *If* $y \in \mathrm{M}_{F,a,D,k}$ *for some* $a \in Y$ *and* $k \in D \setminus \{0\}$*, then* $y \in \mathrm{WEff}(F, D)$.
(b) *If* $\mathrm{M}_{F,a,D,k} = \{y\}$ *for some* $a \in Y$ *and* $k \in D \setminus \{0\}$*, then* $y \in \mathrm{Eff}(F, D)$.
(c) *Take any* $k \in D \setminus \{0\}$*. Then*

$$\begin{aligned} y^0 \in \mathrm{Eff}(F, D + \mathbb{R}_> k) &\iff y^0 \in \mathrm{M}_{F,y^0,D,k}, \\ &\iff \exists a \in Y\colon\ y^0 \in \mathrm{M}_{F,a,D,k}. \\ y^0 \in \mathrm{Eff}(F, D) &\iff \mathrm{M}_{F,y^0,D,k} = \{y^0\}, \\ &\iff \exists a \in Y\colon\ \mathrm{M}_{F,a,D,k} = \{y^0\}. \end{aligned}$$

*For each* $y^0 \in \mathrm{Eff}(F, D)$ *and each* $y^0 \in \mathrm{Eff}(F, D + \mathbb{R}_> k)$*, the problem* $(\mathrm{P}_{F,y^0,D,k})$ *has the optimal value* 0.
(d) *We have*

$$\begin{aligned} \mathrm{Eff}(F, D) &= \bigcup_{a \in Y} \mathrm{Eff}(\mathrm{M}_{F,a,D,k}, D) \quad \textit{for each } k \in D \setminus \{0\}, \\ &= \bigcup_{a \in Y,\ k \in D \setminus \{0\}} \mathrm{Eff}(\mathrm{M}_{F,a,D,k}, D). \end{aligned}$$

***Proof***

(a) and (b) result from Corollary 6.2.23.
(c) follows from Theorem 6.2.29 and Corollaries 6.2.25 and 6.2.23.
(d) For each $k \in D \setminus \{0\}$, the inclusion $\mathrm{Eff}(F, D) \subseteq \bigcup_{a \in Y} \mathrm{Eff}(\mathrm{M}_{F,a,D,k}, D)$ follows from the statement related to the uniqueness of optimal solutions in (c) and by the fact that $\mathrm{Eff}(\{y\}, D) = \{y\}$ for each $y \in Y$. Corollary 6.2.23(b) implies $\bigcup_{a \in Y,\ k \in D \setminus \{0\}} \mathrm{Eff}(\mathrm{M}_{F,a,D,k}, D) \subseteq \mathrm{Eff}(F, D)$. □

We now turn to cones with nonempty interior.

**Proposition 6.2.57** *Assume that* $Y$ *is a topological vector space, that* $D \subset Y$ *is a nontrivial closed convex cone with nonempty interior, and* $F \subseteq Y$.

(a) *If the objective function of some problem* $(\mathrm{P}_{F,a,D,k})$ *with* $a \in Y$ *and* $k \in \mathrm{int}\, D$ *is not bounded below on the feasible set, then* $\mathrm{WEff}(F, D) = \emptyset$.
(b) *Take any* $k \in \mathrm{int}\, D$*. Then*

$$\begin{aligned} y^0 \in \mathrm{WEff}(F, D) &\iff y^0 \in \mathrm{M}_{F,y^0,D,k}, \\ &\iff \exists a \in Y\colon\ y^0 \in \mathrm{M}_{F,a,D,k}. \end{aligned}$$

*For each* $y^0 \in \mathrm{WEff}(F, D)$*, the problem* $(\mathrm{P}_{F,y^0,D,k})$ *has the optimal value* 0.
(c) *Assume* $F \subseteq a + \mathrm{int}\, D$ *for some* $a \in Y$*. Then*

$$\begin{aligned} y^0 \in \mathrm{WEff}(F, D) &\iff y^0 \in \mathrm{M}_{F,a,D,y^0-a}, \\ y^0 \in \mathrm{Eff}(F, D) &\iff \mathrm{M}_{F,a,D,y^0-a} = \{y^0\}. \end{aligned}$$

*For each* $y^0 \in \mathrm{WEff}(F, D)$, *the problem* $(\mathrm{P}_{F,a,D,y^0-a})$ *has the optimal value* 1.

(d) *Assume* $F \subseteq a - \operatorname{int} D$ *for some* $a \in Y$. *Then*

$$y^0 \in \mathrm{WEff}(F, D) \iff y^0 \in \mathrm{M}_{F,a,D,a-y^0},$$
$$y^0 \in \mathrm{Eff}(F, D) \iff \mathrm{M}_{F,a,D,a-y^0} = \{y^0\}.$$

*For each* $y^0 \in \mathrm{WEff}(F, D)$, *the problem* $(\mathrm{P}_{F,a,D,a-y^0})$ *has the optimal value* $-1$.

(e) *We have*

$$\mathrm{Eff}(F, D) = \bigcup_{a\in Y} \mathrm{Eff}(\mathrm{M}_{F,a,D,k}, D) \quad \textit{for each } k \in \operatorname{int} D,$$
$$= \bigcup_{a\in Y,\ k\in \operatorname{int} D} \mathrm{Eff}(\mathrm{M}_{F,a,D,k}, D).$$

***Proof***

(a) follows from Proposition 6.2.52.
(b) follows from Corollaries 6.2.33 and 6.2.23(a).
(c) und (d) result from Theorem 6.2.37.
(e) For each $k \in \operatorname{int} D$, the inclusion $\mathrm{Eff}(F, D) \subseteq \bigcup_{a\in Y} \mathrm{Eff}(\mathrm{M}_{F,a,D,k}, D)$ follows from the statement related to the uniqueness of optimal solutions in Proposition 6.2.56(c) and by the fact that $\mathrm{Eff}(\{y\}, D) = \{y\}$ for each $y \in Y$. Corollary 6.2.23(b) implies $\bigcup_{a\in Y,\ k\in \operatorname{int} D} \mathrm{Eff}(\mathrm{M}_{F,a,D,k}, D) \subseteq \mathrm{Eff}(F, D)$.

□

Furthermore, we get the following in $Y = \mathbb{R}^\ell$.

**Proposition 6.2.58** *Assume* $Y = \mathbb{R}^\ell$, *that* $D \subset Y$ *is a nontrivial closed pointed convex cone, and* $F \subseteq Y$.

(1) *Suppose that* $F$ *or* $F + D$ *is a closed convex set. If some problem* $(\mathrm{P}_{F,a,D,k})$ *with* $a \in Y$ *and* $k \in D \setminus \{0\}$ *which has a nonempty feasible set does not have a nonempty bounded set of optimal solutions* $y$, *then* $\mathrm{Eff}(F, D) = \emptyset$.
(2) *Let* $F$ *be a closed set, and* $k \in D \setminus \{0\}$. *Then the following statements are equivalent to each other.*

    (a) $\mathrm{Eff}(F, D) \neq \emptyset$.
    (b) *There exists* $a \in Y$ *such that* $\mathrm{M}_{F,a,D,k}$ *contains exactly one element.*
    (c) *There exists* $a \in Y$ *such that* $\mathrm{M}_{F,a,D,k}$ *is nonempty and bounded.*
    (d) *There exists* $a \in Y$ *such that* $\mathrm{M}_{F,a,D,k}$ *is nonempty and compact.*

(3) *Assume that* $F + D$ *is closed, and* $k \in D \setminus \{0\}$. *Then the statements* (a)–(c) *in* (2) *are equivalent to each other.*
(4) *Assume that* $F$ *or* $F + D$ *is a closed convex set. Then* $\mathrm{Eff}(F, D) \neq \emptyset$ *is equivalent to the following statement:*

(e) *There exist* $a^0 \in Y$ *and* $k^0 \in D \setminus \{0\}$ *such that* $\mathrm{M}_{F,a^0,D,k^0}$ *is nonempty, and* $\mathrm{M}_{F,a,D,k}$ *is bounded for all* $a \in Y$ *and* $k \in D \setminus \{0\}$.

***Proof***

(1) Assume first that $F$ is closed and convex. Proposition 5.2.1 implies that $\mathrm{M}_{F,a,D,k}$ is closed for all $a \in Y$ and $k \in D \setminus \{0\}$. This yields the assertion for this case because of Proposition 6.2.55.
Assume now that $F + D$ is closed and convex, and that some problem $(\mathrm{P}_{F,a,D,k})$ with $a \in Y$ and $k \in D \setminus \{0\}$ which has a nonempty feasible set does not have a nonempty bounded set of optimal solutions $y$. Because of Corollary 5.1.7, the set $\mathrm{M}_{F+D,a,D,k}$ is empty or unbounded. Moreover, $(\mathrm{P}_{F+D,a,D,k})$ has a nonempty feasible set since $F \subseteq F + D$. Applying the proved part of (1), we get $\mathrm{Eff}(F + D, D) = \emptyset$. By Lemma 6.2.5(d), $\mathrm{Eff}(F, D) = \mathrm{Eff}(F + D, D)$.

The equivalence between (a) and (b) in (2)–(4) follows from Proposition 6.2.53. The other equivalences in (2) result from Proposition 6.2.54.

(3) Obviously, (b) implies (c).
Assume now that (c) holds. Then $\mathrm{cl}\, \mathrm{M}_{F,a,D,k}$ is compact. Theorem 6.2.51 yields $\mathrm{Eff}(\mathrm{cl}\, \mathrm{M}_{F,a,D,k}, D) \neq \emptyset$. By Proposition 5.2.2, $\mathrm{Eff}(\mathrm{M}_{F,a,D,k}, D) = \mathrm{Eff}(\mathrm{cl}\, \mathrm{M}_{F,a,D,k}, D)$. By Corollary 6.2.23(b), $\mathrm{Eff}(\mathrm{M}_{F,a,D,k}, D) \subseteq \mathrm{Eff}(F, D)$. Hence, (a) is fulfilled.
(4) (e) implies (c) for $k = k^0$ and thus $\mathrm{Eff}(F, D) \neq \emptyset$.
Assume now $\mathrm{Eff}(F, D) \neq \emptyset$. Since (a) implies (b), there exist $a^0 \in Y$ and $k^0 \in D \setminus \{0\}$ such that $\mathrm{M}_{F,a^0,D,k^0}$ is nonempty. (1) implies (e). □

**Corollary 6.2.59** *Assume* $Y = \mathbb{R}^\ell$, *that* $D \subset Y$ *is a nontrivial closed pointed convex cone, that* $F \subseteq Y$ *is nonempty, and that* $F$ *or* $F + D$ *is a closed convex set. Then* $\mathrm{Eff}(F, D) \neq \emptyset$ *if and only if each problem* $(\mathrm{P}_{F,a,D,k})$ *with* $a \in Y$ *and* $k \in D \setminus \{0\}$ *which has a nonempty feasible set has some optimal solution* $y \in \mathrm{Eff}(F, D)$.

***Proof*** For each $a \in F$ and $k \in D \setminus \{0\}$, we get $F \cap (a - D + tk) \neq \emptyset$. Consequently, there exist $a \in Y$ and $k \in D \setminus \{0\}$ such that $(\mathrm{P}_{F,a,D,k})$ has a nonempty feasible set. This implies the reverse direction of the assertion.

Assume now $\mathrm{Eff}(F, D) \neq \emptyset$. Take any $a \in Y$ and $k \in D \setminus \{0\}$ for which $(\mathrm{P}_{F,a,D,k})$ has a nonempty feasible set. Then $\mathrm{M}_{F,a,D,k}$ is nonempty and bounded because of Proposition 6.2.58(1). Hence, $\mathrm{cl}\, \mathrm{M}_{F,a,D,k}$ is compact. Theorem 6.2.51 yields $\mathrm{Eff}(\mathrm{cl}\, \mathrm{M}_{F,a,D,k}, D) \neq \emptyset$. If $F + D$ is closed, then Proposition 5.2.2 implies $\mathrm{Eff}(\mathrm{M}_{F,a,D,k}, D) = \mathrm{Eff}(\mathrm{cl}\, \mathrm{M}_{F,a,D,k}, D)$. If $F$ is closed, then $\mathrm{M}_{F,a,D,k}$ is closed because of Proposition 5.2.1. By Corollary 6.2.23(b), $\mathrm{Eff}(\mathrm{M}_{F,a,D,k}, D) \subseteq \mathrm{Eff}(F, D)$. Therefore, $\mathrm{M}_{F,a,D,k}$ contains some $y \in \mathrm{Eff}(F, D)$. □

**Proposition 6.2.60** *Assume* $Y = \mathbb{R}^\ell$, $F \subseteq Y$, *and that* $D \subset Y$ *is a closed pointed convex cone with nonempty interior.*

(1) *For each* $a \in Y$ *and* $k \in \mathrm{int}\, D$, *the set* $\mathrm{M}_{F,a,D,k}$ *is compact and contains some element of* $\mathrm{Eff}(F, D)$ *if one of the following conditions holds:*

(i) *$F$ is nonempty and compact.*
(ii) *$F$ is a nonempty closed set for which there exist a polyhedral cone $H \subset Y$ and $u \in Y$ with $-D \setminus \{0\} \subset \operatorname{int} H$ and $(F - u) \cap \operatorname{int} H = \emptyset$.*

(2) *If $F$ is a closed convex set, then the following statements are equivalent to each other.*

(a) $\operatorname{Eff}(F, D) \neq \emptyset$.
(b) *For each $a \in Y$ and $k \in \operatorname{int} D$, the set $\mathrm{M}_{F,a,D,k}$ is nonempty and compact.*
(c) *There exist $a \in Y$ and $k \in \operatorname{int} D$ such that $\mathrm{M}_{F,a,D,k}$ is nonempty and bounded.*
(d) *For each $a \in Y$ and $k \in \operatorname{int} D$, the set $\mathrm{M}_{F,a,D,k}$ contains some element of* $\operatorname{Eff}(F, D)$.

(3) *Let $F + D$ be a closed convex set. Then the statements* (a)–(d) *in* (2) *are equivalent to each other if the compactness in* (b) *is replaced by boundedness.*

***Proof*** Corollary 4.7.5(d) implies that $\varphi_{a-D,k}$ is finite-valued for each $a \in Y$ and $k \in \operatorname{int} D$. Consequently, $F \cap \operatorname{dom} \varphi_{a-D,k} = F \cap Y = F$ for each $a \in Y$ and $k \in \operatorname{int} D$.

(1) Take any $a \in Y$ and $k \in \operatorname{int} D$.

(i) Because of Proposition 5.2.6, the set $\mathrm{M}_{F,a,D,k}$ is nonempty and compact.
(ii) $\mathrm{M}_{F,a,D,k}$ is nonempty and compact because of Proposition 5.2.20.

Theorem 6.2.51 implies $\operatorname{Eff}(\mathrm{M}_{F,a,D,k}, D) \neq \emptyset$. By Corollary 6.2.23(b), $\operatorname{Eff}(\mathrm{M}_{F,a,D,k}, D) \subseteq \operatorname{Eff}(F, D)$.

(2) (a) and (b) are equivalent to each other because of Proposition 6.2.55. The equivalence of (a) and (c) follows from Proposition 6.2.54. (b) implies $\operatorname{Eff}(\mathrm{M}_{F,a,D,k}, D) \neq \emptyset$ and $\operatorname{Eff}(\mathrm{M}_{F,a,D,k}, D) \subseteq \operatorname{Eff}(F, D)$ for each $a \in Y$ and $k \in \operatorname{int} D$ because of Theorem 6.2.51 and Corollary 6.2.23(b). Thus, (b) implies (d). Obviously, (d) implies (a).

(3) Beside (a), (c) and (d), we consider now the condition:

(b*) For each $a \in Y$ and $k \in \operatorname{int} D$, the set $\mathrm{M}_{F,a,D,k}$ is nonempty and bounded.

We have $\operatorname{Eff}(F + D, D) = \operatorname{Eff}(F, D)$ because of Lemma 6.2.5(d). Applying (2) to $F + D$ instead of $F$, we get the equivalence of (a) to the following condition:

(b') For each $a \in Y$ and $k \in \operatorname{int} D$, the set $\mathrm{M}_{F+D,a,D,k}$ is nonempty and compact.

For each $a \in Y$ and $k \in \operatorname{int} D$, Corollary 5.1.7 yields $\mathrm{M}_{F+D,a,D,k} \cap F = \mathrm{M}_{F,a,D,k}$ and that $\mathrm{M}_{F,a,D,k} \neq \emptyset$ holds if and only if $\mathrm{M}_{F+D,a,D,k}$ is nonempty. Thus, (b') implies (b*). Therefore, (a) implies (b*). Obviously, (b*) implies (c).
Take now any $a \in Y$ and $k \in \operatorname{int} D$ for which $\mathrm{M}_{F,a,D,k}$ is nonempty and bounded. Then $\operatorname{cl} \mathrm{M}_{F,a,D,k}$ is compact. Theorem 6.2.51 yields $\operatorname{Eff}(\operatorname{cl} \mathrm{M}_{F,a,D,k}, D) \neq \emptyset$. Because of Proposition 5.2.2, $\operatorname{Eff}(\mathrm{M}_{F,a,D,k}, D) = \operatorname{Eff}(\operatorname{cl} \mathrm{M}_{F,a,D,k}, D)$. By Corollary 6.2.23(b), $\operatorname{Eff}(\mathrm{M}_{F,a,D,k}, D) \subseteq \operatorname{Eff}(F, D)$.
Hence, (c) implies (a), and (b*) implies (d). Obviously, (d) implies (a). □

Even in the case where $Y = \mathbb{R}^3$, $D = \mathbb{R}^3_+$, $k \in \operatorname{int} D$ and $F$ is a nonempty closed convex set, the existence of optimal solutions to $(\mathrm{P}_{F,a,D,k})$ can depend on $a$. Examples for this statement were mentioned following Proposition 6.2.55.

Proposition 6.2.60(1) implies an existence result for efficient points.

**Corollary 6.2.61** *Assume $Y = \mathbb{R}^\ell$, that $D \subset Y$ is a closed pointed convex cone with nonempty interior and that one of the following conditions is satisfied for the nonempty set $F \subset Y$:*

(i) *$F$ is compact.*
(ii) *$F$ is a closed set for which there exist a polyhedral cone $H \subset Y$ and $u \in Y$ with $-D \setminus \{0\} \subset \operatorname{int} H$ and $(F - u) \cap \operatorname{int} H = \emptyset$.*

*Then* $\mathrm{Eff}(F, D) \neq \emptyset$.

An algorithm for the determination of these efficient points will be given in the next proposition.

If $Y$ is a topological vector space, $F$ is a nonempty compact subset of $Y$, $D \subset Y$, and there exists a function $\varphi \colon Y \to \mathbb{R}$ which is lower semicontinuous and strictly $D$-monotone on $F$, then $\varphi$ attains a minimum on $F$ and each minimizer of $\varphi$ on $F$ belongs to $\mathrm{Eff}(F, D)$. This results from Propositions 3.3.16 and 6.2.16(c). In the case $Y = \mathbb{R}^\ell$ and $\operatorname{qint} D^+ \neq \emptyset$, we can choose any $\varphi \in \operatorname{qint} D^+$. This fact will be used in the subsequent propositions. Here the definition of $\operatorname{qint} D^+$ is related to the dual pair $(\mathbb{R}^\ell, \mathbb{R}^\ell, \langle .,. \rangle)$ with $\langle y, y' \rangle = (y')^\mathrm{T} y$, i.e.,

$$\operatorname{qint} D^+ = \{y' \in \mathbb{R}^\ell \mid \forall d \in D \setminus \{0\} \colon (y')^\mathrm{T} d > 0\}.$$

**Proposition 6.2.62** *Under the assumptions of Corollary 6.2.61, efficient points of $F$ w.r.t. $D$ can be determined by the following algorithm:*

---

*Take any $a \in Y$, $k \in \operatorname{int} D$.*
*Take any $y' \in \operatorname{qint} D^+$, and define $\varphi \colon Y \to \mathbb{R}$ by $\varphi(y) := (y')^\mathrm{T} y$ for each $y \in Y$.*
*Minimize $\varphi$ on $\mathrm{M}_{F,a,D,k}$.*
*The set of minimizers is nonempty. Each minimizer belongs to* $\mathrm{Eff}(F, D)$.

---

*For $k$ and $\varphi$ being fixed, each element of* $\mathrm{Eff}(F, D)$ *can be determined in this way with an appropriate parameter vector $a \in F$.*

*If $F \subseteq a + \operatorname{int} D$ or $F \subseteq a - \operatorname{int} D$ holds, then each element of* $\mathrm{Eff}(F, D)$ *can be determined in the described way with $\varphi$ being fixed by varying $k$ in $F - a$ or $a - F$, respectively.*

*Instead of the chosen function $\varphi$, any function $\varphi \colon Y \to \overline{\mathbb{R}}$ could be used which is finite-valued, lower semicontinuous and strictly $D$-monotone on $\mathrm{M}_{F,a,D,k}$.*

***Proof*** Take any $a \in Y, k \in \operatorname{int} D$, and any function $\varphi \colon Y \to \overline{\mathbb{R}}$ which is finite-valued, lower semicontinuous and strictly $D$-monotone on $\mathrm{M}_{F,a,D,k}$. Then the set $\mathrm{M}_{F,a,D,k}$ is nonempty and compact because of Proposition 6.2.60(1). The function $\varphi$ attains a minimum on $M_{F,a,D,k}$ because of Proposition 3.3.16. The minimizers belong to $\mathrm{Eff}(F, D)$ because of Proposition 6.2.16(c).

Proposition 6.2.56(c) implies that each $y^0 \in \mathrm{Eff}(F, D)$ can be determined in this way with $a = y^0$ even if $k$ and $\varphi$ are fixed. Proposition 6.2.57(c) and (d) yield the statement for varying $k$.

The assumptions guarantee $\mathrm{qint}\, D^+ \neq \emptyset$. If $\varphi$ is defined via an element of $\mathrm{qint}\, D^+$ in the way which is described in the algorithm, then $\varphi$ is finite-valued, lower semicontinuous and strictly $D$-monotone. □

For the case where $F$ or $F + D$ is a closed convex set, the following algorithm determines an efficient element of $F$ or states that $\mathrm{Eff}(F, D)$ is empty.

**Proposition 6.2.63** *Assume $Y = \mathbb{R}^\ell$, that $F \subseteq Y$ is nonempty, and that $D \subset Y$ is a nontrivial closed pointed convex cone. Let $F$ or $F + D$ be a closed convex set.*

*Then the following algorithm can be applied:*

---

*Take any $a \in Y$ and $k \in D \setminus \{0\}$ such that*

$$F \cap (a - D + \mathbb{R}k) \neq \emptyset. \tag{6.2.12}$$

*Solve the problem* $(\mathrm{P}_{F,a,D,k})$.

*If* $M_{F,a,D,k} = \emptyset$, *then* $\mathrm{Eff}(F, D) = \emptyset$.

*If* $M_{F,a,D,k} \neq \emptyset$, *then proceed as follows.*

*Take any $y' \in \mathrm{qint}\, D^+$, and define $\varphi\colon Y \to \mathbb{R}$ by $\varphi(y) := (y')^T y$ for each $y \in Y$.*

*Minimize $\varphi$ on* $\mathrm{cl}\, M_{F,a,D,k}$.

*If the set of minimizers to this problem is empty, then* $\mathrm{Eff}(F, D) = \emptyset$.

*Otherwise, each minimizer belongs to* $\mathrm{Eff}(F, D)$.

---

*Each element of* $\mathrm{Eff}(F, D)$ *can be determined in this way even if $k$ and $\varphi$ are fixed.*

*Instead of the chosen function $\varphi$, any function $\varphi\colon Y \to \overline{\mathbb{R}}$ could be used which is finite-valued, lower semicontinuous and strictly $D$-monotone on* $\mathrm{cl}\, M_{F,a,D,k}$.

*Moreover, the following statements are valid:*

(a) *If $F$ is a closed convex set, then each set* $M_{F,a,D,k}$ *is closed and convex.*
(b) *The condition* (6.2.12) *holds for each $a \in F$ and $k \in D \setminus \{0\}$.*
(c) *If $D$ has a nonempty interior, then* (6.2.12) *is fulfilled for each $a \in Y$ and $k \in \mathrm{int}\, D$.*

***Proof*** (a) follows from Proposition 5.2.1. (c) results from Corollary 2.3.12. (b) is obvious. Hence, there exist parameters for which (6.2.12) holds.

If $M_{F,a,D,k} = \emptyset$, then $\mathrm{Eff}(F, D) = \emptyset$ because of Corollary 6.2.59. Assume now $M_{F,a,D,k} \neq \emptyset$. Take any function $\varphi\colon Y \to \overline{\mathbb{R}}$ which is finite-valued, lower semicontinuous and strictly $D$-monotone on $\mathrm{cl}\, M_{F,a,D,k}$. If $\varphi$ does not attain a minimum on $\mathrm{cl}\, M_{F,a,D,k}$, then Proposition 3.3.16 implies that $\mathrm{cl}\, M_{F,a,D,k}$ is not compact and thus not bounded. We obtain $\mathrm{Eff}(F, D) = \emptyset$ because of Proposition 6.2.58(1). By Proposition 6.2.16(c), minimizers of $\varphi$ on $\mathrm{cl}\ \mathrm{M}_{F,a,D,k}$ are efficient elements of $\mathrm{cl}\ \mathrm{M}_{F,a,D,k}$ w.r.t. $D$. If $F + D$ is closed, then $\mathrm{Eff}(\mathrm{M}_{F,a,D,k}, D) = \mathrm{Eff}(\mathrm{cl}\ \mathrm{M}_{F,a,D,k}, D)$ because of Proposition 5.2.2. This and (a) imply $\mathrm{Eff}(\mathrm{cl}\ \mathrm{M}_{F,a,D,k}, D) = \mathrm{Eff}(\mathrm{M}_{F,a,D,k}, D) \subseteq \mathrm{Eff}(F, D)$ by Corollary 6.2.23(b). The statement related to fixed $k$ and $\varphi$ results from Proposition 6.2.56(c).

The assumptions guarantee $\mathrm{qint}\ D^+ \neq \emptyset$. If $\varphi$ is defined via an element of $\mathrm{qint}\ D^+$ in the way which is described in the algorithm, then $\varphi$ is finite-valued, lower semicontinuous and strictly $D$-monotone. □

For closed convex sets $F$, the following statement can be used for the parameter control in algorithms.

**Corollary 6.2.64** *Assume that $F \subseteq \mathbb{R}^\ell$ is a nonempty closed convex set and $D \subset \mathbb{R}^\ell$ is a nontrivial closed pointed convex cone. Then the set*

$$\{(a, k) \in \mathbb{R}^\ell \times (D \setminus \{0\}) \mid (\mathrm{P}_{F,a,D,k}) \text{ has a feasible solution}\}$$

*is convex. It coincides with*

$$\{(a, k) \in \mathbb{R}^\ell \times (D \setminus \{0\}) \mid \mathrm{M}_{F,a,D,k} \neq \emptyset\}$$

*if* $\mathrm{Eff}(F, D) \neq \emptyset$.

***Proof*** Proposition 5.3.1 yields the first statement. Assume now $\mathrm{Eff}(F, D) \neq \emptyset$. Because of Proposition 6.2.55, each problem $(\mathrm{P}_{F,a,D,k})$ with $(a, k) \in \mathbb{R}^\ell \times (D \setminus \{0\})$ which has a feasible solution has an optimal solution as well. □

We will now investigate a class of scalarizing problems which are defined by elements of the dual space and turn out to be of type $(\mathrm{P}_{F,a,D,k})$. Here we make use of the following assumption:

$$\begin{array}{c}(Y, Y', \langle .,. \rangle) \text{ is a dual pair, } \Gamma \subset Y' \setminus \{0\},\ F \subseteq Y, \text{ and} \\ D := \{y \in Y \mid \forall y' \in \Gamma : \langle y, y' \rangle \geq 0\}.\end{array} \tag{H$\Gamma$}$$

Consider the scalar optimization problem

$$\min_{y \in F} \sup_{y' \in \Gamma} \langle y - q, y' \rangle \tag{P($q$)}$$

with $q \in Y$.

Helbig [228] introduced this problem on a locally convex space $Y$ with the dual pair $(Y, Y^*, \langle .,. \rangle)$ in order to characterize efficient points of $F$ w.r.t. $D$. Because of the following lemma, the general scalarization results which we proved for problems $(\mathrm{P}_{F,a,H,k})$ can be applied to the scalarization (P($q$)).

**Lemma 6.2.65** *Assume* (H$\Gamma$) *and that there exists $k \in Y$ with $\langle k, y' \rangle = 1$ for each $y' \in \Gamma$. Then* (P($q$)) *is equivalent to* $(\mathrm{P}_{F,q,D,k})$.

***Proof*** (P($q$)) can be written as

$$\begin{aligned} &t \to \min \\ &\langle y - q, y' \rangle \leq t \quad \text{for all } y' \in \Gamma \\ &y \in F,\ t \in \mathbb{R}. \end{aligned} \tag{6.2.13}$$

Inequality (6.2.13) is equivalent to the condition that $\langle q - y + tk, y' \rangle \geq 0$ holds for all $y' \in \Gamma$ and thus to $q - y + tk \in D$, i.e., to $y \in q - D + tk$. □

If the assumption related to $k$ in Lemma 6.2.65 is not satisfied, then it is not guaranteed that there exist $a \in Y$, $k \in Y \setminus \{0\}$ and a connected $(-k)$-directionally closed set $C \subseteq Y$ such that (P($q$)) is equivalent to ($P_{F,a,C,k}$) for each $F \subseteq Y$. This is illustrated by the following example.

**Example 6.2.66** Consider the dual pair $(Y, Y, \langle .,. \rangle)$ with $Y = \mathbb{R}^2$ and $\langle y, y' \rangle = (y')^T y$. Take

$$\Gamma := \{(1, 0)^T, (1, -1)^T, (-1, 1)^T\}.$$

Then $D := \{y \in Y \mid \forall y' \in \Gamma\colon \langle y, y' \rangle \geq 0\} = \{(y_1, y_2)^T \in \mathbb{R}^2 \mid y_1 = y_2 \geq 0\}$. Problem (P($q$)) with $q = 0$ becomes

$$\min_{(y_1,y_2)^T \in F} \sup\{y_1,\ y_1 - y_2,\ -(y_1 - y_2)\}. \tag{P(0)}$$

Define $F_1 := \{(y_1, y_2)^T \in \mathbb{R}^2 \mid y_1 - 1 \leq y_2 \leq y_1 + 1\}$, $F_2 := \{(y_1, y_2)^T \in \mathbb{R}^2 \mid y_1 + 1 \leq y_2 \leq y_1 + 3\}$, and $F_3 := \{(y_1, y_2)^T \in \mathbb{R}^2 \mid y_1 - 3 \leq y_2 \leq y_1 - 1\}$ (see Fig. 6.6). For $F = F_1$, (P(0)) has the optimal solution set $-D$. For $F = F_2$, (P(0)) has the optimal solution set $(1, 2)^T - D$. For $F = F_3$, (P(0)) has the optimal solution set $(1, 0)^T - D$. To obtain a contradiction, suppose that there exist $a \in Y$, $k \in Y \setminus \{0\}$ and a connected $(-k)$-directionally closed set $C \subseteq Y$ such that the optimal solutions to (P(0)) coincide with those to ($P_{F,a,C,k}$) for all $F \subseteq Y$. For each $i \in \{1, 2, 3\}$,

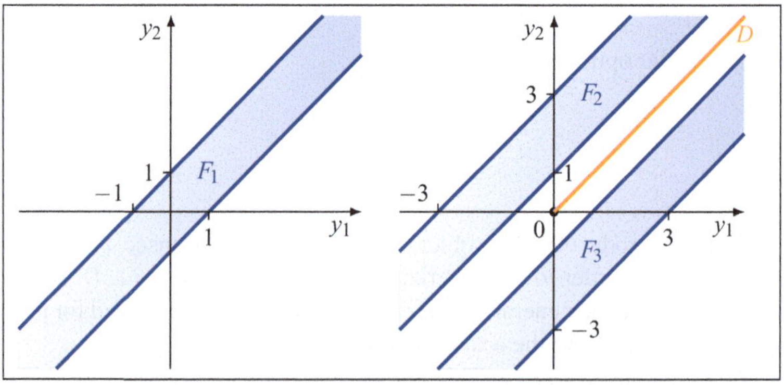


**Fig. 6.6** Sets $D$, $F_1$, $F_2$ and $F_3$ in Example 6.2.66. Adapted from [472, p. 150]

denote by $t_i$ the optimal value of $(\mathrm{P}_{F_i,a,C,k})$. Then $-D = F_1 \cap (a - C + t_1 k)$ because of Theorem 5.1.3(a). Therefore, $-D = a - C + t_1 k$ since $C$ is a connected set and $-D \subset \operatorname{int} F_1$. Furthermore, $(1,2)^T - D = F_2 \cap (a - C + t_2 k) = F_2 \cap ((t_2 - t_1)k - D)$. This implies $(1,2)^T = (t_2 - t_1)k$ since $(1,2)^T - D \subset F_2$. Moreover, $(1,0)^T - D = F_3 \cap (a - C + t_3 k) = F_3 \cap ((t_3 - t_1)k - D)$. Since $(1,0)^T - D \subset F_3$, we get $(1,0)^T = (t_3 - t_1)k$, which contradicts $(1,2)^T = (t_2 - t_1)k$.

In [232], Helbig proved scalarization results using problems (P($q$)) in the more specific context of Lemma 6.2.67(a). Because of Lemma 6.2.67(b), his statements about weakly efficient points of $F$ w.r.t. $D$ [232, Sect. 2] turn out to be special cases of Propositions 6.2.56(a) and 6.2.57(b).

**Lemma 6.2.67** *Assume that $Y$ is a separated locally convex space and that $D \subset Y$ is a nontrivial closed convex cone with nonempty interior. Consider the dual pair $(Y, Y^*, \langle .,. \rangle)$.*

(a) *The cone $D$ can be represented as*

$$D = \{y \in Y \mid \forall y^* \in \Gamma : \langle y, y^* \rangle \geq 0\},$$

*where $\Gamma$ is the set of extreme points of a weak* compact base $B$ for the dual cone $D^+$.*

(b) *The system*

$$\langle k, y^* \rangle = 1 \quad \text{for all } y^* \in \Gamma$$

*has some solution $k \in \operatorname{int} D$. Each problem (P($q$)) with $F \subseteq Y$ and $q \in Y$ is equivalent to $(\mathrm{P}_{F,q,D,k})$.*

***Proof***

(a) was proved in [232].
(b) Because of [272, Lemma 3.3], there exists $k \in Y$ with $\langle k, y^* \rangle > 0$ for each $y^* \in D^+ \setminus \{0\}$ and $B = \{y^* \in D^+ \mid \langle k, y^* \rangle = 1\}$. We get $k \in \operatorname{int} D$ from [4, Lemma 2.17]. Thus, the assertion of part (b) results from Lemma 6.2.65. □

## 6.3 Subsets of the Efficient Point Set and Proper Efficiency

In many vector optimization problems, weakly efficient points can be characterized as minimizers of some extended real-valued functions, whereas efficiency of a point can only be shown if a functional has a unique minimizer, which results in practical problems. Since the weakly efficient point set can be much more comprehensive than the efficient point set, we look for possibilities to guarantee efficiency of solutions. This is the main reason why different notions of proper efficiency came into existence. Properly efficient point sets are subsets of the efficient point set and usually defined in such a way that they can be determined more easily than efficient points in general.

A properly efficient point set can be used for approximating the efficient point set if it is dense in the efficient point set.

Many notions of proper efficiency exist in the literature. We are mainly interested in two general concepts of proper efficiency that arise from the properties of efficient point sets, which we investigated in the previous sections. We will show that different definitions of proper efficiency can result in the same properly efficient point set.

Since each element of $F$ in which some strictly $D$-monotone functional attains its minimum on $F$ is an efficient point of $F$ w.r.t. $D$, such minimal solutions are appropriate for defining proper efficiency.

**Definition 6.3.1** Suppose $Y$ to be a linear space, $F, D \subseteq Y$ with $D \setminus \{0\} \neq \emptyset$ and that $\Phi$ is a nonempty subset of the family of functions $\varphi : F \to \mathbb{R}$ which are strictly $D$-monotone on $F$. Then $y^0 \in F$ is said to be a **$\Phi$-properly efficient point** of $F$ w.r.t. $D$ if there exists a function $\varphi \in \Phi$ that attains its minimum on $F$ in $y^0$.

A special case contained in this definition is proper efficiency according to Bitran and Magnanti [53]. They defined in a topological vector space $Y$ for $F \subseteq Y$ and nontrivial convex cones $D \subset Y$ that $y^0 \in F$ is a properly efficient point of $F$ w.r.t. $D$ if there exists a continuous linear strictly $D$-monotone function $\varphi\colon Y \to \mathbb{R}$ that attains its minimum on $F$ in $y^0$. If $Y$ is the Euclidean space and $D$ the nonnegative orthant in this space, then the proper efficiency by Bitran and Magnanti coincides with the proper efficiency by Schönfeld [406].

Another general concept for proper efficiency is given in the following way:

**Definition 6.3.2** Suppose $Y$ to be a linear space, $F, D \subseteq Y$ with $D \setminus \{0\} \neq \emptyset$ and that $Z$ is a nonempty subset of the family of sets $H \subseteq Y$ with the property $H \supseteq D \setminus \{0\}$. Then $y^0 \in F$ is said to be a **$Z$-properly efficient point** of $F$ w.r.t. $D$ if there exists a set $H \in Z$ with $y^0 \in \mathrm{Eff}(F, H)$.

Because of $H \supseteq D \setminus \{0\}$, each of these properly efficient points is an efficient point of $F$ w.r.t. $D$. If $Y$ is a topological vector space and $H$ is open for each $H \in Z$, then $\mathrm{Eff}(F, H) = \mathrm{WEff}(F, H)$. We have already pointed out the advantage of dealing with weakly efficient point sets.

In a topological vector space $Y$, we get for

$$Z = Z_{\mathrm{He}} := \{H \subset Y \mid H \text{ nontrivial convex cone},\ D \setminus \{0\} \subseteq \operatorname{int} H\}$$

the notion of proper efficiency by Henig [235]; for

$$Z = Z_{\mathrm{NI}} := \{\operatorname{int} H \mid H \subset Y \text{ closed convex set},\ 0 \in \operatorname{bd} H,\ H + (D \setminus \{0\}) \subseteq \operatorname{int} H\},$$

we get the notion of proper efficiency by Nehse und Iwanow [184, 263].

We have to mention that all authors defined their proper efficiency notions under more restrictive assumptions for the space $Y$ and the domination sets $D$. The following notions of proper efficiency in (a) and (c) were originally defined in $Y = \mathbb{R}^\ell$ w.r.t. a nontrivial convex cone $D$ that had to be closed in Benson's definition, which was given in [42].

**Definition 6.3.3** Suppose $Y$ to be a topological vector space and $D, F \subseteq Y$, $D \setminus \{0\} \neq \emptyset$.

(a) $y^0 \in F$ is said to be a **Henig properly efficient point** of $F$ w.r.t. $D$ if $y^0 \in \mathrm{Eff}(F, H)$ for some nontrivial convex cone $H \subset Y$ with $D \setminus \{0\} \subseteq \mathrm{int}\, H$. We will denote the set of these points by He-PEff$(F, D)$.
(b) $y^0 \in F$ is said to be a **Nehse–Iwanow properly efficient point** of $F$ w.r.t. $D$ if $y^0 \in \mathrm{WEff}(F, H)$ for some closed convex set $H \subseteq Y$ with $0 \in \mathrm{bd}\, H$ and $H + (D \setminus \{0\}) \subseteq \mathrm{int}\, H$. We will denote the set of these points by NI-PEff$(F, D)$.
(c) If $0 \in D$, then $y^0 \in F$ is said to be a **Benson properly efficient point** of $F$ w.r.t. $D$ if $\mathrm{cl\,cone}(F + D - y^0) \cap (-D) = \{0\}$. We will denote the set of these points by Be-PEff$(F, D)$.

We define $\text{NI-PEff}_f(S, D) := \{x^0 \in S \mid f(x^0) \in \text{NI-PEff}(f(S), D)\}$ as the set of **Nehse–Iwanow properly efficient solutions** to (VOP).

Note that the assumption $0 \in D$ in (c) guarantees efficiency of the points in Be-PEff$(F, D)$ [463].

The set defined in (a) would not alter if the condition $y^0 \in \mathrm{Eff}(F, H)$ were replaced by $y^0 \in \mathrm{Eff}(F, H \setminus (-H))$. This follows from the facts that $\tilde{H} := (H \setminus (-H)) \cup \{0\}$ is a nontrivial convex cone with $\mathrm{int}\, \tilde{H} = \mathrm{int}\, H$ under the assumptions for $H$ in (a) and that $\mathrm{Eff}(F, H) \subseteq \mathrm{Eff}(F, \tilde{H})$.

The condition $0 \in \mathrm{bd}\, H$ in part (b) of the definition was added by Zălinescu [507, p. 79], who proved that the properly efficient point set of each set $F$ would be empty or $F$ without this condition.

The assumptions for $H$ in part (b) imply $D \setminus \{0\} \subseteq \mathrm{int}\, H$. Zălinescu [507, Theorem 6] proved:

**Lemma 6.3.4** *Suppose $D \subset \mathbb{R}^\ell$ to be a closed convex cone with $\mathrm{int}\, D \neq \emptyset$, $F \subseteq \mathbb{R}^\ell$. Then $y^0 \in$ NI-PEff$(F, D)$ is equivalent to the existence of a closed convex set $H \subseteq \mathbb{R}^\ell$ with $D \setminus \{0\} \subseteq \mathrm{int}\, H$ and $y^0 \in \mathrm{WEff}(F, H)$.*

Lemma 2.1.24 yields:

**Lemma 6.3.5** *Assume $Y$ to be a topological vector space and $D, F \subseteq Y$, $D \setminus \{0\} \neq \emptyset$. Then $y^0 \in$ NI-PEff$(F, D)$ if and only if there exists an open convex set $H \subseteq Y$ with $0 \in \mathrm{bd}\, H$, $\mathrm{cl}\, H + (D \setminus \{0\}) \subseteq H$ and $y^0 \in \mathrm{Eff}(F, H)$.*

Henig's proper efficiency can also be formulated using weakly efficient points w.r.t. $H$.

**Proposition 6.3.6** *Suppose $Y$ to be a topological vector space and $D, F \subseteq Y$ with $D \setminus \{0\} \neq \emptyset$.*

(1) *The following statements are equivalent to each other:*

  (a) $y^0 \in$ He-PEff$(F, D)$.
  (b) $y^0 \in \mathrm{WEff}(F, H)$ *for some nontrivial convex cone $H \subset Y$ with $D \setminus \{0\} \subseteq \mathrm{int}\, H$.*

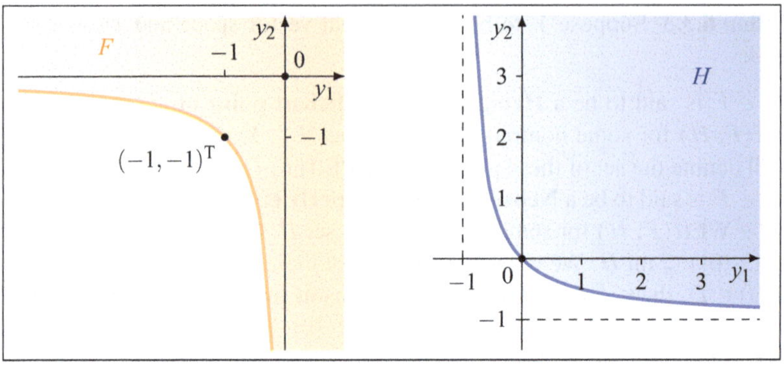


**Fig. 6.7** The sets $F$ and $H$ in Example 6.3.7

(c) $y^0 \in \mathrm{WEff}(F, H)$ *for some nontrivial closed convex cone* $H \subset Y$ *with* $D \setminus \{0\} \subseteq \operatorname{int} H$.

(2) $\text{He-PEff}(F, D) \subseteq \text{NI-PEff}(F, D)$.

(3) $\text{He-PEff}(F, D) = \text{NI-PEff}(F, D) = \mathrm{Eff}(F, D)$ *if* $D$ *is a nontrivial convex cone for which* $D \setminus \{0\}$ *is open.*

***Proof***

(1) (a) implies (b) because of $\mathrm{Eff}(F, H) \subseteq \mathrm{WEff}(F, H)$.
If $H \in Z_{\mathrm{He}}$, then $H^0 := \operatorname{int} H \cup \{0\} \in Z_{\mathrm{He}}$ and $\mathrm{WEff}(F, H) = \mathrm{Eff}(F, H^0)$. Hence, (b) implies (a).
The equivalence between (b) and (c) follows from $\operatorname{int}\operatorname{cl} H = \operatorname{int} H$ and $\mathrm{WEff}(F, H) = \mathrm{WEff}(F, \operatorname{cl} H)$ for convex sets $H$ with nonempty interior.

(2) results from (1).

(3) The definition of $\text{He-PEff}(F, D)$ implies $\text{He-PEff}(F, D) = \mathrm{Eff}(F, D)$ under the given assumptions. The assertion follows from (2) and $\text{NI-PEff}(F, D) \subseteq \mathrm{Eff}(F, D)$.

□

In general, $\text{He-PEff}(F, D) \neq \text{NI-PEff}(F, D)$. This is pointed out by the following example from [263, Remark 2].

**Example 6.3.7** In $Y = \mathbb{R}^2$, consider the set $F := \{(y_1, y_2)^T \in \mathbb{R}^2 \mid y_1 < 0,\ y_2 = \frac{1}{y_1}\} + \mathbb{R}^2_+$. Then $(-1, -1)^T \in \mathrm{Eff}(F, \mathbb{R}^2_+)$ and $(-1, -1)^T \in \mathrm{WEff}(F, H)$ for $H := -(Y \setminus \operatorname{int} F) - (1, 1)^T$. We obtain $\operatorname{int} H \in Z_{\mathrm{NI}}$ for $D = \mathbb{R}^2_+$, hence $y^0 \in \text{NI-PEff}(F, \mathbb{R}^2_+)$, but $\text{He-PEff}(F, \mathbb{R}^2_+) = \emptyset$.

The sets in this example are illustrated by Fig. 6.7.

Obviously, we have:

**Lemma 6.3.8** *Suppose* $Y$ *to be a topological vector space and* $D, F \subseteq Y$ *with* $0 \in D$, $D \setminus \{0\} \neq \emptyset$ *and* $D + D \subseteq D$. *Then* $\text{Be-PEff}(A, D) = \text{Be-PEff}(F, D)$ *for each set* $A \subseteq Y$ *with* $F \subseteq A \subseteq F + D$.

**Proposition 6.3.9** *Suppose $Y = \mathbb{R}^\ell$, $D \subseteq Y$ to be a nontrivial closed pointed convex cone, and $F \subseteq Y$. Then*

(a) He-PEff$(F, D) =$ Be-PEff$(F, D)$.
(b) *If* He-PEff$(F, D) \neq \emptyset$ *and $A$ is closed for some set $A \subseteq Y$ with $F \subseteq A \subseteq F + D$, then* He-PEff$(F, D)$ *is dense in* Eff$(F, D)$.

Proposition 6.3.9 was proved by Henig [235, Theorems 2.1, 5.1 and 5.2] for sets $A = F + B$ with $0 \in B \subseteq D$. Its extension results from Lemma 6.3.8.

Henig's and Benson's proper efficiency coincide also in more general spaces. Part (a) of the following proposition was proved by Guerraggio, Molho and Zaffaroni [200, Theorem 4.2], part (b) results from (a) and [270, Theorem 2.4].

**Proposition 6.3.10** *Suppose that $Y$ is a topological vector space, that $D \subseteq Y$ is a nontrivial closed pointed convex cone, and $F \subseteq Y$. Then*

(a) He-PEff$(F, D) \subseteq$ Be-PEff$(F, D)$.
(b) He-PEff$(F, D) =$ Be-PEff$(F, D)$ *if $Y$ is a normed space and $D$ has a weakly compact base.*

In Sect. 3.2.6 of [192], conditions are given under which He-PEff$(F, D)$ is dense in the efficient point set if $Y$ is a normed space and $D$ is a closed convex cone. Note that density of He-PEff$(F, D)$ in the efficient point set implies density of NI-PEff$(F, D)$ in the efficient point set by Proposition 6.3.6.

A special case of Henig's and Benson's proper efficiency is Geoffrion's proper efficiency [177], which we will discuss in Sect. 7.2 since it is of importance for vector optimization procedures. Engau [157] investigated a slight modification of Geoffrion's proper efficiency for the case that the objective function f consists of a countably infinite number of real-valued functions. Geoffrion's proper efficiency was generalized to the space $C(T)$ of continuous real-valued functions on a compact set $T$ by Winkler [497, 498]. The next two examples contain cones $H$ with $D \setminus \{0\} \subseteq \text{int } H$ for the ordering cone $D$ in $C(T)$ and for its dual cone.

**Example 6.3.11** Let $C[T]$ be the space of continuous real-valued functions on a compact subset $T$ of a separable Banach space. The natural ordering cone in $C[T]$ is given by

$$C[T]_+ := \{y \in C[T] \mid \forall t \in T : y(t) \geq 0\}.$$

It is a closed convex cone with nonempty interior.

For $s \in T$ and $\delta \in \mathbb{R}_>$, we consider the sets

$$D^{s,\delta} := \{y \in C[T] \mid \forall t \in T \setminus \{s\} : y(s) > 0,\ y(s) + \delta y(t) > 0\} \text{ and } D^\delta := \bigcup_{s \in T} D^{s,\delta}.$$

Winkler [498, Lemma 3.1] showed that the sets $D^{s,\delta}$ and $D^\delta$ are open and that the sets $D^{s,\delta} \cup \{0\}$ are nontrivial convex cones. Furthermore, the sets $D^\delta \cup \{0\}$ are cones, and we have $C[T]_+ \setminus \{0\} \subseteq D^\delta$.

**Example 6.3.12** We use the notation introduced in Example 6.3.11. The topological dual space of $C[T]$ is given by the space $M[T]$ of Radon measures on $T$. The dual cone to $C[T]_+$ is the cone of positive Radon measures,

$$M[T]_+ := \{\mu \in M[T] \mid \forall y \in C[T]_+ \colon \int_T y \, d\mu \geq 0\}.$$

$M[T]_+$ has an empty interior. The quasi-interior of $M[T]_+$ is given by

$$\operatorname{qint} M[T]_+ = \Big\{\mu \in M[T] \mid \forall y \in C[T]_+ \setminus \{0\} \colon \int_T y \, d\mu > 0\Big\}.$$

Obviously, $\operatorname{qint} M[T]_+$ is nonempty.

Consider the set

$$C_{\mu,m} := \{y \in C[T] \mid \forall t \in T \colon y(t) + \frac{1}{m} \int_T y \, d\mu \geq 0\}$$

with $\mu \in \operatorname{qint} M[T]_+$ and $m \in \mathbb{R}_>$. Winkler [498, Lemma 3.3] showed that $C_{\mu,m}$ is a convex cone with

$$\operatorname{int} C_{\mu,m} = \{y \in C[T] \mid \forall t \in T \colon y(t) + \frac{1}{m} \int_T y \, d\mu > 0\},$$

where $C[T]_+ \subseteq C_{\mu,m}$ and $C_{\mu,m_1} \setminus \{0\} \subseteq \operatorname{int} C_{\mu,m_2}$ for $m_1 > m_2 > 0$.

The two basic concepts of proper efficiency which we defined, i.e., the definition via a subclass of all strictly $D$-monotone functionals on one hand and via a family of sets $H \supseteq D \setminus \{0\}$ on the other one, could be used to define different notions of proper efficiency and to deduce results for several properly efficient point sets. They are closely related to each other. This will be underscored by the following theorems and corollaries.

**Theorem 6.3.13** *Assume that $Y$ is a linear space, $F, D \subseteq Y$, $D \setminus \{0\} \neq \emptyset$ and that $\varphi : Y \to \mathbb{R}$ is a strictly $D$-monotone functional that attains its minimum on $F$ in $y^0 \in F$.*

*Then $y^0 \in \operatorname{Eff}(F, H)$ for $H := \{y \in Y \mid \varphi(y^0 - y) < \varphi(y^0)\} \supseteq D \setminus \{0\}$.*

*The set $H$ has the following properties depending on properties of $\varphi$.*

(a) *If $Y$ is a topological vector space and $\varphi$ is continuous, then $H$ is open and $\operatorname{cl} H + (D \setminus \{0\}) \subseteq H$.*
*In this case, $\operatorname{Eff}(F, H) = \operatorname{WEff}(F, H)$.*
(b) *If $\varphi$ is convex, then $H$ is convex.*
(c) *If $\varphi$ is linear, then $H \cup \{0\}$ is a convex cone.*
(d) *Define $\bar{\varphi} : Y \to \mathbb{R}$ by $\bar{\varphi}(y) := \varphi(y + y^0)$ for all $y \in Y$.*
*If $\bar{\varphi}$ is sublinear and $\varphi(y^0) = 0$, then $H \cup \{0\}$ is a convex cone.*
(e) *If $0 \in \operatorname{cl}(D \setminus \{0\})$, then $0 \in \operatorname{cl} H \setminus H$.*

***Proof*** For each $y \in H$, we have $\varphi(y^0 - y) < \varphi(y^0)$ and thus $y^0 - y \notin F$. Hence, $F \cap (y^0 - H) = \emptyset$, *i.e.*, $y^0 \in \mathrm{Eff}(F, H)$. The strict $D$-monotonicity of $\varphi$ implies $D \setminus \{0\} \subseteq H$. Define $\bar{\varphi}$ according to (d). We have $H = \{y \in Y \mid \bar{\varphi}(-y) < \bar{\varphi}(0)\}$.

(a) If $\varphi$ is continuous, then $\bar{\varphi}$ is continuous. Thus, $H$ is open.
Take any $y \in \mathrm{cl}\, H$ and $y^1 \in y + (D \setminus \{0\})$. We get $-y \in -y^1 + (D \setminus \{0\})$ and $\bar{\varphi}(-y^1) < \bar{\varphi}(-y) \leq \bar{\varphi}(0)$ because of the strict $D$-monotonicity of $\bar{\varphi}$ and $y \in \mathrm{cl}\, H \subseteq \{\bar{y} \in Y \mid \bar{\varphi}(-\bar{y}) \leq \bar{\varphi}(0)\}$. Hence, $\mathrm{cl}\, H + (D \setminus \{0\}) \subseteq H$.
(b) If $\varphi$ is convex, then $\bar{\varphi}$ is convex, and hence $H$ is convex.
(c) If $\varphi$ is linear, then $H \cup \{0\} = \{y \in Y : \varphi(y) > 0\} \cup \{0\}$ is a convex cone.
(d) $H = \{y \in Y \mid \bar{\varphi}(-y) < 0\}$ yields the assertion.
(e) $0 \in \mathrm{cl}(D \setminus \{0\})$ and $D \setminus \{0\} \subseteq H$ imply $0 \in \mathrm{cl}\, H$. □

If $0 \notin \mathrm{cl}(D \setminus \{0\})$, then the set $H$ in the previous theorem can fail to fulfill $0 \in \mathrm{cl}\, H$.

**Example 6.3.14** In $Y = \mathbb{R}$, take $D := \mathbb{R}_+ \setminus (0, 1)$, $F := [1, 4]$, and $\varphi : Y \to \mathbb{R}$ given by

$$\varphi(y) := \begin{cases} y + \frac{1}{2} & \text{if } y < \frac{1}{2}, \\ 1 & \text{if } \frac{1}{2} \leq y \leq 1, \\ y & \text{if } y > 1. \end{cases}$$

$\varphi$ is continuous, strictly $D$-monotone and attains its minimum on $F$ in $y^0 := 1$.

Note that 0 does not belong to the closure of the set $H := \{y \in Y : \varphi(y^0 - y) < \varphi(y^0)\} = (\frac{1}{2}, \infty)$.

We obtain the following from Theorem 6.2.31 together with Theorems 4.4.4, 4.6.6 and 4.5.5 and Proposition 4.3.5.

**Theorem 6.3.15** *Assume that $Y$ is a topological vector space, $F, D \subseteq Y$, $H$ is a proper closed subset of $Y$ with $0 \in \mathrm{bd}\, H$ and with $H + \mathbb{R}_> k \subseteq \mathrm{int}\, H$ for some $k \in Y \setminus \{0\}$.*

*Suppose $y^0 \in \mathrm{WEff}(F, H)$.*

(a) *There exists a continuous functional $\varphi : Y \to \overline{\mathbb{R}}$ with*

$$\varphi(y^0) = \min_{y \in F} \varphi(y) = 0.$$

*$\varphi_{y^0 - H, k}$ is such a functional, where* $\mathrm{int}\, H = \{y \in Y \mid \varphi_{y^0 - H, k}(y^0 - y) < 0\}$.
(b) *There exists a continuous functional $\varphi : Y \to \overline{\mathbb{R}}$ with*

$$\varphi(y^0 - y^0) = \min_{y \in F} \varphi(y - y^0) = 0.$$

*$\varphi_{-H, k}$ is such a functional, where* $\mathrm{int}\, H = \{y \in Y \mid \varphi_{-H, k}(-y) < 0\}$.

*Moreover, these functionals have the following properties:*

(i) *If $Y = \mathrm{bd}\, H + \mathbb{R}k$, then $\varphi_{y^0 - H, k}$ and $\varphi_{-H, k}$ are finite-valued.*

(ii) *If* $H + D \subseteq H$, *then* $\varphi_{y^0-H,k}$ *and* $\varphi_{-H,k}$ *are D-monotone.*

(iii) *Assume* $H + (D \setminus \{0\}) \subseteq \operatorname{int} H$ *and* $\tilde{F} \subseteq Y$. *If* $\varphi_{y^0-H,k}$ *or* $\varphi_{-H,k}$ *is finite-valued on* $\tilde{F}$ *or on* $\tilde{F} - y^0$, *respectively, then it is strictly D-monotone on this set.*

(iv) $\varphi_{y^0-H,k}$ *and* $\varphi_{-H,k}$ *are convex if* $H$ *is convex.*

(v) $\varphi_{-H,k}$ *is sublinear if* $H$ *is a convex cone.*

**Lemma 6.3.16** *The assumptions for H and k in Theorem 6.3.15 are fulfilled in a topological vector space Y and the constructed functionals are finite-valued if one of the following conditions holds:*

(a) $H$ *is a proper closed convex subset of* $Y$ *with* $0 \in \operatorname{bd} H$, $\operatorname{int} H \neq \emptyset$, $k \in 0^+H$ *and* $Y = H + \mathbb{R}k$.

(b) $H \subset Y$ *is a nontrivial closed convex cone with* $k \in \operatorname{int} H$.

(c) $H$ *is a proper closed subset of* $Y$ *with* $0 \in \operatorname{bd} H$ *and* $k \in \operatorname{int} 0^+H$.

***Proof*** The statement related to (a) follows from Corollary 4.7.10 and Lemma 2.3.15. (b) and Lemma 2.3.14 imply (a). The statement for (c) results from Proposition 4.6.13. □

**Corollary 6.3.17** *Assume that* $Y$ *is a topological vector space,* $F, D \subseteq Y$, *and* $Y = D + \mathbb{R}k$ *for some* $k \in Y \setminus \{0\}$ *with* $\mathbb{R}_{>}k \subseteq D$.

*Then* $y^0 \in$ NI-PEff$(F, D)$ *if and only if* $y^0 \in F$ *is a point in which some continuous convex strictly D-monotone functional* $\varphi\colon Y \to \mathbb{R}$ *attains its minimum on* $F$.

***Proof***

(a) Assume that $H \subseteq Y$ is a closed convex set with $0 \in \operatorname{bd} H$ and $H + (D \setminus \{0\}) \subseteq \operatorname{int} H$.
Then $H + \mathbb{R}_{>}k \subseteq \operatorname{int} H$. We get $Y = H + \mathbb{R}k$ because of $D \setminus \{0\} \subseteq \operatorname{int} H$ and $Y = D + \mathbb{R}k$.
To obtain a contradiction, suppose $k \in -0^+H$. Then $0 = \frac{1}{2}k + \frac{1}{2}(-k) \in \operatorname{int} H$ since $k \in D \setminus \{0\} \subseteq \operatorname{int} H$, $-k = 0 - k \in H$ and $H$ is convex. This contradicts $0 \in \operatorname{bd} H$. Therefore, $k \in 0^+H \setminus (-0^+H)$.
Consequently, the function $\varphi_{-H,k}$ is proper by Proposition 4.7.7, hence finite-valued since $Y = H + \mathbb{R}k$. Then each function $\varphi_{y^0-H,k}$ with $y^0 \in Y$ is also finite-valued because of Proposition 4.3.5.

(b) Suppose $y^0 \in$ NI-PEff$(F, D)$. By Definition 6.3.3(b), $y^0 \in$ WEff$(F, H)$ for some set $H \subseteq Y$ which fulfills the assumptions of (a). By Theorem 6.3.15, $\varphi_{y^0-H,k}$ is continuous, convex, strictly $D$-monotone, and attains the minimal value 0 on $F$ at $y^0$. The function $\varphi_{y^0-H,k}$ is finite-valued because of (a). We get the forward direction of the equivalence with $\varphi = \varphi_{y^0-H,k}$.

The reverse direction of the equivalence follows from Theorem 6.3.13 and Lemma 6.3.5. □

**Corollary 6.3.18** *Assume that $Y$ is a topological vector space, $F, D \subseteq Y$, and $Y = D + \mathbb{R}k$ for some $k \in Y \setminus \{0\}$ with $\mathbb{R}_> k \subseteq D$.*

*Then $y^0 \in$ NI-PEff$(F, D)$ if and only if there exists a closed convex set $H \subseteq Y$ with $0 \in \operatorname{bd} H$ and $H + (D \setminus \{0\}) \subseteq \operatorname{int} H$ such that* (i) *or* (ii) *holds. Note that* (i) *and* (ii) *are equivalent to each other.*

(i) $\varphi_{y^0-H,k}(y^0) = \min_{y \in F} \varphi_{y^0-H,k}(y) = 0.$
(ii) $\varphi_{-H,k}(y^0 - y^0) = \min_{y \in F} \varphi_{-H,k}(y - y^0) = 0.$

*For each $y^0 \in Y$, the functional $\varphi_{y^0-H,k}$ with $H$ having the mentioned properties is continuous, convex, finite-valued, and strictly $D$-monotone.*

***Proof*** The forward direction of the assertion follows from part (b) in the proof of Corollary 6.3.17, where (ii) results from Proposition 4.3.5.

We will now prove the reverse direction. Theorem 6.2.31 implies $y^0 \in$ WEff$(F, H)$ for each minimizer $y^0$ of (i) or (ii). By Definition 6.3.3(b), we have $y^0 \in$ NI-PEff$(F, D)$.

Moreover, the additional statements result from part (a) in the proof of Corollary 6.3.17, Theorem 6.3.15 and Proposition 4.3.5. □

The assumptions of Corollaries 6.3.17 and 6.3.18 are fulfilled if $D \subset Y$ is a nontrivial cone with $k \in \operatorname{core} D$. Because of [506, Proposition 4], the assumptions are also satisfied if $0 \in D$, $D$ is convex, core $D \neq \emptyset$ and $k \in \operatorname{core} 0^+(\operatorname{cl} D)$.

Now we will prove an analogous statement for nonconvex functionals and sets.

**Corollary 6.3.19** *Assume that $Y$ is a topological vector space, $F \subseteq Y$, and $D \subset Y$ is a nontrivial cone with* int $D \neq \emptyset$.

*Then $y^0 \in F$ is a point in which some continuous strictly $D$-monotone functional $\varphi : Y \to \mathbb{R}$ attains its minimum on $F$ if and only if $y^0 \in$ Eff$(F, H)$ for some open set $H \subseteq Y$ with $0 \in \operatorname{bd} H$ and $\operatorname{cl} H + (D \setminus \{0\}) \subseteq H$.*

***Proof*** The first direction of the equivalence results from Theorem 6.3.13.

Assume now $y^0 \in$ Eff$(F, H)$ for some open set $H$ with $0 \in \operatorname{bd} H$ and $\operatorname{cl} H + (D \setminus \{0\}) \subseteq H$. Then we get $y^0 \in$ WEff$(F, \bar{H})$ for $\bar{H} := \operatorname{cl} H$ from Lemma 2.2.23. $\bar{H}$ is a proper closed subset of $Y$, $0 \in \operatorname{bd} \bar{H}$ and $\bar{H} + (D \setminus \{0\}) \subseteq \operatorname{int} \bar{H}$. Theorem 6.3.15 together with Lemma 6.3.16 implies the assertion. □

**Corollary 6.3.20** *Assume that $Y$ is a topological vector space, $F \subseteq Y$, and that $D$ is a proper subset of $Y$ with $D \setminus \{0\} \neq \emptyset$.*

(a) *$y^0 \in F$ is a point for which there exists a continuous sublinear strictly $D$-monotone functional $\varphi : Y \to \mathbb{R}$ with $\varphi(y^0 - y^0) = \min_{y \in F} \varphi(y - y^0) = 0$ if and only if $y^0 \in$ WEff$(F, H)$ for some nontrivial closed convex cone $H \subseteq Y$ with $D \setminus \{0\} \subseteq \operatorname{int} H$, i.e., if and only if $y^0 \in$ He-PEff$(F, D)$.*
(b) *$y^0 \in$ He-PEff$(F, D)$ if and only if there exists a nontrivial closed convex cone $H \subseteq Y$ with $D \setminus \{0\} \subseteq \operatorname{int} H$ such that $\varphi_{-H,k}(y^0 - y^0) = \min_{y \in F} \varphi_{-H,k}(y - y^0) = 0$, where $k \in \operatorname{int} H$ can be chosen arbitrarily. The function $\varphi_{-H,k}$ is finite-valued, continuous, sublinear and strictly $D$-monotone.*

***Proof***

(a) Suppose $y^0 \in F$ is a point for which there exists a continuous sublinear strictly $D$-monotone functional $\varphi : Y \to \mathbb{R}$ with $\varphi(y^0 - y^0) = \min_{y \in F} \varphi(y - y^0) = 0$. Apply Theorem 6.3.13 to $\tilde{\varphi} : Y \to \mathbb{R}$ given by $\tilde{\varphi}(y) := \varphi(y - y^0)$ for all $y \in Y$. Then $\varphi(y) = \tilde{\varphi}(y + y^0)$ for all $y \in Y$, and we get $y^0 \in \mathrm{Eff}(F, H)$ for some open set $H$ for which $\mathrm{cl}\, H + (D \setminus \{0\}) \subseteq H$ holds and $H \cup \{0\}$ is a convex cone. By Lemma 2.2.23, $\mathrm{int}\, \mathrm{cl}\, H = \mathrm{int}\, H$ and hence $y^0 \in \mathrm{WEff}(F, \bar{H})$ for $\bar{H} := \mathrm{cl}\, H$, where $\bar{H}$ is a nontrivial closed convex cone with $D \setminus \{0\} \subseteq \mathrm{int}\, \bar{H}$. The reverse direction follows from Theorem 6.3.15 together with Lemma 6.3.16.

(b) results from the proof of (a) by the construction of the functional in Theorem 6.3.15. □

## 6.4 Bibliographical Notes

In recent years, the interest in efficiency w.r.t. general reference sets has arisen from publications like [25, 164, 209]. Efficiency and weak efficiency for arbitrary domination sets of vector optimization problems were previously investigated in [463]. In this context, Henig's and Benson's proper efficiency were studied, together with proper efficiency defined by other authors. This research was continued in [464], where efficiency was considered in the framework of optima w.r.t. binary relations, motivated by decision making. The results were applied to domination structures in decision making in [465]. In [464], the weakly efficient point set of the vector optimization problem and the efficient point set w.r.t. $\tilde{D} \setminus (-\tilde{D})$ were handled as efficient point sets w.r.t. an appropriate reference set. Properties of general efficient point sets were published in [468]. There, proper efficiency was considered w.r.t. arbitrary domination sets $D$.

Scalarization results with linear functionals can be found for general domination sets in [209, 468]. In [186, 472], functionals $\varphi_{A,k}$ were applied to the scalarization of efficient, weakly efficient and properly efficient points, where the related point sets were characterized by efficient point sets. The studied notions of proper efficiency include Henig's proper efficiency and Nehse–Iwanow proper efficiency. The assumptions in [472] are weaker than those in [186] (see Sect. 4.15). In [472], Weidner pointed out that the scalarization by Pascoletti and Serafini [376] is equivalent to the problem

$$\min\{\varphi_{a-C,k}(y) \mid y \in F \cap \mathrm{dom}\, \varphi_{a-C,k}\},$$

where $F$ is the feasible set in the finite-dimensional linear image space of a vector optimization problem and $C$ is a nontrivial closed convex cone which is the domination set in this space. She extended this approach to arbitrary topological vector spaces and arbitrary sets $C$ and studied its properties. Contrary to [376], sets $C$ were admitted which not coincide with the domination set of the vector optimization problem. It is shown in [472] that many known scalarizations in multicriteria optimization turn out to be special cases of this general optimization problem and thus can be formulated using functionals $\varphi_{A,k}$. In this way, properties of $\varphi_{A,k}$ imply statements for different scalarizations, which will be presented in the next chapter.

Now, we will give more detailed references related to this chapter. Note that the statements in Sects. 6.2.1 and 6.2.2 extend results from [464, 472].

In [464], optimal elements of sets w.r.t. binary relations and sets were investigated under the general assumptions given in this section. In earlier papers, Wallace [459] and Ward [461] had studied optima w.r.t. quasi orders, Hurwicz [250] and Yu [501] had examined optimal elements w.r.t. ordering cones. Chew [96] had considered efficiency in linear spaces w.r.t. additive semigroups. Domination factors according to Definition 6.1.1 were introduced in [48], where efficient points w.r.t. convex sets $D$ with $0 \in D \setminus \operatorname{int} D$ were investigated in $\mathbb{R}^\ell$. The concept of domination structures goes back to Yu [501], but his sets $D_\succ(y)$ coincide with ours only in the case where these sets are cones. While Yu defined a domination structure as a family of cones $D_\succ(y)$, Engau [156] handled it as a mapping. Pre-domination structures were studied by Chen [91] and later in the book by himself et al. [93] for the case where the structure consists of convex cones or of convex sets that contain zero in their boundary.

The first statement of Proposition 6.1.3 as well as parts (a) and (d) and the statements related to cones in this proposition were also given in [192, p. 17f]. The discussion of the condition $D + D \subseteq D$ is taken from [464]. Proposition 6.1.10 can be found in [464, p. 51f] and [490, Proposition 3.2]. Proposition 6.1.14 is given in [490, p. 7].

Propositions 6.2.4(a)–(c) and 6.2.16 were proved in [490]. Under the additional assumptions that $\varphi$ is finite-valued on $F$ and $\operatorname{dom}\varphi = F$, part (a) from Proposition 6.2.16 was shown in [472], part (c) and $\supseteq$ as well as the second statement of part (b) are taken from [468]. Proposition 6.2.16(d) was shown in [468] for finite-valued functions $\varphi$.

Lemmata 6.2.5 and 6.2.6, Theorems 6.2.12 and 6.2.13 were proved in [464]. In [464, 472], also references to previous results for more special sets $D$ are given as well as illustrative examples. Lemma 6.2.5(d) was published in [468]. Examples 6.2.10 and 6.2.11 are taken from [467]. Lemma 6.2.20(a)–(b), Theorem 6.2.21(a)–(d), (f) and (g), Example 6.2.22 and Corollary 6.2.23 can be found in [490].

Theorem 6.2.24 was proved in [490, Theorem 4.5]. It generalizes [472, Theorem 4.3.1] and was also proved in [222, Corollary 4.2].

Corollary 6.2.25 for $D$ being pointed, Theorem 6.2.29 and Corollary 6.2.30 were proved in [490]. In [209], the vector optimization problem is considered under the assumptions that $Y$ is a separated locally convex space, and $D$ is a subset of $Y$ with $0 \notin D$ for which there exists a nontrivial convex cone $K$ with $D + K = D$. There, an analogous statement to (6.2.2) is proved under the additional assumptions that $D$ is closed and $k \in 0^+D \cap \operatorname{cone} D$. A statement for $\operatorname{WEff}(F, D)$ which is analogous to (6.2.3) is proved in [209] under the additional assumptions $\operatorname{cl} D + \mathbb{R}_> k \subseteq \operatorname{int} D$ and $k \in \operatorname{cone}(\operatorname{cl} D)$.

Corollary 6.2.27 and Proposition 6.2.28 are taken from [472, Folgerung 6.2.7 and Satz 6.2.11].

Corollary 6.2.33 was proved in [186, Corollary 3.1]. For $Y = \mathbb{R}^\ell$ and $D = \mathbb{R}^\ell_+$, $\varphi_{y^0-\mathbb{R}^\ell_+,k}$ was previously used by Bernau [49] and by Brosowski [78] to derive scalarization results for the weakly efficient point set, where Bernau assumed $k \in \operatorname{int} \mathbb{R}^\ell_+$ and Brosowski used $k = \mathbf{1}$.

For the special case where $D$ is a closed convex cone in $Y = \mathbb{R}^\ell$, results by Pascoletti and Serafini [376] are equivalent to the last assertion in Corollary 6.2.23(a), to Corollary 6.2.23(c) and to Corollary 6.2.33.

Proposition 6.2.34 and Corollary 6.2.35 can be found in [490]. For a nontrivial closed pointed convex cone $D$ in a Banach space $Y$, the statement

$$\operatorname{Eff}(F, D) = \bigcap_{k\in K} \bigcap_{\varepsilon\in\mathbb{R}_>} \operatorname{Eff}(F, D + \varepsilon k)$$

was proved for $K := \{k \in D \mid \|k\| = 1\}$ in [124, Proposition 2.1]. This statement is equivalent to the first equation in Corollary 6.2.35.

Theorem 6.2.37 was proved in [472, Satz 4.3.2]. The first equation in Theorem 6.2.37 was also formulated by Luc [338, p. 95] under the assumption $F \subseteq a - \operatorname{int} D$ and the additional condition that $D$ is closed. The scalarization statements in Theorems 6.2.36 and 6.2.38 as well as Theorem 6.2.39 for $D$ being pointed were proved in [490].

The second equation in Proposition 6.2.41 is taken from [472, Satz 4.3.4].

Without the statement for topological vector spaces, Proposition 6.2.42 was shown in [490, Proposition 4.11]. Under the assumptions that $D$ is a nontrivial convex cone with $k \in \operatorname{core} D$ and $\varepsilon \in \mathbb{R}_+$, a statement that is equivalent to (6.2.5) was proved in [213, Corollary 24]. In [164, Theorem 4.7], a statement that is equivalent to

$$\operatorname{WEff}(F, D + \varepsilon k) = \{y^0 \in F \mid \forall y \in F \setminus \{y^0\} \colon \varphi_{y^0-D,k}(y) \geq -\varepsilon\}$$

was proved under the assumptions that $Y$ is a topological vector space, $0 \in \operatorname{bd} D \setminus D$, $\operatorname{cl} D + \mathbb{R}_> k \subseteq \operatorname{int} D$ and $\varepsilon \in \mathbb{R}_+$. The "$\supseteq$"-part of (6.2.9) was given under the following special assumptions in [293, Theorem 4.10]: $D$ is a nontrivial pointed convex cone with $\operatorname{vcl}(D) = D$, $k \in \operatorname{core} D$ and $\varepsilon \in \mathbb{R}_>$. Under these assumptions and the additional assumptions that $\operatorname{cone}(F - y^0 + D + \varepsilon k)$ is convex and has a nonempty core, the "$\subseteq$"-part of (6.2.8) was given in [293, Theorem 4.12]. Under the assumptions that $D$ is a nontrivial convex cone with $\operatorname{vcl}(D) = D$, $k \in \operatorname{core} D$ and $\varepsilon \in \mathbb{R}_>$, Kiyani and Soleimani-Damaneh [293, Theorem 4.11] deduced (6.2.11).

Those statements of Sect. 6.2.3 which do not require a topological vector space were proved in [490]. Proposition 6.2.49 is taken from [464]. Propositions 6.2.26(a), 6.2.52, 6.2.53, 6.2.54 and 6.2.55 as well as Corollary 6.2.64 were given in [487, Sect. 8].

Section 6.2.5 is contained in [472], except the statement related to $\operatorname{Eff}(F, D + \mathbb{R}_> k)$ in Proposition 6.2.56. For the space $Y = C(Z)$ of the continuous real-valued functions being defined on a compact Hausdorff space $Z$, the natural ordering cone $D = \{y \in Y \mid \forall z \in Z \colon y(z) \geq 0\}$ of this space, and $k = \mathbf{1} \in Y$, Proposition

6.2.56(b), the second equivalence of Proposition 6.2.57(b) and the statement

$$\forall a \in Y: \bigcup_{a\in Y} \mathrm{Eff}(\mathrm{M}_{F,a,D,k}, D) \subseteq \mathrm{Eff}(F, D),$$

which is part of Proposition 6.2.56(d), were given in [79, Sect. 3]. For closed convex sets $F + D$ and $k \in \operatorname{rint} D$, the statements of Proposition 6.2.58(1), (3) and (4) and of Corollary 6.2.59 can be found in [231, Theorems 2.2 and 2.3] and [229, Theorem 2.10]. The algorithm in Proposition 6.2.63 was formulated by Helbig [231] under the assumptions that $F$ is a closed convex set and $k \in \text{r-int}\, D$, where $\varphi$ was minimized on $\mathrm{M}_{F,a,D,k}$. Most of the results contained in Sect. 6.2.5 were published in [474].

The general approach to proper efficiency at the beginning of Sect. 6.3 was introduced in [186, 472]. Proposition 6.3.6(3) is taken from [463]. Without (d), Theorem 6.3.13 is contained in [186, Theorem 3.3]. Under an assumption that is equivalent to the condition given in Lemma 6.3.16(a), the statement of Theorem 6.3.15 was given in [186, Theorem 3.4]. It was also proved in [186, Theorem 3.5] under an assumption which is sufficient for the condition in Lemma 6.3.16(c). Theorem 6.3.15 and Lemma 6.3.16 extend Theorem 5.1.2 from [472]. Corollaries 6.3.17 and 6.3.19 were given in [186, Corollaries 3.2 and 3.3]. For $Y = \mathbb{R}^\ell$ and $D = \mathbb{R}^\ell_+$, Iwanow und Nehse [263, Theorem 2] proved the statement of Corollary 6.3.17. Zălinescu [507] proved the statement of Corollary 6.3.20 under the assumption that $D$ is a convex cone and $Y$ a separated topological vector space.

# Chapter 7
# Multiobjective Optimization

This chapter focuses on scalarization methods in multiobjective optimization and properties of Geoffrion's properly efficient point set.

In Chap. 6, vector optimization problems in general spaces $Y$ with domination sets $D$ were investigated. Considering $Y = \mathbb{R}^\ell$ and $D = \mathbb{R}^\ell_+$, these vector optimization problems are the multiobjective optimization problems, and we can deduce important statements for the finite-dimensional case. In this chapter, these statements and further results will be applied to scalarization procedures in multiobjective optimization.

The multiobjective optimization problem together with the related optimality notions will be introduced in Sect. 7.1. Section 7.2 deals with Geoffrion's proper efficiency. The properly efficient point set will be described by efficient and weakly efficient point sets w.r.t. cones, especially w.r.t. different types of polyhedral cones. Moreover, existence results and the density in the efficient point set will be proved under rather mild assumptions. The properly efficient point set will be characterized by minimizers of strictly monotone functionals and by minimizers of translation invariant functions. Theorems 7.2.27 and 7.2.28 contain statements about the coincidence of Geoffrion's proper efficiency with Nehse–Iwanow proper efficiency.

In Sect. 7.3, the results of Sect. 7.2—together with those of Chaps. 5 and 6—will be applied to the investigation of several scalarization methods. Beside relationships between their solution sets and optimal point sets of the multiobjective optimization problem, the results for the scalar problems include statements about the existence and uniqueness of their solutions as well as about the parameter control. We list some of these scalarizations, for which modifications and extensions will also be studied:

- the weighted Chebyshev norm minimization [39],
- the minimization of the extended weighted Chebyshev norm by Choo and Atkins [97],
- Wierzbicki's reference point projection [495],
- the $\varepsilon$-constraint method [99, 216],
- the Hurwicz Rule for decision making under uncertainty [249],



C. Tammer and P. Weidner, *Scalarization and Separation by Translation Invariant Functions*, Vector Optimization, https://doi.org/10.1007/978-3-030-44723-6_7

- a method with multiple vectors of weights [472].

The results are of interest for many other scalarization methods since a general framework for their systematic investigation is presented. This framework is based on translation invariant functions.

## 7.1 The Multiobjective Optimization Problem and Optimality Notions

In this chapter, we investigate the **multiobjective optimization problem**

$$\begin{aligned} f_1(s) &\to \min \\ f_2(s) &\to \min \\ &\vdots \\ f_\ell(s) &\to \min \\ s &\in S, \end{aligned} \tag{MOP}$$

where $S_0$ is a nonempty set, $S \subseteq S_0$ and $f_i : S_0 \to \mathbb{R}$ for all $i \in \{1, \dots, \ell\}$, $\ell \geq 2$.

The feasible set in the criterion space $Y := \mathbb{R}^\ell$ is $F := f(S)$ with $f := (f_1, \dots, f_\ell)^T$.

The multiobjective optimization problem is a special case of the vector optimization problem (VOP) defined in Sect. 6.2.1, namely the case with $Y = \mathbb{R}^\ell$ and $D = \mathbb{R}^\ell_+$. We are interested in the sets

$$\mathrm{Eff}(F) := \mathrm{Eff}(F, \mathbb{R}^\ell_+) \quad \text{and} \quad \mathrm{WEff}(F) := \mathrm{WEff}(F, \mathbb{R}^\ell_+)$$

of **Pareto minimal** and **weakly Pareto minimal**, respectively, **points** of $F$ as well as in the sets

$$\mathrm{Eff}_f(S) := \mathrm{Eff}_f(S, \mathbb{R}^\ell_+) \quad \text{and} \quad \mathrm{WEff}_f(S) := \mathrm{WEff}_f(S, \mathbb{R}^\ell_+)$$

of **Pareto minimal** and **weakly Pareto minimal**, respectively, **solutions** to (MOP).

The former notions are illustrated by Fig. 7.1.

The set $\mathrm{Eff}_f(S)$ consists of all $s^0 \in S$ for which the improvement of any objective function value is impossible without some trade-off. In detail:

$s^0 \in \mathrm{Eff}_f(S) \iff (s^0 \in S$, and for all $s \in S$: If there is $i \in \{1, \dots, \ell\}$ with $f_i(s) < f_i(s^0)$, then there exists $j \in \{1, \dots, \ell\}$ with $f_j(s) > f_j(s^0))$.

The set $\mathrm{WEff}_f(S)$ consists of all $s^0 \in S$ for which the simultaneous improvement of all objective function values is impossible. In detail:

$s^0 \in \mathrm{WEff}_f(S) \iff (s^0 \in S$, and there does not exist any $s \in S$ such that $f_i(s) < f_i(s^0)$ for all $i \in \{1, \dots, \ell\})$.

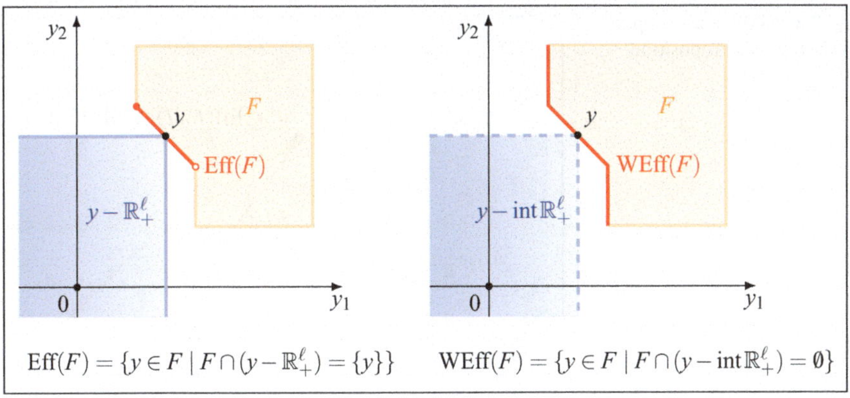


**Fig. 7.1** Pareto minimal and weakly Pareto minimal points of the closed set $F \subset \mathbb{R}^\ell$ with $\ell = 2$

Furthermore, Geoffrion's proper efficiency [177] turns out to be of importance for procedures in multicriteria optimization (see Sect. 7.3). This proper efficiency will be illustrated by Fig. 7.2.

**Definition 7.1.1** $s^0 \in S$ is called a **Geoffrion properly efficient solution** to (MOP) if $\exists K \in \mathbb{R}_> \ \forall s \in S$ with $f_i(s) < f_i(s^0)$ for some $i \in \{1, \ldots, \ell\}$ $\exists j \in \{1, \ldots, \ell\} \setminus \{i\}$:

$$f_i(s^0) - f_i(s) \leq K(f_j(s) - f_j(s^0)).$$

$\mathrm{GEff}_f(S)$ will denote the set of these solutions. $\mathrm{GEff}(F) := f(\mathrm{GEff}_f(S))$ is said to be the set of **Geoffrion properly efficient points** of $F$.

Hence, $\mathrm{GEff}_f(S)$ consists of all $s^0 \in S$ for which the improvement of each objective function value is impossible without some trade-off and such a trade-off is always bounded.

Usually, $\mathrm{Eff}_f(S)$ is the solution set to (MOP) in which the decision maker is interested, and statements about $\mathrm{GEff}_f(S)$ as well as about $\mathrm{WEff}_f(S)$ are useful tools in algorithms. However—depending on the purpose of the objective functions—decision makers can focus on another solution concept. Imagine that each objective function was introduced by another decision maker. If the decision makers do not cooperate, then they can become stuck in a weakly efficient solution which is not efficient. In the case where the decision makers cooperate, they can be interested in properly efficient solutions in order to use the bounded trade-offs in negotiations. For the interesting relationship between Geoffrion's proper efficiency and trade-offs, we recommend the book by Kaliszewski [285].

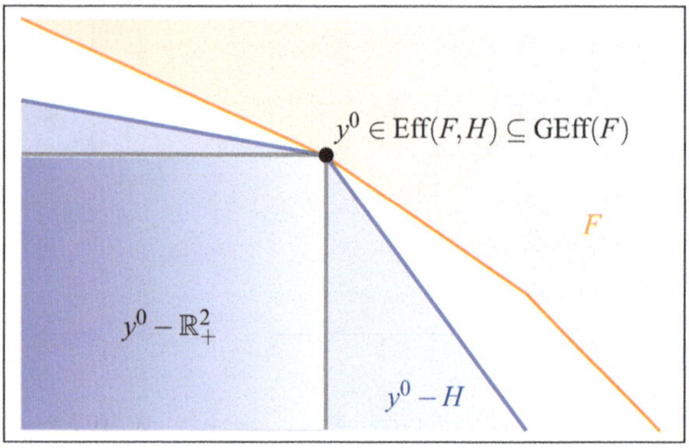


**Fig. 7.2** Geoffrion's proper efficiency (see Proposition 7.2.4)

## 7.2 Geoffrion's Proper Efficiency

This section is devoted to Geoffrion's proper efficiency. Those results which refer to the characterization of Geoffrion's proper efficiency by polyhedral cones and by minimizers of functionals (including Theorem 7.2.22) as well as the existence results are taken from [472].

Benson [42, Theorem 3.2] proved that his proper efficiency notion generalizes Geoffrion's proper efficiency. This implies together with Proposition 6.3.9(a):

**Proposition 7.2.1** $\text{Be-PEff}(F, \mathbb{R}^\ell_+) = \text{He-PEff}(F, \mathbb{R}^\ell_+) = \mathrm{GEff}(F)$.

Obviously, $\mathrm{GEff}(F) \subseteq \mathrm{Eff}(F)$. In Example 6.3.7, the set $\mathrm{GEff}(F)$ is empty, although Pareto minimal points of $F$ exist.

Lemma 6.3.8 yields:

**Proposition 7.2.2** $\mathrm{GEff}(A) = \mathrm{GEff}(F)$ *for each set* $A \subseteq \mathbb{R}^\ell$ *with* $F \subseteq A \subseteq F + \mathbb{R}^\ell_+$.

We get the following from Proposition 6.3.9(b).

**Theorem 7.2.3** *If* $A$ *is closed for some set* $A \subseteq \mathbb{R}^\ell$ *with* $F \subseteq A \subseteq F + \mathbb{R}^\ell_+$ *and* $\mathrm{GEff}(F) \neq \emptyset$*, then* $\mathrm{GEff}(F)$ *is dense in* $\mathrm{Eff}(F)$*, i.e.,* $\mathrm{GEff}(F) \subseteq \mathrm{Eff}(F) \subseteq \mathrm{cl}\,\mathrm{GEff}(F)$.

We will give two constructive proofs of this statement—one in Proposition 7.3.11(h), the other in Proposition 7.3.36. Numerical algorithms cannot distinguish between $\mathrm{GEff}(F)$ and $\mathrm{Eff}(F)$ if the assumptions of Theorem 7.2.3 are fulfilled.

The characterization of Henig's proper efficiency in Definition 6.3.3 and in Proposition 6.3.6(1) results in the following statements.

**Proposition 7.2.4** *The following conditions are equivalent to each other:*

(a) $y \in \mathrm{GEff}(F)$.

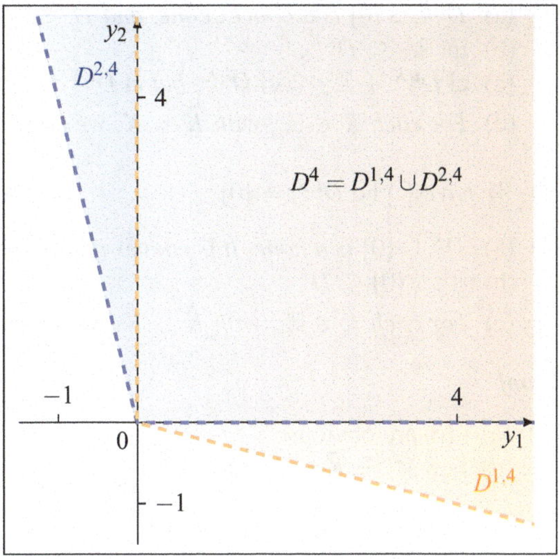


**Fig. 7.3** Set $D^K$ with $K = 4$ in $\mathbb{R}^2$

(b) $y \in \mathrm{Eff}(F, H)$ *for some nontrivial convex cone* $H \subset \mathbb{R}^\ell$ *with* $\mathbb{R}^\ell_+ \setminus \{0\} \subseteq \mathrm{int}\, H$.

(c) $y \in \mathrm{WEff}(F, H)$ *for some nontrivial convex cone* $H \subset \mathbb{R}^\ell$ *with* $\mathbb{R}^\ell_+ \setminus \{0\} \subseteq \mathrm{int}\, H$.

(d) $y \in \mathrm{WEff}(F, H)$ *for some nontrivial closed convex cone* $H \subset \mathbb{R}^\ell$ *with* $\mathbb{R}^\ell_+ \setminus \{0\} \subseteq \mathrm{int}\, H$.

In order to describe Geoffrion properly efficient points as efficient points w.r.t. special cones, we introduce, for each $K \in \mathbb{R}_>$, the sets

$$D^{i,K} := \{y \in \mathbb{R}^\ell \mid y_i > 0,\ y_i + K y_j > 0 \text{ for all } j \in \{1, \dots, \ell\} \setminus \{i\}\}$$

with $i \in \{1, \dots, \ell\}$, and

$$D^K := \bigcup_{i=1}^{\ell} D^{i,K}. \tag{7.2.1}$$

Figure 7.3 illustrates these sets.

First, we investigate properties of the introduced sets. By Proposition 2.2.30, each set $D^{i,K}$ with $i \in \{1, \dots, \ell\}$ and $K \in \mathbb{R}_>$ is the interior of the polyhedral cone

$$\mathrm{cl}\, D^{i,K} = \{y \in \mathbb{R}^\ell \mid y_i \geq 0,\ y_i + K y_j \geq 0 \text{ for all } j \in \{1, \dots, \ell\} \setminus \{i\}\}.$$

Consequently, each of these sets $D^{i,K}$ and each set $D^K$ with $K \in \mathbb{R}_>$ is open.

**Lemma 7.2.5** *Assume* $K \in \mathbb{R}_>$.

(1) *Take any* $i \in \{1, \dots, \ell\}$.

(a) $D^{i,K} \cup \{0\}$ *is a convex cone, and* $D^{i,K} \cap -D^{i,K} = \emptyset$,
(b) $\operatorname{int} \mathbb{R}^\ell_+ \subseteq D^{i,K}$,
(c) $\operatorname{cl} D^{i,K} + \mathbb{R}^\ell_+ \subseteq \operatorname{cl} D^{i,K}$, *but* $\operatorname{cl} D^{i,K} + (\mathbb{R}^\ell_+ \setminus \{0\}) \not\subset D^{i,K}$.
(d) *For each* $\bar{K} \in \mathbb{R}_>$ *with* $\bar{K} > K$, *we have* $D^{i,\bar{K}} \subseteq D^{i,K}$.

(2) *We have* $\bigcap_{i=1}^{\ell} \operatorname{bd} D^{i,K} = \{0\}$.

(3) (a) $D^K \cup \{0\}$ *is a cone. It is convex if and only if* $\ell = 2$ *and* $K \geq 1$.
(b) $\mathbb{R}^\ell_+ \setminus \{0\} \subseteq D^K$.
(c) *For each* $\bar{K} \in \mathbb{R}_>$ *with* $\bar{K} > K$, *we have* $D^{\bar{K}} \subseteq D^K$.

***Proof***

(1) (a) –(c) are obvious.
(d) For $M := \bar{K} > K$, we get $-\frac{1}{M} > -\frac{1}{K}$. Take any $y \in D^{i,M}$. Then $y_i > 0$ and $y_j > -\frac{1}{M} y_i$ for all $j \in \{1, \ldots, \ell\} \setminus \{i\}$. Hence, $y_j > -\frac{1}{K} y_i$ for all $j \in \{1, \ldots, \ell\} \setminus \{i\}$, which yields $y \in D^{i,K}$.

(2) Contrary to the claim, suppose that there exists $y \in \bigcap_{i=1}^{\ell} \operatorname{bd} D^{i,K}$ with $y \neq 0$. Then there is $n \in \{1, \ldots, \ell\}$ with $y_n \neq 0$. From $y \in \operatorname{bd} D^{n,K}$, we obtain $y_n > 0$ and $y_n + K y_m = 0$ for some $m \in \{1, \ldots, \ell\}$. We get $y_m < 0$, a contradiction to $y \in \operatorname{bd} D^{m,K}$.

(3) (a) The cone property is obvious.
For $Y = \mathbb{R}^2$, it can easily be checked that $D^K \cup \{0\}$ is convex if and only if $K \geq 1$ (see Fig. 7.3).
Assume now $\ell > 2$, $K \in \mathbb{R}_>$. We define $y, z \in D^K$ by

$$y_i := \begin{cases} 3K & \text{if } i = 1, \\ -2 & \text{if } i \in \{2, \ldots, \ell\}, \end{cases} \quad \text{and} \quad z_i := \begin{cases} -4 & \text{if } i = 1, \\ 7K & \text{if } i = 2, \\ -6 & \text{if } i \in \{3, \ldots, \ell\}. \end{cases}$$

Then $y + z \notin D^K \cup \{0\}$since
$$(y_1 + z_1) + K(y_3 + z_3) = -5K - 4 < 0,$$
$$(y_2 + z_2) + K(y_3 + z_3) = -K - 2 < 0 \text{ and}$$
$$y_j + z_j < 0 \text{ for all } j \in \{3, \ldots, \ell\}.$$

Thus, $D^K \cup \{0\}$ is not convex.
(b) Take any $z \in \mathbb{R}^\ell_+ \setminus \{0\}$. Then there exists $n \in \{1, \ldots, \ell\}$ such that $z_n > 0$ and $z_j \geq 0$ for all $j \in \{1, \ldots, \ell\}$. Hence, $z \in D^{n,K} \subseteq D^K$.
(c) follows from (1)(d). □

**Proposition 7.2.6** $\operatorname{GEff}(F) = \bigcup_{K>0} \operatorname{Eff}(F, D^K) = \bigcup_{K>0} \bigcap_{i=1}^{\ell} \operatorname{Eff}(F, D^{i,K})$,

*where for all* $K, \bar{K} \in \mathbb{R}_>$ *with* $\bar{K} > K$:

$$\operatorname{Eff}(F, D^K) \subseteq \operatorname{Eff}(F, D^{\bar{K}}) \quad \textit{and}$$
$$\operatorname{Eff}(F, D^{i,K}) \subseteq \operatorname{Eff}(F, D^{i,\bar{K}}) \ \textit{for each } i \in \{1, \ldots, \ell\}.$$

***Proof***

$$
\begin{aligned}
y^0 \in \mathrm{GEff}(F) &\iff y^0 \in F \text{ and} \\
&\quad \exists K > 0\ \forall y \in F \text{ with } y_i^0 > y_i \text{ for some } i \in \{1, \dots, \ell\}: \\
&\quad \exists j \in \{1, \dots, \ell\} \setminus \{i\}: \ y_i^0 - y_i \le K(y_j - y_j^0), \\
&\iff y^0 \in F \text{ and } \exists K > 0\ \forall y \in F: \ y^0 - y \notin D^K, \\
&\iff \exists K > 0: \ y^0 \in \mathrm{Eff}(F, D^K), \\
&\iff y^0 \in \bigcup_{K>0} \mathrm{Eff}(F, D^K).
\end{aligned}
$$

$$\mathrm{Eff}(F, D^K) = \mathrm{Eff}(F, \bigcup_{i=1}^{\ell} D^{i,K}) = \bigcap_{i=1}^{\ell} \mathrm{Eff}(F, D^{i,K}) \text{ for all } K > 0.$$

The results related to $\bar{K}$ follow from Lemma 7.2.5. □

Moreover, we get:

**Proposition 7.2.7** *For each convex cone $H \subseteq \mathbb{R}^\ell$ with $\mathbb{R}^\ell_+ \setminus \{0\} \subseteq \operatorname{int} H$, there exists $K \in \mathbb{R}_>$ with $D^K \subseteq \operatorname{int} H$. Then* $\mathrm{WEff}(F, H) \subseteq \mathrm{Eff}(F, D^K)$.

***Proof*** Assume that $H$ is a convex cone with $\mathbb{R}^\ell_+ \setminus \{0\} \subseteq \operatorname{int} H$. This inclusion implies the following. For each $i \in \{1, \dots, \ell\}$, there exists $t_i > 0$ such that

$$y^i \text{ with } y^i_j := \begin{cases} 1 & \text{if } j = i, \\ -t_i & \text{if } j \neq i, \end{cases}$$

is an element of $H$.

Take any $K > \max_{i=1,\dots,\ell} \frac{1}{t_i}$ and $y^0 \in D^K$. Then there exists $i \in \{1, \dots, \ell\}$ such that $y_i^0 > 0$ and $y_i^0 + K y_j^0 > 0$ for all $j \in \{1, \dots, \ell\} \setminus \{i\}$. Defining $\bar{y} := \frac{1}{y_i^0} y^0$, we get $\bar{y}_j = \frac{1}{y_i^0} y_j^0 > -\frac{1}{K} > -t_i$ for all $j \in \{1, \dots, \ell\} \setminus \{i\}$. Thus, $\bar{y} \in y^i + (\mathbb{R}^\ell_+ \setminus \{0\}) \subseteq H + \operatorname{int} H \subseteq \operatorname{int} H$. We conclude that $y^0 \in \operatorname{int} H$ since $H$ is a cone.

Hence, $D^K \subseteq \operatorname{int} H$. We obtain $\mathrm{WEff}(F, H) \subseteq \mathrm{Eff}(F, D^K)$. □

Geoffrion's proper efficiency is also related to efficiency w.r.t. the convex cones

$$C^p := \{y \in \mathbb{R}^\ell \mid \forall i \in \{1, \dots, \ell\}: \ p y_i + \sum_{\substack{j=1 \\ j \neq i}}^{\ell} y_j \ge 0\} \text{ with } p \in \mathbb{R}_> \tag{7.2.2}$$

(see Fig. 7.4).

For each $p \in \mathbb{R}_>$, Proposition 2.2.30 implies

$$\operatorname{int} C^p = \{y \in \mathbb{R}^\ell \mid \forall i \in \{1, \dots, \ell\}: \ p y_i + \sum_{\substack{j=1 \\ j \neq i}}^{\ell} y_j > 0\}.$$

**Lemma 7.2.8**

(a) $1 \leq p < \bar{p} \implies C^{\bar{p}} \setminus \{0\} \subseteq \operatorname{int} C^p$.
(b) $0 < p < 1 \implies \exists p_1 > 1, p_2 > 1: \ C^{p_2} \subseteq C^p \subseteq C^{p_1}$.
(c) $\forall \bar{p} > 0: \ \mathbb{R}^\ell_+ = \bigcap_{p \geq \bar{p}} C^p, \ \mathbb{R}^\ell_+ \setminus \{0\} = \bigcap_{p \geq \bar{p}} \operatorname{int} C^p$.
(d) $\forall p \in \mathbb{R}_> \ \exists K \in \mathbb{R}_> \ \forall \tilde{K} \geq K: \ D^{\tilde{K}} \subseteq C^p$.
(e) $\forall K \in \mathbb{R}_>: \ C^p \setminus \{0\} \subseteq D^K$ *for* $p = \ell K$.
(f) $\forall p \geq 1: \ \operatorname{int} C^p \subseteq \{y \in \mathbb{R}^\ell \mid \sum_{i=1}^\ell y_i > 0\}$.

***Proof***

(a) Take any $p, \bar{p} \in \mathbb{R}_>$ with $1 \leq p < \bar{p}$ and any $y \in C^{\bar{p}} \setminus \{0\}$. Then $\bar{p} y_i + \sum_{\substack{j=1\\j\neq i}}^{\ell} y_j \geq 0$ for all $i \in \{1, \dots, \ell\}$. To obtain a contradiction, suppose $\sum_{j=1}^{\ell} y_j \leq 0$. We get $0 \leq \bar{p} y_i + \sum_{\substack{j=1\\j\neq i}}^{\ell} y_j = (\bar{p} - 1) y_i + \sum_{j=1}^{\ell} y_j \leq (\bar{p} - 1) y_i$ for all $i \in \{1, \dots, \ell\}$. Hence, $y_i \geq 0$ for all $i \in \{1, \dots, \ell\}$, which contradicts $\sum_{j=1}^{\ell} y_j \leq 0$ because of $y \neq 0$. We conclude that $\sum_{j=1}^{\ell} y_j > 0$. Take any $\{1, \dots, \ell\}$. If $y_i \geq 0$, then $p y_i + \sum_{\substack{j=1\\j\neq i}}^{\ell} y_j \geq \sum_{j=1}^{\ell} y_j > 0$. If $y_i < 0$, then $p y_i + \sum_{\substack{j=1\\j\neq i}}^{\ell} y_j > \bar{p} y_i + \sum_{\substack{j=1\\j\neq i}}^{\ell} y_j \geq 0$. Consequently, $y \in \operatorname{int} C^p$.

(b) Take any $p_0 \in \mathbb{R}_>$ and $y \in C^{p_0}$. Then $p_0 y_i + \sum_{\substack{j=1\\j\neq i}}^{\ell} y_j \geq 0$ for all $i \in \{1, \dots, \ell\}$. Adding these inequalities for $i \neq m \in \{1, \dots, \ell\}$ yields:

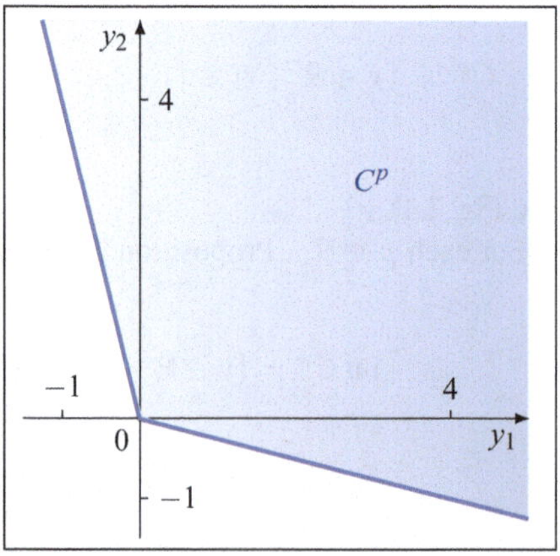


**Fig. 7.4** Cone $C^p$ with $p = 4$ in $\mathbb{R}^2$

$$p_0 \sum_{\substack{j=1\\ j\neq m}}^{\ell} y_j + (\ell-1)y_m + (\ell-2) \sum_{\substack{j=1\\ j\neq m}}^{\ell} y_j \geq 0 \text{ for all } m \in \{1,\ldots,\ell\}.$$

With $p_{00} := \frac{\ell-1}{p_0+(\ell-2)}$, we obtain $p_{00} y_m + \sum_{\substack{j=1\\ j\neq m}}^{\ell} y_j \geq 0$ for all $m \in \{1,\ldots,\ell\}$.

Hence, $y \in C^{p_{00}}$. Consequently, $C^{p_0} \subseteq C^{p_{00}}$.
Take now any $p \in \mathbb{R}_>$ with $0 < p < 1$. Consider first $p_0 := p < 1$. Then $p_2 := p_{00} > 1$.
Consider now $p_1 := p_0 := \frac{\ell-1-(\ell-2)p}{p} > 1$. We get $p_{00} = p$.

(c) Take any $\bar{p} \in \mathbb{R}_>$ and any $y \notin \mathbb{R}^\ell_+$. Then there exists $i \in \{1,\ldots,\ell\}$ with $y_i < 0$. For $p > \max(\bar{p}, -\frac{1}{y_i}\sum_{\substack{j=1\\ j\neq i}}^{\ell} y_j)$, we have $py_i + \sum_{\substack{j=1\\ j\neq i}}^{\ell} y_j < 0$ and hence $y \notin C^p$. Thus, $\bigcap_{p\geq\bar{p}} C^p \subseteq \mathbb{R}^\ell_+$. Since $\mathbb{R}^\ell_+ \subseteq C^p$ for each $p \in \mathbb{R}_>$, we obtain $\bigcap_{p\geq\bar{p}} C^p = \mathbb{R}^\ell_+$. Since $\operatorname{int} C^p \subseteq C^p \setminus \{0\}$ for each $p \in \mathbb{R}_>$, we have $\bigcap_{p\geq\bar{p}} \operatorname{int} C^p \subseteq \mathbb{R}^\ell_+ \setminus \{0\}$. Because $\mathbb{R}^\ell_+ \setminus \{0\} \subseteq \operatorname{int} C^p$ holds for each $p \in \mathbb{R}_>$, we get the assertion.

(d) Take any $p \in \mathbb{R}_>$. Choose $K > \max(\frac{\ell-1}{p},\ p+\ell-2)$. Then

$$p + (\ell-1)(-\frac{1}{K}) > 0 \text{ and } p(-\frac{1}{K}) + 1 + (\ell-2)(-\frac{1}{K}) > 0. \qquad (7.2.3)$$

Take any $y \in D^K$. Then there exists $i \in \{1,\ldots,\ell\}$ such that $y_i > 0$ and $y_j > -\frac{1}{K}y_i$ for all $j \in \{1,\ldots,\ell\} \setminus \{i\}$. This implies by (7.2.3) that $py_i + \sum_{\substack{j=1\\ j\neq i}}^{\ell} y_j > (p + (\ell-1)(-\frac{1}{K}))y_i > 0$, and for each $n \in \{1,\ldots,\ell\} \setminus \{i\}$ :

$$py_n + \sum_{\substack{j=1\\ j\neq n}}^{\ell} y_j = py_n + y_i + \sum_{\substack{j=1\\ j\notin\{n,i\}}}^{\ell} y_j > (p(-\tfrac{1}{K}) + 1 + (\ell-2)(-\tfrac{1}{K}))y_i > 0.$$

We get $py_n + \sum_{\substack{j=1\\ j\neq n}}^{\ell} y_j > 0$ for all $n \in \{1,\ldots,\ell\}$. Therefore, $y \in C^p$.

Consequently, $D^K \subseteq C^p$. By Lemma 7.2.5, $D^{\tilde{K}} \subseteq D^K$ for all $\tilde{K} \geq K$. Thus, $D^{\tilde{K}} \subseteq C^p$ for all $\tilde{K} \geq K$.

(e) Take any $K \in \mathbb{R}_>$, $p := \ell K$ and any $y \in C^p \setminus \{0\}$. Then $py_i + \sum_{\substack{j=1\\ j\neq i}}^{\ell} y_j \geq 0$ for all $i \in \{1,\ldots,\ell\}$.

We get $y_n := \max_{i\in\{1,\ldots,\ell\}} y_i > 0$. Take any $i \in \{1,\ldots,\ell\}$. Then $\sum_{\substack{j=1\\ j\neq i}}^{\ell} y_j \leq (\ell-1)y_n < \ell y_n$ implies $y_i + \frac{\ell}{p} y_n > y_i + \frac{1}{p}\sum_{\substack{j=1\\ j\neq i}}^{\ell} y_j \geq 0$. Hence, $y_n + Ky_i = y_n + \frac{p}{\ell} y_i > 0$. We conclude that $y \in D^K$. Therefore, $C^p \setminus \{0\} \subseteq D^K$.

(f) Take any $p \geq 1$. For each $i \in \{1, \ldots, \ell\}$, $py_i + \sum_{\substack{j=1 \\ j \neq i}}^{\ell} y_j = (p-1)y_i + \sum_{j=1}^{\ell} y_j$.

If we had $y \in \operatorname{int} C^p$ and $\sum_{i=1}^{\ell} y_i \leq 0$, this would imply $y_i > 0$ for each $i \in \{1, \ldots, \ell\}$, which contradicts $\sum_{i=1}^{\ell} y_i \leq 0$. □

Proposition 7.2.7 implies together with Lemma 7.2.8(e) and (c):

**Corollary 7.2.9** *For each convex cone $H \subseteq \mathbb{R}^\ell$ with $\mathbb{R}^\ell_+ \setminus \{0\} \subseteq \operatorname{int} H$, there exists a polyhedral cone $C = C^p$, $p \in \mathbb{R}_>$, with $C \setminus \{0\} \subseteq \operatorname{int} H$. Then $\mathbb{R}^\ell_+ \setminus \{0\} \subseteq \operatorname{int} C$ and $\operatorname{WEff}(F, H) \subseteq \operatorname{Eff}(F, C)$.*

**Proposition 7.2.10** *We have*

$$\operatorname{GEff}(F) = \bigcup_{p>0} \operatorname{Eff}(F, C^p) = \bigcup_{p \geq \bar{p}} \operatorname{Eff}(F, C^p) \text{ for all } \bar{p} \geq 1, \quad \text{where}$$

$$\operatorname{Eff}(F, C^{\bar{p}}) \subseteq \operatorname{Eff}(F, C^p) \text{ for all } p, \bar{p} \in \mathbb{R}_> \text{ with } 1 \leq \bar{p} \leq p, \quad \text{and}$$

$$\forall p \in (0,1) \colon \exists p_1 > 1,\ p_2 > 1 \colon\ \operatorname{Eff}(F, C^{p_1}) \subseteq \operatorname{Eff}(F, C^p) \subseteq \operatorname{Eff}(F, C^{p_2}).$$

***Proof***

(a) The inclusions between the efficient point sets in the second part of the proposition result from Lemma 7.2.8(a) and (b).
(b) Take any $p > 0$. By Lemma 7.2.8(d), there exists $K > 0$ with $\operatorname{Eff}(F, C^p) \subseteq \operatorname{Eff}(F, D^K)$. Proposition 7.2.6 implies $\operatorname{Eff}(F, C^p) \subseteq \operatorname{GEff}(F)$.
(c) Take any $\bar{y} \in \operatorname{GEff}(F)$ and any $\bar{p} \geq 1$. Because of Proposition 7.2.6, there exists $K > \bar{p}$ with $\bar{y} \in \operatorname{Eff}(F, D^K)$. By Lemma 7.2.8(e), $C^p \setminus \{0\} \subseteq D^K$ for $p := \ell K > \bar{p}$. Hence, $\operatorname{Eff}(F, D^K) \subseteq \operatorname{Eff}(F, C^p)$. We conclude that $\bar{y} \in \operatorname{Eff}(F, C^p)$.

□

**Corollary 7.2.11** *The following conditions are equivalent to each other:*

(a) $y \in \operatorname{GEff}(F)$.
(b) $y \in \operatorname{Eff}(F, H)$ *for some nontrivial polyhedral cone* $H \subset \mathbb{R}^\ell$ *with* $\mathbb{R}^\ell_+ \setminus \{0\} \subseteq \operatorname{int} H$.
(c) $y \in \operatorname{WEff}(F, H)$ *for some nontrivial polyhedral cone* $H \subset \mathbb{R}^\ell$ *with* $\mathbb{R}^\ell_+ \setminus \{0\} \subseteq \operatorname{int} H$.

***Proof*** (a) implies (b) by Proposition 7.2.10. (b) implies (c) because of $\operatorname{Eff}(F, H) \subseteq \operatorname{WEff}(F, H)$. (c) yields (a) by Proposition 7.2.4. □

Note that efficient solutions with regard to a polyhedral cone can be interpreted as Pareto minimal solutions with respect to altered objective functions [471].

Proposition 7.2.10 implies the following statement by Podinovskij and Nogin [384, Theorem 2.1.15.].

**Corollary 7.2.12** *For*

$$\Lambda_\varepsilon := \{y \in \mathbb{R}^\ell \mid \forall i \in \{1, \dots, \ell\}\colon (1-(\ell-1)\varepsilon)y_i + \varepsilon \sum_{\substack{j=1\\ j\neq i}}^{\ell} y_j \geq 0\},$$

*we get*

$$\mathrm{GEff}(F) = \bigcup_{\varepsilon \in (0, \frac{1}{\ell})} \mathrm{Eff}(F, \Lambda_\varepsilon).$$

***Proof*** Obviously, $\Lambda_\varepsilon = C^p$ with $p = \frac{1-(\ell-1)\varepsilon}{\varepsilon}$, and $\varepsilon = \frac{1}{p+\ell-1}$. The condition $\varepsilon \in (0, \frac{1}{\ell})$ holds if and only if $p > 1$. □

We will now study proper efficiency as efficiency w.r.t. polyhedral cones that are described in another way than the cones $C^p$. These results will be applied in the scalarization methods of Sect. 7.3. The interdependencies between the polyhedral cones and sets $D^K$ could be used to connect scalarization methods with trade-offs.

Geoffrion's proper efficiency can also be characterized as efficiency w.r.t. the convex cones

$$C(s) := \{y \in \mathbb{R}^\ell \mid \forall i \in \{1, \dots, \ell\}\colon y_i + s^T y \geq 0\} \text{ with } s \in \operatorname{int} \mathbb{R}^\ell_+. \tag{7.2.4}$$

**Lemma 7.2.13** *Assume* $s \in \operatorname{int} \mathbb{R}^\ell_+$.

(a) $\forall m \in \mathbb{R}_>\colon\ \mathbb{R}^\ell_+ \subseteq C(\frac{s}{m}),\ \mathbb{R}^\ell_+ \setminus \{0\} \subseteq \operatorname{int} C(\frac{s}{m})$.
(b) *If* $m > m_0 > 0$, *then* $C(\frac{s}{m}) \setminus \{0\} \subset \operatorname{int} C(\frac{s}{m_0})$.
(c) $\forall m > 1\colon\ C(\frac{s}{m}) \setminus \{0\} \subset D^K$ *for* $K = \frac{m-1}{\sum_{i=1}^{\ell} s_i}$.

***Proof***

(a) is obvious.
(b) Take any $y \in C(\frac{s}{m}) \setminus \{0\}$. Then $y_i + \frac{1}{m} s^T y \geq 0$ for all $i \in \{1, \dots, \ell\}$. If we had $s^T y \leq 0$, then this would imply $y \in \mathbb{R}^\ell_+ \setminus \{0\}$, which contradicts $s^T y \leq 0$. Hence, $s^T y > 0$ and $y_i + \frac{1}{m_0} s^T y > y_i + \frac{1}{m} s^T y \geq 0$ for all $i \in \{1, \dots, \ell\}$. Thus, $y \in \operatorname{int} C(\frac{s}{m_0})$.
(c) Take any $m > 1$ and any $y \in C(\frac{s}{m}) \setminus \{0\}$. Then $\frac{s^T y}{m} + y_i \geq 0$ for all $i \in \{1, \dots, \ell\}$. Therefore, $y_n := \max_{i=1,\dots,\ell} y_i > 0$. We obtain $\frac{s^T y}{m} \leq \frac{1}{m} \sum_{i=1}^{\ell} s_i\, y_n < \frac{1}{m-1} \sum_{i=1}^{\ell} s_i\, y_n$. For all $i \in \{1, \dots, \ell\}$, we have $y_i + \frac{1}{m-1} \sum_{i=1}^{\ell} s_i\, y_n > y_i + \frac{s^T y}{m} \geq 0$ and thus $y_n + \frac{m-1}{\sum_{i=1}^{\ell} s_i} y_i > 0$. Hence, $y \in D^K$ for $K := \frac{m-1}{\sum_{i=1}^{\ell} s_i}$. □

**Proposition 7.2.14** *Assume* $s \in \operatorname{int} \mathbb{R}^\ell_+$. *Then*

$$\mathrm{GEff}(F) = \bigcup_{m > 0} \mathrm{Eff}\left(F, C\left(\frac{s}{m}\right)\right), \quad \textit{where}$$

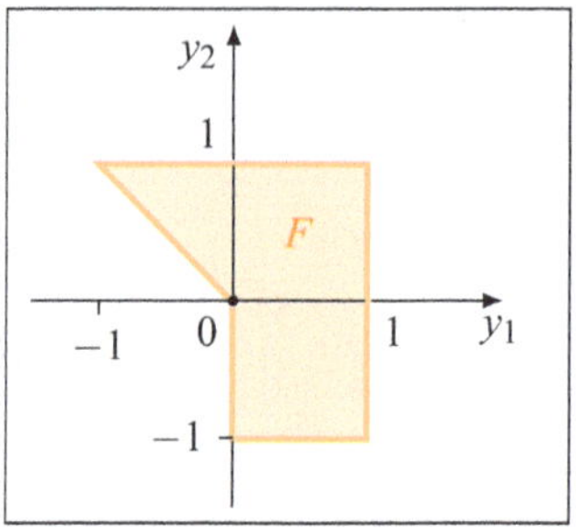


**Fig. 7.5** Set $F$ in Example 7.2.15. Adapted from [472, p. 72]

$$\forall y \in \operatorname{GEff}(F)\ \exists m_0 > 1\ \forall m \geq m_0\colon\ y \in \operatorname{Eff}\left(F, C\left(\frac{s}{m}\right)\right).$$

***Proof*** Take any $s \in \operatorname{int} \mathbb{R}^\ell_+$.

(i) Proposition 7.2.4 and Lemma 7.2.13(a) imply $\bigcup_{m>0} \operatorname{Eff}(F, C(\frac{s}{m})) \subseteq \operatorname{GEff}(F)$.

(ii) Take any $y \in \operatorname{GEff}(F)$. Because of Proposition 7.2.6, there exists $K_0 > 0$ such that $y \in \operatorname{Eff}(F, D^K)$ for all $K \geq K_0$. Define $m_0 := 1 + K_0 \sum_{i=1}^{\ell} s_i > 1$. Take any $m \geq m_0$. Then $C(\frac{s}{m}) \setminus \{0\} \subseteq D^K$ for $K := \frac{m-1}{\sum_{i=1}^{\ell} s_i} \geq K_0$ by Lemma 7.2.13(c). Hence, $\operatorname{Eff}(F, D^K) \subseteq \operatorname{Eff}(F, C(\frac{s}{m}))$. We get $y \in \operatorname{Eff}(F, C(\frac{s}{m}))$. □

Pareto minimality and Geoffrion's proper efficiency coincide for linear vector optimization problems [166]. For such problems, the following statement—which is stronger than Proposition 7.2.14—was formulated by Helbig [227]:

$$\forall s \in \operatorname{int} \mathbb{R}^\ell_+\ \ \exists m_0 \in \mathbb{N}_>\ \ \forall m \geq m_0\colon\ \ \operatorname{WEff}(F, C(\frac{s}{m})) = \operatorname{Eff}(F, C(\frac{s}{m})) = \operatorname{Eff}(F).$$

It is not valid for nonlinear vector optimization problems:

**Example 7.2.15** Consider $F := M_1 \cup M_2$ with $M_1 := \{(y_1, y_2)^T \in \mathbb{R}^2 \mid -1 \leq y_1 \leq 0,\ -y_1 \leq y_2 \leq 1\}$ and $M_2 := \{(y_1, y_2)^T \in \mathbb{R}^2 \mid 0 \leq y_1 \leq 1,\ -1 \leq y_2 \leq 1\}$ (see Fig. 7.5). Then $\operatorname{Eff}(F) = \operatorname{GEff}(F) = \{(0, -1)^T\} \cup \{(y_1, y_2)^T \in \mathbb{R}^2 \mid -1 \leq y_1 < 0,\ y_2 = -y_1\}$, but there does not exist any convex cone $C$ with $\mathbb{R}^\ell_+ \setminus \{0\} \subseteq \operatorname{int} C$ such that $\operatorname{Eff}(F, C) = \operatorname{GEff}(F)$.

**Remark 7.2.16** Kaliszewski [285, Theorem 3.8] proved the following statement:

Assume $Y = \mathbb{R}^\ell$, $F \subset Y$, $y^0 \in \operatorname{Eff}(F)$, and that there exists $\bar{y} \in Y$ with $F \subseteq \bar{y} + \operatorname{int} D$, where $D$ is a polyhedral cone which is not a halfspace, has a nonempty interior and is given by

$$D = \{y \in \mathbb{R}^\ell \mid Wy \geq 0\}$$

with $W = (w^1 \cdots w^r)^T \in \mathbb{R}^{r,\ell}$, $w^i \in \mathbb{R}^\ell$ for all $i \in \{1, \ldots, r\}$. For $\rho \in \mathbb{R}_>$, define

$$D_\rho := \{y \in \mathbb{R}^\ell \mid \forall i \in \{1, \ldots, r\}\colon (w^i)^T y + \rho \sum_{j=1}^{r} (w^j)^T y \geq 0\}.$$

Then $y^0 \in \text{He-PEff}(F, D)$ holds if and only if there exists $\rho \in \mathbb{R}_>$ such that $y^0 \in \text{Eff}(F, D_\rho)$.

For $W = I$ and $s = (\rho, \ldots, \rho)^T \in \mathbb{R}^\ell$, the set $D_\rho$ coincides with $C(s)$ from (7.2.4).

In [206], the formula for $\varphi_{-D_\rho,k}$ is deduced, and the functional is applied to scalarization in multiobjective optimization with a polyhedral domination set.

For each $w \in \operatorname{int} \mathbb{R}^\ell_+$ and $\varepsilon \in \mathbb{R}_>$, define now

$$C_w(\varepsilon) := \{y \in \mathbb{R}^\ell \mid \forall i \in \{1, \ldots, \ell\}\colon w_i y_i + \varepsilon \sum_{j=1}^{\ell} w_j y_j \geq 0\} \quad \text{and} \tag{7.2.5}$$

$$C_{w,\varepsilon} := \{y \in \mathbb{R}^\ell \mid \forall i \in \{1, \ldots, \ell\}\colon w_i y_i + \varepsilon \sum_{j=1}^{\ell} y_j \geq 0\}. \tag{7.2.6}$$

**Lemma 7.2.17** *Assume* $w \in \operatorname{int} \mathbb{R}^\ell_+$.

(a) $\forall \varepsilon \in \mathbb{R}_>\colon\ \mathbb{R}^\ell_+ \subseteq C_w(\varepsilon),\ \mathbb{R}^\ell_+ \setminus \{0\} \subseteq \operatorname{int} C_w(\varepsilon)$.

(b) $\forall \varepsilon \in \mathbb{R}_>\colon\ C_w(\varepsilon) \setminus \{0\} \subseteq \{y \in \mathbb{R}^\ell \mid \sum_{j=1}^{\ell} w_j y_j > 0\}$.

(c) $\forall \varepsilon_0 \in \mathbb{R}_>\ \forall \varepsilon \in (0, \varepsilon_0)\colon\ C_w(\varepsilon) \setminus \{0\} \subseteq \operatorname{int} C_w(\varepsilon_0)$.

(d) $\forall \varepsilon \in \mathbb{R}_>\colon\ C_w(\varepsilon) \setminus \{0\} \subset D^K$ *for* $K = \dfrac{\min_{i=1,\ldots,\ell} w_i}{2\varepsilon \sum_{j=1}^{\ell} w_j}$.

***Proof***

(a) is obvious.

(b) If $\sum_{j=1}^{\ell} w_j y_j \leq 0$ holds for some $y \in C_w(\varepsilon)$, then $y_i \geq 0$ for all $i \in \{1, \ldots, \ell\}$ and hence $y = 0$.

(c) follows from (b).

(d) Take any $\varepsilon \in \mathbb{R}_>$ and $y \in C_w(\varepsilon) \setminus \{0\}$. Then $y_n := \max_{i=1,\ldots,\ell} y_i > 0$. Define $K := \dfrac{\min_{i=1,\ldots,\ell} w_i}{2\varepsilon \sum_{j=1}^{\ell} w_j}$. Take any $i \in \{1, \ldots, \ell\}$. We get $\frac{\varepsilon}{w_i} \sum_{j=1}^{\ell} w_j y_j \leq \frac{\varepsilon}{w_i} \sum_{j=1}^{\ell} w_j y_n < \frac{2\varepsilon}{w_i} \cdot \sum_{j=1}^{\ell} w_j y_n \leq \frac{1}{K} y_n$, which implies $y_i + \frac{1}{K} y_n > y_i + \frac{\varepsilon}{w_i} \sum_{j=1}^{\ell} w_j y_j \geq 0$. Thus, $y_n + K y_i > 0$. We conclude that $y \in D^K$. $\square$

**Proposition 7.2.18** *Assume* $w \in \operatorname{int} \mathbb{R}^\ell_+$. *Then*

$$\mathrm{GEff}(F) = \bigcup_{\varepsilon>0} \mathrm{Eff}(F, C_w(\varepsilon)), \quad \textit{where}$$

$$\forall y \in \mathrm{GEff}(F)\ \exists \varepsilon_0 > 0\ \forall \varepsilon \in (0, \varepsilon_0] \colon\ y \in \mathrm{Eff}(F, C_w(\varepsilon)).$$

***Proof***

(i) Proposition 7.2.4 implies together with Lemma 7.2.17(a): $\bigcup_{\varepsilon>0} \mathrm{Eff}(F, C_w(\varepsilon)) \subseteq \mathrm{GEff}(F)$.

(ii) Take any $y \in \mathrm{GEff}(F)$. Because of Proposition 7.2.6, there exists $K_0 \in \mathbb{R}_>$ such that $y \in \mathrm{Eff}(F, D^K)$ for all $K \geq K_0$. Choose $\varepsilon_0 := \frac{\min_{i=1,\dots,\ell} w_i}{2K_0 \sum_{j=1}^{\ell} w_j} > 0$. Take any $\varepsilon \in (0, \varepsilon_0]$. Define $K := \frac{\min_{i=1,\dots,\ell} w_i}{2\varepsilon \sum_{j=1}^{\ell} w_j} \geq K_0$. We get $C_w(\varepsilon) \setminus \{0\} \subseteq D^K$ by Lemma 7.2.17(d). Thus, $\mathrm{Eff}(F, D^K) \subseteq \mathrm{Eff}(F, C_w(\varepsilon))$. Hence, $y \in \mathrm{Eff}(F, C_w(\varepsilon))$. □

Consider now the sets $C_{w,\varepsilon}$ given by (7.2.6).

**Lemma 7.2.19** *Suppose* $w \in \operatorname{int} \mathbb{R}^\ell_+$.

(a) $\forall \varepsilon \in \mathbb{R}_> \colon\ \mathbb{R}^\ell_+ \subseteq C_{w,\varepsilon},\ \mathbb{R}^\ell_+ \setminus \{0\} \subseteq \operatorname{int} C_{w,\varepsilon}$.

(b) $\forall \varepsilon \in \mathbb{R}_> \colon\ C_{w,\varepsilon} \setminus \{0\} \subseteq \{y \in \mathbb{R}^\ell \mid \sum_{j=1}^{\ell} y_j > 0\}$.

(c) $\forall \varepsilon_0 \in \mathbb{R}_>\ \forall \varepsilon \in (0, \varepsilon_0) \colon\ C_{w,\varepsilon} \setminus \{0\} \subseteq \operatorname{int} C_{w,\varepsilon_0}$.

(d) $\forall \varepsilon \in \mathbb{R}_> \colon\ C_{w,\varepsilon} \setminus \{0\} \subset D^K$ *for* $K = \frac{\min_{i=1,\dots,\ell} w_i}{2\ell\varepsilon}$.

***Proof***

(a) follows from (7.2.6).

(b) If $\sum_{j=1}^{\ell} y_j \leq 0$ holds for some $y \in C_{w,\varepsilon}$, then $y_i \geq 0$ for all $i \in \{1, \dots, \ell\}$ and hence $y = 0$.

(c) follows from (b) and (7.2.6).

(d) Take any $\varepsilon \in \mathbb{R}_>$ and $y \in C_{w,\varepsilon} \setminus \{0\}$. Then $y_n := \max_{i=1,\dots,\ell} y_i > 0$. Choose $K := \frac{\min_{i=1,\dots,\ell} w_i}{2\ell\varepsilon}$. Take any $i \in \{1, \dots, \ell\}$. We get $\frac{\varepsilon}{w_i} \sum_{j=1}^{\ell} y_j \leq \frac{\varepsilon}{w_i} \sum_{j=1}^{\ell} y_n < \frac{2\varepsilon}{w_i} \ell y_n \leq \frac{1}{K} y_n$, which implies $y_i + \frac{1}{K} y_n > y_i + \frac{\varepsilon}{w_i} \sum_{j=1}^{\ell} y_j \geq 0$. Thus, $y_n + K y_i > 0$. We obtain $y \in D^K$. □

**Proposition 7.2.20** *Suppose* $w \in \operatorname{int} \mathbb{R}^\ell_+$. *Then*

$$\operatorname{GEff}(F) = \bigcup_{\varepsilon>0} \operatorname{Eff}(F, C_{w,\varepsilon}), \quad \textit{where}$$

$$\forall y \in \operatorname{GEff}(F)\ \exists \varepsilon_0 > 0\ \forall \varepsilon \in (0, \varepsilon_0]:\ y \in \operatorname{Eff}(F, C_{w,\varepsilon}).$$

***Proof***

(i) Proposition 7.2.4 implies together with Lemma 7.2.19(a): $\bigcup_{\varepsilon>0} \operatorname{Eff}(F, C_{w,\varepsilon}) \subseteq \operatorname{GEff}(F)$.

(ii) Take any $y \in \operatorname{GEff}(F)$. Because of Proposition 7.2.6, there exists $K_0 \in \mathbb{R}_>$ such that $y \in \operatorname{Eff}(F, D^K)$ for all $K \geq K_0$. Choose $\varepsilon_0 := \frac{\min_{i=1,\dots,\ell} w_i}{2\ell K_0} > 0$. Take any $\varepsilon \in (0, \varepsilon_0]$. Define $K := \frac{\min_{i=1,\dots,\ell} w_i}{2\ell\varepsilon} \geq K_0$. By Lemma 7.2.19(d), $C_{w,\varepsilon} \setminus \{0\} \subseteq D^K$. Thus, $\operatorname{Eff}(F, D^K) \subseteq \operatorname{Eff}(F, C_{w,\varepsilon})$. We conclude that $y \in \operatorname{Eff}(F, C_{w,\varepsilon})$. □

Note that the cones $C^p$, $C(s)$, $C_w(\varepsilon)$ and $C_{w,\varepsilon}$ are polyhedral cones of the type

$$\{y \in \mathbb{R}^\ell \mid \forall i \in \{1, \dots, \ell\}:\ u_i y_i + \sum_{\substack{j=1\\ j\neq i}}^{\ell} v_j y_j \geq 0\} \ \text{ with } u, v \in \operatorname{int} \mathbb{R}^\ell_+.$$

Our results imply statements about the existence of properly efficient points.

**Proposition 7.2.21**

(a) *Assume* $u \in \mathbb{R}^\ell$, $K_0, \bar{p}, \varepsilon_0 \in \mathbb{R}_>$, *and* $s, w \in \operatorname{int} \mathbb{R}^\ell_+$. *The following statements are equivalent to each other:*

   (i) *There exists a convex cone* $H \subseteq \mathbb{R}^\ell$ *with* $\mathbb{R}^\ell_+ \setminus \{0\} \subseteq \operatorname{int} H$ *and* $(F - u) \cap (-\operatorname{int} H) = \emptyset$.
   (ii) *There exists a polyhedral cone* $H \subseteq \mathbb{R}^\ell$ *with* $\mathbb{R}^\ell_+ \setminus \{0\} \subseteq \operatorname{int} H$ *and* $(F - u) \cap (-\operatorname{int} H) = \emptyset$.
   (iii) $\exists\, K \geq K_0 :\ (F - u) \cap (-D^K) = \emptyset$.
   (iv) $\exists\, p \geq \bar{p} :\ (F - u) \cap (-C^p) \subseteq \{0\}$.
   (v) $\exists\, m > 0 :\ (F - u) \cap (-C(\frac{s}{m})) \subseteq \{0\}$.
   (vi) $\exists\, \varepsilon \in (0, \varepsilon_0] :\ (F - u) \cap (-C_w(\varepsilon)) \subseteq \{0\}$.

(b) $\operatorname{GEff}(F) \neq \emptyset$ *if and only if* (i) *holds for some* $u \in F$.

(c) *If* $F$ *is nonempty and closed, then* $\operatorname{GEff}(F) \neq \emptyset$ *if and only if* (ii) *holds for some* $u \in \mathbb{R}^\ell$.

***Proof***

(a) The equivalence of (iii) and (iv) follows from Lemma 7.2.8.
(i) implies (iii) because of Propositions 7.2.7 and Lemma 7.2.5. (iii) implies (v) because of Lemma 7.2.13(c). (v) implies (ii) since $C(\frac{s}{m})$ is a polyhedral cone

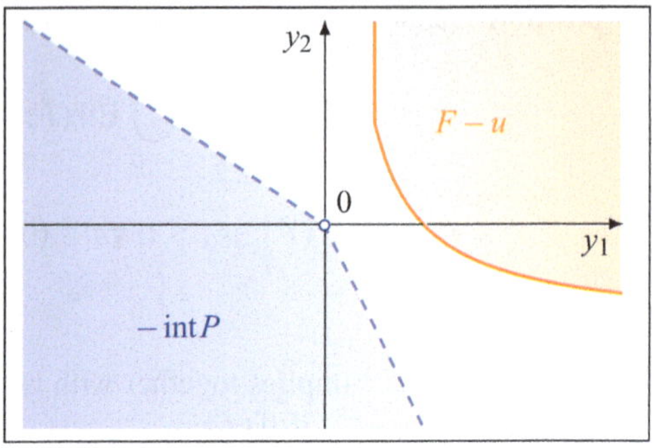


**Fig. 7.6** Illustration of assumption (Sp-ex)

with $\mathbb{R}^\ell_+ \setminus \{0\} \subseteq \operatorname{int} C(\frac{s}{m})$. (ii) yields (i).

(vi) implies (ii). (iii) implies $(F-u) \cap (-D^{\tilde{K}}) = \emptyset$ for each $\tilde{K} > K$ by Lemma 7.2.5, and thus (vi) because of Lemma 7.2.17(d).

(b) results from Proposition 7.2.4.

(c) Assume that $F \neq \emptyset$ is closed and that (ii) is fulfilled for some $u \in \mathbb{R}^\ell$. By Corollary 7.2.9, there exists a polyhedral cone $T$ with $\mathbb{R}^\ell_+ \setminus \{0\} \subseteq \operatorname{int} T$ and $T \setminus \{0\} \subseteq \operatorname{int} H$. Choose some $\bar{y} \in F$. The set $\tilde{F} := F \cap (\bar{y} - T)$ is compact by Lemma 2.2.32. Take any $k \in \operatorname{int} T$. The functional $\varphi_{-T,k}$ is continuous and finite-valued by Proposition 4.7.3(d). Hence, it attains a minimum $t$ on $\tilde{F}$. By (4.6.13) and (4.6.12), $\tilde{F} \cap (-\operatorname{int} T + tk) = \emptyset$, and there exists $y^0 \in \tilde{F}$ with $y^0 \in -T + tk$.

To obtain a contradiction, suppose that $(F - y^0) \cap (-\operatorname{int} T) \neq \emptyset$. Then there exists $y \in F \cap (y^0 - \operatorname{int} T)$. We get $y \in -T + tk - \operatorname{int} T \subseteq -\operatorname{int} T + tk$ since $y^0 \in -T + tk$, and $y \in \bar{y} - T - \operatorname{int} T \subseteq \bar{y} - T$ since $y^0 \in \bar{y} - T$. Thus, $y \in \tilde{F} \cap (-\operatorname{int} T + tk)$, which contradicts $\tilde{F} \cap (-\operatorname{int} T + tk) = \emptyset$. Therefore, the supposition $(F - y^0) \cap (-\operatorname{int} T) \neq \emptyset$ cannot be satisfied. Consequently, (ii) holds for $u := y^0 \in F$. This and (b) imply the assertion. □

For the set $F := \mathbb{R}^2_+ \setminus \{0\}$, which is not closed, we have $\operatorname{GEff}(F) = \emptyset$ even though the condition (i) is fulfilled with $u = (0,0)^T$.

In part (c) of the next theorem, we will use the assumption (see Fig. 7.6):

$$\begin{array}{l}\text{There exist a polyhedral cone } P \subseteq \mathbb{R}^\ell \text{ and } u \in \mathbb{R}^\ell \\ \text{such that } \mathbb{R}^\ell_+ \setminus \{0\} \subseteq \operatorname{int} P \text{ and } (F-u) \cap (-\operatorname{int} P) = \emptyset.\end{array} \qquad \text{(Sp-ex)}$$

By Proposition 7.2.21, (Sp-ex) holds if $\operatorname{GEff}(F) \neq \emptyset$. (Sp-ex) is equivalent to $\operatorname{GEff}(F) \neq \emptyset$ if $F$ is nonempty and closed.

**Theorem 7.2.22**

(1) *We have:*

$$\begin{aligned}\mathrm{GEff}(F) &= \{y^0 \in F \mid \exists \varphi : \mathbb{R}^\ell \to \mathbb{R} \textit{ continuous, sublinear, strictly } \mathbb{R}^\ell_+\textit{-monotone}:\\ &\quad \varphi(y^0 - y^0) = \min\nolimits_{y\in F} \varphi(y - y^0) = 0\}\\ &= \{y^0 \in F \mid \exists \varphi : \mathbb{R}^\ell \to \mathbb{R} \textit{ continuous, sublinear, strictly } \mathbb{R}^\ell_+\textit{-monotone}:\\ &\quad \forall y \in F \setminus \{y^0\} : \varphi(y - y^0) > 0\}.\end{aligned}$$

*In detail:*

(i) $y^0 \in \mathrm{GEff}(F)$ *if and only if there exists a nontrivial closed convex cone* $H \subset \mathbb{R}^\ell$ *with* $\mathbb{R}^\ell_+ \setminus \{0\} \subseteq \mathrm{int}\, H$ *such that* $\varphi_{-H,k}(y^0 - y^0) = \min_{y\in F} \varphi_{-H,k}(y - y^0) = 0$, *where* $k \in \mathrm{int}\, H$ *can be chosen arbitrarily. The function* $\varphi_{-H,k}$ *is continuous, sublinear, finite-valued, and strictly* $\mathbb{R}^\ell_+$*-monotone.*

(ii) *If* $y^0 \in \mathrm{GEff}(F)$, *then there exists* $K \in \mathbb{R}_>$ *such that* $y^0 \in \mathrm{Eff}(F, D^K)$. *For each* $k \in \mathrm{int}\, \mathbb{R}^\ell_+$,

$$\varphi := \sum_{i=1}^{\ell} \varphi_{-\mathrm{cl}\, D^{i,K},k}$$

*is continuous, sublinear, finite-valued, strictly* $\mathbb{R}^\ell_+$*-monotone, and*

$$\varphi(y - y^0) > 0 \textit{ for all } y \in F \setminus \{y^0\}.$$

(2) *Furthermore,*

$$\begin{aligned}\mathrm{GEff}(F) \subseteq \{y^0 \in F \mid \exists \varphi : \mathbb{R}^\ell \to \mathbb{R} &\textit{ continuous, strictly convex,}\\ &\textit{strictly } \mathbb{R}^\ell_+\textit{-monotone}: \ \forall y \in F \setminus \{y^0\} : \ \varphi(y) > \varphi(y^0)\}.\end{aligned}$$

(3) *Assume* (Sp-ex). *Then*

$$\begin{aligned}\mathrm{GEff}(F) &= \{y^0 \in F \mid \exists \varphi \colon \mathbb{R}^\ell \to \mathbb{R} \textit{ continuous, strictly convex,}\\ &\quad \textit{strictly } \mathbb{R}^\ell_+\textit{-monotone}: \ \forall y \in F \setminus \{y^0\} : \ \varphi(y) > \varphi(y^0)\}\\ &= \{y^0 \in F \mid \exists \varphi \colon \mathbb{R}^\ell \to \mathbb{R} \textit{ continuous, convex,}\\ &\quad \textit{strictly } \mathbb{R}^\ell_+\textit{-monotone}: \ \varphi(y^0) = \min_{y\in F} \varphi(y)\}.\end{aligned}$$

***Proof***

(1) (a) Take first any $y^0 \in F$ for which there exists a continuous sublinear strictly $\mathbb{R}^\ell_+$-monotone function $\varphi\colon \mathbb{R}^\ell \to \mathbb{R}$ with $\varphi(y^0 - y^0) = \min_{y\in F} \varphi(y - y^0) = 0$. For $H := \{y \in \mathbb{R}^\ell \mid \varphi(-y) < 0\}$, the set $H \cup \{0\}$ is a convex cone. For each $y \in -\mathbb{R}^\ell_+$, the strict $\mathbb{R}^\ell_+$-monotonicity of $\varphi$ implies $\varphi(-y) > \varphi(0) = 0$. Thus, $H \cup \{0\} \neq \mathbb{R}^\ell$. Define $\varphi_0\colon \mathbb{R}^\ell \to \mathbb{R}$ by $\varphi_0(y) := \varphi(y - y^0)$ for each $y \in \mathbb{R}^\ell$. The function $\varphi_0$ is strictly $\mathbb{R}^\ell_+$-monotone. For each $y \in F$, $\varphi_0(y^0) = \varphi(0) = 0 \leq \varphi(y - y^0) = \varphi_0(y)$. We have $H = \{y \in \mathbb{R}^\ell \mid \varphi_0(y^0 - y) < \varphi_0(y^0)\}$. Theorem 6.3.13 implies $y^0 \in \mathrm{Eff}(F, H)$ and

$\mathbb{R}^\ell_+ \setminus \{0\} \subseteq \operatorname{int} H$. By Proposition 7.2.4, $y^0 \in \operatorname{GEff}(F)$. This yields the first equation in (1) since (i) follows from Corollary 6.3.20.

(b) We will now prove (ii), which implies the second equation in (1). Take now any $y^0 \in \operatorname{GEff}(F)$ and $k \in \operatorname{int} \mathbb{R}^\ell_+$. By Proposition 7.2.6, there exists $K \in \mathbb{R}_>$ with $y^0 \in \operatorname{Eff}(F, D^K)$. Thus, $y^0 \in \operatorname{Eff}(F, D^{i,K})$ for all $i \in \{1, \ldots, \ell\}$. Because of Propositions 4.7.3 and 4.7.1(b), the function $\varphi_{-\operatorname{cl} D^{i,K},k}$ is finite-valued, continuous, and sublinear for all $i \in \{1, \ldots, \ell\}$. By (4.6.13), $(F - y^0) \cap (-D^{i,K}) = \emptyset$ implies $\varphi_{-\operatorname{cl} D^{i,K},k}(y - y^0) \geq 0$ for all $y \in F$ and $i \in \{1, \ldots, \ell\}$. Consequently, $\varphi := \sum_{i=1}^{\ell} \varphi_{-\operatorname{cl} D^{i,K},k}$ is finite-valued, continuous, sublinear, and $\varphi(y - y^0) \geq 0$ for all $y \in F$. Since $\varphi$ is sublinear and finite-valued, we have $\varphi(0) = 0$.
If $\varphi(y - y^0) = 0$ for some $y \in F$, then $\varphi_{-\operatorname{cl} D^{i,K},k}(y - y^0) = 0$ for all $i \in \{1, \ldots, \ell\}$, hence $y - y^0 \in -\operatorname{bd} D^{i,K}$ for all $i \in \{1, \ldots, \ell\}$ by (4.6.14). Lemma 7.2.5(2) yields $\varphi(y - y^0) > 0$ for all $y \in F \setminus \{y^0\}$.
Take any $y^2 \in \mathbb{R}^\ell$ and $y^1 \in y^2 + (\mathbb{R}^\ell_+ \setminus \{0\})$.
For each $i \in \{1, \ldots, \ell\}$, the inclusion $\operatorname{cl} D^{i,K} + \mathbb{R}^\ell_+ \subseteq \operatorname{cl} D^{i,K}$ implies that $\varphi_{-\operatorname{cl} D^{i,K},k}$ is $\mathbb{R}^\ell_+$-monotone by Theorem 4.5.5, and thus $\varphi_{-\operatorname{cl} D^{i,K},k}(y^1) \geq \varphi_{-\operatorname{cl} D^{i,K},k}(y^2)$.
There exists $n \in \{1, \ldots, \ell\}$ with $y^1_n > y^2_n$. Choose $r := \varphi_{-\operatorname{cl} D^{n,K},k}(y^2)$. Then $y^2 \in -\operatorname{bd} D^{n,K} + rk$.
We have $y^1_n > y^2_n$ and $y^1 \geq y^2$, which implies $(-y^1 + rk)_n < (-y^2 + rk)_n$ and

$$(-y^1 + rk)_n + K(-y^1 + rk)_j < (-y^2 + rk)_n + K(-y^2 + rk)_j$$

for all $j \in \{1, \ldots, \ell\} \setminus \{n\}$.
Moreover, $-y^2 + rk \in \operatorname{bd} D^{n,K}$. This yields $-y^1 + rk \notin \operatorname{cl} D^{n,K}$, i.e., $y^1 \notin -\operatorname{cl} D^{n,K} + rk$.
Therefore, $\varphi_{-\operatorname{cl} D^{n,K},k}(y^1) > r = \varphi_{-\operatorname{cl} D^{n,K},k}(y^2)$, which implies $\varphi(y^1) > \varphi(y^2)$. Thus, $\varphi$ is strictly $\mathbb{R}^\ell_+$-monotone.

(2) was proved by Soland [413, Lemma 3].

(3) Take—if they exist—any $y^0 \in F$ and any continuous convex strictly $\mathbb{R}^\ell_+$-monotone function $\varphi\colon \mathbb{R}^\ell \to \mathbb{R}$ with $\varphi(y^0) = \min_{y \in F} \varphi(y)$.
Theorem 6.3.13 implies that $H := \{y \in \mathbb{R}^\ell \mid \varphi(y^0 - y) < \varphi(y^0)\}$ is open and convex, $0 \in \operatorname{bd} H$, $\mathbb{R}^\ell_+ \setminus \{0\} \subseteq H$ and, moreover, $\operatorname{cl} H + (\mathbb{R}^\ell_+ \setminus \{0\}) \subseteq H$ and $y^0 \in \operatorname{Eff}(F, H)$.
Because of (Sp-ex), we have $\mathbb{R}^\ell_+ \setminus \{0\} \subseteq \operatorname{int} P$. By Corollary 7.2.9, there exists a polyhedral cone $T$ with $\mathbb{R}^\ell_+ \setminus \{0\} \subseteq \operatorname{int} T$ and $T \setminus \{0\} \subseteq \operatorname{int} P$. There are $m \in \{1, \ldots, \ell\}$ and vectors $s^j \in \mathbb{R}^\ell$, $j \in \{1, \ldots, m\}$, such that $T = \{y \in \mathbb{R}^\ell \mid \forall j \in \{1, \ldots, m\}\colon \sum_{i=1}^{\ell} s^j_i y_i \geq 0\}$. Since $\mathbb{R}^\ell_+ \setminus \{0\} \subseteq \operatorname{int} T$, we obtain $s^j_i > 0$ for all $i \in \{1, \ldots, \ell\}$ and $j \in \{1, \ldots, m\}$.
Because of Lemma 2.2.32 and (Sp-ex), the set $F \cap (y^0 - T)$ is bounded. Hence,

there exists $w \in \mathbb{R}^\ell$ with $F \cap (y^0 - T) \subseteq w + \operatorname{int} \mathbb{R}^\ell_+$. We get $y^0 \in w + \operatorname{int} \mathbb{R}^\ell_+$.
Take any $n \in \{1, \ldots, \ell\}$.

Define $z^n \in \mathbb{R}^\ell$ by $z^n_j := \begin{cases} w_n & \text{for } j = n, \\ y^0_j & \text{for } j \in \{1, \ldots, \ell\} \setminus \{n\}. \end{cases}$

Then $y^0 - z^n \in \mathbb{R}^\ell_+ \setminus \{0\} \subseteq H$.

Define $u^n \in \mathbb{R}^\ell$ by $u^n_j := \begin{cases} 0 & \text{for } j = n, \\ -1 & \text{for } j \in \{1, \ldots, \ell\} \setminus \{n\}. \end{cases}$

Since $H$ is open, there exists $t_n \in \mathbb{R}_>$ with $(y^0 - z^n) + t_n u^n \in H$.

For $w^n := z^n - t_n u^n \in y^0 - H$, we have $w^n_j = \begin{cases} w_n & \text{for } j = n, \\ y^0_j + t_n & \text{for } j \in \{1, \ldots, \ell\} \setminus \{n\}. \end{cases}$

We now define $K := \max \left( \max\limits_{i=1,\ldots,\ell} \frac{2(y^0_i - w_i)}{t_i}, \max\limits_{\substack{i=1,\ldots,\ell \\ j=1,\ldots,m}} \left(\frac{1}{s^j_i} \sum\limits_{\substack{r=1 \\ r \neq i}}^{\ell} s^j_r + 1\right) \right) > 0.$

For each $y \in D^K$, there exists $i \in \{1, \ldots, \ell\}$ with $y_i > 0$ and $y_i + K y_j > 0$ for all $j \in \{1, \ldots, \ell\} \setminus \{i\}$. For all $j \in \{1, \ldots, m\}$, we get $\sum\limits_{r=1}^{\ell} s^j_r y_r = s^j_i y_i + \sum\limits_{\substack{r=1 \\ r \neq i}}^{\ell} s^j_r y_r > y_i (s^j_i - \frac{1}{K} \sum\limits_{\substack{r=1 \\ r \neq i}}^{\ell} s^j_r) \geq 0$ from the definition of $K$. Thus, $D^K \subseteq T$.

To obtain a contradiction, suppose that there exists $y \in F \cap (y^0 - D^K)$. Then $y^0 - y \in D^K$. Moreover, there is $i \in \{1, \ldots, \ell\}$ such that $y^0_i - y_i > 0$ and $y^0_i - y_i + K(y^0_j - y_j) > 0$ for all $j \in \{1, \ldots, \ell\} \setminus \{i\}$. Because of $y^0 \in w + \operatorname{int} \mathbb{R}^\ell_+$ and $y \in F \cap (y^0 - D^K) \subseteq F \cap (y^0 - T) \subseteq w + \operatorname{int} \mathbb{R}^\ell_+$, we get $w_i < y^0_i$ and $w_i < y_i$. This implies $2w_i < y_i + y^0_i$. Hence, $-y_i < y^0_i - 2w_i$. Therefore, $y^0_i - y_i < 2(y^0_i - w_i)$.

Choose $t := \frac{y^0_i - y_i}{2(y^0_i - w_i)} \in (0, 1)$. We conclude that $y_i = y^0_i + 2t(w_i - y^0_i) < y^0_i + t(w_i - y^0_i)$ since $w_i - y^0_i < 0$. For each $j \in \{1, \ldots, \ell\} \setminus \{i\}$, the inequality $K \geq \frac{2(y^0_i - w_i)}{t_i}$ implies $y^0_j - y_j > -\frac{1}{K}(y^0_i - y_i) \geq -t_i \frac{y^0_i - y_i}{2(y^0_i - w_i)} = -t_i t$, hence $y_j < y^0_j + t t_i = y^0_j + t(w^i_j - y^0_j)$. Thus, $y \in y^0 + t(w^i - y^0) - \operatorname{int} \mathbb{R}^\ell_+$. Since $H$ is convex, $0 \in \operatorname{bd} H$ and $y^0 - w^i \in H$, we have $t(y^0 - w^i) \in H$. Therefore, $y \in y^0 - H - \operatorname{int} \mathbb{R}^\ell_+ = y^0 - (H + \operatorname{int} \mathbb{R}^\ell_+) \subseteq y^0 - H$. We obtain $y \in F \cap (y^0 - H)$, which contradicts $y^0 \in \operatorname{Eff}(F, H)$. Hence, the supposition is wrong, and $y^0 \in \operatorname{Eff}(F, D^K) \subseteq \operatorname{GEff}(F)$. □

**Remark 7.2.23** Because of Proposition 7.2.1, the statement

$$\begin{aligned} \operatorname{GEff}(F) \supseteq \{y^0 \in F \mid \exists \varphi : \mathbb{R}^\ell \to \mathbb{R} \text{ continuous, sublinear, strictly } \mathbb{R}^\ell_+\text{-monotone} : \\ \forall y \in F : \varphi(y - y^0) \geq 0\} \end{aligned}$$

also follows from Proposition 5.5 in [109], where a corresponding assertion for Benson properly efficient points was proved in partially ordered topological vector spaces.

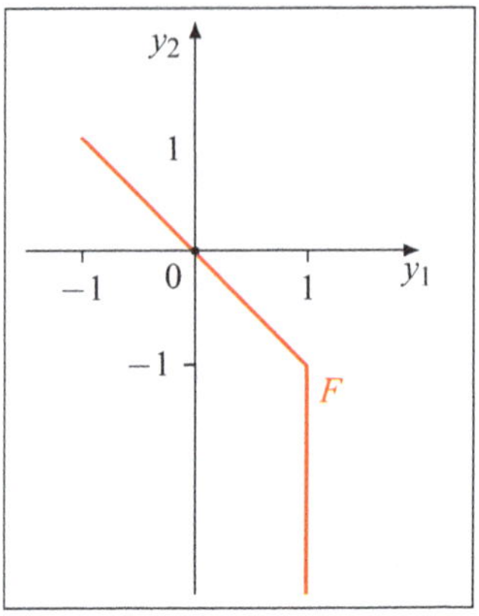


**Fig. 7.7** Set $F$ in Example 7.2.24. Adapted from [472, p. 76]

The inclusion in Theorem 7.2.22(2) cannot be replaced by an equation, which is illustrated by the following example.

**Example 7.2.24** Define $g : \mathbb{R} \to \mathbb{R}$ by

$$g(x) := \begin{cases} e^x - 1 & \text{for } x < 0, \\ x^2 + 2x & \text{for } x \geq 0, \end{cases}$$

and $\varphi : \mathbb{R}^2 \to \mathbb{R}$ by $\varphi((y_1, y_2)^T) := g(y_1) + g(y_2)$.

Then $\varphi$ is continuous, strictly convex and strictly $\mathbb{R}^2_+$-monotone. Consider $F := \{(y_1, y_2)^T \in \mathbb{R}^2 \mid y_1 = 1,\ y_2 \leq -1\} \cup \{(y_1, y_2)^T \in \mathbb{R}^2 \mid y_1 \in [-1, 1],\ y_1 + y_2 = 0\}$ (see Fig. 7.7). We get $\varphi((0, 0)^T) = 0 < \varphi(y)$ for all $y \in F \setminus \{(0, 0)^T\}$, but $\text{GEff}(F) = \emptyset$.

The assumption in Theorem 7.2.22(3) is also not superfluous for the second equation.

**Example 7.2.25** Consider the sets $F$ and $H$ from Example 6.3.7. For $k := (1, 1)^T$ and $\varphi := \varphi_{-H,k}$, we get $\text{Eff}(F) = \text{argmin}_F\, \varphi$. This set coincides with the boundary of $F$ and contains more than one element. $\varphi$ is convex by Theorem 4.4.4, finite-valued and continuous by Theorem 4.6.6, strictly $\mathbb{R}^\ell_+$-monotone by Theorem 4.5.5. This implies together with Corollary 6.3.17

$$\begin{aligned} \text{Eff}(F) = \{y^0 \in F \mid \exists \varphi : \mathbb{R}^\ell \to \mathbb{R} \text{ continuous, convex, strictly } \mathbb{R}^\ell_+\text{-monotone} : \\ \varphi(y^0) = \min_{y \in F} \varphi(y)\}, \end{aligned}$$

although $\text{GEff}(F) = \emptyset$.

We obtain the following from Proposition 6.2.16.

$$\begin{aligned} \text{Eff}(F) \supseteq \{y^0 \in F \mid \exists \varphi : \mathbb{R}^\ell \to \mathbb{R} \text{ continuous, strictly } \mathbb{R}^\ell_+\text{-monotone} : \\ \forall y \in F \setminus \{y^0\} : \varphi(y) > \varphi(y^0)\}. \end{aligned}$$

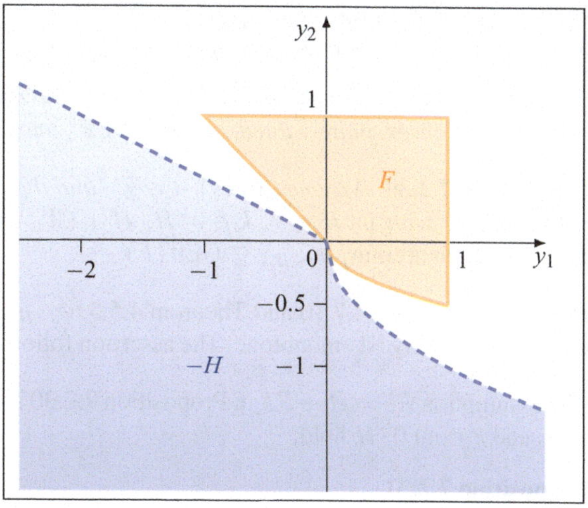


**Fig. 7.8** The sets $F$ and $H$ in Example 7.2.26. Adapted from [472, p. 77]

Luc illustrated by an example [338, p. 85] that the above inclusion cannot be replaced by an equation. As we will show in Example 7.2.26, the elements $y^0$ in this inclusion are not necessarily minimizers of some continuous convex strictly $\mathbb{R}^\ell_+$-monotone functional on $F$.

**Example 7.2.26** In $Y = \mathbb{R}^2$, consider $F := \{(y_1, y_2)^T \in \mathbb{R}^2 \mid -1 \leq y_1 \leq 0,\ -y_1 \leq y_2 \leq 1\} \cup \{(y_1, y_2)^T \in \mathbb{R}^2 \mid 0 \leq y_1 \leq 1,\ -\frac{1}{2}\sqrt{y_1} \leq y_2 \leq 1\}$ and $k := (1, 1)^T$. The set $H := -(\{(y_1, y_2)^T \in \mathbb{R}^2 \mid y_1 \leq 0,\ y_1 + 2y_2 < 0\} \cup \{(y_1, y_2)^T \in \mathbb{R}^2 \mid y_1 > 0,\ y_2 < -\sqrt{y_1}\})$ is open. We have $(0, 0)^T \in \operatorname{bd} H$ and $\mathbb{R}^2_+ \setminus \{(0, 0)^T\} \subseteq H$ (see Fig. 7.8). The function $\varphi_{-\operatorname{cl} H,k}$ is finite-valued. It is continuous and strictly $\mathbb{R}^2_+$-monotone by Theorems 4.6.6 and 4.5.5. Furthermore, $\varphi_{-\operatorname{cl} H,k}((0, 0)^T) = 0 < \varphi_{-\operatorname{cl} H,k}(y)$ for all $y \in F \setminus \{(0, 0)^T\}$.

If there did exist a continuous convex strictly $\mathbb{R}^2_+$-monotone functional $\varphi\colon \mathbb{R}^2 \to \mathbb{R}$ with $\varphi(y) \geq \varphi((0, 0)^T)$ for all $y \in F$, then Theorem 7.2.22(3) would imply $(0, 0)^T \in \operatorname{GEff}(F)$, which is not fulfilled.

Theorem 7.2.22(3) implies together with Corollary 6.3.17:

**Theorem 7.2.27** *If* (Sp-ex) *holds, then* $\operatorname{GEff}(F) = \text{NI-PEff}(F, \mathbb{R}^\ell_+)$.

Thus, we get from Proposition 7.2.21:

**Theorem 7.2.28** *If* $\operatorname{GEff}(F) \neq \emptyset$, *then* $\operatorname{GEff}(F) = \text{NI-PEff}(F, \mathbb{R}^\ell_+)$.

Example 6.3.7 illustrates that $\text{NI-PEff}(F, \mathbb{R}^\ell_+)$ can be nonempty even though $\operatorname{GEff}(F)$ is empty.

Corollary 6.3.18 implies by Theorem 7.2.27:

**Corollary 7.2.29** *Assume* (Sp-ex) *and* $k \in \operatorname{int} \mathbb{R}^\ell_+$.

*Then* $y^0 \in \operatorname{GEff}(F)$ *if and only if there exists a closed convex set* $H \subseteq \mathbb{R}^\ell$ *with* $0 \in \operatorname{bd} H$ *and* $H + (\mathbb{R}^\ell_+ \setminus \{0\}) \subseteq \operatorname{int} H$ *such that* (i) *or* (ii) *holds. Note that* (i) *and* (ii) *are equivalent to each other.*

(i) $\varphi_{y^0-H,k}(y^0) = \min_{y\in F} \varphi_{y^0-H,k}(y) = 0.$
(ii) $\varphi_{-H,k}(y^0 - y^0) = \min_{y\in F} \varphi_{-H,k}(y - y^0) = 0.$

*For each* $y^0 \in \mathbb{R}^\ell$, *the functional* $\varphi_{y^0-H,k}$ *with* $H$ *having the mentioned properties is continuous, convex, finite-valued, and strictly* $\mathbb{R}^\ell_+$*-monotone.*

**Proposition 7.2.30** *Assume* (Sp-ex), $a \in \mathbb{R}^\ell$, *and that* $H$ *is a proper closed convex subset of* $\mathbb{R}^\ell$ *with* $\operatorname{int} H \neq \emptyset$, $k \in 0^+H$, $H + (\mathbb{R}^\ell_+ \setminus \{0\}) \subseteq H + \mathbb{R}_> k$, *and* $\mathbb{R}^\ell = H + \mathbb{R}k$. *Then* $\operatorname{argmin}_F \varphi_{a-H,k} \subseteq \operatorname{GEff}(F)$.

***Proof*** By Corollary 4.7.10 and Theorem 4.5.5, $\varphi_{a-H,k}$ is finite-valued, continuous, convex and strictly $\mathbb{R}^\ell_+$-monotone. The assertion follows from Theorem 7.2.22(3). □

The assumption $\mathbb{R}^\ell = H + \mathbb{R}k$ in Proposition 7.2.30 is satisfied if the other assumptions and $k \in \operatorname{int} 0^+H$ hold.

**Proposition 7.2.31**

(a) *Assume that* $H \subset \mathbb{R}^\ell$ *is a nontrivial convex cone with* $\mathbb{R}^\ell_+ \setminus \{0\} \subset \operatorname{int} H$, $k \in \mathbb{R}^\ell \setminus \{0\}$ *and* $a \in \mathbb{R}^\ell$. *Then* $\operatorname{argmin}_F \varphi_{a-H,k} \subseteq \operatorname{GEff}(F)$.
(b) *For each* $y^0 \in \operatorname{GEff}(F)$, *there exists a nontrivial polyhedral cone* $H \subset \mathbb{R}^\ell$ *with* $\mathbb{R}^\ell_+ \setminus \{0\} \subset \operatorname{int} H$ *such that* $y^0$ *is a unique minimizer of* $\varphi_{y^0-H,k}$ *on* $F$ *for each* $k \in H \setminus \{0\}$.

***Proof***

(a) $\operatorname{argmin}_F \varphi_{a-H,k} \subseteq \operatorname{WEff}(F, H) \subseteq \operatorname{GEff}(F)$ by Corollary 6.2.23 and Proposition 7.2.4.
(b) By Proposition 7.2.10, $y^0 \in \operatorname{GEff}(F)$ implies $y^0 \in \operatorname{Eff}(F, C^p)$ for some $p \in \mathbb{R}_>$. Apply Theorem 6.2.24 to this efficient point set. □

In Sect. 7.3, we will investigate scalar optimization problems that generate Geoffrion properly efficient points.

## 7.3 Scalarization Methods

Most of the procedures in multiobjective decision making—especially interactive methods—use scalarization in auxiliary problems. When choosing a scalarization, different aspects have to be taken into consideration [478]:

1. Does the aim consist in

   - checking the Pareto optimality of a given alternative,
   - determining some Pareto optimal solution,
   - finding the entire Pareto optimal solution set or a set approximating it,
   - a controlled search in the set of Pareto optimal solutions?

Implicitly, one is also interested in whether Pareto optimal solutions of the multi-objective problem exist and whether the Pareto optimal point set differs from the feasible set. The latter question was investigated in [471].
2. If the decision maker is intended to understand the decision making process or even to influence the choice of parameters in the procedure, the parameters in the scalarizing problem should be interpretable in an application-oriented way. The scalarizing problems discussed in this section turn out to have a common geometric interpretation. On this basis, similar properties of different scalarizations can be derived.
3. Do optimal solutions of the scalarizing problems exist and can they be determined in an effective way? Does the nonexistence of an optimal solution imply statements about the Pareto optimal point set?
4. Which parameters in the scalarizing problem can be fixed without changing the solution set?

Depending on the decision process, the above aspects can refer to weakly or properly efficient solutions instead of Pareto optimal solutions (see Sect. 7.1).
In this section, we will only touch on answers to these questions, presenting a very general approach that allows these questions to be studied in a systematic way.

In Chap. 5, we investigated properties of scalar optimization problems $\min_{y\in F} \varphi_{a-H,k}(y)$ and $(\mathrm{P}_{F,a,H,k})$, which were used for characterizing solutions to vector optimization problems in Chap. 6 and in Sect. 7.2. The scalar optimization problem

$$
\begin{aligned}
(\mathrm{P}_{F,a,H,k}):\ & t \to \min \\
& y \in a - H + tk \\
& y \in F,\ t \in \mathbb{R},
\end{aligned}
$$

is now considered in the space $Y = \mathbb{R}^\ell$, $\ell \geq 2$, for $F \subseteq \mathbb{R}^\ell$ under the following assumptions:

$$H \text{ is a proper closed subset of } \mathbb{R}^\ell,\ k \in 0^+H \setminus \{0\},\ a \in \mathbb{R}^\ell.$$

According to Theorem 5.1.4(b), it is equivalent to

$$\min_{y\in F} \varphi_{a-H,k}(y),$$

where $\varphi_{a-H,k}$ is given by

$$\varphi_{a-H,k}(y) = \inf\{t \in \mathbb{R} \mid y \in a - H + tk\} \quad \text{for each } y \in \mathbb{R}^\ell.$$

Here and in the complete chapter, equivalence between optimization problems is to be understood in the sense of Definition 5.1.2.

The parameters $a$ and $k$ of the scalar optimization problem can be interpreted geometrically as a reference point and a direction of improvement, respectively (see Fig. 5.1). Thus, the influence of the parameter choice becomes more obvious.

In this section, we are interested in the optimal point sets $\mathrm{Eff}(F)$, $\mathrm{WEff}(F)$ and $\mathrm{GEff}(F)$ of the multiobjective optimization problem. Many known scalarizations for this case turn out to be special problems $(\mathrm{P}_{F,a,H,k})$, and the statements of the previous chapters can be applied to these scalarizations. Hence, properties of scalarizing problems can be explored in a systematic way, which will be illustrated for different scalarization methods. Although the most frequently used scalarization methods fit in our framework, this is not the case for each scalarization, e.g. not for Kasimbeyli's conic scalarization method [288].

Most of the results in this section are taken from [472]. Some of them were published together with the conceptual ideas in [469]. Detailed references will be given in each subsection. For further details and references related to special scalarization methods, we refer to [89, 132, 136, 272, 354, 400, 426]. There are many applications of the discussed methods in the literature. $(\mathrm{P}_{F,a,H,k})$ extends Pascoletti and Serafini's scalarization in the way described in Sect. 6.4. Pascoletti–Serafini's scalarization was applied, e.g., for developing adaptive scalarization methods by Eichfelder [136] and to linear multiobjective optimization problems in [248].

In Sect. 7.3.1, we start our investigation with a problem $(\mathrm{P}_{F,a,H,k})$ that extends the weighted Chebychev norm minimization. A scalarization which simultaneously makes use of different weight vectors will be introduced in Sect. 7.3.2. It contains the important classical approach of minimizing the weighted sum of objective function values. In Sect. 7.3.3, a scalarization related to the extended weighted Chebyshev norm by Choo and Atkins [97] is studied. Sections 7.3.4 and 7.3.5 deal with scalarizations by Helbig [227] and by Dubov [120], respectively. Wierzbicki's reference point approach [495] is investigated in Sect. 7.3.6, together with some modifications. Section 7.3.7 contains statements about the Hurwicz Rule for decision making under uncertainty. The scalarization method in Sect. 7.3.8 consists in the minimization of one objective function under restrictions to the other objective functions. This includes the $\varepsilon$-constraint method [99]. In Sect. 7.3.9, the weighted sum of objective function values is minimized under restrictions to objective functions. Finally, different aspects of Ester's hyperbola efficiency [158] are discussed in Sect. 7.3.10.

The investigated problems are of type $(\mathrm{P}_{F,a,H,k})$. There sets $H$ with different properties are involved. In Sects. 7.3.1–7.3.9, $H$ will always be a cone, which we will denote by $C$. In Sect. 7.3.7, this cone is not necessarily convex. In the other sections, the cone is polyhedral. In Sects. 7.3.1 and 7.3.8, $H$ coincides with the domination set $\mathbb{R}^{\ell}_{+}$ of the multiobjective optimization problem. In Sect. 7.3.10, we will study a problem $(\mathrm{P}_{F,a,H,k})$ with a strictly convex set $H$. A scalarization in which $H$ is some polyhedral set that is not necessarily a cone will be introduced in Sect. 9.1.3. Up to Sect. 7.3.7 and in Sect. 7.3.10, the considered functional $\varphi_{a-H,k}$ is finite-valued. This is not the case in Sects. 7.3.8 and 7.3.9. Thus, the scalarizing problems in Sects. 7.3.8 and 7.3.9 can have an empty feasible set even though $F$ is nonempty.

### 7.3.1 A Scalarization Related to the Weighted Chebyshev Norm

Consider

$$\min_{y \in F} \max_{i=1,\dots,\ell} w_i(y_i - z_i) \tag{P1}$$

with $w \in \operatorname{int} \mathbb{R}^\ell_+$, $z \in \mathbb{R}^\ell$.

Problem (P1) is equivalent to $(\mathrm{P}_{F,a,C,k})$, and hence to $\min_{y \in F} \varphi_{a-C,k}(y)$, for

$$a = z, \quad C = \mathbb{R}^\ell_+, \quad k_i = \frac{1}{w_i} \text{ for all } i \in \{1, \dots, \ell\}. \tag{7.3.1}$$

Here $C$ coincides with the domination set of the vector optimization problem, and $k \in \operatorname{int} C$.

**Remark 7.3.1** For the understanding of analogous statements in the next sections, we explain in which way the equivalence between (P1) and $(\mathrm{P}_{F,a,C,k})$ was deduced.

Problem (P1) is equivalent to the following optimization problem:

$$\begin{aligned} & t \to \min \\ & w_i(y_i - z_i) \le t \text{ for all } i \in \{1, \dots, \ell\} \\ & y \in F,\ t \in \mathbb{R}. \end{aligned} \tag{7.3.2}$$

Inequality (7.3.2) can be rewritten as $z_i - y_i + t\frac{1}{w_i} \ge 0$ for each $i \in \{1, \dots, \ell\}$, i.e., as $a - y + tk \in C$ with $a$, $C$ and $k$ given by (7.3.1). Thus, (7.3.2) is equivalent to $y \in a - C + tk$.

For $F \subseteq z + \mathbb{R}^\ell_+$, problem (P1) coincides with the minimization of the weighted Chebyshev norm on $F - z$. The **weighted Chebyshev norm** with weight vector $w \in \operatorname{int} \mathbb{R}^\ell_+$ is defined by

$$\|y\| := \max_{i=1,\dots,\ell} w_i |y_i| \text{ for all } y = (y_1, \dots, y_\ell)^{\mathrm{T}} \in \mathbb{R}^\ell.$$

The weighted Chebyshev norm and special problems of type (P1) have often be applied for solving multiobjective optimization problems. The minimization of the weighted Chebyshev norm was used in the interactive procedure STEM for linear vector optimization problems by Benayoun et al. [39]. Zeleny's method of the shifted ideal [511, 512] can be applied together with the weighted Chebyshev norm minimization on $F - z$. Based on the equivalence between $\min\limits_{y \in F} \|y - z\|_\infty$ and the problem $(\mathrm{P}_{F,z,\mathbb{R}^\ell_+,k})$ with $k = \mathbf{1}$ under the assumption $F \subseteq z + \mathbb{R}^\ell_+$, Jahn [266] derived first order conditions which are sufficient for a unique optimal solution to the problem and thus, by [114, Theorem 2], for its Pareto minimality. Problem $(\mathrm{P}_{F,a,\mathbb{R}^\ell_+,k})$ with $k = \mathbf{1}$ was also used by Roy [393, p. 242], who pointed out that it is a linear optimization problem if the multiobjective optimization problem is linear. Brosowski ([80, p. 485], [78, p. 55]) considered $(\mathrm{P}_{F,a,\mathbb{R}^\ell_+,k})$ with $k = \mathbf{1}$ for the scalarization of semiinfinite problems. In this context, Brosowski and Conci [80] formulated an algorithm for finding a Pareto minimal point in the case $\ell \in \{2, 3\}$.

Gembicki [174] introduced $(\mathrm{P}_{F,a,\mathbb{R}^\ell_+,k})$ as Goal Attainment Method, with $a$ as goal vector and $k \in \operatorname{int} \mathbb{R}^\ell_+$ as a vector of weights for the deviations. Disadvantages of this approach and an example from production planning were discussed by Hwang and Masud [251, pp. 97–101].

Michalowski and Szapiro [350, 352] proposed an interactive procedure for the determination of a compromise solution which starts with a worst vector $a$ and an improvement direction $k$. Iteratively, $a$ is altered based on the solution to problem $(\mathrm{P}_{F,a,\mathbb{R}^\ell_+,k})$. This approach was modified in [351], where $k$ is chosen as the difference of the ideal point and $a$ in each step.

The following statements are essential when applying (P1).

**Proposition 7.3.2**

(a) *If the objective function of some problem* (P1) *is not bounded below on* $F$, *then* $\mathrm{WEff}(F) = \emptyset$.
(b) *If* $y$ *is an optimal solution to* (P1), *then* $y \in \mathrm{WEff}(F)$.
(c) *The optimal solution set of each problem* (P1) *is compact and contains some Pareto minimal point of* $F$ *if* $F$ *is a nonempty compact set or if* $F$ *is a nonempty closed set for which there exist* $u \in \mathbb{R}^\ell$ *and a polyhedral cone* $H \subset \mathbb{R}^\ell$ *such that* $\mathbb{R}^\ell_+ \setminus \{0\} \subset \operatorname{int} H$ *and* $(F - u) \cap -\operatorname{int} H = \emptyset$.
(d) *If* $y$ *is the unique optimal solution to* (P1), *then* $y \in \mathrm{Eff}(F)$.
(e) *For each* $w \in \operatorname{int} \mathbb{R}^\ell_+$, *we have:*

$$\begin{aligned} y^0 \in \mathrm{WEff}(F) &\iff y^0 \textit{ solves } (\mathrm{P1}) \textit{ with } z = y^0, \\ &\iff \exists z \in \mathbb{R}^\ell\colon y^0 \textit{ solves } (\mathrm{P1}). \\ y^0 \in \mathrm{Eff}(F) &\iff y^0 \textit{ uniquely solves } (\mathrm{P1}) \textit{ with } z = y^0, \\ &\iff \exists z \in \mathbb{R}^\ell\colon y^0 \textit{ uniquely solves } (\mathrm{P1}). \end{aligned}$$

*The optimal value of the problems* (P1) *with* $z \in \mathrm{WEff}(F)$ *is* 0.
(f) *Assume* $F \subseteq z + \operatorname{int} \mathbb{R}^\ell_+$.

$$\begin{aligned} y^0 \in \mathrm{WEff}(F) \iff & y^0 \textit{ is an optimal solution to } (\mathrm{P1}) \\ & \textit{with } w_i = \frac{1}{y^0_i - z_i} \quad \textit{for all } i \in \{1, \ldots, \ell\}. \\ y^0 \in \mathrm{Eff}(F) \iff & y^0 \textit{ is the unique optimal solution to } (\mathrm{P1}) \\ & \textit{with } w_i = \frac{1}{y^0_i - z_i} \quad \textit{for all } i \in \{1, \ldots, \ell\}. \end{aligned}$$

*The optimal value of these problems* (P1) *with* $y^0 \in \mathrm{WEff}(F)$ *is* 1.

(g) *Assume* $F \subseteq z - \operatorname{int} \mathbb{R}^\ell_+$.

$$y^0 \in \operatorname{WEff}(F) \iff y^0 \text{ is an optimal solution to (P1) with } w_i = \frac{1}{z_i - y_i^0} \text{ for all } i \in \{1, \ldots, \ell\}.$$

$$y^0 \in \operatorname{Eff}(F) \iff y^0 \text{ is the unique optimal solution to (P1) with } w_i = \frac{1}{z_i - y_i^0} \text{ for all } i \in \{1, \ldots, \ell\}.$$

*The optimal value of these problems* (P1) *with* $y^0 \in \operatorname{WEff}(F)$ *is* $-1$.

(h) $\operatorname{Eff}(F) = \bigcup_{z \in \mathbb{R}^\ell} \operatorname{Eff}(\{y \in \mathbb{R}^\ell \mid y \text{ optimal solution to (P1)}\})$ *for all* $w \in \operatorname{int} \mathbb{R}^\ell_+$,

$$= \bigcup_{z \in \mathbb{R}^\ell, w \in \operatorname{int} \mathbb{R}^\ell_+} \operatorname{Eff}(\{y \in \mathbb{R}^\ell \mid y \text{ optimal solution to (P1)}\}).$$

***Proof*** (a), (b) and (d)–(h) follow from Propositions 6.2.56 and 6.2.57. (c) results from Proposition 6.2.60(1). □

Furthermore, Helbig investigated the stability of the scalarization (P1) for $w = \mathbf{1}$ in [226]. Sensitivity results for more general cases were mentioned in Sect. 5.4.

If the assumption $F \subseteq z + \operatorname{int} \mathbb{R}^\ell_+$ in Proposition 7.3.2(f) were replaced by $F \subseteq z + \mathbb{R}^\ell_+$, then there could exist $y^0 \in \operatorname{Eff}(F)$ which does not solve some problem (P1) with parameters $z$ and $w \in \operatorname{int} \mathbb{R}^\ell_+$.

**Example 7.3.3** In $Y = \mathbb{R}^2$, consider $F := \{(y_1, y_2)^T \in \mathbb{R}^2 \mid -1 \le y_1 \le 0,\ y_1^2 \le y_2 \le 1\}$, $z := (-2, 0)^T$, $y^0 := (0, 0)^T$. Then $F \subseteq z + \mathbb{R}^\ell_+$ and $y^0 \in \operatorname{Eff}(F)$. For each $k \in \operatorname{int} \mathbb{R}^2_+$, $y^0$ is not an optimal solution to $(\mathrm{P}_{F,z,\mathbb{R}^2_+,k})$ and hence also not an optimal solution to (P1) with parameters $w \in \operatorname{int} \mathbb{R}^2_+$ and $z$.

Proposition 6.2.60 implies:

**Proposition 7.3.4**

(1) *If* $F$ *is a closed convex set, then the following statements are equivalent to each other.*

  (a) $\operatorname{Eff}(F) \neq \emptyset$.
  (b) *The optimal solution set of each problem* (P1) *is nonempty and compact.*
  (c) *There exist* $z \in \mathbb{R}^\ell$ *and* $w \in \operatorname{int} \mathbb{R}^\ell_+$ *such that the optimal solution set of* (P1) *is nonempty and bounded.*
  (d) *Each problem* (P1) *has some optimal solution which belongs to* $\operatorname{Eff}(F)$.

(2) *Assume that* $F + \mathbb{R}^\ell_+$ *is a closed convex set. Then the statements* (a)–(d) *in* (1) *are equivalent to each other if the compactness in* (b) *is replaced by boundedness.*

The next example illustrates the statements of the above propositions and underscores the following facts:

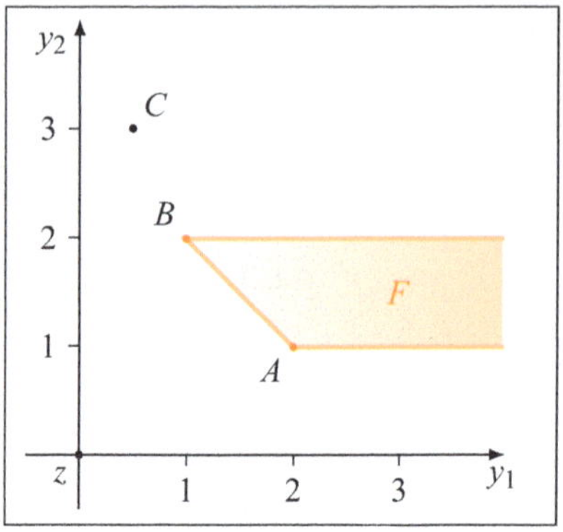


**Fig. 7.9** Set $F$ in Example 7.3.5. Adapted from [473, p. 103]

- Optimal solutions to problems (P1) do not necessarily belong to $\mathrm{Eff}(F)$.
- Problems (P1) can have the same unique optimal solution for an infinite number of parameter choices even though more Pareto minimal points exist.
- A problem (P1) with fixed parameters can contain different Pareto minimal points in its optimal solution set.

**Example 7.3.5** Define $F := \{(y_1, y_2)^T \in \mathbb{R}^2 \mid 1 \leq y_2 \leq 2,\ y_1 + y_2 \geq 3\}$ (see Fig. 7.9), and consider the points $A := (2, 1)$, $B := (1, 2)$ and $C := (0.5, 3)$. Fix $\ell = 2$. Then each problem (P1) has the same optimal solution set as some problem (P1) with $w_1 = 1$. Choose $z := (0, 0)^T$ and $w_1 := 1$. For $w_2 \in (0, 0.5)$, the optimal solution set of (P1) is a line segment which connects $A$ with some point on the line $y_2 = 1$. This segment is as longer as smaller $w_2$ is. For $w_2 = 0.5$, the point $A$ is the unique optimal solution to (P1). For each $w_2 \in (0.5, 2)$, the unique optimal solution to (P1) is some point on the line segment between $A$ and $B$. For each $w_2 \geq 2$, the point $B$ becomes the unique optimal solution to (P1). If $C$ did belong to $F$, then $B$ and $C$ would be the optimal solutions to (P1) for $w_2 = 3$.

Helbig [230] proposed a special case of the algorithm from Proposition 6.2.63 for finding Pareto minimal points. His algorithm uses the scalarization $(\mathrm{P}_{F,a,\mathbb{R}^\ell_+,k})$ with given vectors $a \in \mathbb{R}^\ell$ and $k \in \operatorname{int} \mathbb{R}^\ell_+$ under the assumption that $F$ is a nonempty closed convex set.

The special case of the algorithm in Proposition 6.2.62 which makes use of problems (P1) for the determination of Pareto minimal points was published in [473, Satz 2.3]. In this case, the assumption of Proposition 6.2.62 is equivalent to $F$ being a nonempty closed set with $\mathrm{GEff}(F) \neq \emptyset$.

For a closed convex set $F$ with $\mathrm{WEff}(F) \neq \emptyset$, problems (P1) do not necessarily attain a minimum.

**Example 7.3.6** Choose $F$ like in Example 5.2.19. Then $F = \mathrm{WEff}(F) \neq \emptyset$ is closed and convex, but $\min\limits_{y \in F} \max\limits_{i=1,\ldots,\ell} y_i = \min\limits_{y \in F} y_1$ does not exist, i.e., (P1) does not attain a minimum for $w = (1, 1, 1)^T$ and $z = (0, 0, 0)^T$.

We get from Proposition 7.3.2(c): If $F$ is a closed set with $\mathrm{GEff}(F) \neq \emptyset$, then each problem (P1) attains a minimum. The condition $\mathrm{GEff}(F) \neq \emptyset$ cannot be replaced by $\mathrm{Eff}(F) \neq \emptyset$.

**Example 7.3.7** Take $F := \{(y_1, y_2, y_3)^{\mathrm{T}} \in \mathbb{R}^3 \mid y_1 + y_3 = 0,\ y_2 = -\frac{1}{y_1},\ y_1 > 0\}$. Then $F = \mathrm{Eff}(F) \neq \emptyset$ is closed and $\mathrm{GEff}(F) = \emptyset$. Problem (P1) does not attain a minimum for $w = (1, 1, 1)^{\mathrm{T}}$ and $z = (0, 0, 0)^{\mathrm{T}}$.

**Remark 7.3.8** (*Bibliographical notes related to Sect.* 7.3.1)

Section 7.3.1 is contained in [472], except Example 7.3.5 and the statements related to the compactness of the solution set in Proposition 7.3.2. This content was published without proofs in [473], with the exception of Examples 7.3.6 and 7.3.7. Example 7.3.5 coincides with [473, Beispiel 2.1].

The relationship between Pareto minimality and Chebyshev norm approximation was studied in [72, 113, 114, 173, 268], [272, Sects. 9.4.–9.5.], [365, 398], [426, Chaps. 14–15], [427, 428, 500, 504]. Yu [500] deduced the statement of Proposition 7.3.2(c) related to compact sets $F$ with $z_i = \min_{y \in F} y_i$ for all $i \in \{1, \ldots, \ell\}$. For $F \subseteq z + \mathrm{int}\,\mathbb{R}^\ell_+$, Jahn [271, Corollary 3.1.] proved Proposition 7.3.2(b) and (d) and the equivalences in Proposition 7.3.2(f).

For $F = f(X)$, where $X \subseteq \mathbb{R}^n$ is described by restrictions, $(\mathrm{P}_{F,a,\mathbb{R}^\ell_+,k})$ coincides with problem (1.8) by Bernau [49], who showed Proposition 7.3.2(a) and (b). Germeier [178] considered the special problems (P1) with $z = 0$ or $w = \mathbf{1}$. For $k = \mathbf{1}$, Brosowski [78] proved the related special case of Proposition 7.3.2(c) for a compact set $F$ and that each weakly Pareto minimal point of $F$ solves some problem $(\mathrm{P}_{F,a,\mathbb{R}^\ell_+,k})$. In [226], Helbig proved Proposition 7.3.2(b), (d), (e) and the first two equivalences of Proposition 7.3.4(1) for (P1) with $w = \mathbf{1}$. Helbig [230, Theorem I.2.] also showed that each problem (P1) has an optimal solution if $F$ is a closed convex set with $\mathrm{Eff}(F) \neq \emptyset$.

For $F \subseteq a + \mathrm{int}\,\mathbb{R}^\ell_+$, Steuer and Choo [427, Theorem 4.6] characterized Pareto minimal points as minimizers of $\sum_{i=1}^{\ell} y_i$ on the optimal solution set of some problem (P1).

### 7.3.2 *Multiple Vectors of Weights*

The following scalarizing problem was introduced in [472]:

$$\min_{y \in F} \max_{i=1,\ldots,r} \left((w^i)^T y - z_i\right) \tag{P2}$$

with $r \in \{1, \ldots, \ell\}$, $z \in \mathbb{R}^r$, $W = (w^1 \cdots w^r)^{\mathrm{T}} \in \mathbb{R}^{r,\ell}_+$ and $\mathrm{rank}\, W = r$.

Problem (P2) can be of interest in order to take into consideration different weight vectors $w^i$ of $r$ different decision makers or uncertainties of a decision maker.

It is equivalent to $(\mathrm{P}_{F,a,C,k})$, and hence to problem $\min_{y\in F} \varphi_{a-C,k}(y)$, if the variables are chosen as follows:

$$\begin{aligned} &a \text{ is a solution to the system } Wa = z \text{ of linear equations,}\\ &k \text{ is a solution to the system } Wk = \mathbf{1} \text{ of linear equations,}\\ &C = \{y \in \mathbb{R}^\ell \mid Wy \geq 0\}. \end{aligned} \tag{7.3.3}$$

Because of Lemma 6.2.65, this statement depends on the existence of solutions to the systems $Wk = \mathbf{1}$ and $Wa = z$ of linear equations and not on the additional assumptions $W \in \mathbb{R}_+^{r,\ell}$ and rank $W = r$ for $W \in \mathbb{R}^{r,\ell}$. Of course, the condition rank $W = r$ guarantees the existence of such solutions. Example 6.2.66 underscores that the existence of a vector $k$ with $Wk = \mathbf{1}$ is essential for the equivalence of problem (P2) to some problem $(\mathrm{P}_{F,a,C,k})$ with a closed connected set $C$.

Problem (P1) is a special case of problem (P2). This can be seen by choosing in (P2) $r = \ell$ and, for all $i, j \in \{1, \ldots, \ell\}$,

$$w_j^i = \begin{cases} w_i & \text{if } j = i, \\ 0 & \text{if } j \neq i. \end{cases}$$

In the following sections, we will examine further special cases of (P2).

**Lemma 7.3.9** *Take any problem* (P2), *and define* $k$ *and* $C$ *by* (7.3.3).

(a) $C$ *is a closed convex cone.*
(b) $\operatorname{int} C = \{y \in \mathbb{R}^\ell \mid Wy > 0\} \neq \emptyset$.
(c) $C \setminus \{0\} \not\subseteq \operatorname{int} C$, *and* $C + (C \setminus \{0\}) \not\subseteq \operatorname{int} C$.
(d) $k \in \operatorname{int} C$, $\mathbb{R}^\ell = \operatorname{bd} C + \mathbb{R}k$, *and* $\operatorname{int} C = \operatorname{bd} C + \mathbb{R}_> k$.
(e) $C + \mathbb{R}_+^\ell \subseteq C$.
(f) $C + (\mathbb{R}_+^\ell \setminus \{0\}) \subseteq \operatorname{int} C \iff w^i \in \operatorname{int} \mathbb{R}_+^\ell$ *for all* $i \in \{1, \ldots, r\}$.
(g) $C \cap (-C) = \{0\} \iff r = \ell$.

***Proof***

(a) is obvious.
(b) follows from Proposition 2.2.30.
(c) If $r < \ell$, then there exists $y \neq 0$ with $Wy = 0$.
If $r = \ell$, then the system $Wy = b$ with $b_1 = 0$ and $b_i = 1$ for all $i \in \{2, \ldots, \ell\}$ has a solution $y \neq 0$.
Thus, $C \setminus \{0\} \not\subseteq \operatorname{int} C$. Because of $0 \in C$, we obtain $C + (C \setminus \{0\}) \not\subseteq \operatorname{int} C$.
(d) $Wk = \mathbf{1}$ yields $k \in \operatorname{int} C$. Because of Proposition 4.7.3(d), $\varphi_{C,-k}$ is finite-valued. Theorem 4.6.6 implies the assertion.
(e) and (f) can easily be verified.
(g) $y \in C \cap (-C)$ holds if and only if $Wy = 0$. Clearly, $y = 0$ is the unique solution to $Wy = 0$ if and only if $r = \ell$ holds. □

**Proposition 7.3.10**

(a) *If* $y$ *is an optimal solution to* (P2), *then* $y \in \text{WEff}(F, C)$ *for* $C$ *given by* (7.3.3), *and* $y \in \text{WEff}(F)$.
(b) *If* $y$ *is a unique optimal solution to* (P2), *then* $y \in \text{Eff}(F)$.
(c) *If* $w^i \in \text{int}\,\mathbb{R}^\ell_+$ *for all* $i \in \{1, \ldots, r\}$ *and* $y$ *is an optimal solution to* (P2), *then* $y \in \text{GEff}(F)$.
(d) *For each* $y \in \text{GEff}(F)$, *there exist vectors* $w^i \in \text{int}\,\mathbb{R}^\ell_+$, $i \in \{1, \ldots, \ell\}$, *and* $z \in \mathbb{R}^\ell$ *such that* $y$ *is the unique optimal solution to* (P2) *with* $r = \ell$.
(e) *For each* $y \in \text{Eff}(F)$, *there exist* $W \in \mathbb{R}^{\ell,\ell}_+$ *with* rank $W = \ell$ *and* $z \in \mathbb{R}^\ell$ *such that* $y$ *is a unique optimal solution to* (P2) *with* $r = \ell$. *For each* $y \in \text{WEff}(F)$, *there also exist such a matrix* $W$ *and* $z \in \mathbb{R}^\ell$ *for which* $y$ *is an optimal solution to* (P2) *with* $r = \ell$.

***Proof*** (a) and (b) result from Corollary 6.2.23, $\mathbb{R}^\ell_+ \subseteq C$ and Lemma 6.2.5(b). (c) is implied by Proposition 7.2.31(a). Part (d) follows from Proposition 7.3.11(e), part (e) from Proposition 7.3.2(e). □

For $r = 1$, $z = 0$ and $w = w^1 \in \mathbb{R}^\ell_+ \setminus \{0\}$, the problem (P2) becomes

$$\min_{y \in F} w^T y. \tag{7.3.4}$$

This is the classical linear scalarization. Early papers dealing with (7.3.4) were written by Kuhn and Tucker [313], Arrow, Barankin and Blackwell [10], DaCunha and Polak [106], Geoffrion [177] and Yu [502]. Gass and Saaty [171] and Geoffrion [175, 176] used this problem in a parameter-depending method for solving bicriterial vector optimization problems. Essential properties of (7.3.4) were summarized by Chankong and Haimes [88, pp. 133–136, 141f], who also described its algorithmic use in detail [88, Sect. 6.2].

For $w \in \text{int}\,\mathbb{R}^\ell_+$, the optimal solutions $y$ to (7.3.4) are the properly efficient points defined by Schönfeld [406]. For linear vector optimization problems, the set of these properly efficient points coincides with that defined by Geoffrion [166]. Problem (7.3.4) is often used for finding Pareto minimal points, e.g. by Major [343] and in the interactive methods by Zionts and Wallenius ([515], [251, pp. 150–168]) and by Steuer ([425], [426, Abschn. 9.10.], [251, pp. 226–242]).

Of course, (7.3.4) with varying parameters is not appropriate for approximating the Pareto minimal point set of a nonconvex vector optimization problem. We also have to note that even if $F$ is a polyhedral set, it can happen that (7.3.4) attains a minimum only for one choice of a parameter vector $w$ which fulfills the condition $\sum_{i=1}^{\ell} w_i = 1$ [78, Example 7.3.].

The condition $\sum_{i=1}^{\ell} w_i = 1$ stands for a normalization that does not restrict the solution set.

Section 7.3.2 coincides with [472, Sect. 7.2]. For $W$ being a matrix of positive real numbers, the content of this section was published in [477], except Lemma 7.3.9(c) and Proposition 7.3.10(a), (b) and (e).

### 7.3.3 A Scalarization Related to the Extended Weighted Chebyshev Norm

Consider

$$\min_{y\in F}\ \max_{i=1,\dots,\ell}\ w_i\left(c\,(y_i-g_i)+b\sum_{\substack{j=1\\ j\neq i}}^{\ell}(y_j-g_j)\right) \tag{P3}$$

with $w\in \operatorname{int}\mathbb{R}^{\ell}_{+}$, $g\in\mathbb{R}^{\ell}$, $b,c\in\mathbb{R}_{+}$ and $b+c>0$.

Problem (P3) is equivalent to $(\mathrm{P}_{F,a,C,k})$, and hence to $\min_{y\in F}\varphi_{a-C,k}(y)$, with

$$a=g,$$

$$C=\left\{y\in\mathbb{R}^{\ell}\mid \forall i\in\{1,\dots,\ell\}\colon c\,y_i+b\sum_{\substack{j=1\\ j\neq i}}^{\ell}y_j\geq 0\right\}, \tag{7.3.5}$$

$$k_i=\begin{cases}\dfrac{1}{b\,\ell\max\limits_{j=1,\dots,\ell}w_j} & \text{if } c=b \text{ and } F\subseteq\{y\in\mathbb{R}^{\ell}\mid \sum\limits_{j=1}^{\ell}y_j\geq\sum\limits_{j=1}^{\ell}g_j\},\\[2ex] \dfrac{1}{b\,\ell\min\limits_{j=1,\dots,\ell}w_j} & \text{if } c=b \text{ and } F\not\subseteq\{y\in\mathbb{R}^{\ell}\mid \sum\limits_{j=1}^{\ell}y_j\geq\sum\limits_{j=1}^{\ell}g_j\},\\[2ex] -\dfrac{\ell-2}{b(\ell-1)}\dfrac{1}{w_i}+\dfrac{1}{b(\ell-1)}\sum\limits_{\substack{j=1\\ j\neq i}}^{\ell}\dfrac{1}{w_j} & \text{if } c=0,\\[2ex] s_1\dfrac{1}{w_i}+s_2\sum\limits_{\substack{j=1\\ j\neq i}}^{\ell}\dfrac{1}{w_j} & \text{if } c>0 \text{ and } b\neq c,\end{cases}$$

for all $i\in\{1,\dots,\ell\}$, where

$$s_1:=\frac{c+(\ell-2)b}{c^2+(\ell-2)bc-(\ell-1)b^2},\quad s_2:=\frac{-b}{c^2+(\ell-2)bc-(\ell-1)b^2}.$$

Here $k\in\operatorname{int}C$ holds. For $b=0$, the set $C$ is the nonnegative orthant. For $b\neq 0$, $C$ coincides with $C^p$ given by (7.2.2) for $p=\frac{c}{b}$. Properties of the cone $C$ follow from Lemmata 7.3.9 and 7.2.8.

Each problem (P3) coincides with some problem (P2).

If $c\neq b$, then (P3) is the special case of (P2) with $r=\ell$, $z_i=c\,w_i\,g_i+b\,w_i\sum\limits_{\substack{j=1\\ j\neq i}}^{\ell}g_j$ for all $i\in\{1,\dots,\ell\}$, and, for all $i,j\in\{1,\dots,\ell\}$,

$$w_j^i = \begin{cases} c\,w_i & \text{if } j = i, \\ b\,w_i & \text{if } j \neq i. \end{cases}$$

If $c = b$ and $F \subseteq \{y \in \mathbb{R}^\ell \mid \sum_{j=1}^{\ell} y_j \geq \sum_{j=1}^{\ell} g_j\}$, then (P3) coincides with

$$\min_{y \in F} \left( b\,(\max_{i=1,\ldots,\ell} w_i) \sum_{j=1}^{\ell} (y_j - g_j) \right)$$

and thus with (P2) for $r = 1$, $z_1 = b\,(\max_{i=1,\ldots,\ell} w_i) \sum_{i=1}^{\ell} g_i$ and $w_j^1 = b\,(\max_{i=1,\ldots,\ell} w_i)$ for all $j \in \{1, \ldots, \ell\}$.

If $c = b$ and $F \not\subseteq \{y \in \mathbb{R}^\ell \mid \sum_{j=1}^{\ell} y_j \geq \sum_{j=1}^{\ell} g_j\}$, then (P3) is equivalent to

$$\min_{y \in F} \left( b\,(\min_{i=1,\ldots,\ell} w_i) \sum_{j=1}^{\ell} (y_j - g_j) \right)$$

and hence to (P2) with $r = 1$, $z_1 = b\,(\min_{i=1,\ldots,\ell} w_i) \sum_{i=1}^{\ell} g_i$ and $w_j^1 = b\,(\min_{i=1,\ldots,\ell} w_i)$ for all $j \in \{1, \ldots, \ell\}$.

Problem (P1) is the special case of problem (P3) with $c = 1$ and $b = 0$. The special case with $c = 1$ and $b \neq 0$ will be examined in Sect. 7.3.5.

**Proposition 7.3.11**

(a) *If* $y$ *is an optimal solution to* (P3), *then* $y \in \mathrm{WEff}(F)$.
(b) *If* $y$ *is a unique optimal solution to* (P3), *then* $y \in \mathrm{Eff}(F)$.
(c) *For each* $y \in \mathrm{WEff}(F)$, *there exist* $w \in \operatorname{int} \mathbb{R}^\ell_+$ *and* $g \in \mathbb{R}^\ell$ *such that* $y$ *is an optimal solution to* (P3) *with* $c = 1$ *and* $b = 0$.
*For each* $y \in \mathrm{Eff}(F)$, *there exist* $w \in \operatorname{int} \mathbb{R}^\ell_+$ *and* $g \in \mathbb{R}^\ell$ *such that* $y$ *is the unique optimal solution to* (P3) *with* $c = 1$ *and* $b = 0$.
(d) $y \in \mathrm{GEff}(F)$ *if and only if* $y$ *is an optimal solution to some problem* (P3) *with* $b > 0$ *and* $c > 0$.
(e) *For each* $y^0 \in \mathrm{GEff}(F)$, *there exists* $c_0 > 1$ *such that* $y^0$ *solves the optimization problem* (P3) *with* $b = 1$, $c = c_0$, $w = \mathbf{1}$ *and* $g = y^0$, *where the optimal value is zero. Then* $y^0$ *is the unique optimal solution to each problem* (P3) *with* $b = 1$, $c > c_0$, $w = \mathbf{1}$ *and* $g = y^0$, *and the optimal value of each of these problems is zero.*
(f) *If* $F \subseteq g + (\mathbb{R}^\ell_+ \setminus \{0\})$ *and* $y^0 \in \mathrm{GEff}(F)$, *then there exists* $c_0 > 1$ *such that* $y^0$ *solves the optimization problem* (P3) *with* $b = 1$, $c = c_0$ *and*

$$w_i = \frac{1}{c_0\,(y_i^0 - g_i) + \sum\limits_{\substack{j=1\\ j\neq i}}^{\ell}(y_j^0 - g_j)} \quad \textit{for all } i \in \{1,\dots,\ell\},$$

*where the optimal value is* 1. *Then* $y^0$ *is the unique optimal solution to each problem* (P3) *with* $b = 1$, $c > c_0$ *and*

$$w_i = \frac{1}{c\,(y_i^0 - g_i) + \sum\limits_{\substack{j=1\\ j\neq i}}^{\ell}(y_j^0 - g_j)} \quad \textit{for all } i \in \{1,\dots,\ell\},$$

*with the optimal value* 1.

(g) *For each closed set* $F \subset \mathbb{R}^\ell$ *with* $\mathrm{GEff}(F) \neq \emptyset$ *and each* $w \in \operatorname{int}\mathbb{R}^\ell_+$, $g \in \mathbb{R}^\ell$ *and* $b \in \mathbb{R}_>$, *there exists a real number* $c_0 > b$ *such that* (P3) *with* $c = c_0$ *has an optimal solution. Then each problem* (P3) *with* $c > c_0$ *and the same parameters* $w$, $g$ *and* $b$ *also has an optimal solution, and the optimal point set of each of these problems with* $c > c_0$ *is compact.*

(h) *Assume that* $F \subset \mathbb{R}^\ell$ *is a closed set with* $\mathrm{GEff}(F) \neq \emptyset$ *and* $y^0 \in \mathrm{Eff}(F)$. *Then* $y^0$ *is the limit of each sequence of optimal solutions of the problems*

$$(\mathrm{P}_p): \quad \min_{y\in F}\ \max_{i=1,\dots,\ell}\ (y_i - y_i^0) + \frac{1}{p}\sum_{\substack{j=1\\ j\neq i}}^{\ell}(y_j - y_j^0)$$

*with integers* $p > 1$ *for* $p \to \infty$.

*There exists* $p_0 > 1$ *such that* $(\mathrm{P}_{p_0})$ *has an optimal solution. Each problem* $(\mathrm{P}_p)$ *with* $p > p_0$ *also has an optimal solution.* $y^0$ *is an optimal solution to* $(\mathrm{P}_p)$ *with* $p \geq p_0$ *if and only if the optimal value of* $(\mathrm{P}_p)$ *is zero.*

*The sequence of optimal values of the problems* $(\mathrm{P}_p)$ *with* $p \geq p_0$ *converges to zero for* $p \to \infty$. *It is monotonically increasing. If* $(\mathrm{P}_{\tilde p})$ *with* $\tilde p \geq p_0$ *does not have the optimal value zero, then the optimal values of the problems* $(\mathrm{P}_p)$ *with* $p_0 \leq p \leq \tilde p$ *are strictly monotonically increasing w.r.t. p. If* $(\mathrm{P}_{\tilde p})$ *with* $\tilde p \geq p_0$ *has the optimal value zero, then* $(\mathrm{P}_p)$ *has the optimal value zero for each* $p > \tilde p$.

***Proof***

(a) and (b) result from Proposition 7.3.10.

(c) follows from Proposition 7.3.2(e).

(d) The first direction of the statement follows from (e), the reverse direction from Proposition 7.3.10(c).

(e) Take any $y^0 \in \mathrm{GEff}(F)$. By Proposition 7.2.10, there exists $p_1 > 1$ with $y^0 \in \mathrm{Eff}(F, C^{p_1})$. Consider first $p = p_1$.

   (i) Define $b = 1$, $c = p$, $w = \mathbf{1}$, $C$ by (7.3.5) and $k$ by the formula following (7.3.5). Then $C = C^p$, and $y^0 \in \mathrm{Eff}(F, C)$. Theorem 6.2.24 implies

$\varphi_{y^0-C,k}(y^0) < \varphi_{y^0-C,k}(y)$ for all $y \in F \setminus \{y^0\}$. By Theorem 5.1.3(b), $y^0$ is the unique optimal solution to $(\mathrm{P}_{F,y^0,C,k})$ and thus of (P3) with $g = y^0$. The optimal value of this problem is 0 by Theorem 6.2.24.

(ii) Assume now that $y^0$ is the optimal solution to some problem (P3) with $b = 1, c = c_0 > 1, w = \mathbf{1}$ and $g = y^0$. Then the cone $C$ in (7.3.5) coincides with $C^{p_0}$ for $p_0 = c_0$. By Corollary 6.2.23 and Theorem 5.1.3(b), we have $y^0 \in \mathrm{WEff}(F, C^{p_0})$ for $p_0 = c_0$. Hence, $y^0 \in \mathrm{Eff}(F, C^p)$ for all $p > p_0$ because of Lemma 7.2.8(a). The assertion follows from (i), applied to each $p > p_0$.

(f) Take any $y^0 \in \mathrm{GEff}(F)$. By Proposition 7.2.10, there exists $p_1 > 1$ with $y^0 \in \mathrm{Eff}(F, C^{p_1})$. Consider first $p = p_1$.

(i) Define $b = 1, c = p$ and $C$ by (7.3.5). Then $C = C^p$, and $y^0 \in \mathrm{Eff}(F, C)$. The inclusions $F \subseteq g + (\mathbb{R}^\ell_+ \setminus \{0\})$ and $\mathbb{R}^\ell_+ \setminus \{0\} \subseteq \operatorname{int} C$ imply $F \subseteq g + \operatorname{int} C$. By Theorem 6.2.37, $y^0$ is the unique optimal solution to $(\mathrm{P}_{F,g,C,y^0-g})$ with the optimal value 1. Because of $k = y^0 - g$, we obtain

$$w_i = \frac{1}{c\,k_i + \sum\limits_{\substack{j=1\\ j\neq i}}^{\ell} k_j} = \frac{1}{c\,(y_i^0 - g_i) + \sum\limits_{\substack{j=1\\ j\neq i}}^{\ell} (y_j^0 - g_j)} \quad \text{for all } i \in \{1, \dots, \ell\}.$$

(ii) Assume now that $y^0$ solves problem (P3) with $b = 1, c = c_0 > 1$ and

$$w_i = \frac{1}{c_0\,(y_i^0 - g_i) + \sum\limits_{\substack{j=1\\ j\neq i}}^{\ell} (y_j^0 - g_j)} \quad \text{for all } i \in \{1, \dots, \ell\},$$

i.e., the problem $(\mathrm{P}_{F,g,C^{p_0},y^0-g})$ with $p_0 = c_0$. Theorems 6.2.37 and 5.1.3(b) imply $y^0 \in \mathrm{WEff}(F, C^{p_0})$. Thus, $y^0 \in \mathrm{Eff}(F, C^p)$ for all $p > p_0$ by Lemma 7.2.8(a). Hence, $y^0$ is the unique optimal solution to $(\mathrm{P}_{F,g,C^p,y^0-g})$ because of Theorem 6.2.37. This yields the assertion.

(g) (i) By Proposition 7.2.21, there exist $u \in F$ and $p_1 > 1$ with $(F - u) \cap -C^{p_1} \subseteq \{0\}$. Take any $p > p_1$. In Proposition 5.2.20, choose $z = 0$, $C = -C^{p_1}$ and $H = C^p$. Together with Lemma 7.2.8(a), this implies the existence of an optimal solution to each problem $(\mathrm{P}_{F,a,C^p,k})$ with $a \in \mathbb{R}^\ell$ and $k \in \operatorname{int} C^p$ and the compactness of the optimal point set. For $c = p\,b$, the relationship between $(\mathrm{P}_{F,a,C^p,k})$ and (P3) yields the first part of the assertion.

(ii) Assume now that (P3) with $w \in \operatorname{int} \mathbb{R}^\ell_+$, $g \in \mathbb{R}^\ell$, $b \in \mathbb{R}_>$ and $c = c_0 > b$ has an optimal solution $y$. By Corollary 6.2.23 and Theorem 5.1.3(b), we obtain $y \in \mathrm{WEff}(F, C^{p_0})$ with $p_0 := \frac{c_0}{b} > 1$. This implies $(F - y) \cap -C^p = \{0\}$ for all $p > p_0$ because of Lemma 7.2.8(a). The existence of optimal solutions to each problem (P3) with $c > c_0$ follows by analogous arguments as in (i).

(h) Since $\operatorname{GEff}(F) \neq \emptyset$, Proposition 7.2.21 implies the existence of $p_1 > 1$ and $u \in F$ with $(F - u) \cap (-C^{p_1}) \subseteq \{0\}$. Apply Proposition 7.3.11(g) to $g = y^0$, $b = 1$ and $w = \mathbf{1}$. Thus, we get the existence of an integer $p_0 > 1$ such that $(\mathrm{P}_{p_0})$ has an optimal solution. Moreover, Proposition 7.3.11(g) implies that each problem $(\mathrm{P}_p)$ with $p \geq p_0$ has an optimal solution $y^p$ and an optimal value $t_p$. For each $p \geq p_0$, we have $t_p \leq 0$ since $y^0$ is a feasible solution to $(\mathrm{P}_p)$. Each problem $(\mathrm{P}_p)$ with $p \geq p_0$ is equivalent to $(\mathrm{P}_{F,y^0,C^p,k^p})$ with $k_i^p := \frac{p^2-p}{p^2+(\ell-2)p-(\ell-1)} = \frac{1}{1+\frac{\ell-1}{p}} \in \operatorname{int} C^p$ for all $i \in \{1,\ldots,\ell\}$ and equivalent to $\min_{y \in F} \varphi_{y^0 - C^p, k^p}(y)$. For each $p \geq p_0$, Theorem 4.2.1(f) yields $y^p \in y^0 - C^p + t_p k^p$.
Choose some $p_2 > \max\{p_0, p_1\}$. By Lemma 7.2.8(a), $C^{p_2} \setminus \{0\} \subseteq \operatorname{int} C^{p_1}$. Because of Lemma 2.2.32, applied with $H = -C^{p_1}$, the set $K := F \cap (y^0 - C^{p_2})$ is bounded. Define $M := \{p \in \mathbb{N}_> \mid p \geq p_2\}$. For each $p \in M$, we get $y^p \in y^0 - C^p + t_p k^p \subseteq y^0 - C^p \subseteq y^0 - C^{p_2}$. Hence, the sequence $(y^p)_{p \in M}$ is bounded. It has a cluster point because of Lemma 2.1.49. Let $y^{00}$ be any cluster point of $(y^p)_{p \in M}$. Then $y^{00} \in F$ since $F$ is closed. To obtain a contradiction, suppose $y^{00} \neq y^0$. Since $y^0 \in \operatorname{Eff}(F)$, we get $y^{00} \notin y^0 - \mathbb{R}^\ell_+$. Lemma 7.2.8(c) implies the existence of $\bar{p} \in M$ with $y^{00} \notin y^0 - C^{\bar{p}}$. Since $C^{\bar{p}}$ is closed, there is a neighborhood $U$ of $y^{00}$ with $U \cap (y^0 - C^{\bar{p}}) = \emptyset$. Hence, $U \cap (y^0 - C^p) = \emptyset$ for all $p \geq \bar{p}$, which contradicts the fact that $y^{00}$ is a cluster point of the points $y^p \in y^0 - C^p$. This yields $y^{00} = y^0$. Consequently, $(y^p)_{p \in M}$ has the unique cluster point $y^0$. By Lemma 2.1.49, $y^0$ is the limit of $(y^p)_{p \in M}$.
Take any $p_3, p \in \mathbb{N}$ with $p > p_3 \geq p_0$. We have $y^p \in y^0 - C^p + t_p k^p \subseteq \{y^0 + t_p k^p\} \cup (y^0 - \operatorname{int} C^{p_3} + t_p k^p)$ because of Lemma 7.2.8(a). Consider first the case $y^p \in y^0 - \operatorname{int} C^{p_3} + t_p k^p$. Because of (4.6.13), $F \cap (y^0 - \operatorname{int} C^{p_3} + t_{p_3} k^{p_3}) = \emptyset$. This implies $y^0 - \operatorname{int} C^{p_3} + t_p k^p \not\subseteq y^0 - \operatorname{int} C^{p_3} + t_{p_3} k^{p_3}$ and hence $t_{p_3} k^{p_3} - t_p k^p \notin \operatorname{int} C^{p_3}$. We conclude that $(t_{p_3} - t_p) k^p - t_{p_3}(k^p - k^{p_3}) \notin \operatorname{int} C^{p_3}$. Since $p > p_3$, we obtain $k_i^p > k_i^{p_3}$ for all $i \in \{1,\ldots,\ell\}$, and thus $k^p - k^{p_3} \in \operatorname{int} \mathbb{R}^\ell_+ \subseteq \operatorname{int} C^{p_3}$. Because of $k^p \in \operatorname{int} C^p \subseteq \operatorname{int} C^{p_3}$, we get $t_{p_3} - t_p \leq 0$, i.e., $t_{p_3} \leq t_p$. In the case where $y^p = y^0 + t_p k^p$ holds, we get $y^p \in y^0 - \mathbb{R}^\ell_+$. Then $y^0 \in \operatorname{Eff}(F)$ implies $y^p = y^0$ and thus $t_p = 0 \geq t_{p_3}$. In both cases, $t_p = t_{p_3}$ implies $t_{p_3} = 0$. This proves the monotonicity statements and the statements related to the optimal value zero.
The sequence of optimal values of the problems $(\mathrm{P}_p)$ with $p \geq p_0$ is bounded above by zero. Thus, it has a limit $\varepsilon \leq 0$. To obtain a contradiction, suppose $\varepsilon < 0$. Take any $p \geq p_0$. Then $k^p = \lambda_p \cdot \mathbf{1}$ with $\lambda_p := \frac{1}{1+\frac{\ell-1}{p}}$. By Proposition 4.3.3, $y^p$ is an optimal solution to the problem $(\mathrm{P}_{F,y^0,C^p,\mathbf{1}})$, and this problem has the optimal value $\tilde{t}_p := \lambda_p t_p$. Since $\lambda_p < 1$, we get $\tilde{t}_p < t_p < \varepsilon$. Since $y^p \in y^0 - C^p + \tilde{t}_p \cdot \mathbf{1} = y^0 - C^p + \varepsilon \cdot \mathbf{1} - (\varepsilon - \tilde{t}_p) \cdot \mathbf{1}$, $\varepsilon > \tilde{t}_p$ and $\mathbf{1} \in \operatorname{int} C^p$, we have $y^p \in y^0 + \varepsilon \cdot \mathbf{1} - \operatorname{int} C^p$. This implies $y^p \in y^0 + \varepsilon \cdot \mathbf{1} - \{y \in \mathbb{R}^\ell \mid \sum_{i=1}^{\ell} y_i > 0\}$ by Lemma 7.2.8(f). Hence, the neighborhood $y^0 + \varepsilon \cdot \mathbf{1} - \{y \in \mathbb{R}^\ell \mid \sum_{i=1}^{\ell} y_i < 0\}$ of $y^0$ does not contain any $y^p$ with $p \geq p_0$. This contradicts the fact that $y^0$ is the limit of the optimal solutions $y^p$. Consequently, $\varepsilon = 0$. □

Now, the results of Choo and Atkins [97] are related to those for problem (P3).

Choo and Atkins introduced the **extended weighted Chebyshev norm** $\|\cdot\|_w^\alpha : \mathbb{R}^\ell \to \mathbb{R}$ for $\alpha \in (-\frac{1}{2\ell}, 0)$ and $w \in \operatorname{int} \mathbb{R}^\ell_+$ by

$$\|y\|_w^\alpha := \max_{i=1,\ldots,\ell} w_i \left|(I_\alpha^{-1} y)_i\right|,$$

where the matrix $I_\alpha$ is given by $(I_\alpha)_{ij} := \begin{cases} 1 & \text{if } i = j, \\ \alpha & \text{otherwise.} \end{cases}$

Assuming $F \subseteq g + \operatorname{int} \mathbb{R}^\ell_+$, they proved that $y^0 \in \operatorname{GEff}(F)$ holds if and only if there exist $w \in \operatorname{int} \mathbb{R}^\ell_+$ and $\alpha \in (-\frac{1}{2\ell}, 0)$ such that $y^0$ solves the problem $\min_{y \in F} \|y - g\|_w^\alpha$.

This statement can be deduced from Proposition 7.3.11 in the following way:

For $\alpha \in (-\frac{1}{2\ell}, 0)$, the matrix $I_\alpha^{-1}$ consists of the positive elements

$$(I_\alpha^{-1})_{ij} = \begin{cases} c & \text{if } i = j, \\ b & \text{otherwise,} \end{cases}$$

with $b := \frac{-\alpha}{-\alpha^2(\ell-1)+\alpha(\ell-2)+1}$ and $c := \frac{\alpha(\ell-2)+1}{-\alpha^2(\ell-1)+\alpha(\ell-2)+1}$. The optimization problem $\min_{y \in F} \|y - g\|_w^\alpha$ is equivalent to (P3) because of $F \subseteq g + \operatorname{int} \mathbb{R}^\ell_+$. This implies the reverse direction of the assertion by Proposition 7.3.11(d). The remaining part of the assertion results from Proposition 7.3.11(f) by choosing $c > \max(c_0, \ell + 2)$ and $\alpha = -\frac{1}{\ell-2+c}$.

Kaliszewski [284, Theorem 2.1.] stated the coincidence of $\operatorname{GEff}(F)$ with the union of all optimal solutions to problems (P3) with $b > 0$ and $c = b + 1$ under the assumptions that $F$ is closed and that $F \subseteq g + \operatorname{int} \mathbb{R}^\ell_+$ holds. This statements also follows from Proposition 7.3.11(d) and (f). Based on this result, Kaliszewski proposed an interactive procedure that uses the relationship between $b$ and the constant $K$ in the definition of Geoffrion's proper efficiency for calculating optimal solutions to problems (P3) in order to get a subset of the Pareto minimal point set.

Section 7.3.3 is taken from [472, Sect. 7.4], with the exception of the compactness statement for the optimal point set in Proposition 7.3.11(g). Its content was published under the additional assumptions $b > 0$ and $c > 0$ in [477], except Proposition 7.3.11(a)–(c).

### *7.3.4 A Scalarization by Helbig*

Helbig [227] studied the scalarization $(\mathrm{P}_{F,a,C(\frac{s}{m}),k})$ with the cones introduced in (7.2.4) for linear vector optimization problems.

Consider the problem

$$\min_{y \in F} \max_{i=1,\ldots,\ell} \left( y_i + \frac{1}{m} s^T y - z_i \right) \qquad \text{(P4)}$$

with $s \in \operatorname{int} \mathbb{R}^\ell_+$, $m \in \mathbb{R}_>$ and $z \in \mathbb{R}^\ell$.

It has the same optimal solutions as $(\mathrm{P}_{F,a,C(\frac{s}{m}),k})$, and hence as $\min_{y\in F} \varphi_{a-C(\frac{s}{m}),k}(y)$, if $a \in \mathbb{R}^\ell$ and $k \in \mathbb{R}^\ell$ solve the following systems of linear equations:

$$a \text{ solves } a_i + \frac{1}{m} s^T a = z_i \text{ for each } i \in \{1, \dots, \ell\} \quad \text{and}$$
$$k \text{ solves } k_i + \frac{1}{m} s^T k = 1 \text{ for each } i \in \{1, \dots, \ell\}.$$

This follows immediately from (7.3.3) since (P4) coincides with (P2) in the case where, for all $i, j \in \{1, \dots, \ell\}$,

$$w^i_j = \begin{cases} 1 + \frac{1}{m} s_j & \text{if } j = i, \\ \frac{1}{m} s_j & \text{if } j \neq i. \end{cases}$$

Consequently, Lemma 7.3.9 contains properties of the cones $C(\frac{s}{m})$. Further properties were given in Lemma 7.2.13.

**Proposition 7.3.12**

(a) *If $y$ is an optimal solution to* (P4), *then* $y \in \mathrm{GEff}(F)$.
(b) *For each* $y \in \mathrm{GEff}(F)$ *and* $s \in \operatorname{int} \mathbb{R}^\ell_+$, *there exists* $m_0 \in \mathbb{R}_>$ *such that $y$ is an optimal solution to* (P4) *with* $m = m_0$ *and* $z_i = y_i + \frac{1}{m_0} s^T y$ *for all* $i \in \{1, \dots, \ell\}$. *Then $y$ is a unique optimal solution to each problem* (P4) *with* $m > m_0$, *the same parameter vector $s$ and* $z_i = y_i + \frac{1}{m} s^T y$ *for all* $i \in \{1, \dots, \ell\}$.
(c) *For each closed set* $F \subset \mathbb{R}^\ell$ *with* $\mathrm{GEff}(F) \neq \emptyset$ *and each* $s \in \operatorname{int} \mathbb{R}^\ell_+$ *and* $z \in \mathbb{R}^\ell$, *there exists* $m_0 \in \mathbb{R}_>$ *such that* (P4) *with* $m = m_0$ *has an optimal solution. Then each problem* (P4) *with* $m > m_0$ *and the same parameter vectors $s$ and $z$ also has an optimal solution. The optimal point set of each of these problems with* $m > m_0$ *is compact.*

***Proof***

(a) results from Proposition 7.3.10(c).
(b) Take any $y^0 \in \mathrm{GEff}(F)$, $s \in \operatorname{int} \mathbb{R}^\ell_+$. By Proposition 7.2.14, there exists $m \in \mathbb{R}_>$ with $y^0 \in \mathrm{Eff}(F, C(\frac{s}{m}))$.
    (i) Define $k = k(m)$ as the solution to the following system: $k_i + \frac{1}{m} s^T k = 1$ for all $i \in \{1, \dots, \ell\}$.
    Corollary 6.2.25 implies for $D := C(\frac{s}{m})$ that $\varphi_{y^0-D,k(m)}(y^0) < \varphi_{y^0-D,k(m)}(y)$ for all $y \in F \setminus \{y^0\}$. By Theorem 5.1.3(b), $y^0$ is the unique optimal solution to $(\mathrm{P}_{F,y^0,D,k(m)})$. Thus, $y^0$ is the unique optimal solution to (P4) with $z_i = y^0_i + \frac{1}{m} s^T y^0$ for all $i \in \{1, \dots, \ell\}$.
    (ii) Assume now that $y^0$ is an optimal solution to (P4) with $m = m_0 \in \mathbb{R}_>$ and $z_i = y^0_i + \frac{1}{m_0} s^T y^0$ for all $i \in \{1, \dots, \ell\}$. Then $y^0$ is an optimal solution to $(\mathrm{P}_{F,y^0,C(\frac{s}{m_0}),k(m_0)})$. Corollary 6.2.23 implies together with Theorem 5.1.3(b)

that $y^0 \in \mathrm{WEff}(F, C(\frac{s}{m_0}))$ holds, which results in $y^0 \in \mathrm{Eff}(F, C(\frac{s}{m}))$ for all $m > m_0$ by Lemma 7.2.13(b). The assertion follows from (i).

(c) By Proposition 7.2.21, there exist $u \in F$ and $m_1 \in \mathbb{R}_>$ with $(F - u) \cap (-C(\frac{s}{m_1})) \subseteq \{0\}$.

(i) Take any $m > m_1$. Lemma 7.2.13(b) yields $C(\frac{s}{m}) \setminus \{0\} \subset \mathrm{int}\, C(\frac{s}{m_1})$. In Proposition 5.2.20, choose $z$=0, $C = -C(\frac{s}{m_1})$ and $H = C(\frac{s}{m})$. This implies the existence of an optimal solution to each problem $(\mathrm{P}_{F,a,C(\frac{s}{m}),k})$ with $a \in \mathbb{R}^\ell$ and $k \in \mathrm{int}\, C(\frac{s}{m})$ and the compactness of the optimal point set. Choose $a$ as a solution to the following system: $a_i + \frac{1}{m} s^T a = z_i$ for all $i \in \{1, \ldots, \ell\}$. Then the optimal solutions to $(\mathrm{P}_{F,a,C(\frac{s}{m}),k})$ are those of (P4).

(ii) Assume now that (P4) has an optimal solution $y$ for $s$, $z$ and $m = m_0 \in \mathbb{R}_>$. From Proposition 7.3.10(a), we get $y \in \mathrm{WEff}(F, C(\frac{s}{m_0}))$, which implies $(F - y) \cap -\mathrm{int}\, C(\frac{s}{m_0}) = \emptyset$. Applying (i) to $m_0$ instead of $m_1$ yields the existence of optimal solutions to each problem (P4) with $m > m_0$ and the compactness of the optimal point sets. □

The results of this section can be found in [472, Sect. 7.3], except the uniqueness of optimal solutions in Proposition 7.3.12(b) and the compactness statement in Proposition 7.3.12(c).

### 7.3.5 A Scalarization by Dubov

Consider the scalarization

$$\min_{y \in F} \max_{i=1,\ldots,\ell} w_i \left( y_i + \varepsilon \sum_{\substack{j=1 \\ j \neq i}}^{\ell} y_j - z_i \right) \tag{P5}$$

with $w \in \mathrm{int}\, \mathbb{R}^\ell_+$, $\varepsilon \in \mathbb{R}_+ \setminus \{1\}$, $z \in \mathbb{R}^\ell$.

Problem (P5) is equivalent to $(\mathrm{P}_{F,a,C,k})$, and hence to $\min_{y \in F} \varphi_{a-C,k}(y)$, with

$$a_i = s_1 z_i + s_2 \sum_{\substack{j=1 \\ j \neq i}}^{\ell} z_j \quad \text{for all } i \in \{1, \ldots, \ell\},$$

$$C = \left\{ y \in \mathbb{R}^\ell \mid \forall i \in \{1, \ldots, \ell\}\colon y_i + \varepsilon \sum_{\substack{j=1 \\ j \neq i}}^{\ell} y_j \geq 0 \right\},$$

$$k_i = \frac{s_1}{w_i} + s_2 \sum_{\substack{j=1 \\ j \neq i}}^{\ell} \frac{1}{w_j} \quad \text{for all } i \in \{1, \ldots, \ell\},$$

where $s_1 := \frac{1+(\ell-2)\varepsilon}{1+(\ell-2)\varepsilon-(\ell-1)\varepsilon^2}$, $s_2 := \frac{-\varepsilon}{1+(\ell-2)\varepsilon-(\ell-1)\varepsilon^2}$.

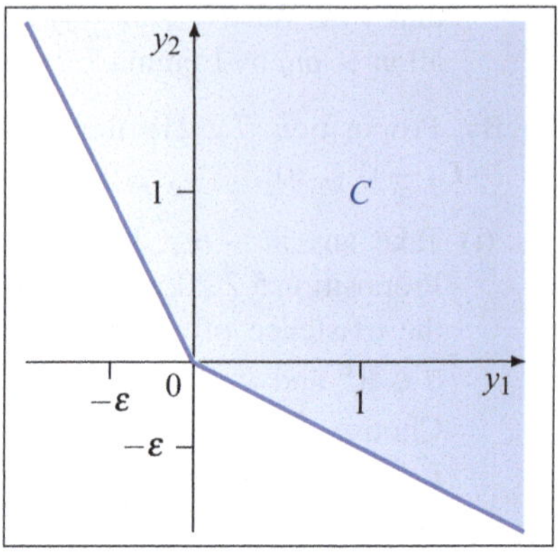


**Fig. 7.10** Set $C$ for problem (P5) with $\varepsilon < 1$ in $\mathbb{R}^2$. Adapted from [472, p. 119]

Problem (P5) is (P3) with $c = 1, b = \varepsilon$ and $z_i = g_i + b \sum_{\substack{j=1 \\ j \neq i}}^{\ell} g_j$ for all $i \in \{1, \ldots, \ell\}$, and thus also (P2) with $r = \ell$, $z_i = w_i\, g_i + \varepsilon\, w_i \sum_{\substack{j=1 \\ j \neq i}}^{\ell} g_j$ for all $i \in \{1, \ldots, \ell\}$ and, for all $i, j \in \{1, \ldots, \ell\}$,

$$w_j^i = \begin{cases} w_i & \text{if } j = i, \\ \varepsilon\, w_i & \text{if } j \neq i. \end{cases}$$

Problem (P1) is the special case of problem (P5) with $\varepsilon = 0$.

For $\varepsilon < 1$, the boundary of the polyhedral cone $C$ in $\mathbb{R}^2$ consists of two rays beginning at the origin, the first one with direction $(1, -\varepsilon)^T$, the second with direction $(-\varepsilon, 1)^T$ (see Fig. 7.10).

**Proposition 7.3.13**

(a) *If* $y$ *is an optimal solution to* (P5), *then* $y \in \mathrm{WEff}(F)$.
(b) *If* $y$ *is a unique optimal solution to* (P5), *then* $y \in \mathrm{Eff}(F)$.
(c) *For each* $y \in \mathrm{WEff}(F)$, *there exist* $w \in \operatorname{int} \mathbb{R}^{\ell}_{+}$ *and* $z \in \mathbb{R}^{\ell}$ *such that* $y$ *is an optimal solution to* (P5) *with* $\varepsilon = 0$.
*For each* $y \in \mathrm{Eff}(F)$, *there exist* $w \in \operatorname{int} \mathbb{R}^{\ell}_{+}$ *and* $z \in \mathbb{R}^{\ell}$ *such that* $y$ *is the unique optimal solution to* (P5) *with* $\varepsilon = 0$.
(d) $y \in \mathrm{GEff}(F)$ *if and only if* $y$ *is an optimal solution to some problem* (P5) *with* $\varepsilon > 0$.
(e) *For each* $y^0 \in \mathrm{GEff}(F)$, *there exists* $\varepsilon_0 \in (0, 1)$ *such that* $y^0$ *solves the optimization problem* (P5) *with* $\varepsilon = \varepsilon_0$, $w = \mathbf{1}$ *and* $z_i = y_i^0 + \varepsilon_0 \sum_{\substack{j=1 \\ j \neq i}}^{\ell} y_j^0$ *for all* $i \in \{1, \ldots, \ell\}$, *where the optimal value is zero. Then* $y^0$ *is the unique optimal solution*

*to each problem* (P5) *with* $\varepsilon \in (0, \varepsilon_0)$, $w = \mathbf{1}$ *and* $z_i = y_i^0 + \varepsilon \sum_{\substack{j=1 \\ j \neq i}}^{\ell} y_j^0$ *for all* $i \in \{1, \ldots, \ell\}$, *and the optimal value of each of these problems is zero.*

(f) *If* $y^0 \in \mathrm{GEff}(F)$ *and* $F \subseteq g + (\mathbb{R}_+^\ell \setminus \{0\})$ *for some* $g \in \mathbb{R}^\ell$, *then there exists* $\varepsilon_0 \in (0, 1)$ *such that* $y^0$ *solves the optimization problem* (P5) *with* $\varepsilon = \varepsilon_0$, $z_i = g_i + \varepsilon_0 \sum_{\substack{j=1 \\ j \neq i}}^{\ell} g_j$ *and* $w_i = \frac{1}{y_i^0 + \varepsilon_0 \sum_{\substack{j=1 \\ j \neq i}}^{\ell} y_j^0 - z_i}$ *for all* $i \in \{1, \ldots, \ell\}$, *where the optimal value is* 1. *Then* $y^0$ *is the unique optimal solution to each problem* (P5) *with* $\varepsilon \in (0, \varepsilon_0)$, $z_i = g_i + \varepsilon \sum_{\substack{j=1 \\ j \neq i}}^{\ell} g_j$ *and* $w_i = \frac{1}{y_i^0 + \varepsilon \sum_{\substack{j=1 \\ j \neq i}}^{\ell} y_j^0 - z_i}$ *for all* $i \in \{1, \ldots, \ell\}$, *with the optimal value* 1.

(g) *For each closed set* $F \subset \mathbb{R}^\ell$ *with* $\mathrm{GEff}(F) \neq \emptyset$ *and each* $w \in \mathrm{int}\, \mathbb{R}_+^\ell$ *and* $z \in \mathbb{R}^\ell$, *there exists a real number* $\varepsilon_0 \in (0, 1)$ *such that* (P5) *with* $\varepsilon = \varepsilon_0$ *has an optimal solution. Each problem* (P5) *with the same parameters* $w$ *and* $z$ *and with* $\varepsilon \in (0, \varepsilon_0)$ *also has an optimal solution, and the optimal point set of each of these problems with* $\varepsilon \in (0, \varepsilon_0)$ *is compact.*

***Proof*** (a), (b), (c) and the reverse direction of the equivalence in (d) follow from Proposition 7.3.11(a)–(d). The other direction of (d) results from (e).

(e) follows from Proposition 7.3.11(e) by choosing, for each $c \geq c_0$, $\varepsilon = \frac{1}{c}$, $w_i = c$ and

$$z_i = y_i^0 + \varepsilon \sum_{\substack{j=1 \\ j \neq i}}^{\ell} y_j^0 \quad \text{for all } i \in \{1, \ldots, \ell\}.$$

Since all $w_i$ have the same value, they can be replaced by 1 without any changes in the optimal solution set and in the optimal value of the problems.

(f) results from Proposition 7.3.11(f) by choosing $\varepsilon_0 = \frac{1}{c_0}$, $\varepsilon = \frac{1}{c}$ and by multiplying each $w_i$ from Proposition 7.3.11(f) with $c$ for each $c \geq c_0$.

(g) follows from Proposition 7.3.11(g). □

Proposition 7.3.11(h) contains a statement about the approximation of $\mathrm{Eff}(F)$ by optimal solutions to problems (P5).

Dubov [120] introduced (P5) with $\varepsilon \in (0, 1)$ and $z = 0$ and proved the Pareto minimality of its solutions for $\varepsilon > 0$. The results of Sect. 7.3.5 are taken from [472, Sect. 7.5], except the compactness in Proposition 7.3.13(g). Its content was published under the additional assumption $\varepsilon > 0$ in [477], with the exception of Proposition 7.3.13(a)–(c).

### 7.3.6 Wierzbicki's Reference Point Projection

Consider the problem

$$\min_{y\in F}\ \max_{i=1,\dots,\ell}\left(w_i(y_i-g_i)+\varepsilon\sum_{j=1}^{\ell}y_j\right) \tag{P6}$$

with $w \in \operatorname{int}\mathbb{R}^{\ell}_{+}$, $\varepsilon \in \mathbb{R}_{>}$, $g \in \mathbb{R}^{\ell}$.

Problem (P6) is equivalent to $(\mathrm{P}_{F,a,C,k})$ , and hence to $\min_{y\in F}\varphi_{a-C,k}(y)$, with

$$a_i = g_i - \frac{\varepsilon\sum\limits_{j=1}^{\ell} g_j}{w_i(1+\varepsilon\sum\limits_{j=1}^{\ell}\frac{1}{w_j})} \quad \text{for all } i \in \{1,\dots,\ell\},$$

$$C = \left\{ y \in \mathbb{R}^{\ell} \mid \forall i \in \{1,\dots,\ell\}\colon w_i\, y_i + \varepsilon\sum_{j=1}^{\ell} y_j \geq 0 \right\},$$

$$k_i = \frac{1}{w_i(1+\varepsilon\sum\limits_{j=1}^{\ell}\frac{1}{w_j})} \quad \text{for all } i \in \{1,\dots,\ell\}.$$

Obviously, $k \in \operatorname{int} C$. The set $C$ coincides with the cone $C_{w,\varepsilon}$ introduced in (7.2.6), the properties of which were investigated in Lemma 7.2.19.

In $\mathbb{R}^2$, the boundary of the polyhedral cone $C$ consists of two rays beginning at the origin, the first one with direction $(-\varepsilon,\ w_1+\varepsilon)^{\mathrm{T}}$, the second with direction $(w_2+\varepsilon,\ -\varepsilon)^{\mathrm{T}}$ (see Fig. 7.11).

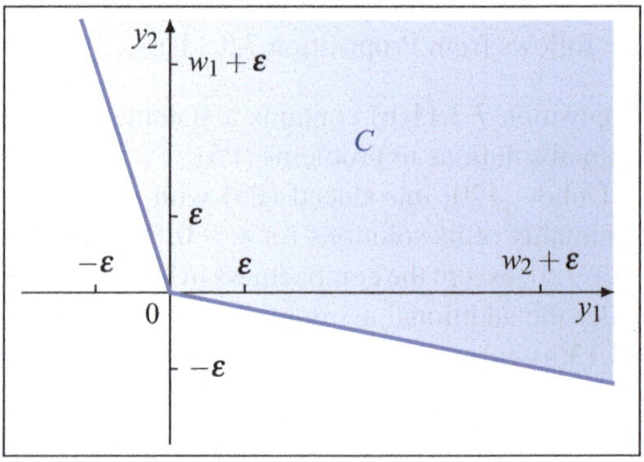


**Fig. 7.11** Set $C$ for problem (P6) in $\mathbb{R}^2$. Adapted from [472, p. 121]

Problem (P6) turns out to be the special case of (P2) with $r = \ell$ and, for all $i, j \in \{1, \dots, \ell\}$, $z_i = w_i\, g_i$ and $w^i_j = \begin{cases} w_i + \varepsilon & \text{if } j = i, \\ \varepsilon & \text{if } j \neq i. \end{cases}$

**Proposition 7.3.14**

(a) *If $y$ is an optimal solution to* (P6), *then* $y \in \mathrm{GEff}(F)$.

(b) *For each $y^0 \in \mathrm{GEff}(F)$, there exists $w_0 \in \mathbb{R}_>$ such that $y^0$ solves the optimization problem* (P6) *with $\varepsilon = 1$, $w_i = w_0$ and $g_i = y^0_i + \frac{1}{w_0} \sum\limits_{j=1}^{\ell} y^0_j$ for all $i \in \{1, \dots, \ell\}$, where the optimal value is zero. Then for each $w > w_0$, $y^0$ is the unique optimal solution to each problem* (P6) *with $\varepsilon = 1$, $w_i = w$ and $g_i = y^0_i + \frac{1}{w} \sum\limits_{j=1}^{\ell} y^0_j$ for all $i \in \{1, \dots, \ell\}$, and the optimal value of each of these problems is zero.*

(c) *For each $y^0 \in \mathrm{GEff}(F)$, there exists $\varepsilon_0 \in (0, 1)$ such that $y^0$ solves the optimization problem* (P6) *with $\varepsilon = \varepsilon_0$, $w_i = 1$ and $g_i = y^0_i + \varepsilon_0 \sum\limits_{j=1}^{\ell} y^0_j$ for all $i \in \{1, \dots, \ell\}$, where the optimal value is zero. Then for each $\varepsilon \in (0, \varepsilon_0)$, $y^0$ is the unique optimal solution to each problem* (P6) *with $w_i = 1$ and $g_i = y^0_i + \varepsilon \sum\limits_{j=1}^{\ell} y^0_j$ for all $i \in \{1, \dots, \ell\}$, and the optimal value of each of these problems is zero.*

(d) *For each closed set $F \subset \mathbb{R}^\ell$ with* $\mathrm{GEff}(F) \neq \emptyset$ *and each $w \in \operatorname{int} \mathbb{R}^\ell_+$, there exists $\varepsilon_0 \in \mathbb{R}_>$ such that each problem* (P6) *with $g \in \mathbb{R}^\ell$ and $\varepsilon = \varepsilon_0$ has an optimal solution. Each problem* (P6) *with the same parameters $w$ and $g$ and with $\varepsilon \in (0, \varepsilon_0)$ also has an optimal solution, and the optimal point set of each of these problems with $\varepsilon \in (0, \varepsilon_0)$ is compact.*

***Proof***

(a) follows from Proposition 7.3.10(c).

(b) results from Proposition 7.3.11(e) by choosing $w_0 = c_0 - 1$ and $w = c - 1$.

(c) follows from (b) when dividing the objective function of (P6) in (b) by $w$.

(d) (i) By Proposition 7.2.21, there exist $u \in F$ and $p > 1$ with $(F - u) \cap -C^p \subseteq \{0\}$. Consider $\varepsilon := \frac{1}{2} \min\limits_{i=1,\dots,\ell} \frac{w_i}{p-1}$. Take any $y \in C_{w,\varepsilon} \setminus \{0\}$, where $C_{w,\varepsilon}$ is defined by (7.2.6). Lemma 7.2.19 yields $\sum\limits_{j=1}^{\ell} y_j > 0$. If $y_i \geq 0$, then $p y_i + \sum\limits_{\substack{j=1 \\ j \neq i}}^{\ell} y_j = (p-1) y_i + \sum\limits_{j=1}^{\ell} y_j > 0$. If $y_i < 0$, then $p y_i + \sum\limits_{\substack{j=1 \\ j \neq i}}^{\ell} y_j = (p-1) y_i + \sum\limits_{j=1}^{\ell} y_j > \frac{w_i}{\varepsilon} y_i + \sum\limits_{j=1}^{\ell} y_j \geq 0$. Consequently, $C_{w,\varepsilon} \setminus \{0\} \subseteq \operatorname{int} C^p$ for $C^p$ given by (7.2.2). Apply Proposition 5.2.20 with $C = -C^p$, $z = 0$ and $H = C_{w,\varepsilon}$. We conclude that the optimal point set of each problem $(\mathrm{P}_{F,a,C_{w,\varepsilon},k})$ with $a \in \mathbb{R}^\ell$ and $k \in \operatorname{int} C_{w,\varepsilon}$ is nonempty and compact. The relationship between (P6) and $(\mathrm{P}_{F,a,C_{w,\varepsilon},k})$ implies the first part of the assertion.

(ii) Assume now that (P6) with $w \in \operatorname{int} \mathbb{R}^{\ell}_{+}$, $g \in \mathbb{R}^{\ell}$ and $\varepsilon = \varepsilon_0$ has an optimal solution $y$. Proposition 7.3.10(a) yields $y \in \mathrm{WEff}(F, C_{w,\varepsilon_0})$. Take any $\varepsilon \in (0, \varepsilon_0)$. By Lemma 7.2.19, $C_{w,\varepsilon} \setminus \{0\} \subseteq \operatorname{int} C_{w,\varepsilon_0}$. Apply Proposition 5.2.20 with $u = y$, $C = -C_{w,\varepsilon_0}$, $z = 0$ and $H = C_{w,\varepsilon}$. We deduce that the optimal point set of each problem $(\mathrm{P}_{F,a,C_{w,\varepsilon},k})$ with $a \in \mathbb{R}^{\ell}$ and $k \in \operatorname{int} C_{w,\varepsilon}$ is nonempty and compact. □

Proposition 7.3.11(h) implies that each Pareto minimal point $y^0$ of a closed set $F$ with $\mathrm{GEff}(F) \neq \emptyset$ is the limit of arbitrary optimal solutions to the problems (P6) with $\varepsilon = \frac{1}{p}$, $w_i = 1 - \frac{1}{p}$ and $g_i = \frac{1}{p-1}\left(py_i^0 + \sum\limits_{\substack{j=1\\ j\neq i}}^{\ell} y_j^0\right)$ for all $i \in \{1, \ldots, \ell\}$, $p \in \mathbb{N}_{>} \setminus \{1\}$, for $p \to \infty$.

The scalarization (P6) has its origin in Wierzbicki's idea to project the reference point $g$ onto the Pareto minimal point set [495]. It was used by Steuer [426, p. 404] for extending the visual interactive algorithm by Korhonen and Laakso [306]. Vlacic et al. [457] discussed its application to group decisions. Originally, the algorithm by Korhonen and Laakso was based on the scalarization (P1), which would be (P6) with $\varepsilon = 0$.

The results in this section related to problem (P6) are contained in [472, Sect. 7.6], except the compactness statement in Proposition 7.3.14(d).

In the decision support system DIDASS [197], the following scalarization is used, which is a modification of (P6) by Wierzbicki [496].

$$\min_{y\in F}\ \max_{i=1,\ldots,\ell}\left(w_i(y_i - g_i) + \varepsilon\sum_{j=1}^{\ell} w_j(y_j - g_j)\right) \tag{P6$'$}$$

with $w \in \operatorname{int} \mathbb{R}^{\ell}_{+}$, $\varepsilon \in \mathbb{R}_{>}$, $g \in \mathbb{R}^{\ell}$.

Problem (P6′) is equivalent to $(\mathrm{P}_{F,a,C,k})$, and hence to $\min_{y\in F} \varphi_{a-C,k}(y)$, with

$$\begin{aligned} a &= g, \\ C &= \left\{ y \in \mathbb{R}^{\ell} \mid \forall i \in \{1, \ldots, \ell\} \colon w_i\, y_i + \varepsilon \sum_{j=1}^{\ell} w_j\, y_j \geq 0 \right\}, \\ k_i &= \frac{1}{w_i\,(1 + \ell\varepsilon)} \quad \text{for all } i \in \{1, \ldots, \ell\}. \end{aligned} \tag{7.3.6}$$

$C$ coincides with the cone $C_w(\varepsilon)$ introduced in (7.2.5), the properties of which were investigated in Lemma 7.2.13.

Problem (P6′) is a special case of (P2), with $r = \ell$, $z_i = w_i\, g_i + \varepsilon \sum\limits_{j=1}^{\ell} w_j g_j$ for all $i \in \{1, \ldots, \ell\}$, and, for all $i, j \in \{1, \ldots, \ell\}$,

$$w^i_j = \begin{cases} (1+\varepsilon)w_i & \text{if } j = i, \\ \varepsilon w_j & \text{if } j \neq i. \end{cases}$$

**Proposition 7.3.15**

(a) *If* $y$ *is an optimal solution to* (P6′)*, then* $y \in \mathrm{GEff}(F)$.

(b) *For each* $y^0 \in \mathrm{GEff}(F)$*, there exists* $\varepsilon_0 \in (0,1)$ *such that* $y^0$ *solves the optimization problem* (P6′) *with* $\varepsilon = \varepsilon_0$*,* $w_i = \frac{1}{\varepsilon_0}$ *for all* $i \in \{1,\dots,\ell\}$ *and* $g = y^0$*, where the optimal value is zero. Then for each* $\varepsilon \in (0,\varepsilon_0)$*,* $y^0$ *is the unique optimal solution to each problem* (P6′) *with* $w_i = \frac{1}{\varepsilon}$ *for all* $i \in \{1,\dots,\ell\}$ *and* $g = y^0$*, and the optimal value of each of these problems is zero.*

(c) *Assume* $F \subseteq g + \mathrm{int}\,\mathbb{R}^\ell_+$ *and* $y^0 \in \mathrm{GEff}(F)$*. Then there exists* $\varepsilon_0 > 0$ *such that* $y^0$ *solves* (P6′) *with* $\varepsilon = \varepsilon_0$ *and* $w_i = \frac{1}{y^0_i - g_i}$ *for all* $i \in \{1,\dots,\ell\}$*. For each* $\varepsilon \in (0,\varepsilon_0)$*,* $y^0$ *is the unique optimal solution to* (P6′) *with* $w_i = \frac{1}{y^0_i - g_i}$ *for all* $i \in \{1,\dots,\ell\}$.

(d) *For each closed set* $F \subset \mathbb{R}^\ell$ *with* $\mathrm{GEff}(F) \neq \emptyset$ *and each* $w \in \mathrm{int}\,\mathbb{R}^\ell_+$*, there exists* $\varepsilon_0 \in \mathbb{R}_>$ *such that* (P6′) *with* $g \in \mathbb{R}^\ell$ *and* $\varepsilon = \varepsilon_0$ *has an optimal solution. Each problem* (P6′) *with the same parameters* $w$ *and* $g$ *and with* $\varepsilon \in (0,\varepsilon_0)$ *also has an optimal solution, and the optimal point set of each of these problems with* $\varepsilon \in (0,\varepsilon_0)$ *is compact.*

***Proof***

(a) follows from Proposition 7.3.10(c).

(b) results from Proposition 7.3.11(e) if $c_0$ is chosen sufficiently large, by taking $\varepsilon_0 = \frac{1}{c_0 - 1}$.

(c) Take any $y^0 \in \mathrm{GEff}(F)$ and $w_i = \frac{1}{y^0_i - g_i}$ for all $i \in \{1,\dots,\ell\}$. Proposition 7.2.18 implies the existence of $\varepsilon > 0$ with $y^0 \in \mathrm{Eff}(F, C_w(\varepsilon))$.

   (i) If we define $C$ by (7.3.6), then $C = C_w(\varepsilon)$ and $y^0 \in \mathrm{Eff}(F, C)$. By Lemma 7.2.17, $\mathbb{R}^\ell_+ \subseteq C$. Thus, $F \subseteq g + \mathrm{int}\,\mathbb{R}^\ell_+ \subseteq g + \mathrm{int}\,C$. Because of Theorem 6.2.37 and Theorem 5.1.3(b), $y^0$ is the unique optimal solution to $(\mathrm{P}_{F,g,C,y^0-g})$ and hence also the unique optimal solution to $(\mathrm{P}_{F,g,C,\lambda(y^0-g)})$ with $\lambda := \frac{1}{1+\ell\varepsilon}$. Problem $(\mathrm{P}_{F,g,C,\lambda(y^0-g)})$ is equivalent to (P6′).

   (ii) Assume now that $y^0$ solves the problem (P6′) with $\varepsilon = \varepsilon_0$ and $w_i = \frac{1}{y^0_i - g_i}$ for all $i \in \{1,\dots,\ell\}$, i.e., the problem $(\mathrm{P}_{F,g,C_w(\varepsilon_0),\lambda(y^0-g)})$ with $\lambda := \frac{1}{1+\ell\varepsilon_0}$. From Proposition 7.3.10(a), we get $y^0 \in \mathrm{WEff}(F, C_w(\varepsilon_0))$. Take any $\varepsilon \in (0,\varepsilon_0)$. By Lemma 7.2.17, $C_w(\varepsilon) \setminus \{0\} \subseteq \mathrm{int}\,C_w(\varepsilon_0)$. Thus, $y^0 \in \mathrm{Eff}(F, C_w(\varepsilon))$. The assertion follows by (i).

(d) (i) Take any $y^0 \in \mathrm{GEff}(F)$ and $w \in \mathrm{int}\,\mathbb{R}^\ell_+$. Proposition 7.2.18 implies the existence of $\varepsilon_1 > 0$ with $(F - y^0) \cap (-C_w(\varepsilon_1)) = \{0\}$. Take any $\varepsilon \in (0,\varepsilon_1)$. In Proposition 5.2.20, choose $C = -C_w(\varepsilon_1)$, $H = C_w(\varepsilon)$, $z = 0$ and $u = y^0$. We obtain that the optimal solution set of each problem $(\mathrm{P}_{F,a,C_w(\varepsilon),k})$ with $a \in \mathbb{R}^\ell$ and $k \in \mathrm{int}\,C_w(\varepsilon)$ is nonempty and compact. The first part of the assertion in (d) follows from the relationship between $(P_{F,a,C_w(\varepsilon),k})$ and (P6′), which we explained in front of Proposition 7.3.15.

(ii) Assume now that (P6′) with $w \in \operatorname{int} \mathbb{R}^\ell_+$, $g \in \mathbb{R}^\ell$ and $\varepsilon_0 \in \mathbb{R}_>$ has an optimal solution $y$. By Proposition 7.3.10(a), we have $y \in \operatorname{WEff}(F, C_w(\varepsilon_0))$. Again, Proposition 5.2.20 yields that the optimal point set of each problem (P6′) with $\varepsilon \in (0, \varepsilon_0)$, $w$ and $g$ is nonempty and compact. □

Under the assumption $F \subseteq g + \operatorname{int} \mathbb{R}^\ell_+$, it was previously pointed out in [115] that $y^0 \in \operatorname{GEff}(F)$ holds if and only if $y^0$ solves (P6′) for some sufficiently small $\varepsilon > 0$ and some $w \in \operatorname{int} \mathbb{R}^\ell_+$.

By Proposition 7.3.11(h), each Pareto minimal point $y^0$ of a closed set $F$ with $\operatorname{GEff}(F) \neq \emptyset$ is the limit of arbitrary optimal solutions to the problems (P6′) with $\varepsilon = \frac{1}{p-1}$, $w_i = 1 - \frac{1}{p}$ for all $i \in \{1, \ldots, \ell\}$, $g = y^0$, $p \in \mathbb{N}_> \setminus \{1\}$, for $p \to \infty$.

We now turn to another modification of problem (P6),

$$\min_{y \in F} \max_{i=1,\ldots,\ell} \left( w_i(y_i - g_i) + \varepsilon \sum_{j=1}^{\ell} (y_j - g_j) \right) \tag{P6''}$$

with $w \in \operatorname{int} \mathbb{R}^\ell_+$, $\varepsilon \in \mathbb{R}_>$, $g \in \mathbb{R}^\ell$.

Problem (P6″) is equivalent to $(P_{F,a,C,k})$, and hence to $\min_{y \in F} \varphi_{a-C,k}(y)$, with

$$\begin{aligned} a &= g, \\ C &= \left\{ y \in \mathbb{R}^\ell \mid \forall i \in \{1, \ldots, \ell\}\colon w_i\, y_i + \varepsilon \sum_{j=1}^{\ell} y_j \geq 0 \right\}, \\ k_i &= \frac{1}{w_i(1 + \varepsilon \sum\limits_{j=1}^{\ell} \frac{1}{w_j})} \quad \text{for all } i \in \{1, \ldots, \ell\}. \end{aligned} \tag{7.3.7}$$

We have $k \in \operatorname{int} C$, and $C$ coincides with the cone $C_{w,\varepsilon}$ introduced in (7.2.6).

Problem (P6″) with parameters $w$, $\varepsilon$ and $g$ coincides with problem (P6) having parameters $w$, $\varepsilon$ and $\bar{g}$ if $\bar{g}_i = g_i + \frac{\varepsilon}{w_i} \sum\limits_{j=1}^{\ell} g_j$ for all $i \in \{1, \ldots, \ell\}$.

Problem (P6″) is a special case of (P2), with $r = \ell$, $z_i = w_i\, g_i + \varepsilon \sum\limits_{j=1}^{\ell} g_j$ for all $i \in \{1, \ldots, \ell\}$, and, for all $i, j \in \{1, \ldots, \ell\}$,

$$w^i_j = \begin{cases} w_i + \varepsilon & \text{if } j = i, \\ \varepsilon & \text{if } j \neq i. \end{cases}$$

**Proposition 7.3.16**

(a) *If* $y$ *is an optimal solution to* (P6″), *then* $y \in \operatorname{GEff}(F)$.

(b) *For each* $y^0 \in \mathrm{GEff}(F)$, *there exists* $w_0 > 1$ *such that* $y^0$ *solves the optimization problem* (P6$''$) *with* $\varepsilon = 1$, $w_i = w_0$ *for all* $i \in \{1, \ldots, \ell\}$ *and* $g = y^0$, *where the optimal value is zero. Then for each* $w > w_0$, $y^0$ *is the unique solution to each problem* (P6$''$) *with* $\varepsilon = 1$, $w_i = w$ *for all* $i \in \{1, \ldots, \ell\}$ *and* $g = y^0$, *and the optimal value of each of these problems is zero.*

(c) *For each* $y^0 \in \mathrm{GEff}(F)$, *there exists* $\varepsilon_0 \in (0, 1)$ *such that* $y^0$ *solves the optimization problem* (P6$''$) *with* $\varepsilon = \varepsilon_0$, $w_i = 1$ *for all* $i \in \{1, \ldots, \ell\}$ *and* $g = y^0$, *where the optimal value is zero. Then for each* $\varepsilon \in (0, \varepsilon_0)$, $y^0$ *is the unique optimal solution to each problem* (P6$''$) *with* $w_i = 1$ *for all* $i \in \{1, \ldots, \ell\}$ *and* $g = y^0$, *and the optimal value of each of these problems is zero.*

(d) *Assume* $F \subseteq g + \mathrm{int}\,\mathbb{R}^\ell_+$ *and* $y^0 \in \mathrm{GEff}(F)$. *Then there exists* $\varepsilon_0 > 0$ *such that* $y^0$ *solves* (P6$''$) *with* $\varepsilon = \varepsilon_0$ *and* $w_i = \frac{1}{y_i^0 - g_i}$ *for all* $i \in \{1, \ldots, \ell\}$. *For each* $\varepsilon \in (0, \varepsilon_0)$, $y^0$ *is the unique optimal solution to* (P6$''$) *with* $w_i = \frac{1}{y_i^0 - g_i}$ *for all* $i \in \{1, \ldots, \ell\}$.

(e) *For each closed set* $F \subset \mathbb{R}^\ell$ *with* $\mathrm{GEff}(F) \neq \emptyset$ *and each* $w \in \mathrm{int}\,\mathbb{R}^\ell_+$, *there exists* $\varepsilon_0 \in \mathbb{R}_>$ *such that each problem* (P6$''$) *with* $g \in \mathbb{R}^\ell$ *and* $\varepsilon = \varepsilon_0$ *has an optimal solution. Each problem* (P6$''$) *with the same parameters* $w$ *and* $g$ *and with* $\varepsilon \in (0, \varepsilon_0)$ *also has an optimal solution, and the optimal point set of each of these problems with* $\varepsilon \in (0, \varepsilon_0)$ *is compact.*

***Proof***

(a) –(c) follow immediately from Proposition 7.3.14.

(d) Take any $y^0 \in \mathrm{GEff}(F)$ and $w_i = \frac{1}{y_i^0 - g_i}$ for all $i \in \{1, \ldots, \ell\}$. Proposition 7.2.20 implies the existence of $\varepsilon > 0$ with $y^0 \in \mathrm{Eff}(F, C_{w,\varepsilon})$.

 (i) If we define $C$ by (7.3.7), then $C = C_{w,\varepsilon}$ and $y^0 \in \mathrm{Eff}(F, C)$. By Lemma 7.2.19, $\mathbb{R}^\ell_+ \subseteq C$. Thus, $F \subseteq g + \mathrm{int}\,\mathbb{R}^\ell_+ \subseteq g + \mathrm{int}\,C$. Because of Theorem 6.2.37 and Theorem 5.1.3(b), $y^0$ is the unique optimal solution to $(\mathrm{P}_{F,g,C,y^0-g})$ and hence also the unique optimal solution to $(\mathrm{P}_{F,g,C,\lambda(y^0-g)})$ with $\lambda := \frac{1}{1+\varepsilon \sum_{j=1}^{\ell} \frac{1}{w_j}}$. $(\mathrm{P}_{F,g,C,\lambda(y^0-g)})$ is equivalent to (P6$''$).

 (ii) Assume now that $y^0$ solves the problem (P6$''$) with $\varepsilon = \varepsilon_0$ and $w_i = \frac{1}{y_i^0 - g_i}$ for all $i \in \{1, \ldots, \ell\}$, i.e., the problem $(\mathrm{P}_{F,g,C_{w,\varepsilon_0},\lambda(y^0-g)})$ with $\lambda := \frac{1}{1+\varepsilon_0 \sum_{j=1}^{\ell} \frac{1}{w_j}}$. From Proposition 7.3.10(a), we get $y^0 \in \mathrm{WEff}(F, C_{w,\varepsilon_0})$. Take any $\varepsilon \in (0, \varepsilon_0)$. By Lemma 7.2.19, $C_{w,\varepsilon} \setminus \{0\} \subseteq \mathrm{int}\,C_{w,\varepsilon_0}$. Thus, $y^0 \in \mathrm{Eff}(F, C_{w,\varepsilon})$. The assertion follows from (i).

(e) Follow the lines in the proof of Proposition 7.3.14(d) and replace (P6) by (P6$''$). □

By Proposition 7.3.11(h), each Pareto minimal point $y^0$ of a closed set $F$ with $\mathrm{GEff}(F) \neq \emptyset$ is the limit of arbitrary optimal solutions to the problems (P6$''$) with $\varepsilon = \frac{1}{p}$, $w_i = 1 - \frac{1}{p}$ for all $i \in \{1, \ldots, \ell\}$, $g = y^0$, $p \in \mathbb{N}_> \setminus \{1\}$, for $p \to \infty$.

Further statements related to problem (P6″) for sets $F$ which are closed and bounded below were given by Kaliszewski ([284, Theorem 2.2], [285, Theorem 4.3]) and—under more restrictive assumptions for $F$—by Steuer and Choo [427], who used it in an interactive procedure.

### 7.3.7 The Hurwicz Rule

The Hurwicz Rule refers to a decision making problem under uncertainty, where different objective functions stand for different scenarios. It was introduced in [249] and has become one of the classical decision rules (see e.g. [88, p. 38]).

The Hurwicz Rule is given by

$$\min_{y\in F} \big(h \max_{i=1,\dots,\ell} y_i + (1-h) \min_{i=1,\dots,\ell} y_i\big) \tag{P7}$$

with $h \in [0, 1]$.

For $h = 1$, (P7) coincides with a special case of (P1).

Problem (P7) is equivalent to $(\mathrm{P}_{F,a,C,k})$, and hence to $\min_{y\in F} \varphi_{a-C,k}(y)$, with

$$\begin{aligned} &a = 0,\ k = \mathbf{1}, \quad \text{and}\\ &C = \{y \in \mathbb{R}^\ell \mid h \min_{i=1,\dots,\ell} y_i + (1-h) \max_{i=1,\dots,\ell} y_i \geq 0\} \end{aligned} \tag{7.3.8}$$

since:

$$\begin{aligned} y \in a - C + tk &\iff -y + a + tk \in C,\\ &\iff h \min_{i=1,\dots,\ell} (-y_i) + (1-h) \max_{i=1,\dots,\ell} (-y_i) + t \geq 0,\\ &\iff h \max_{i=1,\dots,\ell} y_i + (1-h) \min_{i=1,\dots,\ell} y_i \leq t. \end{aligned}$$

We have $C = \mathbb{R}^\ell \setminus (-\operatorname{int} \mathbb{R}^\ell_+)$ for $h = 0$, and $C = \mathbb{R}^\ell_+$ for $h = 1$.

In $\mathbb{R}^2$, the boundary of the cone $C$ consists of two rays beginning at the origin, the first one with direction $(h, h-1)^T$, the second with direction $(h-1, h)^T$ (see Fig. 7.12).

**Lemma 7.3.17** *Assume that $C$ is defined by* (7.3.8) *with $h \in [0, 1]$ and $k = \mathbf{1} \in \mathbb{R}^\ell$.*

(a) *$C$ is a closed cone.*
(b) $\mathbb{R}^\ell_+ \subseteq C \subseteq \mathbb{R}^\ell \setminus (-\operatorname{int} \mathbb{R}^\ell_+)$.
(c) *$C$ is convex if and only if $h = 1$ or ( $Y = \mathbb{R}^2$ and $h \geq \frac{1}{2}$).*
(d) $\operatorname{bd} C = \{y \in \mathbb{R}^\ell \mid h \min_{i=1,\dots,\ell} y_i + (1-h) \max_{i=1,\dots,\ell} y_i = 0\}$,
$\operatorname{int} C = \{y \in \mathbb{R}^\ell \mid h \min_{i=1,\dots,\ell} y_i + (1-h) \max_{i=1,\dots,\ell} y_i > 0\}$.
(e) $k \in \operatorname{int} \mathbb{R}^\ell_+ \subseteq \operatorname{int} C$, $Y = \operatorname{bd} C + \mathbb{R}k$, *and* $\operatorname{int} C = \operatorname{bd} C + \mathbb{R}_> k$.
(f) $C + \mathbb{R}^\ell_+ \subseteq C$.

(g) $C + (\mathbb{R}^\ell_+ \setminus \{0\}) \subseteq \operatorname{int} C \iff Y = \mathbb{R}^2$ *and* $h \in (0, 1)$.
(h) *For each* $i \in \{1, \ldots, \ell\}$, *we have:*
*C does not contain any line in direction* $\mathfrak{e}^i \iff h \in (0, 1]$.

***Proof***

(a) Obviously, $C$ is a cone. Take now any $y \in Y$ with $w := h \min_{i=1,\ldots,\ell} y_i + (1 - h) \max_{i=1,\ldots,\ell} y_i < 0$ and any $\bar{y} \in Y$ with $\|y - \bar{y}\|_2 < -w$. Then $\bar{y}_i < y_i - w$ for all $i \in \{1, \ldots, \ell\}$. We get $\min_{i=1,\ldots,\ell} \bar{y}_i < \min_{i=1,\ldots,\ell} y_i - w$ and $\max_{i=1,\ldots,\ell} \bar{y}_i < \max_{i=1,\ldots,\ell} y_i - w$. Therefore, $h \min_{i=1,\ldots,\ell} \bar{y}_i + (1-h) \max_{i=1,\ldots,\ell} \bar{y}_i < h \min_{i=1,\ldots,\ell} y_i + (1-h) \cdot \max_{i=1,\ldots,\ell} y_i - w = w - w = 0$. Thus, $Y \setminus C$ is an open set. Consequently, $C$ is closed.
(b) follows immediately from the definition of $C$.
(c) In $Y = \mathbb{R}^2$, it is easy to check that $C$ is convex if and only if $h \geq \frac{1}{2}$. For $h = 1$, the set $C = \mathbb{R}^\ell_+$ is convex.
Assume now $Y = \mathbb{R}^\ell$ with $\ell \geq 3$ and $h \in (0, 1)$.
We define vectors $y$ and $z$ by

$$y_i := \begin{cases} h & \text{if } i = 1, \\ -(1-h) & \text{if } i = 2, \\ 0 & \text{if } i > 2, \end{cases}$$

$$z_i := \begin{cases} 0 & \text{if } i = 1 \text{ or } i > 3, \\ -(1-h) & \text{if } i = 2, \\ h & \text{if } i = 3. \end{cases}$$

Then $y \in C$, $z \in C$, but $y + z \notin C$ since
$y_1 + z_1 = h = \max_{i=1,\ldots,\ell} (y_i + z_i)$, $y_2 + z_2 = -2(1-h) = \min_{i=1,\ldots,\ell} (y_i + z_i)$ and

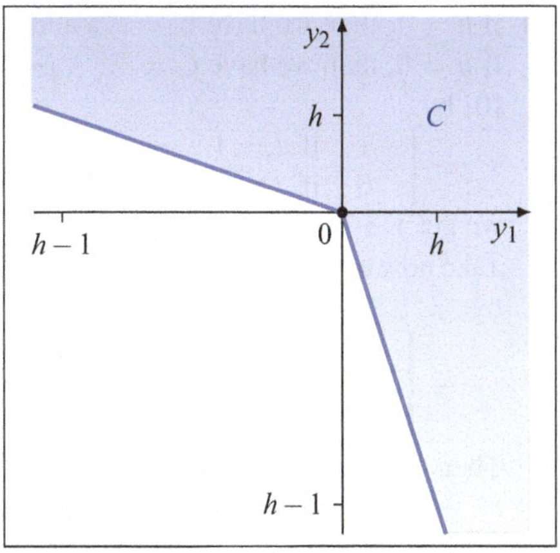


**Fig. 7.12** Set $C$ for problem (P7) in $\mathbb{R}^2$. Adapted from [472, p. 128]

$h(-2(1-h)) + (1-h)h < 0$.
For $h = 0$, choose

$$y_i := \begin{cases} 1 & \text{if } i = 1, \\ -2 & \text{if } i > 1, \end{cases} \quad \text{and} \quad z_i := \begin{cases} 1 & \text{if } i = 2, \\ -2 & \text{if } i \neq 2, \end{cases}$$

in order to get $y \in C$ and $z \in C$ with $y + z \notin C$.
Since $C$ is a cone, it is convex if and only if $C + C \subseteq C$.

(d) We split the proof in two parts.

(i) Take any $y \in Y$ with $z := h \min\limits_{i=1,\dots,\ell} y_i + (1-h) \max\limits_{i=1,\dots,\ell} y_i > 0$ and $\bar{y} \in Y$ with $\|y - \bar{y}\|_2 < z$.
Then $\bar{y}_i > y_i - z$ for all $i \in \{1, \dots, \ell\}$. We get $\min\limits_{i=1,\dots,\ell} \bar{y}_i > \min\limits_{i=1,\dots,\ell} y_i - z$ and $\max\limits_{i=1,\dots,\ell} \bar{y}_i > \max\limits_{i=1,\dots,\ell} y_i - z$. Thus, $h \min\limits_{i=1,\dots,\ell} \bar{y}_i + (1-h) \max\limits_{i=1,\dots,\ell} \bar{y}_i > h \min\limits_{i=1,\dots,\ell} y_i + (1-h) \max\limits_{i=1,\dots,\ell} y_i - z = z - z = 0$. Therefore, $\{y \in Y \mid h \min\limits_{i=1,\dots,\ell} y_i + (1-h) \max\limits_{i=1,\dots,\ell} y_i > 0\} \subseteq \operatorname{int} C$.

(ii) Take any $y \in Y$ with $h \min\limits_{i=1,\dots,\ell} y_i + (1-h) \max\limits_{i=1,\dots,\ell} y_i = 0$ and $t \in \mathbb{R}_>$.
Then $y + tk \in \operatorname{int} C$ and $y - tk \in Y \setminus C$.
We get $\operatorname{bd} C \supseteq \{y \in Y \mid h \min\limits_{i=1,\dots,\ell} y_i + (1-h) \max\limits_{i=1,\dots,\ell} y_i = 0\}$.
Thus, $\operatorname{bd} C = \{y \in Y \mid h \min\limits_{i=1,\dots,\ell} y_i + (1-h) \max\limits_{i=1,\dots,\ell} y_i = 0\}$ by (i) and since $C$ is closed. The second equation in (d) follows from $\operatorname{int} C = C \setminus \operatorname{bd} C$.

(e) The first statement is easy to verify.
(d) implies $\operatorname{bd} C + \mathbb{R}_> k \subseteq \operatorname{int} C$.
Take now any $y \in \operatorname{int} C$. Then $w := h \min\limits_{i=1,\dots,\ell} y_i + (1-h) \max\limits_{i=1,\dots,\ell} y_i > 0$. For $\bar{y} := y - wk$, we get $h \min\limits_{i=1,\dots,\ell} \bar{y}_i + (1-h) \max\limits_{i=1,\dots,\ell} \bar{y}_i = w - w = 0$ and thus $\bar{y} \in \operatorname{bd} C$. Hence, $\operatorname{int} C \subset \operatorname{bd} C + \mathbb{R}_> k$, which yields $\operatorname{int} C = \operatorname{bd} C + \mathbb{R}_> k$. This implies $Y = \operatorname{bd} C + \mathbb{R} k$ since $Y = \operatorname{int} C - \mathbb{R}_> k$ by Lemma 2.3.14.

(f) is obvious.

(g) If $h = 1$, then we have $C = \mathbb{R}^\ell_+$ and thus $C + (\mathbb{R}^\ell_+ \setminus \{0\}) \not\subseteq \operatorname{int} C$.
If $h = 0$, then we have $C = \mathbb{R}^\ell_+ \setminus (-\operatorname{int} \mathbb{R}^\ell_+)$, and defining $y \in C$ and $z \in \mathbb{R}^\ell_+ \setminus \{0\}$ by

$$y_i := \begin{cases} -1 & \text{if } i = 1, \\ 0 & \text{if } i > 1, \end{cases} \quad \text{and} \quad z_i := \begin{cases} 1 & \text{if } i = 1, \\ 0 & \text{if } i > 1, \end{cases}$$

we get $y + z = 0 \notin \operatorname{int} C$.
Take now any $h \in (0, 1)$ and $Y = \mathbb{R}^\ell$ with $\ell \geq 3$. Define $y \in C$ and $z \in \mathbb{R}^\ell_+ \setminus \{0\}$ by

$$y_i := \begin{cases} h & \text{if } i = 1, \\ -(1-h) & \text{if } i = 2, \\ 0 & \text{if } i > 2, \end{cases} \quad \text{and} \quad z_i := \begin{cases} \frac{h}{2} & \text{if } i = 3, \\ 0 & \text{if } i \neq 3. \end{cases}$$

Then $y + z \notin \operatorname{int} C$ since
$h \min\limits_{i=1,\dots,\ell} (y_i + z_i) + (1-h) \max\limits_{i=1,\dots,\ell} (y_i + z_i) = h(-(1-h)) + (1-h)h = 0$.

Consider now $Y = \mathbb{R}^2$ and $h \in (0, 1)$. Take any $y \in C$ and $r \in \mathbb{R}^2_+ \setminus \{0\}$. Then $\min_{i=1,2} (y_i + r_i) > \min_{i=1,2} y_i$ or $\max_{i=1,2} (y_i + r_i) > \max_{i=1,2} y_i$. Hence, $y + r \in \operatorname{int} C$.

(h) If $h = 0$, then $C = \mathbb{R}^\ell \setminus (-\operatorname{int} \mathbb{R}^\ell_+)$ and hence $\{te^i \mid t \in \mathbb{R}\} \subseteq C$ for all $i \in \{1, \dots, \ell\}$.
Consider now $h \in (0, 1]$. Take any $p \in \mathbb{R}^\ell$ and $j \in \{1, \dots, \ell\}$, and choose

$$t < \min \left( \max_{i \in \{1,\dots,\ell\} \setminus \{j\}} p_i \, , \, -\frac{1-h}{h} \max_{i \in \{1,\dots,\ell\} \setminus \{j\}} p_i \right) - p_j .$$

This implies $h \min_{i=1,\dots,\ell} (p_i + te^j_i) + (1-h) \max_{i=1,\dots,\ell} (p_i + te^j_i) \le h(p_j + t) + (1-h) \max_{i \in \{1,\dots,\ell\} \setminus \{j\}} p_i < 0$. Hence, $p + te^j \notin C$. □

**Proposition 7.3.18**

(a) *If* $h = 1$ *and the objective function of* (P7) *is not bounded below on* $F$, *then* $\operatorname{WEff}(F) = \emptyset$.
(b) *If* $y$ *is an optimal solution to* (P7), *then* $y \in \operatorname{WEff}(F)$.
(c) *If* $F$ *is a nonempty compact set and* $h = 1$, *then the optimal solution set of* (P7) *contains some Pareto minimal point of* $F$.
(d) *If* $y$ *is a unique optimal solution to* (P7), *then* $y \in \operatorname{Eff}(F)$.
(e) *Assume* $Y = \mathbb{R}^2$, $h \in (0, 1)$ *and that* $y$ *is an optimal solution to* (P7). *Then* $y \in \operatorname{Eff}(F)$. *Moreover,* $y \in \operatorname{GEff}(F)$ *if* $h \in [\frac{1}{2}, 1)$.
(f) *If* $F$ *is a nonempty closed set and there exists* $u \in \mathbb{R}^\ell$ *with* $F \subseteq u + \mathbb{R}^\ell_+$, *then the optimal solution set of each problem* (P7) *with* $h \in (0, 1]$ *is nonempty and compact.*

***Proof*** (a) and (c) result from Proposition 7.3.2(a) and (c). (b), (d) and the first statement in (e) follow from Theorem 6.2.21(h), (c) and (d), respectively. The second statement in (e) results from Proposition 7.2.31(a). (f) follows from Proposition 5.2.9. □

The statement in Proposition 7.3.18(f) cannot be extended to $h = 0$.

**Example 7.3.19** In $Y = \mathbb{R}^2$, consider $F := \{(y_1, y_2)^T \in \mathbb{R}^2 \mid y_1 > 0, \ y_2 = \frac{1}{y_1}\}$, $a := (0, 0)^T$, $k := (1, 1)^T$ and $C := \mathbb{R}^\ell \setminus (-\operatorname{int} \mathbb{R}^\ell_+)$ (see Fig. 7.13). Then $F \cap (a - C + tk) = \emptyset$ for all $t \le 0$, and $F \cap (a - C + tk) \ne \emptyset$ for all $t > 0$. Thus, $(P_{F,a,C,k})$ does not attain a minimum.

The union of all optimal point sets to problems (P7) is not necessarily dense in $\operatorname{Eff}(F)$.

**Example 7.3.20** Consider $F := \{(y_1, y_2)^T \in \mathbb{R}^2 \mid 0 \le y_1 \le 6, \ y_2 = -\frac{1}{3} y_1 + 4\}$ (see Fig. 7.14). Then $\operatorname{Eff}(F) = \operatorname{GEff}(F) = F$. The optimal point set of (P7) for $h \in [0, \frac{3}{4})$ is $\{(0, 4)^T\}$, for $h = \frac{3}{4}$ the set $\{(y_1, y_2)^T \in F \mid 0 \le y_1 \le 3\}$ and for $h \in (\frac{3}{4}, 1]$ the set $\{(3, 3)^T\}$.

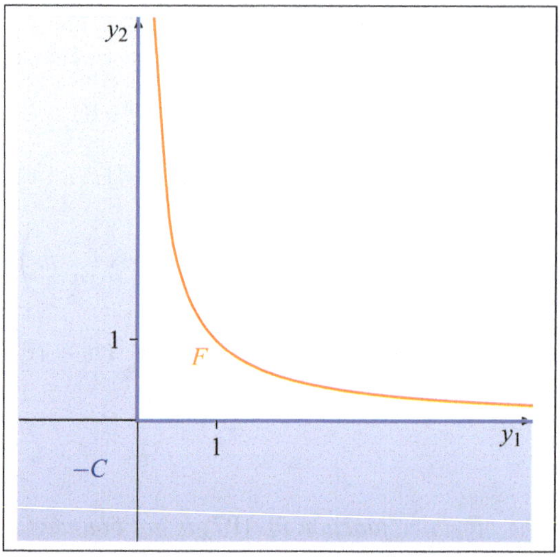


**Fig. 7.13** The sets $F$ and $-C$ in Example 7.3.19. Adapted from [472, p. 82]

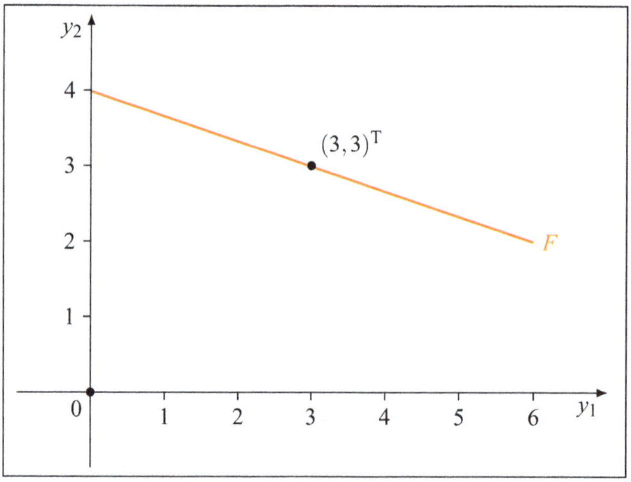


**Fig. 7.14** Set $F$ in Example 7.3.20. Adapted from [472, p. 131]

Section 7.3.7 is taken from [472, Sect. 7.8], except the compactness statement in Proposition 7.3.18(f). Proposition 7.3.18(b), (d), (e) and (f) are contained in [490, Proposition 6.2].

### 7.3.8 The Constraint Method

Consider the scalarizing problem

$$\begin{aligned} & y_j \to \min \\ & y_i \le d_i \text{ for all } i \in \{1, \dots, \ell\} \setminus \{j\} \\ & y \in F \end{aligned} \tag{P8}$$

with $j \in \{1, \dots, \ell\}$ and $d_i \in \mathbb{R}$ for all $i \in \{1, \dots, \ell\} \setminus \{j\}$.

$d_i$ can be a satisfying level for $y_i$ which is given by the decision maker.

Problem (P8) is equivalent to the problem $(\mathrm{P}_{F,a,C,k})$, and hence to $\min_{y \in F} \varphi_{a-C,k}(y)$, for

$$C = \mathbb{R}^\ell_+, \quad k = \mathfrak{e}^j, \quad a_i = \begin{cases} 0 & \text{if } i = j, \\ d_i & \text{if } i \in \{1, \dots, \ell\} \setminus \{j\}. \end{cases}$$

Note that $k \in C \setminus \{0\}$, but $k \notin \operatorname{int} C$.

For any $d \in \mathbb{R}^\ell$, (P8) has the same optimal solutions $y$ as $(\mathrm{P}_{F,d,\mathbb{R}^\ell_+,\mathfrak{e}^j})$ and as

$$\min_{y \in F} \varphi_{d-\mathbb{R}^\ell_+,\mathfrak{e}^j}(y).$$

Then $(\mathrm{P}_{F,d,\mathbb{R}^\ell_+,\mathfrak{e}^j})$ has the optimal value $t_0 - d_j$ if and only if (P8) has the optimal value $t_0$.

**Proposition 7.3.21**

(a) *If $y$ is an optimal solution to* (P8), *then* $y \in \operatorname{WEff}(F)$.

(b) *If $y$ is a unique optimal solution to* (P8), *then* $y \in \operatorname{Eff}(F)$.

(c) *Assume that $F$ is a closed set for which there exist a polyhedral cone $H \subseteq \mathbb{R}^\ell$ and $u \in \mathbb{R}^\ell$ with $\mathbb{R}^\ell_+ \setminus \{0\} \subset \operatorname{int} H$ and $(F - u) \cap (-\operatorname{int} H) = \emptyset$, and that the feasible set of* (P8) *is not empty. Then the optimal solution set of* (P8) *is compact and contains some Pareto minimal point of $F$.*

(d) *If there exists $d \in \mathbb{R}^\ell$ such that $y^0 \in F$ solves* (P8) *for each $j \in \{1, \dots, \ell\}$, then* $y^0 \in \operatorname{Eff}(F)$.

(e) $y^0 \in \operatorname{Eff}(F)$ *is equivalent to each of the following conditions.*

(i) $\forall j \in \{1, \dots, \ell\}$: $y^0$ *solves* (P8) *with* $d_i = y^0_i$ *for all* $i \in \{1, \dots, \ell\} \setminus \{j\}$,

(ii) $\forall j \in \{1, \dots, \ell\}$: $y^0$ *uniquely solves* (P8) *with* $d_i = y^0_i$ *for all* $i \in \{1, \dots, \ell\} \setminus \{j\}$,

(iii) $\exists j \in \{1, \dots, \ell\}$: $y^0$ *uniquely solves* (P8) *with* $d_i = y^0_i$ *for all* $i \in \{1, \dots, \ell\} \setminus \{j\}$.

***Proof***

(a) and (b) follow from Proposition 6.2.56(a) and (b), respectively.
(c) By Proposition 5.2.23, the optimal solution set $\Psi$ of (P8) is nonempty and compact. Theorem 6.2.51 implies $\text{Eff}(\Psi) \neq \emptyset$. By Theorem 6.2.21(b), $\text{Eff}(\Psi) \subseteq \text{Eff}(F)$.
(d) follows from Proposition 6.2.28(a).
(e) Proposition 6.2.28(b) implies

$$\begin{aligned} y^0 \in \text{Eff}(F) &\iff y^0 \text{ solves } (\text{P}_{F,y^0,\mathbb{R}^\ell_+,e^j}) \text{ for all } j \in \{1,\dots,\ell\}, \\ &\iff y^0 \text{ uniquely solves } (\text{P}_{F,y^0,\mathbb{R}^\ell_+,e^j}) \text{ for all } j \in \{1,\dots,\ell\}, \\ &\iff \exists j \in \{1,\dots,\ell\} : y^0 \text{ uniquely solves } (\text{P}_{F,y^0,\mathbb{R}^\ell_+,e^j}). \end{aligned}$$

□

We obtain the following from Proposition 7.3.21(d) and (e).

**Corollary 7.3.22** *In $Y = \mathbb{R}^2$, define $S_j$ as the union of the optimal solution sets of all problems* (P8) *with fixed $j$ and varying real values $d_2$ and $d_1$ for $j = 1$ and $j = 2$, respectively. Then* $\text{Eff}(F) = S_1 \cap S_2$.

A weakly Pareto minimal point is not always an optimal solution to some problem (P8).

**Example 7.3.23** Consider $F$ and $y^0 \in \text{WEff}(F)$ from Example 6.2.32 (see Fig. 6.5). The union of the optimal point sets to all problems (P8) with $j = 1$ is $S_1 = \{(y_1, y_2)^T \in \mathbb{R}^2 \mid y_1 = -2,\ -1 \le y_2 \le 0\} \cup \{(y_1, y_2)^T \in \mathbb{R}^2 \mid y_1 = -1,\ -2 \le y_2 < -1\}$. The union of the optimal point sets to all problems (P8) with $j = 2$ is $S_2 = \{(y_1, y_2)^T \in \mathbb{R}^2 \mid y_2 = -1,\ -2 \le y_1 < -1\} \cup \{(y_1, y_2)^T \in \mathbb{R}^2 \mid y_2 = -2,\ -1 \le y_1 \le 0\}$. Hence, the union of the optimal solutions to problems (P8) is the set $\text{WEff}(F) \setminus \{y^0\}$. We obtain $\text{Eff}(F) = S_1 \cap S_2 = \{(-1, -2)^T, (-2, -1)^T\}$ from Corollary 7.3.22.

**Proposition 7.3.24** *Assume that $F$ is a nonempty closed convex set. For each fixed $j \in \{1, \dots, \ell\}$, we get:*

$\text{Eff}(F) \neq \emptyset$ *if and only if the optimal point set of each problem* (P8) *with a nonempty feasible set is nonempty and compact. Then each of these optimal point sets contains some Pareto minimal point of $F$.*

***Proof*** The first statement follows from Proposition 6.2.55. Assume that (P8) has a nonempty compact optimal point set $\Psi$. Theorem 6.2.51 implies $\text{Eff}(\Psi) \neq \emptyset$. By Theorem 6.2.21(b), $\text{Eff}(\Psi) \subseteq \text{Eff}(F)$. □

In Proposition 7.3.24, the convexity assumption is essential.

**Example 7.3.25** Choose $F$ as in Example 7.3.7, $j := 1$, $d_2 := 0$, $d_3 := 0$. Then $F = \text{Eff}(F)$ is the feasible set of (P8), but (P8) does not have an optimal solution.

The Surrogate Worth Trade-off Method (SWT Method), which was developed by Haimes, Hall and Freedman [216, 217], uses auxiliary problems (P8), where each $d_i$ is the sum of the individual minimum of the i-th objective function on $F$ and some $\varepsilon > 0$. This method is also investigated in depth in [215], [251, pp. 126–144] and [88, Kap. 8]. The auxiliary problem used in the SWT Method was introduced by Cohon and Marks [99] as $\varepsilon$-constraint method.
The Method of Satisfactory Goals, proposed by Benson [46] and explained in [251, pp. 145–149], is also based on (P8).
Problem (P8) with $d_i = y_i^0$ for all $i \in \{1, \dots, \ell\} \setminus \{j\}$ and $y^0 \in F$ was studied in [216–218, 327–329]. Ventcel' [456], Emel'janov et al. [155], Podinovskij and Gavrilov [383], Benson and Morin [45] also worked with (P8). Chankong and Haimes [88, p. 285] gave details about the use of (P8) by Haimes et al., by Giesy [188] and, for approximating the Pareto minimal point set of bicriterial optimization problems, by Payne et al. [377]. They also discussed the applicability of (P8) [88, pp. 128–133,138–144,160ff] and the relationship between the solutions to (P8) and the Pareto minimal points of $F$ [88, Theorem 4.1., Corollary 4.1, Theorem 4.4.]. Essential statements are (b), (d) and the first equivalence of (e) in Proposition 7.3.21. The second equivalence from Proposition 7.3.21(e) was proved by Yu [503, Theorem 3.4.].
Based on calculations for finding Pareto minimal points of several nonconvex multiobjective optimization problems, Sauer [399] stated that the application of (P8) behaved better than those of the scalarizations (P) from Sect. 7.3.10, (7.3.4) and (P1) and better than the penalty scalarization from [494].

Lin [327–330] characterized Pareto minimal points by means of the scalarization

$$\begin{aligned} &y_j \to \min \\ &y_i = d_i \text{ for all } i \in \{1, \dots, \ell\} \setminus \{j\} \\ &y \in F \end{aligned}$$

with $j \in \{1, \dots, \ell\}$ and $d_i \in \mathbb{R}$ for all $i \in \{1, \dots, \ell\} \setminus \{j\}$, which he denoted as Proper Equality Constraint (PEC) Problem.

This problem is equivalent to the problem $(\mathrm{P}_{F,a,\tilde{C},k})$ and also to the problem $(\mathrm{P}_{F,a,C,k})$ for

$$\tilde{C} = \{0\}, \quad C = \mathbb{R}_+ \mathfrak{e}^j, \quad k = \mathfrak{e}^j, \quad a_i = \begin{cases} 0 & \text{if } i = j, \\ d_i & \text{if } i \in \{1, \dots, \ell\} \setminus \{j\}. \end{cases}$$

$(\mathrm{P}_{F,a,C,k})$ is equivalent to the problem $\min_{y \in F} \varphi_{a-C,k}(y)$. This equivalence can be used for deriving statements related to the Proper Equality Constraint Problem.

Section 7.3.8 coincides with [472, Sect. 7.10], except the compactness statement in Proposition 7.3.21(c), Corollary 7.3.22, Proposition 7.3.24 and Example 7.3.25.

### 7.3.9 Constraints Combined with Weight Factors

We now turn to the following extension of scalarization (P8):

$$\begin{aligned} & w^T y \to \min \\ & y_i \le d_i \text{ for all } i \in P \\ & y \in F \end{aligned} \tag{P9}$$

with $P \subsetneq \{1, \dots, \ell\}$ being a nonempty set, $d_i \in \mathbb{R}$ for all $i \in P$, $w \in \mathbb{R}^\ell_+$ and $\sum\limits_{i \notin P} w_i > 0$.

For any $d \in \mathbb{R}^\ell$, (P9) has the same optimal solutions $y$ as $(\mathrm{P}_{F,d,C,k})$, and hence as $\min_{y \in F} \varphi_{d-C,k}(y)$, with

$$\begin{aligned} & k \in \mathbb{R}^\ell \text{ defined by } k_i = \begin{cases} 0 & \text{for all } i \in P, \\ 1 & \text{for all } i \in \{1, \dots, \ell\} \setminus P, \end{cases} \\ & C = \{y \in \mathbb{R}^\ell \mid y_i \ge 0 \text{ for all } i \in P,\ w^T y \ge 0\}. \end{aligned} \tag{7.3.9}$$

**Lemma 7.3.26** *Under the assumptions of problem* (P9)*, consider* $C$ *and* $k$ *given by* (7.3.9).

(a) *$C$ is a closed convex cone.*
(b) $\operatorname{int} C = \{y \in \mathbb{R}^\ell \mid y_i > 0$ *for all* $i \in P,\ w^T y > 0\}$.
(c) $C \cap (-C) = \{0\} \iff P$ *contains exactly* $\ell - 1$ *elements.*
(d) $k \in \operatorname{bd} C$ *and* $C + k \not\subseteq \operatorname{int} C$.
(e) *$C$ does not contain any line in direction $k$.*
(f) $C + \mathbb{R}^\ell_+ \subseteq C$.
(g) $C + (\mathbb{R}^\ell_+ \setminus \{0\}) \not\subseteq \operatorname{int} C$.
(h) $C + (\mathbb{R}^\ell_+ \setminus \{0\}) \subseteq C + \mathbb{R}_> k \iff w \in \operatorname{int} \mathbb{R}^\ell_+$.

***Proof***

(a) and (f) are easy to verify. (b) follows from Proposition 2.2.30(b).
(c) We distinguish two cases.

(i) If $P$ contains exactly $\ell - 1$ elements, then $w_i > 0$ for $i \in \{1, \dots, \ell\} \setminus P$. This implies $C \cap (-C) = \{0\}$.
(ii) Assume now that $P$ contains less than $\ell - 1$ elements. Then there exist $n, j \in \{1, \dots, \ell\} \setminus P$ with $n \neq j$. If $w_n > 0$ and $w_j > 0$, then we can choose

$$y_i := \begin{cases} -w_j & \text{if } i = n, \\ w_n & \text{if } i = j, \\ 0 & \text{if } i \in \{1, \dots, \ell\} \setminus \{n, j\}, \end{cases}$$

as the components of some vector $y \in C \cap (-C)$ with $y \neq 0$.
Otherwise, we can assume $w_n = 0$ without loss of generality. Then we can choose

$$y_i := \begin{cases} 1 & \text{if } i = n, \\ 0 & \text{if } i \in \{1, \ldots, \ell\} \setminus \{n\}, \end{cases}$$

as the components of some vector $y \in C \cap (-C)$ with $y \neq 0$.

(d) $k \in \operatorname{bd} C$ follows from (b). Then $C + k \not\subseteq \operatorname{int} C$ since $0 \in C$.
(e) $p + tk \in C$ for $p \in \mathbb{R}^\ell$ and $t \in \mathbb{R}$ imply $w^{\mathrm{T}}(p + tk) \geq 0$ and thus $w^T p + t \sum\limits_{i \notin P} w_i \geq 0$. This is possible for $t \geq -\frac{w^T p}{\sum\limits_{i \notin P} w_i}$ only. Hence, $C$ does not contain any line in direction $k$.
(g) follows from $C + k \not\subseteq \operatorname{int} C$.
(h) (i) Assume first $w \in \operatorname{int} \mathbb{R}^\ell_+$. Take any $y \in C$ and $r \in \mathbb{R}^\ell_+ \setminus \{0\}$. Then $w^T y \geq 0$ and $w^T r > 0$. This yields $t := \frac{w^T y + w^T r}{\sum\limits_{i \notin P} w_i} > 0$. Define $c \in \mathbb{R}^\ell$ by

$$c_i := \begin{cases} y_i + r_i & \text{for all } i \in P, \\ y_i + r_i - t & \text{for all } i \in \{1, \ldots, \ell\} \setminus P. \end{cases}$$

Thus, $y + r = c + tk$. Because of $c_i \geq 0$ for all $i \in P$ and $w^T c = w^T y + w^T r - t \sum\limits_{i \notin P} w_i = 0$, we get $c \in C$. Hence, $C + (\mathbb{R}^\ell_+ \setminus \{0\}) \subseteq C + \mathbb{R}_> k$.

(ii) Assume now $w \notin \operatorname{int} \mathbb{R}^\ell_+$. Then there exists $j \in \{1, \ldots, \ell\}$ with $w_j = 0$. Define $r \in \mathbb{R}^\ell$ by

$$r_i := \begin{cases} 1 & \text{if } i = j, \\ 0 & \text{if } i \in \{1, \ldots, \ell\} \setminus \{j\}. \end{cases}$$

Then $0 + r \in C + (\mathbb{R}^\ell_+ \setminus \{0\})$, and $w^{\mathrm{T}}(0 + r) = 0$.
Take any $c \in C, t \in \mathbb{R}_>$. Then $c_i + t\, k_i = c_i \geq 0$ for all $i \in P$, and $w^{\mathrm{T}}(c + tk) = w^T c + t \sum\limits_{i \notin P} w_i > 0$. Hence, $C + \mathbb{R}_> k \subseteq \{y \in \mathbb{R}^\ell \mid y_i \geq 0 \text{ for all } i \in P,\ w^T y > 0\}$.
This implies $0 + r \notin C + \mathbb{R}_> k$. Consequently, $C + (\mathbb{R}^\ell_+ \setminus \{0\}) \subseteq C + \mathbb{R}_> k$ is not satisfied. □

**Proposition 7.3.27**

(a) *If $y$ is an optimal solution to* (P9), *then* $y \in \operatorname{WEff}(F)$.
(b) *If $y$ is an optimal solution to* (P9) *with $w \in \operatorname{int} \mathbb{R}^\ell_+$, then* $y \in \operatorname{Eff}(F)$.
(c) $y \in \operatorname{Eff}(F)$ *if and only if there exist a nonempty set $P \subsetneqq \{1, \ldots, \ell\}$, $w \in \mathbb{R}^\ell_+$ with $\sum\limits_{i \notin P} w_i > 0$ and real numbers $d_i$ such that $y$ is the unique optimal solution to* (P9).

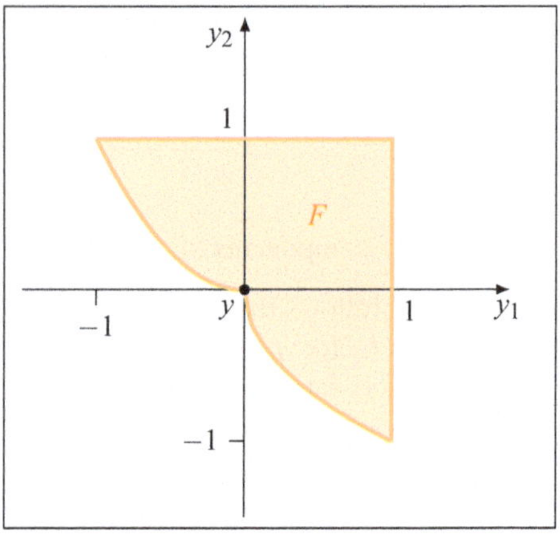


**Fig. 7.15** Set $F$ in Example 7.3.29. Adapted from [472, p. 139]

***Proof*** (a), (b) and the reverse direction of (c) follow from Theorem 6.2.21(h), (d) and (c), respectively. For $y \in \mathrm{Eff}(F)$, special problems (P9) of type (P8) which fulfill the conditions stated in (c) are given in Proposition 7.3.21(e). □

A weakly Pareto minimal point of $F$ does not necessarily solve some problem (P9).

**Example 7.3.28** Consider $F$ and $y^0 := (-1, -1)^{\mathrm{T}} \in \mathrm{WEff}(F)$ given in Example 6.2.32, $y^1 := (-1, -2)^{\mathrm{T}}$, $y^2 := (-2, -1)^{\mathrm{T}}$ (see Fig. 6.5). If $y^0$ is a feasible solution to some problem (P9), then $y^1$ and $y^2$ are also feasible solutions to this problem since $y^1 \leq y^0$ and $y^2 \leq y^0$. If $w_1 > 0$, we get $w^T y^2 = -2w_1 - w_2 < -w_1 - w_2 = w^T y^0$. If $w_2 > 0$, then we have $w^T y^1 = -w_1 - 2w_2 < w^T y^0$. Hence, $y^0$ cannot be an optimal solution to some problem (P9).

A Pareto minimal point of $F$ does not necessarily solve some problem (P9) with $w \in \operatorname{int} \mathbb{R}^{\ell}_{+}$.

**Example 7.3.29** Consider $F := \{y \in \mathbb{R}^2 \mid -1 \leq y_1 \leq 0,\ y_1^2 \leq y_2 \leq 1\} \cup \{y \in \mathbb{R}^2 \mid 0 \leq y_1 \leq 1,\ -\sqrt{y_1} \leq y_2 \leq 1\}$ and $y := (0, 0)^{\mathrm{T}} \in \mathrm{Eff}(F)$ (see Fig. 7.15). $y$ does not solve any problem (P9) with $w \in \operatorname{int} \mathbb{R}^2_{+}$.

We defined (P9) for $P \neq \{1, \ldots, \ell\}$. Without this assumption and admitting only positive weights, we get the problem

$$\begin{aligned} & w^T y \to \min \\ & y \leq d \\ & y \in F \end{aligned} \tag{P9$'$}$$

with $w \in \operatorname{int} \mathbb{R}^{\ell}_{+}$ and $d \in \mathbb{R}^{\ell}$.

There do not exist $a$, $C$ and $k$ such that problem (P9′) is equivalent to the problem ($\mathrm{P}_{F,a,C,k}$) for each set $F \subseteq \mathbb{R}^{\ell}$.

**Lemma 7.3.30** *There do not exist parameters $a \in \mathbb{R}^\ell$, $k \in \mathbb{R}^\ell \setminus \{0\}$ and a nonempty set $C \subseteq \mathbb{R}^\ell$ such that the feasible set of* ($\mathrm{P}_{F,a,C,k}$) *is contained in that of* (P9′) *for each set $F \subseteq \mathbb{R}^\ell$.*

***Proof*** Take any problem (P9′) with fixed parameters $w \in \operatorname{int} \mathbb{R}^\ell_+$ and $d \in \mathbb{R}^\ell$. To obtain a contradiction, suppose that the statement of the lemma is wrong. Then there exist $a \in \mathbb{R}^\ell$, $k \in \mathbb{R}^\ell \setminus \{0\}$ and a nonempty set $C \subseteq \mathbb{R}^\ell$ such that the feasible set of ($\mathrm{P}_{F,a,C,k}$) is contained in that of (P9′) for each set $F \subseteq \mathbb{R}^\ell$. Hence, $(a - C + \mathbb{R}k) \cap F \subseteq (d - \mathbb{R}^\ell_+) \cap F$ for each set $F \subseteq \mathbb{R}^\ell$. This implies

$$C - (a - d) + \mathbb{R}k \subseteq \mathbb{R}^\ell_+. \tag{7.3.10}$$

Take any $c \in C$. There exists $i \in \{1, \dots, \ell\}$ with $k_i \neq 0$. The condition $k_i > 0$ would imply $c_i - (a_i - d_i) - tk_i < 0$ for $t > \frac{c_i - (a_i - d_i)}{k_i}$, which contradicts (7.3.10). Thus, $k_i < 0$. For $t < \frac{(a_i - d_i) - c_i}{-k_i}$, we get $c_i - (a_i - d_i) - tk_i < 0$, which contradicts (7.3.10). □

Nevertheless, problem (P9′) is equivalent to the problem ($\mathrm{P}_{F \cap (d - \mathbb{R}^\ell_+), a, C, k}$), and hence to

$$\min_{y \in F \cap (d - \mathbb{R}^\ell_+)} \varphi_{a - C, k}(y),$$

for $C = \{y \in \mathbb{R}^\ell \mid w^T y \geq 0\}$ if $a$ solves $w^T a = 0$ and $k$ solves $w^T k = 1$. By Lemma 6.2.5(e), each Pareto minimal point of $F \cap (d - \mathbb{R}^\ell_+)$ is also a Pareto minimal point of $F$.

The scalarization (P9′) can be used for the determination of all Pareto minimal points of $F$, even for a fixed $w \in \operatorname{int} \mathbb{R}^\ell_+$. For $w = \mathbf{1}$, this was stated by Wendell and Lee [492], for an arbitrary $w \in \operatorname{int} \mathbb{R}^\ell_+$ by Corley [100]. Guddat et al. [199] used (P9′) for the determination of Pareto minimal points in an interactive procedure.

**Proposition 7.3.31**

(a) *For each $w \in \operatorname{int} \mathbb{R}^\ell_+$, we have:*

$$\begin{aligned} y^0 \in \operatorname{Eff}(F) &\iff y^0 \text{ solves (P9') with } d = y^0, \\ &\iff \exists d \in \mathbb{R}^\ell: \ y^0 \text{ optimal solution to (P9').} \end{aligned}$$

(b) *Assume that $F$ is a nonempty closed set for which there exist $u \in \mathbb{R}^\ell$ and a polyhedral cone $H \subseteq \mathbb{R}^\ell$ such that $\mathbb{R}^\ell_+ \setminus \{0\} \subseteq \operatorname{int} H$ and $(F - u) \cap (-\operatorname{int} H) = \emptyset$. Then each problem* (P9′) *with a nonempty feasible set has an optimal solution.*
(c) *Assume that $F$ is a nonempty closed convex set. Then $\operatorname{Eff}(F) \neq \emptyset$ holds if and only if each problem* (P9′) *with a nonempty feasible set has an optimal solution.*

***Proof***

(a) The forward direction of each equivalence is obvious.
Assume now that $y^0$ solves some problem (P9′). Since $w \in \operatorname{int} \mathbb{R}^\ell_+$, we obtain

$y^0 \in \mathrm{Eff}(\{y \in F \mid y \leq d\}) = \mathrm{Eff}(F \cap (d - \mathbb{R}^\ell_+))$. This yields $y^0 \in \mathrm{Eff}(F)$ by Lemma 6.2.5(e).

(b) By Lemma 2.2.32, the feasible set $Z := F \cap (d - \mathbb{R}^\ell_+)$ of (P9′) is compact. Since $Z \neq \emptyset$, the function $w^T y$ attains a minimum on $Z$.

(c) (a) implies the reverse direction of the equivalence.
Assume now that there exists $y^0 \in \mathrm{Eff}(F)$. Consider any problem (P9′) with a nonempty feasible set $Z$. The set $Z = F \cap (d - \mathbb{R}^\ell_+) \neq \emptyset$ is closed and convex. To obtain a contradiction, suppose that (P9′) does not have any optimal solution. Then $Z$ is not compact and thus not bounded. By Lemma 2.2.31, there exists $y \in 0^+(F \cap (y^0 - \mathbb{R}^\ell_+)) \setminus \{0\}$. Then $y^0 + y \in F \cap (y^0 - \mathbb{R}^\ell_+)$, which contradicts $y^0 \in \mathrm{Eff}(F)$. □

The equivalences in Proposition 7.3.31(a) remain valid if the objective function in (P9′) is replaced by some strictly $\mathbb{R}^\ell_+$-monotone function $\phi(y)$ [413, Theorem 1]. The analogous statement is true for Proposition 7.3.31(b) and (c) if $\phi$ is additionally continuous.

The assumptions of Proposition 7.3.31(b) imply $\mathrm{GEff}(F) \neq \emptyset$ because of Proposition 7.2.21. In the case $\mathrm{GEff}(F) = \emptyset$, a problem (P9′) with a nonempty feasible set does not necessarily have an optimal solution. Even if $F$ is closed, $\mathrm{Eff}(F) \neq \emptyset$ and $d \in F$, the optimal solution set of (P9′) may be empty for each $w \in \operatorname{int} \mathbb{R}^\ell_+$.

**Example 7.3.32** Consider $F$ from Example 7.2.24 (see Fig. 7.7) and $d := (1, -1)^{\mathrm{T}}$. Then $d \in F$. For each $w \in \operatorname{int} \mathbb{R}^\ell_+$, the objective function of (P9′) is not bounded below on the feasible set.

The optimal solution set and the feasible set of problem (P9′) coincide for $d = \bar{y} \in F$ with those of the scalarization

$$
\begin{aligned}
& w^{\mathrm{T}} \varepsilon \to \max \\
& y + \varepsilon = \bar{y} \\
& y \in F,\ \varepsilon \in \mathbb{R}^\ell_+
\end{aligned}
$$

with $w \in \operatorname{int} \mathbb{R}^\ell_+$ and $\bar{y} \in F$.

This scalarization was used by Benson [41], for linear vector optimization problems by Charnes and Cooper [90], for linear vector optimization problems and $w = \mathbf{1}$ by Ecker and Kouada [131].

Section 7.3.9 is taken from [472, Sect. 7.11], except the relationship between the problems (P9′), $(\mathrm{P}_{F\cap(d-\mathbb{R}^\ell_+),a,C,k})$ and $\min_{y \in F\cap(d-\mathbb{R}^\ell_+)} \varphi_{a-C,k}(y)$, and except the results starting with Proposition 7.3.31 and ending with Example 7.3.32.

### *7.3.10 Hyperbola Efficiency*

We will now investigate the scalarizing problem

$$\begin{aligned}
& t \to \min \\
& -w_i\, y_i - z_i + t \geq 0 \text{ for all } i \in \{1, \dots, \ell\} \\
& \prod_{i=1}^{\ell} (-w_i\, y_i - z_i + t) \geq b \\
& y \in F,\ t \in \mathbb{R}
\end{aligned} \tag{P}$$

with parameters $w \in \operatorname{int} \mathbb{R}^{\ell}_{+}$, $z \in \mathbb{R}^{\ell}$ and $b \in \mathbb{R}_{>}$.

(P) has the same set of minimizers $y$ and the same optimal value as

$$\begin{aligned}
& t \to \min \\
& -w_i\, y_i - z_i + t \geq 0 \text{ for all } i \in \{1, \dots, \ell\} \\
& \prod_{i=1}^{\ell} (-w_i\, y_i - z_i + t) = b \\
& y \in F,\ t \in \mathbb{R}.
\end{aligned} \tag{P$'$}$$

Problem (P′) was introduced by Ester [158, 159], who called its optimal solutions $y \in F$ hyperbola efficient points.

The equivalence between (P) and (P′) results from the following lemma.

**Lemma 7.3.33** *Assume $b \in \mathbb{R}_{>}$ and $u_i \in \mathbb{R}_{+}$, $i \in \{1, \dots, \ell\}$, with $\prod_{i=1}^{\ell} u_i > b$. Then there exists $\lambda \in \mathbb{R}_{>}$ such that*

$$\prod_{i=1}^{\ell} (u_i - \lambda) = b.$$

***Proof*** The function $g \colon \mathbb{R} \to \mathbb{R}$ defined by $g(r) := \prod_{i=1}^{\ell} (u_i - r)$ is continuous, attains the value 0, and is not bounded above. Hence, it attains the positive value $b$. The equation $g(r) = b$ implies $r \in \mathbb{R}_{>}$ since otherwise $\prod_{i=1}^{\ell} (u_i - r) \geq \prod_{i=1}^{\ell} u_i > b$. □

Problem (P) coincides with problem $(P_{F,a,H,k})$ if $H \subseteq \mathbb{R}^{\ell}$ and $a$, $k \in \mathbb{R}^{\ell}$ are defined by

$$H := H(w, b) := \left\{ y \in \mathbb{R}^{\ell} \mid \prod_{i=1}^{\ell} (w_i\, y_i + \sqrt[\ell]{b}) \geq b,\ w_i\, y_i + \sqrt[\ell]{b} \geq 0 \text{ for all } i \in \{1, \dots, \ell\} \right\}, \tag{7.3.11}$$

$$a_i := -\frac{z_i + \sqrt[\ell]{b}}{w_i} \text{ and } k_i := \frac{1}{w_i} \text{ for all } i \in \{1, \dots, \ell\}.$$

In $\mathbb{R}^2$, the boundary of $H$ is the curve

$$\left\{ y \in \mathbb{R}^2 \mid \prod_{i=1}^{2} (w_i\, y_i + \sqrt{b}) = b,\ w_i\, y_i + \sqrt{b} \geq 0 \text{ for all } i \in \{1, 2\} \right\},$$

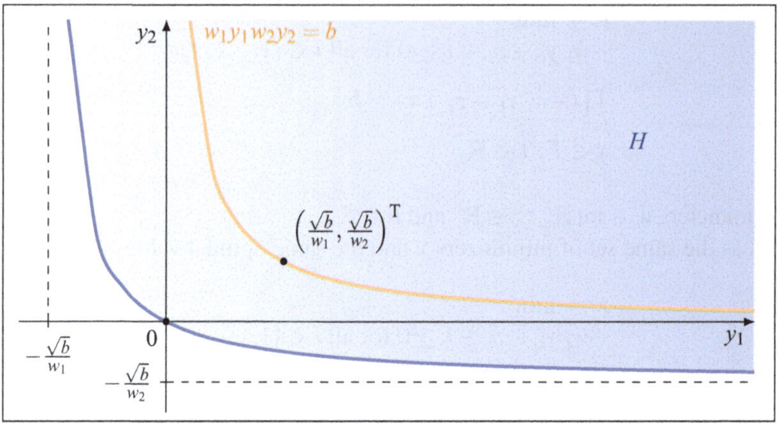


**Fig. 7.16** Set $H$ for hyperbola efficiency in $\mathbb{R}^2$. Adapted from [472, p. 105]

which can be generated shifting that part of the hyperbola $w_1y_1w_2y_2 = b$ which is contained in the nonnegative orthant by $-(\frac{\sqrt{b}}{w_1}, \frac{\sqrt{b}}{w_2})^T$ (see Fig. 7.16).

Fixing the norm of $w$ would not restrict the possible shapes of $H$ since

$$H\left(\frac{w}{\lambda}, b\right) = H(w, \lambda^\ell b) \quad \text{for each } \lambda \in \mathbb{R}_>.$$

The next lemma will be used when studying further properties of the set $H$.

**Lemma 7.3.34**

(a) *Assume $u_i, v_i \in \mathbb{R}_+$ for all $i \in \{1, \dots, \ell\}$. Then*

$$\prod_{i=1}^{\ell}(u_i + v_i) \geq \left(\sqrt[\ell]{\prod_{i=1}^{\ell} u_i} + \sqrt[\ell]{\prod_{i=1}^{\ell} v_i}\right)^\ell. \tag{7.3.12}$$

(b) *Assume $u_i, v_i \in \mathbb{R}_>$ for all $i \in \{1, \dots, \ell\}$ and that not all values $\frac{u_i}{v_i}$ with $i \in \{1, \dots, \ell\}$ coincide. Then*

$$\prod_{i=1}^{\ell}(u_i + v_i) > \left(\sqrt[\ell]{\prod_{i=1}^{\ell} u_i} + \sqrt[\ell]{\prod_{i=1}^{\ell} v_i}\right)^\ell.$$

***Proof***

(a) If $u_i = 0$ or $v_i = 0$ for some $i \in \{1, \dots, \ell\}$, then (7.3.12) is satisfied. Assume now $u_i, v_i \in \mathbb{R}_>$ for all $i \in \{1, \dots, \ell\}$. In this case, (7.3.12) is equivalent to the following statement:
Assume $x_i \in \mathbb{R}_>$ for all $i \in \{1, \dots, \ell\}$. Then

$$\prod_{i=1}^{\ell}(x_i+1) \geq \left(\sqrt[\ell]{\prod_{i=1}^{\ell} x_i}+1\right)^{\ell}. \tag{7.3.13}$$

This equivalence becomes obvious by dividing both sides of (7.3.12) by $\prod_{i=1}^{\ell} v_i$ and defining $x_i = \frac{u_i}{v_i}$. We will now prove (7.3.13), based on the following well-known statement.
Assume $m \in \mathbb{N}_>$, $p_i \in \mathbb{R}_+$ for all $i \in \{1, \dots, m\}$. Then

$$\frac{1}{m}\sum_{i=1}^{m} p_i \geq \left(\prod_{i=1}^{m} p_i\right)^{\frac{1}{m}}. \tag{7.3.14}$$

We have

$$\prod_{i=1}^{\ell}(x_i+1) = 1 + \sum_{j=1}^{\ell} s_j,$$

where $s_j$ stands for the sum of all products with $j$ factors $x_i$, $i \in \{1, \dots, \ell\}$, having different indices. Each sum $s_j$ contains $\binom{\ell}{j}$ summands. Each $x_i$ with $i \in \{1, \dots, \ell\}$ is a factor in $\binom{\ell}{j} - \binom{\ell-1}{j}$ of these summands. For each $j \in \{1, \dots, \ell\}$, (7.3.14) implies

$$\frac{1}{\binom{\ell}{j}} s_j \geq \left(\prod_{i=1}^{\ell} x_i^{\binom{\ell}{j}-\binom{\ell-1}{j}}\right)^{\frac{1}{\binom{\ell}{j}}} = \left(\prod_{i=1}^{\ell} x_i\right)^{\frac{j}{\ell}}. \tag{7.3.15}$$

Consequently,

$$\prod_{i=1}^{\ell}(x_i+1) \geq 1 + \sum_{j=1}^{\ell}\binom{\ell}{j}\left(\prod_{i=1}^{\ell} x_i\right)^{\frac{j}{\ell}} = \left(\sqrt[\ell]{\prod_{i=1}^{\ell} x_i}+1\right)^{\ell}.$$

(b) It is well known that the inequality in (7.3.14) is a strict one if $p_1 = p_2 = \dots = p_m$ does not hold. Following the lines of (a), this fact implies a strict inequality in (7.3.15) and thus in (7.3.13) if $x_1 = x_2 = \dots = x_\ell$ is not satisfied. □

**Lemma 7.3.35** *Under the assumptions of problem* (P), *consider* $H \subseteq \mathbb{R}^\ell$ *and* $k \in \mathbb{R}^\ell$ *given by* (7.3.11).

(a) *$H$ is a closed strictly convex set.*
(b) $\operatorname{int} H = \left\{ y \in \mathbb{R}^\ell \mid \prod_{i=1}^{\ell}(w_i\, y_i + \sqrt[\ell]{b}) > b,\ w_i\, y_i + \sqrt[\ell]{b} \geq 0 \text{ for all } i \in \{1, \dots, \ell\} \right\}$.
(c) $0 \in \operatorname{bd} H$.
(d) $H + (\mathbb{R}^\ell_+ \setminus \{0\}) \subseteq \operatorname{int} H$.
(e) $k \in \mathbb{R}^\ell_+ \setminus \{0\} \subset \operatorname{int} H$ *and* $H + \mathbb{R}_> k \subseteq \operatorname{int} H$.

(f) $\operatorname{int} H = \operatorname{bd} H + \mathbb{R}_{>}k$.
(g) $\mathbb{R}^\ell = \operatorname{bd} H + \mathbb{R}k$.
(h) $H$ *does not be a cone.*
(i) $H + H \nsubseteq H$.
(j) $H \setminus \{0\} \subseteq \{y \in \mathbb{R}^\ell \mid w^T y > 0\}$.
(k) $H \subseteq q + \mathbb{R}^\ell_+$ *with* $q_i := -\frac{1}{w_i}\sqrt[\ell]{b}$ *for all* $i \in \{1, \ldots, \ell\}$.

***Proof***

(b) –(e) are easy to verify.

(a) Obviously, $H$ is closed. Take any $y,\ c \in H$ with $y \neq c$, $s \in (0, 1)$. Then for all $i \in \{1, \ldots, \ell\}$,

$$\begin{aligned} w_i(s\, y_i + (1-s)c_i) + \sqrt[\ell]{b} &= s(w_i\, y_i + \sqrt[\ell]{b}) + (1-s)(w_i\, c_i + \sqrt[\ell]{b}) \\ &\geq s \cdot 0 + (1-s)\cdot 0 = 0. \end{aligned}$$

$$\begin{aligned} \prod_{i=1}^{\ell}(w_i(s\, y_i + (1-s)c_i) + \sqrt[\ell]{b}) &= \prod_{i=1}^{\ell}(s(w_i\, y_i + \sqrt[\ell]{b}) + (1-s)(w_i\, c_i + \sqrt[\ell]{b})) \\ &> \left(\sqrt[\ell]{\prod_{i=1}^{\ell}(s(w_i\, y_i + \sqrt[\ell]{b}))} + \sqrt[\ell]{\prod_{i=1}^{\ell}((1-s)(w_i\, c_i + \sqrt[\ell]{b}))}\right)^\ell \end{aligned}$$

because of Lemma 7.3.34(b).

$$\begin{aligned} &\left(\sqrt[\ell]{\prod_{i=1}^{\ell}(s(w_i\, y_i + \sqrt[\ell]{b}))} + \sqrt[\ell]{\prod_{i=1}^{\ell}((1-s)(w_i\, c_i + \sqrt[\ell]{b}))}\right)^\ell \\ &= \left(s\sqrt[\ell]{\prod_{i=1}^{\ell}(w_i\, y_i + \sqrt[\ell]{b})} + (1-s)\sqrt[\ell]{\prod_{i=1}^{\ell}(w_i\, c_i + \sqrt[\ell]{b})}\right)^\ell \\ &\geq (s\sqrt[\ell]{b} + (1-s)\sqrt[\ell]{b})^\ell = b. \end{aligned}$$

Hence, $s\, y + (1-s)c \in \operatorname{int} H$.

(f) can be proved by means of Lemma 7.3.33.

(g) Corollary 2.3.12 yields $\mathbb{R}^\ell = \mathbb{R}^\ell_+ + \mathbb{R}k$, which implies $\mathbb{R}^\ell \subseteq H + \mathbb{R}k \subseteq \operatorname{bd} H + \mathbb{R}k$ because of (f).

(h) Define $y \in \mathbb{R}^\ell$ by $y_1 := \frac{1}{w_1}$, $y_2 := -\frac{\sqrt[\ell]{b}}{w_2(1+\sqrt[\ell]{b})}$, $y_i := 0$ for all $i \in \{1, \ldots, \ell\} \setminus \{1, 2\}$. Then $\prod_{i=1}^{\ell}(w_i\, y_i + \sqrt[\ell]{b}) = (1+\sqrt[\ell]{b})\left(-\frac{\sqrt[\ell]{b}}{1+\sqrt[\ell]{b}} + \sqrt[\ell]{b}\right)(\sqrt[\ell]{b})^{\ell-2} = (1+\sqrt[\ell]{b})\frac{\sqrt[\ell]{b}\sqrt[\ell]{b}}{1+\sqrt[\ell]{b}} \cdot (\sqrt[\ell]{b})^{\ell-2} = b$ and $y \in H$. For $\bar{y} := (1+b)y$, we get $w_2\,\bar{y}_2 + \sqrt[\ell]{b} = w_2(1+b)y_2 + \sqrt[\ell]{b} = -(1+b)\frac{\sqrt[\ell]{b}}{1+\sqrt[\ell]{b}} + \sqrt[\ell]{b} = \frac{\sqrt[\ell]{b}}{1+\sqrt[\ell]{b}}(\sqrt[\ell]{b} - b) < 0$. Thus, $\bar{y} \notin H$. Hence, $H$ does not be a cone.

(i) $H + H \subseteq H$ would imply that $H$ is a cone because of (a) and (c).

(j) Choose $\varphi(y) := -b + \prod_{i=1}^{\ell}(w_i y_i + \sqrt[\ell]{b})$ for all $y \in \mathbb{R}^\ell$.

Then $\frac{\partial\varphi}{\partial y_i}(y) = w_i \prod_{\substack{j=1\\ j\neq i}}^{\ell}(w_j y_j + \sqrt[\ell]{b})$ for all $i \in \{1, \ldots, \ell\}$.

The supporting hyperplane of the convex set $H$ at the point 0 consists of all vectors $y \in \mathbb{R}^\ell$ with

$$0 = \sum_{i=1}^{\ell} \left( \frac{\partial \varphi}{\partial y_i}(0)(y_i - 0) \right) = \sum_{i=1}^{\ell} \left( w_i (\sqrt[\ell]{b})^{\ell-1} \, y_i \right) = (\sqrt[\ell]{b})^{\ell-1} \, w^T y.$$

Since $\mathbb{R}^\ell_+ \setminus \{0\} \subseteq \operatorname{int} H$, we get $H \subseteq \{y \in \mathbb{R}^\ell \mid w^T y \geq 0\}$. The strict convexity of $H$ and $0 \in H$ imply $H \setminus \{0\} \subseteq \{y \in \mathbb{R}^\ell \mid w^T y > 0\}$. □

The optimization problem (P) has the same set of feasible vectors $y$, the same set of optimal solutions $y$ and the same optimal value as the problem

$$\min_{y \in F} \varphi_{a-H,k}(y)$$

if $a$, $H$ and $k$ are defined by (7.3.11). Because of Lemma 7.3.35, condition $(\mathrm{H2}_{a-H,k})$ from Sect. 4.6 is satisfied. Theorems 4.6.6, 4.5.5 and 4.4.4 imply that $\varphi_{a-H,k}$ is finite-valued, continuous, convex, strictly quasiconvex, and strictly $\mathbb{R}^\ell_+$-monotone.

**Proposition 7.3.36**

(1) *Each optimal solution* $y$ *to* (P) *is a Pareto minimal point of* $F$.
(2) *Assume* (Sp-ex). *If* $y$ *is an optimal solution to* (P), *then* $y \in \operatorname{GEff}(F)$.
(3) *Let* $F$ *be a nonempty closed set and suppose that* (Sp-ex) *holds.*

   (a) *Each problem* (P) *with* $b \in \mathbb{R}_>$, $w \in \operatorname{int} \mathbb{R}^\ell_+$ *and* $z \in \mathbb{R}^\ell$ *has an optimal solution and a compact set of minimizers* $y$.
   (b) *For* $y^0 \in \mathbb{R}^\ell$ *and* $w \in \operatorname{int} \mathbb{R}^\ell_+$, *consider the problems*

$$\begin{aligned} & t \to \min \\ & -w_i y_i + w_i y_i^0 + t \geq 0 \text{ for all } i \in \{1, \ldots, \ell\} \\ & \prod_{i=1}^{\ell} (-w_i y_i + w_i y_i^0 + t) \geq \tfrac{1}{n} \\ & y \in F,\ t \in \mathbb{R} \end{aligned} \qquad (\mathrm{P}_n)$$

   *with* $n \in \mathbb{N}_>$.
   *Each problem* $(\mathrm{P}_n)$ *has an optimal solution and a compact set of minimizers* $y$. *For* $\bar{n} > n$, *each feasible solution* $(y, t)$ *to* $(\mathrm{P}_n)$ *is a feasible solution to* $(\mathrm{P}_{\bar{n}})$. *The sequence of the optimal values of the problems* $(\mathrm{P}_n)$ *is strictly monotonically decreasing for* $n \to \infty$. *It converges to zero if* $y^0 \in \operatorname{Eff}(F)$.
   (c) *Each* $y^0 \in \operatorname{Eff}(F)$ *is, for any* $w \in \operatorname{int} \mathbb{R}^\ell_+$, *the limit of each sequence of optimal solutions* $y^n$ *to the problems* $(\mathrm{P}_n)$ *for* $n \to \infty$.

(4) *If* $F$ *is convex, then* (P) *has at most one minimizer* $y$, *which is the only local minimizer of* (P).

***Proof***

(1) follows from Theorem 6.2.21(d).
(2) results from Theorem 7.2.22(3).

(3) (a) Take any problem (P). Define $a$, $H$ and $k$ by (7.3.11). Since $F$ is not empty, (P) has a feasible solution, i.e., there exist $y^0 \in F$ and $t_0 \in \mathbb{R}$ with $y^0 \in a - H + t_0 k$. Lemma 7.3.35 implies $k \in 0^+ H$ and hence $y \in a - H + t_0 k$ for any $y \in a - H + tk$ with $t \leq t_0$. Consequently, the set of minimizers $y$ of (P) and the optimal value of (P) coincide with that of the problem

$$\varphi_{a-H,k}(y) \to \min$$
$$y \in F \cap (a - H + t_0 k) =: Z.$$

The set $Z = F \cap (a - H + t_0 k) \subseteq F \cap (a - q + t_0 k - \mathbb{R}^\ell_+)$ with $q$ chosen according to Lemma 7.3.35(k) is bounded because of Lemma 2.2.32, thus also compact. Hence, $\varphi_{a-H,k}$ attains a minimum on $Z$ and the set of minimizers $y$ is compact. Therefore, (P) has a nonempty compact set of minimizers $y$.

(b) The first statement results from (a).
Take any $n \in \mathbb{N}_>$. Denote the optimal value of $(\mathrm{P}_n)$ by $t_n$, and let $y^n$ be any minimizer of $(\mathrm{P}_n)$. Take any $\bar{n} \in \mathbb{N}_>$ with $\bar{n} > n$. Then $\frac{1}{n} > \frac{1}{\bar{n}}$, and each feasible solution $(y, t)$ to $(\mathrm{P}_n)$ is a feasible solution to $(\mathrm{P}_{\bar{n}})$. Because of

$$\prod_{i=1}^{\ell}(-w_i y_i^n + w_i y_i^0 + t_n) = \frac{1}{n} > \frac{1}{\bar{n}},$$

the optimal value of $(\mathrm{P}_{\bar{n}})$ is smaller than $t_n$.
Assume now $y^0 \in \mathrm{Eff}(F)$. Then there exists $i \in \{1, \ldots, \ell\}$ with $y_i^n \geq y_i^0$. Since $-w_i y_i^n + w_i y_i^0 + t_n > 0$, we get $t_n > w_i(y_i^n - y_i^0) \geq 0$. Moreover, $(y^0, \sqrt[\ell]{\frac{1}{n}})$ is a feasible solution to $(\mathrm{P}_n)$. This yields the assertion.

(c) Take any $y^0 \in \mathrm{Eff}(F)$ and any $w \in \operatorname{int} \mathbb{R}^\ell_+$. For each $n \in \mathbb{N}_>$, let $(y^n, t_n)$ denote an arbitrary optimal solution to $(\mathrm{P}_n)$. We have $-w_i y_i^n + w_i y_i^0 + t_n \geq 0$ for all $n \in \mathbb{N}_>$, $i \in \{1, \ldots, \ell\}$. Hence, $y_i^n \leq y_i^0 + \frac{1}{w_i} t_n \leq y_i^0 + \frac{1}{w_i}\sqrt[\ell]{\frac{1}{n}} \leq y_i^0 + \frac{1}{w_i}$ for all $n \in \mathbb{N}_>$, $i \in \{1, \ldots, \ell\}$. For each $n \in \mathbb{N}$, we get $y^n \in Q := F \cap \{y \in \mathbb{R}^\ell \mid \forall i \in \{1, \ldots, \ell\}\colon\ y_i \leq y_i^0 + \frac{1}{w_i}\}$. The set $Q$ is bounded because of Lemma 2.2.32. Lemma 2.1.49 implies that the sequence $(y^n)_{n \in \mathbb{N}_>}$ has a cluster point. Let $y^{00}$ be any cluster point of the sequence. Since $F$ is closed, we obtain $y^{00} \in F$. Moreover, $y^{00}$ is the limit of some subsequence $(y^{j_n})_{n \in \mathbb{N}_>}$ of $(y^n)_{n \in \mathbb{N}_>}$, where the sequence of integers $j_n$ is strictly monotonically increasing.
For all $i \in \{1, \ldots, \ell\}$ and $n \in \mathbb{N}_>$, we get $0 \leq -w_i y_i^{j_n} + w_i y_i^0 + t_{j_n}$. Hence,

$$\begin{aligned}
0 &\leq \lim_{n\to\infty}(-w_i y_i^{j_n} + w_i y_i^0 + t_{j_n}) \\
&= -w_i \lim_{n\to\infty} y_i^{j_n} + w_i y_i^0 + \lim_{n\to\infty} t_{j_n} \\
&= -w_i y_i^{00} + w_i y_i^0 + 0 \text{ for all } i \in \{1, \ldots, \ell\}.
\end{aligned}$$

Thus, $y^{00} \leq y^0$. Since $y^0 \in \mathrm{Eff}(F)$, we obtain $y^{00} = y^0$. Consequently, $y^0$ is the only cluster point of $(y^n)_{n\in\mathbb{N}_>}$. By Lemma 2.1.49, $y^0$ is the limit of this sequence.

(4) follows from Theorem 6.2.21(e). □

**Remark 7.3.37** Ester and Schwartz [160] investigated the problem

$$\begin{aligned} & r^T y + s\,t \to \min \\ & -y_i - z_i + g_i t \geq 0 \text{ for all } i \in \{1, \dots, \ell\} \\ & \prod_{i=1}^{\ell}(-y_i - z_i + g_i t) = b \\ & y \in F,\ t \in \mathbb{R} \end{aligned}$$

with $r \in \mathbb{R}^\ell_+$, $s$, $b \in \mathbb{R}_>$, $z \in \mathbb{R}^\ell$ and $g \in \mathbb{R}^\ell_+ \setminus \{0\}$.

They showed [160, Satz 1] that each optimal solution $y$ to such a problem belongs to $\mathrm{Eff}(F)$. This implies part (1) of Proposition 7.3.36.

Further properties of the problems $(\mathrm{P}_n)$ with $n \in \mathbb{R}_>$ and $w = \mathbf{1}$ were stated by Gerstewitz and Iwanow [184], e.g.:

If $\varphi(y, n)$ is defined as the solution $t$ of

$$\prod_{i=1}^{\ell}(-y_i + y_i^0 + t) = \frac{1}{n}, \quad -y_i + y_i^0 + t > 0 \text{ for all } i \in \{1, \dots, \ell\},$$

then the functional $\varphi\colon \mathbb{R}^\ell \times \mathbb{R}_> \to \mathbb{R}$ is finite-valued, continuous, convex, and strictly $\mathbb{R}^{\ell+1}_+$-monotone. They also investigated the influence of alternations in $n$ on the sets $\{y \mid \varphi(y, n) < \varphi(y^0, n)\}$.

**Corollary 7.3.38** *Assume that $F$ is a nonempty closed set with* $\mathrm{GEff}(F) \neq \emptyset$. *Then each problem* (P) *has an optimal solution, and its set of minimizers $y$ is a compact subset of* $\mathrm{GEff}(F)$.

***Proof*** By Proposition 7.2.21, (Sp-ex) is fulfilled if and only if $\mathrm{GEff}(F) \neq \emptyset$. The assertion follows from Proposition 7.3.36(2) and (3a). □

Proposition 7.3.36 illustrates that the properly efficient point set of a nonempty closed set $F$ is dense in the Pareto minimal point set under assumption (Sp-ex), i.e., if $\mathrm{GEff}(F) \neq \emptyset$.

The next example shows that (P) may have an optimal solution even though $\mathrm{GEff}(F)$ is empty.

**Example 7.3.39** Define $F := \{(y_1, y_2)^\mathrm{T} \in \mathbb{R}^2 \mid y_1 = 1,\ y_2 \leq -1\} \cup \{(y_1, y_2)^\mathrm{T} \in \mathbb{R}^2 \mid y_1 \in [-1, 1],\ y_1 + y_2 = 0\}$. Then $\mathrm{GEff}(F) = \emptyset$. Problem (P) with $w_1 = w_2 = b = z_1 = z_2 = 1$ has the unique optimal solution $(0, 0)^\mathrm{T}$. If $z$ is altered into $(-2, 0)^\mathrm{T}$, then $a$ in (7.3.11) becomes $(1, -1)^\mathrm{T}$, and (P) does not attain a minimum (see Fig. 7.17).

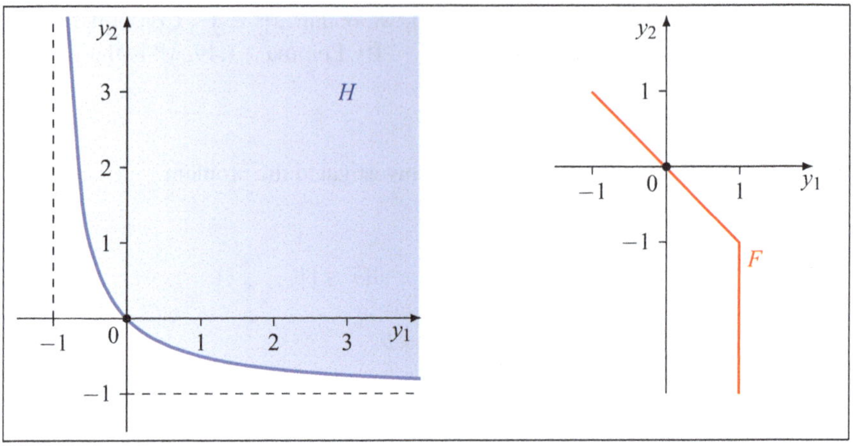


**Fig. 7.17** Sets $H$ and $F$ in Example 7.3.39. Adapted from [472, p. 76]

Conversely, a properly efficient point does not necessarily solve some problem (P).

**Example 7.3.40** Consider $F := \{(y_1, y_2)^T \in \mathbb{R}^2_+ \mid y_1 + y_2 \leq 2\} \cup \{(y_1, y_2)^T \in \mathbb{R}^2 \mid y_2 - 2y_1 \geq 0,\ y_1 + 2y_2 \geq (y_2 - 2y_1)^2,\ y_1 + y_2 \leq 2\} \cup \{(y_1, y_2)^T \in \mathbb{R}^2 \mid y_1 - 2y_2 \geq 0,\ 2y_1 + y_2 \geq (y_1 - 2y_2)^2,\ y_1 + y_2 \leq 2\}$. The blue set in Fig. 7.18 is the set $-D^K$ with $K := 2$, where $D^K$ is defined by (7.2.1). We have $y^0 := (0, 0)^T \in \mathrm{Eff}(F, D^K) \subseteq \mathrm{GEff}(F)$.

To obtain a contradiction, suppose that $y^0$ solves some problem (P) with parameters $w \in \operatorname{int} \mathbb{R}^\ell_+$, $z \in \mathbb{R}^\ell$ and $b \in \mathbb{R}_>$. Let $t_0$ be the optimal value of this problem. Then $-z_i + t_0 \geq 0$ for all $i \in \{1, 2\}$, and $\prod_{i=1}^{2}(-z_i + t_0) = b$. Hence, $y^0 \in \operatorname{bd} B$ for $B := \{(y_1, y_2)^T \in \mathbb{R}^2 \mid -w_i y_i - z_i + t_0 \geq 0$ for all $i \in \{1, 2\}$, $\prod_{i=1}^{2}(-w_i y_i - z_i + t_0) \geq b\}$. The tangent to $B$ at $y^0$ is given by $y_1 = -cy_2$ with $c := \frac{w_2(-z_1+t_0)}{w_1(-z_2+t_0)} > 0$. Therefore, $B$ contains an element $y^1$ of $F \setminus \{y^0\}$ in its interior. The point $y^1$ still remains a feasible solution to (P) if $t$ is chosen smaller than $t_0$ such that $-w_i y_i^1 - z_i + t \geq 0$ for all $i \in \{1, 2\}$ and $\prod_{i=1}^{2}(-w_i y_i^1 - z_i + t) = b$. This contradicts the fact that $t_0$ is the optimal value of (P).

We now turn to the choice of parameters for the problems under consideration. Let $\mathrm{M}_{w,z,b}$ denote the set of optimal solutions $y$ to problem (P) with parameters $w \in \operatorname{int} \mathbb{R}^\ell_+$, $z \in \mathbb{R}^\ell$, and $b \in \mathbb{R}_>$.

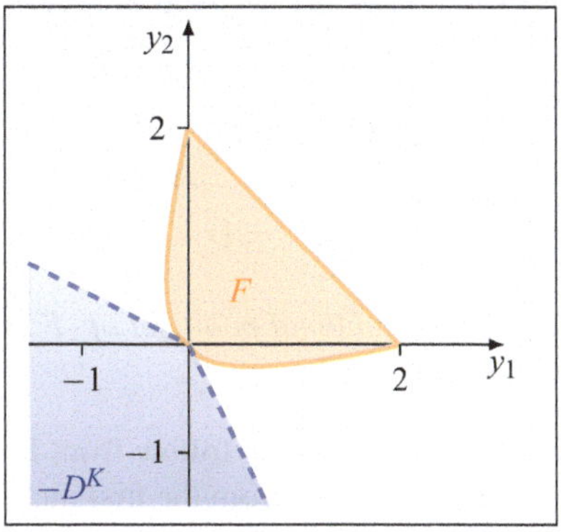


**Fig. 7.18** Sets $F$ and $-D^K$ in Example 7.3.40. Adapted from [472, p. 109]

**Proposition 7.3.41** *Assume* $w \in \operatorname{int} \mathbb{R}^\ell_+$ *and* $b \in \mathbb{R}_>$.

(a) *For each* $z \in \mathbb{R}^\ell$ *and* $\lambda \in \mathbb{R}_>$,

$$\mathrm{M}_{\frac{w}{\lambda},z,b} = \mathrm{M}_{w,\lambda z,\lambda^\ell b}.$$

(b) *For all* $c \in \mathbb{R}$,

$$\bigcup_{z \in \mathbb{R}^\ell} \mathrm{M}_{w,z,b} = \bigcup_{z \in \{y \in \mathbb{R}^\ell \mid y_1 = c\}} \mathrm{M}_{w,z,b}.$$

(c) *We have*

$$\bigcup_{z \in \mathbb{R}^\ell} \mathrm{M}_{w,z,b} = \bigcup_{z \in -\operatorname{int} \mathbb{R}^\ell_+} \mathrm{M}_{w,z,b} = \bigcup_{z \in \operatorname{int} \mathbb{R}^\ell_+} \mathrm{M}_{w,z,b}.$$

***Proof*** (a) is obvious. Consider now (P) with fixed parameters $w$ and $b$. (P) is equivalent to $(\mathrm{P}_{F,a,H,k})$, where $a = a(z)$, $k$ and $H$ are defined by (7.3.11). For all $t,\ s \in \mathbb{R}$, we have:

$$y \in (a - sk) - H + tk \iff y \in a - H + (t - s)k.$$

Consequently, the optimal solutions $y$ to $(\mathrm{P}_{F,a,H,k})$ do not change if $a$ is replaced by $\bar{a} := a - sk$ with $s \in \mathbb{R}$. We have $\bar{a} = a(\bar{z})$ for the corresponding parameter $\bar{z}$ in (P), where $\bar{z} = z + s \cdot \mathbf{1}$.

(b) For $z \in \mathbb{R}^\ell$ and $c \in \mathbb{R}$, take $s := -z_1 + c$. Then $\bar{z}_1 = c$ for $\bar{z} = z + s \cdot \mathbf{1}$.
(c) Take any $z \in \mathbb{R}^\ell$. For $s := -1 - \max\limits_{i=1,\dots,\ell} z_i$, we get $\bar{z} = z + s \cdot \mathbf{1} < 0$. For $s := 1 - \min\limits_{i=1,\dots,\ell} z_i$, we get $\bar{z} = z + s \cdot \mathbf{1} > 0$. □

**Proposition 7.3.42** *Assume that $F$ is a nonempty closed set for which* (Sp-ex) *holds. Then for all* $w \in \operatorname{int} \mathbb{R}^\ell_+$ :

$$\mathrm{Eff}(F) \subseteq \mathrm{cl} \bigcup_{z \in \mathbb{R}^\ell,\, n \in \mathbb{N}} \mathrm{M}_{w,z,\frac{1}{n}}, \quad \textit{especially}$$

$$\mathrm{Eff}(F) \subseteq \mathrm{cl} \bigcup_{z \in W,\, n \in \mathbb{N}} \mathrm{M}_{w,z,\frac{1}{n}}, \quad \textit{where}$$

$$W := \{z \in \mathbb{R}^\ell \mid \exists y \in F \ \forall i \in \{1, \ldots, \ell\} : \ z_i = -w_i y_i\}.$$

*The Pareto minimal point set of* $F$ *can be approximated by varying* $\ell$ *real-valued parameters in* (P).

***Proof*** The inclusions follow from Proposition 7.3.36(3). The last statement of the proposition results from the first inclusion and Proposition 7.3.41(b). □

However, we have to point out the complications for small parameters $b$ in problem (P). If the computer number of $b$ becomes zero, then the scalarizing problem changes from (P) to

$$\min_{y \in F} \max_{i=1,\ldots,\ell} w_i y_i + z_i,$$

and optimal solutions $y$ to this problem are not necessarily Pareto minimal points of $F$. This follows from Sect. 7.3.1.

Finally, we summarize some advantages and disadvantages of using problem (P) for scalarization. In general, the computational effort for solving (P) is higher than that for other scalarizing problems. But each minimizer $y$ of (P) is a Pareto minimal point of $F$. Any choice of the parameters guarantees the existence of an optimal solution to (P) if $F$ is a nonempty closed set with $\mathrm{GEff}(F) \neq \emptyset$. This combination of the two advantages is offered by none of the scalar problems in the previous sections. Each of those problems has the disadvantage that only the uniqueness of the optimal solution corresponds with efficiency or that one parameter has to be chosen sufficiently large or small in order to ensure the existence of optimal solutions.

Sensitivity results for problem (P) were given by Helbig [233]. An interactive procedure for solving multicriteria optimization problems by means of (P) was developed in [491]. In [479], problem (P) was used to calculate trade-off directions between different objective functions in vector optimization.

Section 7.3.10 is taken from [472], except Propositions 7.3.36(4), 7.3.41 and 7.3.42 and the compactness statements in Proposition 7.3.36 and in Corollary 7.3.38. Propositions 7.3.41(b)–(c) and 7.3.42 as well as the results from [472] were published in [475]. Lemma 7.3.34(a) coincides with [184, Lemma 3.1].

# Chapter 8
# Variational Analysis

In this chapter, we recall well-known notions of generalized differentiation for vector-valued functions as well as set-valued maps (for details, see Mordukhovich [359]). These concepts are very important for deriving optimality conditions (see Sects. 12.1, 12.4, 15.3). Under convexity assumptions, corresponding notions and results for the functional $\varphi_{A,k}$ defined by (4.1.11) are already discussed in Sect. 4.10. The focus here is on developing differentiability properties of nonconvex functions (maps).

## 8.1 Generalized Differentiability of Extended Real-Valued Functions

In order to define objects of generalized differentiation, we recall a standard notion of limits for set-valued maps (see the notions in Sect. 8.2). Here, we always understand limits in the sequential sense.

**Definition 8.1.1** Consider a set-valued map $\Gamma : X \rightrightarrows X^*$ acting between a Banach space $X$ and its dual $X^*$. The sequential **Kuratowski Painlevé upper limit** with respect to the norm topology on $X$ and weak*-toplology on $X^*$ is denoted by

$$\operatorname*{Lim\,sup}_{x \to x_0} \Gamma(x),$$

and defined as

$$\operatorname*{Lim\,sup}_{x \to x_0} \Gamma(x) := \{x^* \in X^* \mid \exists \text{ sequences } x_k \to x_0, \text{ and } x_k^* \xrightarrow{w^*} x^*, \\ \text{with } \; x_k^* \in \Gamma(x_k), \quad \text{for all } k = 1, 2, \ldots\},$$

where $\xrightarrow{w^*}$ denotes convergence in the weak*-topology of $X^*$.



C. Tammer and P. Weidner, *Scalarization and Separation by Translation Invariant Functions*, Vector Optimization, https://doi.org/10.1007/978-3-030-44723-6_8

Suppose that $X$ is a Banach space and $\Omega$ a subset of $X$. The **Fréchet normal cone** to $\Omega$ at $x \in \Omega$ is given by

$$\hat{N}(x; \Omega) := \left\{ x^* \in X^* \,\middle|\, \limsup_{u \xrightarrow{\Omega} x} \frac{x^*(u-x)}{\|u-x\|} \leq 0 \right\}. \tag{8.1.1}$$

We extend this definition:

**Definition 8.1.2** (*Generalized normals*) Consider a nonempty subset $\Omega \subset X$ of a Banach space $X$, $x \in \Omega$ and $\varepsilon \in \mathbb{R}_+$. Then,

$$\hat{N}_\varepsilon(x; \Omega) := \left\{ x^* \in X^* \,\middle|\, \limsup_{u \xrightarrow{\Omega} x} \frac{x^*(u-x)}{\|u-x\|} \leq \varepsilon \right\} \tag{8.1.2}$$

is the **set of $\varepsilon$-normals** to $\Omega$ at $x$.

If $x \notin \Omega$, we set $\hat{N}_\varepsilon(x; \Omega) := \emptyset$ for all $\varepsilon \in \mathbb{R}_+$.

In Sect. 12.6, we will study generalized differentiability properties in Asplund spaces because it is possible to show fundamental results like extremal principles in these spaces (see [312, 359, 361]). Furthermore, several calculus rules in generalized differentiation theory presented in this section work in Asplund spaces. For a detailed explanation, see Mordukhovich [359]. This is important for deriving optimality conditions based on limiting subdifferentials in Sect. 12.1.

**Definition 8.1.3** A subset $\Omega$ of an Asplund space $X$ is said to be **closed around** $\overline{x} \in \Omega$ if there is a neighborhood $U$ of $\overline{x}$ such that $\Omega \cap U$ is a closed set.

**Definition 8.1.4** Let $\Omega$ be a subset of an Asplund space $X$ that is closed around $\overline{x} \in \Omega$. The **basic, limiting or Mordukhovich normal cone** to $\Omega$ at $\overline{x} \in \Omega$ is defined by

$$\begin{aligned} N_M(\overline{x}; \Omega) :=\; & \operatorname{Lim\,sup}_{x \to \overline{x}} \hat{N}(x; \Omega) \\ =\; & \left\{ x^* \in X^* \,\middle|\, \exists\, x_k \xrightarrow{\Omega} \overline{x} \;\exists\, x_k^* \xrightarrow{w^*} x^* :\; x_k^* \in \hat{N}(x_k; \Omega) \right\}, \end{aligned} \tag{8.1.3}$$

i.e., it is the sequential Kuratowski Painlevé upper limit of Fréchet normal cones to $\Omega$ at $x$ as $x$ tends to $\overline{x}$.

If $\overline{x} \notin \Omega$, we set $N_M(\overline{x}; \Omega) := \emptyset$.

We would like to mention that the basic (limiting or Mordukhovich) normal cone (8.1.3) is often nonconvex in contrast to (8.1.2). Nevertheless, the basic (limiting or Mordukhovich) normal cone enjoys a full calculus. Both of the cones (8.1.3) and (8.1.1) coincide with the normal cone of convex analysis (see (3.6.2)) in the case that $\Omega$ is convex.

Now, we introduce the basic subdifferential of an extended real-valued function through basic normals to its epigraph.

**Definition 8.1.5** Let $f : X \to \overline{\mathbb{R}}$ be a proper function. The **basic, limiting or Mordukhovich subdifferential** of $f$ at $x^0 \in X$ is given by

$$\partial_M f(x^0) := \begin{cases} \{x^* \in X^* \mid (x^*, -1) \in N_M((x^0, f(x^0)); \operatorname{epi} f)\} & \text{if } f(x^0) \in \mathbb{R} \\ \emptyset & \text{if } f(x^0) \notin \mathbb{R}. \end{cases} \tag{8.1.4}$$

We call the elements of $\partial_M f(x^0)$ **basic or limiting subgradients** of $f$ at $x^0$.

For deriving optimality conditions for constrained optimization problems, it is important to compute the (basic, limiting or Mordukhovich) subdifferential of an indicator function. We get from [359, Proposition 1.79]:

**Proposition 8.1.6** *Consider a nonempty set $A \subset X$ and its indicator function $\iota_A : X \to \overline{\mathbb{R}}$. Then,*

$$\partial_M \iota_A(x^0) = N_M(x^0; A) \textit{ for any } x^0 \in A.$$

We need the following calculus rule given by Mordukhovich in [359, Theorem 3.36 ] for the proof of necessary optimality conditions in Sect. 12.1.

**Lemma 8.1.7** *Suppose that $X$ is an Asplund space, $x^0 \in X$, and that, for each $i \in \{1, 2\}$, $f_i : X \to \overline{\mathbb{R}}$ is a proper function which is lower semicontinuous on* dom $f_i$. *Moreover, assume that $f_1$ or $f_2$ is locally Lipschitz on some neighborhood of $x^0$. Then, the following inclusion holds*

$$\partial_M(f_1 + f_2)(x^0) \subseteq \partial_M f_1(x^0) + \partial_M f_2(x^0).$$

Statements of this kind are especially useful if $f_2$ is the indicator function expressing the restrictions under which $f_1$ has to be minimized.

Furthermore, we introduce concepts of generalized differentiation with full calculus without assuming the convexity of the function $f$.

We now recall the concept of Clarke's subdifferential (see [98]).

**Definition 8.1.8** Suppose that $X$ is a Banach space and that $f : X \to \overline{\mathbb{R}}$ is proper and lower semicontinuous. Assume that $f$ is Lipschitz on some neighborhood of $x^0 \in X$ with $f(x^0) \in \mathbb{R}$.
The **generalized directional derivative** of $f$ at $x^0$ into direction $u \in X$ is given by

$$f^o(x^0; u) := \limsup_{x \to x^0,\ t \to +0} \frac{f(x + tu) - f(x)}{t}.$$

The **Clarke generalized subdifferential** of $f$ at $x^0$ is the set

$$\partial_C f(x^0) = \{x^* \in X^* \mid \forall u \in X \ : \ x^*(u) \le f^o(x^0; u)\}. \tag{8.1.5}$$

Originally, Clarke's subdifferential was defined via Clarke's normal cone to the epigraph of $f$ under the assumption that $f : X \to \overline{\mathbb{R}}$ is a proper lower semicontinuous function on a Banach space $X$ which is Lipschitz on some neighborhood of $x^0 \in X$ with $f(x^0) \in \mathbb{R}$ by

$$\partial_C f(x^0) = \{x^* \in X^* \mid (x^*, -1) \in N_C((x^0, f(x^0)); \operatorname{epi} f)\} \tag{8.1.6}$$

(see [359, Sect. 1.4.3]).

The next assertion is given by Clarke [98, Proposition 2.3.1].

**Proposition 8.1.9** *Consider $f : X \to \overline{\mathbb{R}}$, $X$ being a Banach space. Assume that $f$ is proper and lower semicontinuous on* dom $f$, *$x^0 \in X$ with $f(x^0) \in \mathbb{R}$ and that $f$ is Lipschitz on some neighborhood of $x^0$. Then,*

$$\partial_C(-f)(x^0) = -\partial_C f(x^0).$$

The following subdifferential was introduced by Ioffe [258, 261].

**Definition 8.1.10** Assume that $X$ is a Banach space and $f : X \to \overline{\mathbb{R}}$ is proper and lower semicontinuous. Consider the collection $\mathscr{F}$ of all finite-dimensional subspaces of $X$ and define $f_{\bar{x}+L}$ for each $\bar{x} \in X$ and $L \in \mathscr{F}$ by:

$$f_{\bar{x}+L}(u) := \begin{cases} f(u) & \text{if } \bar{x} \in L, \\ +\infty & \text{otherwise.} \end{cases}$$

Define, for $\varepsilon \in \mathbb{R}_+$,

$$\partial_\varepsilon^- f_{\bar{x}+L}(\bar{x}) := \left\{ x^* \in X^* \mid \forall x \in X \ : \ x^*(x) \le \varepsilon \|x\| + \liminf_{t \to +0} \frac{f_{\bar{x}+L}(\bar{x}+tx) - f_{\bar{x}+L}(\bar{x})}{t} \right\}.$$

The **approximate or Ioffe subdifferential** of $f$ at $x^0 \in X$ is the set

$$\partial_A f(x^0) := \begin{cases} \bigcap_{L \in \mathscr{F}} \operatorname{Lim\,sup}_{(\varepsilon,\bar{x}) \to (+0,x^0)} \partial_\varepsilon^- f_{\bar{x}+L}(\bar{x}) & \text{if } f(x^0) \in \mathbb{R}, \\ \emptyset & \text{if } f(x^0) \notin \mathbb{R}. \end{cases} \tag{8.1.7}$$

The **approximate or Ioffe normal cone** to $S \subset X$ at $x^0 \in S$ is given by

$$N_A(x^0; S) := \mathbb{R}_> \partial_A d(\cdot, S)(x^0).$$

The approximate (Ioffe) normal cone to a set can equivalently be defined as

$$N_A(x^0, S) = \partial_A \iota_S(x^0) \quad \text{for any } \ x^0 \in S.$$

Beneath all the properties of the above subdifferentials, we would like to recall those ones which will be used in our book. The next proposition is shown in [214, Proposition 21.2].

**Proposition 8.1.11** *Let $X$ and $Y$ be Banach spaces, $f$ and $g$ be functions from $X$ to $\overline{\mathbb{R}}$ which are proper and lower semicontinuous on their domains and $x \in$* dom $f$. *Then, it holds that*

(a) $\partial_A f(x) \subseteq \partial_C f(x)$.
(b) *If $f$ is convex and Lipschitz on some neighborhood of $x$, then* $\partial_A f(x) = \partial_C f(x) = \partial f(x)$, *where $\partial f(x)$ is given by (3.6.4).*
(c) *If $f$ attains a local minimum at $x$, then* $0 \in \partial_A f(x)$.
(d) $\partial_A(f + g)(x) \subseteq \partial_A f(x) + \partial_A g(x)$ *for* $x \in \operatorname{dom} f \cap \operatorname{dom} g$ *if $f$ is Lipschitz on some neighborhood of $x$.*

## 8.2 Lipschitz Continuity of Vector-Valued and Set-Valued Functions

First, we introduce certain useful notions and notation related to vector-valued functions as well as to set-valued maps (multifunctions). Let $X$, $Y$ be arbitrary nonempty sets. A function $\Gamma : X \to \mathscr{P}(Y)$ is called a **set-valued map (multifunction)**, and is denoted by $\Gamma : X \rightrightarrows Y$. Provided that $\Gamma : X \rightrightarrows Y$ is a set-valued map, the image of $\Gamma$ at $x \in X$ is a (possibly empty) subset $\Gamma(x)$ of $Y$.

Suppose that $Y$ is a linear topological space and $C$ is a nontrivial, closed, convex cone in $Y$ and $f : X \to Y$.

The **sublevel and strict sublevel** sets of $f$ at $y \in Y$ are $\operatorname{lev}_{f,\leq_C}(y) := \{x \in X \mid f(x) \leq_C y\}$ and $\operatorname{lev}_{f,<_C}(y) := \{x \in X \mid y - f(x) \in \operatorname{int} C\}$, and the **epigraph** of $f$ is $\operatorname{epi} f := \{(x, y) \in X \times Y \mid f(x) \leq_C y\}$. For such a function $f$, we associate the set-valued map $\Gamma_{f,C} : X \rightrightarrows Y$ whose graph is epi $f$. So, $\Gamma_{f,C}(x) = f(x) + C$ for any $x \in X$, $\operatorname{epi} f = \operatorname{epi} \Gamma_{f,C}$, $\operatorname{lev}_{f,\leq_C}(y) = \operatorname{lev}_{\Gamma_{f,C},\leq_C}(y)$, and $\operatorname{lev}_{f,<_C}(y) = \operatorname{lev}_{\Gamma_{f,C},<_C}(y)$ for any $y \in Y$.

We recall Berge's concepts of lower semicontinuity and upper semicontinuity of a set-valued map $\Gamma : X \rightrightarrows Y$ (see [47]).

Assume that $X$, $Y$ are Banach spaces. As usual, we denote a neighborhood of $y \in Y$ by $V(y)$.

**Definition 8.2.1** *(Berge's semicontinuity)* Let $\Gamma : X \rightrightarrows Y$ be a set-valued map, where $X$ and $Y$ are normed spaces.

$\Gamma$ is **Berge-upper semicontinuous (B-u.s.c.) at x** $\in$ **X** if for each open subset $V$ of $Y$ such that $\Gamma(x) \subset V$, there is a neighborhood $U$ of $x$ satisfying $\Gamma(U) \subset V$.

$\Gamma$ is **Berge-lower semicontinuous (B-l.s.c.) at x** $\in$ **X** if for each open subset $V$ of $Y$ such that $\Gamma(x) \cap V \neq \emptyset$ there is a neighborhood $U$ of $x$ satisfying $\Gamma(U) \cap V \neq \emptyset$.

$\Gamma$ is **closed at** $\overline{\mathbf{x}} \in$ **X** if for all $x_n \to \overline{x}$, $y_n \to \overline{y}$, $y_n \in \Gamma(x_n)$ implies that $\overline{y} \in \Gamma(\overline{x})$. $\Gamma$ is **closed** if it is closed at every $x \in X$, i.e., if graph $\Gamma = \{(x, y) \in X \times Y \mid y \in \Gamma(x)\}$ is a closed set.

In the following, a set-valued map $\Gamma : X \rightrightarrows Y$ is called **continuous at x** $\in$ **X** if $F$ is both B-u.s.c. and B-l.s.c. at $x$.

The **epigraph** of a set-valued map $\Gamma$ with respect to a nontrivial, closed, convex and pointed cone $C \subset Y$ is given by

$$\mathrm{epi}\Gamma := \big\{(x, y) \in X \times Y \;\big|\; y \in \Gamma(x) + C\big\}.$$

The **domain of** $\Gamma$ is

$$\mathrm{dom}\,\Gamma := \{x \in X \mid \Gamma(x) \neq \emptyset\},$$

and its **image** is

$$\mathrm{Im}\,\Gamma := \{y \in Y \mid \exists\, x \in X \;:\; y \in \Gamma(x)\}.$$

Usually, a set-valued map $\Gamma : X \rightrightarrows Y$ is identified with its graph,

$$\mathrm{graph}\,\Gamma := \{(x, y) \in X \times Y \mid y \in \Gamma(x)\}.$$

Obviously,

$$\mathrm{dom}\,\Gamma = P_X(\mathrm{graph}\,\Gamma)$$

and

$$\mathrm{Im}\,\Gamma = P_Y(\mathrm{graph}\,\Gamma),$$

where $P_X(\mathrm{graph}\,\Gamma)$ denotes the projection of graph $\Gamma$ onto $X$ and $P_Y(\mathrm{graph}\,\Gamma)$ the projection of graph $\Gamma$ onto $Y$. The image of the set $A \subset X$ through $\Gamma$ is

$$\Gamma(A) := \bigcup_{x \in A} \Gamma(x);$$

so, Im $\Gamma = \Gamma(X)$. The inverse image of the set $B \subseteq \Gamma(X)$ through $\Gamma$ is

$$\Gamma^{-1}(B) := \{x \in X \mid \Gamma(x) \cap B \neq \emptyset\}.$$

$\Gamma^{-1}(B)$ is the image of $B$ through the inverse set-valued map $\Gamma^{-1} : X \rightrightarrows Y$ given by

$$\Gamma^{-1}(y) := \{x \in X \mid y \in \Gamma(x)\};$$

so $\Gamma^{-1}(y) = \Gamma^{-1}(\{y\})$. Obviously, dom $\Gamma^{-1} =$ Im $\Gamma$ and Im $\Gamma^{-1} =$ dom $\Gamma$. There is also another type of the inverse image: $\Gamma^{+1}(B) := \{x \in X \mid \Gamma(x) \subseteq B\}$.

If $\Delta : X \rightrightarrows W$ is another set-valued map, the **composition** of $\Delta$ and $\Gamma$ is the set-valued map $\Delta \circ \Gamma : X \rightrightarrows W$ given by

$$(\Delta \circ \Gamma)(x) := \{z \in W \mid \exists\, y \in \Gamma(x) \text{ with } z \in \Delta(y)\}.$$

We have

$$\mathrm{graph}(\Delta \circ \Gamma) = P_{X \times W}\big(\mathrm{graph}\,\Gamma \times W \cap X \times \mathrm{graph}\,\Delta\big).$$

Suppose that $Y$ is a linear space and $\Gamma, \Gamma_1, \Gamma_2 : X \rightrightarrows Y$. We define the sum $\Gamma_1 + \Gamma_2$ and multiplication by a scalar $\gamma\Gamma$ as the set-valued maps $\Gamma_1 + \Gamma_2, \gamma\Gamma : X \rightrightarrows Y$ according to

$$(\Gamma_1 + \Gamma_2)(x) := \Gamma_1(x) + \Gamma_2(x),$$

$$(\gamma\Gamma)(x) := \gamma \cdot \Gamma(x)$$

with the usual conventions that $A + \emptyset := \emptyset + A := \emptyset$, $\gamma \cdot \emptyset = \emptyset$ for $A \subset Y$.

Now, we study the Lipschitz behavior of vector- and set-valued maps following the explanations in [291, Sect. 3.3]. For more details, see Mordukhovich [359].

We know from real-valued optimization that one has to suppose certain assumptions concerning the objective function as well as the feasible set, particularly in order to prove the stability of solutions with respect to perturbations or rates of convergence in numerical procedures (see Klatte, Kummer [296]). Assuming the classical Lipschitz continuity (see Definition 8.2.2), it is possible to prove such results. An essential characteristic of the classical Lipschitz continuity (Definition 8.2.2) in comparison with the general continuity for single-valued functions is a *linear rate* of continuity quantified by some modulus (Lipschitz constant $L$).

Characterizations of well-posedness and sensitivity analysis for constraint systems as well as variational systems under certain Lipschitz properties are studied in Chap. 4 of the book by Mordukhovich [359]. Moreover, Lipschitz continuity is essential for deriving calculus rules in generalized differentiation (see Sect. 8.3).

**Definition 8.2.2** Let $X$ and $Y$ be real normed spaces, $f : X \to Y$ and $S \subseteq X$ with $S \neq \emptyset$.
$f$ is called **Lipschitzian on S** if there exists a real number $L \in \mathbb{R}_+$ such that $\|f(x^1) - f(x^2)\| \leq L\|x^1 - x^2\|$ for all $x^1, x^2 \in S$. $f$ is called **locally Lipschitzian on S** if for each $x \in S$ there is a neighborhood $U$ of $x$ such that $f$ is Lipschitzian on $U \cap S$. $f$ is **locally Lipschitz continuous** or **Lipschitzian** if $f$ is locally Lipschitzian or Lipschitzian, respectively, on $X$. $L$ is called a **Lipschitz constant**.

The following theorem [339, Theorem 3.1] presents an assertion about Lipschitz properties of $C$-convex vector-valued functions.

**Definition 8.2.3** Let $X$ and $Y$ be real linear spaces, $C$ a nontrivial convex cone in $Y$, $S \subseteq X$ convex. A vector-valued function $f : S \to Y$ is called **$C$-convex**, if for every $x, y \in S$, $\lambda \in [0, 1]$, one has

$$f(\lambda x + (1 - \lambda)y) \in \lambda f(x) + (1 - \lambda)f(y) - C.$$

**Theorem 8.2.4** *Assume that $C \subset \mathbb{R}^\ell$ is a nontrivial convex cone,* $\operatorname{cl} C$ *is pointed, $S \subseteq \mathbb{R}^n$ is a proper convex set. Moreover, suppose that $f : S \to \mathbb{R}^\ell$ is $C$-convex. Then, $f$ is locally Lipschitzian on the relative interior of $S$.*

Borwein [61, Corollary 2.4] derived a more general result for normed spaces $X$, $Y$ and a cone $C \subset Y$ being normal. In the following theorem, we use the notation $[U_Y]_C := (U_Y + C) \cap (U_Y - C)$.

**Theorem 8.2.5** *Suppose that $X$ and $Y$ are normed spaces. Let $f : X \to Y$ be a $C$-convex function. Suppose that $f$ is continuous at some $x_0 \in X$ and the cone $C$ is normal. Then, $f$ is locally Lipschitzian on $X$. Furthermore, if*

$$\forall x \in x_0 + RU_X : \|f(x) - f(x_0)\| \le M, \tag{8.2.1}$$

*for some $R, M > 0$, then*

$$\forall x, x' \in x_0 + rU_X : \left\|f(x) - f(x')\right\| \le \frac{M\gamma}{R} \cdot \frac{R + r}{R - r} \cdot \left\|x - x'\right\|,$$

*where $r \in (0, R)$ and $\gamma := \sup\{\|y\| \mid y \in [U_Y]_C\}$.*

To prove a subdifferential chain rule (Lemma 8.3.6) or the relationship between the coderivative of a vector-valued function and the subdiffential of its scalarization (see (8.3.6) and [359, Theorem 3.28]), we suppose that a scalar function is strictly Lipschitzian in the following sense.

**Definition 8.2.6** Assume that $X$ and $Y$ are normed spaces and $f : X \to Y$ is a single-valued map. $f$ is called **strictly Lipschitzian** at $\overline{x}$ if $f$ is locally Lipschitzian on a neighborhood of $\overline{x}$ and there is a neighborhood $V$ of zero in $X$ such that the sequence with elements

$$y_k := \frac{f(x_k + t_k w) - f(x_k)}{t_k}, \quad k \in \mathbb{N},$$

contains a norm convergent subsequence whenever $w \in V$, $x_k \to \overline{x}$, and $t_k \downarrow 0$.

**Remark 8.2.7** The function $f$ is strictly Lipschitzian at $\overline{x}$ if and only if the sequence $\{\|u_n\|^{-1}(f(x_n + u_n) - f(x_n))\}$ has a norm converging subsequence whenever $\{x_n\} \subset X$ converges to $\overline{x}$, $\{u_n\} \subset X \setminus \{0\}$ converges to $0$ and the sequence $\{\|u_n\|^{-1} u_n\}$ converges in $X$.

**Remark 8.2.8** Suppose that $Y$ is finite-dimensional. Then, the class of strictly Lipschitzian maps coincides with the class of locally Lipschitzian maps $f : X \to \mathbb{R}^\ell$. In [359, Example 1.35], it is shown that this is not the case for $\dim Y = +\infty$. Every map that is strictly differentiable at $\overline{x}$ is strictly Lipschitzian at $\overline{x}$.

For more details regarding the class of strictly Lipschitzian maps (with values in infinite-dimensional spaces), see [359, Sect. 3.1.3].

Next, we study natural extensions of Lipschitz continuity to set-valued maps. These Lipschitz properties are supposed for deriving a generalized differential calculus and optimality conditions (see Sect. 8.3). Particularly, we need an extension of Lipschitz continuity in order to prove necessary optimality conditions in Sect. 12.1.

We recall several notions of Lipschitzian-like set-valued maps. Consider set-valued maps $\Gamma : X \rightrightarrows Y$, where $X$ and $Y$ are normed spaces, with $\operatorname{dom}\Gamma \neq \emptyset$.

**Definition 8.2.9** Let $U \subset X$ and $V \subset Y$ be nonempty sets. A set-valued map $\Gamma : X \rightrightarrows Y$ is called **Lipschitzian-like on $U$ relative to $V$** if there is a real number $L \geq 0$ such that for every $x, u \in U$ it holds that

$$\Gamma(x) \cap V \subset \Gamma(u) + L \|x - u\| B_Y,$$

where $B_Y$ denotes the closed unit ball of $Y$.

**Remark 8.2.10** When $\Gamma(U) \cap V = \emptyset$, then $\Gamma$ is Lipschitzian-like on $U$ relative to $V$. Provided that $\Gamma : X \rightrightarrows Y$ is Lipschitzian-like on $U$ relative to $V$ and $\Gamma(U) \cap V \neq \emptyset$, then $U \subset \operatorname{dom}\Gamma$.

**Definition 8.2.11** Let $\Gamma : X \rightrightarrows Y$ be a set-valued map and $(\overline{x}, \overline{y}) \in \operatorname{graph}\Gamma$. $\Gamma$ is called **locally Lipschitzian-like at $(\overline{x}, \overline{y})$ with modulus $L \geq 0$** if there are neighborhoods $U$ of $\overline{x}$ and $V$ of $\overline{y}$ such that for every $x, u \in U$ it holds that

$$\Gamma(x) \cap V \subset \Gamma(u) + L \|x - u\| B_Y.$$

The infimum of all such moduli $L$ is named the **exact local Lipschitz bound of $\Gamma$ at $(\overline{x}, \overline{y})$** and is denoted by $\operatorname{Lip}\Gamma(\overline{x}, \overline{y})$.

**Remark 8.2.12** Note that the local Lipschitzian-like property is also known as the **pseudo-Lipschitz property** or the **Aubin property** of set-valued maps.

If we suppose $V = Y$ in Definitions 8.2.9 and 8.2.11, we obtain the following definition:

**Definition 8.2.13** Let $U \subset X$ be a nonempty set and $\Gamma : X \rightrightarrows Y$ a set-valued map.

(a) $\Gamma : X \rightrightarrows Y$ is named **locally Lipschitz continuous on $U$** if there is a real number $L \geq 0$ such that for every $x, u \in U$ it holds that

$$\Gamma(x) \subset \Gamma(u) + L \|x - u\| B_Y.$$

(b) $\Gamma$ is called **locally Lipschitzian at $\overline{x}$ with the exact bound $\operatorname{Lip}\Gamma(\overline{x})$** if $V = Y$ in Definition 8.2.11.

**Remark 8.2.14** Note, provided that $\Gamma : X \rightrightarrows Y$ is Lipschitzian on $U$, it holds that $U \cap \operatorname{dom}\Gamma = \emptyset$ or $U \subset \operatorname{dom}\Gamma$.

**Remark 8.2.15** For single-valued maps $f : S \subseteq X \to Y$, all the properties in Definitions 8.2.11 and 8.2.13 coincide with the classical Lipschitz continuity given in Definition 8.2.2.

**Remark 8.2.16** We would like to mention that the local Lipschitzian-like property of $\Gamma : X \rightrightarrows Y$ in Definition 8.2.11 can be considered as a localization of the Lipschitzian behavior not only relative to a point of the domain but also relative to a particular point of the image $\overline{y} \in \Gamma(\overline{x})$.

This leads us to the characterization of local Lipschitzian-like maps in the next theorem in terms of the local Lipschitz continuity of the scalar distance function $d(u, A) := \inf_{a\in A} \|u - a\|$ $(u \in Y,\ A \subset Y)$ (with the convention $\inf \emptyset := +\infty$) to the moving set $\Gamma(x)$ with respect to both variables $(x, y)$.

The next result is shown by Mordukhovich [359, Theorem 1.41].

**Theorem 8.2.17** (Scalarization of the Lipschitzian-like property) *Let $\Gamma : X \rightrightarrows Y$ be a set-valued map and $(\overline{x}, \overline{y}) \in \operatorname{graph}\Gamma$. The following conditions are equivalent:*

(a) *$\Gamma$ is locally Lipschitzian-like at $(\overline{x}, \overline{y})$.*
(b) *The function $\mu : X \times Y \to \overline{\mathbb{R}}$ defined by*

$$\mu(x, y) := d(y, \Gamma(x)) := \inf_{w\in\Gamma(x)} \|y - w\|$$

*(with the convention $\inf \emptyset = +\infty$) is locally Lipschitzian on a neighborhood of $(\overline{x}, \overline{y})$.*

**Remark 8.2.18** Note that the local characteristics in Definitions 8.2.11 and 8.2.13 are stable (robust) with respect to small perturbations of the reference points. These characteristics hold for $\Gamma$ if and only if they hold for the map $\overline{\Gamma} : X \rightrightarrows Y$ with $\overline{\Gamma}(x) := \operatorname{cl}\Gamma(x)$.

Taking into account Definition 8.2.13(a), we conclude that the Lipschitz continuity of $\Gamma$ on $U$ is equivalent to

$$\forall x, u \in U : \qquad \operatorname{haus}(\Gamma(x), \Gamma(u)) \le L\|x - u\|,$$

where $\operatorname{haus}(A_1, A_2)$ is the **Hausdorff distance** (or Pompieu-Hausdorff distance) between $A_1, A_2 \subset Y$ given by

$$\operatorname{haus}(A_1, A_2) := \inf\{\mu \ge 0 \mid A_1 \subset A_2 + \mu B_Y,\ A_2 \subset A_1 + \mu B_Y\}.$$

We would like to mention that the Hausdorff distance provides a **metric** on the space of all nonempty and compact (even closed and bounded) subsets of $Y$. So, when a set-valued map $\Gamma : X \rightrightarrows Y$ is compact-valued, its Lipschitz continuity in the sense of Definition 8.2.13 is equivalent to the Lipschitz continuity of a single-valued map $x \to \Gamma(x)$ from $X$ to the space of all nonempty, compact subsets of $Y$ equipped with the Haussdorff metric (see Definition 8.2.2).

Now, we discuss some relationships between the local Lipschitzian (Definition 8.2.13(b)) and Lipschitzian-like properties (Definition 8.2.9) of set-valued maps. Provided that $\Gamma$ is locally Lipschitzian on a neighborhood of $\overline{x} \in \operatorname{dom}\Gamma$, then it is locally Lipschitzian-like at $(\overline{x}, \overline{y})$ for every $\overline{y} \in \Gamma(\overline{x})$ with

$$\mathrm{Lip}\Gamma(\overline{x}) \geq \sup\{\mathrm{Lip}\Gamma(\overline{x}, \overline{y}) \mid \overline{y} \in \Gamma(\overline{x})\}. \tag{8.2.2}$$

This follows directly from the definitions. Under some additional assumptions concerning $\Gamma$, the converse assertion holds with equality in (8.2.2). The next theorem is proven by Mordukhovich [359, Theorem 1.42] under the assumption that $\Gamma : X \rightrightarrows Y$ is **compact around** $\overline{x} \in \mathrm{dom}\,\Gamma$, i.e., there is a neighborhood $U$ of $\overline{x}$ such that $\mathrm{cl}\,\Gamma(U)$ is compact, that is $\Gamma(U)$ is relatively compact. Moreover, we suppose that $\Gamma$ is **closed at** $\overline{x}$, i.e., for every $y \notin \Gamma(\overline{x})$ there are neighborhoods $U$ of $\overline{x}$ and $V$ of $y$ such that $\Gamma(x) \cap V = \emptyset$ for all $x \in U$. $\Gamma$ is **closed at** $\overline{x}$ in this sense if $\Gamma$ is closed at $\overline{x}$ in the sense of Definition 2.5.7 (ii) in [192]. $\Gamma$ is closed at $\overline{x}$ yields that $\Gamma$ is closed-valued at $\overline{x}$, i.e., $\Gamma(\overline{x})$ is closed.

In the next theorem by Mordukhovich [359, Theorem 1.42], we study the assumptions for the equivalence of the conditions that $\Gamma$ is locally Lipschitzian on a neighborhood of $\overline{x}$ (Definition 8.2.13(b)) and that $\Gamma$ is locally Lipschitzian-like on a neighborhood of $(\overline{x}, \overline{y})$ (Definition 8.2.11).

**Theorem 8.2.19** (Lipschitz continuity of locally compact set-valued maps) *Suppose that $\Gamma : X \rightrightarrows Y$ is closed at $\overline{x} \in \mathrm{dom}\,\Gamma$ and compact around this point.*

*Then, $\Gamma$ is locally Lipschitzian on a neighborhood of $\overline{x}$ if and only if it is locally Lipschitzian-like at $(\overline{x}, \overline{y})$ for every $\overline{y} \in \Gamma(\overline{x})$. In this case, we obtain*

$$\mathrm{Lip}\Gamma(\overline{x}) = \max\{\mathrm{Lip}\Gamma(\overline{x}, \overline{y}) \mid \overline{y} \in \Gamma(\overline{x})\} < +\infty.$$

Now, we consider a map $\Gamma : X \rightrightarrows Y$ acting between Asplund spaces (see Definition 2.1.39) with the image space $Y$ partially ordered by a nontrivial, closed, convex and pointed cone $C \subset Y$. We would like to recall that the **epigraph** of $\Gamma$ with respect to the ordering cone $C$ is given by

$$\mathrm{epi}\,\Gamma = \big\{(x, y) \in X \times Y \;\big|\; y \in \Gamma(x) + C\big\}.$$

The **epigraphical multifunction** of $\Gamma$ denoted by $\mathcal{E}_\Gamma : X \rightrightarrows Y$ is given by

$$\mathcal{E}_\Gamma(x) := \big\{y \in Y \big|\; y \in \Gamma(x) + C\big\} \quad \text{with} \quad \mathrm{graph}\,\mathcal{E}_\Gamma = \mathrm{epi}\,\Gamma. \tag{8.2.3}$$

**Definition 8.2.20** We say that a map $\Gamma : X \rightrightarrows Y$ is **epigraphically Lipschitzian-like** (ELL) around $(\overline{x}, \overline{y})$ with modulus $L \geq 0$ if there are neighborhoods $U$ of $\overline{x}$ and $V$ of $\overline{y}$ with

$$\mathcal{E}_\Gamma(x) \cap V \subset \mathcal{E}_\Gamma(u) + L\|x - u\| B_Y \quad \text{for all} \quad x, u \in U,$$

where $B_Y$ denotes the closed unit ball of $Y$.

This means, a set-valued map $\Gamma$ is ELL at $(\overline{x}, \overline{z})$ if its epigraphical multifunction $\mathcal{E}_\Gamma$ is Lipschitzian-like at that point (Definition 8.2.11). Note that this property is known also as the Aubin property, the pseudo Lipschitz property (compare [359]). In the case of single-valued functions, it agrees with the classical local Lipschitzian behavior. If $V = Y$, it coincides with the standard local Lipschitz property for set-valued maps.

## 8.3 Differentiability of Set-Valued Maps

First, we introduce the **Clarke coderivative** for a set-valued map (see [276], [277] for further approaches).

**Definition 8.3.1** Let $X$ and $Y$ be Banach spaces, $\Gamma : X \rightrightarrows Y$ and $(x, y) \in \operatorname{graph}\Gamma$. Suppose that $\Gamma$ is closed. The *Clarke coderivative* $D^*_C\Gamma(x, y)$ of $\Gamma$ at $(x, y)$ is a set-valued map from $Y^*$ to $X^*$ given by

$$D^*_C\Gamma(x, y)(y^*) := \{x^* \in X^* \mid (x^*, -y^*) \in N_C((x, y); \operatorname{graph}\Gamma)\}.$$

We now turn to the notions of coderivatives and subdifferentials for set-valued maps according to [20–23]. Suppose there is a set-valued map $\Gamma\colon X \rightrightarrows Y$ with its graph

$$\operatorname{graph}\Gamma = \big\{(x, y) \in X \times Y \big|\ y \in \Gamma(x)\big\}$$

acting between Asplund spaces $X$ and $Y$.

**Definition 8.3.2** Suppose there is a set-valued map $\Gamma\colon X \rightrightarrows Y$ between Asplund spaces. Assume that $\operatorname{graph}\Gamma$ is locally closed around $(\overline{x}, \overline{y}) \in \operatorname{graph}\Gamma$. The **(basic, normal, limiting, Mordukhovich) coderivative** map $D^*_M\Gamma(\overline{x}, \overline{y})\colon Y^* \rightrightarrows X^*$ of $\Gamma$ at $(\overline{x}, \overline{y}) \in \operatorname{graph}\Gamma$ is defined by

$$D^*_M\Gamma(\overline{x}, \overline{y})(y^*) := \big\{x^* \in X^* \big|\ x^*(-y^*) \in N_M\big((\overline{x}, \overline{y}); \operatorname{graph}\Gamma\big)\big\}. \quad (8.3.1)$$

If $\Gamma = f\colon X \to Y$ is single-valued, we omit $\overline{y} = f(\overline{x})$ in (8.3.1). Provided that $f\colon X \to Y$ is *strictly differentiable* at $\overline{x}$ (this is automatically fulfilled if $f \in C^1$ around this point), then

$$D^*_M f(\overline{x})(y^*) = \big\{\nabla f(\overline{x})^*(y^*)\big\} \text{ for all } y^* \in Y^*.$$

Moreover, we consider the Fréchet coderivative map of $\Gamma$ at $(\overline{x}, \overline{y}) \in \operatorname{graph}\Gamma$ (see Bao, Mordukhovich [23]) using the Fréchet normal cone defined by (8.1.1).

**Definition 8.3.3** The **Fréchet coderivative** map $\hat{D}^*\Gamma(\overline{x}, \overline{y})\colon Y^* \rightrightarrows X^*$ of $\Gamma$ at $(\overline{x}, \overline{y}) \in \operatorname{graph}\Gamma$ is given by

$$\hat{D}^*\Gamma(\overline{x}, \overline{y})(y^*) := \big\{x^* \in X^* \big|\ (x^*, -y^*) \in \hat{N}\big((\overline{x}, \overline{y}); \operatorname{graph}\Gamma\big)\big\}. \quad (8.3.2)$$

**Remark 8.3.4** We would like to mention (see Bao, Mordukhovich [23]) that both coderivative set-valued maps $D^*_M\Gamma(\overline{x}, \overline{y})\colon Y^* \rightrightarrows X^*$ and $\hat{D}^*\Gamma(\overline{x}, \overline{y})\colon Y^* \rightrightarrows X^*$ are positively homogeneous functions of $y^* \in Y^*$ reducing to the single-valued adjoint derivative linear operator

$$D^*_M f(\overline{x}) = \{\nabla f(\overline{x})^*(y^*)\}, \qquad \hat{D}^* f(\overline{x}) = \{\nabla f(\overline{x})^* y^*\} \tag{8.3.3}$$

(for all $y^* \in Y^*$) under the assumption that $f$ is Fréchet differentiable at $\overline{x}$ for the second equality and strictly differentiable at this point for the first one, i.e.,

$$\lim_{x,u \to \overline{x}} \frac{f(x) - f(u) - \nabla f(\overline{x})(u - x)}{\|x - u\|} = 0,$$

which is automatically fulfilled if $f \in C^1$ around $\overline{x}$.

We define appropriate extensions of the subdifferential notion from extended real-valued functions to vector-valued and set-valued maps with values in partially ordered spaces using coderivatives of the epigraphical multifunction (see (8.2.3)). In [21], the following definition of subdifferentials for set-valued maps $\Gamma : X \rightrightarrows Y$ acting between Asplund spaces $X$ and $Y$ is introduced. It is further developed in [20, 22, 23] (compare the epi-coderivatives for set-valued maps in [514] and references therein).

**Definition 8.3.5** Consider $(\overline{x}, \overline{y}) \in \operatorname{graph}\Gamma$. The **(limiting, basic, normal or Mordukhovich) subdifferential** of $\Gamma$ at $(\overline{x}, \overline{y})$ in direction $y^* \in Y^*$ is defined by

$$\partial_M \Gamma(\overline{x}, \overline{y})(y^*) := D^*_M \mathscr{E}_\Gamma(\overline{x}, \overline{y})(y^*), \tag{8.3.4}$$

where $\mathscr{E}_\Gamma$ is the epigraphical multifunction of $\Gamma$ defined in (8.2.3).

Bao and Mordukhovich pointed out in [23], that the implication

$$D^*_M \mathscr{E}_\Gamma(\overline{x}, \overline{y})(y^*) \neq \emptyset \quad \Longrightarrow \quad -y^* \in N(0; C) \tag{8.3.5}$$

always holds. Therefore, the requirement $-y^* \in N(0; C)$ is redundant for the definitions of the subdifferential for set-valued maps in [20–23], especially in [21, Definition 2.1]. Provided that $\Gamma = f : X \to Y$ is a single-valued vector-valued function, we omit $\overline{y}$ from the notation of subdifferentials. The subdifferential of $f$ at $\overline{x}$ in (8.3.4) turn into

$$\partial_M f(\overline{x})(y^*) = D^*_M \mathscr{E}_f(\overline{x})(y^*).$$

If $\Gamma = f : X \to (-\infty, \infty]$ is an extended real-valued function that is lower semicontinuous on dom $f$, the subdifferential (8.3.4) with $\|y^*\|_* = 1$ coincides with the (basic, limiting, Mordukhovich) subdifferential (see [359] and Definition 8.1.5). The following relationship between the coderivative of a vector-valued function and the subdiffential of its scalarization

$$D^*_M f(\overline{x})(y^*) = \partial_M (y^* \circ f)(\overline{x}) \tag{8.3.6}$$

holds if $f$ is strictly Lipschitzian at $\bar{x}$ in the sense of Definition 8.2.6 (see [359, Theorem 3.28]).

We will not recall all the calculus rules for basic (limiting, Mordukhovich) normal cones and coderivatives. But, we refer to them in Mordukhovich's books [359, 360].

The following calculus rule is shown by Mordukhovich in [359, Theorem 3.41 and Corollary 3.43] and very important for deriving Lagrangian multiplier rules for vector optimization problems in Sect. 12.1.1.

**Lemma 8.3.6** *Let $X$ and $Y$ be Asplund spaces. Suppose that $f : X \to Y$ is strictly Lipschitzian at $x^0$ and $\varphi : Y \to \overline{\mathbb{R}}$ is proper and locally Lipschitzian on a neighborhood of $f(x^0)$. Then, it holds that*

$$\partial_M(\varphi \circ f)(x^0) \subseteq \bigcup_{y^* \in \partial_M \varphi(f(x^0))} \partial_M(y^* \circ f)(x^0).$$

**Remark 8.3.7** The limiting subdifferential $\partial_M$ of the scalarizing functional (4.1.11) for general (nonconvex) sets $A$ is determined in Theorem 12.1.8 (compare the recent papers [19] by Bao, Hillmann and Tammer and [29] by Bao and Tammer).

Furthermore, we introduce the approximate (Ioffe) coderivative and study the relationship between the Clarke coderivative and the approximate (Ioffe) coderivative.

**Definition 8.3.8** Suppose that $\Gamma : X \rightrightarrows Y$, $X$ and $Y$ are Banach spaces and $(x, y) \in \operatorname{graph}\Gamma$. Let $\Gamma$ be closed. Then, the **approximate (Ioffe) coderivative** $D_A^*\Gamma(x, y)$ of $\Gamma$ at $(x, y)$ is a set-valued map from $Y^*$ to $X^*$ defined by

$$D_A^*\Gamma(x, y)(y^*) := \{x^* \in X^* \mid (x^*, -y^*) \in N_A((x, y); \operatorname{graph}\Gamma)\}.$$

The following proposition concerning the relationship between the Clarke coderivative and the approximate (Ioffe) coderivative is shown by Ha in [214, Proposition 21.2].

**Proposition 8.3.9** *Suppose that $X$ and $Y$ are Banach spaces, $\Gamma$ is a set-valued closed map from $X$ to $Y$ and $x \in \operatorname{dom}\Gamma$. Then, $D_A^*\Gamma(x, y)(y^*) \subseteq D_C^*\Gamma(x, y)(y^*)$ for $(x, y) \in \operatorname{graph}\Gamma$ and $y^* \in Y^*$.*

# Chapter 9
# Special Cases and Functionals Related to $\varphi_{A,k}$

In Sect. 9.1, we study the functional $\varphi_{A,k}$ given by (4.1.11) where the involved set $A$ is constructed by block norms or oblique norms. These sets $A$ are described by linear inequalities. The general case for sets $A$ given by linear inequalities was investigated in Sect. 4.8. Now, we characterize efficient elements by means of functionals where oblique norms are involved and weakly efficient elements via scalarization using block norms. Our results are applied to multiobjective d.c. (difference of convex functions) optimization problems (see Sects. 9.1.3 and 9.1.4).

Furthermore, in Sect. 9.2, we discuss several fundamental properties of the directional minimal time function that are important for applications in locational analysis (see Sect. 15.3).

In Sect. 9.3, we present relationships between the nonlinear separating functional $\varphi_{A,k}$ introduced in (4.1.11), and the oriented distance by Hiriart-Urruty.

## 9.1 Functionals $\varphi_{A,k}$ Related to Block Norms and Oblique Norms

The aim of this section is to study the functional $\varphi_{A,k}$ given by (4.1.11) where the involved set $A$ is constructed by block norms or oblique norms. This approach is derived by Tammer and Winkler in [438] and extended by Günther and Popovici in [204]. Employing norms for the construction of the set $A$ in (4.1.11), we get an easy computation and a special structure of the scalarizing functional (4.1.11) that is useful from the theoretical as well as numerical point of view. Furthermore, we derive important monotonicity properties of the scalarizing functional for this special case (see Corollaries 9.1.19 and 9.1.20).

Important properties of the functional $\varphi_{A,k}$ were given in Theorems 4.2.1, 4.4.4, 4.5.5 and 4.6.6. The special case that $A$ is a polyhedral set was investigated in



C. Tammer and P. Weidner, *Scalarization and Separation by Translation Invariant Functions*, Vector Optimization, https://doi.org/10.1007/978-3-030-44723-6_9

Sect. 4.8. We will use the studied properties for deriving the results in this section. These results and the special structure of the functional can be applied in numerical procedures for the determination of efficient solutions (introduced in Chaps. 6 and 7) of vector optimization problems and in algorithms for solving set-valued optimization problems (see Sect. 13.5).

Whereas some of our results also hold in general topological vector spaces, we confine ourselves to $Y = \mathbb{R}^\ell$, $\ell \geq 2$. Especially in Corollary 9.1.1, we use the special structure of $\mathbb{R}^\ell$ in order to get a useful description of the scalarizing functional.

Taking into account the useful properties of the scalarizing functional $\varphi_{A,k}$ in the special case of a polyhedral set $A$, we establish necessary as well as sufficient conditions for solutions of multiobjective optimization problems in Theorems 9.1.22 and 9.1.26. Under certain (explicitly formulated) conditions concerning the structure of the set $A$, we can derive strict monotonicity properties of the scalarizing functional $\varphi_{A,k}$ and improve the possibilities to distinguish efficient elements from weakly efficient ones. The special scalarizing functional is useful for the selection of the most-preferred efficient solution by the decision maker. We will apply these results to multiobjective d.c. optimization problems in Sect. 9.1.4. In due consideration the special structure of the set $A$ in the definition of the scalarizing functional, we preserve the typical d.c. structure during scalarization.

### 9.1.1 Constructions Involving Block Norms and Oblique Norms

Consider the functional $\varphi_{A,k}$ introduced in (4.1.11) by

$$\varphi_{A,k} : \mathbb{R}^\ell \to \overline{\mathbb{R}}, \quad \varphi_{A,k}(y) = \inf\{t \in \mathbb{R} \mid y \in tk + A\}, \tag{9.1.1}$$

where $A$ is a proper, closed subset of $\mathbb{R}^\ell$ and $k \in -0^+A \setminus \{0\}$.

In this section, the shifted set $A$ in the definition of the scalarizing functional (9.1.1) will be constructed using norms. We employ this special structure of the set $A$ in order to derive a characterization of the solutions of multiobjective problems by scalarizing functionals with suitable numerical properties (see Corollary 9.1.1).

First, recall our notations. As already mentioned, we use upper indices in order to distinguish between different elements of a set, for instance $y^1, y^2 \in \mathbb{R}^\ell$, and we use lower indices for the components of a vector: $y = (y_1, \ldots, y_\ell)^T \in \mathbb{R}^\ell$. The basis vectors of $\mathbb{R}^\ell$, i.e., $\mathfrak{e}_i^j = 1$ if $i = j$ and $\mathfrak{e}_i^j = 0$ otherwise, are denoted by $\mathfrak{e}^j$, $j \in \{1, \ldots, \ell\}$. In this section, the scalar product in $\mathbb{R}^\ell$ is termed by the symbol $\langle ., . \rangle$

$$\langle y^1, y^2 \rangle = y_1^1 y_1^2 + \cdots + y_\ell^1 y_\ell^2.$$

The shifted set $A$ in the functional $\varphi_{A,k}$ will now be constructed based on the following set:

$$B_L := \{y \in \mathbb{R}^\ell \mid \langle a^i, y\rangle \le \alpha_i,\ a^i \in \mathbb{R}^\ell,\ \alpha_i \in \mathbb{R},\ i \in \{1, \dots, n\}\} \qquad (9.1.2)$$

with $a^i \in \mathbb{R}^\ell$, $i \in \{1, \dots, n\}$.

From Chaps. 6 and 7, we know that it is important to have certain strict monotonicity properties of the functional (9.1.1). In Corollary 9.1.20 we prove strict monotonicity properties of the functional (9.1.1) under the assumption that an oblique norm (see Definition 9.1.3) is involved in the construction of the functional. In order to derive important properties of oblique norms (see Lemma 9.1.12), we will use the index set $I$ defined by

$$I := \{i \in \{1, \dots, n\} \mid \{y \in \mathbb{R}^\ell \mid \langle a^i, y\rangle = \alpha_i\} \cap B_L \cap \operatorname{int} \mathbb{R}^\ell_+ \neq \emptyset\}. \qquad (9.1.3)$$

The set $I$ is exactly the set of indices $i \in \{1, \dots, n\}$ for which the hyperplanes $\langle a^i, y\rangle = \alpha_i$ are active in the nonnegative orthant (compare Lemma 9.1.11).

Using the index set $I$ given by (9.1.3), we define a set $A_L \subset \mathbb{R}^\ell$ by

$$A_L := \{y \in \mathbb{R}^\ell \mid \langle a^i, y\rangle \le \alpha_i,\ i \in I\}. \qquad (9.1.4)$$

Obviously, the set $A_L$ is a polyhedral set and thus closed. Furthermore, we have

$$\operatorname{int} A_L := \{y \in \mathbb{R}^\ell \mid \langle a^i, y\rangle < \alpha_i,\ i \in I\}.$$

Let $B_L$ and the corresponding set $A_L$ be defined by (9.1.2) and (9.1.4), respectively. Moreover, let vectors $k \in -0^+A_L \setminus \{0\}$ and $w \in \mathbb{R}^\ell$ be given and consider the functional $\varphi_{A,k}$ (see (9.1.1)) with $A = w + A_L$, i.e., we study the functional $\varphi_{w+A_L,k} : \mathbb{R}^\ell \to \overline{\mathbb{R}}$ of type (9.1.1) given by

$$\varphi_{w+A_L,k}(y) = \inf\{t \in \mathbb{R} \mid y \in tk + w + A_L\}, \qquad y \in \mathbb{R}^\ell. \qquad (9.1.5)$$

The functional $\varphi_{w+A_L,k}$ depends on the set $A_L$ and the parameters $k$ and $w$. If $w = 0$, we will often write $\varphi_{A_L,k}$ instead of $\varphi_{0+A_L,k}$.

As a conclusion from the assertions in Sect. 4.8, we get the following result.

**Corollary 9.1.1** *Let $B_L$ given by (9.1.2), $w \in \mathbb{R}^\ell$ be arbitrarily chosen. Consider the corresponding set $A_L$ given by (9.1.4) with the index set (9.1.3), $k \in -0^+A_L \setminus \{0\}$ and the functional $\varphi_{w+A_L,k}$ defined by (9.1.5). Suppose that $\langle a^i, k\rangle \neq 0$ for all $i \in I$.*

*Then, the functional $\varphi_{w+A_L,k}$ defined by formula (9.1.5) has the structure*

$$\varphi_{w+A_L,k}(y) = \max_{i \in I} \frac{\langle a^i, y\rangle - \langle a^i, w\rangle - \alpha_i}{\langle a^i, k\rangle}. \qquad (9.1.6)$$

The formula (9.1.6) will be used in Sect. 13.5 for deriving a descent method for solving set-valued optimization problems.

In order to derive certain monotonicity properties of the functional $\varphi_{w+A_L,k}$ in (9.1.5) with $A_L$ introduced in (9.1.4), we study the case where the involved set $B_L$ in (9.1.2) is given by the unit ball of a suitable polyhedral norm.

It is well known that polyhedral norms or block norms turn out to be useful measures of distances, especially in locational analysis, approximation theory and optimization theory. Block norms can be used in order to approximate arbitrary norms since the set of block norms is dense in the set of all norms (see Ward and Wendell [460]). Moreover, block norms imply piecewise linearity of the corresponding distance measure because of their polyhedral structure. This nice property often leads to effectively solvable (piecewise linear) optimization problems.

In multiobjective optimization, several authors have used norms in order to compute efficient solutions that are closest to the utopia point, compare Yu [500], Steuer and Choo [427], Jahn [269], Lewandowski and Wierzbicki [324] (see Chaps. 6 and 7).

Schandl, Klamroth and Wiecek [403], [404], explore block norms and so-called oblique norms in order to generate efficient solutions of multiobjective optimization problems. They apply a certain norm scalarization concept based on the Minkowski functional (see Definition 2.1.30) generated by block norms and oblique norms and derive necessary as well as sufficient conditions for efficiency.

In contrary to the approach in [403], [404], we are using the scalarizing functional (9.1.5) in this section.

The main aim of this section is to show that it is possible to use the unit ball of a certain block norm or oblique norm $\gamma$ in order to define the set $B_L$ in (9.1.2) and the corresponding set $A_L$ in the definition of the scalarizing functional (9.1.5) and to derive several monotonicity properties of the functional (9.1.5) depending on the choice of the norm $\gamma$ in the definition of the set $B_L := B_\gamma$ in (9.1.2).

Consider a norm $\gamma$ on $\mathbb{R}^\ell$ that is characterized by its (closed) unit ball

$$B_\gamma := \{y \in \mathbb{R}^\ell \mid \gamma(y) \leq 1\}. \tag{9.1.7}$$

Clearly, $B_\gamma$ is a closed, convex, symmetric (i.e., $y \in B_\gamma \Rightarrow -y \in B_\gamma$) set with $0 \in \operatorname{int} B_\gamma$. Reversed, consider a closed, convex, symmetric set $B \subset \mathbb{R}^\ell$ with $0 \in \operatorname{int} B$. Then, $B$ characterizes a norm $\gamma$ defined by

$$\gamma(y) := p_B(y) = \inf\,\{t \in \mathbb{R}_+ \mid y \in tB\}$$

(see Definition 2.1.30 with $A = B \subset \mathbb{R}^\ell$, $\gamma = p_B$). This means that it is no matter whether we consider a norm $\gamma$ given as a function or by its unit ball. In this section, we mainly consider norms given by unit balls.

**Definition 9.1.2** A norm $\gamma$ is called a **block norm**, if its unit ball $B_\gamma$ is polyhedral.

Obviously, the unit ball of a block norm is a polytope. Then, it is the convex hull of finitely many elements of $\mathbb{R}^\ell$. This fact is used by several authors for defining the block norm.

If $B_\gamma$ is described by (9.1.2), then the subsets $F_i := \{y \in B_\gamma \mid \langle a^i, y\rangle = \alpha_i\}$ ($i \in \{1, \ldots, n\}$) are called *faces* of the unit ball $B_\gamma$. Throughout this section, we suppose that the system (9.1.2) contains no redundant inequalities, i.e., that each set $F_i$ consists of at least two different elements.

In case that $\gamma$ is a block norm with a unit ball $B_\gamma$ given by (9.1.2), then

$$\text{int}\, B_\gamma = \{y \in B_\gamma \mid \langle a^i, y\rangle < \alpha_i,\ i \in \{1, \ldots, n\}\}, \tag{9.1.8}$$

$$\text{bd}\, B_\gamma = \{y \in B_\gamma \mid \exists\, i,\ 1 \leq i \leq n : \langle a^i, y\rangle = \alpha_i\}. \tag{9.1.9}$$

This means, bd $B_\gamma$ is equivalently given by the union of all faces of $B_\gamma$.

Schandl, Klamroth and Wiecek introduced the following notion of so-called oblique norms in [403].

**Definition 9.1.3** Consider $\bar{y} \in \mathbb{R}^\ell$. The **reflection set of** $\bar{\mathbf{y}}$ is defined by

$$R(\bar{y}) := \{y \in \mathbb{R}^\ell \mid \forall\, i \in \{1, \ldots, \ell\} : \ |y_i| = |\bar{y}_i|\ \}.$$

A norm $\gamma$ is called **absolute**, if

$$\forall y \in R(\bar{y}) : \quad \gamma(y) = \gamma(\bar{y}). \tag{9.1.10}$$

Furthermore, a block norm $\gamma$ is called **oblique**, if $\gamma$ is absolute and satisfies

$$\forall y \in \mathbb{R}^\ell_+ \cap \text{bd}\, B_\gamma : \quad (y - \mathbb{R}^\ell_+) \cap \mathbb{R}^\ell_+ \cap \text{bd}\, B_\gamma = \{y\}. \tag{9.1.11}$$

The property that a norm is absolute provides a certain geometrical symmetry of the unit ball $B_\gamma$. Indeed, the unit balls of absolute norms are completely described by their positive parts $B_\gamma \cap \mathbb{R}^\ell_+$.

The next lemma is shown in [438, Lemma 2.2].

**Lemma 9.1.4** *Consider the unit ball $B_\gamma$ of an absolute norm $\gamma$. Then,*

$$\forall y \in B_\gamma : \quad \text{conv}\, R(y) \subseteq B_\gamma.$$

***Proof*** Since $\gamma$ is absolute, $R(y) \subseteq B_\gamma$ holds for each $y \in B_\gamma$. Then, the convexity of $B_\gamma$ yields conv $R(y) \subseteq B_\gamma$. □

The following useful property is derived in [438, Lemma 2.3].

**Lemma 9.1.5** *Let the unit ball of an oblique norm $\gamma$ be given by the inequality system $\langle a^i, y\rangle \leq \alpha_i, i \in \{1, \ldots, n\}$. Then, it holds that $\langle a^i, e^j\rangle \neq 0$ for all $i \in \{1, \ldots, n\}$ and for each unit vector $e^j$, $j \in \{1, \ldots, \ell\}$.*

***Proof*** The vectors $a^i$, $i \in \{1, \dots, n\}$, are the normal vectors at the corresponding faces $F_i = \{y \in B_\gamma \mid \langle a^i, y\rangle = \alpha_i\}$ and the assertion follows directly from [403, Lemma 2.9] and [404, Lemma 5]. □

The following examples are given in [438, Examples 2.1 and 2.2].

**Example 9.1.6** The unit ball of the Euclidean norm $\|\cdot\|$ in $\mathbb{R}^\ell$,

$$\|y\| = \sqrt{\langle y, y\rangle} = \sqrt{y_1 y_1 + \cdots + y_\ell y_\ell}, \qquad y = (y_1, \dots, y_\ell)^T \in \mathbb{R}^\ell,$$

is absolute and satisfies the second condition in the definition of oblique norms. However, $\|\cdot\|$ is not a block norm, so $\|\cdot\|$ is absolute but not oblique.

**Example 9.1.7** The unit ball of the maximum norm $\|\cdot\|_\infty$ in $\mathbb{R}^\ell$ is an absolute block norm. We have $(0, y_2, \dots, y_\ell)^T \in ((y_1, \dots, y_\ell)^T - \mathbb{R}^\ell_+) \cap \mathbb{R}^\ell_+ \cap \operatorname{bd} B_{\|\cdot\|_\infty}$ for each $y = (y_1, \dots, y_\ell)^T \in \mathbb{R}^\ell_+ \cap \operatorname{bd} B_{\|\cdot\|_\infty}$, so $\|\cdot\|_\infty$ is not oblique.

**Example 9.1.8** Let $\bar{y} = (\bar{y}_1, \dots, \bar{y}_\ell)^T \in \operatorname{int} \mathbb{R}^\ell_+$, i.e., $\bar{y}_i > 0$ for all $i \in \{1, \dots, \ell\}$. Consider a norm $\gamma$ defined by

$$B_\gamma := \operatorname{conv} R(\bar{y}) = \bigcup \{R(y) \mid y \in (\bar{y} - \mathbb{R}^\ell_+) \cap \mathbb{R}^\ell_+\}.$$

Of course, $B_\gamma$ is closed. The corresponding system of inequalities is given by

$$\langle \mathfrak{e}^i, y\rangle \leq |\bar{y}_i| = \bar{y}_i, \qquad -\langle \mathfrak{e}^i, y\rangle \leq |\bar{y}_i| = \bar{y}_i, \qquad i \in \{1, \dots, \ell\}.$$

This follows easily by $\gamma(y) \leq 1$ if and only if $y \in \operatorname{conv} R(\bar{y})$, respectively if $|y_i| \leq \bar{y}_i$ for all $i \in \{1, \dots, n\}$. Taking into account Lemma 9.1.5, this norm is not oblique: For $i \neq j$ we have $\langle a^i, \mathfrak{e}^j\rangle = \pm\langle \mathfrak{e}^i, \mathfrak{e}^j\rangle = 0$.

**Remark 9.1.9** Consider $\bar{y} = (\bar{y}_1, \dots, \bar{y}_\ell)^T \in \operatorname{int} \mathbb{R}^\ell_+$ and define a norm $\gamma$ by $B_\gamma := \operatorname{conv} R(\bar{y})$. From Example 9.1.8, we know that the corresponding system of inequalities is given by

$$\langle \mathfrak{e}^i, y\rangle \leq \bar{y}_i, \qquad -\langle \mathfrak{e}^i, y\rangle \leq \bar{y}_i, \qquad i \in \{1, \dots, \ell\}.$$

Taking into account Lemma 9.1.5, we get

$$\begin{aligned} A_L &= \{y \in \mathbb{R}^\ell \mid \langle \mathfrak{e}^i, y\rangle \leq \bar{y}_i,\ i \in \{1, \dots, \ell\}\} \\ &= \{y \in \mathbb{R}^\ell \mid y_i \leq \bar{y}_i,\ i \in \{1, \dots, \ell\}\} \\ &= \bar{y} - \mathbb{R}^\ell_+. \end{aligned}$$

This means that the set $A_L$ is given by shifting the negative of the cone $\mathbb{R}^\ell_+$ (see [438, Examples 3.1]).

We set $w = -\bar{y}$ and consider $A_L = \bar{y} - \mathbb{R}^\ell_+$, $k \in \operatorname{int} \mathbb{R}^\ell_+$. Then, we get for all $y \in \mathbb{R}^\ell$

$$\varphi_{w+A_L,k}(y) = \inf\{t \in \mathbb{R} \mid y \in tk - \bar{y} + \bar{y} - \mathbb{R}^\ell_+\} = \inf\{t \in \mathbb{R} \mid y \in tk - \mathbb{R}^\ell_+\} = \varphi_{-\mathbb{R}^\ell_+,k}(y).$$

Accordingly to Proposition 4.13.3, $\|\cdot\|_{-\mathbb{R}^\ell_+,k}$ denotes the norm that is given as the Minkowski functional of the order interval $[-k, k]_{-\mathbb{R}^\ell_+}$ (see Theorem 2.2.10). Taking into account (9.1.6), for all $y \in Y$, it holds that

$$\begin{aligned}\|y\|_{-\mathbb{R}^\ell_+,k} &= \max\{\max_{i\in\{1,\dots,\ell\}} \frac{\langle \mathfrak{e}^i, y\rangle - \langle \mathfrak{e}^i, -\bar{y}\rangle - \bar{y}_i}{\langle \mathfrak{e}^i, -k\rangle}, \max_{i\in\{1,\dots,\ell\}} \frac{\langle -\mathfrak{e}^i, y\rangle - \langle -\mathfrak{e}^i, \bar{y}\rangle - \bar{y}_i}{\langle -\mathfrak{e}^i, k\rangle}\} \\ &= \max\{\max_{i\in\{1,\dots,\ell\}} -\frac{y_i}{k_i}, \max_{i\in\{1,\dots,\ell\}} \frac{y_i}{k_i}\}. \qquad (9.1.12)\end{aligned}$$

Using the special structure of the involved set $A$ in $\varphi_{A,k}$, we get with (9.1.12) a result corresponding to (4.13.2).

The next example is given in [438, Example 2.3].

**Example 9.1.10** Assume that $C$ is a nontrivial, polyhedral, convex, pointed cone with $\mathbb{R}^\ell_+\setminus\{0\} \subset \operatorname{int} C \neq \emptyset$. $C$ can be expressed by

$$C = \{y \in \mathbb{R}^\ell \mid \langle b^i, y\rangle \le 0,\ i \in \{1,\dots,m\}\},$$

where $b^i \in \mathbb{R}^\ell$ ($i \in \{1,\dots,m\}$). Taking into account the assumption $\mathbb{R}^\ell_+\setminus\{0\} \subset \operatorname{int} C \neq \emptyset$, $\mathfrak{e}^j \in C$ implies $b^i \notin \mathbb{R}^\ell_+$ for all $i \in \{1,\dots,m\}$. Because $C$ is assumed to be pointed, we get even $b^i \in -\mathbb{R}^\ell_+$ for all $i \in \{1,\dots,m\}$. In order to construct a norm $\gamma$ with the corresponding unit ball $B_\gamma$, we will use the cone $C$. Then, we prove that this norm $\gamma$ is oblique.

Consider $\bar{y} = (\bar{y}_1,\dots,\bar{y}_\ell)^T \in \operatorname{int}\mathbb{R}^\ell_+$, i.e., $\bar{y}_i > 0$ for all $i \in \{1,\dots,\ell\}$. Analogously to Example 9.1.8, we define

$$\begin{aligned}B_\gamma &:= \bigcup_{y\in(\bar{y}-C)\cap\mathbb{R}^\ell_+} R(y) \\ &= \{y = (y_1,\dots,y_\ell)^T \in \mathbb{R}^\ell_+ \mid \tilde{y} := (|y_1|,\dots,|y_\ell|)^T \in \bar{y} - C\},\end{aligned}$$

where $R(\cdot)$ denotes the reflection set. First, we prove that $B_\gamma$ can be expressed by a corresponding system of inequalities:

$$y \in B_\gamma \iff \langle -a^i, y - \bar{y}\rangle \le 0,\ a^i \in R(b^i),\ i \in \{1,\dots,m\}. \qquad (9.1.13)$$

In order to prove (9.1.13), note that we have $y \in B_\gamma$ if and only if

$$\langle -b^i, (|y_1|,\dots,|y_\ell|)^T\rangle \le -\langle b^i, \bar{y}\rangle, \qquad i \in \{1,\dots,m\}.$$

For each vector $z \in \mathbb{R}^\ell$, we introduce the index set $J_+(z) := \{j \mid 1 \le j \le \ell, z_j \ge 0\}$. Because of $-b^i \in \mathbb{R}^\ell_+$, we get for all $a^i \in R(b^i)$, $i \in \{1,\dots,m\}$, and $y \in \mathbb{R}^\ell$

$$\langle -b^i, (|y_1|, \dots, |y_\ell|)^T \rangle = \langle -\bar{a}^i, y \rangle \geq \langle -a^i, y \rangle,$$

where $\bar{a}^i \in R(b^i)$ and $J_+(-\bar{a}^i) = J_+(y)$. This completes the proof of (9.1.13).

Of course, the norm $\gamma$ is absolute. In order to show obliqueness, we take $\bar{y} \in \mathbb{R}^\ell_+ \cap \operatorname{bd} B_\gamma$ arbitrarily. Suppose the existence of an element $\tilde{y} \in \mathbb{R}^\ell_+ \cap \operatorname{bd} B_\gamma$, $\tilde{y} \neq \bar{y}$, such that $\tilde{y} \in \bar{y} - \mathbb{R}^\ell_+$. We obtain from $\mathbb{R}^\ell_+ \setminus \{0\} \in \operatorname{int} C$ that $\tilde{y} \in \bar{y} - \operatorname{int} C$ and hence $\tilde{y} \in \operatorname{int} B_\gamma$, a contradiction to $\tilde{y} \in \operatorname{bd} B_\gamma$.

Now, we consider the case that the polyhedral set $B_L$ is given by the unit ball $B_\gamma$ (see (9.1.7)) of an absolute block norm $\gamma$, i.e., $B_L = B_\gamma$, and the corresponding set $A_L$ (see (9.1.4)) involved in the definition of the functional $\varphi_{w+A_L,k}$ given by (9.1.5). We state some geometric properties of this set $A_L$ given in [438, Lemma 3.1]. Of course, the properties of the set $A_L$ are important for the properties of the functional $\varphi_{w+A_L,k}$ (see Sect. 9.1.2).

**Lemma 9.1.11** *For each block norm $\gamma$ with unit ball $B_\gamma(= B_L)$, the index set $I$ be defined by (9.1.3) and the corresponding set $A_L$ given by (9.1.4), it holds that $B_\gamma \subseteq A_L$ and $B_\gamma \cap \mathbb{R}^\ell_+ = A_L \cap \mathbb{R}^\ell_+$.*

***Proof*** $I \subseteq \{1, \dots, n\}$ ($I$ according to (9.1.3)) yields $B_\gamma \subseteq A_L$ and $B_\gamma \cap \mathbb{R}^\ell_+ \subseteq A_L \cap \mathbb{R}^\ell_+$.

Conversely, suppose that there is an element $\bar{y} \in A_L \cap \mathbb{R}^\ell_+$ with $\bar{y} \notin B_\gamma \cap \mathbb{R}^\ell_+$. Of course, $\bar{y} \notin B_\gamma$, so we obtain $\gamma(\bar{y}) > 1$ and $\tilde{y} := \bar{y}/\gamma(\bar{y}) \in (\operatorname{bd} B_\gamma) \cap \mathbb{R}^\ell_+$. Because of $\tilde{y} \in \operatorname{bd} B_\gamma$, there is an index $j \in \{1, \dots, n\}$ such that $\langle a^j, \tilde{y} \rangle = \alpha_j$. This implies $\tilde{y} \in B_\gamma \cap \{y \in \mathbb{R}^\ell \mid \langle a^j, y \rangle = \alpha_j\} \cap \mathbb{R}^\ell_+$, i.e., $j \in I$. On the other hand, from $\langle a^j, \bar{y} \rangle = \gamma(\bar{y})\langle a^j, \tilde{y} \rangle = \gamma(\bar{y})\alpha_j > \alpha_j$ and $\bar{y} \in A_L$, we conclude $j \notin I$. □

The next lemma (shown in [438, Lemma 3.2]) gives an important property of the parameters $a^i$, $i \in I$. Indeed, it includes an equivalent definition of the set $A_L$.

**Lemma 9.1.12** *Suppose that $\gamma$ is an absolute block norm with the unit ball*

$$B_\gamma(= B_L) = \{y \in \mathbb{R}^\ell \mid \langle a^i, y \rangle \leq \alpha_i, \ i \in \{1, \dots, n\}\}.$$

*Let the index set $I$ be defined by (9.1.3). Then, $a^i \in \mathbb{R}^\ell_+$ if and only if $i \in I$. If $\gamma$ is an oblique norm, $a^i \in \operatorname{int} \mathbb{R}^\ell_+$ holds if and only if $i \in I$.*

***Proof*** We decompose the proof in three steps:

(1) We prove that $i \in I$ implies $a^i \in \mathbb{R}^\ell_+$: Taking into account the definition of the index set $I$, we obtain for each $i \in I$ some $y^i \in B_\gamma \cap \operatorname{int} \mathbb{R}^\ell_+$ with $\langle a^i, y^i \rangle = \alpha_i$. Detect that $\gamma(y^i) = 1$, since each $y^i$ is contained in some face of $B_\gamma$. For each $j \in \{1, \dots, \ell\}$, we define an element $u^{ij} \in \mathbb{R}^\ell_+$ via $u^{ij}_j := 0$ and $u^{ij}_k := y^i_k$ for $k \neq j$. Obviously, $u^{ij} \neq y^i$ and $u^{ij} \in \operatorname{conv} R(y^i)$ for all $i \in I$ and $j \in \{1, \dots, \ell\}$. Taking into account that $\gamma$ is absolute, we obtain $\gamma(u^{ij}) \leq \gamma(y^i) = 1$ and hence $\langle a^i, u^{ij} \rangle \leq \alpha_i$ for all $i \in I$ and $j \in \{1, \dots, \ell\}$. This implies for all $i \in I$ and $j \in \{1, \dots, \ell\}$

$$0 = \alpha_i - \alpha_i \geq \langle a^i, u^{ij} - y^i \rangle = \langle a^i, -y^i_j \mathfrak{e}^j \rangle = -y^i_j a^i_j.$$

Since $y^i \in \operatorname{int} \mathbb{R}^\ell_+$, we get $a^i_j \geq 0$ for all $i \in I$ and $j \in \{1, \ldots, \ell\}$.

(2) In order to prove the opposite direction, we suppose that $a^i \notin \mathbb{R}^\ell_+$ for some $i \in I$. Then, there is a component $a^i_j < 0$, $1 \leq j \leq \ell$, and some $y \in B_\gamma \cap \operatorname{int} \mathbb{R}^\ell_+$ with $\langle a^i, y \rangle = \alpha_i$. We conclude

$$\langle a^i, y - y_j \mathfrak{e}^j \rangle = \langle a^i, y \rangle - y_j \langle a^i, \mathfrak{e}^j \rangle = \langle a^i, y \rangle - y_j a^i_j > \langle a^i, y \rangle = \alpha_i$$

(remind $y_j > 0$ and $a^i_j < 0$), such that $y - y_j \mathfrak{e}^j \notin B_\gamma$. On the other hand, we get $y - y_j \mathfrak{e}^j \in \operatorname{conv} R(y)$, where $R(y)$ denotes the reflection set of $y$. Then, Lemma 9.1.4 implies $y - y_j \mathfrak{e}^j \in B_\gamma$, in contradiction to $y - y_j \mathfrak{e}^j \notin B_\gamma$.

(3) Finally, we study the case where $\gamma$ is oblique. Again, we consider the element $u^{ij}$ from Part (1) of this proof and verify $u^{ij} \in (y^i - \mathbb{R}^\ell_+) \cap \mathbb{R}^\ell_+$. If $\gamma$ is an oblique norm, we conclude $u^{ij} \notin \operatorname{bd} B_\gamma$ from Definition 9.1.3, i.e., $u^{ij} \in \operatorname{int} B_\gamma$. This yields $\langle a^i, u^{ij} \rangle < \alpha_i$ for all $i \in I$ and $j \in \{1, \ldots, \ell\}$ and finally,

$$0 = \alpha_i - \alpha_i > \langle a^i, u^{ij} - y^i \rangle = \langle a^i, -y^i_j \mathfrak{e}^j \rangle = -y^i_j a^i_j.$$

From $y^i \in \operatorname{int} \mathbb{R}^\ell_+$, we now obtain $a^i_j > 0$ for all $i \in I$ and $j \in \{1, \ldots, \ell\}$.

□

The next example is presented in [438, Example 3.2].

**Example 9.1.13** Consider $\bar{y} = (\bar{y}_1, \ldots, \bar{y}_\ell)^T \in \operatorname{int} \mathbb{R}^\ell_+$ and a polyhedral, convex, pointed cone $C$ with $\mathbb{R}^\ell_+ \setminus \{0\} \subset \operatorname{int} C \neq \emptyset$,

$$C := \{y \in \mathbb{R}^\ell \mid \langle b^i, y \rangle \leq 0,\ i \in \{1, \ldots, m\}\},$$

where $b^i \in \mathbb{R}^\ell$ $(i \in \{1, \ldots, m\})$. We introduce a norm $\gamma$ by $B_\gamma := \bigcup_{y \in (\bar{y} - C) \cap \mathbb{R}^\ell_+} R(y)$. From Example 9.1.10, we know that the corresponding system of inequalities is given by

$$\langle -a^i, y - \bar{y} \rangle \leq 0,\ \ a^i \in R(b^i),\ \ i \in \{1, \ldots, m\}.$$

This implies

$$\begin{aligned} A_L &= \{y \in \mathbb{R}^\ell \mid \langle -a^i, y - \bar{y} \rangle \leq 0,\ a^i \in R(b^i) \cap -\mathbb{R}^\ell_+,\ i \in \{1, \ldots, m\}\} \\ &= \{y \in \mathbb{R}^\ell \mid \langle b^i, \bar{y} - y \rangle \leq 0,\ i \in \{1, \ldots, m\}\} \\ &= \{y \in \mathbb{R}^\ell \mid \bar{y} - y \in C\} \\ &= \bar{y} - C, \end{aligned}$$

i.e., the set $A_L$ is given by shifting the negative of the cone $C$.

The assertions in the following lemma (shown in [438, Lemma 3.3]) are useful for deriving monotonicity as well as strict monotonicity properties of the scalarizing functional (compare Theorems 4.5.5 and Corollary 4.5.6).

**Lemma 9.1.14** *Suppose that $\gamma$ is an absolute block norm with corresponding unit ball $B_\gamma$, $B_L = B_\gamma$ and consider the corresponding set $A_L$ according to (9.1.4). Then, for all $k \in B_\gamma \cap \mathbb{R}^\ell_+$, it holds that*

$$-\mathbb{R}^\ell_+ \subseteq A_L - k \quad \textit{and} \quad -\operatorname{int}\mathbb{R}^\ell_+ \subseteq \operatorname{int} A_L - k.$$

*Furthermore,*

$$-\mathbb{R}^\ell_+ \setminus \{0\} \subseteq \operatorname{int} A_L - k$$

*holds if $\gamma$ is an oblique norm.*

***Proof*** Suppose that $\gamma$ is an absolute block norm, $k \in B_\gamma \cap \mathbb{R}^\ell_+$ and $y \in \mathbb{R}^\ell_+$ arbitrary. By considering $k \in B_\gamma$, we get $\langle a^i, k\rangle \leq \alpha_i$ for $i \in \{1, \ldots, n\}$ and $\langle a^i, y\rangle \geq 0$ for $i \in I$ by Lemma 9.1.12. This yields for all $i \in I$

$$\langle a^i, k - y\rangle = \langle a^i, k\rangle - \langle a^i, y\rangle \leq \langle a^i, k\rangle \leq \alpha_i$$

and therefore $k - y \in A_L$, i.e., $-\mathbb{R}^\ell_+ \subseteq A_L - k$.

If we take $y \in \operatorname{int}\mathbb{R}^\ell_+$ or if $\gamma$ is an oblique norm, we conclude $\langle a^i, y\rangle > 0$ for $i \in I$ and finally $\langle a^i, k - y\rangle < \alpha_i$, $i \in I$. □

**Remark 9.1.15** It is important to mention that Lemma 9.1.14 holds not only for $k \in \operatorname{bd} B_\gamma \cap \mathbb{R}^\ell_+$, but also for $k \in B_\gamma \cap \mathbb{R}^\ell_+$. However, we will use this assertion only for $k \in \operatorname{bd} B_\gamma$.

**Remark 9.1.16** We will apply Lemma 9.1.14 for a characterization of solutions of vector optimization problems by means of the functional $\varphi_{w+A_L,k}$ (see (9.1.5)) with $k \in \operatorname{bd} B_\gamma \cap \operatorname{int}\mathbb{R}^\ell_+$ in Sect. 9.1.3.

### *9.1.2 Monotonicity Properties of the Functional $\varphi_{w+A_L,k}$*

In this section, we derive several important properties of the functional $\varphi_{w+A_L,k}$ like monotonicity and continuity under the assumption that the involved set $B_L$ (in (9.1.2)) is given by the unit ball of an absolute (oblique) block norm $\gamma$.

We establish these properties of $\varphi_{w+A_L,k}$ (see (9.1.5)) taking into account Theorems 4.6.6 and 4.5.5 as well as Proposition 4.6.13. The results given in the following corollaries will be used in Sect. 9.1.3 in order to characterize efficient as well as weakly efficient elements of vector optimization problems.

It is important to mention that the assumptions of Theorem 4.6.6 are fulfilled for $k \in \operatorname{int}\mathbb{R}^\ell_+$, $A = w + A_L$ (where $w \in \mathbb{R}^\ell$, $A_L$ is given by (9.1.4)) under certain assumptions concerning the norm $\gamma$ with the unit ball $B_\gamma = B_L$ included in (9.1.4) (compare the proof of Corollary 9.1.17).

The following properties of the functional $\varphi_{w+A_L,k}$ are shown in [438, Corollary 3.1].

**Corollary 9.1.17** *Consider arbitrary vectors $k \in \operatorname{int} \mathbb{R}^\ell_+$, $w \in \mathbb{R}^\ell$. Suppose that $B_L$ (in (9.1.2)) is the unit ball of an absolute norm $\gamma$. Let the corresponding set $A_L$ be given by (9.1.4) with the index set $I$ defined by (9.1.3) and the functional $\varphi_{w+A_L,k}$ be defined by (9.1.5). Moreover, let $F$ be a proper subset of $\mathbb{R}^\ell$. Then,*

(a) $\varphi_{w+A_L,k}$ *is a finite-valued, continuous, convex functional from $\mathbb{R}^\ell$ onto $\mathbb{R}$.*
(b) $\varphi_{w+A_L,k}(w + A_L) \leq 0$, $\varphi_{w+A_L,k}(w + \operatorname{bd} A_L) = 0$, $\varphi_{w+A_L,k}(w + \operatorname{int} A_L) < 0$.
(c) $F \cap (w + \operatorname{int} A_L) = \emptyset$ *if and only if* $\varphi_{w+A_L,k}(F) \geq 0$.
(d) $\varphi_{w+A_L,k}$ *is $D$-monotone for each set $D \subset \mathbb{R}^\ell$ with* $\operatorname{bd} A_L - D \subseteq A_L$. $\varphi_{w+A_L,k}$ *is strictly $D$-monotone for each set $D \subset \mathbb{R}^\ell$ with* $\operatorname{bd} A_L - (D \setminus \{0\}) \subseteq \operatorname{int} A_L$.
(e) $\varphi_{w+A_L,k}$ *is subadditive if* $\operatorname{bd} A_L + \operatorname{bd} A_L + w \subseteq A_L$.

***Proof*** We operate Theorem 4.6.6 choosing $A$ as $w + A_L$ and Theorem 4.5.5 for $D \subset \mathbb{R}^\ell$. Of course, $A_L$ has a nonempty interior. The inclusion $w + A_L - \mathbb{R}_{>}k \subseteq w + \operatorname{int} A_L$ follows directly from $a^i \in \mathbb{R}^\ell_+ \setminus \{0\}$ for $i \in I$, $\langle a^i, y \rangle > 0$ for $i \in I$ and $y \in \operatorname{int} \mathbb{R}^\ell_+$. Taking into account Proposition 4.6.11, this inclusion is equivalent to $w + A_L - \operatorname{int} \mathbb{R}^\ell_+ \subseteq w + A_L$ such that we obtain $k \in -\operatorname{int} 0^+(w + A_L)$ from Proposition 4.6.18. So, the assumptions of Proposition 4.6.13 are fulfilled. From Proposition 4.6.13, we get that $\varphi_{w+A_L,k}$ is finite-valued and continuous. Obviously, $w + A_L$ is convex. By Theorem 4.4.4(f), $\varphi_{w+A_L,k}$ is a convex functional such that (a) holds. The conditions (b) and (c) follow from Theorems 4.4.4 and 4.6.6. Furthermore, we get (d) from Corollary 4.5.6. In order to prove (e), we consider the equivalence of $\operatorname{bd} A_L + w + \operatorname{bd} A_L \subseteq A_L$ and $(w + \operatorname{bd} A_L) + (w + \operatorname{bd} A_L) \subseteq w + A_L$. The second inclusion is exactly the condition for subadditivity in Theorem 4.4.4(h) taking into account Lemma 4.4.7. □

**Remark 9.1.18** For assertion (c) in Corollary 9.1.17, we need $\operatorname{int} A_L \neq \emptyset$. In our setting, this is trivially fulfilled by the construction of the set $A_L$.

The following monotonicity properties of the functional $\varphi_{w+A_L,k}$ are shown in [438, Corollaries 3.2 and 3.3]. In the following corollaries, we construct the set $A_L$ using block norms as well as oblique norms such that the functional $\varphi_{A,k}$ has different monotonicity properties depending from the construction of $A_L$. So, we improve the possibility to distinguish weakly efficient elements from efficient ones (see Sect. 9.1.3).

**Corollary 9.1.19** *Suppose that $B_L$ in (9.1.2) is the unit ball of an absolute block norm $\gamma$. Furthermore, let $k \in \operatorname{int} \mathbb{R}^\ell_+$ and $w \in \mathbb{R}^\ell$ be arbitrarily chosen. We consider the corresponding set $A_L$ given by (9.1.4) with the index set $I$ defined by (9.1.3) and the functional $\varphi_{w+A_L,k}$ defined by (9.1.5).*

*Then, $\varphi_{w+A_L,k}$ is $\mathbb{R}^\ell_+$-monotone and strictly* $(\operatorname{int} \mathbb{R}^\ell_+)$*-monotone.*

***Proof*** Consider $y^1 \in \text{bd}\, A_L$, $y^2 \in \mathbb{R}^\ell_+$ and $y^3 \in \text{int}\, \mathbb{R}^\ell_+$, arbitrarily chosen. Taking into account the definition of the set $A_L$ and Lemma 9.1.12, we obtain for all $i \in I$ the inequalities $\langle a^i, y^1 \rangle \leq \alpha_i$, $\langle a^i, y^2 \rangle \geq 0$ and $\langle a^i, y^3 \rangle > 0$. This implies $\langle a^i, y^1 - y^2 \rangle \leq \alpha_i$ and $\langle a^i, y^1 - y^3 \rangle < \alpha_i$ for all $i \in I$, i.e., $y^1 - y^2 \in A_L$ and $y^1 - y^3 \in \text{int}\, A_L$, respectively. So we have $\text{bd}\, A_L - \mathbb{R}^\ell_+ \subseteq A_L$ and $\text{bd}\, A_L - \text{int}\, \mathbb{R}^\ell_+ \subseteq \text{int}\, A_L$, respectively. We now apply part (d) of Corollary 9.1.17 which implies the monotonicity assertions. □

**Corollary 9.1.20** *Suppose that $B_L$ in (9.1.2) is the unit ball of an oblique block norm $\gamma$. Let $k \in \text{int}\, \mathbb{R}^\ell_+$ and $w \in \mathbb{R}^\ell$ be arbitrarily chosen. Furthermore, consider the corresponding set $A_L$ given by (9.1.4) with the index set $I$ defined by (9.1.3) and the functional $\varphi_{w+A_L,k}$ defined by (9.1.5). Then, $\varphi_{w+A_L,k}$ is strictly $\mathbb{R}^\ell_+$-monotone.*

***Proof*** We proceed as in the proof of Corollary 9.1.19 with $y^1 \in \text{bd}\, A_L$ and $y^2 \in \mathbb{R}^\ell_+ \setminus \{0\}$. Taking into account the definition of the set $A_L$ and Lemma 9.1.12, we obtain for all $i \in I$ the inequalities $\langle a^i, y^1 \rangle \leq \alpha_i$ and $\langle a^i, y^2 \rangle > 0$. This implies $\langle a^i, y^1 - y^2 \rangle < \alpha_i$ for all $i \in I$, i.e., $y^1 - y^2 \in \text{int}\, A_L$. So, we get $\text{bd}\, A_L - (\mathbb{R}^\ell_+ \setminus \{0\}) \subseteq \text{int}\, A_L$. Now, we apply part (d) of Corollary 9.1.17 such that we obtain the mentioned monotonicity property. □

**Remark 9.1.21** We get the following properties of the functional $\varphi_{w+A_L,k}$ from Corollaries 9.1.17, 9.1.19, 9.1.20 and Corollary 9.1.1. Let $B_L$ in (9.1.2) be the unit ball of an absolute (oblique) block norm $\gamma$. Let $k \in \text{int}\, \mathbb{R}^\ell_+$ and $w \in \mathbb{R}^\ell$ be arbitrarily chosen. Consider the corresponding set $A_L$ given by (9.1.4) and the functional $\varphi_{w+A_L,k}$ defined by (9.1.5).

Then, $\varphi_{w+A_L,k}$ is a finite-valued, continuous, convex, $\mathbb{R}^\ell_+$-monotone (strictly $\mathbb{R}^\ell_+$-monotone) functional with

$$\varphi_{w+A_L,k}(y) = \max_{i \in I} \frac{\langle a^i, y \rangle - \langle a^i, w \rangle - \alpha_i}{\langle a^i, k \rangle}. \tag{9.1.14}$$

### *9.1.3 Characterization of Solutions to Multiobjective Optimization Problems*

In Sect. 9.1, we studied a special case (9.1.5) of the nonlinear scalarizing functional $\varphi_{A,k}$ given by (4.1.11), where $A$ in the definition of the functional is given by a system of linear inequalities such that $A$ is a polyhedral set. We will use the properties of the scalarizing functional in this special case (see Corollary 9.1.17) in order to characterize efficiency via scalarization by means of the functional $\varphi_{w+A_L,k}(y) = \inf\{t \in \mathbb{R} \mid \; y \in tk + w + A_L\}$, $y \in \mathbb{R}^\ell$, introduced in (9.1.5) with $A_L = \{y \in \mathbb{R}^\ell \mid \; \langle a^i, y \rangle \leq \alpha_i, \; i \in I\}$ ($\alpha_i \in \mathbb{R}$), $k \in \text{int}\, \mathbb{R}^\ell_+$, $w \in \mathbb{R}^\ell$, the unit ball $B_L$ (in (9.1.2)) of an absolute (oblique) block norm $\gamma$ and the index set $I = \{i \in \{1, \ldots, n\} \mid \{y \in \mathbb{R}^\ell \mid \langle a^i, y \rangle = \alpha_i\} \cap B_\gamma \cap \text{int}\, \mathbb{R}^\ell_+ \neq \emptyset\}$.

By considering a special polyhedral set $A_L$ in the definition of the functional (9.1.5) that is useful from the theoretical as well as from the numerical point of view, we present characterizations of efficient and weakly efficient elements of multiobjective optimization problems on the base of a scalarization by means of the functional (9.1.5). In Sect. 9.1.4, we will apply our results for the case of multiobjective d.c. problems.

The following characterization of solutions to multiobjective optimization problems (see Sect. 6.2) by means of the functional $\varphi_{A_L,k}$ is shown in [438, Theorem 3.2].

**Theorem 9.1.22** (Necessary conditions) *Assume that $F \subseteq \operatorname{int} \mathbb{R}^\ell_+$ is a nonempty set.*

(a) *If $\bar{y} \in \operatorname{WEff}(F, \mathbb{R}^\ell_+)$, then there is a unit ball $B_\gamma (= B_L)$ of a block norm $\gamma$ and the corresponding set $A_L$ given by (9.1.4) with the index set I defined by (9.1.3), such that for $w = 0$ and for each $k \in \operatorname{bd} B_L \cap \operatorname{int} \mathbb{R}^\ell_+$ and the functional $\varphi_{A_L,k}$ defined by (9.1.5), it holds that*

$$\forall\, y \in F : \quad \varphi_{A_L,k}(y) \geq \varphi_{A_L,k}(\bar{y}) = 0.$$

(b) *If $\bar{y} \in \operatorname{Eff}(F, \mathbb{R}^\ell_+)$, then there is a unit ball $B_\gamma (= B_L)$ of a block norm $\gamma$ and the corresponding set $A_L$ given by (9.1.4) with the index set I defined by (9.1.3), such that for $w = 0$ and for each $k \in \operatorname{bd} B_L \cap \operatorname{int} \mathbb{R}^\ell_+$ and the functional $\varphi_{A_L,k}$ defined by (9.1.5), it holds that*

$$\forall\, y \in F \backslash \{\bar{y}\} : \quad \varphi_{A_L,k}(y) > \varphi_{A_L,k}(\bar{y}) = 0.$$

(c) *If $\bar{y} \in \operatorname{Eff}(F, C)$ for some convex, polyhedral cone $C$ with $\mathbb{R}^\ell_+ \backslash \{0\} \subset \operatorname{int} C$, then we can chose the norm $\gamma$ in (b) to be oblique.*

***Proof*** Consider a nonempty set $F \subseteq \mathbb{R}^\ell_+$.

(a) Take $\bar{y} \in \operatorname{WEff}(F, \mathbb{R}^\ell_+)$ and define a norm $\gamma$ by

$$B_\gamma (= B_L) := \operatorname{conv} R(\bar{y}),$$

where $R(\bar{y})$ denotes the reflection set of $\bar{y}$ (see Definition 9.1.3). From Remark 9.1.9, we obtain $A_L = \bar{y} - \mathbb{R}^\ell_+$ as the corresponding set $A_L$ for the scalarization functional. Then, $\bar{y} \in \operatorname{WEff}(F, \mathbb{R}^\ell_+)$ is equivalent to $\emptyset = F \cap (\bar{y} - \operatorname{int} \mathbb{R}^\ell_+) = F \cap \operatorname{int} A_L$, and by Corollary 9.1.17(c), equivalent to $\varphi_{A_L,k}(y) \geq 0$ for all $y \in F$ and arbitrary fixed $k \in \operatorname{bd} B_L \cap \operatorname{int} \mathbb{R}^\ell_+$. Obviously, $\bar{y} \in \operatorname{bd} A_L$, such that Corollary 9.1.17(b) implies $\varphi_{A_L,k}(\bar{y}) = 0$ and consequently, $\varphi_{A_L,k}(y) \geq \varphi_{A_L,k}(\bar{y})$ for all $y \in F$ and an arbitrary fixed element $k \in B_L \cap \operatorname{int} \mathbb{R}^\ell_+$.

(b) We consider the sets $B_\gamma$ $(= B_L)$, $A_L$ and the functional $\varphi_{A_L,k}$ as in part (a) of this proof. Suppose $\bar{y} \in \operatorname{Eff}(F, \mathbb{R}^\ell_+)$ and $\varphi_{A_L,k}(y) = \varphi_{A_L,k}(\bar{y})$ for some $y \in F$, $y \neq \bar{y}$. Then, $\varphi_{A_L,k}(y) = 0$, and Corollary 9.1.17(b) yields

$y \in \operatorname{bd} A_L = \bar{y} - \operatorname{bd} \mathbb{R}^\ell_+$. We obtain $y \in \bar{y} - \mathbb{R}^\ell_+ \setminus \{0\}$, a contradiction to $\bar{y} \in \operatorname{Eff}(F, \mathbb{R}^\ell_+)$. Then, $\bar{y} \in \operatorname{Eff}(F, \mathbb{R}^\ell_+)$ yields $\varphi_{A_L,k}(y) > \varphi_{A_L,k}(\bar{y})$ for all $y \in F \setminus \{\bar{y}\}$ and an arbitrary fixed element $k \in \operatorname{bd} B_L \cap \operatorname{int} \mathbb{R}^\ell_+$.

(c) Consider $\bar{y} \in \operatorname{Eff}(F, C)$. Employing the cone $C$, we define a norm $\gamma$ by

$$B_\gamma \ (= B_L) := \{y = (y^1, \dots, y^\ell)^T \in \mathbb{R}^\ell \mid (|y^1|, \dots, |y^\ell|)^T \in \bar{y} - C\}.$$

As shown in Example 9.1.10, $\gamma$ is oblique. Moreover, from Example 9.1.13 we conclude that the corresponding set $A_L$ in (9.1.4) has the form $A_L = \bar{y} - C$. We now consider the functional $\varphi_{A_L,k} : \mathbb{R}^\ell \to \mathbb{R}$ ($w = 0$) in (9.1.5) with $A_L = \bar{y} - C$. For $\bar{y} \in \operatorname{Eff}(F, C)$, it holds that $\emptyset = F \cap (\bar{y} - C \setminus \{0\})$ and so $\emptyset = F \cap (\bar{y} - \operatorname{int} C) = F \cap \operatorname{int} A_L$. By Corollary 9.1.17(c), we obtain $\varphi_{A_L,k}(y) \geq 0$ for all $y \in F$ and arbitrary fixed $k \in \operatorname{bd} B_L \cap \operatorname{int} \mathbb{R}^\ell_+$. Of course, we have $\bar{y} \in \operatorname{bd} A_L$, such that Corollary 9.1.17(b) implies $\varphi_{A_L,k}(\bar{y}) = 0$ and consequently, $\varphi_{A_L,k}(y) \geq \varphi_{A_L,k}(\bar{y})$ for all $y \in F$ and an arbitrary fixed element $k \in B_L \cap \operatorname{int} \mathbb{R}^\ell_+$.
We now act as in part (b) of the proof: Suppose $\varphi_{A_L,k}(y) = \varphi_{A_L,k}(\bar{y})$ for some $y \in F$, $y \neq \bar{y}$. Then, $\varphi_{A_L,k}(y) = 0$ and Corollary 9.1.17(b) yields $y \in \operatorname{bd} A_L = \bar{y} - \operatorname{bd} C$. We obtain $y \in \bar{y} - C \setminus \{0\}$, a contradiction to $\bar{y} \in \operatorname{Eff}(F, C)$. So, $\bar{y} \in \operatorname{Eff}(F, C)$ yields $\varphi_{A_L,k}(y) > \varphi_{A_L,k}(\bar{y})$ for all $y \in F \setminus \{\bar{y}\}$ and an arbitrary fixed element $k \in B_L \cap \operatorname{int} \mathbb{R}^\ell_+$. □

**Remark 9.1.23** In the statements of Theorem 9.1.22, we have taken $w = 0$. Hence, the shift of the set $A_L$ starts always in $\bar{y}$, which cannot be canceled. For $w \neq 0$, we probably can state only inequalities of the type $\varphi_{w+A_L,k}(y - \bar{y}) \geq 0$. Obviously, this is not the same, because $\varphi_{w+A_L,k}$ is not homogeneous and not subadditive in general, too.

**Remark 9.1.24** We have supposed that $k$ belongs to $\operatorname{bd} B_L$. Since we only use $k \in \operatorname{int} \mathbb{R}^\ell_+$, this is a rather technical assumption.

**Remark 9.1.25** In [285, Theorem 3.7], Kaliszewski characterizes efficient elements of a certain set with respect to polyhedral cones by some inconsistency assertions: Consider a polyhedral cone

$$C := \{y \in \mathbb{R}^\ell \mid \langle -b_i, y \rangle \geq 0, \ b_i \in \mathbb{R}^\ell, \ i \in \{1, \dots, m\}\}.$$

Then, Kaliszewski shows in [285]: $\bar{y} \in \operatorname{Eff}(F, C)$ implies that the system

$$\langle b_i, y - \hat{y} \rangle < \langle b_i, \bar{y} - \hat{y} \rangle, \qquad i \in \{1, \dots, m\}\},$$

where $F \subset \hat{y} + C$, is inconsistent for all $y \in F$.

This characterization, indeed, is a consequence of our Theorem 9.1.22(c): We take $A_L = \bar{y} - C$ and consider $\hat{F} = F - \hat{y}$. Then, we obtain $\hat{F} \subset C$ which is a weaker condition than $\hat{F} \subset \mathbb{R}^\ell_+$, but it suffices for our purposes. Theorem 9.1.22(c) implies

$$\forall\, y \in F : \qquad \varphi_{w+A_L,k}(y - \hat{y}) \geq \varphi_{w+A_L,k}(\bar{y} - \hat{y}).$$

By formula (9.1.6) with $w = 0$, $\alpha_i = 0$, we obtain for all $y \in F$

$$\max_{i=1,\dots,m} \frac{\langle b_i, y - \hat{y}\rangle}{\langle b_i, k\rangle} \geq \max_{i=1,\dots,m} \frac{\langle b_i, \bar{y} - \hat{y}\rangle}{\langle b_i, k\rangle}.$$

Then, for each $y \in F$, there are $i^* \in I$ and

$$\frac{\langle b_{i^*}, y - \hat{y}\rangle}{\langle b_{i^*}, k\rangle} \geq \max_{i=1,\dots,m} \frac{\langle b_i, \bar{y} - \hat{y}\rangle}{\langle b_i, k\rangle} \geq \frac{\langle b_{i^*}, \bar{y} - \hat{y}\rangle}{\langle b_{i^*}, k\rangle},$$

that is $\langle b_{i^*}, y - \hat{y}\rangle \geq \langle b_{i^*}, \bar{y} - \hat{y}\rangle$. This means, $\langle b_i, y - \hat{y}\rangle < \langle b_i, \bar{y} - \hat{y}\rangle$ for all $i = 1, \dots, m$ is not possible.

The following characterization of solutions of vector optimization problems by means of the functional $\varphi_{w+A_L,k}$ is shown in [438, Theorem 3.3].

**Theorem 9.1.26** (Sufficient conditions) *Assume that $B_\gamma (= B_L)$ is the unit ball of an absolute block norm $\gamma$, $A_L$ is the corresponding set given by (9.1.4) with the index set $I$ defined by (9.1.3) and $k \in \operatorname{bd} B_L \cap \operatorname{int} \mathbb{R}^\ell_+$ arbitrary.*

*Suppose that $F \subseteq \operatorname{int} \mathbb{R}^\ell_+$ is a nonempty set. Then, we have*

(a) *If $\bar{y} \in F$ satisfies the relation*

$$\forall\, y \in F: \quad \varphi_{w+A_L,k}(y) \geq \varphi_{w+A_L,k}(\bar{y}),$$

*where $\varphi_{w+A_L,k}$ with $w = \bar{y} - k$ is given by (9.1.5), then $\bar{y} \in \operatorname{WEff}(F, \mathbb{R}^\ell_+)$.*

(b) *If $\bar{y} \in F$ satisfies the relation*

$$\forall\, y \in F\backslash\{\bar{y}\}: \quad \varphi_{w+A_L,k}(y) > \varphi_{w+A_L,k}(\bar{y}),$$

*where $\varphi_{w+A_L,k}$ with $w = \bar{y} - k$ is given by (9.1.5), then $\bar{y} \in \operatorname{Eff}(F, \mathbb{R}^\ell_+)$.*

(c) *Suppose that $\gamma$ is an oblique norm. If $\bar{y} \in F$ satisfies the relation*

$$\forall\, y \in F: \quad \varphi_{w+A_L,k}(y) \geq \varphi_{w+A_L,k}(\bar{y}),$$

*where $\varphi_{w+A_L,k}$ with $w = \bar{y} - k$ is given by (9.1.5), then $\bar{y} \in \operatorname{Eff}(F, \mathbb{R}^\ell_+)$.*

***Proof*** Suppose that $F \subseteq \mathbb{R}^\ell_+$ is a nonempty set.

(a) Consider an absolute block norm $\gamma$ with unit ball $B_\gamma = B_L$ and $k \in \operatorname{bd} B_L \cap \operatorname{int} \mathbb{R}^\ell_+$ arbitrary. Furthermore, suppose that $\bar{y} \in F$ satisfies

$$\forall\, y \in F: \quad \varphi_{w+A_L,k}(y) \geq \varphi_{w+A_L,k}(\bar{y}) \tag{9.1.15}$$

with $w = \bar{y} - k$. Now, we transform $\varphi_{w+A_L,k}$ into $\varphi_{A_L,k}$ and vice versa by

$$\begin{aligned}\varphi_{w+A_L,k}(y) &= \inf\{t \in \mathbb{R} \mid y \in tk + w + A_L\}\\ &= \inf\{t \in \mathbb{R} \mid y - w \in tk + A_L\}\\ &= \varphi_{A_L,k}(y - w)\\ &= \varphi_{A_L,k}(y - \bar{y} + k).\end{aligned}$$

This means that we can write inequality (9.1.15) as

$$\varphi_{A_L,k}(y - \bar{y} + k) \geq \varphi_{A_L,k}(\bar{y} - \bar{y} + k) = \varphi_{A_L,k}(k)$$

or equivalently, because of $k \in \operatorname{bd} B_L \cap \operatorname{int} \mathbb{R}^\ell_+ \subset \operatorname{bd} A_L$ and $\varphi_{A_L,k}(\operatorname{bd} A_L) = 0$ (recall $w = 0$),

$$\varphi_{A_L,k}(y - \bar{y} + k) \geq 0. \tag{9.1.16}$$

Corollary 9.1.17(c) implies $y - \bar{y} + k \notin \operatorname{int} A_L$ for all $y \in F$. From Lemma 9.1.14, we obtain $-\operatorname{int} \mathbb{R}^\ell_+ \subseteq \operatorname{int} A_L - k$, such that we get $y - \bar{y} \notin -\operatorname{int} \mathbb{R}^\ell_+$ for all $y \in F$, which implies $\bar{y} \in \operatorname{WEff}(F, \mathbb{R}^\ell_+)$.

(b) We proceed as in part (a) of this proof, but now the inequality (9.1.16) is strongly satisfied,

$$\varphi_{A_L,k}(y - \bar{y} + k) > 0. \tag{9.1.17}$$

Corollary 9.1.17(c) implies $y - \bar{y} + k \notin \operatorname{int} A_L$ for all $y \in F$. Taking into account Corollary 9.1.17(b), we can enhance this inclusion to $y - \bar{y} + k \notin A_L$ for all $y \in F$. Lemma 9.1.14 says $-\mathbb{R}^\ell_+ \subseteq A_L - k$, such that we obtain $y - \bar{y} \notin -\mathbb{R}^\ell_+ \setminus \{0\}$ for all $y \in F \setminus \{\bar{y}\}$, which implies $\bar{y} \in \operatorname{Eff}(F, \mathbb{R}^\ell_+)$.

(c) We proceed as in part (a) of this proof. Lemma 9.1.14 even says $-\mathbb{R}^\ell_+ \setminus \{0\} \subseteq \operatorname{int} A_L - k$ if $\gamma$ is an oblique norm. Then, we obtain $y - \bar{y} \notin -\mathbb{R}^\ell_+ \setminus \{0\}$ for all $y \in F \setminus \{\bar{y}\}$. Hence, $\bar{y} \in \operatorname{Eff}(F, \mathbb{R}^\ell_+)$. □

Theorems 9.1.22 and 9.1.26 have the same structure: We proved that solutions to a scalarized problem where a block norm is involved in the construction of the scalarizing functional are (only) weakly efficient elements, while efficiency can be shown either for unique solutions to the scalarized problems or for solutions to the scalarized problems where oblique norms are involved.

We would like to mention that the assumption $F \subset \operatorname{int} \mathbb{R}^\ell_+$ can be satisfied easily if $F$ is bounded from below, i.e., if there exists an element $b \in \mathbb{R}^\ell_+$ with $y \in b + \mathbb{R}^\ell_+$ for all $y \in F$. Then, the original optimization problem is equivalent to the problem of finding some efficient elements in $F + b + k$ for some $k \in \operatorname{int} \mathbb{R}^\ell_+$.

### 9.1.4 Application in Multiobjective D.C. Optimization

In this section, we apply our results from Sects. 9.1.2 and 9.1.3 in order to give a useful characterization of efficient elements of nonconvex multiobjective d.c. (difference of convex functions) optimization problems via scalarization.

We now study a multiobjective d.c. optimization problem, where the objective function $f : S \to \mathbb{R}^\ell$ is given by

$$f(x) = g(x) - h(x) = \begin{pmatrix} g_1(x) - h_1(x) \\ \vdots \\ g_\ell(x) - h_\ell(x) \end{pmatrix}, \qquad x \in S.$$

Suppose that $S$ is a nonempty, convex subset of a topological vector space, the functions $g_i, h_i : S \to \mathbb{R} \cup \{+\infty\}$ are continuous and convex. We introduce a multiobjective d.c. optimization problem by

$$\mathrm{Eff}((g - h)(S), \mathbb{R}^\ell_+). \qquad (P_{CD})$$

This means that we are looking for elements belonging to the set

$$\{g(x) - h(x) \mid x \in S, (g - h)(S) \cap (g(x) - h(x) - \mathbb{R}^\ell_+\setminus\{0\}) = \emptyset\}.$$

Dealing with scalar d.c. problems, a significant question is to obtain lower bounds for the objective function values. Such lower bounds are very important for solving global optimization problems by branch-and-bound techniques. Applying polyhedral branching strategies, it is important to generate sharp lower bounds for subproblems in d.c. form at each stage in the resolution of the d.c. optimization problem. Moreover, lower bounds are useful in order to check feasibility: If a lower bound for $\min_{x\in S}\{g(x) - h(x)\}$ is strictly positive, then the infeasibility of the problem

$$\min\{f(x) \mid g(x) - h(x) \leq 0, x \in S\}$$

follows.

The aim of this section is to characterize the solutions to the multiobjective d.c. problem using a suitable scalarization by means of the functional $\varphi_{w+A_L,k}$ defined in (9.1.5). From the numerical point of view, it is important to preserve the typical d.c. structure during the scalarization such that it is possible to apply well-known results from scalar d.c. optimization, compare Horst and Thoai [245], Horst and Tuy [246], especially duality assertions, compare Martinez-Legaz and Volle [348], or Lagrangian decomposition methods, see Carrizosa [84].

Taking into account Definition 8.2.3 (with $C = \mathbb{R}^\ell_+$), a function $g : S \to \mathbb{R}^\ell$ is called $\mathbb{R}^\ell_+$ *-convex* if

$$g(\lambda x^1 + (1-\lambda)x^2) \in \lambda g(x^1) + (1-\lambda)g(x^2) - \mathbb{R}^\ell_+$$

holds for all $x^1, x^2 \in S$, $\lambda \in [0, 1]$. Obviously, if $g(x) = (g_1(x), \dots, g_\ell(x))^T$ is a vector-valued function with convex functions $g_i : S \to \mathbb{R}$, $i \in \{1, \dots, \ell\}$, then $g$ is $\mathbb{R}^\ell_+$-convex.

We now show that convexity is preserved under scalarization by means of monotone functionals (see [438, Lemma 4.1]). This fact guarantees the useful d.c. structure after a scalarization.

**Lemma 9.1.27** *Consider a $\mathbb{R}^\ell_+$-convex function $g : S \to \mathbb{R}^\ell$. Moreover, suppose that $\varphi : \mathbb{R}^\ell \to \mathbb{R}$ is a $\mathbb{R}^\ell_+$-monotone, convex functional. Then, $\varphi \circ g$ is convex.*

***Proof*** For all $x^1, x^2 \in S$ and for all $\lambda \in (0, 1)$ it holds that

$$\begin{aligned}\varphi(g(\lambda x^1 + (1-\lambda)x^2)) &\le \varphi(\lambda g(x^1) + (1-\lambda)g(x^2)) \\ &\le \lambda\varphi(g(x^1)) + (1-\lambda)\varphi(g(x^2)),\end{aligned}$$

which means convexity of $\varphi \circ g$ taking into account Theorem 3.4.4. □

Next, we will show necessary and sufficient conditions for solutions to ($\mathrm{P}_{CD}$) using the functional (9.1.5) defined by

$$\varphi_{w+A_L,k}(y) = \inf\{t \in \mathbb{R} \mid y \in tk + w + A_L\}, \qquad y \in \mathbb{R}^\ell.$$

We deduce a lower bound for scalarized objective function values of feasible elements of ($\mathrm{P}_{CD}$) applying Theorem 9.1.22 in the following theorem (see [438, Theorem 4.1]). For this intention, it is important that the lower bound has the d.c. structure, too. So, it would be possible to apply known results from scalar d.c. optimization, especially duality assertion for real-valued d.c. problems.

**Theorem 9.1.28** (Necessary condition for solutions of d.c. optimization problems) *We consider the problem ($\mathrm{P}_{CD}$) and $\bar{y} = g(\bar{x}) - h(\bar{x}) \in \mathrm{Eff}((g-h)(S), \mathbb{R}^\ell_+)$. Furthermore, assume $(g-h)(S) \subseteq \mathrm{int}\,\mathbb{R}^\ell_+$. Then, there is a unit ball $B_\gamma = B_L$ of a block norm $\gamma$ and the corresponding set $A_L$ given by (9.1.4) with the index set $I$ defined by (9.1.3), such that the functional $\varphi_{w+A_L,k}$ defined by (9.1.5) with $k = \bar{y} = g(\bar{x}) - h(\bar{x})$ and $\bar{w} = -\bar{y}$ fulfills*

$$\varphi_{\bar{w}+A_L,k}(g(\bar{x})) - \varphi_{\bar{w}+A_L,k}(h(\bar{x})) < \varphi_{\bar{w}+A_L,k}(g(x) - h(x))$$

*for all $x \in S$ with $g(x) - h(x) \neq g(\bar{x}) - h(\bar{x})$.*

***Proof*** Consider $\bar{y} = g(\bar{x}) - h(\bar{x}) \in \mathrm{Eff}((g-h)(S), \mathbb{R}^\ell_+)$. Under the assumption $(g-h)(S) \subseteq \mathrm{int}\,\mathbb{R}^\ell_+$, from Theorem 9.1.22(b), we obtain for all $y \in (g-h)(S)\setminus\{\bar{y}\}$

$$\varphi_{w+A_L,k}(\bar{y}) < \varphi_{w+A_L,k}(y), \tag{9.1.18}$$

where $\varphi_{w+A_L,k}$ defined by (9.1.5) is constructed by a norm $\gamma$ with $B_\gamma(= B_L) :=$ conv $R(\bar{y})$, an arbitrary element $k \in \operatorname{bd} B_L \cap \operatorname{int} \mathbb{R}^\ell_+$ and $w = 0$. The set $A_L$ is given by $\bar{y} - \mathbb{R}^\ell_+$ (see Remark 9.1.9). We put $k = \bar{y}$ (that is in fact an element of $\operatorname{bd} B_L \cap \operatorname{int} \mathbb{R}^\ell_+$) and obtain

$$\begin{aligned}\varphi_{-\bar{y}+A_L,\bar{y}}(y) &= \max_{i\in I} \frac{\langle a^i, y\rangle + \langle a^i, \bar{y}\rangle - \alpha_i}{\langle a^i, \bar{y}\rangle} \\ &= \max_{i\in I} \frac{\langle a^i, y\rangle - \alpha_i}{\langle a^i, \bar{y}\rangle} + 1 \\ &= \varphi_{0+A_L,\bar{y}}(y) + 1\end{aligned}$$

(compare (9.1.6)), where the set $I$ is the index set defined in (9.1.3). Hence, inequality (9.1.18) remains valid if we write $\varphi_{-\bar{y}+A_L,\bar{y}}$ instead of $\varphi_{0+A_L,\bar{y}}$. Replacing $y$ by $g(x) - h(x)$, we conclude

$$\varphi_{-\bar{y}+A_L,\bar{y}}(g(\bar{x}) - h(\bar{x})) < \varphi_{-\bar{y}+A_L,\bar{y}}(g(x) - h(x))$$

for all $x \in S$ with $g(x) - h(x) \neq g(\bar{x}) - h(\bar{x})$. Moreover, $\varphi_{-\bar{y}+A_L,\bar{y}}$ is subadditive due to Corollary 9.1.17(e). Taking into account

$$\operatorname{bd} A_L + \operatorname{bd} A_L - \bar{y} = \bar{y} - \operatorname{bd} \mathbb{R}^\ell_+ + \bar{y} - \operatorname{bd} \mathbb{R}^\ell_+ - \bar{y} \subseteq -\mathbb{R}^\ell_+ + \bar{y} = A_L$$

and applying the subadditivity of $\varphi_{\bar{w}+A_L,k}$, we get

$$\varphi_{\bar{w}+A_L,k}(g(x) - h(x)) > \varphi_{\bar{w}+A_L,k}(g(\bar{x}) - h(\bar{x})) \geq \varphi_{\bar{w}+A_L,k}(g(\bar{x})) - \varphi_{\bar{w}+A_L,k}(h(\bar{x}))$$

for all $x \in S$ with $g(x) - h(x) \neq g(\bar{x}) - h(\bar{x})$. □

**Remark 9.1.29** In Theorem 9.1.28, we derived a lower bound for optimal solutions to the scalarized problem by means of the functional $\varphi_{w+A_L,k}$, where a block norm $\gamma$ is involved and $k \in \operatorname{bd} B_\gamma \cap \operatorname{int} \mathbb{R}^\ell_+$. Because of Theorem 9.1.26(a), we conclude that any solution to the scalarized problem is a weakly efficient element of the multiobjective problem ($\mathrm{P}_{CD}$).

The assertion in the following theorem is derived in [438, Theorem 4.2].

**Theorem 9.1.30** (Sufficient condition for solutions of d.c. optimization problems) *Consider the functional $\varphi_{w+A_L,k}$ defined by (9.1.5), where $B_\gamma(= B_L)$ is the unit ball of an oblique block norm $\gamma$, $A_L$ is the corresponding set given by (9.1.4) with the index set $I$ defined by (9.1.3), $k \in \operatorname{bd} B_\gamma \cap \operatorname{int} \mathbb{R}^\ell_+$ arbitrary and $w = g(\bar{x}) - h(\bar{x}) - k$, $\bar{x} \in S$. Furthermore, suppose that $\bar{x} \in S$ is a solution to the scalarized problem*

$$\varphi_{w+A_L,k}(g(x) - h(x)) \to \min, \qquad x \in S.$$

*Then, $g(\bar{x}) - h(\bar{x}) \in \operatorname{Eff}((g - h)(S), \mathbb{R}^\ell_+)$.*

***Proof*** The assertion of the theorem follows from Theorem 9.1.26(c). □

**Remark 9.1.31** In Theorem 9.1.28, we deduced a lower bound in d.c structure for the objective function values of the scalarized multiobjective d.c. problem. Using these results, it is possible to apply techniques from scalar d.c. optimization, see Horst and Thoai [245], and Horst and Tuy [246]. Especially, the Lagrangian decomposition method given by Carrizosa [84], is applicable. The special structure of the scalarizing functional where the shifted set is defined via block norms with polyhedral unit balls is very useful for the application of polyhedral branching schemes.

Moreover, taking into account Theorem 9.1.28, under certain additional assumptions, it is possible to apply duality assertions for real-valued d.c. problems investigated by Martinez-Legaz and Volle [348], in order to get lower bounds for the d.c. problem.

## 9.2 Directional Minimal Time Functions

In this section, we study the directional minimal time function introduced by Nam and Zălinescu [366]. The directional minimal time function is related to the nonlinear separating functional $\varphi_{A,-k}$ given by (4.1.11) and has interesting applications in locational analysis (see Sect. 15.3).

Throughout this section, we suppose that $Y$ is a real normed vector space. Furthermore, let $k \in Y \setminus \{0\}$ and $A \subseteq Y$ be a nonempty closed set. The directional minimal time function $\varphi^+_{A,-k} : Y \to \mathbb{R} \cup \{+\infty\}$ with direction $k$ and a target set $A$ is introduced by

$$\varphi^+_{A,-k}(y) := \inf\{t \geq 0 \mid y + tk \in A\}. \tag{9.2.1}$$

Taking into account $\inf \emptyset = +\infty$, it holds that $\operatorname{dom} \varphi^+_{A,-k} = \{y \in Y \mid \varphi^+_{A,-k}(y) \in \mathbb{R}\}$. The following results concerning the domain and epigraph of $\varphi^+_{A,-k}$ and the relationship to the separating functional $\varphi_{A,-k}$ given by (4.1.11) under the assumption $k \in 0^+A$ are shown by Nam and Zălinescu in [366, Proposition 2.1].

**Proposition 9.2.1** *Suppose that $Y$ is a real normed vector space, $k \in Y \setminus \{0\}$ and $A \subseteq Y$ is a nonempty closed set. Then,*

$$\operatorname{dom} \varphi^+_{A,-k} = \{y \in Y \mid \varphi^+_{A,-k} \in \mathbb{R}\} = -\mathbb{R}_+k + A.$$

*Furthermore, if $k \in 0^+A$, then*

$$\operatorname{dom} \varphi^+_{A,-k} = \{y \in Y \mid \varphi^+_{A,-k} \in \mathbb{R}\} = -\mathbb{R}k + A = -\mathbb{R}_+k + A, \tag{9.2.2}$$

$$\operatorname{epi} \varphi^+_{A,-k} = \{(y,t) \in Y \times \mathbb{R} \mid t \geq 0,\ y + tk \in A\} \tag{9.2.3}$$

*and*

$$\forall y \in Y: \; \varphi^{+}_{A,-k}(y) = \max\{\varphi_{A,-k}(y), 0\}. \tag{9.2.4}$$

***Proof*** It holds that

$$\text{dom}\varphi^{+}_{A,-k} = \{y \in Y \mid \varphi^{+}_{A,-k} \in \mathbb{R}\} = \{y \in Y \mid y + tk \in A \text{ for some } t \geq 0\}$$

$$= \{y \in Y \mid y \in A - tk \;\text{ for some }\; t \geq 0\} = -\mathbb{R}_{+}k + A.$$

Now, we suppose $k \in 0^{+}A$. For any $y \in A - \mathbb{R}k$, there are $a \in A$ and $\lambda \in \mathbb{R}$ such that $y = a - \lambda k$. If $\lambda \geq 0$, then $y \in A - \mathbb{R}_{+}k = \{y \in Y \mid \varphi^{+}_{A,-k} \in \mathbb{R}\}$. On the other hand, if $\lambda < 0$, we get $y = a + (-\lambda)k \in A$ because of $k \in 0^{+}A$ and so, $y \in A \subseteq \{y \in Y \mid \varphi^{+}_{A,-k} \in \mathbb{R}\}$.
We get (9.2.2) taking into account that $k \in 0^{+}A$ implies $A = A + \mathbb{R}_{+}k$ and $A - \mathbb{R}k = A + \mathbb{R}k = (A + \mathbb{R}_{+}k) \cup (A - \mathbb{R}_{+}k) = A \cup (A - \mathbb{R}_{+}k) = A - \mathbb{R}_{+}k$.
The proof of (9.2.3) is easy.

In order to prove (9.2.4), we show $\varphi^{+}_{A,-k}(y) \leq \max\{\varphi_{A,-k}(y), 0\}$ for all $y \in Y$. The inequality $\varphi^{+}_{A,-k}(y) \geq \max\{\varphi_{A,-k}(y), 0\}$ for all $y \in Y$ is obvious. Let $\max\{\varphi_{A,-k}(y), 0\} < \lambda$. Taking into account the definition of $\varphi_{A,-k}$, there are $t \in (-\infty, \lambda)$ such that $y \in -tk + A$. Then,

$$y + \lambda k = y + tk + (\lambda - t)k \in A + \mathbb{R}_{+}k = A,$$

and so $\varphi^{+}_{A,-k}(y) \leq \lambda$. This yields $\varphi^{+}_{A,-k}(y) \leq \max\{\varphi_{A,-k}(y), 0\}$. □

The following result is given by Nam and Zălinescu in [366, Proposition 2.4].

**Proposition 9.2.2** *Assume that $k \in Y \setminus \{0\}$ and $A \subseteq Y$ is a nonempty closed set. For any $\alpha \in \mathbb{R}$ with $\alpha \geq 0$, it holds that*

$$\text{lev}_{\varphi_{A,\ k},\leq}(\alpha) = -[0, \alpha]\, k + A, \tag{9.2.5}$$

$$\text{lev}_{\varphi_{A,-k},<}(\alpha) = -[0, \alpha)\, k + A \tag{9.2.6}$$

*and*

$$\varphi^{+}_{A,-k}(y) = 0 \iff y \in A. \tag{9.2.7}$$

*Furthermore, $\varphi^{+}_{A,-k}$ is a proper and lower semicontinuous functional.*

***Proof*** Obviously, (9.2.5) and (9.2.6) hold. If we suppose $y \in A$, then $y + 0k \in A$ and so $\varphi^{+}_{A,-k}(y) \leq 0$ such that we get $\varphi^{+}_{A,-k}(y) = 0$. Suppose now $\varphi^{+}_{A,-k}(y) = 0$, then $y + \varphi^{+}_{A,-k}(y)k = y \in A$, i.e., (9.2.7) holds. $\varphi^{+}_{A,-k}$ is proper because $A \subseteq \{y \in Y \mid \varphi^{+}_{A,-k}(y) \in \mathbb{R}\}$. Because of the closedness of the sublevel set $\{y \in Y \mid \varphi^{+}_{A,-k}(y) \leq \alpha\}$ for each $\alpha \in \mathbb{R}$ ($\text{lev}_{\varphi_{A,-k},\leq}(\alpha)$ is empty for $\alpha < 0$), the functional $\varphi^{+}_{A,-k}$ is lower semicontinuous taking into account Proposition 3.3.12. □

**Remark 9.2.3** Corresponding properties are shown for the separating functional $\varphi_{A,-k}$ given by (4.1.11) in Theorem 4.4.4. Since $\varphi^+_{A,-k}$ is lower semicontinuous, in the case $k \in 0^+A$, we get that $\varphi^+_{A,-k}$ and $\varphi_{A,-k}$ given by (4.1.11) coincide on a neighborhood of $y$ for any $y \in Y$ with $\varphi^+_{A,-k} > 0$. This tells that the properties of the directional minimal time function $\varphi^+_{A,-k}$ shown for elements $y \in Y$ with $\varphi^+_{A,-k} > 0$ can be extended to corresponding properties of the nonlinear separating functional $\varphi_{A,-k}$.

## 9.2.1 *Fundamental Properties*

The convexity of the function $\varphi^+_{A,-k}$ is discussed in the following proposition (see [366, Proposition 2.6]).

**Proposition 9.2.4** *Assume that $k \in Y \setminus \{0\}$ and $A \subseteq Y$ is a nonempty closed set. Then, $\varphi^+_{A,-k}$ is convex if and only if $A$ is convex.*

***Proof*** First, assume that $A$ is convex. Choose arbitrary $y^1, y^2 \in \{y \in Y \mid \varphi^+_{A,-k} \in \mathbb{R}\}$ and $\lambda \in (0, 1)$, fixed. Then, we have

$$y^1 + \varphi^+_{A,-k}(y^1)k \in A \quad \text{and} \quad y^2 + \varphi^+_{A,-k}(y^2)k \in A.$$

Because of the convexity of $A$, this yields

$$\lambda y^1 + (1-\lambda)y^2 + \left(\lambda\varphi^+_{A,-k}(y^1) + (1-\lambda)\varphi^+_{A,-k}(y^2)\right) k \in A.$$

So, we can conclude

$$\varphi^+_{A,-k}(\lambda y^1 + (1-\lambda)y^2) \le \lambda\varphi^+_{A,-k}(y^1) + (1-\lambda)\varphi^+_{A,-k}(y^2)$$

and we get the convexity of $\varphi^+_{A,-k}$ taking into account Theorem 3.4.4. Conversely, if $\varphi^+_{A,-k}$ is convex, then $\{y \in Y \mid \varphi^+_{A,-k}(y) \le 0\} = A$ is convex. □

The translation invariance along $k$ of $\varphi^+_{A,-k}$ in the following proposition (see [366, Proposition 2.8]) is very useful for deriving results concerning continuity properties in Corollary 9.2.6 and Lipschitz continuity in Proposition 9.2.9.

**Proposition 9.2.5** *Assume that $k \in Y \setminus \{0\}$ and $A \subseteq Y$ is a nonempty closed set. Consider $\overline{y} \in \operatorname{dom} \varphi^+_{A,-k} = \{y \in Y \mid \varphi^+_{A,-k} \in \mathbb{R}\}$. For all $0 \le \lambda \le \varphi^+_{A,-k}(\overline{y})$, $\varphi^+_{A,-k}$ is translation invariant along $k$, i.e.,*

$$\varphi^+_{A,-k}(\overline{y} + \lambda k) = \varphi^+_{A,-k}(\overline{y}) - \lambda. \tag{9.2.8}$$

*For $\lambda > 0$, assume further $\overline{y} - \gamma k \notin A$ for every $\gamma \in (0, \lambda]$. Then,*

$$\varphi^+_{A,-k}(\overline{y} - \lambda k) = \varphi^+_{A,-k}(\overline{y}) + \lambda.$$

*Furthermore, for all* $y \in \operatorname{dom} \varphi^+_{A,-k}$ *and* $t \geq 0$*, it holds that*

$$\varphi^+_{A,-k}(y + tk) \geq \max\{\varphi^+_{A,-k}(y) - t, 0\} \tag{9.2.9}$$

*and, if* $k \in 0^+A$

$$\varphi^+_{A,-k}(y + tk) = \max\{\varphi^+_{A,-k}(y) - t, 0\}. \tag{9.2.10}$$

***Proof*** It holds that $\overline{y} + \lambda k + (\varphi^+_{A,-k}(\overline{y}) - \lambda)k \in A$ since $\overline{y} + \varphi^+_{A,-k}(\overline{y})k \in A$. This yields

$$\varphi^+_{A,-k}(\overline{y} + \lambda k) \leq \varphi^+_{A,-k}(\overline{y}) - \lambda \in \mathbb{R}.$$

We will prove that

$$\varphi^+_{A,-k}(\overline{y}) \leq \lambda + \varphi^+_{A,-k}(\overline{y} + \lambda k). \tag{9.2.11}$$

Indeed, for $t := \varphi^+_{A,-k}(\overline{y} + \lambda k)(\in \mathbb{R})$, it holds that

$$\overline{y} + \lambda k + tk \in A.$$

This implies $\varphi^+_{A,-k}(\overline{y}) \leq \lambda + t$ such that (9.2.11) follows. From (9.2.11), we get

$$\varphi^+_{A,-k}(\overline{y}) - \lambda \leq \varphi^+_{A,-k}(\overline{y} + \lambda k),$$

hence (9.2.8) holds.
We now prove $\varphi^+_{A,-k}(\overline{y} - \lambda k) = \varphi^+_{A,-k}(\overline{y}) + \lambda$ under the assumption $\overline{y} - \gamma k \notin A$ for every $\gamma \in (0, \lambda]$.

Let $\overline{w} := \overline{y} - \lambda k$. Then, $\overline{y} = \overline{w} + \lambda k$. From (9.2.11), we get

$$\varphi^+_{A,-k}(\overline{w}) = \varphi^+_{A,-k}(\overline{y} - \lambda k) = \varphi^+_{A,-k}(\overline{y}) + \lambda < +\infty.$$

This yields $\overline{w} \in \{y \in Y \mid \varphi^+_{A,-k} \in \mathbb{R}\}$. It follows that $0 < \lambda \leq \varphi^+_{A,-k}(\overline{w})$ in this case. Indeed, consider $\eta := \varphi^+_{A,-k}(\overline{w})$. Then, $\overline{w} + \eta k = \overline{y} - \lambda k + \eta k = \overline{y} - (\lambda - \eta)k \in A$. Hence, $\lambda - \eta \notin (0, \lambda]$, such that $\lambda \leq \eta = \varphi^+_{A,-k}(\overline{w})$. Using (9.2.8), we obtain $\varphi^+_{A,-k}(\overline{y}) = \varphi^+_{A,-k}(\overline{w}) - \lambda = \varphi^+_{A,-k}(\overline{y} - \lambda k) - \lambda$, and so $\varphi^+_{A,-k}(\overline{y} - \lambda k) = \varphi^+_{A,-k}(\overline{y}) + \lambda$ holds.
From (9.2.11), the inequality in (9.2.9) follows. Finally, we will show that the equality in (9.2.9) holds if $k \in 0^+A$. If $\varphi^+_{A,-k}(y) = +\infty$, the assertion is obvious. We consider the case $\varphi^+_{A,-k}(y) > t$, then it holds that

$$y + tk + (\varphi^+_{A,-k}(y) - t)k = y + \varphi^+_{A,-k}(y)k \in A.$$

This yields $\varphi^+_{A,-k}(y + tk) \le \varphi^+_{A,-k}(y) - t$, and the equation in (9.2.9) holds. Furthermore, in the case $\varphi^+_{A,-k}(y) \le t$ we get

$$y + tk = y + \varphi^+_{A,-k}(y)k + (t - \varphi^+_{A,-k}(y))k \in A + 0^+A = A,$$

such that $\varphi^+_{A,-k}(y + tk) = 0$, i.e., the equation in (9.2.9) holds in this case too. □

The following assertion is given in [366, Corollary 2.9] and will be used for deriving necessary and sufficient conditions for the Lipschitz continuity of $\varphi^+_{A,-k}$ in Proposition 9.2.9.

**Corollary 9.2.6** *Assume that $k \in Y \setminus \{0\}$ and $A \subseteq Y$ is a nonempty closed set. Suppose that there are $\overline{y} \in A$ and $\overline{t} > 0$ with $\overline{y} + \overline{t}k \in A$ but $[\overline{y}, \overline{y} + \overline{t}k] \not\subseteq A$. Then, there is an element $t \in [0, \overline{t})$ such that $\varphi^+_{A,-k}$ is finite-valued and not continuous at $\overline{y} + tk$.*

***Proof*** Under the given assumptions, there is an element $t_0 \in (0, \overline{t})$ such that $\overline{y} + t_0k \notin A$. We consider the largest interval $I \subseteq [0, \overline{t}]$ containing $t_0 \in I$ such that $\overline{y} + sk \notin A$ for every $s \in I$, we may suppose that $I = (0, \overline{t})$. This implies that for every $\lambda \in (0, \overline{t})$, we obtain $\overline{y} + \overline{t}k - \gamma k \notin A$ for each $\gamma \in (0, \lambda]$. Taking into account Proposition 9.2.5, we have

$$\varphi^+_{A,-k}(\overline{y} + \overline{t}k - \lambda k) = \varphi^+_{A,-k}(\overline{y} + \overline{t}k) + \lambda = \lambda,$$

and so

$$\lim_{t \to +0} \varphi^+_{A,-k}(\overline{y} + tk) = \lim_{\lambda \to \overline{t}-0} \varphi^+_{A,-k}(\overline{y} + \overline{t}k - \lambda k) = \overline{t} \ne \varphi^+_{A,-k}(\overline{y}) = 0.$$

This means that $\varphi^+_{A,-k}$ is not continuous at $\overline{y}$, i.e., there is an element $t \in [0, \overline{t})$ such that $\varphi^+_{A,-k}$ is finite-valued and not continuous at $\overline{y} + tk$. □

Furthermore, the following assertions (see [366, Proposition 2.12]) are important for deriving Lipschitz properties of $\varphi^+_{A,-k}$ in the next section.

**Proposition 9.2.7** *Suppose that $A$ and $B$ are nonempty, closed subsets of $Y$ and $k \in Y \setminus \{0\}$. Then,*

(a) $A \subseteq B \implies \forall y \in Y: \ \varphi^+_{B,-k}(y) \le \varphi^+_{A,-k}(y)$.
(b) $\forall y^1, y^2 \in Y: \ \varphi^+_{A+B,-k}(y^1 + y^2) \le \varphi^+_{A,-k}(y^1) + \varphi^+_{B,-k}(y^2)$.

***Proof*** First, we show (a). Now, let be $t := \varphi^+_{A,-k}(y) (\in \mathbb{R})$. Then,

$$y + tk \in A \subseteq B.$$

This yields $\varphi^+_{B,-k}(y) \le t = \varphi^+_{A,-k}(y)$. If $\varphi^+_{A,-k}(y) = +\infty$, the assertion is obvious.

In order to show (b), let $s := \varphi^+_{A,-k}(y^1)$ and $t := \varphi^+_{B,-k}(y^2)$. Then,

$$y^1 + sk \in A \quad \text{and} \quad y^2 + tk \in B,$$

such that we obtain $y^1 + y^2 + (s+t)k \in A + B$ and

$$\varphi^+_{A+B,-k}(y^1 + y^2) \le s + t = \varphi^+_{A,-k}(y^1) + \varphi^+_{B,-k}(y^2)$$

for all $y^1, y^2 \in Y$. If $\varphi^+_{A,-k}(y^1) = +\infty$ or $\varphi^+_{B,-k}(y^2) = +\infty$, the assertion is obvious. □

### 9.2.2 Lipschitz Continuity

In this section, we present Lipschitz properties of the directional minimal time function that are important for deriving optimality conditions in Sect. 15.3. The first result is shown in [366, Proposition 4.1].

**Proposition 9.2.8** *Assume that $A \subseteq Y$ is a nonempty closed set and $k \in \operatorname{int} 0^+A$. Then, $\varphi^+_{A,-k}$ is globally Lipschitzian with Lipschitz constant*

$$L := \inf\{r^{-1} \mid r > 0,\ B(k,r) \subseteq 0^+A\} = \frac{1}{d(k, \operatorname{bd} 0^+A)},$$

*where $d(k, \operatorname{bd} 0^+A)$ is the distance from $k$ to $\operatorname{bd} 0^+A$ (see Definition 2.1.28).*

***Proof*** First, we prove that $\operatorname{dom} \varphi^+_{A,-k} = \{y \in Y \mid \varphi^+_{A,-k} \in \mathbb{R}\} = Y$. Indeed,

$$\operatorname{dom} \varphi^+_{A,-k} = \{y \in Y \mid \varphi^+_{A,-k}(y) \in \mathbb{R}\} = A - \mathbb{R}k = A + [0^+A - \mathbb{R}k] = A + Y = Y$$

because of $k \in \operatorname{int} 0^+A$. Furthermore, from Proposition 9.2.7(b) we obtain for all $y^1, y^2 \in Y$

$$\varphi^+_{A+0^+A,-k}(y^1) \le \varphi^+_{A,-k}(y^2) + \varphi^+_{0^+A,-k}(y^1 - y^2).$$

This yields

$$\varphi^+_{A,-k}(y^1) \le \varphi^+_{A,-k}(y^2) + \varphi^+_{0^+A,-k}(y^1 - y^2)$$

and

$$\varphi^+_{A,-k}(y^1) - \varphi^+_{A,-k}(y^2) \le \varphi^+_{0^+A,-k}(y^1 - y^2). \tag{9.2.12}$$

We fix $r > 0$ such that $B(k,r) \subseteq 0^+A$, i.e., $B(0,r) \subseteq 0^+A - k$. Because $0^+A$ is a cone, we obtain for $y \notin 0^+A$:

$$\varphi^+_{0^+A,-k}(y) = \inf\{t \ge 0 \mid y + tk \in 0^+A\} = \inf\{t > 0 \mid y \in t(0^+A - k)\} \le \inf\{t > 0 \mid y \in tB(0,r)\} = \|r^{-1}y\|.$$

Taking into account $\varphi^+_{0^+A,-k}(y) = 0 \le r^{-1}\|y\|$ for $y \in 0^+A$, we get for all $y \in Y$:

$$\varphi^+_{0^+A,-k}(y) \le r^{-1}\|y\|.$$

From (9.2.12), it follows that

$$\varphi^+_{A,-k}(y^1) - \varphi^+_{A,-k}(y^2) \le \varphi^+_{0^+A,-k}(y^1 - y^2) \le L\|y^1 - y^2\|$$

and finally

$$|\varphi^+_{A,-k}(y^2) - \varphi^+_{A,-k}(y^1)| \le L\|y^1 - y^2\|$$

such that the proof is completed. □

The following necessary and sufficient condition for the property that $\varphi^+_{A,-k}$ is finite-valued and Lipschitzian goes back to [366, Proposition 4.2].

**Proposition 9.2.9** *Assume that $A \subseteq Y$ is a nonempty closed set. Then,*

$$\varphi^+_{A,-k} \text{ is finite-valued and Lipschitzian} \iff k \in \operatorname{int} 0^+A.$$

***Proof*** (a) Assume that $\varphi^+_{A,-k}$ is finite-valued and Lipschitzian. Then, $k \in 0^+A$. Indeed, if we suppose $k \notin 0^+A$, there is an element $\overline{y} \in A$ and $t > 0$ with $\overline{y} + tk \notin A$. Taking into account Proposition 9.2.1 we have $\overline{y} + tk \in Y = \{y \in Y \mid \varphi^+_{A,-k} \in \mathbb{R}\} = -\mathbb{R}_+k + A$. Then, there is an element $s > 0$ such that $\overline{y} + tk + sk \in A$. Applying Corollary 9.2.6 for $\overline{t} := t + s$, we get for some $t_0 \in [0, \overline{t})$ such that $\varphi^+_{A,-k}$ is not continuous at $\overline{y} + t_0k$, a contradiction to our assumption. Therefore, $k \in 0^+A$.
In order to see that $k \in \operatorname{int} 0^+A$, we consider the Lipschitz constant $L \ge 0$ of $\varphi^+_{A,-k}$, i.e., for all $x, y \in Y$:

$$|\varphi^+_{A,-k}(x) - \varphi^+_{A,-k}(y)| \le L\|x - y\|.$$

We proceed with showing $k + \frac{1}{L}B_Y \in 0^+A$. Indeed, if we fix an element $w \in \frac{1}{L}B_Y$, we get by using Proposition 9.2.5 and $k \in 0^+A$ for any $y \in A$ and $t > 0$:

$$\varphi^+_{A,-k}(y + t(k + w)) = \varphi^+_{A,-k}(y + tw + tk) = \max\{\varphi^+_{A,-k}(y + tw) - t, 0\} \le \max\{\varphi^+_{A,-k}(y) + tL\|w\| - t, 0\} \le 0.$$

This implies $y + t(k + w) \in A$ and so $k + w \in 0^+A$, i.e., $k \in \operatorname{int} 0^+A$.
(b) Proposition 9.2.8 yields the converse implication. □

### 9.2.3 *Conjugate and Subgradients*

In this section, we compute the Fenchel conjugate as well as the subdifferential of the directional minimal time function $\varphi^+_{A,-k}$.

We present the Fenchel conjugate of the directional minimal time function $\varphi^+_{A,-k}$ introduced by (9.2.1) in terms of the support function $\sigma_A$ of $A$ defined on $Y^*$. The next result concerning the conjugate of $\varphi^+_{A,-k}$ is shown in [366, Proposition 3.1].

**Proposition 9.2.10** *Assume that $k \in Y \setminus \{0\}$ and $A \subseteq Y$ is a nonempty closed set. Then, the function $\varphi^+_{A,-k}$ is a proper lower semicontinuous functional and its conjugate is*

$$(\varphi^+_{A,-k})^*(y^*) = \begin{cases} \sigma_A(y^*) & \text{if } y^*(-k) \leq 1 \\ +\infty & \text{otherwise.} \end{cases}$$

***Proof*** The lower semicontinuity of $\varphi^+_{A,-k}$ is shown in Proposition 9.2.2 as well as that $\varphi^+_{A,-k}$ is proper. Taking into account the definition of the Fenchel conjugate (see Definition 3.6.4), we get

$$\varphi^+_{A,-k}(y^*) = \sup\{y^*(y) - \varphi^+_{A,-k}(y) \mid y \in \operatorname{dom} \varphi^+_{A,-k}\} = \sup\{y^*(y) - t \mid t \geq 0,\, y + tk \in A\}$$

$$= \sup\{y^*(a) + t(y^*(-k) - 1) \mid t \geq 0,\ a \in A\} = \sup\{y^*(a) \mid a \in A\} + \sup\{t(y^*(-k) - 1) \mid t \geq 0\}.$$

Then, the formula for the conjugate of $\varphi^+_{A,-k}$ follows. □

Furthermore, in [366, Theorem 3.2] the Fenchel subdifferential of $\varphi^+_{A,-k}$ is computed under the assumption that $A$ is convex.

**Theorem 9.2.11** *Assume that $k \in Y \setminus \{0\}$ and $A \subseteq Y$ is a nonempty, closed, convex set. Consider $\overline{y} \in \{y \in Y \mid \varphi^+_{A,-k} \in \mathbb{R}\}$. Then,*

$$\partial \varphi^+_{A,-k}(\overline{y}) = \begin{cases} N(\overline{y}; A) \cap \{y^* \in Y^* \mid y^*(k) \geq -1\} & \text{if } \overline{y} \in A \\ N(\overline{y} + \varphi^+_{A,-k}(\overline{y})k; A) \cap \{y^* \in Y^* \mid y^*(k) = -1\} & \text{otherwise.} \end{cases} \tag{9.2.13}$$

*Furthermore, if $\overline{y} \in A$ and $k \in 0^+A \setminus \{0\}$, then*

$$\partial \varphi^+_{A,-k}(\overline{y}) = N(\overline{y}; A) \cap \{y^* \in Y^* \mid -1 \leq y^*(k) \leq 0\}. \tag{9.2.14}$$

***Proof*** Let $y^* \in \partial \varphi^+_{A,-k}(\overline{y})$. Then, $\varphi^+_{A,-k}(\overline{y}) + (\varphi^+_{A,-k})^*(y^*) = y^*(\overline{y})$. With Proposition 9.2.10, we get

$$y^*(k) \geq -1 \quad \text{and} \quad \varphi^+_{A,-k}(\overline{y}) + \sigma_A(y^*) = y^*(\overline{y}).$$

Because of $\overline{y} + \varphi^+_{A,-k}(\overline{y})k \in A$ it holds that

$$y^*(\overline{y}) \geq \varphi^+_{A,-k}(\overline{y}) + y^*(\overline{y} + \varphi^+_{A,-k}(\overline{y})k).$$

This implies

$$\varphi^+_{A,-k}(\overline{y})\left(1 + y^*(k)\right) \leq 0.$$

Since $\varphi^+_{A,-k}(\overline{y}) \geq 0$ and $1 + y^*(k) \geq 0$, we obtain

$$\varphi^+_{A,-k}(\overline{y})\left(1 + y^*(k)\right) = 0.$$

If $\overline{y} \in A$, then $\sigma_A(y^*) = y^*(\overline{y})$, and so $y^* \in N(\overline{y}; A)$. If $\overline{y} \notin A$, then $\varphi^+_{A,-k}(\overline{y}) > 0$ and so $1 + y^*(k) = 0$. This yields

$$\sigma_A(y^*) = y^*(\overline{y} + \varphi^+_{A,-k}(\overline{y})k),$$

and so $y^* \in N(\overline{y} + \varphi^+_{A,-k}(\overline{y})k; A)$. Hence, $\subseteq$ holds in (9.2.13).

Inversely, if $\overline{y} \in A$ and $y^* \in N(\overline{y}; A) \cap \{y^* \in Y^* \mid y^*(k) \geq -1\}$, then

$$\varphi^+_{A,-k}(\overline{y}) \left(1 + y^*(k)\right) = 0 \quad \text{and} \quad \sigma_A(y^*) = y^*(\overline{y}).$$

This yields $\varphi^+_{A,-k}(\overline{y}) + (\varphi^+_{A,-k})^*(y^*) = y^*(\overline{y})$ and furthermore, $y^* \in \partial\varphi^+_{A,-k}(\overline{y})$. If $\overline{y} \notin A$ and $y^* \in N(\overline{y} + \varphi^+_{A,-k}(\overline{y})k; A) \cap \{y^* \in Y^* \mid y^*(k) = -1\}$, then

$$\varphi^+_{A,-k}(\overline{y}) \left(1 + y^*(k)\right) = 0 \quad \text{and} \quad \sigma_A(y^*) = y^*(\overline{y} + \varphi^+_{A,-k}(\overline{y})k).$$

Therefore, $\varphi^+_{A,-k}(\overline{y}) + (\varphi^+_{A,-k})^*(y^*) = y^*(\overline{y})$, such that we get again $y^* \in \partial\varphi^+_{A,-k}(\overline{y})$.

Under the assumption $k \in 0^+A$, we conclude $y^*(k) \leq 0$ for all $y^* \in N(\overline{y}; A)$. Hence, assertion (9.2.14) follows. □

**Remark 9.2.12** An extension of the directional minimal time function given by (9.2.1) related to set optimization with respect to variable domination structures is given in the paper by Köbis, Tam Le, Tammer and Yao in [299].

**Remark 9.2.13** Applications of the directional minimal time function introduced in (9.2.1) are given in Sect. 15.3 where this functional (9.2.1) is involved in the formulation of location problems.

## 9.3 Relationship to the Hiriart-Urruty Oriented Distance

In this section, we discuss the relationship between the functional $\varphi_{A,k}$ given by (4.1.11) and the oriented distance introduced by Hiriart-Urruty in [241], [240].

**Definition 9.3.1** Assume $A$ to be a proper subset of a Banach space $Y$, $d$ to be the distance function in $Y$ according to Definition 2.1.28. The **oriented distance function** $\Delta_A : Y \to \mathbb{R}$ is given as

$$\Delta_A(y) := d(y, A) - d(y, Y \setminus A). \tag{9.3.1}$$

**Remark 9.3.2** If $\operatorname{int} A = \emptyset$, then Hiriart-Urruty's function coincides with the distance $d$ to $A$, i.e., $\Delta_A(y) = d(y, A)$ for all $y \in Y$. If, moreover, $A$ is convex, we have $\Delta_A = \Delta_{\operatorname{cl} A}$.

Note that oriented distance functions are just the so-called *signed distance functions* which are widely used e.g. in image processing and in solving differential equations [373].

Zaffaroni [505] showed the following properties of $\Delta_A$:

**Proposition 9.3.3** *Suppose that $A$ is a proper subset of a Banach space $Y$. Then,*

$$\Delta_A(y) < 0 \Leftrightarrow y \in \operatorname{int} A,$$
$$\Delta_A(y) = 0 \Leftrightarrow y \in \operatorname{bd} A,$$
$$\Delta_A(y) > 0 \Leftrightarrow y \in Y \setminus \operatorname{cl} A,$$

*and*

*(a) $\Delta_A$ is Lipschitzian with Lipschitz constant* 1.
*(b) If $A$ is convex, then $\Delta_A$ is convex.*
*(c) If $Y \setminus A$ is convex, then $\Delta_A$ is concave.*
*(d) If $A$ is a cone, then $\Delta_A$ is positively homogeneous.*
*(e) If $A$ is a nontrivial, closed, convex cone, then $\Delta_A$ is $(-A)$-monotone.*
*(f) $\Delta_{Y\setminus A} = -\Delta_A$.*
*(g) Suppose that $A$ is convex and has a nonempty interior. Then, for each $x \in \operatorname{bd} A$*

$$\partial \Delta_A(x) \subseteq N(x; A) \setminus \{0\},$$

*where $N(x; A)$ is the normal cone to $A$ at $x$ (see (3.6.2)).*

Similarly to [207, Theorem 6.15] (compare also [104]), we will show that the oriented distance function $\Delta_{-C}$ coincides with the functional $\varphi_{-C,k}$ under the additional assumption that $Y$ is equipped with the norm $\|\cdot\|_{C,k}$ which is the Minkowski functional of the order interval $[-k, k]_C$ according to Theorem 2.2.10.

**Lemma 9.3.4** *Suppose that $Y$ is a topological vector space, that $C \subset Y$ is a pointed, closed, convex cone with nonempty interior, $k \in \operatorname{int} C$. If the distance function $d$ refers to the norm $\|\cdot\|_{C,k}$, it follows that*

$$\forall y \in Y,\ \forall A \subset Y: \quad d(y, A) = \inf\{t > 0 \mid y \in t[-k, k]_C + A\}.$$

*Furthermore, we have*

$$\forall t > 0: \quad t[-k, k]_C + C = -tk + C. \tag{9.3.2}$$

**Theorem 9.3.5** *Suppose that $Y$ is a topological vector space and $C \subset Y$ is a pointed, closed, convex cone with nonempty interior, $k \in \operatorname{int} C$. Then,*

$$\varphi_{-C,k} = \Delta_{-C},$$

*where $\Delta_{-C} : Y \to \mathbb{R}$ is defined by (9.3.1) with distance functions induced by the norm $\|\cdot\|_{C,k}$.*

***Proof*** (a) Suppose that $y \in -C$. Then, $d(y, -C) = 0$ and

$$\begin{aligned}
\Delta_{-C}(y) &= d(y,-C) - d(y, Y\setminus(-C)) = -d(y, Y\setminus(-C)) = -d(0, -y + Y\setminus(-C)) \\
&= -\inf\{t > 0 \mid 0 \in t[-k,k]_C - y + (Y\setminus(-C))\} \\
&= -\inf\{t > 0 \mid 0 \in -tk - y + (Y\setminus(-C))\} \\
&= -\inf\{t > 0 \mid -tk \in y - (Y\setminus(-C))\} \\
&= -\inf\{t \in \mathbb{R} \mid -tk \in y - (Y\setminus(-C))\} \\
&= \sup\{t \in \mathbb{R} \mid tk \in y - (Y\setminus(-C))\} \\
&= \sup\{t \in \mathbb{R} \mid tk \notin y + C\} \\
&= \inf\{t \in \mathbb{R} \mid tk \in y + C\} = \varphi_{-C,k}(y).
\end{aligned}$$

(b) Suppose that $y \notin -C$. Then, $0 \in Y\setminus(y+C)$, i.e., $d(y, Y\setminus(-C)) = d(0, Y\setminus(y+C)) = 0$, and taking into account (9.3.2) we get

$$\begin{aligned}
\Delta_{-C}(y) &= d(y,-C) - d(y, Y\setminus(-C)) = d(y,-C) \\
&= \inf\{t > 0 \mid y \in t[-k,k]_C - C\} \\
&= \inf\{t > 0 \mid y \in tk - C\} \\
&= \inf\{t \in \mathbb{R} \mid y \in tk - C\} = \varphi_{-C,k}(y).
\end{aligned}$$

The equation $\inf\{t > 0 \mid y \in tk - C\} = \inf\{t \in \mathbb{R} \mid y \in tk - C\}$ follows because $y \notin -C$.

□

**Remark 9.3.6** Relationships in the sense of inclusion between the functional $\varphi_{A,k}$ defined by (4.1.11), the scalarizing functional by Hiriart-Urruty (see (9.3.1)) and the functional by Graña Drummond and Svaiter [196] are studied by Bouza, Quintana and Tammer in [69].

For a given $w^*-$compact generator $G$, the Graña Drummond-Svaiter functional associated to $G$ is defined by

$$\Psi_G(y) := \sigma_G(y) = \max\{y^*(y) \mid y^* \in G\}, \tag{9.3.3}$$

where we say that a set $G \subseteq Y$ is a generator of a convex cone $C \subset Y$ ($Y$ is a topological vector space) if it satisfies the following properties:

(i) $G$ is convex,
(ii) $0 \notin \operatorname{cl} G$,
(iii) $C = \operatorname{cone}(G)$.

Comparing the classes of functionals (4.1.11), (9.3.1) and (9.3.3), Bouza, Quintana and Tammer have shown in [69] that the class of functionals (4.1.11) is minimal in the sense of inclusion. Furthermore, a new (and larger) class of scalarizing functionals that are not necessarily convex, but rather quasidifferentiable and positively homogeneous is introduced in [69]. There are relationships to certain set relations (see Sect. 10.1).

# Chapter 10
# Set-Valued Optimization Problems

Set optimization has become a very important field in optimization theory as well as in various applications, especially in welfare economics, mathematical finance, optimization under uncertainty and medical image processing. These topics are discussed in the books [272, 291] (see also the references therein) and in many papers (for example, see [9, 24, 27, 103, 112, 165, 207, 211, 212, 220, 237, 278, 290, 292, 295, 297, 301, 303, 314–316, 318–320, 342, 410, 513]). In particular, certain concepts of robustness for dealing with uncertainties in scalar as well as vector optimization can be described using approaches from set-valued optimization (see [133, 253–255, 291, 295]). In the applications mentioned above, it is necessary to assume a feasible element to be associated with a whole set of objective function values instead of just one vector. This means, set optimization deals with the process of obtaining minimal sets, where the objective map to be optimized is set-valued (see [272]). In order to obtain optimal solutions of a set optimization problem, one uses set relations (see Sect. 10.1). We are interested in characterizing certain set relations by using the nonlinear separating functional (4.1.11) and corresponding extensions. Such characterizations of set relations via scalarization are important for deriving numerical methods for solving set-valued optimization problems, see Sect. 13.5. Recently, there are many papers dealing with these extensions of the functional (4.1.11) related to set relations in order to characterize solutions of set-valued optimization problems.

The aim of this chapter is to show that it is possible to give characterizations of solutions to set-valued optimization problems by means of the nonlinear scalarizing functional $\varphi_{A,k}$ in (4.1.11). First, we introduce set relations (Sect. 10.1) and define solution concepts for set-valued optimization problems based on these set relations (Sect. 10.2). Hernández and Rodríguez-Marín [237] introduced a functional $\Phi_k$ that can be considered as an extension of the functional $\varphi_{A,k}$ in (4.1.11) related to set-valued optimization. In Sect. 10.3.1, we discuss the properties of the functional $\Phi_k$, especially boundedness and monotonicity. A characterization of the solutions of



C. Tammer and P. Weidner, *Scalarization and Separation by Translation Invariant Functions*, Vector Optimization, https://doi.org/10.1007/978-3-030-44723-6_10

set-valued optimization problems via scalarization by means of the functional $\Phi_k$ is given in Sect. 10.4.2. In Sect. 10.3.2, we will see that it is possible to derive a complete characterization of solutions to set-valued optimization problems using the functional $\varphi_{A,k}$ that has very useful properties (see Chap. 4) and nice geometrical interpretations.

## 10.1 Set Relations

In this section, we introduce set relations (compare Sect. 2.1.2) that we will use in Sect. 10.2 in order to define solution concepts for set-valued optimization problems (see [291] for an overview).

We assume that $Y$ is a topological vector space partially ordered by a nontrivial, convex, closed, pointed cone $C \subset Y$.

**Definition 10.1.1** The set of all subsets of $Y$ is called **power set** of $Y$ and is denoted by $\mathscr{P}(Y)$.

The important concepts of the lower (upper) set less relation are introduced by Kuroiwa [314–316] (see Figs. 10.1, 10.2).

**Definition 10.1.2** (*Lower (upper) set less relation*) Suppose that $C \subset Y$ is a nontrivial, closed, convex and pointed cone and $A, B \in \mathscr{P}(Y)$ are arbitrary proper sets. The lower set less relation $\preceq^l_C$ is defined by

$$A \preceq^l_C B :\iff A + C \supseteq B \iff \forall b \in B\ \exists a \in A \text{ such that } a \leq_C b$$

and the upper set less relation $\preceq^u_C$ is defined by

$$A \preceq^u_C B :\iff A \subseteq B - C \iff \forall a \in A\ \exists b \in B \text{ such that } a \leq_C b.$$

In order to distinguish between vector minimum and maximum problems, we use the following notations in this chapter:

$$\text{Min}(B, C) := \text{Eff}(B, C),$$

$$\text{Max}(B, C) := \text{Eff}(B, -C),$$

where $C \subset Y$ is a nontrivial, closed, convex and pointed cone and $B \in \mathscr{P}(Y)$ is an arbitrary proper set.

**Remark 10.1.3** We recall several important relationships and assertions concerning the lower and upper set less relation (see [291, Remark 2.6.11]). Of course,

$$A \preceq^l_C B \iff A + C \supseteq B + C.$$

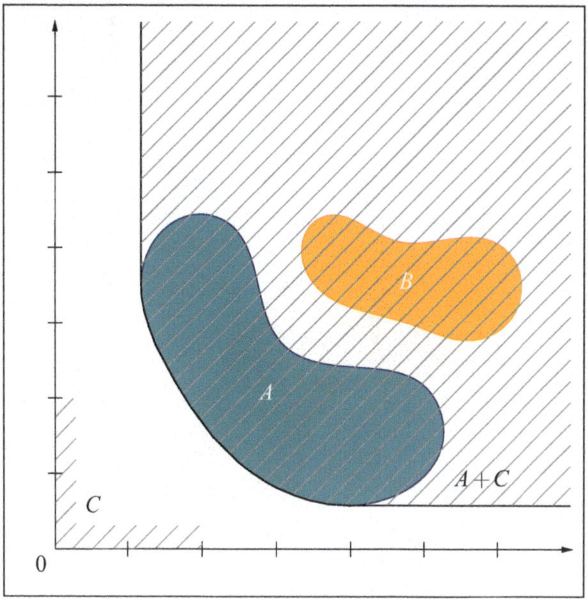


**Fig. 10.1** $A \preccurlyeq^l_C B$ $(B \subseteq A + C)$

Moreover,

$$A \preceq^u_C B \iff A - C \subseteq B - C$$

and

$$A \preceq^l_C B \text{ and } B \preceq^l_C A \iff A + C = B + C.$$

Since $C$ is a pointed, closed, convex cone, it holds that $\mathrm{Min}(A + C, C) = \mathrm{Min}(A, C)$ and $\mathrm{Min}(B + C, C) = \mathrm{Min}(B, C)$ such that the following implication holds

$$A \preceq^l_C B \text{ and } B \preceq^l_C A \implies \mathrm{Min}(A, C) = \mathrm{Min}(B, C).$$

If we additionally assume that $A \subseteq \mathrm{Min}(A, C) + C$ and $B \subseteq \mathrm{Min}(B, C) + C$ (domination property, see [43, 338]), we get

$$\mathrm{Min}(A, C) = \mathrm{Min}(B, C) \iff A + C = B + C$$

such that

$$A \preceq^l_C B \text{ and } B \preceq^l_C A \iff \mathrm{Min}(A, C) = \mathrm{Min}(B, C).$$

Moreover, we similarly conclude

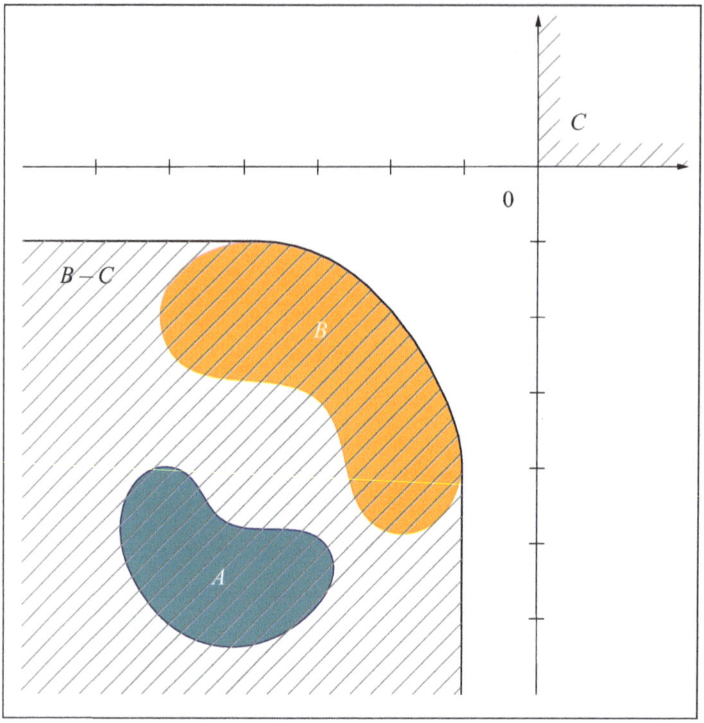


**Fig. 10.2** $A \preccurlyeq^u_C B \; (A \subseteq B - C)$

$$A \preceq^u_C B \text{ and } B \preceq^u_C A \quad \Longleftrightarrow \quad A - C = B - C$$

and since $\operatorname{Max}(A - C, C) = \operatorname{Max}(A, C)$ and $\operatorname{Max}(B - C, C) = \operatorname{Max}(B, C)$, the following implication holds

$$A \preceq^u_C B \text{ and } B \preceq^u_C A \Longrightarrow \operatorname{Max}(A, C) = \operatorname{Max}(B, C).$$

If we additionally assume that $A \subseteq \operatorname{Max}(A, C) - C$ and $B \subseteq \operatorname{Max}(B, C) - C$, we get

$$\operatorname{Max}(A, C) = \operatorname{Max}(B, C) \Longleftrightarrow A - C = B - C$$

and, finally,

$$A \preceq^u_C B \text{ and } B \preceq^u_C A \Longleftrightarrow \operatorname{Max}(A, C) = \operatorname{Max}(B, C).$$

Furthermore, we will consider the set less relation $\preceq^s_C$ that was independently introduced by Young [499] and Nishnianidze [370] (see Eichfelder, Jahn [147]) for comparing sets:

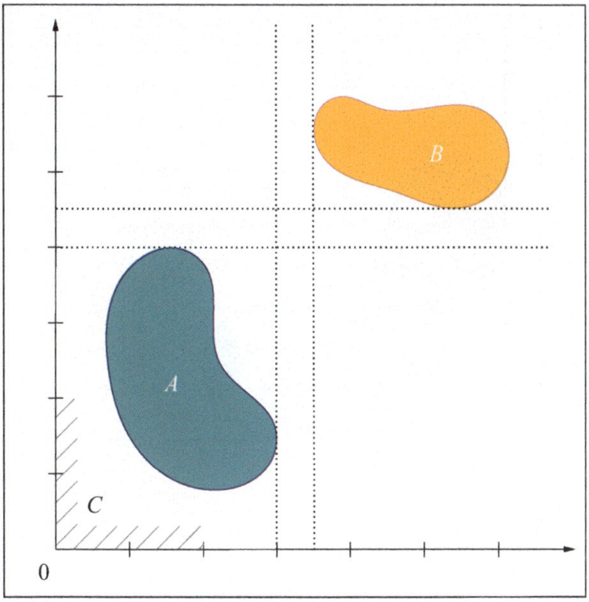


**Fig. 10.3** $A \preccurlyeq_C^c B$ $(\forall a \in A\ \forall b \in B{:}a \leq_C b)$

**Definition 10.1.4** (*Set less relation*) Let $C \subset Y$ be a nontrivial, closed, convex and pointed cone. Moreover, let $A, B \in \mathscr{P}(Y)$ be arbitrarily chosen proper sets. The set less relation is defined by

$$A \preceq_C^s B{:} \iff A \subseteq B - C \text{ and } A + C \supseteq B.$$

The certainly less relation $\preceq_C^c$ (Kuroiwa [314–316], compare Eichfelder, Jahn [147]), introduced in the following definition (see Fig. 10.3), is useful in interval analysis.

**Definition 10.1.5** (*Certainly less relation* $\preceq_C^c$) Suppose that $C \subset Y$ is a nontrivial, closed, convex and pointed cone. Let $A, B \in \mathscr{P}(Y)$ be arbitrary proper sets. The **certainly less relation** $\preceq_C^c$ is defined by

$$A \preceq_C^c B{:} \iff (A = B) \text{ or } (A \neq B, \forall a \in A\ \forall b \in B{:}a \leq_C b).$$

Furthermore, we introduce the possibly less relation $\preceq_C^p$ (Kuroiwa [315, 316]) (see Fig. 10.4).

**Definition 10.1.6** (*Possibly less relation* $\preceq_C^p$) Let $A, B \in \mathscr{P}(Y)$ be arbitrary proper sets and $C \subset Y$ a nontrivial, closed, convex and pointed cone. The possibly less relation $\preceq_C^p$ is defined by

$$A \preceq_C^p B{:} \iff (\exists a \in A, \exists b \in B{:}a \leq_C b).$$

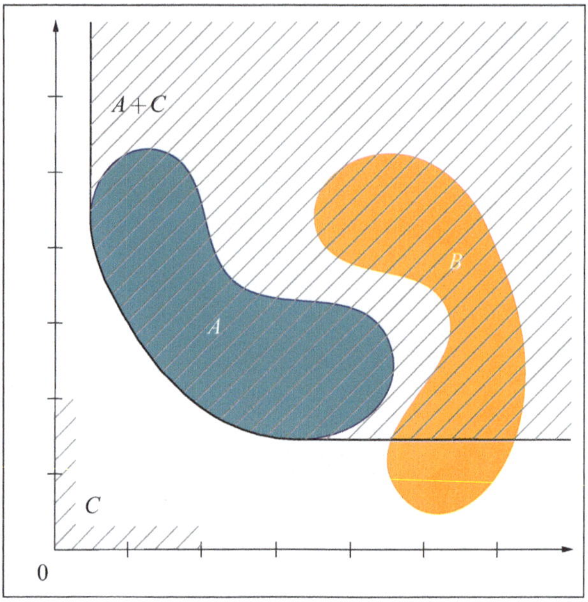


**Fig. 10.4** $A \preccurlyeq^p_C B$ ($\exists a \in A, \exists b \in B{:}a \leq_C b$)

**Remark 10.1.7** The following relationships between the certainly less relation, the possibly less relation and the lower set less relation are explained in [291, Remark 2.6.14].

Obviously, $A \preceq^c_C B$ yields

$$\exists a \in A \quad \text{such that} \quad \forall b \in B{:}a \leq_C b. \tag{10.1.1}$$

Furthermore, (10.1.1) yields $A \preceq^l_C B$ (compare Definition 10.1.2) such that

$$A \preceq^c_C B \Longrightarrow A \preceq^l_C B$$

and $A \preceq^l_C B$ yields

$$\exists a \in A,\ \exists b \in B \text{ such that } a \leq_C b. \tag{10.1.2}$$

We obtain

$$A \preceq^c_C B \quad \Longrightarrow \quad A \preceq^l_C B \quad \Longrightarrow \quad A \preceq^p_C B \tag{10.1.3}$$

taking into account Definition 10.1.6.

**Remark 10.1.8** The following relationships between the certainly less relation, the possibly less relation and the upper set less relation are discussed in [291, Remark 2.6.15].

$A \preceq_C^c B$ implies

$$\exists b \in B \quad \text{such that} \quad \forall a \in A: a \leq_C b \tag{10.1.4}$$

and (10.1.4) implies $A \preceq_C^u B$ (see Definition 10.1.2) such that we get the implication

$$A \preceq_C^c B \Longrightarrow A \preceq_C^u B.$$

Moreover, $A \preceq_C^u B$ yields

$$\exists a \in A, \ \exists b \in B \text{ such that } a \leq_C b, \tag{10.1.5}$$

such that we can conclude

$$A \preceq_C^c B \quad \Longrightarrow \quad A \preceq_C^u B \quad \Longrightarrow \quad A \preceq_C^p B$$

taking into account Definition 10.1.6.

Moreover, we introduce the minmax less relation $\preceq_C^m$ for proper sets $A$, $B$ belonging to

$$\mathcal{F} := \{A \in \mathcal{P}(Y) \mid \text{Min}(A, C) \neq \emptyset \text{ and } \text{Max}(A, C) \neq \emptyset\}.$$

For instance, in a topological vector space $Y$, for every proper compact set in $\mathcal{P}(Y)$, minimal and maximal elements exist.

**Definition 10.1.9** (*Minmax less relation*) Consider $A, B \in \mathcal{F}$. Suppose that $C \subset Y$ is a nontrivial, closed, convex and pointed cone. The minmax less relation $\preceq_C^m$ is given by

$$A \preceq_C^m B :\Longleftrightarrow \text{Min}(A, C) \preceq_C^s \text{Min}(B, C) \text{ and } \text{Max}(A, C) \preceq_C^s \text{Max}(B, C).$$

**Definition 10.1.10** (*Minmax certainly less relation*) Consider $A, B \in \mathcal{F}$. Suppose that $C \subset Y$ is a nontrivial, closed, convex and pointed cone. The *minmax* certainly less relation $\preceq_C^{mc}$ is given by

$$A \preceq_C^{mc} B :\Longleftrightarrow (A = B) \text{ or } (A \neq B, \text{Min}(A, C) \preceq_C^c \text{Min}(B, C) \text{ and } \text{Max}(A, C) \preceq_C^c \text{Max}(B, C)).$$

The minmax certainly nondominated relation $\prec_C^{mn}$ is introduced by Jahn and Ha in [275].

**Definition 10.1.11** (*Minmax certainly nondominated relation*) Suppose that $C \subset Y$ is a nontrivial, closed, convex and pointed cone. Consider $A, B \in \mathcal{P}(Y)$. The *minmax certainly nondominated relation* $\preceq_C^{mn}$ is given by

$$A \preceq_C^{mn} B :\Longleftrightarrow (A = B) \; or \; (A \neq B, \operatorname{Max}(A, C) \preceq_C^s \operatorname{Min}(B, C)).$$

**Remark 10.1.12** Recently, in Köbis, Thanh, Tammer and Yao [300], several set relations equipped with a variable domination structure that is not necessarily given by a cone-valued map, are introduced. These set relations are used in order to define corresponding solution concepts. Furthermore, necessary optimality conditions for these solutions are derived in [300].

## 10.2 Solution Concepts in Set-Valued Optimization

In Sect. 10.1, we introduced the set relations $\preceq_C^l$, $\preceq_C^u$, $\preceq_C^s$, $\preceq_C^m$, $\preceq_C^{mc}$ and $\preceq_C^{mn}$. Under the assumption that $\preceq_C$ denotes one of these set relations, we introduce the following general concept of nondominated elements of a system of nonempty sets.

Throughout this section, we suppose that $Y$ is a topological vector space and $C \subset Y$ a nontrivial, convex, closed and pointed cone. Let $\mathscr{P}(Y)$ be the power set of $Y$. Furthermore, let $A, \bar{A} \in \mathscr{P}(Y)$ be proper sets and $\mathscr{A}$ be a system of nonempty subsets of $Y$.

**Definition 10.2.1** We say that $\bar{A} \in \mathscr{A}$ is a **nondominated element of** $\mathscr{A}$ w.r.t. $\preceq_C$ if

$$A \preceq_C \bar{A}, \; A \in \mathscr{A} \implies \bar{A} \preceq_C A.$$

Analogously, we introduce optimal solutions with respect to $\preceq_C$ to the following set-valued optimization problem:

$$\preceq_C -\operatorname*{Opt}_{x \in S} \Gamma(x), \qquad (SP - \preceq_C)$$

where we suppose that $S$ is a subset of $X$, $X$ is a linear space, $\Gamma : X \rightrightarrows Y$.

According to the definition of nondominated elements with respect to a binary relation (see Definition 2.1.42), we introduce nondominated solutions of problem $(SP - \preceq_C)$ w.r.t. $\preceq_C$ in the pre-image space $X$.

**Definition 10.2.2** (*Nondominated solutions of* $(SP - \preceq_C)$ *w.r.t. the preorder* $\preceq_C$) We say that $\bar{x} \in S$ is a nondominated solution of problem $(SP - \preceq_C)$ w.r.t. $\preceq_C$ if

$$\Gamma(x) \preceq_C \Gamma(\bar{x}) \text{ for some } x \in S \quad \Longrightarrow \quad \Gamma(\bar{x}) \preceq_C \Gamma(x).$$

**Remark 10.2.3** If we use the set relation $\preceq_C^l$ introduced in Definition 10.1.2 in the formulation of the solution concept (Definition 10.2.2) (with other words, we consider the set-valued problem $(SP - \preceq_C^l)$), we recognize that this solution concept is based on comparisons among sets of (minimal) nondominated points of values of $\Gamma$ (see Definition 6.0.1). On the other hand, if we use the upper set less relation $\preceq_C^u$ (Definition 10.1.2) in the formulation of the solution concept, i.e., if we consider the problem $(SP - \preceq^u_C)$, we observe that this solution concept is based on comparisons of (maximal) nondominated points of values of $\Gamma$ (see Definition 6.0.1).

We define the set-relation $\approx^l$ by

$$A \approx^l B :\Longleftrightarrow A \preceq^l_C B \;\text{ and }\; B \preceq^l_C A. \tag{10.2.1}$$

It is easy to check that the set-relation $\approx^l$ is an equivalence relation on $\mathscr{P}(Y) \setminus \{\emptyset\}$. Furthermore, we consider an order on the family of equivalence classes that are denoted by $[\cdot]^l$.

Of course,

$$A \in [B]^l \quad \Longleftrightarrow \quad A + C = B + C$$

(compare Definition 10.1.2 and Remark 10.1.3).

## 10.3 Characterization of Set Relations Using the Functional $\varphi_{A,k}$

The aim of this section is to show the usefulness of the nonlinear scalarizing functional given by (4.1.11) in order to characterize set relations (see Sect. 10.1). In the literature, many authors have extended this nonlinear scalarizing functional in order to apply it to set-valued optimization problems, where the solution concept is based on set relations introduced in Sect. 10.1 (see [9, 112, 165, 207, 210–212, 220, 237, 292, 297, 301, 318–320, 342, 410, 513]).

However, in Sect. 10.3.2, we will show that set relations can be completely characterized using the nonlinear separating functional $\varphi_{A,k}$ given by (4.1.11).

### *10.3.1 The Scalarization Method by Hernández and Rodríguez-Marín*

We present the scalarization method by Hernández and Rodríguez-Marín [237] in this section. For further recent developments on the field of extensions related to the set-valued case, see the paper by Gutiérrez, Jiménez, Miglierina, Molho [207] and references therein.

In this section, we suppose that $Y$ is a topological vector space ordered by a nontrivial, convex, closed, pointed cone $C \subset Y$ with nonempty interior, $k \in -\operatorname{int} 0^+C$. Note that $k \in -\operatorname{int} 0^+C = -\operatorname{int} C$ under the given assumptions concerning $C$ (see Lemma 2.2.21(b)).

For $d \in Y$, $k \in -\operatorname{int} 0^+C$ and $A = d + C$, the functional $\varphi_{A,k} : Y \to \overline{\mathbb{R}}$ in (4.1.11) is given by

$$\varphi_{d+C,k}(y) = \inf\{t \in \mathbb{R} \mid y \in tk + d + C\}. \tag{10.3.1}$$

If we replace the single element $d \in Y$ in (10.3.1) by a set $D \in \mathscr{P}(Y)$, $D \neq \emptyset$, and suppose that $D + C$ is closed, $k \in -\operatorname{int} 0^+C$, we study the functional $\varphi_{D+C,k} : Y \to \overline{\mathbb{R}}$ given by

$$\varphi_{D+C,k}(y) := \inf\{t \in \mathbb{R} \mid y \in tk + D + C\}, \quad y \in Y. \tag{10.3.2}$$

Of course, if we set $A = D + C$, the functional $\varphi_{D+C,k}$ coincides with the functional $\varphi_{A,k}$ given by (4.1.11).

Note, under the given assumptions concerning $C$, $D$ and $k$, the functional in (10.3.2) takes the values in $\mathbb{R} \cup \{-\infty\}$, i.e., $\varphi_{D+C,k} : Y \to \mathbb{R} \cup \{-\infty\}$.

Furthermore, for any $y \in Y$, it holds that

$$\varphi_{D+C,k}(y) = \inf_{d \in D} \varphi_{d+C,k}(y). \tag{10.3.3}$$

Consider the nontrivial, convex, closed, pointed cone $C \subset Y$ (with nonempty interior). In Definition 2.2.14, we introduced $C$-proper subsets $D$ of $Y$, i.e., $D$ is a nonempty subset of $Y$ and $D + C \neq Y$. We now introduce the family of all $C$-proper subsets $D$ of $Y$ with the additional assumption that $D + C$ is closed, i.e., $D$ is $C$-closed in the sense of Definition 2.2.16:

$$\mathscr{P}_C(Y) := \{D \subset Y \mid D \text{ is a } C\text{-proper set and } D + C \text{ is closed}\}.$$

Furthermore, we introduce the notation

$$\mathscr{P}_C^0(Y) := \{D \subset Y \mid D \text{ is a } C\text{-proper set}\}.$$

**Remark 10.3.1** The assertion

$$D \in \mathscr{P}_C^0(Y) \iff \forall y \in Y : \varphi_{D+C,k}(y) > -\infty$$

is shown by Hernández and Rodríguez-Marín in [237, Lemma 2.16] under the assumption $k \in -\operatorname{int} C$ $(= -\operatorname{int} 0^+C)$.

Following Hernández and Rodríguez-Marín [237], we discuss an extension of the functional $\varphi_{D+C,k} : Y \to \mathbb{R} \cup \{-\infty\}$ given by (10.3.2) to a function $\Phi_k : \mathscr{P}_C(Y) \times \mathscr{P}_C^0(Y) \longrightarrow \mathbb{R} \cup \{+\infty\}$ (see Definition 10.3.8) in this section.

The results in this section are derived by Hernández and Rodríguez-Marín [237]. Using a certain scalarization technique where the nonlinear scalarizing functional (4.1.11) is involved, Hernández and Rodríguez-Marín have shown a characterization of solutions of set-valued optimization problems where the solution concept is given by the lower set less relation $\preceq_C^l$ for proper sets $A, B \in \mathscr{P}(Y)$ introduced in Definition 10.1.2

$$A \preceq^l_C B \Longleftrightarrow A + C \supseteq B$$

(see Sect. 10.4.2). For a characterization of other set relations and corresponding solution concepts, see Sects. 10.3.2 and 10.4.

**Remark 10.3.2** Important properties of the functional $\varphi_{D+C,k}$ in (10.3.2) are already shown in Proposition 6.2.41.

For applications as well as from the theoretical point of view, it is important to find sufficient conditions such that the scalarizing functional $\varphi_{D+C,k}$ is finite-valued, continuous, convex and monotone.

**Lemma 10.3.3** *Let $C \subset Y$ be a nontrivial, convex, closed, pointed cone with nonempty interior and $D \in \mathcal{P}_C(Y)$. Suppose that $k \in -\operatorname{int} 0^+C$. Then,*

(a) *$\varphi_{D+C,k}(\cdot)$ is finite-valued and continuous.*
(b) *If $D + C$ is convex, then $\varphi_{D+C,k}(\cdot)$ is convex.*
(c) *$\varphi_{D+C,k}(\cdot)$ is $(-\operatorname{int} C)$-monotone as well as strictly $(-\operatorname{int} C)$-monotone.*

***Proof***

(a) From Lemma 2.2.21(b), we have $k \in -\operatorname{int} 0^+C = -\operatorname{int} C$ under the given assumptions concerning $C$. For $k \in -\operatorname{int} C$ and since $C \subseteq 0^+(D + C)$, it holds that $k \in -\operatorname{int} 0^+(D + C)$. So, the assumptions of Proposition 4.6.13 are fulfilled and we get that $\varphi_{D+C,k}(\cdot)$ is finite-valued and continuous.
(b) If $D + C$ is convex, then $\varphi_{D+C,k}(\cdot)$ is convex because of Theorem 4.4.4(f).
(c) Note that $(Y \setminus \operatorname{int}(D + C)) - \operatorname{int} C \subseteq Y \setminus \operatorname{int}(D + C)$ holds. Indeed, if we suppose that $(Y \setminus \operatorname{int}(D + C)) - \operatorname{int} C \subseteq Y \setminus \operatorname{int}(D + C)$ does not hold, then there are elements $y \in Y$ with $y \notin \operatorname{int}(D + C)$ and $c \in \operatorname{int} C$ such that $y - c \in \operatorname{int}(D + C)$. This yields $y \in c + \operatorname{int}(D + C) \subseteq \operatorname{int} C + \operatorname{int}(D + C) \subseteq \operatorname{int}(D + C)$ in contradiction to $y \notin \operatorname{int}(D + C)$. So, we have $-k \in 0^+(Y \setminus (\operatorname{int}(D + C)))$. From Corollary 4.5.6, we get that $\varphi_{Y\setminus(\operatorname{int}(D+C)),-k}$ is $(\operatorname{int} C)$-monotone as well as strictly $(\operatorname{int} C)$-monotone. This implies that $\varphi_{D+C,k} = -\varphi_{Y\setminus(\operatorname{int}(D+C)),-k}$ is $(-\operatorname{int} C)$-monotone as well as strictly $(-\operatorname{int} C)$-monotone.

□

Analogously to the proof of the properties (4.4.5), (4.6.13), (4.6.14) of the functional $\varphi_{A,k}$ given by (4.1.11) in Theorems 4.4.4, 4.6.6, we get the corresponding assertions concerning $\varphi_{D+C,k}$ in the following lemma:

**Lemma 10.3.4** *Let $D \in \mathcal{P}_C(Y)$, $k \in -\operatorname{int} 0^+C$. Then, for any $y \in Y$ and for all $t \in \mathbb{R}$, it holds that*

(a) $\operatorname{lev}_{\varphi_{D+C,k},<}(t) = tk + D + \operatorname{int} C,$
(b) $\operatorname{lev}_{\varphi_{D+C,k},\leq}(t) = tk + D + C,$
(c) $\operatorname{lev}_{\varphi_{D+C,k},=}(t) = tk + \operatorname{bd}(D + C),$
(d) $\operatorname{lev}_{\varphi_{D+C,k},\geq}(t) = Y \setminus \{y \in Y \mid y \in tk + D + \operatorname{int} C\},$
(e) $\operatorname{lev}_{\varphi_{D+C,k},>}(t) = Y \setminus \{y \in Y \mid y \in tk + D + C\}.$

Taking into account the properties of the functional $\varphi_{D+C,k}$ in (10.3.2), we immediately get the following assertions for sets belonging to $\mathscr{P}_C(Y)$ (compare Hernández and Rodríguez-Marín [237, Proposition 2.18]).

**Proposition 10.3.5** *Suppose that $A, B \in \mathscr{P}_C(Y)$, $k \in -\operatorname{int} 0^+C$ and $y \in Y$. Then, the following assertions hold*

(a) $\varphi_{A+C,k}(y) = \inf_{a\in A} \varphi_{a+C,k}(y)$.

(b) $A \preceq^l_C B \implies \varphi_{A+C,k}(y) \le \varphi_{B+C,k}(y)$,
$A \in [B]^l \implies \varphi_{A+C,k}(y) = \varphi_{B+C,k}(y)$.

***Proof*** Under the assumptions $A \in \mathscr{P}_C(Y)$, $k \in -\operatorname{int} 0^+C$, we obtain $\varphi_{A+C,k}(y) = \inf_{a\in A} \varphi_{a+C,k}(y)$ from the definition of the functional $\varphi_{A+C,k}$ in (10.3.2) (see Sect. 4.3), i.e., (a) holds. Assertion (b) follows from the definition of the functional $\varphi_{A+C,k}$ and corresponding monotonicity properties derived in Theorem 4.5.5. □

**Remark 10.3.6** Further characterizations of the set relation $\preceq^l_C$ are given in Theorem 10.3.18 and in Corollary 10.3.38.

**Remark 10.3.7** Sufficient conditions concerning the attainment property

$$\varphi_{A+C,k}(y) = \inf_{a\in A} \varphi_{a+C,k}(y) = \min_{a\in A} \varphi_{a+C,k}(y)$$

are derived in Sect. 5.2, compare also Remark 10.3.21.

In order to compare sets, we study a very useful extension of the nonlinear separating functional (4.1.11) to a functional $\Phi_k : \mathscr{P}_C(Y) \times \mathscr{P}^0_C(Y) \to \mathbb{R} \cup \{+\infty\}$ (see [237, Definition 3.1]). Note that the functional $\varphi_{A+C,k}$ is finite-valued for $A \in \mathscr{P}_C(Y)$ and $k \in -\operatorname{int} 0^+C$ (see Lemma 10.3.3).

Already in Proposition 6.2.41, a motivation for the construction of the functional $\Phi_k$ and the relationship to the functional $\varphi_{A+C,k}$ including important properties are given.

**Definition 10.3.8** Let $k \in -\operatorname{int} 0^+C$ be a fixed element. Then, the **scalarizing functional by Hernández and Rodríguez-Marín** $\Phi_k : \mathscr{P}_C(Y) \times \mathscr{P}^0_C(Y) \longrightarrow \mathbb{R} \cup \{+\infty\}$ is given as

$$\Phi_k(A, B) := \sup_{b\in B}\{\varphi_{A+C,k}(b)\}. \tag{10.3.4}$$

**Remark 10.3.9** Taking into account the definition of $\Phi_k(A, B)$ and Proposition 10.3.5(a), we get

$$\Phi_k(A, B) = \sup_{b\in B} \inf_{a\in A} \varphi_{a+C,k}(b)$$

for $A \in \mathscr{P}_C(Y)$ and $B \in \mathscr{P}^0_C(Y)$. This means that $\Phi_k(A, B)$ is defined using the supremum and infimum of the nonlinear separating functional (4.1.11) over certain sets $A$ and $B$. We will discuss further corresponding scalarizations in Sect. 10.3.2.

**Remark 10.3.10** It is important to mention that the supremum in (10.3.4) is attained, i.e., $\Phi_k(A, B) = \max_{b\in B}\{\varphi_{A+C,k}(b)\}$, if $B$ is a compact set taking into account the continuity of the functional $\varphi_{A+C,k}$ under the given assumptions (see Lemma 10.3.3(a)). Furthermore, in [237, Proposition 3.4], it is shown that the supremum in (10.3.4) is attained if $B$ is a $C$-compact set (see Definition 2.2.16).

Taking into account Lemma 10.3.4(b) and Definition 10.1.2, we observe

$$\forall\, A, B \in \mathscr{P}_C(Y),\ t \in \mathbb{R}:\quad \Phi_k(A, B) \le t \iff B \subseteq tk + A + C \iff tk + A \preceq^l_C B. \tag{10.3.5}$$

In the following proposition (see [237, Proposition 3.2]), we mention an equivalent formulation of the functional (10.3.4) using (10.3.5).

**Proposition 10.3.11** *Suppose that $A, B \in \mathscr{P}_C(Y)$ and $k \in -\operatorname{int} 0^+C$. Furthermore, assume that $\Phi_k(A, B) < +\infty$. Then, the functional $\Phi_k : \mathscr{P}_C(Y)^2 \longrightarrow \mathbb{R} \cup \{+\infty\}$ can be expressed as follows:*

$$\Phi_k(A, B) = \inf\{t \in \mathbb{R} \mid B \subseteq tk + A + C\} = \inf\{t \in \mathbb{R} \mid tk + A \preceq^l_C B\}. \tag{10.3.6}$$

***Proof*** Let $t \in \mathbb{R}$ and $\Phi_k(A, B) \le t$. Then,

$$\sup_{b\in B}\{\varphi_{A+C,k}(b)\} \le t$$

and

$$\forall b \in B : \quad \varphi_{A+C,k}(b) \le t.$$

From Lemma 10.3.4(b), we get

$$B \subseteq \operatorname{lev}_{\varphi_{A+C,k},\le}(t) = tk + A + C.$$

This yields, together with (10.3.5) and the assumption $\Phi_k(A, B) < +\infty$

$$\Phi_k(A, B) = \inf\{t \in \mathbb{R} \mid B \subseteq tk + A + C\} = \inf\{t \in \mathbb{R} \mid tk + A \preceq^l_C B\}.$$

The proof is completed. □

**Remark 10.3.12** Formula (10.3.6) in Proposition 10.3.11 shows the relationship between the functional $\Phi_k$ given by (10.3.4) and the scalarizing functional introduced by Hamel and Löhne in [220] ($\Psi_k(A, B) := \inf\{t \in \mathbb{R} \mid tk + A \preceq^l_C B\}$). In the definition of the scalarizing functional $\Psi_k(A, B)$, a set relation is involved. But, for deriving optimality conditions and numerical procedures, one needs a calculation of the subdifferential of the scalarizing functional. We know the structure of the subdifferential of the nonlinear scalarizing functional (4.1.11) (see Sect. 4.10

and Theorem 12.1.8) that is involved in the definition of (10.3.4) by Hernández and Rodríguez-Marín such that this functional is very useful.

The following characterization of the $C$-boundedness of a set $A \in \mathscr{P}_C(Y)$ (see Definition 2.2.16) by the boundedness of the functional $\Phi_k$ is shown in [237, Lemma 3.5].

**Lemma 10.3.13** *Let $A \in \mathscr{P}_C(Y)$, $k \in -\operatorname{int} 0^+C$ be given. Then,*

$$A \text{ is } C\text{–bounded} \iff \Phi_k(C, A) = \sup_{a \in A} \inf_{c \in C} \varphi_{c+C,k}(a) < +\infty.$$

***Proof*** "$\Rightarrow$": Suppose that $A$ is $C$-bounded and consider a neighborhood of $0 \in Y$

$$U := k + \operatorname{int} C.$$

Then, there is a real number $\alpha > 0$ with $A \subseteq \alpha(k + \operatorname{int} C) + C \subseteq \alpha k + C + C$, compare Definition 2.2.16. Hence, for each $a \in A$ it holds that $\varphi_{C+C,k}(a) \leq \alpha$ taking into account Lemma 10.3.4(b). This implies $\Phi_k(C, A) = \sup_{a \in A} \varphi_{C+C,k}(a) \leq \alpha$.

"$\Leftarrow$": Now, we assume that $\Phi_k(C, A) < +\infty$. Then, there is a real number $r \in \mathbb{R}$ with $\Phi_k(C, A) < r$. This means, for each $a \in A$ it holds that

$$\varphi_{C+C,k}(a) < r.$$

Because of Lemma 10.3.4(a), the inequality given above can be written as $a \in rk + \operatorname{int} C$, and this means that

$$A \subseteq rk + \operatorname{int} C. \tag{10.3.7}$$

Consider a neighborhood $U'$ of $0 \in Y$. If we chose $\lambda > 0$ such that $rk \in \lambda U'$, we get from (10.3.7) that

$$A \subseteq \lambda U' + C,$$

i.e., $A$ is $C$-bounded. □

The following assertion concerning the $C$-boundedness of a set $B \in \mathscr{P}_C(Y)$ (see Definition 2.2.16) and the property $\Phi_k(A, B) < +\infty$ (for $A, B \in \mathscr{P}_C(Y)$) is given in [237, Theorem 3.6].

**Theorem 10.3.14** *Suppose that $A, B \in \mathscr{P}_C(Y)$ is a $C$-bounded set and $k \in -\operatorname{int} 0^+C$. Then,*

$$B \text{ is } C\text{–bounded} \iff \Phi_k(A, B) = \sup_{b \in B} \inf_{a \in A} \varphi_{a+C,k}(b) < +\infty.$$

***Proof*** "$\Rightarrow$": Assume that $B$ is $C$-bounded. For every $b \in B$, we get

$$\varphi_{A+C,k}(b) = \inf_{a \in A} \{\varphi_{a+C,k}(b)\} \tag{10.3.8}$$

because of (10.3.3). Obviously, it holds that $\varphi_{a+C,k}(b) = \varphi_{0+C,k}(b-a)$ and $\varphi_{0+C,k}(b-a) = \varphi_{C+C,k}(b-a)$. This implies that the equality (10.3.8) can be written as

$$\varphi_{A+C,k}(b) = \inf_{a \in A}\{\varphi_{C+C,k}(b-a)\}.$$

Then, we get

$$\varphi_{A+C,k}(b) \leq \varphi_{C+C,k}(b-a_0) \quad \text{for all} \quad b \in B$$

for any fixed $a_0 \in A$. Hence,

$$\Phi_k(A, B) \leq \sup_{b \in B}\{\varphi_{C+C,k}(b-a_0)\} = \Phi_k(C, B-a_0).$$

Taking into account Lemma 10.3.13, it follows that $\Phi_k(C, B-a_0) < +\infty$ since $B$ is $C$-bounded. Hence, $\Phi_k(A, B) < +\infty$.

"$\Leftarrow$": Conversely, we assume $\Phi_k(A, B) = r$ with $r \in \mathbb{R}$. Then, $\varphi_{A+C,k}(b) \leq r$ for each $b \in B$. Letting $\alpha > 0$, we have $\varphi_{A+C,k}(b) < r + \alpha$ and with Lemma 10.3.4(a), it holds that $b \in (r+\alpha)k + A + C$ for all $b \in B$. Therefore,

$$B \subseteq (r+\alpha)k + A + C. \tag{10.3.9}$$

Consider a neighborhood $U$ of $0 \in Y$. Because $A$ is $C$-bounded, the set $(r+\alpha)k + A$ is $C$-bounded too. Hence, there is a real number $\lambda > 0$ such that

$$(r+\alpha)k + A \subseteq \lambda U + C.$$

This inclusion implies

$$B \subseteq \lambda U + C$$

taking into account (10.3.9), i.e., $B$ is $C$-bounded. □

For deriving characterizations of solutions of set-valued optimization problems (where the solution concept is given by the lower set less relation $\preceq^l_C$ (see Definition 10.1.2)) by means of the functional $\Phi_k$ given by (10.3.4) in Sect. 10.4.2, it is important to have certain monotonicity properties w.r.t. $\preceq^l_C$ of the functional $\Phi_k$. These monotonicity properties in Definition 10.3.15 are a natural extension of the monotonicity studied in Sect. 4.5 to the case where the comparison of sets is involved.

**Definition 10.3.15** Consider $\mathscr{S} \subseteq \mathscr{P}_C(Y)$. A functional $\varphi : \mathscr{P}_C(Y) \longrightarrow \mathbb{R}$ is called $\preceq^l_C$**-increasing** on $\mathscr{S}$ if

$$A, B \in \mathscr{S},\ A \preceq^l_C B \quad \Longrightarrow \quad \varphi(A) \leq \varphi(B).$$

Furthermore, $\varphi : \mathscr{P}_C(Y) \to \mathbb{R}$ is called $\preceq^l_C$**-decreasing** on $\mathscr{S}$ if

$$A, B \in \mathscr{S}, \ A \preceq^l_C B \quad \Longrightarrow \quad \varphi(A) \geq \varphi(B).$$

The next result (see [237, Theorem 3.8(v)]) will be used to prove characterizations of solutions of set-valued optimization problems via scalarization in Theorem 10.4.8.

**Theorem 10.3.16** *Suppose that $D \in \mathscr{P}_C(Y)$, $k \in -\operatorname{int} 0^+C$. Then, $\Phi_k(\cdot, D)$ is $\preceq^l_C$-increasing on $\mathscr{P}_C(Y)$.*

***Proof*** The assertion follows directly from Proposition 10.3.5(b). □

Moreover, we have the following properties (see [237, Theorem 3.8]):

**Theorem 10.3.17** *For $B, D, F \in \mathscr{P}_C(Y)$, $k \in -\operatorname{int} 0^+C$, the following assertions hold*

(a) *If $B \in [D]^l$, then $\Phi_k(D, \cdot) = \Phi_k(B, \cdot)$.*
(b)
$$B \preceq^l_C F \quad \Longrightarrow \quad \sup_{b \in B} \inf_{d \in D} \varphi_{d+C,k}(b) \geq \sup_{y \in F} \inf_{d \in D} \varphi_{d+C,k}(y),$$

*i.e., $\Phi_k(D, B) \geq \Phi_k(D, F)$ and $\Phi_k(D, \cdot)$ is $\preceq^l_C$-decreasing on $\mathscr{P}_C(Y)$.*
(c) *If $B \in [F]^l$, then $\Phi_k(\cdot, B) = \Phi_k(\cdot, F)$.*
(d) *If $B \in [D]^l$, then $\Phi_k(D, B) = \Phi_k(B, D)$.*

***Proof***

(a) The result follows directly from Proposition 10.3.5(b).
(b) Consider $B, D, F \in \mathscr{P}_C(Y)$. If $B \preceq^l_C F$ (this means $F \subseteq B + C$), we get that for each $\overline{y} \in F$ there is an element $\overline{b} \in B$ such that

$$\varphi_{D+C,k}(\overline{y}) \leq \varphi_{D+C,k}(\overline{b})$$

since $\varphi_{D+C,k}(\cdot)$ is $-C$-monotone (see Lemma 10.3.3(c)). This yields

$$\sup_{y \in F} \varphi_{D+C,k}(y) \leq \sup_{b \in B} \varphi_{D+C,k}(b).$$

From Proposition 10.3.5(a), we get

$$\sup_{y \in F} \inf_{d \in D} \varphi_{d+C,k}(y) \leq \sup_{b \in B} \inf_{d \in D} \varphi_{d+C,k}(b),$$

Taking into account the definition of the functional $\Phi_k$, we get $\Phi_k(D, F) \leq \Phi_k(D, B)$.
(c) The assertion (c) follows directly from (b).
(d) Obviously, (a) and (c) imply (d). □

Moreover, for $A, B \in \mathscr{P}_C(Y)$, the following assertions hold (see Hernández and Rodríguez-Marín [237, Theorem 3.10]).

**Theorem 10.3.18** *Suppose that $A, B \in \mathscr{P}_C(Y)$, $k \in -\operatorname{int} 0^+C$. Then, the following assertions hold:*

(a) $\Phi_k(A, A) = 0$,
(b) $A \in [B]^l \implies \sup_{b \in B} \inf_{a \in A} \varphi_{a+C,k}(b) = \sup_{a \in A} \inf_{b \in B} \varphi_{b+C,k}(a) = 0$,
(c) $A \preceq^l_C B \iff \sup_{b \in B} \inf_{a \in A} \varphi_{a+C,k}(b) \leq 0$.

***Proof***

(a) Taking into account $A \in [A + C]^l$, we get from Theorem 10.3.17(a), (d) that $\Phi_k(A, A) = \Phi_k(A, A + C)$. So it is sufficient to show $\Phi_k(A, A + C) = 0$. We have $\Phi_k(A, A + C) \leq 0$ because of $(A + C) \subseteq 0k + A + C$ in (10.3.5). Furthermore, since $A$ is a $C$-proper set, we get $\operatorname{bd}(A + C) \neq \emptyset$. This means that there is an element $y \in \operatorname{bd}(A + C)$. Because of Lemma 10.3.4(c), it holds that $\varphi_{A+C,k}(y) = 0$. So, we get $0 = \varphi_{A+C,k}(y) = \inf_{a \in A} \varphi_{a+C,k}(y) \leq \sup_{y \in A+C} \inf_{a \in A} \varphi_{a+C,k}(y) = \Phi_k(A, A + C) \leq 0$, i.e., $\Phi_k(A, A + C) = 0$.
(b) The assertion (a) and Theorem 10.3.17(d) imply (b).
(c) Assertion (c) follows from Theorem 10.3.17(b) and Proposition 10.3.11.

□

**Remark 10.3.19** It is important to mention that a characterization of the lower set less relation (see Definition 10.1.2) by means of the functional $\Phi_k$ introduced in (10.3.4) is given in Theorem 10.3.18(c). Further characterizations of the lower (and upper) set less relation using the nonlinear separating functional (4.1.11) are presented in Sect. 10.3.2. A slight extension of Theorem 10.3.18 is given in Theorem 10.3.34.

We will use the results of this section for a characterization of solutions to set-valued optimization problems via scalarization in Sect. 10.4.2.

### *10.3.2 A Complete Characterization of Set Relations by $\varphi_{A,k}$*

In Theorem 10.3.18(c) of the previous section, we have seen that the lower set less relation $\preceq^l_C$ (Definition 10.1.2) can be characterized by the nonlinear separating functional (4.1.11). Now, we explain how the nonlinear separating functional $\varphi_{A,k}$ given by (4.1.11) can be used in order to give a complete characterization of several set relations (see Sect. 10.1). We will see that this functional plays a key role in set optimization.

The results of this section concerning the nonlinear separating functional (4.1.11) are derived by Köbis and Tammer [301] and extended by Köbis and Köbis in [297].

As set relations, we consider the lower set less relation and the upper set less relation introduced in Definition 10.1.2 and the set less relation given in Definition 10.1.4.

In Sect. 10.3.1, we have shown that the functional $\Phi_k$ given by (10.3.4) is very useful to characterize the lower set less relation. Now, we study a general approach using the nonlinear separating functional $\varphi_{-C,k}$ (see (4.1.11) with $A = -C$, $C \subset Y$ is a nontrivial, convex, closed and pointed cone) to characterize this set relation. Our approach even allows us to present a full characterization of other set relations known from the literature, especially the upper set less relation (see Definition 10.1.2) and the set less relation (Definition 10.1.4). Of course, the functional $\varphi_{-C,k}$ in (4.1.11) is involved in the construction of the functional $\Phi_k$ in (10.3.4) too. The functional (4.1.11) has nice geometrical interpretations and useful continuity properties that are important for deriving optimality conditions (see Chap. 4) and for developing numerical procedures (see Sect. 13.5).

For proper sets $A, B \in \mathscr{P}(Y)$, $Y$ a separated locally convex space, $C \subset Y$ a nontrivial, convex, closed cone, Jahn [273] has shown that

$$A \subseteq B - C \iff \forall\, l \in C^+ \setminus \{0\} : \ \sup_{a\in A} l(a) \leq \sup_{b\in B} l(b) \tag{10.3.10}$$

and

$$A \subseteq B + C \iff \forall\, l \in C^+ \setminus \{0\} : \ \inf_{b\in B} l(b) \leq \inf_{a\in A} l(a)$$

under the assumption that the sets $B - C$ and $B + C$ are closed and convex. As usual, $C^+$ denotes the continuous dual cone of $C$.

The approach presented in this section allows us to give a full characterization of the relations $A \subseteq B - C$ and $A \subseteq B + C$ using the functional $\varphi_{-C,k}$ without any convexity assumptions on the considered sets.

Throughout this section, let the following assumption $(H_{A,B,C})$ be satisfied.

**Assumption** ($\mathbf{H_{A,B,C}}$). Let $Y$ be a topological vector space and $C \subset Y$ a nontrivial, convex, closed and pointed cone. Furthermore, let $A, B \in \mathscr{P}(Y)$ be proper sets and

$$k \in C_{A,B} := \{k \in C \setminus \{0\} \mid A \subseteq \operatorname{dom} \varphi_{-C,k} \text{ and } B \subseteq \operatorname{dom} \varphi_{-C,k}\}.$$

In Sect. 10.1, we introduced set relations, where the ordering cone $C$ is attached pointwise to the considered nonempty sets $A, B \subseteq Y$. In the next corollary, we show useful properties of the functional $\varphi_{-C,k}$ given by (4.1.11) with respect to the sets $A + C$ and $B - C$.

The following result is shown in [297, Corollary 2.2].

**Corollary 10.3.20** *Suppose that assumption* $(H_{A,B,C})$ *is fulfilled. Then, it holds that*

$$\sup_{b\in B} \varphi_{-C,k}(b) = \sup_{y\in B-C} \varphi_{-C,k}(y)$$

$$\inf_{a\in A} \varphi_{-C,k}(a) = \inf_{y\in A+C} \varphi_{-C,k}(y)\,.$$

***Proof*** We show the first assertion, the second one can be proven in the same way.

Consider arbitrary elements $k\in C_{A,B}$, $c\in C$ and $b \in B$. Using the $C$-monotonicity of the functional $\varphi_{-C,k}$ (see Theorem 4.5.5 with $A = -C$), we get

$$\varphi_{-C,k}(b) \geq \varphi_{-C,k}(b-c).$$

This yields that

$$\sup_{b\in B} \varphi_{-C,k}(b) \geq \sup_{y\in B-C} \varphi_{-C,k}(y).$$

Furthermore, $\sup_{b\in B} \varphi_{-C,k}(b) \leq \sup_{y\in B-C} \varphi_{-C,k}(y)$ follows from the definition of the supremum and $0 \in C$ (such that $B \subseteq B - C$). □

**Remark 10.3.21** (*Attainment Property*) In some of the following assertions, especially in Theorem 10.3.25, we suppose that $\inf_{b\in B} \varphi_{-C,k}(a-b)$ is attained for all $a \in A$, and in Theorem 10.3.34 we suppose that $\inf_{a\in A} \varphi_{-C,k}(a-b)$ is attained for all $b \in B$ (similar attainment assumptions are involved in Corollaries 10.3.41, 10.4.2, 10.4.4, 10.4.5, as well as in Attainment Properties 10.4.1 and 10.4.3). Sufficient conditions for such attainment properties are derived in Sect. 5.2, compare also Remark 10.3.7. Assertions concerning the existence of solutions of certain optimization problems known from the literature can be applied to derive sufficient conditions for these attainment properties. Especially, the classical Weierstrass Theorem and corresponding extensions (compare Zeidler [510, Proposition 9.13]) give sufficient conditions for such existence results. Taking into account that the functional $\varphi_{-C,k}$ is lower semi-continuous and convex if $C \subset Y$ is a nontrivial, closed, convex cone and $k \in C \setminus \{0\}$ (compare Theorem 4.4.4), we get that the attainment property for $\inf_{a\in A} \varphi_{-C,k}(a-b)$ (with $b \in B$ fixed) is fulfilled if $A$ is a proper, closed, bounded, convex subset of a reflexive Banach space.

First, we consider the upper set less relation and study corresponding characterizations by means of the functional $\varphi_{-C,k}$ (see (4.1.11) with $A = -C$, $C \subset Y$).

In the following theorem (see [297, Theorem 3.1] and [301, Theorem 2]), we describe a relationship between the upper set less relation and the nonlinear scalarizing functional $\varphi_{-C,k}$.

**Theorem 10.3.22** *Suppose that assumption ($H_{A,B,C}$) is fulfilled. Then, it holds that*

$$A \subseteq B - C \implies \forall\, k \in C_{A,B} : \ \sup_{a\in A} \varphi_{-C,k}(a) \leq \sup_{b\in B} \varphi_{-C,k}(b).$$

***Proof*** Consider $k \in C_{A,B}$ arbitrary, but fixed and suppose that $A \subseteq B - C$. By the definition of the functional (4.1.11), we get $\sup_{a\in A} \varphi_{-C,k}(a) \leq \sup_{y\in B-C} \varphi_{-C,k}(y)$. This inequality yields

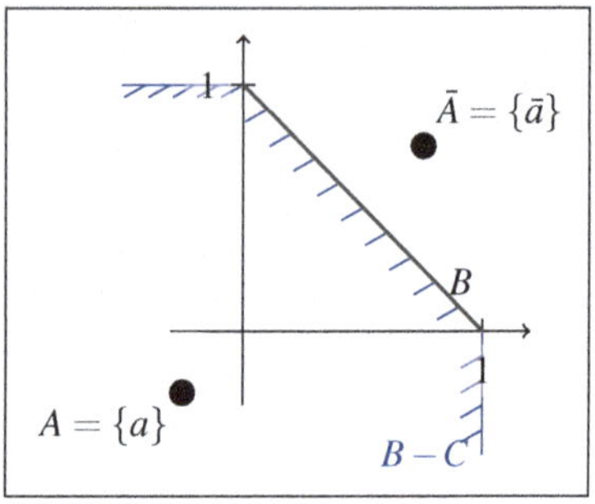


**Fig. 10.5** The sets $A$, $\bar{A}$, $B$ and $B - C$ in Example 10.3.23. Reprinted from [297, Fig. 3]

$$\sup_{a \in A} \varphi_{-C,k}(a) \leq \sup_{y \in B-C} \varphi_{-C,k}(y) = \sup_{b \in B} \varphi_{-C,k}(b)$$

taking into account Corollary 10.3.20 and the proof is completed. □

The following examples (see [297]) show that the converse implication of the assertion in Theorem 10.3.22 is not fulfilled in general.

**Example 10.3.23** ([297, Example 3.2]) Consider $Y := \mathbb{R}^2$,

$$A := \{(-1/4, -1/4)^T\} =: \{a\},$$

$$\bar{A} := \{(3/4, 3/4)^T\} =: \{\bar{a}\},$$

$B := \{(s, 1 - s)^T \mid s \in [0, 1]\}, k := (k_1, k_2)^T, k_1,\ k_2 > 0$, and $C = \mathbb{R}^2_+$ (see Fig. 10.5). Obviously, it holds that

$$\bar{A} \not\subseteq B - C\,.$$

Then, it holds that

$$\sup_{b \in B} \varphi_{-C,k}(b) = \sup_{s \in [0,1]} \max\left\{\frac{s}{k_1}, \frac{1-s}{k_2}\right\} = \max\left\{\frac{1}{k_1}, \frac{1}{k_2}\right\} > 0$$

and

$$\sup_{\tilde{a} \in A} \varphi_{-C,k}(\tilde{a}) = \varphi_{-C,k}(a) = -\frac{1}{4}\min\left\{\frac{1}{k_1}, \frac{1}{k_2}\right\} < 0 < \sup_{b \in B} \varphi_{-C,k}(b)$$

corresponding to $A \subseteq B - C$ as well as

$$\sup_{\tilde{a} \in \bar{A}} \varphi_{-C,k}(\tilde{a}) = \varphi_{-C,k}(\bar{a}) = \frac{3}{4}\max\left\{\frac{1}{k_1}, \frac{1}{k_2}\right\} < \sup_{b \in B} \varphi_{-C,k}(b).$$

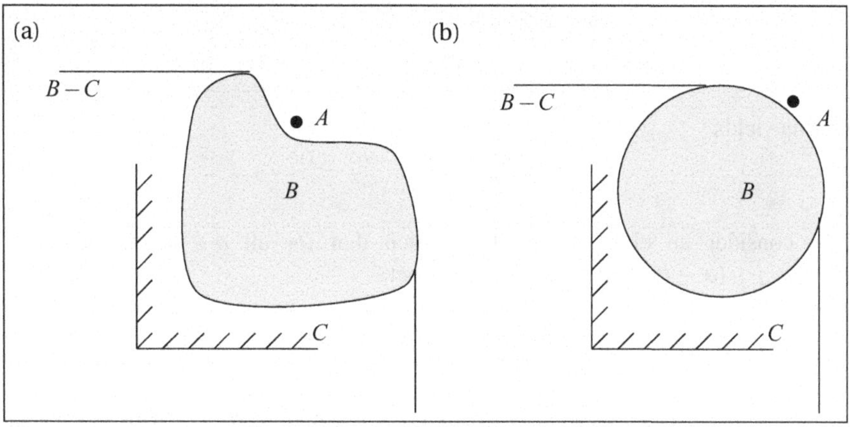


**Fig. 10.6** $B - C$ and $A = \{a\}$ in Example 10.3.24(a),(b). Reprinted from [301, Fig. 2]

**Example 10.3.24** ([301, Example 1]) Consider $C = \mathbb{R}^2_+$, $A = \{a\}$ and the set $B$ in Fig. 10.6(a). Then, we have that $\varphi_{-C,k}(a) \le \sup_{b\in B} \varphi_{-C,k}(b)$ for every $k \in \operatorname{int} C$, but obviously $A \not\subseteq B - C$. The same holds if the set $B - C$ is convex, as Fig. 10.6(b) illustrates.

The following theorem ([297, Theorem 3.3] and [301, Theorem 3]) verifies how to characterize the inclusion $A \subseteq B - C$ for two arbitrary nonempty sets $A, B \subseteq Y$ by using the nonlinear separating functional $\varphi_{-C,k}$.

**Theorem 10.3.25** *Let assumption ($H_{A,B,C}$) be fulfilled. Then, it holds that*

$$A \subseteq B - C \implies \forall k \in C_{A,B} : \sup_{a\in A} \inf_{b\in B} \varphi_{-C,k}(a - b) \le 0 . \tag{10.3.11}$$

*Furthermore, assume that there is an element $k^0 \in C_{A,B}$ such that for all $a \in A$, $\inf_{b\in B} \varphi_{-C,k^0}(a - b)$ is attained, then,*

$$\sup_{a\in A} \inf_{b\in B} \varphi_{-C,k^0}(a - b) \le 0 \implies A \subseteq B - C . \tag{10.3.12}$$

***Proof*** Suppose $A \subseteq B - C$. This is equivalent to

$$\forall\, a \in A\ \exists\, b \in B :\ a \in b - C$$

and

$$\forall\, a \in A\ \exists\, b \in B :\ a - b \in -C .$$

Taking into account Theorem 4.4.4, we get for $k \in C_{A,B}$

$$\begin{aligned} &\forall\, a \in A \;\exists\, b \in B: \; a - b \in -C \\ \iff &\forall\, a \in A \;\exists\, b \in B: \; \varphi_{-C,k}(a-b) \le 0, \end{aligned}$$

and this yields

$$\sup_{a \in A} \inf_{b \in B} \varphi_{-C,k}(a-b) \le 0\,.$$

Now, consider an element $k^0 \in C_{A,B}$ such that for all $a \in A$ the infimum $\inf_{b \in B} \varphi_{-C,k^0}(a-b)$ is attained. Suppose that

$$\sup_{a \in A} \inf_{b \in B} \varphi_{-C,k^0}(a-b) \le 0, \tag{10.3.13}$$

but $A \not\subseteq B - C$. Then, there is an element $\bar{a} \in A$ with $\bar{a} \notin B - C$. This means that for all $b \in B$, it holds that $\bar{a} - b \notin -C$. With Theorem 4.4.4, we obtain

$$\exists\, \bar{a} \in A\, \forall b \in B: \varphi_{-C,k^0}(\bar{a}-b) > 0 \quad \Longrightarrow \quad \exists\, \bar{a} \in A \text{ with } \inf_{b \in B} \varphi_{-C,k^0}(\bar{a}-b) > 0\,.$$

Since the last infimum is attained by assumption, we conclude that

$$\sup_{\bar{a} \in A} \inf_{b \in B} \varphi_{-C,k^0}(\bar{a}-b) > 0\,,$$

in contradiction to (10.3.13). □

**Remark 10.3.26** It is important to mention that $A \subseteq B - C$ also implies

$$\sup_{k \in C_{A,B}} \sup_{a \in A} \inf_{b \in B} \varphi_{-C,k}(a-b) \le 0$$

because of Theorem 10.3.25.

**Remark 10.3.27** Under the assumption that there is an element $k^0 \in C_{A,B}$ such that

$$\inf_{b \in B} \varphi_{-C,k^0}(a-b)$$

is attained for all $a \in A$, one concludes from Theorem 10.3.25

$$\begin{aligned} A \subseteq B - C \iff &\sup_{a \in A} \inf_{b \in B} \varphi_{-C,k^0}(a-b) \le 0 \\ \iff &\sup_{k \in C_{A,B}} \sup_{a \in A} \inf_{b \in B} \varphi_{-C,k}(a-b) \le 0. \end{aligned} \tag{10.3.14}$$

**Example 10.3.28** ([297, Example 3.5]) Again, we look at the setting of Example 10.3.23 with fixed $k_1, k_2 > 0$.

(a) For $\bar{A} = \{\bar{a}\} = (3/4, 3/4)^T$, it holds that

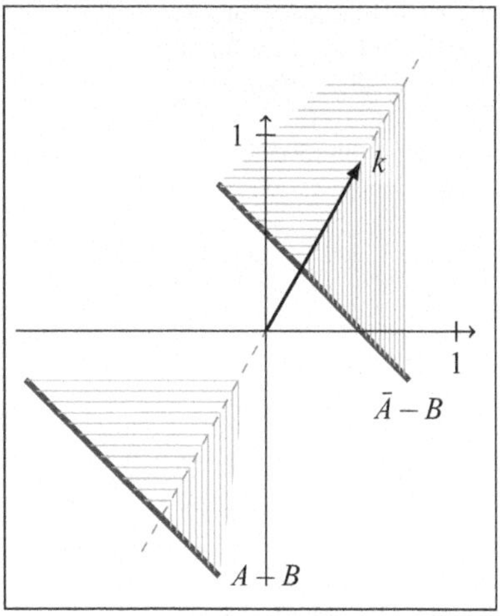


**Fig. 10.7** Illustration of Example 10.3.28. Reprinted from [297, Fig. 4]

$$\sup_{\tilde{a}\in\bar{A}}\inf_{b\in B}\varphi_{-C,k}(\tilde{a}-b) = \min_{s\in[0,1]}\max\left\{\frac{3/4-s}{k_1}, \frac{s-1/4}{k_2}\right\} > 0,$$

see Fig. 10.7.

(b) For $A = \{a\} = (-1/4, -1/4)^T$, one gets

$$\sup_{\tilde{a}\in A}\inf_{b\in B}\varphi_{-C,k}(\tilde{a}-b) = \min_{s\in[0,1]}\max\left\{\frac{-s-1/4}{k_1}, \frac{s-5/4}{k_2}\right\} < 0.$$

These relations indicate $\bar{A} \nsubseteq B - C$ and $A \subseteq B - C$.

**Example 10.3.29** ([301, Example 2]) Again, we consider Example 10.3.24 with $C = \mathbb{R}^2_+$. In Fig. 10.8, we can see that $\inf_{b\in B}\varphi_{-C,k}(a-b) > 0$. Because of Theorem 10.3.25, this is equivalent to $\{a\} \nsubseteq B - C$, as supposed here.

**Example 10.3.30** ([301, Example 3]) We consider Example 10.3.29 with $C = \mathbb{R}^2_+$, but now we suppose that $A = \{a\} \subseteq B - C$. In Fig. 10.9, it is possible to see that $\inf_{b\in B}\varphi_{-C,k}(a-b) \leq 0$. In this example, the attainment property ($\inf_{b\in B}\varphi_{-C,k^0}(a-b) = \min_{b\in B}\varphi_{-C,k^0}(a-b)$) is satisfied because of the compactness of $A = \{a\}$ and $B$. So, the converse $\sup_{a\in A}\inf_{b\in B}\varphi_{-C,k^0}(a-b) \leq 0$ implies $A \subseteq B - C$ is possible as well.

From Theorem 10.3.25, we get the following characterization of the upper set less relation directly by means of the functional $\varphi_{-C,k}$ (see [301, Corollary 1]).

**Corollary 10.3.31** *Suppose that assumption ($H_{A,B,C}$) is fulfilled. Then, it holds that*

$$A \preceq^u_C B \implies \forall\, k \in C_{A,B} : \sup_{a\in A}\inf_{b\in B}\varphi_{-C,k}(a-b) \leq 0.$$

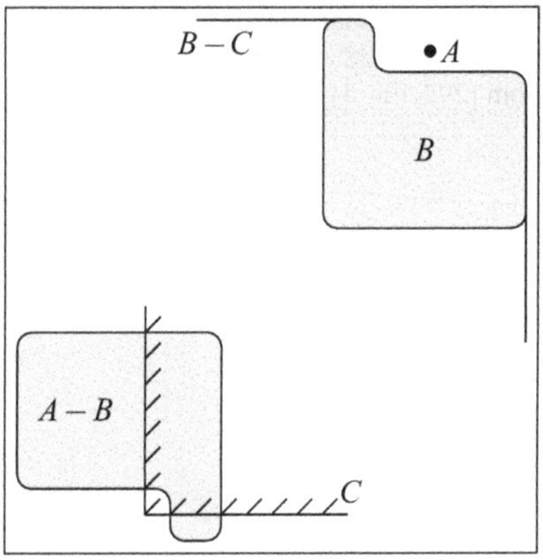


**Fig. 10.8** Illustration of sets $A = \{a\}$ and $B$, where for all $k \in C_{A,B}$, $\inf_{b\in B} \inf\{t \in \mathbb{R} | a - b \in tk - C\} > 0$. Reprinted from [301, Fig. 3]

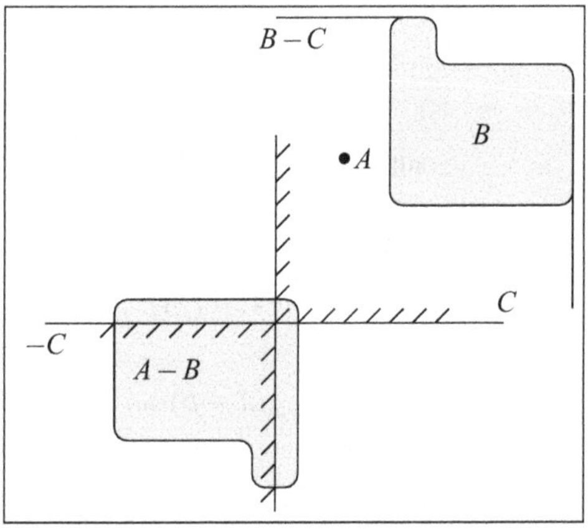


**Fig. 10.9** Illustration of Example 10.3.30. Reprinted from [301, Fig. 4]

*If there is a vector* $k^0 \in C_{A,B}$ *such that* $\inf_{b\in B} \varphi_{-C,k^0}(a-b)$ *is attained for all* $a \in A$, *the converse holds as well:*

$$\begin{aligned} A \preceq^u_C B &\iff \sup_{a\in A} \inf_{b\in B} \varphi_{-C,k^0}(a-b) \le 0 \\ &\iff \sup_{k\in C_{A,B}} \sup_{a\in A} \inf_{b\in B} \varphi_{-C,k}(a-b) \le 0 \,. \end{aligned} \tag{10.3.15}$$

Under convexity assumptions concerning the set $B - C$, we will show how the characterization of the set relation $A \subseteq B - C$ by means of the functional $\varphi_{-C,k}$ can be simplified. We look at a half space $\widetilde{C}^w \subset Y$ defined by

$$\widetilde{C}^w := \{y \in Y \mid w(y) \ge 0\} \tag{10.3.16}$$

with $w \in C^+ \setminus \{0\}$, where $C^+$ is the continuous dual cone of a nontrivial, closed, convex cone $C$ in a separated locally convex space $Y$. Now, we obtain the following assertion shown in [297, Theorem 3.6].

**Theorem 10.3.32** *Suppose that $Y$ is a separated locally convex space and assumption $(H_{A,B,C})$ is satisfied. Let $A$, $B$ be two nonempty, closed subsets of $Y$. Consider*

$$\widetilde{C}^w_{A,B} := \{k \in \widetilde{C}^w \setminus \{0\} \mid A \subseteq \operatorname{dom} \varphi_{-\widetilde{C}^w,k} \text{ and } B \subseteq \operatorname{dom} \varphi_{-\widetilde{C}^w,k}\}$$

*(where $\widetilde{C}^w$ is defined in (10.3.16)). Assume that for all $w \in C^+ \setminus \{0\}$, there is an element $k_w \in \operatorname{int} \widetilde{C}^w_{A,B}$ with*

$$\sup_{a\in A} \varphi_{-\widetilde{C}^w,k_w}(a) \leq \sup_{b\in B} \varphi_{-\widetilde{C}^w,k_w}(b). \tag{10.3.17}$$

*Moreover, assume that $B - C$ is closed and convex.*

*Then, it holds that*

$$A \subseteq B - C.$$

***Proof*** For every $w \in C^+ \setminus \{0\}$ and for some $k_w \in \operatorname{int} \widetilde{C}^w_{A,B}$ it holds that

$$\begin{aligned}
\sup_{a\in A} \varphi_{-\widetilde{C}^w,k_w}(a) &= \sup_{a\in A} \inf\{t \in \mathbb{R} \mid a \in tk_w - \widetilde{C}^w\} \\
&= \sup_{a\in A} \inf\{t \in \mathbb{R} \mid w(a) \leq w(tk_w)\} \\
&= \sup_{a\in A} \inf\left\{t \in \mathbb{R} \mid \frac{w(a)}{w(k_w)} \leq t\right\} \\
&= \sup_{a\in A} \frac{w(a)}{w(k_w)} \\
&= \frac{1}{w(k_w)} \sup_{a\in A} w(a)
\end{aligned}$$

(because $k_w \in \operatorname{int} \widetilde{C}^w_{A,B}$, we have $w(k_w) > 0$).

Analogously,

$$\begin{aligned}
\sup_{b\in B} \varphi_{-\widetilde{C}_w,k_w}(b) &= \sup_{b\in B} \inf\{t \in \mathbb{R} | b \in tk_w - \widetilde{C}_w\} \\
&= \sup_{b\in B} \inf\{t \in \mathbb{R} \mid w(b) \leq w(tk_w)\} \\
&= \sup_{b\in B} \inf\left\{t \in \mathbb{R} \mid \frac{w(b)}{w(k_w)} \leq t\right\} \\
&= \sup_{b\in B} \frac{w(b)}{w(k_w)}
\end{aligned}$$

$$= \frac{1}{w(k_w)} \sup_{b \in B} w(b).$$

Taking into account the above calculation, one can see that the assertion in (10.3.17) is equivalent to

$$\forall\, w \in C^+ \setminus \{0\}\ \exists\, k_w \in \operatorname{int} \widetilde{C}^w_{A,B} : \quad \frac{1}{w(k_w)} \sup_{a \in A} w(a) \leq \frac{1}{w(k_w)} \sup_{b \in B} w(b)$$

and this yields

$$\forall\, w \in C^+ \setminus \{0\} : \quad \sup_{a \in A} w(a) \leq \sup_{b \in B} w(b). \tag{10.3.18}$$

The rest of the proof is analog to a corresponding proof by Jahn in [273]. Assume that $A \not\subseteq B - C$, i.e., there is an element $a \in A$ with $a \notin B - C$. Applying a classical separation argument to the convex set $B - C$ and the element $a$, we get the existence of a linear, continuous functional $w \in Y^* \setminus \{0\}$ with

$$\forall\, y \in B - C : \quad w(a) > \alpha \geq w(y), \tag{10.3.19}$$

such that we get

$$\sup_{a \in A} w(a) > \alpha \geq \sup_{y \in B - C} w(y). \tag{10.3.20}$$

In order to show that $w \in C^+ \setminus \{0\}$, we suppose that $w \notin C^+ \setminus \{0\}$. This means that there is an element $c \in C$ with $w(c) < 0$. Thus, taking into account the separating inequality (10.3.19),

$$\alpha \geq w(b - \lambda c) = w(b) - \lambda w(c) \overset{\lambda \to +\infty}{\rightarrow} +\infty$$

for arbitrary $b \in B$ and $\lambda > 0$, a contradiction.

From the inequality (10.3.20), we obtain

$$\sup_{a \in A} w(a) > \alpha \geq \sup_{y \in B - C} w(y) = \sup_{b \in B} w(b) - \sup_{c \in C} w(c) = \sup_{b \in B} w(b),$$

in contradiction to (10.3.18). □

Now, we consider the lower set less relation (see Definition 10.1.2) and its characterization by means of the functional $\varphi_{-C,k}$.

In Theorems 10.3.33 and 10.3.34 (see [297, Theorems 3.7, 3.8] and [301, Theorems 4, 5]), we show how the set relation $B \subseteq A + C$ is characterized by the functional $\varphi_{-C,k}$. We refrain from giving the proofs of the following Theorems 10.3.33 and 10.3.34, since they can be deduced in a similar way as the proofs of Theorems 10.3.22 and 10.3.25. The following Theorem 10.3.33 is presented in [297, Theorem 3.7] and [301, Theorem 4].

**Theorem 10.3.33** *Let assumption* $(H_{A,B,C})$ *be fulfilled. Then, the following implication holds*

$$B \subseteq A + C \implies \forall\, k \in C_{A,B} : \inf_{a \in A} \varphi_{-C,k}(a) \leq \inf_{b \in B} \varphi_{-C,k}(b).$$

The following Theorem 10.3.34 (see [297, Theorems 3.8] and [301, Theorems 5]) is a slight extension of Theorem 10.3.18(c) and [237, Theorem 3.10].

**Theorem 10.3.34** *Let assumption* $(H_{A,B,C})$ *be fulfilled. Then, it holds that*

$$B \subseteq A + C \quad \implies \quad \forall\, k \in C_{A,B} : \sup_{b \in B} \inf_{a \in A} \varphi_{-C,k}(a - b) \leq 0.$$

*Moreover, if there is an element* $k^0 \in C_{A,B}$ *such that* $\inf_{a \in A} \varphi_{-C,k^0}(a - b)$ *is attained for all* $b \in B$*, then, it holds that*

$$\sup_{b \in B} \inf_{a \in A} \varphi_{-C,k^0}(a - b) \leq 0 \quad \implies \quad B \subseteq A + C\,.$$

**Remark 10.3.35** Sufficient conditions for the attainment property

$$\inf_{a \in A} \varphi_{-C,k^0}(a - b) = \min_{a \in A} \varphi_{-C,k^0}(a - b)$$

in Theorem 10.3.34 are discussed in Remark 10.3.21.

**Remark 10.3.36** The set inclusion $B \subseteq A + C$ in Theorem 10.3.34 also yields

$$\sup_{k \in C_{A,B}} \ \sup_{b \in B} \ \inf_{a \in A} \varphi_{-C,k}(a - b) \leq 0.$$

**Remark 10.3.37** If there is an element $k^0 \in C_{A,B}$ such that $\inf_{a \in A} \varphi_{-C,k^0}(a - b)$ is attained for all $b \in B$, then, from Theorem 10.3.34, it follows that

$$\begin{aligned} B \subseteq A + C &\iff \sup_{b \in B} \ \inf_{a \in A} \varphi_{-C,k^0}(a - b) \leq 0 \\ &\iff \sup_{k \in C_{A,B}} \ \sup_{b \in B} \ \inf_{a \in A} \varphi_{-C,k^0}(a - b) \leq 0. \end{aligned}$$

Theorem 10.3.34 directly yields the following characterization for the lower set less relation (see [301, Corollary 2]).

**Corollary 10.3.38** *Suppose that assumption* $(H_{A,B,C})$ *is fulfilled. Then, it holds that*

$$A \preceq^l_C B \quad \implies \quad \forall\, k \in C_{A,B} : \sup_{b \in B} \ \inf_{a \in A} \varphi_{-C,k}(a - b) \leq 0\,.$$

*If there is an element* $k^0 \in C_{A,B}$ *such that for all* $b \in B$ *the infima* $\inf_{a \in A} \varphi_{-C,k^0}(a - b)$ *are attained, the inverse holds as well. In particular:*

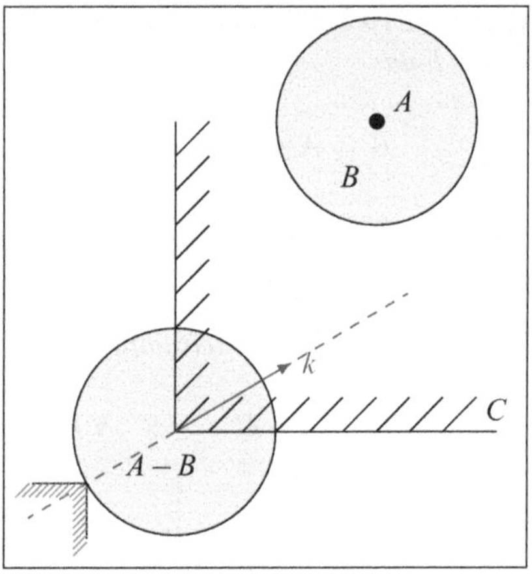


**Fig. 10.10** Illustration of the set $A - B = B - A$, where $A = \{a\} \subseteq B$. Reprinted from [301, Fig. 5]

$$A \preceq^l_C B \iff \sup_{b \in B} \inf_{a \in A} \varphi_{-C,k^0}(a-b) \le 0$$
$$\iff \sup_{k \in C_{A,B}} \sup_{b \in B} \inf_{a \in A} \varphi_{-C,k}(a-b) \le 0.$$

**Example 10.3.39** ([297, Example 3.10]) We study again the sets considered in Example 10.3.23 and $k_1, k_2 > 0$. Obviously, it holds that

$$A \preceq^l_C B \text{ and } \bar{A} \not\preceq^l_C B.$$

From Corollary 10.3.38, we get

$$\begin{aligned} \sup_{b \in B} \inf_{\tilde{a} \in \bar{A}} \varphi_{-C,k}(\tilde{a} - b) &= \max_{s \in [0,1]} \max \left\{ \frac{3/4 - s}{k_1}, \frac{s - 1/4}{k_2} \right\} > 0 \\ \sup_{b \in B} \inf_{\tilde{a} \in A} \varphi_{-C,k}(\tilde{a} - b) &= \max_{s \in [0,1]} \max \left\{ \frac{-s - 1/4}{k_1}, \frac{s - 5/4}{k_2} \right\} < 0, \end{aligned} \tag{10.3.21}$$

see Example 10.3.28 and Fig. 10.7.

The converse assertion also holds, i.e., $A \preceq^l_C B$ and $\bar{A} \not\preceq^l_C B$ follow from (10.3.21).

**Example 10.3.40** ([301, Example 4]) Set $C = \mathbb{R}^2_+$ and consider the nonempty subsets $\{a\} = A \subset B$ of $Y$ in Fig. 10.10. Then, $A \subseteq B - C$ and $A \subseteq B + C$. Because of the symmetry of $B$ and $A$ being a singleton, it holds that $A - B = B - A$. By Theorems 10.3.25 and 10.3.34, we conclude

$$\forall\, k \in C_{A,B} : \inf_{b \in B} \varphi_{-C,k}(a - b) \le 0 \text{ and } \inf_{b \in B} \varphi_{-C,k}(b - a) \le 0.$$

In the following, we give a characterization of the *set less relation* (see Definition 10.1.4) by means of the functional $\varphi_{-C,k}$. The next result goes back to [297, Corollary 3.11] and [301, Corollary 3].

**Corollary 10.3.41** *Let assumption $(H_{A,B,C})$ be fulfilled. Then, it holds that*

$$A \preceq^s_C B \implies \left[\forall\, k \in C_{A,B} :\ \sup_{a\in A}\ \inf_{b\in B} \varphi_{-C,k}(a-b) \le 0 \text{ and } \sup_{b\in B}\ \inf_{a\in A} \varphi_{-C,k}(a-b) \le 0\right].$$

*Furthermore, suppose additionally that there is an element $k^0 \in C_{A,B}$ such that for all $a \in A$ the infima $\inf_{b\in B} \varphi_{-C,k^0}(a-b)$ are attained and an element $k^1 \in C_{A,B}$ such that for all $b \in B$ the infima $\inf_{a\in A} \varphi_{-C,k^1}(a-b)$ are attained. Then,*

$$\left[\sup_{a\in A}\ \inf_{b\in B} \varphi_{-C,k^0}(a-b) \le 0 \text{ and } \sup_{b\in B}\ \inf_{a\in A} \varphi_{-C,k^1}(a-b) \le 0\right] \implies A \preceq^s_C B$$

*and, moreover,*

$$\begin{aligned} A \preceq^s_C B \iff & \Big[\forall\, k \in C_{A,B} :\ \sup_{a\in A}\ \inf_{b\in B} \varphi_{-C,k}(a-b) \le 0 \\ & \quad \text{and } \forall\, k \in C_{A,B} :\ \sup_{b\in B}\ \inf_{a\in A} \varphi_{-C,k}(a-b) \le 0\Big] \\ \iff & \Big[\sup_{k\in C_{A,B}}\ \sup_{a\in A}\ \inf_{b\in B} \varphi_{-C,k}(a-b) \le 0 \\ & \quad \text{and } \sup_{k\in C_{A,B}}\ \sup_{b\in B}\ \inf_{a\in A} \varphi_{-C,k}(a-b) \le 0\Big] \\ \iff & \left[\max\left(\sup_{k\in C_{A,B}}\ \sup_{a\in A}\ \inf_{b\in B} \varphi_{-C,k}(a-b),\ \sup_{k\in C_{A,B}}\ \sup_{b\in B}\ \inf_{a\in A} \varphi_{-C,k}(a-b)\right) \le 0\right]. \end{aligned}$$

***Proof*** The results directly follow from Theorems 10.3.25 and 10.3.34. □

**Example 10.3.42** (*Weighted Sum Scalarization*) ([297, Corollary 3.12] and [301, Example 5]) Suppose $Y := \mathbb{R}^\ell$ and let $A$, $B$ be nonempty, closed, bounded subsets in $\mathbb{R}^\ell$, $a \in A$, $b \in B$, $w := (w_1, \dots, w_\ell)^T$ with $w_i > 0$, $i \in \{1, \dots, \ell\}$, $C := \{y \in \mathbb{R}^\ell \mid w^T y \ge 0\}$ (notice, $C$ is a convex cone, but $C$ is not pointed) and $k := (k_1, \dots, k_\ell)^T \in \operatorname{int} C$. Then, it holds that (see (4.8.2))

$$\begin{aligned} \varphi_{-C,k}(a-b) &= \inf\{t \in \mathbb{R} \mid (a-b) \in tk - C\} \\ &= \inf\{t \in \mathbb{R} \mid w^T(a-b) \le w^T(tk)\} \\ &= \inf\{t \in \mathbb{R} \mid w^T(a-b) \le t \cdot (w^T k)\} \\ &= \inf\{t \in \mathbb{R} \mid \frac{1}{w^T k} \cdot \sum_{i=1}^{\ell} w_i(a_i - b_i) \le t\} \quad \text{since } k \in \operatorname{int} C \end{aligned}$$

$$= \frac{1}{w^T k} \cdot \sum_{i=1}^{\ell} w_i (a_i - b_i),$$

such that we get

$$\begin{aligned}\sup_{a\in A} \inf_{b\in B} \varphi_{-C,k}(a-b) &= \sup_{a\in A} \inf_{b\in B} \frac{1}{w^T k} \cdot \sum_{i=1}^{\ell} w_i (a_i - b_i) \\ &= \sup_{a\in A} \frac{1}{w^T k} \cdot \sum_{i=1}^{\ell} w_i a_i - \sup_{b\in B} \frac{1}{w^T k} \cdot \sum_{i=1}^{\ell} w_i b_i \\ &= \frac{1}{w^T k} \cdot \Big(\sup_{a\in A} \sum_{i=1}^{\ell} w_i a_i - \sup_{b\in B} \sum_{i=1}^{\ell} w_i b_i\Big).\end{aligned}$$

Using these calculations, from Theorems 10.3.25 and 10.3.34 and taking into account the above definitions of $C$ and $k$ and different weights $w_i > 0,\ i \in \{1, \dots, \ell\}$, we conclude

$$\begin{aligned}A \subseteq B - C &\iff \forall\, k \in \operatorname{int} C:\ \frac{1}{w^T k} \sup_{a\in A} \sum_{i=1}^{\ell} w_i a_i \le \frac{1}{w^T k} \sup_{b\in B} \sum_{i=1}^{\ell} w_i b_i \\ &\iff \sup_{a\in A} \sum_{i=1}^{\ell} w_i a_i \le \sup_{b\in B} \sum_{i=1}^{\ell} w_i b_i \text{ and, analogously,} \\ B \subseteq A + C &\iff \forall\, k \in \operatorname{int} C:\ \frac{1}{w^T k} \inf_{a\in A} \sum_{i=1}^{\ell} w_i a_i \le \frac{1}{w^T k} \inf_{b\in B} \sum_{i=1}^{\ell} w_i b_i \\ &\iff \inf_{a\in A} \sum_{i=1}^{\ell} w_i a_i \le \inf_{b\in B} \sum_{i=1}^{\ell} w_i b_i.\end{aligned}$$

We considered elements $k \in \operatorname{int} C$ in this example in order to exclude division by zero for this in fact algorithmic example. Furthermore, the attainment of the infima and suprema, respectively, is implicitly required. In the next example, we also need these two assumptions.

**Example 10.3.43** (*Natural Ordering*) ([297, Corollary 3.13] and [301, Example 6]) Again, consider $Y := \mathbb{R}^\ell$, $C := \mathbb{R}^\ell_+$ and $k := (k_1, \dots, k_\ell)^T \in \operatorname{int} C$. Then, it holds that (see (4.8.1))

$$\begin{aligned}\varphi_{-C,k}(a-b) &= \inf\{t \in \mathbb{R} \mid (a-b) \in tk - C\} \\ &= \inf\{t \in \mathbb{R} \mid (a-b) - tk \in -C\} \\ &= \inf\{t \in \mathbb{R} \mid \forall\, i \in \{1, \dots, \ell\}:\ (a-b)_i - tk_i \le 0\}\end{aligned}$$

$$= \inf\{t \in \mathbb{R} \mid \forall\, i \in \{1, \dots, \ell\} : \frac{(a-b)_i}{k_i} \le t\}$$
$$= \sup_{i=1,\dots,\ell} \frac{(a-b)_i}{k_i}.$$

Taking into account these calculations, Theorems 10.3.25 and 10.3.34, the above definitions of $C$ and $k$, we get

$$A \subseteq B - C \iff \forall\, k \in \text{int}\mathbb{R}^\ell_+ : \sup_{a\in A} \inf_{b\in B} \sup_{i=1,\dots,\ell} \frac{(a-b)_i}{k_i} \le 0,$$
$$B \subseteq A + C \iff \forall\, k \in \text{int}\mathbb{R}^\ell_+ : \sup_{b\in B} \inf_{a\in A} \sup_{i=1,\dots,\ell} \frac{(a-b)_i}{k_i} \le 0.$$

**Remark 10.3.44** A detailed discussion of the characterization of further set relations (see Sect. 10.1) by means of the functional $\varphi_{-C,k}$ is given in the paper by Köbis and Köbis in [297].

**Remark 10.3.45** Recently, Bao and Tammer [30, Sect. 3] are using the nonlinear scalarizing functional $\varphi_{A,k}$ given by (4.1.11) where the hypergraphical (epigraphical) set of $A$ with respect to $C$ is involved to characterize the upper (lower) set less relation $\preceq^u_C$ ($\preceq^l_C$), see Definition 10.1.2. The results in [30] are derived under the assumption that $Y$ is a real linear space.

## 10.4 Characterization of Solutions to Set-Valued Optimization Problems

### *10.4.1 Characterization of Nondominated Elements*

In Sects. 10.1 and 10.2, we introduced set relations and corresponding solution concepts in set optimization. Now, we will consider the set relations $\preceq^u_C$, $\preceq^l_C$, and $\preceq^s_C$ introduced in Definitions 10.1.2 and 10.1.4 in order to give a characterization of the corresponding solution concepts via scalarization.

Suppose that assumption ($H_{A,B,C}$) (introduced in Sect. 10.3.2) is fulfilled. Let $\mathscr{A}$ be a system of nonempty subsets of $Y$ and let $\preceq_C$ be a set relation. Recall (see Definition 10.2.1) that $\bar{A}$ is a nondominated element of $\mathscr{A}$ w.r.t. $\preceq_C$ if

$$A \preceq_C \bar{A},\ A \in \mathscr{A} \implies \bar{A} \preceq_C A.$$

**Attainment Property 10.4.1** Consider $\bar{A} \in \mathscr{A}$, suppose that for all $A \in \mathscr{A}$ there is an element

(i) $k^0 = k^0(A) \in C_{A,\bar{A}}$ such that the infima $\inf_{\bar{a}\in\bar{A}} \varphi_{-C,k^0}(a-\bar{a})$ are attained for every $a \in A$;
(ii) $k^1 = k^1(A) \in C_{A,\bar{A}}$ such that the infima $\inf_{a\in A} \varphi_{-C,k^1}(\bar{a}-a)$ are attained for all $\bar{a} \in \bar{A}$.

The assertions in the following corollary are derived in [297, Corollary 4.2] and [301, Corollary 4].

**Corollary 10.4.2** *Consider a family of nonempty subsets $\mathscr{A}$ of $Y$ and $\bar{A} \in \mathscr{A}$. Moreover, assume that the Attainment Property 10.4.1 is satisfied. Then, $\bar{A}$ is a nondominated element of $\mathscr{A}$ w.r.t. $\preceq_C^u$ if and only if there does not exist any $A \in \mathscr{A}$ with*

$$\sup_{a\in A}\inf_{\bar{a}\in\bar{A}} \varphi_{-C,k^0}(a-\bar{a}) \le 0 \text{ and}$$
$$\sup_{\bar{a}\in\bar{A}}\inf_{a\in A} \varphi_{-C,k^1}(\bar{a}-a) > 0.$$

***Proof***

$$\begin{aligned}
&\bar{A} \text{ is a nondominated element of } \mathscr{A} \text{ w.r.t. } \preceq_C^u\\
\iff\ & \left(A \in \mathscr{A} : A \preceq_C^u \bar{A} \implies \bar{A} \preceq_C^u A\right)\\
\iff\ & \Big(A \in \mathscr{A} : \ \sup_{a\in A}\inf_{\bar{a}\in\bar{A}} \varphi_{-C,k^0}(a-\bar{a}) \le 0\\
&\quad \implies \sup_{\bar{a}\in\bar{A}}\inf_{a\in A} \varphi_{-C,k^1}(\bar{a}-a) \le 0\Big)\\
\iff\ & \nexists A \in \mathscr{A} : \Big(\sup_{a\in A}\inf_{\bar{a}\in\bar{A}} \varphi_{-C,k^0}(a-\bar{a}) \le 0\\
&\quad \text{and } \sup_{\bar{a}\in\bar{A}}\inf_{a\in A} \varphi_{-C,k^1}(\bar{a}-a) > 0\Big).
\end{aligned}$$

□

**Attainment Property 10.4.3** For $A, \bar{A} \in \mathscr{A}$, suppose that for all $A \in \mathscr{A}$ there is an element

(i) $k^2 = k^2(A) \in C_{A,\bar{A}}$ such that the infima $\inf_{a\in A} \varphi_{-C,k^2}(a-\bar{a})$ are attained for every $\bar{a} \in \bar{A}$;
(ii) $k^3 = k^3(A) \in C_{A,\bar{A}}$ such that the infima $\inf_{\bar{a}\in\bar{A}} \varphi_{-C,k^3}(\bar{a}-a)$ are attained for all $a \in A$.

The assertion in the following corollary (derived in [297, Corollary 4.4(a)] and [301, Corollary 5]) can be established in a similar manner as Corollary 10.4.2.

**Corollary 10.4.4** *Consider a family of nonempty subsets $\mathcal{A}$ of $Y$. $\bar{A} \in \mathcal{A}$ is a nondominated element of $\mathcal{A}$ w.r.t. $\preceq^l_C$ if Attainment Property 10.4.3 is fulfilled and if there does not exist any $A \in \mathcal{A}$ with*

$$\sup_{\bar{a}\in\bar{A}} \inf_{a\in A} \varphi_{-C,k^2}(a-\bar{a}) \leq 0 \text{ and}$$
$$\sup_{a\in A} \inf_{\bar{a}\in\bar{A}} \varphi_{-C,k^3}(\bar{a}-a) > 0.$$

Finally, we are able to present a characterization of nondominated elements of a system of sets w.r.t. the set less relation. This result (see [297, Corollary 4.4(b)] and [301, Corollary 6]) is deduced from Corollaries 10.4.2 and 10.4.4.

**Corollary 10.4.5** *Let $\mathcal{A}$ be a family of nonempty subsets of $Y$. $\bar{A} \in \mathcal{A}$ is a nondominated element of $\mathcal{A}$ w.r.t. $\preceq^s_C$ if Attainment Properties 10.4.1 and 10.4.3 are fulfilled and if there does not exist any $A \in \mathcal{A}$ such that*

$$\sup_{a\in A} \inf_{\bar{a}\in\bar{A}} \varphi_{-C,k^0}(a-\bar{a}) \leq 0 \text{ and } \sup_{\bar{a}\in\bar{A}} \inf_{a\in A} \varphi_{-C,k^2}(a-\bar{a}) \leq 0 \text{ and}$$
$$\left[\sup_{\bar{a}\in\bar{A}} \inf_{a\in A} \varphi_{-C,k^1}(\bar{a}-a) > 0 \text{ or } \sup_{a\in A} \inf_{\bar{a}\in\bar{A}} \varphi_{-C,k^3}(\bar{a}-a) > 0\right].$$

**Remark 10.4.6** In this section, we discussed a unifying approach for representing set relations via nonlinear scalarization by means of the functional (4.1.11). The characterization of set relations via scalarization can be used for developing optimality conditions and corresponding numerical procedures for solving set-valued optimization problems (see Sect. 13.5).

**Remark 10.4.7** Characterizations of nondominated elements of $\mathcal{A}$ w.r.t. $\preceq^u_C$ ($\preceq^l_C$) via scalarization by means of the functional (4.1.11) where thc hypergraphical multifunction (epigraphical multifunction) is involved are derived by Bao and Tammer [30, Sect. 4].

### 10.4.2 Characterization by Hernández and Rodríguez-Marín

We now give a characterization of solutions of set-valued optimization problems (see Definition 10.2.2) via scalarization by means of the functional $\Phi_k$ (see (10.3.4)) introduced in Definition 10.3.8. In this section, we suppose that $Y$ is a topological vector space partially ordered by a nontrivial, convex, closed, pointed cone $C \subset Y$ with nonempty interior and $k \in -\operatorname{int} 0^+C$ is a fixed element.

Now, we consider a set-valued optimization problem given by

$$\preceq^l_C -\operatorname*{Opt}_{x\in S} \Gamma(x), \qquad (SP - \preceq^l_C)$$

where we assume that $S$ is a subset of $X$, $X$ is a linear space, $\Gamma : X \rightrightarrows Y$ and where the solution concept is given with respect to the lower set less relation $\preceq^l_C$ (Definition 10.1.2). In this section, we suppose that the following assumption $(A_\Gamma)$ is fulfilled.

Assumption $(\mathbf{A}_\Gamma)$. $\Gamma(x)$ belongs to $\mathscr{P}_C(Y)$ and $\Gamma(x)$ is $C$-bounded (see Definition 2.2.16) for all $x \in S$.

Recall that the function $\Phi_k : \mathscr{P}_C(Y) \times \mathscr{P}^0_C(Y) \longrightarrow \mathbb{R} \cup \{+\infty\}$ is given by (see (10.3.4))

$$\Phi_k(B, D) = \sup_{d \in D}\{\varphi_{B+C,k}(d)\} \quad \text{for} \quad (B, D) \in \mathscr{P}_C(Y) \times \mathscr{P}^0_C(Y).$$

Denote the set of all functions from $\mathscr{P}_C(Y)$ to $\mathbb{R}$ by $\mathscr{M}(\mathscr{P}_C(Y), \mathbb{R})$ and recall that $[\cdot]^l$ denotes a family of equivalence classes (see (10.2.1)).

The following assertions are shown by Hernández and Rodríguez-Marín [237, Theorem 4.1 and Corollary 4.3] and can be considered as a certain kind of a separation theorem.

**Theorem 10.4.8** *We consider the problem $(SP - \preceq^l_C)$ and suppose that $(A_\Gamma)$ is fulfilled.*

*Then, $x^0 \in S$ is a nondominated solution of $(SP - \preceq^l_C)$ if and only if there is an $\preceq^l_C$-increasing (on $\mathscr{P}_C(Y)$) functional $\varphi \in \mathscr{M}(\mathscr{P}_C(Y), \mathbb{R})$ with*

(a) $x \in S,\ \Gamma(x) \in [\Gamma(x^0)]^l \implies \varphi(\Gamma(x)) = 0.$
(b) $x \in S,\ \Gamma(x) \notin [\Gamma(x^0)]^l \implies \varphi(\Gamma(x)) > 0.$
(c) $D \in \mathscr{P}_C(Y),\ D \preceq^l_C \Gamma(x^0) \implies \varphi(D) \leq 0.$

***Proof*** Suppose that $x^0$ is a nondominated solution of $(SP - \preceq^l_C)$. We consider a fixed element $k \in -\operatorname{int} 0^+C$. The functional

$$\varphi(\cdot) := \Phi_k(\cdot, \Gamma(x^0)) \in \mathscr{M}(\mathscr{P}_C(Y), \mathbb{R})$$

is $\preceq^l_C$-increasing on $\mathscr{P}_C(Y)$ taking into account Theorem 10.3.16. We prove that $\varphi$ satisfies the assertions (a), (b) and (c). The assertions (a) and (c) are conclusions from Theorem 10.3.18(b) and (c), respectively.

We now prove that (b) holds. Since $x^0$ is a nondominated solution of $(SP - \preceq^l_C)$, for each $x \in S$ with $\Gamma(x) \notin [\Gamma(x^0)]^l$, we get $\Gamma(x) \not\preceq^l_C \Gamma(x^0)$. Therefore,

$$\varphi(\Gamma(x)) = \Phi_k(\Gamma(x), \Gamma(x^0)) > 0$$

by Theorem 10.3.18(c).

Conversely, we suppose that (a), (b) and (c) hold for some $\varphi \in \mathscr{M}(\mathscr{P}_C(Y), \mathbb{R})$ that is $\preceq^l_C$-increasing on $\mathscr{P}_C(Y)$. We show that $x^0$ is a nondominated solution of $(SP - \preceq^l_C)$. Indeed, if we assume that $x^0$ is not a nondominated solution of $(SP - \preceq^l_C)$, then there is an element $x' \in S$ with $\Gamma(x') \notin [\Gamma(x^0)]^l$ and

$$\Gamma(x') \preceq^l_C \Gamma(x^0).$$

Therefore, we get $\varphi(\Gamma(x')) > 0$ by (b). Furthermore, it follows $\varphi(\Gamma(x')) \leq 0$ by (c), a contradiction. □

From Theorem 10.4.8, we obtain well-known characterizations of solutions of a vector optimization problem by means of $C$-monotone scalarizing functionals (see Sect. 6.2, Proposition 6.2.16 and Theorem 6.2.24).

Now, we consider the following vector optimization problem (see Chap. 6). Let $f : S \longrightarrow Y$ be a single-valued map, $C$ a nontrivial, pointed, closed, convex cone in the topological vector space $Y$, $S \subseteq X$, $X$ a linear space. We are using the solution concept introduced in Sect. 6.2.1, see also Definition 6.0.1 and consider the problem

$$C - \operatorname*{Opt}_{x \in S} f(x). \tag{10.4.1}$$

In the assertion of the following corollary (given in [237, Corollary 4.3] as a conclusion of Theorem 10.4.8), we are dealing with (strictly) $C$-monotone functionals $\varphi : Y \to \mathbb{R}$ in the sense of Definition 3.7.1. This result is derived in Corollary 6.2.25 in terms of the functional $\varphi_{-C,k}$ in the image space $Y$ and it is well known in vector optimization.

**Corollary 10.4.9** *We consider the problem (10.4.1). Then, $f(x^0) \in \mathrm{Eff}(f(S), C)$ if and only if there is a continuous, $C$-monotone functional $\varphi : Y \to \mathbb{R}$ with*

(a) $\varphi(f(x^0)) = 0$,
(b) $x \in S,\ f(x) \neq f(x^0) \implies \varphi(f(x)) > 0$,
(c) $y \in f(x^0) - C \implies \varphi(y) \leq 0$.

**Remark 10.4.10** A corresponding result for weakly efficient elements $f(x^0) \in \mathrm{WEff}(f(S), C)$ is shown in [237, Corollary 4.4]. We already derived corresponding properties in Proposition 6.2.17 for the nonlinear separating functionals of type (4.1.11).

At the end of this section, we prove sufficient conditions for nondominated solutions of the set-valued optimization problem $(SP - \preceq^l_C)$ using the functional $\Phi_k(B, \cdot)$ (see (10.3.4)). In order to derive these assertions, we recall the following notations.

**Definition 10.4.11** A net $\{D_\alpha \mid \alpha \in I\}$ of proper sets from $\mathscr{P}(Y)$ is called $\preceq^l_C$-**decreasing**, if $D_\alpha \preceq^l_C D_\beta$ for each $\alpha, \beta \in I$ with $\beta < \alpha$.

Let $x^0 \in S$ be given, then we introduce

$$L(x^0) := \{x \in S \mid \Gamma(x) \preceq^l_C \Gamma(x^0)\}.$$

**Definition 10.4.12** Consider elements $x^0 \in X$, $k \in -\operatorname{int} 0^+C$ and the set-valued map $\Gamma$. If for any $\{x_\alpha\} \subset L(x^0)$ and $t \in \mathbb{R}$ with $\{\Gamma(x_\alpha)\}$ is $\preceq^l_C$-decreasing and $\Phi_k(\Gamma(x^0), \Gamma(x_\alpha)) = t$ for all $\alpha$ there is an element $\overline{x} \in S$ such that $\Gamma(\overline{x}) \preceq^l_C \Gamma(x^0) + tk$, the set-valued map $\Gamma$ is called **$\Phi_k$-stable at** $x^0$ .

Hernández and Rodríguez-Marín [237, Theorem 5.3] have shown the following existence result for nondominated solutions of $(SP - \preceq^l_C)$ in the sense of Definition 10.2.2 under the assumption that $\Gamma$ is $\Phi_k$-stable. Furthermore, the $C$-compactness of $\bigcup_{x \in L(x_0)} \Gamma(x)$ in the sense of Definition 2.2.16 is supposed.

**Theorem 10.4.13** *Consider the set-valued optimization problem $(SP - \preceq^l_C)$, $x^0 \in S$ and $k \in -\operatorname{int} 0^+C$. Suppose that $\Gamma(x)$ belongs to $\mathscr{P}_C(Y)$ for all $x \in S$ and $\Gamma(x^0)$ is $C$-bounded, $\Gamma$ is $\Phi_k$-stable at $x^0$ and $\bigcup_{x \in L(x_0)} \Gamma(x)$ is $C$-compact. Then, there is a nondominated solution $x (\in L(x^0))$ of $(SP - \preceq^l_C)$.*

Moreover, the next existence result is shown by Hernández and Rodríguez-Marín [237, Theorem 5.4] using Cantor's intersection theorem and a nonlinear scalarization technique by means of the functional $\Phi_k$ given by (10.3.4).

**Theorem 10.4.14** *Consider the set-valued problem $(SP - \preceq^l_C)$, $k \in -\operatorname{int} 0^+C$. Suppose that $(X, d)$ is a complete metric space, $S \subset X$, $\Gamma(x)$ is $C$-bounded for all $x \in X$ and the set $L(x)$ is closed for each $x \in S$. Furthermore, assume that $x^0 \in S$ and that the following conditions are fulfilled:*

(a) *$\bigcup_{x \in L(x^0)} \Gamma(x)$ is $C$-bounded,*
(b) *there is a $C$-bounded set $B \in \mathscr{P}_C(Y)$ such that for a certain $k \in -\operatorname{int} 0^+C$*

$$x^1, x^2 \in S,\ \Gamma(x^1) \preceq^l_C \Gamma(x^2) \quad \Longrightarrow \quad d(x^1, x^2) \le \Phi_k(B, \Gamma(x^1)) - \Phi_k(B, \Gamma(x^2)).$$

*Then, there is a nondominated solution $\overline{x}$ $(\in L(x^0))$ of $(SP - \preceq^l_C)$.*

***Proof*** Assume that $x^0$ is not a nondominated solution of $(SP - \preceq^l_C)$. Then, by Definition 10.2.2, there is an element $x \in L(x^0)$ with

$$\Gamma(x) \preceq^l_C \Gamma(x^0) \quad \text{and} \quad \Gamma(x^0) \not\preceq^l_C \Gamma(x).$$

Taking into account assumption (b), we get

$$0 < \Phi_k(B, \Gamma(x)) - \Phi_k(B, \Gamma(x^0)).$$

Furthermore, since $\bigcup_{x \in L(x^0)} \Gamma(x)$ and $\Gamma(x^0)$ are $C$-bounded sets, we get by Theorem 10.3.14

$$0 < \sup_{x \in L(x_0)} \{\Phi_k(B, \Gamma(x))\} - \Phi_k(B, \Gamma(x^0)) < +\infty. \tag{10.4.2}$$

Now, we introduce $v(z) := \sup_{x \in L(z)} \{\Phi_k(B, \Gamma(x))\}$ for $z \in S$. Because of (10.4.2), we conclude that there is an element $x^1 \in L(x^0)$ with

$$v(x^0) - \Phi_k(B, \Gamma(x^1)) \le 2^{-1}.$$

In the case that $x^1 \in L(x^0)$ and $x^1$ is not a nondominated solution of $(SP - \preceq^l_C)$, there is an element $x_2 \in L(x^1)$ such that

$$v(x^1) - \Phi_k(B, \Gamma(x^2)) \le 2^{-2}.$$

We obtain a sequence $\{x_n\}_{n\in\mathbb{N}}$ from $L(x^0)$ in this way such that for each $n \in \mathbb{N}$ it holds that $L(x_{n+1}) \subseteq L(x_n)$.

Furthermore, it follows

$$\{\operatorname{diam} L(x_n)\} \to 0 \tag{10.4.3}$$

since the set $L(x_n)$ is closed for each $n \in \mathbb{N}$ and

$$v(x_n) - \Phi_k(B, \Gamma(x_{n+1})) \le 2^{-(n+1)}. \tag{10.4.4}$$

Indeed, if we suppose that (10.4.3) does not hold, then there is a real number $\delta > 0$ with $\operatorname{diam} L(x_n) > \delta$ for each $n \in \mathbb{N}$. Then, there is an element $z' \in L(x_n)$ such that $d(z', x_n) > \frac{\delta}{2}$ by the triangle inequality of $d(\cdot, \cdot)$ for fixed $n \in \mathbb{N}$. Taking into account assumption (b), we have

$$\frac{\delta}{2} < d(z', x_n) \le \Phi_k(B, \Gamma(z')) - \Phi_k(B, \Gamma(x_n)). \tag{10.4.5}$$

Moreover, it holds that

$$\Phi_k(B, \Gamma(z')) - \Phi_k(B, \Gamma(x_n)) \le v(x_n) - \Phi_k(B, \Gamma(x_n)).$$

Because of $x_n \in L(x_{n-1})$, we get $v(x_n) \le v(x_{n-1})$. Therefore, we have

$$\frac{\delta}{2} < d(z', x_n) \le v(x_{n-1}) - \Phi_k(B, \Gamma(x_n)) \le 2^{-n}$$

since (10.4.4) and (10.4.5). For $n \to +\infty$, the above inequality yields $\delta < 0$ and we get a contradiction.

Taking into account that the set $L(x_n)$ is closed for every $n \in \mathbb{N}$, $(X, d)$ is a complete metric space, $L(x_{n+1}) \subseteq L(x_n)$ and $\operatorname{diam} L(x_n) \to 0$, we conclude from Cantor's intersection theorem

$$\bigcap_{n\in\mathbb{N}} L(x_n) = \{\overline{x}\}.$$

Therefore, $\overline{x} \in L(x^0)$ and $L(\overline{x}) = \{\overline{x}\}$. This means that $\overline{x}$ is a nondominated solution of $(SP - \preceq^l_C)$. This completes the proof. □

From Theorem 10.4.14, we get the following existence result for solutions of vector optimization problems (see Chap. 6) as a special case. This result is shown in [237, Corollary 5.5]).

**Corollary 10.4.15** *We consider the vector optimization problem*

$$C - \operatorname*{Opt}_{x \in S} f(x),$$

*where* $f : S \to Y$, $C$ *is a nontrivial, pointed, closed, convex cone in the topological vector space* $Y$, $S \subseteq X$, $X$ *a linear space. Suppose that* $(X, d)$ *is a complete metric space,* $S \subset X$ *and for each* $x \in S$ *the set* $L(x)$ *is closed. Moreover, we assume that* $x^0 \in S$ *and:*

(a) $f(L(x^0))$ *is* $C$*-bounded,*
(b) *there is a* $C$*-bounded set* $B \in \mathscr{P}_C(Y)$ *such that for a certain* $k \in -\operatorname{int} 0^+C$

$$x^1, x^2 \in S,\ f(x^1) \in f(x^2) - C \implies d(x^1, x^2) \leq \varphi_{B+C,k}(f(x^1)) - \varphi_{B+C,k}(f(x^2)).$$

*Then, there is an element* $\overline{x} \in L(x^0)$ *such that* $f(\overline{x}) \in \operatorname{Eff}(f(S), C)$.

**Remark 10.4.16** Under similar assumptions to those supposed in Theorems 10.4.13 and 10.4.14, Kuroiwa [314, Theorems 4.1 and 4.2] derived existence results for nondominated solutions of $(SP - \preceq^l_C)$. But, the assumptions in Theorems 10.4.13 and 10.4.14 are weaker in comparison with the assumptions in [314, Theorems 4.1 and 4.2].

# Chapter 11
# Vector Optimization with Variable Domination Structures

Vector optimization with variable domination structures is a growing up and expanding field of applied mathematics that deals with optimization problems where the domination structure is given by a set-valued map acting between abstract or finite-dimensional spaces. Interesting and important applications of vector optimization with variable domination structure arise in economics, psychology, capability of human behavior, in portfolio management, location theory and radiotherapy treatment in medicine (see Bao, Mordukhovich [26], Bao, Mordukhovich, Soubeyran [27], Chen, Huang, Yang [93], Engau [156], Eichfelder [138, 141–144], Eichfelder, Kasimbeyli [148–151], Le Thanh Tam [321] and Hang, Yang, Chan [247]). Especially, Eichfelder [138, 141, 144] explained that variable domination structures are motivated by certain applications in medical image registration because a variable domination structure better reflects the problem structure than a corresponding problem with respect to a fixed domination structure. For these problems, Wacker proposed a special solution procedure in [458].

We have seen that nondominated points of a set w.r.t. a relation $\succ$ can be described via the domination structure or via the pre-domination structure in Sect. 6.1. Now, in Sect. 11.1, we give a detailed discussion of solution concepts for problems with variable domination structures based on the (pre-) domination structures introduced in Sect. 6.1. We present certain modifications of the nonlinear scalarizing functional (4.1.11) in Sect. 11.2 and show characterizations of solutions to vector optimization problems with variable domination structure by means of the nonlinear scalarizing functional (4.1.11) as well as corresponding modifications in Sect. 11.3. Furthermore, we introduce several concepts for approximate solutions to vector optimization problems with variable domination structures in Sect. 11.4.1 and show corresponding characterizations by means of nonlinear functionals in Sect. 11.4.2. These results are very useful for further research on the field of vector optimization with variable domination structure, especially, for deriving optimality conditions, duality assertions and numerical procedures.

The main aim of this chapter is to characterize different solution concepts of vector optimization problems with variable domination structure via scalarization using

 
C. Tammer and P. Weidner, *Scalarization and Separation by Translation Invariant Functions*, Vector Optimization, https://doi.org/10.1007/978-3-030-44723-6_11

several nonlinear functionals that will be introduced in Sect. 11.2. These functionals are modifications of the nonlinear scalarizing functional (4.1.11).

## 11.1 Vector Optimization Problems with Variable Domination Structure (VVDS)

An interesting application of vector optimization with variable domination structures is given in the theory of consumer demand in economics by John [282, 283].

**Example 11.1.1** John presented in [282, 283] (compare Eichfelder [143]) a local and a global theory in order to explain consumer behavior. In the local approach, it is supposed that the consumer faces a nonempty set of admissible alternatives $F \subset \mathbb{R}^\ell$ from which he is allowed to choose. A local preference only requires that the consumer is able to rank alternatives in a small neighborhood of a given commodity bundle relative to that bundle in contrast to the global approach. This concept can be incorporated by an economical comparative function $g : \mathbb{R}^\ell \to \mathbb{R}^\ell$ such that $y$ in a neighborhood of $\overline{y}$ is interpreted to be better than $\overline{y}$ if and only if $g(\overline{y})(y - \overline{y}) < 0$. The choice set assigned to $F \subset \mathbb{R}^\ell$ in the local theory is characterized by (see [283])

$$\mathcal{F}(F) := \{\overline{y} \in F \mid \forall y \in F : g(\overline{y})(y - \overline{y}) \geq 0\}.$$

This leads us to a pre-domination structure $\tilde{D}_{\succ} : \mathbb{R}^\ell \to \mathscr{P}(\mathbb{R}^\ell)$ (see Sect. 6.1) defined by (for $\overline{y} \in F$):

$$\tilde{D}_{\succ}(\overline{y}) := \{b \in \mathbb{R}^\ell \mid g(\overline{y})(b) \geq 0\}.$$

In the case that the consumer is interested in an alternative $\overline{y} \in F$ such that

$$\forall y \in F : \quad g(\overline{y})(y - \overline{y}) \geq 0,$$

then $y - \overline{y} \in \tilde{D}_{\succ}(\overline{y})$ for all $y \in F$, i.e., $F \subseteq \{\overline{y}\} + \tilde{D}_{\succ}(\overline{y})$.

Moreover, if the consumer is searching for alternatives $\overline{y} \in F$ such that

$$\forall y \in F \setminus \{\overline{y}\} : \quad g(\overline{y})(\overline{y} - y) < 0,$$

then for all $y \in F \setminus \{\overline{y}\}$,

$$\overline{y} - y \notin \tilde{D}_{\succ}(\overline{y}).$$

This means that the consumer is searching for alternatives $\overline{y} \in F$ such that

$$F \cap (\overline{y} - \tilde{D}_{\succ}(\overline{y})) = \{\overline{y}\},$$

i.e., the consumer is looking for nondominated-like points of a vector optimization problem with variable domination structure in the sense of Definition 11.1.3 (compare (11.1.4)) in the following section.

Let $Y$ be a topological vector space, $F \subset Y$ and $C \subset Y$ is a nontrivial, closed, convex and pointed cone in $Y$. In vector optimization, recall that $\overline{y} \in F$ is an efficient point of $F$ with respect to (the fixed ordering cone) $C$ if the following holds:

$$F \cap (\overline{y} - (C \setminus \{0\})) = \emptyset, \tag{11.1.1}$$

(see Sect. 6.1, Definition 6.0.1).

In the case of a constant domination structure, equivalently, we can write (11.1.1) in the form

$$\forall y \in F: \qquad \overline{y} \notin y + (C \setminus \{0\}). \tag{11.1.2}$$

Now, we consider an asymmetric or antisymmetric relation $\succ$ described via the domination structure $D_{\succ} : Y \to \mathscr{P}(Y)$ and we are looking for elements $\overline{y} \in F$ such that

$$\forall y \in F: \qquad \overline{y} \notin y + (D_{\succ}(y) \setminus \{0\}) \tag{11.1.3}$$

(see Lemma 6.1.15). The concept in (11.1.3) was initiated by Yu [502], the so-called nondominated elements (see Definition 11.1.2, compare also Definition 2.1.42 and Sect. 6.4 for bibliographical notes).

Furthermore, we consider an asymmetric or antisymmetric relation $\succ$ described via the pre-domination structure $\tilde{D}_{\succ} : Y \to \mathscr{P}(Y)$ and we are looking for elements $\overline{y} \in F$ such that

$$F \cap (\overline{y} - (\tilde{D}_{\succ}(\overline{y}) \setminus \{0\})) = \emptyset \tag{11.1.4}$$

(see Lemma 6.1.15). Solutions $\overline{y} \in F$ with the property (11.1.4) are called nondominated-like elements of $F$ with respect to $\tilde{D}_{\succ} : Y \to \mathscr{P}(Y)$. Nondominated-like elements of $F$ with respect to $\tilde{D}_{\succ} : Y \to \mathscr{P}(Y)$ are introduced by Chen, Huang, Yang [93, Definition 1.13] under the name nondominated-like minimal points (see Chen, Yang [94, Definition 3.1]). We present this solution concept in Definition 11.1.3.

A nondominated-like element $\overline{y}$ of a set $F$ is an element which is not dominated by another point $y$ of $F$ with respect to the associated set $\tilde{D}_{\succ}(\overline{y})$ at this point $\overline{y}$. In the definition of nondominated-like elements, a domination set is a set associated to the nondominated-like element. But, for a nondominated element $\overline{y}$, a domination set $D_{\succ}(y)$ is a set associated to another point $y$. Important properties of these elements can be found in [93, 94, 138, 141–143, 502].

Moreover, using $\hat{D}_{\succ} : Y \rightarrow \mathscr{P}(Y)$, we are looking for elements $\overline{y} \in F$ with

$$\forall y \in F : \quad F \cap (\overline{y} - (\hat{D}_{\succ}(y) \backslash \{0\})) = \emptyset. \tag{11.1.5}$$

Elements $\overline{y} \in F$ with the property given by (11.1.5) are called completely nondominated elements of $F$ with respect to $\hat{D}_{\succ} : Y \rightarrow \mathscr{P}(Y)$, (see [94, Definition 3.1] and [93, Definition 1.11] under a different name (nondominated minimal points)). We will introduce the solution concept of completely nondominated elements in Definition 11.1.4.

Yu initiated the concept of nondominated elements to vector optimization problems with variable domination structure based on general domination set-valued maps in [502] (see Definition 6.1.1 and Sect. 6.4). Vector optimization problems with respect to a cone-valued map $D$ are studied intensively by Eichfelder in [138, 141–143] and Eichfelder, Ha [146], where corresponding solution concepts, characterizations by scalarization methods, optimality conditions and numerical procedures are presented. Eichfelder [139, 143] gives a very detailed overview of solution concepts for vector optimization problems with variable domination structure and presents a complete characterization of these solution concepts (see also Eichfelder, Kasimbeyli [148] and Eichfelder, Gerlach [145]).

In Sect. 6.1, corresponding to the decision maker's strict or weak preference relation $\succ$, the set $\mathrm{ND}(F, \succ)$ of decision outcomes that are optimal for the decision maker is introduced. Now, we introduce and discuss several concepts of solutions to problems with variable domination structure taking into account Lemma 6.1.15.

First, accordingly to (11.1.3) with the domination structure $D_{\succ}$, we give the definition of (weakly) nondominated elements (see Lemma 6.1.15). The following concept of nondominated elements was initiated by Yu in [502] (compare Eichfelder [143] and Sect. 6.1).

**Definition 11.1.2** Let $Y$ be a topological vector space, $F \subset Y$. Suppose that the domination structure $D_{\succ}$ is associated to $D : Y \rightarrow \mathscr{P}(Y)$ and $D(y)$ is a proper, closed, pointed set with $0 \in D(y)$ for all $y \in Y$.

(a) $\overline{y} \in F$ is a nondominated element of the set $F$ with respect to $D : Y \rightarrow \mathscr{P}(Y)$, if no $y \in F$ exists with

$$\overline{y} \in y + (D(y) \setminus \{0\}),$$

or equivalently,

$$\overline{y} \notin \bigcup_{y \in F} (y + (D(y) \setminus \{0\})).$$

(b) Suppose that $\operatorname{int} D(y) \neq \emptyset$ for all $y \in F$. $\overline{y} \in F$ is a weakly nondominated element of the set $F$ with respect to $D : Y \rightarrow \mathscr{P}(Y)$, if no $y \in F$ exists with

$$\overline{y} \in y + \operatorname{int} D(y),$$

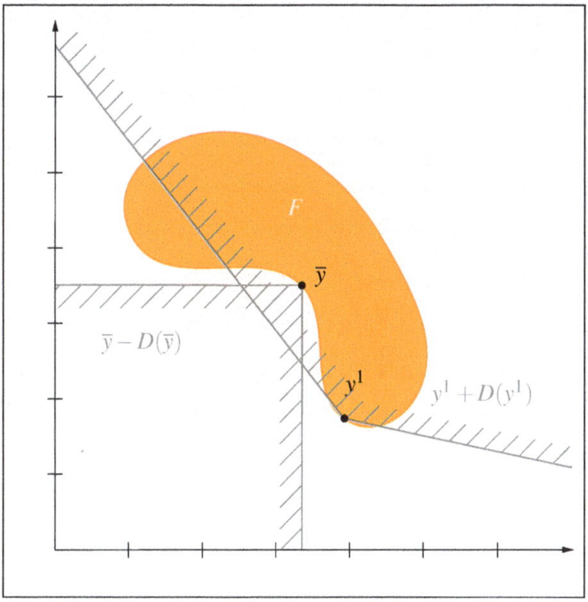


**Fig. 11.1** $\overline{y} \in \mathrm{M}(F, D(\cdot))$, $\overline{y} \notin \mathrm{N}(F, D(\cdot))$

or equivalently,

$$\overline{y} \notin \bigcup_{y \in F} (y + \operatorname{int} D(y)).$$

Accordingly to the set of nondominated elements $\mathrm{ND}(F, \succ)$ in Sect. 6.1, we denote the set of all (weakly) nondominated elements of $F$ with respect to $D(\cdot)$ by $\mathrm{N}(F, D(\cdot))$ ($\mathrm{WN}(F, D(\cdot))$, respectively) (see Figs. 11.1 and 11.2).

Moreover, accordingly to (11.1.4), we introduce another solution concept, namely nondominated-like elements of a set $F \subset Y$ with respect to $\tilde{D} : Y \to \mathscr{P}(Y)$, see (11.1.4) and [143].

**Definition 11.1.3** Let $Y$ be a topological vector space, $F \subset Y$. Suppose that the pre-domination structure $\tilde{D}_{\succ}$ is associated to $\tilde{D} : Y \to \mathscr{P}(Y)$ and $\tilde{D}(y)$ is a proper, closed, pointed set with $0 \in \tilde{D}(y)$ for all $y \in Y$.

(a) $\overline{y} \in F$ is a nondominated-like element of the set $F$ with respect to $\tilde{D} : Y \to \mathscr{P}(Y)$, if no $y \in F$ exists with

$$\overline{y} \in y + (\tilde{D}(\overline{y}) \setminus \{0\}),$$

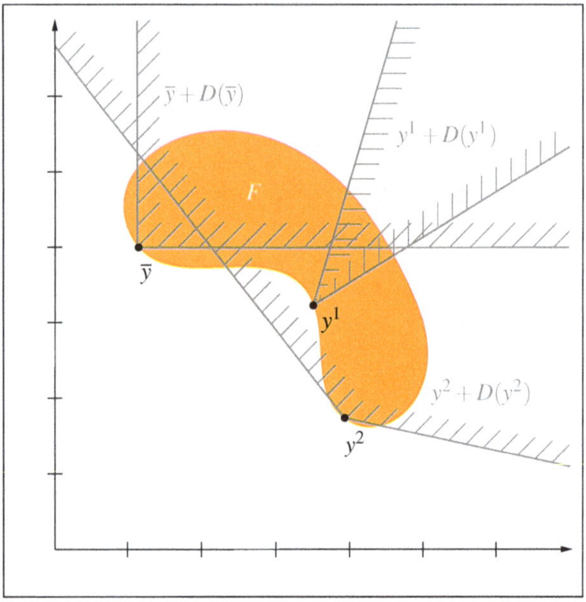


**Fig. 11.2** $\overline{y} \in N(F, D(\cdot))$, $y^1 \notin N(F, D(\cdot))$

or equivalently

$$F \cap (\overline{y} - \tilde{D}(\overline{y})) = \{\overline{y}\}.$$

(b) Suppose that $\operatorname{int} \tilde{D}(\overline{y}) \neq \emptyset$ for $\overline{y} \in F$. $\overline{y} \in F$ is a weakly nondominated-like element of the set $F$ with respect to $\tilde{D} : Y \to \mathscr{P}(Y)$, if no $y \in F$ exists with

$$\overline{y} \in y + \operatorname{int} \tilde{D}(\overline{y}),$$

or equivalently

$$F \cap (\overline{y} - \operatorname{int} \tilde{D}(\overline{y})) = \emptyset.$$

The set of all (weakly) nondominated-like elements of $F$ with respect to $\tilde{D}(\cdot)$ is denoted by $\mathrm{M}(F, \tilde{D}(\cdot))$ ($\mathrm{WM}(F, \tilde{D}(\cdot))$, respectively) (compare Fig. 11.1).

Furthermore, we will use the solution concept of completely nondominated elements introduced by Chen, Huang, Yang [93, Definition 1.11].

**Definition 11.1.4** Let $Y$ be a topological vector space, $F \subset Y$. Suppose that $\hat{D}_{\succ}$ is associated to $\hat{D} : Y \to \mathscr{P}(Y)$ and $\hat{D}(y)$ is a proper, closed, pointed set with $0 \in \hat{D}(y)$ for all $y \in Y$.

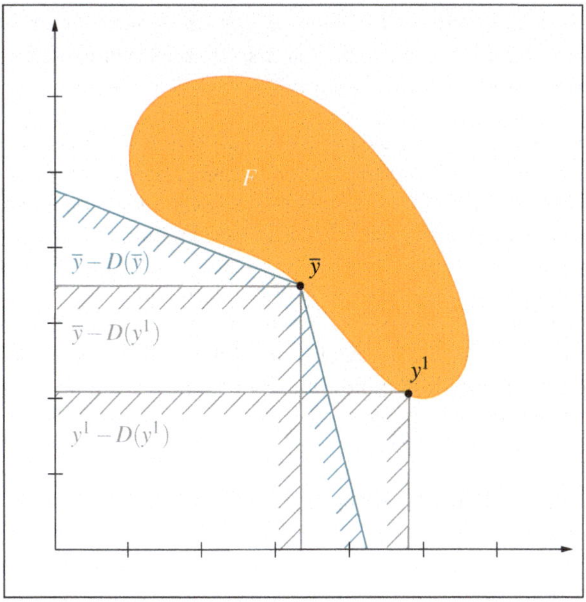


**Fig. 11.3** $\overline{y} \in MZ(F, D)$

(a) $\overline{y} \in F$ is a completely nondominated element of the set $F$ with respect to $\hat{D} : Y \to \mathscr{P}(Y)$, if

$$\forall y \in F : \qquad F \cap (\overline{y} - (\hat{D}(y) \backslash \{0\})) = \emptyset.$$

(b) Suppose that int $\hat{D}(\overline{y}) \neq \emptyset$ for $\overline{y} \in F$. The element $\overline{y} \in F$ is a weakly completely nondominated element of the set $F$ with respect to $\hat{D} : Y \to \mathscr{P}(Y)$, if

$$\forall y \in F : \qquad F \cap (\overline{y} - \text{int } \hat{D}(\overline{y})) = \emptyset.$$

The set of all (weak) completely nondominated elements of $F$ with respect to $\hat{D}(\cdot)$ is denoted by MZ$(F, \hat{D}(\cdot))$ (WMZ$(F, \hat{D}(\cdot))$, respectively) (see Fig. 11.3).

Important relationships between nondominated elements of $F$ w.r.t. $D : Y \to \mathscr{P}(Y)$ (Definition 11.1.2) and nondominated-like elements of $F$ w.r.t. $\tilde{D} : Y \to \mathscr{P}(Y)$ (Definition 11.1.3) are presented in the next lemma. The following assertions are derived by Eichfelder in [143, Lemma 2.11] under the assumption that $D(y) = \tilde{D}(y)$ is a nontrivial, convex, pointed cone for all $y \in Y$.

**Lemma 11.1.5**

(a) *If $\overline{y} \in F$ is a nondominated-like element of the set $F$ w.r.t. $D : Y \to \mathscr{P}(Y)$ and $D(y) \subseteq D(\overline{y})$ for all $y \in F$, then $\overline{y}$ is also a nondominated element of $F$ w.r.t. $D : Y \to \mathscr{P}(Y)$.*

(b) *If* $\overline{y} \in F$ *is a nondominated element of the set* $F$ *w.r.t.* $D : Y \to \mathscr{P}(Y)$ *and* $D(\overline{y}) \subseteq D(y)$ *for all* $y \in F$*, then* $\overline{y}$ *is also a nondominated-like element of* $F$ *w.r.t.* $D : Y \to \mathscr{P}(Y)$.

**Remark 11.1.6** Of course, in the case that for all $y \in Y$ it holds that $D = D(y) = \tilde{D}(y) = \hat{D}(y)$, where $D \subseteq Y$ is a fixed set, the concepts of nondominated elements, nondominated-like elements and completely nondominated elements coincide with the concept of efficient elements of $F$ w.r.t. $D$ (see Definition 6.0.1).

In Sect. 11.3, we present characterizations of nondominated, nondominated-like and completely nondominated elements of vector optimization problems with variable domination structures by means of extensions of the nonlinear separating functional (4.1.11). These extensions will be introduced in the next section.

## 11.2 Scalarizing Functionals Related to Domination Structures

In this section, we introduce and discuss extensions of the scalarizing functional (4.1.11) that we will apply in Sect. 11.3 to give characterizations of solutions to vector optimization problems with variable domination structures via scalarization using these functionals. Furthermore, we will use these extensions of (4.1.11) for deriving variational principles and necessary conditions for approximate solutions of vector optimization problems with variable domination structures in Chap. 12 as well as in numerical methods for solving vector optimization problems with variable domination structures in Sect. 13.6.

We derive the results of this section under certain assumptions concerning $D : Y \to \mathscr{P}(Y)$ involved in the definition of a separating functional. The set-valued map $D : Y \to \mathscr{P}(Y)$ defines the variable domination structure in Definition 6.1.1 as well as in this chapter. Concerning $D : Y \to \mathscr{P}(Y)$, we suppose in this section the following assumption.

**Assumption (A).** Let $Y$ be a topological vector space and $k \in Y \setminus \{0\}$. Suppose $D : Y \to \mathscr{P}(Y)$ and $D(y)$ is a proper, closed, pointed set with $0 \in D(y)$ for all $y \in Y$.

If not otherwise stated in this section, we suppose that (A) holds.

**Remark 11.2.1** The pointedness of $D(y)$ and $0 \in D(y)$ for all $y \in Y$ implies $0 \in \mathrm{bd}(y)$ for all $y \in Y$. Indeed, if we suppose that $0 \notin \mathrm{bd}\, D(y)$, then $0 \in \mathrm{int}\, D(y)$. Because of $0 \in \mathrm{int}\, D(y)$ there is a sufficiently small neighborhood $U(0)$ of $0$ such that $0 \neq u \in U(0) \subset D(y)$ and $-u \in U(0) \subset D(y)$. This means $u \neq 0$, $u \in D(y)$ and $u \in -D(y)$, a contradiction to the pointedness of $D(y)$.

Furthermore, for some assertions we suppose partially stronger assumptions than in (A) concerning $D : Y \to \mathscr{P}(Y)$, especially convexity and cone properties.

**Assumption** ($A_{cone}$). Let $Y$ be a topological vector space. Suppose $D : Y \to \mathscr{P}(Y)$ and $D(y)$ is a nontrivial, closed, convex cone for all $y \in Y$.

We study the functional $\zeta_{D(\cdot),k(\cdot)}$ given by

$$\zeta_{-D(\cdot),k(\cdot)} : Y \times Y \to \overline{\mathbb{R}}, \qquad \zeta_{-D(\cdot),k(\cdot)}(y, z) := \inf\{t \in \mathbb{R} \mid y \in tk(z) - D(z)\}. \tag{11.2.1}$$

Under the assumptions ($A_{cone}$), $\operatorname{int} D(z) \neq \emptyset$, $k : Y \to Y$ and $k(z) \in \operatorname{int} D(z)$ for all $z \in Y$, the functional (11.2.1) was introduced by Chen, Yang, Yu in [95] (see also Chen, Huang, Yang [93]).

**Remark 11.2.2** In the special case $D(z) = C$ and $k(z) = k$ for any $z \in Y$, $C \subset Y$ is a nontrivial, closed, convex, pointed cone and $k \in \operatorname{int} C$ is fixed, the functional (11.2.1) coincides with the functional $\varphi_{A,k}$ given by (4.1.11) with $A = -C$ and $k \in \operatorname{int} C$.

**Remark 11.2.3** For the special case $k(z) = k$ for each $z \in Y$, the functional $\zeta_{-D(\cdot),k(\cdot)}$ in (11.2.1) reduces to the functional

$$\zeta_{-D(\cdot),k} : Y \times Y \to \overline{\mathbb{R}}, \qquad \zeta_{-D(\cdot),k}(y, z) := \inf\{t \in \mathbb{R} \mid y \in tk - D(z)\}. \tag{11.2.2}$$

The functional (11.2.2) was studied by Chen and Yang in [94] under the additional assumption $k \in \bigcap_{z \in Y} \operatorname{int} D(z)$.

Properties of the functionals (11.2.1) and (11.2.2) will be discussed in Sect. 11.2.1.

Now, we consider the functional

$$\varphi_{a-D(\cdot),k} : Y \to \overline{\mathbb{R}}, \quad \varphi_{a-D(\cdot),k}(y) := \inf\{t \in \mathbb{R} \mid y \in tk + a - D(y)\}, \tag{11.2.3}$$

where $a \in Y$.

**Remark 11.2.4** If we set $y = z$ in (11.2.2), then we get the functional (11.2.3) with $a = 0$.

Under the assumptions that ($A_{cone}$) holds and $k \in \left(\bigcap_{z \in Y} D(z)\right) \setminus \{0\}$, the functional (11.2.3) was studied by Eichfelder in [139, Section 5.2.1], [143, Section 5.2.1]; see also [141]. We discuss Eichfelder's extension of the functional (4.1.11) in Sect. 11.2.2. The functional (11.2.3) will be applied for a characterization of non-dominated elements of a set $F \subset Y$ with respect to a variable domination structure (see Sect. 11.3). Moreover, we use the functional (11.2.3) for deriving variational principles for vector optimization problems with variable domination structures in Sect. 12.3.2 and necessary conditions for approximate solutions of vector optimization problems with variable domination structures in Sect. 12.4.1.

Furthermore, under the assumptions (A) and $k \in 0^+D(z)$ for all $z \in Y$ (without convexity assumptions concerning $D$ (see Soleimani, Tammer [416])), we consider the case $D(z) = D(w)$ ($w \in Y$ fixed) for all $z \in Y$ such that the functional (11.2.2) has the following form:

$$\varphi_{-D(w),k} : Y \to \overline{\mathbb{R}}, \quad \varphi_{-D(w),k}(y) := \inf\{t \in \mathbb{R} \mid y \in tk - D(w)\}. \tag{11.2.4}$$

For fixed $w \in Y$, this functional coincides with the functional (4.1.11) and will be discussed in Sect. 11.2.1.

**Remark 11.2.5** If we put $D(y) = D(w)$ ($w \in Y$ fixed) for all $y \in Y$ and $a = 0$ in (11.2.3), then the functional (11.2.3) coincides with the functional (11.2.4).

**Remark 11.2.6** Of course, if we put $k(z) = k \in Y \setminus \{0\}$ and $D(z) = D(w)$ ($w \in Y$ fixed) for all $z \in Y$, then the functional (11.2.1) coincides with the functional (11.2.4).

In Sect. 11.4.2, we will use the functional (11.2.4) for a characterization of approximate solutions of vector optimization problems with variable domination structures. Furthermore, the functional (11.2.4) plays an important role for deriving optimality conditions for vector optimization problems with variable domination structures in Sect. 12.4.2.

Under the assumption (A), an additional convexity assumption concerning the set-valued map $D$ and for a set $A \subset Y$ we consider the functional

$$\varphi_{A+D(\cdot)} : Y \times Y \to \overline{\mathbb{R}}, \quad \varphi_{A+D(\cdot)}(y, k) := \inf\{t \in \mathbb{R} \mid y + tk \in A + D(y)\}, \tag{11.2.5}$$

(see Bouza and Tammer [70, 71]). We will discuss the properties of this functional in Sect. 11.2.3 and we will use it for a characterization of solutions to vector optimization problems with variable domination structure in Sect. 11.3. Moreover, the functional (11.2.5) is useful in numerical procedures for generating solutions of vector optimization problems with variable domination structure (see Sect. 13.6).

### *11.2.1 The Separating Functional $\zeta_{-D(\cdot),K(\cdot)}$*

Unless otherwise stated, we suppose that $(A_{cone})$ holds in this section.

The functional $\zeta_{-D(\cdot),k(\cdot)} : Y \times Y \to \overline{\mathbb{R}}$ (see (11.2.1)) is given by

$$\zeta_{-D(\cdot),k(\cdot)}(y, z) = \inf\{t \in \mathbb{R} \mid y \in tk(z) - D(z)\}.$$

In order to show the attainment property, we consider the special case $k(z) = k \in Y$ for all $z \in Y$ and the corresponding functional (11.2.2) given by

$$\zeta_{-D(\cdot),k}(y, z) = \inf\{t \in \mathbb{R} \mid y \in tk - D(z)\}$$

under the assumption $(A_{cone})$ and $k \in \bigcap_{z\in Y} \operatorname{int} D(z)$.

In order to prove the attainment property of the functional $\zeta_{-D(\cdot),k}$ given by (11.2.2) in Theorem 11.2.9, we start with two helpful lemmata (see [94, Lemma 2.1] and [94, Lemma 2.2]) that immediately follow from our results derived in Sect. 2.1.2, especially from Lemma 2.3.13.

**Lemma 11.2.7** *Suppose that* $(A_{cone})$ *holds and* $k \in \operatorname{int}\left(\bigcap_{z\in Y} D(z)\right)$. *Then, it holds that*

$$Y = \bigcup \left\{ tk - \operatorname{int}\left( \bigcap_{z\in Y} D(z) \right) \mid t \in \mathbb{R}_> \right\}.$$

***Proof*** The result follows from Lemma 2.3.13. □

**Lemma 11.2.8** *Suppose that* $(A_{cone})$ *holds,* $k \in \operatorname{int}\left(\bigcap_{z\in Y} D(z)\right)$. *For* $t \in \mathbb{R}$ *and* $z \in Y$ *we consider the set* $S_t(z) := tk - D(z)$. *Then, the following assertions are true:*

(a) *For each* $y \in Y$, *there is an element* $t \in \mathbb{R}$ *such that* $y \in S_t(z)$ *for all* $z \in Y$.
(b) *If* $y \in S_t(z)$ *is fulfilled for some* $t \in \mathbb{R}$ *and* $z \in Y$, *then*

$$\forall \mu > t: \qquad y \in \mu k - \operatorname{int} D(z) \subseteq \mu k - D(z).$$

(c) *For all* $y, z \in Y$, *there is a real number* $t \in \mathbb{R}$ *such that* $y \notin S_t(z)$.
(d) *Consider* $y \in Y$. *If* $y \notin S_t(z)$ *for some* $t \in \mathbb{R}$ *and* $z \in Y$, *then,*

$$\forall \mu < t: \qquad y \notin S_\mu(z).$$

***Proof***

(a) Taking into account Lemma 11.2.7, for each $y \in Y$, there is an element $t \in \mathbb{R}_+ \setminus \{0\}$ such that $y \in tk - \operatorname{int}\left(\bigcap_{z\in Y} D(z)\right)$ and so

$$\forall z \in Y: \qquad y \in tk - \operatorname{int} D(z).$$

(b) Let $\mu > t$ and let $y \in S_t(z)$ for some $z \in Y$. Then, we obtain

$$\mu k - y = (\mu - t)k + tk - y \in \operatorname{int}\left( \bigcap_{z\in Y} D(z) \right) + D(z) \subseteq \operatorname{int} D(z).$$

This yields

$$y \in \mu k - \operatorname{int} D(z) \subseteq \mu k - D(z).$$

(c) Suppose that there are elements $y^0, z^0 \in Y$ such that, for all $t \in \mathbb{R}$, $y^0 \in S_t(z^0)$. From (b), we obtain

$$\forall t \in \mathbb{R}: \qquad y^0 \in tk - \operatorname{int} D(z^0).$$

This means $\{tk - y^0 \mid t \in \mathbb{R}\} \subseteq \operatorname{int} D(z^0)$, which is equivalent to $\{-tk - y^0 \mid t \in \mathbb{R}\} \subseteq \operatorname{int} D(z^0)$.

For $k \in \operatorname{int}\left(\bigcap_{z\in Y} D(z)\right) \subseteq \operatorname{int} D(z)$ for each $z \in Y$, we get from Lemma 11.2.7 that for all $z \in Y$ it holds that

$$Y = \bigcup \{tk - \operatorname{int} D(z) \mid t \in \mathbb{R}_+ \setminus \{0\}\}.$$

Hence, for each $y \in Y$, there are elements $c \in \operatorname{int} D(z^0)$ and $\alpha \in \mathbb{R}_+ \setminus \{0\}$ such that $-y = \alpha k - c$. Then, we get

$$y = -\alpha k + c = (-\alpha k - y^0) + c + y^0 \subseteq \operatorname{int} D(z^0) + \operatorname{int} D(z^0) + y^0 = y^0 + \operatorname{int} D(z^0).$$

This means $Y \subseteq y^0 + \operatorname{int} D(z^0)$, a contradiction to $D(z^0) \neq Y$.

(d) Consider $y \notin S_t(z)$ for some $t \in \mathbb{R}$, $y \in Y$. If we suppose that for some $\mu < t$ it holds that $y \in S_\mu(z)$, then we get $y \in S_t(z)$ from (b), a contradiction.

□

Now, we are able to present the following important attainment property (see [94, Proposition 2.1]) under the assumption that $D(\cdot)$ is a cone-valued map.

**Theorem 11.2.9** *Suppose that* $(A_{cone})$ *holds and* $k \in \operatorname{int}\left(\bigcap_{z\in Y} D(z)\right)$. *Then, the functional* $\zeta_{-D(\cdot),k}$ *given by (11.2.2) is finite-valued and the infimum in (11.2.2) is attained, i.e.,*

$$\zeta_{-D(\cdot),k}(y, z) = \min\{t \in \mathbb{R} \mid y \in tk - D(z)\}.$$

***Proof*** For any $y, z \in Y$, we define

$$L := \{t \in \mathbb{R} \mid y \in tk - D(z)\}.$$

Now, it is sufficient to prove that $L$ is a lower bounded and closed subset in $\mathbb{R}$. Assume that $\{t_n\} \subset L$ and $t_n \to \bar{t}$ for $n \to +\infty$. From the definition of $L$, we get $t_n k - y \in D(z)$ for all $n$. Taking into account the closedness of $D(z)$, we get $\bar{t}k - y \in D(z)$. This yields that $\bar{t} \in L$ and so $L$ is closed.

Suppose that, for each $r \in \mathbb{R}$, there is a real number $t_r \in \mathbb{R}$ such that $t_r < r$ and $y \in t_r k - D(z)$. Taking into account Lemma 11.2.8(c), there is a real number $\lambda > 0$ such that $y \notin \lambda k - D(z)$. Because of Lemma 11.2.8(d), we get $y \notin \mu k - D(z)$ for all $\mu < \lambda$, a contradiction. Hence, $L$ is lower bounded. Together with the closedness of $L$, the proof is complete. □

Now, we study a slight modification of the functional (11.2.2) defined by Chen and Yang in [94], especially concerning the assumptions for the set-valued map $D$. We replace the assumption $(A_{cone})$ by the assumption (A). Under the assumption (A) and $k \in Y\setminus\{0\}$ such that $D(z) + \mathbb{R}_> k \subseteq \operatorname{int} D(z)$ for all $z \in Y$, we use the functional $\zeta_{-D(\cdot),k}(y, z) : Y \times Y \to \overline{\mathbb{R}}$ (see (11.2.2)):

$$\zeta_{-D(\cdot),k}(y,z) := \inf\{t \in \mathbb{R} \mid y \in tk - D(z)\}. \tag{11.2.6}$$

The next result is shown by Chen and Yang in [94, Lemma 2.3] under the stronger assumptions $(A_{cone})$ and $k \in \operatorname{int}\left(\bigcap_{z\in Y} D(z)\right)$ using the attainment property given in Theorem 11.2.9. We derive the result for set-valued maps $D(\cdot)$ without assuming convexity and cone properties.

The assertions (a), (b), (c) of the next lemma for arbitrary fixed elements $z \in Y$ immediately follow from corresponding conditions in Theorem 4.4.4 concerning the sublevel sets.

**Lemma 11.2.10** *Let assumption* (A) *be fulfilled and assume that* $k \in Y\setminus\{0\}$ *such that* $D(z) + \mathbb{R}_{>}k \subseteq \operatorname{int} D(z)$ *for all* $z \in Y$. *Consider an arbitrary fixed element* $z \in Y$. *Then, for each* $t \in \mathbb{R}$, *the following assertions hold*

(a) $\operatorname{lev}_{\zeta_{-D(\cdot),k},<}(t) = \{(y,z) \in Y \times Y \mid y \in tk - \operatorname{int} D(z)\}$,
(b) $\operatorname{lev}_{\zeta_{-D(\cdot),k},\leq}(t) = \{(y,z) \in Y \times Y \mid y \in tk - D(z)\}$,
(c) $\operatorname{lev}_{\zeta_{-D(\cdot),k},=}(t) = \{(y,z) \in Y \times Y \mid y \in tk - \operatorname{bd} D(z)\}$,
(d) $\operatorname{lev}_{\zeta_{-D(\cdot),k},\geq}(t) = (Y \times Y) \setminus \{(y,z) \in (Y \times Y) \mid y \in tk - \operatorname{int} D(z)\}$,
(e) $\operatorname{lev}_{\zeta_{-D(\cdot),k},>}(t) = (Y \times Y) \setminus \{(y,z) \in (Y \times Y) \mid y \in tk - D(z)\}$.

***Proof***

(a) The following implication holds:

$$\lambda \in \mathbb{R},\ y \in \lambda k - D(z) \quad \Longrightarrow \quad \forall \mu > \lambda :\ \ y \in \mu k - \operatorname{int} D(z). \tag{11.2.7}$$

Indeed, for $y \in \lambda k - D(z)$ and $\mu > \lambda$ it holds that

$$y - \mu k = y - \lambda k + (\lambda - \mu)k \in -D(z) + (\lambda - \mu)k \subseteq -\operatorname{int} D(z) \tag{11.2.8}$$

since $D(z) + \mathbb{R}_{>}k \subseteq \operatorname{int} D(z)$.
If we suppose that $\zeta_{-D(\cdot),k}(y,z) < t$, then $y \in \lambda k - D(z)$ for (at least one) $\lambda < t$. This yields $y \in tk - \operatorname{int} D(z)$, taking into account (11.2.7).
Now, suppose that $y \in tk - \operatorname{int} D(z)$. Therefore, there is an element $c^1 \in \operatorname{int} D(z)$ such that

$$y = tk - c^1. \tag{11.2.9}$$

Since $c^1 \in \operatorname{int} D(z)$, there is a real number $\gamma > 0$ such that $c^1 - \gamma k \in D(z)$. This yields, together with (11.2.9),

$$y = (t - \gamma)k - (c^1 - \gamma k)$$

and

$$y \in (t - \gamma)k - D(z).$$

Hence, $\zeta_{-D(\cdot),k}(y,z) \leq (t - \gamma) < t$.

(b) Suppose that $\zeta_{-D(\cdot),k}(y,z) \le t$ and consider a neighborhood $U$ of $tk$. Then, there is a real number $\lambda > t$ with $\lambda k \in U$. Because of (11.2.7), $\zeta_{-D(\cdot),k}(y,z) \le t$ yields

$$\forall \lambda > t: \qquad y \in \lambda k - D(z)$$

and so $\lambda k \in y + D(z)$. On the other hand, we have $\lambda k \in U$ such that

$$tk \in \mathrm{cl}(y + D(z)) = y + D(z)$$

and

$$y \in tk - D(z).$$

Now, suppose $y \in tk - D(z)$. Then, we get $\zeta_{-D(\cdot),k}(y,z) \le t$ from the definition of $\zeta_{-D(\cdot),k}$.

(d) We get (d) from (a).
(c) Assertion (c) follows from (b) and (d).
(e) (b) implies (e).

□

For many applications, monotonicity properties and subadditivity of the scalarizing functional are important. Under the assumption that the map $D(\cdot)$ is linear, these properties are shown in the following theorem ([94, Theorem 2.1]).

**Theorem 11.2.11** *Suppose that* $(A_{cone})$ *holds and* $k \in \mathrm{int}\left(\bigcap_{z\in Y} D(z)\right)$. *Furthermore, assume that* $D(\cdot)$ *is linear. Then, the functional* $\zeta_{-D(\cdot),k}$ *is finite-valued and*

(a) $\zeta_{-D(\cdot),k}$ *is positively homogeneous and subadditive on* $Y \times Y$,
(b) $\zeta_{-D(\cdot),k}$ *is convex on* $Y \times Y$,
(c) *Consider some* $z \in Y$, *fixed. Then,* $\zeta_{-D(\cdot),k}(\cdot, z)$ *is strictly* $\mathrm{int}\, D(z)$*-monotone, i.e., for* $y^1, y^2, z \in Y$, $y^1 \in y^2 + \mathrm{int}\, D(z)$ *implies* $\zeta_{-D(\cdot),k}(y^2,z) < \zeta_{-D(\cdot),k}(y^1,z)$.

*Moreover, if* $D(\cdot)$ *is continuous, then*

(d) $\zeta_{-D(\cdot),k}$ *is lower semicontinuous.*

***Proof*** Under the given assumptions, the functional $\zeta_{-D(\cdot),k}$ is finite-valued (see Theorem 11.2.9).

(a) In order to show that $\zeta_{-D(\cdot),k}$ is positively homogeneous on $Y \times Y$, let $\lambda > 0$. For all $(y,z) \in Y \times Y$, it holds that

$$\begin{aligned}\zeta_{-D(\cdot),k}(\lambda y, \lambda z) &= \min\{t \in \mathbb{R} \mid \lambda y \in tk - D(\lambda z)\} = \lambda \min\{\frac{t}{\lambda} \in \mathbb{R} \mid y \in \frac{t}{\lambda}k - \frac{1}{\lambda}D(\lambda z)\} \\ &= \lambda \min\{\alpha \in \mathbb{R} \mid y \in \alpha k - D(z)\} = \lambda \zeta_{-D(\cdot),k}(y,z).\end{aligned}$$

This means that $\zeta_{-D(\cdot),k}$ is positively homogeneous on $Y \times Y$.
In order to show the subadditivity of $\zeta_{-D(\cdot),k}$, consider $(y^1, z^1), (y^2, z^2) \in Y \times Y$ and $t_1, t_2 \in \mathbb{R}$. From Lemma 11.2.10(c), we get that

$$\zeta_{-D(\cdot),k}(y^1, z^1) = t_1 \Longleftrightarrow y^1 \in t_1 k - \text{bd}\, D(z^1),$$

$$\zeta_{-D(\cdot),k}(y^2, z^2) = t \Longleftrightarrow y^2 \in t_2 k - \text{bd}\, D(z^2).$$

Since $D(\cdot)$ is linear, we get

$$y^1 + y^2 \in (t_1 + t_2)k - \text{bd}\, D(z^1) - \text{bd}\, D(z^2) \subseteq (t_1 + t_2)k - D(z^1 + z^2).$$

This yields, together with Lemma 11.2.10(b),

$$\zeta_{-D(\cdot),k}(y^1 + y^2, z^1 + z^2) \leq t_1 + t_2 = \zeta_{-D(\cdot),k}(y^1, z^1) + \zeta_{-D(\cdot),k}(y^2, z^2).$$

This means that $\zeta_{-D(\cdot),k}$ is subadditive on $Y \times Y$.

(b) Let $(y^1, z^1)$, $(y^2, z^2) \in Y \times Y$ and $\lambda \in (0, 1)$. From Theorem 11.2.9, we get

$$\lambda y^1 \in \zeta_{-D(\cdot),k}(\lambda y^1, \lambda z^1)k - D(\lambda z^1),$$

$$(1 - \lambda)y^2 \in \zeta_{-D(\cdot),k}((1 - \lambda)y^2, (1 - \lambda)z^2)k - D((1 - \lambda)z^2).$$

This yields

$$\lambda y^1 + (1 - \lambda)y^2 \in (\zeta_{-D(\cdot),k}(\lambda y^1, \lambda z^1) + \zeta_{-D(\cdot),k}((1 - \lambda)y^2, (1 - \lambda)z^2))k - D(\lambda z^1) - D((1 - \lambda)z^2),$$

such that

$$\lambda y^1 + (1 - \lambda)y^2 \in (\lambda\zeta_{-D(\cdot),k}(y^1, z^1) + (1 - \lambda)\zeta_{-D(\cdot),k}(y^2, z^2))k - D(\lambda z^1 + (1 - \lambda)z^2), \tag{11.2.10}$$

taking into account that $\zeta_{-D(\cdot),k}$ is positively homogeneous (see (a)) and $D(\cdot)$ is supposed to be linear. From the definition of $\zeta_{-D(\cdot),k}$ and (11.2.10), we get

$$\zeta_{-D(\cdot),k}(\lambda y^1 + (1 - \lambda)y^2, \lambda z^1 + (1 - \lambda)z^2) \leq \lambda\zeta_{-D(\cdot),k}(y^1, z^1) + (1 - \lambda)\zeta_{-D(\cdot),k}(y^2, z^2).$$

This means (together with Theorem 3.4.4) that $\zeta_{-D(\cdot),k}$ is convex on $Y \times Y$.

(c) Consider $z \in Y$, fixed. In order to show the strict $\text{int}\, D(z)$-monotonicity of $\zeta_{-D(\cdot),k}(\cdot, z)$, let $y^1 \in y^2 + \text{int}\, D(z)$. Put $t = \zeta_{-D(\cdot),k}(y^1, z)$. From Theorem 11.2.9, we get

$$y^2 \in y^1 - \text{int}\, D(z) \subseteq tk - \text{int}\, D(z) - \text{int}\, D(z) \subseteq tk - \text{int}\, D(z).$$

With Lemma 11.2.10(a), we have

$$\zeta_{-D(\cdot),k}(y^2, z) < t = \zeta_{-D(\cdot),k}(y^1, z).$$

(d) Suppose that $D(\cdot)$ is continuous on $Y$. Taking into account Theorem 11.2.9, it holds that

$$\zeta_{-D(\cdot),k}(y,z) = \min\{t \in \mathbb{R} \mid y \in tk - D(z)\} = \min\{t \in \mathbb{R} \mid tk \in D(z) + y\},$$

i.e., $\zeta_{-D(\cdot),k}(y,z)$ is the marginal function for the scalar optimization problem

$$\min\ t \qquad \text{subject to} \qquad tk \in D(z) + y.$$

Because of the continuity of the set-valued map $D(\cdot) + y$, the functional $\zeta_{-D(\cdot),k}$ is lower semicontinuous (see [14, Proposition 19]). □

**Remark 11.2.12** Under the assumptions that $Y$ is a Hausdorff locally convex space, $(A_{cone})$, $\operatorname{int} D(z) \neq \emptyset$, $k : Y \to Y$ and $k(z) \in \operatorname{int} D(z)$ for all $z \in Y$, Chen, Yang, Yu studied (in [95]) the more general functional (11.2.1) given by

$$\zeta_{-D(\cdot),k(\cdot)} : Y \times Y \to \overline{\mathbb{R}}, \qquad \zeta_{-D(\cdot),k(\cdot)}(y,z) = \inf\{t \in \mathbb{R} \mid y \in tk(z) - D(z)\}.$$

In [95, Proposition 2.4], Chen, Yang, Yu proved that $\zeta_{-D(\cdot),k(\cdot)}(\cdot, z)$ is positively homogenous and strictly $\operatorname{int} D(z)$- monotone for $z \in Y$, fixed (compare also Chen, Huang, Yang [93, Proposition 1.55]).

The continuity of $\zeta_{-D(\cdot),k}(\cdot, z)$ for fixed $z \in Y$ in the following theorem is derived in [418, Theorem 4.4].

**Theorem 11.2.13** *Suppose that assumption* (A) *holds and* $k \in Y\backslash\{0\}$ *such that* $D(z) + \mathbb{R}_{>}k \subseteq \operatorname{int} D(z)$ *for all* $z \in Y$*. Then,* $\zeta_{-D(\cdot),k}(\cdot, z)$ *is continuous for each arbitrary fixed* $z \in Y$*.*

***Proof*** Suppose that $Y$ is a normed space. For more general spaces (linear topological spaces), we proceed in the same way using nets (see Definition 2.1.45) instead of sequences.

Consider an arbitrary but fixed element $z \in Y$. We write $\zeta_{-D(z),k}(y) := \zeta_{-D(\cdot),k}(y,z)$ for fixed $z \in Y$. Now, we show that the sublevel set

$$\operatorname{lev}_{\zeta_{-D(z),k},\leq}(t) = \{y \in Y \mid \zeta_{-D(z),k}(y) \leq t\}$$

is closed for any $t \in \mathbb{R}$. Indeed, we suppose that $\{y^n\}$ is a sequence with $y^n \to y^0$ and $y^n \in \operatorname{lev}_{\zeta_{-D(z),k},\leq}(t)$, $y^0 \in Y$. We prove that the limit point of this sequence belongs to the set $\operatorname{lev}_{\zeta_{-D(z),k},\leq}(t)$ such that $\operatorname{lev}_{\zeta_{-D(z),k},\leq}(t)$ is a closed set. Because of $y^n \in \operatorname{lev}_{\zeta_{-D(z),k},\leq}(t)$, it holds that $\zeta_{-D(\cdot),k}(y^n, z) \leq t$. By Lemma 11.2.10, we have $y^n \in tk - D(z)$ such that $tk - y^n \in D(z)$ holds.

Taking into account that $D(z)$ is a closed set, the limit point of the sequence $tk - y^n \to tk - y^0$ also belongs to $D(z)$ and $y^0 \in tk - D(z)$. From Lemma 11.2.10, we obtain $\zeta_{-D(\cdot),k}(y^0, z) \leq t$. This means that $\operatorname{lev}_{\zeta_{-D(z),k},\leq}(t)$ is a closed set for any $t \in \mathbb{R}$ and $\zeta_{-D(\cdot),k}(\cdot, z)$ is lower semicontinuous for any $z \in Y$ taking into account Proposition 3.3.12.

In order to show that $\zeta_{-D(\cdot),k}(\cdot, z)$ is upper semicontinuous, we prove that

$$\text{lev}_{\zeta_{-D(z),k},\geq}(t) = \{y \in Y \mid \zeta_{-D(z),k}(y) \geq t\}$$

is a closed set for any $t \in \mathbb{R}$. For this, we suppose that $\{y^n\}$ is a sequence with $y^n \to y^0$ and $y^n \in \text{lev}_{\zeta_{-D(z),k},\geq}(t)$. Since $y^n \in \text{lev}_{\zeta_{-D(z),k},\geq}(t)$, we obtain $\zeta_{-D(\cdot),k}(y^n, z) \geq t$. By Lemma 11.2.10, we get

$$y^n \notin tk - \text{int}\, D(z) \Rightarrow tk - y^n \notin \text{int}\, D(z).$$

This yields $tk - y^n \in (\text{int}\, D(z))^c$. Since int $D(z)$ is an open set, its complement $(\text{int}\, D(z))^c$ is closed and includes all the limit points. Therefore $tk - y^0 \in (\text{int}\, D(y))^c$ and this means

$$tk - y^0 \notin \text{int}\, D(z) \quad \Longrightarrow \quad y^0 \notin tk - \text{int}\, D(z).$$

We obtain $\zeta_{-D(\cdot),k}(y^0, z) \geq t$ from Lemma 11.2.10 and this yields that $\text{lev}_{\zeta_{-D(z),k},\geq}(t)$ is a closed set and $\zeta_{-D(\cdot),k}(\cdot, z)$ is upper semicontinuous.

Taking into account that $\zeta_{-D(\cdot),k}(\cdot, z)$ is also lower semicontinuous, $\zeta_{-D(\cdot),k}(\cdot, z)$ is continuous. □

The convexity of $\zeta_{-D(\cdot),k}(\cdot, z)$ for fixed $z \in Y$ in the following theorem is shown in [418, Theorem 4.5].

**Theorem 11.2.14** *Suppose that assumption (A) holds and $k \in Y \setminus \{0\}$ such that $D(z) + \mathbb{R}_{>}k \subseteq \text{int}\, D(z)$ for all $z \in Y$ and additionally assume that $D(z)$ is a convex set for all $z \in Y$. Then, $\zeta_{-D(\cdot),k}(\cdot, z)$ is convex for all fixed $z \in Y$.*

***Proof*** Consider an arbitrary but fixed element $z \in Y$. Suppose that $\lambda \in (0, 1)$ and $y^1, y^2 \in Y$ such that $\zeta_{-D(\cdot),k}(y^1, z) = t_1$ and $\zeta_{-D(\cdot),k}(y^2, z) = t_2$. From Lemma 11.2.10, we obtain

$$\zeta_{-D(\cdot),k}(y^1, z) = t_1 \Longrightarrow y^1 \in t_1 - D(z),$$

$$\zeta_{-D(\cdot),k}(y^2, z) = t_2 \Longrightarrow y^2 \in t_2 - D(z),$$

i.e., there are elements $c, d \in D(z)$ such that $y^1 = t_1 k - c$ and $y^2 = t_2 k - d$ and

$$\begin{aligned}
&\lambda y^1 + (1 - \lambda) y^2 \\
&= \lambda t_1 k - \lambda c + (1 - \lambda) t_2 k - (1 - \lambda) d \\
&= \big(\lambda t_1 + (1 - \lambda) t_2\big) k - \big(\lambda c + (1 - \lambda) d\big).
\end{aligned}$$

Taking into account $c, d \in D(z)$ and the convexity of $D(z)$, we obtain $\lambda c + (1 - \lambda) d \in D(z)$ and hence,

$$\lambda y^1 + (1 - \lambda) y^2 \in (\lambda t_1 + (1 - \lambda) t_2) k - D(z).$$

By Lemma 11.2.10(c), we get

$$\zeta_{-D(\cdot),k}(\lambda y^1 + (1-\lambda)y^2, z) \leq \lambda t_1 + (1-\lambda)t_2,$$

$$\zeta_{-D(\cdot),k}(\lambda y^1 + (1-\lambda)y^2, z) \leq \lambda \zeta_{-D(\cdot),k}(y^1, z) + (1-\lambda)\zeta_{-D(\cdot),k}(y^2, z)$$

and thus $\zeta_{-D(\cdot),k}(\cdot, z)$ is convex taking into account Theorem 3.4.4. This completes the proof. □

We will use the functional $\zeta_{-D(\cdot),k}$ for deriving necessary conditions for approximate solutions of vector optimization problems with variable domination structure in Sect. 12.4.2.

Furthermore, we study the case where, for all $z \in Y$, it holds that $D(z) = D(w)$ ($w \in Y$ fixed) and $k(z) = k$ in the definition of the functional $\zeta_{-D(\cdot),k(\cdot)}$ given by (11.2.1). This leads us to the functional (11.2.4) that we will use for the characterization of three different kinds of approximate solutions to vector optimization problems with variable domination structure in Sect. 11.4.1 and for deriving necessary conditions for approximative solutions in Sect. 12.4.2.

Under the assumptions (A) and $k \in 0^+D(y)$ for all $y \in Y$, for fixed $w \in Y$, we use the functional $\varphi_{-D(w),k} : Y \to \overline{\mathbb{R}}$ given by (11.2.4):

$$\varphi_{-D(w),k}(y) = \inf\{t \in \mathbb{R} \mid y \in tk - D(w)\}.$$

We get the following assertions analogously to corresponding properties in Chap. 4.

**Corollary 11.2.15** *Suppose that assumption (A) holds and assume that* $k \in 0^+D(y)$ *for all* $y \in Y$. *If* $y, w \in Y$ *and* $\varphi_{-D(w),k}(y) = t_1$, *then for any* $t_2 \geq t_1$,

$$y \in t_2 k - D(w). \tag{11.2.11}$$

***Proof*** By $k \in 0^+D(y)$, $y \in t_1 k - D(w)$ and $t_2 - t_1 \geq 0$, we obtain

$$y \in t_1 k - D(w) = t_2 k - [(t_2 - t_1)k + D(w)] \subseteq t_2 k - D(w).$$ □

The following corollary shows us that the functional (11.2.4) is continuous and finite-valued under certain assumptions supposed in Proposition 4.6.18.

**Corollary 11.2.16** *Suppose that assumption* (A) *holds. If there is a nontrivial, closed, convex cone* $C \subset Y$ *such that* $k \in \operatorname{int} C$ *and* $D(w) = D(w) + C$ *for a fixed element* $w \in Y$, *then* $\varphi_{-D(w),k}$ *in* (11.2.4) *is continuous and finite-valued for* $w \in Y$.

***Proof*** The assertion follows from Propositions 4.6.18(c) and 4.6.13. □

In the special case when $D(\cdot)$ in (A) is a convex cone-valued map, we have the following corollary (compare [94] and Theorem 11.2.9 for the functional $\varphi_{-D(\cdot),k}$).

**Corollary 11.2.17** *Suppose that assumption* $(A_{cone})$ *is fulfilled and* $k \in \bigcap_{y\in Y} \operatorname{int} D(y)$. *Then,* $\varphi_{-D(w),k}$ *in (11.2.4) is continuous and finite-valued for each* $w \in Y$.

***Proof*** Under our assumptions, $D(w)$ is a closed, convex cone such that $D(w) + D(w) = D(w)$ for all $w \in Y$. By Corollary 11.2.16, $\varphi_{-D(w),k}$ is continuous and finite-valued for each $w \in Y$. □

The following corollary collects some important properties of the functional $\varphi_{-D(w),k}$ in (11.2.4) that immediately follow from Theorem 4.6.6.

**Corollary 11.2.18** *Suppose that assumption* (A) *holds, and additionally that* $D(y) + \mathbb{R}_{>}k \subseteq \operatorname{int} D(y)$ *for all* $y \in Y$. *Then, for each* $t \in \mathbb{R}$ *and for any fixed* $w \in Y$, *we have the following properties:*

(a) $\operatorname{lev}_{\varphi_{-D(w),k},<}(t) = tk - \operatorname{int} D(w)$,
(b) $\operatorname{lev}_{\varphi_{-D(w),k},\leq}(t) = tk - D(w)$,
(c) $\operatorname{lev}_{\varphi_{-D(w),k},=}(t) = tk - \operatorname{bd} D(w)$,
(d) $\operatorname{lev}_{\varphi_{-D(w),k},\geq}(t) = Y \setminus \{y \in Y \mid y \in tk - \operatorname{int} D(w)\}$,
(e) $\operatorname{lev}_{\varphi_{-D(w),k},>}(t) = Y \setminus \{y \in Y \mid y \in tk - D(w)\}$.

***Proof*** The assertions immediately follow from Theorem 4.6.6 for fixed $w \in Y$ with $A = -D(w)$. □

In the following, we will see that the scalarizing functional (11.2.4) is lower semicontinuous, subadditive, homogenous, monotone and continuous under certain assumptions. These properties are important for deriving optimality conditions and generalizations of Ekeland's variational principle for vector optimization problems with a variable domination structure in Chap. 12.

**Corollary 11.2.19** *Suppose that assumption (A) and* $k \in 0^+D(y)$ *for all* $y \in Y$ *hold,* $B \subset Y$ *and* $w \in Y$.

*Then,* $\varphi_{-D(w),k}$ *in* (11.2.4) *is lower semicontinuous.*

*It holds that* $\varphi_{-D(w),k}(y + tk) = \varphi_{-D(w),k}(y) + t$ *for all* $t \in \mathbb{R}$, $y \in Y$ *and* $\varphi_{-D(w),k}$ *is* $B$*-monotone if* $D(w) + B \subseteq D(w)$.

$\varphi_{-D(w),k}$ *in* (11.2.4) *is positively homogeneous if and only if* $D(w)$ *is a cone.*

*Furthermore, if* $\varphi_{-D(w),k}$ *in* (11.2.4) *is proper, then it is subadditive if and only if* $D(w) + D(w) \subseteq D(w)$ *and* $\varphi_{-D(w),k}$ *is convex if and only if* $D(w)$ *is a convex set.*

***Proof*** The proof follows with Theorems 4.4.4 and 4.5.5 for $A = -D(w)$. □

Moreover, under a stronger assumption, we get continuity and strict monotonicity properties of the scalarization functional (11.2.4) (compare Corollary 4.6.6). These properties will be used in Sect. 11.4.2 for a characterization of approximate optimal solutions of vector optimization problems with a variable domination structure and also in Sect. 12.4.2 for deriving necessary optimality conditions for vector optimization problems with a variable domination structure.

**Corollary 11.2.20** *Suppose that* assumption *(A) holds, and additionally that* $D(y) + \mathbb{R}_{>}k \subseteq \operatorname{int} D(y)$ *for all* $y \in Y$ *and* $B \subset Y$. *Then, for each* $w \in Y$, *the following properties hold for* $\varphi_{-D(w),k}$ *defined in* (11.2.4)*:*

(a) $\varphi_{-D(w),k}$ *is continuous.*
(b) *If* $\varphi_{-D(w),k}$ *is finite-valued, then* $\varphi_{-D(w),k}$ *is* $B$*-monotone* $\Leftrightarrow D(w) + B \subseteq D(w) \Leftrightarrow \operatorname{bd} D(w) + B \subseteq D(w)$ *and,* $\varphi_{-D(w),k}$ *is strictly* $B$*-monotone* $\Leftrightarrow D(w) + (B\backslash\{0\}) \subseteq \operatorname{int} D(w) \Leftrightarrow \operatorname{bd} D(w) + (B\backslash\{0\}) \subseteq \operatorname{int} D(w)$.

***Proof*** The results follow from Theorem 4.6.6, Lemma 4.4.7 and Corollary 4.5.6 for $A = -D(w)$. □

The following result is important for a characterization of approximate solutions of vector optimization problems with variable domination structure via scalarization by means of the functional $\varphi_{-D(w),k}$ in Sect. 11.4.2.

**Lemma 11.2.21** *Suppose that* assumption *(A) holds, and additionally that* $D(y) + \mathbb{R}_{>}k \subseteq \operatorname{int} D(y)$ *for all* $y \in Y$. *Then, for all* $w \in Y$ *it holds that*

$$\varphi_{-D(w),k}(0) = 0.$$

***Proof*** Taking into account Corollary 11.2.18(c), it holds that

$$\varphi_{-D(w),k}(y) = t \Leftrightarrow y \in tk - \operatorname{bd} D(w). \tag{11.2.12}$$

This yields, for $y = 0$,

$$tk \in \operatorname{bd} D(w). \tag{11.2.13}$$

If we suppose $t > 0$, we get $tk \in \operatorname{int} D(w)$ since $D(w) + \mathbb{R}_{>}k \subseteq \operatorname{int} D(w)$ and $0 \in D(w)$, a contradiction to (11.2.13). So, we obtain $t \leq 0$. Furthermore, if we assume that $t < 0$, we get the existence of a real number $\lambda < 0$ with $0 \in \lambda k - D(w) \subseteq -\operatorname{int} D(w)$, a contradiction to the pointedness of $D(w)$. □

## *11.2.2 The Separating Functional* $\varphi_{a-D(\cdot),k}$

In this section, we discuss the extension (11.2.3) of the functional (4.1.11), where we replace the set $A \subset Y$ in (4.1.11) by a cone-valued map $D : Y \rightarrow \mathscr{P}(Y)$ (see Eichfelder [139, 143]). We suppose that assumption ($A_{cone}$) holds and additionally that $D(y)$ is pointed for all $y \in Y$.

**Assumption** ($A'_{cone}$). Let $Y$ be a topological vector space. Suppose $D : Y \rightarrow \mathscr{P}(Y)$, where $D(y)$ is a nontrivial, closed, pointed, convex cone for all $y \in Y$. Furthermore, we assume that $\emptyset \neq F \subset Y$, $a \in Y$ and $k \in \left(\bigcap_{y \in F} D(y)\right) \backslash\{0\}$.

Under assumption ($A'_{cone}$), we study properties of the functional $\varphi_{a-D(\cdot),k} : Y \to \overline{\mathbb{R}}$ (see (11.2.3)) given by

$$\varphi_{a-D(\cdot),k}(y) = \inf\{t \in \mathbb{R} \mid y \in tk + a - D(y)\} \quad (y \in Y).$$

First, we mention the following attainment property that follows immediately from Theorem 11.2.9 (see Eichfelder [143, Corollary 5.10]).

**Corollary 11.2.22** *Suppose that* $\emptyset \neq F \subset Y, a \in Y$ *and* $k \in \operatorname{int}\left(\bigcap_{y\in F} D(y)\right)\setminus\{0\}$. *Then,*

$$\varphi_{a-D(\cdot),k}(y) = \min\{t \in \mathbb{R} \mid y \in tk + a - D(y)\} \quad (y \in Y).$$

The following assertions by Eichfelder [143, Section 3.1.1] are very important for the examination of the convexity of the functional (11.2.3). In the following results, we are using a constant cone $C \subset Y$.

**Definition 11.2.23** Assume that $C$ is a nontrivial convex cone in a topological vector space $Y$, $X$ is a linear space and $S \subseteq X$ is convex. A set-valued map $D : S \to \mathscr{P}(Y)$ is called $C$-convex, if for all $x^1, x^2 \in S$ and $\lambda \in [0, 1]$

$$\lambda D(x^1) + (1-\lambda)D(x^2) \subseteq D(\lambda x^1 + (1-\lambda)x^2) + C. \tag{11.2.14}$$

The following result is shown by Eichfelder in [143, Lemma 3.3].

**Lemma 11.2.24** *Assume that* $C$ *is a nontrivial convex cone in a topological vector space* $Y$, $X$ *is a linear space and* $S \subseteq X$ *is algebraically open and convex. Let* $D : S \to \mathscr{P}(Y)$ *be a* $C$*-convex cone-valued map. Then,*

$$\forall x \in S : \quad D(x) + C = D(S) + C. \tag{11.2.15}$$

***Proof*** Since $D$ is a cone-valued map, we obtain $\lambda D(x) = D(x)$ and $0 \in D(x)$ for all $x \in S$ and all $\lambda > 0$. This yields

$$D(x^2) \subseteq D(x^2) + \lambda D(x^1) - \lambda D(x^2) \subseteq D(\lambda x^1 + (1-\lambda)x^2) + C \tag{11.2.16}$$

for all $x^1, x^2 \in S$, $\lambda \in (0, 1)$ taking into account (11.2.14).

Consider an element $\overline{x} \in \operatorname{core} S = S$ and choose $x \in S$ arbitrarily. Then, there is a real number $\mu > 0$ with

$$\overline{x} + \mu(\overline{x} - x) \in S.$$

Taking into account (11.2.16), we get

$$D(x) \subseteq D\left(\frac{1}{1+\mu}(\overline{x} + \mu(\overline{x} - x)) + \frac{\mu}{1+\mu}x\right) + C = D(\overline{x}) + C$$

such that

$$D(S) \subseteq D(\overline{x}) + C. \tag{11.2.17}$$

Since $S$ is algebraically open, there is some $\alpha > 0$ with

$$x + \alpha(x - \overline{x}) \in S.$$

Again, from the inclusion (11.2.16) we get

$$D(\overline{x}) \subseteq D\left(\frac{1}{1+\alpha}(x + \alpha(x - \overline{x})) + \frac{\alpha}{1+\alpha}\overline{x}\right) + C = D(x) + C,$$

and this implies

$$D(S) + C \subseteq D(\overline{x}) + C \subseteq D(x) + C \subseteq D(S) + C$$

taking into account (11.2.17) and since $C$ is a nontrivial convex cone. □

The following theorem is shown by Eichfelder [143, Theorem 3.4] using Lemma 11.2.24 for a cone-valued map $D : S \to \mathcal{P}(Y)$, where $S \subseteq X$ is algebraically open and convex. Under different assumptions concerning $D$, a corresponding result is shown in Theorem 11.2.36.

**Theorem 11.2.25** *Assume that $C$ is a nontrivial convex cone in a topological vector space $Y$, $X$ is a linear space and $S \subseteq X$ is algebraically open and convex. Suppose that $D : S \to \mathcal{P}(Y)$ is a cone-valued map with $D(x)$ a convex cone for all $x \in S$ and suppose*

$$C \subseteq \bigcap_{x \in S} D(x).$$

*If $D$ is $C$-convex in the sense of Definition 11.2.23, then $D$ is a constant cone-valued map, i.e., $D(x) = K$ for all $x \in S$ and for some convex cone $K \subset Y$.*

***Proof*** Consider $x \in S$. Then,

$$D(x) + C \subseteq D(x) + D(x) \subseteq D(x) \subseteq D(x) + C, \tag{11.2.18}$$

taking into account the assumptions on $C$, since $D(x)$ is a nontrivial convex cone and $0 \in C$. Formula (11.2.18) means $D(x) = D(x) + C$. Applying Lemma 11.2.24, this yields for all $x \in S$

$$D(x) = D(x) + C = D(S) + C = \bigcup_{x \in S}(D(x) + C) = \bigcup_{x \in S} D(x) = D(S) =: K.$$

□

**Remark 11.2.26** For a characterization of nondominated elements of a set $F \subset Y$ (see Sect. 11.3), it is important to know something about certain properties of the functional (11.2.3) like convexity. In order to check the convexity of $\varphi_{a-D(\cdot),k}$ for $D : Y \to \mathscr{P}(Y)$ (compare Eichfelder [139, Section 5.2.1]), we consider arbitrary elements $y^1, y^2 \in Y$ and $\lambda \in (0, 1)$ and we have to show that

$$\varphi_{a-D(\cdot),k}(\lambda y^1 + (1-\lambda)y^2) \leq \lambda\varphi_{a-D(\cdot),k}(y^1) + (1-\lambda)\varphi_{a-D(\cdot),k}(y^2). \quad (11.2.19)$$

Setting $t^i := \varphi_{a-D(\cdot),k}(y^i)$ $(i = 1, 2)$ and assuming $k \in \operatorname{int}\left(\bigcap_{y\in F} D(y)\right)$, we get

$$a + t^1k - y^1 \in D(y^1) \quad \text{and} \quad a + t^2k - y^2 \in D(y^2). \quad (11.2.20)$$

To check (11.2.19) means to check that

$$\varphi_{a-D(\cdot),k}(\lambda y^1 + (1-\lambda)y^2) \leq \lambda t^1 + (1-\lambda)t^2,$$

i.e., that

$$a + (\lambda t^1 + (1-\lambda)t^2)k - (\lambda y^1 + (1-\lambda)y^2) \in D(\lambda y^1 + (1-\lambda)y^2),$$

what is equivalent to

$$\lambda(a + t^1k - y^1) + (1-\lambda)(a + t^2 - y^2) \in D(\lambda y^1 + (1-\lambda)y^2). \quad (11.2.21)$$

Because of (11.2.20), we get

$$\lambda(a + t^1k - y^1) + (1-\lambda)(a + t^2 - y^2) \in \lambda D(y^1) + (1-\lambda)D(y^2). \quad (11.2.22)$$

Taking into account (11.2.22) and (11.2.21), we obtain that convexity of $\varphi_{a-D(\cdot),k}$ is given if

$$\lambda D(y^1) + (1-\lambda)D(y^2) \subseteq D(\lambda y^1 + (1-\lambda)y^2) \quad \text{for all} \quad \lambda \in (0, 1),\ y^1, y^2 \in Y,$$

this means, if $D(y)$ is a convex cone for all $y \in Y$. This is equivalent to $D(\cdot)$ being constant, taking into account Theorem 11.2.25. So, it is important to mention that the functional $\varphi_{a-D(\cdot),k}$ is not convex in general, especially if $D : Y \to \mathscr{P}(Y)$ is cone-valued. Furthermore, a discussion of the convexity of the functional $\varphi_{A+D(\cdot)}$ (see (11.2.5)) is given in Theorems 11.2.36, 11.2.40 and in Remark 11.2.38.

In Sect. 11.3, we apply the functional (11.2.3) for a characterization of solutions of vector optimization problems with variable domination structure.

### 11.2.3 The Separating Functional $\varphi_{A+D(\cdot)}$

In this section, we discuss the extension (11.2.5) of the nonlinear separation functional (4.1.11) studied by Bouza and Tammer in [70, 71]. Unless otherwise stated, we suppose in this section:

**Assumption** (A′). Let $Y$ be a Banach space. Suppose $D : Y \to \mathcal{P}(Y)$, where $D(y)$ is a proper, closed, convex and pointed set, $0 \in D(y)$, $D(y) \neq \{0\}$ for all $y \in Y$ and $A \subset Y$, $A \neq \emptyset$.

Now, under these assumptions, we use the functional $\varphi_{A+D(\cdot)} : Y \times (Y \setminus \{0\}) \to \overline{\mathbb{R}}$ on the product space $Y \times (Y \setminus \{0\})$ introduced in (11.2.5):

$$\varphi_{A+D(\cdot)}(y, k) = \inf\{t \in \mathbb{R} \mid y + tk \in A + D(y)\},$$

where $y \in Y$ and $k \in Y \setminus \{0\}$ (see Fig. 11.4). Again, we are using the convention $\inf \emptyset = +\infty$.

We introduce a concept of boundedness of the set $A$ in order to get assertions concerning the finiteness of $\varphi_{A+D(\cdot)}$.

**Definition 11.2.27** A subset $A$ of a Banach space $Y$ is said to be **bounded** if there is a real number $M > 0$ such that $A \subset \mathring{B}_{\|\cdot\|}(0, M)$.

The following result concerning the finiteness of the functional $\varphi_{A+D(\cdot)}$ is shown in [70, Proposition 3.2].

**Proposition 11.2.28** *Suppose that Assumption (A′) is fulfilled. Consider the functional* $\varphi_{A+D(\cdot)} : Y \times (Y \setminus \{0\}) \to \overline{\mathbb{R}}$, *given by* (11.3.10). *Then, the following assertions hold:*

(a) $\varphi_{A+D(\cdot)}(y, k) \leq 0$, *for all* $y \in A$, $k \in D(y) \setminus \{0\}$.
(b) *Assume that either*

(A′1) $A + D(y)$ *is a bounded set,*

*or*

(A′2) $\dim(Y) < +\infty$ *and* $A$ *bounded.*

*Then, for all* $y \in A$ *and* $k \in D(y) \setminus \{0\}$, *it holds that* $\varphi_{A+D(\cdot)}(y, k) \in \mathbb{R}$.

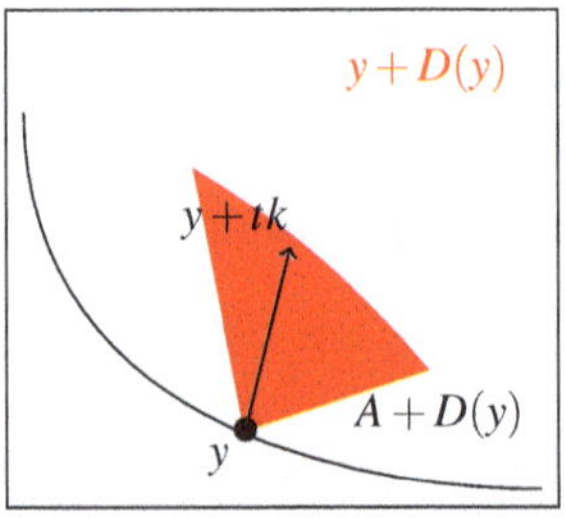


**Fig. 11.4** Construction of $\varphi_{A+D(\cdot)}(y, k) = \inf\{t \in \mathbb{R} \mid y + tk \in A + D(y)\}$ given in (11.2.5)

***Proof***

(a) Under our assumptions, we have $0 \in D(y)$. Because of $y \in A$, we get $y \in A + D(y)$. Hence, $\varphi_{A+D(\cdot)}(y,k) \leq 0 < +\infty$ for all $k \in D(y) \setminus \{0\}$, $y \in A$.

(b) By (a), it is sufficient to show that $\varphi_{A+D(\cdot)}(y,k) > -\infty$ for all fixed elements $y \in A$, $k \in D(y) \setminus \{0\}$. We define the set $T(y,k) := \{t \in \mathbb{R} \mid y + tk \in A + D(y)\}$ and consider the following two cases.

Case 1: Assume that (A'1) holds. Then, it is clear that the set $A + D(y) - y$ is bounded and, since $k \neq 0$, $T(y,k)$ is bounded. Hence, $\varphi_{A+D(\cdot)}(y,k) \in \mathbb{R}$ for all $y \in A$ and all $k \in D(y) \setminus \{0\}$.

Case 2: We assume that (A'2) holds, i.e., $A$ is bounded and $\dim(Y) < +\infty$. We take $\|\cdot\|$ as the Euclidean norm and consider a fixed element $y \in A$ and an arbitrary element $k \in D(y) \setminus \{0\}$. Then, the set $\{z \in A \mid z \in y + tk - D(y) \text{ for some } t \in \mathbb{R}\}$ is bounded since $A$ is bounded.
Assume that there are fixed elements $y \in A$ and $k \in D(y) \setminus \{0\}$ with $\varphi_{A+D(\cdot)}(y,k) = -\infty$. Then, there are sequences $\{t_n\}$, $t_n \to -\infty$, and $\{d_n\}$, $d_n \in D(y) \setminus \{0\}$, with

$$y + t_n k - d_n \in A \quad (n = 1, 2, \ldots).$$

Note that $\{t_n k - d_n\}$ is a bounded sequence since $A$ is bounded. This yields

$$\|d_n\| \to +\infty, \tag{11.2.23}$$

since $t_n \to -\infty$. W.l.o.g. we suppose that $\|k\| = 1$. We now define

$$u_n := \frac{d_n}{\|d_n\|} \in Y,$$

i.e., $d_n = \alpha_n u_n$ where $\alpha_n := \|d_n\| > 0$. Therefore, $\|u_n\| = \|k\| = 1$ and, $\alpha_n \to +\infty$ because of (11.2.23).
Taking into account our assumptions, we have $0 \in D(y)$ and $\alpha_n u_n = d_n \in D(y)$. Since $\alpha_n > 1$ for $n$ large enough, $u_n$ can be written as a convex combination of these two elements ($\alpha_n u_n$ and 0) of $D(y)$ as follows:

$$u_n = \frac{1}{\alpha_n}(\alpha_n u_n) + \left(1 - \frac{1}{\alpha_n}\right) 0.$$

Of course, $u_n \in D(y)$ since $D(y)$ is a convex set.
Moreover, for each limit point $u$ of the sequence $\{u_n\}$, we have $u \in D(y)$ because $D(y)$ is closed. Now, let $M > 0$ be an upper bound of $\{\| - t_n k + d_n\|^2\}$. Then,

$$\begin{aligned} M \geq \| - t_n k + d_n\|^2 &= t_n^2 - 2t_n\alpha_n \langle k, u_n\rangle + \alpha_n^2 \\ &= (t_n + \alpha_n)^2 - 2t_n\alpha_n(1 + \langle k, u_n\rangle) \geq -2t_n\alpha_n(1 + \langle k, u_n\rangle). \end{aligned}$$

Therefore,

$$M \geq -2t_n\alpha_n(1 + \langle k, u_n\rangle), \tag{11.2.24}$$

i.e., $\{-2t_n\alpha_n(1 + \langle k, u_n\rangle)\}$ is upper bounded. Furthermore, we have $(-2t_n)\alpha_n \to +\infty$ since $t_n \to -\infty$ and $\alpha_n \to +\infty$. Because of $\|u_n\| = \|k\| = 1$, by the Cauchy-Schwartz inequality, $(1 + \langle k, u_n\rangle) \geq 0$. Then, we get

$$\lim_{n\to+\infty} (1 + \langle k, u_n\rangle) = 0$$

because of (11.2.24), i.e., $\langle k, u\rangle = -1$. Taking into account $\|k\| = \|u\| = 1$, we get $u = -k \in -D(y) \setminus \{0\}$, a contradiction to the pointedness of $D(y)$ for all $y \in Y$.

So, there does not exist a sequence $\{t_n\}$, $t_n \in T(y, k)$ with $t_n \to -\infty$ and hence $\varphi_{A+D(\cdot)}(y, k) > -\infty$ for all fixed $y \in A$ and for all $k \in D(y) \setminus \{0\}$.

□

**Remark 11.2.29** (A′2) guarantees the finiteness of the functional $\varphi_{A+D(\cdot)}$ at $(y, k) \in A \times (D(y) \setminus \{0\})$ if $D$ is a cone-valued map.

Now, we show some essential properties of the separating functional $\varphi_{A+D(\cdot)}$ given in (11.2.5).

Assume that $X$, $Y$ are Banach spaces. We introduce $[a, z] := \{u = \alpha u + (1 - \alpha)z \mid \alpha \in [0, 1]\}$. As usual, we denote a neighborhood of $y \in Y$ by $V(y)$.

Here, we will use Berge's concepts of lower semicontinuity and upper semicontinuity (see Definition 8.2.1).

Recall that a functional $\varphi : X \to \overline{\mathbb{R}}$ ($X$ is a normed space) is lower semicontinuous (respectively upper semicontinuous) at $x^0 \in \operatorname{dom}\varphi$ if $\liminf_{x\to x^0} \varphi(x) \geq \varphi(x^0)$ ($\limsup_{x\to x^0} \varphi(x) \leq \varphi(x^0)$, respectively) (see Proposition 3.3.6).

The following sufficient condition for the lower semicontinuity of the functional $\varphi_{A+D(\cdot)}$ is shown in [70, Proposition 3.8].

**Proposition 11.2.30** *Suppose that Assumption (A') is fulfilled and $A + D(y)$ is convex for all $y \in Y$. Let $(\overline{y}, \overline{k}) \in A \times (Y \setminus \{0\})$ be fixed and assume that the map $A + D(\cdot)$ is closed at $\overline{y}$ in the sense of Definition 8.2.1, and that $\varphi_{A+D(\cdot)}(\overline{y}, \overline{k})$ is finite. Then, the functional $\varphi_{A+D(\cdot)} : Y \times (Y \setminus \{0\}) \to \overline{\mathbb{R}}$ is l.s.c. at $(\overline{y}, \overline{k})$.*

***Proof*** Consider $(\overline{y}, \overline{k}) \in A \times (Y \setminus \{0\})$. We suppose that $\varphi_{A+D(\cdot)}(\overline{y}, \overline{k})$ is finite, but $\varphi_{A+D(\cdot)}$ is not l.s.c. at $(\overline{y}, \overline{k})$. Taking into account Proposition 3.3.6, there are sequences $\{y_n\}$, $y_n \in Y$, $\{k_n\}$, $k_n \in Y$ and a real number $\varepsilon > 0$ such that $(y_n, k_n) \to (\overline{y}, \overline{k})$ and for some $n$

$$\varphi_{A+D(\cdot)}(y_n, k_n) < \varphi_{A+D(\cdot)}(\overline{y}, \overline{k}) - \varepsilon,$$

i.e., there is a real number $t_n$ such that

$$t_n < \varphi_{A+D(\cdot)}(\overline{y}, \overline{k}) - \varepsilon \tag{11.2.25}$$

and

$$y_n + t_n k_n \in A + D(y_n).$$

Because of $\overline{y} \in A$, we have $\varphi_{A+D(\cdot)}(\overline{y}, \overline{k}) \le 0$. With (11.2.25), we get

$$t_n < \varphi_{A+D(\cdot)}(\overline{y}, \overline{k}) - \varepsilon < 0. \tag{11.2.26}$$

Defining $\alpha_n := \frac{\varphi_{A+D(\cdot)}(\overline{y},\overline{k})-\varepsilon}{t_n}$, it holds that $\alpha_n \in (0, 1)$ because of (11.2.26).

Under our assumptions, it holds that $0 \in D(y)$. Hence, $\overline{y} \in A$ implies that $\overline{y} \in A + D(y)$ for all $y \in Y$. The convexity of $A + D(y)$ for all $y \in Y$ yields

$$\alpha_n(y_n + t_n k_n) + (1 - \alpha_n)\overline{y} \in A + D(y_n). \tag{11.2.27}$$

From (11.2.25) and (11.2.27), we get

$$\alpha_n y_n + (\varphi_{A+D(\cdot)}(\overline{y}, \overline{k}) - \varepsilon)k_n + (1 - \alpha_n)\overline{y} \in A + D(y_n). \tag{11.2.28}$$

Because of

$$\lim_{n\to+\infty} \alpha_n y_n + (1 - \alpha_n)\overline{y} = \overline{y},$$

$$\lim_{n\to+\infty} (\varphi_{A+D(\cdot)}(\overline{y}, \overline{k}) - \varepsilon)k_n = (\varphi_{A+D(\cdot)}(\overline{y}, \overline{k}) - \varepsilon)\overline{k}$$

and $A + D(y)$ is closed at $\overline{y}$, we obtain

$$\overline{y} + (\varphi_{A+D(\cdot)}(\overline{y}, \overline{k}) - \varepsilon)\overline{k} \in A + D(\overline{y}),$$

from (11.2.28). This is a contradiction to the definition of $\varphi_{A+D(\cdot)}$. This means, $\varphi_{A+D(\cdot)}$ is l.s.c. at $(\overline{y}, \overline{k})$. □

**Remark 11.2.31** The assumption that $A + D(y)$ is convex for all $y \in Y$ in Proposition 11.2.30 can be substituted by the assumption that there are a bounded set $U$ and a neighborhood $V$ of $\overline{y}$, such that for all $y \in V$, $A + D(y) \subset U$. As in the proof of Proposition 11.2.30, we again consider a sequence of the type $\{y_n + t_n k_n\}$, $y_n + t_n k_n \in A + D(y_n)$, where $t_n = \varphi_{A+D(\cdot)}(\overline{y}, \overline{k}) - \varepsilon_n$. It holds that $\varepsilon_n \ge \varepsilon > 0$ because of (11.2.25). $\varepsilon_n$ is bounded since $A + D(y)$ is bounded. Therefore, after taking a convergent subsequence it holds that $\varepsilon_n \to \varepsilon^* \ge \varepsilon > 0$. Taking the limit, we get $\overline{y} + (\varphi_{A+D(\cdot)}(\overline{y}, \overline{k}) - \varepsilon)\overline{k} \in A + D(\overline{y})$, a contradiction to the definition of the functional (11.2.5).

**Remark 11.2.32** For a fixed element $\overline{y} \in Y$, the lower semicontinuity of $\varphi_{A+D(\cdot)}(\overline{y}, \cdot)$ as a functional of $k$ at $\overline{k} \in Y \setminus \{0\}$ can be easily obtained, if $A + D(y)$ is closed and starshaped with respect to $\overline{y}$ (that is, if $\overline{y} + k \in D(y)$, then $\overline{y} + \alpha k \in D(y)$ for all $\alpha \in [0, 1]$).

The upper semicontinuity of $\varphi_{A+D(\cdot)} : Y \times (Y \setminus \{0\}) \to \overline{\mathbb{R}}$ in the next proposition is shown in [70, Proposition 3.11].

**Proposition 11.2.33** *Suppose that Assumption (A') is fulfilled. Moreover, let $Y = \mathbb{R}^\ell$ and $\overline{y} \in Y, \overline{k} \in Y \setminus \{0\}$. Assume that:*

(a) *$A + D(\cdot)$ is a B-l.s.c. map at $\overline{y}$ (in the sense of Definition 8.2.1).*
(b) *$A + D(y)$ is a convex set for all $y \in Y$.*
(c) *$\varphi_{A+D(\cdot)} : Y \times (Y \setminus \{0\}) \to \overline{\mathbb{R}}$ (given by (11.2.5)) is finite at $(\overline{y}, \overline{k})$.*
(d) *There is a real number $\varepsilon_0 > 0$ such that*

$$\forall t \in (\varphi_{A+D(\cdot)}(\overline{y}, \overline{k}), \varphi_{A+D(\cdot)}(\overline{y}, \overline{k}) + \varepsilon_0) : \quad \overline{y} + t\overline{k} \in \operatorname{int}(A + D(\overline{y})).$$

*Then, $\varphi_{A+D(\cdot)} : Y \times (Y \setminus \{0\}) \to \overline{\mathbb{R}}$ is u.s.c. at $(\overline{y}, \overline{k})$.*

***Proof*** By contradiction, assume that $\varphi_{A+D(\cdot)}$ is finite but not u.s.c. at $(\overline{y}, \overline{k})$. Then, there are a real number $\varepsilon > 0$ and sequences $\{y_n\},\ y_n \in Y,\ y_n \to \overline{y}$ and $\{k_n\},\ k_n \in Y, k_n \to \overline{k}$, such that for some $n$,

$$\varphi_{A+D(\cdot)}(y_n, k_n) > \varphi_{A+D(\cdot)}(\overline{y}, \overline{k}) + \varepsilon.$$

W.l.o.g., we suppose that $\varepsilon < \varepsilon_0$.
Consider $z = \overline{y} + t\overline{k},\ t \in \big[\varphi_{A+D(\cdot)}(\overline{y}, \overline{k}) + \varepsilon/2,\ \varphi_{A+D(\cdot)}(\overline{y}, \overline{k}) + \varepsilon\big)$. We obtain

$$z \in \operatorname{int}(A + D(\overline{y}))$$

because of (d). Furthermore, note that for the elements of the sequence $\{z_n\}$ with $z_n := y_n + tk_n$, it holds that $z_n \notin A + D(y_n)$ since otherwise the following (contradicting) inequality would be true:

$$\varphi_{A+D(\cdot)}(\overline{y}, \overline{k}) + \varepsilon < \varphi_{A+D(\cdot)}(y_n, k_n) \le t < \varphi_{A+D(\cdot)}(\overline{y}, \overline{k}) + \varepsilon.$$

Because of (b), $A + D(y_n)$ is convex and $z_n \notin A + D(y_n)$, we get from a classical separation argument (see [31, Theorem 2.4.4]), that there is a sequence $\{p_n\}$, $p_n \in Y^*$ (we have $Y^* = Y$ under the given assumptions), $p_n \neq 0,\ n = 1, 2, \ldots$ such that

$$\forall u \in A + D(y_n) : \quad \langle p_n, u \rangle \le \langle p_n, z_n \rangle.$$

W.l.o.g., we suppose that $\|p_n\| = 1$, and hence that $p_n \to p$.
Taking into account the B-l.s.c. of $A + D(\cdot)$ at $\overline{y}$ (supposed in (a)), if $u \in A + D(\overline{y})$, there is a sequence $\{u_n\}$ with $u_n \to u$, such that $u_n \in A + D(y_n)$. This yields

$$\forall u \in A + D(\overline{y}) : \qquad \langle p, u \rangle \le \langle p, z \rangle. \tag{11.2.29}$$

Since $z \in \operatorname{int}(A + D(\overline{y}))$, for some $\varepsilon > 0$ it holds that $u = z + \varepsilon p \in A + D(\overline{y})$. Therefore,

$$\langle p, z\rangle + \varepsilon\|p\|^2 \leq \langle p, z\rangle,$$

because of (11.2.29). We obtain $p = 0$, a contradiction to $\|p\| = \lim_{n\to+\infty}\|p_n\| = 1$. This means that $\varphi_{A+D(\cdot)}$ is an u.s.c. functional at $(\overline{y}, \overline{k})$. □

The following lemma will be useful for the analysis of Algorithm 1 in Sect. 13.6.1 and is shown in [70, Lemma 3.12].

**Lemma 11.2.34** *Assume that $Y = \mathbb{R}^\ell$, $\overline{y} \in Y$ and $D : Y \to \mathscr{P}(Y)$, where $D(y)$ is a proper, convex, closed, pointed set with $0 \in D(y)$, $D(y) \neq \{0\}$ for all $y \in Y$. If $D$ is B-l.s.c. at $\overline{y} \in Y$ and $d \in \operatorname{int} D(\overline{y})$, then there is a neighborhood $V(\overline{y})$ of $\overline{y}$ and a positive number $\delta$ such that for all $y \in V(\overline{y})$, it holds that $\mathring{B}_{\|\cdot\|}(d, \delta) \subset \operatorname{int} D(y)$.*

***Proof*** Suppose $\dim(Y) = \ell$ and define $E := \{e_1, \ldots, e_\ell, -e_1, \ldots, -e_\ell\}$, where $e_i$ is the vector whose i-th component is 1 and the other components are 0.

Let $d \in \operatorname{int} D(\overline{y})$. In order to simplify the notations, w.l.o.g. we suppose that

$$\forall e \in E: \quad d + e \in D(\overline{y}). \tag{11.2.30}$$

Let $\varepsilon > 0$ be fixed and consider the open ball $\mathring{B}_{\|\cdot\|}(d + e, \varepsilon)$. Taking into account (11.2.30), $\mathring{B}_{\|\cdot\|}(d + e, \varepsilon)$ satisfies

$$D(\overline{y}) \cap \mathring{B}_{\|\cdot\|}(d + e, \varepsilon) \neq \emptyset.$$

Then, the B-l.s.c. of $D$ at $\overline{y}$ yields that there is a neighborhood $V_\varepsilon^e(\overline{y})$ of $\overline{y}$ such that for all $y \in V_\varepsilon^e(\overline{y})$

$$D(y) \cap \mathring{B}_{\|\cdot\|}(d + e, \varepsilon) \neq \emptyset. \tag{11.2.31}$$

We define

$$V_\varepsilon := \bigcap_{e\in E} V_\varepsilon^e(\overline{y}). \tag{11.2.32}$$

$V_\varepsilon$ is an open neighborhood of $\overline{y}$ for all $\varepsilon > 0$ since $E$ is finite.

We now show the existence of $\delta > 0$ small enough, such that for all $y \in Y$ with $\|y - \overline{y}\| < \delta$, it holds that $d \in \operatorname{int} D(y)$. By contradiction, assume that there are sequences $\{y_n\}$, $y_n \in Y$, $y_n \to \overline{y}$ and $\{d_n\}$, $d_n \in Y$, $d_n \to d$, such that $d_n \notin D(y_n)$. W.l.o.g. we suppose that $y_n \in V_{1/n}$, for $n = 1, 2, \ldots$ where $V_{1/n}$ is defined in (11.2.32) for $\varepsilon = 1/n$.

By (11.2.31), we get

$$D(y_n) \cap \mathring{B}_{\|\cdot\|}\left(d + e, \frac{1}{n}\right) \neq \emptyset$$

since $V_{1/n} \subset V_{1/n}^e(\overline{y})$ for all $e \in E$. Let $\hat{d}_n(e) \in D(y_n) \cap \mathring{B}_{\|\cdot\|}(d + e, 1/n)$. Taking into account that $D(y_n)$ is closed and convex and $d_n \notin D(y_n)$, there is a sequence

$\{p_n\}$, $p_n \in Y^*$ (recall that $Y^* = Y$ since $Y$ is a finite-dimensional space), $\|p_n\| = 1$ which separates $d_n$ from $\hat{d}_n(e) \in D(y_n)$ ([31, Theorem 2.4.4]). This means, for all $e \in E$:

$$\langle p_n, d_n \rangle > \langle p_n, \hat{d}_n(e) \rangle.$$

Because of $\|p_n\| = 1$, there is a convergent subsequence. So, we can suppose that $p_n \to p$. We get that

$$\langle p, d \rangle \geq \langle p, d + e \rangle$$

since $\hat{d}_n(e) \to d + e$ for all $e \in E$.

From the definition of $E$ and the above inequality, it follows $\langle p, e \rangle = 0$ for all $e \in E$, which yields $p_i = 0$ for $i \in \{1, \ldots, \ell\}$. Therefore, $p = 0$, in contradiction to the fact that $\|p_n\| = 1$. Hence, there is a real number $\delta > 0$ such that for all $y \in \mathring{B}_{\|\cdot\|}(\overline{y}, \delta)$ it holds that

$$d \in \operatorname{int} D(y).$$

□

Now, the convexity of $\varphi_{A+D(\cdot)}$ is analyzed.

Under the assumption that $A \subset Y$ is a proper, closed, convex set and $D : Y \to \mathcal{P}(Y)$, where $D(y)$ is a pointed, convex and closed set with $0 \in D(y)$, $D(y) \neq \{0\}$ for all $y \in Y$, Bouza and Tammer [70, Proposition 3.13] have shown that a sufficient condition ensuring the convexity of the functional $\varphi_{A+D(\cdot)} : Y \times (Y \setminus \{0\}) \to \overline{\mathbb{R}}$ defined by (11.2.5) is that $D(\cdot)$ satisfies

$$\alpha D(y^1) + (1 - \alpha) D(y^2) \subseteq D(\alpha y^1 + (1 - \alpha) y^2) \tag{11.2.33}$$

for all $\alpha \in [0, 1]$ and for all $y^1, y^2 \in Y$. Taking into account the following theorem and the closedness of $D(y)$ for all $y \in Y$, it follows that $D(\cdot)$ is a constant map.

In Theorem 11.2.36, we will see that $D(\cdot)$ is constant under the closedness assumption concerning $D(\cdot)$ and (11.2.33).

**Definition 11.2.35** A proper set $D \subseteq Y$ is called **radiant** if $\lambda D \subseteq D$ for all $\lambda \in [0, 1]$.

The following assertion is shown by Doagooei, Le and Tammer [116].

**Theorem 11.2.36** *Let $D : Y \to \mathcal{P}(Y)$ be given and $D(y)$ is a proper set with $0 \in D(y)$, $D(y) \neq \{0\}$ for all $y \in Y$ and $\alpha D(y^1) + (1 - \alpha) D(y^2) \subseteq D(\alpha y^1 + (1 - \alpha) y^2)$ for all $\alpha \in [0, 1]$ and for all $y^1, y^2 \in Y$. Then,*

(a) *$D(y)$ is a radiant set for all $y \in Y$.*
(b) *$\operatorname{cl} D : Y \to \mathcal{P}(Y)$, defined by $\operatorname{cl} D(y) := \operatorname{cl}(D(y))$ for all $y \in Y$, is constant.*

***Proof***

(a) Let $y \in Y$ and $\lambda \in [0, 1]$ be arbitrary. Taking into account the assumptions $0 \in D(y)$ and (11.2.33), we get

$$\lambda D(y) \subseteq \lambda D(y) + (1 - \lambda) D(y) \subseteq D(y).$$

(b) Consider $x, u \in Y$, $\lambda \in [0, 1)$ arbitrarily, and put $y := \dfrac{u - \lambda x}{1 - \lambda}$. Under the given assumptions concerning $D(\cdot)$, we have

$$\lambda D(x) \subseteq \lambda D(x) + (1 - \lambda)D(y) \subseteq D(\lambda x + (1 - \lambda)y),$$

i.e., $\lambda D(x) \subseteq D(u)$ for $y = \dfrac{u - \lambda x}{1 - \lambda}$. If $\lambda$ goes to 1 from the left, we get $D(x) \subseteq \operatorname{cl} D(u)$, and hence $\operatorname{cl} D(x) \subseteq \operatorname{cl} D(u)$. Because $x$ and $u$ are arbitrarily chosen, we also get $\operatorname{cl} D(u) \subseteq \operatorname{cl} D(x)$. This means that $\operatorname{cl} D(\cdot)$ is a constant map.

□

**Remark 11.2.37** A corresponding result was shown by Eichfelder in [143, Theorem 3.4] (see Theorem 11.2.25 in Sect. 11.2.2) under different assumptions concerning the map $D$. Eichfelder supposed that $S \subseteq X$ is algebraically open and convex and $D : S \to \mathcal{P}(Y)$ is a cone-valued map.

**Remark 11.2.38** If $D : Y \to \mathcal{P}(Y)$ is a closed cone-valued map, then condition $\alpha D(y^1) + (1 - \alpha)D(y^2) \subseteq D(\alpha y^1 + (1 - \alpha)y^2)$ in (11.2.33) holds if and only if $D(y) = C$ for all $y \in Y$, where $C \subset Y$ is a (fixed) cone.

The next example (see Doagooei, Le and Tammer [116]). shows that a set-valued map $D : Y \to \mathcal{P}(Y)$ with $0 \in D(y)$, $D(y) \neq \{0\}$ for all $y \in Y$ and $\alpha D(y^1) + (1 - \alpha)D(y^2) \subseteq D(\alpha y^1 + (1 - \alpha)y^2)$ for all $\alpha \in [0, 1]$ and for all $y^1, y^2 \in Y$ is not necessarily constant.

**Example 11.2.39** Let $D : \mathbb{R} \rightrightarrows \mathbb{R}$ be given by

$$D(x) := [0, 1) \text{ if } x < 0 \text{ and } D(x) := [0, 1] \text{ if } x \geq 0.$$

Of course, $D(\cdot)$ satisfies the assumptions of Theorem 11.2.36 but it is not constant.

The next result is derived by Doagooei, Le and Tammer in [116].

**Theorem 11.2.40** *Let $A \subset Y$ be a proper, convex set and $D : Y \to \mathcal{P}(Y)$. Assume that $\overline{k} \in Y \setminus \{0\}$ and the map $D(\cdot)$ satisfies*

$$\alpha D(y^1) + (1 - \alpha)D(y^2) \subseteq D(\alpha y^1 + (1 - \alpha)y^2)$$

*for all $\alpha \in [0, 1]$ and for all $y^1, y^2 \in Y$. Consider the functional $\varphi_{A+D(\cdot)} : Y \times (Y \setminus \{0\}) \to \overline{\mathbb{R}}$ defined by (11.2.5). Then, $\varphi_{A+D(\cdot)}(\cdot, \overline{k})$ is convex.*

***Proof*** Let $\overline{k} \in Y \setminus \{0\}$ be fixed. Because of Definition 3.4.1, we have to prove that the epigraph of $\varphi_{A+D(\cdot)}(\cdot, \overline{k})$ (see (3.2.4)) is convex. We consider $\operatorname{epi}_< \varphi_{A+D(\cdot)}(\cdot, \overline{k}) := \{(y, \xi) \in Y \times \mathbb{R} \mid \varphi_{A+D(\cdot)}(y, \overline{k}) < \xi\}$. Furthermore, let $(y^1, \xi_1)$, $(y^2, \xi_2)$ be arbitrary elements of $\operatorname{epi}_< \varphi_{A+D(\cdot)}(\cdot, \overline{k})$.
Obviously, $\varphi_{A+D(\cdot)}(y^1, \overline{k}) < \xi_1$ and $\varphi_{A+D(\cdot)}(y^2, \overline{k}) < \xi_2$. By definition of $\varphi_{A+D(\cdot)}$

$(y^1, \overline{k})$, there exits $t_1 \in \mathbb{R}, t_1 < \xi_1$ such that $y^1 + t_1\overline{k} \in A + D(y^1)$. Analogously, we get the existence of $t_2 \in \mathbb{R}$, $t_2 < \xi_2$ such that $y^2 + t_2\overline{k} \in A + D(y^2)$. Suppose $\alpha \in [0, 1]$. Because of the convexity of $A$ and the property $\alpha D(y^1) + (1 - \alpha)D(y^2) \subseteq D(\alpha y^1 + (1 - \alpha)y^2)$ for all $\alpha \in [0, 1]$ and for all $y^1, y^2 \in Y$, we get

$$\begin{aligned} \alpha y^1 + (1 - \alpha)y^2 + (\alpha t_1 + (1 - \alpha)t_2)\overline{k} &\in A + \alpha D(y^1) + (1 - \alpha)D(y^2) \\ &\subseteq A + D(\alpha y^1 + (1 - \alpha)y^2). \end{aligned} \tag{11.2.34}$$

This yields $(\alpha y^1 + (1 - \alpha)y^2, \alpha\xi_1 + (1 - \alpha)\xi_2) \in \text{epi}_< \varphi_{A+D(\cdot)}(\cdot, \overline{k})$, i.e., $\text{epi}_< \varphi_{A+D(\cdot)}(\cdot, \overline{k})$ is convex. Then, $\text{epi}\, \varphi_{A+D(\cdot)}(\cdot, \overline{k})$ is convex too (see [508, Theorem 2.1.1]). Taking into account Definition 3.4.1, $\varphi_{A+D(\cdot)}(\cdot, \overline{k})$ is convex. □

Now, we study the quasiconvexity of the functional $\varphi_{A+D(\cdot)}(y, \cdot)$. This result is shown in [70, Proposition 3.13].

**Proposition 11.2.41** *Suppose that Assumption (A′) is fulfilled and, additionally, $A \subset Y$ is a proper, convex and closed set. Consider the functional $\varphi_{A+D(\cdot)} : Y \times (Y \setminus \{0\}) \to \overline{\mathbb{R}}$ defined by* (11.2.5).

*Then, for all fixed elements $y \in A$, $\varphi_{A+D(\cdot)}(y, \cdot)$ is a quasiconvex functional of $k$ ($k \in D(y) \setminus \{0\}$).*

***Proof*** We look at a fixed element $y \in A$ and $k^1, k^2 \in Y \setminus \{0\}$ arbitrarily chosen. Taking into account Proposition 11.2.28, we have

$$\forall\, k \in Y \setminus \{0\} : \quad \varphi_{A+D(\cdot)}(y, k) \leq 0, \tag{11.2.35}$$

since $D(y)$ is pointed with $0 \in D(y)$ and $y \in A$. We get

$$\varphi_{A+D(\cdot)}(y, k^1), \varphi_{A+D(\cdot)}(y, k^2) < +\infty.$$

Because of the definition of $\varphi_{A+D(\cdot)}$, there are two sequences $\{t_n^1\} \in \mathbb{R}$ and $\{t_n^2\} \in \mathbb{R}$ such that $t_n^i \to \varphi_{A+D(\cdot)}(y, k^i)$, $i = 1, 2$.

Using (11.2.35), in the case $\varphi_{A+D(\cdot)}(y, k^i) = 0$ for some $i$, $i = 1, 2$, we obtain

$$\varphi_{A+D(\cdot)}(y, \alpha k^1 + (1 - \alpha)k^2) \leq 0 = \max\{\varphi_{A+D(\cdot)}(y, k^1), \varphi_{A+D(\cdot)}(y, k^2)\}.$$

We now assume that $\varphi_{A+D(\cdot)}(y, k^1), \varphi_{A+D(\cdot)}(y, k^2) < 0$ and, therefore, w.l.o.g. that $t_n^i < 0$, for $i = 1, 2$. From the convexity of $A$ and $D(y)$ for all $y \in A$, we conclude that $A + D(y)$ is a convex set for all $y \in A$ too. Because of $y + t_n^i k^i \in A + D(y)$ ($i = 1, 2$) and $y \in A + D(y)$, we get

$$\forall \alpha \in [0, 1] : \; y + \alpha t_n^i k^i \in A + D(y).$$

Taking into account $t_n^i < 0$, $i = 1, 2$, it holds that

$$\frac{\max\{t_n^1, t_n^2\}}{t_n^i} \in [0, 1].$$

Then,

$$y + \max\{t_n^1, t_n^2\}k^i \in A + D(y)$$

for $i = 1, 2$ and for all $y \in A$.

From the convexity of $A + D(y)$ for all $y \in A$, we conclude

$$y + \max\{t_n^1, t_n^2\}(\alpha k^1 + (1 - \alpha)k^2) \in A + D(y)$$

for all $\alpha \in [0, 1]$.

Therefore, $\varphi_{A+D(\cdot)}(y, \alpha k^1 + (1 - \alpha)k^2) \leq \max\{t_n^1, t_n^2\}$ and, taking limits as $n \to +\infty$, we get

$$\varphi_{A+D(\cdot)}(y, \alpha k^1 + (1 - \alpha)k^2) \leq \max\{\varphi_{A+D(\cdot)}(y, k^1), \varphi_{A+D(\cdot)}(y, k^2)\},$$

and the proof is complete. □

As already mentioned, monotonicity properties of scalarizing functionals are very important for characterizations of solutions to vector optimization problems (see Chap. 6). Now, we discuss the homogeneity and certain monotonicity properties of $\varphi_{A+D(\cdot)}(\cdot, k)$ as a functional of $y \in Y$, for fixed $k \in Y \setminus \{0\}$.

**Definition 11.2.42** $\varphi_{A+D(\cdot)} : Y \times (Y \setminus \{0\}) \to \overline{\mathbb{R}}$ is called a **D(·)-decreasing functional in y ∈ Y** if, for all $d \in D(y)$, the inequality $\varphi_{A+D(\cdot)}(y + d, k) \leq \varphi_{A+D(\cdot)}(y, k)$ is satisfied for all $k \in Y \setminus \{0\}$.

Under the assumption that for all $y \in Y$, $\operatorname{int} D(y) \neq \emptyset$, $\varphi_{A+D(\cdot)}$ is called a **strictly D(·)-decreasing functional in y ∈ Y** if, for all $d \in \operatorname{int} D(y)$, the strict inequality $\varphi_{A+D(\cdot)}(y + d, k) < \varphi_{A+D(\cdot)}(y, k)$ is fulfilled for all $k \in Y \setminus \{0\}$.

The following result concerning the homogeneity and monotonicity of $\varphi_{A+D(\cdot)}(\cdot, k)$ as a functional of $y \in Y$, for fixed $k \in Y \setminus \{0\}$ is shown in [70, Proposition 3.16].

**Proposition 11.2.43** *Suppose that Assumption (A') is fulfilled. Moreover, suppose that $A \subset Y$ is a proper closed set. Consider the functional $\varphi_{A+D(\cdot)} : Y \times (Y \setminus \{0\}) \to \overline{\mathbb{R}}$ defined by* (11.2.5).

(a) *Let $A$ be a cone and suppose that $D(\lambda y) = \lambda D(y)$ for all $y \in Y$ and $\lambda > 0$. Then, for all $\overline{k} \in Y \setminus \{0\}$ and for all $y \in Y$ with $\varphi_{A+D(\cdot)}(y, \overline{k}) \in [-\infty, +\infty)$, it holds that $\varphi_{A+D(\cdot)}(\lambda y, \overline{k}) = \lambda \varphi_{A+D(\cdot)}(y, \overline{k})$.*
(b) *Assume that $k \in Y \setminus \{0\}$ is fixed, $\varphi_{A+D(\cdot)}$ is finite-valued and that $d + D(y) \subseteq D(y + d)$ for all $y \in Y$ and for all $d \in D(y)$. Then, $\varphi_{A+D(\cdot)}$ is a $D(\cdot)$-decreasing functional in $y$.*

(c) *Assume that* int $D(y) \neq \emptyset$ *and* $k \in D(y) \setminus \{0\}$ *for all* $y \in Y$. *Suppose that the assumptions of (b) are fulfilled and, additionally, that* $D(y) \subseteq D(y + \lambda d)$ *for all* $y \in Y$, $d \in D(y)$ *and* $\lambda \geq 0$. *Then,* $\varphi_{A+D(\cdot)}(\cdot, k)$ *is strictly* $D(\cdot)$*-decreasing in* $y$.

(d) *Assume that* $\varphi_{A+D(\cdot)}$ *is finite-valued. For all* $y \in Y$, *we suppose that* $D(y + \lambda \bar{k}) = D(y)$ *for all* $\bar{k} \in D(y) \setminus \{0\}$ *and* $\lambda \in \mathbb{R}$. *Then, the translation invariance along* $\bar{k}$

$$\varphi_{A+D(\cdot)}(y + \lambda \bar{k}, \bar{k}) = \varphi_{A+D(\cdot)}(y, \bar{k}) - \lambda$$

*is satisfied for all* $y \in Y$, $\bar{k} \in D(y) \setminus \{0\}$ *and* $\lambda \in \mathbb{R}$.

***Proof***

(a) First, we assume that $\varphi_{A+D(\cdot)}(y, \bar{k}) \in \mathbb{R}$. Moreover, we suppose that $\lambda > 0$. Taking into account the definition of $\varphi_{A+D(\cdot)}$, there are sequences $\{\varepsilon_n\}$, $\varepsilon_n \in \mathbb{R}$, $\{y_n\}$, $y_n \in A$ and $\{d_n\}$, $d_n \in D(y)$ such that $\varepsilon_n \to 0$, $y = y_n - (\varphi_{A+D(\cdot)}(y, \bar{k}) + \varepsilon_n)\bar{k} + d_n$. For $\lambda > 0$, we obtain

$$\lambda y = \lambda y_n - \lambda(\varphi_{A+D(\cdot)}(y, \bar{k}) + \varepsilon_n)\bar{k} + \lambda d_n.$$

Our assumptions guarantee that $\lambda d_n \in D(\lambda y)$ and $\lambda y_n \in A$, therefore

$$\varphi_{A+D(\cdot)}(\lambda y, \bar{k}) \leq \lambda(\varphi_{A+D(\cdot)}(y, \bar{k}) + \varepsilon_n).$$

Taking the limit for $n \to +\infty$, we get

$$\varphi_{A+D(\cdot)}(\lambda y, \bar{k}) \leq \lambda \varphi_{A+D(\cdot)}(y, \bar{k}). \tag{11.2.36}$$

Furthermore, for $\lambda > 0$ we consider sequences $\{\varepsilon_n^\lambda\}$, $\varepsilon_n^\lambda \in \mathbb{R}$, $\varepsilon_n^\lambda \to 0$, $\{d_n^\lambda\}$, $d_n^\lambda \in D(\lambda y)$ and $\{y_n^\lambda\}$, $y_n^\lambda \in A$ such that

$$\lambda y = y_n^\lambda - (\varphi_{A+D(\cdot)}(\lambda y, \bar{k}) + \varepsilon_n^\lambda)\bar{k} + d_n^\lambda.$$

With $\lambda > 0$, we get

$$y = \left(\frac{y_n^\lambda}{\lambda} - \frac{(\varphi_{A+D(\cdot)}(\lambda y, \bar{k}) + \varepsilon_n^\lambda)}{\lambda}\bar{k} + \frac{d_n^\lambda}{\lambda}\right). \tag{11.2.37}$$

Under the given assumptions it holds that $\frac{d_n^\lambda}{\lambda} \in D(y)$ and $\frac{y_n^\lambda}{\lambda} \in A$. This yields together with (11.2.37)

$$\varphi_{A+D(\cdot)}(y, \bar{k}) \leq \frac{\varphi_{A+D(\cdot)}(\lambda y, \bar{k})}{\lambda} + \frac{\varepsilon_n^\lambda}{\lambda}.$$

For $n \to +\infty$, we get

$$\varphi_{A+D(\cdot)}(y, \bar{k}) \le \frac{\varphi_{A+D(\cdot)}(\lambda y, \bar{k})}{\lambda}. \tag{11.2.38}$$

Because of (11.2.36) and (11.2.38), we obtain

$$\varphi_{A+D(\cdot)}(\lambda y, \bar{k}) = \lambda \varphi_{A+D(\cdot)}(y, \bar{k})$$

for $\lambda > 0$.

Now, we consider the case $\varphi_{A+D(\cdot)}(y, \bar{k}) = -\infty$ for some $y \in Y$. Then, there are sequences $\{t_n\},\ t_n \in \mathbb{R},\ \{y_n\},\ y_n \in A$, and $\{d_n\},\ d_n \in D(y)$ such that $t_n \to -\infty$ and

$$y = y_n - t_n\bar{k} + d_n.$$

For $\lambda > 0$, we get $\lambda y = \lambda y_n - \lambda t_n\bar{k} + \lambda d_n$, where $\lambda y_n \in A$ and $\lambda d_n \in D(\lambda y)$. Because of $t_n \to -\infty$ and $\lambda > 0$ is fixed, we have $\lambda t_n \to -\infty$. This implies $\varphi_{A+D(\cdot)}(\lambda y, \bar{k}) = -\infty$.

(b) We look at $y \in Y,\ k \in Y \setminus \{0\}$ and $d \in D(y)$. Because of the definition of $\varphi_{A+D(\cdot)}$, there are sequences $\{t_n\},\ t_n \in \mathbb{R}, t_n \to \varphi_{A+D(\cdot)}(y, k), \{y_n\},\ y_n \in A$ and $\{d_n\},\ d_n \in D(y)$, such that

$$y + t_n k = y_n + d_n.$$

Of course, this implies

$$y + d + t_n k = y_n + d_n + d.$$

Since $d \in D(y)$ yields $d + D(y) \subseteq D(y + d)$ and $d_n \in D(y)$, we obtain $d_n + d \in D(y + d)$. Then,

$$y + d + t_n k \in A + D(y + d)$$

and

$$\varphi_{A+D(\cdot)}(y + d, k) \le t_n. \tag{11.2.39}$$

For $n \to +\infty$, we get from (11.2.39)

$$\varphi_{A+D(\cdot)}(y + d, k) \le \varphi_{A+D(\cdot)}(y, k),$$

i.e., $\varphi_{A+D(\cdot)}(\cdot, k)$ is $D(\cdot)$-decreasing in $y$.

(c) First, for $\lambda > 0$ and $k \in D(y)$ for all $y \in Y$ we will show that

$$\varphi_{A+D(\cdot)}(y + \lambda k, k) \le \varphi_{A+D(\cdot)}(y, k) - \lambda. \tag{11.2.40}$$

Indeed, because $\varphi_{A+D(\cdot)}$ is finite-valued, there are sequences $\{t_n\},\ t_n \in \mathbb{R}, t_n \to \varphi_{A+D(\cdot)}(y,k), \{y_n\},\ y_n \in A$, and $\{d_n\},\ d_n \in D(y)$ with

$$y + t_n k = y_n + d_n,$$

or, equivalently,

$$y + \lambda k + (t_n - \lambda)k = y_n + d_n.$$

Taking into account the assumption $D(y) \subseteq D(y + \lambda k)$, we get

$$y + \lambda k + (t_n - \lambda)k \in A + D(y + \lambda k).$$

This yields

$$\varphi_{A+D(\cdot)}(y + \lambda k, k) \le t_n - \lambda.$$

For $n \to +\infty$, we obtain (11.2.40).

We now take $d \in \operatorname{int} D(y)$. Then, there is $\varepsilon > 0$ such that $d - \varepsilon k \in \operatorname{int} D(y)$. Because of the monotonicity of $\varphi_{A+D(\cdot)}$, we obtain

$$\varphi_{A+D(\cdot)}(y + d - \varepsilon k, k) \le \varphi_{A+D(\cdot)}(y,k). \tag{11.2.41}$$

Taking into account (11.2.40) and the monotonicity condition (11.2.41), we get

$$\varphi_{A+D(\cdot)}(y+d,k) = \varphi_{A+D(\cdot)}(y+d-\varepsilon k+\varepsilon k,k) \le \varphi_{A+D(\cdot)}(y+d-\varepsilon k,k) - \varepsilon \le \varphi_{A+D(\cdot)}(y,k) - \varepsilon$$

$$< \varphi_{A+D(\cdot)}(y,k),$$

i.e., we have shown the strict monotonicity of $\varphi_{A+D(\cdot)}(\cdot,k)$ in $y$.

(d) Suppose that $D(y + \lambda\overline{k}) = D(y)$ for all $y \in Y, \overline{k} \in D(y) \setminus \{0\}$ and $\lambda \in \mathbb{R}$. Analogously to the proof of (c), we obtain for all $y \in Y, \overline{k} \in D(y) \setminus \{0\}$ and $\lambda \in \mathbb{R}$

$$\varphi_{A+D(\cdot)}(y + \lambda\overline{k}, \overline{k}) \le \varphi_{A+D(\cdot)}(y, \overline{k}) - \lambda. \tag{11.2.42}$$

Then,

$$\varphi_{A+D(\cdot)}(y - \lambda\overline{k}, \overline{k}) = \varphi_{A+D(\cdot)}(y + (-\lambda)\overline{k}, \overline{k}) \le \varphi_{A+D(\cdot)}(y, \overline{k}) - (-\lambda) = \varphi_{A+D(\cdot)}(y, \overline{k}) + \lambda$$

and setting $y := y + \lambda\overline{k}$,

$$\varphi_{A+D(\cdot)}(y, \overline{k}) \le \varphi_{A+D(\cdot)}(y + \lambda\overline{k}, \overline{k}) + \lambda. \tag{11.2.43}$$

From (11.2.42) and (11.2.43), we get the desired result. □

**Remark 11.2.44** If $D(y)$ is a convex cone for all $y \in Y$, the assumption of Proposition 11.2.43(b) is $D(y) \subseteq D(y + d)$ for all $d \in D(y)$. This is equivalent to

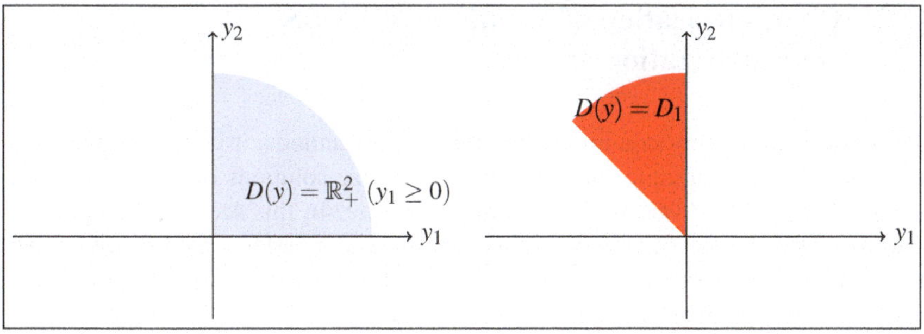


**Fig. 11.5** The domination map $D$ involving the cones $\mathbb{R}^2_+$ and $D_1$ in Example 11.2.46

the assumption of (c). Furthermore, the equality between both sets must be fulfilled in (d).

In the following examples, we will present non-constant set-valued maps which fulfill the hypotheses of Proposition 11.2.43. These examples are given in [70, Example 3.18, Example 3.19]. We start with the homogeneity.

**Example 11.2.45** Consider a bi-linear operator $B : Y \times Y \to \mathbb{R}$ and let the domination structure be given by $D(y) = \{z \in Y : B(y, z) \leq 0\}$. We obtain $D(y) = D(\lambda y)$ for all $\lambda > 0$.

Of course, if $D(y + d) = D(y)$ for all $d \in D(y)$, the conditions assumed in Proposition 11.2.43(b)–(d) hold. Taking into account the close relationship between them, the next example presents a domination structure for which this stronger property is fulfilled.

**Example 11.2.46** Let $D_1 := \left\{(r\cos(\theta), r\sin(\theta)) \mid r \geq 0, \theta \in \left[\frac{\pi}{2}, \frac{5\pi}{4}\right]\right\}$

$$D(y) := \begin{cases} \mathbb{R}^2_+ & \text{if } y_1 \geq 0 \\ D_1 & \text{otherwise.} \end{cases}$$

The two cones which define the variable domination structure in Example 11.2.46 are shown in Fig. 11.5.

For the case $y_1 \geq 0$, and $d \in D(y)$, it holds that $d \geq 0$ and therefore, $d_1 + y_1 \geq 0$. So, we have $D(y + d) = \mathbb{R}^2_+ = D(y)$. In the other case, i.e., if $y_1 < 0$, $d \in D(y)$ yields $d_1 \leq 0$. Hence, $y_1 + d_1 < 0$ and again $D(y + d) = D(y)$.

An algorithm for computing weakly nondominated-like elements of a geometric vector optimization problem with variable domination structure based on the properties of the functional $\varphi_{A+D(\cdot)} : Y \times (Y \setminus \{0\}) \to \overline{\mathbb{R}}$ presented in this section is given in Sect. 13.6.

## 11.3 Characterization of Solutions to VVDS via Scalarization

For deriving optimality conditions and numerical methods, it is very important to use scalarization techniques in order to characterize solutions of vector optimization problems with a variable domination structure. In this section, we apply the scalarizing functional (11.2.3) for a characterization of nondominated elements. Furthermore, a characterization of nondominated-like elements to vector optimization problems with variable domination structure by means of the functionals (11.2.4) and (11.2.5) is presented. Later on in Sect. 11.4.2, we will use the functional (11.2.4) for characterizations of three types of (approximate) solutions to vector optimization problems with a variable domination structure.

In order to apply the nonlinear separating functional (11.2.3) for a characterization of nondominated and nondominated-like elements to vector optimization problems with a variable domination structure, we suppose (compare Sect. 11.2.2) that the maps $D : Y \to \mathcal{P}(Y)$ and $\tilde{D} : Y \to \mathcal{P}(Y)$ are cone-valued and additionally we assume that $\emptyset \neq F \subset Y, a \in Y$.

In the following, we give characterizations of nondominated elements of a set $F \subset Y$ with respect to a variable domination structure by means of the nonlinear scalarizing functional (11.2.3) introduced in Sect. 11.2.

We use the functional (11.2.3) for $a := \overline{y} \in Y$ and $k \in \left(\bigcap_{y\in F} D(y)\right) \setminus\{0\}$:

$$\varphi_{\overline{y}-D(\cdot),k}(y) = \inf\{t \in \mathbb{R} \mid y \in tk + \overline{y} - D(y)\} \quad \text{for all } y \in Y. \tag{11.3.1}$$

The following characterization of nondominated elements in the sense of Definition 11.1.2 is shown by Eichfelder in [139, Theorem 5.11], [143, Theorem 5.11].

**Theorem 11.3.1** (Characterization of nondominated elements via scalarization)
*Let $Y$ be a topological vector space. Suppose $D : Y \to \mathcal{P}(Y)$, where $D(y)$ is a nontrivial, closed, convex cone for all $y \in Y$.*

*(a) Let $k \in \left(\bigcap_{y\in F} D(y)\right) \setminus \{0\}$. Then, $\overline{y} \in \mathrm{N}(F, D(\cdot))$ if and only if*

$$\varphi_{\overline{y}-D(\cdot),k}(y) > \varphi_{\overline{y}-D(\cdot),k}(\overline{y}) = 0 \quad \textit{for all} \quad y \in F \setminus \{\overline{y}\}. \tag{11.3.2}$$

*(b) Let $k \in \operatorname{int}\left(\bigcap_{y\in F} D(y)\right)$. Then, $\overline{y} \in \mathrm{WN}(F, D(\cdot))$ if and only if*

$$\varphi_{\overline{y}-D(\cdot),k}(y) \geq \varphi_{\overline{y}-D(\cdot),k}(\overline{y}) = 0 \quad \textit{for all} \quad y \in F. \tag{11.3.3}$$

***Proof***

(a) $[\Longrightarrow]$ : Consider $\overline{y} \in \mathrm{N}(F, D(\cdot))$. Suppose that for $\overline{y}$, (11.3.2) is not fulfilled. Then, there is some $t \in \mathbb{R}$, $t \leq 0$ such that for some $y \in F \setminus \{\overline{y}\}$ it holds that

$$\overline{y} + tk - y \in D(y).$$

This yields

$$\overline{y} \in y + (D(y) \setminus \{0\}),$$

taking into account that $tk \in -D(y)$ and $D(y)$ is a convex cone for each $y \in Y$. This is a contradiction to $\overline{y} \in \mathrm{N}(F, D(\cdot))$.

$[\Longleftarrow]$ : Suppose that (11.3.2) is fulfilled for $\overline{y}$. If there is some $y \in F \setminus \{\overline{y}\}$ with $\overline{y} \in y + D(y)$, then

$$\overline{y} + 0k - y \in D(y).$$

This means $\varphi_{\overline{y}-D(\cdot),k}(y) \leq 0$, a contradiction to (11.3.2).

(b) $[\Longrightarrow]$ : Consider $\overline{y} \in \mathrm{WN}(F, D(\cdot))$. Suppose that for $\overline{y}$, (11.3.3) is not fulfilled. Then, there is some $t \in \mathbb{R}$, $t < 0$ such that for some $y \in F \setminus \{\overline{y}\}$, it holds that

$$\overline{y} + tk - y \in D(y).$$

This yields

$$\overline{y} \in y + \operatorname{int} D(y)$$

since $tk \in -\operatorname{int} D(y)$ and $D(y)$ is a convex cone for each $y \in Y$. This is a contradiction to $\overline{y} \in \mathrm{WN}(F, D(\cdot))$.

$[\Longleftarrow]$ : Suppose that (11.3.3) is fulfilled for $\overline{y}$. If there is some $y \in F$ with $\overline{y} \in y + \operatorname{int} D(y)$, then this implies that there is some $t < 0$ such that

$$\overline{y} + tk - y \in D(y)$$

and hence $\varphi_{\overline{y}-D(\cdot),k}(y) < 0$, a contradiction to (11.3.3). □

Differently to the characterization of nondominated elements using the functional (11.3.1) in Theorem 11.3.1, a characterization of nondominated-like elements $\overline{y}$ to $F$ with respect to a variable domination structure in the sense of Definition 11.1.3(a) is given in the following theorem using a scalarization by means of the functional $\varphi_{a-K,k}$, where $a \in Y$, $K$ is a cone in $Y$ with certain properties related to $\tilde{D}(\overline{y})$ and $k \in \operatorname{int} K$ (compare (4.1.11) with $A = a - K$):

$$\varphi_{a-K,k}(y) := \inf\{t \in \mathbb{R} \mid y \in tk + a - K\} \qquad (y \in Y).$$

This result is shown by Eichfelder in [139, 143, Theorem 5.13].

**Theorem 11.3.2** (Characterization of nondominated-like elements via scalarization) *Let $Y$ be a topological vector space. Suppose $\tilde{D} : Y \to \mathscr{P}(Y)$, where $\tilde{D}(y)$ is a nontrivial, closed, convex cone for all $y \in Y$, $F \subset Y$ and let $\overline{y} \in F$.*

(a) *If $\overline{y} \in \mathrm{M}(F, \tilde{D}(\cdot))$, then for any nontrivial, closed, convex, pointed cone $K \subseteq \tilde{D}(\overline{y})$ and any $k \in K \setminus \{0\}$ it holds that*

$$\varphi_{\overline{y}-K,k}(y) > \varphi_{\overline{y}-K,k}(\overline{y}) = 0 \quad \textit{for all} \quad y \in F \setminus \{\overline{y}\}. \tag{11.3.4}$$

(b) *If for* $a \in Y$, *a nontrivial, closed, convex, pointed cone* $K$ *with* $\tilde{D}(\overline{y}) \subseteq K$ *and* $k \in K \setminus \{0\}$ *it holds that*

$$\varphi_{a-K,k}(y) > \varphi_{a-K,k}(\overline{y}) \quad \textit{for all} \quad y \in F \setminus \{\overline{y}\}, \tag{11.3.5}$$

*then* $\overline{y} \in \mathrm{M}(F, \tilde{D}(\cdot))$.

***Proof***

(a) Suppose $\overline{y} \in \mathrm{M}(F, \tilde{D}(\cdot))$, but (11.3.4) does not hold. Then, there are $t \in \mathbb{R}$, $t \leq 0$ and $y \in F \setminus \{\overline{y}\}$ such that $\overline{y} + tk - y \in K$. This implies

$$y \in \overline{y} - (K \setminus \{0\}) \subseteq \overline{y} - (\tilde{D}(\overline{y}) \setminus \{0\})$$

since $tk \in -K$ and $K$ is a convex cone. So, we have a contradiction to $\overline{y} \in \mathrm{M}(F, \tilde{D}(\cdot))$.

(b) Put $\overline{t} := \varphi_{a-K,k}(\overline{y}) = \inf\{t \in \mathbb{R} \mid \overline{y} \in tk + a - K\}$. Then, we get

$$\overline{y} \in \overline{t}k + a - K$$

taking into account the definition of $\varphi_{a-K,k}$ and Corollary 11.2.22. Suppose that $\overline{y}$ is not a nondominated-like element of $F$ with respect to the set-valued map $\tilde{D}$, then there is an element $y \in F \setminus \{\overline{y}\}$ with $y \in \overline{y} - \tilde{D}(\overline{y})$ and this implies

$$a + \overline{t}k - y = a + \overline{t}k - \overline{y} + (\overline{y} - y) \in K + \tilde{D}(\overline{y}) \subseteq K.$$

This means that $\varphi_{a-K,k}(y) \leq \varphi_{a-K,k}(\overline{y})$, a contradiction. □

A corresponding result can be shown for weakly nondominated-like elements of $F$ with respect to the set-valued map $\tilde{D}$ in the sense of Definition 11.1.3(b). The next result is shown by Eichfelder in [139, 143, Theorem 5.13].

**Theorem 11.3.3** (Characterization of weakly nondominated-like elements) *Let* $Y$ *be a topological vector space. Suppose* $\tilde{D} : Y \to \mathcal{P}(Y)$, *where* $\tilde{D}(y)$ *is a nontrivial, closed, convex cone for all* $y \in Y$, $F \subset Y$ *and let* $\overline{y} \in F$.

(a) *Suppose* $\operatorname{int} \tilde{D}(\overline{y}) \neq \emptyset$. *If* $\overline{y} \in \mathrm{WM}(F, \tilde{D}(\cdot))$, *then for any nontrivial, closed, convex, pointed cone* $K \subseteq \tilde{D}(\overline{y})$ *and any* $k \in \operatorname{int} K$ *it holds that*

$$\varphi_{\overline{y}-K,k}(y) \geq \varphi_{\overline{y}-K,k}(\overline{y}) = 0 \quad \textit{for all} \quad y \in F. \tag{11.3.6}$$

(b) *Suppose* $\operatorname{int} \tilde{D}(\overline{y}) \neq \emptyset$. *If for* $a \in Y$, *a nontrivial closed, convex, pointed cone* $K$ *with* $\tilde{D}(\overline{y}) \subseteq K$ *and* $k \in \operatorname{int} K$ *it holds that*

$$\varphi_{a-K,k}(y) \geq \varphi_{a-K,k}(\overline{y}) \quad \textit{for all} \quad y \in F \setminus \{\overline{y}\}, \tag{11.3.7}$$

*then* $\overline{y} \in \text{WM}(F, \tilde{D}(\cdot))$.

***Proof***

(a) Consider $\overline{y} \in \text{WM}(F, \tilde{D}(\cdot))$, but (11.3.6) does not hold. Then, there are $t \in \mathbb{R}$, $t < 0$ and $y \in F \setminus \{\overline{y}\}$ such that $\overline{y} + tk - y \in K$. This implies

$$y \in \overline{y} - \text{int}\, K \subseteq \overline{y} - \text{int}\, \tilde{D}(\overline{y})$$

since $tk \in -\,\text{int}\, K$ and $K$ is a convex cone. So we have a contradiction to $\overline{y}$ is a weakly nondominated-like element of $F$ with respect to $\tilde{D}$.

(b) Put $\overline{t} := \varphi_{a-K,k}(\overline{y}) = \inf\{t \in \mathbb{R} \mid \overline{y} \in tk + a - K\}$. Then, we get

$$\overline{y} \in \overline{t}k + a - K,$$

taking into account the definition of $\varphi_{a-K,k}$ and Corollary 11.2.22. Suppose that $\overline{y}$ is not a weakly nondominated-like element of $F$ with respect to the set-valued map $\tilde{D}$. Then, there is an element $y \in F \setminus \{\overline{y}\}$ with $y \in \overline{y} - \text{int}\, \tilde{D}(\overline{y})$ and this implies

$$a + \overline{t}k - y = a + \overline{t}k - \overline{y} + (\overline{y} - y) \in K + \text{int}\, \tilde{D}(\overline{y}) \subseteq \text{int}\, K.$$

This means that there is a real number $\varepsilon > 0$ such that $a + (\overline{t} - \varepsilon)k - y \in K$, i.e., $\varphi_{a-K,k}(y) \leq \overline{t} - \varepsilon$, a contradiction to $\overline{t} = \varphi_{a-K,k}(\overline{y}) > \overline{t} - \varepsilon$.

□

From Theorems 11.3.2 and 11.3.3, one gets the following assertion by Eichfelder in [139, 143, Corollary 5.14].

**Corollary 11.3.4** *Let $Y$ be a topological vector space. Suppose $\tilde{D} : Y \to \mathcal{P}(Y)$, where $\tilde{D}(y)$ is a nontrivial, closed, convex cone for all $y \in Y$, $F \subset Y$ and let $\overline{y} \in F$.*

(a) *Let $K = \tilde{D}(\overline{y})$ and $k \in \text{int}\, K$, then $\overline{y} \in \text{M}(F, \tilde{D}(\cdot))$ if and only if*

$$\varphi_{\overline{y}-K,k}(y) > \varphi_{\overline{y}-K,k}(\overline{y}) = 0 \quad \textit{for all} \quad y \in F \setminus \{\overline{y}\}. \tag{11.3.8}$$

(b) *Let $K = \tilde{D}(\overline{y})$ and $k \in \text{int}\, K$, then $\overline{y} \in \text{WM}(F, \tilde{D}(\cdot))$ if and only if*

$$\varphi_{\overline{y}-K,k}(y) \geq \varphi_{\overline{y}-K,k}(\overline{y}) = 0 \quad \textit{for all} \quad y \in F \setminus \{\overline{y}\}. \tag{11.3.9}$$

**Remark 11.3.5** In Theorems 11.4.29 and 11.4.30, we will show corresponding characterizations of approximately nondominated-like elements $y_\varepsilon \in F$ via scalarization by means of the nonlinear functional $\varphi_{-\tilde{D}(y_\varepsilon),k}$.

Furthermore, we present characterizations of weakly nondominated-like and nondominated-like elements in the sense of Definition 11.1.3 by means of the functional $\varphi_{A+\tilde{D}(\cdot)}$ introduced in Sect. 11.2.3 (see (11.2.5)). Unless otherwise stated, we assume in the following again (compare Sect. 11.2.3) that Assumption (A') is fulfilled for $\tilde{D} : Y \to \mathscr{P}(Y)$, i.e., $Y$ is a Banach space, $\tilde{D} : Y \to \mathscr{P}(Y)$, where $\tilde{D}(y)$ is a proper, closed, convex, pointed, set, $0 \in \tilde{D}(y)$, $\tilde{D}(y) \neq \{0\}$ for all $y \in Y$ and $A \subset Y$, $A \neq \emptyset$.

We consider the functional $\varphi_{A+\tilde{D}(\cdot)} : Y \times (Y \setminus \{0\}) \to \overline{\mathbb{R}}$ given by (see (11.2.5)):

$$\varphi_{A+\tilde{D}(\cdot)}(y, k) = \inf\{t \in \mathbb{R} \mid y + tk \in A + \tilde{D}(y)\}, \tag{11.3.10}$$

where $y \in Y$ and $k \in Y \setminus \{0\}$.

In the following, we study the functional $\varphi_{A+\tilde{D}(\cdot)}$ and use it for characterizations of weakly nondominated-like and nondominated-like elements of $A \subset Y$ with respect to $\tilde{D} : Y \to \mathscr{P}(Y)$. Again, we use the convention $\varphi_{A+\tilde{D}(\cdot)}(y, k) = +\infty$ if $\{y + tk \mid t \in \mathbb{R}\} \cap (A + \tilde{D}(y)) = \emptyset$.

Now, we will apply the functional $\varphi_{A+\tilde{D}(\cdot)}$ given by (11.3.10) in order to derive necessary conditions for nondominated-like elements of $A$ with respect to the variable domination map $\tilde{D}(\cdot)$. This result is shown in [70, Lemma 3.4].

**Theorem 11.3.6** *Suppose that Assumption (A') is fulfilled. Furthermore, consider $\overline{y} \in Y$. Assume that either $\tilde{D}(\overline{y})$ is a cone or $A$ is a convex set.*
*If $\overline{y} \in \mathrm{M}(A, \tilde{D}(\cdot))$, then*

$$\forall k \in \tilde{D}(\overline{y}) \setminus \{0\} : \quad \varphi_{A+\tilde{D}(\cdot)}(\overline{y}, k) = 0.$$

***Proof*** We consider $\overline{y} \in \mathrm{M}(A, \tilde{D}(\cdot))$ and $k \in \tilde{D}(\overline{y}) \setminus \{0\}$. Taking into account Proposition 11.2.28(a), we obtain $\varphi_{A+\tilde{D}(\cdot)}(\overline{y}, k) \leq 0$ for $\overline{y} \in A$ and for all $k \in \tilde{D}(\overline{y}) \setminus \{0\}$, since $\tilde{D}(\overline{y})$ is pointed and $\overline{y} \in A$. Assume that $\varphi_{A+\tilde{D}(\cdot)}(\overline{y}, k) < 0$. Then, there are $y \in A,\ y \neq \overline{y}$ and $t < 0$ such that

$$\overline{y} + tk = y + c \text{ for some } c \in \tilde{D}(\overline{y}). \tag{11.3.11}$$

Now, we study two cases.

Case 1: Suppose that $\tilde{D}(\overline{y})$ is a cone. Then, $-tk \in \tilde{D}(\overline{y}) \setminus \{0\}$ since $t < 0$. Because $\tilde{D}(\overline{y})$ is convex and pointed, we get $-tk + c \in \tilde{D}(\overline{y}) \setminus \{0\}$ for $c \in \tilde{D}(\overline{y})$. This implies, together with (11.3.11),

$$\overline{y} \in y + \tilde{D}(\overline{y}) \setminus \{0\},$$

a contradiction to the assumption $\overline{y} \in \mathrm{M}(A, \tilde{D}(\cdot))$.

Case 2: Now, we suppose that $A$ is convex. We take $k, c, 0 \in \tilde{D}(\overline{y})$. Because the convexity of $\tilde{D}(\overline{y})$, we can chose $\alpha \in (0, 1)$ (depending from $t < 0$) such that $\alpha(-tk + c) \in \tilde{D}(\overline{y}) \setminus \{0\}$. Moreover, since $A$ is convex and $y \in A$, we obtain

$$\alpha y + (1 - \alpha)\overline{y} \in A.$$

Since $\alpha \neq 0$, we get $\alpha y + (1-\alpha)\overline{y} \neq \overline{y}$. Taking into account (11.3.11), it holds that

$$\overline{y} = \alpha y + (1-\alpha)\overline{y} + \alpha(\overline{y} - y) = [\alpha y + (1-\alpha)\overline{y}] + \alpha(-tk + c) \in (\alpha y + (1-\alpha)\overline{y}) + (\tilde{D}(y) \setminus \{0\}).$$

This implies that $\overline{y}$ is not a nondominated-like element of $A$, a contradiction. □

In the next theorem, we derive sufficient conditions for the property that the element $\overline{y} := y + \varphi_{A+\tilde{D}(\cdot)}(y,k)k - c \in A$ ($y \in A$, $k \in \tilde{D}(y) \setminus \{0\}$, $c \in \tilde{D}(y)$) is a weakly nondominated-like element of $A$. This result is shown in [70, Proposition 3.5].

**Theorem 11.3.7** *Suppose that Assumption (A') is fulfilled. Let $A \subset Y$, $\tilde{D} : Y \to \mathscr{P}(Y)$, $y \in A$, $k \in \tilde{D}(y) \setminus \{0\}$ and $\overline{y} = y + \varphi_{A+\tilde{D}(\cdot)}(y,k)k - c$ for some $c \in \tilde{D}(y)$. Assume that*

(a) *For all $\overline{k} \in \tilde{D}(y) \setminus \{0\}$, $\operatorname{int} \tilde{D}(y - \overline{k}) \subseteq \operatorname{int} \tilde{D}(y)$.*
(b) *$\tilde{D}(y)$ is a nontrivial, pointed, closed and convex cone for each $y \in Y$.*

*If $\overline{y} \in A$, then $\overline{y} \in \mathrm{WM}(A, \tilde{D}(\cdot))$.*

***Proof*** Under our assumptions, $\tilde{D}(y)$ is pointed and $y \in A$. Applying Proposition 11.2.28(a), we get

$$\varphi_{A+\tilde{D}(\cdot)}(y,k) \leq 0.$$

Furthermore, since $\tilde{D}(y)$ is a pointed, convex cone for each $y \in Y$, we obtain

$$-\varphi_{A+\tilde{D}(\cdot)}(y,k)k + c \in \tilde{D}(y) \setminus \{0\} \text{ for } k, c \in \tilde{D}(y).$$

Suppose that $\overline{y} = y + \varphi_{A+\tilde{D}(\cdot)}(y,k)k - c$ is not a weakly nondominated-like point. Then, there is $\hat{y} \in A$ such that

$$\overline{y} - \hat{y} \in \operatorname{int} \tilde{D}(\overline{y}). \tag{11.3.12}$$

Taking into account $-\varphi_{A+\tilde{D}(\cdot)}(y,k)k + c \in \tilde{D}(y) \setminus \{0\}$ and assumption (a), the following condition holds

$$\operatorname{int} \tilde{D}(y - (-\varphi_{A+\tilde{D}(\cdot)}(y,k)k + c)) \subseteq \operatorname{int} \tilde{D}(y),$$

i.e.,

$$\operatorname{int} \tilde{D}(\overline{y}) \subseteq \operatorname{int} \tilde{D}(y). \tag{11.3.13}$$

With (11.3.12) and (11.3.13) , we have $\overline{y} - \hat{y} \in \operatorname{int} \tilde{D}(y)$. Then, we conclude for some $\varepsilon > 0$ small enough

$$\overline{y} - \hat{y} - \varepsilon k \in \tilde{D}(y).$$

For $\overline{y} = y + \varphi_{A+\tilde{D}(\cdot)}(y,k)k - c$, we obtain

$$y + (\varphi_{A+\tilde{D}(\cdot)}(y,k) - \varepsilon)k \in \hat{y} + \tilde{D}(y) \subseteq A + \tilde{D}(y).$$

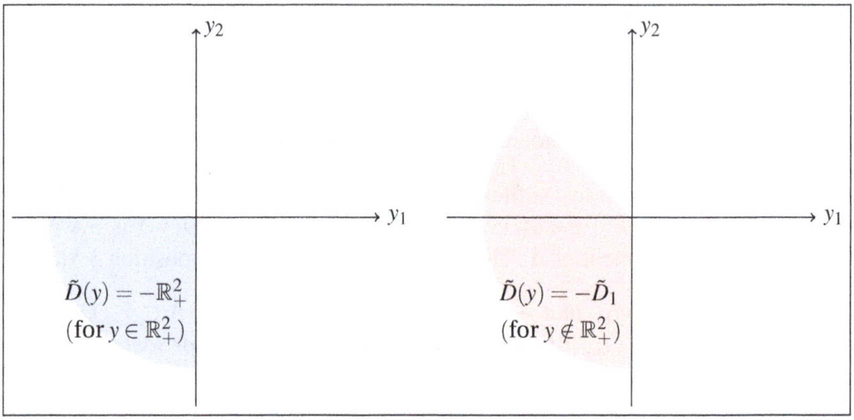


**Fig. 11.6** The cones $-\mathbb{R}^2_+$ and $-\tilde{D}_1$ defining the domination map $\tilde{D}$ in Example 11.3.9

Therefore, $\varphi_{A+\tilde{D}(\cdot)}(y,k) \leq \varphi_{A+\tilde{D}(\cdot)}(y,k) - \varepsilon$, a contradiction. □

**Remark 11.3.8** The proof of Theorem 11.3.7 is based on the fulfillment of the following two conditions for all $y \in Y$:

- $0 \in \tilde{D}(y)$.
- $\forall \lambda \geq 0: \quad \lambda k + c \in \tilde{D}(y)$ .

Because the vectors $k, c$ are not known in advance, these two conditions have to be supposed for all $k, c \in \tilde{D}(y)$. These conditions are fulfilled if and only if $\tilde{D}(y)$ is a convex cone. This means, unless the point $y$ is given, the sufficient condition in Theorem 11.3.7 has not much practical use for vector optimization problems with respect to domination maps $\tilde{D} : Y \to \mathscr{P}(Y)$ which are not cone-valued.

Although the assumptions in Theorem 11.3.7 are strong, there are non constant maps fulfilling (a) in Theorem 11.3.7, as the next example ([70, Example 3.7]) will show.

**Example 11.3.9** Consider $\tilde{D}_1 := \{z \in \mathbb{R}^2 \mid z_2 \geq -z_1, z_1 \geq 0\}$ and define the domination map by (see Fig. 11.6):

$$\tilde{D}(y) := \begin{cases} -\mathbb{R}^2_+, & \text{if } y \in \mathbb{R}^2_+, \\ -\tilde{D}_1, & \text{otherwise.} \end{cases}$$

If $y \in \mathbb{R}^2_+$, then $\tilde{D}(y) = -\mathbb{R}^2_+$. We obtain $\tilde{D}(y-k) = -\mathbb{R}^2_+$ for all $k \in -\mathbb{R}^2_+ \setminus \{0\}$, since $y - k \in \mathbb{R}^2_+$. This implies $\operatorname{int} \tilde{D}(y-k) \subseteq \operatorname{int} \tilde{D}(y)$ for all $k \in \tilde{D}(y) \setminus \{0\}$. Moreover, if $y \notin \mathbb{R}^2_+$, then $\tilde{D}(y) = -\tilde{D}_1$. Because of $-\mathbb{R}^2_+ \subseteq -\tilde{D}_1$, it holds that $\operatorname{int}(-\mathbb{R}^2_+) \subseteq \operatorname{int}(-\tilde{D}_1)$ such that we have $\operatorname{int} \tilde{D}(y-k) \subseteq \operatorname{int}(-\tilde{D}_1) = \operatorname{int} \tilde{D}(y)$ for all $k \in \tilde{D}(y) \setminus \{0\}$, i.e., condition (a) of Theorem 11.3.7 is fulfilled.

## 11.4 Approximate Solutions to VVDS

In optimization theory, it is well known that one needs certain compactness assumptions in order to show existence results for solutions of optimization problems. Such compactness assumption are not satisfied for many optimization problems. So, one has to deal with approximate solutions of general optimization problems under weak assumptions (without compactness or convexity conditions). We can show several assertions without any compactness assumptions for approximate solutions (see Sections 12.3.2, 12.4). Furthermore, if we derive numerical algorithms for solving optimization problems, then these algorithms usually generate approximate solutions which are close to the exact solutions. Now, we will introduce approximate solutions of vector optimization problems with a variable domination structure. In the literature, many authors deal with different concepts for approximate solutions with respect to a constant domination structure, see [111, 185, 317, 337, 367, 422, 432, 442, 453–455, 493] for different definitions, concepts and properties of these elements.

### *11.4.1 Concepts of Approximate Solutions*

The results of this section are derived by Soleimani in [414, 415] and by Soleimani, Tammer in [417]. In this section, we again suppose that assumption (A) is fulfilled, where no convexity assumptions concerning the set-valued map $D$ are involved (compare Sect. 11.2):

**Assumption (A)** Let $Y$ be a topological vector space. Suppose $D : Y \to \mathscr{P}(Y)$, where $D(y)$ is a proper, closed, pointed set with $0 \in D(y)$ for all $y \in Y$. Furthermore, we suppose that the set-valued maps $\tilde{D} : Y \to \mathscr{P}(Y)$ and $\hat{D} : Y \to \mathscr{P}(Y)$ have the same properties.

Suppose that $\varepsilon \geq 0$ and $k \in Y \setminus \{0\}$. We define $\varepsilon k$-nondominated, $\varepsilon k$-nondominated-like elements and $\varepsilon k$-completely nondominated points with respect to a variable domination structure. Furthermore, we define weakly $\varepsilon k$-nondominated-like (nondominated) elements and completely nondominated points. For sure, there is no difference between (weakly) $\varepsilon k$-nondominated, (weakly) $\varepsilon k$-nondominated-like elements and (weak) $\varepsilon k$-completely nondominated points in the case of constant domination structure. As already mentioned in Sect. 11.1 for exact solutions, this statement cannot be true for (approximate) solutions of vector optimization problems with a variable domination structure and all these three definitions describe different elements. This will be shown by several examples.

First, we introduce a concept of approximate nondominated elements to vector optimization problems with a variable domination structure (see Soleimani and Tammer [417, Definition 1]).

**Definition 11.4.1** Suppose that assumption (A) holds and additionally that $k \in 0^+D(y) \setminus \{0\}$ for all $y \in Y$, $F \subset Y$, $\varepsilon \geq 0$ and $y_\varepsilon \in F$.

(a) $y_\varepsilon$ is said to be an $\varepsilon k$-nondominated element of $F$ with respect to the domination map $D : Y \to \mathcal{P}(Y)$ if no $y \in F$ exists such that

$$y_\varepsilon - \varepsilon k \in y + (D(y)\setminus\{0\}).$$

(b) Suppose that int $D(y) \neq \emptyset$ for all $y \in F$. $y_\varepsilon$ is said to be a weakly $\varepsilon k$-nondominated element of $F$ with respect to the domination map $D : Y \to \mathcal{P}(Y)$ if no $y \in F$ exists such that

$$y_\varepsilon - \varepsilon k \in y + \text{int}\, D(y).$$

We denote the set of all $\varepsilon k$-nondominated elements of $F$ with respect to the domination map $D : Y \to \mathcal{P}(Y)$ by $\varepsilon k - \text{N}(F, D(\cdot))$. Furthermore, we denote the set of all weakly $\varepsilon k$-nondominated elements of $F$ with respect to the domination map $D$ by $\varepsilon k - \text{WN}(F, D(\cdot))$.

**Remark 11.4.2** If $\varepsilon = 0$, then the above concepts in Definition 11.4.1 coincide with the usual definitions of nondominated elements, see Definition 11.1.2. Moreover, for the case $D(y) = D$ for all $y \in Y$, where $D$ is a proper, closed, pointed set with $0 \in D$ and $D \neq \{0\}$ and $\varepsilon = 0$, the concepts in Definition 11.4.1 coincides with the concepts in Definitions 6.0.1 and 6.2.2, respectively.

**Remark 11.4.3** For the special case $D(y) = D$ for all $y \in Y$, where $D$ is a proper, closed, pointed set with $0 \in D$ and $D \neq \{0\}$, the concepts in Definition 11.4.1 coincide with the corresponding concepts for problems with fixed ordering cones, see [185, 317, 337, 432].

If $\varepsilon = 0$, then all these definitions coincide with the usual definitions of nondominated points, see Definition 11.1.2.

Furthermore, we introduce a concept of approximate nondominated-like elements to vector optimization problems with a variable domination structure (see Soleimani and Tammer [417, Definition 2]).

**Definition 11.4.4** Suppose that Assumption (A) holds for $\tilde{D} : Y \to \mathcal{P}(Y)$ and additionally that $k \in 0^+\tilde{D}(y)\setminus\{0\}$ for all $y \in Y$, $F \subset Y$, $\varepsilon \geq 0$ and $y_\varepsilon \in F$.

(a) $y_\varepsilon$ is said to be an $\varepsilon k$-nondominated-like element of $F$ with respect to $\tilde{D} : Y \to \mathcal{P}(Y)$ if

$$F \cap \Big(y_\varepsilon - \varepsilon k - (\tilde{D}(y_\varepsilon)\setminus\{0\})\Big) = \emptyset.$$

(b) Suppose that int $\tilde{D}(y_\varepsilon) \neq \emptyset$. $y_\varepsilon$ is said to be a weakly $\varepsilon k$-nondominated-like element of $F$ with respect to $\tilde{D} : Y \to \mathcal{P}(Y)$ if

$$F \cap (y_\varepsilon - \varepsilon k - \text{int}\, \tilde{D}(y_\varepsilon)) = \emptyset.$$

We denote the set of all $\varepsilon k$-nondominated-like elements of $F$ with respect to the domination map $\tilde{D} : Y \to \mathscr{P}(Y)$ by $\varepsilon k - \mathrm{M}(F, \tilde{D}(\cdot))$. Furthermore, we denote the set of all weakly $\varepsilon k$-nondominated-like elements of $F$ with respect to the domination map $\tilde{D}$ by $\varepsilon k - \mathrm{WM}(F, \tilde{D}(\cdot))$.

**Remark 11.4.5** If $\varepsilon = 0$, then the concepts in Definition 11.4.4 coincide with the usual definitions of nondominated-like elements, see Definition 11.1.3. Moreover, for the case $\tilde{D}(y) = D$ for all $y \in Y$, where $D$ is a proper, closed, pointed set with $0 \in D$ and $D \neq \{0\}$ and $\varepsilon = 0$, the concepts in Definition 11.4.4 coincides with the concepts in Definitions 6.0.1 and 6.2.2, respectively.

**Remark 11.4.6** For the special case $\tilde{D}(y) = D$ for all $y \in Y$, where $D$ is a proper, closed, pointed set with $0 \in D$ and $D \neq \{0\}$, the concepts in Definition 11.4.4 coincide with the corresponding concepts for problems with fixed ordering cones, see [185, 317, 337, 432].

We now introduce a concept of approximate completely nondominated elements to vector optimization problems with a variable domination structure (see Soleimani and Tammer [417, Definition 3]).

**Definition 11.4.7** Let Assumption (A) be fulfilled for the set-valued map $\hat{D} : Y \to \mathscr{P}(Y)$ and suppose additionally that $k \in 0^+\hat{D}(y)\backslash\{0\}$ for all $y \in Y$, $F \subset Y$, $\varepsilon \geq 0$ and $y_\varepsilon \in F$.

(a) $y_\varepsilon$ is said to be an $\varepsilon k$-completely nondominated element of $F$ with respect to the domination map $\hat{D} : Y \to \mathscr{P}(Y)$ if

$$\forall y, y' \in F : \qquad (y_\varepsilon - \varepsilon k - (\hat{D}(y)\backslash\{0\})) \cap \{y'\} = \emptyset.$$

(b) Suppose that $\operatorname{int} \hat{D}(y) \neq \emptyset$ for all $y \in F$. $y_\varepsilon$ is said to be a weakly $\varepsilon k$-completely nondominated point of $F$ with respect to the domination map $\hat{D} : Y \to \mathscr{P}(Y)$ if

$$\forall y, y' \in F : \qquad (y_\varepsilon - \varepsilon k - \operatorname{int} \hat{D}(y)) \cap \{y'\} = \emptyset.$$

We denote the set of all $\varepsilon k$-completely nondominated points of $F$ with respect to the domination map $\hat{D} : Y \to \mathscr{P}(Y)$ by $\varepsilon k - \mathrm{MZ}(F, \hat{D}(\cdot))$.

Moreover, we denote the set of all weakly $\varepsilon k$-completely nondominated points of $F$ with respect to the domination map $\hat{D}$ by $\varepsilon k - \mathrm{WMZ}(F, \hat{D}(\cdot))$.

The following example is discussed in [417, Example 1].

**Example 11.4.8** Let $\varepsilon = \frac{1}{100}$ and $k = (1, 0)^T$. Also, suppose that

$$F := \left\{(y_1, y_2)^T \in \mathbb{R}^2 \mid y_1 + y_2 \geq 2, \; y_1 \geq 0, \; 0 \leq y_2 \leq 2\right\}$$

and

$$D((y_1, y_2)^T) := \begin{cases} \mathbb{R}^2_+ & \text{if } y_2 = 0 \\ \operatorname{cone\,conv} \{(2, 0)^T, (y_1, y_2)^T\} & \text{otherwise} \end{cases}$$

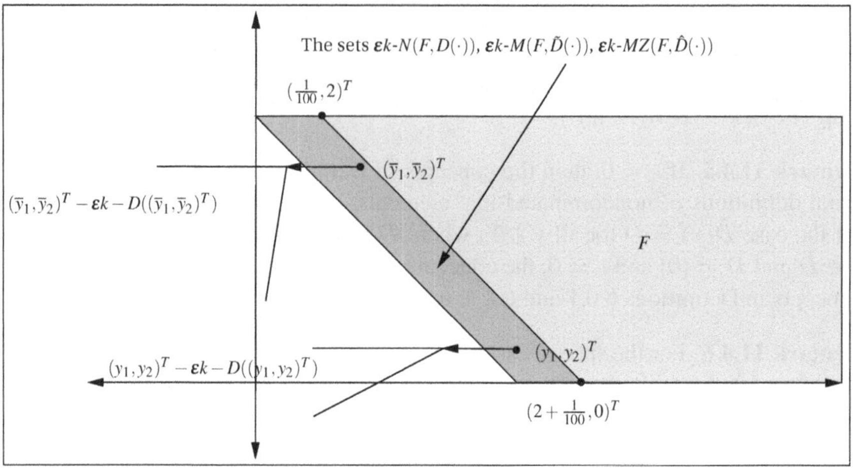


**Fig. 11.7** Illustration of Example 11.4.8. Adapted from [417, Figure 1]

and $D = \tilde{D} = \hat{D}$. Of course, $k \in 0^+D(y)$ for all $y \in F$. Then, $\{(y_1, y_2)^T \in F \mid y_1 + y_2 \le 2 + \frac{1}{100}\}$ are $\varepsilon k$-nondominated, $\varepsilon k$-completely nondominated point and also $\varepsilon k$-nondominated-like elements and the sets of all these points coincide. Moreover,

$$\left\{(y_1, y_2)^T \in F \mid y_1 + y_2 \le 2 + \frac{1}{100}\right\} \cup \left\{(y_1, 0)^T\right\}$$

describes the sets of weakly $\varepsilon k$-nondominated, $\varepsilon k$-completely nondominated point and also weakly $\varepsilon k$-nondominated-like elements (see Fig. 11.7).

As already mentioned, in the case of constant domination structure, (weakly) $\varepsilon k$-nondominated elements and (weakly) $\varepsilon k$-nondominated-like elements coincide, but the following examples show that this is not true when we are dealing with variable domination structures.

The next example is studied in [417, Example 2].

**Example 11.4.9** Let $\varepsilon = \frac{1}{100}$ and $k = (1, 0)^T$. Furthermore, consider

$$F := \left\{(y_1, y_2)^T \in \mathbb{R}^2 \mid y_1 + y_2 \ge -1, \quad y_1 \le 0, \quad y_2 \le 0\right\}.$$

and

$$D((y_1, y_2)^T) := \begin{cases} \{(d_1, d_2)^T \in \mathbb{R}^2 \mid d_1 \ge 0,\ d_2 \le 0\} & \text{for } (y_1, y_2)^T = (-1, 0)^T \\ \mathbb{R}^2_+ & \text{otherwise} \end{cases}$$

and $D = \tilde{D} = \hat{D}$. Of course, $k \in 0^+D(y)$ for all $y \in F$. Then,

$$\left\{(y_1, y_2)^T \in F \mid y_1 + y_2 \le -1 + \varepsilon\right\}$$

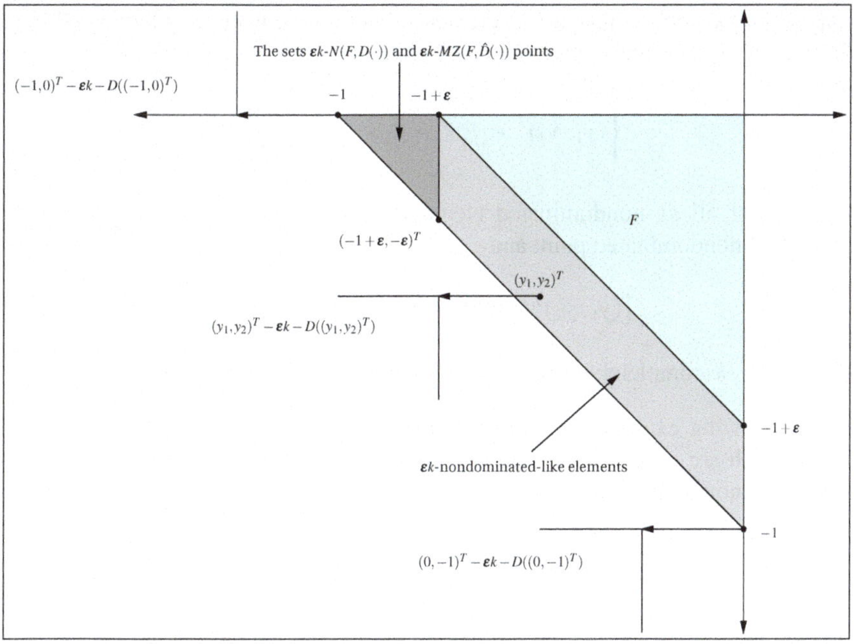


**Fig. 11.8** Illustration of Example 11.4.9. Adapted from [417, Figure 2]

are $\varepsilon k$-nondominated-like, but only the elements of the set

$$\{(y_1, y_2)^T \in F \mid y_1 < -1 + \varepsilon\} \cup \{(-1 + \varepsilon, 0)^T\}$$

are $\varepsilon k$-nondominated and $\varepsilon k$-completely nondominated points (see Fig. 11.8).

The next example is studied in [417, Example 3].

**Example 11.4.10** Let $\varepsilon = \frac{1}{100}$ and $k = (1, 1)^T$. We consider

$$F := \{(y_1, y_2)^T \in \mathbb{R}^2 \mid y_1 + y_2 \geq -1, \quad y_1 \leq 0, \quad y_2 \leq 0\}$$

and

$$D((y_1, y_2)^T) := \begin{cases} \{(d_1, d_2)^T \in \mathbb{R}^2 | \, d_2 \geq 0, \; d_1 + d_2 \geq -1\} & \text{for } (y_1, y_2)^T = (-1, 0)^T \\ \mathbb{R}^2_+ & \text{otherwise} \end{cases}$$

and $D = \tilde{D} = \hat{D}$. Again, we have $k \in 0^+ D(y)$ for all $y \in F$. Then,

$$\left\{(y_1, y_2)^T \in F \mid y_1 + y_2 \leq -\frac{98}{100}, \; y_1 \neq -1\right\}$$

is the set of $\varepsilon k$-nondominated-like elements, but $(-1, 0)$ is not nondominated-like. However, $(-1, 0)$ is an $\varepsilon k$-nondominated point and

$$\left\{(y_1, y_2)^T \in F \mid y_1 + y_2 \leq -\frac{98}{100}\right\}$$

is the set of all $\varepsilon k$-nondominated elements. Obviously, $(-1, 0)^T$ is not an $\varepsilon k$-completely nondominated point and

$$\left\{(y_1, y_2)^T \in F \mid -1 < y_2 < -1 + \varepsilon\right\}$$

is the set of $\varepsilon k$-completely nondominated points (see Fig. 11.9).

In the following example (see [417, Example 4]), we show that there are some points which are $\varepsilon k$-nondominated and also $\varepsilon k$-nondominated-like but are not $\varepsilon k$-completely nondominated points.

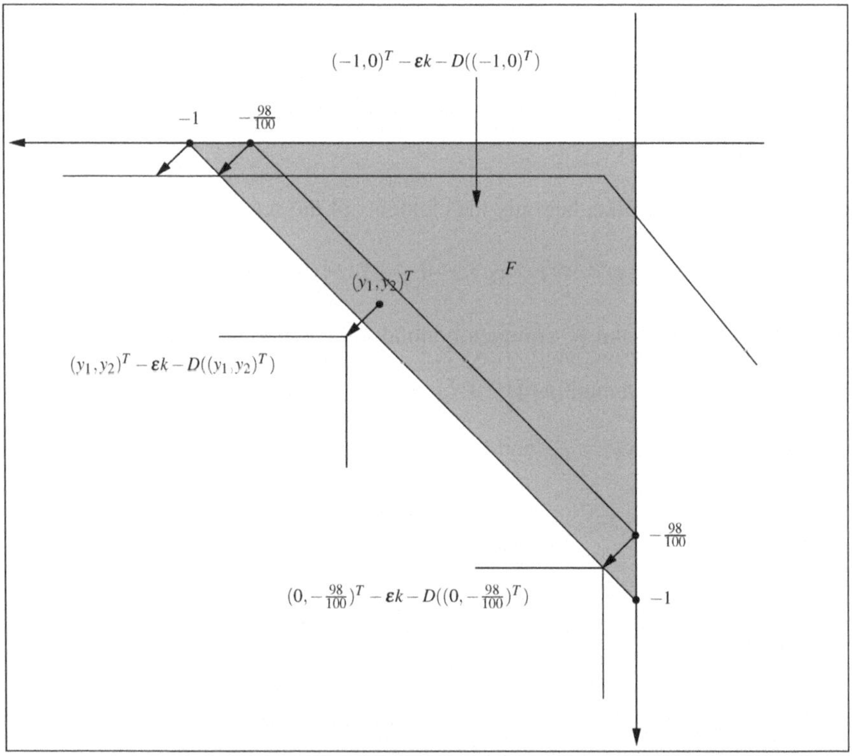


**Fig. 11.9** Illustration of Example 11.4.10. Adapted from [417, Figure 3]

**Example 11.4.11** Let $\varepsilon = \frac{1}{100}$ and $k = (1, 0)^T$. We consider

$$F := \{(y_1, y_2)^T \in \mathbb{R}^2 \mid y_1 + y_2 \geq -1, \quad y_1 \leq 0, \quad y_2 \leq 0\}$$

and

$$D((y_1, y_2)^T) = \begin{cases} \{(d_1, d_2)^T \in \mathbb{R}^2 \mid d_2 \geq 0,\ d_1 + d_2 \geq 0\} & \text{for } (y_1, y_2)^T = (0, 0)^T \\ \mathbb{R}^2_+ & \text{otherwise} \end{cases}$$

and $D = \tilde{D} = \hat{D}$. Of course, $k \in 0^+ D(y)$ for all $y \in F$. Then,

$$\left\{\{(y_1, y_2)^T \in F \mid y_1 + y_2 \leq -\frac{99}{100}\right\}$$

is the set of $\varepsilon k$-nondominated-like and $\varepsilon k$-nondominated points. However, only points of the set

$$\left\{(y_1, y_2)^T \in F \mid y_1 + y_2 < -\frac{99}{100}\right\}$$

are $\varepsilon k$-completely nondominated points and the points

$$\left\{(y_1, y_2)^T \in F \mid y_1 + y_2 = -\frac{99}{100}\right\}$$

are not $\varepsilon k$-completely nondominated points. This shows that there are some points which are both $\varepsilon k$-nondominated and $\varepsilon k$-nondominated-like but not $\varepsilon k$-completely nondominated points (see Fig. 11.10).

Definitions of locally $\varepsilon k$-nondominated, locally $\varepsilon k$-nondominated-like and locally $\varepsilon k$-completely nondominated elements are similar. We just need to substitute $F$ with $F \cap U$ in Definitions 11.4.1, 11.4.4 and 11.4.7 where $U$ is a neighborhood of the point in consideration.

In Lemma 11.1.5 (see [143, Lemma 2.11]), we presented relationships between exact nondominated and nondominated-like solutions to vector optimization problems with variable domination structures. In the following, we will show relationships between different kinds of approximately optimal elements ($\varepsilon k$-nondominated, $\varepsilon k$-nondominated-like and $\varepsilon k$-completely nondominated points) to vector optimization problems with respect to variable domination structures.

First, we will show relationships between (weakly) $\varepsilon k$-completely nondominated points of the set $F$ with respect to domination map $\hat{D} : Y \to \mathcal{P}(Y)$ and (weakly) $\varepsilon k$-nondominated elements of $F$ with respect to $D$ (compare [417, Theorem 4]).

**Theorem 11.4.12** *Suppose that assumption (A) holds for $D = \hat{D} : Y \to \mathcal{P}(Y)$ and, additionally, $k \in Y \setminus \{0\}$ such that $k \in 0^+ D(y)$ for all $y \in Y$, $\emptyset \neq F \subset Y$ and $\varepsilon \geq 0$. Then, the following assertions hold:*

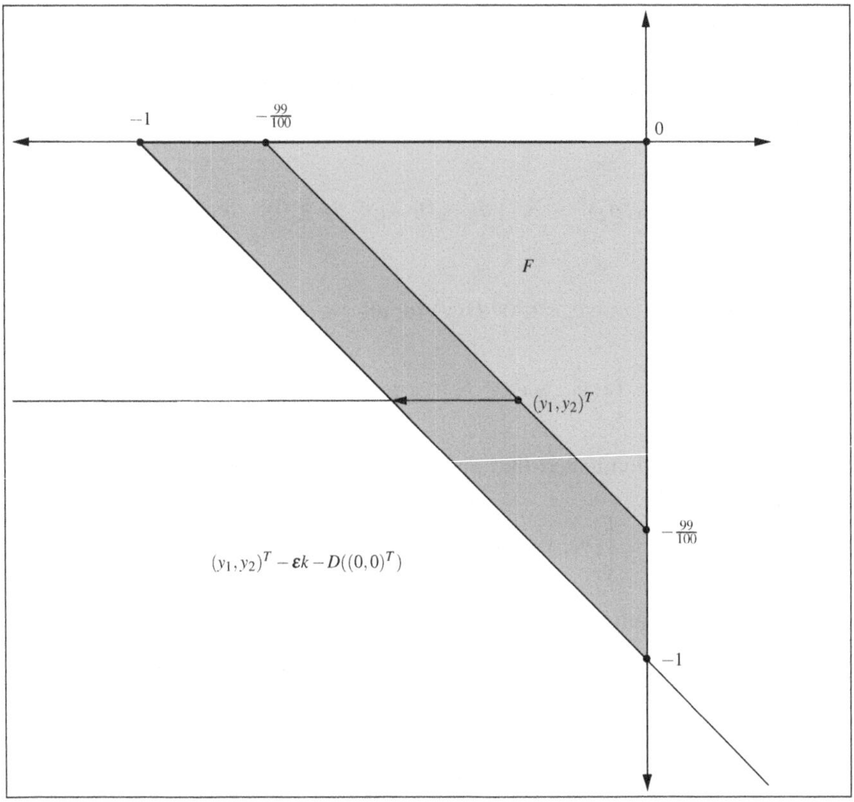


**Fig. 11.10** Illustration of Example 11.4.11. Adapted from [417, Figure 4]

(a)

$$y_\varepsilon \in \varepsilon k - \mathrm{MZ}(F, D(\cdot)) \implies y_\varepsilon \in \varepsilon k - \mathrm{N}(F, D(\cdot)).$$

(b)

$$y_\varepsilon \in \varepsilon k - \mathrm{N}(F, D(\cdot)) \implies y_\varepsilon \in \varepsilon k - \mathrm{MZ}(F, D(\cdot))$$

*if* $D(y) = D(y')$ *for all* $y, y' \in F$. *This means that each approximately nondominated element is an approximate completely nondominated point if the domination map is constant for all* $y \in F$.

(c) *Assume that* $\operatorname{int} D(y) \neq \emptyset$ *for all* $y \in F$, *then*

$$y_\varepsilon \in \varepsilon k - \mathrm{WMZ}(F, D(\cdot)) \implies y_\varepsilon \in \varepsilon k - \mathrm{WN}(F, D(\cdot)).$$

(d) *Suppose that* $\operatorname{int} D(y) \neq \emptyset$ *for all* $y \in F$, *then*

$$y_\varepsilon \in \varepsilon k - \mathrm{WN}(F, D(\cdot)) \implies y_\varepsilon \in \varepsilon k - \mathrm{WMZ}(F, D(\cdot))$$

*if* int $D(y) = \text{int } D(y')$ *for all* $y, y' \in F$.

***Proof***

(a) Obviously, the assertion follows from Definitions 11.4.1(a), 11.4.4(a).
(b) Assume that $y_\varepsilon \in \varepsilon k - \text{N}(F, D(\cdot))$, then

$$\forall y \in F: \quad y_\varepsilon - \varepsilon k \notin y + (D(y)\backslash\{0\}).$$

Because of $D(y) = D(y')$ for all $y, y' \in F$, we obtain

$$\forall y, y' \in F: \quad y_\varepsilon - \varepsilon k \notin y + (D(y')\backslash\{0\}),$$

i.e., $(y_\varepsilon - \varepsilon k - (D(y')\backslash\{0\})) \cap \{y\} = \emptyset$ for all $y, y' \in F$ and therefore $y_\varepsilon \in \varepsilon k - \text{MZ}(F, D(\cdot))$.
(c) The assertion follows from Definition 11.4.1(b) and Definition 11.4.7(b).
(d) Let $y_\varepsilon \in \varepsilon k - \text{WN}(F, D(\cdot))$, then it holds that

$$\forall y \in F: \quad y_\varepsilon - \varepsilon k \notin y + \text{int } D(y).$$

If int $D(y) = \text{int } D(y')$ for all $y, y' \in F$, we get

$$\forall y, y' \in F: \quad y_\varepsilon - \varepsilon k \notin y + \text{int } D(y').$$

This implies $(y_\varepsilon - \varepsilon k - \text{int } D(y')) \cap \{y\} = \emptyset$ for all $y, y' \in F$ and $y_\varepsilon \in \varepsilon k - \text{WMZ}(F, D(\cdot))$. □

The following examples (compare [417, Examples 5, 6]) illustrate that the condition $D(y) = D(y')$ for all $y, y' \in F$ in Theorem 11.4.12(b) is not a necessary condition and is just a sufficient condition.

**Example 11.4.13** Consider Example 11.4.9. Obviously, $(-1, 0)^T \in \varepsilon k - \text{N}(F, D(\cdot))$ and also $(-1, 0)^T \in \varepsilon k - \text{MZ}(F, D(\cdot))$ but $D(y') = D(y)$ does not hold for all $y, y' \in F$.

**Example 11.4.14** Consider Example 11.4.10. It is easy to see that $(-1, 0)^T \in \varepsilon k - \text{N}(F, D(\cdot))$ but $(-1, 0)^T \notin \varepsilon k - \text{MZ}(F, D(\cdot))$. It is obvious that $\{(d_1, d_2)^T \in \mathbb{R}^2 | d_2 \geq 0,\ d_1 + d_2 \geq -1\}$ is not a subset of $\mathbb{R}^2_+$.

In the following theorem ([417, Theorem 5]), we discuss the relationships between the sets of (weakly) $\varepsilon k$-completely nondominated points and (weakly) $\varepsilon k$-nondominated-like elements with respect to the domination map $D = \tilde{D} = \hat{D} : Y \to \mathcal{P}(Y)$. We already mentioned that in the case of vector optimization problems with constant domination structures, these sets coincide. However, in vector optimization problems with variable domination structures, Examples 11.4.9 and 11.4.11 illustrated that there are some approximately nondominated-like elements which are not approximate completely nondominated points.

**Theorem 11.4.15** *Assume that assumption (A) holds, $D = \tilde{D} = \hat{D} : Y \to \mathscr{P}(Y)$ and additionally $k \in Y\backslash\{0\}$ such that $k \in 0^+D(y)$ for all $y \in Y$, $\emptyset \neq F \subset Y$ and $\varepsilon \geq 0$. Then, the following assertions hold:*

(a)
$$y_\varepsilon \in \varepsilon k - \mathrm{MZ}(F, D(\cdot)) \implies y_\varepsilon \in \varepsilon k - \mathrm{M}(F, D(\cdot)).$$

(b)
$$y_\varepsilon \in \varepsilon k - \mathrm{M}(F, D(\cdot)) \implies y_\varepsilon \in \varepsilon k - \mathrm{MZ}(F, D(\cdot))$$

*if $D(y) \subseteq D(y_\varepsilon)$ for all $y \in F$.*

(c) *Suppose that* $\mathrm{int}\, D(y) \neq \emptyset$ *for all $y \in F$, then*
$$y_\varepsilon \in \varepsilon k - \mathrm{WMZ}(F, D(\cdot)) \implies y_\varepsilon \in \varepsilon k - \mathrm{WM}(F, D(\cdot)).$$

(d) *Suppose that* $\mathrm{int}\, D(y) \neq \emptyset$ *for all $y \in F$, then*
$$y_\varepsilon \in \varepsilon k - \mathrm{WM}(F, D(\cdot)) \implies y_\varepsilon \in \varepsilon k - \mathrm{WMZ}(F, D(\cdot))$$

*if* $\mathrm{int}\, D(y) \subseteq \mathrm{int}\, D(y_\varepsilon)$ *for all $y \in F$.*

***Proof***

(a) The assertion directly follows from Definition 11.4.4(a) and Definition 11.4.7(a).
(b) Assume $y_\varepsilon \in \varepsilon k - \mathrm{M}(F, D(\cdot))$, then $(y_\varepsilon - \varepsilon k - (D(y_\varepsilon)\backslash\{0\})) \cap F = \emptyset$. Since $D(y) \subseteq D(y_\varepsilon)$ for all $y \in F$, hence
$$\forall y, y' \in F: \quad (y_\varepsilon - \varepsilon k - (D(y)\backslash\{0\})) \cap \{y'\} \subseteq (y_\varepsilon - \varepsilon k - (D(y_\varepsilon)\backslash\{0\})) \cap \{y'\} = \emptyset,$$
i.e., $y_\varepsilon \in \varepsilon k - \mathrm{MZ}(F, D(\cdot))$.
(c) Obviously, the assertion follows from Definition 11.4.4(b) and Definition 11.4.7(b).
(d) Taking into account $\mathrm{int}\, D(y) \subseteq \mathrm{int}\, D(y_\varepsilon)$ for all $y \in F$ and $y_\varepsilon \in \varepsilon k - \mathrm{WM}(F, D(\cdot))$, we get
$$\forall y, y' \in F: \quad (y_\varepsilon - \varepsilon k - \mathrm{int}\, D(y)) \cap \{y'\} \subseteq (y_\varepsilon - \varepsilon k - \mathrm{int}\, D(y_\varepsilon)) \cap \{y'\} = \emptyset,$$
i.e., $(y_\varepsilon - \varepsilon k - \mathrm{int}\, D(y)) \cap \{y'\} = \emptyset$ for all $y, y' \in F$ and $y_\varepsilon \in \varepsilon k - \mathrm{WMZ}(F, D(\cdot))$.

□

The following examples ([417, Examples 7, 8]) show that the condition $D(y) \subseteq D(y_\varepsilon)$ for all $y \in F$ in Theorem 11.4.15(b) is not a necessary condition.

**Example 11.4.16** Consider Example 11.4.9. Obviously, $(-1, 0)^T \in \varepsilon k - \mathrm{MZ}(F, D(\cdot))$ and $(-1, 0)^T \in \varepsilon k - \mathrm{M}(F, D(\cdot))$, however $D(y) \subseteq D(y_\varepsilon)$ does not hold for all $y \in F$.

**Example 11.4.17** Consider Example 11.4.9. It is easy to see that $(0, -1)^T \in \varepsilon k - \mathrm{M}(F, D(\cdot))$, however $(0, -1)^T \notin \varepsilon k - \mathrm{MZ}(F, D(\cdot))$. Obviously, $\{(d_1, d_2)^T \in \mathbb{R}^2 | d_1 \geq 0,\ d_2 \leq 0\}$ is not a subset of $\mathbb{R}^2_+$.

In the next theorem ([417, Theorem 6]), we present relationships between (weakly) $\varepsilon k$-nondominated elements of the set $F$ and (weakly) $\varepsilon k$-nondominated-like elements of $F$ with respect to $D$. From this theorem, we can conclude that the sets of $\varepsilon k$-nondominated-like and $\varepsilon k$-nondominated elements do coincide for vector optimization problems with constant domination structures. Eichfelder [139, 143] studied a corresponding result for exact solutions of vector optimization problems with respect to variable domination structures.

**Theorem 11.4.18** *Suppose that assumption (A) holds, $D = \tilde{D} : Y \to \mathcal{P}(Y)$ and additionally $k \in Y\backslash\{0\}$ such that $k \in 0^+ D(y)$ for all $y \in Y$, $\emptyset \neq F \subset Y$ and $\varepsilon \geq 0$. Then, the following assertions hold:*

(a)
$$y_\varepsilon \in \varepsilon k - \mathrm{M}(F, D(\cdot)) \implies y_\varepsilon \in \varepsilon k - \mathrm{N}(F, D(\cdot))$$
*if $D(y) \subseteq D(y_\varepsilon)$ for all $y \in F$.*

(b)
$$y_\varepsilon \in \varepsilon k - \mathrm{N}(F, D(\cdot)) \implies y_\varepsilon \in \varepsilon k - \mathrm{M}(F, D(\cdot))$$
*if $D(y_\varepsilon) \subseteq D(y)$ for all $y \in F$.*

(c) *Suppose that* $\mathrm{int}\, D(y) \neq \emptyset$ *for all $y \in F$, then*
$$y_\varepsilon \in \varepsilon k - \mathrm{WM}(F, D(\cdot)) \implies y_\varepsilon \in \varepsilon k - \mathrm{WN}(F, D(\cdot))$$
*if* $\mathrm{int}\, D(y) \subseteq \mathrm{int}\, D(y_\varepsilon)$ *for all $y \in F$.*

(d) *Suppose that* $\mathrm{int}\, D(y) \neq \emptyset$ *for all $y \in F$, then*
$$y_\varepsilon \in \varepsilon k - \mathrm{WN}(F, D(\cdot)) \implies y_\varepsilon \in \varepsilon k - \mathrm{WM}(F, D(\cdot))$$
*if* $\mathrm{int}\, D(y_\varepsilon) \subseteq \mathrm{int}\, D(y)$ *for all $y \in F$.*

***Proof***

(a) Taking into account $D(y) \subseteq D(y_\varepsilon)$ and Theorem 11.4.15(b), we have $y_\varepsilon \in \varepsilon k - \mathrm{MZ}(F, D(\cdot))$. From Theorem 11.4.12(a), we obtain $y_\varepsilon \in \varepsilon k - \mathrm{N}(F, D(\cdot))$.
(b) Because of $y_\varepsilon \in \varepsilon k - \mathrm{N}(F, D(\cdot))$ and $D(y_\varepsilon) \subseteq D(y)$ for all $y \in F$, we obtain the implication
$$\forall y \in F: \quad y_\varepsilon - \varepsilon k \notin y + (D(y)\backslash\{0\}) \implies y_\varepsilon - \varepsilon k \notin y + (D(y_\varepsilon)\backslash\{0\}),$$
i.e., $(y_\varepsilon - \varepsilon k - D(y_\varepsilon)\backslash\{0\}) \cap F = \emptyset$ and $y_\varepsilon \in \varepsilon k - \mathrm{M}(F, D(\cdot))$.

(c) Since $D(y) \subseteq D(y_\varepsilon)$ and by Theorem 11.4.15(d), we get that $y_\varepsilon \in \varepsilon k - \mathrm{WM}(F, D(\cdot))$ implies $y_\varepsilon \in \varepsilon k - \mathrm{WMZ}(F, D(\cdot))$. With Theorem 11.4.12(c), it follows that $y_\varepsilon \in \varepsilon k - \mathrm{WN}(F, D(\cdot))$.
(d) Similar to the proof of (b) by considering $\operatorname{int} D(y_\varepsilon) \subseteq \operatorname{int} D(y)$ for all $y \in F$, we get the assertion. □

The following examples ([417, Examples 9, 10]) show that the condition $D(y) \subseteq D(y_\varepsilon)$ for all $y \in F$ in Theorem 11.4.18(a) is a sufficient condition but it is not a necessary condition.

**Example 11.4.19** We consider Example 11.4.9. Of course, $(-1, 0)^T \in \varepsilon k - \mathrm{M}(F, D(\cdot))$ and also $(-1, 0)^T \in \varepsilon k - \mathrm{N}(F, D(\cdot))$, however $D(y) \subseteq D(y_\varepsilon)$ does not hold for all $y \in F$.

**Example 11.4.20** We consider Example 11.4.9. Of course, $(0, -1)^T \in \varepsilon k - \mathrm{M}(F, D(\cdot))$, however $(0, -1)^T \notin \varepsilon k - \mathrm{N}(F, D(\cdot))$. Obviously, $\mathbb{R}^2_+$ is not a subset of $\{(d_1, d_2)^T \in \mathbb{R}^2 | d_1 \geq 0,\ d_2 \leq 0\}$.

The following examples (compare [417, Examples 11, 12]) show that the condition $D(y_\varepsilon) \subseteq D(y)$ for all $y \in F$ in Theorem 11.4.18(b) is a sufficient condition but it is not a necessary condition.

**Example 11.4.21** We consider Example 11.4.9. Obviously, $(-1, 0)^T$ is an $\varepsilon k$-nondominated element and also it is an $\varepsilon k$-nondominated-like element, however $D(y_\varepsilon) \subseteq D(y)$ does not hold for all $y \in F$.

**Example 11.4.22** We consider Example 11.4.10. From Example 11.4.10, it is known that $(-1, 0)^T$ is an $\varepsilon k$-nondominated element, however not an $\varepsilon k$-nondominated-like element.

Obviously, $\{(d_1, d_2)^T \in \mathbb{R}^2 \mid d_2 \geq 0,\ d_1 + d_2 \geq -1\}$ is not a subset of $\mathbb{R}^2_+$.

### *11.4.2 Characterization of Approximate Solutions via Scalarization*

In scalarization methods for vector optimization problems, we replace a vector optimization problem by a suitable scalar optimization problem to characterize the optimal elements (see Chap. 6). The results presented in this section are shown by Soleimani in [414].

In this section, we characterize $\varepsilon k$-optimal elements of vector optimization problems with variable domination structure (see Sect. 11.4.1) by scalarization via nonlinear functionals introduced and discussed in Sect. 11.2.1 (see (11.2.4)). Using these results, we show that an approximate solution of the original vector optimization problem with variable domination structure can be characterized by certain scalarizations and vice versa.

The results in this section are derived again under the assumption (A) introduced in Sect. 11.4.1 concerning the set-valued maps $D : Y \to \mathscr{P}(Y)$, $\tilde{D} : Y \to \mathscr{P}(Y)$ and $\hat{D} : Y \to \mathscr{P}(Y)$ with an additional (stronger) assumption $D(y) + \mathbb{R}_{>}k \subseteq \operatorname{int} D(y)$ ($\tilde{D}(y) + \mathbb{R}_{>}k \subseteq \operatorname{int} \tilde{D}(y)$, $\hat{D}(y) + \mathbb{R}_{>}k \subseteq \operatorname{int} \hat{D}(y)$, respectively) for any $y \in Y, k \in Y \setminus \{0\}$.

Under these assumptions, for $w \in Y$, we use the functional $\varphi_{-\hat{D}(w),k} : Y \to \mathbb{R}$ (see (11.2.4) and Sect. 11.2.1):

$$\varphi_{-\hat{D}(w),k}(y) = \inf\{t \in \mathbb{R} \mid y \in tk - \hat{D}(w)\}.$$

In the following theorem ([414, Theorem 5.1]), we show that (weakly) $\varepsilon k$-completely nondominated points of the set $F$ (see Sect. 11.4.1) can be characterized by scalarization via the functional $\varphi_{-\hat{D}(w),k}$.

**Theorem 11.4.23** *Suppose that (A) holds, $k \in Y \setminus \{0\}$ with $\hat{D}(y) + \mathbb{R}_{>}k \subseteq \operatorname{int} \hat{D}(y)$ for any $y \in Y$, $F \subset Y$, and $\varepsilon \geq 0$. Then,*

(a)

$$y_\varepsilon \in \varepsilon k - \mathrm{MZ}(F, \hat{D}(\cdot)) \implies \forall w \in F,\ \forall y \in F : \varphi_{-\hat{D}(w),k}(0) \leq \varphi_{-\hat{D}(w),k}(y - y_\varepsilon) + \varepsilon.$$

(b)

$$y_\varepsilon \in \varepsilon k - \mathrm{WMZ}(F, \hat{D}(\cdot)) \implies \forall w \in F,\ \forall y \in F : \varphi_{-\hat{D}(w),k}(0) \leq \varphi_{-\hat{D}(w),k}(y - y_\varepsilon) + \varepsilon.$$

***Proof***

(a) Assume $y_\varepsilon \in \varepsilon k - \mathrm{MZ}(F, \hat{D}(\cdot))$. Suppose that there are elements $w, y \in F$ such that $\varphi_{-\hat{D}(w),k}(y - y_\varepsilon) + \varepsilon < \varphi_{-\hat{D}(w),k}(0) =: t$. Taking into account Lemma 11.2.21, we obtain $\varphi_{-\hat{D}(w),k}(0) = 0$, i.e., $t = 0$.
Because of $\varphi_{-\hat{D}(w),k}(y - y_\varepsilon) + \varepsilon < \varphi_{-\hat{D}(w),k}(0) = 0$, there is a real number $\gamma > 0$ such that $\varphi_{-\hat{D}(w),k}(y - y_\varepsilon) = -\gamma - \varepsilon$ and by Corollary 11.2.18(c), it holds that

$$(-\gamma - \varepsilon)k + y_\varepsilon - y = c^1 \in \hat{D}(w), \text{ i.e., } y_\varepsilon - y - \varepsilon k \in \hat{D}(w) + \gamma k. \tag{11.4.1}$$

Because of $\gamma > 0$ and $\hat{D}(w) + \mathbb{R}_{>}k \subseteq \operatorname{int} \hat{D}(w)$, we get $y_\varepsilon - \varepsilon k - y \in \operatorname{int} \hat{D}(w)$ and so $y \in y_\varepsilon - \varepsilon k - \operatorname{int} \hat{D}(w)$, i.e.,

$$(y_\varepsilon - \varepsilon k - (\hat{D}(w) \setminus \{0\})) \cap F \neq \emptyset,$$

a contradiction to the assumption.
(b) The proof is similar to the proof of assertion (a). □

In Theorem 11.4.23, we have shown that each (weak) $\varepsilon k$-completely nondominated point of the set $F$ can be characterized by a scalarization by means of the functional

$\varphi_{-\hat{D}(w),k}$. Now, we show sufficient conditions for (weak) $\varepsilon k$-completely nondominated points of the set $F$ using a scalarization by means of $\varphi_{-\hat{D}(w),k}$.

The assertions of the next theorem were shown in [414, Theorem 5.2].

**Theorem 11.4.24** *Suppose that (A) holds, $k \in Y\backslash\{0\}$ with $\hat{D}(y) + \mathbb{R}_{>}k \subseteq \operatorname{int} \hat{D}(y)$ for any $y \in Y$, $F \subset Y$, and $\varepsilon \geq 0$. Then,*

(a)

$$y_\varepsilon \in F \text{ and } \forall w \in F,\ \forall y \in F : \varphi_{-\hat{D}(w),k}(0) < \varphi_{-\hat{D}(w),k}(y - y_\varepsilon) + \varepsilon \implies y_\varepsilon \in \varepsilon k - \mathrm{MZ}(F, \hat{D}(\cdot)).$$

(b)

$$y_\varepsilon \in F \text{ and } \forall w \in F,\ \forall y \in F : \varphi_{-\hat{D}(w),k}(0) \leq \varphi_{-\hat{D}(w),k}(y - y_\varepsilon) + \varepsilon \implies y_\varepsilon \in \varepsilon k - \mathrm{WMZ}(F, \hat{D}(\cdot)).$$

***Proof***

(a) Assume that $\varphi_{-\hat{D}(w),k}(0) < \varphi_{-\hat{D}(w),k}(y - y_\varepsilon) + \varepsilon$ for all $w, y \in F$ but $y_\varepsilon \notin \varepsilon k - \mathrm{MZ}(F, \hat{D}(\cdot))$, then there are elements $y^1, y^2 \in F$ such that $y^1 \in y_\varepsilon - \varepsilon k - (\hat{D}(y^2)\backslash\{0\})$, i.e., $y^1 - y_\varepsilon \in -\varepsilon k - (\hat{D}(y^2)\backslash\{0\})$. By Corollary 11.2.18(b), it holds that $\varphi_{-\hat{D}(y^2),k}(y^1 - y_\varepsilon) \leq -\varepsilon$. This implies $\varphi_{-\hat{D}(y^2),k}(y^1 - y_\varepsilon) + \varepsilon \leq 0$. From Lemma 11.2.21 we get $\varphi_{-\hat{D}(y^2),k}(y^1 - y_\varepsilon) + \varepsilon \leq 0 = \varphi_{-\hat{D}(y^2),k}(0)$, a contradiction to the assumption.

(b) Assume that $\varphi_{-\hat{D}(w),k}(0) \leq \varphi_{-\hat{D}(w),k}(y - y_\varepsilon) + \varepsilon$ for all $w, y \in F$ but $y_\varepsilon \notin \varepsilon k - \mathrm{WMZ}(F, \hat{D}(\cdot))$, then there are elements $y^1, y^2 \in F$ such that $y^1 \in y_\varepsilon - \varepsilon k - \operatorname{int} \hat{D}(y^2)$, i.e., $y^1 - y_\varepsilon \in -\varepsilon k - \operatorname{int} \hat{D}(y^2)$. Taking into account Corollary 11.2.18(a), we get $\varphi_{-\hat{D}(y^2),k}(y^1 - y_\varepsilon) < -\varepsilon$. This implies $\varphi_{-\hat{D}(y^2),k}(y^1 - y_\varepsilon) + \varepsilon < 0 = \varphi_{-\hat{D}(y^2),k}(0)$ taking into account Lemma 11.2.21, a contradiction to our assumption.

□

By Theorems 11.4.23 and 11.4.24, we characterized approximate completely nondominated points by scalarizations via nonlinear functionals $\varphi_{-D(w),k}$ given by (11.2.4). Furthermore, we can use this method in order to derive a characterization of approximate nondominated elements. In the next theorem ([414, Theorem 5.3]), we show that each (weakly) $\varepsilon k$-nondomianted element of the set $F$ is characterized by suitable scalar optimization problems using nonlinear functionals $\varphi_{-D(w),k}$ given by (11.2.4).

**Theorem 11.4.25** *Suppose that (A) holds, $k \in Y\backslash\{0\}$ with $D(y) + \mathbb{R}_{>}k \subseteq \operatorname{int} D(y)$ for any $y \in Y$, $F \subset Y$, and $\varepsilon \geq 0$. Then,*

(a)

$$y_\varepsilon \in \varepsilon k - \mathrm{N}(F, D(\cdot)) \implies \forall w \in F :\ \varphi_{-D(w),k}(0) \leq \varphi_{-D(w),k}(w - y_\varepsilon) + \varepsilon.$$

(b)

$$y_\varepsilon \in \varepsilon k - \mathrm{WN}(F, D(\cdot)) \implies \forall w \in F : \varphi_{-D(w),k}(0) \le \varphi_{-D(w),k}(w - y_\varepsilon) + \varepsilon.$$

***Proof***

(a) Assume that $y_\varepsilon \in \varepsilon k - \mathrm{N}(F, D(\cdot))$ and there is an element $w \in F$ such that $\varphi_{-D(w),k}(w - y_\varepsilon) + \varepsilon < \varphi_{-D(w),k}(0) =: t$. Because of Lemma 11.2.21, we have $\varphi_{-D(w),k}(0) = 0$, i.e., $t = 0$.
Since $\varphi_{-D(w),k}(w - y_\varepsilon) + \varepsilon < \varphi_{-D(w),k}(0) = 0$, there is a real number $\gamma > 0$ such that $\varphi_{-D(w),k}(w - y_\varepsilon) = -\gamma - \varepsilon$ and by Corollary 11.2.18(c),

$$w - y_\varepsilon \in (-\varepsilon - \gamma)k - \mathrm{bd}\, D(w). \tag{11.4.2}$$

This yields, together with $\gamma > 0$ and $D(w) + \mathbb{R}_{>}k \subseteq \mathrm{int}\, D(w)$, that $w - y_\varepsilon \in -\varepsilon k - \mathrm{int}\, D(y)$, i.e., $w \in y_\varepsilon - \varepsilon k - \mathrm{int}\, D(y)$, and

$$y_\varepsilon - \varepsilon k \in w + (D(w)\backslash\{0\}),$$

which is a contradiction to the assumption that $y_\varepsilon$ is an $\varepsilon k$-nondominated element of the set $F$.

(b) The proof is analog to the proof of (a). □

In Theorem 11.4.25, we have shown necessary conditions for (weakly) $\varepsilon k$-nondominated elements of the set $F$ using a nonlinear scalararization by means of the functional $\varphi_{-D(w),k}$. Now, we give sufficient conditions for (weakly) $\varepsilon k$-nondominated elements of the set $F$ applying a scalarization by means of the functional $\varphi_{-D(w),k}$. The assertion of the next theorem is shown in [414, Theorem 5.4].

**Theorem 11.4.26** *Suppose that (A) holds, $k \in Y\backslash\{0\}$ with $D(y) + \mathbb{R}_{>}k \subseteq \mathrm{int}\, D(y)$ for any $y \in Y$, $F \subset Y$, and $\varepsilon \ge 0$.*

(a) *Assume that $y_\varepsilon \in F$. Then,*

$$\forall w \in F : \varphi_{-D(w),k}(0) < \varphi_{-D(w),k}(w - y_\varepsilon) + \varepsilon \implies y_\varepsilon \in \varepsilon k - \mathrm{N}(F, D(\cdot)).$$

(b) *Assume that $y_\varepsilon \in F$. Then,*

$$\forall w \in F : \varphi_{-D(w),k}(0) \le \varphi_{-D(w),k}(w - y_\varepsilon) + \varepsilon \implies y_\varepsilon \in \varepsilon k - \mathrm{WN}(F, D(\cdot)).$$

***Proof***

(a) Assume that $\varphi_{-D(w),k}(0) < \varphi_{-D(w),k}(w - y_\varepsilon) + \varepsilon$ for all $w \in F$ but $y_\varepsilon \notin \varepsilon k - \mathrm{N}(F, D(\cdot))$, then there is an element $w \in F$ such that $w \in y_\varepsilon - \varepsilon k - (D(w)\backslash\{0\})$. This means (for fixed $w \in F$)

$$y_\varepsilon - \varepsilon k \in w + (D(w)\backslash\{0\}).$$

By Corollary 11.2.18(b) and because of Lemma 11.2.21, it holds that $\varphi_{-D(w),k}(w - y_\varepsilon) + \varepsilon \leq 0 = \varphi_{-D(w),k}(0)$, a contradiction to our assumption.

(b) The proof is analog to the proof of Theorem 11.4.24(b). □

**Remark 11.4.27** In Theorems 11.4.25 and 11.4.26, we derived a characterization of $\varepsilon k$-nondominated elements by scalarization via nonlinear functionals $\varphi_{-D(w),k}$ given by (11.2.4). However, in Sect. 12.3.2, Theorem 12.3.28, we will prove a variational principle of Ekeland's type for $\varepsilon k$-nondominated solutions using the functional $\varphi_{\overline{y}-D(\cdot),k}$ given by (11.2.3), see also Lemmata 12.3.23 and 12.3.24 for further results about a corresponding scalarization.

**Remark 11.4.28** In the special case, when $\varepsilon = 0$ and $D$ is a cone-valued map, where $D(y)$ is a nontrivial, closed and convex cone for all $y \in Y$, corresponding characterizations of (weakly) nondominated elements to vector optimization problems with variable domination structures based on a scalarization by means of the functional (11.2.3) are given in Theorem 11.3.1, see Sect. 11.3 and Eichfelder [143].

In the following theorem, we characterize approximate nondominated-like elements $y_\varepsilon$ of $F$ with respect to the domination map $\tilde{D} : Y \to \mathscr{P}(Y)$ by scalarization via nonlinear functionals $\varphi_{-\tilde{D}(y_\varepsilon),k}$. We show that each (weakly) $\varepsilon k$-nondominated-like element $y_\varepsilon$ of the set $F$ can be characterized by scalarization using the functional $\varphi_{-\tilde{D}(y_\varepsilon),k}$. The next theorem is shown in [414, Theorem 5.5].

**Theorem 11.4.29** *Suppose that (A) holds, $k \in Y\backslash\{0\}$ with $\tilde{D}(y) + \mathbb{R}_{>}k \subseteq \operatorname{int} \tilde{D}(y)$ for any $y \in Y$, $F \subset Y$, and $\varepsilon \geq 0$. Then,*

(a)

$$y_\varepsilon \in \varepsilon k - \mathrm{M}(F, \tilde{D}(\cdot)) \implies \forall y \in F : \varphi_{-\tilde{D}(y_\varepsilon),k}(0) \leq \varphi_{-\tilde{D}(y_\varepsilon),k}(y - y_\varepsilon) + \varepsilon.$$

(b)

$$y_\varepsilon \in \varepsilon k - \mathrm{WM}(F, \tilde{D}(\cdot)) \implies \forall y \in F : \varphi_{-\tilde{D}(y_\varepsilon),k}(0) \leq \varphi_{-\tilde{D}(y_\varepsilon),k}(y - y_\varepsilon) + \varepsilon.$$

***Proof***

(a) Assume that $y_\varepsilon \in \varepsilon k - \mathrm{M}(F, \tilde{D}(\cdot))$ and there is an element $y \in F$ such that $\varphi_{-\tilde{D}(y_\varepsilon),k}(y - y_\varepsilon) + \varepsilon < \varphi_{-\tilde{D}(y_\varepsilon),k}(0) =: t$. Because of Lemma 11.2.21, we have $\varphi_{-\tilde{D}(y_\varepsilon),k}(0) = 0$, i.e., $t = 0$. Since $\varphi_{-\tilde{D}(y_\varepsilon),k}(y - y_\varepsilon) + \varepsilon < 0$, there is a real number $\gamma > 0$ such that $\varphi_{-\tilde{D}(y_\varepsilon),k}(y - y_\varepsilon) = -\gamma - \varepsilon$. Taking into account Corollary 11.2.18(c), we get

$$y - y_\varepsilon \in (-\gamma - \varepsilon)k - \operatorname{bd} \tilde{D}(y_\varepsilon) \tag{11.4.3}$$

and

$$y_\varepsilon - \varepsilon k - y \in \operatorname{bd} \tilde{D}(y_\varepsilon) + \gamma k.$$

Because of $\gamma > 0$ and $\tilde{D}(y_\varepsilon) + \mathbb{R}_> k \subseteq \operatorname{int} \tilde{D}(y_\varepsilon)$, we obtain $y_\varepsilon - \varepsilon k - y \in \operatorname{int} \tilde{D}(y_\varepsilon)$ and

$$(y_\varepsilon - \varepsilon k - (\tilde{D}(y_\varepsilon)\backslash\{0\})) \cap F \neq \emptyset,$$

a contradiction to the assumption.

(b) The assertion follows in the same way like in the proof of (a). □

Theorem 11.4.29 tells us that each (weakly) $\varepsilon k$-nondominated-like element $y_\varepsilon$ of the set $F$ can be characterized by scalarization using the functional $\varphi_{-\tilde{D}(y_\varepsilon),k}$. In the following, we show sufficient conditions for (weakly) $\varepsilon k$-nondominated-like elements of the set $F$ using a suitable scalarization. The assertion of the next theorem is shown in [414, Theorem 5.6].

**Theorem 11.4.30** *Suppose that (A) holds, $k \in Y\backslash\{0\}$ with $\tilde{D}(y) + \mathbb{R}_> k \subseteq \operatorname{int} \tilde{D}(y)$ for any $y \in Y$, $F \subset Y$, and $\varepsilon \geq 0$.*

(a) *Assume that $y_\varepsilon \in F$, then*

$$\forall y \in F: \ \varphi_{-\tilde{D}(y_\varepsilon),k}(0) < \varphi_{-\tilde{D}(y_\varepsilon),k}(y - y_\varepsilon) + \varepsilon \quad \Longrightarrow \quad y_\varepsilon \in \varepsilon k - \mathrm{M}(F, \tilde{D}(\cdot)).$$

(b) *Assume that $y_\varepsilon \in F$, then*

$$\forall y \in F: \ \varphi_{-\tilde{D}(y_\varepsilon),k}(0) \leq \varphi_{-\tilde{D}(y_\varepsilon),k}(y - y_\varepsilon) + \varepsilon \quad \Longrightarrow \quad y_\varepsilon \in \varepsilon k - \mathrm{WM}(F, \tilde{D}(\cdot)).$$

***Proof***

(a) Assume that $\varphi_{-\tilde{D}(y_\varepsilon),k}(0) < \varphi_{-\tilde{D}(y_\varepsilon),k}(y - y_\varepsilon) + \varepsilon$ for all $y \in F$ but $y_\varepsilon \notin \varepsilon k - \mathrm{M}(F, \tilde{D}(\cdot))$. Then, there is an element $y \in F$ such that $y \in y_\varepsilon - \varepsilon k - (\tilde{D}(y_\varepsilon)\backslash\{0\})$, i.e., $y - y_\varepsilon \in -\varepsilon k - (\tilde{D}(y_\varepsilon)\backslash\{0\})$.
By Corollary 11.2.18(b) and because of Lemma 11.2.21, we obtain $\varphi_{-\tilde{D}(y_\varepsilon),k}(y - y_\varepsilon) + \varepsilon \leq 0 = \varphi_{-\tilde{D}(y_\varepsilon),k}(0) = 0$, a contradiction to the assumption.

(b) The assertion follows similar to the proof of Theorem 11.4.24(a). □

**Remark 11.4.31** For the special case $\varepsilon = 0$, corresponding characterizations of (weakly) nondominated-like elements are shown in Corollary 11.3.4 under the assumption that $\tilde{D}(y)$ is a nontrivial, closed, convex cone for all $y \in Y$.

**Remark 11.4.32** We will use the results presented in Theorems 11.4.29 and 11.4.30 concerning a characterization of $\varepsilon k$-nondominated-like elements via scalarization in order to prove a variational principle of Ekeland's type for $\varepsilon k$-nondominated-like solutions (see Theorem 12.3.20).

# Chapter 12
# Variational Methods in Topological Vector Spaces

In this chapter, we explain that the application of the functional $\varphi_{A,k}$ defined by (4.1.11) (or its modifications) is an important tool for deriving necessary optimality conditions in vector optimization (Sect. 12.1), duality assertions (Sect. 12.2), minimal point theorems and variational principles (Sect. 12.3), necessary conditions for approximate solutions of vector optimization problems with respect to variable domination structure (Sect. 12.4), existence results for solutions of vector variational inequalities (Sect. 12.5) and many other results in optimization theory and functional analysis. Furthermore, we derive certain relationships to the Extremal Principle by Kruger and Mordukhovich (Sect. 12.6). For the proofs of these results, the separation properties of the functional $\varphi_{A,k}$ as well as the monotonicity, convexity, continuity properties and the translation invariance, as shown in Chap. 4, are essential.

## 12.1 Lagrange Multiplier Rules

The aim of this section is to prove necessary optimality conditions for solutions of vector optimization problems (VOP) (see Sect. 6.1) using the functional $\varphi_{A,k}$ defined by (4.1.11) and certain concepts of generalized differentiation. The separation properties of $\varphi_{A,k}$ shown in Separation Theorem 4.11.3 as well as the structure of the subdifferential of $\varphi_{A,k}$ presented in Theorem 4.10.3 are important for the proofs of the assertions.

There are many papers where Lagrangian multiplier rules are shown for vector optimization problems under certain differentiability or convexity (or convexity-like) assumptions (see Jahn [272]). Necessary conditions for weakly efficient solutions in nonsmooth vector optimization where the objective function takes its values in finite-dimensional spaces are derived by Clarke [98], Minami [356], Craven [102] and Miettinen [354, 355]. Lagrangian multiplier rules for weakly efficient solutions of vector optimization problems on Banach spaces are presented by El Abdouni and



C. Tammer and P. Weidner, *Scalarization and Separation by Translation Invariant Functions*, Vector Optimization, https://doi.org/10.1007/978-3-030-44723-6_12

Thibault in [154]. Furthermore, Lagrange multiplier rules in a general context are derived by Dutta and Tammer [130], by Durea and Tammer [129], by Durea, Dutta and Tammer [123–125] and by Durea and Dutta [122]. For recent results concerning necessary optimality conditions for vector optimization problems even in the case that the ordering cone has an empty interior, we refer to the papers by Bao and Tammer [28] and Durea, Strugariu and Tammer [127] and references therein.

The results in this section are shown by Tammer and Zălinescu in [441, Sect. 5].

In order to formulate the vector optimization problem, we assume

$(H_{X,Y,C})$ $X$ and $Y$ are Asplund spaces and $C \subset Y$ is a nontrivial, closed, convex, pointed cone.

Furthermore, we suppose $f : X \to Y$, $S \subseteq X$ and consider the vector minimization problem (VOP) (see Chap. 6)

$$C-\operatorname*{Opt}_{x\in S} f(x).$$

The solution concepts for the vector optimization problem (VOP) are described in Chap. 6.

In this section, we do not suppose convexity assumptions concerning the objective function $f$ and the feasible set $S$ in (VOP). For deriving Lagrange multiplier rules for properly and weakly efficient solutions of the nonconvex problem (VOP), we will use a scalarization by means of the nonlinear scalarizing functional (4.1.11) as well as the limiting (Mordukhovich) subdifferential and corresponding calculus (see Mordukhovich [359]). In Sect. 12.4.2, we will use a more general generic approach concerning the subdifferential for deriving optimality conditions.

We are interested in developing the Lagrangian necessary conditions for the vector optimization problem (VOP) and also try to demonstrate how Lagrangian multipliers can be interpreted as subgradients of the nonlinear functional (4.1.11). In the Lagrangian theory of vector optimization, one of the most important approaches is scalarization, where the vector optimization problem is converted into an equivalent scalar optimization problem and then the usual techniques of scalar optimization are applied.

For example, we look at the case where the function $f$ is a $C$-convex function, $S$ a closed, convex subset of $X$, $\operatorname{int} C \neq \emptyset$ and let $x_0 \in S$ be a weakly efficient solution for (VOP). Then, by using a classical separation argument for convex sets, it is easy to show that there is an element $y^* \in C^+ \setminus \{0\}$ such that $x_0$ also solves the following scalar optimization problem (SP)

$$\inf_{x\in S} y^*(f(x)).$$

Then, we get the well-known necessary optimality condition in the form

$$0 \in \partial y^*(f(x_0)) + N(x_0; S),$$

where $\partial y^*(f(x_0))$ is the (Fenchel) subdifferential of $y^* \circ f$ at $x_0$ (see Definition 3.6.4) and $N(x_0; S)$ denotes the normal cone in the sense of convex analysis to $S$ at $x_0$ (see (3.6.2)).

The converse of this fact is also true. But when the function $f$ is not convex, such a classical scalarization through linear continuous functionals is not possible since the convex separation results are no longer available. This leads us to an application of the nonlinear scalarization scheme by means of the functional (4.1.11).

Under the assumption that the ordering cone has a nonempty interior, we will show the existence of nontrivial, positive multipliers $y^* \in C^+ \setminus \{0\}$. Unfortunately, this procedure does not work for the important situation where $\text{int}\, C = \emptyset$. An overview on results for the case $\text{int}\, C = \emptyset$ is given by Durea, Strugariu and Tammer in [127], compare also Bao and Tammer [28].

### 12.1.1 Necessary Conditions for Properly Efficient Solutions

In this section, we develop Lagrangian multiplier rules for (VOP) using the nonconvex scalarization scheme by means of the functional (4.1.11). Moreover, we will also demonstrate that the Lagrangian multipliers are in fact subgradients of the convex functional (4.1.11) (for $A$ convex) that are generated by these nonlinear scalarization schemes.

The aim of this section is to derive necessary optimality conditions for solutions to the vector optimization problem (VOP) based on the calculus for the limiting subdifferential $\partial_M$ (Definition 8.1.5) and the limiting normal cone $N_M$ (Definition 8.1.4), see Mordukhovich [359].

**Remark 12.1.1** Assume that $A \subset Y$ is proper, closed and convex, $C \subseteq Y$ is a cone, satisfying the assumption $A - (C \setminus \{0\}) = \text{int}\, A$ and $A \neq Y$. Then, the following assertions hold:

(i) $C$ is pointed, that is, $C \cap (-C) \subseteq \{0\}$. Indeed, assume that $c \in C \cap (-C) \setminus \{0\}$. Take $a \in \text{bd}\, A$ ($\subseteq A$). Such an element $a$ exists because $A \neq Y$. Then, $a' := a - c \in \text{int}\, A \subset A$, and so $a = a' - (-c) \in \text{int}\, A$, a contradiction.

(ii) If $k \in C \setminus \{0\}$, then $A - \mathbb{R}_{>}k \subseteq \text{int}\, A$. Hence, by Theorem 4.6.6, especially (4.6.11), $\varphi_{A,k}$ is continuous on $\text{dom}\, \varphi_{A,k}$,

$$\text{dom}\, \varphi_{A,k} = \mathbb{R}k + A = \mathbb{R}k + \text{int}\, A = \text{int}(\text{dom}\, \varphi_{A,k}) \text{ is an open set}$$

and $\text{bd}\, A = \{y \in Y \mid \varphi_{A,k}(y) = 0\}$.

In the proof of the following necessary condition for Nehse–Iwanow properly efficient solutions of (VOP) (see Definition 6.3.3), we use the separating functional $\varphi_{A,k}$ defined by (4.1.11) and corresponding properties given in Theorem 4.6.6. We formulate the necessary conditions for properly efficient solutions of (VOP) using the limiting (Mordukhovich) subdifferential $\partial_M$ introduced in Definition 8.1.5 and the corresponding calculus. Furthermore, we suppose that the function $f : X \to Y$ is strictly Lipschitzian in the sense of Definition 8.2.6 in the following theorem shown by Tammer and Zălinescu in ([441, Theorem 5.4]).

**Theorem 12.1.2** *Assume that $(H_{X,Y,C})$ holds. Let $k \in C \setminus \{0\}$ be fixed and $Y = C + \mathbb{R}k$. Furthermore, assume that $f : X \to Y$ is strictly Lipschitzian, $S$ is a closed subset of $X$ and $x_0 \in S$. If $x_0$ is a properly efficient solution to (VOP) with $f(x_0) \in$ NI-PEff$(f(S), C)$, then there is an element $y^* \in \operatorname{qint} C^+$ with $y^*(k) = 1$ such that*

$$0 \in \partial_M y^*(f(x_0)) + N_M(x_0; S). \tag{12.1.1}$$

*Moreover, if $f$ is strictly differentiable at $x_0$ then*

$$(f'(x_0))^*(y^*) \in -N_M(x_0; S). \tag{12.1.2}$$

***Proof*** Assume that $x_0$ is a properly efficient solution for (VOP) with $f(x_0) \in$ NI-PEff$(f(S), C)$. Then, there is a closed, convex set $H \subset Y$ that satisfies the assumption $0 \in \operatorname{bd} H$, $H + (C \setminus \{0\}) \subseteq \operatorname{int} H$ and $H \neq Y$ such that $f(S) \cap (f(x_0) - \operatorname{int} H) = \emptyset$. This set $H$ and the element $k \in C \setminus \{0\}$ are used for the construction of the nonlinear functional $g : X \to \overline{\mathbb{R}}$ given by

$$g(x) := \varphi_{-H,k}(f(x) - f(x_0)) + \iota_S(x).$$

By Corollary 6.3.18, $x_0$ is a minimum point of $g$ over $X$.

From Lemma 4.6.10 and Proposition 4.6.11, we know that $H + (C \setminus \{0\}) \subseteq \operatorname{int} H$ is equivalent to $H + \mathbb{R}_{>}k \subseteq \operatorname{int} H$. As mentioned in Remark 12.1.1(ii), $\operatorname{dom} \varphi_{-H,k}$ is open because $k \in C \setminus \{0\}$ and $\varphi_{-H,k}(0) = 0$ because $0 \in \operatorname{bd} H$. Note that $k \notin 0^+(-H)$; otherwise we get the contradiction $0 = (0 + k) - k \in -H - (0, +\infty)k \subseteq \operatorname{int}(-H)$. So, we have $k \in 0^+H \setminus (-0^+H)$ such that the assumptions of Proposition 4.7.9 are fulfilled. Since $0 \in H \subset \operatorname{dom} \varphi_{-H,k}$, by Proposition 4.7.9(f) we have that $\varphi_{-H,k}$ is Lipschitzian on a neighborhood of 0. Furthermore, $\varphi_{-H,k}$ is convex (see Theorem 4.4.4). It follows that $0 \in \partial_M g(x_0)$ (see [359, Prop. 1.114]). Since $f$ is strictly Lipschitzian and $\varphi_{-H,k}$ is Lipschitzian on a neighborhood of 0, the function $x \mapsto \varphi_{-H,k}(f(x) - f(x_0))$ is Lipschitzian on a neighborhood of $x_0$. Moreover, we have that $\iota_S$ is a lower semicontinuous function since $S$ is a closed subset of $X$. Applying Lemmata 8.1.7, 8.3.6 and Proposition 8.1.6, we get

$$0 \in \partial_M y^*(f(x_0)) + N_M(x_0; S)$$

for some $y^* \in \partial_M \varphi_{-H,k}(0) = \partial \varphi_{-H,k}(0)$, $\varphi_{-H,k}$ being convex and finite and Lipschitzian on a neighborhood of 0. From Theorem 4.10.3, we obtain $y^* \in \operatorname{qint} C^+$ and $y^*(k) = 1$, and so $y^* \neq 0$ such that (12.1.1) is fulfilled.

If $f$ is strictly differentiable at $x_0$, then $\partial_M f(x_0) = \{f'(x_0)\}$ and we conclude that (12.1.2) holds. □

**Remark 12.1.3** In the proof of Theorem 12.1.2, we can see that the Lagrange multipliers $y^* \in \operatorname{qint} C^+$ are nothing else than the subgradients of the nonlinear separating functional $\varphi_{-H,k}$ at the point 0 ($y^* \in \partial_M \varphi_{-H,k}(0) = \partial \varphi_{-H,k}(0)$).

For weakly efficient solutions of (VOP), we have the following result ([441, Theorem 5.5]).

**Theorem 12.1.4** *Assume that* $(H_{X,Y,C})$ *holds. Let* $k \in \operatorname{int} C$ *be fixed. Furthermore, assume that* $f : X \to Y$ *is strictly Lipschitzian,* $S$ *is a closed subset of* $X$ *and* $x_0 \in S$. *If* $x_0$ *is a weakly efficient solution to (VOP), then there is an element* $y^* \in C^+ \setminus \{0\}$ *with* $y^*(k) = 1$ *such that (12.1.1) holds. Moreover, if* $f$ *is strictly differentiable at* $x_0$ *then (12.1.2) holds.*

***Proof*** If $\operatorname{int} C \neq \emptyset$ and $x_0$ is a weakly efficient solution for (VOP), by Theorem 6.2.31, we have that $x_0$ is a minimum point of the corresponding scalar problem with the scalarizing functional $\varphi_{-C,k}$ with $k \in \operatorname{int} C$. From Proposition 4.7.9 and Proposition 4.7.1(b), we get that $\varphi_{-C,k}$ is Lipschitzian on a neighborhood of 0 and sublinear. The rest of the proof is similar to the proof of Theorem 12.1.2. □

**Remark 12.1.5** If $X$ is an Asplund space and $g : X \to \overline{\mathbb{R}}$ is finite and Lipschitzian on a neighborhood of $x_0 \in S \subset X$ with $S$ closed, the following well-known relations

$$\partial_C g(x_0) = \operatorname{cl}_{w^*}(\operatorname{conv} \partial_M g(x_0)) \text{ and } N_C(x_0; S) = \operatorname{cl}_{w^*}(\operatorname{conv} N_M(x_0; S))$$

hold (see [359, Th. 3.57]), where $\partial_C g(x_0)$ and $N_C(x_0; S)$ represent Clarke's subdifferential of $g$ at $x_0$ (see (8.1.6)) and Clarke's normal cone of $S$ at $x_0$, respectively. Here, we are using the fact that Clarke's normal cone $N_C(x_0; S)$ to $S$ at $x_0 \in S$ is described via the weak*-closure of the cone spanned on the generalized subdifferential of the distance function:

$$N_C(x_0; S) = \operatorname{cl}_{w^*}(\mathbb{R}_+ \partial_C d(x_0, S)). \tag{12.1.3}$$

In the hypotheses of Theorem 12.1.2 from (12.1.1) we get the necessary optimality condition

$$\exists y^* \in C^+ \setminus \{0\} \;:\; 0 \in \partial_C y^*(f(x_0)) + N_C(x_0; S) \tag{12.1.4}$$

in terms of Clarke's subdifferential and Clarke's normal cone. However, the optimality condition given by (12.1.1) is sharper than the condition given by (12.1.4).

**Remark 12.1.6** Note that Theorems 12.1.2 and 12.1.4 remain valid when the limiting (Mordukhovich) subdifferential $\partial_M$ is replaced by another subdifferential $\partial$ which meets conditions (H1)–(H4) in Durea and Tammer [129] (compare also Sect. 12.4.2). In such a situation, Theorem 12.1.4 corresponds to Lagrangian necessary condition for weakly efficient solutions in [129, Th. 3.1] (compare also [130, Th. 3.2] for the case $\dim Y < +\infty$). In Theorem 12.1.2, we have established the result for properly efficient solutions without assuming $\operatorname{int} C \neq \emptyset$.

**Remark 12.1.7** Recently, necessary optimality conditions for vector optimization problems using a penalization method and a scalarization by means of the nonlinear scalarizing functional (4.1.11) are given in [128].

In the proofs of Theorems 12.1.2, 12.1.4, we supposed that the set involved in the concept of proper minimality is closed and convex in order to ensure the existence of subgradients and to apply Theorem 4.10.3. For more general solution concepts without assuming the convexity of the involved set, we can use the following result concerning the subdifferential of the nonlinear scalarizing functional shown by Bao and Tammer in [29, Theorem 3.1].

We derive the formula for the limiting subdifferential of the nonlinear separating functional (4.1.11). Here, we do not assume that the set $A$ is convex. From Theorem 4.4.4(f), we know that the nonlinear separating functional (4.1.11) is not convex if $A$ is not convex. When $A$ has an empty interior, (4.1.11) is not Lipschitz. Therefore, we use the limiting generalized differentiation to compute the limiting subdifferential of the nonlinear separating functional.

**Theorem 12.1.8** (Subdifferentials of nonlinear separating functional in the non-convex case) *Let $Y$ be an Asplund space, $A$ be a proper and closed set in $Y$, and $k \in 0^+A \setminus \{0\}$. Consider the nonlinear separating functional $\varphi_{-A,k}$ and $y \in \operatorname{dom} \varphi_{-A,k}$. Then, the limiting subdifferential of $\varphi_{-A,k}$ at $(\overline{y}, \overline{t}) \in \operatorname{epi} \varphi_{-A,k}$ is given by (for $\lambda \in \mathbb{R}$)*

$$\partial_M \varphi_{-A,k}(\overline{y})(\lambda) = H_\lambda(k) \cap \big( - N_M(\overline{t}k - \overline{y}; A) \big), \tag{12.1.5}$$

*where $H_\lambda(k) := \{y^* \in Y^* \mid y^*(k) = \lambda\}$.*

***Proof*** We look at a set-valued map $F : \mathbb{R} \rightrightarrows Y$ defined by $F(t) := tk - A$. Note that $\operatorname{graph} F^{-1} = \operatorname{epi} \varphi_{-A,k}$. First, we will prove that $\operatorname{epi} \varphi_{-A,k} \subseteq \operatorname{graph} F^{-1}$. Indeed, for any $(-u, t) \in \operatorname{epi} \varphi_{-A,k}$, we have $\varphi_{-A,k}(tk - u - tk) \leq t$. Taking into account Theorem 4.4.4, we know that $\varphi_{-A,k}(-u - tk) + t \leq t$ and so $\varphi_{-A,k}(-u - tk) \leq 0$. This implies $-u - tk \in -A$. Put $y = u + tk \in A$. Then, we have $(-u, t) = (tk - y, t) \in \operatorname{graph} F^{-1}$. In order to prove the converse inclusion, we fix an arbitrary pair $(t, tk - y) \in \operatorname{graph} F$ implying that $y \in A$ and $(tk - y, t) \in \operatorname{graph} F$. Again from Theorem 4.4.4, we get $\varphi_{-A,k}(-y) \leq 0$ and $\varphi_{-A,k}(tk - y) = t + \varphi_{-A,k}(-y) \leq t$ clearly verifying that $(tk - y, t) \in \operatorname{epi} \varphi_{-A,k}$. Since $(tk - y, t)$ was arbitrary in $\operatorname{graph} F^{-1}$, we get $\operatorname{graph} F^{-1} \subseteq \operatorname{epi} \varphi_{-A,k}$.

Furthermore, we introduce a constant set-valued map $\mathscr{A} : \mathbb{R} \rightrightarrows Y$ with $\mathscr{A}(t) := -A$ for all $t \in \mathbb{R}$ and a strictly differentiable function $\kappa : \mathbb{R} \to Y$ with $\kappa(t) := tk$ to describe the set-valued map $F$ as a sum of two maps $F := \kappa + \mathscr{A}$. Now, we fix an arbitrary pair $(\overline{t}, \overline{y}) \in \operatorname{graph} F$, i.e., $(\overline{y}, \overline{t}) \in \operatorname{epi} \varphi_{-A,k}$. We employ the sum rule for coderivatives with equality from [359, Theorem 1.62] to $\kappa$ and $\mathscr{A}$ such that for every $y^* \in Y^*$, it holds that

$$D_M^* F(\overline{t}, \overline{y})(-y^*) = -y^*(k) + D_M^* \mathscr{A}(\overline{t}, \overline{y} - \overline{t}k)(-y^*). \tag{12.1.6}$$

Because of graph $\mathscr{A} = \mathbb{R} \times (-A)$, [359, Proposition 1.2] yields

$$N_M((\bar{t}, \bar{y} - \bar{t}k); \operatorname{graph} \mathscr{A}) = N_M(\bar{t}; \mathbb{R}) \times N_M(\bar{y} - \bar{t}k; -A) = \{0\} \times N_M(\bar{y} - \bar{t}k; -A),$$

where $N_M((\bar{t}, \bar{y} - \bar{t}k); \operatorname{graph} \mathscr{A})$ stands for the limiting normal cone to graph $\mathscr{A}$ at $(\bar{t}, \bar{y} - \bar{t}k)$ introduced in Definition 8.1.4. Taking into account Definition 8.3.2 for coderivatives, we obtain

$$y^* \in N_M(\bar{y} - \bar{t}k; -A) \text{ and } D_M^* \mathscr{A}(\bar{y} - \bar{t}k; -A)(-y^*) = \{0\}.$$

Now, we obtain from (12.1.6) that $y^* \in N_M(\bar{y} - \bar{t}k; -A) = -N_M(\bar{t}k - \bar{y}; A)$ and $D_M^* F(\bar{t}, \bar{y})(-y^*) = -y^*(k)$. Reformulating the latter equality, we get

$$\begin{aligned} &(-y^*(k), y^*) \in N_M((\bar{t}, \bar{y}); \operatorname{graph} F) \Leftrightarrow (y^*, -y^*(k)) \in N_M((\bar{y}, \bar{t}); \operatorname{graph} F^{-1}) \\ \Longleftrightarrow\ &(y^*, -y^*(k)) \in N_M((\bar{y}, \bar{t}); \operatorname{epi} \varphi_{-A,k}) \Longleftrightarrow y^* \in D_M^* \mathscr{E}_{\varphi_{-A,k}}(\bar{y}, \bar{t})(y^*(k)). \end{aligned}$$

From Definition 8.3.5, we conclude

$$\partial_M \varphi_{-A,k}(\bar{y}, \bar{t})(\lambda) = D_M^* \mathscr{E}_{\varphi_{-A,k}}(\bar{y}, \bar{t})(\lambda) = H_\lambda(k) \cap (-N_M(\bar{t}k - \bar{y}; A)) \qquad (12.1.7)$$

where $H_\lambda(k) = \{y^* \in Y^* \mid y^*(k) = \lambda\}$. This completes the proof. □

**Remark 12.1.9**

(a) Theorem 12.1.8 improves the results in [19] where the formulas were derived via bd $(A)$ instead of $A$. Furthermore, Theorem 12.1.8 provides subdifferentials of the scalarization functionals at pairs in the epigraph which are essential for proving the equivalence between the SNC property of the set $A$ and the SNEC property of the scalarization functional $\varphi_{-A,k}$ (see Bao and Tammer [29]).

(b) When $A = -C$ is convex, the limiting normal cone to $C$ reduces to the normal cone of convex analysis (see (3.6.2)). Therefore, the limiting subdifferential of $\varphi_{-C,k}$ reduces to the known formulas in Theorem 4.10.3 (with less restrictive assumptions imposed on $C$). Especially, we obtain for $H_\lambda(k) := \{y^* \in Y^* \mid y^*(k) = \lambda\}$, $\lambda \in \mathbb{R}$:

$$\begin{aligned} \partial \varphi_{-C,k}(\bar{y}) &= H_1(k) \cap \big( - N(\varphi_{C,k}(\bar{y})k - \bar{y}; C)\big) \\ &= H_1(k) \cap \{y^* \in Y^* \mid \forall y \in C: \ y^*(y + \bar{y}) - \varphi_{-C,k}(\bar{y}) \geq 0\} \\ &= H_1(k) \cap \{y^* \in Y^* \mid y^*(\bar{y}) = \varphi_{-C,k}(\bar{y})\}, \end{aligned}$$

where the second equality holds for any convex set $C$ and the last equality is valid for any convex cone $C$.

### 12.1.2 Application to Approximation Problems

We consider a class of vector-valued control approximation problems (where the image space of the objective function is $\mathbb{R}^\ell$) in which each component of the objective function is a sum of two terms, a locally Lipschitzian function and a vector of a power of norm terms. The feasible set is not necessarily convex. An important special case of such problems are (nonconvex) vector-valued location problems with forbidden regions. Necessary conditions for weakly efficient solutions of these problems were derived using the results from Sect. 12.1.1.

In our proofs it is possible to use the special structure of the subdifferential of a power of the norm and Theorem 12.1.2.

We study a general vector-valued control approximation problem and suppose that

$(H_{X,U_i,Y})$ $(X, \|\cdot\|_X)$, $(U_i, \|\cdot\|_i)$ $(i \in \{1, \dots, \ell\})$ are Asplund spaces, $Y{=}\mathbb{R}^\ell$.
$(H_S)$ $S \subseteq X$ is a nonempty closed set.
$(H_C)$ $C \subset \mathbb{R}^\ell$ is a nontrivial, closed, convex, pointed cone with nonempty interior and $C + \mathbb{R}^\ell_+ \subseteq C$.

Furthermore, assume

$(H_P)$ $x \in X, a^i \in U_i, \alpha_i \geq 0$, $\beta_i \geq 1, A_i \in L(X, U_i), (i = 1, \dots, \ell), f_1{:}X{\to}\mathbb{R}^\ell$ is a locally Lipschitzian cost function.

As usual, $\|\cdot\|_*$ denotes the dual norm to $\|\cdot\|_X$, and $\|\cdot\|_{i^*}$ the dual norm to $\|\cdot\|_i$. We would like to mention that the dual norm $\|\cdot\|_*$ to $\|\cdot\|_X$ is defined by

$$\|x^*\|_* := \sup_{\|x\|_X=1} |\, x^*(x) \,| \quad (x^* \in X^*).$$

Under the assumptions $(H_{X,U_i,Y})$, $(H_S)$, $(H_C)$ and $(H_P)$, we consider the following (nonconvex) *vector control approximation problem*

$$C\text{-}\operatorname*{Opt}_{x \in S} f(x), \qquad (P_{app})$$

where

$$f(x) := f_1(x) + \begin{pmatrix} \alpha_1 \|A_1(x) - a^1\|_1^{\beta_1} \\ \cdots \\ \alpha_\ell \|A_\ell(x) - a^\ell\|_\ell^{\beta_\ell} \end{pmatrix}$$

is the vector-valued objective function. We will derive necessary conditions for weakly efficient solutions of the vector control approximation problem ($\mathrm{P}_{app}$) using Theorem 12.1.4.

In order to prove this result, we need the following assertion (cf. [508], Corollary 2.4.16) concerning the subdifferential of the norm (in terms of the Fenchel subdifferential $\partial$ of a convex function (see (3.6.4))):

**Lemma 12.1.10** *Let $X$ be a Banach space. Then,*

$$\partial \|x\| = \begin{cases} \{x^* \in L(X,\mathbb{R}) \mid x^*(x) = \|x\|,\ \|x^*\|_* = 1\} & \text{if } x \neq 0, \\ \{x^* \in L(X,\mathbb{R}) \mid \|x^*\|_* \leq 1\} & \text{if } x = 0, \end{cases}$$

*and for $\beta > 1$,*

$$\partial\left(\frac{1}{\beta}\|\cdot\|^\beta\right)(x) = \{x^* \in L(X,\mathbb{R}) \mid \|x^*\|_* = \|x\|^{\beta-1},\ x^*(x) = \|x\|^\beta\}.$$

In the next theorem, we will derive necessary conditions for weakly efficient solutions of ($P_{app}$).

**Theorem 12.1.11** *Consider a weakly efficient solution $x_0$ of the approximation problem ($P_{app}$). Then, there is a functional $y^* \in C^+ \setminus \{0\}$ such that*

$$0 \in \Big\{\partial_M y^*(f_1(x_0)) + \sum_{i=1}^{\ell} \alpha_i\beta_i A_i^* y_i^* M_{0i} \;\Big|\; M_{0i} \in L(U_i,\mathbb{R}),\ M_{0i}(A_i(x_0) - a^i) = \|A_i(x_0) - a^i\|_i^{\beta_i},\ \|M_{0i}\|_{i^*} \leq 1 \text{ if } \beta_i = 1 \text{ and } A_i(x_0) = a^i, \|M_{0i}\|_{i^*} = \|A_i(x_0) - a^i\|_i^{\beta_i - 1} \text{otherwise } (i \in \{1,\ldots,\ell\})\Big\} + N_M(x_0; S).$$

***Proof*** Assume $x_0$ is a weakly efficient solution to ($P_{app}$). Then, from Theorem 12.1.4, we get the existence of $y^* \in C^+ \setminus \{0\}$ and

$$0 \in \partial_M y^*(f(x_0)) + N_M(x_0; S). \tag{12.1.8}$$

Furthermore, we have

$$\partial_M y^*(f(x_0)) = \partial_M y^*\left(f_1(\cdot) + \begin{pmatrix} \alpha_1\|A_1(\cdot) - a^1\|_1^{\beta_1} \\ \cdots \\ \alpha_\ell\|A_\ell(\cdot) - a^\ell\|_\ell^{\beta_\ell} \end{pmatrix}\right)(x_0).$$

The rule of sums for Mordukhovich subdifferentials (Lemma 8.1.7) yields the inclusion

$$\partial_M y^*(f(x_0)) \subseteq \partial_M y^*(f_1(x_0)) + \sum_{i=1}^{\ell} \alpha_i y_i^* \partial\|A_i(\cdot) - a^i\|_i^{\beta_i}(x_0). \tag{12.1.9}$$

Applying Lemma 12.1.10, relation (12.1.9) implies

$$\partial y^*(f(x_0)) \subseteq \partial_M y^*(f_1(x_0)) + \sum_{i=1}^{\ell} \alpha_i A_i^* y_i^* \partial(\|u\|_i^{\beta_i}) \mid_{u=A_i(x_0)-a^i}$$

$$= \Big\{ \partial_M y^*(f_1(x_0)) + \sum_{i=1}^{\ell} \alpha_i \beta_i A_i^* y_i^* M_{0i} \mid M_{0i} \in L(U_i, \mathbb{R}),$$

$$M_{0i}(A_i(x_0) - a^i) = \|A_i(x_0) - a^i\|_i^{\beta_i}, \ \|M_{0i}\|_{i*} \leq 1 \text{ if } \beta_i = 1 \text{ and } A_i(x_0) = a^i,$$

$$\|M_{0i}\|_{i*} = \|A_i(x_0) - a^i\|_i^{\beta_i - 1} \text{otherwise } (i \in \{1, \dots, \ell\}) \Big\}.$$

Then, together with (12.1.8), we get the desired relation. □

## 12.2 Duality for Vector Optimization Problems

Duality assertions have a long history in different fields of mathematics. Especially, the dual variational principles by Dirichlet and Thompson or simply the pair of dual programs in linear optimization are well-known examples.

The idea is to study, in addition to a given (primal) optimization problem

$$f(x) \to \inf_{x \in S}, \qquad (P^0)$$

where $f : X \to \mathbb{R}$, $S \subseteq X$, $X$ a linear space, with infimal value $\alpha$, a corresponding dual problem

$$f^*(u) \to \sup_{u \in S^*} \qquad (D^0)$$

with supremal value $\beta$ such that the weak duality assertion $\beta \leq \alpha$ holds.

There are several reasons for the introduction of a useful dual problem:

- Under certain additional assumptions, the dual problem $(D^0)$ has the same optimal value as the given primal optimization problem $(P^0)$, but the dual problem $(D^0)$ could be solved with other methods of variational analysis or numerical procedures.
- An objective function value of an approximate solution of $(P^0)$ arise an estimation of the infimal value $\alpha$ from above, whereas an objective function value of an approximate solution of $(D^0)$ is an estimation of $\alpha$ from below, so that one gets intervals containing $\alpha$. These estimations are very useful for deriving stopping criteria in numerical methods.
- Taking into account the Lagrange method, saddle points, equilibrium points of two person games and shadow prices in economics, one can see that optimal dual variables often have a certain meaning or interpretation for the original problem.

Obviously, the listed advantages require a carefully chosen dual problem. However, the mentioned arguments are motivation enough, to look for useful dual problems in vector optimization too. Many books and (survey) papers are dedicated to this aim (see Jahn [272], Bot, Grad and Wanka [66], Bot [65], Luc [338], Göpfert, Riahi, Tammer and Zălinescu [192], Hernández, Löhne, Rodríguez-Marín and Tammer [236], Löhne and Tammer [335] and Löhne [333, 334] and references therein).

To construct a dual problem to a given vector optimization problem, one has several approaches, especially Conjugation, Lagrangian and Axiomatic Duality (see the book by Luc [338]), but there seems to be no unified concept to dualization in vector optimization. One of the difficulties in the existing concepts is the fact that the set of solutions to a vector optimization problem is in general not a singleton. Very important for the development of duality theory in vector optimization is the definition of infimum (or supremum) of a set equipped with a partial order. Pallaschke and Rolewicz [374] and Nakayama [364] give an interesting discussion of these aspects.

In the literature, there are at least three main approaches that are used for overcoming the difficulties that accrue when generalizing duality assertions well known in scalar optimization theory to the vector-valued case.

(1) The first one is a scalarization by means of linear or nonlinear functionals involved in the formulation of the dual problem (see Schönfeld [406], Breckner [73], Jahn [267, 272], Gerstewitz [181], Göpfert, Riahi, Tammer and Zălinescu [192], Bot and Wanka [54], Bot, Grad and Wanka [66] and Bot [65]). Scalarization concepts and corresponding duality assertions from scalar optimization are applied in order to construct suitable dual problems to a given vector optimization problem, to prove duality statements or to solve the dual problem.
(2) The second approach to duality in vector optimization is predicated on the observation that a dual vector optimization problem is naturally set-valued (see Khan, Tammer and Zălinescu [291], Tanino, Sawaragi [445], Corley [101], Tanino [443, 444], Löhne and Tammer [335], Luc [338], Nakayama [364], Tammer [430], Dolecki and Malivert [118], Pallaschke and Rolewicz [374], Song [419], Löhne [333, 334]). In these books and papers, duality assertions for vector optimization problems are shown without a scalarization "from the beginning" such that the dual problem becomes set-valued.
Already in the paper by Tanino [444], the set-valued structure of the primal and dual vector optimization problem is observed: Tanino presents a weak duality assertion by embedding the primal problem into a family (depending on perturbation parameters) of set-valued optimization problems and by employing an extension of Fenchel's inequality. Furthermore, Tanino shows a strong duality statement using the relationship between a map and its biconjugate.
Moreover, a set-valued approach in combination with Lagrangian techniques and perturbations of marginal relations is used in the paper by Dolecki and Malivert [118]. Dolecki and Malivert prove duality statements for general vector optimization problems where the solution concept is described by a transitive, translation invariant relation.
(3) A third approach to develop a dual problem is predicated on solution concepts with respect to the supremum and infimum in the sense of a vector lattice (see Löhne [333–335]). Pallaschke and Rolewicz [374] develop a duality theory for

objective functions with values in vector lattices using corresponding notions of infimum and supremum in the sense of utopia minima (maxima). But, these infima and suprema may not be solutions in the sense of vector optimization (see Definition 6.0.1) because they may not belong to the set of objective function values.

Because the assumption that the objective function has its values in a vector lattice is too restrictive for vector optimization, solution concepts on the basis of infimal (supremal) sets are introduced by Nieuwenhuis in [369]. Using these solution concepts, Nieuwenhuis [369] and Tanino [443, 444] derived corresponding duality statements. These infimal sets are closely related to weakly efficient points of vector optimization problems (see Definition 6.2.2). In [118], Dolecki and Malivert extended these concepts to infimal sets being closely related to other kinds of efficiency too.

In [335], the approach to duality in vector optimization is characterized by an embedding of the the image space of the vector optimization problem into a complete lattice without linear structure and the lattice structure is consequently used.

The results in this section are related to the first approach based on a scalarization method. Using a suitable scalarization for deriving duality assertions in vector optimization, one often gets a beneficial structure of the dual problem. We discuss the application of the functional $\varphi_{A,k} : Y \to \overline{\mathbb{R}}$ defined by (4.1.11) in order to prove duality statements for nonconvex vector optimization problems. Without convexity assumptions, this method was introduced by Gerstewitz [181], Gerstewitz and Iwanow [184] and Göpfert and Gerth [191] (compare the book by Göpfert, Riahi, Tammer and Zălinescu [192]). Furthermore, in the book by Bot, Grad and Wanka [66, Sect. 4.4], duality assertions for vector optimization problems via general scalarization methods are presented.

The most important properties of the functional $\varphi_{A,k}$ (with $A \subset Y, k \in 0^+A \setminus \{0\}$) for deriving duality assertions are the following:

- strict monotonicity,
- continuity.

These properties of the functional $\varphi_{A,k}$ are shown in Theorems 4.4.4, 4.5.5 and 4.6.6 depending on the properties of the set $A \subset Y$ and of the element $k \in Y \setminus \{0\}$ in the definition of the functional $\varphi_{A,k}$.

We will use the scalarization method by means of the nonlinear separating functional $\varphi_{A,k}$ in order to characterize (properly) efficient elements of the primal vector optimization problem (see Definition 6.3.3(b)) and in the construction of the dual problem. The advantage of using the functional $\varphi_{A,k}$ in the construction of the dual problem is the possibility for the decision maker to chose $A$ and $k$ corresponding to the applications (see sets $A$ of acceptable risk in Sect. 15.1). In the duality assertions by Göpfert, Riahi, Tammer and Zălinescu [192], general strictly $C$-monotone and continuous functions are used, but here we are using (in the construction of the dual problem) the functional $\varphi_{A,k}$ under certain assumptions concerning $A \subset Y$ and $k \in Y \setminus \{0\}$.

The aim of this section is to prove duality theorems for nonconvex vector optimization problems based on scalarization by means of the functional $\varphi_{A,k}$ given by (4.1.11) using its monotonicity as well as continuity properties.

### *12.2.1 Lagrange-Duality for Nonconvex Problems*

Throughout this section, we consider the (primal) vector optimization problem (see (VOP) in Chap. 6) in the following form:

$$C-\operatorname*{Opt}_{x\in S} f(x) \quad \text{with} \quad S := \{x \in M \mid g(x) \in V\} \neq \emptyset, \tag{VOP}$$

where $f : M \to Y$, $\emptyset \neq M \subseteq X$, $g : M \to Z$, $V$ is a nonempty set in $Z$, $M$ is a nonempty set in $X$ and $X$, $Y$, $Z$ are topological vector spaces, $Y$ partially ordered by a nontrivial, closed, convex, pointed cone $C \subset Y$.

For $k \in C \setminus \{0\}$ with $Y = C + \mathbb{R}k$, we introduce a family $\mathscr{A}(C, k)$ of proper, closed, convex sets $A \subset Y$ with

$$A - (C \setminus \{0\}) \subseteq \operatorname{int} A. \tag{12.2.1}$$

For $A \in \mathscr{A}(C, k)$, we consider the functional (4.1.11)

$$\varphi_{A,k} : Y \to \overline{\mathbb{R}}, \quad \varphi_{A,k}(y) = \inf\{t \in \mathbb{R} \mid y \in tk + A\}. \tag{12.2.2}$$

Under the given assumptions concerning $A \in \mathscr{A}(C, k)$, we obtain that $\varphi_{A,k}$ is finite-valued, convex and continuous (Corollary 4.7.10). From Corollary 4.5.6, it follows that $\varphi_{A,k}$ is strictly $C$-monotone.

In order to formulate a dual problem for (VOP), we denote the dual feasible set by

$$M^* := \{L : Y \to \overline{\mathbb{R}} \mid \forall u \in V \ : \ L(u) \leq 0\}.$$

Furthermore, we denote the set of finite-valued, convex, continuous, strictly $C$-monotone functionals $\varphi : Y \to \mathbb{R}$ by $M_s$.

Using functionals $\varphi \in M_s$, we define the dual image set

$$Y_D := \left\{y \in Y \mid \exists \varphi \in M_s,\, \exists L \in M^* \ : \ \varphi(y) \leq \inf_{x\in M} \{\varphi(f(x)) + L(g(x))\} > -\infty\right\}. \tag{12.2.3}$$

We formulate the strong (direct/converse) duality statements for Nehse–Iwanow properly efficient solutions of the vector minimum problem (VOP) (see Definition 6.3.3(b)) and maximal points of $Y_D$ in (D) (in the sense of Definition 6.0.1). In order to distinguish between the primal (minimum-) and dual (maximum-) vector optimization problem, we use the following notations in this section:

$$\mathrm{Min}(f(S), C) := \mathrm{Eff}(f(S), C),$$

$$\text{NI-PMin}(f(S), C) := \text{NI-PEff}(f(S), C),$$

$$\mathrm{Max}(Y_D, C) := \mathrm{Eff}(Y_D, -C).$$

Taking into account these notations as well as the notations and assumptions for (VOP), the primal problem is given by

$$\text{Determine} \quad \text{NI-Min}(f(S), C) \tag{P}$$

and the dual problem to (P) is formulated as

$$\text{Determine} \quad \mathrm{Max}(Y_D, C). \tag{D}$$

For elements $y \in Y_D$, we say that $\varphi$ is the corresponding functional to $y$ if the conditions in the definition of $Y_D$ are fulfilled for $y$ and $\varphi$ ($\varphi$ corresponds to $y$ according to the definition of $Y_D$).

**Lemma 12.2.1** (Weak duality) *The pair* $(P), (D)$ *satisfies*

$$f(S) \cap (Y_D - (C \setminus \{0\})) = \emptyset.$$

***Proof*** Assume that there are $x \in S$ and $y \in Y_D$ with the corresponding functional $\varphi \in M_s$ (strictly $C$-monotone) such that $f(x) \in y - C \setminus \{0\}$. Using the functional $\varphi$, we get

$$\varphi(f(x)) < \varphi(y)$$

because $\varphi$ is strictly $C$-monotone, taking into account (12.2.1) and Theorem 4.5.5.

On the other hand, for $y \in Y_D$ with the corresponding $\varphi$ and $L$, we derive from the definition of $Y_D$

$$\forall x \in S: \quad \varphi(y) \leq \inf_{x \in M} \{\varphi(f(x)) + L(g(x))\} \leq \varphi(f(x) + L(g(x)) \leq \varphi(f(x)), \tag{12.2.4}$$

a contradiction. □

Moreover, we can prove strong (direct and converse) duality assertions for Nehse–Iwanow properly efficient solutions of (P) (see Definition 6.3.3(b)) and elements of $\mathrm{Max}(Y_D, C)$ in (D).

**Theorem 12.2.2** (Strong (direct) duality) *Suppose* $Y = C + \mathbb{R}k$ *for some* $k \in C \setminus \{0\}$. *Consider the problems (P) and (D) and an element* $f(\overline{x}) \in \text{NI-PMin}(f(S), C)$. *Then, there is an element* $\overline{y} \in \mathrm{Max}(Y_D, C)$ *with* $\overline{y} = f(\overline{x})$, *i.e.,* $f(\overline{x})$ *is a maximal point of* $Y_D$ *in (D) with respect to* $C$.

***Proof*** Let $f(\overline{x}) \in \text{NI-PMin}(f(S), C)$ be a properly efficient element of the primal problem (P). Taking into account Corollary 6.3.18, there is a closed convex set $\tilde{H} \subset Y$ with $0 \in \text{bd}\,\tilde{H}$ and $\tilde{H} + (C \setminus \{0\}) \subseteq \text{int}\,\tilde{H}$ such that $\overline{x}$ is a solution to the corresponding scalarized problem:

$$\varphi_{f(\overline{x})-\tilde{H},k}(f(\overline{x})) = \min_{x\in S} \varphi_{f(\overline{x})-\tilde{H},k}(f(x)).$$

Setting $\tilde{A} := f(\overline{x}) - \tilde{H}$, we get $\tilde{A} \in \mathscr{A}(C, k)$. The functional $\varphi_{\tilde{A},k}$ is finite-valued, continuous, convex and strictly $C$-monotone (see Corollaries 4.7.10 and 4.5.6).

We put

$$L(u) := \begin{cases} 0, & \text{if } u \in V, \\ +\infty & \text{otherwise.} \end{cases}$$

Using the functional $\varphi_{\tilde{A},k}$, we get

$$\begin{aligned} \inf_{x\in M} \{\varphi_{\tilde{A},k}(f(x)) + L(g(x))\} &= \inf\big\{ \inf_{x\in S} \varphi_{\tilde{A},k}(f(x)) + L(g(x))\,,\ \inf_{x\in M\setminus S} \varphi_{\tilde{A},k}(f(x)) + L(g(x))\big\} \\ &= \inf_{x\in S} \varphi_{\tilde{A},k}(f(x)) = \varphi_{\tilde{A},k}(f(\overline{x})). \end{aligned}$$

This yields $f(\overline{x}) \in Y_D$ with $\varphi_{\tilde{A},k}$ corresponds to $f(\overline{x})$ according to the definition of $Y_D$. From the weak duality assertion in Lemma 12.2.1, we can conclude that $f(\overline{x})$ is a maximal point of $Y_D$ in (D) with respect to $C$. □

Next, we complete the study of (P) and (D) with a converse duality assertion. At first, we prove that elements of $\text{Max}(Y_D, C)$ can be characterized via scalarization.

**Lemma 12.2.3** *Suppose $y^0 \in Y_D$. Then, $y^0 \in \text{Max}(Y_D, C)$ if and only if $\varphi(y) \le \varphi(y^0)$ for all $y \in Y_D$, where $\varphi$ corresponds to $y$ according to the definition of $Y_D$ in (D).*

***Proof***

1. If $y^0 \in \text{Max}(Y_D, C)$ and $\overline{y} \in y^0 + (C \setminus \{0\})$, then $\overline{y} \notin Y_D$. Hence, for all $\varphi \in M_s$ and $L \in M^*$ it holds that

$$\varphi(\overline{y}) > \inf_{x\in M} \{\varphi(f(x)) + L(g(x))\}. \tag{12.2.5}$$

Then, we get for all $y \in Y_D$ with the corresponding $\varphi$ and $L$,

$$\begin{aligned} \varphi(y^0) &= \inf\{\varphi(y) \mid y \in y^0 + (C \setminus \{0\})\} \\ &\ge \inf_{x\in M} \{\varphi(f(x)) + L(g(x))\} \\ &\ge \varphi(y), \end{aligned}$$

taking into account the continuity and strict $C$-monotonicity of $\varphi$ as claimed, where the first inequality follows from the strict inequality in (12.2.5) above and the second inequality follows from the definition of $Y_D$.

2. Suppose $\varphi(y) \leq \varphi(y^0)$ for all $y \in Y_D$, where $\varphi$ corresponds to $y$. If $y^0 \in Y_D$ but $y^0 \notin \mathrm{Max}(Y_D, C)$, then there is $y \in Y_D$ such that $y \in y^0 + (C \setminus \{0\})$. Therefore, $\varphi(y) > \varphi(y^0)$ for all strictly $C$-monotone functionals $\varphi$, a contradiction to our assumption.

□

Additionally, a strong converse duality statement holds.

**Theorem 12.2.4** (Strong converse duality) *Suppose $Y = \mathbb{R}k - C$ for some $k \in C \setminus \{0\}$. Consider the problems $(P)$, $(D)$. Furthermore, suppose that $P := f(S) + C$ is closed and for all continuous, convex, strictly $C$-monotone functionals $\varphi : Y \to \mathbb{R}$ with $\inf\{\varphi(f(x)) \mid x \in S\} > -\infty$, there is an element $x_0 \in S$ with $\inf\{\varphi(f(x)) \mid x \in S\} = \varphi(f(x_0))$. Then, for $\overline{y} \in \mathrm{Max}(Y_D, C)$, there is an element $f(\overline{x}) \in$ NI-PMin$(f(S), C)$ with $\overline{y} = f(\overline{x})$.*

***Proof*** Let $\overline{y}$ be an element of $\mathrm{Max}(Y_D, C)$ with the corresponding functional $\varphi \in M_s$. Suppose $\overline{y} \notin P$. Then, there is an open convex neighborhood $U$ of $\overline{y}$ with $U \cap P = \emptyset$. Hence,

$$(U - C) \cap P = \emptyset.$$

Otherwise, there would exist elements $u \in U$, $k \in C$ such that $u - k \in P = f(S) + C$, which means that $u \in f(S) + C + k = P$, a contradiction.

Now, we study the set $B := U - C$. Of course, $B$ is open and convex and $B - C = B$, and so $\mathrm{cl}\, B - C \subseteq \mathrm{cl}(B - C) \subseteq \mathrm{cl}\, B$. For $A_1 := \mathrm{cl}(B - C)$ and $k \in C \setminus \{0\}$ the assumptions of Separation Theorem 4.11.3 are fulfilled, such that we get from Theorem 4.11.3

$$\forall p \in P,\ \forall u \in B : \quad \varphi_{A_1,k}(p) \geq 0 > \varphi_{A_1,k}(u),$$

where $\varphi_{A_1,k}$ is finite-valued, continuous, convex and $C$-monotone. We have $\overline{y} \in B$, since $0 \in C$, so there is a number $\gamma$ with

$$\forall\, p \in P : \quad \varphi_{A_1,k}(\overline{y}) < \gamma < 0 \leq \varphi_{A_1,k}(p). \tag{12.2.6}$$

From (12.2.4) in the proof of the weak duality assertion (Lemma 12.2.1), we get for the strictly $C$-monotone functional $\varphi \in M_s$ corresponding to $\overline{y}$ according to the definition of $Y_D$,

$$\forall\, p \in P : \quad \varphi(\overline{y}) \leq \varphi(p).$$

Taking $\lambda \in (0, 1)$, we introduce $\varphi^\lambda := \lambda\varphi + (1 - \lambda)\varphi_{A_1,k}$. Then, $\varphi^\lambda$ is strictly $C$-monotone, convex and continuous, and

$$\begin{aligned}
\varphi^\lambda(\overline{y}) &= \varphi_{A_1,k}(\overline{y}) + \lambda(\varphi(\overline{y}) - \varphi_{A_1,k}(\overline{y})),\\
\forall x \in S : \varphi^\lambda(f(x)) &= \lambda(\varphi(f(x))) + (1 - \lambda)(\varphi_{A_1,k}(f(x))),\\
&\geq \lambda\varphi(f(x)) \geq \lambda\varphi(\overline{y}),
\end{aligned}$$

taking into account $f(x) \in P$. It holds that $\varphi_{A_1,k}(\overline{y}) < \gamma$ because of (12.2.6). Now, we choose $\lambda$ sufficiently small. Then, $\varphi^\lambda(\overline{y}) < \gamma$. From $\gamma < 0$, it follows that $\gamma < \lambda\varphi(\overline{y})$, such that

$$\forall x \in S: \quad \varphi^\lambda(\overline{y}) < \gamma < \lambda\varphi(\overline{y}) \leq \varphi^\lambda(f(x))$$

holds. This yields $\inf\{\varphi^\lambda(f(x)) \mid x \in S\} \geq \gamma$. Because of our assumptions, we have the existence of an element $x_0 \in S$ such that

$$\varphi^\lambda(\overline{y}) < \gamma \leq \inf\{\varphi^\lambda(f(x)) \mid x \in S\} = \varphi^\lambda(f(x_0)). \tag{12.2.7}$$

This means that $f(x_0)$ is a solution to a scalarized problem, where the scalarizing functional $\varphi^\lambda$ is strictly $C$-monotone, convex and continuous. Taking into account Corollary 6.3.17, we get $f(x_0) \in$ NI-PMin$(f(S), C)$ and Theorem 12.2.2 yields $f(x_0) \in Y_D$ with $\varphi^\lambda$ corresponds to $f(x_0)$ according to the definition of $Y_D$. Applying Lemma 12.2.3, it follows $\varphi^\lambda(f(x_0)) \leq \varphi^\lambda(\overline{y})$, a contradiction to (12.2.7). So, we have $\overline{y} \in P = f(S) + C$.

From the weak duality assertion (Lemma 12.2.1), we get $\overline{y} \in f(S)$. This means that there is an element $\overline{x} \in S$ with $\overline{y} = f(\overline{x})$.

Finally, taking into account the definition of the set $Y_D$, we get, for $\overline{y} \in Y_D$ with the corresponding $\varphi$, and $L$,

$$\varphi(f(\overline{x})) = \varphi(\overline{y}) \leq \inf_{x\in M}\{\varphi(f(x)) + L(g(x)\} \leq \{\varphi(f(x)) + L(g(x)\} \leq \varphi(f(x))$$

for all $x \in S$. This means that $\overline{x}$ is a solution to a scalarized problem with a convex and continuous scalarizing functional $\varphi$ that is strictly $C$-monotone. From Corollary 6.3.17 we get $f(\overline{x}) \in$ NI-PMin$(f(S), C)$. □

**Remark 12.2.5** The procedure in the proofs of the duality assertions in this section based on a scalarization by means of the nonlinear separating functional (4.1.11) can be used for deriving other approaches to duality for vector optimization problems, especially conjugation.

**Remark 12.2.6** Duality assertions based on conjugation and applying nonlinear strictly monotone functionals are shown by Bot, Grad, Wanka in [66].

## *12.2.2 Geometric Duality for Convex Problems*

Using the separating functional (4.1.11) and the corresponding properties derived in Chap. 4, it is possible to show geometric duality assertions for convex vector optimization problems in finite-dimensional spaces. The results in this section are derived by Heyde in [239].

In [239], *geometric duality relationships* between the epigraph of a proper, lower semicontinuous, convex function and the epigraph of its conjugate are studied (see

(3.2.4)). We consider a vector-valued objective function $f : \mathbb{R}^n \to \mathbb{R}^\ell$, a feasible set $S \subseteq \mathbb{R}^n$ and a nontrivial, closed, convex cone $C \subset \mathbb{R}^\ell$ that is not a linear subspace of $\mathbb{R}^\ell$. Suppose that $f$ is $C$-convex, i.e., for all $x^1, x^2 \in \mathbb{R}^n$, $\lambda \in [0, 1]$ it holds that

$$f(\lambda x^1 + (1 - \lambda)x^2) \in \lambda f(x^1) + (1 - \lambda) f(x^2) - C.$$

Under these assumptions, we study the following vector optimization problem (VOP) in this section:

$$C\text{-}\operatorname*{Opt}_{x \in S} f(x). \tag{VOP}$$

The aim of this section is to derive geometric duality relationships for the upper closed extended *primal image set* (discussed in [335])

$$P := \mathrm{cl}(f(S) + C) \qquad (\subseteq \mathbb{R}^\ell) \tag{12.2.8}$$

of the vector optimization problem (VOP) and a corresponding *dual image set* $D \subseteq \mathbb{R}^\ell$ (see (12.2.13)). Of course, the set $P$ is convex (since $f$ is $C$-convex) and $P$ is closed.

For deriving the results, we construct a linear transformation $T : \mathbb{R}^\ell \to \mathbb{R}^\ell$ and a proper, convex function $h : \mathbb{R}^{\ell-1} \to \overline{\mathbb{R}}$ such that

$$P = T(\mathrm{epi}\, h). \tag{12.2.9}$$

Let $k \in$ r-int $C$ (where r-int $C$ denotes the relative interior of $C$) and $e^1, \cdots, e^{\ell-1}$ be vectors in $\mathbb{R}^\ell$ such that $e^1, \cdots, e^{\ell-1}, k$ are linearly independent. Using these vectors, we introduce the nonsingular matrix

$$T := (e^1, \cdots, e^{\ell-1}, k) \tag{12.2.10}$$

and put

$$E := (e^1, \cdots, e^{\ell-1}).$$

Furthermore, we consider the functional $\varphi_{-C,k} : \mathbb{R}^\ell \to \overline{\mathbb{R}}$ (see (4.1.11) with $A = -C$ and $k \in$ r-int $C$)

$$\varphi_{-C,k}(y) = \inf\{t \in \mathbb{R} \mid y \in tk - C\}.$$

Now, we introduce a certain scalarization of the vector optimization problem (VOP) by means of a functional $h : \mathbb{R}^{\ell-1} \to \overline{\mathbb{R}}$ where the functional $\varphi_{-C,k}$ is involved:

$$h(z) := \inf_{x \in S} \varphi_{-C,k}(f(x) - Ez) \qquad (z \in \mathbb{R}^{\ell-1}). \tag{12.2.11}$$

In Theorems 4.4.4 and 4.5.5, we have shown essential properties of the functional $\varphi_{-C,k}$, especially that the functional is lower semicontinuous, sublinear, dom $\varphi_{-C,k} =$

$\mathbb{R}k - C \neq \emptyset$ and $C$-monotone. These properties of $\varphi_{-C,k}$ are important tools for deriving the geometric duality relationships in this section.

In the next lemma, we give an equivalent formulation of the functional $h : \mathbb{R}^{\ell-1} \to \overline{\mathbb{R}}$. This assertion is shown in [239, Lemma 5.2] and will be used later on.

**Lemma 12.2.7** *For $z \in \mathbb{R}^{\ell-1}$, it holds that*

$$h(z) = \inf \left\{ t \in \mathbb{R} \mid T\begin{pmatrix} z \\ t \end{pmatrix} \in P \right\} = \inf_{y \in P} \varphi_{-C,k}(y - Ez).$$

***Proof*** Taking into account the definition of the functional $\varphi_{-C,k}$, we obtain

$$\begin{aligned} h(z) &= \inf_{x \in S} \varphi_{-C,k}(f(x) - Ez) \\ &= \inf\{t \in \mathbb{R} \mid tk - f(x) + Ez \in C,\ x \in S\} \\ &= \inf \left\{ t \in \mathbb{R} \mid T\begin{pmatrix} z \\ t \end{pmatrix} \in f(S) + C \right\} \\ &= \inf\{t \in \mathbb{R} \mid T\begin{pmatrix} z \\ t \end{pmatrix} \in P\} \\ &= \inf\{t \in \mathbb{R} \mid Ez + tk \in y + C,\ y \in P\} \\ &= \inf_{y \in P} \varphi_{-C,k}(y - Ez). \end{aligned}$$

□

Moreover, we define for $z \in \mathbb{R}^{\ell-1}$

$$R(z) := \{t \in \mathbb{R} \mid T\begin{pmatrix} z \\ t \end{pmatrix} \in P\}.$$

We obtain that $R(z)$ is closed for every $z \in \mathbb{R}^{\ell-1}$ taking into account the closedness of $P$. Moreover, $R(z)$ is an upper set, i.e., $t \in R(z)$ and $t' \geq t$ imply $t' \in R(z)$ because of the definition of $P$ and $k \in C$. This yields for $z \in \mathbb{R}^{\ell-1}$

$$t \geq h(z) \iff t \in R(z) \iff T\begin{pmatrix} z \\ t \end{pmatrix} \in P$$

and

$$\operatorname{epi} h = T^{-1}(P) \quad \text{or} \quad P = T(\operatorname{epi} h). \tag{12.2.12}$$

Therefore, $h$ is a lower semicontinuous convex function taking into account that $P$ is a closed, convex set and $T$ is linear and continuous. Furthermore, $h$ is proper if and only if $P$ is proper, i.e., $\emptyset \neq P \neq \mathbb{R}^{\ell}$.

A point $(z, t) \in \operatorname{epi} h$ is an efficient element of epi $h$ with respect to

$$K := \{(z, t) \in \mathbb{R}^{\ell-1} \times \mathbb{R} \mid z = 0,\ t \geq 0\}$$

(see Definition 6.0.1) if and only if $t = h(z)$.

Now, the optimality properties of the transformed points $T\begin{pmatrix} z \\ h(z) \end{pmatrix}$ in $P$ with respect to $C$ will be analyzed. For this reason, we introduce the following solution concept (compare Tasset [446]).

**Definition 12.2.8** Suppose that r-int $C \neq \emptyset$. A point $y \in P$ is called **relatively efficient** in $P$ with respect to $C$ if $P \cap (y - \text{r-int}\, C) = \emptyset$.

We denote the set of all relatively efficient points in $P$ with respect to $C$ by r-Eff$(P, C)$.

In [239, Proposition 5.4], the following characterization of r-Eff$(P, C)$ is shown. For the proof of the result, the monotonicity properties of the functional $\varphi_{-C,k}$ (see Theorem 4.5.5 with $A = -C$ in (4.1.11)) are essential.

**Proposition 12.2.9**

$$y \in \text{r-Eff}(P, C) \iff \exists z \in \mathbb{R}^{\ell-1} \textit{ such that } y = T\begin{pmatrix} z \\ h(z) \end{pmatrix}.$$

***Proof*** Suppose that $y = T\begin{pmatrix} z \\ h(z) \end{pmatrix} = Ez + h(z)k$ and $y \notin$ r-Eff$(P, C)$. Then, there is an element $y^0 \in P$ with $y - y^0 \in$ r-int $C$. From $y - y^0 \in$ r-int $C$ and $k \in C$, we get the existence of some $\lambda > 1$ such that $(1-\lambda)k + \lambda(y - y^0) \in C$, i.e., $y - y^0 - tk \in C$ for $t := \frac{(\lambda-1)}{\lambda} > 0$. From $y - y^0 - tk \in C$ and $y = Ez + h(z)k$, we get

$$Ez + h(z)k - y^0 - tk \in C,$$

i.e.,

$$y^0 - Ez \in (h(z) - t)k - C.$$

Taking into account the monotonicity properties of the functional $\varphi_{-C,k}$ shown in Theorem 4.5.5, we obtain

$$\varphi_{-C,k}(y^0 - Ez) \leq h(z) - t < h(z),$$

in contradiction to $h(z) = \inf_{y \in P} \varphi_{-C,k}(y - Ez)$.

Now, suppose $\overline{y} \in$ r-Eff$(P, C)$ and $\begin{pmatrix} \overline{z} \\ \overline{t} \end{pmatrix} := T^{-1}\overline{y}$, i.e., $\overline{y} = E\overline{z} + \overline{t}k$. In order to show $\overline{t} = h(\overline{z})$, we suppose by contraposition that

$$\inf_{y \in P} \varphi_{-C,k}(y - E\overline{z}) = h(\overline{z}) < \overline{t}.$$

This implies that there is some $y \in P$ with $\varphi_{-C,k}(y - E\overline{z}) < \overline{t}$. Furthermore, taking into account the definition of $\varphi_{-C,k}$ in (4.1.11) with $A = -C$ and $k \in$ r-int $C$, there are elements $t \in \mathbb{R}$ with $y - E\overline{z} \in tk - C$ and $t < \overline{t}$. Since $C$ is a convex cone and $k \in$ r-int $C$, we get

$$\overline{y} - y = E\overline{z} + \overline{t}k - y = (tk - y + E\overline{z}) + (\overline{t} - t)k \in C + \text{r-int}\, C \subseteq \text{r-int}\, C,$$

in contradiction to $\overline{y} \in \text{r-Eff}(P, C)$. □

For deriving the duality assertions, the following characterization of the conjugate of $h$ is important. In the next proposition (see [239, Proposition 5.5]), we denote the inverse of the transpose $T^T$ of $T$ by $T^{-T}$. For the proof of the next proposition, the computation of the conjugate of the functional $\varphi_{-C,k}$ (given in Theorem 4.10.3) is important.

**Proposition 12.2.10** *For each $w \in \mathbb{R}^{\ell-1}$, the conjugate of $h$ is given by*

$$h^*(w) = \begin{cases} (\langle c^*(-w), f\rangle + \iota_S)^*(0) & \text{if } c^*(-w) \in C^+, \\ +\infty & \text{if } c^*(-w) \notin C^+, \end{cases}$$

*where $c^*(w) : \mathbb{R}^{\ell} \to \mathbb{R}$ is given by $c^*(w) := T^{-T}\begin{pmatrix} w \\ 1 \end{pmatrix}$.*

***Proof*** Taking into account the definition of the function $h$, it holds that

$$h(z) = \inf_{x\in\mathbb{R}^{\ell}} \Phi(x, z) \quad \text{with} \quad \Phi(x, z) := \varphi_{-C,k}(f(x) - Ez) + \iota_S(x)$$

and so, because of [508, Theorem 2.6.1],

$$h^*(w) = \Phi^*(0, w).$$

In order to calculate $\Phi^*$, we apply [508, Theorem 2.8.10] (it is easy to see that the assumptions of this theorem are fulfilled in our finite-dimensional setting). With [508, Theorem 2.8.10], we get

$$\Phi^*(0, w) = \min\{(\langle c^*, f(x) - Ez\rangle + \iota_S(x))^*(0, w) + \varphi^*_{-C,k}(c^*) \mid c^* \in C^+\}.$$

Furthermore, because of Theorem 4.10.3, it holds that

$$\varphi^*_{-C,k}(c^*) = \begin{cases} 0, & \text{if } c^* \in C^+,\ \langle c^*, k\rangle = 1 \\ +\infty & \text{otherwise}. \end{cases}$$

Then, we have

$$\begin{aligned} h^*(w) &= \min\{(\langle c^*, f(x) - Ez\rangle + \iota_S(x))^*(0, w) \mid c^* \in C^+,\ \langle c^*, k\rangle = 1\} \\ &= \min\{(\langle c^*, f\rangle + \iota_S)^*(0) + \langle -c^*, E\rangle^*(w) \mid c^* \in C^+,\ \langle c^*, k\rangle = 1\} \\ &= \min\{(\langle c^*, f\rangle + \iota_S)^*(0) \mid c^* \in C^+,\ \langle c^*, k\rangle = 1,\ E^T c^* = -w\} \\ &= \begin{cases} (\langle c^*(-w), f\rangle + \iota_S)^*(0), & \text{if } c^*(-w) = T^{-T}\begin{pmatrix} -w \\ 1 \end{pmatrix} \in C^+ \\ +\infty & \text{otherwise}. \end{cases} \end{aligned}$$

□

Equation (12.2.12) shows that the extended primal image set $P \subseteq \mathbb{R}^\ell$ of the convex vector optimization problem (VOP) can be transformed into the epigraph of a suitable function $h$, where the functional $\varphi_{-C,k}$ is involved in the definition of $h$ (see (12.2.11)). Now, the dual problem with the dual image set $D$ will be derived by an appropriate transformation of the epigraph of $h^*$.

Heyde [239] introduced the following *dual image set* $D \subseteq \mathbb{R}^\ell$

$$D := \left\{ \begin{pmatrix} w \\ s \end{pmatrix} \mid w \in \mathbb{R}^{\ell-1},\ s \in \mathbb{R},\ c^*(w) \in C^+,\ s \leq -(\langle c^*(w), f\rangle + \iota_S)^*(0) \right\}. \tag{12.2.13}$$

Then, it holds that

$$D = -\operatorname{epi} h^*.$$

This means that the nonlinear scalarizing functional $\varphi_{-C,k}$ is involved in the construction of the dual image set due to the definition of $h$ in (12.2.11) and the corresponding calculation of $h^*$ in Proposition 12.2.10.

**Remark 12.2.11** The right hand side of the inequality in the definition of $D$ (see (12.2.13))

$$-(\langle c^*(w), f\rangle + \iota_S)^*(0) = \inf_{x \in S} \langle c^*(w), f(x)\rangle$$

is the optimal value of a scalarization of the primal problem by means of the linear functional $c^*(w) \in C^+$.

**Definition 12.2.12** Consider $y^* \in \mathbb{R}^\ell \setminus \{0\}$ and $\alpha \in \mathbb{R}$. We introduce a hyperplane in $\mathbb{R}^\ell$ by

$$H_{y^*,\alpha} := \{y \in \mathbb{R}^\ell \mid \langle y^*, y\rangle = \alpha\}.$$

$H_{y^*,\alpha}$ is called **supporting hyperplane to a set B** $\subset \mathbb{R}^\ell$ if $B \cap H_{y^*,\alpha} \neq \emptyset$ and $B \subset \{y \in \mathbb{R}^\ell \mid \langle y^*, y\rangle \leq \alpha\}$. Let $B \subseteq \mathbb{R}^\ell$ be a convex set. A convex subset $\overline{B} \subseteq B$ is called a **constructor** of $B$ if the following implication holds

$$(y^1, y^2 \in B,\ \lambda \in (0,1),\ \lambda y^1 + (1-\lambda)y^2 \in \overline{B}) \implies y^1, y^2 \in \overline{B}.$$

We say that a constructor $\overline{B}$ of $B$ is **proper** if $\emptyset \neq \overline{B} \neq B$. Moreover, a set $\hat{B} \subseteq B$ is called an **exposed constructor** of $B$ if there is a supporting hyperplane $H_{y^*,\alpha}$ to $B$ with

$$\hat{B} = H_{y^*,\alpha} \cap B.$$

Each exposed constructor of a convex set $B$ is a proper constructor of $B$ as well if $\dim B = \ell$. The converse assertion holds for polyhedral convex sets.

Consider now

$$K := \{(x,t) \in \mathbb{R}^n \times \mathbb{R} \mid x = 0,\ t \geq 0\}$$

and a proper, lower semicontinuous, convex function $g : \mathbb{R}^n \to \overline{\mathbb{R}}$.

In Heyde [239], it is shown how an inclusion-reversing one-to-one map between $K$-efficient exposed constructors of the epigraph of $g$ and the epigraph of the Legendre-Fenchel conjugate $g^*$ can be received.

**Definition 12.2.13** A proper constructor is called **$K$-efficient** if all of its points are efficient with respect to $K$.

Exposed constructors are obtained by supporting hyperplanes and so it is important to recall some properties of supporting hyperplanes to epi $g$ . In the following assertions (see [239, Lemma 3.1]), the Fenchel subdifferential $\partial$ given by (3.6.4) is used.

**Lemma 12.2.14** *Let $g : \mathbb{R}^n \to \overline{\mathbb{R}}$ be a proper, lower semicontinuous, convex function. Consider* epi $g \subset \mathbb{R}^n \times \mathbb{R}$ *and supporting hyperplanes to* epi $g$ *defined for $(x^*, s) \in \mathbb{R}^n \times \mathbb{R}$ and $\alpha \in \mathbb{R}$.*

(i) *If $H_{(x^*,s),\alpha}$ is a supporting hyperplane to* epi $g$, *then $s \leq 0$.*
(ii) *If $H_{(x^*,s),\alpha}$ is a supporting hyperplane to* epi $g$, *then*

$$H_{(x^*,s),\alpha} \cap \operatorname{epi} g$$

*is $K$-efficient in* epi $g$ *if and only if $s < 0$.*
(iii) *The hyperplane $H_{(x^*,-1),\alpha}$ is a supporting hyperplane to* epi $g$ *if and only if $\partial g^*(x^*) \neq \emptyset$ and $\alpha = g^*(x^*)$.*

The next proposition is shown in [239, Proposition 3.2].

**Proposition 12.2.15** *Let $g : \mathbb{R}^n \to \overline{\mathbb{R}}$ be a proper, lower semicontinuous, convex function. A subset $\hat{B} \subseteq$* epi $g$ *is a $K$-efficient exposed constructor of* epi $g$ *if and only if there are elements $x^* \in \mathbb{R}^n$ with $g^*(x^*) < +\infty$ and $\partial g^*(x^*) \neq \emptyset$ with*

$$\hat{B} = \{(x, g(x)) \in \mathbb{R}^n \times \mathbb{R} \mid x^* \in \partial g(x)\}. \tag{12.2.14}$$

*Furthermore, $\hat{B}^* \subseteq$* epi $g^*$ *is a $K$-efficient exposed constructor of* epi $g^*$ *if and only if there are elements $x \in \mathbb{R}^n$ with $g(x) < +\infty$ and $\partial g(x) \neq \emptyset$ with*

$$\hat{B}^* = \{(x^*, g^*(x^*)) \in \mathbb{R}^n \times \mathbb{R} \mid x^* \in \partial g(x)\}. \tag{12.2.15}$$

***Proof*** Taking into account the Young-Fenchel inequality, it holds that

$$\begin{aligned}\{(x, g(x)) \in \mathbb{R}^n \times \mathbb{R} \mid x^* \in \partial g(x)\} &= \{(x, t) \in \mathbb{R}^n \times \mathbb{R} \mid t = g(x),\ \langle x^*, x\rangle = g(x) + g^*(x^*)\} \\ &= \{(x, t) \in \mathbb{R}^n \times \mathbb{R} \mid t \geq g(x),\ \langle x^*, x\rangle - t = g^*(x^*)\} \\ &= H_{(x^*,-1),g^*(x^*)} \cap \operatorname{epi} g.\end{aligned}$$

Therefore, the first assertion follows by Lemma 12.2.14. Because of $x^* \in \partial g(x)$ if and only if $x \in \partial g^*(x^*)$ for a proper, lower semicontinuous function $g$, we can show the second assertion analogously. □

Consider the set-valued map $\Psi : \mathbb{R}^{n+1} \rightrightarrows \mathbb{R}^{n+1}$ defined by

$$\Psi(\hat{B}^*) := \bigcap_{(x^*, g^*(x^*)) \in \hat{B}^*} \{(x, g(x)) \in \mathbb{R}^n \times \mathbb{R} \mid x^* \in \partial g(x)\}. \tag{12.2.16}$$

The following assertion is shown by Heyde in [239, Theorem 3.3].

**Theorem 12.2.16** *Let $g : \mathbb{R}^n \to \overline{\mathbb{R}}$ be a proper, lower semicontinuous, convex function. The map $\Psi$ defined by (12.2.16) is an inclusion reversing one-to-one map between $K$-efficient exposed constructors of* epi $g^*$ *and $K$-efficient exposed constructors of* epi $g$. *The corresponding inverse map is given by*

$$\Psi^*(\hat{B}) := \bigcap_{(x, g(x)) \in \hat{B}} \{(x^*, g^*(x^*)) \in \mathbb{R}^n \times \mathbb{R} \mid x^* \in \partial g(x)\}. \tag{12.2.17}$$

*Proof*

1° By definition, $\Psi$ is inclusion reversing.

2° We will prove that $\Psi(\hat{B}^*)$ is a $K$-efficient exposed constructor of epi $g$. Furthermore, we will show $\Psi^*(\Psi(\hat{B}^*)) = \hat{B}^*$ if $\hat{B}^*$ is a $K$-efficient exposed constructor of epi $g^*$.

If $\hat{B}^*$ is a $K$-efficient exposed constructor of epi $g^*$, we get from Proposition 12.2.15 that there is some $\overline{x} \in \mathbb{R}^n$ with $g(\overline{x}) < +\infty$ and $\partial g(\overline{x}) \neq \emptyset$ and

$$\hat{B}^* = \{(x^*, g^*(x^*)) \in \mathbb{R}^n \times \mathbb{R} \mid x^* \in \partial g(\overline{x})\}.$$

If $x^* \in \mathbb{R}^n$ with $g^*(x^*) < +\infty$ and $\partial g^*(x^*) \neq \emptyset$, then

$$\{(x, g(x)) \in \mathbb{R}^n \times \mathbb{R} \mid x^* \in \partial g(x)\}$$

is a $K$-efficient exposed constructor of epi $g$. Because the intersection of exposed constructors is an exposed constructor again if it is nonempty (compare Webster [462, Theorem 2.6.17]), $\Psi(\hat{B}^*)$ is a $K$-efficient exposed constructor of epi $g$ whenever $\Psi(\hat{B}^*)$ is nonempty. This is true because $(\overline{x}, g(\overline{x})) \in \Psi(\hat{B}^*)$.

Furthermore, we have

$$\begin{aligned} \hat{B}^* &= \{(x^*, g^*(x^*)) \in \mathbb{R}^n \times \mathbb{R} \mid x^* \in \partial g(\overline{x})\} \\ &\supseteq \bigcap_{(x, g(x)) \in \Psi(\hat{B}^*)} \{(x^*, g^*(x^*)) \in \mathbb{R}^n \times \mathbb{R} \mid x^* \in \partial g(x)\} = \Psi^*(\Psi(\hat{B}^*)). \end{aligned}$$

Now, we show that $\hat{B}^* \subseteq \Psi^*(\Psi(\hat{B}^*))$. If we suppose to the contrary that there is some $(x^*, g^*(x^*)) \in \hat{B}^*$ with $(x^*, g^*(x^*)) \notin \Psi^*(\Psi(\hat{B}^*))$, we get that there is some $(x, g(x)) \in \Psi(\hat{B}^*)$ such that $x^* \notin \partial g(x)$, a contradiction to $(x^*, g^*(x^*)) \in \hat{B}^*$.

3° Next, we prove that $\Psi^*(\hat{B})$ is a $K$-efficient exposed constructor of epi $g^*$. Moreover, we will show $\Psi(\Psi^*(\hat{B})) = \hat{B}$ if $\hat{B}$ is a $K$-efficient exposed constructor of epi $g$. If $\hat{B}$ is a $K$-efficient exposed constructor of epi $g$, we get from Proposition 12.2.15 that there is some $\overline{x}^* \in \mathbb{R}^n$ with $g^*(\overline{x}^*) < +\infty$ and $\partial g^*(\overline{x}^*) \neq \emptyset$ and

$$\hat{B} = \{(x, g(x)) \in \mathbb{R}^n \times \mathbb{R} \mid \overline{x}^* \in \partial g(x)\}.$$

If $x \in \mathbb{R}^n$ with $g(x) < +\infty$ and $\partial g(x) \neq \emptyset$, then

$$\{(x^*, g^*(x^*)) \in \mathbb{R}^n \times \mathbb{R} \mid x^* \in \partial g(x)\}$$

is a $K$-efficient exposed constructor of epi $g^*$. Taking into account that the intersection of exposed constructors is an exposed constructor again if it is nonempty (Webster [462, Theorem 2.6.17]), $\Psi^*(\hat{B})$ is a $K$-efficient exposed constructor of epi $g^*$ whenever $\Psi^*(\hat{B})$ is nonempty. This assertion is true since $(\overline{x}^*, g^*(\overline{x}^*)) \in \Psi^*(\hat{B})$.

Furthermore, we have

$$\begin{aligned}\hat{B} &= \{(x, g(x)) \in \mathbb{R}^n \times \mathbb{R} \mid \overline{x}^* \in \partial g(x)\}\\ &\supseteq \bigcap_{(x^*, g^*(x^*)) \in \Psi^*(\hat{B})} \{(x, g(x)) \in \mathbb{R}^n \times \mathbb{R} \mid x^* \in \partial g(x)\} = \Psi(\Psi^*(\hat{B})).\end{aligned}$$

Next, we prove that $\hat{B} \subseteq \Psi(\Psi^*(\hat{B}))$. Contrary, suppose that there is some $(x, g(x)) \in \hat{B}$ with $(x, g(x)) \notin \Psi(\Psi^*(\hat{B}))$. Then, we get that there is some $(x^*, g^*(x^*)) \in \Psi^*(\hat{B})$ such that $x^* \notin \partial g(x)$, a contradiction to $(x, g(x)) \in \hat{B}$. □

We consider a map $\hat{\Psi}$ given by

$$\hat{\Psi}(\hat{B}^*) := T(\Psi(-\hat{B}^*)).$$

Taking into account Theorem 12.2.16, the map $\hat{\Psi}$ defines an inclusion reversing one-to-one map between $K$-maximal exposed constructors of the dual image set $D = -\operatorname{epi} h^*$ (see (12.2.13)) and relatively $C$-efficient exposed constructors of $P = T(\operatorname{epi} h) = \operatorname{cl}(f(S) + C)$ with the inverse map $\hat{\Psi}^{-1} = -\Psi^* \circ T^{-1}$.

This means that we have a *geometric duality relationship between the upper closed extended primal image set* $P$ *given in* (12.2.8) and the *dual image set* $D$ *introduced in* (12.2.13):

For $y, v \in \mathbb{R}^\ell$, we consider

$$g(y, v) := (v_1, \ldots, v_{\ell-1}, 1)T^{-1}y - v_\ell$$

and

$$H(v) := \{y \in \mathbb{R}^\ell \mid g(y, v) = 0\} \quad \text{and} \quad H^*(y) := \{v \in \mathbb{R}^\ell \mid g(y, v) = 0\}.$$

Then, Heyde [239] has shown

$$\begin{aligned}
\hat{\Psi}(\hat{B}^*) = T(\Psi(-\hat{B}^*)) &= \bigcap_{(x^*, h^*(x^*)) \in -\hat{B}^*} T\left(\{(z, h(z)) \in \mathbb{R}^\ell \mid x^* \in \partial h(z)\}\right) \\
&= \bigcap_{(x^*, h^*(x^*)) \in -\hat{B}^*} T\left(\{(z, t) \in \mathbb{R}^\ell \mid t = h(z),\ \langle x^*, z\rangle = h(z) + h^*(x^*)\}\right) \\
&= \bigcap_{(x^*, h^*(x^*)) \in -\hat{B}^*} T\left(\{(z, t) \in \operatorname{epi} h \mid \langle x^*, z\rangle = t + h^*(x^*)\}\right) \\
&= \bigcap_{v \in \hat{B}^*} T\left(\{(z, t) \in \operatorname{epi} h \mid t - v_\ell + (v_1, \ldots, v_{\ell-1})z = 0\}\right) \\
&= \bigcap_{v \in \hat{B}^*} T\left(\left\{(z, t) \in \operatorname{epi} h \mid (v_1, \ldots, v_{\ell-1}, 1)\begin{pmatrix} z \\ t \end{pmatrix} - v_\ell = 0\right\}\right) \\
&= \bigcap_{v \in \hat{B}^*} \{y \in P \mid g(y, v) = 0\} \\
&= \bigcap_{v \in \hat{B}^*} H(v) \cap P.
\end{aligned}$$

Furthermore, we have

$$\begin{aligned}
\hat{\Psi}^{-1}(\hat{B}) = -\Psi^*(T^{-1}(\hat{B})) &= \bigcap_{(z, h(z)) \in T^{-1}(\hat{B})} -\{(x^*, f^*(x^*)) \in \mathbb{R}^\ell \mid x^* \in \partial h(z)\} \\
&= \bigcap_{(z, h(z)) \in T^{-1}(\hat{B})} -\{(x^*, s) \in \mathbb{R}^\ell \mid s = h^*(x^*),\ \langle x^*, z\rangle = h(z) + h^*(x^*)\} \\
&= \bigcap_{(z, h(z)) \in T^{-1}(\hat{B})} -\{(x^*, s) \in \operatorname{epi} h^* \mid -\langle x^*, z\rangle + h(z) + s = 0\} \\
&= \bigcap_{y \in \hat{B}} -\{(x^*, s) \in \operatorname{epi} h^* \mid -(x^{*T}, 1)T^{-1}y + s = 0\} \\
&= \bigcap_{y \in \hat{B}} \{v \in D \mid g(y, v) = 0\} = \bigcap_{y \in \hat{B}} H^*(y) \cap D.
\end{aligned}$$

**Remark 12.2.17** In this geometric duality relationship, we observe the influence of the nonlinear scalarizing functional $\varphi_{-C,k}$:

The functional $\varphi_{-C,k}$ is involved in the assertion

$$\hat{\Psi}(\hat{B}^*) = \bigcap_{v \in \hat{B}^*} H(v) \cap P$$

via the functional $h$ (see (12.2.11)) taking into account the condition $P = T(\operatorname{epi} h)$ for the primal image set $P$ (see (12.2.12)).

Furthermore, $\varphi_{-C,k}$ is involved in the assertion

$$\hat{\Psi}^{-1}(\hat{B}) = \bigcap_{y \in \hat{B}} H^*(y) \cap D$$

via the conjugate $h^*$ (see Proposition 12.2.10) taking into account the condition $D = -\operatorname{epi} h^*$ for the dual image set $D$.

Note that convex, continuous, strictly $C$-monotone functionals are used in the construction of the dual image set $Y_D$ in (12.2.3) too. Furthermore, the functional (4.1.11) is also an important tool in the proof of the duality assertions in Sect. 12.2.1.

**Remark 12.2.18** These geometric duality relationships between $P$ and $D$ are used by Heyde in [239, Example 5.8] for deriving geometric duality relations for linear vector optimization problems with a general ordering cone.

## 12.3 Minimal Point Theorems and Variational Principles of Ekeland's Type

Ekeland's variational principle is an assertion about the existence of an exact solution to a perturbed optimization problem in a neighborhood of an approximate solution to the original problem without convexity and compactness assumptions. It is a very useful tool to solve problems in optimization, variational analysis, optimal control theory, game theory, nonlinear equations and dynamical systems.

Let $(X, d)$ be a metric space and $\varphi : X \to \mathbb{R} \cup \{+\infty\}$ be a function.

Already Bishop and Phelps [52, 381] as well as Ekeland [152, 153], Turinici [451, 452], Dancs et al. [107] (see also [192, 195]), observed that this variational principle is nothing else than a minimal point theorem for a partial order on $X \times \mathbb{R}$ defined by

$$\left(x', r'\right) \preceq (x, r) \quad :\Longleftrightarrow \quad r' + d\left(x', x\right) \leq r$$

for $\left(x', r'\right), (x, r) \in X \times \mathbb{R}$. This relation can be applied to epigraphs of functions $\varphi : X \to \mathbb{R} \cup \{+\infty\}$, such that it generates a partial order on $X$ by

$$x' \preceq_X x \quad \Longleftrightarrow \quad \varphi\left(x'\right) + d\left(x', x\right) \leq \varphi\left(x\right).$$

We would like to recall the variational principle initiated by Ekeland in 1972 (see [152] and [153, Theorem 1]).

**Theorem 12.3.1** (Ekeland's variational principle)*([153, Theorem 1]) Let $(X, d)$ be a complete metric space and let $\varphi : X \to \mathbb{R} \cup \{+\infty\}$ be lower semicontinuous and bounded from below. Consider $\varepsilon > 0$ and $x_0 \in X$ such that $\varphi(x_0) \leq \inf \varphi + \varepsilon$. Then, for every $\lambda > 0$, there is an element $\overline{x} \in X$ with $f(\overline{x}) \in \mathbb{R}$ such that*

$$\varphi(\overline{x}) + \lambda^{-1}\varepsilon d(\overline{x}, x_0) \leq \varphi(x_0), \qquad d(\overline{x}, x_0) \leq \lambda, \tag{12.3.1}$$

*and*

$$\forall x \in X \setminus \{\overline{x}\} : \quad \varphi(x) + \lambda^{-1}\varepsilon d(\overline{x}, x) > \varphi(\overline{x}). \tag{12.3.2}$$

**Remark 12.3.2** Furthermore, all assertions of Theorem 12.3.1 hold for functions $\varphi : S \to \mathbb{R}$, where $S \subset X$ is a closed set.

**Remark 12.3.3** The assertion (12.3.2) in Theorem 12.3.1 says that $\overline{x}$ is an exact solution to the perturbed minimization problem

$$\inf_{x \in X} \{\varphi(x) + \lambda^{-1}\varepsilon d(\overline{x}, x)\}.$$

The aim of this section is to show minimal point theorems and corresponding variational principles in a more general setting concerning the involved spaces using the functional (4.1.11). We replace $\mathbb{R}$ by a separated l.c.s. $Y$ with a nontrivial, convex cone $C \subseteq Y$ generating a preorder on $Y$. For the proof of variational principles in such a general setting (for $f : X \to Y$, a complete metric space $X$ and a separated l.c.s. $Y$), it is useful to apply minimal point theorems on product spaces $X \times Y$. Dealing with such product spaces $X \times Y$, where we do not suppose that the involved sets are convex, we will see that the application of the nonlinear scalarizing functional (4.1.11) is very important. Göpfert, Tammer and Zălinescu derived corresponding results in this direction in a series of papers [193–195, 431, 433, 436, 439], compare also the books [192, Sect. 3.10] and [291, Chap. 10]. In [195] and [192, Sect. 3.10], it is explained that there are two different kinds of minimal point theorems. Considering a set $\mathscr{A} \subseteq X \times Y$ and a partial order $\preceq$ on $X \times Y$. Under different assumptions concerning boundedness and closedness of the involved set $\mathscr{A}$, it is possible to show the following types of minimal point theorems:

*Authentic minimal point theorem*: There is a (efficient) point $(\bar{x}, \bar{v})$ of $\mathscr{A}$ such that

$$(x, v) \in \mathscr{A}, \ (x, v) \preceq (\bar{x}, \bar{v}) \quad \Longrightarrow \quad (x, v) = (\bar{x}, \bar{v}).$$

*Nonauthentic minimal point theorem*: There is a point $(\bar{x}, \bar{v}) \in \mathscr{A}$ such that

$$(x, v) \in \mathscr{A}, \ (x, v) \preceq (\bar{x}, \bar{v}) \quad \Longrightarrow \quad x = \bar{x}.$$

Of course, in order to prove the weaker assertion in a nonauthentic minimal point theorem (see Theorem 12.3.4 and [192, 195]), one needs less stringent assumptions. However, as stated in [195], it is enough to prove Ekeland's type variational principles for vector-valued functions in Theorem 12.3.7, see [195]. Especially, since the preorder induced by the cone $C$ on $Y$ is used to generate the order $\preceq$ on $X \times Y$, the cone $C$ needs additional properties in order to show an authentic minimal point theorem in Theorem 12.3.10 and a corresponding variational principle in Theorem 12.3.11.

For related material, compare also the books [192, 252, 291] and references therein.

### 12.3.1 Variational Principles for Vector Optimization Problems

We now derive a minimal point theorems and a corresponding variational principles of Ekeland's type using the functional (4.1.11) in the proof of the assertions. Suppose that $(X, d)$ is a complete metric space, $Y$ is a separated l.c.s., $C \subset Y$ is a nontrivial, closed, convex cone, and $k \in C \setminus (-C)$. In the following assertions, $P_X$ and $P_Y$ denote the projections of $X \times Y$ onto $X$ and $Y$, respectively; so $P_X(x, y) = x$ for every $(x, y) \in X \times Y$.

Suppose that $\mathcal{A} \subset X \times Y$ is a nonempty set. For $x \in X$ and $y \in Y$, we consider the sets

$$\mathcal{A}_x := \{y \in Y \mid (x, y) \in \mathcal{A}\}, \quad \mathcal{A}_y := \{x \in X \mid (x, y) \in \mathcal{A}\}.$$

Obviously, $P_X(\mathcal{A}) = \{x \in X \mid \mathcal{A}_x \neq \emptyset\}$.

Using an element $k \in C \setminus (-C)$, we introduce a preorder on $X \times Y$, denoted by $\preceq_k$:

$$(x^1, y^1) \preceq_k (x^2, y^2) \iff y^1 + kd(x^1, x^2) \in y^2 - C. \tag{12.3.3}$$

Note that $\preceq_k$ is reflexive and transitive; if $(x^1, y^1) \preceq_k (x^2, y^2)$ and $(x^2, y^2) \preceq_k (x^1, y^1)$, then $x^1 = x^2$. If $C$ is pointed, then $\preceq_k$ is antisymmetric, too.

Furthermore, if $(X, \|\cdot\|)$ is a normed space (n.v.s.), then $\preceq_k$ is determined by the convex cone $C_k := \{(x, y) \in X \times Y \mid \|x\| \, k \in y - C\}$. Furthermore, if $C$ is pointed, then $C_k$ is pointed.

The first result is a nonauthentic minimal point theorem in the product space $X \times Y$ that we prove using a scalarization by means of the separating functional (4.1.11). This result is shown in [192, Theorem 3.10.12].

**Theorem 12.3.4** ((Nonauthentic) Minimal point theorem) *Assume that the following conditions hold:*

(A) *$(X, d)$ is a complete metric space, $Y$ is a separated l.c.s., $C \subset Y$ is a nontrivial, closed, convex cone, and $k \in C \setminus (-C)$. Let $\mathcal{A} \subset X \times Y$ be a nonempty set and $(x_0, y_0) \in \mathcal{A}$ be fixed.*

(B) *The following boundedness assumption is fulfilled: There is an element $\tilde{t} \in \mathbb{R}$ such that*

$$P_Y(\mathcal{A}) \cap \big(y_0 - \tilde{t}k - (C \setminus \{0\})\big) = \emptyset.$$

(C) *The set $A^r := \{x \in X \mid \exists\, y \in y_0 + rk - C, \ (x, y) \in \mathcal{A}\}$ is closed for every $r \in \mathbb{R}$.*

(D) *Every function* $\psi : Y \to \overline{\mathbb{R}}$ *that is* $C$*-monotone, proper, lower semicontinuous and sublinear, attains its infimum on* $\mathcal{A}_x - y_0$ *for every* $x \in P_X(\mathcal{A})$.

*Then, there is an element* $(\bar{x}, \bar{y}) \in \mathcal{A}$ *such that*

$$\bar{y} + kd(\bar{x}, x_0) \in y_0 - C \tag{12.3.4}$$

*and*

$$(x', y') \in \mathcal{A},\ y' + kd(x', \bar{x}) \in \bar{y} - C \implies \begin{cases} x' = \bar{x} \text{ and} \\ y' \in (\bar{y} - C) \setminus (\bar{y} - (0, +\infty)\, k - C)\,. \end{cases} \tag{12.3.5}$$

*If* $k \in \operatorname{int} C$, *assumption* (D) *can be replace by:*

(E) *Every function* $\psi : Y \to \mathbb{R}$ *that is* $C$*-monotone, continuous and sublinear, attains its infimum on* $\mathcal{A}_x - y_0$ *for every* $x \in P_X(\mathcal{A})$,
*and (12.3.5) by*

$$(x', y') \in \mathcal{A},\ y' + kd(x', \bar{x}) \in \bar{y} - C \implies x' = \bar{x},\ y' \notin \bar{y} - \operatorname{int} C. \tag{12.3.6}$$

***Proof*** We look at the function

$$\xi : X \to \overline{\mathbb{R}}, \quad \xi(x) := \begin{cases} \inf\{\varphi_{-C,k}(y - y_0) \mid y \in \mathcal{A}_x\} & \text{for } x \in P_X(\mathcal{A}), \\ +\infty & \text{for } x \notin P_X(\mathcal{A}), \end{cases}$$

where $\varphi_{-C,k}$ is defined by

$$\varphi_{-C,k}(y) = \inf\{t \in \mathbb{R} \mid y \in tk - C\}$$

(see (4.1.11) with $A = -C$). From Propositions 4.7.3 and 4.7.1(b), we know that the functional $\varphi_{-C,k}$ is proper, sublinear, $C$-monotone and lower semicontinuous. Since $\varphi_{-C,k}$ is $C$-monotone (Corollary 4.5.8), we obtain that $\xi$ is bounded from below from (B). Indeed, assume that for each $\tilde{t} \in \mathbb{R}$ there is an element $x \in X$ such that $\xi(x) < -\tilde{t}$. Then, there is an element $y \in Y$ such that $(x, y) \in \mathcal{A}$ and $\varphi_{-C,k}(y - y_0) < -\tilde{t}$, therefore $y - y_0 \in -\tilde{t}k - tk - C$ for some $t > 0$, a contradiction to (B).

Furthermore, we show that the function $\xi$ is also lower semicontinuous. Indeed, let $r, r' \in \mathbb{R}$ be such that $r < r'$. We prove

$$A^r \subseteq \{x \in X \mid \xi(x) \le r\} \subseteq A^{r'}. \tag{12.3.7}$$

If $x \in A^r$, there is an element $y \in y_0 + rk - C$ with $(x, y) \in \mathcal{A}$, hence $\xi(x) \le \varphi_{-C,k}(y - y_0) \le \varphi_{-C,k}(rk) = r$. Now, let $x \in X$ be such that $\xi(x) \le r$. Because of $r < r'$, there is an element $y \in \mathcal{A}_x$ such that $\varphi_{-C,k}(y - y_0) \le r'$ taking into account the definition of $\xi$. Therefore, $y \in y_0 + r'k - C$ by Theorem 4.4.4, (4.4.5). This yields $x \in A^{r'}$. Hence, (12.3.7) holds.

We can conclude

$$\{x \in X \mid \xi(x) \le r\} \subseteq \bigcap_{r'>r} A^{r'} \subseteq \bigcap_{r'>r} \{x \in X \mid \xi(x) \le r'\}$$
$$= \{x \in X \mid \xi(x) \le r\}.$$

Under the assumption (C), $A^{r'}$ is closed for every $r' \in \mathbb{R}$, it follows that $\{x \in X \mid \xi(x) \le r\}$ is closed for every $r \in \mathbb{R}$. Taking into account Proposition 3.3.12, we get that $\xi$ is lower semicontinuous.

Now, it is possible to apply Ekeland's variational principle (Theorem 12.3.1) and we get the existence of an element $\bar{x} \in X$ such that

$$\xi(\bar{x}) + d(\bar{x}, x_0) \le \xi(x_0) \le \varphi_{-C,k}(y_0 - y_0) = 0 \tag{12.3.8}$$

and

$$\forall x \in X,\ x \ne \bar{x}: \quad \xi(x) + d(x, \bar{x}) > \xi(\bar{x}). \tag{12.3.9}$$

From (D), we get the existence of an element $\bar{y} \in \mathcal{A}_{\bar{x}}$ such that $\xi(\bar{x}) = \varphi_{-C,k}(\bar{y} - y_0)$. By (12.3.8), we obtain that $\bar{y} - y_0 + kd(\bar{x}, x_0) \in -C$ taking into account (4.2.2). This means, (12.3.4) holds.

Now, consider $(x', y') \in \mathcal{A}$ with $y' + kd(x', \bar{x}) \in \bar{y} - C$. Then, it holds that

$$\xi(x') + d(x', \bar{x}) \le \varphi_{-C,k}(y' - y_0) + d(x', \bar{x}) = \varphi_{-C,k}(y' - y_0 + kd(x', \bar{x}))$$
$$\le \varphi_{-C,k}(\bar{y} - y_0) = \xi(\bar{x}).$$

Therefore, we preserve that $x' = \bar{x}$ from (12.3.9). If $y' \in \bar{y} - (0, +\infty)k - C$, then $y' = \bar{y} - tk - k$ for some $t > 0$ and $k \in C$. This yields

$$\xi(\bar{x}) = \xi(x') \le \varphi_{-C,k}(y' - y_0) = \varphi_{-C,k}(\bar{y} - y_0 - tk - k) \le \varphi_{-C,k}(\bar{y} - y_0 - tk)$$
$$= \varphi_{-C,k}(\bar{y} - y_0) - t < \xi(\bar{x}),$$

a contradiction. Hence, assertion (12.3.5) holds.

Finally, we suppose $k \in \operatorname{int} C$, then we additionally get from Proposition 4.7.3 that $\varphi_{-C,k}$ is finite-valued and continuous. Furthermore, it holds that $\operatorname{int} C \subseteq (0, +\infty)k + C$. The conclusion immediately follows. □

**Remark 12.3.5** The result in Theorem 12.3.4 is not an authentic minimal point theorem. Furthermore, if condition (C) is replaced by the condition

($\tilde{C}$) $\{x \in X \mid \exists\, y \in rk - C,\ (x, y) \in \mathcal{A}\}$ is closed for every $r \in \mathbb{R}$,

then the assertion in (12.3.4) must be replaced (when $\operatorname{int} C \ne \emptyset$) by the weaker one

$$\bar{y} + kd(\bar{x}, x_0) \notin y_0 + \operatorname{int} C.$$

**Remark 12.3.6** In the assertions of Theorem 12.3.4, we also have that $d(\bar{x}, x_0) \leq \tilde{t}$. Indeed, since $\xi(\bar{x}) \geq \tilde{t}$, we get $-\tilde{t} + d(\bar{x}, x_0) \leq 0$ from (12.3.8), i.e., our assertion.

We get the following variational principle of Ekeland's type for vector optimization problems (shown in [192, Corollary 3.10.14]) by applying Theorem 12.3.4 to the epigraph of a vector-valued function.

**Theorem 12.3.7** (Variational principle of Ekeland's type I) *Assume that the following conditions are fulfilled:*

(A′) *$(X, d)$ is a complete metric space, $Y$ is a separated l.c.s., $C \subset Y$ is a nontrivial, closed, convex, pointed cone, and $k \in C \setminus (-C)$ and $f : X \to Y$.*
(B′) *$f(X) \cap (f(x_0) - \varepsilon k - (C \setminus \{0\})) = \emptyset$ for some $\varepsilon > 0$.*
(C′) *$\{x \in X \mid f(x) \in f(x_0) + rk - C\}$ is closed for every $r \in \mathbb{R}$.*

*Then, for every $\lambda > 0$, there is an element $\bar{x} \in X$ such that*

(a) $f(\bar{x}) + \lambda^{-1}\varepsilon k d(\bar{x}, x_0) \in f(x_0) - C$,
(b) $d(\bar{x}, x_0) \leq \lambda$,
(c) $f(x) + \lambda^{-1}\varepsilon k d(x, \bar{x}) \in f(\bar{x}) - C \implies x = \bar{x}$.

***Proof*** We consider $\mathscr{A} = \operatorname{epi} f$ and put $y_0 = f(x_0)$. Obviously, the hypotheses of Theorem 12.3.4 are satisfied for $\tilde{t} = \varepsilon$. From Theorem 12.3.4, we get that there is an element $(\bar{x}, \bar{y}) \in \mathscr{A}$ such that (12.3.4) and (12.3.5) hold with $\lambda^{-1}\varepsilon d$ instead of $d$. Assertion (a) follows from (12.3.4) and Remark 12.3.6.

In order to show (b), we assume that $d(\bar{x}, x_0) = \lambda + t$ with $t > 0$ and obtain that for some $c \in C$,

$$\begin{aligned} f(\bar{x}) = f(x_0) - \lambda^{-1}\varepsilon k d(\bar{x}, x_0) - c &= f(x_0) - \varepsilon k - (\lambda^{-1}\varepsilon t k + c) \\ &\in f(x_0) - \varepsilon k - (C \setminus \{0\}), \end{aligned}$$

a contradiction. We obtain $d(\bar{x}, x_0) \leq \lambda$.

Finally, let $x \in X$ be such that $f(x) + \lambda^{-1}\varepsilon k d(x, \bar{x}) \in f(\bar{x}) - C$ $(\in \bar{y} - C)$. Taking $y = f(x)$, we get $x = \bar{x}$ from (12.3.5). □

**Remark 12.3.8** The assumption (B′) in Theorem 12.3.7 is a (weak) boundedness assumption. Corresponding (stronger) boundedness assumptions are supposed in assumption (A5) in Theorems 12.3.16, 12.3.20 and 12.3.28 in order to prove extensions of Ekeland's variational principle for vector optimization problems with variable domination structure.

**Remark 12.3.9** The assumption (C′) in Theorem 12.3.7 is a certain kind of lower semicontinuity of functions $f : X \to Y$. If the condition

" (C′) $\{x \in X \mid f(x) \in f(x_0) + rk - C\}$ is closed for every $r \in \mathbb{R}$ "

in Theorem 12.3.7 is replaced by

" $\{x \in X \mid f(x) \in rk - C\}$ is closed for every $r \in \mathbb{R}$ ", then (a) must be replaced by

$$\text{“ } f(\bar{x}) + \lambda^{-1}\varepsilon k d(\bar{x}, x_0) \notin f(x_0) + \operatorname{int} C \text{ ”}$$

(if $\operatorname{int} C \neq \emptyset$).

Corresponding assumptions concerning the lower semicontinuity of $f : X \to Y$ for vector optimization problems with variable domination structure are formulated in assumption (A6) in Theorem 12.3.16 and in assumption (A6′) in Theorem 12.3.20 in order to prove extensions of Ekeland's variational principle for problems with variable domination structure.

Furthermore, we obtain another minimal point theorem in slightly different conditions using the nonlinear separating functional (4.1.11) and embedding $C \setminus \{0\}$ in the interior of another nontrivial cone. This result is derived in [192, Theorem 3.10.15].

**Theorem 12.3.10** ((Authentic) Minimal point theorem) *Assume that the following conditions are fulfilled:*

(A″) *$(X, d)$ is a complete metric space, $Y$ is a separated l.c.s., $C \subset Y$ is a nontrivial, closed, convex cone, $k \in C \setminus (-C)$. Furthermore, there is a nontrivial, closed, convex cone $B \subset Y$ with $C \setminus \{0\} \subset \operatorname{int} B$. Let $\mathcal{A} \subset X \times Y$ be a nonempty set and $(x_0, y_0) \in \mathcal{A}$ be fixed.*
(B″) *$P_Y(\mathcal{A}) \cap (\widetilde{y} - \operatorname{int} B) = \emptyset$ for some $\widetilde{y} \in Y$.*
(C″) *$\mathcal{A} \subset X \times Y$ satisfies the condition: For every $\preceq_k$-decreasing sequence $\{(x_n, y_n)\} \subset \mathcal{A}$ with $x_n \to x \in X$, there is an element $y \in Y$ such that $(x, y) \in \mathcal{A}$ and $(x, y) \preceq_k (x_n, y_n)$ for every $n \in \mathbb{N}$.*

*Then, there is an element $(\overline{x}, \overline{y}) \in \mathcal{A}$, such that*

$$(\overline{x}, \overline{y}) \preceq_k (x_0, y_0)$$

*and*

$$(x', y') \in \mathcal{A},\ (x', y') \preceq_k (\bar{x}, \bar{y}) \implies (x', y') = (\bar{x}, \bar{y}).$$

***Proof*** Let $\varphi_{-B,k}$ be given by (4.1.11) with $A = -B$. From Theorem 4.6.6 and Propositions 4.7.3, 4.7.1(b), we get that $\varphi_{-B,k}$ is a finite-valued, continuous, sublinear function for which (4.6.12), (4.6.13), and (4.6.14) hold with $A$ replaced by $-B$. Furthermore, (4.2.2) with $A = -B$ holds. Moreover, $\varphi_{-B,k}$ is strictly $C$-monotone (Proposition 4.7.3), i.e., if $y_2 - y_1 \in C \setminus \{0\}$, then $\varphi_{-B,k}(y_1) < \varphi_{-B,k}(y_2)$. For $(x, y) \in \mathcal{A}$, it holds that $\varphi_{-B,k}(y - \widetilde{y}) \geq 0$. Indeed, if $\varphi_{-B,k}(y - \widetilde{y}) < 0$ for some $(x, y) \in \mathcal{A}$, there is an element $\lambda > 0$ such that $y - \widetilde{y} \in -\lambda k - B$. Therefore,

$$y \in \widetilde{y} - (\lambda k + B) \subset \widetilde{y} - (B + \operatorname{int} B) \subset \widetilde{y} - \operatorname{int} B,$$

a contradiction to (B″). Because of $0 \leq \varphi_{-B,k}(y - \widetilde{y}) \leq \varphi_{-B,k}(y) + \varphi_{-B,k}(-\widetilde{y})$, we get that $\varphi_{-B,k}$ is bounded from below on $P_Y(\mathcal{A})$.

Now, we construct a sequence $\{(x_n, y_n)\}_{n \geq 0} \subset \mathcal{A}$ in the following way: Having $(x_n, y_n) \in \mathcal{A}$, we chose $(x_{n+1}, y_{n+1}) \in \mathcal{A}$, $(x_{n+1}, y_{n+1}) \preceq_k (x_n, y_n)$, with

$$\varphi_{-B,k}(y_{n+1}) \leq \inf\{\varphi_{-B,k}(y) \mid (x, y) \in \mathscr{A},\ (x, y) \preceq_k (x_n, y_n)\} + 1/(n+1).$$

Obviously, the sequence $\{(x_n, y_n)\}$ is $\preceq_k$-decreasing such that

$$\forall n, p \in \mathbb{N}_> : \quad y_{n+p} + kd(x_{n+p}, x_n) \leq_C y_n,$$

and

$$\forall n, p \in \mathbb{N}_> : \quad d(x_{n+p}, x_n) \leq \varphi_{-B,k}(y_n) - \varphi_{-B,k}(y_{n+p}) \leq 1/n.$$

This yields that $\{x_n\}$ is a Cauchy sequence in the complete metric space $(X, d)$, and so $(x_n)$ is convergent to some $\bar{x} \in X$. Taking into account (C$''$), there is an element $\overline{y} \in Y$ such that $(\overline{x}, \overline{y}) \in \mathscr{A}$ and $(\bar{x}, \bar{y}) \preceq_k (x_n, y_n)$ for every $n \in \mathbb{N}$. Now, we show that $(\overline{x}, \overline{y})$ is the desired element. Indeed, $(\bar{x}, \bar{y}) \preceq_k (x_0, y_0)$. Assume that $(x', y') \in \mathscr{A}$ is such that $(x', y') \preceq_k (\bar{x}, \bar{y})$ ($\preceq_k (x_n, y_n)$ for every $n \in \mathbb{N}$). Thus, $\varphi_{-B,k}(y') + d(x', \overline{x}) \leq \varphi_{-B,k}(\overline{y})$. We conclude

$$\forall n \geq 1 : \quad d(x', \bar{x}) \leq \varphi_{-B,k}(\bar{y}) - \varphi_{-B,k}(y') \leq \varphi_{-B,k}(y_n) - \varphi_{-B,k}(y') \leq 1/n.$$

We obtain that $d(x', \bar{x}) = \varphi_{-B,k}(\bar{y}) - \varphi_{-B,k}(y') = 0$. Therefore, $x' = \overline{x}$. If $y' \neq \overline{y}$, we get $\overline{y} - y' \in C \setminus \{0\}$ from $y' \leq_C \overline{y}$. This yields $\varphi_{-B,k}(y') < \varphi_{-B,k}(\bar{y})$, a contradiction. Hence, $(x', y') = (\bar{x}, \bar{y})$. □

Note that when $C$ and $k$ are as in Theorem 12.3.10, we get another variant of Ekeland's variational principle in the following theorem (see [192, Corollary 3.10.16]).

**Theorem 12.3.11** (Variational principle of Ekeland's type II) *Assume that the following conditions are fulfilled:*

(A$'''$) *$(X, d)$ is a complete metric space, $Y$ is a separated l.c.s., $C \subset Y$ is a nontrivial, closed, convex cone, $k \in C \setminus (-C)$. There is a nontrivial convex cone $B \subset Y$ such that $C \setminus \{0\} \subset \operatorname{int} B$ and $f : X \to Y$.*
(B$'''$) *$f(X) \cap (\widetilde{y} - B) = \emptyset$ for some $\widetilde{y} \in Y$.*
(C$'''$) *$\{x' \in X \mid f(x') + kd(x', x) \in f(x) - C\}$ is closed for every $x \in X$.*

*Then, for every $x_0 \in X$ there is an element $\bar{x} \in X$ such that*

(a) $f(\bar{x}) + kd(\bar{x}, x_0) \in f(x_0) - C$,
(b) $\forall x \in X : f(x) + kd(x, \bar{x}) \in f(\bar{x}) - C \implies x = \bar{x}$.

***Proof*** Consider $\mathscr{A} = \operatorname{graph} f \subset X \times Y$. We show that assumption (C$''$) in Theorem 12.3.10 holds. Indeed, let $\{(x_n, y_n)\} \subset \mathscr{A}$ be a $\preceq_k$-decreasing sequence with $x_n \to x \in X$. Of course, since $y_n = f(x_n)$, we have $x_{n+p} \in \{x \in X \mid f(x) + kd(x, x_n) \in f(x_n) - C\}$ for all $n, p \in \mathbb{N}$. Since this set is closed (because of assumption (C$'''$)), it also contains the limit $x$ of the sequence $\{x_{n+p}\}_{p\in\mathbb{N}}$. Therefore, we obtain $(x, y) \preceq_k (x_n, y_n)$ for every $n$, where $y = f(x)$. Obviously, $(x, y) \in \mathscr{A}$. So, the assumptions of Theorem 12.3.10 are satisfied and we get the existence of $\bar{x} \in X$ such that $(\bar{x}, f(\bar{x})) \in \operatorname{graph} f$ fulfills the assertions of the theorem, i.e., (a) and (b) hold. □

**Remark 12.3.12** Again, the assumption (B‴) in Theorem 12.3.11 is a (weak) boundedness assumption. Corresponding (stronger) boundedness assumptions are supposed in assumption (A5) in Theorems 12.3.16, 12.3.20 and 12.3.28 in order to prove extensions of Ekeland's variational principle for vector optimization problems with variable domination structure.

**Remark 12.3.13** The assumption (C‴) in Theorem 12.3.11 is a certain kind of lower semicontinuity of functions $f : X \to Y$.

Notice also that the assumption (C‴) in Theorem 12.3.11 can be replaced by the assumption that $C$ is closed in the direction $k$ and for every sequence $\{x_n\}$, with $x_n \to x$ and $\{f(x_n)\}$ decreasing (with respect to $C$), it holds that $f(x) \in f(x_n) - C$ for all $n \in \mathbb{N}$. $C$ is closed in the direction $k$ if $C \cap (y - \mathbb{R}_+k)$ is closed for every $y \in C$.

Corresponding assumptions to (C‴) in Theorem 12.3.11 concerning the lower semicontinuity of $f : X \to Y$ for vector optimization problems with variable domination structure are formulated in assumption (A6) in Theorem 12.3.16, in assumption (A6′) in Theorem 12.3.20 and in assumption (A6″) in Theorem 12.3.28 in order to prove corresponding variational principle of Ekeland's type.

### *12.3.2 Variational Principles for VVDS*

The aim of this section is to show variational principles of Ekeland's type for $\varepsilon k$-nondominated-like solutions and $\varepsilon k$-nondominated solutions of vector optimization problems with respect to variable domination structures (see Definition 6.1.1 and Sect. 11.4.1) applying extensions of the nonlinear separating functional (4.1.11). These extensions are discussed in Sect. 11.2. We will apply these results for deriving necessary conditions for approximate solutions of (VVOP) in Sect. 12.4.

The results in this section are shown by Bao, Eichfelder, Soleimani and Tammer in [18].

Throughout this section, we suppose

(A1) $X$ is a Banach space, $S \subseteq X$ a closed set, $Y$ a topological vector space, $f : X \to Y$ is a vector-valued function and $\varepsilon \geq 0$.

(A2) Let $D : Y \to \mathscr{P}(Y)$ be a set-valued map with $0 \in \text{bd}\, D(y)$ and $D(y)$ is closed for all $y \in Y$. $k \in Y \setminus \{0\}$ satisfies $D(y) + \mathbb{R}_+k \subseteq D(y)$ for all $y \in Y$. Furthermore, we suppose that the set-valued map $\tilde{D} : Y \to \mathscr{P}(Y)$ has the same properties.

In this section, under the assumptions (A1) and (A2), we consider the following vector optimization problem with respect to a variable domination structure:

$$D(\cdot)\text{–}\operatorname*{Opt}_{x \in S} f(x), \qquad \text{(VVOP)}$$

where we are looking for two different kinds of approximate solutions: $\varepsilon k$-nondominated-like solutions with the domination map $\tilde{D}$ (Definition 12.3.14) and $\varepsilon k$-nondominated solutions with the domination map $D$ (Definition 12.3.22). These concepts are based on the following binary relations on $Y$ with respect to the variable domination structure $D$, denoted by $\preceq_1$, and $\preceq_2$, respectively:

$$y^1 \preceq_1 y^2 \iff y^1 \in y^2 - \tilde{D}(y^2) \tag{12.3.10}$$

and

$$y^1 \preceq_2 y^2 \iff y^2 \in y^1 + D(y^1) \tag{12.3.11}$$

(see Sect. 11.1). When $D(y) \equiv K$ for some fixed set $K$ in $Y$ and for all $y \in Y$, the two binary relations defined in (12.3.10) and (12.3.11) are identical.

In [414], a variational principle for $\varepsilon k$-nondominated-like solutions of (VVOP) is given. Following [18], we now show variational principles of Ekeland's type for $\varepsilon k$-nondominated as well as for $\varepsilon k$-nondominated-like solutions of (VVOP) under weaker assumptions in comparison with corresponding results in [414]. In vector optimization with fixed domination structure, there is no difference between $\varepsilon k$-nondominated and $\varepsilon k$-nondominated-like solutions but in the case of vector optimization problems with variable domination structure (see Chap. 11) these solution concepts are different. Generalizations of Ekeland's variational principle for vector optimization with fixed domination structure are given in Sect. 12.3.1, compare [440] and references therein.

In Definition 11.4.4 (see (12.3.10)), we have introduced approximate solutions of a vector optimization problem with variable domination structure in the image space $Y$. In the following, we formulate a corresponding definition in the pre-image space using the notations $\varepsilon k - \mathrm{M}(F, \tilde{D}(\cdot))$ and $\varepsilon k - \mathrm{WM}(F, \tilde{D}(\cdot))$ introduced in Definition 11.4.4 with $F = f(S)$.

**Definition 12.3.14** Consider problem (VVOP) and $\varepsilon \geq 0$. Then,

(a) $x_\varepsilon \in S$ is called an $\varepsilon k$-nondominated-like solution of (VVOP) with respect to the variable domination structure $\tilde{D}(\cdot)$ if there is no element $y \in f(S) = \bigcup_{x \in S}\{f(x)\}$ such that $y + \varepsilon k \preceq_1 f(x_\varepsilon)$, i.e.,

$$f(x_\varepsilon) \in \varepsilon k - \mathrm{M}(f(S), \tilde{D}(\cdot)).$$

(b) Assume that int $\tilde{D}(f(x_\varepsilon)) \neq \emptyset$. $x_\varepsilon \in S$ is called a weakly $\varepsilon k$-nondominated-like solution of (VVOP) with respect to $\tilde{D}(\cdot)$ if

$$f(x_\varepsilon) \in \varepsilon k - \mathrm{WM}(f(S), \tilde{D}(\cdot)).$$

If $\varepsilon = 0$, the notions in Definition 12.3.14 coincide with the usual definitions of nondominated-like solutions and weakly nondominated-like solutions (see Definition 11.1.3 for the definition in the image space). The sets of $\varepsilon k$-nondominated-like and weakly $\varepsilon k$-nondominated-like solutions are denoted by $\varepsilon k$-$\mathrm{M}_f(S, \tilde{D}(\cdot))$ and $\varepsilon k$-$\mathrm{WM}_f(S, \tilde{D}(\cdot))$, respectively. Furthermore, if $\varepsilon = 0$, we also write $\mathrm{M}_f(S, \tilde{D}(\cdot))$ and $\mathrm{WM}_f(S, \tilde{D}(\cdot))$.

**Remark 12.3.15** In Theorems 12.3.16, 12.3.20 and 12.3.28, we will apply a scalarization by means of an extension of the nonlinear scalarizing functional (4.1.11) and Ekeland's variational principle for scalar optimization problems (see Theorem 12.3.1) in order to show a variational principle for (weakly) $\varepsilon k$-nondominated-like solutions (and $\varepsilon k$-nondominated solutions) of (VVOP). So, it is important to formulate the boundedness (see assumption (A5) in Theorems 12.3.16, 12.3.20, 12.3.28) as well as the lower semicontinuity assumption (see assumption (A6) in Theorem 12.3.16 and assumption (A6′) in Theorem 12.3.20) in such a way that the scalarized function has corresponding properties as supposed in Theorem 12.3.1.

Corresponding assumptions concerning the lower semicontinuity in variational principles for vector optimization problems with fixed domination structure are supposed in assumption (C′) in Theorem 12.3.7 and in assumption (C‴) in Theorem 12.3.11 (compare Remarks 12.3.9 and 12.3.13).

The next theorem (shown in [414, Theorem 5.1]) presents a generalization of Ekeland's variational principle (Theorem 12.3.1) for $\varepsilon k$-nondominated-like solutions of (VVOP) under the assumptions that $f : X \to Y$ is bounded from below (assumption (A5)) and that $f$ has a certain lower semicontinuity property (assumption (A6)).

**Theorem 12.3.16** (Variational principle of Ekeland's type for $\varepsilon k$-nondominated-like solutions) *We consider problem* (VVOP), $\overline{x} \in \varepsilon k-\mathrm{M}_f(S, \tilde{D}(\cdot))$ *and put* $\overline{y} := f(\overline{x})$. *In addition to* (A1) *and* (A2), *we suppose:*

(A3) $\tilde{D}(y)$ *are proper, closed, pointed sets with* $0 \in \tilde{D}(y)$, $\mathrm{int}\, \tilde{D}(y) \neq \emptyset$ *satisfying* $\tilde{D}(y) + \mathbb{R}_{>}k \subseteq \mathrm{int}\, \tilde{D}(y)$ *and* $\tilde{D}(y) + \tilde{D}(y) \subseteq \tilde{D}(y)$ *for all* $y \in f(S)$.
(A4) *There is a cone-valued map* $B : Y \to \mathcal{P}(Y)$ *with* $k \in \mathrm{int}\, B(y)$ *and* $\tilde{D}(y) + (B(y) \setminus \{0\}) \subseteq \tilde{D}(y)$ *for all* $y \in f(S)$.
(A5) *There is an element* $\underline{y} \in Y$ *with* $f(S) \subseteq \underline{y} + \tilde{D}(\overline{y})$.
(A6) *The sets*

$$M(y, t) := \{x \in S \mid f(x) \in tk - \tilde{D}(y)\}$$

*are closed in* $X$ *for all* $y \in f(S)$ *and* $t \in \mathbb{R}$.
(A7) $\tilde{D}(y) \subseteq \tilde{D}(\overline{y})$ *for all* $y \in f(S)$.

*Then, there is an element* $x_\varepsilon \in S$ *with*

(a) $x_\varepsilon \in \varepsilon k-\mathrm{WM}_f(S, B(\cdot))$, *i.e.*, $\big(f(x_\varepsilon) - \varepsilon k - \mathrm{int}\, B(f(x_\varepsilon))\big) \cap f(S) = \emptyset$.
(b) $\|x_\varepsilon - \overline{x}\| \leq \sqrt{\varepsilon}$.
(c) $x_\varepsilon \in \mathrm{WM}_{f_{\varepsilon k}}(S, B(\cdot))$, *where* $f_{\varepsilon k}(x) := f(x) + \sqrt{\varepsilon}\|x - x_\varepsilon\|k$.

***Proof*** See Theorem 5.1 in [414]. □

**Remark 12.3.17** Assumption (A5) in Theorem 12.3.16 is a (stronger) boundedness assumptions in comparison with the boundedness assumption (B′) in Theorem 12.3.7 or (B‴) in Theorem 12.3.11.

**Remark 12.3.18** Assumption (A6) in Theorem 12.3.16 is a certain kind of lower semicontinuity of functions $f : X \to Y$. According to the discussion in Remark 12.3.9 concerning assumption (C′) in Theorem 12.3.7, we get the assertion "(a) $x_\varepsilon \in \varepsilon k - \mathrm{WM}_f(S, B(\cdot))$" under the assumptions of Theorem 12.3.16 including the lower semicontinuity of $f : X \to Y$ supposed in (A6).

**Remark 12.3.19** Observe that Theorem 12.3.16 can be shown by using the nonlinear separating function $\varphi_{-\tilde{D}(v),k}$ defined in (11.2.4) with $v = \overline{y} := f(\overline{x})$:

$$\varphi_{-\tilde{D}(\overline{y}),k}(y) = \inf \left\{ t \in \mathbb{R} \mid y \in tk - \tilde{D}(\overline{y}) \right\}$$

and the corresponding properties of this scalarization function derived in Theorem 4.6.6. In this case, we assume instead of $(A3)$ the assumption $(A3^*)$ and instead of $(A4)$, the assumption $(A4^*)$ given by

$(A3^*)$ $\tilde{D}(\overline{y})$ is a proper, closed, pointed set with $0 \in \tilde{D}(\overline{y})$, $\operatorname{int} \tilde{D}(\overline{y}) \neq \emptyset$ satisfying satisfying $\mathbb{R}k - \tilde{D}(\overline{y}) = Y$.

$(A4^*)$ There is a cone-valued map $B : Y \to \mathscr{P}(Y)$ with $k \in \operatorname{int} B(\overline{y})$, $\tilde{D}(\overline{y}) + (B(\overline{y}) \setminus \{0\}) \subseteq \tilde{D}(\overline{y})$ and $B(f(x)) \subseteq B(\overline{y})$ for all $x \in S$ with $\|x - \overline{x}\| \leq \sqrt{\varepsilon}$.

Furthermore, taking into account that $\tilde{D}(y)$ for $y \in f(S) \setminus \{\overline{y}\}$ does not play any role in the definition of the scalarization function $\varphi_{-\tilde{D}(\overline{y}),k}$, we have to suppose the closedness of all the sets

$$M(\overline{y}, t) := \left\{ x \in S \mid f(x) \in tk - \tilde{D}(\overline{y}) \right\}$$

for all $t \in \mathbb{R}$. With other words, we do not need the closedness of the sets $M(y, t)$ in assumption (A6) when $y \neq \overline{y}$ and $y \in f(S)$.

This discussion leads us to a variational principle for $\varepsilon k$-nondominated-like solutions of (VVOP) under a weaker condition concerning lower semicontinuity (assumption (A6′)) in comparison with assumption (A6) in Theorem 12.3.16.

The following theorem is shown in [18, Theorem 3.8].

**Theorem 12.3.20** (Variational principle of Ekeland's type for $\varepsilon k$-nondominated-like solutions) *We consider problem* (VVOP), $\overline{x} \in \varepsilon k - \mathrm{M}_f(S, \tilde{D}(\cdot))$ *and put* $\overline{y} := f(\overline{x})$. *In addition to* (A1) *and* (A2), *suppose that the following conditions hold:*

(A3′) $\overline{D} := \tilde{D}(\overline{y})$ *is a proper, closed, pointed set with* $0 \in \tilde{D}(\overline{y})$, $\operatorname{int} \overline{D} \neq \emptyset$ *and* $Y = \mathbb{R}k - \overline{D}$.

(A4′) *There is a cone-valued map* $B : Y \to \mathscr{P}(Y)$ *with* $k \in \operatorname{int} \overline{B}$, *where* $\overline{B} := B(\overline{y})$, $\overline{D} + (\overline{B} \setminus \{0\}) \subseteq \operatorname{int} \overline{D}$, *and* $B(f(x)) \subseteq \overline{B}$ *for all* $x \in S$ *with* $\|x - \overline{x}\| \leq \sqrt{\varepsilon}$.

(A5) *There is an element* $\underline{y} \in Y$ *with* $f(S) \subseteq \underline{y} + \overline{D}$.

(A6′) *The sets*

$$M(t) := \left\{ x \in S \mid f(x) \in tk - \overline{D} \right\}$$

*are closed in* $X$ *for all* $t \in \mathbb{R}$.

*Then, there is an element* $x_\varepsilon \in S$ *such that*

(a′) $x_\varepsilon \in \varepsilon k\text{-}\mathrm{M}_f(S, B(\cdot))$, *i.e.*, $\big(f(x_\varepsilon) - \varepsilon k - (B(f(x_\varepsilon)) \setminus \{0\})\big) \cap f(S) = \emptyset$.
(b) $\|x_\varepsilon - \overline{x}\| \leq \sqrt{\varepsilon}$.
(c′) $x_\varepsilon \in \mathrm{M}_{f_{\varepsilon k}}(S, B(\cdot))$, *where* $f_{\varepsilon k}(x) := f(x) + \sqrt{\varepsilon}\|x - x_\varepsilon\|k$ $(x \in X)$.

***Proof*** Consider $\overline{x} \in \varepsilon k\text{-}\mathrm{M}_f(S, \tilde{D}(\cdot))$ and $\overline{y} = f(\overline{x})$. We apply the nonlinear separating functional (4.1.11) with $A = -\overline{D}$

$$\varphi_{-\overline{D},k}(y) = \inf\left\{t \in \mathbb{R} \mid y \in tk - \overline{D}\right\}.$$

Under the assumptions of this theorem, we get the following properties of $\varphi_{-\overline{D},k}$ from Theorem 4.6.6 and Proposition 4.6.13:

- $\varphi_{-\overline{D},k}(y + tk) = \varphi_{-\overline{D},k}(y) + t$ for all $y \in Y$ and for all $t \in \mathbb{R}$ due to (A3′) by (4.2.2).
- $\varphi_{-\overline{D},k}$ is finite-valued and continuous due to (A3′) and (A4′) by Proposition 4.7.10.
- $\varphi_{-\overline{D},k}$ is strictly $\overline{B}$-monotone (and thus $\overline{B}$-monotone), i.e.,

$$y_2 - y_1 \in \overline{B} \setminus \{0\} \implies \varphi_{-\overline{D},k}(y_1) < \varphi_{-\overline{D},k}(y_2)$$

 due to (A4′) by Corollary 4.5.6.

We show that $\overline{x}$ is an $\varepsilon$-minimal solution to some scalar optimization problem. Indeed, set $g(x) := f(x) - f(\overline{x})$. Of course, $g(\overline{x}) = 0$. For $\overline{x} \in \varepsilon k\text{-}\mathrm{M}_f(S, \tilde{D}(\cdot))$ (see Definition 11.4.4), we obtain

$$\begin{aligned}
&(f(x) - f(\overline{x}) + \varepsilon k) \notin -\tilde{D}(f(\overline{x})), \ \text{for all } x \in S \text{ with } f(x) \neq f(\overline{x}) \\
\Longleftrightarrow &(g(x) + \varepsilon k) \notin 0 - \overline{D}, \ \text{for all } x \in S \text{ with } g(x) \neq 0 \\
\Longrightarrow &\varphi_{-\overline{D},k}(g(x) + \varepsilon k) = \varphi_{-\overline{D},k}(g(x)) + \varepsilon \geq 0, \ \text{for all } x \in S.
\end{aligned}$$

The last implication holds because of the translation invariance along $k$ and the strict $\overline{B}$-monotonicity of $\varphi_{-\overline{D},k}$ and $\varphi_{-\overline{D},k}(0) = 0$, which holds because of (A2) and the pointedness of $\overline{D}$ in (A3′). This yields (together with $\varphi_{-\overline{D},k}(g(\overline{x})) = \varphi_{-\overline{D},k}(0) = 0$)

$$\inf_{x \in S} \varphi_{-\overline{D},k}(g(x)) + \varepsilon \geq \varphi_{-\overline{D},k}(g(\overline{x})), \tag{12.3.12}$$

i.e., $\overline{x}$ is an $\varepsilon$- minimal solution to the composition $\varphi_{-\overline{D},k} \circ g$ over $S$.

The validity of (A5) and (A6′) yields the boundedness from below and the lower semicontinuity of $\varphi_{-\overline{D},k} \circ g$, respectively. Now, we apply Ekeland's variational principle in Theorem 12.3.1 to $\varphi_{-\overline{D},k} \circ g$ and the $\varepsilon$-minimal solution $\overline{x}$ of $\inf_{x \in S} \varphi_{-\overline{D},k}(g(x))$. Then, we obtain the existence of some $x_\varepsilon \in S$ with

($\tilde{a}$) $\varphi_{-\overline{D},k}(g(x_\varepsilon)) \leq \varphi_{-\overline{D},k}(g(\overline{x})) = 0$;
($\tilde{b}$) $\|x_\varepsilon - \overline{x}\| \leq \sqrt{\varepsilon}$;

($\tilde{c}$) $\varphi_{-\overline{D},k}(g(x)) + \sqrt{\varepsilon}\|x - \overline{x}\| > \varphi_{-\overline{D},k}(g(x_\varepsilon))$ for all $x \in S$ and $x \neq x_\varepsilon$.

Of course, (b) holds.

Furthermore, we prove that $x_\varepsilon$ fulfills the conditions (a′) and (c′) in this theorem. By contradiction, we suppose that (a′) does not hold, i.e., $x_\varepsilon \notin \varepsilon k - \mathrm{M}_f(S, B(\cdot))$. Then, taking into account Definition 11.4.4, we get the existence of $x \in S$ with

$$\begin{aligned} & f(x) \in f(x_\varepsilon) - \varepsilon k - (B(f(x_\varepsilon)) \setminus \{0\}) \\ \Longleftrightarrow\ & f(x) - f(\overline{x}) + \varepsilon k \in (f(x_\varepsilon) - f(\overline{x})) - (B(f(x_\varepsilon)) \setminus \{0\}) \\ \Longleftrightarrow\ & g(x) + \varepsilon k \in g(x_\varepsilon) - (B(f(x_\varepsilon)) \setminus \{0\}) \overset{(A4')}{\subseteq} g(x_\varepsilon) - (\overline{B} \setminus \{0\}). \end{aligned}$$

From the last inclusion, taking into account the strict $\overline{B}$-monotonicity of $\varphi_{-\overline{D},k}$, we obtain that

$$\varphi_{-\overline{D},k}(g(x_\varepsilon)) > \varphi_{-\overline{D},k}(g(x) + \varepsilon k) = \varphi_{-\overline{D},k}(g(x)) + \varepsilon \geq \inf_{x \in S} \varphi_{-\overline{D},k}(g(x)) + \varepsilon \geq \varphi_{-\overline{D},k}(g(\overline{x})), \tag{12.3.13}$$

where the last inequality ($\geq \varphi_{-\overline{D},k}(g(\overline{x}))(= 0)$ ) holds due to (12.3.12). So, we get a contradiction to ($\tilde{a}$) such that (a′) in this theorem is true.

Finally, we have to show that condition (c′) holds. By contradiction, we suppose that $x_\varepsilon \notin \mathrm{M}_{f_{\varepsilon k}}(S, B(\cdot))$. Then, there is some $x \in S$ and $x \neq x_\varepsilon$ such that we get from the (strict) $\overline{B}$-monotonicity of $\varphi_{-\overline{D},k}$

$$\begin{aligned} & f(x) + \sqrt{\varepsilon}\|x - x_\varepsilon\|k \in f(x_\varepsilon) - B(f(x_\varepsilon)) \\ \Longleftrightarrow\ & f(x) - f(\overline{x}) + \sqrt{\varepsilon}\|x - x_\varepsilon\|k \in f(x_\varepsilon) - f(\overline{x}) - B(f(x_\varepsilon)) \\ \Longleftrightarrow\ & g(x) + \sqrt{\varepsilon}\|x - x_\varepsilon\|k \in g(x_\varepsilon) - B(f(x_\varepsilon)) \overset{(A4')}{\subseteq} g(x_\varepsilon) - \overline{B} \\ \Longrightarrow\ & \varphi_{-\overline{D},k}\left(g(x) + \sqrt{\varepsilon}\|x - x_\varepsilon\|k\right) = \varphi_{-\overline{D},k}(g(x)) + \sqrt{\varepsilon}\|x - x_\varepsilon\| \leq \varphi_{-\overline{D},k}(g(x_\varepsilon)), \end{aligned}$$

in contradiction to ($\tilde{c}$). So, (c′) holds and the proof is completed. □

**Remark 12.3.21** Theorem 12.3.20 improves the result in Theorem 12.3.16 because some of the assumptions are weaker than the corresponding assumptions in Theorem 12.3.16. In [18, Theorem 3.12], a variational principle for $\varepsilon k$-nondominated-like solutions is shown without assuming that the involved sets have a nonempty interior like in (A3′) and (A4′).

Our next results are derived for (weakly) $\varepsilon k$-nondominated solutions of (VVOP) with respect to $D(\cdot)$. In Definition 11.4.1 (see (12.3.11)), we have introduced approximate nondominated solutions of a vector optimization problem with variable domination structure in the image space $Y$. In the following, we formulate a corresponding definition in the pre-image space using the notations $\varepsilon k - \mathrm{N}(F, D(\cdot))$ and $\varepsilon k - \mathrm{WN}(F, D(\cdot))$ introduced in Definition 11.4.1 with $F = f(S)$.

**Definition 12.3.22** We consider problem (VVOP) and $\varepsilon \geq 0$. Then,

(a) $x_\varepsilon \in S$ is called an $\varepsilon k$-nondominated solution of (VVOP) with respect to the variable domination structure $D(\cdot)$ if there is no element $y \in f(S)$ such that $y \preceq_2 f(x_\varepsilon) - \varepsilon k$, i.e.,

$$f(x_\varepsilon) \in \varepsilon k - \mathrm{N}(f(S), D(\cdot)).$$

(b) Assume that $\operatorname{int} D(f(x)) \neq \emptyset$ for all $x \in S$. $x_\varepsilon \in S$ is called a weakly $\varepsilon k$-nondominated solution to (VVOP) with respect to $D(\cdot)$ if

$$f(x_\varepsilon) \in \varepsilon k - \mathrm{WN}(f(S), D(\cdot)).$$

The sets of all the $\varepsilon k$-nondominated and weakly $\varepsilon k$-nondominated solutions of (VVOP) are denoted by $\varepsilon k\text{-}\mathrm{N}_f(S, D(\cdot))$ and $\varepsilon k\text{-}\mathrm{WN}_f(S, D(\cdot))$, respectively. In the case $\varepsilon = 0$, these nondominated solution concepts coincide with the definitions of nondominated solutions (compare Definition 11.1.2) and we use the notations $\mathrm{N}_f(S, D(\cdot))$ and $\mathrm{WN}_f(S, D(\cdot))$.

In order to show the variational principle for nondominated solutions of (VVOP), we apply the functional $\varphi_{\overline{y}-D(\cdot),k} : Y \mapsto \overline{\mathbb{R}}$ (with $\overline{y} \in Y$ and $k \in Y \setminus \{0\}$) given by (11.3.1) (see Sect. 11.3)

$$\varphi_{\overline{y}-D(\cdot),k}(y) = \inf\{t \in \mathbb{R} \mid y \in tk + \overline{y} - D(y)\} \text{ for all } y \in Y. \tag{12.3.14}$$

The next lemma characterizes (approximate) solutions using the functional (12.3.14). Part (c) of the following lemma was proven under slightly stronger assumptions on $D$ and for a similar but different scalarization functional in [411, Theorems 4.3 and 4.4]. Parts (a) and (b) of the following lemma were shown in [143, Theorem 5.11] for cone-valued maps. Obviously, (b) also follows from (a) and (c).

Note that in [414] a different scalarization method is used in order to give a characterization of approximate solutions to a vector optimization problem with variable domination structure via scalarization.

The following lemma is shown in [18, Lemma 4.2].

**Lemma 12.3.23** *Suppose that* (A1) *holds and* $D : Y \to \mathcal{P}(Y)$ *is a set-valued map, where* $D(y)$ *is closed for each* $y \in Y$ *and fulfills the following conditions for some* $k \in Y \setminus \{0\}$*:*

(C1) $(-\infty, 0)k \cap D(y) = \emptyset$ *and* $0 \in \operatorname{bd} D(y)$ *for all* $y \in f(S)$.
(C2) $D(y) + \mathbb{R}_> k \subseteq \operatorname{int} D(y)$ *for all* $y \in f(S)$.

*Consider the functional* $\varphi_{\overline{y}-D(\cdot),k} : Y \to \overline{\mathbb{R}}$ *defined for some* $\overline{y} \in f(S)$*. Then,*

(a) *Under condition* (C1) *only, it holds that*

$$\varphi_{\overline{y}-D(\cdot),k}(\overline{y}) = 0. \tag{12.3.15}$$

(b) *Let* $\overline{x} \in S$ *and* $\overline{y} = f(\overline{x})$. *Then,*

$$\overline{x} \in \mathrm{WN}_f(S, D(\cdot)) \iff \inf_{y \in f(S)} \varphi_{\overline{y}-D(\cdot),k}(y) \geq 0. \tag{12.3.16}$$

(c) *Let* $\varepsilon \geq 0$, $\overline{x} \in S$, *and* $\overline{y} = f(\overline{x})$. *Then,*

$$\overline{x} \in \varepsilon k - \mathrm{WN}_f(S, D(\cdot)) \iff \inf_{y \in f(S)} \varphi_{\overline{y}-D(\cdot),k}(y) \geq -\varepsilon. \tag{12.3.17}$$

***Proof*** We consider $\varphi_{\overline{y}-D(\cdot),k}(y)$ for $y \in f(S)$. Because (b) follows from (c) for $\varepsilon = 0$, we show only (a) and (c).

(a) For $y = \overline{y}$, we get $\varphi_{\overline{y}-D(\cdot),k}(\overline{y}) = \inf\{t \in \mathbb{R} \mid tk \in D(\overline{y})\}$. Since $0 \in \mathrm{bd}\, D(y)$ for all $y \in f(S)$ and $(-\infty, 0)k \cap D(y) = \emptyset$, we obtain $\varphi_{\overline{y}-D(\cdot),k}(\overline{y}) = 0$.

(c) Suppose $\varphi_{\overline{y}-D(\cdot),k}(y) \geq -\varepsilon$ for all $y \in f(S)$ but $\overline{x} \notin \varepsilon k$-$\mathrm{WN}_f(S, D(\cdot))$. Then, there is an element $y \in f(S)$ with $\overline{y} - \varepsilon k - y \in \mathrm{int}\, D(y)$. Hence, there is a real number $t < 0$ with

$$(\overline{y} - y) + (t - \varepsilon)\, k \in D(y),$$

i. e., $\overline{y} + (t - \varepsilon)\, k - y \in D(y)$ and therefore, $\varphi_{\overline{y}-D(\cdot),k}(y) \leq t - \varepsilon < -\varepsilon$, a contradiction.

Now, suppose $\overline{x} \in \varepsilon k$-$\mathrm{WN}_f(S, D(\cdot))$ but assume $\exists t \in \mathbb{R}$, $t < -\varepsilon$ and $y \in f(S)$ with

$$\overline{y} + t\, k - y \in D(y).$$

Since $D(y) + (-t - \varepsilon)\, k \in \mathrm{int}\, D(y)$ by (C2), we obtain

$$\overline{y} - \varepsilon k \in y + D(y) + (-t - \varepsilon)\, k \subseteq y + \mathrm{int}\, D(y)$$

in contradiction to $\overline{x} \in \varepsilon k$-$\mathrm{WN}_f(S, D(\cdot))$. □

The following assertion is shown in [18, Lemma 4.3].

**Lemma 12.3.24** *Suppose that* (A1) *holds and* $D : Y \to \mathscr{P}(Y)$ *is a set-valued map, where* $D(y)$ *is closed for each* $y \in Y$. *Moreover, suppose that* $D$ *fulfills the following condition for some* $k \in Y \setminus \{0\}$:

$$\forall y \in f(S): \quad D(y) + \mathbb{R}_{>}k \subseteq D(y) \setminus \{0\}. \tag{12.3.18}$$

*Let* $\varepsilon \geq 0$, $\overline{x} \in S$, *and* $\overline{y} = f(\overline{x})$. *Then,*

$$\overline{x} \in \varepsilon k - \mathrm{N}_f(S, D(\cdot)) \implies \inf_{y \in f(S)} \varphi_{\overline{y}-D(\cdot),k}(y) \geq -\varepsilon. \tag{12.3.19}$$

***Proof*** The result follows analogously to the proof of Lemma 12.3.23(c), using the condition (12.3.18) instead of (C2). □

In the following lemma (see [18, Lemma 4.4]), we show a sufficient condition for the lower semicontinuity of $\varphi_{\overline{y}-D(\cdot),k} \circ f$.

**Lemma 12.3.25** *Suppose that* (A1) *and* (A2) *hold. Consider problem* (VVOP), $\overline{x} \in X$, $\overline{y} = f(\overline{x})$, *and the scalarizing functional* $\varphi_{\overline{y}-D(\cdot),k}$. *Suppose that the set-valued map* $D : Y \to \mathscr{P}(Y)$ *satisfies the following condition:*

(C3) *$D$ has a closed graph over $f(X)$ in the following sense: $y_n \in f(X)$ and $v_n \in D(y_n)$ for all $n \in \mathbb{N}$ and $\{(y_n, v_n)\} \to (y_*, v_*)$ as $n \to +\infty$ implies $y_* \in f(X)$ and $v_* \in D(y_*)$.*

*Then,* $(\varphi_{\overline{y}-D(\cdot),k} \circ f)(\cdot)$ *is a lower semicontinuous functional on $S$ if $f$ is a continuous function on $S$.*

***Proof*** We suppose that $f$ is a continuous function on $S$. For the proof of the lower semicontinuity of $\varphi_{\overline{y}-D(\cdot),k}$ on $S$, it is sufficient to show that the sublevel set

$$\operatorname{lev}_{\varphi_{\overline{y}-D(\cdot),k} \circ f, \leq}(t) = \left\{x \in S \mid \varphi_{\overline{y}-D(\cdot),k}(f(x)) \leq t\right\}$$

is closed in $X$ for all $t \in \mathbb{R}$. For $t \in \mathbb{R}$ arbitrarily, but fixed, and any sequence $\{x_n\}$ in $\operatorname{lev}_{\varphi_{\overline{y}-D(\cdot),k} \circ f, \leq}(t)$ (and thus in $S$) such that $x_n \to x_*$ as $n \to +\infty$, we get $\varphi_{\overline{y}-D(\cdot),k}(f(x_n)) \leq t$ by the definition of $\operatorname{lev}_{\varphi_{\overline{y}-D(\cdot),k} \circ f, \leq}(t)$. Now, we conclude

$$\overline{y} + tk - f(x_n) \in D(f(x_n)).$$

Because of the continuity of $f$ on $S$, the sequence $(y_n, v_n) \in \operatorname{graph} D$ with $y_n := f(x_n)$ and $v_n := \overline{y} + tk - f(x_n)$ converges to $(f(x_*), \overline{y} + tk - f(x_*))$. From (C3), we obtain

$$\overline{y} + tk - f(x_*) \in D(f(x_*))$$

and so

$$\varphi_{\overline{y}-D(\cdot),k}(f(x_*)) \leq t \tag{12.3.20}$$

taking into account the definition of $\varphi_{\overline{y}-D(\cdot),k}$ in (12.3.14). Formula (12.3.20) implies $x_* \in \operatorname{lev}_{\varphi_{\overline{y}-D(\cdot),k} \circ f, \leq}(t)$ and thus the closedness of the set $\operatorname{lev}_{\varphi_{\overline{y}-D(\cdot),k} \circ f, \leq}(t)$. The proof is complete because of Proposition 3.3.12. □

**Remark 12.3.26** In the assumption (C3) in Lemma 12.3.25, it is supposed that $D$ has a closed graph over $f(X)$, which is related to the upper semicontinuity of $D$. Eichfelder [140] has shown that any cone-valued map $D : Y \rightrightarrows Y$, $Y$ a reflexive Banach space, with $D(y^0)$ a closed convex cone for $y^0 \in Y$, is upper semicontinuous at $y^0$ if and only if there is a neighborhood $U$ of $y^0$ such that $D(y) \subseteq D(y^0)$ for all $y \in U$. Upper semicontinuous maps with closed images have a closed graph. The converse assertion is true if the images are compact.

The following assertion is shown in [18, Lemma 4.6].

**Lemma 12.3.27** *We look at the problem* (VVOP) *and the functional* $\varphi_{\overline{y}-D(\cdot),k}$ *given in (12.3.14). Suppose that the assumptions* (A1) *and* (A2) *are fulfilled.*

*Furthermore, suppose that* $f$ *satisfies the following boundedness condition:*

(A5) *There is an element* $\underline{y} \in Y$ *with* $f(S) \subseteq \underline{y} + D(\underline{y})$.

*Moreover, assume that the set-valued map* $D : Y \to \mathscr{P}(Y)$ *satisfies for* $\underline{y}$ *from assumption (A5):*

(C4) $D(y) + D(\underline{y}) \subseteq D(\underline{y})$ *for all* $y \in f(S)$ *(which holds provided that* $D(y) \subseteq D(\underline{y})$ *for all* $y \in f(S)$ *and* $D(\underline{y})$ *is a convex cone). Assume that there is a cone* $D$ *with* $k \in \operatorname{int} D$ *and* $D(\underline{y}) + \operatorname{int} D \subseteq D(\underline{y})$.

*Then,* $\varphi_{\overline{y}-D(\cdot),k} \circ f$ *is bounded from below on* $S$.

***Proof*** We consider the element $\underline{y}$ given by assumption (A5). Because of assumption (C4) and Proposition 4.6.13 with $A = D(\underline{y}) - \overline{y}$, there is an element $\underline{t} \in \mathbb{R}$ with

$$\overline{y} + \underline{t}k - \underline{y} \notin D(\underline{y}). \tag{12.3.21}$$

Now, suppose that $f$ is bounded from below on $S$ by $\underline{y}$ with respect to $D(\underline{y})$, but $\varphi_{\overline{y}-D(\cdot),k} \circ f$ is not bounded from below on $S$. The assumption that $f$ is bounded from below on $S$ by $\underline{y}$ ensures that

$$-\underline{y} \in -f(x) + D(\underline{y}). \tag{12.3.22}$$

Since we supposed that $\varphi_{\overline{y}-D(\cdot),k} \circ f$ is not bounded from below on $S$, we find some $x \in S$ such that $\varphi_{\overline{y}-D(\cdot),k}(f(x)) < \underline{t}$. Taking into account (11.3.1) and (A2), we obtain

$$\overline{y} + \underline{t}k - f(x) \in D(f(x)). \tag{12.3.23}$$

From (12.3.22), (12.3.23) and taking into account (C4), we obtain

$$\overline{y} + \underline{t}k - \underline{y} \in D(f(x)) + D(\underline{y}) \subseteq D(\underline{y}),$$

a contradiction to (12.3.21) such that $\varphi_{\overline{y}-D(\cdot),k} \circ f$ is lower bounded on $S$. This completes the proof. □

Using the previous results, we are ready to prove an extension of Ekeland's theorem for $\varepsilon k$-nondominated solutions of vector optimization problems with a variable domination structure. This assertion is shown in [18, Theorem 4.7].

**Theorem 12.3.28** (Variational principle of Ekeland's type for $\varepsilon k$-nondominated solutions) *Let problem* (VVOP) *be given,* $\overline{x} \in \varepsilon k\text{-N}_f(S, D(\cdot))$, *and put* $\overline{y} := f(\overline{x})$. *In addition to the standing assumptions* (A1) *and* (A2), *we suppose that the following conditions hold:*

(A5) *There is an element* $\underline{y} \in Y$ *with* $f(S) \subseteq \underline{y} + D(\underline{y})$.
(A6‴) $f$ *is continuous on* $X$.

*Furthermore, suppose that the conditions* (C1), *(12.3.18),* (C3) *and* (C4) *from Lemmata 12.3.23, 12.3.24, 12.3.25 and 12.3.27 are satisfied for some* $k \in Y \setminus \{0\}$.

*Then, there is an element* $x_\varepsilon \in S$ *such that*

(a) $\varphi_{\overline{y}-D(\cdot),k}(f(x_\varepsilon)) + \sqrt{\varepsilon}\,\|\overline{x} - x_\varepsilon\| \le \varphi_{\overline{y}-D(\cdot),k}(f(\overline{x})) \le \inf_{x\in S} \varphi_{\overline{y}-D(\cdot),k}(f(x)) + \varepsilon$,
(b) $\|\overline{x} - x_\varepsilon\| \le \sqrt{\varepsilon}$,
(c) $x_\varepsilon \in S$ *is an exact solution to the scalar problem*

$$\inf_{x\in S} \varphi_{\overline{y}-D(\cdot),k}(f(x)) + \sqrt{\varepsilon}\|x - x_\varepsilon\|.$$

***Proof*** We consider $\overline{x} \in \varepsilon k\text{-N}_f(S, D(\cdot))$, $\overline{y} := f(\overline{x})$, and $\varphi_{\overline{y}-D(\cdot),k}$. Under the supposed conditions (C1) and (12.3.18) and taking into account Lemmata 12.3.23 and 12.3.24, we obtain

$$0 = \varphi_{\overline{y}-D(\cdot),k}(\overline{y}) \le \inf_{y\in f(S)} \varphi_{\overline{y}-D(\cdot),k}(y) + \varepsilon.$$

This means that $\overline{y}$ is an $\varepsilon$-minimal solution to $\varphi_{\overline{y}-D(\cdot),k} \circ f$ over $S$. The functional $\varphi_{\overline{y}-D(\cdot),k} \circ f$ is lower semicontinuous and bounded from below on $S$ because of Lemmata 12.3.25 and 12.3.27 under the assumptions of this theorem. So, all assumptions of Theorem 12.3.1 are satisfied. Hence, from Theorem 12.3.1 for $\lambda = \sqrt{\varepsilon}$, we obtain the existence of $x_\varepsilon \in S$ with

(a) $\varphi_{\overline{y}-D(\cdot),k}(f(x_\varepsilon)) \le \varphi_{\overline{y}-D(\cdot),k}(f(x_\varepsilon)) + \sqrt{\varepsilon}\,\|\overline{x} - x_\varepsilon\| \le \varphi_{\overline{y}-D(\cdot),k}(f(\overline{x})) \le \inf_{x\in S} \varphi_{\overline{y}-D(\cdot),k}(f(x)) + \varepsilon$.
(b) $\|\overline{x} - x_\varepsilon\| \le \sqrt{\varepsilon}$.
(c) $\varphi_{\overline{y}-D(\cdot),k}(f(x)) + \sqrt{\varepsilon}\,\|x - x_\varepsilon\| > \varphi_{\overline{y}-D(\cdot),k}(f(x_\varepsilon))$ for all $x \in S$ and $x \ne x_\varepsilon$.

This completes the proof. □

**Remark 12.3.29** In Theorem 12.3.28, the assumptions imposed on the set-valued map $D$ are weaker than the assumptions in the variational principles for $\varepsilon k$-nondominated-like elements in Theorems 12.3.16 and 12.3.20, however, the assertions in Theorem 12.3.28 are weaker too.

## 12.4 Necessary Conditions for Approximate Solutions of VVDS

### 12.4.1 Conditions Based on the Limiting Subdifferential Calculus

In this section, we use the variational principle presented in Theorem 12.3.28 in order to derive necessary conditions for approximate solutions of the problem (VVOP) (introduced in Sect. 12.3.2) based on the subdifferential calculus by Mordukhovich [359] (see Sect. 8.3 and compare also Sect. 12.1.1 for necessary conditions for (exact) properly and weakly efficient solutions of vector optimization problems with constant domination structure). Here, we work again with the limiting (Mordukhovich) subdifferential $\partial_M$ and limiting (Mordukhovich) normal cone $N_M$ (see Sect. 8.3, Definitions 8.1.5 and 8.1.4).

In order to formulate a qualification condition, we suppose that $Y$ is an Asplund space and consider a functional $\varphi : Y \to \mathbb{R} \cup \{+\infty\}$ and a point $\overline{y} \in \operatorname{dom} \varphi$. The set

$$\partial^{\infty}\varphi(\overline{y}) := \left\{ y^* \in Y^* \mid (y^*, 0) \in N_M((\overline{y}, \varphi(\overline{y})); \operatorname{epi} \varphi) \right\}$$

is the *singular subdifferential* of $\varphi$ at $\overline{y}$, and its elements are *singular subgradients* of $\varphi$ at this point.

Note, if $\varphi$ is locally Lipschitzian on a neighborhood of $\overline{y}$, then $\partial^{\infty}\varphi(\overline{y}) = \{0\}$. Furthermore, if $\varphi$ is strictly Lipschitz continuous at $\overline{y}$, in particular it is $C^{1,1}$ (where $C^{1,1}$ is the space of all functions with locally Lipschitzian first Fréchet derivative), then $\partial_M\varphi(\overline{y}) = \{\nabla\varphi(\overline{y})\}$.

For deriving necessary optimality conditions, we need a certain kind of regularity assumption. Let $X$ and $Y$ be Asplund spaces, $\varphi : Y \to \mathbb{R} \cup \{+\infty\}$ and $f : X \to Y$. The pair of functions $(\varphi, f)$ satisfies a *qualification condition* if

$$\partial^{\infty}\varphi(\overline{y}) \cap \ker \partial_M \langle \cdot, f \rangle(\overline{x}) = \{0\}, \tag{12.4.1}$$

where $\ker \partial_M \langle \cdot, f \rangle(\overline{x}) := \{y^* \in Y^* \mid 0 \in \partial_M y^*(f(\overline{x}))\}$ (see Mordukhovich [359]).

In the following theorem ([18, Theorem 5.5]), we give necessary conditions for approximately nondominated solutions of (VVOP) in the sense of Definition 12.3.22.

**Theorem 12.4.1** *Consider problem* (VVOP) *introduced in Sect. 12.3.2,* $\overline{x} \in \varepsilon k\text{-}\mathrm{N}_f(S, D(\cdot))$ *and* $\overline{y} := f(\overline{x})$. *Suppose that X and Y are Asplund spaces,* $f : X \to Y$ *is strictly Lipschitzian, S is a closed subset of X, and* $D : Y \to \mathscr{P}(Y)$ *is a set-valued map. In addition to the standing assumptions* (A1), (A2), *suppose that all assumptions* (A5), (A6‴), (C1), *(12.3.18),* (C3) *and* (C4) *from Theorem 12.3.28 are satisfied. Furthermore, assume that the pair of functions* $(\varphi_{\overline{y}-D(\cdot),k}, f)$ *satisfies the qualification condition (12.4.1) for all* $x \in S$ *with* $\|x - \overline{x}\| \leq \sqrt{\varepsilon}$. *Then, there is an element* $x_\varepsilon \in S$ *with*

(a′) $\varphi_{\overline{y}-D(\cdot),k}(f(x_\varepsilon)) + \sqrt{\varepsilon}\,\|\overline{x} - x_\varepsilon\| \leq \varphi_{\overline{y}-D(\cdot),k}(f(\overline{x}))$,

(b) $\|\overline{x} - x_\varepsilon\| \leq \sqrt{\varepsilon}$,
(c) $y^* \in \partial_M \varphi_{\overline{y}-D(\cdot),k}(f(x_\varepsilon))$ *such that* $0 \in \partial_M(y^* \circ f)(x_\varepsilon) + N_M(x_\varepsilon; S) + \sqrt{\varepsilon} B_{X^*}$.

***Proof*** From Theorem 12.3.28, we get the existence of an element $x_\varepsilon \in S$ such that it satisfies (a′), (b) and it is an exact solution to a scalar minimization problem over $S$ with the objective function $h : X \to \overline{\mathbb{R}}$:

$$h(x) := (\varphi_{\overline{y}-D(\cdot),k} \circ f)(x) + \sqrt{\varepsilon}\, \|x - x_\varepsilon\| \quad \text{for all } x \in X.$$

Taking into account [359, Proposition 5.1], we obtain

$$0 \in \partial_M h(x_\varepsilon) + N_M(x_\varepsilon; S). \tag{12.4.2}$$

Furthermore, Lemma 12.3.25 yields that $\varphi_{\overline{y}-D(\cdot),k} \circ f$ is lower-semicontinuous on a neighborhood of $x_\varepsilon$. Taking into account Lemma 8.1.7 for the lower semicontinuous functional $\varphi_{\overline{y}-D(\cdot),k} \circ f$ and the Lipschitz continuous function $\|\cdot\|$, we obtain

$$\partial_M h(x_\varepsilon) \subseteq \partial_M\big(\varphi_{\overline{y}-D(\cdot),k} \circ f\big)(x_\varepsilon) + \partial_M(\sqrt{\varepsilon}\, \|\cdot - x_\varepsilon\|)(x_\varepsilon). \tag{12.4.3}$$

With Lemma 8.3.6, we get

$$\partial_M(\varphi_{\overline{y}-D(\cdot),k} \circ f)(x_\varepsilon) \subseteq \bigcup \big\{\partial_M(y^* \circ f)(x_\varepsilon) \mid y^* \in \partial_M \varphi_{\overline{y}-D(\cdot),k}(f(x_\varepsilon))\big\}. \tag{12.4.4}$$

From (12.4.2), (12.4.3), (12.4.4) and taking into account the subdifferential of the norm $\partial_M \|\cdot - x_\varepsilon\| (x_\varepsilon) = B_{X^*}$, we find an element $y^* \in \partial_M \varphi_{\overline{y}-D(\cdot),k}(f(x_\varepsilon))$ with

$$0 \in \partial_M(y^* \circ f)(x_\varepsilon) + N_M(x_\varepsilon; S) + \sqrt{\varepsilon} B_{X^*}.$$

This completes the proof. □

In the following corollary ([18, Corollary 5.6]), we derive a necessary condition for (exact) nondominated solutions of (VVOP).

**Corollary 12.4.2** *Let problem* (VVOP) *be given,* $\overline{x} \in \mathrm{N}_f(S, D(\cdot))$, *i.e.,* $\overline{x}$ *is a nondominated solution to problem* (VVOP), *and* $\overline{y} := f(\overline{x})$. *Suppose that* $X$ *and* $Y$ *are Asplund spaces and* $f : X \to Y$ *is strictly Lipschitzian. In addition to the standing assumptions* (A1), (A2), *we suppose that all assumptions* (A5), (A6‴), (C1), *(12.3.18),* (C3) *and* (C4) *from Theorem 12.3.28 are satisfied for some* $k \in Y \setminus \{0\}$. *Furthermore, suppose that the pair of functions* $(\varphi_{\overline{y}-D(\cdot),k}, f)$ *fulfills the qualification condition (12.4.1) at* $\overline{x}$. *Then, for any* $\lambda > 0$, *there is an element* $y^* \in \partial_M(\varphi_{\overline{y}-D(\cdot),k}(f(\overline{x}))$ *with*

$$0 \in \partial_M(y^* \circ f)(\overline{x}) + N_M(\overline{x}; S) + \lambda B_{X^*}. \tag{12.4.5}$$

***Proof*** Because of $\overline{x} \in \mathrm{N}_f(S, D(\cdot))$, i.e., $\overline{x}$ is a $0k$-nondominated solution to (VVOP), it is also an $\varepsilon k$-nondominated solution to (VVOP) with $\varepsilon = \lambda^2 > 0$ for all $\lambda > 0$ and a weakly nondominated solution to (VVOP). Taking into account Theorem 12.4.1, the only point $x_\varepsilon \in S$ which fulfills condition (a′)

$$\varphi_{\overline{y}-D(\cdot),k}(f(x_\varepsilon)) + \lambda \|\overline{x} - x_\varepsilon\| \le \varphi_{\overline{y}-D(\cdot),k}(\overline{y}) \text{ with } \overline{y} = f(\overline{x})$$

is $\overline{x}$ since Lemma 12.3.23(a),(b) yields $\varphi_{\overline{y}-D(\cdot),k}(\overline{y}) = 0$ and $\varphi_{\overline{y}-D(\cdot),k}(f(x_\varepsilon)) \ge 0$. Because $\overline{x}$ satisfies Theorem 12.4.1(c), we find $y^* \in \partial_M \varphi_{\overline{y}-D(\cdot),k}(f(\overline{x}))$ with

$$0 \in \partial_M(y^* \circ f)(\overline{x}) + N_M(\overline{x}; S) + \lambda B_{X^*},$$

i.e., (12.4.5) holds. This completes the proof. □

**Remark 12.4.3** The necessary conditions for nondominated solutions of problem (VVOP) obtained in this section are different from those in [26, 146].

**Remark 12.4.4** Recently, in the paper [19] by Bao, Hillmann and Tammer, the subdifferential of the scalarizing functional (4.1.11) for general (nonconvex) sets $A$ is determined and necessary conditions for solutions of vector optimization problems with general (fixed and variable) domination sets are derived. These results are applied to derive necessary optimality conditions for nonconvex vector-valued location problems.

### *12.4.2 Conditions Based on a Generic Approach to Subdifferentials*

The aim of this section is to derive necessary conditions for approximately nondominated solutions, approximately nondominated-like solutions and approximately completely nondominated solutions of vector optimization problems with variable domination structures by using nonlinear separating functionals of the type (4.1.11) and their subdifferentials. Bao and Mordukhovich [26] have shown necessary conditions for nondominated points of sets and nondominated solutions of vector optimization problems with variable domination structures and general geometric constraints, applying methods of variational analysis and generalized differentiation by Mordukhovich (see Sect. 8.3). Furthermore, in Theorem 12.4.1 (see Bao, Eichfelder, Soleimani and Tammer [18]), we have shown necessary conditions for approximately nondominated solutions of vector optimization problems with variable domination structures in Asplund spaces using the subdifferential calculus by Mordukhovich and the Ekeland's type variational principle for (VVOP), given in Theorem 12.3.28.

In order to formulate our assertions for several subdifferentials, we introduce a generic approach to subdifferentials that includes many well-known subdifferentials. We will give necessary conditions for approximately nondominated-like solutions of vector optimization problems with variable domination structures. For this purpose, we will use an extension of the nonlinear separating functional (4.1.11). Moreover, we give necessary conditions for approximately completely nondominated elements. The results of this section are derived by Soleimani and Tammer in [418].

As usual, for a set $S \subset X$, we denote the indicator function of $S$ by $\iota_S$ ($\iota_S(x) = 0$ if $x \in S$ and $\iota_S(x) = +\infty$ if $x \notin S$).

We derive necessary conditions for approximately completely nondominated elements and nondominated-like solutions of vector optimization problems with variable domination structures using the following generic approach to subdifferentials (see [118, 129, 259]):

Let $\mathcal{X}$ be a class of Banach spaces which contains the class of finite-dimensional normed vector spaces. By an abstract subdifferential $\partial$, we mean a map which associates to every lower semicontinuous function $h : X \in \mathcal{X} \to \mathbb{R} \cup \{+\infty\}$ and to every $x \in X$ a (possibly empty) subset $\partial h(x) \subset X^*$. Let $X, Y \in \mathcal{X}$ and denote by $\mathcal{F}(X, Y)$ a class of functions acting between $X$ and $Y$ having the property that by left composition with a lower semicontinuous functional $h : Y \to \mathbb{R} \cup \{+\infty\}$, the resulting function is still lower semicontinuous.
In the following, we work with the next properties of the abstract subdifferential $\partial$:

(H1) If $h$ is convex and lower semicontinuous, then $\partial h(x)$ coincides with the Fenchel subdifferential.

(H2) If $x$ is a local minimum point for $h$, then $0 \in \partial h(x)$; $\partial h(u) = \emptyset$ if $h(u) \notin \mathbb{R}$.

Note that (H1) and (H2) are very natural requirements for any subdifferential.

(H3) If $\varphi : Y \to \overline{\mathbb{R}}$ is convex and $\psi \in \mathcal{F}(X, Y)$, then for every $x$,

$$\partial(\varphi \circ \psi)(x) \subseteq \bigcup_{y^* \in \partial\varphi(\psi(x))} \partial(y^* \circ \psi)(x).$$

(H4) If $\varphi : Y \to \overline{\mathbb{R}}$ is convex, $\psi \in \mathcal{F}(X, Y)$, and $S \subset X$ is a closed set containing $x$, then

$$\partial(\varphi \circ \psi + \iota_S)(x) \subseteq \partial(\varphi \circ \psi)(x) + \partial\iota_S(x).$$

(H5) If $h$ is convex and $g : X \to \mathbb{R}$ is locally Lipschitzian, then for every $x \in X$ with $h(x) \in \mathbb{R}$ and $g(x) \in \mathbb{R}$

$$\partial(h + g)(x) \subseteq \partial h(x) + \partial g(x).$$

As usual, for a closed set $S \subset X$ the set $\partial\iota_S(x)$ is denoted by $N_\partial(x; S)$ and is called the set of normal directions to $S$ at $x \in S$ with respect to $\partial$.

The properties (H3), (H4) and (H5) are "exact calculus rules" for sums and for composition, and as examples of subdifferentials with these properties we can mention:

- the limiting (or Mordukhovich) subdifferential (see Definition 8.1.5) when $\mathcal{X}$ is the class of Asplund spaces, $Y$ is finite-dimensional and $\mathcal{F}(X, Y)$ is the class of Lipschitzian functions from $X$ into $Y$ (see [357, 359] and Sect. 8.3);

- the approximate (or Ioffe) subdifferential (see Definition 8.1.10) when $\mathscr{X}$ is the class of Banach spaces and $\mathscr{F}(X, Y)$ is the class of strongly compact Lipschitzian functions from $X$ into $Y$ (see [258, 260] and Sect. 8.3).

Furthermore, it is important to mention that Clarke's subdifferential satisfies (H5) on Banach spaces (see [98] and Definition 8.1.8).

We will use the following assumptions in this section.

(A) $X, Y$ are Banach spaces, $S \subset X$ is a closed set in $X$, $f \in \mathscr{F}(X, Y)$ and $\varepsilon \geq 0$.
(B) For the set-valued map $D : Y \to \mathscr{P}(Y)$, $D(y)$ is closed, pointed, $0 \in D(y)$ and $\operatorname{int} D(y) \neq \emptyset$ for all $y \in Y$. $k \in Y \setminus \{0\}$ satisfies $D(y) + \mathbb{R}_{>}k \subseteq \operatorname{int} D(y)$ for all $y \in Y$. Furthermore, we suppose that the set-valued maps $\tilde{D} : Y \to \mathscr{P}(Y)$ and $\hat{D} : Y \to \mathscr{P}(Y)$ have the same properties.

Under assumptions (A) and (B), we consider again the vector optimization problem (VVOP) with respect to a variable domination structure (see Sect. 12.3.2):

$$D(\cdot)\text{–}\operatorname*{Opt}_{x \in S} f(x),$$

where we are looking for three different kinds of approximate solutions: $\varepsilon k$-nondominated-like solutions with the domination map $\tilde{D}$ (see Definition 12.3.14), $\varepsilon k$-nondominated solutions with the domination map $D$ (Definition 12.3.22) and $\varepsilon k$-completely nondominated solutions with the domination map $\hat{D}$ (Definition 12.4.5).

In this section, we suppose that the assumptions (A) and (B) are fulfilled. As already mentioned in Sect. 11.1, the solution concepts for vector optimization problems with variable domination structure (nondominated-like elements, nondominated elements and completely nondominated elements) in the image space are given by three different (domination) relations. Here, we are dealing with corresponding solution concepts in the pre-image space. The concept of approximately nondominated-like solutions is introduced in Definition 12.3.14 and the concept of approximately nondominated solutions is given by Definition 12.3.22. Now, we will introduce approximately completely nondominated solutions (compare Definition 11.4.7 in the image space $Y$). In the following, we formulate a corresponding definition in the pre-image space using the notations $\varepsilon k - \mathrm{MZ}(F, \hat{D}(\cdot))$, introduced in Definition 11.4.1 with $F = f(S)$.

**Definition 12.4.5** Let assumptions (A) and (B) be fulfilled, $\varepsilon \geq 0$ and consider (VVOP).

$x_\varepsilon \in S$ is said to be an $\varepsilon k$-completely nondominated solution to the problem (VVOP) with respect to the domination map $\hat{D} : Y \to \mathscr{P}(Y)$ if

$$f(x_\varepsilon) \in \varepsilon k - \mathrm{MZ}(F, \hat{D}(\cdot)).$$

We denote the set of $\varepsilon k$-completely nondominated solutions of (VVOP) by $\varepsilon k$-$MZ_f(S, \hat{D}(\cdot))$. For $\varepsilon = 0$, we also write $\mathrm{MZ}_f(S, \hat{D}(\cdot))$.

According to Definitions 12.3.14, 12.3.22 and 12.4.5, sets of $\varepsilon k$-completely nondominated, $\varepsilon k$-nondominated and $\varepsilon k$-nondominated-like solutions of vector optimization problems with constant domination structures coincide.

In this section, using certain extensions of the nonlinear separating functional (4.1.11) and its properties, we will characterize approximately nondominated-like solutions as well as approximately completely nondominated solutions of vector optimization problems with variable domination structures. We will derive necessary conditions for approximately nondominated-like solutions as well as approximately completely nondominated solutions of vector optimization problems with variable domination structures (see Theorems 12.4.8 and 12.4.15) using these characterizations in Theorems 12.4.6 and 12.4.13.

We suppose that the conditions (A) and (B) are fulfilled for $\tilde{D} : Y \to \mathscr{P}(Y)$ and consider $\overline{x} \in X$. Set $\overline{y} := f(\overline{x})$ and $\overline{D} := \tilde{D}(f(\overline{x}))$. In order to derive necessary optimality conditions for approximately nondominated-like solutions of (VVOP), we use the functional $\varphi_{-\tilde{D}(w),k}$ (see (11.2.4)), $w := \overline{y}$, in the following form

$$\varphi_{\overline{y}-\overline{D},k}(y) := \varphi_{-\overline{D},k}(y - f(\overline{x})) = \inf\{t \in \mathbb{R} \mid y \in tk + f(\overline{x}) - \overline{D}\}. \qquad (12.4.6)$$

In the following theorem ([418, Theorem 3.4]), we present a characterization of approximately nondominated-like solutions of (VVOP) by using a scalarization by means of $\varphi_{\overline{y}-\overline{D},k} : Y \to \overline{\mathbb{R}}$ given in (12.4.6). For similar assertions based on different scalarizations and corresponding characterizations of approximately nondominated solutions and approximately completely nondominated solutions, see Sect. 11.4.2.

**Theorem 12.4.6** *Let the assumptions (A) and (B) be fulfilled, let $\overline{x} \in S$ be an $\varepsilon k$-nondominated-like solution to (VVOP) with respect to $\tilde{D}$. Set $\overline{y} := f(\overline{x})$ and $\overline{D} := \tilde{D}(f(\overline{x}))$. Consider $\varphi_{\overline{y}-\overline{D},k}$ introduced in (12.4.6). Then, it holds that*

$$\varphi_{\overline{y}-\overline{D},k}(f(\overline{x})) \leq \inf_{x \in S} \varphi_{\overline{y}-\overline{D},k}(f(x)) + \varepsilon.$$

***Proof*** From Lemma 11.2.21, be get $\varphi_{\overline{y}-\overline{D},k}(\overline{y}) = 0$. Now by contradiction, suppose that there is an element $x \in S$ such that $\varphi_{\overline{y}-\overline{D},k}(f(x)) + \varepsilon < \varphi_{\overline{y}-\overline{D},k}(\overline{y}) = 0$. Then, there is a real number $\gamma > 0$ with $\varphi_{\overline{y}-\overline{D},k}(f(x)) + \varepsilon + \gamma = 0$ and $\varphi_{\overline{y}-\overline{D},k}(f(x)) = -\varepsilon - \gamma$. By Corollary 11.2.18(c), we get for $y = f(x)$

$$(-\varepsilon - \gamma)k + \overline{y} - f(x) \in \operatorname{bd} D(\overline{y}) \implies \overline{y} - \varepsilon k - y \in \operatorname{bd} D(\overline{y}) + \gamma k \subset \operatorname{int} \overline{D}.$$

This means that $(\overline{y} - \varepsilon k - (\overline{D} \setminus \{0\})) \cap f(S) \neq \emptyset$, a contradiction to the approximate nondominated-like property of $\overline{x}$ and hence

$$\varphi_{\overline{y}-\overline{D},k}(f(\overline{x})) \leq \inf_{x \in S} \varphi_{\overline{y}-\overline{D},k}(f(x)) + \varepsilon. \qquad \square$$

In the next lemma ([418, Lemma 3.6]), we study the boundedness of the scalarizing functional $\varphi_{\overline{y}-\overline{D},k} \circ f$ in (12.4.6).

**Lemma 12.4.7** *Let assumptions (A) and (B) be fulfilled. We consider the problem (VVOP) with $\overline{x} \in S$. Put $\overline{y} := f(\overline{x})$ and $\overline{D} := \tilde{D}(f(\overline{x}))$. Furthermore, consider the functional $\varphi_{\overline{y}-\overline{D},k}$ given by (12.4.6). Suppose that $\overline{D} + \overline{D} \subseteq \overline{D}$. If $f(S) \subseteq y + \overline{D}(\overline{y})$ for an element $y \in Y$ with $\varphi_{\overline{y}-\overline{D},k}(y) > -\infty$, then $\varphi_{\overline{y}-\overline{D},k} \circ f$ is bounded from below.*

***Proof*** Under the given assumption $\overline{D} + \overline{D} \subseteq \overline{D}$, $\varphi_{\overline{y}-\overline{D},k}$ is $\overline{D}$-monotone, taking into account Theorem 4.5.5(a). Furthermore, the $\overline{D}$-monotonicity of $\varphi_{\overline{y}-\overline{D},k}$ and $f(S) \subseteq y + \overline{D}$ yields

$$\forall x \in S: \qquad \varphi_{\overline{y}-\overline{D},k}(f(x)) \geq \varphi_{\overline{y}-\overline{D},k}(y) > -\infty,$$

i.e., $\varphi_{\overline{y}-\overline{D},k} \circ f$ is bounded from below. □

In the next theorem ([418, Theorem 3.7]), we derive necessary conditions for approximately nondominated-like solutions of (VVOP) in the sense of Definition 12.3.14.

**Theorem 12.4.8** *Let the assumptions (A) and (B) be fulfilled and consider problem (VVOP), $\overline{x} \in \varepsilon k\text{-M}_f(S, \tilde{D}(\cdot))$. Put $\overline{y} := f(\overline{x})$ and $\overline{D} := \tilde{D}(f(\overline{x}))$. Furthermore, consider the functional $\varphi_{\overline{y}-\overline{D},k}$ given by (12.4.6). Assume that $\overline{D}$ is a convex set with $\overline{D} + \overline{D} \subseteq \overline{D}$.*

*Suppose that $f \in \mathscr{F}(X, Y)$ is locally Lipschitzian and $f(S) \subseteq y + \overline{D}$ for an element $y \in Y$ with $\varphi_{\overline{y}-\overline{D},k}(y) > -\infty$. Consider an abstract subdifferential $\partial$ for that (H1) – (H5) are satisfied. Then, there are elements $x_\varepsilon \in S$ and $y^* \in \partial\varphi_{\overline{y}-\overline{D},k}(f(x_\varepsilon))$ such that*

$$0 \in \partial(y^* \circ f)(x_\varepsilon) + N_\partial(x_\varepsilon; S) + \sqrt{\varepsilon} B_{X^*}. \tag{12.4.7}$$

***Proof*** We consider $\overline{x} \in \varepsilon k$ - $M_f(S, \tilde{D}(\cdot))$. Taking into account Theorem 12.4.6, we get $\varphi_{\overline{y}-\overline{D},k}(f(\overline{x})) \leq \inf_{x \in S} \varphi_{\overline{y}-\overline{D},k}(f(x)) + \varepsilon$. Hence, $\overline{x}$ is an approximate solution to the scalar problem with the objective functional $\varphi_{\overline{y}-\overline{D},k} \circ f$. Applying Theorem 4.6.6, we get that $(\varphi_{\overline{y}-\overline{D},k} \circ f)$ is lower semicontinuous because $f \in \mathscr{F}(X, Y)$. Moreover, $\varphi_{\overline{y}-\overline{D},k} \circ f$ is bounded from below because of Lemma 12.4.7. This implies that the assumptions of the scalar Ekeland's variational principle (Theorem 12.3.1) are satisfied.

Because of Theorem 12.3.1, there is an element $x_\varepsilon \in S$ such that it fulfills the conditions of Theorem 12.3.1, i.e., $x_\varepsilon$ is an exact solution to a minimization problem with the feasible set $S$ and the scalar objective function $h : X \to \overline{\mathbb{R}}$ over $S$ with

$$h(x) := (\varphi_{\overline{y}-\overline{D},k} \circ f)(x) + \sqrt{\varepsilon}\, \|x - x_\varepsilon\| \quad \text{for all } x \in X.$$

Under the conditions (H2) and (H4), we get

$$0 \in \partial h(x_\varepsilon) + N_\partial(x_\varepsilon; S). \tag{12.4.8}$$

From Theorems 4.4.4 and 4.6.6, we obtain that the functional $\varphi_{\overline{y}-\overline{D},k}$ is convex and continuous under the given assumptions. Taking into account that $f$ is locally Lipschitzian and $\varphi_{\overline{y}-\overline{D},k}$ is convex and continuous (and hence locally Lipschitzian, see

[381, Proposition 1.6]), $\varphi_{\overline{y}-\overline{D},k} \circ f$ is locally Lipschitzian too. This yields, together with the convexity of $\|\cdot\|$ and (H5), that

$$\partial h(x_\varepsilon) \subseteq \partial\big(\varphi_{\overline{y}-\overline{D},k} \circ f\big)(x_\varepsilon) + \partial(\sqrt{\varepsilon}\,\|\cdot - x_\varepsilon\|)(x_\varepsilon). \tag{12.4.9}$$

With (H3), we conclude

$$\partial(\varphi_{\overline{y}-\overline{D},k} \circ f)(x_\varepsilon) \subseteq \bigcup \big\{\partial(y^* \circ f)(x_\varepsilon) \mid y^* \in \partial\varphi_{\overline{y}-\overline{D},k}(f(x_\varepsilon))\big\}. \tag{12.4.10}$$

Taking into account the convexity of the norm and (H1), we conclude $\partial \|\cdot - x_\varepsilon\| (x_\varepsilon) = B_{X^*}$, and by (12.4.8), (12.4.9), (12.4.10), we find $y^* \in \partial\varphi_{\overline{y}-\overline{D},k}(f(x_\varepsilon))$ satisfying

$$0 \in \partial(y^* \circ f)(x_\varepsilon) + N_\partial(x_\varepsilon; S) + \sqrt{\varepsilon}B_{X^*}.$$

The proof is complete. □

**Remark 12.4.9** Taking into account Theorem 4.10.3, we get the existence of $y^* \in Y^*$ such that the conditions in (4.10.3) hold for $y^*$ in Theorem 12.4.8.

Under certain additional assumptions concerning the domination map $D$, the following corollary ([418, Corollary 3.9]) gives a necessary condition of approximately nondominated solutions of (VVOP) in the sense of Definition 12.3.22.

**Corollary 12.4.10** *Let assumptions (A) and (B) be satisfied. We consider problem* (VVOP) *with* $\overline{x} \in \varepsilon k\text{-N}_f(S, D(\cdot))$. *Put* $\overline{y} := f(\overline{x})$ *and* $\overline{D} := D(f(\overline{x}))$. *Moreover, consider the functional* $\varphi_{\overline{y}-\overline{D},k}$ *given by (12.4.6). Assume that* $\overline{D}$ *is a convex set,* $\overline{D} + \overline{D} \subseteq \overline{D}$ *and* $\overline{D} \subseteq D(f(x))$ *for all* $x \in S$.

*Furthermore, assume that* $f \in \mathcal{F}(X, Y)$ *is locally Lipschitzian and* $f(S) \subseteq y + \overline{D}$ *for an element* $y \in Y$ *with* $\varphi_{\overline{y}-\overline{D},k}(y) > -\infty$.

*We consider an abstract subdifferential* $\partial$ *such that (H1)–(H5) are fulfilled. Then, there are elements* $x_\varepsilon \in S$ *and* $y^* \in \partial(\varphi_{\overline{y}-\overline{D},k}(f(x_\varepsilon))$ *such that*

$$0 \in \partial(y^* \circ f)(x_\varepsilon) + N_\partial(x_\varepsilon; S) + \sqrt{\varepsilon}B_{X^*}. \tag{12.4.11}$$

***Proof*** Let $\overline{x} \in \varepsilon k\text{-N}_f(S, D(\cdot))$. Because of $D(f(\overline{x})) \subseteq D(f(x))$ for all $x \in S$ and Theorem 11.4.18, $\overline{x}$ is an approximately nondominated-like solution to (VVOP), i.e., $\overline{x} \in \varepsilon k\text{-M}_f(S, D(\cdot))$. By applying Theorem 12.4.8, we get the assertion. □

**Remark 12.4.11** If $\overline{x}$ is an approximately nondominated-like solution to (VVOP) and $D(f(x)) \subseteq D(f(\overline{x}))$ for all $x \in S$, then by Theorem 11.4.18, $\overline{x}$ is also an approximately nondominated solution to (VVOP) and all the results about optimality conditions for approximately nondominated solutions given in Theorem 12.4.1 and Corollary 12.4.2 (see Bao, Eichfelder, Soleimani and Tammer [18]) can be used also for approximately nondominated-like solutions.

Furthermore, we give necessary conditions for approximately completely nondominated solutions of (VVOP) in the sense of Definition 12.4.5 of vector optimization problems with variable domination structures.

**Remark 12.4.12** By Theorems 11.4.12 and 11.4.15, all necessary conditions for approximately nondominated-like and approximately nondominated solutions presented before also hold for approximately completely nondominated solutions of (VVOP) under the assumptions of these theorems.

Let assumptions (A) and (B) be fulfilled and $\overline{x} \in X$. In order to derive necessary conditions for approximately completely nondominated solutions of vector optimization problems with variable domination structures, we use the functional $\zeta_{f(\overline{x})-\hat{D}(\cdot),k} : Y \times Y \to \overline{\mathbb{R}}$ (see (11.2.2) and (11.2.6)):

$$\zeta_{f(\overline{x})-\hat{D}(\cdot),k}(y, z) = \inf\{t \in \mathbb{R} \mid y \in tk + f(\overline{x}) - \hat{D}(z)\}. \tag{12.4.12}$$

In the following theorem ([418, Theorem 4.7]), we derive a characterization of approximately completely nondominated solutions of (VVOP) using a scalarization by means of $\zeta_{f(\overline{x})-\hat{D}(\cdot),k} : Y \times Y \to \overline{\mathbb{R}}$ given in (12.4.12).

**Theorem 12.4.13** *Let assumptions* (A) *and* (B) *be fulfilled and* $\overline{x} \in S$ *be an* $\varepsilon k$-*completely nondominated solution to (VVOP). Then, for all* $\omega \in S$,

$$\zeta_{f(\overline{x})-\hat{D}(\cdot),k}(f(\overline{x}), f(\omega)) \leq \inf_{x\in S} \zeta_{f(\overline{x})-\hat{D}(\cdot),k}(f(x), f(\omega)) + \varepsilon. \tag{12.4.13}$$

***Proof*** Suppose that $\omega$ is an arbitrary but fixed element of $S$ and put $\overline{y} = f(\overline{x})$. Now, we prove that $\zeta_{f(\overline{x})-\hat{D}(\cdot),k}(f(\overline{x}), f(\omega)) = 0$. Assume $\zeta_{f(\overline{x})-\hat{D}(\cdot),k}(f(\overline{x}), f(\omega)) = \overline{t}$.

From Lemma 11.2.10, we obtain

$$tk + \overline{y} - \overline{y} \in \hat{D}(f(\omega)) \implies \overline{t}k \in \hat{D}(f(\omega)).$$

Because of $0 \in \text{bd}\, \hat{D}(f(\omega))$, we conclude $\overline{t} \leq 0$. If $\zeta_{f(\overline{x})-\hat{D}(\cdot),k}(f(\overline{x}), f(\omega)) < 0$, then $\overline{t} < 0$ and $-\overline{t} > 0$. By $0 \in \text{bd}\, \hat{D}(f(\omega))$ and $\hat{D}(f(\omega)) + \mathbb{R}_+k \subseteq \hat{D}(f(\omega))$, we get $-\overline{t}k \in \hat{D}(f(\omega))$ and $\overline{t}k \in \hat{D}(f(\omega)) \cap (-\hat{D}(f(\omega)))$ and we arrive at a contradiction because we supposed $\hat{D}(y)$ is a pointed set for all $y \in Y$. This means that $\overline{t} = 0$. Now, we show that $\zeta_{f(\overline{x})-\hat{D}(\cdot),k}(f(\overline{x}), f(\omega)) \leq \inf_{x\in S} \zeta_{f(\overline{x})-\hat{D}(\cdot),k}(f(x), f(\omega)) + \varepsilon$. By contradiction, suppose that (12.4.13) does not hold and there is an element $x \in S$ such that $\zeta_{f(\overline{x})-\hat{D}(\cdot),k}(f(x), f(\omega)) + \varepsilon < \zeta_{f(\overline{x})-\hat{D}(\cdot),k}(f(\overline{x}), f(\omega)) = 0$. This implies that there is a real number $\beta > 0$ such that $\zeta_{f(\overline{x})-\hat{D}(\cdot),k}(f(x), f(\omega)) = -\varepsilon - \beta$. From Lemma 11.2.10, we obtain

$$(-\varepsilon - \beta)k + \overline{y} - f(x) \in \hat{D}(f(\omega))$$

and this yields

$$\overline{y} - \varepsilon k - f(x) \in \hat{D}(f(\omega)) + \beta k \subset \hat{D}(f(\omega)) \setminus \{0\}.$$

This yields that there is an element $\omega \in S$ such that

$$(\overline{y} - \varepsilon k - (\hat{D}(f(\omega)) \setminus \{0\}) \cap f(S) \neq \emptyset$$

and $\overline{y} \notin \varepsilon k\text{-}\mathrm{MZ}_f(S, \hat{D}(\cdot))$, a contradiction because we supposed that $\overline{x}$ is an $\varepsilon k$-completely nondominated solution to (VVOP). Hence,

$$\zeta_{f(\overline{x})-\hat{D}(\cdot),k}(f(\overline{x}), f(\omega)) \leq \inf_{x \in S} \zeta_{f(\overline{x})-\hat{D}(\cdot),k}(f(x), f(\omega)) + \varepsilon$$

for all $x, \omega \in S$. □

In the following assertions, we use Theorem 12.4.13 and Ekeland's variational principle (Theorem 12.3.1) in order to derive necessary conditions for approximately completely nondominated solutions of vector optimization problems with variable domination structures. To proceed in this way, we need a certain boundedness condition with respect to the scalarization by means of the functional (12.4.12) formulated in the following definition.

**Definition 12.4.14** We consider $\overline{x} \in X$ and the functional $\zeta_{f(\overline{x})-\hat{D}(\cdot),k} : Y \times Y \to \overline{\mathbb{R}}$ given by (12.4.12). $f : X \to Y$ is called **bounded from below on $S$** with respect to $\zeta_{f(\overline{x})-\hat{D}(\cdot),k}$ if for all $\omega \in S$, there is a real number $\alpha > -\infty$ with

$$\inf_{x \in S} \zeta_{f(\overline{x})-\hat{D}(\cdot),k}(f(x), f(\omega)) > \alpha.$$

In the next theorem ([418, Theorem 4.8]), we derive a necessary condition for approximately completely nondominated solutions of a vector optimization problem with variable domination structure.

**Theorem 12.4.15** *Let assumptions* (A) *and* (B) *be satisfied. Assume that $\hat{D}(y)$ is a convex set for all $y \in Y$.*

*We consider problem $(VVOP)$ with $\overline{x} \in \varepsilon k - \mathrm{MZ}_f(S, \hat{D}(\cdot))$, the functional $\zeta_{f(\overline{x})-\hat{D}(\cdot),k}$ given by (12.4.12) and set $\overline{y} = f(\overline{x})$.*

*Furthermore, we suppose that $f \in \mathscr{F}(X, Y)$ is locally Lipschitzian and bounded from below in the sense of Definition 12.4.14 on S with respect to $\zeta_{f(\overline{x})-\hat{D}(\cdot),k}$ given in (12.4.12). Consider the abstract subdifferential $\partial$ for that (H1) – (H5) are satisfied. Then, for all $\omega \in S$, there are elements $x_\omega \in S$ and $y^*_\omega \in \partial(\zeta_{f(\overline{x})-\hat{D}(\cdot),k}(f(x_\omega), f(\omega))$ such that*

$$0 \in \partial(y^*_\omega \circ f)(x_\omega) + N_\partial(x_\omega; S) + \sqrt{\varepsilon} B_{X^*}. \tag{12.4.14}$$

***Proof*** Consider $\overline{x} \in \varepsilon k - \mathrm{MZ}_f(S, \hat{D}(\cdot))$. Using Theorem 12.4.13, we get for all $\omega \in S$

$$\zeta_{f(\overline{x})-\hat{D}(\cdot),k}(f(\overline{x}), f(\omega)) \leq \inf_{x \in S} \zeta_{f(\overline{x})-\hat{D}(\cdot),k}(f(x), f(\omega)) + \varepsilon.$$

Hence, $\overline{x}$ is an approximate solution to the scalar problem with the objective functional $\zeta_{f(\overline{x})-\hat{D}(\cdot),k}(\cdot, f(\omega))$ for all $\omega \in S$. Because of Theorem 11.2.13 and $f \in \mathscr{F}(X, Y)$, we get that $\zeta_{f(\overline{x})-\hat{D}(\cdot),k}(\cdot, f(\omega))$ is lower semicontinuous for all $\omega \in S$. Furthermore, $\zeta_{f(\overline{x})-\hat{D}(\cdot),k}(\cdot, f(\omega))$ is bounded from below for all $\omega \in S$. Taking into account the scalar Ekeland's variational principle (Theorem 12.3.1), for all $\omega \in S$ there is an element $x_\omega \in S$ such that it fulfills the conditions of Theorem 12.3.1 and it is an exact solution to a scalar optimization problem with the objective function $\overline{h} : X \to \overline{\mathbb{R}}$ over $S$ with

$$\overline{h}(x) := \zeta_{f(\overline{x})-\hat{D}(\cdot),k}(f(x), f(\omega)) + \sqrt{\varepsilon}\,\|x - x_\omega\| \quad \text{for all } x \in X.$$

From (H2) and (H4), we get

$$0 \in \partial\overline{h}(x_\omega) + N_\partial(x_\omega; S). \tag{12.4.15}$$

Under the given assumptions, the functional $\zeta_{f(\overline{x})-\hat{D}(\cdot),k}(\cdot, f(\omega))$ is convex (Theorem 11.2.14) and continuous (Theorem 11.2.13). Since $f \in \mathscr{F}(X, Y)$ is supposed to be locally Lipschitzian and $\zeta_{f(\overline{x})-\hat{D}(\cdot),k}(\cdot, f(\omega))$ is convex and continuous and hence locally Lipschitzian, the composition $\zeta_{f(\overline{x})-\hat{D}(\cdot),k}(f(\cdot), f(\omega))$ is also locally Lipschitzian. This yields together with the convexity of the norm $\|\cdot\|$ and (H5), that

$$\partial\overline{h}(x_\omega) \subseteq \partial\big(\zeta_{f(\overline{x})-\hat{D}(\cdot),k}(f(\cdot), f(\omega))\big)(x_\omega) + \partial(\sqrt{\varepsilon}\,\|\cdot - x_\omega\|)(x_\omega). \tag{12.4.16}$$

Furthermore, by (H3), we derive

$$\partial(\zeta_{f(\overline{x})-\hat{D}(\cdot),k}(f(\cdot), f(\omega)))(x_\omega) \subseteq \bigcup\big\{\partial(y^*_\omega \circ f)(x_\omega) \mid y^*_\omega \in \partial\zeta_{f(\overline{x})-\hat{D}(\cdot),k}(f(x_\omega), f(\omega))\big\}. \tag{12.4.17}$$

Taking into account the convexity of the norm and (H1), we get $\partial\,\|\cdot - x_\omega\|\,(x_\omega) = B_{X^*}$ and by (12.4.15), (12.4.16), (12.4.17), we find $y^*_\omega \in \partial\zeta_{f(\overline{x})-\hat{D}(\cdot),k}(f(x_\omega), f(\omega))$ satisfying

$$0 \in \partial(y^*_\omega \circ f)(x_\omega) + N_\partial(x_\omega; S) + \sqrt{\varepsilon}B_{X^*}$$

and the proof is complete. □

## 12.5 Existence of Solutions to Vector Variational Inequalities

Vector Variational Inequalities (VVI) have been of great interest in the academic and professional communities ever since the path-breaking paper by Giannessi [187]. Following the paper by Chen and Hou [92], we present an existence result for solutions of generalized VVI where the nonlinear separating functional (4.1.11) plays in important role in the proof.

In this section, let $Y$ be a Hausdorff topological vector space and let $C$ be a nontrivial, closed, convex cone in $Y$ with $\operatorname{int} C \neq \emptyset$. Furthermore, let $X$ be another Hausdorff topological vector space. As usual, we denote the set of all linear continuous maps from $X$ to $Y$ by $L(X, Y)$. The value of a linear map $l \in L(X, Y)$ at $x$ is denoted by $l(x)$.

Let $K$ be a subset of $X$ and let $T : K \to L(X, Y)$.

First, we introduce the following *Vector Variational Inequalitie* (VVI): Find $y \in K$, such that

$$\forall x \in K : \qquad T(y)(x - y) \notin -C \setminus \{0\}. \tag{12.5.1}$$

**Remark 12.5.1** A necessary optimality condition for a weakly efficient solution to a vector optimization problem (VOP) (see Sect. 6.1) can be formulated under certain differentiability assumptions in the form of a VVI (cf. Jahn [272, Theorems 7.3 and 7.6]).

Furthermore, the problem: Find $y \in K$ such that

$$\forall x \in K : \qquad T(y)(x - y) \notin -\operatorname{int} C \tag{12.5.2}$$

is called *Weak Vector Variational Inequality* (WVVI).

In order to formulate Generalized Vector Variational-Like Inequalities (GVVLI), we additionally consider a Hausdorff topological vector space $Z$. Furthermore, let $K \subseteq X$ and $E \subseteq Z$ be nonempty sets. Suppose that $\eta : K \times K \to K$ is a vector-valued function, $V : K \to \mathscr{P}(E)$ a set-valued map, and $H : X \times Z \to L(X, Y)$ a continuous map.

In this section, we study the *Generalized Vector Variational-Like Inequality* (GVVLI): Find $y \in K, z \in V(y)$ with

$$\forall x \in K : \qquad H(y, z)(\eta(x, y)) \notin -\operatorname{int} C. \tag{12.5.3}$$

In Theorem 12.5.5, we will show an existence result for solutions of GVVLI using the functional $\varphi_{A,k}$ defined by (4.1.11) with $A = -C$ and $k \in \operatorname{int} C$:

$$\varphi_{-C,k}(y) = \inf\{t \in \mathbb{R} \mid y \in tk - C\}.$$

Using the fixed point theorem by Fan and Browder and a continuous selection theorem, we will show an existence result for solutions of GVVLI in Theorem 12.5.5.

**Theorem 12.5.2** (Fixed point theorem by Fan and Browder) *Let $X$ be a nonempty compact convex subset of a Hausdorff topological vector space and $S : X \to \mathscr{P}(X)$ a set-valued map, such that for each $x \in X$ the image $S(x)$ is nonempty and convex, whereas for each $y \in X$, the pre-image $S^{-1}(y)$ is open in $X$. Then, $S$ has a fixed point.*

Continuous selection theorems are given in the book by Aubin and Frankowska[15]. In the following, we mention two selection theorems.

**Theorem 12.5.3** (Continuous selection theorem) *Let $X$ be a compact metric space, $Y$ a Banach space, and $\Gamma : X \to \mathscr{P}(Y)$ a lower semicontinuous map in the sense* *of Berge (see Definition 8.2.1). Furthermore, suppose that the images $\Gamma(\lambda)$ are nonempty and convex for all $\lambda \in \Lambda$. Then, there is a continuous selection $f : X \to Y$ with*

$$\forall x \in X: \qquad f(x) \in \operatorname{cl} \Gamma(x).$$

The following continuous selection theorem is a special case of the more general result by Park [375, Theorem 6] (see [34, Lemma 1] and [35, Theorem 2.1]).

**Theorem 12.5.4** (Continuous selection theorem) *Let $X$ be a compact Hausdorff space, $Y$ a topological vector space, $V : X \to \mathscr{P}(Y)$ a set-valued map with nonempty convex values and open fibers $V^{-1}(y) := \{x \in X \mid y \in V(x) \subseteq Y\}$ for all $y \in Y$. Then, $V$ has a continuous selection $f : X \to Y$ (with $\forall x \in X$: $f(x) \in V(x)$ ).*

The following existence result is shown by Chen and Hou in [92, Theorem 7] using the nonlinear separating functional (4.1.11) and its properties.

**Theorem 12.5.5** *Assume that $K \subset X$ is a nonempty, compact and convex set and $E \subset Z$ is a nonempty, closed and convex set. Consider a map $\eta : K \times K \to K$ with $\eta(x, x) = 0$ for all $x \in K$. Furthermore, let $V : K \to \mathscr{P}(E)$ be a set-valued map with $V(k) \neq \emptyset$ and convex for all $k \in K$ and let $H : K \times E \to L(X, Y)$ be a map. Assume that the following conditions are fulfilled:*

(a) *$\eta$ is continuous,*
(b) *$V$ is upper semi-continuous and $V^{-1}(l) := \{x \in K \mid l \in V(x)\}$ is an open set for all $l \in Z$,*
(c) *$H$ is a continuous map,*
(d) *$\exists k \in \operatorname{int} C$ such that the functional*

$$\varphi_{-C,k}(H(x, z)(\eta(w, x)))$$

*is quasiconvex with respect to $w \in K$.*

*Then, the Generalized Vector Variational-Like Inequality (GVVLI) has at least one solution.*

***Proof*** Under the given assumptions and from the continuous selection theorem (Theorem 12.5.3), it follows that there is a continuous map $f : K \to Z$ such that $f(x) \in V(x)$ for all $x \in K$. Consider

$$\mu(x, z, w) := H(x, z)(\eta(w, x)) \quad (x \in K, z \in E, w \in K). \tag{12.5.4}$$

$\mu$ is a continuous map from $K \times E \times K$ into $Y$ and $\varphi_{-C,k} \circ \mu$ is quasiconvex with respect to $w$ because of (d).

We now define for $n = 1, 2, \ldots$ set-valued maps $F_n : K \to \mathscr{P}(K)$ by

$$F_n(x) := \{w \in K \mid (\varphi_{-C,k} \circ \mu)(x, f(x), w) - \inf_{u \in K} (\varphi_{-C,k} \circ \mu)(x, f(x), u) < \frac{1}{n}\} \quad (x \in K). \tag{12.5.5}$$

Taking into account that $K$ is compact and $\varphi_{-C,k}$ is continuous (Proposition 4.7.3), we get, for all $x \in K$, that $(\varphi_{-C,k} \circ \mu)(x, f(x), w)$ is a continuous function with respect to $w$. Furthermore, it follows that $(\varphi_{-C,k} \circ \mu)(x, f(x), K)$ is a compact subset of $\mathbb{R}$ because of the compactness of $K$. Hence, $F_n(x)$ is nonempty for all $n$.

In order to apply the Fan-Browder fixed point theorem (Theorem 12.5.2), we will show the following assertions:

(I) For each $n$ and $x \in K$, $F_n(x)$ is a convex set.

Consider $w_1, w_2 \in F_n(x)$ and $\lambda \in (0, 1]$. Then, $w_1, w_2 \in K$ and there is an element $a_i > 0$, $(i = 1, 2)$ with

$$(\varphi_{-C,k} \circ \mu)(x, f(x), w_i) = \inf_{u \in K} (\varphi_{-C,k} \circ \mu)(x, f(x), u) + \frac{1}{n} - a_i, \quad i = 1, 2.$$

Set $a_0 := \inf\{a_1, a_2\}$ and $t_0 := \inf_{u \in K} (\varphi_{-C,k} \circ \mu)(x, f(x), u) + \frac{1}{n} - a_0$. Then, we obtain

$$(\varphi_{-C,k} \circ \mu)(x, f(x), w_i) \le t_0, \quad i = 1, 2.$$

Therefore,

$$w_i \in \mathrm{lev}_{(\varphi_{-C,k} \circ \mu)(x, f(x), \cdot), \le}(t_0) = \{u \in K \mid (\varphi_{-C,k} \circ \mu)(x, f(x), u) \le t_0\}.$$

Because of assumption $(d)$, $(\varphi_{-C,k} \circ \mu)(x, f(x), \cdot)$ is quasiconvex, and hence $\mathrm{lev}_{(\varphi_{-C,k} \circ \mu)(x, f(x), \cdot), \le}(t_0)$ is a convex set such that

$$\lambda w_1 + (1 - \lambda) w_2 \in \mathrm{lev}_{(\varphi_{-C,k} \circ \mu)(x, f(x), \cdot), \le}(t_0).$$

This yields

$$(\varphi_{-C,k} \circ \mu)(x, f(x), \lambda w_1 + (1 - \lambda) w_2) \le t_0,$$

such that we obtain

$$(\varphi_{-C,k} \circ \mu)(x, f(x), \lambda w_1 + (1 - \lambda) w_2) - \inf_{u \in K} (\varphi_{-C,k} \circ \mu)(x, f(x), u) \le \frac{1}{n},$$

and hence $\lambda w_1 + (1 - \lambda) w_2 \in F_n(x)$, i.e., $F_n(x)$ is a convex set for each $n$ and $x \in K$.

Furthermore, for the application of the Fan-Browder fixed point theorem (Theorem 12.5.2) we show the statement:

(II) The set $F_n^{-1}(w)$ is open for every $n$ and $w \in K$.

For all $w \in K$, it holds that

$$\begin{aligned} F_n^{-1}(w) &= \{x \in K \mid w \in F_n(x)\} \\ &= \{x \in K \mid (\varphi_{-C,k} \circ \mu)(x, f(x), w) - \inf_{u \in K} (\varphi_{-C,k} \circ \mu)(x, f(x), u) < \frac{1}{n}\}. \end{aligned}$$

By contradiction, suppose that (II) is not true. Then, there are an element $y \in F_n^{-1}(w)$ and a net $\{x_\alpha\}$ such that $x_\alpha \to y$ and $x_\alpha \notin F_n^{-1}(w)$ for every $\alpha$, namely $w \notin F_n(x_\alpha)$. Taking into account the definition of $F_n(x)$ in (12.5.5), we get

$$(\varphi_{-C,k} \circ \mu)(x_\alpha, f(x_\alpha), w) - \inf_{u \in K} (\varphi_{-C,k} \circ \mu)(x_\alpha, f(x_\alpha), u) \geq \frac{1}{n}.$$

Because of the continuity of the function $(x, w) \to (\varphi_{-C,k} \circ \mu)(x, f(x), w)$ and $K$ is compact, we conclude that the function $x \to \inf_{u \in K} (\varphi_{-C,k} \circ \mu)(x, f(x), u)$ is continuous, and so we obtain

$$(\varphi_{-C,k} \circ \mu)(x_\alpha, f(x_\alpha), w) - \inf_{u \in K} (\varphi_{-C,k} \circ \mu)(x_\alpha, f(x_\alpha), u)$$

$$\longrightarrow (\varphi_{-C,k} \circ \mu)(y, f(y), w) - \inf_{u \in K} (\varphi_{-C,k} \circ \mu)(y, f(y), u) \geq \frac{1}{n},$$

in contradiction to $w \in F_n(y)$. So, we get that $F_n^{-1}(w)$ is open.

Since the statements (I) and (II) hold, all assumptions of the Fan-Browder fixed point theorem (Theorem 12.5.2) are fulfilled. From the Fan-Browder fixed point theorem, we get the existence of $x_n \in K$ such that

$$x_n \in F_n(x_n), \qquad n = 1, 2, \ldots.$$

We suppose $x_n \to y \in K$ and $f(x_n) \to f(y) \in V(y)$ since $K$ is compact. From the definition of $F_n$ in (12.5.5), we obtain that for each $n$ it holds that

$$(\varphi_{-C,k} \circ \mu)(x_n, f(x_n), x_n) < \inf_{u \in K} (\varphi_{-C,k} \circ \mu)(x_n, f(x_n), u) + \frac{1}{n},$$

and for $n \to +\infty$

$$(\varphi_{-C,k} \circ \mu)(y, f(y), y) \leq \inf_{u \in K} (\varphi_{-C,k} \circ \mu)(y, f(y), u).$$

Because of $\eta(x, x) = 0$ and $\varphi_{-C,k}(0) = 0$ (see (4.6.14)), we obtain

$$\inf_{u \in K} (\varphi_{-C,k} \circ \mu)(y, f(y), u) \geq 0.$$

This means

$$\forall u \in K : \qquad (\varphi_{-C,k} \circ \mu)(y, f(y), u) \geq 0.$$

Under the given assumptions, $\varphi_{-C,k}$ is strictly int $C$-monotone (see Proposition 4.7.3) such that

$$\forall u \in K : \quad \mu(y, f(y), u) \notin -\operatorname{int} C.$$

Put $z := f(y) \in V(y)$, we get

$$\forall u \in K : \quad H(y, z)(\eta(u, y)) \notin -\operatorname{int} C.$$

The proof is completed. □

**Remark 12.5.6** Recently, vector equilibrium problems are studied by Gutiérrez, Novo, Ródenas-Pedregosa and Tanaka in [213]: Consider a linear space $Y$, an arbitrary nonempty decision set $S$, a vector-valued bifunction $f : S \times S \to Y$ and a subset $D \subset Y$ with core $D \neq \emptyset$ and $0 \notin$ core $D$. Then, the vector equilibrium problem is given as follows: Find $\overline{x} \in S$ such that

$$\forall x \in S : \quad f(\overline{x}, x) \notin -D \setminus \{0\}. \tag{VEP}$$

An element $\overline{x} \in S$ is called a weakly efficient solution of problem (VEP) if

$$\forall x \in S : \quad f(\overline{x}, x) \notin -\operatorname{core} D.$$

The set of all weakly efficient solutions of problem (VEP) is denoted by $\mathrm{WE}_f(S, D)$.

In order to give a characterization of the set $\mathrm{WE}_f(S, D)$, the following scalar equilibrium problem is studied in [213]: Find $\overline{x} \in S$ such that

$$\forall x \in S : \quad (\varphi_{-D,k} \circ f)(\overline{x}, x) \geq 0, \tag{12.5.6}$$

where $\varphi_{-D,k}$ is given by (4.1.11) with $A = -D$ and $k \in Y \setminus \{0\}$. The set of all $\overline{x}$ satisfying (12.5.6) is denoted by $\mathrm{Sol}(\varphi_{-D,k} \circ f, S)$.

Consider $\mathrm{vcl}_k(D) = \{y \in Y \mid \forall \lambda > 0 \ \exists \lambda' \in [0, \lambda] \text{ such that } y + \lambda' k \in D\}$. In [213, Theorem 20], the authors prove for all $k \in Y \setminus \{0\}$ with $\mathrm{vcl}_k(D) + \mathbb{R}_{>}k =$ core $D$:

$$\mathrm{WE}_f(S, D) = \mathrm{Sol}(\varphi_{-D,k} \circ f, S).$$

Note, that $Y$ is supposed to be a linear space in [213]. This characterization of weakly efficient solutions to the vector equilibrium problem (VEP) is shown under a weak closedness assumption and using corresponding properties of $\varphi_{-D,k}$ (compare Chap. 4).

**Remark 12.5.7** Recent developments on the field of vector variational inequalities are presented in the book [7] by Ansari, Köbis and Yao.

## 12.6 The Extremal Principle by Kruger and Mordukhovich

The extremal principle by Kruger and Mordukhovich [311, 312, 357, 358] is a very deep assertion concerning the existence of linear continuous functionals with certain separation properties. It plays a fundamental role for deriving optimality conditions and variational principles for not necessarily convex optimization problems. Recent developments and extensions of the extremal principle are given by Bui and Kruger in [81, 82].

A general concept of set extremality is presented in the books by Mordukhovich [359] (see Mordukhovich [357, 358], Kruger, Mordukhovich [312], Kruger [311]). The relationships between extremal principles and separation properties of nonconvex sets are studied as well as corresponding applications for deriving optimality conditions in constrained optimization are explained in [357, 359]. We will elucidate these fundamental results in this section, particularly several versions of the extremal principle. Bao, Mordukhovich [20, 24] studied applications of these extremal principles for deriving a subdifferential variational principle for set-valued maps.

We would like to recall the concept of local extremal points and extremal systems (see [359, Definition 2.1]).

**Definition 12.6.1** We consider nonempty subsets $\Omega_1, \ldots, \Omega_n$ $(n \geq 2)$ of a Banach space $X$ and a common point $\overline{x}$ of these sets. This point $\overline{x}$ is called a **local extremal point** of the set system $\{\Omega_1, \ldots, \Omega_n\}$ if there are sequences $\{a_{ik}\} \subset X, i = 1, \ldots, n$, and a neighborhood $V$ of $\overline{x}$ such that $a_{ik} \overset{k\to+\infty}{\to} 0$ and for all large $k \in \mathbb{N}$

$$\bigcap_{i=1}^{n} (\Omega_i - a_{ik}) \cap V = \emptyset.$$

We call $\{\Omega_1, \cdots, \Omega_n, \overline{x}\}$ an **extremal system** in $X$ in this case.

**Remark 12.6.2** We consider two closed sets $\Omega_1$ and $\Omega_2$ in the Banach space $X$. A point $\bar{x} \in \Omega_1 \cap \Omega_2$ is called *locally extremal* for the set system $\{\Omega_1, \Omega_2\}$ if there is a neighborhood $V$ of $\bar{x}$ such that for any $\varepsilon > 0$, there is an element $a \in \varepsilon B_X$ with

$$\Omega_1 \cap (\Omega_2 + a) \cap V = \emptyset. \tag{12.6.1}$$

We would like to mention that the condition $\Omega_1 \cap \Omega_2 = \{\overline{x}\}$ does not necessarily imply that $\overline{x}$ is a local extremal point of $\{\Omega_1, \Omega_2\}$.

Mordukhovich [359, Definition 2.5] described the following three fundamental versions of the extremal principle in Banach spaces and explained that they can be viewed as a certain kind of local separation of nonconvex sets around extremal points. In the next definition, we are using the set of $\varepsilon$-normals $\hat{N}_\varepsilon(x; \Omega)$ (see Definition 8.1.2), the Fréchet normal cone $\hat{N}(x; \Omega)$ (see (8.1.1)) and the limiting normal cone $N_M(x; \Omega)$ (see Definition 8.1.4) (with $X$ a Banach space, $\Omega \subset X$, $x \in \Omega$).

**Definition 12.6.3** We consider an extremal system $\{\Omega_1, \cdots, \Omega_n, \overline{x}\}$ in $X$. Then,

(a) $\{\Omega_1, \cdots, \Omega_n, \overline{x}\}$ fulfills the $\varepsilon$-**extremal principle** if for every $\varepsilon > 0$ there are $x_j \in \Omega_j \cap (\overline{x} + \varepsilon B_X)$ and $x_j^* \in X^*$ such that

$$x_j^* \in \hat{N}_\varepsilon(x_j; \Omega_j) \quad j = 1, \ldots, n, \tag{12.6.2}$$

$$x_1^* + \cdots + x_n^* = 0, \qquad \|x_1^*\|_* + \cdots + \|x_n^*\|_* = 1. \tag{12.6.3}$$

(b) $\{\Omega_1, \cdots, \Omega_n, \overline{x}\}$ fulfills the **approximate extremal principle** if for each $\varepsilon > 0$ there are $x_j \in \Omega_j \cap (\overline{x} + \varepsilon B_X)$ and $x_j^* \in X^*$ and

$$x_j^* \in \hat{N}(x_j; \Omega_j) + \varepsilon B_{X^*} \quad j = 1, \ldots, n, \tag{12.6.4}$$

such that (12.6.3) holds.

(c) $\{\Omega_1, \cdots, \Omega_n, \overline{x}\}$ fulfills the **exact extremal principle** if there are basic normals

$$x_j^* \in N_M(\overline{x}; \Omega_j), \quad j = 1, \ldots, n, \tag{12.6.5}$$

such that (12.6.3) holds.

The $\varepsilon$- extremal principle according to (a), (b) or (c) holds in the space $X$ if it holds for each extremal system $\{\Omega_1, \cdots, \Omega_n, \overline{x}\}$ in $X$, where all the sets $\Omega_j$ $(j = 1, \cdots, n)$ are closed around $\overline{x}$.

**Remark 12.6.4** Because of $\hat{N}(x; \Omega) + \varepsilon B_{X^*} \subset \hat{N}_\varepsilon(x; \Omega)$, it is easy to see that the $\varepsilon$-extremal principle is a conclusion from the approximate extremal principle for every extremal system $\{\Omega_1, \cdots, \Omega_n, \overline{x}\}$ in a Banach space $X$.

**Remark 12.6.5** We would like to mention that the statements in the extremal principles yield necessary conditions for local extremal points of set systems and can be treated as generalized Euler equations in an abstract geometric setting. Moreover, the extremal principles are connected to local separation of nonconvex sets. Actually, if we study the exact extremal principle for two sets $\Omega_1$ and $\Omega_2$, then (12.6.3) and (12.6.5) reduce to the condition that there is an element $x^* \in X^* \setminus \{0\}$ with

$$x^* \in N_M(\overline{x}; \Omega_1) \cap (-N_M(\overline{x}; \Omega_2)). \tag{12.6.6}$$

We also get such a kind of extremal principle using the nonlinear separating functionals given by (4.1.11) and their properties. Applying (4.1.11) and the properties given in Propositions 4.7.3 and 4.7.1(b) to an extremal system $\{\Omega_1, \Omega_2, \overline{x}\}$ and taking into account the properties of the subgradients of the nonlinear separating functional (see Theorem 4.10.3), we get a condition corresponding to (12.6.6) (see Theorem 12.6.6).

In general, the optimality conditions are expressed in terms of generalized differentiation objects such as subdifferentials and normal cones (see Sects. 12.1 and 8.3). We do not consider here a specific subdifferential, but we work with a generic concept (see Sect. 12.4.2): Let $Y$ be a Banach space. We consider an operator $\partial$ which associates with every $h : Y \to \mathbb{R} \cup \{+\infty\}$ and every $y \in \operatorname{dom} h$ a subset $\partial h(y) \subset Y^*$ such that the following axioms are satisfied:

(A1) If $h$ is convex and lower semicontinuous, then $\partial h$ coincides with the Fenchel subdifferential, i.e., $\partial h(y) = \{y^* \in Y^* \mid \forall z \in Y : \; y^*(z - y) + h(y) \leq h(z)\}$.

(A2) If $S \subset Y$ is a nonempty and closed set, $h$ is locally Lipschitzian and $\overline{y}$ is a minimum point for $h$ over $S$, then

$$0 \in \partial h(\overline{y}) + N_{\partial}(\overline{y}; S),$$

where $N_{\partial}(\overline{y}; S) := \partial \iota_S(\overline{y})$ and $\iota_S$ is the indicator function of $S$.

Note that these axioms are fulfilled by several important well-known subdifferentials on suitable classes of Banach spaces. Once again, we avoid technical details, but we mention the subdifferentials of Mordukhovich, Michel-Penot, Ioffe, and for details we refer the reader to [359].

Under special assumptions, we show that it is possible to derive corresponding results to (12.6.6) by using the extremal principle in a Banach space $Y$ for two sets $\Omega_1, \Omega_2$ as well as by using Separation Theorem 4.11.4. We consider the formulation of the extremal principle for a general subdifferential $\partial$ satisfying (A1) and (A2) and the corresponding normal cone $N_{\partial}$. Of course, we suppose special structures concerning the sets $\Omega_1, \Omega_2$, however these structures are given for many optimization problems.

**Theorem 12.6.6** *Assume that $S \subset Y$ is a proper set and $C \subset Y$ is a closed, convex, pointed cone with $k \in \operatorname{int} C$ and $\operatorname{cone} S \cap (-C) = \{0\}$. Consider a system $\{\Omega_1, \Omega_2, 0\}$ with $\Omega_1 = \operatorname{cone} S$ and $\Omega_2 = -C$. Suppose that $\Omega_1$ is a proper, closed set. Furthermore, assume that $\partial$ satisfies (A1) and (A2). Then, there is an element $y^* \in C^+$, $y^*(k) = 1$ with*

$$-y^* \in N_{\partial}(0; \Omega_1) \cap (-N_{\partial}(0; \Omega_2)).$$

***Proof*** Under the given assumptions, we have $\Omega_1 \cap \Omega_2 = \{0\}$ and $k \in \operatorname{int} C$. This means that the assumptions of Separation Theorem 4.11.4 are fulfilled. Consider the functional $\varphi_{A,k}$ given by (4.1.11) with $A = \Omega_2 = -C$. For $A = -C$ and $D = \Omega_1$, we get

$$\forall y \in \Omega_1 : \quad \varphi_{-C,k}(y) \geq 0$$

from Theorem 4.11.4 and $\varphi_{-C,k}$ is a finite-valued continuous sublinear functional (see Theorem 4.6.6 and Proposition 4.7.1(b)).

Since $0 \in \Omega_1$ and $\varphi_{-C,k}(0) = 0$, one gets that 0 is a minimum for the scalar function $\varphi_{-C,k}$ over the closed set $\Omega_1$. The functional $\varphi_{-C,k}$ is continuous and sublinear and so locally Lipschitzian. From (A2), one gets

$$0 \in \partial\varphi_{-C,k}(0) + N_\partial(0, \Omega_1).$$

Since $\varphi_{-C,k}$ is continuous and convex (see Proposition 4.7.3), from (A1) and Theorem 4.10.3, we conclude that there is an element $y^* \in \partial\varphi_{-C,k}(0)$ with $y^* \in C^+$, $y^*(k) = 1$ and

$$-y^* \in N_\partial(0; \Omega_1). \tag{12.6.7}$$

Taking into account $C^+ = -N_\partial(0; C)$ (see Sect. 8.3), for the closed, convex, pointed cone $C$, one deduces that $y^* \in -N_\partial(0; C) = N_\partial(0; \Omega_2)$ as desired. □

**Remark 12.6.7** If we suppose that $\Omega_1$ and $\Omega_2$ are convex sets, (12.6.6) coincides with

$$\forall\, x_1 \in \Omega_1,\ \forall x_2 \in \Omega_2: \qquad x^*(x_1) \le x^*(x_2),$$

i.e., we get the classical separation property for two convex sets.

Mordukhovich [359, Theorem 2.20] has shown the following essential assertion in an Asplund space $X$.

**Theorem 12.6.8** *Assume that $X$ is an Asplund space. Then, the approximate extremal principle holds in $X$.*

**Remark 12.6.9** Because of Remark 12.6.4, we obtain that the $\varepsilon$-extremal principle holds in an Asplund space $X$ taking into account Theorem 12.6.8 (see [359, Theorem 2.20]).

If we consider a system $\{\Omega_1, \Omega_2\}$, a locally extremal point $\bar{x} \in \Omega_1 \cap \Omega_2$ is given by (12.6.1). Then, we obtain the following result from Theorem 12.6.8 (see [23]).

**Theorem 12.6.10** (The extremal principle) *Suppose that $X$ is an Asplund space. Let $\bar{x}$ be a local extremal point of the set system $\{\Omega_1, \Omega_2\}$, where both $\Omega_1$ and $\Omega_2$ are closed around $\bar{x}$. Then, for every $\varepsilon > 0$ there are*

$$x_j \in \Omega_j \cap (\bar{x} + \varepsilon B_X) \quad \textit{and} \quad x_j^* \in \hat{N}(x_j; \Omega_j) \quad j = 1, 2,$$

*such that the conditions*

$$1 - \varepsilon \le \|x_1^*\|_* + \|x_2^*\|_* \le 1 + \varepsilon \qquad \textit{and} \qquad \|x_1^* + x_2^*\|_* \le \varepsilon$$

*are fulfilled.*

As mentioned in Remark 12.6.5, the extremal principle can be viewed as a nonconvex variational counterpart of the classical separation principle for convex sets, similar to the separation theorems for nonconvex sets (see Sect. 4.11). In fact, the extremal principle plays an essential role in nonlinear variational analysis, nonlinear functional analysis and nonsmooth optimization analogical to classical convex separation theorems in convex analysis and convex optimization; see the books by Mordukhovich [359]] for more details and substantial applications.

# Chapter 13
# Algorithms for the Solution of Optimization Problems

For generating solutions of vector optimization problems via algorithms based on a scalarization, the monotonicity properties of the scalarizing functionals are important. In this chapter, we present Benson's Outer Approximation Algorithm that uses a scalarization by means of the nonlinear separating functional (4.1.11). Furthermore, we present proximal-point algorithms as well as an adaptive algorithm for solving vector optimization problems where the nonlinear separating functional (4.1.11) is involved. We show that a scalarization by means of the functional (4.1.11) is useful for deriving an algorithm for solving set-valued optimization problems. Finally, we derive algorithms for solving vector optimization problems with variable domination structure using the extension of the functional (4.1.11) given by (11.2.5) introduced and studied in Sect. 11.2.

## 13.1 The Benson Algorithm for Linear Vector Optimization Problems

*Benson's Outer Approximation Algorithm* is very popular for solving linear vector optimization problems (see Löhne [334] and references therein). For deriving this algorithm, a scalarization by means of the nonlinear separating functional $\varphi_{A,k}$ given by (4.1.11) plays a crucial role. Also for an extension of Benson's algorithm to convex vector optimization problems, the functional $\varphi_{A,k}$ is an essential tool in the proofs (see Ehrgott, Shao, Schöbel [134]).

The Outer Approximation Algorithm is a finite algorithm for generating the set of all efficient extreme points in the image set (set of objective function values over the feasible set) of a linear vector optimization problem (LVP). Because it works in the image set rather than in the pre-image (decision) set of problem (LVP), the Outer Approximation Algorithm has several advantages over decision-set-based algorithms, especially, it is easier for the decision maker to compare the objective function values.



C. Tammer and P. Weidner, *Scalarization and Separation by Translation Invariant Functions*, Vector Optimization, https://doi.org/10.1007/978-3-030-44723-6_13

We consider a linear vector optimization problem, where the solution concept is given by Definition 6.0.1 with respect to the ordering cone $C = \mathbb{R}^\ell_+$:

$$\text{Determine} \quad \text{Eff}(P(S), C), \tag{LVP}$$

where $P$ is a $\mathbb{R}^{\ell,n}$-matrix,

$$S := \{x \in \mathbb{R}^n \mid Ax \geq b\},$$

$A$ is a $\mathbb{R}^{m,n}$-matrix and $b \in \mathbb{R}^m$ and $P(S) := \bigcup_{x \in S} Px$.

In the sequel, we are using the symbols $\leq, \geq$ for the partial orderings induced by the natural ordering cones in the spaces $\mathbb{R}^\ell$, $\mathbb{R}^m$ as well as $\mathbb{R}^n$. We suppose that the feasible set $S$ is nonempty and compact.

For each $i = 1, 2, \ldots, \ell$, we consider

$$y_i^0 := \max\{y_i \mid y = (y_1, \ldots, y_\ell)^T \in P(S)\}. \tag{13.1.1}$$

Let $\hat{y} \in \mathbb{R}^\ell$ be such that $\hat{y} \in y^0 + \text{int}\, C$ and consider

$$\hat{Y} := \{y \in \mathbb{R}^\ell \mid Px \leq y \leq \hat{y} \text{ for some } x \in S\}. \tag{13.1.2}$$

The following result is shown by Benson in [44, Theorem 2.4] and gives a sufficient condition for the property that a certain element $\overline{y}$ belongs to the set of weakly efficient elements of $\hat{Y}$ with respect to $C$.

**Theorem 13.1.1** *Consider $y \in \hat{y} - C$ with $y \notin \hat{Y}$ and $\overline{p} \in \text{int}\, \hat{Y}$. Assume that $\overline{y}$ is the unique point on the boundary of $\hat{Y}$ that belongs to the line segment connecting $y$ and $\overline{p}$. Then, $\overline{y} \in \text{WEff}(\hat{Y}, C)$.*

Let $Y^*(y^*)$ be the optimal solution set to the problem

$$\min_{y \in \hat{Y}} \langle y^*, y \rangle \tag{$\text{P}_{y^*}$}$$

for some $y^* \in C \setminus \{0\}$. Note, $\hat{Y}$ given by (13.1.2) is the feasible set in $(\text{P}_{y^*})$.

Taking into account $\hat{y} \in Px + \text{int}\, C$ for all $x \in S$ and the definition of $\hat{Y}$, we get that $\ell$ of the $(\ell - 1)$-dimensional faces of $\hat{Y}$ are given by the sets

$$F_\nu := \{y \in \hat{Y} \mid y_\nu = \hat{y}_\nu\},$$

for $\nu = 1, 2, \ldots, \ell$. It is well known that for each $\nu = 1, 2, \ldots, \ell$, either $F_\nu \subseteq \text{WEff}(\hat{Y}, C)$ or r-int $F_\nu \cap \text{WEff}(\hat{Y}, C) = \emptyset$ (see [503]). This yields that r-int $F_\nu \cap \text{WEff}(\hat{Y}, C) = \emptyset$ for each $\nu = 1, 2, \ldots, \ell$ because of $\hat{y} \in F_\nu$ and $\hat{y} \notin \text{WEff}(\hat{Y}, C)$. This implies that $\overline{y}$ in Theorem 13.1.1 lies in some face $F \subseteq \text{WEff}(\hat{Y}, C)$ of $\hat{Y}$ with $F \neq F_\nu$ for each $\nu = 1, 2, \ldots, \ell$. Any such face $F$ coincides with the optimal solution set $Y^*(y^*)$ of problem $(\text{P}_{y^*})$ for some $y^* \in C \setminus \{0\}$ (see [503]). The result in the following Theorem 13.1.2 explains how to find such a face $F$ and to give the representation of it.

A main tool for the proof of Theorem 13.1.2 is a scalarization of (LVP) by means of the nonlinear separating functional $\varphi_{A,k}$ with $A = \overline{y} - C, \overline{y} \in \mathbb{R}^\ell$ and $k \in \operatorname{int} \mathbb{R}^\ell_+$. We formulate this scalarized problem in the form (5.0.1) and, equivalently (see Theorem 5.1.4) as

$$\min t \qquad (\text{LVP}(\varphi_{\overline{y}-C,k}))$$

subject to

$$Px \in tk + \overline{y} - C, \qquad (13.1.3)$$

$$Ax \geq b. \qquad (13.1.4)$$

Furthermore, in Theorem 13.1.2, we use the duality between problem (LVP($\varphi_{\overline{y}-C,k}$)) and the corresponding dual problem (see Ehrgott, Shao, Schöbel [134]) given by

$$\max \langle b, u\rangle - \langle \overline{y}, w\rangle \qquad (\text{DLVP}(\varphi_{\overline{y}-C,k}))$$

$$\text{subject to} \quad A^T u - P^T w = 0,$$

$$\langle k, w\rangle = 1,$$

$$u \in \mathbb{R}^m, \ u \geq 0, \ w \in \mathbb{R}^\ell, \ w \geq 0.$$

The following result is shown by Benson in [44, Theorem 2.5] using the classical duality between the scalar linear problems (LVP($\varphi_{\overline{y}-C,k}$)) and (DLVP($\varphi_{\overline{y}-C,k}$)).

**Theorem 13.1.2** *Suppose that $\overline{y} \in \text{WEff}(\hat{Y}, C)$ and consider an optimal solution $(u^{*T}, w^{*T})$ of the dual linear problem (DLVP ($\varphi_{\overline{y}-C,k}$)) to the problem (LVP ($\varphi_{\overline{y}-C,k}$)), where $u^* \in \mathbb{R}^m$ and $w^* \in \mathbb{R}^\ell$ correspond to the constraints (13.1.4) and (13.1.3) of problem (LVP ($\varphi_{\overline{y}-C,k}$)), respectively. Then, it holds that $w^* \geq 0$, $w^* \neq 0$, and $\overline{y}$ appertains to the weakly efficient face $Y^*(w^*)$ of $\hat{Y}$, where $Y^*(w^*)$ is given by*

$$Y^*(w^*) = \{y \in \hat{Y} \mid \langle w^*, y\rangle = \langle b, u^*\rangle\}.$$

***Proof*** Consider $Y^C := \{y \in \mathbb{R}^\ell \mid y \geq Px \text{ for some } x \in S\}$. We denote the optimal objective function value of problem (LVP($\varphi_{\overline{y}-C,k}$)) with $\overline{y} = y$ by $g(y)$ for each $y \in Y^C$. Taking into account $\overline{y} \in \text{WEff}(\hat{Y}, C)$, it is easy to see that $g(\overline{y}) = 0$ (compare Theorem 6.2.31). Hence, we know from the duality theory for linear optimization problems that the dual linear problem (DLVP($\varphi_{\overline{y}-C,k}$)) to problem (LVP($\varphi_{\overline{y}-C,k}$)) also has an optimal objective function value equal to $g(\overline{y}) = 0$.

For an optimal solution $(u^{*T}, w^{*T})$ to problem (DLVP($\varphi_{\overline{y}-C,k}$)) it holds that $w^* \geq 0$, $w^* \neq 0$ taking into account the constraint $\langle k, w\rangle = 1$ in (DLVP($\varphi_{\overline{y}-C,k}$)).

Moreover, because the optimal objective function value of problem (DLVP($\varphi_{\overline{y}-C,k}$)) is equal to 0, we obtain

$$\langle \overline{y}, w^* \rangle = \langle b, u^* \rangle. \tag{13.1.5}$$

Because of $w^* \geq 0$ and $w^* \neq 0$, the optimal solution set $Y^*(w^*)$ for problem (P$_{w^*}$) corresponds to a weakly efficient face of $\hat{Y}$ (see Yu [503]). This means together with (13.1.5), the proof of the theorem is completed if we show that $\overline{y}$ is an optimal solution to problem (P$_{w^*}$). In order to show that $\overline{y}$ is an optimal solution to problem (P$_{w^*}$), we can consider the following reformulation of (P$_{w^*}$) (taking into account the definition of $\hat{Y}$)

$$\min \langle w^*, y \rangle \tag{P$_{w^*}$}$$

$$\text{subject to} \quad -y \geq -\hat{y},$$

$$-Px + y \geq 0,$$

$$Ax \geq b.$$

This means that we are looking for elements $(x^T, y^T)$ solving (P$_{w^*}$).

Furthermore, we consider the dual linear problem to (P$_{w^*}$) given by

$$\max -\langle \hat{y}, s \rangle + \langle b, q \rangle \tag{D$_{w^*}$}$$

$$\text{subject to} \quad -s + r = w^*,$$

$$-P^T r + A^T q = 0,$$

$$s \in \mathbb{R}^\ell, s \geq 0, \; r \in \mathbb{R}^\ell, r \geq 0, \; q \in \mathbb{R}^m, q \geq 0.$$

Taking into account that $(u^{*T}, w^{*T})$ is an optimal solution to problem (DLVP($\varphi_{\overline{y}-C,k}$)), the element

$$(s^T, r^T, q^T) = (0^T, w^{*T}, u^{*T})$$

is feasible in (D$_{w^*}$) with the objective function value $\langle b, u^* \rangle$.

Consider an optimal solution $(x^0, t^0)$ to problem (LVP($\varphi_{\overline{y}-C,k}$)). Because of $g(\overline{y}) = 0$, we get that $t^0 = 0$, $Px^0 \in \overline{y} - C$, $Ax^0 \geq b$. This yields, together with $\overline{y} \leq \hat{y}$, that $(x^T, y^T) = (x^{0T}, \overline{y}^T)$ is a feasible element for (P$_{w^*}$) with an objective function value $\langle w^*, \overline{y} \rangle$. By (13.1.5), we know that $\langle w^*, \overline{y} \rangle = \langle b, u^* \rangle$ holds. Because $(0^T, w^{*T}, u^{*T})$ is a feasible element to (D$_{w^*}$) with the objective function value $\langle b, u^* \rangle$, this yields that $(x^T, y^T) = (x^{0T}, \overline{y}^T)$ is an optimal solution to problem (P$_{w^*}$) taking into account the duality theory for scalar linear optimization problems. Hence, $\overline{y}$ is an optimal solution to the reformulation of problem (P$_{w^*}$) as

$$\min_{y \in \hat{Y}} \langle w^*, y \rangle.$$

□

**Remark 13.1.3** In Theorem 13.1.2, it is explained how to find a weakly efficient face of $\hat{Y}$ using the nonlinear scalarizing functional $\varphi_{A,k}$ given by (4.1.11) with $A = \overline{y} - C$, $\overline{y} \in \mathbb{R}^\ell$ and $k \in \operatorname{int} C$. So, Theorem 13.1.2 procures the background for constructing certain linear inequality cuts required in *Benson's Outer Approximation Algorithm* for problem (LVP) presented in the following.

For deriving the algorithm, we need the following assertions given by Benson in [44]. Set

$$\beta := \min_{y \in \hat{Y}} \langle k, y \rangle,$$

where $k \in \operatorname{int} C$. Taking into account [44, Proposition 2.2], $\beta$ is a finite number.

Furthermore, let $w^0 = \hat{y}$ and let $w^\nu \in \mathbb{R}^\ell$ be defined (for each $\nu = 1, 2 \ldots, \ell$) by

$$w_i^\nu := \begin{cases} \hat{y}_i & \text{if } i \neq \nu \\ \beta + \hat{y}_\nu - \langle k, \hat{y} \rangle & \text{if } i = \nu, \end{cases} \tag{13.1.6}$$

$i = 1, 2, \ldots, \ell$.

In Algorithm 13.1.6 (Benson's outer approximation algorithm for problem (LVP)), an initial simplex containing $\hat{Y}$ is constructed. The theoretical background of this construction is given in the following assertions (see Benson in [44, Theorems 2.2 and 2.3]).

**Theorem 13.1.4** *Let $w^\nu$ ($\nu \in \{0, 1, \ldots, \ell\}$) be given by (13.1.6). The convex hull $\mathcal{S}$ of*

$$V(\mathcal{S}) := \{w^\nu \mid \nu \in \{0, 1, \ldots, \ell\}\} \tag{13.1.7}$$

*is a $\ell$-dimensional simplex with the vertex set $V(\mathcal{S})$, and $\hat{Y} \subseteq \mathcal{S}$.*

In Benson's Outer Approximation Algorithm, the following representation of the simplex $\mathcal{S}$ is used in the initialization step.

**Theorem 13.1.5** *The inequality representation of the $\ell$-dimensional simplex $\mathcal{S}$ defined in Theorem 13.1.4 using $k \in \operatorname{int} C$ is given by*

$$\mathcal{S} = \{y \in \mathbb{R}^\ell \mid \hat{y} \geq y, \quad \langle k, y \rangle \geq \beta\}. \tag{13.1.8}$$

In the following algorithm, we choose a point $\overline{p} \in \operatorname{int} \hat{Y}$. Benson's outer approximation algorithm starts with the $\ell$-dimensional simplex $\mathcal{S}$ introduced in Theorem 13.1.4 and iteratively generates a finite number of nonempty, compact polyhedral sets $\mathcal{S}^j$, $j = 0, 1, 2, \ldots, L$, with $\mathcal{S} = \mathcal{S}^0 \supseteq \mathcal{S}^1 \supseteq \ldots \supseteq \mathcal{S}^{L-1} \supseteq \mathcal{S}^L = \hat{Y}$. In *iteration j* of Benson's outer approximation algorithm, a vertex $y^j$ of $\mathcal{S}^j$ is chosen such that $y^j \notin \hat{Y}$. Then, the unique point $\overline{y}^j \in \operatorname{bd} \hat{Y}$ that lies on the line segment connecting $\overline{p}$ and $y^j$ is computed. Theorem 13.1.1 shows that $\overline{y}^j \in \operatorname{WEff}(\hat{Y}, C)$. This assertion is important for the proof of the finiteness of Benson's outer approximation algorithm (see [44, Theorem 3.1]).

---

**Algorithm 13.1.6** *(Benson's Outer approximation algorithm [44])*

---

**Input:** $P$ *($\mathbb{R}^{\ell,n}$-matrix), $A$ ($\mathbb{R}^{m,n}$-matrix) and $b \in \mathbb{R}^m$.*

**Initialization**
*Construct the vertex set $V(\mathscr{S}^0)$ given by (13.1.7) and the inequality representation of $\mathscr{S}^0$ given by (13.1.8).*
*Store the vertex set $V(\mathscr{S}^0)$.*
*Store the inequality representation of $\mathscr{S}^0$.*
*Compute a point $\overline{p} \in \operatorname{int} \hat{Y}$.*
*Set $j = 0$ and go to iteration loop.*

**Iteration loop**
*Choose $j$, $j > 0$.*
***if** for each $y \in V(\mathscr{S}^j)$, $y \in \hat{Y}$ is satisfied, **then***
**STOP:** $\hat{Y} = \mathscr{S}^j$.
***else** choose any $y^j \in V(\mathscr{S}^j)$ such that $y^j \notin \hat{Y}$.*
*Find the unique value $\lambda^j$ of $0 < \lambda < 1$, such that $\lambda y^j + (1 - \lambda)\overline{p} \in \operatorname{bd} \hat{Y}$, and set $\overline{y}^j := \lambda^j y^j + (1 - \lambda^j)\overline{p}$.*
*Set $\mathscr{S}^{j+1} := \mathscr{S}^j \cap \{y \in \mathbb{R}^\ell \mid \langle w^j, y\rangle \geq \langle b, u^j\rangle\}$, where $(u^{jT}, w^{jT})$ is any optimal solution to the dual linear problem (DLVP ($\varphi_{\overline{y}-C,k}$)) with $\overline{y} = \overline{y}^j$ (see Theorem 13.1.2).*
*Using $V(\mathscr{S}^j)$ and the definition of $\mathscr{S}^{j+1}$, determine $V(\mathscr{S}^{j+1})$. Set $j := j + 1$.*
***end if***

---

The finiteness of the Outer Approximation Algorithm (Algorithm 13.1.6) is shown by Benson in [44, Theorem 3.1]. Furthermore, the Outer Approximation Algorithm terminates with generating all extreme elements of $\mathrm{Eff}(P(S), C)$ (see Benson in [44, Theorem 3.2]).

A free solver (BENSOLVE) developed by Löhne and Weißing [336], is available. Further information can be found at

www.bensolve.org.

A discussion of Benson's algorithm and recent developments are given by Löhne in [334, 368], see also references therein.

**Remark 13.1.7** Recently, Niebling and Eichfelder [368] developed a new branch-and-bound-based algorithm for smooth nonconvex multiobjective optimization problems with convex constraints. In this algorithm, techniques from the $\alpha$BB method well known in scalar global optimization and outer approximation techniques from multiobjective convex optimization involving the nonlinear scalarizing functional $\varphi_{A,k}$ given by (4.1.11) are combined.

## 13.2 Proximal-Point Algorithms for Optimization Problems

The (approximate) solution of an optimization problem is closely related to the problem of finding zero elements of maximal monotone operators (cf. Rockafellar [388]). One of the most important theoretical schemes to find zero elements of maximal monotone operators is the proximal-point algorithm (Rockafellar [388]). Its relevance is (mostly) a theoretical one and the success of its implementation depends on the effectiveness of the methods used to solve the involved subproblems. However, for special structures, and especially for approximation problems where it is possible to compute the (classical Fenchel) subdifferential, the proximal-point algorithm works effectively (see [40, 192, 256, 353, 437]). In the literature, there are many papers with possible extensions and generalizations of the proximal-point algorithm (see [83, 262]). Recently, some attention was focused on proximal-point algorithms for vector optimization problems (see Bonnel, Iusem, Svaiter [56], Ceng, Yao [85] and Graña Drummond, Svaiter [196]). In Durea, Strugariu [126], a proximal-point algorithm for vector optimization problems is derived from the classical one via scalarization by means of functionals of the type $\varphi_{A,k}$ given by (4.1.11).

### *13.2.1 Rockafellar's Algorithm for Convex Problems*

Assume that $(X, \langle\cdot,\cdot\rangle)$ is a real Hilbert space. We identify the topological dual of $X$ with the space $X$ itself. The classical proximal-point algorithm (PPA) for generating approximate solutions to the optimization problem

$$\min_{x\in X} f(x), \tag{P}$$

where $f : X \to \mathbb{R}\cup\{+\infty\}$ is a convex l.s.c. function, is based on the Moreau–Yosida regularization of the initial function and generates a sequence $\{x_n\} \subset X$ given by

$$x_{n+1} = \operatorname{argmin}_{x\in X}\left(f(x) + \frac{\lambda_n}{2}\|x - x_n\|^2\right), \tag{13.2.1}$$

where $\{\lambda_n\}$ is a bounded sequence of positive real numbers and $x_0$ is arbitrarily chosen.

Rockafellar [388] has shown that $\{x_n\}$ is bounded and weakly converges towards the minimum of $f$ if $f$ admits a minimum.

### *13.2.2 A Proximal-Point Algorithm for Vector Optimization Problems*

In this section, we present a proximal-point algorithm for vector optimization problems derived by Durea and Strugariu in [126] using a scalarization technique by means of the functional $\varphi_{A,k}$ (see (4.1.11)).

Suppose that $X$ is a Hilbert space, $S$ is a nonempty, closed, convex subset of $X$, $Y$ is a Banach space, $k \in Y \setminus \{0\}$ and $B \subset Y$ is a proper, closed, convex set with $0 \in \operatorname{bd} B$, $B + \mathbb{R}_{>}k \subseteq \operatorname{int} B$, $Y = \mathbb{R}k - \operatorname{bd} B$ (see the assumptions in Theorem 4.6.6) and $B + B \subseteq B$. It is important to mention that these assumptions concerning $B$ do not imply that $B$ is a cone or a shifted cone. We consider general vector optimization problems where the objective function $f$ takes values in the space $Y$, i.e., $f : X \to Y$.

Under these assumptions, we study the problem

$$B - \operatorname*{Opt}_{x \in S} f(x) \tag{13.2.2}$$

for finding weakly efficient solutions of $f(S)$ w.r.t. $B$ (see Chap. 6, Definition 6.2.2).

In order to apply the proximal-point algorithm known from scalar optimization (see Sect. 13.2.1), we use the scalarizing functional $\varphi_{-B,k}$ given by (4.1.11) with $A = -B$.

In the next corollary, we summarize the properties of the scalarizing functional, which are important for deriving the algorithm.

**Corollary 13.2.1** *We consider the vector optimization problem (13.2.2) under the assumptions mentioned above and the functional $\varphi_{-B,k} : Y \to \overline{\mathbb{R}}$ given by*

$$\varphi_{-B,k}(y) = \inf\{t \in \mathbb{R} \mid y \in tk - B\}. \tag{13.2.3}$$

*The functional $\varphi_{-B,k}$ is finite-valued, convex, strictly* (int $B$)*-monotone, $B$-monotone and for every $t \in \mathbb{R}$ it holds that*

$$\{y \in Y \mid \varphi_{-B,k}(y) \leq t\} = tk - B, \qquad \{y \in Y \mid \varphi_{-B,k} < t\} = tk - \operatorname{int} B.$$

***Proof*** From Theorem 4.6.6, we get that the functional $\varphi_{-B,k}$ is finite-valued, the convexity of $\varphi_{-B,k}$ follows from Theorem 4.4.4. Furthermore, Theorem 4.5.5 yields that $\varphi_{-B,k}$ is $B$-monotone and from Corollary 4.5.10, we obtain that $\varphi_{-B,k}$ is strictly (int $B$)-monotone. The properties $\{y \in Y \mid \varphi_{-B,k}(y) \leq t\} = tk - B$ and $\{y \in Y \mid \varphi_{-B,k} < t\} = tk - \operatorname{int} B$ follow from Theorem 4.4.4. □

The next lemma (see [126]) shows the relationship between minimal solutions of the composite function $\varphi_{-B,k} \circ f$ over the feasible set $S$ and weakly efficient solutions to problem (13.2.2) in the sense of Definition 6.2.2.

**Lemma 13.2.2** *If $\overline{x} \in S$ is a minimal point of $\min_{x \in S} \varphi_{-B,k} \circ f(x)$, then it is a weakly efficient solution to problem (13.2.2), i.e., $\overline{x} \in \mathrm{WEff}_f(S, B)$.*

***Proof*** Since $\overline{x}$ is a minimal point of $\min_{x \in S} \varphi_{-B,k} \circ f(x)$, we get that $\overline{x}$ is a weakly efficient solution of $f$ over $S$ with respect to $B$ (see Theorem 6.2.31). □

Furthermore, for deriving a proximal-point algorithm, we assume

(A1) $f : X \to Y$ is $B$-convex (see Sect. 8.2), i.e., for every $x^1, x^2 \in X$ with $f(x^1), f(x^2) \in Y$ and every $\lambda \in [0, 1]$,

$$f(\lambda x^1 + (1-\lambda)x^2) \in \lambda f(x^1) + (1-\lambda) f(x^2) - B.$$

(A2) The set $\{x \in X \mid f(x) \in tk - B\}$ is closed for every $t \in \mathbb{R}$.

**Remark 13.2.3** The assumption (A2) corresponds to the lower semicontinuity supposed in the variational principles shown in Sect. 12.3.2.

Using the indicator function $\iota_S : X \to \mathbb{R} \cup \{+\infty\}$ of $S$ (see Definition 3.2.1), we formulate the constrained scalar optimization problem $\min_{x \in S} \varphi_{-B,k} \circ f(x)$ as an unconstrained scalar optimization problem $\min_{x \in X} (\varphi_{-B,k} \circ f(x) + \iota_S(x))$. Now, we consider the proximal-point algorithm for $(\varphi_{-B,k} \circ f + \iota_S) : X \to \mathbb{R} \cup \{+\infty\}$.

Take a bounded sequence of positive real numbers $\{\lambda_n\}$. Let $\{\varepsilon_n\}$ be a sequence of positive real numbers such that $\sum_{n=0}^{+\infty} \varepsilon_n < +\infty$. Consider a starting point $x_0 \in X$ arbitrarily chosen. At every step, take $\tilde{x}_{n+1}$ defined by

$$\tilde{x}_{n+1} := \operatorname{argmin}_{x \in X} ((\varphi_{-B,k} \circ f)(x) + \iota_S(x) + \frac{\lambda_n}{2} \|x_n - x\|^2) \tag{13.2.4}$$

$$= \operatorname{argmin}_{x \in X} (\varphi_{-B,k}(f(x) + \frac{\lambda_n}{2} \|x_n - x\|^2 k) + \iota_S(x)).$$

The last equality holds because of the translation invariance along $k$ of the functional $\varphi_{-B,k}$ (compare (4.2.2)). Subsequently, consider $x_{n+1} \in X$ such that $\|x_{n+1} - \tilde{x}_{n+1}\| < \varepsilon_n$.

In the next theorem (see [126]), we will see that it is possible to show a convergence result for the sequence generated by the proximal iteration (13.2.4) using the convexity and monotonicity properties of the functional $\varphi_{-B,k}$.

**Theorem 13.2.4** *Assume that* (A1) *and* (A2) *are fulfilled. Suppose that $\varphi_{-B,k} \circ f + \iota_S$ admits a minimum point on $X$. Then, the sequence $\{x_n\}$ generated by (13.2.4) with $\|x_{n+1} - \tilde{x}_{n+1}\| < \varepsilon_n$ is well defined. Furthermore, the sequence $\{x_n\}$ is bounded and $\{x_n\}$ is weakly convergent to a weakly efficient solution $\overline{x}$ to problem (13.2.2), i.e., $\overline{x} \in \mathrm{WEff}_f(S, B)$.*

***Proof*** We show that for the scalarization $\varphi_{-B,k} \circ f$ of the vector-valued objective function $f$ the assumptions of the PPA (see Sect. 13.2.1) are fulfilled:

(a) Convexity: Take $x^1, x^2 \in X$ and $\lambda \in (0, 1)$. By the definition of $\varphi_{-B,k}$, for every $\varepsilon > 0$ one has

$$f(x^1) \in ((\varphi_{-B,k} \circ f(x^1)) + \varepsilon)k - B$$

and

$$f(x^2) \in ((\varphi_{-B,k} \circ f(x^2)) + \varepsilon)k - B.$$

This yields

$$\lambda f(x^1) + (1-\lambda) f(x^2) \in \big(\lambda(\varphi_{-B,k} \circ f)(x^1) + (1-\lambda)(\varphi_{-B,k} \circ f)(x^2) + \varepsilon\big)k - B.$$

This yields together with the $B$-convexity of $f$ supposed by (A1) and the assumptions concerning $B$

$$f(\lambda x^1 + (1-\lambda)x^2) \in \lambda f(x^1) + (1-\lambda) f(x^2) - B$$

$$\subseteq \big(\lambda(\varphi_{-B,k} \circ f)(x^1) + (1-\lambda)(\varphi_{-B,k} \circ f)(x^2) + \varepsilon\big)k - B.$$

By the definition of the functional $\varphi_{-B,k}$, we get

$$\varphi_{-B,k}(f(\lambda x^1 + (1-\lambda)x^2)) \leq \lambda\varphi_{-B,k}(f(x^1)) + (1-\lambda)\varphi_{-B,k}(f(x^2)) + \varepsilon.$$

For $\varepsilon \to 0$ we obtain the convexity of $\varphi_{-B,k} \circ f$. The convexity of $\varphi_{-B,k} \circ f + \iota_S : X \to \mathbb{R} \cup \{+\infty\}$ follows immediately.

(b) Lower semicontinuity: By (4.4.5), we get for any $t \in \mathbb{R}$

$$\{x \in X \mid (\varphi_{-B,k} \circ f)(x) \leq t\} = \{x \in X \mid f(x) \in tk - B\}.$$

By (A2), it follows that the sublevel sets of $\varphi_{-B,k} \circ f$ are closed and so $\varphi_{-B,k} \circ f : X \to \mathbb{R}$ is lower semicontinuous. Furthermore, $\varphi_{-B,k} \circ f + \iota_S : X \to \mathbb{R} \cup \{+\infty\}$ is lower semicontinuous.

So, the assumptions of Rockafellar's PPA are fulfilled. Taking into account Rockafellar's results in [388], the sequence $\{x_n\}$ generated by (13.2.4) with $\|x_{n+1} - \tilde{x}_{n+1}\| < \varepsilon_n$ is bounded. Furthermore, $\{x_n\}$ is weakly convergent to one of the minima of $\varphi_{-B,k} \circ f + \iota_S$, say $\overline{x}$. From Theorem 6.2.31, we get that $\overline{x}$ is a weakly efficient solution of $f$ over $S$ with respect to $B$, hence the conclusion holds. □

## 13.3 Proximal-Point Algorithms for Control Approximation Problems

In this section, we derive a proximal-point algorithm (analogously to the approach in [192, Sect. 4.2]) based on a scalarization by means of the functional $\varphi_{A,k}$ defined in (4.1.11) for solving vector-valued approximation problems (see problem ($P_{app}$) in Sect. 12.1.2).

For deriving the algorithm, we use monotonicity, convexity and continuity properties of the functional $\varphi_{A,k}$.

In order to formulate the problem we suppose that

(H1) $x \in X, a^i \in Y, \beta_i \geq 1, \; A^i \in L(X, Y)$ $(i \in \{1, \dots, \ell\})$; ($L(X, Y)$ denotes the space of linear continuous operators from X to Y); $\| \cdot \|_i$ are norms in $Y$ $(i \in \{1, \dots, \ell\})$.

(H2) $S_j \subset X$ $(j \in \{1, \dots, m\})$ are closed and convex sets, $S_0 \subseteq X$ is a closed linear subspace, and $S = \bigcap_{j=0}^{m} S_j$ is nonempty and bounded. Furthermore, we assume that a suitable constraint qualification is satisfied.

(H3) One of the following conditions (H3-1) or (H3-2) is satisfied:

(H3-1) $(H, \langle \cdot, \cdot \rangle)$ is a Hilbert space and $X = Y = H$;
(H3-2) $X = \mathbb{R}^s$ and $Y = \mathbb{R}^s$.

(H4) $C \subset \mathbb{R}^\ell$ is a nontrivial, closed, convex, pointed cone with $C + (\mathbb{R}^\ell_+ \setminus \{0\}) \subset \operatorname{int} C$ and $k \in \operatorname{int} C$.

Under the assumptions (H1), (H2), (H3) and (H4), we consider the following **vector control approximation problem** (see ($P_{app}$) in Sect. 12.1.2):

$$C - \operatorname*{Opt}_{x \in S} f(x), \tag{$P_{app}$}$$

where

$$f(x) := \begin{pmatrix} \|A^1(x) - a^1\|_1^{\beta_1} \\ \cdots \\ \|A^\ell(x) - a^\ell\|_\ell^{\beta_\ell} \end{pmatrix}. \tag{13.3.1}$$

Many authors have studied generalizations of the well-known Fermat-Weber problem (especially, vector-valued approximation problems) from the theoretical and computational point of view (see Michelot and Lefebvre [353], Idrissi, Lefebvre, and Michelot [257], Tammer [434, 435]). For recent developments, see Günther [203] and references therein.

**Remark 13.3.1** We consider a scalarization of problem ($P_{app}$), which is an extension of a problem studied by Idrissi, Lefebvre, and Michelot [256, 257]. This class of problems contains many practically important special cases such as approximation, location, and optimal control problems, perturbed linear programming problems, and surrogate problems for large linear programming problems with empty feasible sets, especially:

- **multiobjective location problems** if $A^i = I$ $(i \in \{1, \dots, \ell\})$,
- **multiobjective optimal control problems (optimal regulator problems)** of the form $(\ell = 2, \beta_i = p)$:

$$C - \mathrm{Opt}_{u \in U}\, f(u),$$

where

$$f(u) := \begin{pmatrix} \|A(u) - r\|^p \\ \|u\|^p \end{pmatrix} \tag{13.3.2}$$

and $U \subset H$.

We introduce a suitable scalarization of the vector control approximation ($\mathrm{P}_{app}$) by means of the functional $\varphi_{A,k}$ defined in (4.1.11), derive a proximal-point algorithm for the scalarized problem and we present an interactive algorithm for the vector control approximation problem using a surrogate parametric optimization problem, taking into account stability results of this special parametric optimization problem.

Under the given assumptions it is easy to see that the vector-valued objective function $f : X \longrightarrow \mathbb{R}^\ell$ in (13.3.1) is $(\mathbb{R}^\ell_+)$-convex; i.e., for all $x_1,\ x_2 \in X$, $\mu \in [0, 1]$, we have $\mu f(x_1) + (1 - \mu) f(x_2) \in f(\mu x_1 + (1 - \mu)x_2) + \mathbb{R}^\ell_+$ (compare assumption (A1) in Sect. 13.2.1 with $B = \mathbb{R}^\ell_+$).

Under the assumptions (H1), (H2), (H3) and (H4) and taking into account Theorem 6.2.31, we know that $f(x^0) \in \mathrm{WEff}(f(S), C)$ if and only if $x^0$ solves the real-valued optimization problem

$$\varphi_{-C,k}(f(x) - f(x^0)) := \varphi_{-C,k}\left( \begin{pmatrix} \|A^1(x) - a^1\|_1^{\beta_1} \\ \cdots \\ \|A^\ell(x) - a^\ell\|_\ell^{\beta_\ell} \end{pmatrix} - f(x^0) \right) \longrightarrow \min_{x \in S}, \tag{$\mathrm{P}'(\varphi_{-\mathrm{C,k}})$}$$

where $\varphi_{-C,k}$ is given by (4.1.11) with $A = -C$ and $k \in \operatorname{int} C$.

Using the *indicator functions* $\iota_{S_j}$ of $S_j$ defined by $\iota_{S_j}(x) = 0$, if $x \in S_j$ and $\iota_{S_j}(x) = +\infty$ otherwise, the problem ($\mathrm{P}'(\varphi_{-C,k})$) is equivalent to the following unconstrained minimization problem:

$$F(x) := \varphi_{-C,k}\left( \begin{pmatrix} \|A^1(x) - a^1\|_1^{\beta_1} \\ \cdots \\ \|A^\ell(x) - a^\ell\|_\ell^{\beta_\ell} \end{pmatrix} - f(x^0) \right) + \sum_{j=0}^{m} \iota_{S_j}(x) \rightarrow \min_{x \in X}. \tag{$\mathrm{P}(\varphi_{-\mathrm{C,k}})$}$$

$B_{\|\cdot\|_{i*}}$ denotes the closed unit ball associated with the dual norm of the norm $\|\cdot\|_i$ $(i \in \{1, \dots, \ell\})$. As usual, $P_{\mathscr{S}}(e)$ denotes the projection of $e$ onto a set $\mathscr{S}$. Furthermore, $N(x^0; S_j)$ denotes the normal cone in the sense of Convex Analysis to $S_j$ at $x^0 \in X$ (see (3.6.2)) defined by

$$N(x^0; S_j) = \begin{cases} \{x^* \in X^* \mid \quad \forall x' \in S_j : \langle x^*, x' - x^0 \rangle \le 0\} : & \text{if } x^0 \in S_j \\ \emptyset & : \text{otherwise.} \end{cases}$$

Moreover, $S_0^{\perp} := \{x^* \in X^* \mid \forall x \in S_0 : \langle x^*, x\rangle = 0\}$ denotes the orthogonal complement to $S_0$.

We derive optimality conditions for problem ($P_{app}$) using the sub-differential calculus for convex functions (see (3.6.4) and Sect. 8.3). These optimality conditions are useful for an application of Spingarn's proximal-point algorithm (see [40, 256, 353, 421, 437]) for solving problem ($P_{app}$). We present two proximal-point algorithms with different subspaces in the formulation of the algorithm for solving the scalar problem ($P(\varphi_{-C,k})$).

The following approach is inspired by papers of Michelot and Lefebvre [353] and Idrissi, Lefebvre, and Michelot [256]. We consider the *unconstrained minimization problem* ($P(\varphi_{-C,k})$):

$$F(x) = \varphi_{-C,k}(f(x) - f(x^0)) + \sum_{j=0}^{m} \iota_{S_j}(x) \to \min_{x\in X}. \tag{13.3.3}$$

Obviously, the functional $F(\cdot)$ is convex under the given assumptions. In order to derive a primal-dual algorithm based on subdifferential calculus, we assume in the following that any suitable constraint qualification (generalized Slater condition, stability, etc.) is satisfied.

Hence,

$$0 \in \partial(\varphi_{-C,k}(f(\cdot) - f(x^0))(x^0) + N(x^0; S_j) \tag{13.3.4}$$

is necessary and sufficient for the optimality of $x^0$ in (13.3.3). By estimating the subdifferential of the functional $\varphi_{-C,k}(f - f(x^0))$ in (13.3.3), the condition (13.3.4) is equivalent to the following optimality conditions (13.3.5), (13.3.6), (13.3.7):

$$q_i \in \partial(\,\|A^i(x^0) - a^i\|_i^{\beta_i}), \qquad i \in \{1, \ldots, \ell\}, \tag{13.3.5}$$

$$r_j \in N(x^0; S_j), \qquad j \in \{1, \ldots, m\}, \tag{13.3.6}$$

$$\sum_{i=1}^{\ell} y_i^* q_i + \sum_{j=1}^{m} r_j \in S_0^{\perp}, \tag{13.3.7}$$

where

$$y^* \in \partial\varphi_{-C,k}(0) \qquad (y^* = (y_1^*, \ldots, y_\ell^*)^T). \tag{13.3.8}$$

From Theorem 4.10.3, we know $y^* \in \partial\varphi_{-C,k}(0) = \{y^* \in C^+ \mid y^*(k) = 1\}$.

### 13.3.1 Algorithm I for Scalar Problems

For deriving the first algorithm (analogously to the approach in [192, Sect. 4.2]) for solving (P($\varphi_{-C,k}$)) under the assumptions (H1), (H2), (H3), and (H4), we assume additionally that the inverse operator $(A^{iT})^{-1}$ of the adjoint operator $A^{iT}$ to $A^i$ exists for all $i \in \{1, \dots, \ell\}$.

In order to reformulate the optimality conditions (13.3.5)–(13.3.7) in a more practical way, we introduce a Hilbert space $E$, suitable subspaces $\mathscr{A}$, $\mathscr{B}$, and a set-valued map $T$. This will be done in different ways. For the first algorithm, we will use the following notation:

We consider the Hilbert space

$$E := (X)^{\ell+m}$$

with the inner product

$$\langle u \, , \, w \rangle := \sum_{i=1}^{\ell+m} \langle u_i \, , \, w_i \rangle$$

for $u = (u_1, \dots, u_{\ell+m})$ and $w = (w_1, \dots, w_{\ell+m}) \in E$,
and for $y^* \in \{y^* \in C^+ \mid y^*(k) = 1\}$ ($y^* = (y_1^*, \dots, y_\ell^*)$) the subspaces

$$\mathscr{A}(y^*) := \left\{ y \in E \mid y = (y_1^* y_1, \dots, y_\ell^* y_\ell, y_{\ell+1}, \dots, y_{\ell+m});\ y_j = x,\ j \in \{1, \dots, \ell+m\};\ x \in S_0 \right\},$$

$$\mathscr{B}(y^*) := \left\{ p \in E \mid p = (p_1, \dots, p_{\ell+m}),\ \sum_{i=1}^{\ell} y_i^* p_i + \sum_{j=1}^{m} p_{\ell+j} \in S_0^{\perp} \right\},$$

and the set-valued map $T$ defined on $E$ by

$$T(y) := \prod_{i=1}^{\ell+m} T^i(x) = (T^1(x), \dots, T^{\ell+m}(x)) \tag{13.3.9}$$

where

$$\begin{aligned} T^i(x) &:= \partial(\| A^i(x) - a^i \|_i^{\beta_i}), \qquad i \in \{1, \dots, \ell\}, \\ T^{\ell+j}(x) &:= N(x; S_j), \qquad j \in \{1, \dots, m\}. \end{aligned}$$

using these notations, the optimality conditions (13.3.5)–(13.3.7) can be reformulated as

$$\text{Find} \quad (y^0, p^0) \in \mathscr{A}(y^*) \times \mathscr{B}(y^*) \quad \text{such that} \quad p^0 \in T(y^0). \tag{13.3.10}$$

The proof of the following lemma concerning the properties of the subspaces $\mathscr{A}(y^*)$ and $\mathscr{B}(y^*)$ is analogously to the proof of a corresponding result in [192, Sect. 4.2] for different subspaces.

**Lemma 13.3.2** *Suppose that $y^* \in \{y^* \in C^+ \mid y^*(k) = 1\}$ is fixed. $\mathscr{A}(y^*)$ and $\mathscr{B}(y^*)$ are closed linear subspaces of $E$ with $\mathscr{A}(y^*) \perp \mathscr{B}(y^*)$. Additionally, the subspaces $\mathscr{A}(y^*)$ and $\mathscr{B}(y^*)$ are complementary; i.e., $E = \mathscr{A}(y^*) \oplus \mathscr{B}(y^*)$.*

***Proof***

(a) For all $y \in \mathscr{A}(y^*)$ and $p \in \mathscr{B}(y^*)$ it holds that

$$\langle y, p\rangle = \sum_{i=1}^{\ell} \langle y_i^* x, p_i\rangle + \sum_{j=1}^{m} \langle x, p_{\ell+j}\rangle = \sum_{i=1}^{\ell} \langle x, y_i^* p_i\rangle + \sum_{j=1}^{m} \langle x, p_{\ell+j}\rangle = \langle x, \sum_{i=1}^{\ell} y_i^* p_i + \sum_{j=1}^{m} p_{\ell+j}\rangle = 0,$$

because $x \in S_0$ and $\sum_{i=1}^{\ell} y_i^* p_i + \sum_{j=1}^{m} p_{\ell+j} \in S_0^{\perp}$.

(b) It holds that $\mathscr{A}(y^*) \cap \mathscr{B}(y^*) = \{0\}$. Indeed, since $0 \in S_0$, we get $y = 0 = (0, \dots, 0) \in \mathscr{A}(y^*)$. Furthermore for $p = (0, \dots, 0)$, it holds that $\sum_{i=1}^{\ell} y_i^* p_i + \sum_{j=1}^{m} p_{\ell+j} = 0$ and so $p = (0, \dots, 0) \in S_0^{\perp}$, i.e., $p = (0, \dots, 0) \in \mathscr{B}(y^*)$.

(c) We show $\mathscr{B}(y^*) = \mathscr{A}(y^*)^{\perp}$. It holds that $\mathscr{B}(y^*) \subseteq \mathscr{A}(y^*)^{\perp}$. Indeed, let $p \in \mathscr{B}(y^*)$. Then, $\sum_{i=1}^{\ell} y_i^* p_i + \sum_{j=1}^{m} p_{\ell+j} \in S_0^{\perp}$ and so for all $x \in S_0$ it holds that $\langle x, \sum_{i=1}^{\ell} y_i^* p_i + \sum_{j=1}^{m} p_{\ell+j}\rangle = 0$. This yields $\sum_{i=1}^{\ell} \langle x, y_i^* p_i\rangle + \sum_{j=1}^{m} \langle x, p_{\ell+j}\rangle = 0$ for all $x \in S_0$ such that $\sum_{i=1}^{\ell} \langle y_i, y_i^* p_i\rangle + \sum_{j=1}^{m} \langle y_i, p_{\ell+j}\rangle = 0$ for all $y \in \mathscr{A}(y^*)$. Taking into account the definition of the inner product in $E$, we get $\langle y, p\rangle = 0$ for all $y \in \mathscr{A}(y^*)$, i.e., $p \in \mathscr{A}(y^*)^{\perp}$.
Moreover, we show $\mathscr{A}(y^*)^{\perp} \subseteq \mathscr{B}(y^*)$. Let $p \in \mathscr{A}(y^*)^{\perp}$, then $\langle p, y\rangle = 0$ for all $y \in \mathscr{A}(y^*)$. So we get, with the definition of the inner product in $E$ for all $y \in \mathscr{A}(y^*)$, and for all $x \in S_0$,

$$\langle y, p\rangle = \sum_{i=1}^{\ell} \langle y_i^* y_i, p_i\rangle + \sum_{j=1}^{m} \langle y_i, p_{\ell+j}\rangle = \sum_{i=1}^{\ell} \langle y_i^* x, p_i\rangle + \sum_{j=1}^{m} \langle x, p_{\ell+j}\rangle =$$
$$= \sum_{i=1}^{\ell} \langle x, y_i^* p_i\rangle + \sum_{j=1}^{m} \langle x, p_{\ell+j}\rangle = \langle x, \sum_{i=1}^{\ell} y_i^* p_i + \sum_{j=1}^{m} p_{\ell+j}\rangle = 0.$$

This means $\sum_{i=1}^{\ell} y_i^* p_i + \sum_{j=1}^{m} p_{\ell+j} \in S_0^{\perp}$ and finally $p \in \mathscr{B}(y^*)$.

So we have shown that the subspaces $\mathscr{A}(y^*)$ and $\mathscr{B}(y^*)$ are complementary, i.e., $E = \mathscr{A}(y^*) \oplus \mathscr{B}(y^*)$. □

**Remark 13.3.3** Of course, the set-valued map $T$ introduced in (13.3.9) is a maximal monotone set-valued map, since $T$ is composed of subdifferentials in the sense of Convex Analysis. For a numerical procedure, we need the condition (13.3.10) in the form $0 \in T(y^0) - p^0$, but the new multi-function $\widetilde{T}(y, p) := T(y) - p$ is not maximal monotone. Hence, we apply an idea by Spingarn [421] in the following.

Using the results by Spingarn [421], the problem (13.3.10) is equivalent to: Find $(y^0, p^0) \in \mathscr{A}(y^*) \times \mathscr{B}(y^*)$ such that

$$0 \in T_{\mathscr{A}(y^*)}(y^0 + p^0), \tag{13.3.11}$$

where $T_{\mathscr{A}(y^*)}$ is the *partial inverse* of $T$ with respect to $\mathscr{A}(y^*)$. The set-valued map $T_{\mathscr{A}(y^*)}$ is characterized by its graph:

$$\big\{(y_{\mathscr{A}(y^*)} + p_{\mathscr{B}(y^*)}, p_{\mathscr{A}(y^*)} + y_{\mathscr{B}(y^*)}) \mid p \in T(y),\ p = p_{\mathscr{A}(y^*)} + p_{\mathscr{B}(y^*)},\ y = y_{\mathscr{A}(y^*)} + y_{\mathscr{B}(y^*)}\big\},$$

where $v_{\mathscr{A}(y^*)}$ and $v_{\mathscr{B}(y^*)}$ are the projections of $v$ onto $\mathscr{A}(y^*)$ and $\mathscr{B}(y^*)$, respectively. Note that $T_{\mathscr{A}(y^*)}$ is maximal monotone if and only if $T$ is maximal monotone. Putting

$$z = z_{\mathscr{A}(y^*)} + z_{\mathscr{B}(y^*)}$$

and

$$z_{\mathscr{A}(y^*)} = y, \qquad z_{\mathscr{B}(y^*)} = p,$$

problem (13.3.11) has the following form:

$$\text{Find} \quad z^0 \in E \quad \text{such that} \quad 0 \in T_{\mathscr{A}(y^*)}(z^0). \tag{13.3.12}$$

Now, we apply the proximal-point algorithm (compare Michelot, Lefebvre [353], Idrissi, Lefebvre, Michelot [256], [192, Sect. 4.2]) in order to solve (13.3.12). This algorithm is given by

$$z^{n+1} = (I + c_n T_{\mathscr{A}(y^*)})^{-1}(z^n) \qquad n = 1, 2, \ldots, \tag{13.3.13}$$

with an arbitrary starting point $z^1$, and a sequence of positive real numbers $c_n$. Provided that $\{c_n\}$ is bounded away from zero (we set $c_n = 1$), then either $\{z^n\}$ converges weakly to a zero of $T_{\mathscr{A}(y^*)}$, or $\|z^n\| \to +\infty$ and $T_{\mathscr{A}(y^*)}$ has no zeros (see [388]).

Analogously to [192, Sect. 4.2]), we derive the following algorithm for solving $(P(\varphi_{-C,k}))$, $k \in \operatorname{int} C$, in the special case $\beta_i = 1$ $(i \in \{1, \ldots, \ell\})$.

---

**Algorithm 13.3.4** *(Proximal-Point Algorithm I for solving $(P(\varphi_{-C,k}))$)*

---

**Input:** *$A^i \in L(X, Y)$, $\|\cdot\|_i$, $\beta_i = 1$, $(i \in \{1, \ldots, \ell\})$, $S_j \subset X$ $(j \in \{0, \ldots, m\})$, ordering cone $C \subset \mathbb{R}^\ell$, $k \in \operatorname{int} C$, choose $\overline{y}^* \in \{y^* \in C^+ \mid y^*(k) = 1\}$, maximal number $n_{max}$ of iterations*

**Initialization**

$n := 1$

*Choose*

$$x^1 \in S_0 \quad \textit{and} \quad p^1 \in E \quad \textit{with} \ \sum_{i=1}^{\ell} \overline{y}_i^* p_i^1 + \sum_{j=1}^{m} p_{\ell+j}^1 \in S_0^{\perp}. \tag{13.3.14}$$

**Iteration loop**
***while*** *$n \le n_{max}$* ***do***
*Compute $x^{n+1}$ and $p^{n+1}$ from*

$$x^n - x^{n+1} + p_i^{n+1} = P_{M_i}(x^n + p_i^n - \left(A^i\right)^{-1} a^i), \ i \in \{1, \dots, \ell\}, \tag{13.3.15}$$

$$x^{n+1} + p_{\ell+j}^n - p_{\ell+j}^{n+1} = P_{S_j}(x^n + p_{\ell+j}^n), \qquad j \in \{1, \dots, m\}, \tag{13.3.16}$$

*with*

$$\sum_{i=1}^{\ell} \overline{y}_i^* p_i^{n+1} + \sum_{j=1}^{m} p_{\ell+j}^{n+1} \in S_0^{\perp} \qquad \textit{and} \qquad x^{n+1} \in S_0. \tag{13.3.17}$$

$n := n + 1$
***end while***
**Output:** $x^n$*, an approximate solution of* $(P(\varphi_{-C,k}))$.

---

In Algorithm 13.3.4, we first calculate the projections on $M_i$ and $S_j$ in the right-hand side of the Eqs. (13.3.15) and (13.3.16). Then, we solve the linear equations (13.3.15)–(13.3.16) such that the relations (13.3.17) are fulfilled. The solution of the linear equations (13.3.15)–(13.3.16) is very easy:

Putting for $1 \le i \le \ell - 1$ and $1 \le j \le m$,

$$\begin{aligned}
b_i^n &= P_{M_i}\left(x^n + p_1^n - \left(A^i\right)^{-1} a^1\right) - P_{M_i}\left(x^n + p_{i+1}^n - \left(A^{i+1}\right)^{-1} a^{i+1}\right), \\
b_{\ell+j-1}^n &= P_{M_i}\left(x^n + p_1^n - \left(A^i\right)^{-1} a^1\right) + P_{S_j}\left(x^n + p_{\ell+j}^n\right) - x^n + p_{\ell+j}^n, \\
b_{\ell+m}^n &= P_{M_i}\left(x^n + p_1^n - \left(A^i\right)^{-1} a^1\right),
\end{aligned}$$

we immediately get the solutions

$$p_i^{n+1} = p_1^{n+1} - b_{i-1}^n, \qquad i \in \{2, \dots, \ell + m\}, \tag{13.3.18}$$

where $p_1^{n+1}$ is settled by (13.3.17):

$$(\ell + m)\, \overline{y}_i^* p_1^{n+1} - \sum_{i=1}^{\ell} \overline{y}_i^* b_i^n - \sum_{j=1}^{m} b_{\ell+j}^n \in S_0^{\perp} \quad \text{and} \quad (x^{n+1} =)\, x^n + p_1^{n+1} - b_{\ell+m}^n \in S_0. \tag{13.3.19}$$

**Remark 13.3.5**

(1) Only if the subspace $S_0$ of $X$ is concretely given, the determination of $p_1^{n+1}$ from the relations (13.3.19) is possible. Especially, provided that $S_0 = X = H$, we obtain from (13.3.19)

$$p_1^{n+1} = \frac{1}{\ell + m} \sum_{i=1}^{\ell+m} b_i^n \quad \text{and} \quad x^{n+1} = x^n + p_1^{n+1} - b_{\ell+m}^n. \tag{13.3.20}$$

(2) Furthermore, another possibility for the solution of the Eqs. (13.3.15)–(13.3.16) is explained in the following:
Using the settings

$$\widetilde{p}^n = y^n - y^{n+1} + p^{n+1},$$

$$\widetilde{y}^n = p^n - p^{n+1} + y^{n+1},$$

which implies

$$\widetilde{y}^n + \widetilde{p}^n = y^n + p^n, \qquad y^{n+1} = (\widetilde{y}^n)_{\mathscr{A}(y^*)}, \qquad p^{n+1} = (\widetilde{p}^n)_{\mathscr{B}(y^*)},$$

we get

$$\widetilde{p}_1^n, \ldots, \widetilde{p}_\ell^n$$

from (13.3.15),

$$\widetilde{y}_{\ell+1}^n, \ldots, \widetilde{y}_{\ell+m}^n$$

from (13.3.16), and the remaining components from

$$\widetilde{p}_{\ell+j}^n = x^n + p_{\ell+j}^n - \widetilde{y}_{\ell+j}^n \ (1 \le j \le m) \text{ and } \widetilde{y}_i^n = x^n + p_i^n - \widetilde{p}_i^n \ (1 \le i \le \ell).$$

Finally, we get the solution vectors

$$p^{n+1} = (p_1^{n+1}, \ldots, p_{\ell+m}^{n+1}) \text{ and } y^{n+1} = (x^{n+1}, \ldots, x^{n+1})$$

by calculating the projections $(\widetilde{p}^n)_{\mathscr{B}(y^*)}$ and $(\widetilde{y}^n)_{\mathscr{A}(y^*)}$ from $\widetilde{p}^n$ and $\widetilde{y}^n$ onto the subspaces $\mathscr{B}(y^*)$ and $\mathscr{A}(y^*)$, respectively.

**Remark 13.3.6** In the case $\beta_i > 1$ ($i \in \{1, \ldots, \ell\}$), we compute the iteration (13.3.15) using the structure of the subdifferential in (13.3.5).

## *13.3.2 Algorithm II for Scalar Problems*

In the previous section, we derived Algorithm 13.3.4, but this algorithm has some disadvantages. Especially, it is supposed that the maps $A^i$ is invertible, and some operator equation has to be solved if $A^i \not\equiv I$ ($I$ = identity). Analogously to [192, Sect. 4.2]), we define the subspaces $\mathscr{A}(y^*)$ and $\mathscr{B}(y^*)$ in a different way than in Sect. 13.3.1 and obtain another specification of Spingarn's algorithm under weaker assumptions. The different idea for deriving the proximal-point algorithm is to include the maps $A^i$ in the definition of the subspaces $\mathscr{A}(y^*)$ and $\mathscr{B}(y^*)$.

Now, we derive a proximal-point algorithm (Algorithm 13.3.7) for the scalar control approximation problem (P($\varphi_{-C,k}$)) under the assumptions (H1), (H2), (H3-2) and (H4) (compare [192, Sect. 4.2]).

In contrast to the assumptions for Algorithm 13.3.4, we now consider a control approximation problem (P($\varphi_{-C,k}$)) (without assuming that the operators $A^i$ are invertible) in finite-dimensional spaces.

We consider the scalarized problem (P($\varphi_{-C,k}$)), but now under the assumptions (H1), (H2), (H3-2) and (H4):

$$F(x) = \varphi_{-C,k}\left(\begin{pmatrix} \|A^1(x) - a^1\|_1^{\beta_1} \\ \cdots \\ \|A^\ell(x) - a^\ell\|_\ell^{\beta_\ell} \end{pmatrix} - f(x^0)\right) + \sum_{j=1}^{m} \iota_{S_j}(x) \to \min_{x \in X},$$

where $\varphi_{-C,k}$ is given by (4.1.11) with $A = -C$ and $k \in \operatorname{int} C$.

For

$$y^* \in \partial\varphi_{-C,k}(0) \qquad (y^* = (y_1^*, \ldots, y_\ell^*)^T), \tag{13.3.21}$$

we obtain $y^* \in \partial\varphi_{-C,k}(0) = \{y^* \in C^+ \mid y^*(k) = 1\}$ taking into account Theorem 4.10.3.

Now, the space $E$ is given by

$$E := \mathbb{R}^{s_1} \times \mathbb{R}^{s_2} \times \cdots \times \mathbb{R}^{s_{\ell+m}}, \tag{13.3.22}$$

where $s_1 = \cdots = s_{\ell+m} = s$. For $y^* \in \{y^* \in C^+ \mid y^*(k) = 1\}$, we define the operator $\mathscr{S}_{y^*} : E \to \mathbb{R}^s$ by

$$\mathscr{S}_{y^*}(e) := \sum_{i=1}^{\ell} y_i^* A^{iT} e_i + \sum_{j=1}^{m} e_{\ell+j}, \quad e = (e_1, \ldots, e_{\ell+m}) \in E .$$

Furthermore, the subspaces are given by

$$\mathscr{A}_2(y^*) := \{y \in E \mid y = \underbrace{(y_1^* A^1 x, y_2^* A^2 x, \ldots, y_\ell^* A^\ell x, x, \ldots, x)}_{\ell+m},\ x \in \mathbb{R}^s\}, \tag{13.3.23}$$

$$\mathscr{B}_2(y^*) := \{p \in E \mid \mathscr{S}_{y^*}(p) = 0\} . \tag{13.3.24}$$

The subspace $\mathscr{B}_2(y^*)$ is the orthogonal complement of $\mathscr{A}_2(y^*)$. Indeed, with $u \in E$ and $a \in \mathscr{A}_2(y^*)$, it follows

$$\langle a, u \rangle = \sum_{i=1}^{\ell} \langle y_i^* A^i x, u_i \rangle + \sum_{j=1}^{m} \langle x, u_{\ell+j} \rangle = \langle x, \mathscr{S}_{y^*}(u) \rangle . \tag{13.3.25}$$

For $u \in \mathscr{B}_2(y^*)$, the right-hand side in (13.3.25) is zero. Hence, (13.3.25) implies $\mathscr{B}_2(y^*) \subseteq \mathscr{A}_2(y^*)^\perp$. Provided that $u \in \mathscr{A}_2(y^*)^\perp$, then (13.3.25) has to be zero for all $a \in \mathscr{A}_2(y^*)$ and hence for all $x \in \mathbb{R}^s$, i.e., the right part in the inner product has to be zero, so we obtain $u \in \mathscr{B}_2(y^*)$ and $\mathscr{A}_2(y^*)^\perp \subseteq \mathscr{B}_2(y^*) \subseteq \mathscr{A}_2(y^*)^\perp$, $\mathscr{A}_2(y^*)^\perp = \mathscr{B}_2(y^*)$. Because of the closedness of $\mathscr{A}_2(y^*)$ and $\mathscr{B}_2(y^*)$ it follows that

$$E = \mathscr{A}_2(y^*) \oplus \mathscr{B}_2(y^*) . \tag{13.3.26}$$

Again, the set-valued map $T : E \rightrightarrows E$ is given by

$$\begin{aligned} \tilde{p} \in T(\tilde{y}) &\Leftrightarrow \tilde{p}_i \in T_i(\tilde{y}_i), \quad i \in \{1, \dots, \ell+m\}, \quad \tilde{p}, \tilde{y} \in E . \\ T_i(y_i) &:= \partial(\|y_i - a^i\|_i^{\beta_i}), \; i \in \{1, \dots, \ell\}, \\ T_{\ell+j}(x) &:= N(x; S_j) \qquad j \in \{1, \dots, m\} . \end{aligned}$$

Analogously to [192, Sect. 4.2]), we compute the projection onto the subspaces $\mathscr{A}_2(y^*)$ and $\mathscr{B}_2(y^*)$ and get the following algorithm for solving the control approximation problem $(\mathrm{P}(\varphi_{-C,k}))$, $k \in \operatorname{int} C$, under the assumptions (H1), (H2), (H3-2) and (H4).

---

**Algorithm 13.3.7** *(Proximal-Point Algorithm II for solving* $(\mathrm{P}(\varphi_{-C,k}))$*)*

---

**Input:** $A^i \in L(X, Y)$, $\|\cdot\|_i$, $\beta_i = 1$ *(*$i \in \{1, \dots, \ell\}$*)*, $S_j \subset X$ *(*$j \in \{0, \dots, m\}$*), ordering cone* $C \subset \mathbb{R}^\ell$, $k \in \operatorname{int} C$. *Choose* $\overline{y}^* \in \{y^* \in C^+ \mid y^*(k) = 1\}$, $\varepsilon > 0$.

**Initialization**

$n := 1$

*Choose* $x^1 \in \mathbb{R}^s$ *and* $p^1 \in \mathscr{B}_2(\overline{y}^*)$, *set* $(p^1 + y^1)_i = p_i^1 + \overline{y}_i^* A^i x^1$ *and* $p_{\ell+j}^1 + y_{\ell+j}^1 = p_{\ell+j}^1 + x^1$.

*Calculate* $B^{-1} := \left(mI + \sum_{i=1}^{\ell} \overline{y}_i^* A^{iT} A^i\right)^{-1}$.

**Proximal Step**

*Set* $b_i := y_i^n + p_i^n - a^i$ *for every* $i \in \{1, \dots, \ell\}$.

*if* $\beta_i = 1$ ***then***

*set*

$$\tilde{p}_i^n := \begin{cases} b_i \text{ if } \|b_i\| < 1 \\ P_{B_{\|\cdot\|_{i*}}}(b_i) \text{ otherwise.} \end{cases}$$

***end if***

*Set* $\tilde{p}^n_{\ell+j} = p^n_{\ell+j} + y^n_{\ell+j} - P_{S_j}(p^n_{\ell+j} + y^n_{\ell+j})$ *for every* $j \in \{1, \dots, m\}$.

**Iteration loop**
*Calculate:*

$$\bar{p}^n := B^{-1}\underbrace{\Big(\sum_{i=1}^{\ell} \overline{y}_i^* A^{iT}\tilde{p}_i^n + \sum_{j=1}^{m} \tilde{p}_{\ell+j}\Big)}_{=:o_1}.$$

$x^{n+1} := x^n - \bar{p}^n$.
$p^{n+1} + y^{n+1} := \tilde{p}^n + (\overline{y}_1^* A^1(x^n - 2\bar{p}^n), \dots, \overline{y}_\ell^* A^\ell(x^n - 2\bar{p}^n), x^n - 2\bar{p}^n, \dots, x^n - 2\bar{p}^n)$.
***if*** $\|o_1\| + \|\bar{p}^n\| = \|\mathscr{S}_{y^*}(\tilde{p})\| + \|\bar{p}^n\| < \varepsilon$ *and* $\|(p^{n+1} + y^{n+1}) - (p^n + y^n)\| < \varepsilon$ *for* $\varepsilon > 0$ ***then***
***STOP***
***else***
*set* $n := n + 1$ *and return to Proximal Step*
***end if***

---

**Remark 13.3.8** If $\beta_i > 1$ and $\|\cdot\|_i$ is the sum, maximum, or Euclidean norm, then a special construction of $\tilde{p}_i^n$ is given in [437].

Finally, we remark that the general convergence results for Spingarn's algorithm (compare [421]) also hold for our application of the proximal-point algorithm.

### 13.3.3 An Algorithm for Vector-Valued Control Approximation Problems

In the previous sections, we derived Algorithm 13.3.4 and Algorithm 13.3.7 for solving the scalar approximation problem ($\text{P}(\varphi_{-C,k})$) under certain assumptions. We now consider the vector-valued control approximation problem ($\text{P}_{app}$) under the assumptions (H1), (H2), (H3) and (H4):

$$\left.\begin{array}{l} C-\operatorname{Opt}_{x\in S} f(x) \quad \text{with} \\[2ex] f(x) = \begin{pmatrix} \|A^1(x) - a^1\|_1^{\beta_1} \\ \cdots \\ \|A^\ell(x) - a^\ell\|_\ell^{\beta_\ell} \end{pmatrix}. \end{array}\right\} \tag{13.3.27}$$

Taking into account Corollary 6.3.20, we get that for each element $f(x^0) \in \operatorname{PEff}(f(S), C)$ there is a parameter $k \in \operatorname{int} C$ such that $x^0$ solves the scalar optimization problem ($\text{P}(\varphi_{-C,k})$).

In order to solve the vector-valued problem (13.3.27) under the assumptions ((H1), (H2), (H3-1), and (H4)) or ((H1), (H2), (H3-2), and (H4)), we derive an interactive procedure based on a scalarization by means of the functional $\varphi_{-C,k}$ and Algorithms 13.3.4 and 13.3.7 for solving the corresponding scalar approximation problems $(\mathrm{P}(\varphi_{-C,k}))$.

---

**Algorithm 13.3.9** *(Algorithm III for generating approximate solutions of* $(\mathrm{P}_{app})$*)*

---

**Input:** $A^i \in L(X, Y), \|\cdot\|_i, \beta_i \geq 1\, (i \in \{1, \dots, \ell\}), S_j \subset X\, (j \in \{0, \dots, m\})$, ordering cone $C \subset \mathbb{R}^\ell$, choose $k \in \operatorname{int} C$ and $\overline{y}^* \in \{y^* \in C^+ \mid y^*(k) = 1\}$.

**Initial Step**
Compute an approximate solution $(x^0, p^0)$ of $(\mathrm{P}(\varphi_{-C,k}))$ with Algorithm 13.3.4 (or 13.3.7).
 **if** $(x^0, p^0)$ is accepted by the decision maker **then**
  **STOP**
 **else** go to Initialization.
 **end if**

**Initialization**
Put $j := 0$, $t_0 := 0$. Choose $\hat{y}^*(\neq \overline{y}^*) \in \{y^* \in C^+ \mid y^*(k) = 1\}$. Go to Iteration loop.

**Iteration loop**
Choose $t_{j+1}$ with $t_j < t_{j+1} \leq 1$ and compute an approximate solution $(x^{j+1}, p^{j+1})$ of $(\mathrm{P}(\varphi_{-C,k}))$ using the **Algorithm 13.3.4** (or **13.3.7**) with $y^{*j} = \overline{y}^* + t_{j+1}(\hat{y}^* - \overline{y}^*)$ and $(x^j, p^j)$ as starting point.
 **if** an approximate solution cannot be found for $t > t_j$ **then** go to Input
 **else** go to Interactive Step.
 **end if**

**Interactive Step**
The point $(x^{j+1}, p^{j+1})$ is to be evaluated by the decision maker.
 **if** $(x^{j+1}, p^{j+1})$ is accepted by the decision maker **then**
  **STOP**
 **else**
  **if** $t_{j+1} = 1$ **then** go to Input.
  **else** set $j = j + 1$ and return to Iteration loop.
  **end if**
 **end if**

---

## 13.4 An Adaptive Solution Procedure for Multiobjective Optimization Problems

Following Eichfelder [135–137], we present an adaptive algorithm based on a scalarization by means of a special case of the nonlinear separating functional (4.1.11) and well-known Lagrange multiplier rules for continuously differentiable functions in this section.

In order to introduce the Lagrange multiplier rule, we consider the scalar optimization problem

$$\min f(x) \tag{P}$$

subject to the constraints

$$g(x) \in K_{\mathbb{R}^r},$$

$$h(x) = 0_q,$$

$$x \in S,$$

where $K_{\mathbb{R}^r} \subset \mathbb{R}^r$ is a closed, convex cone, $\hat{S} \subset \mathbb{R}^n$ is a nonempty open set, $S \subset \hat{S}$ is a closed, convex set and $f : \hat{S} \to \mathbb{R}$, $g : \hat{S} \to \mathbb{R}^r$ and $h : \hat{S} \to \mathbb{R}^q$ are continuously differentiable.

Let $L : \hat{S} \times K^+_{\mathbb{R}^r} \times \mathbb{R}^q \to \mathbb{R}$ be the classical Lagrange function:

$$L(x, \mu, \xi) := f(x) - \mu^T g(x) - \xi^T h(x). \tag{13.4.1}$$

Provided that $x$ is a feasible element and $(\mu, \xi) \in K^+_{\mathbb{R}^r} \times \mathbb{R}^q$ elements with

$$\forall s \in S : \quad \nabla_x L(x, \mu, \xi)^T (s - x) \geq 0$$

and

$$\mu^T g(x) = 0,$$

we say that $\mu$ and $\xi$ are *Lagrange multipliers* to the point $x \in \hat{S}$.

Consider a multiobjective optimization problem with the objective function

$$f(x) := \begin{pmatrix} f_1(x) \\ \cdots \\ f_\ell(x) \end{pmatrix},$$

where $f_i : \mathbb{R}^n \to \mathbb{R}^\ell$ $(i \in \{1, \ldots, \ell\})$. Furthermore, consider the constraint set

$$S := \{x \in \mathbb{R}^n \mid g_s(x) \geq 0 \ (s \in \{1, \ldots, r\}),\ h_p(x) = 0 \ (p \in \{1, \ldots, q\})\}.$$

In order to formulate the multiobjective optimization problem, let $C = \mathbb{R}^\ell_+$ be the ordering cone and consider

$$C - \operatorname*{Opt}_{x \in S} f(x). \tag{13.4.2}$$

Now, the aim is to generate an equidistant approximation of the set of efficient elements to this problem by solving scalarized problems where the scalarization is given by means of the functional $\varphi_{A,k}$. We study the special case of $\varphi_{A,k}$ (cf. (4.1.11)), where $C = \mathbb{R}^\ell_+, a \in \mathbb{R}^\ell, A := a - C$ and $k \in C \setminus \{0\}$ (see Chaps. 5 and 6). From Theorem 6.2.31, we know that we obtain weakly minimal solutions of (13.4.2) by solving the following scalarization of (13.4.2) using the nonlinear separating functional $\varphi_{a-C,k}$ (see Sect. 7.3):

$$\min\ t \tag{$\mathrm{P}_{\varphi_{\mathrm{a-C,k}}}$}$$

subject to the constraint

$$f(x) \in tk + a - C,$$

$$g_s(x) \geq 0\ (s \in \{1, \ldots, r\}),$$

$$h_p(x) = 0\ (p \in \{1, \ldots, q\}),$$

$$t \in \mathbb{R}, \quad x \in \mathbb{R}^n.$$

In the following, we study a special case of (13.4.2), namely the bicriteria problem ($\ell = 2$) with the objective function

$$f_{bi}(x) := \begin{pmatrix} f_1(x) \\ f_2(x) \end{pmatrix},$$

and

$$C - \operatorname*{Opt}_{x \in S} f_{bi}(x). \tag{13.4.3}$$

By variation of the parameter $a \in \mathbb{R}^\ell$ in $\mathrm{P}(\varphi_{a-C,k})$, certain approximate solutions of the bicriteria problem can be generated. In Eichfelder [135–137], an adaptive parameter choice is proposed in the scalarization approach using sensitivity results derived in [136, Chap. 3]. In order to avoid arbitrarily distributed approximation points, a procedure for controlling the choice of the parameter $k \in C \setminus \{0\}$ is developed in Eichfelder's algorithm.

The theoretical foundation of this procedure is given in the following assertions by Eichfelder [136, Theorem 2.21] (compare also [136, Theorem 4.8]).

**Theorem 13.4.1** *Consider a hyperplane* $H := \{y \in \mathbb{R}^\ell | b^T y = \beta\}$ *with* $b \in \mathbb{R}^\ell \setminus \{0\}$ *and* $\beta \in \mathbb{R}$. *Assume that* $(\bar{t}, \bar{x})$ *is a solution of the scalarized problem* $(\mathrm{P}_{\varphi_{a-C,k}})$ *for* *the parameters* $a \in \mathbb{R}^\ell$ *and* $k \in C \setminus \{0\}$ *with* $b^T k \neq 0$. *This means that there is an element* $\bar{c} \in C$ *with*

$$a + \bar{t}k - f(\bar{x}) = \bar{c}.$$

*Then, there is a parameter* $a' \in H$ *and some* $t' \in \mathbb{R}$ *such that* $(t', \bar{x})$ *is an optimal solution of* $(\mathrm{P}_{\varphi_{a'-C,k}})$ *with*

$$a' + t'k - f(\bar{x}) = 0.$$

Using the Lagrange function (see (13.4.1)), we get an extended version of Theorem 13.4.1 (see Eichfelder [136, Lemma 2.26]).

**Theorem 13.4.2** *Suppose that* $C$ *is a nontrivial, closed, convex, pointed cone in* $\mathbb{R}^\ell$, $k \in C \setminus \{0\}$, $K_{\mathbb{R}^r} \subset \mathbb{R}^r$ *is a closed, convex cone,* $\hat{S} \subset \mathbb{R}^n$ *is a nonempty open set,* $S \subset \hat{S}$ *is a closed, convex set and* $f : \hat{S} \to \mathbb{R}^\ell$, $g : \hat{S} \to \mathbb{R}^r$ *and* $h : \hat{S} \to \mathbb{R}^q$ *are continuously differentiable.*

*Consider the scalarized problem* $(\mathrm{P}_{\varphi_{a-C,k}})$ *and assume that* $(\bar{t}, \bar{x})$ *is an optimal solution of* $(\mathrm{P}_{\varphi_{a-C,k}})$. *Suppose that there are Lagrange multipliers* $(\lambda, \mu, \xi) \in C^+ \times K^+_{\mathbb{R}^r} \times \mathbb{R}^q$ *to the point* $(\bar{t}, \bar{x})$. *According to Theorem 13.4.1, there are a parameter* $a' \in H$ *and some* $t' \in \mathbb{R}$ *such that* $(t', \bar{x})$ *is an optimal solution of* $(\mathrm{P}_{\varphi_{a'-C,k}})$ *and* $a' + t'k = f(\bar{x})$.

*Then, we obtain that* $(\lambda, \mu, \xi)$ *are Lagrange multipliers to the point* $(t', \bar{x})$ *for the problem* $(\mathrm{P}_{\varphi_{a'-C,k}})$, *too.*

In Chaps. 6 and 7, we have seen that it is possible to characterize (weakly) minimal elements of vector optimization problems using a scalarization by means of the functional $\varphi_{A,k}$ given by (4.1.11). We consider the special case $\ell = 2$ and the problem (13.4.3) with $C = \mathbb{R}^2_+$. In order to develop an effective procedure for finding representatives of the set of minimal elements of vector optimization problems by solving $(\mathrm{P}_{\varphi_{a^0-C,k}})$, Eichfelder [136, Theorem 4.8] has shown the following result.

**Theorem 13.4.3** *Let* $b \in \mathbb{R}^2 \setminus \{0\}$, $\beta \in \mathbb{R}$ *and* $k \in C \setminus \{0\}$ *with* $b^T k \neq 0$. *Suppose that* $(t^0, x^0)$ *is a minimal solution of the scalarized problem* $(\mathrm{P}_{\varphi_{a^0-C,k}})$ *(with* $\ell = 2$*) for a parameter* $a^0 \in H = \{y \in \mathbb{R}^2 \mid b^T y = \beta\}$ *with*

$$a^0 + t^0 k - f(x^0) = c^0$$

*and* $c^0 \in C \setminus \{0\}$.

*Consider*

$$\bar{t} := \frac{b^T f(\bar{x}) - \beta}{b^T k} \quad \text{and} \quad \bar{a} := f(\bar{x}) - \bar{t}k,$$

*and*

$$\tilde{t}^0 := \frac{b^T f(x^0) - \beta}{b^T k} \quad \text{and} \quad \tilde{a}^0 := f(x^0) - \tilde{t}^0 k.$$

*Then, there is no* $\bar{x} \in S$, $f(\bar{x}) \in \mathrm{Eff}(f(S), C)$ *with*

$$\overline{a} = a^0 + \lambda(\tilde{a}^0 - a^0) \quad \textit{for some} \quad \lambda \in (0, 1).$$

**Remark 13.4.4** Theorem 13.4.3 explains that the line segment with endpoints $a^0$ and $\tilde{a}^0$ on the hyperplane $H$, i.e., all parameters $a$ with $a = a^0 + \lambda(\tilde{a}^0 - a^0) \in H$ for $\lambda \in (0, 1)$, can be ignored without loosing the possibility that all efficient points of (13.4.3) can be generated by solving ($\mathrm{P}_{\varphi_{a-C,k}}$) for appropriate parameters $a \in H$.

Using information concerning the Lagrange multipliers (see Theorem 13.4.2), we get the following algorithm based on a scalarization by means of the functional $\varphi_{A,k}$ (see [135–137]).

---

**Algorithm 13.4.5** *(Adaptive solution procedure for a bicriteria optimization problem (13.4.3), Eichfelder [136, Algorithm 1])*

---

**Input:** *Choose* $k \in \mathbb{R}^2_+$ *with* $k_1 > 0$, *and predefine the desired distance* $\alpha \in \mathbb{R}$, $\alpha > 0$, *between the approximation points. Choose a hyperplane*

$$H = \{y \in \mathbb{R}^2 \mid b^T y = \beta\}$$

*by setting* $b \in \mathbb{R}^2$ *with* $b^T k \neq 0$ *and* $\beta \in \{0, 1\}$.

*Provide* $M^1 \in \mathbb{R}$ *with*

$$M^1 > f_2(x) - f_1(x)\frac{k_2}{k_1} \textit{ for all } x \in S.$$

**Initialization**

*Solve* ($\mathrm{P}_{\varphi_{\tilde{a}^1-C,k}}$) for $\tilde{a}^1 := (0, M^1)^T$. Let $(\tilde{t}^1, x^1)$ be the minimal solution of ($\mathrm{P}_{\varphi_{\tilde{a}^1-C,k}}$) and $\mu^1 \in \mathbb{R}^2_+$ the Lagrange multiplier to the constraint $\tilde{a}^1 + tk - f(x) \in \mathbb{R}^2_+$. Calculate

$$t^1 := \frac{b^T f(x^1) - \beta}{b^T k} \text{ and } a^1 := f(x^1) - t^1 k.$$

Set $r^1 := 0 \in \mathbb{R}^2$, $j := 1$.

Solve $\min_{x \in S} f_2(x)$ with the minimal solution $\hat{x}$ and calculate

$$\hat{t} := \frac{b^T f(\hat{x}) - \beta}{b^T k} \quad \text{and} \quad \hat{a} := f(\hat{x}) - \hat{t}k.$$

Set $w := \hat{a} - a^1$.

**Iteration loop**

Determine $a^{j+1}$ by

**if** $r^j = 0 \in \mathbb{R}^2$ **then**

set

$$a^{j+1} := a^j + \frac{\alpha}{\|w + ((-\mu^j)^T w)k\|} w. \tag{13.4.4}$$

**else**

**if** ($r_1^j = 0, r_2^j > 0$ and $\frac{k_1 b_2}{b^T k} > 0$) or ($r_1^j > 0, r_2^j = 0$ and $\frac{k_2 b_2}{b^T k} < 0$) **then**

calculate

$$\tilde{t}^j := \frac{b^T f(x^j) - \beta}{b^T k} \quad \text{and} \quad \tilde{a}^j := f(x^j) - \tilde{t}^j k$$

and put

$$a^{j+1} := \tilde{a}^j + \frac{\alpha}{\|w + ((-\mu^j)^T w)k\|} w.$$

**else**

**if** ($r_1^j = 0, r_2^j > 0$ and $\frac{k_1 b_2}{b^T k} \leq 0$) or ($r_1^j > 0, r_2^j = 0$ and $\frac{k_2 b_2}{b^T k} \geq 0$) **then**

set

$$a^{j+1} := \tilde{a}^j + \left(1 + \frac{\alpha}{\|r^j\|}\right)\left(r^j - \frac{b^T r^j}{b^T k} k\right).$$

**end if**

**Lagrange Step**

Set $j := j + 1$.

**if** $a^j = a^1 + \lambda w$ for a $\lambda \in [0, 1]$, **then** solve ($P_{\varphi_{a^j - C,k}}$) with the minimal solution $(t^j, x^j)$ and Lagrange multiplier $\mu^j$ to the constraint $a^j + tk - f(x) \in \mathbb{R}^2_+$, set $r^j := a^j + t^j k - f(x^j)$ and go to Iteration loop.

**else**

**STOP**

**end if**

**Output:** The set $\text{Eff}_f := \{x^1, \cdots, x^{j-1}, \hat{x}\}$ is an approximation of the set of weakly efficient solutions $\text{WEff}_f(S, C)$ (see Definition 6.2.2) in (13.4.3).

---

**Remark 13.4.6** In the **Input**, on has to provide a value $M^1 \in \mathbb{R}$ with

$$M^1 > f_2(x) - f_1(x)\frac{k_2}{k_1} \text{ for all } x \in S.$$

Such an element exists if there are solutions of the problems

$$\max_{x \in S} f_2(x)$$

and

$$\min_{x \in S} f_1(x).$$

Then, we can pick

$$M^1 > \max_{x \in S} f_2(x) - \min_{x \in S} f_1(x)\frac{k_2}{k_1}.$$

**Remark 13.4.7** The adaptive algorithm generates an approximation of the set of weakly efficient solutions of the bicriteria optimization problem (13.4.3).

**Remark 13.4.8** Since the $\varepsilon$-constraint method (see Sect. 7.3.8) is a special case of $(\mathrm{P}_{\varphi_{a-C,k}})$, one gets a corresponding algorithm based on the $\varepsilon$-constraint method (see Eichfelder [136, Sect. 4.2.2]).

In Sect. 15.2 we will use the adaptive algorithm presented in this section for solving vector-valued optimal control problems with PDE-constraints.

**Remark 13.4.9** Recently, the nonlinear scalarizing functional $\varphi_{A,k}$ given by (4.1.11) is used to compute the search direction in every iteration of a *trust-region algorithm* for solving multiobjective heterogeneous optimization problems by Thomann and Eichfelder in [447, 448].

## 13.5 A Descent Method for Set-Valued Optimization Problems

In this section, we will present a descent method developed by Köbis and Köbis in [297] for solving set-valued optimization problems using a scalarization by means of the functional $\varphi_{A,k}$ given by (4.1.11). Especially, we apply the results given in Sect. 10.3.2 for deriving the descent method. Under convexity assumptions, a descent method is derived by Jahn in [273, 274] using a scalarization by means of linear continuous functionals. Without any convexity assumptions concerning the set-valued optimization problem, the following descent method is derived in [297] using the nonlinear separating functional (4.1.11).

We study a set-valued optimization problem like introduced in Sect. 10.2. Consider a set-valued mapping $\Gamma : \mathbb{R}^n \rightrightarrows \mathbb{R}^\ell$, $C := \mathbb{R}^\ell_+$, and a binary relation $\preceq_C$. $\preceq_C$ denotes one of the set relations $\preceq^s_C$, $\preceq^l_C$, $\preceq^u_C$, $\preceq^m_C$, $\preceq^{mc}_C$ and $\preceq^{mn}_C$ (see Sect. 10.1). In this section, we are looking for (minimal) nondominated solutions w.r.t. the binary relation $\preceq_C$ in the sense of Definition 10.2.2 of the problem

$$\preceq_C - \operatorname*{Opt}_{x\in\mathbb{R}^n} \Gamma(x). \tag{13.5.1}$$

Specifically, we consider the set-valued optimization problem (13.5.1) w.r.t. the upper set less relation $\preceq_C^u$ (see Definition 10.1.2), where $C := \mathbb{R}^\ell_+$.

For two points $x^1, x^2 \in \mathbb{R}^n$, Corollary 10.3.31 then states that

$$\Gamma(x^1) \preceq_C^u \Gamma(x^2) \iff \sup_{k\in C_{\Gamma(x^1),\Gamma(x^2)}} \sup_{y^1\in\Gamma(x^1)} \inf_{y^2\in\Gamma(x^2)} \varphi_{-C,k}(y^1 - y^2) \leq 0, \tag{13.5.2}$$

where $\varphi_{-C,k}$ is defined by (4.1.11) with $A = -C$, $C = \mathbb{R}^\ell_+$, $k \in \operatorname{int} C$ under a certain attainment property concerning $\inf_{y^2\in\Gamma(x^2)} \varphi_{-C,k}(y^1 - y^2)$.

For Algorithm 13.5.1, it is important to have a possibility to calculate the functional $\varphi_{-C,k}$ in an easy way. This can be done using the results derived in Sect. 9.1, especially using formula (9.1.6) for the case that $C$ is constructed by a system of linear inequalities.

Now, it is our goal to obtain an improvement of feasible points by means of the equivalence (13.5.2) using the results derived in Sect. 10.3.2.

---

**Algorithm 13.5.1** *(Descent algorithm for generating an approximation of a nondominated solution of the set optimization problem* (13.5.1) *by Köbis, Köbis [297, Algorithm 4.5])*

---

**Input:** $\Gamma : \mathbb{R}^n \rightrightarrows \mathbb{R}^\ell$, *ordering cone* $C$ *with induced set relation* $\preceq_C$, $k \in C \setminus \{0\}$ *fulfilling the attainment property, starting point* $x^0 \in \mathbb{R}^n$, *maximal number* $i_{max}$ *of iterations, maximal number of search directions* $n_s$, *maximal number* $p_{max}$ *of iterations for the determination of the step size, initial step size* $h_0$ *and minimum step size* $h_{\min}$

**Initialization**
$i := 0$, $h := h_0$
*choose* $n_s$ *points* $\tilde{x}^1, \ldots, \tilde{x}^{n_s}$ *on the unit sphere around* $0_{\mathbb{R}^n}$

**Iteration loop**
***while*** $i \leq i_{max}$ ***do***

*check* $\Gamma(x^i + h\tilde{x}^j) \preceq_C \Gamma(x^i)$ *for every* $j \in \{1, \ldots, n_s\}$ *by evaluating the extremal term (e.g.,* $\sup_{a\in A} \inf_{b\in B} \varphi_{-C,k}(a - b)$ *for* $A = \Gamma(x^i + h\tilde{x}^j)$ *and* $B = \Gamma(x^i)$, *when* $\preceq_C = \preceq_C^u$ *(compare (13.5.2))). Choose the index* $n_0$ *with the smallest function value* extremal$_{\min}$.

***if*** $\text{extremal}_{\min} < 0$ ***then***
  $x^{i+1} := x^i + h\tilde{x}^{n_0}$ *% new iteration point*
  $j := 1$
  ***while*** $\Gamma(x^i + (j+1)h\tilde{x}^{n_0}) \preceq_C \Gamma(x^i + jh\tilde{x}^{n_0})$ *and* $j \leq j_{max}$ ***do***
    $j := j + 1$
    $x^{i+1} := x^{i+1} + h\tilde{x}^{n_0}$ *% new iteration point*
  ***end while***
***else***
  $h := h/2$
  ***if*** $h \leq h_{\min}$ ***then***
    ***STOP*** $x := x^i$
  ***end if***
***end if***
$i := i + 1$
***end while***

**Output:** *An approximation* $x$ *of a nondominated solution of the set optimization problem* (13.5.1).

---

**Remark 13.5.2** Note that for the Descent Method in the algorithm given above, we do not need any convexity assumptions on the sets in consideration. A numerical example (see Example 13.5.3) for solving a set-valued optimization problem where the objective map $\Gamma$ has nonconvex images is given by Köbis and Köbis in [297, Example 4.6].

**Example 13.5.3** We define $\Delta_t := 2\pi/40$ and $\mathcal{T} := \{j \cdot \Delta_t,\ j = 0, \ldots, 40\}$ and the set-valued mapping $\Gamma\colon \mathbb{R}^2 \rightrightarrows \mathbb{R}^2$

$$\Gamma(x) := \left\{ \begin{pmatrix} x_1^2 + x_2^2 \cdot \sin(2t) \\ x_2^2 + x_1^2 \cdot \cos(3t) \end{pmatrix} \;\middle|\; t \in \mathcal{T} \right\}$$

where $x = (x_1, x_2)^T$. This example is taken such that the unique nondominated solution is attained at $x = 0_{\mathbb{R}^2}$.

Algorithm 13.5.1 is applied to the set optimization problem (13.5.1) with the starting point $x^0 := (5, 6)^T$ using the natural ordering cone $C := \mathbb{R}^2_+$ and the solution concept based on the upper set less relation $\preceq^u_C$. Furthermore, the initial and minimal step lengths $h_0 := 2.1$ and $h_{\min} := 10^{-3}$ are used.

The attainment property is trivially satisfied for this discrete example such that we can use any $k \in C \setminus \{0\}$ in order to get equivalence in (10.3.14). In this numerical example, $k := (1, 1)^T$ and $n_s := 4$ search directions are taken (compass-search).

The numerical results are illustrated in Fig. 13.1. The iterates $x^i \in \mathbb{R}^2$ (in the diagrams at the left side) are shown with their corresponding function values (in the diagrams at the right side). The algorithm performed 25 main iterations and the

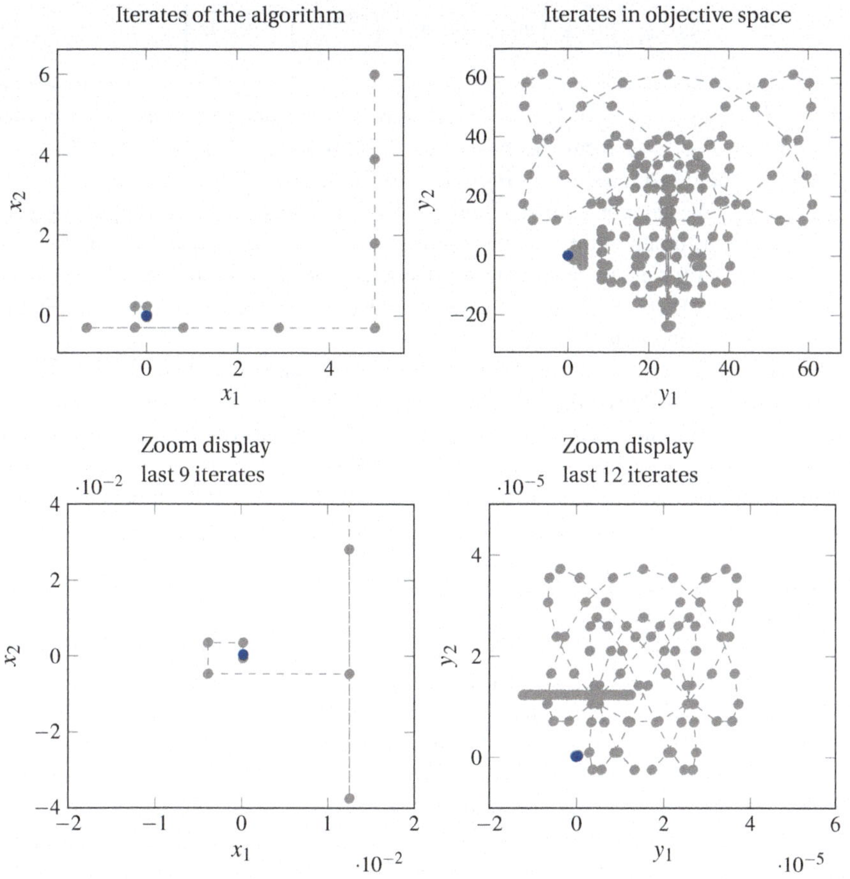


**Fig. 13.1** Numerical results for Example 13.5.3. Reprinted from [297, Fig. 7]

objective function $\Gamma$ was evaluated 148 times for this setup. This is an appropriate measure of computational effort for realistic problems. The algorithm terminates at $x^{25} \approx 10^{-4} \cdot (1.953, 4.395)^T$ for the chosen minimal step length $h_{\min}$. Obviously, this is within a ball of radius $h_{\min}$ around the actual minimum.

The next example is given in [297, Example 4.7].

**Example 13.5.4** We consider a set-valued extension $\Gamma\colon \mathbb{R}^2 \rightrightarrows \mathbb{R}^2$ of the linear-quadratic objective function

$$f(x_1, x_2) = \begin{pmatrix} x_1^2 + x_2^2 \\ 2(x_1 + x_2) \end{pmatrix}.$$

The values are clustered on a circle around $f$ similar to Example 13.5.3:

$$\Gamma(x_1, x_2) = \left\{ f(x_1, x_2) + \frac{1}{4} \begin{pmatrix} \sin(t) \\ \cos(t) \end{pmatrix} \mid t \in \bar{\mathcal{T}} \right\},$$

where $\bar{\mathcal{T}} = \left\{\frac{2\pi}{14} \cdot i \mid i = 0, 1, \ldots, 13\right\}$. It is well known that functions with a similar structure as $f(x_1, x_2)$ form the basis of the Markowitz stock model [345]. This model may be considered as an important example for a large class of real-world applications. We do not consider convex objective sets in this example, only to simplify the argumentation that the optimal solutions are adjusted along the line $x_1 = x_2$, $x_1 \leq 0$, with images forming discretized circles around the Pareto front $\{(f_1, f_2) : f_2 \leq 0, f_1 = f_2^2/8\}$. Again, we consider the upper set less relation and the natural ordering cone $C = \mathbb{R}^2_+$.

Numerical experiments are discussed in [297, Example 4.7]. In the pre-image space (argument values), the average error is $4.7 \times 10^{-5}$ for 15 experiments (15 different starting points). On the other hand, in the image space (calculated as distance of the center points from the Pareto front), the average error is even $1.59 \cdot 10^{-9}$. Furthermore, the average number of steps is 26.4 with an average number of function evaluations of 260.2 (for a detailed discussion, see [297, Example 4.7]).

**Remark 13.5.5** In the literature, there are several algorithms for solving set-valued optimization problems (see Jahn [274] and references therein). Jahn [274] proposes a descent method that generates approximations of nondominated elements of set-valued optimization problems under convexity assumptions on the considered sets. In [274], the set less relation is characterized by means of linear functionals. The method presented above (see [297]) is a similar descent method for obtaining approximations of nondominated elements of set-valued optimization problems. In these methods, several set relations are characterized by the nonlinear scalarizing functional $\varphi_{A,k}$ given by (4.1.11), where $A$ is assumed to be a proper convex cone. Since the nonlinear functional $\varphi_{A,k}$ is used, no convexity assumptions concerning the involved operator $\Gamma$ are needed. The approaches in Jahn [274] and given above [297] all rely on set relations where the involved domination structure is given by cones.

The Jahn–Graef–Younes method, which was introduced in [279] for vector optimization problems, selects efficient elements of a set of finitely many elements. Its advantage is that this method reduces the numerical effort by excluding elements which cannot be efficient for a given set. In Köbis, Kuroiwa and Tammer [298], this method is extended to the set-valued case in order to obtain nondominated elements of a family of finitely many sets. Furthermore, the involved sets are compared by means of set relations (see Sect. 10.1) applying the scalarizing results from Sect. 10.3.2, [297, 301] to evaluate $A \preceq B$ by using the nonlinear scalarizing functional (4.1.11).

Recently, characterizations of approximate nondominated elements of set-valued optimization problems based on a nonlinear scalarization technique and corresponding numerical methods for finding all approximate nondominated elements are given in [205].

## 13.6 Algorithms for Solving VVDS

### 13.6.1 An Algorithm for Generating Weakly Nondominated-Like Elements of Sets

Vector optimization problems with variable domination structures and corresponding solution concepts are introduced and discussed in Chap. 11. Numerical procedures for generating nondominated elements of these problems (see Definition 11.1.2) are derived by Eichfelder in [142, 144]. Especially, a Jahn–Graef–Younes method (see [279]) for computing nondominated elements is developed by Eichfelder in [144, Algorithm 4].

In this section, we present an algorithm derived by Bouza and Tammer in [70, Algorithm 1] that can be used for solving geometric vector optimization problems with variable domination structure (see Sect. 11.1), i.e., to find weakly nondominated-like elements according to Definition 11.1.3(b).

In this section, we consider the problem:

$$\text{Determine} \quad \mathrm{WM}(A, \tilde{D}(\cdot)),$$

where $Y = \mathbb{R}^\ell$, $A \subset Y$, $\tilde{D} : Y \to \mathscr{P}(Y)$ is a set-valued map such that $\tilde{D}(y)$ is a nontrivial convex, closed, pointed cone with $\operatorname{int} \tilde{D}(y) \neq \emptyset$ for all $y \in Y$.

As a measure which determines whether a point is an element of $\mathrm{WM}(A, \tilde{D}(\cdot))$ or not, we will use the scalarization defined in (11.2.5) by the function $\varphi_{A+\tilde{D}(\cdot)} : Y \times (Y \setminus \{0\}) \to \overline{\mathbb{R}}$,

$$\varphi_{A+\tilde{D}(\cdot)}(y, k) = \inf\{t \in \mathbb{R} \mid y + tk \in A + \tilde{D}(y)\}.$$

In the following, $\mathbb{Z}$ represents the set of integers.

We now introduce an algorithm ([70, Algorithm 1]) for generating weakly nondominated-like elements of $A$ with respect to the domination map $\tilde{D}(\cdot)$.

---

**Algorithm 13.6.1** *(Algorithm for generating elements of* $\mathrm{WM}(A, \tilde{D}(\cdot))$*)*

---

**Input:** $A \subset Y$, $\tilde{D} : Y \to \mathscr{P}(Y)$.

**Initialization**

$n := 0$, *choose* $y_0 \in A$.

**Iteration loop**

*if*

$$\inf_{\|k\|=1} \varphi_{A+\tilde{D}(\cdot)}(y_n, k) = 0, \tag{13.6.1}$$

***then STOP***

***else***

*take* $y_n \in A$, $k_n \in \tilde{D}(y_n)$ *such that* $\varphi_{A+\tilde{D}(\cdot)}(y_n, k_n) < 0$ *and* $\|k_n\| = 1$. *Find* $j_n \in \mathbb{Z}$, $d_n \in \tilde{D}(y_n) \setminus \{0\}$ *such that*

$$j_n := \sup\{j \in \mathbb{Z} \mid y_n - 2^j k_n - d_n \in A,\ d_n \in \tilde{D}(y_n)\}. \tag{13.6.2}$$

$$y_{n+1} := y_n - 2^{j_n} k_n - d_n, \tag{13.6.3}$$

$n := n + 1$.

***end if***

---

**Remark 13.6.2** Provided that the hypothesis of Propositions 11.2.28, 11.2.30 and 11.2.33 are satisfied for $\overline{y} = y_n$ and for all $\overline{k} \in Y$ with $\|\overline{k}\| = 1$, the function $\varphi_{A+\tilde{D}(\cdot)}$ is continuous for all feasible elements of the auxiliary problem (13.6.1). Hence, the infimum in (13.6.1) is attained.

In the case that $\|\cdot\|$ represents the Euclidean norm, the set of feasible solutions of the auxiliary model is not convex. Nevertheless, another norm $\|\cdot\|$ such that $\{k \in Y \mid \|k\| = 1\}$ is convex can be considered. If the assumptions of Proposition 11.2.41(b) are satisfied, the objective function is quasiconvex.

The following characterization of the limit points of the sequence generated by Algorithm 13.6.1 is given in [70, Proposition 4.2].

**Proposition 13.6.3** *Assume that* $Y = \mathbb{R}^\ell$. *Suppose that the domination map is given by* $\tilde{D} : Y \to \mathscr{P}(Y)$, *where* $\tilde{D}(y)$ *is a nontrivial, pointed, convex and closed cone with* $\operatorname{int} \tilde{D}(y) \neq \emptyset$ *for all* $y \in Y$. *Moreover, assume that* $\tilde{D}(\cdot)$ *is a B-l.s.c. and closed map. Provided that* $A \subset Y$ *is a proper, convex, closed set with nonempty interior and the sequence* $\{y_n\}$ *generated by Algorithm 13.6.1 converges to* $\overline{y}$, *then* $\overline{y} \in \mathrm{WM}(A, \tilde{D}(\cdot))$.

***Proof*** Consider the sequence $\{y_n\}$ generated by Algorithm 13.6.1 and its limit $\overline{y}$. Assume that $\overline{y} \notin \mathrm{WM}(A, \tilde{D}(\cdot))$. Hence, there is an element $y^0 \in A$ with $\overline{y} = y^0 + \overline{d}$ for a certain $\overline{d} \in \operatorname{int} \tilde{D}(\overline{y})$. We chose $\delta > 0$ such that $\mathring{B}_Y(\overline{d}, \delta) \subset \operatorname{int} \tilde{D}(\overline{y})$. Taking into account that $A$ is a convex set with nonempty interior, there is an element $\hat{y} \in \operatorname{int} A \cap \mathring{B}_Y(y^0, \delta)$. Then, we obtain

$$\overline{y} - \hat{y} - \overline{d} = y^0 - \hat{y} \in \mathring{B}_Y(0, \delta).$$

Therefore,

$$\overline{y} - \hat{y} \in \mathring{B}_Y(\overline{d}, \delta) \subset \operatorname{int} \tilde{D}(\overline{y}),$$

i.e., there is an element $\hat{d} \in \operatorname{int} \tilde{D}(\overline{y})$ with $\overline{y} = \hat{y} + \hat{d}$. Since $\hat{y} \in \operatorname{int} A$, we find $\alpha > 0$ such that

$$\mathring{B}_Y(\hat{y}, \alpha) \subset A. \tag{13.6.4}$$

We consider

$$y_n - (\hat{y} - \alpha k_n) = (-y_{n+1} + y_n) + (y_{n+1} - \overline{y}) + \overline{y} - \hat{y} + \alpha k_n, \tag{13.6.5}$$
$$= 2^{j_n} k_n + d_n + (y_{n+1} - \overline{y}) + \hat{d} + \alpha k_n \tag{13.6.6}$$
$$= (2^{j_n} + \alpha) k_n + \hat{d} + d_n + (y_{n+1} - \overline{y}), \tag{13.6.7}$$

where $\{j_n\}$, $j_n \in \mathbb{Z}$, $\{d_n\}$, $d_n \in \tilde{D}(y_n) \setminus \{0\}$ and $\{k_n\}$, $k_n \in Y \setminus \{0\}$, $\|k_n\| = 1$, are the sequences generated by Algorithm 13.6.1 for defining $y_{n+1}$, see (13.6.3).

We note that $\hat{d} \in \operatorname{int} \tilde{D}(\overline{y})$ and $\tilde{D}$ is a B-l.s.c. map. Taking into account Lemma 11.2.34, there is a neighborhood $V$ of $\overline{y}$ and a real number $\varepsilon > 0$ such that for all $y \in V$, $\mathring{B}_Y(\hat{d}, \varepsilon) \subset \tilde{D}(y)$. In particular, since $y_n \to \overline{y}$, we get $\mathring{B}_Y(\hat{d}, \varepsilon) \subset \tilde{D}(y_n)$ for $n$ large enough. Since $y_{n+1} - \overline{y} \to 0$, we get

$$\hat{d} + y_{n+1} - \overline{y} \in \tilde{D}(y_n)$$

for $n$ large enough. Because $\tilde{D}(y_n)$ is a convex cone, $\hat{d} + y_{n+1} - \overline{y} \in \tilde{D}(y_n)$ and $d_n \in \tilde{D}(y_n)$ imply $\hat{d} + y_{n+1} - y + d_n \in \tilde{D}(y_n)$. Then, we derive from (13.6.7) and (13.6.4):

$$y_n - (2^{j_n} + \alpha) k_n = (\hat{y} - \alpha k_n) + (\hat{d} + y_{n+1} - \overline{y} + d_n) \in A + \tilde{D}(y_n).$$

Because of the definition of $j_n$ in Eq. (13.6.2), it holds that

$$2^{j_n} + \alpha \leq 2^{j_n + 1}.$$

We obtain

$$0 < \alpha < 2^{j_n}. \tag{13.6.8}$$

Since $y_n \to \overline{y}$ (and so of course $y_{n+1} \to \overline{y}$) by (13.6.3), we get

$$-2^{j_n} k_n - d_n \to 0. \tag{13.6.9}$$

Therefore,

$$\lim_{n \to +\infty} \|2^{j_n} k_n + d_n\|^2 = 0,$$

that is equivalent to

$$\lim_{n \to +\infty} (2^{j_n} - \|d_n\|)^2 + 2^{j_n + 1} \|d_n\| \left( \langle k_n, \frac{d_n}{\|d_n\|} \rangle + 1 \right) = 0,$$

taking into account $\|k_n\| = 1$.

Because both terms are nonnegative, we have

$$\lim_{n\to+\infty}(2^{j_n} - \|d_n\|)^2 = \lim_{n\to+\infty} 2^{j_n+1}\|d_n\| \left(\langle k_n, \frac{d_n}{\|d_n\|}\rangle + 1\right) = 0.$$

Here, the second limit yields that either

$$\left(\left\langle k_n, \frac{d_n}{\|d_n\|}\right\rangle + 1\right) \to 0 \tag{13.6.10}$$

or

$$2^{j_n+1}\|d_n\| \to 0. \tag{13.6.11}$$

Assume that (13.6.10) holds, then due to $\{k_n\}$ and $\{\frac{d_n}{\|d_n\|}\}$ are bounded, we can consider that $k_n \to \overline{k}$ and $\frac{d_n}{\|d_n\|} \to \overline{d}$. Of course, $\|\overline{k}\| = \|\overline{d}\| = 1$ holds, and hence $\overline{k} \neq 0, \overline{d} \neq 0$. So, we get

$$\overline{k},\ \overline{d} \in \tilde{D}(y) \setminus \{0\}$$

since $\tilde{D}(\overline{y})$ is closed.

Together with (13.6.10), we get

$$\overline{k} = -\overline{d} \in \tilde{D}(\overline{y}) \setminus \{0\},$$

a contradiction to the pointedness of $\tilde{D}(\overline{y})$.
This means that $2^{j_n}\|d_n\| \to 0$ holds.

Taking into account $\lim_{n\to+\infty}(2^{j_n} - \|d_n\|)^2 = 0$, we derive

$$\lim_{n\to+\infty} 2^{j_n} = \lim_{n\to+\infty} \|d_n\| = 0.$$

Because $\alpha$ is fixed, for $n$ large enough, (13.6.8) is not fulfilled. Therefore, $\overline{y} \in \mathrm{WM}(A, \tilde{D}(\cdot))$. □

**Remark 13.6.4** It is important to mention that the condition that the variable domination mapping $\tilde{D} : Y \to \mathcal{P}(Y)$ is cone-valued cannot be relaxed. Indeed, $\tilde{D}(y) + \tilde{D}(y) \subseteq \tilde{D}(y)$ for all $y \in Y$ is needed. Because $\tilde{D}(y)$ is pointed and convex for all $y \in Y$, this implies that $\tilde{D}(y)$ is a convex cone for all $y \in Y$.

Now, we show the convergence of the sequence $\{y_n\}$ generated by Algorithm 13.6.1 ([70, Theorem 4.4]).

**Theorem 13.6.5** *Assume that*

- $Y = \mathbb{R}^\ell$.
- $\tilde{D} : Y \to \mathcal{P}(Y)$ *is a B-l.s.c. closed map with* $\tilde{D}(y)$ *is a nontrivial, pointed, convex and closed cone with* $\operatorname{int} \tilde{D}(y) \neq \emptyset$ *for all* $y \in Y$.

- *There is a nontrivial, closed, convex and pointed cone $\mathscr{C}$ such that $\tilde{D}(y) \subset \mathscr{C}$ for all $y \in Y$.*
- *$A \subset Y$ is convex and $\operatorname{int} A \neq \emptyset$.*

*Provided that the sequence $\{y_n\}$ generated by Algorithm 13.6.1 has an accumulation point $y$, then $y_n \to y$ and $y \in \mathrm{WM}(A, \tilde{D}(\cdot))$.*

***Proof*** If there is a closed, convex and pointed cone $\mathscr{C}$ such that $\tilde{D}(y) \subset \mathscr{C}$ for all $y \in Y$, then $\{y_n\}$ is a decreasing sequence with respect to $\mathscr{C}$. $\{y_n\}$ converges since it has a convergent subsequence. From Proposition 13.6.3, we conclude that the limit is a weakly nondominated-like element. □

The following assertion concerning the convergence of Algorithm 13.6.1 is shown in [70, Corollary 4.5].

**Corollary 13.6.6** *Provided that $A \subset Y$ is bounded and that the assumptions of Theorem 13.6.5 are fulfilled, then Algorithm 13.6.1 converges.*

***Proof*** The assertion follows from Theorem 13.6.5, since $y_n \in A$ and $A$ is bounded. □

**Remark 13.6.7** Even though the algorithm is simple, at each iteration one has to determine whether certain points are elements of $A$ or not. This means that the procedure can only be recommended in cases when this step is not too expensive. Nevertheless, in practical cases $A$ may have a more complex structure. If $A$ is the image of a certain function $f : S \to Y$ over $S$, in general there does not exist an effective algorithm which establishes whether $y \in A$ or not. In the following section, Algorithm 13.6.1 is adapted to this case.

### *13.6.2 An Algorithm for Computing Stationary Points of VVDS*

In this section, we suppose that $X$ and $Y$ are finite-dimensional spaces and that the variable domination in $Y$ is defined by a set-valued mapping $\tilde{D} : Y \to \mathscr{P}(Y)$, where $\tilde{D}(y)$ is a nontrivial, convex, closed and pointed cone with $\operatorname{int} \tilde{D}(y) \neq \emptyset$ for all $y \in Y$. Assume that $S$ is a closed subset of $X$ and $f : X \to Y$ is a $C^1$-differentiable function.

The aim of this section is to present results and an algorithm derived by Bouza and Tammer in [70, Sect. 5, Algorithm 2]. We consider the vector optimization problem with respect to a variable domination structure (see (VVOP) in Sect. 12.3.2)

$$\tilde{D}(\cdot) - \operatorname*{Opt}_{x \in S} f(x). \tag{13.6.12}$$

Now, we are looking for (weakly) nondominated-like solutions of Problem (13.6.12) in the sense of Definition 12.3.14 with $\varepsilon = 0$, i.e., for solutions

$\overline{x} \in \mathrm{M}_f(S, \tilde{D}(\cdot))$. Recall that $\overline{x} \in \mathrm{M}_f(S, \tilde{D}(\cdot))$ if for all $x \in S$, $f(x) \notin f(\overline{x}) - (\tilde{D}(f(\overline{x})) \setminus \{0\})$ (see the corresponding Definition 11.1.3 in the image space).

Of course, Problem (13.6.12) is related to the problem of finding weakly nondominated-like elements, i.e., elements of $\mathrm{WM}(A, \tilde{D}(\cdot))$ (or nondominated-like elements, i.e., elements of $\mathrm{M}(A, \tilde{D}(\cdot))$) studied in Sect. 13.6.1 because it is equivalent to finding elements of $\mathrm{WM}(f(S), \tilde{D}(\cdot))$ (or $\mathrm{M}(f(S), \tilde{D}(\cdot))$). As already mentioned in Remark 13.6.7, Algorithm 13.6.1 is not an effective algorithm. Nevertheless, the structure of the set $A$ can be used. Assume that $f \in C^1$ and $w \in S - x$, then $f(x + w)$ can be approximated as follows

$$f(x + w) \approx f(x) + \nabla f(x)w$$

and the set $A$ as the image of this linear function. This means

$$A(x) \approx \{f(x) + \nabla f(x)w \mid w \in S - x\}.$$

Now, we generate weakly nondominated-like elements of the approximate linearized set $A(x)$, i.e., the set $\mathrm{WM}(A(x), \tilde{D}(\cdot))$. The weakly nondominated-like elements of $A(x)$ will be used as the next iteration point. The stopping criterion is $\nabla f(x)w \notin -\operatorname{int} \tilde{D}(x)$ which means that $x$ is a weakly stationary point in the following sense:

**Definition 13.6.8** Suppose $f \in C^1$. An element $x \in S$ is a **weakly stationary point** of Problem (13.6.12) if $\nabla f(x)w \notin -\operatorname{int} \tilde{D}(f(x))$ for all $w \in S - x$.

As already mentioned in [33], weakly nondominated-like elements are also weakly stationary such that we suppose that by generating the set $\mathrm{WM}(A(x), \tilde{D}(\cdot))$ at each step, Algorithm 13.6.1 estimates a sequence that either stops after finitely many steps or the accumulation points of the generated sequence are stationary points. We consider the scalarizing function (11.2.5) with $A = A(x)$ ($x \in S$ is given)

$$\varphi_{A+\tilde{D}(\cdot)}(f(x), k) = \inf\{t \in \mathbb{R} \mid tk \in \nabla f(x)w + \tilde{D}(f(x)),\ w \in S - x\}.$$

Obviously, the linear term $\nabla f(x)$ may lead to the unboundedness of $\varphi_{A+\tilde{D}(\cdot)}$. This difficulty can be overcome by adding $\|w\|^2/2$. Nevertheless, even with this new term, the resulting function is not convex in $k$. As shown in Proposition 11.2.41, $\varphi_{A+\tilde{D}(\cdot)}(f(x), \cdot)$ is only a quasiconvex function. This means, the solution of the approximate problem

$$\inf_{w \in S-x} \|w\|^2/2 + \inf_{\|k\|=1} \varphi_{A+\tilde{D}(\cdot)}(y, k)$$

is not simple. Furthermore, a point of the form $f(x_n) - d = f(x_{n+1})$ for some $d \in \tilde{D}(f(x_n))$ must be found. This is the reason that we consider domination maps $\tilde{D} : Y \to \mathscr{P}(Y)$ for that it easy to verify $d \in \tilde{D}(y)$ and its scalarizing function enjoys very useful properties.

Now, we suppose that

(E1) $l : \mathbb{R}^\ell \times \mathbb{R}^\ell \to \mathbb{R}$ is a continuous function.
(E2) For all $y \in Y$, $l(y, \cdot)$ is a convex and positively homogeneous function.
(E3) The set $\{l(y, z) \mid y \in \mathbb{R}^\ell,\ z \in \mathbb{R}^\ell,\ \|z\| = 1\}$ is bounded.
(E4) The cone $\tilde{D}(y)$, $y \in \mathbb{R}^\ell$, is defined as

$$\tilde{D}(y) := \{z \in \mathbb{R}^\ell \mid l(y, z) \le 0\}$$

and

$$\operatorname{int} \tilde{D}(y) = \{z \in \mathbb{R}^\ell \mid l(y, z) < 0\} \neq \emptyset \text{ for all } y \in Y.$$

(E5) $f \in C^1$.

$\tilde{D}$ is a closed map since $l : \mathbb{R}^\ell \times \mathbb{R}^\ell \to \mathbb{R}$ is a continuous function.

Considering this representation of the cone $\tilde{D}(y)$, we define the auxiliary function

$$\theta(x, w) := l(f(x), -\nabla f(x)w) \quad (x, w \in X).$$

If $S$ is unbounded, then $\theta(x, w)$ can be unbounded and so $\inf_{w \in S-x} \theta(x, w) = -\infty$ such that the iterative step will be unsuccessful.

We consider the problem

$$\inf_{w \in S-x} \frac{\|w\|^2}{2} + \theta(x, w)$$

(see ($P_n$) in the following Algorithm 13.6.9).

In this framework, the following algorithm for generating stationary points of problem (13.6.12) is introduced in [70, Algorithm 2].

---

**Algorithm 13.6.9** *(Algorithm for computing stationary points of problems with variable domination structure)*

---

**Input:** $S \subseteq X$, $f : X \to Y$, $\tilde{D} : Y \to \mathscr{P}(Y)$.

**Initialization**

$n := 0$

*choose* $x_0 \in S$, $\sigma \in (0, 1)$.

**Iteration loop**

*Find a solution* $w_n$ *of the problem*

$$\arg \min_{w \in S-x_n} \|w\|^2/2 + l(f(x_n), -\nabla f(x_n)w) \tag{$P_n$}$$

*if* $w_n = 0$

*then STOP*

*else*

*find* $j_n \in \mathbb{Z}$ *such that*

$$j_n := \sup\{j \in \mathbb{Z} \mid f(x_n) + \sigma 2^j \nabla f(x_n) w_n - f(x_n + 2^j w_n) \in \tilde{D}(f(x_n))\}.$$

$$x_{n+1} := x_n + 2^{j_n} w_n, \ n = n + 1.$$

*end if*

---

In the following proposition ([70, Proposition 5.2]), we show that Algorithm 13.6.9 is well defined.

**Proposition 13.6.10**

(a) *The problem* $(\mathrm{P}_n)$ *has a unique optimal solution* $w_n$.
(b) $w_n = 0$ *is a unique optimal solution of* $(\mathrm{P}_n)$ *if and only if* $x_n$ *is a weakly stationary point of (13.6.12).*
(c) *There is an element* $M > 0$ *such that* $\|w_n\| \leq M \|\nabla f(x_n)\|$ *for all* $n = 1, 2, \ldots$.
(d) *If* $w_n \neq 0$, *then there is an element* $j \in \mathbb{Z}$ *such that* $f(x_n) + 2^j \nabla f(x_n) w_n - f(x_n + 2^j w_n) \in \tilde{D}(f(x_n))$.

***Proof***

(a) Since $l$ is positively homogeneous, $\|w\|^2/2 + l(f(x_n), -\nabla f(x_n) w) \to +\infty$ if $\|w\| \to +\infty$. This yields the existence of a solution of $(\mathrm{P}_n)$. Furthermore, $\|w\|^2/2$ is a strictly convex function and $l(f(x_n), -\nabla f(x_n) w)$ is convex in $w$ due to (E2). This means, the objective function of $(\mathrm{P}_n)$ is strictly convex and, therefore, the minimum is also unique.
(b) Obviously, 0 is a feasible solution of problem $(\mathrm{P}_n)$, therefore

$$\lambda_n := \min_{w \in S - x_n} \|w\|^2/2 + l(f(x_n), -\nabla f(x_n) w) \leq 0. \tag{13.6.13}$$

(i) If $x_n$ is a weakly stationary point, we get $-\nabla f(x_n) w \notin \operatorname{int} \tilde{D}(f(x_n))$ for all $w \in S - x_n$, i.e.,

$$l(f(x_n), -\nabla f(x_n) w) \geq 0 \text{ for all } w \in S - x_n.$$

Then, for all $w \in S - x_n$, we obtain

$$\|w\|^2/2 + l(F(x_n), -\nabla F(x_n) w) \geq 0.$$

From this inequality, together with (13.6.13), we obtain that $\lambda_n = 0$.
(ii) Now, we show that $\lambda_n < 0$ if $x_n$ is not a weakly stationary point. Then, for $w \in S - x_n$, it holds that $-\nabla f(x_n) w \in \operatorname{int} \tilde{D}(f(x_n))$. Because of (E4), it

holds that

$$l(f(x_n), -\nabla f(x_n)w) < 0.$$

Since $l$ is a positively homogeneous function in the second component, for some $\alpha > 0$ small enough, it holds that

$$\|\alpha w\|^2/2 + l(f(x_n), -\nabla f(x_n)\alpha w) = \alpha^2\|w\|^2/2 + \alpha l(f(x_n), -\nabla f(x_n)w) < 0.$$

Furthermore, $\alpha w \in S - x_n$ for all $\alpha \in [0, 1]$ since $S$ is convex. Then, as demanded

$$\lambda_n \le \alpha\|w\|^2/2 + l(f(x_n), -\nabla f(x_n)\alpha w) < 0.$$

Taking into account (a) and (b), it holds that $\lambda_n = 0$ if and only if $x_n$ is a weakly stationary point. Because the objective function of $(P_n)$ stated at $w = 0$ coincides with the optimal value and, by (a), $(P_n)$ has a unique solution, we derive that $w_n = 0$ is its unique solution if and only if $x_n$ is a weakly stationary point.

(c) Taking into account (13.6.13), we obtain for a solution $w_n$ of $(P_n)$

$$\|w_n\|^2/2 \le -l(f(x_n), -\nabla f(x_n)w_n) = |-l(f(x_n), -\nabla f(x_n)w_n)|.$$

Since $l$ is a continuous, positively homogeneous function, we get

$$\|w_n\|^2/2 \le |-l(f(x_n), -\nabla f(x_n)w_n)| \le \max_{\|r\|=1} |l(f(x_n), r)|\|\nabla f(x_n)w_n\|.$$

Because of (E3), $l(f(x_n), r)$ is bounded for all $r$ with $\|r\| = 1$. Therefore, for

$$M = 2\max_{\|r\|=1} |l(f(x_n), r)|,$$

we get

$$\|w_n\|^2 \le M\|\nabla f(x_n)w_n\| \le M\|\nabla f(x_n)\|\,\|w_n\|$$

and, hence

$$\|w_n\| \le M\|\nabla f(x_n)\|.$$

(d) Note that $w_n \neq 0$ means that $\nabla f(x_n)w_n \in -\tilde{D}(f(x_n))$. By considering the Taylor expansion of $f(x_n + 2^j w_n)$, we obtain

$$f(x_n) + 2^j\nabla f(x_n)w_n + o(2^j) = f(x_n + 2^j w_n)$$

and, equivalently, for $\sigma \in (0, 1)$,

$$f(x_n) + 2^j\sigma\nabla f(x_n)w_n - f(x_n + 2^j w_n) = -2^j(1-\sigma)\nabla f(x_n)w_n - o(2^j). \tag{13.6.14}$$

Since $-\nabla f(x_n)w_n \in \operatorname{int} \tilde{D}(f(x_n))$ and $0 < \sigma < 1$, we obtain

$$-(1-\sigma)\nabla f(x_n)w_n \in \operatorname{int} \tilde{D}(f(x_n)).$$

So, there is $j \in \mathbb{Z}$ such that

$$-(1-\sigma)\nabla f(x_n)w_n - \frac{o(2^j)}{2^j} \in \operatorname{int} \tilde{D}(f(x_n)) \subset \tilde{D}(f(x_n)),$$

i.e.,

$$-2^j(1-\sigma)\nabla f(x_n)w_n - o(2^j) \in \operatorname{int} \tilde{D}(f(x_n)) \subset \tilde{D}(f(x_n))$$

and the assertion follows with (13.6.14).

□

Now, we obtain a sufficient condition for the convergence of $\{f(x_n)\}$ in the following proposition ([70, Proposition 5.3]).

**Proposition 13.6.11** *Assume that assumptions (E1)–(E5) are fulfilled and that there is a closed, convex and pointed cone $\mathscr{C} \subset Y$ such that $\tilde{D}(f(x)) \subset \mathscr{C}$ for all $x \in S$. Provided that $\overline{x}$ is an accumulation point of the sequence $\{x_n\}$ generated by Algorithm 13.6.9, we obtain $f(x_n) \to f(\overline{x})$.*

***Proof*** Note that $f(\overline{x})$ is a limit point of $\{f(x_n)\}$ since $f$ is a continuous function and $\overline{x}$ is an accumulation point of the sequence $\{x_n\}$. Taking into account the definition of Algorithm 13.6.9, $f(x_n) + \sigma 2^{j_n}\nabla f(x_n)w_n - f(x_{n+1})$ and $-\nabla f(x_n)w_n$ are elements of $\tilde{D}(f(x_n))$. Since $\tilde{D}(f(x_n))$ is a convex cone, we obtain

$$\forall \alpha > 0: \quad f(x_n) + (\sigma 2^{j_n} - \alpha)\nabla f(x_n)w_n - f(x_{n+1}) \in \tilde{D}(f(x_n)). \tag{13.6.15}$$

Put $\alpha := \sigma 2^{j_n} \geq 0$, we get

$$f(x_n) - f(x_{n+1}) \in \tilde{D}(f(x_n)) \subset \mathscr{C}.$$

This yields that $\{f(x_n)\}$ is a decreasing sequence with an accumulation point. Taking into account that $\mathscr{C}$ is a closed, convex and pointed cone and $\mathscr{C} \neq \{0\}$, since $\{0\} \neq \tilde{D}(f(x)) \subset \mathscr{C}$ for all $x \in S$, we get

$$f(x_n) \to f(\overline{x}).$$

□

In the following, we analyze certain properties of the accumulation points of the sequence generated by Algorithm 13.6.9. We start with some preliminary results derived in [70, Lemma 5.4].

**Lemma 13.6.12** *Assume that $\{x_n\}$ is a subsequence of points generated by Algorithm 13.6.9 which converges to $\overline{x}$. Then, the sequence $\{w_n\}$ of solutions of the associated scalar problems ($\mathrm{P}_n$ ) has a convergent subsequence $\{w_{n_i}\}$, and if $w_{n_i} \to \overline{w}$, it* *holds that* $\overline{w} \in S - \overline{x}$*. Furthermore, if* $l(f(\overline{x}), -\nabla f(\overline{x})\overline{w}) = 0$*, then* $\overline{x}$ *is a weakly stationary point of* $f$ *over* $S$*.*

***Proof*** Since $\{x_n\}$ converges to $\overline{x}$ and $f \in C^1$, we get that $\|w_n\|$ is bounded from Proposition 13.6.10(c). Therefore, it has an accumulation point. For simplicity, we suppose that the whole sequence $\{w_n\}$ converges to $\overline{w}$. Because $S$ is closed, the mapping $\mathcal{S} : X \to \mathcal{P}(X)$, defined by $\mathcal{S}(x) := S - x$ is closed. Then, it holds that $\overline{w} \in S - \overline{x}$ taking into account $x_n \to \overline{x}$, $w_n \to \overline{w}$ and $w_n \in S - x_n$.

In order to prove the second assertion, note that, as $x_n \to \overline{x}$ and $w_n \to \overline{w}$, by the continuity of $\nabla f(x)$ we obtain

$$\lim_{n\to+\infty} l(f(x_n), -\nabla f(x_n)w_n) = l(f(\overline{x}), -\nabla f(\overline{x})\overline{w}) = 0.$$

Now, we suppose that $\overline{x}$ is not a weakly stationary point. Then, there is an element $w \in S - \overline{x}$ such that

$$l(f(\overline{x}), -\nabla f(\overline{x})w) < 0.$$

Analogously to the proof of Proposition 13.6.10(c), we can suppose that

$$\frac{\|w\|^2}{2} + l(f(\overline{x}), -\nabla f(\overline{x})w) < 0.$$

We consider $(w + (x - x_n))$. Of course, $(w + (x - x_n))$ is an element of $S - x_n$ since $w \in S - x$. Because $w_n$ is the solution of $(\mathrm{P}_n)$, we obtain

$$\frac{\|w_n\|^2}{2} + l(f(x_n), -\nabla f(x_n)w_n) \le \frac{\|w + (x - x_n)\|^2}{2} + l(f(x_n), -\nabla f(x_n) + (x - x_n)).$$

For $n \to +\infty$, we derive

$$0 \le \frac{\|\overline{w}\|^2}{2} + l(f(\overline{x}), -\nabla f(\overline{x})\overline{w}) \le \frac{\|w\|^2}{2} + l(f(\overline{x}), -\nabla f(\overline{x})w) < 0.$$

This contradiction yields that $\overline{x}$ is a weakly stationary point. □

The following assertion is shown in [70, Corollary 5.5].

**Corollary 13.6.13** *Consider a subsequence* $\{x_n\}$ *of points generated by Algorithm 13.6.9 converging to* $\overline{x}$*. Let* $w_n$ *be the solution of* $(\mathrm{P}_n)$*. Assume that* $w_n \to \overline{w}$*. If* $\nabla f(\overline{x})\overline{w} = 0$*, then* $\overline{x}$ *is a weakly stationary point of* $f$ *over* $S$*.*

***Proof*** Note that $l(f(x), 0) = 0$. Hence, $l(f(\overline{x}), -\nabla f(\overline{x})\overline{w}) = 0$. Then, the result follows directly from Lemma 13.6.12. □

Finally, we get the following assertion that every accumulation point of the sequence generated by Algorithm 13.6.9 is a weakly stationary point ([70, Theorem 5.6]).

**Theorem 13.6.14** *Suppose that (E1)–(E5) are fulfilled and that $\tilde{D}(f(x)) \subset \mathscr{C}$ for all $x \in S$, where $\mathscr{C}$ is a closed, convex and pointed cone in $Y$. Then, every accumulation point of the sequence $\{x_n\}$ generated by Algorithm 13.6.9 is a weakly stationary point.*

***Proof*** Let $\overline{x}$ be an accumulation point of the sequence $\{x_n\}$ generated by Algorithm 13.6.9. Consider a subsequence $\{x_n\}$ generated by Algorithm 13.6.9, such that $x_n \to \overline{x}$. We have $f(x_n) \to f(\overline{x})$, as already shown in Proposition 13.6.11. Furthermore, by Lemma 13.6.12, w.l.o.g., we can suppose that $w_n \to \overline{w}$ for a subsequence $\{w_n\}$ of solutions of the associated scalar problems ($\mathrm{P}_n$). Moreover, for $\sigma \in (0, 1)$, we have

$$f(x_n) + \sigma 2^{j_n} \nabla f(x_n) w_n - f(x_{n+1}) \in \tilde{D}(f(x_n)). \tag{13.6.16}$$

We now consider two cases:
*Case 1:* If the sequence $\{2^{j_n}\}$ is upper bounded, then $\{2^{j_n}\}$ is bounded since it is nonnegative. Let $\{2^{j_n}\} \to t^*$ be a convergent subsequence, and taking the limit for $n \to +\infty$ in (13.6.16), we get

$$t^* \nabla f(\overline{x}) \overline{w} \in \tilde{D}(f(\overline{x})). \tag{13.6.17}$$

Taking into account the definition of Algorithm 13.6.9, we obtain

$$-\nabla f(x_n) w_n \in \tilde{D}(f(x_n))$$

and, because $\tilde{D}(f(x_n))$ is a cone,

$$-2^{j_n} \nabla f(x_n) w_n \in \tilde{D}(f(x_n)).$$

For $n \to +\infty$, we get

$$-t^* \nabla f(\overline{x}) \overline{w} \in \tilde{D}(f(\overline{x})).$$

Because of (13.6.17) and the last inclusion, we obtain

$$-t^* \nabla f(\overline{x}) \overline{w} \in \tilde{D}(f(\overline{x})) \cap -\tilde{D}(f(\overline{x})).$$

Because $\tilde{D}(f(\overline{x}))$ is pointed, we get

$$t^* \nabla f(\overline{x}) \overline{w} = 0.$$

In the following, we again consider two cases:
*Case (1.a)* If $t^* > 0$, then $\nabla f(\overline{x}) \overline{w} = 0$. With Corollary 13.6.13, we get that $\overline{x}$ is a weakly stationary point.
*Case (1.b)* We now suppose that $2^{j_n} \to 0$ and fix $q \in \mathbb{N}$. Then, there is an element $n_q \in \mathbb{N}$, such that

$$\forall n > n_q : \quad f(x_n) + \sigma 2^q \nabla f(x_n) w_n - f(x_n + 2^q w_n) \notin \tilde{D}(f(x_n)),$$

or, equivalently,

$$l(f(x_n), f(x_n) + \sigma 2^q \nabla f(x_n) w_n - f(x_n + 2^q w_n)) \geq 0.$$

Taking into account that $l$ is positively homogeneous, we obtain

$$l\left(f(x_n), \frac{f(x_n) - f(x_n + 2^q w_n)}{2^q} + \sigma \nabla f(x_n) w_n\right) \geq 0.$$

For $q \to +\infty$ and $n \to +\infty$, we obtain

$$l(f(\overline{x}), (\sigma - 1) \nabla f(\overline{x}) \overline{w}) \geq 0$$

and,

$$l(f(\overline{x}), -\nabla f(\overline{x}) \overline{w}) \geq 0 \tag{13.6.18}$$

since $l$ is positively homogeneous.
Furthermore, by Proposition 13.6.10, either $w_n = 0$ or $l(f(x_n), -\nabla f(x_n) w_n) < 0$. Then, we obtain

$$l(f(\overline{x}), -\nabla f(\overline{x}) \overline{w}) \leq 0.$$

Together with (13.6.18), the following equality is fulfilled

$$l(f(\overline{x}), \nabla f(\overline{x}) w) = 0$$

and, by Lemma 13.6.12, this yields that $\overline{x}$ is a weakly stationary point.
*Case 2*: Now, we suppose that $\{2^{j_n}\}$ is unbounded. For a fixed $\varepsilon > 0$, we consider (13.6.15) for $\alpha > 0$ such that $\{2^{j_n} - \alpha\}$ is bounded and $2^{j_n} - \alpha > \varepsilon$. Then, we get

$$f(x_n) + (\sigma 2^{j_n} - \alpha) \nabla f(x_n) w_n - f(x_{n+1}) \in \tilde{D}(f(x_n)).$$

For $n \to +\infty$, and following the ideas used Case 1, we conclude

$$(2^{j_n} - \alpha) \nabla f(\overline{x}) \overline{w} = 0.$$

Because of $2^{j_n} - \alpha > \varepsilon > 0$, we get the result as in Case (1.a). □

**Remark 13.6.15** Especially, if the domination map $\tilde{D}(\cdot)$ is a Bishop-Phelps cone of the form $\{z \in Y \mid \|z\| \leq a(y)\}$ for all $y \in Y$ and $a \in Y^*$, the natural representation of $l(y, z)$ is $\|y\| - a(x)$. If $f \in C^1$, the results again are apparently valid.

We finish this section with an illustration of Algorithm 13.6.9 applied to a numerical example ([70, Example 5.8]).

**Example 13.6.16** We look at the problem (13.6.12) with $f(x) := (x+1, x^2+1)^T$, $S := [0,1]$ and the domination map $\tilde{D} : Y \to \mathscr{P}(Y)$, $Y = \mathbb{R}^2$,

$$\tilde{D}(y) := \{z \in Y \mid z_1 \geq 0,\ y_2 z_1 - y_1 z_2 \leq 0\}.$$

For generating nondominated-like elements, we note that for fixed $\overline{x} \in [0,1]$, it is clear that $f(x) = f(\overline{x})$ if and only if $\overline{x} = x$. Hence, $\overline{x}$ is a nondominated-like element if and only if $\overline{x}$ is the unique solution of the system

$$(\overline{x}+1, (\overline{x})^2+1) - (x+1, x^2+1) \in \tilde{D}(f(\overline{x})).$$

This means

$$\overline{x} - x \geq 0 \tag{13.6.19}$$

$$((\overline{x})^2+1)(\overline{x}-x) - (\overline{x}+1)\left((\overline{x})^2 - x^2\right) \leq 0. \tag{13.6.20}$$

From (13.6.20), we obtain that either $\overline{x} = x$ or

$$(\overline{x})^2 + 1 - (\overline{x}+1)(\overline{x}+x) \leq 0. \tag{13.6.21}$$

Obviously, (13.6.21) is a linear inequality in $x$. Taking into account $\overline{x}+1 > \overline{x} \geq 0$, (13.6.21) is equivalent to proving that the inequality holds for $x = \overline{x}$ (compare (13.6.19)). Moreover, as the case $\overline{x} = x$ was already analyzed, because of the strict monotonicity of linear functions, the equality can be dropped. This means that

$$(\overline{x})^2 + 1 - 2(\overline{x}+1)\overline{x} < 0 \tag{13.6.22}$$

is to be solved. The solutions of $-(\overline{x})^2 + 1 - 2\overline{x} = 0$ are $\overline{x} = -1 \pm \sqrt{2}$. Hence, (13.6.22) holds for all $\overline{x} \in [0,1] \cap \left((\sqrt{2}-1, +\infty) \cup [-\infty, -1-\sqrt{2})\right) = (\sqrt{2}-1, 1)$, i.e., $\overline{x} \in (\sqrt{2}-1, 1)$ are not nondominated-like elements and for

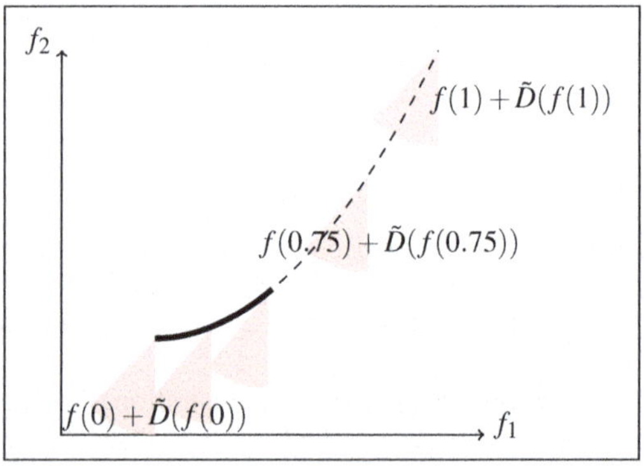


**Fig. 13.2** $f([0,1])$ and its nondominated-like elements with respect to $\tilde{D}(\cdot)$

$\overline{x} \in [0, \sqrt{2} - 1]$, the unique solution of (13.6.19)–(13.6.20) is $x = \overline{x}$. Therefore, we conclude that the set of nondominated-like elements is given by $[0, \sqrt{2} - 1]$, see Fig. 13.2. The values of $f(x)$, $x \in S$, are given by the dashed curve whose intrinsic part corresponds with the nondominated-like elements of $f$ with respect to the domination structure given by $\tilde{D}(\cdot)$. The domination cones for $x = 1$ and $x = 0.75$ show that the objective function values of these points are not nondominated-like elements while at $x = 0.4$, $x = 0.2$, $x = 0$, the condition for nondominated-like elements holds.

The algorithm is implemented in MatLab R2012 and run on an Intel(R) Atom(TM) CPU N270 at 1.6 GHz. The generated solutions are

$$x = 0.3152,\ 0.3893,\ 0.3955,\ 0.3822,\ 0.2647\ 0.2838,\ 0.4111,\ 0.4136,\ 0.3292,\ 0.3944$$

using ten randomly generated starting points in $[0, 1]$. In the algorithm, at most 4 iterations were needed and the largest cputime was 0.9828 s.

# Chapter 14
# Optimization Under Uncertainty

Robust optimization is a very active field of research. In this chapter, we show that the nonlinear scalarizing functional (4.1.11) can be considered in order to describe many concepts of robustness and stochastic optimization well known from scalar optimization under uncertainty as special cases of a scalarization by means of (4.1.11). Based on this unified approach to robustness and stochastic optimization, it is possible to derive new concepts of robustness in scalar optimization under uncertainty. Moreover, we explain that the well-studied properties of the functional (4.1.11) (see Chap. 4) allow the establishment of useful relationships to multiobjective optimization problems.

## 14.1 Problem Formulation

The results presented in this chapter are derived by Klamroth, Köbis, Schöbel and Tammer in [294], compare also [253, 254, 295] for further results.

To study optimization problems under uncertainty is very important because most real world optimization problems are contaminated with uncertain data. The concept of robustness gives one way of dealing with such optimization problems where uncertainties are involved: One allows different scenarios for the input parameters and looks for a solution that works well in every uncertain scenario instead of assuming that all data are known. In the books by Ben-Tal, L. El Ghaoui and Nemirovski [36] and Kouvelis and Yu [309], very interesting overviews on results and applications for the most prominent concepts for robust optimization are presented. Recently, new concepts of robustness were introduced, especially the concept of light robustness by Fischetti and Monaci [163] or the concept of recovery-robustness by Liebchen et al. in [326]. Furthermore, a scenario-based approach is suggested in Goerigk and Schöbel [189] and the concept of $\varepsilon$-constraint robustness in [294]. In the approaches mentioned above, the optimization problem under uncertainty is replaced by a deter-



C. Tammer and P. Weidner, *Scalarization and Separation by Translation Invariant Functions*, Vector Optimization, https://doi.org/10.1007/978-3-030-44723-6_14

ministic formulation, called the *robust counterpart* of the optimization problem under uncertainty. A robust solution hedges against the worst case that may happen and no probability distribution of the uncertain data is needed in robust optimization.

If the probability distribution is known, it is possible to formulate a stochastic optimization problem where the expected value of a certain objective or cost function is going to be optimized. In stochastic optimization, a solution is desired which is good on average. The field of stochastic optimization is another prominent way of dealing with optimization problems under uncertainty. An introduction is given in the book by Birge and Louveaux [51].

In this chapter, we connect different concepts of robustness and of stochastic programming that have been studied basically different in the literature, in a very general and unifying framework using the nonlinear scalarizing functionals (4.1.11). Under the assumption that the set of scenarios is finite, Klamroth, Köbis, Schöbel and Tammer [294] have shown that many optimization problems under uncertainty have a (deterministic) counterpart in this framework. Our discussion is based on two closely related approaches:

- We use the nonlinear scalarizing functional (4.1.11) and corresponding properties presented in Chap. 4 to describe the robust counterparts of optimization problems under uncertainty.
- We relate solutions of scalar optimization problems defined by the functional (4.1.11) (and likewise of the respective optimization problems under uncertainty) to the set of (weakly) efficient elements to a *multiobjective counterpart* using results from Chaps. 6 and 7.

These approaches provide the basement for a careful analysis of the interrelations and the differences between established concepts in robust optimization as well as stochastic optimization. Furthermore, the multiobjective counterpart can support the decision making process by preparing additional trade-off information between alternative efficient solutions.

The connection between scalar optimization models under uncertainty and an associated (deterministic) multiobjective counterpart problem were discussed by several authors for specific robustness concepts, see Kouvelis and Yu [309], Perny et al. [380], Kouvelis and Sayin [308, 401]. Kouvelis and Sayin focus on two concepts for robustness (strict robustness and deviation robustness), compare Sects. 14.2 and 14.3. Furthermore, Ogryczak [372] studied the relationships between uncertain scalar optimization and deterministic multiple objective optimization. Concerning stochastic optimization, Gast [172] introduced a multiobjective counterpart for a two-stage stochastic programming problem and studied the interrelation of stochastic programming models and the concept of recoverable robustness (see also Stiller [429]). Hites et al. [243] give a critical view and a qualitative description of the similarities and differences between the two modeling paradigms. They conclude that a multiobjective counterpart can in general not be used to represent a scalar optimization problem under uncertainty from a modeling point of view. Nevertheless, there certainly is a strong relation from a theoretical point of view (see Sect. 14.9). Our approaches show interesting and useful properties of alternative solutions for scalar optimization problems under uncertainty.

In the following sections, we derive robust counterpart problems to scalar optimization problems under uncertainty as special cases of a general scalarization by means of the nonlinear scalarizing functional given by (4.1.11). As discussed in Sect. 7.3, the scalarization method by means of the functional (4.1.11) includes, for instance, weighted-sums, Chebyshev- and $\varepsilon$-constraint-scalarization. We show that the scalarization by means of (4.1.11) includes a variety of different models from robust optimization and stochastic optimization as special cases. We get a unified approach for different types of optimization problems with uncertainties with our methods.

Let $Y$ be a topological vector space and let $\mathcal{F}$, $B$ be proper subsets of $Y$. We assume that

$$k \in 0^+ B \setminus \{0\}. \tag{14.1.1}$$

We study the functional (4.1.11) with $A = -B$ and denote it by $\varphi_{-B,k} : Y \to \overline{\mathbb{R}}$:

$$\varphi_{-B,k}(y) := \inf\{t \in \mathbb{R} | y \in tk - B\} \tag{14.1.2}$$

and use its properties shown in Chap. 4.

In order to describe several robust counterpart problems to optimization problems under uncertainty, we consider the problem

$$\varphi_{-B,k}(y) \to \inf_{y \in \mathcal{F}} . \tag{$\mathrm{P}_{k,B,\mathcal{F}}$}$$

In this section, we show that many different concepts for scalar optimization problems under uncertainty can be described by means of the functional (14.1.2) and problem ($\mathrm{P}_{k,B,\mathcal{F}}$) by choosing $B$, $k$ and $\mathcal{F}$ in different ways. For our results, the monotonicity and continuity properties of $\varphi_{-B,k}$ studied in Chap. 4 are useful. Furthermore, several important properties of the functional $\varphi_{-B,k}$, for example the translation invariance along $k$ and sublinearity, shown in Chap. 4, will be applied.

In Sect. 14.9, based on the interpretation of scalar optimization problems under uncertainty by means of the nonlinear scalarizing functional (14.1.2), we will formulate multiobjective counterparts with the corresponding sets of (weakly) efficient elements in order to characterize optimal solutions for the considered scalar optimization problems under uncertainty.

In this chapter, we show how several concepts of robustness can be expressed using the nonlinear scalarizing functional $\varphi_{-B,k}$. Furthermore, new concepts for robustness (see [294]) using the nonlinear scalarization approach by varying the parameters $B$ and $k$ are derived. Especially, we introduce the $\varepsilon$-constraint robust counterpart problem (compare Sect. 7.3.8).

In order to explain the relationships between the efficient points of a deterministic multiobjective optimization problem (see Chaps. 6 and 7) and minimal solutions of problem ($\mathrm{P}_{k,B,\mathcal{F}}$), we use the properties of the functional $\varphi_{-B,k}$ shown in Chap. 4.

First, we formulate an optimization problem under uncertainty. Consistently, in this section, let $\mathcal{U} := \{\xi_1, \dots, \xi_\ell\}$ be the uncertainty set, i.e., $\xi \in \mathcal{U}$ can take on $\ell$ different values. Here, $\xi$ could be real numbers or real vectors. Moreover, let $f : \mathbb{R}^n \times \mathcal{U} \to \mathbb{R}$, $F_i : \mathbb{R}^n \times \mathcal{U} \to \mathbb{R}$, $i \in \{1, \dots, m\}$. We define a *scalar optimization problem under uncertainty* as a parametrized optimization problem

$$(Q(\xi), \xi \in \mathcal{U}), \tag{14.1.3}$$

where for a given $\xi \in \mathcal{U}$ the optimization problem $(Q(\xi))$ is specified by

$$\begin{aligned} & f(x, \xi) \to \min \\ & \text{subject to } F_i(x, \xi) \le 0, \ i \in \{1, \dots, m\}, \\ & x \in \mathbb{R}^n. \end{aligned} \tag{$Q(\xi)$}$$

It is not known which value $\xi \in \mathcal{U}$ is going to be realized at the time the optimization problem under uncertainty (14.1.3) has to be solved. $\hat{\xi} \in \mathcal{U}$ is called the *nominal value*, i.e., the value of $\xi$ that we believe is true today. Furthermore, $(Q(\hat{\xi}))$ is called the *nominal problem*. Consistently, we suppose that the minima of the problems to be proposed in the following sections exist.

## 14.2 Strict Robustness

The first robustness concept that we will study is called *strict robustness* here. This concept was first mentioned by Soyster [420], later formalized and analyzed by Ben-Tal, El Ghaoui, and Nemirovski in several publications (see e.g. [37] for early contributions and [36] for an overview on results). The concept of strict robustness comprised the idea that the worst possible objective function value is minimized in order to get a solution that is "good enough" even in the worst case scenario. Moreover, constraints have to be fulfilled for each scenario $\xi \in \mathcal{U}$. The *strictly robust counterpart* of the optimization problem $(Q(\xi), \xi \in \mathcal{U})$ including uncertainties is specified by

$$\begin{aligned} & \rho_{RC}(x) = \max_{\xi \in \mathcal{U}} f(x, \xi) \to \min \\ & \text{subject to } \forall \xi \in \mathcal{U} : \ F_i(x, \xi) \le 0, \ i \in \{1, \dots, m\}, \\ & x \in \mathbb{R}^n. \end{aligned} \tag{$RC$}$$

A feasible solution of $(RC)$ is called *strictly robust* and the set of strictly robust solutions is designated by

$$\mathfrak{A} := \{x \in \mathbb{R}^n \mid \forall \xi \in \mathcal{U} : \ F_i(x, \xi) \le 0, \ i \in \{1, \dots, m\}\}. \tag{14.2.1}$$

Now, we will show how $(RC)$ can be embodied using the nonlinear scalarizing functional $\varphi_{-B,k}$ given by (14.1.2). The following theorem in shown in [294, Theorem 3].

**Theorem 14.2.1** *We consider*

$$\mathfrak{A}_1 := \mathfrak{A}, \tag{14.2.2}$$

$$B_1 := \mathbb{R}^\ell_+, \tag{14.2.3}$$

$$k_1 = 1_\ell := (1, \dots, 1)^T, \tag{14.2.4}$$

$$\mathscr{F}_1 = \{(f(x, \xi_1), \dots, f(x, \xi_\ell))^T \mid x \in \mathfrak{A}_1\}. \tag{14.2.5}$$

*Then, condition (14.1.1) is fulfilled for $k = k_1$, $B = B_1$ and with $\mathscr{F} = \mathscr{F}_1$, problem $(\mathrm{P}_{k,B,\mathscr{F}})$ is equivalent to problem $(RC)$ in the following sense:*

$$\begin{aligned}\min\{\varphi_{-B_1,k_1}(y) \mid y \in \mathscr{F}_1\} &= \varphi_{-B_1,k_1}(y^*)\\ &= \min\{\rho_{RC}(x) \mid x \in \mathfrak{A}_1\}\\ &= \rho_{RC}(x^*),\end{aligned}$$

*where $y^* = (f(x^*, \xi_1), \dots, f(x^*, \xi_\ell))^T$.*

***Proof*** Taking into account $B_1 + \mathbb{R}_+ k_1 = \mathbb{R}^\ell_+ + \mathbb{R}_+ 1_\ell \subseteq \mathbb{R}^\ell_+ = B_1$, the condition $k_1 \in 0^+ B_1 \setminus \{0\}$ (see (14.1.1)) is fulfilled. Because of $k_1 \in \operatorname{int} \mathbb{R}^\ell_+$ and $B_1 = \mathbb{R}^\ell_+$ is closed, the infimum in the definition of $\varphi_{-B_1,k_1}$ is finite and attained and we replace the infimum by a minimum:

$$\begin{aligned}\min_{y\in\mathscr{F}_1} \varphi_{-B_1,k_1}(y) &= \min_{y\in\mathscr{F}_1} \min\{t \in \mathbb{R} \mid y \in tk_1 - B_1\}\\ &= \min_{y\in\mathscr{F}_1} \min\{t \in \mathbb{R} \mid y - tk_1 \in -B_1\}\\ &= \min_{x\in\mathfrak{A}_1} \min\{t \in \mathbb{R} \mid (f(x, \xi_1), \dots, f(x, \xi_\ell))^T - t(1, \dots, 1)^T \leq 0_\ell\}\\ &= \min_{x\in\mathfrak{A}_1} \min\{t \in \mathbb{R} \mid (f(x, \xi_1), \dots, f(x, \xi_\ell))^T \leq t(1, \dots, 1)^T\}\\ &= \min\{\max_{\xi\in\mathscr{U}} f(x, \xi) \mid x \in \mathfrak{A}_1\}\\ &= \min\{\rho_{RC}(x) \mid x \in \mathfrak{A}_1\}.\end{aligned}$$

□

Because we have chosen $k_1 = 1_\ell$, every objective function $f(x, \xi)$, $\xi \in \mathscr{U}$, is handled in the same way, i.e., no objective function is preferred to another one.

**Remark 14.2.2** Taking into account Propositions 4.7.3 and 4.7.1(b), the functional $\varphi_{-B_1,k_1}$ is continuous, finite-valued, $\mathbb{R}^\ell_+$-monotone, strictly $(\operatorname{int} \mathbb{R}^\ell_+)$-monotone and sublinear since $B_1$ is a nontrivial, closed, convex cone and $k_1 \in \operatorname{int} B_1$.

**Remark 14.2.3** Obviously, the concept of strict robustness corresponds to the Chebyshev scalarization (see Sect. 7.3.1) with the origin as reference point as a special case of functional (14.1.2) (compare Sect. 7.3). Theorem 14.2.1 shows that $(RC)$ can be interpreted as a max-ordering problem as studied in multiobjective optimization (see Sect. 7.3.1). Kouvelis and Sayin [308, 401] considered this relationship too. In [308, 401], this relationship was used to determine the nondominated set of discrete bicriteria optimization problems.

## 14.3 Deviation Robustness

Now, we will discuss another important concept of robustness, named *deviation robustness* or *min max regret robustness*. Here, the function to be minimized is $\max_{\xi\in\mathcal{U}}(f(x,\xi)-f^*(\xi))$, where $f^*(\xi)\in\mathbb{R}$ is the optimal value of problem $(Q(\xi))$ for the parameter $\xi\in\mathcal{U}$ in contrast to the concept of strict robustness. This concept of robustness is often used in applications, especially in scheduling or locational analysis, usually if no uncertainties are involved in the constraints, see [309] for interesting applications. We look at the *deviation-robust counterpart* of (14.1.3) given by

$$\begin{aligned}&\rho_{dRC}(x)=\max_{\xi\in\mathcal{U}}(f(x,\xi)-f^*(\xi))\to\min\\&\text{subject to }\forall\xi\in\mathcal{U}:\ F_i(x,\xi)\le 0,\ i\in\{1,\dots,m\},\\&x\in\mathbb{R}^n.\end{aligned}\tag{$dRC$}$$

We consider the vector

$$f^*:=(f^*(\xi_1),\dots,f^*(\xi_\ell))^T \tag{14.3.1}$$

consisting of the individual minimizers for the particular scenarios. This vector can be interpreted as an *ideal solution vector*. The relationship to the ideal point of a multiobjective counterpart of problem (14.1.3) will be studied later but should already be mentioned here. The following theorem ([294, Theorem 4]) shows the relationship between problems $(P_{k,B,\mathcal{F}})$ and $(dRC)$.

**Theorem 14.3.1** *We look at*

$$\mathfrak{A}_2:=\mathfrak{A}, \tag{14.3.2}$$

$$B_2:=\mathbb{R}^\ell_+-f^*, \tag{14.3.3}$$

$$k_2:=1_\ell, \tag{14.3.4}$$

$$\mathcal{F}_2=\{(f(x,\xi_1),\dots,f(x,\xi_\ell))^T\mid x\in\mathfrak{A}_2\}. \tag{14.3.5}$$

*For* $k = k_2$, $B = B_2$, *condition (14.1.1) is fulfilled and with* $\mathscr{F} = \mathscr{F}_2$, *problem* $(\mathrm{P}_{k,B,\mathscr{F}})$ *is equivalent to problem (dRC) in the following sense:*

$$\begin{aligned}\min\{\varphi_{-B_2,k_2}(y) \mid y \in \mathscr{F}_2\} &= \varphi_{-B_2,k_2}(y^*) \\ &= \min\{\rho_{dRC}(x) \mid x \in \mathfrak{A}_2\} \\ &= \rho_{dRC}(x^*),\end{aligned}$$

*where* $y^* = (f(x^*, \xi_1), \ldots, f(x^*, \xi_\ell))^T$.

***Proof*** Because $B_2 + \mathbb{R}_+ k_2 = (\mathbb{R}^\ell_+ - f^*) + \mathbb{R}_+ 1_\ell \subseteq \mathbb{R}^\ell_+ - f^* = B_2$, condition $k_2 \in 0^+ B_2 \setminus \{0\}$ (see (14.1.1)) is fulfilled. Furthermore,

$$\begin{aligned}\min_{y \in \mathscr{F}_2} \varphi_{-B_2,k_2}(y) &= \min_{y \in \mathscr{F}_2} \min\{t \in \mathbb{R} \mid y \in tk_2 - B_2\} \\ &= \min_{x \in \mathfrak{A}_2} \min\{t \in \mathbb{R} \mid (f(x,\xi_1),\ldots,f(x,\xi_\ell))^T - (f^*(\xi_1),\ldots,f^*(\xi_\ell))^T - t1_\ell \leq 0_\ell\} \\ &= \min_{x \in \mathfrak{A}_2} \min\{t \in \mathbb{R} \mid (f(x,\xi_1),\ldots,f(x,\xi_\ell))^T - (f^*(\xi_1),\ldots,f^*(\xi_\ell))^T \leq t1_\ell\} \\ &= \min\{\max_{\xi \in \mathscr{U}} (f(x,\xi) - f^*(\xi)) \mid x \in \mathfrak{A}_2\} \\ &= \min\{\rho_{dRC}(x) \mid x \in \mathfrak{A}_2\}.\end{aligned}$$

□

Substituted, we could put $\widetilde{B_2} := B_1 = \mathbb{R}^\ell_+$ and we could minimize $\varphi_{-B_1,k_1}$ over the set $\widetilde{\mathscr{F}_2} := \{(f(x,\xi_1),\ldots,f(x,\xi_\ell))^T - (f^*(\xi_1),\ldots,f^*(\xi_\ell))^T \mid x \in \mathfrak{A}_2\}$. Hence, one can find that (dRC) is a shifted version of (RC), i.e., whenever we consider a finite set of uncertainties and $f^*$ is known beforehand, both counterpart problems (strict robustness and deviation robustness) can be solved within the same complexity. But, (dRC) is usually harder to solve than (RC) for general uncertainty sets.

**Remark 14.3.2** Taking into account Theorems 4.4.4, 4.5.5 and 4.6.6, we obtain that the functional $\varphi_{-B_2,k_2}$ is continuous, finite-valued, convex, $\mathbb{R}^\ell_+$-monotone and strictly $(\mathrm{int}\,\mathbb{R}^\ell_+)$-monotone. Note that we cannot apply Propositions 4.7.3 and 4.7.1 since $B_2$ is not a cone.

**Remark 14.3.3** The concept of deviation robustness can be described by the Chebyshev scalarization (see Sect. 7.3.1) similar to the case of strict robustness (not with the origin as reference point but with the ideal point $f^*$ defined in (14.3.1) as reference point). Once again, this shows the close relationship between these two robustness concepts, compare Sect. 7.3.1 and Kouvelis and Sayin [308, 401].

## 14.4 Reliable Robustness

In many applications, it is difficult to find a point $x$ that fulfills all constraints $F_i(x,\xi) \leq 0$ $(i \in \{1,\ldots,m\})$ for all $\xi \in \mathscr{U}$, or, from the practical point of view, it is not useful for $x$ to fulfill the constraints at the cost of minimality of $f$.

Therefore, we study the concept of reliable robustness, where the constraints are allowed to differ from the original optimization problem. Now, we allow the constraints to fulfill a certain infeasibility tolerance $\delta_i \in \mathbb{R}_+$ in order to score soft constraints $F_i(x, \xi) \leq \delta_i$ instead of having strong restrictions $F_i(x, \xi) \leq 0$ for all $\xi \in \mathcal{U}$. However, the original restrictions for the nominal value $\hat{\xi}$ should be satisfied, i.e., $F_i(x, \hat{\xi}) \leq 0,\ i \in \{1, \ldots, m\}$. With this intention, the *reliably robust counterpart* of (14.1.3) introduced by Ben-Tal and Nemirovski in [38], is given by

$$\begin{aligned} &\rho_{rRC}(x) = \max_{\xi \in \mathcal{U}} f(x, \xi) \to \min \\ &\text{subject to } F_i(x, \hat{\xi}) \leq 0,\ i \in \{1, \ldots, m\}, \\ &\forall \xi \in \mathcal{U} :\ F_i(x, \xi) \leq \delta_i,\ i \in \{1, \ldots, m\}, \\ &x \in \mathbb{R}^n. \end{aligned} \tag{$rRC$}$$

We call a feasible solution of $(rRC)$ *reliably robust.* If $\delta_i = 0$ for all $i \in \{1, \ldots, m\}$, the reliably robust counterpart problem $(rRC)$ is equivalent to the strictly robust optimization problem $(RC)$. Therefore, strict robustness is a special case of reliable robustness.

In [294, Theorem 5], the following relationship between $(\mathrm{P}_{k,B,\mathcal{F}})$ and $(rRC)$ is shown.

**Theorem 14.4.1** *We look at*

$$\mathfrak{A}_3 := \{x \in \mathbb{R}^n \mid F_i(x, \hat{\xi}) \leq 0,\ \forall \xi \in \mathcal{U} :\ F_i(x, \xi) \leq \delta_i,\ i \in \{1, \ldots, m\}\}, \tag{14.4.1}$$

$$B_3 := \mathbb{R}^\ell_+, \tag{14.4.2}$$

$$k_3 := 1_\ell, \tag{14.4.3}$$

$$\mathcal{F}_3 = \{(f(x, \xi_1), \ldots, f(x, \xi_\ell))^T \mid x \in \mathfrak{A}_3\}. \tag{14.4.4}$$

*For* $k = k_3,\ B = B_3$, *condition (14.1.1) is fulfilled and with* $\mathcal{F} = \mathcal{F}_3$, *problem* $(\mathrm{P}_{k,B,\mathcal{F}})$ *is equivalent to problem* $(rRC)$ *in the following sense:*

$$\begin{aligned} &\min\{\varphi_{-B_3,k_3}(y) \mid y \in \mathcal{F}_3\} \\ &= \varphi_{-B_3,k_3}(y^*) \\ &= \min\{\rho_{rRC}(x) \mid x \in \mathfrak{A}_3\} \\ &= \rho_{rRC}(x^*), \end{aligned}$$

*where* $y^* = (f(x^*, \xi_1), \ldots, f(x^*, \xi_\ell))^T$.

Taking into account that $\rho_{rRC} = \rho_{RC}$ and only the set $\mathcal{F}_3$ of feasible points differs from the concept of strict robustness, we relinquish the proof of Theorem 14.4.1.

**Remark 14.4.2** Because of $\varphi_{-B_3,k_3} = \varphi_{-B_1,k_1}$, we obtain that the functional $\varphi_{-B_3,k_3}$ is continuous, finite-valued, $\mathbb{R}^\ell_+$-monotone, strictly ($\operatorname{int} \mathbb{R}^\ell_+$)-monotone and sublinear from Propositions 4.7.3 and 4.7.1.

**Remark 14.4.3** Similar to the concept of strict robustness, the concept of reliable robustness is described by the Chebyshev scalarization (see Sect. 7.3.1) with the origin as reference point and on the basis of a relaxed feasible set, as a special case of functional (14.1.2).

## 14.5 Light Robustness

The concept of light robustness is useful when we examine a certain variation of the constraints $F_i(x, \xi) \leq \delta_i$ (where $F_i,\ \delta_i,\ i \in \{1, \ldots, m\}$, are defined as in Sect. 14.4) and these tolerances are to be minimized. This concept was introduced by Fischetti and Monaci in [163] for linear problems with the $\Gamma$-uncertainty set (the $\Gamma$-uncertainty set was introduced by Bertsimas and Sim [50]) and generalized to common optimization problems under uncertainty by Schöbel [405]. Suppose that $\varphi_*$ is the optimal value of the nominal problem $(Q(\hat{\xi}))$. We assume that $\varphi_*$ is positive. Furthermore, we suppose that the nominal value $f(x, \hat{\xi})$ is bounded by $(1+\gamma)\varphi_*$, where $\gamma \geq 0$. A solution of the *lightly robust counterpart* of (14.1.3)

$$
\begin{aligned}
&\rho_{lRC}(\delta, x) = \sum_{i=1}^{m} w_i \delta_i \ \rightarrow \ \min \\
&\text{subject to } F_i(x, \hat{\xi}) \leq 0,\ i \in \{1, \ldots, m\}, \\
&f(x, \hat{\xi}) \leq (1+\gamma)\varphi_*, \\
&\forall \xi \in \mathcal{U}:\ F_i(x, \xi) \leq \delta_i,\ i \in \{1, \ldots, m\}, \\
&x \in \mathbb{R}^n, \\
&\delta_i \in \mathbb{R}_+,\ i \in \{1, \ldots, m\},
\end{aligned}
\qquad (lRC)
$$

where $w_i \geq 0,\ i \in \{1, \ldots, m\},\ \sum_{i=1}^m w_i = 1$, is called *lightly robust*.

The equivalence of $(\mathrm{P}_{k,B,\mathcal{F}})$ and $(lRC)$ in the following theorem is shown in [294, Theorem 6].

**Theorem 14.5.1** *We consider*

$$B_4 := \{(\delta_1, \ldots, \delta_m)^T \mid \sum_{i=1}^{m} w_i\delta_i \geq 0,\ \delta_i \in \mathbb{R},\ i \in \{1, \ldots, m\}\}, \tag{14.5.1}$$

$$k_4 := 1_m, \tag{14.5.2}$$

$$\begin{aligned}\mathcal{F}_4 =&\{(\delta_1, \ldots, \delta_m)^T \mid \exists x \in \mathbb{R}^n :\ F_i(x, \hat{\xi}) \leq 0,\ f(x, \hat{\xi}) \leq (1+\gamma)\varphi_*,\\ &\forall \xi \in \mathcal{U} :\ F_i(x, \xi) \leq \delta_i,\ \delta_i \in \mathbb{R}_+,\ i \in \{1, \ldots, m\}\}.\end{aligned} \tag{14.5.3}$$

*For $k = k_4$, $B = B_4$, condition (14.1.1) is fulfilled and with $\mathcal{F} = \mathcal{F}_4$, problem $(\mathrm{P}_{k,B,\mathcal{F}})$ is equivalent to problem $(lRC)$ in the following sense:*

$$\begin{aligned}&\min\{\varphi_{-B_4,k_4}(y) \mid y \in \mathcal{F}_4\}\\ &= \varphi_{-B_4,k_4}(y^*)\\ &= \min\{\rho_{lRC}(\delta) \mid \delta \in \mathcal{F}_4\}\\ &= \rho_{lRC}(\delta^*),\end{aligned}$$

*where $y^* = \delta^* = (\delta_1^*, \ldots, \delta_m^*)^T$.*

***Proof*** Taking into account $B_4 + \mathbb{R}_+ k_4 = \{(\delta_1, \ldots, \delta_m)^T \in \mathbb{R}^m \mid \sum_{i=1}^m w_i\delta_i \geq 0\} + \mathbb{R}_+ 1_m \subseteq B_4$, we conclude that $k_4 \in 0^+B_4 \setminus \{0\}$ (see (14.1.1)) is fulfilled in $\mathbb{R}^m$. Furthermore,

$$\begin{aligned}\min_{y\in\mathcal{F}_4} \varphi_{-B_4,k_4}(y) &= \min_{y\in\mathcal{F}_4} \min\{t \in \mathbb{R} \mid y \in tk_4 - B_4\}\\ &= \min_{y\in\mathcal{F}_4} \min\{t \in \mathbb{R} \mid y - tk_4 \in -B_4\}\\ &= \min_{\delta\in\mathcal{F}_4} \min\{t \in \mathbb{R} \mid \sum_{i=1}^{m} w_i(\delta_i - t) \leq 0\}\\ &= \min_{\delta\in\mathcal{F}_4} \min\{t \in \mathbb{R} \mid \sum_{i=1}^{m} w_i\delta_i \leq t\sum_{i=1}^{m} w_i\}\\ &= \min\{\sum_{i=1}^{m} w_i\delta_i \mid \delta \in \mathcal{F}_4\}\\ &= \min\{\rho_{lRC}(\delta) \mid \delta \in \mathcal{F}_4\}.\end{aligned}$$

□

**Remark 14.5.2** Since $B_4$ is a nontrivial, closed, convex cone with $k_4 \in \operatorname{int} B_4$, Propositions 4.7.3 and 4.7.1 yield that $\varphi_{-B_4,k_4}$ is continuous, finite-valued, $\mathbb{R}^m_+$-monotone, strictly ($\operatorname{int}\mathbb{R}^m_+$)-monotone and sublinear.

**Remark 14.5.3** We can interpret the concept of light robustness studied in this section as a weighted sum approach with upper bound constraints (in the dimension given by the number of constraints of the optimization problem with uncertainties), including the original restrictions in the weighted objective function.

## 14.6 Stochastic Optimization

In stochastic optimization (see Birge and Louveaux [51] and Shapiro et al. [408]), a certain knowledge of the probability distribution of the uncertainties is supposed. So, stochastic optimization models have an other conceptual framework than robust optimization models. In this section, we study two-stage stochastic optimization models (for early references, see Beale [32], Dantzig [108] and Tintner [450]). When the realization of the random data is known, two-stage stochastic optimization problems allow a later correction of a solution $x$ selected in stage 1 of the decision process by a recourse action $u$. Taking into account that we supposed that the scenario set $\mathcal{U}$ is finite, each scenario $\xi_j \in \mathcal{U}$ has an associated probability $p_j \geq 0$, $j \in \{1, \dots, \ell\}$, $\sum_{j=1}^{\ell} p_j = 1$. The *two-stage stochastic counterpart* is formulated as

$$\begin{aligned} &\rho_{SP}(x) = \mathbb{E}[Q(x, \xi)] = \sum_{j=1}^{\ell} p_j Q(x, \xi_j) \to \min \\ &\text{subject to } x \in X. \end{aligned} \tag{14.6.1}$$

In (14.6.1), $X$ denotes the feasible set of the first-stage problem. For example, this feasible set could be defined based on the nominal scenario as $X := \{x \in \mathbb{R}^n \mid F_i(x, \hat{\xi}) \leq 0,\ i \in \{1, \dots, m\}\}$, or as the set of strictly robust solutions $X = \mathfrak{A}$ (see (14.2.1)). We study the problem to minimize the expectation of the overall cost $Q(x, \xi)$ that involves an optimal recourse action $u$ ( for given $x \in X$ and known $\xi \in \mathcal{U}$), i.e., an optimal solution of the second-stage problem

$$\begin{aligned} &Q(x, \xi) = \min f(x, u, \xi) \\ &\text{subject to } u \in \mathcal{G}(x, \xi). \end{aligned} \tag{14.6.2}$$

In (14.6.2), the second-stage objective function $f(x, u, \xi)$ and the feasible set $\mathcal{G}(x, \xi)$ of the second-stage problem are both parametrized with respect to the stage 1 solution $x \in X$ and the scenario $\xi \in \mathcal{U}$.

In consideration of the scalar optimization problem involving uncertainties (14.1.3), we can suppose that the objective function $f$ in (14.1.3) depends both on the first-stage and the second-stage variables, i.e., on the nominal cost and the cost of the recourse action. Therefore, we discuss the following specification of problem (14.6.1):

$$\begin{aligned}
&\rho_{SP}(x,u) = \sum_{j=1}^{\ell} p_j f(x,u_j,\xi_j) \to \min \\
&\text{subject to } \forall \xi_j \in \mathcal{U}: \; F_i(x,\xi_j) - \delta_j(u_j) \le 0, \; i \in \{1,\dots,m\}, \\
&x \in \mathbb{R}^n, \\
&u_j \in \mathcal{G}(x,\xi_j), \; j \in \{1,\dots,\ell\},
\end{aligned} \tag{SP}$$

with adjustments $\delta_j : \mathbb{R}^{\bar{n}} \to \mathbb{R}$ depending on the second-stage decisions $u_j \in \mathbb{R}^{\bar{n}}$, $j \in \{1,\dots,\ell\}$.

**Remark 14.6.1** We obtain a static model as a special case in which the second-stage variables $u \in \mathbb{R}^{\bar{n}^\ell}$ can be omitted if we put $\mathcal{G}(x,\xi) = \emptyset$ in the two-stage stochastic programming formulation $(SP)$. Because this model plays a special role in the comparison of the different concepts for robustness and stochastic optimization in Sect. 14.9, we mention it here for the sake of completeness.

$$\begin{aligned}
&\rho_{sSP}(x) = \sum_{j=1}^{\ell} p_j f(x,\xi_j) \to \min \\
&\text{subject to } \forall \xi_j \in \mathcal{U}: \; F_i(x,\xi_j) \le 0, \; i \in \{1,\dots,m\}, \\
&x \in \mathbb{R}^n.
\end{aligned} \tag{sSP}$$

The next theorem is shown in [294, Theorem 7] and presents the relationships between problems $(\mathrm{P}_{k,B,\mathcal{F}})$ and $(SP)$.

**Theorem 14.6.2** *Consider*

$$\begin{aligned}
\mathfrak{A}_5 :=&\{(x,u) := (x,u_1,\dots,u_\ell) \in \mathbb{R}^{n\times\bar{n}^\ell} \mid \forall \xi_j \in \mathcal{U}: \\
&F_i(x,\xi_j) - \delta_j(u_j) \le 0, \; i \in \{1,\dots,m\}\},
\end{aligned} \tag{14.6.3}$$

$$B_5 := \{(y_1,\dots,y_\ell)^T \mid \sum_{j=1}^{\ell} p_j y_j \ge 0, \; y_j \in \mathbb{R}, \; j \in \{1,\dots,\ell\}\}, \tag{14.6.4}$$

$$k_5 := 1_\ell, \tag{14.6.5}$$

$$\mathcal{F}_5 =\{(f(x,u_1,\xi_1),\dots,f(x,u_\ell,\xi_\ell))^T \mid (x,u) \in \mathfrak{A}_5\}. \tag{14.6.6}$$

*For $k = k_5$, $B = B_5$, condition (14.1.1) is fulfilled and with $\mathcal{F} = \mathcal{F}_5$, problem $(\mathrm{P}_{k,B,\mathcal{F}})$ is equivalent to problem $(SP)$ in the following sense:*

$$\begin{aligned}
&\min\{\varphi_{-B_5,k_5}(y) \mid y \in \mathcal{F}_5\} \\
&\quad = \varphi_{-B_5,k_5}(y^*) \\
&\quad = \min\{\rho_{SP}(x,u) \mid (x,u) \in \mathfrak{A}_5\} \\
&\quad = \rho_{SP}(x^*,u^*),
\end{aligned}$$

*where* $y^* = (f(x^*, u_1^*, \xi_1), \ldots, f(x^*, u_\ell^*, \xi_\ell))^T$.

***Proof*** Because of $B_5 + \mathbb{R}_+ k_5 = \{(y_1, \ldots, y_\ell)^T \in \mathbb{R}^\ell \mid \sum_{j=1}^{\ell} p_j y_j \geq 0\} + \mathbb{R}_+ 1_m \subseteq B_5$, $k_5 \in 0^+ B_5 \setminus \{0\}$ (see (14.1.1)) is satisfied. Furthermore,

$$\begin{aligned}
\min_{y \in \mathcal{F}_5} \varphi_{-B_5,k_5}(y) &= \min_{y \in \mathcal{F}_5} \min\{t \in \mathbb{R} \mid y \in tk_5 - B_5\} \\
&= \min_{y \in \mathcal{F}_5} \min\{t \in \mathbb{R} \mid y - tk_5 \in -B_5\} \\
&= \min_{y \in \mathcal{F}_5} \min\{t \in \mathbb{R} \mid \sum_{j=1}^{\ell} p_j(y_j - t) \leq 0\} \\
&= \min_{y \in \mathcal{F}_5} \min\{t \in \mathbb{R} \mid \sum_{j=1}^{\ell} p_j y_j \leq t \underbrace{\sum_{j=1}^{\ell} p_j}_{=1}\} \\
&= \min\{\sum_{j=1}^{\ell} p_j y_j \mid y \in \mathcal{F}_5\} \\
&= \min\{\rho_{SP}(x,u) \mid (x,u) \in \mathfrak{A}_5\}.
\end{aligned}$$

□

**Remark 14.6.3** By Propositions 4.7.3 and 4.7.1, the functional $\varphi_{-B_5,k_5}$ is continuous, finite-valued, $\mathbb{R}^\ell_+$-monotone, strictly $(\operatorname{int} \mathbb{R}^\ell_+)$-monotone and sublinear since $B_5$ is a nontrivial, closed, convex cone with $k_5 \in \operatorname{int} B_5$.

**Remark 14.6.4** Comparable with the case of light robustness, the two-stage stochastic optimization problem formulated above can be interpreted as a weighted sum approach, nevertheless, in this case with a relaxed feasible set. Gast [172] observed this relation in the multiobjective context too. The feasible set is in fact identical to the set of strictly robust solutions $\mathfrak{A}$ (and not relaxed) (see (14.2.1)) in the special case of the static model $(sSP)$.

## 14.7 $\varepsilon$-Constraint Robustness

As shown in Sect. 7.3, the nonlinear scalarizing functional (14.1.2) describes many scalarization methods known in multiobjective optimization, for instance the weighted Chebyshev scalarization or weighted sum scalarization. In the previous sections, we have seen that these scalarizations can be considered in the context of robustness. Furthermore, one can derive new concepts for robustness that are well matched to specific requirements of a decision maker. As an example, we will use the well-known *ε-constraint scalarization* (see Sect. 7.3.8 and [132, 136, 215, 218]) in order to illustrate this fact. We will analyze which type of robust counterpart is defined by this scalarization.

A comparable similar analysis can be done for every diversification of the parameters $B$, $k$ and for every feasible set $\mathcal{F}$ in $(\mathrm{P}_{k,B,\mathcal{F}})$, i.e., several other possibilities for defining a robust counterpart of an optimization problem under uncertainty may be derived.

First, we recall the ε-constraint scalarization introduced in Sect. 7.3.8 using the notations of this section. For this purpose, let some $j \in \{1, \ldots, \ell\}$ and some real values $\varepsilon_q \in \mathbb{R}$, $q = 1, \ldots, \ell, q \neq j$ be given. Now, the ε-constraint scalarization is given by

$$k_6 = (k_{61}, \ldots, k_{6\ell})^T \text{ where } k_{6q} = \begin{cases} 1 \text{ for } q = j, \\ 0 \text{ for } q \neq j, \end{cases} \tag{14.7.1}$$

$$B_6 := \mathbb{R}^\ell_+ - \bar{b}, \text{ with } \bar{b} = (\bar{b}_1, \ldots, \bar{b}_\ell)^T, \ \bar{b}_q = \begin{cases} 0 & \text{for } q = j, \\ \varepsilon_q & \text{for } q \neq j, \end{cases} \tag{14.7.2}$$

$$\mathcal{F}_6 = \{(f(x, \xi_1), \ldots, f(x, \xi_\ell))^T \mid x \in \mathfrak{A}\}. \tag{14.7.3}$$

Considering these parameters, the functional $\varphi_{-B_6,k_6} : \mathbb{R}^\ell \to \overline{\mathbb{R}}$ describes the ε-constraint-method (see Sect. 7.3.8, Eichfelder [136] and Haimes, Lasdon, Wismer [218]). Then, the following assertion concerning the relationship between the problems $(\mathrm{P}_{k,B,\mathcal{F}})$ and $(\varepsilon RC)$ (see [294, Theorem 8]) holds.

**Theorem 14.7.1** *We consider* $\varepsilon := (\varepsilon_1, \ldots, \varepsilon_\ell)^T \in \mathbb{R}^\ell$ *and* $j \in \{1, \ldots, \ell\}$. *Then, for* $k = k_6$, $B = B_6$, *condition (14.1.1) is fulfilled and with* $\mathcal{F} = \mathcal{F}_6$, *problem* $(\mathrm{P}_{k,B,\mathcal{F}})$ *is equivalent to*

$$\begin{aligned} &\rho_{\varepsilon RC}(x) = f(x, \xi_j) \to \min \\ &\textit{subject to } \forall \xi \in \mathcal{U} : \ F_i(x, \xi) \leq 0, \ i \in \{1, \ldots, m\}, \\ &x \in \mathbb{R}^n, \\ &f(x, \xi_q) \leq \varepsilon_q, \ q \in \{1, \ldots, \ell\}, \ q \neq j. \end{aligned} \tag{$\varepsilon RC$}$$

***Proof*** Because of $B_6 + \mathbb{R}_+ k_6 \subseteq B_6$, condition $k_6 \in 0^+ B_6 \setminus \{0\}$ (see (14.1.1)) is fulfilled. Furthermore,

$$\begin{aligned}\min_{y\in\mathcal{F}_6}\varphi_{-B_6,k_6}(y) &= \min_{y\in\mathcal{F}_6}\min\{t\in\mathbb{R}\mid y\in tk_6-B_6\}\\ &= \min_{y\in\mathcal{F}_6}\min\{t\in\mathbb{R}\mid y-tk_6\in -B_6\}\\ &= \min_{x\in\mathfrak{A}}\min\{t\in\mathbb{R}\mid f(x,\xi_j)\leq t,\ f(x,\xi_q)\leq\varepsilon_q,\ q\in\{1,\dots,\ell\},\ q\neq j\}\\ &= \min\{\rho_{\varepsilon RC}(x)\mid x\in\mathbb{R}^n,\ \forall\xi\in\mathcal{U}:\ F_i(x,\xi)\leq 0,\ i\in\{1,\dots,m\},\\ &\qquad f(x,\xi_q)\leq\varepsilon_q,\ q\in\{1,\dots,\ell\},\ q\neq j\}.\end{aligned}$$

□

Theorem 14.7.1 indicates that problem ($\varepsilon RC$) can be formulated using the nonlinear scalarizing functional $\varphi_{-B,k}$. We call ($\varepsilon RC$) the **$\varepsilon$-constraint robust counterpart** of (14.1.3). Now, we explain its meaning for robust optimization. In contrast to the other robustness concepts, the parameter $k_6$ symbolizes that only a single objective function is minimized, i.e., the decision maker selects one particular objective function that is to be minimized subject to the constraints that were also used in strict and deviation-robustness. Moreover, the former objective functions $f(x,\xi_q)$, $q\in\{1,\dots,\ell\}$, $q\neq j$, are moved to the constraints. This concept is useful if a solution is required with a given nominal quality for every scenario $\xi_q$, $q=1,\dots,\ell, q\neq j$ while finding the best possible for the remaining scenarios $j$. One important question arises when applying this concept: Can a decision maker be sure how to pick the upper bounds $\varepsilon_l$ for these constraints? It is evident that the set of feasible solutions of ($\varepsilon RC$) may be empty or the objective function value of $f(x,\xi_j)$ may become undesirable if the bounds $\varepsilon_l$ are chosen too small. Furthermore, if the bounds $\varepsilon_j$ are chosen too large, the quality decreases for the other scenarios.

It is important to mention that we could have included the constraint $f(x,\xi_q)\leq\varepsilon_q$, $q\in\{1,\dots,\ell\}$, $q\neq j$ in the set of feasible elements $\widetilde{\mathcal{F}}_6$, and we would have obtained

$$\widetilde{\mathcal{F}}_6=\{(f(x,\xi_1),\dots,f(x,\xi_\ell))^T\mid x\in\mathbb{R}^n:\ f(x,\xi_q)\leq\varepsilon_q,\ q\in\{1,\dots,\ell\},\ q\neq j,$$

$$\forall\xi\in\mathcal{U}:\ F_i(x,\xi)\leq 0,\ i\in\{1,\dots,m\}\}.$$

Then, we could have used $\widetilde{B_6}=\mathbb{R}^\ell_+$ instead of $B_6$, but the set of feasible elements $\widetilde{\mathcal{F}}_6$ would have been smaller and possibly harder to deal with.

**Remark 14.7.2** Because condition (14.1.1) is satisfied, Theorem 4.4.4 yields that $\varphi_{-B_6,k_6}:\mathbb{R}^\ell\to\overline{\mathbb{R}}$ is lower semi-continuous, convex and proper. Furthermore, $\varphi_{-B_6,k_6}$ is strictly ($\operatorname{int}\mathbb{R}^\ell_+$)-monotone on $\operatorname{dom}\varphi_{-B_6,k_6}$ applying Theorem 4.5.5 (with $A=-B_6$ and $k=k_6$).

## 14.8 Comparison of the Concepts for Robust Optimization

The introduced concepts of robustness and stochastic programming, and the corresponding parameters $B$, $k$ in the formulation of the functional $\varphi_{-B,k}$ as well as the corresponding sets $\mathcal{F}$ are summarized in the following table (presented in [294]).

| Robustness Concept | $B$ | $k$ | $\mathcal{F}$ |
|---|---|---|---|
| Strict | $\mathbb{R}^\ell_+$ | $1_\ell$ | $\{(f(x,\xi_1),\dots,f(x,\xi_\ell))^T \mid x \in \mathfrak{A}_1\}$ |
| Deviation | $\mathbb{R}^\ell_+ - f^*$ | $1_\ell$ | $\{(f(x,\xi_1),\dots,f(x,\xi_\ell))^T \mid x \in \mathfrak{A}_2\}$ |
| Reliable | $\mathbb{R}^\ell_+$ | $1_\ell$ | $\{(f(x,\xi_1),\dots,f(x,\xi_\ell))^T \mid x \in \mathfrak{A}_3\}$ |
| Light | $B_4$ | $1_m$ | $\mathcal{F}_4$, see below |
| Stochastic | $B_5$ | $1_\ell$ | $\{(f(x,u_1,\xi_1),\dots,f(x,u_\ell,\xi_\ell))^T \mid (x,u) \in \mathfrak{A}_5\}$ |
| $\varepsilon$-constraint | $\mathbb{R}^\ell_+ - \bar{b}$ | $k_q = \begin{cases} 1 \text{ for } q=j \\ 0 \text{ for } q \neq j \end{cases}$ | $\{(f(x,\xi_1),\dots,f(x,\xi_\ell))^T \mid x \in \mathfrak{A}_6\}$ |

Here,

$$f^* = (f^*(\xi_1),\dots,f^*(\xi_\ell))^T, \text{ where } f^*(\xi) \in \mathbb{R} \text{ is the optimal value of problem } (Q(\xi)),$$

$$B_4 = \{(\delta_1,\dots,\delta_m)^T \in \mathbb{R}^m \mid \sum_{i=1}^m w_i\delta_i \geq 0\},\ w_i \geq 0,\ i \in \{1,\dots,m\},\ \sum_{i=1}^m w_i = 1,$$

$$B_5 = \{(y_1,\dots,y_\ell)^T \in \mathbb{R}^\ell \mid \sum_{j=1}^\ell p_j y_j \geq 0\},\ p_j \geq 0,\ j \in \{1,\dots,\ell\},\ \sum_{j=1}^\ell p_j = 1,$$

$$\bar{b} = (\bar{b}_1,\dots,\bar{b}_\ell)^T, \text{ where } \bar{b}_q = \begin{cases} 0 & \text{for } q = j, \\ \varepsilon_q & \text{for } q \neq j, \end{cases}, (j \in \{1,\dots,\ell\}, \text{ cf. } (14.7.2)),$$

$$\mathcal{F}_4 = \{(\delta_1,\dots,\delta_m)^T \mid \exists x \in \mathbb{R}^n : F_i(x,\hat{\xi}) \leq 0,\ f(x,\hat{\xi}) \leq (1+\gamma)\varphi_*,\ \forall \xi \in \mathcal{U} : F_i(x,\xi) \leq \delta_i,$$
$$\delta_i \in \mathbb{R}_+,\ i \in \{1,\dots,m\}\}.$$

Furthermore, the following sets $\mathfrak{A}$ are used:

$$\mathfrak{A}_1 = \mathfrak{A},$$
$$\mathfrak{A}_2 = \mathfrak{A},$$
$$\mathfrak{A}_3 = \{x \in \mathbb{R}^n \mid F_i(x,\hat{\xi}) \leq 0,\ \forall \xi \in \mathcal{U} : F_i(x,\xi) \leq \delta_i,\ i \in \{1,\dots,m\}\},$$
$$\mathfrak{A}_5 = \{(x,u) := (x,u_1,\dots,u_\ell) \in \mathbb{R}^{n \times \bar{n}^\ell} \mid \forall \xi_j \in \mathcal{U} : F_i(x,\xi_j) - \delta_j(u_j) \leq 0,\ i \in \{1,\dots,m\}\},$$
$$\mathfrak{A}_6 = \{x \in \mathbb{R}^n \mid x \in \mathfrak{A},\ f(x,\xi_q) \leq \varepsilon_q,\ q \in \{1,\dots,\ell\},\ q \neq j\},\ j \in \{1,\dots,\ell\}.$$

In the special case of static stochastic programming $(sSP)$, we have $\mathfrak{A}_5 = \mathfrak{A}$ and $\mathcal{F}_5 = \mathcal{F}_1$.

Important properties of the nonlinear scalarizing functional $\varphi_{-B,k}$ defined by (14.1.2) that were used for the description of the robust and stochastic optimization counterparts are summarized in the following corollary (as conclusions of Propositions 4.7.3 and 4.7.1) (see [294, Corollary 2]).

**Corollary 14.8.1** *The following properties are fulfilled for $i = 1, 2, 3, 5$ ($i = 1$: strict robustness, $i = 2$: deviation robustness, $i = 3$: reliable robustness, $i = 5$: stochastic programming): The corresponding functional $\varphi_{-B_i,k_i}$ is continuous, finite-*

*valued, convex,* $\mathbb{R}^{\ell}_{+}$*-monotone and strictly* $(\operatorname{int}\mathbb{R}^{\ell}_{+})$*-monotone, and the following properties hold:*

$$\forall y \in \mathcal{F}_i,\ \forall r \in \mathbb{R}: \ \varphi_{-B_i,k_i}(y) \le r \iff y \in rk_i - B_i, \tag{P1}$$

$$\forall y \in \mathcal{F}_i,\ \forall r \in \mathbb{R}: \ \varphi_{-B_i,k_i}(y + rk_i) = \varphi_{-B_i,k_i}(y) + r. \tag{P2}$$

$$\forall y \in \mathcal{F}_i,\ \forall r \in \mathbb{R}: \ \varphi_{-B_i,k_i}(y) = r \iff y \in rk_i - \operatorname{bd} B_i, \tag{P3}$$

$$\forall y \in \mathcal{F}_i,\ \forall r \in \mathbb{R}: \ \varphi_{-B_i,k_i}(y) < r \iff y \in rk_i - \operatorname{int} B_i. \tag{P4}$$

*For* $i = 1, 3, 5$, $\varphi_{-B_i,k_i}$ *is even sublinear. For* $i = 4$ *(light robustness), the properties (P1)–(P4) are fulfilled, and* $\varphi_{-B_4,k_4}$ *is continuous, sublinear,* $\mathbb{R}^{m}_{+}$*-monotone and strictly* $(\operatorname{int}\mathbb{R}^{m}_{+})$*-monotone.*

The next result is shown in [294, Corollary 3].

**Corollary 14.8.2** *The functional* $\varphi_{-B_6,k_6} : \mathbb{R}^{\ell} \to \overline{\mathbb{R}}$ *is lower semicontinuous, convex, proper,* $\mathbb{R}^{\ell}_{+}$*-monotone and strictly* $(\operatorname{int}\mathbb{R}^{\ell}_{+})$*-monotone on* $\operatorname{dom}\varphi_{-B_6,k_6}$*, and the properties (P1) and (P2) from Corollary 14.8.1 hold for* $i = 6$.

**Remark 14.8.3** We derived the $\mathbb{R}^{\ell}_{+}$-monotonicity, continuity, translation invariance along $k_i$ and convexity of the functional $\varphi_{-B_i,k_i}$ for i=1, ..., 5. These important properties appear in the theory of risk measures (see Sect. 15.1.1) too. Moreover, the functional $\varphi_{-B_i,k_i}$ is sublinear (see Proposition 4.7.1(b)) for $i = 1, 3, 4, 5$.

## 14.9 Multiobjective Counterpart Problems and Relations to Scalar Robust Optimization and Stochastic Programming

In this section, we replace a scalar optimization problem under uncertainty $(Q(\xi), \xi \in \mathcal{U})$, as introduced in (14.1.3) by its multiobjective (deterministic) counterpart. Here, the main idea is that every scenario $\xi_l \in \mathcal{U}, l = 1, \dots, \ell$ brings its own objective function $h_l(x) := f(x, \xi_l)$, the only exception being the case of light robustness where the roles of objective and constraints are conversed. We follow the discussion of different robustness concepts above, then the multiobjective counterparts formulated in this section can be characterized with respect to the solution set $\mathfrak{A}$, i.e., the way in which the (uncertain) constraints are handled. In the case of stochastic programming, we focus on the static model $(sSP)$ in order to simplify the following analysis.

Consider

$$h(x) := \begin{pmatrix} h_1(x) \\ \vdots \\ h_\ell(x) \end{pmatrix} := \begin{pmatrix} f(x,\xi_1) \\ \vdots \\ f(x,\xi_\ell) \end{pmatrix}. \tag{14.9.1}$$

We recall from (14.2.1) (see also Sect. 14.8) that

$$\mathfrak{A}_1 = \mathfrak{A}_2 = \mathfrak{A}_5 = \mathfrak{A} = \{x \in \mathbb{R}^n \mid \forall \xi \in \mathcal{U} : F_i(x, \xi) \leq 0, \ i \in \{1, \ldots, m\}\}.$$

Then, the *multiobjective strictly robust counterpart* to $(Q(\xi), \xi \in \mathcal{U})$ is given by

$$\mathbb{R}^\ell_+ - \text{Opt}_{x \in \mathfrak{A}_1} h(x), \qquad (RC')$$

where $h(\mathfrak{A}_1) = \mathcal{F}_1$ (see (14.2.5)). Here, we are looking for elements of $\text{Eff}(\mathcal{F}_1, \mathbb{R}^\ell_+)$ and $\text{WEff}(\mathcal{F}_1, \mathbb{R}^\ell_+)$.

Comparable, recall from (14.4.1) that

$$\mathfrak{A}_3 := \{x \in \mathbb{R}^n \mid F_i(x, \hat{\xi}) \leq 0, \ \forall \xi \in \mathcal{U} : F_i(x, \xi) \leq \delta_i, \ i \in \{1, \ldots, m\}\}.$$

We consider the *multiobjective reliably robust counterpart* to $(Q(\xi), \xi \in \mathcal{U})$ as

$$\mathbb{R}^\ell_+ - \text{Opt}_{x \in \mathfrak{A}_3} h(x), \qquad (rRC')$$

where $h(\mathfrak{A}_3) = \mathcal{F}_3$ (see (14.4.4)). Here, we are looking for elements of $\text{Eff}(\mathcal{F}_3, \mathbb{R}^\ell_+)$ and $\text{WEff}(\mathcal{F}_3, \mathbb{R}^\ell_+)$.

We now introduce a multiobjective counterpart that corresponds to the lightly robust counterpart $(lRC)$. Let $\mathcal{F}_4$ be given by (14.5.3), i.e.,

$$\begin{aligned}\mathcal{F}_4 =\{(\delta_1, \ldots, \delta_m)^T \mid \exists x \in \mathbb{R}^n : \ & F_i(x, \hat{\xi}) \leq 0, \ f(x, \hat{\xi}) \leq (1+\gamma)\varphi_*, \ \forall \xi \in \mathcal{U} : F_i(x, \xi) \leq \delta_i, \\ & \delta_i \in \mathbb{R}, \ i \in \{1, \ldots, m\}\},\end{aligned}$$

where $\varphi_* > 0$ is the optimal value of the nominal problem (compare $(lRC)$).

We consider the *multiobjective lightly robust counterpart* to $(Q(\xi), \xi \in \mathcal{U})$ by

$$\mathbb{R}^m_+ - \text{Opt}_{x \in \mathfrak{A}_4} h(x), \qquad (lRC')$$

where we are looking for elements of $\text{Eff}(\mathcal{F}_4, \mathbb{R}^m_+)$ and $\text{WEff}(\mathcal{F}_4, \mathbb{R}^m_+)$.

We conclude that problem $(\text{P}_{k_i, B_i, \mathcal{F}_i})$, $i = 1, 2, 5, 6$, $((\text{P}_{k_3, B_3, \mathcal{F}_3})$, $(\text{P}_{k_4, B_4, \mathcal{F}_4})$, respectively) is a scalarization of the multiobjective counterpart $(RC')$ $((rRC')$, $(lRC')$, respectively), using Corollary 6.2.23 together with Corollaries 14.8.1 and 14.8.2.

The following corollary ([294, Corollary 4]) holds due to the monotonicity properties of $\varphi_{-B_i, k_i}$, $i = 1, \ldots, 6$.

**Corollary 14.9.1** *For* $i = 1, 2, 5, 6$ *(*$i = 1$*: strict robustness,* $i = 2$*: deviation robustness,* $i = 5$*: static stochastic programming,* $i = 6$*:* $\varepsilon$*-constraint robustness), we conclude*

$$[\forall y \in \mathcal{F}_i \setminus \{y^*\} : \varphi_{-B_i,k_i}(y^*) < \varphi_{-B_i,k_i}(y)] \Longrightarrow y^* \in \mathrm{Eff}(h(\mathfrak{A}_1), \mathbb{R}^\ell_+),$$
$$[\forall y \in \mathcal{F}_i : \varphi_{-B_i,k_i}(y^*) \leq \varphi_{-B_i,k_i}(y)] \Longrightarrow y^* \in \mathrm{WEff}(h(\mathfrak{A}_1), \mathbb{R}^\ell_+).$$

*Concerning reliably robustness ($i = 3$), it holds that*

$$[\forall y \in \mathcal{F}_3 \setminus \{y^*\} : \varphi_{-B_3,k_3}(y^*) < \varphi_{-B_3,k_3}(y)] \Longrightarrow y^* \in \mathrm{Eff}(h(\mathfrak{A}_3), \mathbb{R}^\ell_+),$$
$$[\forall y \in \mathcal{F}_3 : \varphi_{-B_3,k_3}(y^*) \leq \varphi_{-B_3,k_3}(y)] \Longrightarrow y^* \in \mathrm{WEff}(h(\mathfrak{A}_3), \mathbb{R}^\ell_+).$$

*For light robustness ($i = 4$), we conclude*

$$[\forall y \in \mathcal{F}_4 \setminus \{y^*\} : \varphi_{-B_4,k_4}(y^*) < \varphi_{-B_4,k_4}(y)] \Longrightarrow y^* \in \mathrm{Eff}(\mathcal{F}_4, \mathbb{R}^m_+),$$
$$[\forall y \in \mathcal{F}_4 : \varphi_{-B_4,k_4}(y^*) \leq \varphi_{-B_4,k_4}(y)] \Longrightarrow y^* \in \mathrm{WEff}(\mathcal{F}_4, \mathbb{R}^m_+).$$

Problem $(\mathrm{P}_{k_i,B_i,\mathcal{F}_i})$, $i = 1, 2, 5, 6$, with the nonlinear scalarizing functional $\varphi_{-B_i,k_i}$ is a scalarization of the multiobjective counterpart $(RC^{'})$. Furthermore, $(\mathrm{P}_{k_1,B_1,\mathcal{F}_1})$ ($(\mathrm{P}_{k_2,B_2,\mathcal{F}_2})$, $(\mathrm{P}_{k_5,B_5,\mathcal{F}_5})$, $(\mathrm{P}_{k_6,B_6,\mathcal{F}_6})$, respectively) is equivalent to the optimization problem $(RC)$ ($(dRC)$, $(sSP)$, $(\varepsilon RC)$, respectively). So, we get:

If $x^*$ solves $(RC)$ ($(dRC)$, $(sSP)$, $(\varepsilon RC)$, respectively), then $y^* = (f(x^*, \xi_1), \ldots, f(x^*, \xi_\ell))^T$ is weakly efficient for $(RC^{'})$. If $x^*$ is the unique solution of problem $(RC)$ ($(dRC)$, $(sSP)$, $(\varepsilon RC)$, respectively), then $y^* = (f(x^*, \xi_1), \ldots, f(x^*, \xi_\ell))^T$ is an efficient element for $(RC^{'})$.

**Remark 14.9.2** Thus, the set of (weakly) efficient elements of the multiobjective strictly robust counterpart $(RC^{'})$ incorporates optimal solutions of the (scalar) strictly robust counterpart $(RC)$ (which are generated by Chebyshev scalarization with the origin as reference point, see Sect. 7.3.3), the deviation-robust counterpart $(dRC)$ (Chebyshev scalarization with the ideal point as reference point), the static stochastic programming equivalent $(sSP)$ (weighted sums scalarization), and the $\varepsilon$-constraint robust counterpart $(\varepsilon RC)$ ($\varepsilon$-constraint scalarization), see Sect. 7.3.8.

Similarly, corresponding results also hold for the other concepts of robustness: $(\mathrm{P}_{k_3,B_3,\mathcal{F}_3})$ is equivalent to the reliably robust counterpart $(rRC)$. Hence, we get: If $x^*$ solves $(rRC)$, then $y^* = (f(x^*, \xi_1), \ldots, f(x^*, \xi_\ell))^T$ is weakly efficient for $(rRC^{'})$. If $x^*$ is the unique solution of problem $(rRC)$, then $y^* = (f(x^*, \xi_1), \ldots, f(x^*, \xi_\ell))^T$ is efficient for $(rRC^{'})$. Regarding the concept of light robustness, we have: If $\delta^* = (\delta^*_1, \ldots, \delta^*_m)^T$ is a (unique) solution of the lightly robust counterpart $(lRC)$, then $\delta^*$ is (efficient) weakly efficient for the corresponding multiobjective counterpart $(lRC^{'})$.

**Remark 14.9.3** In the paper [302] by Köbis and Tammer, a new definition of robustness for uncertain vector optimization problems equipped with a variable domination structure, scalarization results and corresponding algorithms for computing robust solutions are derived.

# Chapter 15
# Further Applications

In this chapter, we show that coherent risk measures in Mathematical Finance can be formulated using functionals of type (4.1.11) such that it is possible to use the results from Chap. 4 in order to derive corresponding properties for coherent risk measures. Furthermore, we study the relationship between coherent risk measures and a strictly robust counterpart problem of an optimization problem under uncertainty. The benefit function and shortage function in Mathematical Economics are functions of type (4.1.11). Furthermore, we consider a vector-valued optimal control problem with PDE-constraints and apply the scalarization technique by means of the functional (4.1.11) for deriving characterizations of solutions to this vector-valued optimal control problem that are useful for corresponding adaptive algorithms (see Sect. 13.4). Finally, we use a modification of the functional (4.1.11) discussed in Sect. 9.2 for the formulation of location problems and present necessary optimality conditions for solutions of these (in general nonconvex) location problems.

## 15.1 Mathematical Economics

### *15.1.1 Coherent Risk Measures*

Risk measures play an important role in mathematical finance (see Föllmer, Schied [169]), where *risk measures* are understood as a certain amount of money (capital that secures almost surely the solvency of an enterprise in the case of losses). Analogously, the reserves of insurance companies that ensure the survival in the case of a large number of insurance events can be understood as risk measures.

Assume that $\Omega$ is a fixed set of elementary events (scenarios) supplied with a probability measure $P$ and a field $\Sigma$ of measurable sets. Then, the result of an investment in the future is considered as a random variable $y : \Omega \to \mathbb{R}$. If no investment will be done, then the random variable is equal to zero, positive outcomes are *wins* and negative ones are *losses*. $Y$ denotes a linear space of random variables (of financial



C. Tammer and P. Weidner, *Scalarization and Separation by Translation Invariant Functions*, Vector Optimization, https://doi.org/10.1007/978-3-030-44723-6_15

positions). In terms of mathematical finance, a *financial position* is described by the corresponding payoff profile, i.e., by a functional $y : \Omega \to \mathbb{R}$ with $y$ belonging to a certain class of financial positions. Related to the applications in mathematical finance, we suppose that the linear space of financial positions $Y$ belongs to the class $\mathcal{Y}$ of bounded functions containing the constants. For $w \in \Omega$, $y(w)$ is the discounted net worth of the position at the end of the time period if the scenario $w$ is realized.

An important question for decision makers in mathematical finance is: How to find a quantification of an existing risk such that the solvency is almost sure on one hand and on the other, that not too much capital is engaged? The aim of the decision maker is to quantify the risk of a financial position $y$, i.e., he has to find a function $\mu : Y \to \overline{\mathbb{R}}$ which assigns a real number or the values $-\infty$, $+\infty$ to every financial position $y \in Y$ (see Chap. 3). A classical measure of risk is the variance (in the classical Markowitz model, see [345]). However, in modern approaches in mathematical finance, the portfolio managers consider other measures of risk such as coherent risk measures. We will see that a very natural way to express risk measures is the description by means of functions of type (4.1.11).

Artzner, Delbaen, Eber and Heath [11] introduced axioms to be satisfied for a set $D \subset Y$ of random variables corresponding to *acceptable investments*:

$$(A1)\ \{y \in Y \mid \forall\, w \in \Omega :\ y(w) \geq 0\} \subseteq D,\ D \cap \{y \in Y \mid \forall\, w \in \Omega\colon y(w) < 0\} = \emptyset,$$

$$(A2)\ D \text{ is a cone,}$$

$$(A3)\ D + D \subseteq D.$$

In the case $Y = L^p(\Omega, \Sigma, P)$ for some probability space $(\Omega, \Sigma, P)$ and $p \geq 1$ in the above axioms, "$\forall w \in \Omega$" is to understood as " for $P$-almost all $w \in \Omega$ ".

Of course, if the axioms (A2) and (A3) are fulfilled, then $D$ is a convex cone.

The subsets $D$ of $Y$ satisfying the axioms (A1)–(A3) of acceptable investments can be employed in order to introduce a strict preference relation $\succ$ on $Y$ (see Sect. 6.1): $y^1$ is strictly preferred to $y^2$ (changing from $y^2$ to $y^1$ is an acceptable risk), if $y^1 - y^2$ is an element of $D \setminus \{0\}$, i.e.,

$$y^1 \succ y^2 \iff y^1 - y^2 \in D \setminus \{0\}.$$

Formulated in terms of finance, Axiom (A1) says that every investment with almost sure nonnegative results will be accepted and every investment with almost sure negative results is not acceptable. Furthermore, the cone property in Axiom (A2) means that every nonnegative multiple of an acceptable investment is again acceptable. The convexity attribute in Axiom (A3) says that merging two acceptable investments together, again leads to an acceptable investment. However, the axioms (A2) and (A3) are not useful in some applications. Especially, if the investor does not want to lose more than a certain amount of money. In this case, Föllmer and Schied [168, 169] replace the axioms (A2) and (A3) by a convexity axiom.

A function of the type $\varphi_{A,k}$ in (4.1.11) is used in mathematical finance in order to express a *risk measure* (for example a valuation of risky investments) with respect to an acceptance set $D \subset Y$ and $k \in Y \setminus \{0\}$. Coherent risk measures in form of (4.1.11) are introduced by Artzner, Delbean, Eber and Heath in [11]. The relationships to scalarization methods in vector optimization are mentioned in [219, 238]. Recently, Noyan and Rudolf [371] consider models in multiobjective decision making under uncertainties, especially multiobjective optimization problems with stochastic preference constraints (compare also [332]). It is typical that these problems involve an univariate stochastic preference relation, often based on stochastic dominance, or a scalarization method, for instance a coherent risk measure such as conditional value-at-risk. Noyan and Rudolf adapt certain nonlinear scalarization techniques well known in deterministic multiobjective optimization (see Chap. 7) into a stochastic context by introducing the class of min-biaffine functions. Corresponding optimization problems in finite probability spaces with multivariate stochastic benchmarking constraints based on min-biaffine scalarizations are studied. Furthermore, duality assertions, optimality conditions, and a cut generation method for solving these problems are presented in [371].

The following attributes of *coherent risk measures* $\mu : Y \to \overline{\mathbb{R}}$ are supposed in the papers by Artzner, Delbaen, Eber and Heath [11] and Rockafellar, Uryasev and Zabarankin [390]:

$(P1)$ $\mu(y + tk) = \mu(y) - t$,

$(P2)$ $\mu(0) = 0$ and $\mu(\lambda y) = \lambda\mu(y)$ for all $y \in Y$ and $\lambda > 0$,

$(P3)$ $\mu(y^1 + y^2) \le \mu(y^1) + \mu(y^2)$ for all $y^1, y^2 \in Y$,

$(P4)$ $\mu(y^1) \le \mu(y^2)$ if $y^1 \ge y^2$.

**Remark 15.1.1** The following interpretation (in the scope of Finance) of the properties (P1)–(P4) is important.

In financial terms, the function $\mu$ represents the minimal amount of capital that one has to add such that the whole investment is acceptable.

The translation invariance along $k$ in (P1) says that the risk would be palliated by an additional safe investment with a corresponding amount, particularly, it holds that

$$\mu(y + \mu(y)k) = 0.$$

Furthermore, the positive homogeneity of the risk measure in (P2) says that double risk must be secured by double risk capital.

The subadditivity in (P3) says that a diversification of risk should be remunerated.

Finally, the monotonicity of the risk measure in (P4) says that higher risk needs more risk capital.

Obviously, a risk measure may be negative. Then, the risk measure can be interpreted as a maximal amount of cash that could be given away such that the reduced capital remains acceptable.

Examples of coherent risk measures are the *Average Value at Risk* (see [169, Sect. 4.4, Definition 4.43] and Example 15.1.4) and the *Worst-Case Risk Measure* (see [169, Example 4.8] and Example 15.1.3).

**Example 15.1.2** (*Value at Risk*) We consider a fixed set $\Omega$ of scenarios and a probability measure $P$ on $(\Omega, \Sigma)$. A financial position $y : \Omega \to \mathbb{R}$ is considered to be acceptable if the probability of a loss is bounded by a given level $\lambda \in (0, 1)$, i.e., if $P[y < 0] \leq \lambda$. The corresponding monetary risk measure $V@R_\lambda$, the so-called *Value at Risk* at level $\lambda$ is given by

$$V@R_\lambda(y) := \inf\{m \in \mathbb{R} \mid P[m + y < 0] \leq \lambda\}.$$

Formulated in terms of finance, the Value at Risk $V@R_\lambda$ is the smallest amount of capital which, if added to the financial position $y$ and invested in the risk-free asset, keeps the probability of a negative outcome below the level $\lambda$.

The Value at Risk $V@R_\lambda$ is positively homogeneous but in general it is not convex (see Föllmer and Schied [169, Examples 4.11, 4.41 and Definition 4.40]). This means that the Value at Risk is not a coherent risk measure.

**Example 15.1.3** (*Worst-Case Risk Measure*) Let $Y$ be the linear space of all bounded measurable functions on some measurable space $(\Omega, \Sigma)$. The *Worst-Case Risk Measure* $\rho_{max}$ (see Föllmer and Schied [169, Example 4.8]) is given by

$$\forall y \in Y : \ \rho_{max}(y) := -\inf_{w \in \Omega} y(w). \tag{15.1.1}$$

In (15.1.1), it is to see that the value $\rho_{max}(y)$, $y \in Y$, is the least upper bound for the potential loss which occur in any scenario $w \in \Omega$. It is known that the corresponding acceptance set $D$ is given by the nontrivial convex cone of all nonnegative functions in $Y$ such that the Worst-Case Risk Measure $\rho_{max}$ is a coherent risk measure on $Y$ (see [169, Example 4.8]).

In the following example, we discuss the Average Value at risk (or Conditional Value at Risk), see [169, Sect. 4.4, Definition 4.43].

**Example 15.1.4** (*Average Value at Risk*) The *Average Value at Risk* at level $\lambda \in (0, 1]$ of a financial position $y \in Y$ is defined by

$$AV@R_\lambda(y) := \frac{1}{\lambda}\int_0^\lambda V@R_\gamma(y)d(\gamma).$$

The Average Value at Risk (that is defined in terms of Value at Risk) is a coherent risk measure whereas the Value at Risk is not a coherent risk measure (see Föllmer, Schied [169] and Sect. 15.1.1).

**Remark 15.1.5** We discussed functionals with the properties (P1), (P2), (P3) and (P4) in Chap. 4. The sublevel set $\text{lev}_{\mu,\leq}(0) =: D$ of $\mu$ to the level 0 is a convex cone and corresponds to the acceptance set $D$.

In [11, 390], it is shown that a coherent risk measure admits a representation as

$$\mu(y) = \inf\{t \in \mathbb{R} \mid y + tk \in D\}, \tag{15.1.2}$$

where $k \in D \setminus \{0\}$.

For the relationship between the function and its sublevel sets, see (4.4.5). Formula (15.1.2) leads us to the following formulation of a coherent risk measure $\mu$ in terms of the nonlinear scalarizing functional (4.1.11) (see [219, 238] and references therein):

$$\mu(-y) = \inf\{t \in \mathbb{R} \mid -y + tk \in D\} = \inf\{t \in \mathbb{R} \mid y \in tk - D\} = \varphi_{-D,k}(y) = \varphi_{D,-k}(-y). \tag{15.1.3}$$

The last equation in formula (15.1.3) follows from Proposition 4.3.9. If we consider the functional $\varphi_{A,k}$ in (4.1.11) with $A = -D$ and $D$ is a convex cone (see (A2) and (A3)), we get $k \in -0^+A \setminus \{0\}$ (see Lemma 2.2.21). If $D$ is a proper, closed subset of a topological vector $Y$, then all assumptions of Theorem 4.4.4 (for $A = -D$) are fulfilled such that we can use the properties of the functional $\varphi_{A,k}$ in Theorem 4.4.4 for deriving corresponding properties of coherent risk measures $\mu$. Especially, we know from Theorem 4.4.4 that the functional $\varphi_{A,k} : Y \to \overline{\mathbb{R}}$ is lower semicontinuous and so $\mu : Y \to \overline{\mathbb{R}}$ is lower semicontinuous.

Taking into account the results in Chap. 4, especially Theorems 4.2.1, 4.4.4 and 4.6.6, it is possible to see that a lower semicontinuous functional $\mu : Y \to \overline{\mathbb{R}}$ with

$$D = \{y \in Y \mid \mu(y) \leq 0\}$$

such that the properties (P1)–(P4) are fulfilled for $\mu$ can be defined by (15.1.2) with closed acceptance sets $D$ satisfying axioms (A1)–(A3) (compare Artzner, Delbaen, Eber and Heath [11], Hamel [219] and Heyde [238]). This means, for a functional $\mu : Y \to \overline{\mathbb{R}}$ given by (15.1.2) that is lower semicontinuous on dom $\mu$, the properties (P1)–(P4) are fulfilled if and only if $\mu$ is defined by (15.1.2) for a closed set $D = \{y \in Y \mid \mu(y) \leq 0\}$ satisfying (A1)–(A3).

Furthermore, as mentioned above, in financial mathematics (see Föllmer, Schied [168, 169]) the axioms (A2) and (A3) are replaced sometimes by a convexity assumption concerning the acceptance set $D$. Then, one has corresponding results between convex acceptance sets and convex risk measures (compare Theorem 4.4.4(f)).

Recently, extensions of the functional (15.1.2) are introduced by Baes and Munari in [17] and Baes, Koch-Medina and Munari in [16]. In these papers, a one-period economy with the initial date $t = 0$ and a terminal date $t = 1$ is considered. The space $Y$ of all financial positions at the terminal date $t = 1$ is supposed to be a separated locally convex space, partially ordered by a reflexive, antisymmetric, transitive relation induced by a nontrivial, closed, convex and pointed cone $C := \{y \in Y \mid y \geq 0\}$. As usual, the associated continuous dual cone is denoted by $C^+$. An element $y \in C$

is called *strictly positive* if $y^*(y) > 0$ for all $y^* \in C^+ \setminus \{0\}$ and the set of all strictly positive elements is denoted by $C_{++}$. Again, in the space $Y$ of financial positions, we consider the subset $D$ of acceptable positions from the point of view of financial regulators. From the point of view of risk or portfolio managers, $D$ denotes the set of desirable positions.

Furthermore, we introduce the finite-dimensional subspace $\mathscr{K} \subseteq Y$ of elements representing the payoffs of a finite number of financial assets that are used to push unacceptable positions into the acceptance set $D$. Let the space $\mathscr{K}$ be endowed with the relative topology induced by $Y$. Each payoff in $\mathscr{K}$ carries a certain price that is represented by a linear functional $c^* : Y \to \mathbb{R}$. If the capital manager is interested to know at which cost a certain financial position can be made acceptable, he can study the functional (risk measure) $\mu_{D,\mathscr{K}} : Y \to \overline{\mathbb{R}}$ defined by

$$\mu_{D,\mathscr{K}}(y) := \inf\{c^*(k) \mid k \in \mathscr{K},\ y + k \in D.\} \tag{15.1.4}$$

In [16, 17], the (set-valued) *optimal set mapping* $\text{opt} : Y \to \mathscr{P}(\mathscr{K})$ associated to the parametric optimization problem (15.1.4) given by

$$\text{opt}(y) := \{k \in \mathscr{K} \mid y + k \in D,\ c^*(k) = \mu_{D,\mathscr{K}}(y)\} \tag{15.1.5}$$

is studied. Any element of $\text{opt}(y)$ is referred to an optimal payoff for the financial position $y$.

In [16], Baes, Koch-Medina and Munari study the qualitative robustness for optimal set mappings and show that the lower semicontinuity (see Definition 8.2.1) of $\text{opt} : Y \to \mathscr{P}(\mathscr{K})$ is typically not satisfied. The lower semicontinuity of opt is always satisfied if polyhedral risk measures are involved. But, it is important to mention that the lower semicontinuity of the optimal set mapping opt is not given in general, even if the reference risk measure is convex. Baes, Koch-Medina and Munari present counterexamples to lower semicontinuity of opt in [16]. However, the lower semicontinuity of opt can be often restored if one considers portfolios that are close to being optimal. Examples of optimal set mappings on a finite-dimensional space $Y$ that are not lower semicontinuous or do not admit a continuous selection are presented in the paper by Baes and Munari in [17]. In financial terms, this shows that risk minimization problems with a convex risk measure defined on a finite-dimensional space $Y$ need not allow for a robust way to select optimal portfolios of hedging assets.

Furthermore, Baes, Koch-Medina and Munari [16] show the existence and stability of optimal portfolios of eligible assets under the assumption that $\mu_{D,\mathscr{K}}$ given in (15.1.4) takes only finite values and is continuous. Sufficient conditions for the finiteness and continuity of $\mu_{D,\mathscr{K}}$ are given in the following result ([16, Proposition 2.4]), where $\ker(c^*)$ denotes the kernel of the linear functional $c^*$.

**Proposition 15.1.6** *Assume $D + \ker(c^*) \neq Y$ for a linear functional $c^* : Y \to \mathbb{R}$. Then, $\mu_{D,\mathscr{K}}$ in (15.1.4) is finite-valued and continuous if any of the following conditions hold:*

(a) $\text{int} C \cap \mathscr{K} \neq \emptyset$,

(b) *D is convex and* $C_{++} \cap \mathscr{K} \neq \emptyset$,
(c) *D is a convex cone and* $\operatorname{int} D \cap \mathscr{K} \neq \emptyset$.

**Remark 15.1.7**

(i) The assumption in (a) that the ordering cone $C$ in $Y$ has a nonempty interior is strong. This assumption is fulfilled for the natural ordering cone of $Y = \mathbb{R}^\ell$, but this condition is typically not fulfilled in an infinite-dimensional setting. Especially, the natural ordering cone of $Y = L^p$ has an empty interior.
(ii) In Proposition 4.7.3(d) and Corollary 4.7.10, we have shown that the functional $\varphi_{A,k}$ in (4.1.11) is finite-valued and continuous under corresponding assumptions concerning $A = D$ and $k$.

In the paper [16], Baes, Koch-Medina and Munari consider a special case of the finite-dimensional subspace $\mathscr{K}$. They suppose that the subspace $\mathscr{K}$ of $Y$ is spanned by the $\ell$ payoffs (frictionless and liquid assets)

$$S^1 = (S_0^1, S_1^1), \ldots, S^\ell = (S_0^\ell, S_1^\ell), \tag{15.1.6}$$

which stand for the eligible assets. The quantity $S_0^i \in \mathbb{R}$ represents the price of one unit of the i-th asset at time 0 and $S_1^i \in Y$ represents the payoff of one unit of the same asset at time 1 for each $i \in \{1, \ldots, \ell\}$. The subspace of $Y$ spanned by the above payoffs is denoted by $\mathscr{K}$, i.e.,

$$\mathscr{K} := \operatorname{span}(S_1^1, \ldots, S_1^\ell).$$

In the literature, the subspace $\mathscr{K}$ is called the space of eligible payoffs. Each $k \in \mathscr{K}$ can be viewed as the payoff of a suitable portfolio of eligible assets. We suppose that the *Law of One Price* holds, i.e., every portfolio generating the same payoff has the same initial price. Then, one can study a pricing functional $\tilde{c}^* : \mathscr{K} \to \mathbb{R}$ defined by

$$\tilde{c}^*(k) := \sum_{i=1}^{\ell} \lambda_i S_0^i \quad \text{for any } \lambda \in \mathbb{R}^\ell \text{ such that } k = \sum_{i=1}^{\ell} \lambda_i S_1^i. \tag{15.1.7}$$

Baes, Koch-Medina and Munari [16] pointed out that the minimal amount of capital that needs to be raised and invested in a portfolio of eligible payoffs to pass the prescribed acceptability test is represented by the risk measure $\mu_{D,\mathscr{K}} : \mathscr{K} \to \overline{\mathbb{R}}$ (see (15.1.4)) given by

$$\tilde{\mu}_{D,\mathscr{K}}(y) = \inf\{\tilde{c}^*(k) \mid k \in \mathscr{K},\ y + k \in D\}. \tag{15.1.8}$$

From the definition of $\tilde{\mu}_{D,\mathscr{K}}$, we get a natural interpretation of the quantity $\tilde{\mu}_{D,\mathscr{K}}$ as a capital requirement. Baes, Koch-Medina and Munari explain that the risk measures introduced by Artzner et al. [11] (see (P1)–(P4) and the functional $\varphi_{D,k}$ in

(4.1.11)) can be considered as a prototype of capital requirement functionals given by (15.1.4) and correspond to the simplest form of management action, i.e., raising capital and investing it in a single eligible asset.

In what follows, we study the relationship between coherent risk measures and the strictly robust counterpart (*RC*) of an optimization problem under uncertainty (see Sect. 14.2) using the notations introduced there.

Now, the following interpretation of coherent risk measures is potential (see [294]): Consider the case $Y = \mathbb{R}^\ell$ (there are $\ell$ states of the future), $B = \mathbb{R}^\ell_+$ (indeed, this is the smallest set that satisfies the axioms (A1), (A2), (A3) in Sect. 15.1.1) and $k = 1_\ell$. Then,

$$\mu(y) = \varphi_{-B,k}(-y) = \max_{\xi \in \mathcal{U}}(-f(x,\xi)) = -\min_{\xi \in \mathcal{U}} f(x,\xi)$$

is a coherent risk measure. Especially, the risk measure $\max_{\xi \in \mathcal{U}}(-f(x,\xi))$ is the objective function of the strictly robust counterpart (*RC*) (see Sect. 14.2) with negative values of $f$. Since $\mu(y) = -\min_{\xi \in \mathcal{U}} f(x,\xi)$, negative payments $f$ of an investment in the future result in a positive risk measure, and positive payments result in a negative risk measure. This appears very suitable since negative payments (losses) are much more riskier than investments with only positive payments (bonds). It is important to mention that this interpretation can analogously be employed for other concepts of robustness (supposing that the set $B$ in the definition of the functional (14.1.2) satisfies the assumptions (A1), (A2) and (A3) for acceptable investments in this section).

Quaranta and Zaffaroni in [386] studied interrelations between robustness and coherent risk measures. In [386], the Average Value at Risk (which is a coherent risk measure, see Example 15.1.4) of a portfolio of shares is minimized using concepts of robust optimization.

### 15.1.2 Utility Theory

A function of type (4.1.11) is studied by Bonnisseau and Crettez [57, 58] and by Luenberger in [341, Definition 2.1], [340] under the name of *benefit function*. The benefit function converts preferences of a decision maker into a certain measure that has several useful properties and is a powerful tool in economics, especially in analyzing welfare issues.

In [57, 58], [341], two (main) principles of decision making are presented: The primal concept concerns allocations (see Definition 15.1.17), while the dual concerns prices. Both concepts intersect in an equilibrium problem involving both allocations and prices. The primal optimization problem is expressed in terms of a simple measure of individual welfare associated with changes in allocation. So, the primal problem is related to the benefit. The benefit function assigns, relative to a given utility level, a certain benefit to any bundle of goods. The primal optimization principle is formulated as follows: If an allocation is an optimal solution, then it maximizes the total benefit—defined as the sum of the individual benefit functions relative to the utility levels at the solution.

The following notions are introduced by Luenberger in [340, 341].

We consider a total of $\ell$ commodities. Then, a *commodity bundle* is a vector $y = (y_1, ..., y_\ell)^T \in \mathbb{R}^\ell$ representing the amount of commodity $i$ in the bundle (compare Example 11.1.1). The cost of a commodity bundle is measured by the price vector $p = (p_1, ..., p_\ell)^T$.

The decision maker (consumer) selects a commodity bundle $y$ for consumption under the restriction that this choice belongs to a certain *set of consumption possibilities* $D \subset \mathbb{R}^\ell$. An important special case is $D = \mathbb{R}^\ell_+$. In general, we suppose the following attributes of the set of consumption possibilities:

(CP1) $D$ is bounded from below, i.e., $\exists\, \overline{y} \in \mathbb{R}^\ell$ with $D \subseteq \overline{y} + \mathbb{R}^\ell_+$.
(CP2) $D$ is a proper, closed and convex set.
(CP3) $\operatorname{int} D \neq \emptyset$.

The preferences of the decision maker are described by an *utility function* $u : \mathbb{R}^\ell \to \mathbb{R}$, where $u$ is a continuous function on $D$. Assume that $k \in 0^+D \setminus \{0\}$ is a given reference consumption possibility.

**Definition 15.1.8** Let $u : \mathbb{R}^\ell \to \mathbb{R}$ be an utility function. A bundle $k \in \mathbb{R}^\ell$ is said to be ***u*-good** if, for any $y \in D$, it holds that

$$\forall t > 0 : \quad y + tk \in D \ \text{ and } \ u(y + tk) > u(y).$$

Based on this standard description of the consumer's preferences, Luenberger [341, Definition 2.1] introduced the benefit function.

**Definition 15.1.9** The **benefit function** $b^u_{D,k} : D \to \overline{\mathbb{R}}$ associated with a certain utility function $u : \mathbb{R}^\ell \to \mathbb{R}$, a given utility level $u_0 \in \mathbb{R}$ and the set of consumption possibilities $D \subset \mathbb{R}^\ell$ is given by

$$b^u_{D,k}(y) := \sup\{t \in \mathbb{R} \mid y \in tk + D,\ u(y - tk) \geq u_0\}, \tag{15.1.9}$$

where $D$ is a proper, closed and convex subset of $\mathbb{R}^\ell$, $k \in 0^+D \setminus \{0\}$, $y \in D$, $u : \mathbb{R}^\ell \to \mathbb{R}$ a continuous utility function.

The case when $k = (1, \ldots, 1)^T$ was introduced earlier by Bonnisseau and Cornet [57].

As in Chap. 4, we define $\sup \emptyset = -\infty$. Of course, the functional $b^u_{D,k}$ is a functional of type (4.1.11).

In economic notions, $b^u_{D,k}(y)$ is the amount of good $k$ that the decision maker would give up to shift from utility level $u$ to the bundle represented by the vector $y$. Luenberger [341] explained that the Lagrange multipliers $p$ of the benefit maximization problem serve as prices. Then, the decision maker is looking for a pair $(a, p)$, where $a$ is an allocation (see Definition 15.1.17) and $p$ is a price vector such that $(a, p)$ satisfies a generalized saddlepoint property (Lagrange equilibrium, see Definition 15.1.19).

**Remark 15.1.10** The assumption $k \in 0^+D \setminus \{0\}$ is equivalent to the *free disposal assumption* $D = D + C$ for some nontrivial, closed, convex cone $C \subset Y$ with $k \in C$ (see Proposition 4.2.6(c)). The free disposal assumption is an important assumption in production theory but in optimization theory too (see Sect. 4.6).

**Remark 15.1.11** Under the assumption $k \in Y \setminus \{0\}$, $D + \mathbb{R}_{>}k \subseteq \operatorname{int} D$, the supremum in (15.1.9) is attained because the set of consumption possibilities $D$ is closed and bounded from below, $y \in D$, $u : \mathbb{R}^\ell \to \mathbb{R}$ is continuous (see Luenberger [341]).

**Remark 15.1.12** We get important properties of the functional $b^u_{D,k}$ given by (15.1.9) analogously to corresponding properties of the functional (4.1.11) in Theorems 4.4.4 and 4.5.5.

Luenberger stated in [341, Proposition 2.1]:

**Corollary 15.1.13** *Let $u : \mathbb{R}^\ell \to \mathbb{R}$ be an utility function. Suppose that $D$ is a proper closed subset of $\mathbb{R}^\ell$, $k \in 0^+D \setminus \{0\}$, $y \in D$. Then, the benefit function $b^u_{D,k}$ introduced in (15.1.9) has the following properties*

(a) *If $k$ is $u$-good, then*

$$u(y) = u_0 \implies b^u_{D,k}(y) = 0.$$

(b) *If $y \in \operatorname{int} D$, then*

$$b^u_{D,k}(y) = 0 \implies u(y) = u_0.$$

Furthermore, the following corollary is shown by Luenberger in [341, Proposition 2.2].

**Corollary 15.1.14** *Let $u : \mathbb{R}^\ell \to \mathbb{R}$ be an utility function. Suppose that $D$ is a proper closed subset of $\mathbb{R}^\ell$, $k \in 0^+D \setminus \{0\}$, $y \in D$. Then, the benefit function introduced in (15.1.9) has the following properties*

(a) *If $y \in D$, $\alpha \in \mathbb{R}$ and $y + \alpha k \in D$, then $b^u_{D,k}(y + \alpha k) = b^u_{D,k}(y) + \alpha$.*
(b)

$$u(y) \geq u_0 \implies b^u_{D,k}(y) \geq 0.$$

Assertion (a) describes the translation invariance along $k$ (see (4.2.2)). The benefit function $b^u_{D,k}$ given by (15.1.9) is closely related to the *expenditure function*. It is important to mention that the function $b^u_{D,k}$ in (15.1.9) and the function $e^u$ in (15.1.10) are dual in a certain sense. The expenditure function $e^u$ (see Luenberger [341, Definition 2.4]) is defined for a given utility function $u : \mathbb{R}^\ell \to \mathbb{R}$ by

$$e^u(p) := \inf_{y \in D} \{p^T y \mid u(y) \geq u_0\}, \tag{15.1.10}$$

where $p \in \mathbb{R}^\ell$ is a price vector.

The relationship between the expenditure function and the benefit function are to be seen in the following characterization of the expenditure function (see [341, Proposition 2.3]):

**Proposition 15.1.15** *Assume that $u : \mathbb{R}^\ell \to \mathbb{R}$ is an utility function, $k \in 0^+ D \setminus \{0\}$ a reference vector, $p \in \mathbb{R}^\ell$ a price vector with $p^T k > 0$. Then,*

$$e^u(p) = \inf_{y \in D} \{p^T y \mid b^u_{D,k}(y) \geq 0\}.$$

In the next result (see [341, Proposition 2.4]), one can see that the expenditure function can be described as an other infimum problem.

**Proposition 15.1.16** *Suppose that $u : \mathbb{R}^\ell \to \mathbb{R}$ is an utility function, $k \in 0^+ D \setminus \{0\}$ a reference vector, $p \in \mathbb{R}^\ell$ a price vector with $p^T k > 0$. Then,*

$$e^u(p) = \inf_{y \in D} \{p^T y - b^u_{D,k}(y) p^T k\}. \tag{15.1.11}$$

In many applications, one normalizes prices by $p^T k = 1$. Then, (15.1.11) reduces to

$$e^u(p) = \inf_{y \in D} \{p^T y - b^u_{D,k}(y)\}.$$

Taking into account the relationship between *primal* variables $y \in D$ (commodity bundles belonging to the set of consumption possibilities) and *dual* variables $p \in \mathbb{R}^\ell$ (price vectors used to measure the cost of commodity bundles), it is to be expected that the Lagrange multipliers associated with the benefit maximization problem will provide appropriate prices. This causal relationship is described in the concept of the *benefit equilibrium* that is a natural modification of the competitive equilibrium where the utility is replaced by the individual benefit.

We formulate the equilibrium in the framework of an economy: We look at $\ell$ commodities and $n$ individuals. Each individual $j \in \{1, \ldots, n\}$ has a set of consumption possibilities $D_j \subset \mathbb{R}^\ell$ and a continuous utility function $u_j : D_j \to \mathbb{R}$. Moreover, we consider the so-called transformation set $\mathcal{T} \subset \mathbb{R}^\ell$.

**Definition 15.1.17** An **allocation** is a collection of $n$ commodity bundles $a = (y^1, \ldots, y^n)$ where each $y^j \in D_j \subset \mathbb{R}^\ell$, $j \in \{1, \ldots, n\}$. If $\sum_{j=1}^n y^j \in \mathcal{T}$ ($\mathcal{T} \subset \mathbb{R}^\ell$ is a transformation set), an allocation $a$ is called feasible.

Suppose that $k \in \mathbb{R}^\ell_+ \setminus \{0\}$ is a given reference bundle. For all $j \in \{1, \ldots, n\}$, let $D_j \subset \mathbb{R}^\ell$ be sets of consumption possibilities, $u_j : D_j \to \mathbb{R}$ be continuous utility functions. The $n$ benefit functions (compare (15.1.9))

$$b^{u_j}_{D_j,k} : D_j \to \overline{\mathbb{R}} \quad (j \in \{1, \ldots, n\})$$

are defined in terms of this reference bundle $k$ and correspond to each of the $n$ utility functions $u_j$.

Consider $n$ commodity bundles $\overline{y}^j \in D_j$. Corresponding to an allocation $\overline{a} = (\overline{y}^1, \dots, \overline{y}^n)$, there are $\overline{u}_j := u_j(\overline{y}^j)$, $j \in \{1, \dots, n\}$.

In [341, Definition 3.7] and [341, Definition 3.8], the benefit equilibrium and the Lagrange equilibrium are introduced.

**Definition 15.1.18** We consider $\ell$ commodities and $n$ individuals, the sets of consumption possibilities $D_j \subset \mathbb{R}^\ell$, $n$ commodity bundles $\overline{y}^j \in D_j$, a continuous utility function $u_j : D_j \to \mathbb{R}$, $\overline{u}_j := u(\overline{y}^j)$ for each individual $j \in \{1, \dots, n\}$ and let $k \in \mathbb{R}^\ell_+ \setminus \{0\}$ be a given reference bundle. We consider an allocation $\overline{a} := (\overline{y}^1, \dots, \overline{y}^n)$ and a price vector $\overline{p} \in \mathbb{R}^\ell \setminus \{0\}$. Let $\mathcal{T} \subset \mathbb{R}^\ell$ be a transformation set. The pair $(\overline{a}, \overline{p})$ is called **benefit equilibrium** if

(i) $\overline{a} = (\overline{y}^1, \dots, \overline{y}^n)$ is feasible, i.e., $\sum_{j=1}^n \overline{y}^j \in \mathcal{T}$.
(ii) For each $j$ and all $y^j \in D_j$, it holds that

$$\overline{p}^T y^j \leq \overline{p}^T \overline{y}^j \quad \Longrightarrow \quad b^{\overline{u}_j}_{D_j,k}(y^j) \leq b^{\overline{u}_j}_{D_j,k}(\overline{y}^j).$$

(iii) For $\overline{z} := \sum_{j=1}^n \overline{y}^j$, it holds that

$$\forall\, z \in \mathcal{T}: \quad \overline{p}^T \overline{z} \geq \overline{p}^T z.$$

**Definition 15.1.19** We consider $\ell$ commodities and $n$ individuals, the sets of consumption possibilities $D_j \subset \mathbb{R}^\ell$, $n$ commodity bundles $\overline{y}^j \in D_j$, a continuous utility function $u_j : D_j \to \mathbb{R}$, $\overline{u}_j := u(\overline{y}^j)$ for each individual $j \in \{1, \dots, n\}$ and let $k \in \mathbb{R}^\ell_+ \setminus \{0\}$ be a given reference bundle. We consider an allocation $\overline{a} := (\overline{y}^1, \dots, \overline{y}^n)$ and a price vector $\overline{p} \in \mathbb{R}^\ell \setminus \{0\}$ with $\overline{p}^T k = 1$. Let $\mathcal{T} \subset \mathbb{R}^\ell$ be a transformation set. The pair $(\overline{a}, \overline{p})$ is called **Lagrange equilibrium** if

(i) $\overline{a}$ is feasible, i.e., $\sum_{j=1}^n \overline{y}^j \in \mathcal{T}$.
(ii) For each $j \in \{1, \dots, n\}$ and all $y^j \in D_j$, it holds that

$$b^{\overline{u}_j}_{D_j,k}(\overline{y}^j) - \overline{p}^T \overline{y}^j \geq b^{\overline{u}_j}_{D_j,k}(y^j) - \overline{p}^T y^j.$$

(iii) For $\overline{z} := \sum_{j=1}^n \overline{y}^j$, it holds that

$$\forall\, z \in \mathcal{T}: \quad \overline{p}^T \overline{z} \geq \overline{p}^T z.$$

Luenberger [341] presented the following economic interpretation of a Lagrange equilibrium: The price vector $\overline{p}$ can be employed to measure the cost of commodity bundles. Because $\overline{p}$ is normalized by $\overline{p}^T k = 1$, this cost is expressed in benefit units. At a Lagrange equilibrium (see Definition 15.1.19), each commodity bundle $\overline{y}^j$ of individual $j$ ($j = 1, \dots, n$) maximizes the net of benefits minus cost. Suppose that the prices are given. Then, each individual (faced with these prices) would be satisfied with the choice $\overline{y}^j$ even without a budget constraint—as long as the individual had to give up benefit equal to the cost. This means that the individuals are in an equilibrium with respect to an unconstrained benefit maximization problem. In Definitions 15.1.18 and 15.1.19, the third condition is a condition of profit maximization.

Moreover, Luenberger [341, Definition 4.1] introduced the shortage function (see Definition 15.1.20). This shortage function associated to production technologies can be considered as the counterpart to the benefit function given by (15.1.9) associated to consumer preferences. The shortage functions of enterprises can be combined with the benefit functions of individuals.

The technology of an enterprise is described by a *production possibility set* $B \subset \mathbb{R}^\ell$ and we suppose that $B$ is proper and closed. The enterprise can select any element from $B$ as its net output vector $y \in \mathbb{R}^\ell$.

**Definition 15.1.20** For $y \in \mathbb{R}^\ell$, the **shortage function** $\sigma_{B,k} : \mathbb{R}^\ell \to \overline{\mathbb{R}}$ associated to the closed production possibility set $B \subset \mathbb{R}^\ell$ and $k \in -0^+B \setminus \{0\}$ is given by

$$\sigma_{B,k}(y) := \inf\{t \in \mathbb{R} \mid y \in tk + B\}. \tag{15.1.12}$$

As in Chap. 4, we are using the convention $\inf \emptyset = +\infty$. It is clear that the shortage function in (15.1.12) coincides with the functional (4.1.11).

**Remark 15.1.21** The assumption $k \in -0^+B \setminus \{0\}$ is equivalent to the free disposal assumption $B = B - C$ for some nontrivial, closed, convex cone $C \subset Y$ with $k \in C$ (see Proposition 4.2.6(c)).

The shortage function can be interpreted as the shortest distance in the direction $k \in -0^+B \setminus \{0\}$ of an input-output vector $y$ from the upper boundary of the production possibility set $B$. This means that $\sigma_{B,k}(y)$ measures the amount in units of the reference vector $k \in -0^+B \setminus \{0\}$ by which any netput vector $y$ is short of reaching the production possibility set $B$. If $\sigma_{B,k}(y) > 0$ and one is trying to produce $y$, then this leads to shortages.

Using our results in Theorems 4.2.1, 4.4.4 and 4.5.5, we get, e.g., the following properties of the shortage function.

**Corollary 15.1.22** *Let $B \subset \mathbb{R}^\ell$ be a closed production possibility set, $k \in -0^+B \setminus \{0\}$ a reference vector and $\sigma_{B,k} : \mathbb{R}^\ell \to \mathbb{R}$ the shortage function associated to $B$ and $k$. Then, the shortage function $\sigma_{B,k}$ introduced in (15.1.12) has the following properties*

(a) $\sigma_{B,k}(y + \alpha k) = \sigma_{B,k}(y) + \alpha$.
(b) *If $B$ is convex, then $\sigma_{B,k}$ is convex.*
(c) *If $B$ is a convex cone, then $\sigma_{B,k}$ is sublinear.*
(d) *Assume that $H \subseteq \mathbb{R}^\ell$ and $B - H \subseteq B$, then $\sigma_{B,k}$ is $H$-monotone.*

Chambers, Chung and Färe [86, 87] introduced a certain measure (the directional distance) as a generalization of Luenberger's shortage function that defines how much outputs may be increased and inputs decreased. There are relationships to

Shephard's distance function. Shephard [409] introduced a distance function in a general production framework. Especially, Shephard's input distance function is a measure for the largest radial contraction of an input vector consistent with remaining technically feasible (compare Aparicio and Pastor [8]). Shephard's input distance function is very useful in production theory and there are certain dual relationships concerning the cost function.

Consider the production function $f$ of an enterprise. Then, the shortage function is closely related to $f$. If the output quantity is $q$ and the input vector is $z$, then the feasible set is given by $q \le f(z)$. Now, we define the netput vector $y := (-z, q)^T$. Furthermore, consider the reference vector $k$ corresponding to the output, that is, $k = \overline{k} = (0, 1)^T$. The corresponding shortage function is equal to the minimal amount $s$ that can be subtracted from $q$ and maintain feasibility. This means that it is the smallest $s$ such that

$$q - s \le f(z).$$

Taking into account the definition of $\sigma_{B,\overline{k}}$, we obtain

$$\sigma_{B,\overline{k}}(-z, q) = q - f(z).$$

This equation shows that the shortage function can be found directly from the production function $f$.

It is interesting to note that the shortage function in (15.1.12) is the natural precursor to the *profit function* $\pi : \mathbb{R}^\ell \to \mathbb{R}$ related to the transformation set $\mathscr{T} \subset \mathbb{R}^\ell$:

$$\pi(p) := \sup\{p^T y \mid y \in \mathscr{T}\}, \tag{15.1.13}$$

where $p \in \mathbb{R}^\ell$ is a price vector. Furthermore, (15.1.12) and (15.1.13) are dual problems (see [341, Proposition 4.1]).

The following result concerning the relationships between the shortage function (15.1.12) and the profit function (15.1.13) is given in [341, Proposition 4.1].

**Proposition 15.1.23** *Assume $p \in \mathbb{R}^\ell$ is a price vector, $k \in -0^+B \setminus \{0\}$ a reference vector and $p^T k > 0$. Suppose that $B \subset \mathbb{R}^\ell$ is a closed production possibility set. Then,*

$$\pi(p) = \sup_{y \in \mathbb{R}^\ell} \{p^T y - \sigma_{B,k}(y) p^T k\}.$$

**Remark 15.1.24** In the paper [76] by Briec, Kerstens and Lesourd, the authors introduce the *efficiency improvement possibility function* (a variation of the shortage function, see Definition 15.1.20), discuss its axiomatic attributes in the context of the Markowitz efficient frontier, and make a link to the *indirect mean-variance utility function*.

Briec [75] explains that the benefit function enjoys many important and useful attributes in the context of game theory. He pointed out that the benefit function can be adapted to compare the mixed profiles of a game. Briec suggests a dual framework and gives a characterization of Nash equilibriums in terms of the benefit function inspired by Luenberger's approach.

Under the assumption that the production set is convex, Briec and Lesourd [77] mention that Luenberger's shortage function and the directional distance function by Chambers, Chung and Färe [86] appear as special cases of some Hölder metric distance function defined using the Hölder norm $\|\cdot\|_p$, $p \in [1, +\infty]$ (compare Sect. 4.13 for further results concerning the relationships between the functional (4.1.11) and norms).

**Remark 15.1.25** Recently, Andriamasy, Briec and Solonandrasana [6] discuss models in production theory with an upper semilattice structure endowed with certain translation homothetic attributes. In their paper, the authors suggest an approach allowing to take into consideration both semilattice structure and translation homothetic properties in productivity measurement using the concept of the so-called Max-Plus-convexity.

## 15.2 A Vector-Valued Optimal Control Problem with PDE-Constraints

Leugering and Schiel [323] derived a regularized nonlinear scalarization methods for vector optimization problems with PDE-constraints. They develop these methods taking into account the properties of the separating functional (4.1.11) that are useful for a characterization of solutions to vector optimization problems with PDE-constraints. In this section, we present ideas given by Leugering and Schiel [323]. However, we are using different solution concepts and modify several assertions in comparison with [323].

### *15.2.1 Problem Formulation*

Suppose that $Y$ and $Z$ are real Banach spaces partially ordered by nontrivial, closed, convex, pointed and reproducing cones $C \subset Y$ and $K \subset Z$. Recall that a cone $C \subset Y$ is called reproducing if $C - C = Y$ (see Definition 2.2.2). In this section, we do not suppose that the ordering cones $C$ and $K$ have a nonempty interior such that it is possible to deal with natural ordering cones in $L^p$ spaces, i.e., with the cones

$$L^p(\Omega)_+ := \{f \in L^p(\Omega) \mid f(x) \geq 0 \quad \text{for a. e.} \quad x \in \Omega\}$$

for $p \in (1, +\infty]$, where $\Omega \subset \mathbb{R}^n$ is a bounded domain. Furthermore, consider a space $U$ of controls, a given set $U_0 \subset U$ of admissible controls and a subset $X_0$ of the state space $X$.

Now, we introduce an abstract vector-valued control problem. Suppose that $J : X \times U \to Y$ is a vector-valued cost function and $A : X \times U \to X^*$ is a possibly nonlinear operator, especially a differential operator that is associated to PDE-constraints, $F : X \times U \to Z$ and $f \in X^*$. The objective function $J$ is to be minimize subject to the constraints

$$A(x, u) = f, \tag{15.2.1}$$

$$F(x, u) \in K, \tag{15.2.2}$$

$$x \in X_0, \quad u \in U_0. \tag{15.2.3}$$

Furthermore, we introduce the set

$$S := \{(x, u) \in X_0 \times U_0 \mid (x, u) \text{ fulfills } (15.2.1), (15.2.2)\}.$$

Using these notations and the solution concept introduced in Sect. 6.2.1, we formulate the abstract vector-valued control problem as

$$C - \operatorname*{Opt}_{(x,u)\in S} J(x, u). \tag{P}$$

In the problem (P), $x$ is the state variable whereas $u$ is the control variable.

**Definition 15.2.1** A pair $(x, u)$ belonging to the set $S$ is called **admissible** to (P).

We suppose that $S$ is nonempty.

**Example 15.2.2** In this example, we consider a vector-valued control problem with nonlinear PDE-constraints. Suppose that $\Omega \subset \mathbb{R}^n$ is an open and bounded set with Lipschitz boundary bd$\Omega$. Furthermore, suppose $U = Y$. The objective function $J : X \times U \to Y$ is given by $J(x, u) = u$. With this objective function $J$ we study the vector-valued control problem

$$C - \operatorname*{Opt}_{(x,u)\in S} J(x, u) \tag{P$_{\text{PDE}}$}$$

where the set $S$ is given by

$$-\sum_{i=1}^{n} \frac{\partial}{\partial w_i}\left(\left|\frac{\partial x}{\partial w_i}\right|^{p-2} \frac{\partial x}{\partial w_i}\right) + \lambda |x|^{p-2} x = f \quad \text{in } \Omega,$$

$$\frac{\partial x}{\partial \mathbf{z}_p} = u \text{ on } \mathrm{bd}\Omega,$$

$$x \geq 0 \text{ a.e. in } \Omega,$$

$$u \in U_0$$

and $p \in [2, +\infty)$, $\frac{1}{p} + \frac{1}{q} = 1$, $\lambda \in (0, +\infty)$, $U = Y := L^q(\mathrm{bd}\Omega)$, $X_0 = X := W^{1,p}(\Omega)$, $w \in \Omega$, $Z := L^p(\Omega)$, $C := L^q(\mathrm{bd}\Omega)_+$, $K := L^p(\Omega)_+$ and $f \in L^q(\Omega)$. $\mathbf{z}$ denotes the outer unit normal to bd$\Omega$. Furthermore, we suppose that

$$U_0 := \{u \in L^q(\text{bd}\Omega)_+ \mid \|u\|_{L^q(\text{bd}\Omega)} \leq 1\}$$

is a closed, convex and bounded set of controls.

In problem ($P_{PDE}$), we are looking for controls $u \in \text{Eff}(J(S), C)$ in a pointwise sense such that the constraints of the problem ($P_{PDE}$) are fulfilled. It is important to mention that for problem ($P_{PDE}$) the assumptions for an existence result for solutions (see Kogut [304, Theorem 5.2, Lemma 5.3]) are fulfilled.

**Remark 15.2.3** In Example 15.2.2, we do not use the $p$-Laplacian involved in [323, Example 2.4]. The $p$-Laplacian is given by $\Delta_p x := \sum_{i=1}^n \frac{\partial}{\partial w_i}\left(\left|\frac{\partial x}{\partial w_i}\right|^{p-2} \frac{\partial x}{\partial w_i}\right)$ and the co-normal-derivative by $\frac{\partial x}{\partial \mathbf{z}_p} := \sum_{i=1}^n \left|\frac{\partial x}{\partial w_i}\right|^{p-2} \frac{\partial x}{\partial w_i}\mathbf{z}_i$, where $\mathbf{z}$ denotes the outer unit normal to $\text{bd}\Omega$. The $p$-Laplacian is a nonlinear partial differential operator (apart from the case $p = 2$ which leads to the standard Laplacian).

### 15.2.2 *Scalarization Technique by Means of the Functional* $\varphi_{A,k}$

The aim of this section is to solve the vector-valued control problem (P) using the nonlinear scalarizing functional (4.1.11). We suppose that the image space $Y$ of the objective function in (P) is partially ordered by a nontrivial, closed, pointed, convex and reproducing ordering cone $C$. Now, we study vector-valued control problems in a Banach space $Y$ and we focus our attention to the case $\text{int}\, C = \emptyset$.

In Chap. 5, Theorem 5.1.4, we discussed the relationships between the scalar optimization problem (5.0.1), where the functional $\varphi_{a-H,k}$ (see (4.1.11) with $A = a - H$) is involved, and the scalarized problem ($P_{F,a,H,k}$). The problem ($P_{F,a,C,k}$) was introduced by Pascoletti, Serafini [376] for the case $Y = \mathbb{R}^\ell$ (compare also Sect. 6.4).

Now, we write the scalar problem ($P_{F,a,H,k}$) in terms of a scalarization of the vector-valued control problem (P) with $F := J(S)$, $a \in Y$, $H = C$ and $k \in C \setminus \{0\}$.

Let $k \in C \setminus \{0\}$ and $a \in Y$ be fixed elements.

$$\inf\ t \qquad (P_S)$$

subject to

$$A(x, u) = f,$$

$$F(x, u) \in K,$$

$$a + tk - J(x, u) \in C,$$

$$(t, x, u) \in \mathbb{R} \times X_0 \times U_0.$$

The elements $k \in C \setminus \{0\}$ and $a \in Y$ in problem $(\mathrm{P}_S)$ are arbitrary chosen parameters.

**Remark 15.2.4** In Theorem 5.1.4, it is shown that the problem $(\mathrm{P}_S)$ is equivalent to a problem of type (5.0.1):

$$\min_{(x,u)\in S} \varphi_{a-C,k}(J(x,u)). \tag{15.2.4}$$

**Definition 15.2.5** Let the parameters $a \in Y$ and $k \in C \setminus \{0\}$ be arbitrary chosen. A triplet $(t, x, u) \in \mathbb{R} \times S$ with $a + tk - J(x, u) \in C$ is called **admissible** to $(\mathrm{P}_S)$.

The idea of the following assertion goes back to Leugering, Schiel [323, Proposition 3.1]. We derive a corresponding result for admissible elements of (P) and $(\mathrm{P}_S)$ in the sense of Definitions 15.2.1 and 15.2.5.

**Theorem 15.2.6** *Suppose* $\mathrm{int}C \neq \emptyset$, *chose fixed parameters* $(a, k) \in Y \times \mathrm{int}C$ *and consider the scalar optimization problem* $(\mathrm{P}_S)$. *Then, the vector-valued control problem* (P) *has admissible pairs if and only if the scalar problem* $(\mathrm{P}_S)$ *has admissible triplets.*

***Proof*** Suppose that $(x, u)$ is an admissible pair of (P), i.e., $(x, u) \in S$. Because of $k \in \mathrm{int}C$, there is a real number $\varepsilon > 0$ with

$$k + \varepsilon(a - J(x, u)) \in C,$$

such that

$$\frac{1}{\varepsilon}k + a - J(x, u) \in C$$

taking into account that $C$ is a cone. This yields that $(\frac{1}{\varepsilon}, x, u)$ is admissible for the scalarized problem $(\mathrm{P}_S)$.

Conversely, let $(t, x, u)$ be an admissible triplet of $(\mathrm{P}_S)$. Obviously, then $(x, u) \in S$ and $(x, u)$ is an admissible point of (P). □

In the following theorem, we use $\mathscr{C}$-efficient solutions of (P) (see [124]). In the literature, several concepts of approximate solutions to vector optimization problems are introduced, see [111, 185, 317, 337, 367, 422, 432, 442, 453–455, 493]. Here, we will use a slightly modified version of these definitions that is useful especially in the case where the ordering cone has an empty interior (compare [124]):

**Definition 15.2.7** ($\mathscr{C}$-efficient solution) Consider the problem (P), $k \in C$, $\|k\|_Y = 1$. A pair $(\overline{x}, \overline{u}) \in S$ is called **$\mathscr{C}$-efficient solution** of (P) if

$$J(S) \cap (J(\overline{x}, \overline{u}) - C - \varepsilon k) = \emptyset \tag{15.2.5}$$

holds for all $\varepsilon > 0$. The set of $\mathscr{C}$-efficient solutions of (P) is denoted by $\mathrm{Eff}_J^{\mathscr{C}}(S, C)$.

**Remark 15.2.8** In difference to the concepts in [185, 317, 337, 432], we suppose in Definition 15.2.7 that (15.2.5) holds for all $\varepsilon > 0$. This slightly modified definition is very useful for a characterization of the set $\mathrm{Eff}_J(S, C)$ in (P) in the sense of Definition 6.0.1 via $\mathscr{C}$-efficiency in the sense of Definition 15.2.7 (see Corollary 6.2.35) for the case of arbitrary Banach spaces with $\mathrm{int}C = \emptyset$. Note that $\mathrm{Eff}_J^{\mathscr{C}}(S, C)$ is related to the set $\mathrm{Eff}_J(S, C + \mathbb{R}_{>}k)$. Efficient point sets w.r.t. $C + \mathbb{R}_{>}k$ were investigated in Chap. 6.

In the next theorem, we are dealing with unique solutions of $(\mathrm{P}_S)$ in the following sense:

**Definition 15.2.9** Consider problem $(\mathrm{P}_S)$ with the parameters $k \in C \setminus \{0\}$ and $a \in Y$. An admissible triplet $(\overline{t}, \overline{x}, \overline{u}) \in \mathbb{R} \times S$ to $(\mathrm{P}_S)$ is called **unique solution** of $(\mathrm{P}_S)$ if $\overline{t} < t$ for all admissible triplets $(t, x, u)$ to $(\mathrm{P}_S)$ with $J(x, u) \neq J(\overline{x}, \overline{u})$.

The ideas in the following assertion about the relationships between elements of the set $\mathrm{Eff}_J(S, C)$ in (P) and solutions of the problem $(\mathrm{P}_S)$ are given by Leugering, Schiel [323, Theorem 3.5]. We derive the results for admissible elements of (P) and $(\mathrm{P}_S)$ in the sense of Definitions 15.2.1 and 15.2.5 under different assumptions.

**Theorem 15.2.10** *Consider the problems* (P) *and* $(\mathrm{P}_S)$. *Then, the following assertions hold:*

(a) *Suppose that* $\mathrm{int}C \neq \emptyset$. *Let* $k \in \mathrm{int}C$ *be chosen arbitrarily. If* $(\overline{x}, \overline{u}) \in \mathrm{Eff}_J(S, C)$, *then* $(0, \overline{x}, \overline{u}) \in \mathbb{R} \times X \times U$ *is a solution of the scalar problem* $(\mathrm{P}_S)$ *with the parameters* $k \in \mathrm{int}C$ *and* $\overline{a} := J(\overline{x}, \overline{u})$.
(b) *Suppose that* $\mathrm{int}C \neq \emptyset$. *Let* $k \in \mathrm{int}C$ *and* $\overline{a} := J(\overline{x}, \overline{u})$. *Then,* $J(\overline{x}, \overline{u})$ *is an efficient element of* $J(S)$ *with respect to* $C$, *i.e.,* $J(\overline{x}, \overline{u}) \in \mathrm{Eff}(J(S), C)$ *if and only if* $(\overline{t}, \overline{x}, \overline{u}) \in \mathbb{R} \times S$ *is a unique solution of* $(\mathrm{P}_S)$ *with the parameters* $k$ *and* $\overline{a}$.
(c) *Let* $a \in Y$ *and* $k \in C$ *with* $\|k\|_Y = 1$ *be chosen arbitrarily. If* $(\overline{t}, \overline{x}, \overline{u}) \in \mathbb{R} \times S$ *is a solution of problem* $(\mathrm{P}_S)$ *with the parameters* $a$ *and* $k$, *then* $(\overline{x}, \overline{u})$ *is a* $\mathscr{C}$*-efficient solution of* (P).
(d) *Suppose that* $\mathrm{int}C \neq \emptyset$. *Let* $a \in Y$ *and* $k \in \mathrm{int}\,C$ *be chosen arbitrarily. If* $(\overline{t}, \overline{x}, \overline{u}) \in \mathbb{R} \times S$ *is a solution of* $(\mathrm{P}_S)$ *with the parameters* $a$ *and* $k$, *then* $(\overline{x}, \overline{u}) \in \mathrm{WEff}_J(S, C)$ *and* $a + \overline{t}k - J(\overline{x}, \overline{u}) \in \mathrm{bd}C$.

***Proof***

(a) Suppose that $(\overline{x}, \overline{u}) \in \mathrm{Eff}_J(S, C)$. Then, we obtain that $(\overline{x}, \overline{u}) \in S$ is a solution of the scalar problem (15.2.4) with $k \in \mathrm{int} C$ and $a = \overline{a} := J(\overline{x}, \overline{u})$ from Corollary 6.2.25. Since $\overline{a} - J(\overline{x}, \overline{u}) = 0$, we get $\overline{a} - J(\overline{x}, \overline{u}) = 0 \in C$, i.e., $(0, \overline{x}, \overline{u})$ is an admissible element of $(\mathrm{P}_S)$.
Suppose that $(0, \overline{x}, \overline{u})$ is not a global solution of $(\mathrm{P}_S)$. Then, there is a triplet $(t, x, u) \in (-\infty, 0) \times S$ with

$$J(\overline{x}, \overline{u}) + tk - J(x, u) \in C$$

and so, taking into account the pointedness of $C$

$$J(\overline{x}, \overline{u}) - J(x, u) \in -tk + C \subseteq C \setminus \{0\},$$

a contradiction to $(\overline{x}, \overline{u}) \in \mathrm{Eff}_J(S, C)$.

(b) The assertion follows with Corollary 6.2.25.

(c) In order to prove (c), we suppose by contradiction that $(\overline{x}, \overline{u})$ is not a $\mathscr{C}$-efficient solution of (P). Then, there are $\varepsilon > 0$ and a pair $(x, u) \in S$ satisfying

$$J(x, u) - J(\overline{x}, \overline{u}) \in -C - \varepsilon k.$$

This means that there is an element $c \in C$ with

$$J(x, u) - J(\overline{x}, \overline{u}) = -c - \varepsilon k. \tag{15.2.6}$$

Since $(\overline{t}, \overline{x}, \overline{u}) \in \mathbb{R} \times S$ is a solution of problem $(\mathrm{P}_S)$, we get $a + \overline{t}k - J(\overline{x}, \overline{u}) \in C$ and therefore there is an element $\overline{c} \in C$ such that

$$a + \overline{t}k - J(\overline{x}, \overline{u}) = \overline{c}. \tag{15.2.7}$$

From (15.2.6) and (15.2.7), we get

$$a + (\overline{t} - \varepsilon)k - J(x, u) = \overline{c} + c \in C.$$

This yields that $(\overline{t} - \varepsilon, x, u)$ is an admissible triplet to $(\mathrm{P}_S)$ and has a value $\overline{t} - \varepsilon$ that is smaller than the minimal value $\overline{t}$, a contradiction.

(d) The assertion follows with Corollary 6.2.23. □

**Remark 15.2.11** It is important to mention that we do not suppose $\mathrm{int}\, C \neq \emptyset$ in Theorem 15.2.10(c) whereas we need the assumption $\mathrm{int}\, C \neq \emptyset$ in assertions (a), (b) and (d) of Theorem 15.2.10. This means that the assertion (c) is important for dealing with vector-valued optimal control problems where $\mathrm{int}\, C = \emptyset$ for the ordering cone $C$ of $Y$, especially for the natural ordering cone in $L^p$.

**Remark 15.2.12** Consider problem (P). Taking into account [124, Proposition 2.1], it is possible to use Theorem 15.2.10(c) to compute elements of $\mathrm{Eff}\, J((S), C)$.

In order to approximate the set $\mathrm{Eff}\, J((S), C)$ of efficient elements of (P) by solving ($\mathrm{P}_S$) for different pairs of parameters $(a, k) \in Y \times C$, we employ the adaptive procedure by Eichfelder [136] presented in Sect. 13.4. The following theorem is shown by Leugering, Schiel in [323, Theorem 3.6] for the case of arbitrary Banach spaces with $\mathrm{int} C = \emptyset$ by computing $\mathscr{C}$-efficient solutions of (P) for $k \in C$ with $\|k\|_Y = 1$ taking into account Theorem 15.2.10(c).

**Theorem 15.2.13** *Consider the problems* (P) *and* ($\mathrm{P}_S$). *Furthermore, for arbitrary* $y^* \in Y^* \setminus \{0\}$ *and* $\alpha \in \mathbb{R}$, *we consider a hyperplane* $H_\alpha \subset Y$ *defined by*

$$H_\alpha := \{y \in Y \mid y^*(y) = \alpha\}.$$

*Let* $k \in C$ *be chosen such that* $y^*(k) \neq 0$.

*Then, the following assertions hold true.*

(a) *Consider* $(\overline{x}, \overline{u}) \in \mathrm{Eff}_J(S, C)$. *Then, there are a parameter* $\overline{a} \in H_\alpha$ *and a real number* $\overline{t} \in \mathbb{R}$ *such that* $(\overline{t}, \overline{x}, \overline{u})$ *is a solution of* ($\mathrm{P}_S$) *satisfying* $\overline{a} + \overline{t}k - J(\overline{x}, \overline{u}) = 0$.
(b) *Suppose that* $(\overline{t}, \overline{x}, \overline{u})$ *is a solution of* ($\mathrm{P}_S$) *satisfying* $\overline{a} + \overline{t}k - J(\overline{x}, \overline{u}) \in C \setminus \{0\}$ *for some* $\overline{a} \in H_\alpha$. *Then, there are a parameter* $\hat{a} \in H_\alpha$ *and a real number* $\hat{t} \in \mathbb{R}$ *with* $\hat{a} + \hat{t}k - J(\overline{x}, \overline{u}) = 0$ *such that* $(\hat{t}, \overline{x}, \overline{u})$ *is also a solution of* ($\mathrm{P}_S$) *with parameter* $\hat{a}$.

***Proof*** Let $y^* \in Y^*$, $\alpha \in \mathbb{R}$ and $k \in C$ be given such that $y^*(k) \neq 0$. Consider $\overline{t} := \frac{y^*(J(\overline{x}, \overline{u})) - \alpha}{y^*(k)}$ and $\overline{a} := J(\overline{x}, \overline{u}) - \overline{t}k$. It is easy to see that $\overline{a} \in H_\alpha$ and $\overline{a} + \overline{t}k - J(\overline{x}, \overline{u}) = 0$.

(a) Let $(\overline{x}, \overline{u}) \in \mathrm{Eff}_J(S, C)$. Suppose that $(\overline{t}, \overline{x}, \overline{u})$ is not a solution of ($\mathrm{P}_S$). Then, there is an admissible triple $(t', x', u') \in \mathbb{R} \times S$ with $t' < \overline{t}$ and it holds that

$$J(\overline{x}, \overline{u}) + (t' - \overline{t})k - J(x', u') = \overline{a} + t'k - J(x', u') \in C.$$

This is equivalent to

$$J(x', u') \in J(\overline{x}, \overline{u}) - C - (\overline{t} - t')k \subseteq J(\overline{x}, \overline{u}) - (C \setminus \{0\})$$

since $\overline{t} - t' > 0$, $k \neq 0$ and $C$ is a convex, pointed cone. This yields $(x', u') \in S$ and $J(x', u') \in J(\overline{x}, \overline{u}) - (C \setminus \{0\})$, a contradiction to $(\overline{x}, \overline{u}) \in \mathrm{Eff}_J(S, C)$.
(b) Let $(\overline{t}, \overline{x}, \overline{u})$ be a solution of ($\mathrm{P}_S$) satisfying

$$\overline{c} := \overline{a} + \overline{t}k - J(\overline{x}, \overline{u}) \in C \setminus \{0\} \tag{15.2.8}$$

for some $\overline{a} \in H_\alpha$.
Now, we define

$$\hat{t} := \frac{y^*(J(\overline{x}, \overline{u})) - \alpha}{y^*(k)} \tag{15.2.9}$$

and

$$\hat{a} := J(\overline{x}, \overline{u}) - \hat{t}k. \tag{15.2.10}$$

Of course,

$$\hat{a} + \hat{t}k - J(\overline{x}, \overline{u}) = 0 \tag{15.2.11}$$

because of (15.2.10). Furthermore, from (15.2.9), we get $\hat{t}y^*(k) = y^*(J(\overline{x}, \overline{u})) - \alpha$, i.e., $y^*(\hat{t}k) = y^*(J(\overline{x}, \overline{u})) - \alpha$, such that $\alpha = y^*(J(\overline{x}, \overline{u}) - \hat{t}k)$. This means $\hat{a} = J(\overline{x}, \overline{u}) - \hat{t}k \in H_\alpha$.

Suppose now that $(\hat{t}, \overline{x}, \overline{u})$ is not a solution of $(\mathrm{P}_S)$ with parameter $\hat{a}$. Then, there is an admissible triplet $(t', x', u')$ to $(\mathrm{P}_S)$ with parameter $\hat{a}$ such that $t' < \hat{t}$. Since $(t', x', u')$ is an admissible triplet to $(\mathrm{P}_S)$ with parameter $\hat{a}$, it holds that

$$c' := \hat{a} + t'k - J(x', u') \in C. \tag{15.2.12}$$

Equations (15.2.10) and (15.2.8) yield

$$\hat{a} = J(\overline{x}, \overline{u}) - \hat{t}k = \overline{a} + (\overline{t} - \hat{t})k - \overline{c},$$

such that we get with (15.2.12)

$$\overline{a} + (\overline{t} - \hat{t} + t')k - J(x', u') = \overline{c} + c' \in C.$$

This implies that $(\overline{t} - \hat{t} + t', x', u')$ is admissible for $(\mathrm{P}_S)$ with parameter $\overline{a}$. Moreover, it holds that $\overline{t} - \hat{t} + t' < \overline{t}$ and so $(\overline{t}, \overline{x}, \overline{u})$ is not a solution of $(\mathrm{P}_S)$ with parameter $\overline{a}$, a contradiction. □

### *15.2.3 Regularization*

The inequality constraints $a + tk - J(x, u) \in C$ and $F(x, u) \in K$ are hard to be verified directly in infinite-dimensional spaces. Therefore, in [323] a regularization of $(\mathrm{P}_S)$ is proposed by using a technique introduced by Kogut, Manzo and Nechay in [304]. The scalar optimization problem $(\mathrm{P}_S)$ is replaced by another optimization problem without these constraints. The violation of the restrictions will be penalized by high values of the objective (cost) function.

To proceed in this way, for fixed $k \in C \setminus \{0\}$, $a \in Y$, the inequality constraints are combined by the operator $T_{k,a,t} : X \times U \to Y \times Z$ for all $(x, u) \in X \times U$ defined by

$$T_{k,a,t}(x, u) := (a + tk - J(x, u), F(x, u)).$$

Then, a triplet $(t, x, u) \in \mathbb{R} \times S$ is admissible for $(\mathrm{P}_S)$ if and only if $T_{k,a,t}(x, u) \in C \times K$. Obviously,

$$C_0 := C \times K$$

is a nontrivial, closed, convex, pointed and reproducing cone in $Y_0 := Y \times Z$ if $C$ and $K$ have these properties too. Equipped with the product topology, $Y_0$ is a Banach space and the duality map is given by

$$(y^*, z^*)((y, z)) = y^*(y) + z^*(z)$$

for all $(y, z) \in Y_0$ and for all $(y^*, z^*) \in Y^* \times Z^* =: Y_0^*$. Furthermore, we need the continuous dual cone to $C_0$ given by

$$C_0^+ = \{y^* \in Y_0^* \mid \forall\, y \in C_0 : \; y^*(y) \geq 0\}.$$

Taking into account the Hahn–Banach-Theorem we have $T_{k,a,t}(x, u) \in C_0$ if and only if

$$y^*(T_{k,a,t}(x, u)) \geq 0$$

holds for all $y^* \in C_0^+$.

In order to penalize the violation of the condition $T_{k,a,t}(x, u) \in C_0$, we introduce a certain cost function. To do this, let $\sigma > 0$ and $\mu : \mathbb{R} \to [0, +\infty)$ be a monotone decreasing function that is strictly monotone decreasing on $(-\infty, 0]$ and $\mu(0) = 0$. For example, we can choose $\mu(x) = \max\{0, -x\}$. Furthermore, $\mu$ is supposed to be lower semicontinuous. Consequently,

$$\sup_{(c^*,k^*)\in C_0^+,\|(c^*,k^*)\|=1} \mu((c^*, k^*)(T_{k,a,t}(x, u)))$$

is always nonnegative and equal to zero if $T_{k,a,t}(x, u) \in C_0$.

For all $\sigma > 0$, the penalized cost function is defined by

$$J_\sigma(t, x, u) := t + \frac{1}{\sigma} \sup_{(c^*,k^*)\in C_0^+,\|(c^*,k^*)\|=1} \mu((c^*, k^*)(T_{k,a,t}(x, u))). \tag{15.2.13}$$

This leads us to the following scalar optimization problem without inequality constraints

$$\inf J_\sigma(t, x, u) \tag{P$_\sigma$}$$

subject to

$$A(x, u) = f \text{ in } \Omega,$$

$$(t, x, u) \in \mathbb{R} \times X_0 \times U_0.$$

For this problem, all triplets $(t, x, u) \in \mathbb{R} \times X_0 \times U_0$ with $A(x, u) = f$ are admissible to (P$_\sigma$). If the problem (P$_S$) has admissible points, then for every $\sigma > 0$ the set of admissible points of problem (P$_\sigma$) is nonempty.

In the following, we suppose (see Kogut, Manzo and Nechay [304, Sect. 4]) *Assumption (A)*:

1. $U$ is the topological dual space to some separable Banach space and $U_0 \subset U$ is bounded and sequentially weak* closed.
2. $X_0 \subset X$ is weakly closed.
3. $C \subset Y$ and $K \subset Z$ are closed, convex, pointed and reproducing cones.
4. The operator $A$ is coercive, i.e., for all bounded sets $V \subset U$ it holds that

$$\lim_{\|x\| \to +\infty} \inf_{u \in V} \frac{A(x, u)(x)}{\|x\|} = +\infty.$$

5. The operator $A$ fulfills the following monotonicity condition: For all sequences $\{(x_n, u_n)\} \subseteq X \times U$ with $x_n \rightharpoonup x \in X$, $u_n \rightharpoonup^* u \in U$, $A(x_n, u_n) \rightharpoonup x^* \in X^*$ for $n \to +\infty$ and

$$\limsup_{n \to +\infty} A(x_n, u_n)(x_n) \leq x^*(x)$$

it follows

$$A(x, u) = x^*.$$

6. The functions $F$ and $J$ are weakly sequentially continuous, i.e., for all $\{(x_n, u_n)\} \subseteq X \times U$ with $x_n \rightharpoonup x \in X$ and $u_n \rightharpoonup^* u \in U$ the sequences $\{J(x_n, u_n)\}$ and $\{F(x_n, u_n)\}$ converge weakly to $J(x, u)$ and $F(x, u)$, respectively.

In the next theorem, we present existence results for the problems $(\mathrm{P}_\sigma)$ and (P) shown by Kogut et al. in [304, Theorem 5.2, Lemma 7.3], compare [323, Theorem 4.2].

**Theorem 15.2.14** *Suppose that assumption (A) is fulfilled. Then, the scalar problem* $(\mathrm{P}_\sigma)$ *has a solution for each* $\sigma > 0$ *and each l.s.c., monotone decreasing function* $\mu : \mathbb{R} \to [0, +\infty)$ *that is strictly monotone decreasing on* $(-\infty, 0]$ *and* $\mu(0) = 0$.

*Moreover, suppose that each* $C$*-monotone decreasing sequence in* $Y$ *weakly converges in* $Y$ *if and only if it has a lower bound with respect to* $C$*. Then, for the vector-valued control problem* (P)*, it holds that* $\mathrm{WEff}(J(S), C) \neq \emptyset$.

Furthermore, by Kogut et al. in [304, Lemma 5.3], the existence of weakly $C$-efficient solutions for a regularization of problem (P) is shown.

For the proof of necessary optimality conditions, the Lipschitz continuity of the penalized function $J_\sigma$ defined by (15.2.13) is important. In the following theorem, we improve the result in [323, Theorem 4.4].

**Theorem 15.2.15** *Assume that (A) is fulfilled. Consider the scalar problem* $(\mathrm{P}_\sigma)$ *and suppose that* $F$ *and* $J$ *are Lipschitz continuous with the same Lipschitz constant* $L$*. Furthermore, we assume that* $\mu$ *is Lipschitz continuous with the Lipschitz constant* $L_\mu$*. Then, the penalized cost function*

$$(x, u) \to J_\sigma(t, x, u)$$

*given by (15.2.13) is Lipschitz continuous on $X_0 \times U_0$ with the Lipschitz constant $\frac{2}{\sigma}LL_\mu$ for each $t \in \mathbb{R}$, each $\sigma > 0$, each $a \in Y$ and each $k \in C \setminus \{0\}$.*

***Proof*** Let $t \in \mathbb{R}$, $\sigma > 0$, $a \in Y$ and $k \in C \setminus \{0\}$ be arbitrarily chosen. Consider $(x_1, u_1), (x_2, u_2) \in X_0 \times U_0$. Furthermore, let $\varepsilon > 0$ be arbitrarily chosen and $w_1^* := (y_1^*, k_1^*)$, $w_2^* := (y_2^*, k_2^*) \in C_0^+$, $\|(y_i^*, k_i^*)\| = \|y_i^*\| + \|k_i^*\| = 1$ $(i = 1, 2)$ chosen such that the following inequality holds for $i = 1, 2$:

$$w_i^*(T_{k,a,t}(x_i, u_i)) + \varepsilon \geq \sup_{w^* \in C_0^+, \|w^*\|=1} w^*(T_{k,a,t}(x_i, u_i)).$$

Without loss of generality, suppose that

$$w_1^*(T_{k,a,t}(x_1, u_1)) \geq w_2^*(T_{k,a,t}(x_2, u_2)).$$

Hence,

$$\begin{aligned}
|w_1^*(T_{k,a,t}(x_1, u_1)) - w_2^*(T_{k,a,t}(x_2, u_2))| &= w_1^*(T_{k,a,t}(x_1, u_1)) - w_2^*(T_{k,a,t}(x_2, u_2)) \\
&\leq w_1^*(T_{k,a,t}(x_1, u_1)) - w_1^*(T_{k,a,t}(x_2, u_2)) + \varepsilon \\
&\leq |w_1^*(T_{k,a,t}(x_1, u_1) - T_{k,a,t}(x_2, u_2))| + \varepsilon \\
&\leq \|w_1^*\| \, \|T_{k,a,t}(x_1, u_1) - T_{k,a,t}(x_2, u_2)\| + \varepsilon \\
&\leq (\|J(x_2, u_2) - J(x_1, u_1)\| + \|F(x_1, u_1) \\
&\qquad -F(x_2, u_2)\|) + \varepsilon \\
&\leq 2L\|(x_1, u_1) - (x_2, u_2)\| + \varepsilon.
\end{aligned}$$

Then, we get

$$\begin{aligned}
&|J_\sigma(t, x_1, u_1) - J_\sigma(t, x_2, u_2)| \\
&= \frac{1}{\sigma}\left| \sup_{w^* \in C_0^+, \|w^*\|=1} \mu(w^*(T_{k,a,t}(x_1, u_1))) - \sup_{w^* \in C_0^+, \|w^*\|=1} \mu(w^*(T_{k,a,t}(x_2, u_2)))\right| \\
&\leq \frac{1}{\sigma}\left(\left|\mu(w_1^*(T_{k,a,t}(x_1, u_1))) - \mu(w_2^*(T_{k,a,t}(x_2, u_2)))\right| + L_\mu \varepsilon\right) \\
&\leq \frac{2}{\sigma} LL_\mu \|(x_1, u_1) - (x_2, u_2)\| + \frac{2}{\sigma} L_\mu \varepsilon.
\end{aligned}$$

Taking into account that $\varepsilon > 0$ is arbitrarily chosen, the proof is completed. □

**Remark 15.2.16** Recently, in Jadambaa, Khan, Lopez and Sama [264], interesting new developments concerning conical regularization for multiobjective optimization problems, especially for PDE-constrained multiobjective optimal control problems, are presented (compare also [265] for a survey on the topic).

## 15.3 Locational Analysis

In this section, we present results given by Nam and Zălinescu [366]. They discussed directional minimal time functions (see (9.2.1)) that specify the minimal time for a vector to reach an object following its given direction. Important properties of directional minimal time functions are studied in Sect. 9.2. These functions are related to the functional (4.1.11). Furthermore, Nam and Zălinescu [366] derived generalized differentiation properties of this class of functions. Their methods allow the study of certain location problems where sets are involved and to derive necessary optimality conditions for these location problems.

We consider location problems where the distances are described by directional minimal time functions (see (9.2.1)). Suppose that $Y$ is a real normed vector space, $k \in Y \setminus \{0\}$ and $A \subseteq Y$ is a nonempty closed set. The directional minimal time function with direction $k$ and a target set $A$ is defined in (9.2.1) by

$$\varphi^{+}_{A,-k}(y) = \inf\{t \geq 0 \mid y + tk \in A\}. \tag{15.3.1}$$

Of course, this class of functions is similar to the class of separating functionals (4.1.11) that we discuss in this book. In this section, we are using generalized differentiation properties of the functional (15.3.1) and apply these properties for deriving necessary optimality conditions for facility location problems. We study the following location problem that is a generalization of the well-known Fermat–Torricelli problem: Consider a finite number of nonempty closed target sets $A_i \subseteq Y$ for $i \in \{1, \dots, n\}$ and $n$ elements $k_i \in Y \setminus \{0\}$ for $i \in \{1, \dots n\}$. Furthermore, a nonempty, closed, constraint set $A_0$ is given. The location problem is to find an element $\overline{y} \in A_0$ such that the total times for the vectors to reach the target sets are minimal. This leads us to the following facility location problem:

$$\text{minimize} \quad f(y) := \sum_{i=1}^{n} \varphi^{+}_{A_i,-k_i}(y) \quad \text{subject to} \quad y \in A_0. \tag{$\text{P}_L$}$$

Because the functions involved in problem $(\text{P}_L)$ are not differentiable in general, one needs certain tools of variational analysis for these functions (see Sect. 3.6). We apply such methods in order to solve $(\text{P}_L)$.

Obviously,

$$\operatorname{dom} f = \bigcap_{i=1}^{n} \operatorname{dom} \varphi^{+}_{A_i,-k_i} = \bigcap_{i=1}^{n} (A_i - \mathbb{R}_+ k_i).$$

The following assertion concerning the existence of a minimal solution of $(\text{P}_L)$ is shown in [366, Proposition 5.1]).

**Proposition 15.3.1** *Consider the location problem* $(\text{P}_L)$ *and assume* $\operatorname{dom} f \cap A_0 \neq \emptyset$. *Then,* $(\text{P}_L)$ *has an optimal solution if one of the following conditions is fulfilled:*

(a) *At least one of the given sets $A_i$ $(i = 0, 1, \cdots, n)$ is compact.*
(b) *$Y$ is a reflexive Banach space, the sets $A_i$ $(i = 0, 1, \cdots, n)$ are convex and at least one of these sets is bounded.*

***Proof*** Because of $\operatorname{dom} f \cap A_0 \neq \emptyset$ we have $\alpha := \inf\{f(x) \mid x \in A_0\} \neq +\infty$.

Suppose that (a) is satisfied. If $A_0$ is compact, then there is a minimal solution of $(\mathrm{P}_L)$, taking into account the classical Weierstrass theorem because $f$ is lower semicontinuous (see Proposition 9.2.2).

Assume now, w.l.o.g., that $A_1$ is compact. Let a minimizing sequence $\{y_i\} \subseteq A_0$ be given, i.e., $f(y_i) \to \alpha$ as $i \to +\infty$. Hence, $\varphi^+_{A_1,-k_1}(y_i) < \alpha + 1$ for sufficiently large $i$. Put $t_i := \varphi^+_{A_1,-k_1}(y_i)$, then it holds that

$$y_i + t_i k_1 \in A_1.$$

Because of the compactness of $A_1$, it follows that $\{y_i\}$ has a convergent subsequence to $\overline{y} \in A_0$. Because $f$ is lower semicontinuous (Proposition 9.2.2),

$$f(\overline{y}) \leq \liminf_{i \to +\infty} f(y_i) = \alpha.$$

This yields that $\overline{y}$ is a minimal solution of $(\mathrm{P}_L)$.

Assume now that (b) is fulfilled. It is possible to use a similar argument like before, taking into account that $f$ is weakly lower semicontinuous (see [392]), since it is convex and lower semicontinuous. Furthermore, every closed, bounded, convex set in a reflexive Banach space is weakly sequentially compact. □

In the next theorem, we present necessary conditions for minimal solutions of $(\mathrm{P}_L)$ (shown in [366, Theorem 5.3]) assuming that the underlying spaces are Asplund spaces (see Definition 2.1.39). For deriving optimality conditions, we will use Mordukhovich's normal cone $N_M(y; A_i)$ to $A_i$ at $y \in A_i$ (see Definition 8.1.4) and Mordukhovich's subdifferential $\partial_M$ (see Definition 8.1.5). We define (for every $y \in Y$)

$$I(y) := \{i \in \{1, \dots, n\} \mid y \in A_i\},$$

and

$$J(y) := \{i \in \{1, \dots, n\} \mid y \notin A_i\}.$$

Furthermore, we introduce the *projection* from $y \in \operatorname{dom}\varphi^+_{A_i,-k_i}$ to $A_i$ with respect to the directional minimal time function $\varphi^+_{A_i,-k_i}$ $(i \in \{1, \dots, n\})$ as

$$\Pi_{k_i}(y; A_i) := y + \varphi^+_{A_i,-k_i}(y) k_i \in A_i.$$

**Theorem 15.3.2** *Suppose that $Y$ is an Asplund space and consider the location problem $(\mathrm{P}_L)$. Assume that $\overline{y} \in A_0$ is a minimal solution of $(\mathrm{P}_L)$ and $\varphi^+_{A_i,-k_i}(\cdot)$ are locally Lipschitzian on a neighborhood $U$ of $\overline{y}$ for all $i \in \{1, \dots, n\}$. Then, there are elements $y_i^* \in Y^*$ with*

(a) $y_i^* \in N_M(\hat{y}_i; A_i)$, *where* $\hat{y}_i := \Pi_{k_i}(\overline{y}; A_i)$ *for* $i \in \{1, \dots, n\}$.
(b) $y_i^*(-k_i) \leq 1$ *for all* $i \in I(\overline{y})$ *and* $y_j^*(k_j) = -1$ *for all* $j \in J(\overline{y})$.
(c) $-\sum_{i=1}^n y_i^* \in N_M(\overline{y}; A_0)$.

***Proof*** Of course, $\overline{y} \in A_0$ is a minimal solution of $(P_L)$ if and only if it is a solution of the following unconstrained location problem:

$$\min_{y \in Y}(f(y) + \iota_{A_0}(y)), \tag{15.3.2}$$

where $\iota_{A_0}(\cdot)$ is the indicator function of the set $A_0$ given by $\iota_{A_0}(y) = 0$ if $y \in A_0$, and $+\infty$ otherwise. Under the assumption that each function $\varphi^+_{A_i,-k_i}(\cdot)$ is locally Lipschitzian on a neighborhood $U$ of $\overline{y}$ and since each $\varphi^+_{A_i,-k_i}(\cdot)$ is a proper lower semicontinuous function (see Proposition 9.2.2), it is possible to apply the limiting sum rule (see Lemma 8.1.7) such that we obtain

$$0 \in \partial_M \left(f(\cdot) + \iota_{A_0}(\cdot)\right)(\overline{y}) \subseteq \sum_{i=1}^n \partial_M \varphi^+_{A_i,-k_i}(\overline{y}) + N_M(\overline{y}; A_0).$$

This means that there are elements $y_i^* \in \partial_M \varphi^+_{A_i,-k_i}(\overline{y})$ with

$$-\sum_{i=1}^n y_i^* \in N_M(\overline{y}; A_0),$$

i.e., (c) holds.

Furthermore, we get the assertions (a) and (b) from the properties of the limiting (Mordukhovich) subdifferential of $\varphi^+_{A_i,-k_i}$ (see Sect. 3.6 and Theorem 9.2.11). □

Moreover, in the convex case, it is possible to formulate necessary and sufficient optimality conditions for solutions of $(P_L)$ (see [366, Theorem 5.4]) using the (Fenchel) normal cone introduced in (3.6.2).

**Theorem 15.3.3** *Suppose that* $Y$ *is a normed linear space and consider the location problem* $(P_L)$. *Assume that* $A_i$ $(i \in \{1, \dots, n\})$ *and* $A_0$ *are convex sets. Furthermore, assume that there is an element* $y \in \operatorname{dom} f \cap A_0$ *at which all functions* $\varphi^+_{A_i,-k_i}(\cdot)$ *are continuous. If* $\overline{y} \in A_0$ *is a minimal solution of* $(P_L)$, *then there are elements* $y_i^* \in Y^*$ $(i \in \{1, \dots, n\})$ *with*

(a) $y_i^* \in N(\hat{y}_i; A_i)$, *where* $\hat{y}_i := \Pi_{k_i}(\overline{y}; A_i)$ *for* $i \in \{1, \dots, n\}$.
(b) $y_i^*(-k_i) \leq 1$ *for all* $i \in I(\overline{y})$ *and* $y_j^*(k_j) = -1$ *for all* $j \in J(\overline{y})$.
(c) $-\sum_{i=1}^n y_i^* \in N(\overline{y}; A_0)$.

*Furthermore, if* $\overline{y} \in A_0$ *fulfills (a), (b) and (c), then* $\overline{y}$ *is a minimal solution of the location problem* $(P_L)$.

***Proof*** Taking into account that $f$ is a convex function and $A_0$ is a convex set, we can apply the Fenchel subdifferential calculus well known in convex analysis (especially the sum rule) and Theorem 9.2.11. Following the line of the proof of Theorem 15.3.2, we get that $\overline{y} \in A_0$ is a minimal solution of the location problem ($\text{P}_L$) if and only if

$$0 \in \partial \left( f(\cdot) + \iota_{A_0}(\cdot) \right) (\overline{y}) = \sum_{i=1}^{n} \partial \varphi^{+}_{A_i,-k_i}(\overline{y}) + N(\overline{y}; A_0),$$

where $\partial$ denotes the Fenchel subdifferential (see Definition 3.6.4). □

# Correction to: Scalarization and Separation by Translation Invariant Functions

**Correction to:**
**C. Tammer and P. Weidner, *Scalarization and Separation by Translation Invariant Functions*, Vector Optimization,**
**https://doi.org/10.1007/978-3-030-44723-6**

The following was revised in the original version of this book:

1. In the original version author names were mentioned inadvertently on the chapter opening pages. The names were removed.
2. Missing crosslinks in the references of Chapters 5 and 7 were fixed.

The erratum book has been updated with the changes.

---

The updated version of the book can be found at https://doi.org/10.1007/978-3-030-44723-6



C. Tammer and P. Weidner, *Scalarization and Separation by Translation Invariant Functions*, Vector Optimization, https://doi.org/10.1007/978-3-030-44723-6_16

# References


1. Adan, M., Novo, V.: Weak efficiency in vector optimization using a closure of algebraic type under cone-convexlikeliness. Eur. J. Oper. Res. **149**, 641–653 (2003)
2. Adan, M., Novo, V.: Proper efficiency in vector optimization on linear spaces. J. Optim. Theory Appl. **121**, 515–540 (2004)
3. Aliprantis, C.D., Border, K.C.: Infinite Dimensional Analysis, 3rd edn. Springer, Berlin (2006)
4. Aliprantis, C.D., Tourky, R.: Cones and Duality. American Mathematical Society, Providence (2007)
5. Alt, H.W.: Lineare Funktionalanalysis [Linear Functional Analysis], 5th edn. Springer, Berlin (2006)
6. Andriamasy, R., Briec, W., Solonandrasana, B.: Tropical production technologies. Pacific J. Optim. **13**(4), 683–706 (2017)
7. Ansari, Q.H., Köbis, E., Yao, J.C.: Vector Variational Inequalities and Vector Optimization. Springer, Berlin (2018)
8. Aparicio, J., Pastor, J.T.: A general input distance function based on opportunity costs. Adv. Decis. Sci. (Hindawi Publishing Corporation), Article ID 505241 (2011). https://doi.org/10.1155/2011/505241
9. Araya, Y.: Four types of nonlinear scalarizations and some applications in set optimization. Nonlinear Anal. **75**, 3821–3835 (2012)
10. Arrow, K.J., Barankin, E.W., Blackwell, D.: Admissible points of convex sets. In: Kuhn, H.W., Tucker, A.W. (eds.) Contributions to the Theory of Games, pp. 87–91. Princeton University Press, Princeton (1953)
11. Artzner, P., Delbaen, F., Eber, J.M., Heath, D.: Coherent measures of risk. Math. Finance **9**, 203–228 (1999)
12. Attouch, H.: Variational Convergence for Functions and Operators. Pitman, Boston (1984)
13. Aubin, J.P.: Optima and Equilibria. Springer, Berlin (1993)
14. Aubin, J.P., Ekeland, I.: Applied Nonlinear Analysis. Wiley, New York (1984)
15. Aubin, J.P., Frankowska, H.: Set Valued Analysis. Birkhäuser, Boston (1990)
16. Baes, M., Koch-Medina, P., Munari, C.: Existence, uniqueness and stability of optimal portfolios of eligible assets. Research Report, Department of Mathematics, ETH Zurich, Switzerland, arXiv: 1702.01936 (2018)
17. Baes, M., Munari, C.: A continuous selection for optimal portfolios under convex risk measures does not always exist. Math. Meth. Oper. Res. (2019). https://doi.org/10.1007/s00186-019-00681-x




C. Tammer and P. Weidner, *Scalarization and Separation by Translation Invariant Functions*, Vector Optimization, https://doi.org/10.1007/978-3-030-44723-6

18. Bao, T.Q., Eichfelder, G., Soleimani, B., Tammer, C.: Ekeland's variational principle for vector optimization with variable ordering structure. J. Convex Anal. **24**(2), 393–415 (2017)
19. Bao, T.Q., Hillmann, M., Tammer, C.: Subdifferentials of nonlinear scalarization functions and applications. J. Nonlinear Convex Anal. **18**(4), 589–605 (2017)
20. Bao, T.Q., Mordukhovich, B.S.: Existence of minimizers and necessary conditions in set-valued optimization with equilibrium constraints. Appl. Math. **52**, 452–562 (2007)
21. Bao, T.Q., Mordukhovich, B.S.: Variational principles for set-valued mappings with applications to multiobjective optimization. Control Cybernet. **36**(3), 531–562 (2007)
22. Bao, T.Q., Mordukhovich, B.S.: Necessary conditions for super minimizers in constrained multiobjective optimization. J. Global Optim. **43**(4), 533–552 (2009)
23. Bao, T.Q., Mordukhovich, B.S.: Relative Pareto minimizers for multiobjective problems: existence and optimality conditions. Math. Program. A **122**(2), 301–347 (2010)
24. Bao, T.Q., Mordukhovich, B.S.: Set-valued optimization in welfare economics. In: Kusuoka, S., Maruyama, T. (eds.) Advances in Mathematical Economics, pp. 113–153. Springer, Tokyo (2010)
25. Bao, T.Q., Mordukhovich, B.S.: Extended Pareto optimality in multiobjective problems. In: Ansari, Q.H., Yao, J.C. (eds.) Recent Developments in Vector Optimization, pp. 467–515. Springer, Berlin (2012)
26. Bao, T.Q., Mordukhovich, B.S.: Necessary nondomination conditions in set and vector optimization with variable ordering structures. J. Optim. Theory Appl. **162**(2), 350–370 (2014)
27. Bao, T.Q., Mordukhovich, B.S., Soubeyran, A.: Variational analysis in psychological modeling. J. Optim. Theory Appl. **164**(1), 290–315 (2015)
28. Bao, T.Q., Tammer, C.: Lagrange necessary conditions for Pareto minimizers in Asplund spaces and applications. Nonlinear Anal. **75**(3), 1089–1103 (2012)
29. Bao, T.Q., Tammer, C.: Subdifferentials and snc property of scalarization functionals with uniform level sets and applications. J. Nonlinear Var. Anal. **2**(3), 355–378 (2018)
30. Bao, T.Q., Tammer, C.: Scalarization functionals with uniform level sets in set optimization. J. Optim. Theory Appl. **182**(1), 310–335 (2019)
31. Bazaraa, M.S., Sherali, H.D., Shetty, C.M.: Nonlinear Programming, 3rd edn. Wiley, New York (2006)
32. Beale, E.: On minimizing a convex function subject to linear inequalities. J. R. Stat. Soc. Ser. B **17**, 173–184 (1955)
33. Bello Cruz, J.Y., Bouza Allende, G.: A steepest descent-like method for variable order vector optimization problems. J. Optim. Theory Appl. **162**(2), 371–391 (2014)
34. Ben-El-Mechaiekh, H., Deguire, P., Granas, A.: Points fixes et coïncidences pour les applications multivoques (applications de Ky Fan) [Fixed points and coincidences for set-valued maps (applications by Ky Fan)]. C. R. Acad. Sci. Paris Sér. I Math. **295**(4), 337–340 (1982)
35. Ben-El-Mechaiekh, H., Deguire, P., Granas, A.: Points fixes et coïncidences pour les fonctions multivoques (applications de type $\varphi$ et $\varphi^*$) II [Fixed points and coincidences for set-valued maps (applications to type $\varphi$ and $\varphi^*$) II]. C. R. Acad. Sci. Paris Sér. I Math. **295**(5), 381–384 (1982)
36. Ben-Tal, A., El Ghaoui, L., Nemirovski, A.: Robust Optimization. Princeton University Press, Princeton (2009)
37. Ben-Tal, A., Nemirovski, A.: Robust convex optimization. Math. Oper. Res. **23**(4), 769–805 (1998)
38. Ben-Tal, A., Nemirovski, A.: Robust solutions of linear programming problems contaminated with uncertain data. Math. Program. A **88**(3), 411–424 (2000)
39. Benayoun, R., Larichev, O.I., DeMontgolfier, J., Tergny, J.: Linear programming with multiple objective functions. The method of constraints. Automat. Remote Control **32**(8), 1257–1264 (1971)
40. Benker, H., Hamel, A., Tammer, C.: A proximal point algorithm for control approximation problems. I. Theoretical background. Math. Methods Oper. Res. **43**(3), 261–280 (1996)
41. Benson, H.P.: Existence of efficient solutions for vector maximization problems. J. Optim. Theory Appl. **26**(4), 569–580 (1978)

42. Benson, H.P.: An improved definition of proper efficiency for vector maximization with respect to cones. J. Math. Anal. Appl. **71**, 232–241 (1979)
43. Benson, H.P.: On a domination property for vector maximization with respect to cones. J. Optim. Theory Appl. **39**(1), 125–132 (1983)
44. Benson, H.P.: An outer approximation algorithm for generating all efficient extreme points in the outcome set of a multiple objective linear programming problem. J. Global Optim. **13**(1), 1–24 (1998)
45. Benson, H.P., Morin, T.L.: The vector maximization problem: proper efficiency and stability. SIAM J. Appl. Math. **32**(1), 64–72 (1977)
46. Benson, R.G.: Interactive multiple criteria optimization using satisfactory goals. Ph.D. Thesis, University of Iowa (1975)
47. Berge, D.: Topological Spaces. Macmillan Co., New York (1963)
48. Bergstresser, K., Charnes, A., Yu, P.L.: Generalization of domination structures and non-dominated solutions in multicriteria decision making. J. Optim. Theory Appl. **18**(1), 3–13 (1976)
49. Bernau, H.: Interactive methods for vector optimization. In: Brosowski, B., Martensen, E. (eds.) Optimization in Mathematical Physics, pp. 21–36. Peter Lang, Frankfurt a. M. (1987)
50. Bertsimas, D., Sim, M.: The price of robustness. Oper. Res. **52**(1), 35–53 (2004)
51. Birge, J.R., Louveaux, F.: Introduction to Stochastic Programming, 2nd edn. Springer, New York (2011)
52. Bishop, E., Phelps, R.R.: The support functionals of a convex set. Proc. Sympos. Pure Math. **VII**, 27–35 (1963)
53. Bitran, G.R., Magnanti, T.L.: The structure of admissible points with respect to cone dominance. J. Optim. Theory Appl. **29**(4), 573–614 (1979)
54. Boţ, R.I., Wanka, G.: An analysis of some dual problems in multiobjective optimization. I, II. Optimization **53**(3), 281–300, 301–324 (2004)
55. Bonnans, J.F., Shapiro, A.: Perturbation Analysis of Optimization Problems. Springer, New York (2000)
56. Bonnel, H., Iusem, A.N., Svaiter, B.F.: Proximal methods in vector optimization. SIAM J. Optim. (electronic) **15**(4), 953–970 (2005)
57. Bonnisseau, J.M., Cornet, B.: Existence of equilibria when firms follow bounded losses pricing rules. J. Math. Econ. **17**, 119–147 (1988)
58. Bonnisseau, J.M., Crettez, B.: On the characterization of efficient production vectors. Econ. Theory **31**, 213–223 (2007)
59. Bonsall, F.F.: Sublinear functionals and ideals in partially ordered vector spaces. Proc. London Math. Soc. **4**(3), 402–418 (1954)
60. Borwein, J., Zhu, Q.J.: Techniques of Variational Analysis. Springer, New York (2005)
61. Borwein, J.M.: Continuity and differentiability properties of convex operators. Proc. London Math. Soc. **44**(3), 420–444 (1982)
62. Borwein, J.M., Lewis, A.S.: Partially finite convex programming, Part I: quasi relative interiors and duality theory. Math. Progr. A **57**, 15–48 (1992)
63. Borwein, J.M., Lewis, A.S.: Convex analysis and nonlinear optimization: theory and examples, 2nd edn. Springer, New York (2005)
64. Borwein, J.M., Vanderwerff, J.D.: Convex Functions - Constructions, Characterizations and Counterexamples. Cambridge University Press, New York (2010)
65. Boţ, R.I.: Conjugate Duality in Convex Optimization. Springer, Berlin (2010)
66. Boţ, R.I., Grad, S.M., Wanka, G.: Duality in Vector Optimization. Springer, Berlin (2009)
67. Bourbaki, N.: Espaces vectoriels topologiques. Chapitres 1–2 [Topological vector spaces, Chaps. 1–2]. Act. Sci. Ind. 1189. Hermann, Paris (1953)
68. Bourbaki, N.: Espaces vectoriels topologiques. Chapitres 3–5 [Topological vector spaces, Chaps. 3–5]. Act. Sci. Ind. 1229. Hermann, Paris (1955)
69. Bouza, G., Quintana, E., Tammer, C.: A unified characterization of nonlinear scalarizing functionals in optimization. Vietnam J. Math. **47**(3), 683–713 (2019)

70. Bouza, G., Tammer, C.: Nonlinear scalarizing functionals for computing minimal points under variable ordering structures. Appl. Anal. Optim. **1**(1), 67–97 (2017)
71. Bouza Allende, G., Tammer, C.: Scalar functions for computing minimizers under variable order structures. Presentation at Congreso Latino-Iberoamericano de Investigacion Operativa, Rio de Janeiro (2012)
72. Bowman, V.J.: On the relationship of the Tchebycheff norm and the efficient frontier of multiple-criteria objectives. In: Thiriez, H., Zionts, S. (eds.) Multiple Criteria Decision Making, pp. 76–86. Springer, Berlin (1976)
73. Breckner, W.W.: Dualität bei Optimierungsaufgaben in halbgeordneten topologischen Vektorräumen. I [Duality for optimization problems in partially ordered topological vector spaces]. Math., Rev. Anal. Numér. Théor. Approx. **1**, 5–35 (1972)
74. Breckner, W.W., Trif, T.: Convex Functions and Related Functional Equations. Cluj University Press, Cluj (2008)
75. Briec, W.: Benefit function and duality in finite normal form games. Int. Game Theory Rev. **9**(3), 495–513 (2007)
76. Briec, W., Kerstens, K., Lesourd, J.B.: Single-period Markowitz portfolio selection, performance gauging, and duality: a variation on the Luenberger shortage function. J. Optim. Theory Appl. **120**(1), 1–27 (2004)
77. Briec, W., Lesourd, J.B.: Metric distance function and profit: some duality results. J. Optim. Theory Appl. **101**(1), 15–33 (1999)
78. Brosowski, B.: A criterion for efficiency and some applications. In: Brosowski, B., Martensen, E. (eds.) Optimization in Mathematical Physics, pp. 37–59. Peter Lang, Frankfurt a. M. (1987)
79. Brosowski, B.: A recursive procedure for the solution of linear and nonlinear vector optimization problems. In: Eschenauer, H.A., Thierauf, G. (eds.) Discretization Methods and Structural Optimization - Procedures and Applications, pp. 95–101. Springer, Berlin (1989)
80. Brosowski, B., Conci, A.: On vector optimization and parametric programming. Segundas Jornadas Latino Americanas De Matematica Aplicada **II**, 483–495 (1983)
81. Bui, H.T., Kruger, A.Y.: About extensions of the extremal principle. Vietnam J. Math. **46**(2), 215–242 (2018)
82. Bui, H.T., Kruger, A.Y.: Extremality, stationarity and generalized separation of collections of sets. J. Optim. Theory Appl. **182**(1), 211–264 (2019)
83. Burachik, R.S., Iusem, A.N.: A generalized proximal point algorithm for the nonlinear complementarity problem. RAIRO Oper. Res. **33**(4), 447–479 (1999)
84. Carrizosa, E.: An optimal bound for d.c. programs with convex constraints. Math. Methods Oper. Res. **54**(1), 47–51 (2001)
85. Ceng, L.C., Yao, J.C.: Approximate proximal methods in vector optimization. Eur. J. Oper. Res. **183**(1), 1–19 (2007)
86. Chambers, R., Chung, Y., Färe, R.: Benefit and distance functions. J. Econ. Theory **70**, 407–419 (1996)
87. Chambers, R., Chung, Y., Färe, R.: Profit, directional distance functions, and nerlovian efficiency. J. Optim. Theory Appl. **98**(3), 351–364 (1998)
88. Chankong, V., Haimes, Y.Y.: Multiobjective Decision Making - Theory and Methodology. North-Holland, New York (1983)
89. Chankong, V., Haimes, Y.Y.: Optimization-based methods for multiobjective decision-making: an overview. Large Scale Syst. **5**(1), 1–33 (1983)
90. Charnes, A., Cooper, W.W.: Management Models and Industrial Application of Linear Programming, Vol. 1. Wiley, New York (1961)
91. Chen, G.y.: Existence of solutions for a vector variational inequality: an extension of the Hartmann-Stampacchia theorem. J. Optim. Theory Appl. **74**, 445–456 (1992)
92. Chen, G.y., Hou, S.H.: Existence of solutions for vector variational inequalities. In: Giannessi, F. (ed.) Vector Variational Inequalities and Vector Equilibria, pp. 73–86. Kluwer Academic Publishers, Dordrecht (2000)
93. Chen, G.y., Huang, X., Yang, X.: Vector Optimization. Springer, Berlin (2005)

94. Chen, G.Y., Yang, X.Q.: Characterizations of variable domination structures via nonlinear scalarization. J. Optim. Theory Appl. **112**(1), 97–110 (2002)
95. Chen, G.Y., Yang, X.Q., Yu, H.: A nonlinear scalarization function and generalized quasi-vector equilibrium problems. J. Global Optim. **32**(4), 451–466 (2005)
96. Chew, K.L.: Maximal points with respect to cone dominance in Banach spaces and their existence. J. Optim. Theory Appl. **44**(1), 1–53 (1984)
97. Choo, E.U., Atkins, D.R.: Proper efficiency in nonconvex multicriteria programming. Math. Ops. Res. **8**(3), 467–470 (1983)
98. Clarke, F.H.: Optimization and Nonsmooth Analysis. Wiley, New York (1983)
99. Cohon, J.L., Marks, D.H.: A review and evaluation of multiobjective programming techniques. Water Resour. Res. **11**(7), 208–220 (1975)
100. Corley, H.W.: A new scalar equivalence for Pareto optimization. IEEE Trans. Autom. Control **25**(4), 829–830 (1980)
101. Corley, H.W.: Duality theory for maximizations with respect to cones. J. Math. Anal. Appl. **84**, 560–568 (1981)
102. Craven, B.D.: Nonsmooth multiobjective programming. Numer. Funct. Anal. Optim. **10**(1–2), 49–64 (1989)
103. Crespi, G.P., Ginchev, I., Rocca, M.: First-order optimality conditions in set-valued optimization. Math. Methods Oper. Res. **63**(1), 87–106 (2006)
104. Crespi, G.P., Ginchev, I., Rocca, M.: Points of efficiency in vector optimization with increasing-along-rays property and Minty variational inequalities. In: Konnov, I.V., Luc, D.T., Rubinov, A.M. (eds.) Generalized Convexity and Related Topics, pp. 209–226. Springer, Berlin (2006)
105. Crouzeix, J.P.: Conjugacy in quasiconvex analysis. In: Auslender, A. (ed.) Convex Analysis and Its Applications, pp. 66–99. Springer, Berlin (1977)
106. Da Cunha, N.O., Polak, E.: Constrained minimization under vector-valued criteria in finite dimensional spaces. J. Math. Anal. Appl. **19**, 103–124 (1967)
107. Dancs, S., Hegedüs, M., Medvegyev, P.: A general ordering and fixed-point principle in complete metric space. Acta Sci. Math. **46**(1–4), 381–388 (1983)
108. Dantzig, G.: Linear programming under uncertainty. Manag. Sci. **1**, 197–206 (1955)
109. Dauer, J.P., Saleh, O.A.: A characterization of proper minimal points as solutions of sublinear optimization problems. J. Math. Anal. Appl. **178**, 227–246 (1993)
110. Debreu, G.: Theory of Value. Wiley, New York (1959)
111. Dentcheva, D., Helbig, S.: On variational principles, level sets, well-posedness, and $\epsilon$-solutions in vector optimization. J. Optim. Theory Appl. **89**(2), 325–349 (1996)
112. Dhingra, M., Lalitha, C.S.: Well-setness and scalarization in set optimization. Optim. Lett. **10**(8), 1657–1667 (2016)
113. Dinkelbach, W.: Über einen Lösungsansatz zum Vektormaximumproblem [On an approach for solving the vector maximum problem]. In: Beckmann, M. (ed.) Unternehmensforschung heute, pp. 1–13. Springer, Berlin (1971)
114. Dinkelbach, W., Dürr, W.: Effizienzaussagen bei Ersatzprogrammen zum Vektormaximumproblem [Efficiency statements for surrogate problems related to the vector maximum problem]. Oper. Res. Verfahr. **12**, 69–77 (1972)
115. Dinkelbach, W., Isermann, H.: On decision making under multiple criteria and under incomplete information. In: Cochrane, J.L., Zeleny, M. (eds.) Multiple Criteria Decision Making. University of South Carolina Press, Columbia (1973)
116. Doagooei, A.R., Le, T.T., Tammer, C.: Convexity in the framework of variable domination structures and applications in optimization. Accepted for publication in: J. Nonlinear Convex Anal. (2019)
117. Dolecki, S., Greco, G.H.: Niveloids. Topol. Methods Nonlinear Anal. **5**, 1–22 (1995)
118. Dolecki, S., Malivert, C.: General duality in vector optimization. Optimization **27**(1–2), 97–119 (1993)
119. Drapeau, S., Kupper, M.: Risk preferences and their robust representation. Math. Oper. Res. **38**(1), 28–62 (2013)

120. Dubov, J.A.: Ustojčivost' optimal'nych po Pareto vektor'nych ocenok i $\epsilon$-ravnomernye rešenija [Stability of Pareto optimal vectors and $\epsilon$-regular solutions]. Avtom. Telem. **6**, 139–146 (1981)
121. Durea, M.: Estimations of the Lagrange multipliers' norms in set-valued optimization. Pacific J. Optim. **2**, 487–499 (2006)
122. Durea, M., Dutta, J.: Lagrange multipliers for Pareto minima in general Banach spaces. Pacific J. Optim. **4**, 447–463 (2008)
123. Durea, M., Dutta, J., Tammer, C.: Bounded sets of Lagrange multipliers for vector optimization problems in infinite dimension. J. Math. Anal. Appl. **348**, 589–606 (2008)
124. Durea, M., Dutta, J., Tammer, C.: Lagrange multipliers for $\epsilon$-Pareto solutions in vector optimization with nonsolid cones in Banach spaces. J. Optim. Theory Appl. **145**(1), 196–211 (2010)
125. Durea, M., Dutta, J., Tammer, C.: Stability properties of KKT points in vector optimization. Optimization **60**(7), 823–838 (2011)
126. Durea, M., Strugariu, R.: Some remarks on proximal point algorithm in scalar and vectorial cases. Nonlinear Funct. Anal. Appl. **15**(2), 307–319 (2010)
127. Durea, M., Strugariu, R., Tammer, C.: Scalarization in geometric and functional vector optimization revisited. J. Optim. Theory Appl. **159**(3), 635–655 (2013)
128. Durea, M., Strugariu, R., Tammer, C.: On some methods to derive necessary and sufficient optimality conditions in vector optimization. J. Optim. Theory Appl. **175**(3), 738–763 (2017)
129. Durea, M., Tammer, C.: Fuzzy necessary optimality conditions for vector optimization problems. Optimization **58**, 449–467 (2009)
130. Dutta, J., Tammer, C.: Lagrangian conditions for vector optimization in Banach spaces. Math. Meth. Oper. Res. **64**, 521–541 (2006)
131. Ecker, J.G., Kouada, I.A.: Finding efficient points for linear multiple objective programs. Math. Program. **8**(3), 375–377 (1975)
132. Ehrgott, M.: Multicriteria Optimization, 2nd edn. Springer, Berlin (2005)
133. Ehrgott, M., Ide, J., Schöbel, A.: Minmax robustness for multi-objective optimization problems. Eur. J. Oper. Res. **239**, 17–31 (2014)
134. Ehrgott, M., Shao, L., Schöbel, A.: An approximation algorithm for convex multi-objective programming problems. J. Global Optim. **50**(3), 397–416 (2011)
135. Eichfelder, G.: An adaptive scalarization method in multiobjective optimization. SIAM J. Optim. **19**(4), 1694–1718 (2008)
136. Eichfelder, G.: Adaptive Scalarization Methods in Multiobjective Optimization. Springer, Berlin (2008)
137. Eichfelder, G.: Scalarizations for adaptively solving multi-objective optimization problems. Comput. Optim. Appl. **44**(2), 249–273 (2009)
138. Eichfelder, G.: Optimal elements in vector optimization with a variable ordering structure. J. Optim. Theory Appl. **151**(2), 217–240 (2011)
139. Eichfelder, G.: Variable ordering structures in vector optimization. Habilitation Thesis, Universität Erlangen-Nürnberg, Germany (2011)
140. Eichfelder, G.: Cone-valued maps in optimization. Appl. Anal. **91**(10), 1831–1846 (2012)
141. Eichfelder, G.: Variable ordering structures in vector optimization. In: Ansari, Q.H., Yao, J.C. (eds.) Recent Developments in Vector Optimization, pp. 95–126. Springer, Berlin (2012)
142. Eichfelder, G.: Numerical procedures in multiobjective optimization with variable ordering structures. J. Optim. Theory Appl. **162**(2), 489–514 (2014)
143. Eichfelder, G.: Variable Ordering Structures in Vector Optimization. Springer, Heidelberg (2014)
144. Eichfelder, G.: Vector optimization in medical engineering. In: Rassias, T.M., Pardalos, P. (eds.) Mathematics Without Boundaries, pp. 181–215. Springer, New York (2014)
145. Eichfelder, G., Gerlach, T.: Characterization of properly optimal elements with variable ordering structures. Optimization **65**(3), 571–588 (2016)
146. Eichfelder, G., Ha, T.X.D.: Optimality conditions for vector optimization problems with variable ordering structures. Optimization **62**(5), 597–627 (2013)

147. Eichfelder, G., Jahn, J.: Vector optimization problems and their solution concepts. In: Ansari, Q.H., Yao, J.C. (eds.) Recent Developments in Vector Optimization, pp. 1–27. Springer, Berlin (2012)
148. Eichfelder, G., Kasimbeyli, R.: Properly optimal elements in vector optimization with variable ordering structures. J. Global Optim. **60**(4), 689–712 (2014)
149. Eichfelder, G., Pilecka, M.: Set approach for set optimization with variable ordering structures Part I: set relations and relationship to vector approach. J. Optim. Theory Appl. **171**(3), 931–946 (2016)
150. Eichfelder, G., Pilecka, M.: Set approach for set optimization with variable ordering structures Part II: scalarization approaches. J. Optim. Theory Appl. **171**(3), 947–963 (2016)
151. Eichfelder, G., Pilecka, M.: Ordering structures and their applications. In: Rassias, T. (ed.) Applications of Nonlinear Analysis, pp. 265–304. Springer, Cham (2018)
152. Ekeland, I.: On the variational principle. J. Math. Anal. Appl. **47**, 324–353 (1974)
153. Ekeland, I.: Nonconvex minimization problems. Bull. Amer. Math. Soc. **1**, 443–474 (1979)
154. El Abdouni, B., Thibault, L.: Lagrange multipliers for Pareto nonsmooth programming problems in Banach spaces. Optimization **26**(3–4), 277–285 (1992)
155. Emel'janov, S.V., Borisov, V.I., Malevič, A.A., Čerkashin, A.: Modeli i metody vektornoj optimizacii [Models and methods of vector optimization]. Izv. AN SSSR, TK. **1973**(6), 386–448 (1973)
156. Engau, A.: Variable preference modeling with ideal-symmetric convex cones. J. Global Optim. **42**(2), 295–311 (2008)
157. Engau, A.: Definition and characterization of Geoffrion proper efficiency for real vector optimization with infinitely many criteria. J. Optim. Theory Appl. **165**(2), 439–457 (2015)
158. Ester, J.: Dialogverfahren als wichtiges Hilfsmittel zur Lösung mehrkriterialer Optimierungsaufgaben [Interactive procedures as an important tool for the solution of multicriteria optimization problems]. Habilitation Thesis, TH Karl–Marx–Stadt, Germany (1979)
159. Ester, J.: Systemanalyse und mehrkriterielle Entscheidung [System analysis and multicriterial decision]. Verlag Technik, Berlin (1987)
160. Ester, J., Schwartz, B.: Ein erweitertes Effizienztheorem [An extended efficiency theorem]. Math. Operationsforsch. Statist. Ser. Optim. **14**, 331–342 (1983)
161. Fenchel, W.: Convex cones, sets, and functions. Lecture Notes. Princeton University, Princeton (1953)
162. Filipović, D., Kupper, M.: Monotone and cash-invariant convex functions and hulls. Insur. Math. Econ. **41**, 1–16 (2007)
163. Fischetti, M., Monaci, M.: Light robustness. In: Ahuja, R.K., Moehring, R., Zaroliagis, C. (eds.) Robust and Online Large-Scale Optimization, pp. 61–84. Springer, Berlin (2009)
164. Flores-Bazán, F., Hernández, E.: A unified vector optimization problem: complete scalarizations and applications. Optimization **60**(12), 1399–1419 (2011)
165. Flores-Bazán, F., Jiménez, B.: Strict efficiency in set-valued optimization. SIAM J. Control Optim. **48**(2), 881–908 (2009)
166. Focke, J.: Vektormaximumproblem und parametrische Optimierung [Vector maximum problem and parametric optimization]. Math. Operationsforsch. Statist. **4**, 365–369 (1973)
167. Focke, J.: Strenge lineare Ungleichungen bezüglich Kegelhalbordnungen [Strict linear inequalities related to cone orders]. Math. Operationsforsch. Statist. **6**(6), 881–900 (1975)
168. Föllmer, H., Schied, A.: Convex measures of risk and trading constraints. Financ. Stoch. **6**, 429–447 (2002)
169. Föllmer, H., Schied, A.: Stochastic Finance. Walter de Gruyter, Berlin (2004)
170. Frittelli, M., Rosazza Gianin, E.: Putting orders in risk measures. J. Bank. Financ. **26**, 1473–1486 (2002)
171. Gass, S., Saaty, T.: The computational algorithm for the parametric objective function. Nav. Res. Log. Quart. **2**, 39–45 (1955)
172. Gast, S.: Network flow problems with uncertain input data in the context of supply chain management applications. Ph.D. Thesis, Universität Wuppertal, Germany (2010)

173. Gearhart, W.B.: Compromise solutions and estimation of the noninferior set. J. Optim. Theory Appl. **28**, 29–47 (1979)
174. Gembicki, F.: Performance and sensitivity optimization: a vector index approach. Ph.D. Thesis, Case Western Research University (1974)
175. Geoffrion, A.M.: Strictly concave parametric programming, Part I: basic theory. Manag. Sci. **13**, B244–B253 (1967)
176. Geoffrion, A.M.: Strictly concave parametric programming, Part II: additional theory and computational considerations. Manag. Sci. **13**, B359–B370 (1967)
177. Geoffrion, A.M.: Proper efficiency and the theory of vector maximization. J. Math. Anal. Appl. **22**, 618–630 (1968)
178. Germeier, J.B.: Einführung in die Theorie der Operationsforschung [Introduction to the theory of operations research]. Akademie-Verlag, Berlin (1974)
179. Gerstenhaber, M.: Theory of convex polyhedral cones. In: Koopmans, T.C. (ed.) Activity Analysis of Production and Allocation, pp. 298–316. Wiley, New York (1951)
180. Gerstewitz (Tammer), C.: Eigentliche Effizienz in der Vektoroptimierung und ihre Anwendung in der Dualitätstheorie [Proper efficiency in vector optimization and its application in duality theory]. In: Vorträge zu Problemen der Polyoptimierung und der mehrkriteriellen Entscheidungstheorie I, pp. 21–31. TU Karl-Marx-Stadt, Germany (1983)
181. Gerstewitz (Tammer), C.: Nichtkonvexe Dualität in der Vektoroptimierung [Nonconvex duality in vector optimization]. Wiss. Z. TH Leuna-Merseburg **25**(3), 357–364 (1983)
182. Gerstewitz (Tammer), C.: Beiträge zur Dualitätstheorie der nichtlinearen Vektoroptimierung [Contributions to duality theory in nonlinear vector optimization]. Ph.D. Thesis, Technische Hochschule Leuna-Merseburg, Germany (1984)
183. Gerstewitz (Tammer), C.: Nichtkonvexe Trennungssätze und deren Anwendung in der Theorie der Vektoroptimierung [Nonconvex separation theorems and their application in vector optimization theory]. Seminarberichte der Sektion Mathematik der Humboldt-Universität Berlin **80**, 19–31 (1986)
184. Gerstewitz (Tammer), C., Iwanow, E.: Dualität für nichtkonvexe Vektoroptimierungsprobleme [Duality for nonconvex vector optimization problems]. Wiss. Z. TH Ilmenau **31**(2), 61–81 (1985)
185. Gerth (Tammer), C.: Näherungslösungen in der Vektoroptimierung [Approximate solutions in vector optimization]. Seminarberichte der Sektion Mathematik der Humboldt-Universität Berlin **90**, 67–76 (1987)
186. Gerth (Tammer), C., Weidner, P.: Nonconvex separation theorems and some applications in vector optimization. J. Optim. Theory Appl. **67**(2), 297–320 (1990)
187. Giannessi, F.: Theorems of alternative, quadratic programs and complementarity problems. In: Cottle, R.W., Giannessi, F., Lions, J.-L. (eds.) Variational Inequalities and Complementarity Problems, pp. 151–186. Wiley, Chichester (1980)
188. Giesy, D.P.: Calculation of Pareto-optimal solutions to multiple-objective problems using threshold-of-acceptability constraints. IEEE Trans. Autom. Control **23**(6), 1114–1115 (1978)
189. Goerigk, M., Schöbel, A.: A scenario-based approach for robust linear optimization. In: Marchetti-Spaccamela, A., Segal, M. (eds.) Theory and Practice of Algorithms in (Computer) Systems, pp. 139–150. Springer, Berlin (2011)
190. Göpfert, A.: Mathematische Optimierung in allgemeinen Vektorräumen [Mathematical optimization in general vector spaces]. Teubner, Leipzig (1973)
191. Göpfert, A., Gerth (Tammer), C.: Über die Skalarisierung und Dualisierung von Vektoroptimierungsproblemen [On scalarization and duality in vector optimization]. Z. Anal. Anwendungen **5**(4), 377–384 (1986)
192. Göpfert, A., Riahi, H., Tammer, C., Zălinescu, C.: Variational Methods in Partially Ordered Spaces. Springer, New York (2003)
193. Göpfert, A., Tammer, C.: A new maximal point theorem. Z. Anal. Anwendungen **14**(2), 379–390 (1995)
194. Göpfert, A., Tammer, C., Zălinescu, C.: A new minimal point theorem in product spaces. Z. Anal. Anwendungen **18**(3), 767–770 (1999)

195. Göpfert, A., Tammer, C., Zălinescu, C.: On the vectorial Ekeland's variational principle and minimal points in product spaces. Nonlinear Anal., Ser. A: Theory Methods **39**(7), 909–922 (2000)
196. Graña Drummond, L.M., Svaiter, B.F.: A steepest descent method for vector optimization. J. Comput. Appl. Math. **175**(2), 395–414 (2005)
197. Grauer, M., Lewandowski, A., Wierzbicki, A.: DIDASS - theory, implementation and experiences. In: Grauer, M., Wierzbicki, A. (eds.) Interactive Decision Analysis, pp. 22–30. Springer, Berlin (1984)
198. Grothendieck, A.: Topological Vector Spaces, 3rd edn. Gordon and Breach Science Publishers, New York (1992)
199. Guddat, J., GuerraVasquez, F., Tammer, K., Wendler, K.: Multiobjective and Stochastic Optimization Based on Parametric Optimization. Akademie-Verlag, Berlin (1985)
200. Guerraggio, A., Molho, E., Zaffaroni, A.: On the notion of proper efficiency in vector optimization. J. Optim. Theory Appl. **82**, 1–21 (1994)
201. Gunawardena, J.: An introduction to idempotency. In: Gunawardena, J. (ed.) Idempotency, pp. 1–49. Cambridge University Press, Cambridge (1998)
202. Gunawardena, J.: From max-plus algebra to nonexpensive mappings: a nonlinear theory for discrete event systems. Theor. Comput. Sci. **293**, 141–167 (2003)
203. Günther, C.: On generalized-convex constrained multi-objective optimization and application in location theory. Ph.D. Thesis, Martin-Luther-Universität Halle-Wittenberg, Germany (2018)
204. Günther, C., Popovici, N.: New algorithms for discrete vector optimization based on the Graef-Younes method and cone-monotone sorting functions. Optimization **67**(7), 975–1003 (2018)
205. Gutiérrez, C., Huerga, L., Köbis, E., Tammer, C.: Approximate solutions of set-valued optimization problems using set-criteria. Appl. Anal. Optim. **1**(3), 501–519 (2017)
206. Gutiérrez, C., Huerga, L., Novo, V.: Nonlinear scalarization in multiobjective optimization with a polyhedral ordering cone. Intl. Trans. Oper. Res. **25**, 763–779 (2018)
207. Gutiérrez, C., Jiménez, B., Miglierina, E., Molho, E.: Scalarization in set optimization with solid and nonsolid ordering cones. J. Global Optim. **61**(3), 525–552 (2015)
208. Gutiérrez, C., Jiménez, B., Novo, V.: A unified approach and optimality conditions for approximate solutions of vector optimization problems. SIAM J. Optim. (electronic) **17**(3), 688–710 (2006)
209. Gutiérrez, C., Jiménez, B., Novo, V.: Improvement sets and vector optimization. Eur. J. Oper. Res. **223**, 304–311 (2012)
210. Gutiérrez, C., Jiménez, B., Novo, V.: Nonlinear scalarizations of set optimization problems with set orderings. In: Hamel, A.H., Heyde, F., Löhne, A., Rudloff, B., Schrage, C. (eds.) Set Optimization and Applications-the State of the Art, pp. 43–63. Springer, Berlin (2015)
211. Gutiérrez, C., Jiménez, B., Novo, V., Thibault, L.: Strict approximate solutions in set-valued optimization with applications to the approximate Ekeland variational principle. Nonlinear Anal. **73**(12), 3842–3855 (2010)
212. Gutiérrez, C., Miglierina, E., Molho, E., Novo, V.: Pointwise well-posedness in set optimization with cone proper sets. Nonlinear Anal. **75**, 1822–1833 (2012)
213. Gutiérrez, C., Novo, V., Ródenas-Pedregosa, J.L., Tanaka, T.: Nonconvex separation functional in linear spaces with applications to vector equilibria. SIAM J. Optim. **26**(4), 2677–2695 (2016)
214. Ha, T.X.D.: Optimality conditions for several types of efficient solutions of set-valued optimization problems. In: Pardalos, P., Rassias, T.M., Khan, A.A. (eds.) Nonlinear Analysis and Variational Problems, pp. 305–324. Springer, New York (2010)
215. Haimes, Y.Y.: Risk Modeling, Assessment, and Management, 4th edn. Wiley, Hoboken (2016)
216. Haimes, Y.Y., Hall, W.A.: Multiobjectives in water resources systems analysis: the surrogate worth trade-off method. Water Resour. Res. **10**(4), 615–624 (1974)
217. Haimes, Y.Y., Hall, W.A., Freedman, H.T.: Multiobjective Optimization in Water Resources Systems: The Surrogate Worth Trade-off Method. Elsevier, Amsterdam (1975)

218. Haimes, Y.Y., Lasdon, L.S., Wismer, D.A.: On a bicriterion formulation of the problems of integrated system identification and system optimization. IEEE Trans. Syst. Man Cybern. **SMC-1**(3), 296–297 (1971)
219. Hamel, A.: From real- to set-valued coherent risk measures. In: Hamel, A., Henrion, R. (eds.) Risikomaße und ihre Anwendungen [Risk measures and their applications], pp. 10–22, Report No.19, Martin-Luther-Universität Halle-Wittenberg, Germany (2004)
220. Hamel, A., Löhne, A.: Minimal element theorems and Ekeland's principle with set relations. J. Nonlinear Convex Anal. **7**(1), 19–37 (2006)
221. Hamel, A.H.: Translative sets and functions and their applications to risk measure theory and nonlinear separation. Preprint Serie D 21/2006, IMPA, Rio de Janeiro (2006)
222. Hamel, A.H.: Translative functions. Draft (2011)
223. Hamel, A.H.: A very short history of directional translative functions. Draft, Yeshiva University, New York (2012)
224. Hamel, A.H.: Monetary measures of risk. arXiv: 1812.04354v1 (2018)
225. Hamel, A.H., Heyde, F., Löhne, A., Rudloff, B., Schrage, C.: Set optimization - a rather short introduction. In: Hamel, A.H., Heyde, F., Löhne, A., Rudloff, B., Schrage, C. (eds.) Set Optimization and Applications - The State of the Art, pp. 65–141. Springer, Berlin (2015)
226. Helbig, S.: Parametric optimization with a bottleneck objective and vector optimization. In: Jahn, J., Krabs, W. (eds.) Recent Advances and Historical Development of Vector Optimization, pp. 146–159. Springer, Berlin (1987)
227. Helbig, S.: Experiences with interactive methods for vector optimization problems. Abstract of a presentation at the conference "300 Jahre Mathematische Gesellschaft in Hamburg", Hamburg, Germany (1989)
228. Helbig, S.: A scalarization for vector optimization problems in locally convex spaces. Z. Angew. Math. Mech. **69**(4), T89–T91 (1989)
229. Helbig, S.: Stability and stability sets in non-linear vector optimization. Preprint, Universität Frankfurt a. M., Frankfurt a. M. (1989)
230. Helbig, S.: An algorithm for quadratic vector optimization problems. Z. Angew. Math. Mech. **70**(6), T751–T753 (1990)
231. Helbig, S.: An interactive algorithm for nonlinear vector optimization. Appl. Math. Optim. **22**, 147–151 (1990)
232. Helbig, S.: On the connectedness of the set of weakly efficient points of a vector optimization problem in locally convex spaces. J. Optim. Theory Appl. **65**(2), 257–270 (1990)
233. Helbig, S.: On quantitative and qualitative stability of a scalarization method, I. In: Guddat, J., Kummer, B., Jongen, H.T., Nožička, F. (eds.) Parametric Optimization and Related Topics. Peter Lang, Frankfurt a. M. (1993)
234. Helbig, S.: Quantitative and qualitative stability of a scalarization method, II. In: Brosowski, B., Ester, J., Helbig, S., Nehse, R. (eds.) Multicriteria Decision, pp. 19–42. Peter Lang, Frankfurt a. M. (1993)
235. Henig, M.I.: Proper efficiency with respect to cones. J. Optim. Theory Appl. **36**(3), 387–407 (1982)
236. Hernández, E., Löhne, A., Rodríguez-Marín, L., Tammer, C.: Lagrange duality, stability and subdifferentials in vector optimization. Optimization **62**(3), 415–428 (2013)
237. Hernández, E., Rodríguez-Marín, L.: Nonconvex scalarization in set optimization with set-valued maps. J. Math. Anal. Appl. **325**(1), 1–18 (2007)
238. Heyde, F.: Coherent risk measures and vector optimization. In: Küfer, K.H., Rommelfanger, H., Tammer, C., Winkler, K. (eds.) Multicriteria Decision Making and Fuzzy Systems, pp. 3–12. Shaker, Aachen (2006)
239. Heyde, F.: Geometric duality for convex vector optimization problems. J. Convex Anal. **20**(3), 813–832 (2013)
240. Hiriart-Urruty, J.B.: New concepts in nondifferentiable programming. Mém. Soc. Math. France **60**, 57–85 (1979)
241. Hiriart-Urruty, J.B.: Tangent cones, generalized gradients and mathematical programming in Banach spaces. Math. Oper. Res. **4**(1), 79–97 (1979)

242. Hiriart-Urruty, J.B., Lemaréchal, C.: Convex Analysis and Minimization Algorithms. I. Springer, Berlin (1993)
243. Hites, R., De Smet, Y., Risse, N., Salazar-Neumann, M., Vincke, P.: About the applicability of MCDA to some robustness problems. Eur. J. Oper. Res. **174**, 322–332 (2006)
244. Holmes, R.B.: Geometric Functional Analysis and Its Applications. Springer, New York (1975)
245. Horst, R., Thoai, N.V.: DC programming: overview. J. Optim. Theory Appl. **103**(1), 1–43 (1999)
246. Horst, R., Tuy, H.: Global Optimization, 2nd edn. Springer, Berlin (1993)
247. Huang, N.J., Yang, X.Q., Chan, W.K.: Vector complementarity problems with a variable ordering relation. Eur. J. Oper. Res. **176**(1), 15–26 (2007)
248. Huong, N.T.T., Yen, N.D.: The Pascoletti-Serafini scalarization scheme and linear vector optimization. J. Optim. Theory Appl. **162**, 559–576 (2014)
249. Hurwicz, L.: A class of criteria for decision–making under ignorance. Cowles Commission Discussion Paper, Statistics, No. 356, Yale University (1951)
250. Hurwicz, L.: Programming in linear spaces. In: Arrow, K.J., Hurwicz, L., Uzawa, H. (eds.) Studies in Linear and Nonlinear Programming, pp. 38–102. Stanford University Press, Stanford (1958)
251. Hwang, C.L., Masud, A.S.: Multiple Objective Decision Making - Methods and Applications. Springer, Berlin (1979)
252. Hyers, D.H., Isac, G., Rassias, T.M.: Topics in Nonlinear Analysis and Applications. World Scientific, Singapore (1997)
253. Ide, J., Köbis, E.: Concepts of efficiency for uncertain multi-objective optimization problems based on set order relations. Math. Methods Oper. Res. **80**(1), 99–127 (2014)
254. Ide, J., Köbis, E., Kuroiwa, D., Schöbel, A., Tammer, C.: The relationship between multi-objective robustness concepts and set-valued optimization. Fixed Point Theory Appl. **2014**, Article no. 83 (2014)
255. Ide, J., Schöbel, A.: Robustness for uncertain multi-objective optimization: a survey and analysis of different concepts. OR Spectrum **38**(1), 235–271 (2016)
256. Idrissi, H., Lefebvre, O., Michelot, C.: A primal dual algorithm for a constrained Fermat-Weber problem involving mixed gauges. Recherche Operationelle **22**, 313–330 (1988)
257. Idrissi, H., Loridan, P., Michelot, C.: Approximation of solutions for location problems. J. Optim. Theory Appl. **56**, 127–143 (1988)
258. Ioffe, A.: Approximate subdifferentials and applications. I: The finite dimensional theory. Trans. Amer. Math. Soc. **281**, 389–416 (1984)
259. Ioffe, A.: Metric regularity and subdifferential calculus. Russ. Math. Surv. **55**, 501–558 (2000)
260. Ioffe, A.D.: Approximate subdifferentials and applications II. Mathematika **33**(1), 111–128 (1986)
261. Ioffe, A.D.: Approximate subdifferentials and applications 3: the metric theory. Mathematika **36**(1), 1–38 (1989)
262. Iusem, A., Otero, R.G.: Inexact versions of proximal point and augmented Lagrangian algorithms in Banach spaces. Numer. Funct. Anal. Optim. **22**(5–6), 609–640 (2001)
263. Iwanow, E., Nehse, R.: O sobstvenno ėffektivnich rešenijach mnogokriterial'nych zadač [On properly efficient solutions of multicriterial problems]. Wiss. Z. TH Ilmenau **30**(5), 55–60 (1984)
264. Jadamba, B., Khan, A.A., López, R., Sama, M.: Conical regularization for multi-objective optimization problems. J. Math. Anal. Appl. **479**(2), 2056–2075 (2019)
265. Jadamba, B., Khan, A.A., Sama, M., Tammer, C.: Regularization methods for scalar and vector control problems. In: Khan, A., Köbis, E., Tammer, C. (eds.) Variational Analysis and Set Optimization, pp. 296–321. CRC Press, Boca Raton (2019)
266. Jahn, J.: The HAAR condition in vector optimization. In: Fandel, G., Gal, T. (eds.) Multiple Criteria Decision Making - Theory and Application, pp. 128–134. Springer, Berlin (1980)
267. Jahn, J.: Duality in vector optimization. Math. Program. **25**, 343–353 (1983)

268. Jahn, J.: Neuere Entwicklungen in der Vektoroptimierung [Recent developments in vector optimization]. In: Steckhan, H., Bühler, W., Jäger, K.E., Schneeweiß, C., Schwarze, J. (eds.) Operations Research Proceedings 1983, pp. 511–519. Springer, New York (1984)
269. Jahn, J.: Scalarization in vector optimization. Math. Program. **29**(2), 203–218 (1984)
270. Jahn, J.: A characterization of properly minimal elements of a set. SIAM J. Control Optim. **23**, 649–656 (1985)
271. Jahn, J.: Parametric approximation problems arising in vector optimization. J. Optim. Theory Appl. **54**(3), 503–516 (1987)
272. Jahn, J.: Vector Optimization, 2nd edn. Springer, Berlin (2011)
273. Jahn, J.: Vectorization in set optimization. J. Optim. Theory Appl. **150**(2), 209–236 (2013)
274. Jahn, J.: A derivative-free descent method in set optimization. Comput. Optim. Appl. **60**(2), 393–411 (2015)
275. Jahn, J., Ha, T.X.D.: New order relations in set optimization. J. Optim. Theory Appl. **148**(2), 209–236 (2011)
276. Jahn, J., Khan, A.A.: Existence theorems and characterizations of generalized contingent epiderivatives. J. Nonlinear Convex Anal. **3**(3), 315–330 (2002)
277. Jahn, J., Khan, A.A.: Generalized contingent epiderivatives in set-valued optimization: optimality conditions. Numer. Funct. Anal. Optim. **23**(7–8), 807–831 (2002)
278. Jahn, J., Khan, A.A., Zeilinger, P.: Second-order optimality conditions in set optimization. J. Optim. Theory Appl. **125**(2), 331–347 (2005)
279. Jahn, J., Rathje, U.: Graef-Younes method with backward iteration. In: Küfer, K.H., Rommelfanger, H., Tammer, C., Winkler, K. (eds.) Multicriteria Decision Making and Fuzzy Systems, pp. 75–81. Shaker, Aachen (2006)
280. Jaschke, S., Küchler, U.: Coherent risk measures and good-deal bounds. Financ. Stoch. **5**, 181–200 (2001)
281. Jofré, A., Jourani, A.: Characterizations of the free disposal condition for nonconvex economies on infinite dimensional commodity spaces. SIAM J. Optim. **25**(1), 699–712 (2015)
282. John, R.: The concave nontransitive consumer. J. Global Optim. **20**, 297–308 (2001)
283. John, R.: Local and global consumer preferences. In: Konnov, I.V., Luc, D.T., Rubinov, A.M. (eds.) Generalized Convexity and Related Topics, pp. 315–325. Springer, Berlin (2006)
284. Kaliszewski, I.: Generating nested subsets of efficient solutions. In: Jahn, J., Krabs, W. (eds.) Recent Advances and Historical Development of Vector Optimization, pp. 173–182. Springer, Berlin (1987)
285. Kaliszewski, I.: Quantitative Pareto Analysis by Cone Separation Technique. Kluwer Academic Publishers, Boston (1994)
286. Kantorowitsch, L.W., Akilow, G.P.: Funktionalanalysis in normierten Räumen [Functional analysis in normed spaces]. Akademie-Verlag, Berlin (1978)
287. Kasimbeyli, R.: A nonlinear cone separation theorem and scalarization in nonconvex vector optimization. SIAM J. Optim. **20**(3), 1591–1619 (2010)
288. Kasimbeyli, R.: A conic scalarization in multi-objective optimization. J. Global Optim. **56**, 279–297 (2013)
289. Kelley, J.L., Namioka, I.: Linear Topological Spaces. Springer, New York (1963)
290. Khan, A.A., Raciti, F.: A multiplier rule in set-valued optimisation. Bull. Aust. Math. Soc. **68**(1), 93–100 (2003)
291. Khan, A.A., Tammer, C., Zălinescu, C.: Set-Valued Optimization - An Introduction with Applications. Springer, Berlin (2015)
292. Khoshkhabar-amiranloo, S., Khorram, E., Soleimani-Damaneh, M.: Nonlinear scalarization functions and polar cone in set optimization. Optim. Lett. **11**(3), 521–535 (2017)
293. Kiyani, E., Soleimani-Damaneh, M.: Approximate proper efficiency on real linear vector spaces. Pacific J. Optim. **10**(4), 715–734 (2014)
294. Klamroth, K., Köbis, E., Schöbel, A., Tammer, C.: A unified approach for different concepts of robustness and stochastic programming via non-linear scalarizing functionals. Optimization **62**(5), 649–671 (2013)

295. Klamroth, K., Köbis, E., Schöbel, A., Tammer, C.: A unified approach to uncertain optimization. Eur. J. Oper. Res. **260**(2), 403–420 (2017)
296. Klatte, D., Kummer, B.: Nonsmooth Equations in Optimization. Kluwer Academic Publishers, Dordrecht (2002)
297. Köbis, E., Köbis, M.: Treatment of set order relations by means of a nonlinear scalarization functional: a full characterization. Optimization **65**(10), 1805–1827 (2016)
298. Köbis, E., Kuroiwa, D., Tammer, C.: Generalized set order relations and their numerical treatment. Appl. Anal. Optim. **1**(1), 45–66 (2017)
299. Köbis, E., Le, T.T., Tammer, C., Yao, J.C.: A new scalarizing functional in set optimization with respect to variable domination structures. Appl. Anal. Optim. **1**(2), 301–326 (2017)
300. Köbis, E., Le, T.T., Tammer, C., Yao, J.C.: Necessary conditions for solutions of set optimization problems with respect to variable domination structures. Pure Appl. Funct. Anal. **4**(2), 317–343 (2019)
301. Köbis, E., Tammer, C.: Characterization of set relations by means of a nonlinear scalarization functional. In: Le Thi, H.A., Pham Dinh, T., Nguyen, N.T. (eds.) Modelling, Computation and Optimization in Information Systems and Management Sciences, pp. 491–503. Springer, Cham (2015)
302. Köbis, E., Tammer, C.: Robust vector optimization with a variable domination structure. Carpathian J. Math. **33**(3), 343–351 (2017)
303. Köbis, E., Tammer, C., Yao, J.C.: Optimality conditions for set-valued optimization problems based on set approach and applications in uncertain optimization. J. Nonlinear Convex Anal. **18**(6), 1001–1014 (2017)
304. Kogut, P.I., Manzo, R., Nechay, I.V.: Generalized efficient solutions to one class of vector optimization problems in Banach spaces. Aust. J. Math. Anal. Appl. **7**(1), Article 13, 27 (2010)
305. Kolmogorov, A.N., Fomin, S.V.: Reelle Funktionen und Funktionalanalysis [Real-valued functions and functional analysis]. VEB Deutscher Verlag der Wissenschaften, Berlin (1975)
306. Korhonen, P., Laakso, J.: A visual interactive method for solving the multiple criteria problem. Eur. J. Oper. Res. **24**(2), 277–287 (1986)
307. Köthe, G.: Topologische lineare Räume [Topological vector spaces]. I. Springer, Berlin (1960)
308. Kouvelis, P., Sayin, S.: Algorithm robust for the bicriteria discrete optimization problem. Ann. Oper. Res. **147**, 71–85 (2006)
309. Kouvelis, P., Yu, G.: Robust Discrete Optimization and Its Applications. Kluwer Academic Publishers, Dordrecht (1997)
310. Krasnosel'skiĭ, M.A.: Positive Solutions of Operator Equations. P. Noordhoff Ltd., Groningen (1964)
311. Kruger, A.Y.: On extremality of set systems (in Russian). Dokl. Nats. Akad. Nauk Belarusi **42**(1), 24–28 (1998)
312. Kruger, A.Y., Mordukhovich, B.S.: Extremal points and the Euler equation in nonsmooth optimization problems (in Russian). Dokl. Akad. Nauk BSSR **24**(8), 684–687 (1980)
313. Kuhn, H.W., Tucker, A.W.: Nonlinear programming. In: Neyman, J. (ed.) Proceedings of the Second Berkeley Symposium on Mathematical Statistics and Probability, pp. 481–492. University of California Press, Berkeley and Los Angeles (1951)
314. Kuroiwa, D.: Existence Theorems of Set Optimization with Set-Valued Maps. Manuscript, Shimane University, Matsue (1997)
315. Kuroiwa, D.: Natural criteria of set-valued optimization. Shimane University (Japan), Departmental Bulletin Paper, Vol. 1031, pp. 85–90 (1998)
316. Kuroiwa, D.: Some duality theorems of set-valued optimization with natural criteria. In: Tanaka, T. (ed.) Proceedings of the International Conference on Nonlinear Analysis and Convex Analysis, pp. 221–228. World Scientific, Kyoto (1999)
317. Kutateladze, S.S.: Convex $\varepsilon$-programming. Dokl. Akad. Nauk SSSR **245**(5), 1048–1050 (1979)
318. Kuwano, I., Tanaka, T., Yamada, S.: Characterization of nonlinear scalarizing functions for set-valued maps. In: Akashi, S., Takahashi, W., Tanaka, T. (eds.) Nonlinear Analysis and Optimization, pp. 193–204. Yokohama Publishers, Yokohama (2009)

319. Kuwano, I., Tanaka, T., Yamada, S.: Inherited properties of nonlinear scalarizing functions for set-valued maps. In: Akashi, S., Takahashi, W., Tanaka, T. (eds.) Nonlinear Analysis and Convex Analysis, pp. 161–177. Yokohama Publishers, Yokohama (2010)
320. Kuwano, I., Tanaka, T., Yamada, S.: Unified scalarization for sets and set-valued Ky Fan minimax inequality. J. Nonlinear Convex Anal. **11**, 513–525 (2010)
321. Le, T.T.: Multiobjective approaches based on variable ordering structures for intensity problems in radiotherapy treatment. Revista Investigación Operacional **39**(3), 426–448 (2018)
322. Leontief, W.W.: The Structure of the American Economy, 1919–1939, 2nd edn. Oxford University Press, New York (1951)
323. Leugering, G., Schiel, R.: Regularized nonlinear scalarization for vector optimization problems with PDE-constraints. GAMM-Mitt. **35**(2), 209–225 (2012)
324. Lewandowski, A., Wierzbicki, A.: Aspiration based decision analysis. Part I: theoretical and methodological backgrounds. Working Paper WP-88-03, International Institute for Applied Systems Analysis, Laxemburg, Austria (1988)
325. Li, G.H., Li, S.J., You, M.X.: Relationships between the oriented distance functional and a nonlinear separation functional. J. Math. Anal. Appl. 1109–1117 (2018)
326. Liebchen, C., Lübbecke, M., Möhring, R.H., Stiller, S.: The concept of recoverable robustness, linear programming recovery, and railway applications. In: Ahuja, R.K., Möhring, R., Zaroliagis, C. (eds.) Robust and Online Large-Scale Optimization. Springer, Berlin (2009)
327. Lin, J.G.: Maximal vectors and multi-objective optimization. J. Optim. Theory Appl. **18**(1), 41–64 (1976)
328. Lin, J.G.: Multiple objective problems: Pareto–optimal solution by method of proper equality constraints. IEEE Trans. Autom. Control **AC–21**(5), 641–650 (1976)
329. Lin, J.G.: Proper equality constraints and maximization of index vectors. J. Optim. Theory Appl. **20**(2), 215–244 (1976)
330. Lin, J.G.: Multiple–objective optimization by a multiplier method of proper equality constraints. Part I: theory. IEEE Trans. Autom. Control **24**(4), 567–573 (1979)
331. Lippold, G.: Dualitätssätze für Optimierungsprobleme in topologischen linearen Räumen [Duality approaches for optimization problems in topological vector spaces]. Math. Operationsforsch. Stat. **6**, 913–928 (1975)
332. Liu, X., Küçükyavuz, S., Noyan, N.: Robust multicriteria risk-averse stochastic programming models. Ann. Oper. Res. **259**(1–2), 259–294 (2017)
333. Löhne, A.: Duality in vector optimization with infimum and supremum. In: Ansari, Q.H., Yao, J.C. (eds.) Recent Developments in Vector Optimization, pp. 61–94. Springer, Berlin (2012)
334. Löhne, A.: Vector Optimization with Infimum and Supremum. Springer, Berlin (2012)
335. Löhne, A., Tammer, C.: A new approach to duality in vector optimization. Optimization **56**(1–2), 221–239 (2007)
336. Löhne, A., Weißing, B.: Bensolve - a free vector linear program solver. www.bensolve.org (2015)
337. Loridan, P.: $\epsilon$-solutions in vector minimization problems. J. Optim. Theory Appl. **43**(2), 265–276 (1984)
338. Luc, D.T.: Theory of Vector Optimization. Springer, Berlin (1989)
339. Luc, D.T., Tan, N.X., Tinh, P.N.: Convex vector functions and their subdifferential. Acta Math. Vietnam. **23**(1), 107–127 (1998)
340. Luenberger, D.G.: Benefit functions and duality. J. Math. Econ. **21**, 461–481 (1992)
341. Luenberger, D.G.: New optimality principles for economic efficiency and equilibrium. J. Optim. Theory Appl. **75**(2), 221–264 (1992)
342. Maeda, T.: On optimization problems with set-valued objective maps: existence and optimality. J. Optim. Theory Appl. **153**(2), 263–279 (2012)
343. Major, D.C.: Benefit cost ratios for projects in multiple objective investment programs. Water Resour. Res. **5**(6), 1174–1178 (1969)
344. Makarov, V.L., Levin, M.J., Rubinov, A.M.: Mathematical Economic Theory: Pure and Mixed Types of Economic Mechanisms. North-Holland, Amsterdam (1995)
345. Markowitz, H.: Portfolio selection. J. Financ. **7**(1), 77–91 (1952)

346. Martinez-Legaz, J.E.: Level sets and the minimal time function of linear control processes. Numer. Funct. Anal. Optim. **9**, 105–129 (1987)
347. Martinez-Legaz, J.E., Rubinov, A.M., Singer, I.: Downward sets and their separation and approximation properties. J. Global Optim. **23**, 111–137 (2002)
348. Martínez-Legaz, J.E., Volle, M.: Duality in D.C. programming: the case of several D.C. constraints. J. Math. Anal. Appl. **237**(2), 657–671 (1999)
349. Mas-Colell, A., Whinston, M.D., Green, J.R.: Microeconomic Theory. Oxford University Press, New York (1995)
350. Michalowski, W.: Use of the displaced worst compromise in interactive multiobjective programming. IEEE Trans. Syst. Man Cybern. **18**(3), 472–477 (1988)
351. Michalowski, W., Szapiro, T.: Generating solutions in MCDM problems using dual frame reference. Working Paper 89–23, School of Business, Carleton University (1989)
352. Michalowski, W., Szapiro, T.: A procedure for worst outcomes displacement in multiple criteria decision making. Comput. Oper. Res. **16**(3), 195–206 (1989)
353. Michelot, C., Lefebvre, O.: A primal-dual algorithm for the Fermat-Weber problem involving mixed gauges. Math. Program. **39**, 319–335 (1987)
354. Miettinen, K.: Nonlinear Multiobjective Optimization. Kluwer Academic Publishers, Boston (1999)
355. Miettinen, K., Mäkelä, M.M.: Tangent and normal cones in nonconvex multiobjective optimization. In: Haimes, Y.Y., Steuer, R.E. (eds.) Research and Practice in Multiple Criteria Decision Making, pp. 114–124. Springer, Berlin (2000)
356. Minami, M.: Weak Pareto-optimal necessary conditions in a nondifferentiable multiobjective program on a Banach space. J. Optim. Theory Appl. **41**(3), 451–461 (1983)
357. Mordukhovich, B.S.: Maximum principle in the problem of time optimal response with nonsmooth constraints: PMM **40**(6), 1014–1023 (1976). J. Appl. Math. Mech. **40**(6), 960–969 (1976)
358. Mordukhovich, B.S.: Generalized differential calculus for nonsmooth and set-valued mappings. J. Math. Anal. Appl. **183**(1), 250–288 (1994)
359. Mordukhovich, B.S.: Variational Analysis and Generalized Differentiation, Vol. I: Basic Theory, Vol. II: Applications. Springer, Berlin (2006)
360. Mordukhovich, B.S.: Variational Analysis and Applications. Springer, Cham (2018)
361. Mordukhovich, B.S., Shao, Y.H.: Nonsmooth sequential analysis in Asplund spaces. Trans. Amer. Math. Soc. **348**(4), 1235–1280 (1996)
362. Moreau, J.J.: Fonctionnelles convexes [Convex functions]. Séminaire sur les équations aux dérivées partielles, Collège de France (1967)
363. Nachbin, L.: A theorem of the Hahn-Banach type for linear transformations. Trans. Amer. Math. Soc. **68**, 28–46 (1950)
364. Nakayama, H.: Duality in multi-objective optimization. In: Gal, T., Stewart, T., Hanne, T. (eds.) Multicriteria Decision Making, pp. 3–29. Kluwer Academic Publishers, Boston (1999)
365. Nakayama, H.J.: On the components in interactive multiobjective programming methods. In: Grauer, M., Thompson, A.P., Wierzbicki, A.P. (eds.) Plural Rationality and Interactive Decision Processes. Springer, Berlin (1984)
366. Nam, N.M., Zălinescu, C.: Variational analysis of directional minimal time functions and applications to location problems. Set-Valued Var. Anal. **21**(2), 405–430 (2013)
367. Németh, A.B.: Between Pareto efficiency and Pareto $\epsilon$-efficiency. Optimization **20**(5), 615–637 (1989)
368. Niebling, J., Eichfelder, G.: A branch-and-bound-based algorithm for nonconvex multiobjective optimization. SIAM J. Optim. **29**(1), 794–821 (2019)
369. Nieuwenhuis, J.W.: Supremal points and generalized duality. Math. Operationsforsch. Statist. Ser. Optim. **11**(1), 41–59 (1980)
370. Nishnianidze, Z.G.: Fixed points of monotone multivalued operators (in Russian). Soobshch. Akad. Nauk Gruzin. SSR **114**(3), 489–491 (1984)
371. Noyan, N., Rudolf, G.: Optimization with stochastic preferences based on a general class of scalarization functions. Oper. Res. **66**(2), 463–486 (2018)

372. Ogryczak, W.: Multiple criteria optimization and decisions under risk. Control Cybern. **31**, 975–1003 (2002)
373. Osher, S., Fedkiw, R.: Level Set Methods and Dynamic Implicit Surfaces. Springer, New York (2003)
374. Pallaschke, D., Rolewicz, S.: Foundations of Mathematical Optimization. Kluwer Academic Publishers, Dordrecht (1997)
375. Park, S.: Continuous selection theorems in generalized convex spaces. Numer. Funct. Anal. Optim. **20**, 567–583 (1999)
376. Pascoletti, A., Serafini, P.: Scalarizing vector optimization problems. J. Optim. Theory Appl. **42**, 499–524 (1984)
377. Payne, H.J., Polak, E., Collins, D.C., Miesel, W.S.: An algorithm for bicriteria optimization based on the sensitivity function. IEEE Trans. Autom. Control **AC–20**, 546–548 (1975)
378. Peleg, B.: Efficiency prices for optimal consumption plans. J. Math. Anal. Appl. **29**, 83–90 (1970)
379. Peressini, A.L.: Ordered Topological Vector Spaces. Harper & Row Publishers, New York (1967)
380. Perny, P., Spanjaard, O., Storme, L.X.: A decision-theoretic approach to robust optimization. Ann. Oper. Res. **147**, 317–341 (2006)
381. Phelps, R.R.: Convex Functions, Monotone Operators and Differentiability, 2nd edn. Springer, Berlin (1993)
382. Pirlot, M., Vincke, P.: Semiorders. Springer Science + Business Media, Dordrecht (1997)
383. Podinovskij, V.V., Gavrilov, V.M.: Optimizacija po posledovatel'no primenjaemych kriterijam [Optimization by successively applied criteria]. Sov. Radio, Moskva (1975)
384. Podinovskij, V.V., Nogin, V.D.: Pareto-optimal'nye rešenija mnogokriterial'nych zadač [Pareto optimal solutions of multicriterial problems]. Nauka, Moskva (1982)
385. Qiu, J.H., He, F.: A general vectorial Ekeland's variational principle with a p-distance. Acta Math. Sin. Engl. Ser. **29**(9), 1655–1678 (2013)
386. Quaranta, A., Zaffaroni, A.: Robust optimization of conditional value at risk and portfolio selection. J. Bank. Financ. **32**, 2046–2056 (2008)
387. Radner, R.: A note on maximal points of convex sets in $\ell_\infty$. In: Lecam, L., Neyman, J. (eds.) Proceedings of the Fifth Berkeley Symposium on Mathematical Statistics and Probability. Vol. 1: Statistics, pp. 351–354. University of California Press, Berkeley and Los Angeles (1967)
388. Rockafellar, R.T.: Augmented Lagrangians and applications of the proximal point algorithm in convex programming. Math. Oper. Res. **1**(2), 97–116 (1976)
389. Rockafellar, R.T.: Convex Analysis. Princeton University Press, Princeton (1997)
390. Rockafellar, R.T., Uryasev, S.: Conditional value-at-risk for general loss distributions. J. Bank. Financ. **26**, 1443–1471 (2002)
391. Rockafellar, R.T., Uryasev, S., Zabarankin, M.: Deviation measures in risk analysis and optimization. Research Report 2002-7, ISE Dept., University of Florida (2002)
392. Rockafellar, R.T., Wets, R.J.B.: Variational Analysis. Springer, Berlin (1998)
393. Roy, B.: Problems and methods with multiple objective functions. Math. Program. **1**, 239–265 (1971)
394. Rubinov, A., Gasimov, R.: Scalarization and nonlinear scalar duality for vector optimization with preferences that are not necessarily a pre-order relation. J. Global Optim. **29**(4), 455–477 (2004)
395. Rubinov, A.M.: Sublinear operators and their applications. Uspehi Mat. Nauk **32**(4(196)), 113–174, 287 (1977)
396. Rubinov, A.M., Singer, I.: Best approximation by normal and conormal sets. J. Approx. Theory **107**, 212–243 (2000)
397. Rubinov, A.M., Singer, I.: Topical and sub-topical functions, downward sets and abstract convexity. Optimization **50**(5–6), 307–351 (2001)
398. Sakawa, M., Mori, N.: Interactive multiobjective decisionmaking for nonconvex problems based on the weighted Tchebycheff norm. Large Scale Syst. **5**(1), 69–82 (1983)

399. Sauer, H.: Vergleichende Untersuchungen zu Lösungsverfahren für nichtlineare Probleme der Vektoroptimierung [Parallel investigation of solution procedures for nonlinear vector optimization problems]. Ph.D. Thesis, TH Ilmenau (1984)
400. Sawaragi, Y., Nakayama, H., Tanino, T.: Theory of Multiobjective Optimization. Academic Press, Orlando (1985)
401. Sayin, S., Kouvelis, P.: The multiobjective discrete optimization problem: a weighted min-max two-stage optimization approach and a bicriteria algorithm. Manag. Sci. **51**, 1572–1581 (2005)
402. Schaefer, H.: Topological Vector Spaces. Springer, New York (1971)
403. Schandl, B., Klamroth, K., Wiecek, M.M.: Using block norms in bicriteria optimization. In: Haimes, Y.Y., Steuer, R.E. (eds.) Research and Practice in Multiple Criteria Decision Making, pp. 149–160. Springer, Berlin (2000)
404. Schandl, B., Klamroth, K., Wiecek, M.M.: Introducing oblique norms into multiple criteria programming. J. Global Optim. **23**(1), 81–97 (2002)
405. Schöbel, A.: Generalized light robustness and the trade-off between robustness and nominal quality. Math. Methods Oper. Res. **80**(2), 161–191 (2014)
406. Schönfeld, P.: Some duality theorems for the non-linear vector maximum problem. Unternehmensforschung **14**, 51–63 (1970)
407. Schrage, C.: Algebraische Trennungsaussagen [Algebraic separation statements]. Diploma Thesis, Martin-Luther-Universität Halle-Wittenberg, Germany (2005)
408. Shapiro, A., Dentcheva, D., Ruszczyński, A.: Lectures on Stochastic Programming – Modeling and Theory. Society for Industrial and Applied Mathematics (SIAM) and Mathematical Programming Society (MPS), Philadelphia, PA (2009)
409. Shephard, R.: Cost and Production Functions. Princeton University Press, Princeton (1953)
410. Shimizu, A., Nishizawa, S., Tanaka, T.: Optimality conditions in set-valued optimization using nonlinear scalarization methods. In: Akashi, S., Takahashi, W., Tanaka, T. (eds.) Proceedings of the 4th International Conference on Nonlinear Analysis and Convex Analysis, pp. 565–574. Yokohama Publishers, Yokohama (2007)
411. Shukla, P., Braun, M.: Indicator based search in variable orderings: theory and algorithms. In: Purshouse, R.C., Fleming, P.J., Fonseca, C.M., Greco, S., Shaw, J. (eds.) Evolutionary Multi-Criterion Optimization, pp. 66–80. Springer, Berlin (2013)
412. da Silva, M.R.: Nonnegative order iterates of Bernstein polynomials and their limiting semi-group. Portugal. Math. **42**(3), 225–248 (1983/84)
413. Soland, R.M.: Multicriteria optimization: a general characterization of efficient solutions. SECISION - J. Amer. Inst. Decis. Sci. **10**(1), 26–38 (1979)
414. Soleimani, B.: Characterization of approximate solutions of vector optimization problems with a variable order structure. J. Optim. Theory Appl. **162**(2), 605–632 (2014)
415. Soleimani, B.: Vector optimization problems with variable ordering structures. Ph.D. Thesis, Martin-Luther-Universität Halle-Wittenberg, Germany (2014)
416. Soleimani, B., Tammer, C.: Scalarization of approximate solutions of vector optimization problems with variable order structure based on nonlinear scalarization. Report No 3. Institut für Mathematik, Martin-Luther-Universität Halle-Wittenberg, Germany (2013)
417. Soleimani, B., Tammer, C.: Concepts for approximate solutions of vector optimization problems with variable order structures. Vietnam J. Math. **42**(4), 543–566 (2014)
418. Soleimani, B., Tammer, C.: Optimality conditions for approximate solutions of vector optimization problems with variable ordering structures. Bull. Iranian Math. Soc. **42**(7), 5–23 (2016)
419. Song, W.: A generalization of Fenchel duality in set-valued vector optimization. Math. Methods Oper. Res. **48**(2), 259–272 (1998)
420. Soyster, A.: Convex programming with set-inclusive constraints and applications to inexact linear programming. Oper. Res. **21**, 1154–1157 (1973)
421. Spingarn, J.: Partial inverse of a monotone operator. Appl. Math. Optim. **10**, 247–265 (1983)
422. Staib, T.: On two generalizations of Pareto minimality. J. Optim. Theory Appl. **59**(2), 289–306 (1988)

423. Sterna-Karwat, A.: Continuous dependence of solutions on a parameter in a scalarization method. J. Optim. Theory Appl. **55**(3), 417–434 (1987)
424. Sterna-Karwat, A.: Lipschitz and differentiable dependence of solutions on a parameter in a scalarization method. J. Aust. Math. Soc. Ser. A **42**, 353–364 (1987)
425. Steuer, R.E.: An interactive multiple objective linear programming procedure. In: Starr, M.K., Zeleny, M. (eds.) Multiple Criteria Decision Making, pp. 225–239. North-Holland, Amsterdam (1977)
426. Steuer, R.E.: Multiple Criteria Optimization: Theory, Computation, and Application. Wiley, New York (1986)
427. Steuer, R.E., Choo, E.U.: An interactive weighted Tchebycheff procedure for multiple objective programming. Math. Program. **26**, 326–344 (1983)
428. Steuer, R.E., Wood, E.F.: On the 0–1 implementation of the Tchebycheff solution approach: a water quality illustration. Large Scale Syst. **10**(1), 243–255 (1986)
429. Stiller, S.: Extending concepts of reliability: network creation games, real-time scheduling, and robust optimization. Ph.D. Thesis, TU Berlin (2009)
430. Tammer, C.: Lagrange-Dualität in der Vektoroptimierung [Lagrange duality in vector optimization]. Wiss. Z. TH Ilmenau **37**(3), 71–88 (1991)
431. Tammer, C.: A generalization of Ekeland's variational principle. Optimization **25**(2–3), 129–141 (1992)
432. Tammer, C.: Stability results for approximately efficient solutions. OR Spektrum **16**(1), 47–52 (1994)
433. Tammer, C.: A variational principle and a fixed point theorem. In: Henry, J., Yvon, J.P. (eds.) System Modelling and Optimization, pp. 248–257. Springer, Berlin (1994)
434. Tammer, C.: Necessary conditions for approximately efficient solutions of vector approximation problems. In: Florenzano, M., Guddat, J., Jimenez, M., Jongen, H.T., Lagomasino, G., Marcellan, F. (eds.) Approximation and Optimization in the Caribbean II, pp. 651–663. Peter Lang, Frankfurt a. M. (1995)
435. Tammer, C.: Approximate solutions of vector-valued control-approximation problems. Stud. Locat. Anal. **10**, 151–162 (1996)
436. Tammer, C.: Multiobjective optimal control problems. In: Schmidt, W.H., Heier, K., Bittner, L., Bulirsch, R. (eds.) Variational Calculus, Optimal Control and Applications, pp. 97–106. Birkhäuser, Basel (1998)
437. Tammer, C., Gergele, M., Patz, R., Weinkauf, R.: Standortprobleme in der Landschaftsgestaltung [Location problems in landscaping]. In: Habenicht, W., Scheubrein, B.R. (eds.) Multi-Criteria- und Fuzzy-Systeme in Theorie und Praxis, pp. 261–286. Deutscher Universitätsverlag, Wiesbaden (2003)
438. Tammer, C., Winkler, K.: A new scalarization approach and applications in multicriteria d.c. optimization. J. Nonlinear Convex Anal. **4**(3), 365–380 (2003)
439. Tammer, C., Zălinescu, C.: Vector variational principles for set-valued functions. Optimization **60**(7), 839–857 (2011)
440. Tammer, C., Zălinescu, C.: Vector variational principles for set-valued functions. In: Ansari, Q.H., Yao, J.C. (eds.) Recent Developments in Vector Optimization, pp. 367–415. Springer, Berlin (2012)
441. Tammer, C., Zălinescu, C.: Lipschitz properties of the scalarization function and applications. Optimization **59**, 305–319 (2010)
442. Tanaka, T.: Approximately efficient solutions for vector optimization problems. Sūrikaisekikenkyūsho Kōkyūroku **899**, 159–166 (1995)
443. Tanino, T.: On supremum of a set in a multidimensional space. J. Math. Anal. Appl. **130**(2), 386–397 (1988)
444. Tanino, T.: Conjugate duality in vector optimization. J. Math. Anal. Appl. **167**(1), 84–97 (1992)
445. Tanino, T., Sawaragi, Y.: Duality theory in multiobjective programming. J. Optim. Theory Appl. **27**(4), 509–529 (1979)

446. Tasset, T.N.: Lagrange multipliers for set-valued functions when ordering cones have empty interior. Ph.D. Thesis, University of Colorado at Boulder (2010)
447. Thomann, J., Eichfelder, G.: Representation of the Pareto front for heterogeneous multiobjective optimization. http://www.optimization-online.org/DB_HTML/2019/08/7350.html (2019)
448. Thomann, J., Eichfelder, G.: A trust-region algorithm for heterogeneous multiobjective optimization. SIAM J. Optim. **29**(2), 1017–1047 (2019)
449. van Tiel, J.: Convex Analysis. Wiley, Chichester (1984)
450. Tintner, G.: Stochastic linear programming with applications to agricultural economics. In: Proceedings of the Second Symposium in Linear Programming, Washington, D. C., 1955, pp. 197–228. National Bureau of Standards, Washington, D. C. (1955)
451. Turinici, M.: Maximality principles and mean value theorems. An. Acad. Brasil. Ciênc. **53**(4), 653–655 (1981)
452. Turinici, M.: Variational principles on semi-metric structures. Libertas Math. **20**, 161–171 (2000)
453. Vályi, I.: Approximate solutions of vector optimization problems. In: Sydow, A., Thoma, M., Vichnevetsky, R. (eds.) Systems Analysis and Simulation 1985: Theory and Foundations, pp. 246–250. Akademie-Verlag, Berlin (1985)
454. Vályi, I.: On duality theory related to approximate solutions of vector-valued optimization problems. In: Demyanov, V., Pallaschke, D. (eds.) Nondifferentiable Optimization: Motivations and Applications, pp. 150–162. Springer, Berlin (1985)
455. Vályi, I.: Approximate saddle-point theorems in vector optimization. J. Optim. Theory Appl. **55**(3), 435–448 (1987)
456. Ventcel', E.S.: Issledovanie operacij [Operations research]. Sov. Radio, Moskva (1972)
457. Vlacic, L., Wierzbicki, A., Matic, B.: Aggregation procedures for hierarchically grouped decision attributes with application to control system performance evaluation. In: Jahn, J., Krabs, W. (eds.) Recent Advances and Historical Development of Vector Optimization, pp. 285–310. Springer, Berlin (1987)
458. Wacker, M.: Multikriterielle Optimierung bei Registrierung medizinischer Daten [Multicriterial optimization for medical data registration]. Diploma Thesis, Universität Erlangen-Nürnberg, Erlangen, Germany (2008)
459. Wallace, A.D.: A fixed point theorem. Bull. Amer. Math. Soc. **51**, 413–416 (1945)
460. Ward, J.E., Wendell, R.E.: Using block norms for location modeling. Oper. Res. **33**(5), 1074–1090 (1985)
461. Ward, L.E.: Partially ordered topological spaces. Proc. Amer. Math. Soc. **5**, 144–161 (1954)
462. Webster, R.: Convexity. Oxford University Press, New York (1994)
463. Weidner, P.: Vergleichende Darstellung von Optimalitätsbegriffen und Dualitätsansätzen in der Vektoroptimierung [Unified description of optimality notions and duality approaches in vector optimization]. Diploma Thesis, Martin-Luther-Universität Halle-Wittenberg, Germany (1983)
464. Weidner, P.: Charakterisierung von Mengen effizienter Elemente in linearen Räumen auf der Grundlage allgemeiner Bezugsmengen [Characterization of efficient element sets in linear spaces based on general reference sets]. Ph.D. Thesis, Martin-Luther-Universität Halle-Wittenberg, Germany (1985)
465. Weidner, P.: Dominanzmengen und Optimalitätsbegriffe in der Vektoroptimierung [Domination sets and optimality notions in vector optimization]. Wiss. Z. TH Ilmenau **31**(2), 133–146 (1985)
466. Weidner, P.: Extensions of the feasible range leaving invariant the efficient point set. Seminarberichte der Sektion Mathematik der Humboldt-Universität Berlin **85**, 137–147 (1986)
467. Weidner, P.: The influence of neglecting feasible points and of varying preferences on the efficient point set. Wiss. Z. TH Ilmenau **33**(6), 181–188 (1987)
468. Weidner, P.: On the characterization of efficient points by means of monotone functionals. Optimization **19**(1), 53–69 (1988)

469. Weidner, P.: An approach to different scalarizations in vector optimization. Wiss. Z. TH Ilmenau **36**(3), 103–110 (1990)
470. Weidner, P.: Comparison of six types of separating functionals. In: Sebastian, H.J., Tammer, K. (eds.) System Modelling and Optimization, pp. 234–243. Springer, Berlin (1990)
471. Weidner, P.: Complete efficiency and interdependencies between objective functions in vector optimization. Z. Oper. Res. **34**, 91–115 (1990)
472. Weidner, P.: Ein Trennungskonzept und seine Anwendung auf Vektoroptimierungsverfahren [A separation concept and its application to vector optimization procedures]. Habilitation Thesis, Martin-Luther-Universität Halle-Wittenberg, Germany (1990)
473. Weidner, P.: Verallgemeinerte Tschebyscheff-Normen in der Vektoroptimierung [Extended Chebyshev norms in vector optimization]. Wiss. Z. TH Ilmenau **37**(3), 101–114 (1991)
474. Weidner, P.: On the determination of efficient points with respect to cones. In: Černý, M., Glückaufova, D., Loula, D. (eds.) Multicriteria Decision Making: Methods - Algorithms - Applications, pp. 197–210. The Institute of Economics of the Czechoslovak Academy of Sciences, Prague (1992)
475. Weidner, P.: Advantages of hyperbola efficiency. In: Brosowski, B., Ester, J., Helbig, S., Nehse, R. (eds.) Multicriteria Decision, pp. 123–137. Peter Lang, Frankfurt a. M. (1993)
476. Weidner, P.: Separation of nonconvex sets. In: Guddat, J., Kummer, B., Jongen, H.T., Nožička, F. (eds.) Parametric Optimization and Related Topics. Peter Lang, Frankfurt a. M. (1993)
477. Weidner, P.: The influence of proper efficiency on optimal solutions of scalarizing problems in multicriteria optimization. OR Spektrum **16**(4), 255–260 (1994)
478. Weidner, P.: Problems in scalarizing multicriteria approaches. In: Köksalan, M., Zionts, S. (eds.) Multiple Criteria Decision Making in the New Millenium, pp. 199–209. Springer, Berlin (2001)
479. Weidner, P.: Tradeoff directions and dominance sets. In: Tanino, T., Tanaka, T., Inuiguchi, M. (eds.) Multi-Objective Programming and Goal-Programming, pp. 275–280. Springer, Berlin (2003)
480. Weidner, P.: A unified approach to extended-real-valued functions. Research Report, HAWK Hildesheim/Holzminden/Göttingen, Germany. arXiv: 1605.03456 (2015)
481. Weidner, P.: Functions with uniform level sets. Research Report, HAWK Hildesheim/Holzminden/Göttingen, Germany. arXiv: 1606.00714 (2016)
482. Weidner, P.: Functions with uniform sublevel sets and scalarization in linear spaces. Research Report, HAWK Hildesheim/Holzminden/Göttingen, Germany. arXiv: 1608.04091 (2016)
483. Weidner, P.: Gerstewitz functionals on linear spaces and functionals with uniform sublevel sets. J. Optim. Theory Appl. **173**(3), 812–827 (2017)
484. Weidner, P.: Lower semicontinuous functionals with uniform sublevel sets. Optimization **66**(4), 491–505 (2017)
485. Weidner, P.: Construction of functions with uniform sublevel sets. Optim. Lett. **12**(1), 35–41 (2018)
486. Weidner, P.: Extended real-valued functions - a unified approach. Revista Investigación Operational **39**(3), 303–325 (2018)
487. Weidner, P.: Minimization of Gerstewitz functionals extending a scalarization by Pascoletti and Serafini. Optimization **67**(7), 1121–1141 (2018)
488. Weidner, P.: Functions with uniform sublevel sets, epigraphs and continuity. In: Khan, A., Köbis, E., Tammer, C. (eds.) Variational Analysis and Set Optimization, pp. 92–111. CRC Press, Boca Raton (2019)
489. Weidner, P.: A new tool for the investigation of extended real-valued functions. Optim. Lett. **13**(7), 1651–1661 (2019)
490. Weidner, P.: Scalarization in vector optimization with arbitrary domination sets. Optimization (2019). https://doi.org/10.1080/02331934.2019.1672071
491. Weidner, P., Hakanen, J., Miettinen, K., Mäkelä, M.M.: Experiences with a multicriteria optimization procedure using hyperbola efficiency. Reports of the Department of Mathematical Information Technology, Series B. Scientific Computing, No. B 17/2003, University of Jyväskylä, Finland (2003)

492. Wendell, R.E., Lee, D.N.: Efficiency in multiple objective optimization problems. Math. Program. **12**, 406–414 (1977)
493. White, D.J.: Epsilon efficiency. J. Optim. Theory Appl. **49**(2), 319–337 (1986)
494. Wierzbicki, A.P.: Penalty methods in solving optimization problems with vector performance criteria. Technical Report 12/1974, Inst. of Aut. Control, Techn. University Warsaw (1974)
495. Wierzbicki, A.P.: The use of reference objectives in multiobjective optimization. In: Fandel, G., Gal, T. (eds.) Multiple Criteria Decision Making - Theory and Application, pp. 468–486. Springer, Berlin (1980)
496. Wierzbicki, A.P.: On the completeness and constructiveness of parametric characterizations to vector optimization problems. OR Spektrum **8**(2), 73–87 (1986)
497. Winkler, K.: Aspekte mehrkriterieller Optimierung $C(T)$–wertiger Abbildungen [Aspects of multicriterial optimization with $C(T)$–valued mappings]. Ph.D. Thesis, Martin-Luther-Universität Halle-Wittenberg, Germany (2003)
498. Winkler, K.: Geoffrion proper efficiency in an infinite dimensional space. Optimization **53**(4), 355–368 (2004)
499. Young, R.C.: The algebra of many-valued quantities. Math. Ann. **104**(1), 260–290 (1931)
500. Yu, P.L.: A class of solutions for group decision problems. Manag. Sci. **19**, 936–946 (1973)
501. Yu, P.L.: Introduction to domination structures in multicriteria decision problems. In: Cochrane, J.L., Zeleny, M. (eds.) Multicriteria Decision Making, pp. 249–261. University of South Carolina Press, Columbia (1973)
502. Yu, P.L.: Cone convexity, cone extreme points and nondominated solutions in decision problems with multiobjectives. J. Optim. Theory Appl. **14**(3), 319–377 (1974)
503. Yu, P.L.: Multiple-Criteria Decision Making: Concepts, Techniques, and Extensions. Plenum Press, New York (1985)
504. Yu, P.L., Leitmann, G.: Compromise solutions, domination structures, and Salukvadze's solution. J. Optim. Theory Appl. **13**, 362–378 (1974)
505. Zaffaroni, A.: Degrees of efficiency and degrees of minimality. SIAM J. Control Optim. **42**, 1071–1086 (2003)
506. Zălinescu, C.: On a class of convex sets. Comment. Math. Univ. Carolin. **27**(3), 543–549 (1986)
507. Zalinescu, C.: On two notions of proper efficiency. In: Brosowski, B., Martensen, E. (eds.) Optimization in Mathematical Physics, pp. 77–86. Peter Lang, Frankfurt a. M. (1987)
508. Zălinescu, C.: Convex Analysis in General Vector Spaces. World Scientific Publishing Co., Inc., River Edge (2002)
509. Zeidler, E.: Vorlesungen über nichtlineare Funktionalanalysis III - Variationsmethoden und Optimierung [Lectures on nonlinear functional analysis III - variational methods and optimization]. Teubner, Leipzig (1978)
510. Zeidler, E.: Nonlinear Functional Analysis and Its Applications. Part I: Fixed-Point Theorems. Springer, New York (1986)
511. Zeleny, M.: A concept of compromise solutions and the method of the displaced ideal. Comput. Oper. Res. **1**(3–4), 479–496 (1974)
512. Zeleny, M.: Linear Multiobjective Programming. Springer, Berlin (1974)
513. Zheng, X.Y.: Well-posedness for set optimization problems. Nonlinear Anal. **71**, 3769–3778 (2009)
514. Zheng, X.Y., Ng, K.F.: The Lagrange multiplier rule for multifunctions in Banach spaces. SIAM J. Optim. (electronic) **17**(4), 1154–1175 (2006)
515. Zionts, S., Wallenius, J.: An interactive programming method for solving the multiple criteria problem. Manag. Sci. **22**, B652–B663 (1976)

# Index





C. Tammer and P. Weidner, *Scalarization and Separation by Translation Invariant Functions*, Vector Optimization, https://doi.org/10.1007/978-3-030-44723-6

The manufacturer's authorised representative in the EU is Springer Nature Customer Service Centre GmbH, Europaplatz 3, 69115 Heidelberg, Germany. If you have any concerns regarding our products, please contact ProductSafety@springernature.com

Printed and bound by CPI Group (UK) Ltd, Croydon, CR0 4YY
05/10/2026
02228126-0019